I B	II B	III A	IV A	V A	VI A	VII A	O
							2 **He** Helium 4.00260
		5 **B** Boron 10.81	6 **C** Carbon 12.011	7 **N** Nitrogen 14.0067	8 **O** Oxygen 15.9994	9 **F** Fluorine 18.99840	10 **Ne** Neon 20.179
		13 **Al** Aluminum 26.98154	14 **Si** Silicon 28.086	15 **P** Phosphorus 30.97376	16 **S** Sulfur 32.06	17 **Cl** Chlorine 35.453	18 **Ar** Argon 39.948
28 **Ni** Nickel 58.70	29 **Cu** Copper 63.546	30 **Zn** Zinc 65.38	31 **Ga** Gallium 69.72	32 **Ge** Germanium 72.59	33 **As** Arsenic 74.9216	34 **Se** Selenium 78.96	35 **Br** Bromine 79.904
46 **Pd** Palladium 106.4	47 **Ag** Silver 107.868	48 **Cd** Cadmium 112.40	49 **In** Indium 114.82	50 **Sn** Tin 118.69	51 **Sb** Antimony 121.75	52 **Te** Tellurium 127.60	53 **I** Iodine 126.9045
78 **Pt** Platinum 195.09	79 **Au** Gold 196.9665	80 **Hg** Mercury 200.59	81 **Tl** Thallium 204.37	82 **Pb** Lead 207.2	83 **Bi** Bismuth 208.9804	84 **Po** Polonium (209)	85 **At** Astatine (210)

(Group columns 36 Kr Krypton 83.80; 54 Xe Xenon 131.30; 86 Rn Radon (222) in the O group)

63 **Eu** Europium 151.96	64 **Gd** Gadolinium 157.25	65 **Tb** Terbium 158.9254	66 **Dy** Dysprosium 162.50	67 **Ho** Holmium 164.9304	68 **Er** Erbium 167.26	69 **Tm** Thulium 168.9342	70 **Yb** Ytterbium 173.04	71 **Lu** Lutetium 174.97
95 **Am** Americium (243)	96 **Cm** Curium (247)	97 **Bk** Berkelium (247)	98 **Cf** Californium (251)	99 **Es** Einsteinium (254)	100 **Fm** Fermium (257)	101 **Md** Mendelevium (258)	102 **No** Nobelium (255)	103 **Lr** Lawrencium (260)

CHEMICAL PRINCIPLES

ROBERT S. BOIKESS
Douglass College, Rutgers University

EDWARD EDELSON
New York University

HARPER & ROW, PUBLISHERS
New York   Hagerstown   San Francisco   London

Credits for photographs appearing on chapter opening pages:

Chapter 1, *Herwig, Stock, Boston;* Chapter 2, *Granger;* Chapter 3, *Pinney, Monkmeyer;* Chapter 4, *Wide World;* Chapter 5, *Kalman, Stock, Boston;* Chapter 6, *Granger;* Chapter 8, *Computer Graphics by Wilma K. Olson, Photo by Robert Boikess;* Chapter 9, *Albertson, Stock, Boston;* Chapter 10, *Virginia Phelps;* Chapter 11, *Urban, Stock, Boston;* Chapter 12, *Fundamental Photographs, Granger;* Chapter 13, *Wide World;* Chapter 14, *National Park Service;* Chapter 15, *Sunkist Growers, Inc.;* Chapter 16, *Union Carbide;* Chapter 17, *Beckwith Studios;* Chapter 18, *Reed, Animals, Animals;* Chapter 19, *Granger,* Chapter 20, *Granger;* Chapter 21, *DPI;* Chapter 22, *Franken, Stock, Boston.*

Credits for photographs appearing on the cover:

Front Cover

TOP: Earthrise from the moon (NASA)

MIDDLE: *Top, Left,* Alchemist's symbol for phosphorus; Tin-lead-tellurium polycrystalline ingot for semiconductor diode laser (E. Sperko, *The Changing Challenge,* General Motors); *Top, Center,* Computer enhanced thermogram of human female (Sochurek, Woodfin Camp); *Top, Right,* Double exposure of laser-excited I_2 and photon echo from the I_2 displayed on an oscilloscope (Richard G. Brewer, IBM); *Bottom,* Sequence of sunset from the earth (Mark Berghash)

BOTTOM: *Left,* Photomicrograph of vapor-deposited silicon surface (Bell Laboratories); *Right,* Molecular model of DNA

Back Cover

TOP: Orion Nebula (Naval Photographic Center)

MIDDLE: *Top,* Trees (Mark Berghash); *Bottom,* Computer enhancement of aerial infrared photograph (Spatial Data Systems)

BOTTOM: *Left,* E. Coli Bacteriophage magnified 90,000 × (A. K. Klein Schmidt); *Right,* Computer Graphic of a π molecular orbital of diacetylene (*Orbital and Electron Density Diagrams* by Andrew Streitwieser, Jr. and Peter H. Owens, Macmillan Publishing Co., Inc., 1973.

Sponsoring Editor: Wayne E. Schotanus/John A. Woods
Project Editor: Lois Lombardo
Designer: Gayle Jaeger
Production Supervisor: Marion A. Palen
Photo Researcher: Myra Schachne
Compositor: York Graphic Services, Inc.
Printer: The Murray Printing Company
Binder: Halliday Lithograph Corporation
Art Studio: J & R Technical Services Inc.
Cover Designer: Mark Berghash, 20/20 Services, Inc.

CHEMICAL PRINCIPLES
Copyright © 1978 by Robert S. Boikess and Edward Edelson

Library of Congress Cataloging in Publication Data

Boikess, Robert S
    Chemical principles.

    Includes index.
    1. Chemistry.   I. Edelson, Edward, joint author. II. Title.
QD31.2.B63   540   77-17023
ISBN 0-06-040807-3

For Karen and Phyllis,
    Bruce, Anne, Daniel, and Noah
Who gave help, good cheer, advice, and incentive.

CONTENTS

SELECTED TABLES

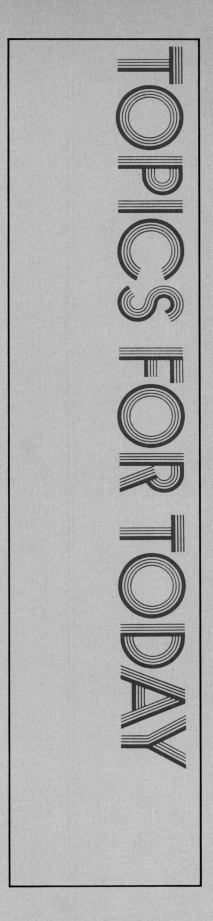

xiii

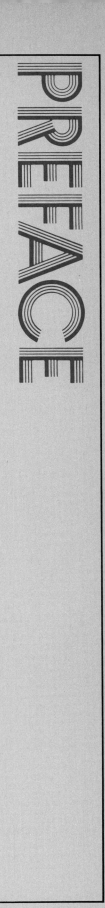

A chemistry textbook must meet the needs of both students and teachers. One reason for writing this book was a feeling that many texts fail to do this. This textbook aims to meet those needs by being both comprehensible and comprehensive. In the long process of writing it, our goal has always been to combine clarity of exposition with completeness of presentation. Our objective is to present the basic principles of chemistry in a way that can be easily read, understood, and appreciated by students without oversimplification or omission.

Most students study chemistry not because of an abstract interest in chemical principles, but rather because these principles are used in their chosen fields of study. Throughout the book, we show the practical applications of the principles we present. We believe that students will master chemical ideas and develop chemical skills much more readily when they can see the purpose and value of those ideas and skills. Accordingly, we have not only included many such discussions in the body of the text but have also interspersed, within chapters, a number of boxed topics of current interest to illustrate the importance and usefulness of basic chemical principles.

Chemical Principles is intended for a full-year general chemistry course for students majoring in any science from agriculture to zoology, including those whose career goals are in the health professions, in engineering, and in the environmental sciences. Our approach does not use calculus, but it does require some familiarity with simple algebra, including facility with exponents and logarithms. An appendix covering mathematical procedures is included.

In choosing the level and sequence of presentation, we have been helped greatly by extensive consultations with many teachers of general chemistry. The thoughtful comments and criticisms of many reviewers who have carefully read the manuscript, as well as student feedback from classroom testing of much of the material, have had a major influence on the book.

We have chosen an approach that allows the teacher a great deal of flexibility in selecting the order in which material is presented. The table of contents is only one of many possible sequences. We have tried to write chapters or small groups of chapters as self-contained units. For example, although the topic of thermodynamics is presented in the text before the topic of equilibrium, it is quite feasible for the teacher to reverse that sequence. A beginning chapter which can be drawn upon throughout the course presents such basic topics as the writing of chemical formulas and chemical equations, the naming of compounds, and the classification of compounds and reactions. Early discussions of stoichiometry and of solutions permit the coordination of meaningful laboratory work with the material presented in class.

The opening chapters introduce some basic principles of chemistry, and so permit the presentation of the properties of substances as more than a mere catalog of unrelated observations. The description of the chemical and physical properties of substances also helps in the understanding of the basic principles and how they are formulated. Accordingly, we have included substantial amounts of descriptive material, not in separate chapters relegated to the end of the book but as part of the presentation of chemical principles. For example, an early and extensive chapter on the nonmetals is used to reinforce the discussions of molecular structure, thermochemistry, and oxidation states. Those concepts are, in turn, used as the basis for the presentation and correlation of much of the chemistry of the nonmetals. The in-depth descriptions of the biological and technological interest of the nonmetals is typical of our approach to descriptive chemistry. Similarly, in the discussion of oxidation-reduction we incorporate descriptions of biological oxidations and of corrosion. The discussion of kinetics includes descriptions of enzymes, the role of homogeneous catalysts in atmospheric pollution, and the role of heterogeneous catalysts in industry. The discussion of acids and bases incorporates a description of blood and biological buffers, and so on.

Problem-solving skills are developed by the use of numerous illustrative examples that are incorporated in the body of the text. We have also included more than 1100 end-of-chapter exercises, with answers to some selected ones given in an appendix. This comprehensive set of exercises provides a review guide that allows the student to test understanding of the various concepts that are presented and to gain facility in solving numerical problems. More than one exercise is included for each concept, to offer the student many self-testing opportunities and to give the teacher a wide choice of materials to assign. We have also included extensive tables of data to assist the teacher in formulating additional exercises or examination questions.

Because of the growing acceptance of the International System of Units (SI), it is important that students be introduced to this system of measurements. In *Chemical Principles,* SI units are used to the full extent that is compatible with chemical clarity. Although non-SI units are not used extensively, appropriate exercises and discussions are included to enable the student to master the use of scientific information that is presented in these units. Some units that are technically not SI units are permissible in this system. Also some non-SI units are widely used by chemists because of their convenience. Accordingly, we use the liter as a unit of volume and the atmosphere as a unit of pressure. However, the milliliter is not used; instead, we use the equivalent cubic centimeter. Similarly, the torr and the millimeter of mercury, which are not SI units, are not used. The joule, rather than the calorie, is utilized in energy measurements, and the mole is defined and used as the unit of amount. We generally use the kelvin as the unit of temperature and the nanometer as the unit for atomic distances.

Supporting materials for this book are available to assist both the student and the teacher. For the teacher we have prepared an *Instructor's Manual* in which we analyze in detail every section of each chapter to help in the development of alternate sequences of presentation. In addition, all the end-of-chapter exercises are analyzed in terms of the topics they cover and the level of difficulty of each exercise.

A *Study Guide and Glossary* that provides valuable assistance in the use and understanding of the text, in the development of skills, and in the delineation of learning objectives has been written by Professors Daniel L. Reger and Edward E. Mercer of the Department of Chemistry, University of South Carolina and Professor Robert S. Boikess.

Chemical Principles in the Laboratory, written by Professors Lester R. Morss and Robert S. Boikess of the Department of Chemistry, Rutgers University, is a laboratory manual that accompanies the text. The manual reinforces many of the principles presented in the text, developing them in greater depth and describing a number of instructive and interesting experimental procedures. The order of experiments and of topics discussed in the manual coincide with those of this book.

The authors owe a great debt to the large number of persons who have been involved in the planning, writing, and production of this book. Their comments and suggestions have provided invaluable help at every stage of the work, from the initial preparation of the general outline and approach to the final revisions. We hope that their help has enabled us to achieve our goal of combining conceptual rigor, a high standard of chemistry, a balanced approach, and a readable style to give both students and teachers what they want and need for a general chemistry course.

Robert S. Boikess
Edward Edelson

It has been our good fortune to have had the assistance of many skilled people during the entire course of the development of this book. Their efforts have had a major impact, and we are grateful to them for their time, their patience, and their willingness to help.

More than thirty professors of chemistry from around the country helped us to develop the general approach, the conceptual level, and the table of contents. Many reviewers assisted us further in preparing an order of presentation to satisfy the needs of those who teach chemistry. Substantial portions of the manuscript were read and many valuable suggestions for improvement were made by a number of people. We especially thank Professors Lawrence Epstein, University of Pittsburgh; Henry Heikkinen, University of Maryland; Marvin Kientz, California State College, Sonoma; Paul Kimmel, East Brunswick High School; Daniel C. Krezenski, Joliet Junior College; Richard Lynde, Montclair State College; Edward E. Mercer, University of South Carolina; John P. Mitchell, Tarrant County Junior College; Gardiner Myers, University of Florida, Gainesville; Robert Niedzielski, University of Toledo; William Plucknett, University of Kentucky; and Daniel L. Reger, University of South Carolina.

Many colleagues at Rutgers helped with the revision of the original manuscript. We would especially like to thank Professors Ken Breslauer, Gregory Herzog, Stephan Isied, Omar Khalil, John Krenos, Stanley Mandeles, Lester Morss, Wilma Olson, Donald Shombert, George Strauss, Joan Valentine, and Carol Venanzi. Several students at Rutgers provided valuable help—Susan Barnes and Carol Oken, with a rare combination of chemical understanding and artistic skill, prepared the sketches for the art; Deborah Plick and Kim Hodgson solved all the exercises.

We have been fortunate indeed to have been able to work with the staff at Harper & Row. Without John A. Woods, the first sponsoring editor, the book would never have been written. The project was his conception, and he initiated and organized it most skillfully. When John heeded the advice of Horace Greeley and went west, Wayne E. Schotanus helped us through many deadlines to the end. Lois Lombardo, the project editor, was of immense help. Her professionalism and dedication above and beyond the call of duty are in large part responsible for transforming our manuscript into a finished book. The handsome appearance of the book is due to the skill and talent of the designer, Gayle Jaeger.

Finally we should like to thank all our teachers and students from whom we have learned.

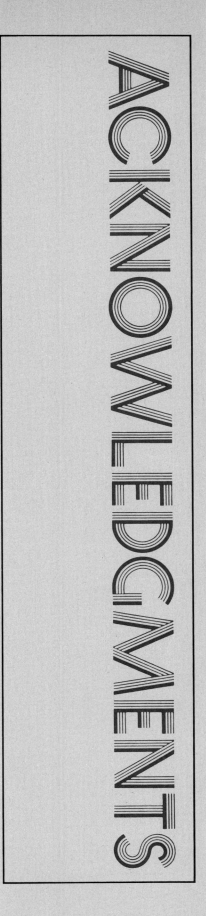

ACKNOWLEDGMENTS

xvii

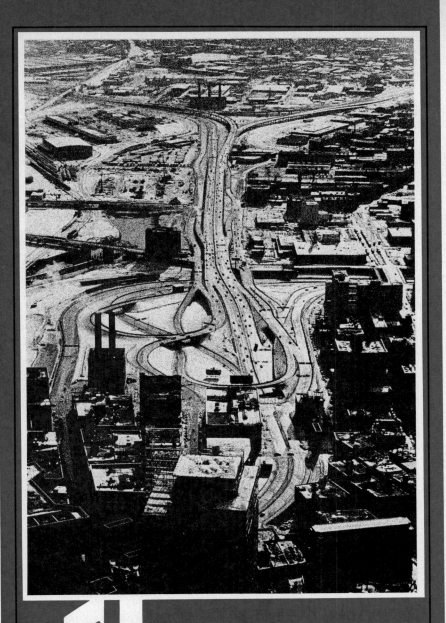

THE SCIENCE OF CHEMISTRY

1.1 THE NATURE OF SCIENCE; THE SCIENTIFIC METHOD

Science is an organized body of knowledge that is based on a method of looking at the world. The method of science is unique. It requires that any explanation of what is seen be based on the results of experiments and observations. These experiments and observations must be verifiable by anyone who has the time and means needed to repeat them.

This requirement has many important consequences. Because scientific theories are subject to testing and retesting, they must be put forth precisely enough to avoid ambiguity. Because a new experiment may be performed at any time to produce new data, theory is regarded as always being open to revision. Because an experiment must be verifiable, it is preferable to express results in a way that makes direct comparison between experiments possible. Whenever possible, measurements, rather than broad descriptions, are preferred.

The scientific way of examining phenomena is rare in human history. Most societies, including many now in existence, have preferred methods that do not require experimentation and measurement. Many societies view the world as a mysterious place, governed by forces so vast and changeable that they are beyond human understanding. Even societies which believe that a rational understanding of these forces is possible usually do not demand that an explanation be verified by experiments. In ancient Greece, for example, philosophers developed many theories that, in modified form, are accepted today. But the Greeks generally were content with theorizing. It is only in the modern era of Western history, which began in Europe around A.D. 1500, that theories are regarded as unacceptable unless they are supported by observation and experimentation. The result of this requirement is the creation of a society in which the relatively small number of persons who have mastered the scientific method have achieved an amazingly thorough transformation of our planet.

The scientific method is not easily summarized. No single set of rules can describe the great variety of scientific activities. But an idealized form of the scientific method can be described. Often, the beginning point is the development of a hypothesis about a specific phenomenon. The hypothesis may be tested by observation, by experimentation, or by a combination of the two, depending on the phenomenon that is being studied. A chemist, for example, can perform an experiment by attempting to carry out a given chemical reaction in the laboratory under varying conditions. By contrast, an astronomer can only observe the distant stars. He or she cannot experiment by changing them.

A hypothesis that is supported by observation or experiment may lead to a unifying explanation for many different phenomena. This kind of unifying explanation often is called a law of nature. Any law is subject to revision as new observations and experiments are done.

Science is ordinarily regarded as being done outside the influence of personality, economics, or social structure. In such a framework, the first step in science will be an experiment or an observation; theorizing begins only after data are gathered. Studies of the way that science is actually done indicate that this picture is rarely accurate. Science is influenced by personality, money, and social forces. Scientific research generally begins with a quest for a specific piece of new knowledge, chosen by a combination of social, economic, and personal influences. The role of intuition (for lack of a better word to describe a mental process that is poorly understood) in the formulation of theories by scientists is dramatically important. Hunches, guesses, and even dreams have played a part in some of the greatest scientific discoveries. There is also a strong esthetic element in scientific discovery. One of the highest forms of praise for a scientific experi-

THE SCIENTIFIC METHOD

"When I think of formal scientific method an image sometimes comes to mind of an enormous juggernaut, a huge bulldozer—slow, tedious, lumbering, laborious, but invincible. It takes twice as long, five times as long, maybe a dozen times as long as informal mechanic's technique, but you know in the end you're going to *get* it. There's no fault-isolation problem in motorcycle maintenance can stand up to it. When you've hit a really tough one, tried everything, racked your brains and nothing works, and you know that this time Nature has really decided to be difficult, you say, 'Okay, Nature, that's the end of the *nice* guy,' and you crank up the formal scientific method. . . ."

". . . . That part of the formal scientific method called experimentation is sometimes thought of by romantics as all of science itself because that's the only part with much visible surface. They see lots of test tubes and bizarre equipment and people running around making discoveries. They do not see the experiment as part of a larger intellectual process and so they often confuse experiments with demonstrations, which look the same. A man conducting a gee-whiz science show with fifty thousand dollars' worth of Frankenstein equipment is not doing anything scientific if he knows beforehand what the results of his effort are going to be. A motorcycle mechanic, on the other hand, who honks the horn to see if the battery works is informally conducting a true scientific experiment. He is testing a hypothesis by putting the question to nature. The TV scientist who mutters sadly, 'The experiment is a failure; we have failed to achieve what we hoped for,' is suffering mainly from a bad scriptwriter. An experiment is never a failure solely because it fails to achieve predicted results. An experiment is a failure only when it also fails adequately to test the hypothesis in question, when the data it produces don't prove anything one way or another."

Robert M. Pirsig, *Zen and the Art of Motorcycle Maintenance*

ment or hypothesis is to call it "elegant," which implies not only admiration for its effectiveness but also appreciation of its beauty.

In the early stages of scientific discovery, the major effort is to describe phenomena and to classify them by their characteristics. Later, measurement—quantification—replaces the qualitative descriptions. Still later, a mass of quantitative data can be described by a few concise statements or mathematical equations that are called laws—always with the understanding that these laws are tentative and subject to modification. In some cases, it may be possible to construct a theory that explains many different laws by a few general principles. In biology, the theory of evolution is such a unifying principle, as the atomic and molecular theory of matter is in chemistry.

Theories and laws are always subject to refinement and modification as new observations are made. For example, Newton's theory of gravitation eventually was refined and modified by Einstein's theories, which are themselves constantly being studied with a view toward modification.

NEW THEORIES FOR OLD

Every scientific theory, however well established, is subject to revision as new data are gathered. As this book is being written in early 1977, major theories in both biology and physics have been challenged because of new data. By the time you read this book, these theories may no longer be accepted, and new theories may have replaced them.

In biology, studies of genes on the molecular level have led to a challenge of a major part of the theory of evolution, one of the cornerstones of modern biology. As set forth by Charles Darwin and elaborated by other biologists, the theory of evolution proposes that natural selection—"the survival of the fittest"—acts to increase the incidence of genes that make an organism more suited to a given environment. According to classical Darwinian theory, evolution occurs primarily because individuals and species that are better suited to a given environment have a greater chance of surviving and reproducing. Thus, the genes of these individuals and species are transmitted with a higher frequency than the genes of less fit individuals and species.

However, recent studies have led several molecular geneticists to propose that pure chance plays an extremely important role in evolution. According to these studies, the rate of transmission of genes from generation to generation is determined by chance alone, not by natural selection. (It is important to note that the theory of evolution itself is not being challenged; the mechanism by which evolution occurs is being debated.) As this is being written, more data are being gathered, both in support of and in opposition to the new interpretation of evolution. The evaluation of these data eventually may resolve the issue.

In February 1977, a research group at Cambridge University in England announced the determination of the complete molecular structure of the genes of a virus called Phi X174. Study of the virus's genetic structure raised questions about another theory, which states that one gene can govern the production of only one protein (Chapter 22). It has been found that the same section of the genetic material of Phi X174 can be read in two different ways, so more information is packed into the relatively small amount of genetic material than might have been expected. The finding raised the possibility that the same sort of multiple interpretation might be possible in organisms with a much greater amount of genetic material. If this is so, theories about the amount of information coded into genetic material will have to be fundamentally altered.

In physics, experiments at the Swiss Institute for Nuclear Research indicate that a subatomic particle called the muon decayed in a way that could not be accounted for by existing theory. It appeared that the muon experiments could be explained by proposing the existence of a new force of nature whose effects had never before been seen in particle decay processes. The existence of such a new force had been postulated by Stephen Weinberg, a physicist at Harvard University, and others, who described it as being much weaker than any of the previously known forces. As this book is being written, workers in a number of laboratories are attempting to repeat the Swiss experiments. The success or failure of these attempts will play an important role in determining whether the new theory is accepted or rejected.

In all three of these examples, the lesson is the same. All "truths" in science are provisional and are subject to change in the light of new evidence from observation or experiment.

1.2 CHEMISTRY AS A PHYSICAL SCIENCE

For convenience, science is divided into a number of separate disciplines. However, the division never is as clear-cut as it might appear to be. All the sciences are related and draw on each other.

Chemistry is defined as the branch of science that deals with the composition, structure, properties, and transformations of matter. Chemistry tends to study matter from an atomic and molecular point of view. Therefore, chemistry is closely related to physics, which can be defined in the most general way as the study of matter and energy and the interaction between them. Chemistry is also related to biology, which has as one of its major concerns the chemistry of living organisms. The relationship of chemistry to other sciences, such as astronomy and geology, has become more evident in recent years with the rise of such disciplines as astrochemistry and geochemistry.

In a hierarchy of the sciences, chemistry often is placed between physics and biology. Physics is viewed as taking the most general approach. It deals in all ways with matter, energy, and the interactions between them. Chemistry uses the information gathered by physics about the nature of matter and energy to study the properties and interactions of substances from a somewhat different point of view. Chemistry depends heavily on an understanding of the atom, which it regards as a basic unit of matter, to explain many interactions between substances. Biology uses the findings of both physics and chemistry to study living organisms.

Some working definitions are required for the discussions that lie ahead. **Matter** is defined as anything possessing mass and occupying space. **Energy** is defined as the ability to do work. **Kinetic energy** is the work that a moving object is capable of performing as it is brought to rest. **Potential energy** is the capability of doing work that matter has by virtue of its position in a force field. For example, a boulder at the top of a hill is capable of doing work because of its position in the earth's gravitational field. **Radiant energy** is more difficult to define in simple terms. For the moment, we can define radiant energy as a combined electric field and magnetic field oscillating in space. More intuitively, we can define radiant energy as anything moving at the speed of light.

We shall discuss all these subjects at length later. But let us look more closely at the last definition of radiant energy, which is one of the major accomplishments of twentieth-century science. Until the beginning of the twentieth century, it was believed that matter and energy were completely different entities. One difference seemed obvious: Matter has mass and energy does not. In 1905, however, Albert Einstein said that radiant energy has a mass equivalent; indeed, that radiant energy is simply matter traveling at the speed of light. The equivalence of matter and radiant energy has been established, and it has had a profound effect on scientific laws.

Before Einstein, there were two separate laws of conservation. One law said that matter could be transformed into other forms of matter but could not be lost.

The other law said that energy could be transformed into other forms of energy but could not be lost. Now there is a single law which says that mass-energy is the entity that is conserved. Einstein's equation linking energy and mass:

$$E = mc^2$$

where E is the energy, m is the mass, and c is the speed of light, expresses the equivalence of energy and matter. (It should be noted that the transformation of matter into energy and vice versa is not important in ordinary chemical reactions.)

Matter in Chemistry

The matter that chemists deal with can be classified in several different ways. For example, a distinction is drawn between **homogeneous** material, which has the same properties throughout, and **heterogeneous** material, which has parts with visibly different properties. A lump of pure gold is a homogeneous material, while the rock in which the gold was found is a heterogeneous material.

A **pure substance** is a homogeneous material that has a definite chemical composition throughout. There are two kinds of pure substances. One kind can be decomposed into two or more different substances by simple chemical change. The other kind cannot be decomposed by simple chemical change. Ordinary table salt is a pure substance of the first kind. It can be decomposed into sodium and chlorine by an appropriate process. Gold is a pure substance of the second kind. It cannot be decomposed by chemical change. A pure substance that cannot be decomposed is called an **element**. Pure substances that can be decomposed by simple chemical changes are called **compounds**. There are many millions of compounds but just over 100 elements.

Most of the materials that we see in the world around us are not pure substances. There are **mixtures**, which consist of a number of different substances. A bowl of vegetable soup is a mixture; so is a candy bar. There are **solutions,** which are homogeneous combinations of different substances. The chief difference between a mixture and a solution is that any sample of a solution has the same composition, while the composition of a mixture is not the same throughout. Solutions can be gaseous, liquid, or solid. Carbon dioxide in water is a liquid solution, air is a gaseous solution, and sterling silver, an alloy of copper and silver, is a solid solution.

A distinction must also be drawn between a mixture and a compound. The elements making up a compound cannot be recovered without a chemical change. The substances making up a mixture or a solution can. The substances making up a mixture or a solution need not be elements. For example, we can prepare a solution by dissolving salt, a compound, in water, another compound. In addition, the substances making up a mixture or a solution can be combined in varying proportions. The elements in a compound have fixed proportions.

All substances have characteristics, or properties, by which we identify them. The **chemical properties** of a substance are those that describe the way in which it can undergo change, either alone or in interactions with other substances, to form different materials. Such changes are called chemical reactions. We can list the chemical properties that are characteristic of any substance. Iron combines readily with oxygen to form the compound called rust. Gold is inert to nitric acid, while copper is not. Sulfur combines with silver to form a black compound that we call tarnish. And so on, almost endlessly.

Physical properties are characteristics by which we can describe a substance—color, density, hardness, melting point, and the like. Thus, we can describe gallium as a white metal, soft enough to be cut with a knife, which has a density 5.91 times greater than that of water, a melting point of 29.78°C, and a boiling point of 1983°C. A distinction usually is drawn between the **intensive properties** of a substance, those properties found in any sample of the substance, and the **extensive properties**, which are characteristics of a specific sample. Shape, size, length, and weight are extensive properties.

1.3 FROM ALCHEMY TO CHEMISTRY

Modern chemistry was born out of alchemy. There was much more to alchemy than the mere quest for the ability to transform base metals into gold. Alchemy was a complex mixture of philosophy, astrology, mysticism, magic, real chemistry, and other ingredients, including the belief that all forms of matter have a soul in common. While alchemists were interested primarily in proving the truth of their philosophical system, their patrons had more practical ends in mind—for example, the discovery of the philosophers' stone, which would transmute any base metal into gold. From the modern point of view, alchemy was valuable because it led to the development of laboratory procedures and to the discovery of new substances.

Modern chemistry emerged from alchemy in the seventeenth century. The most important event in the transformation was the publication in 1661 of Robert Boyle's book, *The Sceptical Chymist,* which rejected the alchemists' belief that the properties of all substances were derived from a mixture of four ''elements'' (earth, water, fire, and air) and four ''principles'' (hot, cold, dry, and moist). This philosophy went back to Aristotle. Boyle replaced it with a new philosophy, which described an element as any substance that cannot be broken down into a simpler substance. His description is essentially the modern definition of an element. But Boyle did not move into the modern world at a single bound. While he put forth many other ideas that remain basically unchanged, such as the distinction between an acid and a base and the definition of a compound, Boyle clung to a number of the old ideas. For example, he seems to have retained a lingering belief in the transmutation of elements.

The transition from alchemy to chemistry took many years. One of the hindrances to progress was a theory put forward by a German physician, Georg Ernst Stahl (1660–1734), to explain combustion. Stahl's theory, proposed as the seventeenth century ended, said that all flammable substances contain a material called phlogiston. When a substance burned, the theory said, phlogiston escaped into the air, leaving behind something called calyx.

An alchemist's laboratory, as seen in a seventeenth-century engraving. Several assistants are tending a large still in the middle of the laboratory. A smaller still is operating at right. A young assistant grinds plant material in a mortar with a pestle while the alchemist consults a book of alchemical lore. (*Granger*)

ALCHEMY INTO CHEMISTRY

Alchemy was a mixture of mystical, speculative thought and practical laboratory techniques of chemistry and metallurgy. In addition to their main goal, the transmutation of base metals into silver and gold, alchemists also sought to prepare an elixir to prolong life indefinitely. Some alchemists also sought a substance that would dissolve all other substances, although they never said how such a universal solvent could be contained.

While alchemy generally is identified with medieval times, it flourished in Hellenistic Egypt. Some 80 different kinds of apparatus, many of which are used in modified form by chemists today, are mentioned in Greek treatises of the third and fourth centuries. They include furnaces, beakers, pestles and mortars, filters, stills, and stirring rods. Alchemy was a popular study in Islam. Moslem alchemists received much of their information from the earlier Greek writings and, in turn, passed their alchemical lore along to medieval Europe.

A number of chemical and metallurgical achievements of alchemists were absorbed into modern chemistry. The word ''gas'' was invented by an alchemist, J. B. van Helmont (1577–1644), who derived it from the Greek word *khaos.* Van Helmont recognized a number of gases, including hydrogen, chlorine, carbon monoxide, and methane, although he did not isolate any of them.

Another alchemist, J. R. Glauber, (1604–1668), prepared hydrochloric acid by heating salt with oil of vitriol, or sulfuric acid. A compound produced in this reaction is sodium sulfate, called *sal mirable* by Glauber. He described it as a powerful medicine for many ailments. It is still used in medicine and is now called Glauber's salt.

Alchemists placed great importance on the development of furnaces, because they believed that the base metals were impure forms of gold, and could be converted to gold by some sort of roasting, subliming, or distilling operation. They placed almost no value on quantitative measurements. Balances or other weighing devices were hardly to be found in an alchemist's laboratory, since changes of weight had little significance in alchemy. Concepts of temperature were also vague. The seventeenth-century German alchemist J. D. Mylius, for example, recognized only four degrees of heat: the heat of a human body, the heat of June sunshine, the heat of a calcining body and the heat of fusion.

It took the better part of a century to replace the phlogiston theory with something closer to a correct description of combustion. As late as 1774, when the British scientist Joseph Priestley (1733–1804) heated mercury to produce a reddish powder and then heated the powder to liberate a gas in which things burned more brightly than they did in air, he described his experiment in terms of phlogiston. Priestley believed that the gas permitted brighter burning because it had no phlogiston, and so he called it ''dephlogisticated air.''

The ''father of modern chemistry,'' the French scientist Antoine Lavoisier (1743–1794), finally disproved the phlogiston theory. Lavoisier showed that after a substance burns, the products of combustion weigh more than the original substance did. Lavoisier concluded that it was absurd to believe that phlogiston was being liberated by combustion, since phlogiston would have to weigh less than nothing to explain his results. Lavoisier proposed that in combustion, the burning substance combines with a portion of the air to form new compounds. Priestley had called that portion of the air dephlogisticated air. Lavoisier called it oxygen, a name derived from the Greek words for ''acid-former.'' Lavoisier's naming of the gas was not quite correct, since it was based on his belief that all acids contain oxygen. But the name, and Lavoisier's explanation of combustion, are still accepted.

Another major step toward modern chemistry was taken in the first decade of the nineteenth century when the English chemist John Dalton postulated that all elements are made up of atoms. Dalton used the word element in its modern sense, and he pictured atoms as tiny, indestructible units that could combine with other atoms to form ''compound atoms,'' or molecules. Dalton proposed that each element has its own kind of atom, that the atoms of different elements differ in essentially nothing but their masses, and that atoms combine in definite proportions. Dalton went on to determine the relative weights of atoms of several elements, including oxygen, hydrogen, and carbon. A new era had begun.

1.4 CHEMISTRY IN THE MODERN WORLD

At the height of the cold war in the 1950s, steel production was a matter of great concern and national pride to both the United States and the Soviet Union. It was the stated aim of the Soviet Union to overtake the United States in total steel production. Some U.S. observers feared that such an event would signal a major shift in the balance of power.

The Soviet Union did, in fact, surpass the United States in steel production in the 1960s. However, that milestone caused no major upheaval. The reason was chemistry. In the last half of the twentieth century, steel production yielded to chemical production as an indicator of industrial expertise and power. The clothes we wear, the paint on the wall, the furniture in our rooms are all likely to be products of the chemical industry. Our automobiles contain many components produced by the chemical industry. Even the food we eat is emulsified, colored, preserved, and flavored by products of the chemical industry.

We live in a world of synthetic chemicals, and in recent years we have become aware that these chemicals may not be entirely beneficial. The production of nitrogenous fertilizers provides a good example of the mixture of good and ill provided by modern chemistry. Millions of tons of fertilizer are produced annually by modern versions of the nitrogen-fixing method developed before World War I by Fritz Haber and Karl Bosch, two German chemists. This fertilizer enables farmers to achieve the high yields that are needed to feed an increasing population. Without ample supplies of artificial fertilizers, the ''Green Revolution,'' based on new, high-yielding strains of food grains, would be impossible.

However, many scientists have become concerned about the possible long-term ecological effects of fixing large amounts of nitrogen—that is, of converting the inert nitrogen that makes up 80% of air to a biologically active form. When nitrogenous fertilizer is washed from a field into a lake or stream, it can serve as a nutrient for species of microscopic algae, which can multiply until these clear waters are clogged with a green scum. The resulting ''eutrophication'' can accelerate the natural aging process by which a lake is slowly filled in by accumulating organic matter.

Nuclear energy, which provides a growing portion of the nation's energy, can also be regarded as a mixed blessing. Nuclear energy could not have been developed without chemical technology. For example, a chemical process for producing the compound uranium hexafluoride was essential for the enrichment process by which first the atomic bomb and later fuel elements for nuclear reactors were manufactured. Without nuclear power, the shortage of such fossil fuels as oil and natural gas would impose a crippling burden on industrialized societies. Yet there are pressing issues in nuclear energy: the need for long-term storage of radioactive wastes, the possible use of plutonium to make illicit nuclear weapons, and unresolved issues of reactor safety.

There are many such examples. Phonograph records enable us to hear either Mozart or rock music at our convenience. But phonograph records are made of polyvinyl chloride, produced from vinyl chloride, which can cause liver cancer and other diseases in industrial workers. Antibiotics, many of them synthesized in the laboratory, have virtually eliminated infectious diseases as a major cause of death in the developed nations. But overuse of antibiotics is threatening to create infectious agents that are resistant to the antibiotics on which we must rely.

The perception of the relative balance between benefits and problems caused by progress in chemistry changes with time. Modern insecticides increased the food supply and added to personal comfort, so they were used on a large scale. Today, we balance these benefits against the harm done by insecticides to birds, fishes, and useful insects. We have entered a new and complex era that requires a better understanding of the multitude of chemical reactions, both natural and man-made, that shape the world we live in.

1.5 CHEMISTRY IN OTHER SCIENTIFIC DISCIPLINES

We mentioned that all the sciences are interrelated. In a sense, the division between sciences is an artificial construct, because all the sciences borrow information and methods from each other. Chemistry contributes greatly to other scientific fields.

Biology began as an almost purely descriptive science. A large part of modern biology is the effort to describe the processes of life in purely chemical terms. Molecular biology, which has grown to a dominant position in recent years, is almost pure chemistry, the complex chemistry of living organisms.

Geology also depends heavily on chemistry. The chemical processes that form the rocks and soil of the earth are studied intensively by geologists. Atmospheric studies rely heavily on chemistry. The recent controversy about the effects of aerosol propellants on the upper atmosphere was the result of information gained from studies of atmospheric chemistry. Ecology, a cross-disciplinary field whose importance has been recognized only recently, is another science in which chemistry plays a major role. Pollution can be described as the addition of unwanted chemicals to the environment. A substance cannot be described as unwanted unless its chemistry is known.

However, it must be kept in mind that the study of chemistry does not have to be justified only by examples of practical applications.

1.6 MEASUREMENTS AND UNITS IN CHEMISTRY

We mentioned in Section 1.1 that science relies essentially on quantitative data—numerical information. At the heart of any quantitative examination of our surroundings is the performance of operations called measurement.

Measurement is nothing more than counting. When we measure the value of a physical quantity, we either count that quantity or we count a ratio between two examples of the quantity. This counting usually is done with the help of some sort of measuring instrument.

Measurement probably is as old as civilization. There are records of meas-

TABLE 1.1
Units of Length

Unit	Origin	Equivalent in Meters (m)
cubit	the length from the tip of the middle finger to the elbow	0.46 m
inch	a thumb breadth or the length of three barleycorns, from the middle of the ear of barley	0.0254 m
foot	12 inches or a man's foot	0.3048 m
yard	3 feet or the length from the nose to the thumb of King Henry I of England	0.9144 m
furlong	220 yards or the length of a furrow in a common field	201.2 m
mile	1760 yards; originally 1000 Roman double-step paces	1609.3 m

urements that were made as far back as 6000 B.C. The first measurements probably were those of length, which illustrate the counting aspect of measurement well. If you want to measure the length of a room, a simple way is to pace it off, counting the number of times you place one foot in front of the other as you go the length of the room. You can measure other rooms in the same way, comparing the lengths of different rooms by comparing the number of feet counted in pacing off each room. In each case, you are using your foot as the common standard of comparison, so it is easy to compare the lengths of different rooms.

But a problem arises in trying to learn the length of a room that you cannot measure yourself. Suppose a friend agrees to pace off his room and compare its length to that of your room. If your room's length is 16 of your feet and the length of your friend's room is 15 of his feet, you have only part of the answer. You cannot compare the lengths of your rooms unless you know the relative lengths of your feet.

In spite of such problems, most early units of measurement and many that are still used are based on some dimension of the human body. Table 1.1 lists some units of length. It also lists their equivalence in meters, an important unit of length in modern science.

The Metric System

A table listing all the units of length that have been used around the world would go on for many pages. Such a table would be essential for any sort of international communication without an agreement to standardize measurements. There is such an agreement, in which virtually every major country has adopted the same units of length and many other basic physical quantities. The agreed-upon unit of length is the meter, and the other agreed-upon units are part of what originally was called the metric system.

The metric system was developed in France shortly after the French Revolution. It has two important features. One of them is an attempt to relate the units of measurement to natural phenomena. For example, the meter was originally defined as one ten-millionth of the length of a line running from the North Pole to the equator and passing through Paris.

The second important feature of the metric system concerns the relationship between different units for the same quantity. All these units are related decimally—that is, by powers of 10. This relationship usually is indicated by a prefix.

For example, the meter, the kilometer, and the centimeter are units of length in the metric system. One kilometer is 1000 meters and one centimeter is the

INCHING TOWARD STANDARDIZATION

The difficulty of achieving uniformity in weights and measures is illustrated by the long, complex process of negotiation by which the English-speaking countries agreed on a common inch and a common pound. Both are defined in terms of metric units, but the definitions differed until 1959.

Starting in 1893, the official inch in the United States was defined as 2.540005 cm. However, American industry became dissatisfied with the complicated conversion problem, so the American Standards Association in 1933 informally adopted a 2.54-cm inch for general use in industry. In Britain, meanwhile, the inch was defined as 2.539996 cm, slightly less than either of the American inches. The British value was adopted in 1922, when the bronze bar that had served as the national standard yard since 1878 was found to be shrinking. The Canadian inch was slightly different from the British and American inches, at 2.54 cm.

As for the pound, its value in Britain was equal to 0.453592338 kg. The value in the United States was 0.4535924277 kg, while the Canadian pound was 0.45359237 kg.

On July 1, 1959, a standard inch and a standard pound went into effect in all three countries, as well as in Australia, New Zealand, and South Africa. All agreed to adopt the Canadian inch of exactly 2.54 cm, and to set the pound equal to the Canadian value of 0.45359237 kg.

However, agreement was found to be impossible over a standard value for the gallon. The United States gallon is defined as equal to 231 cubic inches, while the British imperial gallon is the volume of 10 pounds of water under a fixed set of conditions. The imperial gallon is equal to 1.20094 United States gallons. The U.S. National Bureau of Standards said that "the standard United States gallon and the imperial gallon are so substantially different that a compromise international gallon was not practicable."

And despite the agreement on the inch, the U.S. Coast and Geodetic Survey said that it would continue to base all its maps and reference points on an inch of 2.540005 cm.

The agreement to standardize the inch and the pound was described as a step toward an ultimate adoption of the metric system by the English-speaking nations. Britain started to go metric in 1965, and Canada committed itself to metrication shortly afterward. The United States also is committed to metrication, although on an uncertain timetable.

hundredth part of a meter. You can see that conversion between these units of length requires no more than a shift in the decimal point. The simplicity of this relationship is evident by comparison with the units listed in Table 1.1. To convert inches into feet, we must multiply by 12. To convert feet into yards, we multiply by 3. To convert yards into miles, we multiply by 1760. By comparison, we can convert centimeters into meters and meters into kilometers by using the appropriate powers of 10.

The metric system has been used universally by scientists for some time. As science developed and new phenomena were investigated, different units were introduced. It was not uncommon at one time for different scientists to be using different units derived from the metric system for the same quantity. For example, among the units used to express a quantity of electric current were the ampere, the international ampere, the statampere, and the abampere. Listings such as Table 1.1, giving the relationship between these units became essential. One such table in a commonly used chemical handbook has more than 3000 entries and is far from complete.

The SI

To avoid confusion, attempts to extend and improve the metric system have been made since 1875. The result is a system, completed in 1969, which has been adopted by many nations and international scientific bodies, including the one that represents chemists. This system is called the International System of Units, or SI (from the initials of its French name). The use of SI is designed to bring order out of the chaos that previously existed among scientific units of measurement. However, the adoption of the SI has been and continues to be a slow process.

One reason is that old habits are hard to break, in science as in other fields. Another reason is that some units of the SI are not regarded as completely satisfactory by all scientists. Often SI units represent compromises between conflicting requirements of several disciplines. In chemistry, certain SI units are inconvenient, so the older non-SI units remain in almost universal use. In Chapter 4, for example, we shall see that the SI unit of pressure, the pascal, is particularly inconvenient for chemists. We shall use primarily the older unit for pressure, the atmosphere, which most chemists still use.

Finally, since a large body of chemical literature was written before the adoption of the SI, it is impossible to ignore the old units. Even today, much scientific literature includes non-SI units. We shall use the SI units as much as is consistent with current practice and convenience.

The SI is partially what is called a coherent system of units. A coherent system starts with a set of base units, that is, a set of independent physical quantities. There are seven such base units in SI. The units for all other physical quantities in a coherent system are derived by multiplication and division of the base units, without using any numerical factors, not even powers of 10. Table 1.2 lists the names and symbols of the seven SI base units.

The decision to set the number of base units at seven is arbitrary and is based to a large extent on convenience. By using seven units, it is relatively easy to carry out dimensional analysis, a technique by which relationships between physical quantities can be discerned from the relationships between their units.

TABLE 1.2
SI Base Units

Physical Quantity	SI Unit	Symbol for Unit
length	meter	m
mass	kilogram	kg
time	second	s
amount	mole	mol
thermodynamic temperature	kelvin	K
electric current		
luminous intensity	ampere	A
	candela	cd

The first five units in Table 1.2 are used in almost every branch of chemistry. The sixth, the unit for electric current, has more limited use. The seventh, the unit for luminous intensity, is rarely used in chemistry.

With the exception of the kilogram, the SI unit for mass, all the base units are defined very precisely in terms of appropriate natural phenomena. The kilogram is defined as the mass of a specific sample of platinum-iridium alloy that is kept at the International Bureau of Weights and Measures at Sèvres in France. However, a way of defining the kilogram in terms of a natural phenomenon is being sought.

The SI also includes a set of prefixes that are used to form decimal multiples and decimal fractions of the SI units. The units formed in this way are not coherent units, since they are derived by multiplying or dividing the base units by numerical factors. But they are SI units and are used quite correctly as part of the SI.

Table 1.3 lists SI prefixes, their meanings, and their symbols. You will note that prefixes are not provided for every power of 10. Allowed fractions smaller than 10^{-3} and allowed multiples larger than 10^3 all must have exponents that are divisible by 3.

The symbol for an SI prefix is placed before the symbol for an SI base unit or an SI-derived unit. The exception to this rule is the base unit of mass, the kilogram, which already has a prefix. For historical reasons, prefixes are added to the gram instead. Thus, one milligram is 1/1000 of a gram and 1/1 000 000 of a kilogram.

The combination of a prefix and a unit is regarded as a single symbol for a unit that is a decimal multiple or fraction of the original unit.

For example, the SI base unit for length is the meter, whose symbol is m. A centimeter, whose symbol is cm, is one-tenth (10^{-1}) of a meter. A micrometer, whose symbol is μm, is one-millionth (10^{-6}) of a meter. A kilometer, symbol km, is 1000 (10^3) meters, and a gigameter, symbol Gm, is one billion (10^9) meters.

As an example of a derived unit, we shall consider the physical quantity of area. Since area is determined by multiplying length $\times$ length, the unit for area is meter $\times$ meter, or square meter. Such a unit is written more conveniently by using an exponent; rather than writing square meter, we write meter^2 or, using the symbol, m^2. Exponents that are used with units have the same meaning as exponents used with a number or any other symbol. Thus, m^3 means meter $\times$ meter $\times$ meter, or m^3, while m^{-1} means 1/m, which can be read as ''per meter.''

The selection of the proper unit for a given measurement is quite important. The square meter (m^2), for example, is a convenient unit in which to express the area of a house or a farm. However, the square meter is too small a unit in which to express the area of a country conveniently. We are free, within the set of SI units, to use the unit square kilometer (km^2), which means kilometer $\times$ kilometer. For smaller areas, such as the page of this book, a more convenient unit is the square centimeter (cm^2). You should note that the square centimeter and the square kilometer are not regarded as coherent units based on the meter because they include numerical factors. But, they still are part of the SI and are perfectly valid.

TABLE 1.3
SI Prefixes

Fraction	Prefix	Symbol	Multiple	Prefix	Symbol
10^{-1}	deci	d	10	deka	da
10^{-2}	centi	c	10^2	hecto	h
10^{-3}	milli	m	10^3	kilo	k
10^{-6}	micro	μ	10^6	mega	M
10^{-9}	nano	n	10^9	giga	G
10^{-12}	pico	p	10^{12}	tera	T

TABLE 1.4
SI-Derived Units

Physical Quantity	SI Unit	Symbol of Unit	Definition of SI Unit
energy	joule	J	$kg\ m^2\ s^{-2}$
force	newton	N	$kg\ m\ s^{-2}$ (or $J\ m^{-1}$)
pressure	pascal	Pa	$kg\ m^{-1}\ s^{-2}$ (or $N\ m^{-2}$)
power	watt	W	$kg\ m^2\ s^{-3}$ (or $J\ s^{-1}$)
frequency	hertz	Hz (or s^{-1})	s^{-1}
electric charge	coulomb	C	$A\ s$
electric potential difference	volt	V	$kg\ m^2\ s^{-3}\ A^{-1}$ (or $J\ s^{-1}\ A^{-1}$)

Units for all other physical quantities can be derived coherently from the seven SI base units. Table 1.4 lists some physical quantities of interest to the chemist and the SI units for these quantities.

All the units in Table 1.4 have special names. Decimal multiples and fractions of these units may be formed by the use of SI prefixes. Many other units that do not have special names are also used by chemists. These units can always be derived coherently from the SI base units. However, in some cases it is more convenient to derive units from SI-prefixed units.

For example, the coherent SI unit of volume is the m^3, or cubic meter. One m^3 is a much larger volume than that normally encountered by most chemists. Therefore, chemists usually use the cm^3 or, in many cases, the dm^3. Formally, the dm^3 is called the cubic decimeter. However, the chemist has been using the dm^3 as a common unit of volume for many years and has become accustomed to calling it the liter, whose symbol is l. The liter is accepted in SI, and is allowed as an acceptable nickname for the dm^3. We shall use the liter extensively as a unit of volume in this book.

Conversion of Units

Even using SI units, it is often necessary to convert a value for a physical quantity measured in one unit into a value measured in another unit. In addition, since many non-SI units are still in use, we must also be able to convert between such units. Conversion between SI units only requires multiplication by a power of 10. Conversion between non-SI units may require numerical factors that are not powers of 10; these numerical factors must be obtained from a table of conversion factors.

The use of conversion factors can be regarded as purely an arithmetic technique. However, to avoid errors, it is helpful to understand exactly what is done when these conversions are performed.

The first question to ask in performing a conversion is: Which of the two units represents the larger physical quantity? The answer will show whether the conversion requires a multiplication by a number larger than 1 or smaller than 1.

For example, suppose your height is 183 centimeters, and you are told to express it in meters. A meter is a greater length than a centimeter. Therefore, the numerical value of your height in meters will be smaller than it is in centimeters. Thus, you must reduce 183 to a smaller value by multiplying by a number less than one.

The next step is to find the relationship between the two units. The prefix ''c'' in the symbol cm for centimeter tells us that 1 meter = 100 centimeters. This equation is the basis of all conversions between meters and centimeters. We can rewrite the equation in two ways:

$$\frac{1\ meter}{100\ centimeters} = 1 \quad \text{or} \quad \frac{100\ centimeters}{1\ meter} = 1$$

The fraction on the left side of each equation is a conversion factor. These conversion factors are used to convert between the two units, centimeters and meters. To convert from one unit to another, the given quantity is multiplied by the appropriate conversion factor. Note that this operation does not change the value of the quantity, because each conversion factor is equal to 1.

The final step is to decide which conversion factor should be used for a given conversion. There are several lines of reasoning that can give the correct answer. For the conversion of 183 centimeters to meters, you could keep in mind the fact that the numerical value in meters must be smaller than the numerical value in centimeters. Or, you could realize that the numerator of the conversion factor must be the desired unit and the denominator must be the original unit, which becomes clear if the units are written in each step of the calculation. By either line of reasoning, the first conversion factor is the correct one, and the desired conversion is:

$$183 \text{ cm} \times \frac{1 \text{ m}}{100 \text{ cm}} = 1.83 \text{ m}$$

It can be seen that the units of cm cancel, leaving only m, the desired unit.

EXAMPLE 1.1

What is the mass in grams and in milligrams of a 50-kg student?

SOLUTION

The relationship between grams (g) and kilograms (kg) is:

1 kg = 1000 g

Since a gram is a smaller mass than a kilogram, the mass of the student in grams must have a larger numerical value than the mass of the student in kilograms. The appropriate conversion factor is therefore 1000 g / 1 kg and:

$$50 \text{ kg} \times \frac{1000 \text{ g}}{1 \text{ kg}} = 50\ 000 \text{ g}$$

The units of kilograms (kg) cancel, leaving only grams (g).

The relationship between grams and milligrams is:

1 g = 1000 mg

and conversion from kilograms (kg) to milligrams (mg) can be carried out in one step by using two conversion factors:

$$50 \text{ kg} \times \frac{1000 \text{ g}}{1 \text{ kg}} \times \frac{1000 \text{ mg}}{1 \text{ g}} = 50\ 000\ 000 \text{ mg}$$

Again, it can be noted that all the units except milligrams (mg) cancel when the appropriate conversion factors are used. The use of conversion factors in the solution of chemistry problems will be developed more fully in Chapter 3.

To find relationships between derived SI units, we start with the relationship between the base unit and the unit with the prefix. The operation that converts the base unit to the derived unit is then carried out on both of these quantities. For example, suppose we want to find the relationship between cubic meters (m^3) and cubic centimeters (cm^3). The relationship between the meter and the centimeter is 1 m = 100 cm. We must raise both equivalent quantities in this equation to the third power:

$$(1 \text{ m})^3 = (100 \text{ cm})^3$$

or

$$1 \text{ m}^3 = 1\ 000\ 000 \text{ cm}^3$$

TABLE 1.5
Non-SI Units

Physical Quantity	Name of Unit	Symbol of Unit	Definition[a]
length	angstrom	Å	10^{-10} m
length	inch	in	2.54×10^{-2} m
mass	pound	lb	0.453 592 kg
force	dyne	dyn	10^{-5} N
pressure	atmosphere	atm	101 325 Pa
pressure	torr[b]	Torr	$\dfrac{101\ 325}{760}$ Pa
pressure	millimeters[b] of mercury	mmHg	13.5951×9.80665 Pa
energy	erg	erg	10^{-7} J
energy	calorie	cal	4.184 J

[a] These definitions can be changed into equations by setting the unit equal to the definition. Thus, 1 in = 2.54×10^{-2} m.
[b] These two units have essentially the same value.

Table 1.5 lists some non-SI units that are still likely to be encountered in chemistry, and their definitions in terms of SI units. It should be noted that some of these units are related to SI units by powers of 10. They are not included in SI because the power of 10 is not divisible by three, as required for an SI prefix. Nondecimal conversion factors are required for other units in Table 1.5. You will also find that SI prefixes are used with these non-SI units. Thus, 1 kcal = 1000 cal. Although the calorie is not an SI unit, the prefix k has its SI meaning. Similarly 1 ml = 1000 liters. Since 1 ml = 1 cm³ we shall prefer the latter SI unit.

One conversion that is often required involves temperature. In SI, the unit of temperature is the kelvin (K), which is used for the thermodynamic temperature scale. It is the only temperature scale that scientists should use, but in practice the Celsius temperature scale is also used. The temperature interval in both scales is the same: 1 K = 1°C. But the thermodynamic temperature T is 273.15 K higher than the Celsius temperature t:

$$T = t + 273.15 \text{ K}$$

Thus, to convert from degrees Celsius (°C) to thermodynamic temperature in kelvin (K), add 273.15 to the Celsius value. We shall discuss this point further in Section 4.3.

To convert between the units listed in Table 1.5 and SI units, we use the definitions listed in the table and the method outlined in Example 1.1. We must also include any necessary nondecimal factors. For example, to express 50 kg in pounds, we first find from Table 1.5 that 1 lb = 0.454 kg. Therefore, the conversion factor is 1 lb/0.454 kg, and:

$$50 \text{ kg} = 50 \text{ kg} \times \frac{1 \text{ lb}}{0.454 \text{ kg}} = 110 \text{ lb}$$

Significant Figures*

An expression of the magnitude of a physical quantity has two parts, a number and a unit. So far we have discussed the units. Let us now turn our attention to the numbers.

---

*A more detailed treatment of many of the arithmetic procedures used in general chemistry can be found in Appendix IV.

Almost no measurement of a physical quantity is exact. Generally, there is some error in the counting process, and this error affects the numerical value of the measurement. The number used to express the magnitude of a physical quantity should also indicate the margin of error.

The simplest way to indicate the margin of error in a measurement is by writing more or fewer digits in the number. The number of digits is called the number of **significant figures.** The last digit written in the number is understood to be uncertain.

Thus, if we say that a sample has a mass of 8 kg, we take it to mean that the mass of the sample is between 7.5 kg and 8.5 kg. If the mass is said to be 8.0 kg, we take it to mean that the mass is between 7.95 kg and 8.05 kg. If the mass is said to be 8.01 kg, we take it to mean that the mass is between 8.005 kg and 8.015 kg. In other words, the number of digits in the numerical part of the expression gives us information about the limits of precision of the measurement. We can say that the expression ''8 kg'' has one significant figure. The expression ''8.0 kg'' has two significant figures, and the expression ''8.01 kg'' has three significant figures.

There are some exceptions to this rule. Usually, a zero is a significant figure; there is a difference between writing ''8 kg'' and ''8.0 kg.'' However, the number of significant figures in whole numbers that have one or more zeros to the left of the decimal point and no zeros to the right of the decimal point can be ambiguous.

Consider the margin of error in a measurement of a distance such as 400 km. We do not know if the distance lies between 350 km and 450 km (one significant figure), between 395 km and 405 km (two significant figures) or between 399.5 km and 400.5 km (three significant figures). However, there is no ambiguity if we write 400.0 km. We know that the distance is between 399.95 and 400.05 km, since there are four significant figures.

Zeros that serve only to fix the position of the decimal point in numbers less than 1 are not counted as significant figures. Thus, there are two significant figures in 0.00023, while there are three significant figures in 0.000230, and there are six significant figures in 1.00023.

Ambiguity about significant figures can be avoided by using **exponential notation.** Exponential notation is a method of writing numbers that is especially useful for dealing with very large or very small numbers. In exponential notation, a number is expressed as the product of an ordinary number and a power of 10, in the form

$$N \times 10^x$$

where N is the ordinary number and x is the exponent, or power to which 10 is raised. If the decimal point is moved one space to the left, x is increased by one. If the decimal point is moved one space to the right, x is decreased by one. For example, $25\ 486 = 2.5486 \times 10^4$ and $0.000147 = 1.47 \times 10^{-4}$. Note that in numbers less than 1, the exponent, x, is negative, while in numbers greater than 1, x is positive.

In chemistry, N is written with only one integer to the left of the decimal point. The number of significant figures is one more than the number of integers to the right of the decimal point. Thus, the number 400 can be written as 4×10^2 to show one significant figure, as 4.0×10^2 to show two significant figures, and as 4.00×10^2 to show three significant figures.

When arithmetic operations are performed with numbers that represent measurements, the number of significant figures in the result can be determined by following these rules:

1. In addition or subtraction, the number of significant figures to the right of the decimal point in the final sum or difference is the lowest number of significant figures in any of the original numbers. The following examples illustrate this rule:

$$
\begin{array}{r}
12.234 \\
+\ 2.34 \\
\hline
14.57
\end{array}
\qquad
\begin{array}{r}
4.083 \\
+0.1 \\
\hline
4.2
\end{array}
\qquad
\begin{array}{r}
4 \times 10^2 \\
-30 \\
\hline
4 \times 10^2
\end{array}
\qquad
\begin{array}{r}
4.00 \times 10^2 \\
-3 \\
\hline
3.97 \times 10^2
\end{array}
$$

2. In multiplication and division, the final product or quotient cannot have more significant figures than are in the original number with the least number of significant figures. The following examples illustrate this rule:

$$4.3 \times 3.2 = 14 \qquad \frac{14.1}{13.8} = 1.02 \qquad 6.789 \times 0.000032 = 0.00022$$

During these operations, an extra significant figure usually is carried until the calculations are completed. The answer is then rounded off. However, in such a calculation as 3×15.99876, it is simpler to round off the original number before performing the calculation, since the answer will have only one significant figure. Care should always be taken to express the result of a calculation with the correct number of significant figures, since significant figures carry an implication about the precision of the measurements.

Measurements are so central to chemistry that care must be taken to express both parts of any measurement correctly. The correct number of significant figures tells us about the error limits of the measurement. The units that are used make the numerical value of the measurement meaningful.

EXERCISES

* 1.1[a] Classify each of the following statements as a hypothesis, an observation, a law, or a theory:
(a) Matter is composed of very small particles called atoms, whose properties determine the properties of the matter.
(b) The interior of the planet Saturn is made of green cheese.
(c) The moon is not made of green cheese.
(d) There is a force of attraction between opposite electric charges.
(e) Fossil records always show the development of invertebrates before vertebrates.
(f) Coffee is more expensive than tea.
1.2 Classify the following substances as elements, compounds, solutions, or mixtures:
(a) water, (b) earth, (c) fire, (d) very clean air, (e) air above a city, (f) peanut butter, (g) sugar, (h) whiskey, (i) oxygen.
1.3 Classify the following as intensive or extensive properties:
(a) volume, (b) density, (c) boiling point, (d) temperature, (e) mass, (f) energy,

(g) melting point.
1.4 Examine your surroundings and try to list 10 different nonsynthetic substances that you can see.
1.5 For each of the units listed below, suggest a physical quantity for which the unit is appropriate, indicate whether the unit is an SI or non-SI unit, and if it is an SI unit indicate whether or not it is a coherent unit:
(a) m/s, (b) km/hour, (c) minute^{-1}, (d) g/m^3, (e) kg/m^3 (f) atm, (g) kPa.
1.6 Suggest a convenient SI unit for the following quantities and indicate the corresponding coherent unit: (a) the distance from the earth to the moon, (b) the capacity of a can, (c) the mass of a diamond any of us is likely to encounter, (d) the area of the space inside this O, (e) the velocity of a snail.
1.7 Find your height measured in (a) meters, (b) centimeters, (c) feet, (d) inches, (e) yards, (f) cubits.
1.8 The density of water is 1 g/cm^3. Express the density of water in the following units: (a) kg/cm^3, (b) g/m^3, (c) kg/m^3, (d) g/liter.
1.9 The mass of Phobos, the larger of Mars' two moons, is 2.7×10^{16} kg. Express this mass using three other SI mass units.
1.10 A diver at a depth of 50 m is

subjected to a pressure of 6 atm. Express this pressure using three other units.
1.11 Dieters commonly encounter a unit called the ''calorie.'' This calorie is related to the calorie that is used as the unit of energy by: 1 food calorie = 1000 energy calories. A table of food composition lists the ''calorie content'' of 1 cup of roasted peanuts as 805 ''calories.'' Convert this value to joules.
1.12 The speed limit on a highway is posted as 50 mph. Find the speed limit expressed in SI base units.
1.13 Express all the data in today's weather forecast in SI units.
1.14 Express the following numbers in exponential notation: (a) 54 321, (b) 1 000 001, (c) 0.000401, (d) 0.0000010, (e) 765.32.
1.15 Use exponential notation to indicate the number of significant figures in the following statements:
(a) Fifty million Frenchmen can't be wrong.
(b) ''For every ten jokes, thou hast got an hundred enemies'' (Laurence Sterne, 1713–1768).
(c) The diameter of a cesium atom is three one hundred millionths of a centimeter.
(d) ''I see one-third of a nation ill-

---

[a] The answers to exercises whose numbers are in italics can be found in Appendix VII. The asterisk indicates an exercise that is more challenging than average.

housed . . ." (Franklin D. Roosevelt).
(e) One hundred dollars is a high price for that bicycle.

1.16 Perform the following calculations and report the results with the correct number of significant figures:

(a) $12.346 + 0.21 + 1.02$,
(b) $1.21 + 8.45 + 0.34$,
(c) $(123)(0.0037)$,
(d) $(1234)(4321)/5332$, (e) $424 - 25$.

1.17 There are exactly 60 seconds in a minute, there are exactly 60 minutes in an hour, there are exactly 24 hours in a mean solar day, and there are 365.24 solar days in a solar year. Find the number of seconds in a solar year. Be sure to give your answer with the correct number of significant figures.

Acid Amalgam Ammoniac Salt Antimony Arsenic-Sulphur Copper

Gold Hydrochloric Acid Iron Iron Filings Iron Vitriol Gravel

Borax Brass Brimstone Glass Lead Lead Sulphate

Lime Magnesia Mercury Metal Lime Minium Nickel

2

THE LANGUAGE OF CHEMISTRY

When you explore new territory, it is helpful to know the language. You probably know a little of the language of chemistry. The aim of this chapter is to give you a wider and deeper knowledge of the symbols, numbers, and names used by chemists. Sentences in English are made up of words. The ''sentences'' of chemistry are equations similar to those you have encountered in algebra, and are made up of symbols and numbers. The language of chemistry is designed to convey the maximum amount of information in the most orderly, precise, and concise way. By learning how chemists communicate, you will take an important first step toward learning how chemistry is done.

2.1 ATOMS, MOLECULES, AND FORMULAS

The world is a complicated place, composed of a bewildering array of materials that the untrained mind cannot possibly classify. Historically, one of the major achievements of chemistry has been to simplify this complexity, to show that the tremendous variety of animal, vegetable, and mineral substances on earth can be explained as different combinations of a relatively small number of basic ingredients.

These basic ingredients are the **chemical elements.** There are not very many of them. To date, 106 elements have been identified; 89 of them have been found to occur naturally on earth or in the atmosphere, and the rest are synthetic. Every object on earth—the table on which this book rests, the book itself, even the reader of the book—can be described as a collection of substances made up of these elements.

A sample of a chemical element can be divided and subdivided only to a point. The division and subdivision of such a sample can go on until one has a quantity far too small for the human senses to distinguish. But sooner or later, one arrives at the smallest quantity of an element that retains the identity of that element. **The basic unit of an element is an atom.** The chemical behavior of a single atom may not differ significantly from the behavior of a billion atoms of that element. But if you break an atom into pieces—a feat that became possible only a few years ago—you are left with entities of a completely different sort.

It is almost impossible to appreciate the minute size of this basic unit, the atom. A 1-cm cube of iron would fit comfortably on the tip of your thumb. This tiny cube contains 8.2×10^{22} atoms of iron (82 000 000 000 000 000 000 000 atoms). That is more than all the grains of sand on all the beaches of earth. The thumb on which the cube rests contains a substantially larger number of atoms. The individual to which the thumb is attached contains even more atoms.

Most substances do not exist as single atoms in nature. An exception to this rule is the family of elements known as noble gases, found in low concentrations in the atmosphere, which exist as single atoms. But we usually find that **atoms join together to form molecules** (Figure 2.1). The smallest molecules consist of only two atoms. The largest contain many thousands. Some elements normally exist as two-atom molecules. Among these are hydrogen, nitrogen, oxygen, fluorine, chlorine, bromine, and iodine. Two-atom molecules in which both atoms are of the same element are called *homonuclear diatomic molecules*. Some elements normally consist of molecules containing more than two atoms; molecules of sulfur contain eight atoms, and molecules of phosphorus contain four atoms, for example.

The largest molecules are those produced by living organisms. Some of the largest of these molecules are in viruses, which are on the threshold of life. A virus

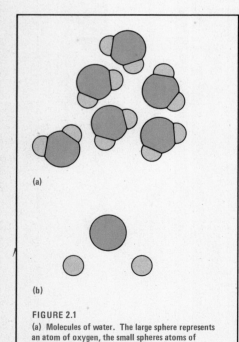

(a)

(b)

FIGURE 2.1
(a) Molecules of water. The large sphere represents an atom of oxygen, the small spheres atoms of hydrogen. The smallest drop of water contains billions of such molecules. (b) A molecule of water has been split. It is no longer a molecule of water, but rather two separate hydrogen atoms and a single oxygen atom.

typically consists of a long molecule, called a nucleic acid, with a protein coat. If the coat is stripped away, the long, stringy nucleic acid molecule can be seen under extremely high magnification in an electron microscope. This molecule may contain half a million atoms, but it is insignificantly small by the standards of everyday life. You would need 10^{15} (one million billion) individual molecules of this size to make a 1-cm cube.

You may be momentarily bewildered by the miniscule size and enormous quantities of atoms and molecules. But atoms and molecules are enormously helpful in chemistry. They allow us to describe any substance in terms of a few fundamental units of structure. If there were no such fundamental units, the chemical properties of a substance could vary with the size of a sample of that substance. This idea was in fact intellectually defensible for a long time, in the early days of modern chemistry. Chemists at all levels should be grateful that the properties of a pure substance do not vary as the quantity of the substance changes, because that makes classroom exercises, laboratory experiments, and life much simpler.

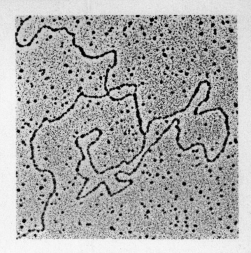

A molecule of nucleic acid, containing many tens of thousands of atoms. Nucleic acids and other molecules created by living organisms are the largest that are known to exist. (*Louise Tsi Chow, Coldspring Harbor Laboratories*)

Symbols and Formulas

To simplify work with atoms and molecules, a system of symbols is used. Each element is represented by a symbol of one or two letters. These symbols usually consist of the first letter of the English or Latin name* of the element; this letter is capitalized. Since the names of many elements start with the same letter, a second letter, uncapitalized, is added when necessary to distinguish one element from another. Thus, the symbol for hydrogen is H; for helium He; for boron B; for beryllium Be. Examples of symbols derived from Latin names are listed in Table 2.1.

An element's symbol can mean several things. It can be used as an abbreviation in a sentence. In more formal scientific writing, a symbol can mean one atom of the element. If we write, "Among the compounds containing Fe are . . .," the symbol Fe stands for any quantity of iron. But the symbol Fe means not only the element iron but also a single atom of iron, the fundamental unit of structure in any quantity of iron.

Symbols are used to indicate the elements in molecules. Most molecules are made up of atoms of different elements. If we want to discuss the molecule made of one atom of oxygen and one atom of carbon, we could call it carbon monoxide. We could also call it CO, which is simpler. Similarly, it is easier to write "HCN"

TABLE 2.1
Elemental Symbols from Latin Names

English Name	Symbol	Latin Name
antimony	Sb	stibium
copper	Cu	cuprum
gold	Au	aurum
iron	Fe	ferrum
lead	Pb	plumbum
mercury	Hg	hydrargentum
potassium	K	kalium
silver	Ag	argentum
sodium	Na	natrium
tin	Sn	stannum

*The symbol W, for element 74, wolfram, (called tungsten in English) is an exception. It comes from *wolfrumb*, the early German word for "wolf turnip," because the metal is found in turniplike lumps.

than to write "hydrogen cyanide," the name given to a molecule containing one atom of hydrogen, one atom of carbon, and one atom of nitrogen. The combination of symbols representing the atoms in a molecule of a compound is called a *formula*. For example, CO is the formula for a molecule of the compound carbon monoxide and HCN is the formula for a molecule of the compound hydrogen cyanide.

Most molecules contain more than one atom of a given element, so we need a way to specify the number of atoms of any element in a molecule. This is done by using *subscripts*, small numerals that are written below and to the right of symbols in a formula. The subscript indicates the number of atoms of that element in the molecule.

For example, a molecule of the oxygen in the air we breathe consists of two joined atoms of oxygen. The formula for this molecule is O_2. We can also find small traces of atomic oxygen, whose symbol is O. A third form of oxygen that is important in atmospheric studies is ozone, a three-atom molecule whose formula is O_3. Again, the formula for hydrazine, an important rocket fuel, is N_2H_4. A single molecule of hydrazine contains two atoms of nitrogen and four atoms of hydrogen. You will not find the subscript 1 in a chemical formula. If a molecule has only one atom of an element, that symbol has no subscript, as in CO, the formula for carbon monoxide, or O, the formula for atomic oxygen.

Numerals sometimes are written to the left of chemical formulas. These numerals are called *coefficients*. They specify the relative number of molecules taking part in a chemical reaction. You have no trouble distinguishing between coefficients and subscripts. In the formula Cl_2, the numeral 2 is a subscript that indicates that we have a two-atom molecule of chlorine. But if we write 2Cl, we have two separate atoms of chlorine (Figure 2.2). Similarly, CO_2 is one molecule of carbon dioxide, while 2CO is two molecules of carbon monoxide. To repeat: **a coefficient** (a numeral written to the left of a chemical symbol) **indicates the number of separate particles. A subscript** (a numeral written to the right of and below a symbol) **indicates the number of atoms of a given element in a single molecule.**

FIGURE 2.2
The subscript 2 used with the symbol Cl indicates one molecule of Cl_2, left. The coefficient 2 used with the symbol Cl indicates two Cl atoms, right.

2.2 THE LAW OF CONSTANT COMPOSITION

One of the major controversies in late eighteenth-century chemistry raged between Claude-Louis Berthollet (1749–1822), one of the leading chemists of the time, and Joseph-Louis Proust (1754–1826), a relative unknown. On the basis of experimental work, Berthollet believed that the composition of a pure substance could vary from sample to sample. Proust, on the other hand, maintained that **all samples of a pure substance contain the same elements in the same proportions.** To defend his idea, Proust offered much data obtained in many years of experimentation. It took more than a decade for Proust to convince other chemists that he was right. The idea championed by Proust is now called **the law of constant composition.**

As we shall see in Chapter 5, this law has been thoroughly verified by modern studies of the atomic structure of matter.

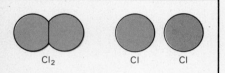

Berthollet.

Claude-Louis Bethollet (1749–1822), born of poor parents and trained as a physician at the University of Turin, was one of the first to accept Lavoisier's theories. He met Napoleon on a trip to Egypt in 1798, taught him chemistry and was made a senator and a count. Among Berthollet's achievements were the determination of the composition of ammonia and the introduction of chlorine as a bleaching agent. (*Culver*)

EXAMPLE 2.1
Analysis of a sample of pure table salt, sodium chloride, whose mass is 3.20 g shows that the sample contains 1.26 g of sodium and 1.94 g of chlorine. How much sodium does 451 g of table salt contain?
SOLUTION
The analysis shows that the fraction of sodium in the sample of sodium chloride is 1.26/3.20. The law of constant composition requires that any sample of pure sodium chloride have the same fractional content, 1.26/3.20, of

sodium. Therefore, 451 g of sodium chloride contains

$$451 \times \frac{1.26}{3.20} = 177 \text{ g sodium}$$

The best way to carry out this calculation is to include all the units:

$$\text{mass sodium} = 451 \text{ g sodium chloride} \times \frac{1.26 \text{ g sodium}}{3.20 \text{ g sodium chloride}}$$
$$= 177 \text{ g sodium}$$

2.3 THE LAW OF MULTIPLE PROPORTIONS

John Dalton (1766–1844) is remembered primarily for his atomic theory, which we shall discuss in Chapter 5. He also made one of the most important observations in the early investigation of the chemical composition of substances.

In Dalton's day, the fact that two elements can form different compounds under different conditions was already known. We have already mentioned an example: Carbon and oxygen can form either CO, carbon monoxide, or CO_2, carbon dioxide, depending on conditions. Studying the composition of such compounds, Dalton found a relatively simple relationship between the masses of the elements. On the basis of his experiments, he formulated what is now called the law of multiple proportions, which states **If two elements unite to form more than one compound, the different masses of one element which combine with a fixed mass of the other element are in a ratio of small whole numbers.**

The compounds of carbon and oxygen illustrate the law of multiple proportions. An analysis of 1.00 g of carbon monoxide shows that it consists of 0.57 g of oxygen and 0.43 g of carbon. The ratio of oxygen to carbon is 1.33; put another way, 1.33 g of oxygen combine with 1 g of carbon. An analysis of 1.00 g of carbon dioxide shows that it consists of 0.73 g of oxygen and 0.27 g of carbon. The ratio of oxygen to carbon here is 2.67, which means that 2.67 g of oxygen combine with 1 g of carbon. In these two compounds, the ratio of the masses of oxygen that combine with 1 g of carbon is 2.67:1.33, or 2:1. The law of multiple proportions requires this result for samples of any size.

Joseph-Louis Proust (1754–1826) was the son of an apothecary and established himself in that business while he did research in chemistry. In addition to helping to establish the validity of the law of definite proportions, Proust also pioneered in the study of sugars, distinguishing several different kinds. He spent a great part of his life in Spain to avoid the upheavals that followed the French Revolution but returned to France to end his career. (*Edgar Fahs Smith Collection*)

EXAMPLE 2.2

Analysis of a sample of MnO shows that the ratio of Mn to O by mass, expressed in integers, is 55:16. Calculate the ratio by mass of Mn to O in MnO_2 and Mn_2O_7.

SOLUTION

The law of constant composition requires that any sample of MnO have the same ratio by mass of Mn to O, 55:16. Suppose our sample consists of a single molecule, containing one atom of Mn and one atom of O. The ratio of the mass of the Mn atom to the mass of the O atom must be 55:16. For the sake of convenience, it is possible to assign a mass of 55 to a single atom of Mn and a mass of 16 to a single atom of O, if we use a suitably small unit of mass. We can then say that in MnO_2, which contains two O atoms for each Mn atom, the ratio of Mn to O will be 55:(16 × 2) = 55:32. In Mn_2O_7, the ratio by mass of Mn to O will be (2 × 55):(16 × 7) = 110:112 = 55:56). (Note that in these compounds, the masses of O that combine with a mass of 55 Mn have the ratio 16:32:56 = 2:4:7.)

EXAMPLE 2.3

Analysis of still another compound of manganese and oxygen reveals a ratio of Mn to O of 55:24. What is the simplest formula of this compound?

John Dalton (1766–1844), the son of a weaver, was a practicing Quaker. His chemical studies began while he was teaching at a Quaker school. He first became interested in meteorology, which he studied with instruments he built himself. Weather observations led to studies of the composition of air and then of other gases. Dalton published his atomic theory in 1803. His Quaker principles made him refuse many proffered honors, but he did accept a pension from King William IV. (*Granger*)

SOLUTION

In Example 2.2, a mass of 55 was arbitrarily assigned to Mn and a mass of 16 was assigned to oxygen, for convenience. In the compound we are now considering, one atom of Mn, represented by 55, is combined with a mass equivalent to $1\frac{1}{2}$ atoms of O. This combination suggests the formula $MnO_{1.5}$. However, subscripts in formulas must always be whole numbers, since atoms cannot be subdivided. If we multiply the subscripts in the tentative formula by the same factor, the mass ratio will remain the same. The smallest factor that will convert 1.5 to a whole number is 2, and so we multiply by 2. The correct formula is therefore Mn_2O_3.

A more extended example of the law of multiple proportions is found in the oxides of nitrogen (Table 2.2). It can be seen from the table that the ratio of the masses of oxygen that combine with 28 g of nitrogen is $16:32:48:64:80$, or $1:2:3:4:5$.

TABLE 2.2
The Law of Multiple Proportions

Compound	Formula	Mass of Nitrogen	Mass of Oxygen	Relative Mass of Oxygen
nitrous oxide	N_2O	28	16	1
nitric oxide	NO	28	32	2
dinitrogen trioxide	N_2O_3	28	48	3
nitrogen dioxide	NO_2	28	64	4
dinitrogen pentoxide	N_2O_5	28	80	5

The law of multiple proportions was particularly important at the time it was introduced because it promoted acceptance of Dalton's atomic theory. Acceptance did not always come readily. Some scientists steadfastly refused to accept the existence of atoms. They considered the atom to be a useful concept, but they would not believe that anything as small as an atom could exist. It took the experiments of twentieth-century physicists to produce the evidence that finally convinced the diehards that atoms are real things.

2.4 CHEMICAL NOMENCLATURE

Numerals and symbols are not enough for a full description of the materials and substances of chemistry. We also need names for these compounds.

The names used in chemistry have varied origins. Some are the result of a logical system based on chemical composition. But many originate from simple custom based on long usage. Many familiar substances have names of the second kind, so-called *common names*. The older common names have nothing to do with chemical composition, but they remain in use by sheer force of habit. After all, could we call H_2O anything but water? Who would use ''sodium chloride'' to refer to table salt? Baking soda, washing soda, aspirin—these are only a few of the compounds that are rarely called by their chemical names.

Common names have one major disadvantage: They convey a chemical meaning only to someone who already knows the meaning. You know what ''water'' is chemically, but a person who speaks only French might not. And since common names arose haphazardly, there is no system to help us memorize them.

Fortunately, most of the substances used by chemists are named systematically—that is, their names are derived from their composition according to fixed rules. By learning a relatively small number of rules, it is possible to name a great

ATOMS OBSERVED

The achievements of twentieth-century microscopy have banished any remaining doubts about the existence of atoms by producing images of single atoms. In the 1950s, an instrument called the field ion microscope was used to create an image of the atomic structure of the tip of a needle. The needle is placed in a tube, and its temperature is lowered to about 20 K. A gas such as helium is introduced into the tube, and a high positive voltage is applied to the tip of the needle. The combination of extreme cold and the high voltage causes the atoms of the gas to become electrically charged and to recoil from the atoms in the tip of the needle. The gas atoms strike a fluorescent screen, creating a pattern that is identical to the atomic structure of the tip of the needle. Magnifications of up to 750 000 are possible with the field ion microscope.

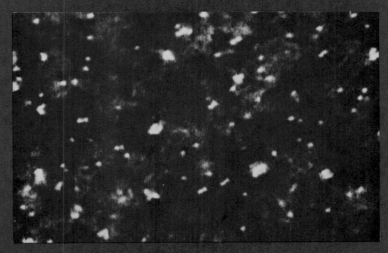

In 1976, physicists at the University of Chicago went one step further and produced motion pictures of the motion of single atoms. Using a scanning electron microscope of special design, the physicists were able to get images of single uranium atoms placed on a specimen of carbon 500 nm thick. The motion pictures show the uranium atoms as blurred bright spots against the dark background of the carbon, magnified about 5.5 million times. Images taken in sequence show the slow thermal motion of the uranium atoms.

many substances from their chemical formulas, and to obtain chemical formulas from names.

The study of the rules used to name substances is called **chemical nomenclature.** Learning chemical nomenclature is like learning to spell: You must learn both the rules and the exceptions. As you might expect, historical custom creates most of the exceptions. Usually, it is the more common compounds that have irregular chemical names. The irregularities exist because the rules of nomenclature developed over long periods of time. Chemists meet periodically to simplify the rules and eliminate as many irregularities as possible, but that work is far from complete.

The basic guideline in chemical nomenclature is to give each chemical compound a name that is unique and that distinguishes it from all other compounds. Two things must be done to meet this need. First, the name must specify the composition of the substance, the number and types of atoms it contains. The second requirement arises from the fact that two substances may have the same type and number of atoms and still be different, because the atoms are arranged differently in the molecules. The names must distinguish between such substances.

The rules of nomenclature that we will study are primarily concerned with specifying the composition of substances. To arrive at the correct name of a compound, you must know more than its formula. You must also have a good understanding of how chemical elements behave, how they are classified, and how they combine to form compounds.

Classifying the Elements

All of the elements, the 89 that occur in nature and the 17 created by scientists, are classified by an arrangement called the **periodic table** (see inside front cover). The periodic table is the most important tool ever devised for understanding the elements. We shall discuss it in detail in Chapter 6, but a brief introduction here will help you understand the rules of chemical nomenclature.

All the elements can be divided roughly into three types (Figure 2.3): the metals, the semimetals, and the nonmetals. This is not a rigid, precise division; sometimes elements are not easily assigned to one group or another. In making the assignment, it is necessary to consider not only chemical behavior but also the physical structure and electrical properties of an element.

The electrical properties of an element are particularly important in making the classification. All elements that are good conductors of electricity and whose conductivity decreases as the temperature rises are defined as **metals.** If an element is a poor conductor of electricity, and if its conductivity increases as its temperature rises, it is defined as a **semimetal.** (Semimetals include silicon and germanium. They are also called **semiconductors**, and modern electronics is based on their characteristics.) Any element that does not conduct electricity is a **nonmetal.**

As for physical state, almost all metals are solids at room temperature. The one exception to this rule is mercury, which is a liquid at room temperature. The semimetals are also solids at room temperature, although their solid structure differs in important ways from that of the metals. Some nonmetals are gases at room temperature (H_2, N_2, O_2, F_2, Cl_2). Most of the others are solids that usually do not resemble one another. Carbon can exist as a hard, transparent solid called *diamond* or a soft, black solid called *graphite*. Iodine, another nonmetal, is a solid whose color can vary from violet to black, and which is composed of I_2 molecules. Sulfur, which is commonly composed of S_8 molecules, is a yellow solid. The only nonmetal that is a liquid at room temperature is bromine, which is composed of Br_2 molecules; its color is an intense red-brown. The nonmetallic elements that occupy the column at the far right of the periodic table are all monoatomic gases: helium (He), neon (Ne), argon (Ar), krypton (Kr), xenon (Xe), and radon (Rn), called the noble gases.

It is important to note that each of these groups occupies a specific area of the periodic table. The nonmetals are found toward the top and right of the periodic table (hydrogen is an exception). The most important of the nonmetals include the halogens (fluorine, chlorine, bromine, and iodine), oxygen, sulfur, selenium, nitrogen, phosphorus, carbon, and the gases occupying the far-right column in the periodic table, headed by helium.

The semimetals occupy a diagonal band starting at the upper left of the periodic table. They include, in addition to silicon and germanium, arsenic, tellurium, and boron. There are a number of elements that are close to the

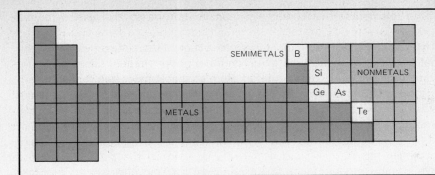

semimetals in the periodic table whose properties make their classification as metals or nonmetals rather arbitrary. An example is phosphorus, which exists in several different forms whose characteristics differ. *White phosphorus,* the most common form, consists of P_4 molecules, is very reactive, does not conduct electricity, and acts as a typical nonmetal. *Red phosphorus* is not as vigorously nonmetallic. Its structure is similar to that of a nonmetal. It is usually a nonconductor, but does conduct electricity if it contains arsenic as an impurity. *Black phosphorus* looks like a metal and conducts electricity.

When you have identified the nonmetals and semimetals, you will be left with the majority of elements, occupying the greater part of the periodic table. These are the metals.

Writing Formulas of Compounds

The three-part classification of metal, semimetal, and nonmetal is the starting point for the rules that govern the naming of compounds and the way that formulas for compounds are written. You may not know these rules, but you probably have used them already, since most of us follow the rules in writing about common substances, even without special training in chemistry. You would never write the formula for water as OH_2; if you happen to use the chemical name for table salt you almost certainly will call it "sodium chloride," not "chloride sodium." You can arrive at the correct formula and name for many compounds by following a relatively small number of rules. These rules work especially well for binary compounds (those containing two elements), and they are useful for all compounds. The rules are:

1. In a compound containing a metal and one or more nonmetals, the metal is written and named first. *Example:* NaCl, sodium chloride.

2. In binary compounds containing two nonmetals or semimetals, the nonmetals have the same order from left to right in the formula as they do in the periodic table. The one exception to this rule is oxygen, which comes last in a formula even when combined with Cl, Br, or I, which are to its right in the periodic table. *Examples:* NF_3, P_4O_6, SiO_2, ClO_2, BrO_2, I_2O_5, and OF_2.

3. When binary compounds contain nonmetals that are in the same column of the periodic table, the element closest to the bottom of the table is written first. *Examples:* SO_2, SeS_2, ClF_3, $BrCl_5$, SiC.

4. In a binary compound of hydrogen, the symbol for hydrogen is written *first* when hydrogen is combined with elements from columns VIA and VIIA of the periodic table and *last* when hydrogen is combined with elements from columns IA through VA. *Examples:* LiH, B_2H_6, CaH_2, NH_3, H_2S, HBr.

Valences and Oxidation Numbers

During the nineteenth century, as chemists were investigating the composition of substances and developing the chemical formulas for those substances, it became clear that each element had a characteristic capacity for combining with other elements. This combining capacity was called the *valence* of the element.

The valence of an element is a small integer. That integer is either the number of hydrogen atoms with which an atom of the element can combine or the number of hydrogen atoms that an atom of the element can replace when forming a compound. The valence of oxygen in H_2O is 2, since the oxygen atom combines with two hydrogen atoms. The valence of bromine in HBr is 1, since the bromine atom combines with one hydrogen atom. The valence of magnesium in MgO is 2; if we compare MgO with H_2O, we see that the Mg atom has replaced two hydrogen atoms.

When more modern ideas about chemical bonding and atomic structure were developed, the use of valence as an expression for the combining power of the elements became outmoded—or, to be more precise, the concept of valence was replaced and extended. We now use **oxidation numbers.** These usually are integers that can have either positive or negative signs. The oxidation number of an element conveys the same information about the element's combining power as the valence does, but the oxidation number gives additional information about chemical behavior. As we shall see, oxidation numbers are related to the electrical charge on atoms. We shall discuss the meaning of oxidation numbers at length in Chapter 7. For the moment, we shall use oxidation numbers to help us determine the formulas and names of many compounds of metals and nonmetals.

The basic rule for all compounds is that the sum of the positive oxidation numbers must equal the sum of the negative oxidation numbers. To give this rule practical value, you must memorize some facts about oxidation numbers of some elements in their compounds.

1. Any element in column IA of the periodic table always has an oxidation number of $+1$. Any element in column IIA always has an oxidation number of $+2$. Any element in column IIIA generally has an oxidation number of $+3$. Elements in column IVA generally have oxidation numbers of $+4$. There are exceptions to some of these rules. Some of the heavier elements in columns IIIA and IVA can have more than one oxidation number. For instance, Tl can have an oxidation number of $+1$ or $+3$; Sn and Pb can have oxidation numbers of either $+2$ or $+4$.

2. When they combine with metals, elements in column VA usually have oxidation numbers of -3, elements in column VIA have oxidation numbers of -2, and elements in column VIIA usually have oxidation numbers of -1. The usual oxidation number of oxygen is -2.

3. Hydrogen has an oxidation number of $+1$ when combined with most nonmetals and -1 when combined with metals.

Table 2.3 shows how oxidation numbers can serve as guides to correct formulas.

TABLE 2.3
Formulas from Oxidation Numbers

Element	Oxidation Number	Element	Oxidation Number	Formula
K	$+1$	Br	-1	KBr
Ca	$+2$	F	-1	CaF_2
Al	$+3$	O	-2	Al_2O_3
C	$+4$	S	-2	CS_2
Mg	$+2$	N	-3	Mg_3N_2

Many of the elements near the center of the table in columns IB, IVB–VIIB, and VIII usually have more than one oxidation number, so more than one binary compound can be formed by one of these elements and a nonmetal. The most common oxidation numbers for these elements are $+2$ and $+3$, but there are more complex cases. Manganese, for instance, can have oxidation numbers of +2, +3, +4, +5, +6, and +7. You should be certain that you have all the necessary facts when you write formulas for compounds of these elements.

Polyatomic Groups

There are some groups of atoms that act as a single unit in chemical reactions. You will encounter such groups of atoms often. These polyatomic groups, as they are called, are always found in combination with other elements or groups of elements. Although the oxidation numbers of the atoms in a molecule add up to zero, the sum of the oxidation numbers of atoms in a polyatomic group is not zero. A polyatomic group carries an electrical charge, either positive or negative. This charge is the sum of the oxidation numbers of the atoms of the polyatomic group.

An atom or group of atoms that carries an electrical charge is called an ion. An ion with a positive charge is called a **cation.** An ion with a negative charge is called an **anion.** The charge of an ion (+ or −) is written as a superscript after the formula of the ion. A monoatomic ion, containing a single atom, carries a charge that is identical to the atom's oxidation number. Some important polyatomic groups and their ionic charges are listed in Table 2.4.

You will note that all of these groups are anions, with the sole exception of ammonium, NH_4^+. The anionic groups tend to form compounds with hydrogen and with metals.

The formulas of compounds containing polyatomic groups are written to emphasize the fact that these groups are independent units. If the formula of a substance contains more than one unit of any polyatomic group, the formula for the group is written in parentheses, and the subscript is written outside the parentheses. This means that the subscript applies to the group as a unit. Some examples are $Ba(HSO_4)_2$, $(NH_4)_2CO_3$, $Ca_3(PO_4)_2$.

TABLE 2.4
Important Polyatomic Groups

Formula	Name
HCO_3^-	bicarbonate
CO_3^{2-}	carbonate
$C_2H_3O_2^-$	acetate
CN^-	cyanide
NO_3^-	nitrate
NO_2^-	nitrite
NH_4^+	ammonium
$H_2PO_4^-$	dihydrogen phosphate
HPO_4^{2-}	hydrogen phosphate
PO_4^{3-}	phosphate
OH^-	hydroxide
O_2^{2-}	peroxide
O_2^-	superoxide
HSO_4^-	hydrogen sulfate
SO_4^{2-}	sulfate
$S_2O_3^{2-}$	thiosulfate
HSO_3^-	hydrogen sulfite
SO_3^{2-}	sulfite
ClO_4^-	perchlorate
ClO_3^-	chlorate
ClO_2^-	chlorite
ClO^-	hypochlorite
MnO_4^-	permanganate
CrO_4^{2-}	chromate
$Cr_2O_7^{2-}$	dichromate

HOW COMPOUNDS ARE NAMED

The International Union of Pure and Applied Chemistry (IUPAC) consists of national organizations representing chemists in more than 40 countries. The United States member is the National Academy of Sciences. The governing body of IUPAC is a council of delegates that meets every two years, together with commissions and task forces on specific questions of importance to chemists. The responsibilities of IUPAC include standardization of the table of atomic weights and the establishment of definitive rules of nomenclature for all classes of chemical compounds.

In organic chemistry, for example, IUPAC's Commission on the Nomenclature of Organic Chemistry first met in Geneva in 1892, proposing rules that were revised and extended at meetings in Liège (1930), Lucerne (1936), and Rome (1938). An extensive revision and updating of the rules took place after World War II, and the current rules were issued in 1957.

These rules provide specific names for a large number of compounds. For example, the name of the saturated unbranched acyclic hydrocarbon with one carbon atom is specified as methane, while the member of the same family with 132 carbon atoms is named dotriacontahectane.

The rules for nomenclature of inorganic chemistry also were adopted in 1957, after a number of meetings in which comments and criticisms were considered. These rules include procedures for naming the most complex compounds. For example, the name of the compound $[Fe(CN)_2(CH_3NC)_4]$ is specified to be dicyanotetrakis (methyl isocyanide) iron(II), while the name of the compound $Na(UO_2)_3[Zn(H_2O)_6](C_2H_3O_2)_9$ is designated as hexaaquazinc sodium triuranyl(VI) nonacetate.

IUPAC acknowledges that its rules are an attempt to achieve "the best compromises" among differing points of view and various national usages. Periodic revisions of the rules of nomenclature are made by the commissions of IUPAC in light of new developments.

Some Rules of Chemical Nomenclature

In naming compounds, we use the same information about oxidation numbers and the same classification of elements into metals, nonmetals, and semimetals that we use when writing formulas. There are a few relatively simple rules, which are complicated by exceptions handed down from the past. The rules are:

1. In a binary (two-element) compound that contains (a) a metal or semimetal that commonly displays only one oxidation number in any of its compounds and (b) a nonmetal, we write the name of the metal first, followed by the stem of the name of the nonmetal with the suffix *ide* added. For example, LiF is lithium fluoride, $CaCl_2$ is calcium chloride, Al_2O_3 is aluminum oxide, Li_3N is lithium nitride, Ag_2S is silver sulfide. KI is potassium iodide. Table 2.5 lists the stems of the names of some elements.

TABLE 2.5
Stems Used in Naming Compounds of Nonmetals

Element	Stem	Element	Stem
C	carb	S	sulf
N	nitr	F	fluor
P	phosph	Cl	chlor
As	arsen	Br	brom
O	ox	I	iod

2. With metals or nonmetals that can have more than one oxidation number, a roman numeral is written in parentheses after the name of the metal to give its oxidation number in the specific compound. *Examples:* FeO, iron(II) oxide; Fe_2O_3, iron(III) oxide; SnS, tin(II) sulfide; SnS_2, tin(IV) sulfide; $TiCl_2$, titanium(II) chloride; $TiCl_4$, titanium(IV) chloride.*

3. Binary compounds of hydrogen and a nonmetal are named in the same way as compounds of a metal and a nonmetal, with *hydrogen* substituted for the name of the metal. *Hydrogen* comes first, followed by the stem of the nonmetal and the suffix *ide*. Thus, HCl is hydrogen chloride, H_2S is hydrogen sulfide, and HI is hydrogen iodide. The exceptions are hydrogen compounds that have common names, such as methane (CH_4), silane (SiH_4), ammonia (NH_3), water (H_2O), and phosphine (PH_3).

Some of these binary compounds are usually found dissolved in water; their common names are then different. For instance, HF in water is hydrofluoric acid, HCl in water is hydrochloric acid, HBr in water is hydrobromic acid, and HI in water is hydriodic acid.

4. Names for compounds of a metal with a polyatomic group have the name of the metal followed by the name of the polyatomic group. If the metal has more than one oxidation number, the correct oxidation number is given by a roman numeral in parentheses. *Examples:* $Cr(OH)_3$, chromium(III) hydroxide; Ag_2CO_3, silver carbonate; KCN, potassium cyanide; $Na_2Cr_2O_7$, sodium dichromate; $CuSO_4$, copper(II) sulfate; $Mg(HSO_4)_2$, magnesium hydrogen sulfate; NaH_2PO_4, sodium dihydrogen phosphate; $LiC_2H_3O_2$, lithium acetate.

5. Binary compounds containing two nonmetals have the name of the first element in the formula followed by the stem of the name of the second element with the suffix *ide*. The number of atoms of each element in the compound is indicated by a prefix. Some of these prefixes are:

mono = 1	tetra = 4
di = 2	penta = 5
tri = 3	hexa = 6

(The prefix *mono* is often omitted; no prefix means that the compound contains only one atom of the element.) *Examples:* ICl, iodine chloride; NO_2, nitrogen dioxide; Cl_2O, dichlorine monoxide; PCl_3, phosphorus trichloride; CF_4, carbon tetrafluoride; N_2O_5, dinitrogen pentoxide; P_4O_6, tetraphosphorus hexoxide.

---

*There is an older method that uses different suffixes for metals to indicate oxidation numbers. In this method, if a metal has only two common oxidation numbers the suffix *ous* indicates the lower number and the suffix *ic* indicates the higher number. Thus, $CoCl_2$ would be called cobaltous chloride and $CoCl_3$ would be called cobaltic chloride; $FeCl_2$ would be ferrous chloride and $FeCl_3$ would be ferric chloride; $SnCl_2$ is stannous chloride and $SnCl_4$ is stannic chloride. This system is more complicated and less precise than the modern method agreed upon by members of the International Union of Pure and Applied Chemistry (IUPAC). Fortunately, the older method is being abandoned.

Some compounds of two nonmetals have common names that are frequently used and must be memorized. Important examples are N_2O, nitrous oxide, and NO, nitric oxide.

EXAMPLE 2.4

Write the formula and name for a binary compound of each of the following pairs of elements: (a) lithium and bromine, (b) calcium and iodine, (c) aluminum and sulfur, (d) carbon and chlorine, (e) hydrogen and fluorine.

SOLUTION

The formulas are written by using the known oxidation numbers of the elements and following the rule that the sum of the oxidation numbers in a substance must be zero. The names are written by examining the formulas and following the preceding rules. The specific rule used is given in parentheses:

(a) $LiBr$ lithium bromide (rule 1)
(b) CaI_2 calcium iodide (rule 1)
(c) Al_2S_3 aluminum sulfide (rule 1)
(d) CCl_4 carbon tetrachloride (rule 5)
(e) HF hydrogen fluoride (rule 3)

EXAMPLE 2.5

Name the following: (a) $MnSO_4$, (b) $V(CN)_4$, (c) $Ba(HCO_3)_2$, (d) N_2O_5.

SOLUTION

The names are written by following the appropriate rules. If it is necessary to find the oxidation number of a metal in a substance, use the fact that the sum of the oxidation numbers must be zero.

(a) By rule 2, the name is manganese(II) sulfate. The roman numeral is necessary because manganese displays more than one oxidation state. Since the sulfate ion has a charge of -2, the Mn atom in the compound has an oxidation number of $+2$.

(b) By rule 2, the name is vanadium(IV) cyanide. The oxidation number of vanadium is found to be $+4$ because the compound has four cyanide groups, each with a charge of -1.

(c) By rule 4, the name is barium bicarbonate. No roman numeral is needed because barium displays only one oxidation state.

(d) By rule 5, the name is dinitrogen pentoxide.

2.5 BALANCED CHEMICAL EQUATIONS

Giving compounds proper formulas and names is only one part of chemical language. We also need a concise and accurate method of describing how substances are converted into other substances by chemical reactions. This is done by writing balanced chemical equations.

A balanced chemical equation is a precise, nonverbal statement of the conversion of one or more substances, called **starting materials** or **reactants**, into one or more new substances, called **products.**

Chemical equations have most of the characteristics of the algebraic equations used in mathematics, with chemical formulas replacing the x and y of algebra. The most noticeable difference is that the chemical equation usually has an arrow instead of an equal sign. The arrow and the equal sign have the same meaning, to a point. They both say that the quantities on one side of the equation equal the quantities on the other side of the equation. The arrow also indicates the direction of chemical change. By tradition, the formulas of the reactants appear to the left of the arrow, with the formulas of the products to the right.

Ordinary chemical processes obey the **law of conservation of mass**; the total mass of the reactants must equal the total mass of the products. Furthermore, transmutation of elements, the conversion of one element to another, does not occur in an ordinary chemical process. The same number and type of atoms that appear on one side of a chemical equation must appear on the other side; all the atoms of all the elements found in the reactants must be found in the products of the chemical reaction.

To write a balanced chemical equation, you need a knowledge of the reactants, of the products, and of arithmetic, the last to ensure that the equation is indeed balanced.

As an example, we shall examine one of the important reactions that occurs in the pollution control devices of automobiles: the conversion of carbon monoxide to carbon dioxide by reaction with oxygen. This change can be written

$$CO + O_2 \longrightarrow CO_2$$

The formula for oxygen is written as O_2 because oxygen occurs as a diatomic molecule in air. As it is written, the expression has three atoms of oxygen on the left and only two on the right. Obviously, this expression is not balanced. It does designate the reactants and products of the equation. But it can be read to say that one molecule of CO reacts with one molecule of O_2 to produce one molecule of CO_2. This cannot be so, since it violates the law of conservation of mass. We cannot balance an equation by adding new products; the only product of the reaction is known to be CO_2. We must balance the number of oxygen and carbon atoms on each side of the expression without changing either the reactants or the product.

We obtain a balanced equation by using *coefficients,* numerals that precede the formulas in the equation. An equation is balanced by answering one question: What are the smallest coefficients that will give the same numbers of each type of atom on each side of the equation? Many equations are balanced simply by inspection. Others require careful calculation. There is no simple rule covering all equations.

In the unbalanced expression

$$CO + O_2 \longrightarrow CO_2$$

we could start by assuming that the right side is deficient in oxygen atoms, and that perhaps two molecules of CO_2 are formed. The expression would then read

$$CO + O_2 \longrightarrow 2CO_2$$

But this expression is not balanced. It has four oxygen atoms and two carbon atoms on the right side, while there are three oxygen atoms and one carbon atom on the left. It can be balanced by assuming that two CO molecules take part in the reaction. The equation

$$2CO + O_2 \longrightarrow 2CO_2$$

then reads that two molecules of CO can combine with one molecule of O_2 to form two molecules of CO_2. This equation is consistent with conservation of mass, since it has the same numbers of each type of atom on each side of the equation. As you will see in Chapter 3, this equation also gives a great deal of information about the mass relationships of the substances in this reaction.

Sometimes several steps are needed to write correct equations, as in the case of the chemical reaction describing what happens when propane gas burns. Combustion is the rapid reaction of a substance with oxygen, usually liberating large amounts of energy in the form of heat. Propane is C_3H_8. It combines with O_2 to form two products of combustion, CO_2 and H_2O. The reactants and products are

$$C_3H_8 + O_2 \longrightarrow CO_2 + H_2O$$

This expression is not balanced. Coefficients are needed for some or all of the formulas. A useful first step in balancing this expression is to select the substance in the process that has the largest number of atoms, in this case C_3H_8. Add coefficients to balance the numbers of atoms in this compound on both sides of the expression. This gives

$$C_3H_8 + O_2 \longrightarrow 3CO_2 + 4H_2O$$

Another step is needed because the number of oxygen atoms is unbalanced; there are two oxygen atoms on the left and ten on the right. The oxygen atoms can be balanced by multiplying the O_2 on the left by 5, giving a balanced equation:

$$C_3H_8 + 5O_2 \longrightarrow 3CO_2 + 4H_2O$$

Now consider the equation that describes the thermite reaction, by which aluminum reacts with various metal oxides at high temperatures. The thermite reaction is ignited by a magnesium fuse and generates an immense amount of heat. It is used for welding and, in wartime, for incendiary bombs. The products of the reaction are aluminum oxide and the free metal from the metallic oxide, so the thermite reaction can be used industrially to prepare pure metals from oxides. The reaction for the production of manganese by this method is

$$Al + MnO_2 \longrightarrow Al_2O_3 + Mn$$

The most convenient way to obtain a correct equation is to start by considering the formula for aluminum oxide. There are two atoms of Al on the right and only one on the left, so we place a coefficient of 2 before the Al on the left. Balancing the oxygen atoms is not as easy. The oxygen-containing molecule on the left, MnO_2, has two oxygen atoms per molecule, while the oxygen-containing molecule on the right, Al_2O_3, has three oxygen atoms. A balance can be achieved by using a fractional coefficient, $\frac{3}{2}$, on the left, giving $\frac{3}{2}MnO_2$, to balance the oxygen atoms, and using the same fractional coefficient on the right to balance the manganese atoms:

$$2Al + \tfrac{3}{2}MnO_2 \longrightarrow Al_2O_3 + \tfrac{3}{2}Mn$$

Equations with fractional coefficients are used occasionally. But it is usually more convenient to multiply both sides of such equations by the denominator of the fraction (in this case, 2), to give an equation with whole-number coefficients:

$$4Al + 3MnO_2 \longrightarrow 2Al_2O_3 + 3Mn$$

Every chemical equation must be balanced. It is also desirable that a chemical equation be *net*—that is, the equation should contain only substances that undergo change in the chemical reaction. Writing the formulas for substances that do not undergo change in the reaction merely clutters the equation without giving information about the reaction. In this respect, the chemical equation is analogous to an algebraic equation, which is also reduced to its simplest form. An equation written $2x + y = 6 + y$ would automatically be changed to $2x = 6$. Similarly, if hydrogen that is mixed with helium burns to form water, it is not proper to write

$$2H_2 + O_2 + He \longrightarrow 2H_2O + He$$

The correct equation is

$$2H_2 + O_2 \longrightarrow 2H_2O$$

An ability to correlate and interpret chemical equations is essential for an understanding of chemistry and for the solution of chemistry problems. In addition, equations transmit a great deal of information about chemical reactions, in a concise form.

EXAMPLE 2.6

Convert the following expressions into chemical equations:

(a) $H_2O_2 \rightarrow H_2O + O_2$, (b) $XeF_6 + H_2O \rightarrow XeO_3 + HF$, (c) $P + Cl_2 \rightarrow PCl_3$, (d) $P + Cl_2 \rightarrow PCl_5$.

SOLUTION

(a) First balance the O atoms:

$$2H_2O_2 \longrightarrow 2H_2O + O_2$$

Both the H and the O atoms are balanced by this first step, which is sufficient.

(b) First balance the F atoms:

$$XeF_6 + H_2O \longrightarrow XeO_3 + 6HF$$

Next balance the O atoms:

$$XeF_6 + 3H_2O \longrightarrow XeO_3 + 6HF$$

This is the desired result.

(c) First balance the Cl atoms:

$$P + 3Cl_2 \longrightarrow 2PCl_3$$

Next balance the P atoms:

$$2P + 3Cl_2 \longrightarrow 2PCl_3$$

This is the desired equation.

(d) First balance the Cl atoms:

$$P + 5Cl_2 \longrightarrow 2PCl_5$$

Then balance the P atoms for the desired equation:

$$2P + 5Cl_2 \longrightarrow 2PCl_5$$

2.6 IMPORTANT CLASSES OF COMPOUNDS

Most of the compounds that you will encounter in beginning chemistry can be grouped into a relatively small number of classes, where the substances in each class have many properties in common. This method of classification is quite useful, because it lets us describe the essential properties of many compounds in a few words. The classification is based on the way atoms are held together in compounds, and on the behavior of compounds when dissolved in water.

Salts

A salt can be defined as a compound of a metal and a nonmetal, or of a metal with a negative polyatomic group, such as those in Table 2.4. Compounds that have an ammonium group (NH_4^+) instead of a metal are also classified as salts. Some salts are NaCl, $Mg(HSO_4)_2$, $KMnO_4$, $KAl(SO_4)_2$, NH_4Cl, and $(NH_4)_3PO_4$.

A salt is an ionic solid at room temperature. We can describe most salts by saying that they each have two ionic components: (a) a cation, which can be a polyatomic group such as ammonium or a monoatomic metal ion such as Na^+, K^+, Ca^{2+}, or Mn^{3+}, and (b) an anion, which can be a negative polyatomic group such as those listed in Table 2.4 or a monoatomic ion such as F^-, Cl^-, S^{2-}, or N^{3-}. This description helps to determine the structural features of solid salts and their behavior when dissolved in water.

A salt is held together by the attractive forces of its negative and positive ions. For example, NaCl is composed of Na^+ cations and Cl^- anions. Any amount of NaCl contains equal numbers of Na^+ and Cl^- ions. If the numbers were not equal, sodium chloride in bulk would have a detectable net electrical charge. No such charge is detectable. Another salt, $Ca(HSO_4)_2$, contains Ca^{2+} cations and HSO_4^- anions. It is obvious that any quantity of $Ca(HSO_4)_2$ must contain half as many Ca^{2+} as HSO_4^- ions, in order to be electrically neutral. Again, ammonium chloride, NH_4Cl, consists of equal numbers of NH_4^+ (ammonium) cations and Cl^- (chloride) anions, whereas $(NH_4)_3PO_4$ contains three times as many NH_4^+ cations as PO_4^{3-} (phosphate) anions. A complex salt such as $KAl(SO_4)_2$ consists of K^+, Al^{3+}, and SO_4^{2-} in the ratio of $1:1:2$; the total negative charge of -4 is balanced by a total positive charge of $+4$.

The fact that solid salts exist in ionic form creates a possible ambiguity in the interpretation of the formula of a salt. The formula NaCl, for example, might be regarded as representing a single "molecule" of sodium chloride. But solid sodium chloride does not consist of NaCl molecules. It consists of Na^+ and Cl^- ions in a lattice that has each Na^+ ion surrounded by six Cl^- ions and each Cl^- ion surrounded by six Na^+ ions, as shown in Figure 2.4. The formula of a salt defines the smallest whole-number ratio of ions that make up the salt, but it does not describe a molecule that is a discrete unit of structure. This does not affect the way we write the formulas of salts, but it does affect our interpretation of those formulas.

A solid salt consists of ions in close association. When the salt dissolves in water, the ions are separated. Substances that exist as ions in solution are called **electrolytes.** The existence of electrolytes is reflected in chemical notation. When sodium chloride dissolves in water, the correct formula is no longer NaCl but $Na^+ + Cl^-$. When ammonium phosphate dissolves in water, its formula is no longer written as $(NH_4)_3PO_4$ but as $3NH_4^{2+} + PO_4^{3-}$. These formulas treat the component ions of the salts as independent entities, which is approximately what they behave like in water solution. Salts are called *strong electrolytes*, because they usually separate completely into ions in water.

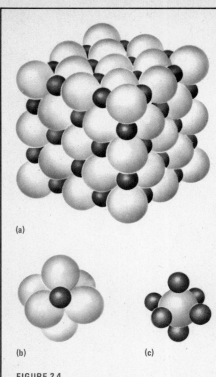

(a)

(b)   (c)

FIGURE 2.4

(a) A sodium chloride crystal consists of a lattice of sodium ions and chloride ions. (b) Each sodium ion is surrounded by six chloride ions. (c) Each chloride ion is surrounded by six sodium ions.

Acids

By the classic definition, an acid is a compound that is a source of H^+ ions. This is a useful introduction, although the definition is rather narrow and not quite correct, as you will see in Chapter 15.

An acid usually is a compound of hydrogen and a nonmetal or a negative polyatomic group. The word *acid* usually is part of its name. Unlike salts, acids usually are not aggregates of ions. An acid may be a gas (HCl, hydrogen chloride), a liquid (H_2SO_4, sulfuric acid), or a solid ($H_2C_2O_4$, oxalic acid). Like salts, acids tend to form ions when they dissolve in water. When a substance separates into ions, it is said to *dissociate*. Some acids dissociate completely; they are called *strong acids*. Table 2.6 lists some common strong acids and the ions they form (the positive ions in the table are listed as H^+ for convenience). As in the case of

TABLE 2.6
Common Strong Acids

Formula	Name	Ions in Water
HNO_3	nitric acid	$H^+ \ NO_3^-$
H_2SO_4	sulfuric acid	$H^+ \ HSO_4^-$
$HClO_4$	perchloric acid	$H^+ \ ClO_4^-$
HCl	hydrochloric acid	$H^+ \ Cl^-$
HBr	hydrobromic acid	$H^+ \ Br^-$
HI	hydriodic acid	$H^+ \ I^-$

salts, the formulas for strong acids dissolved in water are written to reflect their complete dissociation into ions. For example, the formula for nitric acid is HNO_3. When nitric acid dissolves in water, we write $H^+ + NO_3^-$.

Most acids dissociate only partially when dissolved in water. These are called *weak acids;* they are weak electrolytes. We do not write a formula that shows dissociation into ions because the major part of the dissolved weak acid is the undissociated molecule. Very few acids are classified as strong. If you do not know that an acid is strong, assume that it is weak.

Bases

A useful, if limited, definition describes a base as a compound that is a source of OH^- ions in water solution. By this definition, a compound of a cation and the OH^- (hydroxide) anion is a base.

Bases resemble salts in many ways. They, too, are ionic solids that dissociate into ions when dissolved in water. Bases that contain a cation and OH^- generally dissociate completely in water; they are classified as strong bases. Some strong bases are NaOH (sodium hydroxide), KOH (potassium hydroxide), and $Ba(OH)_2$ (barium hydroxide).

Compounds that do not contain hydroxide ions are defined as bases if they produce OH^- ions by reaction with water. The best known of these compounds is ammonia, NH_3, which reacts with water to produce hydroxide ions by the reaction

$$NH_3 + H_2O \rightleftharpoons NH_4^+ + OH^-$$

This reaction proceeds only partially, so ammonia is classified as a weak base. (The double arrow in the equation indicates that the reaction proceeds in both directions. We can form ammonia and water by starting with ammonium ions and hydroxide ions.)

Nonelectrolytes

Compounds containing only nonmetals usually exist as discrete molecules, rather than as collections of ions. These compounds usually do not dissociate into ions when they dissolve in water. They are called nonelectrolytes.

Most of the compounds that you will study in organic chemistry are nonelectrolytes. You usually can recognize organic molecules by their formulas. Organic compounds always contain carbon and either hydrogen or a halogen, one of the elements from column VIIA of the periodic table. An organic compound may also contain nitrogen, oxygen, phosphorus, or sulfur. The simpler organic compounds usually do not contain metals.

Many nonelectrolytes will not dissolve appreciably in water. Oil, for example, is a mixture of nonelectrolytes, and it is well known that oil and water don't mix. But some nonelectrolytes will dissolve in water, although they will not dissociate into ions. Sugar is one such nonelectrolyte. Another is ethyl alcohol, C_2H_5OH. Vodka is nothing more than a half-and-half mixture of ethyl alcohol and water. Alcohol does not dissociate into ions when it dissolves in water.

Oxides

An oxide is a binary compound of any element with oxygen, when the oxygen has an oxidation number of -2. Almost every element forms at least one oxide. The properties of oxides vary widely. An oxide may have properties resembling those of a salt, of an acid, of a base, or of a nonelectrolyte.

The oxides of metals with only one valence, such as those in columns IA and IIA of the periodic table, are ionic solids resembling salts. However, when these oxides dissolve in water they form hydroxide ions, behaving like strong bases. These oxides of metals are called *basic oxides*.

When sodium oxide, Na_2O, dissolves in water, the reaction

$$Na_2O + H_2O \longrightarrow 2Na^+ + 2OH^-$$

takes place, forming the same solution that is formed when NaOH dissolves in water. And when barium oxide, BaO, dissolves in water, we have the same solution as when barium hydroxide, $Ba(OH)_2$, dissolves:

$$BaO + H_2O \longrightarrow Ba^{2+} + 2OH^-$$

Oxides of nonmetals are not ionic compounds. But when they dissolve in water, they often produce the same solution that is obtained by dissolving a related acid in water. The compounds that produce these solutions are called *acidic oxides*. One of them is N_2O_5, which dissolves in water to produce the same solution that is produced by nitric acid, HNO_3:

$$N_2O_5 + H_2O \longrightarrow 2H^+ + 2NO_3^-$$

The basic oxides behave as though they were bases with the water removed. The acidic oxides behave as though they were acids with water removed. They are often called *anhydrides* of bases or acids (from *anhydrous*, meaning ''without water'').

Oxides of metals that have more than one oxidation number are ionic solids. Such metal oxides may have complex reactions in water. As a rough generalization, we can say that an oxide produces acidic solutions when the metal has one of its high oxidation numbers and basic solutions when the metal has one of its low oxidation numbers. For example, manganese has an oxidation number of $+7$ in Mn_2O_7, which dissolves in water to give an acidic solution:

$$Mn_2O_7 + H_2O \longrightarrow 2H^+ + 2MnO_4^-$$

But manganese has an oxidation number of $+2$ in MnO, a compound that behaves like a basic oxide in water. Many oxides of metals are very insoluble in water. One example is iron(III) oxide, commonly known as rust. These oxides are not classified by their behavior in water but by the chemical reactions they undergo.

2.7 IMPORTANT TYPES OF CHEMICAL REACTIONS

What seems to be a bewildering variety of chemical reactions can be simplified somewhat by a classification that places reactions in specific categories. Many of the reactions studied in chemistry can be placed into one of the following classes:

1. Addition or combination reactions: Two substances combine to form one:

$$2Na + Cl_2 \longrightarrow 2NaCl$$
$$P_4 + 3O_2 \longrightarrow P_4O_6$$
$$H_2O + C_2H_4 \longrightarrow C_2H_5OH$$
$$NO_2 + NO_2 \longrightarrow N_2O_4$$

2. Decomposition reactions: One compound breaks into two or more compounds or elements. These reactions are the reverse of addition or combination reactions.

$$CaCO_3 \longrightarrow CaO + CO_2$$
$$N_2O_4 \longrightarrow 2NO + O_2$$
$$Mg(OH)_2 \longrightarrow MgO + H_2O$$

3. Displacement reactions: Substances exchange parts. There are many different types of displacement reactions. One of the most important, often called *metathesis*, is the exchange of ions by two ionic compounds, with the anion of one compound joining the cation of the other compound and vice versa. A generalized formula for this exchange is

$$AB + CD \longrightarrow AD + CB$$

This kind of displacement reaction is especially important for compounds in aqueous solution. We shall discuss them in more detail in Section 2.8. One such reaction is

$$AgNO_3 + NaI \longrightarrow AgI + NaNO_3$$

Many other reactions resemble metathesis, although they do not necessarily involve ionic substances. Some of them are:

(a) Hydrolysis, the reaction of a substance or ion with water. We can write the formula for water as HOH, to emphasize the fact that water is a source of H^+ and OH^- ions. By doing this, we can predict the products of a hydrolysis reaction. The OH^- anion combines with the positive portion of the compound that is hydrolyzed. This positive portion may be a cation or an atom with a positive oxidation number. The H^+ cation combines with the negative portion of the compound, which may be an anion or an atom with a negative oxidation number:

$$AlCl_3 + 3HOH \longrightarrow Al(OH)_3 + 3HCl$$
$$LiH + HOH \longrightarrow LiOH + H_2$$
$$PBr_3 + 3HOH \longrightarrow P(OH)_3 + 3HBr$$

If these reactions are carried out with a large excess of water, so that the products are in aqueous solution when the reaction is completed, the formulas for the strong acids and bases, such as LiOH and HBr, are written to show dissociation into ions.

(b) Acid-base reactions. Generations of chemists have learned the catchword: Acid plus base gives water plus salt. This is true of bases that contain hydroxide. The reaction can be regarded as a simple switch of partners between compounds, as in these examples:

$$HCN + NaOH \longrightarrow NaCN + H_2O$$
$$2HF + Ca(OH)_2 \longrightarrow CaF_2 + 2H_2O$$

In Chapter 15, where more general definitions of acids and bases are given, you will see that acid-base reactions are not necessarily metatheses.

2.8 NET IONIC REACTIONS

Until now, we have written chemical equations in a form that conveys information only about the relative quantities of the reactants and products of reactions. We can write formulas in a way that conveys more information, especially about the physical state of the reactants and products. It is better to write formulas that give the physical state of substances, because this information is often quite important in determining what happens in a reaction. The rate at which a chemical process proceeds, the heat produced by the process, and other variables can depend on whether each reactant is a solid, a liquid, or a gas, or is in aqueous solution.

Information about the state of a reactant is given by a letter in parentheses after the formula. The symbol (l) indicates that a substance is a liquid, as in $Br_2(l)$, $H_2SO_4(l)$, $C_2H_5OH(l)$. The symbol (s) indicates that a substance is a solid, as in $NaCl(s)$, $Ba(OH)_2(s)$, $AgI(s)$. The symbol (g) indicates that a substance is a gas, as in $H_2(g)$, $NH_3(g)$, $HCl(g)$.

For some reactions, we need information about conditions and the state of reactants and products at these conditions to write the correct state symbols. For example, at 500 K, carbon monoxide and water form carbon dioxide and hydrogen:

$$CO(g) + H_2O(g) \longrightarrow CO_2(g) + H_2(g)$$

TABLE 2.7
Common Gases at Room Temperature

Name	Formula	Name	Formula
hydrogen	H_2	carbon monoxide	CO
nitrogen	N_2	carbon dioxide	CO_2
oxygen	O_2	sulfur dioxide	SO_2
fluorine	F_2	nitrous oxide	N_2O
chlorine	Cl_2	nitric oxide	NO
hydrogen chloride	HCl	nitrogen dioxide	NO_2
hydrogen bromide	HBr	methane	CH_4
hydrogen iodide	HI	ethane	C_2H_6
hydrogen sulfide	H_2S	propane	C_3H_8
ammonia	NH_3	butane	C_4H_{10}

elements from column O: He, Ne, Ar, Kr, and Xe

In this expression, H_2O is followed by (g) because water boils at 373 K, which means that it is a gas at 500 K. If the reaction occurred at 350 K, the equation would have H_2O(l) instead.

Reactions often occur at room temperature, which is assumed to be 298 K or 25 °C. If the conditions of a reaction are not specified, we can assume that the reaction occurs at or near room temperature. Therefore, we must know the state of a substance at room temperature to assign the correct state symbol for such a reaction.

Most substances are solids at room temperature, a fact you can verify by looking around you. All metals except mercury, all semimetals, and some nonmetals such as iodine, phosphorus, sulfur, and carbon are solids at room temperature. So are all salts and most other compounds that contain metals, as well as most organic compounds and many compounds of semimetals and nonmetals. A relatively small number of common substances are gases at room temperature. They are listed in Table 2.7.

Some organic compounds are liquids at room temperature. Among them are carbon tetrachloride, CCl_4; chloroform, $CHCl_3$; and benzene, C_6H_6. Bromine, Br_2, is also a liquid at room temperature.

By using state symbols, we can write chemical equations for changes in state. One example of a change in state is the change from liquid to gas that occurs when water boils:

$$H_2O(l) \longrightarrow H_2O(g)$$

A different change in state occurs when water is cooled below 273 K, and it becomes the solid called ice:

$$H_2O(l) \longrightarrow H_2O(s)$$

A change in state from solid to gas occurs when dry ice, solid carbon dioxide, is kept at room temperature:

$$CO_2(s) \longrightarrow CO_2(g)$$

State symbols can also be used to indicate whether a substance is in aqueous solution. In this text, we shall adopt the convention that **any substance or ion whose formula is not followed by a state symbol is assumed to be in aqueous solution.** There is another convention that uses the state symbol (aq) to indicate that a substance is dissolved in water. Since general chemistry deals primarily with substances in aqueous solution, we shall use the convention that allows us to use fewer symbols.

Omission now is important. The meaning of

$$C_2H_5OH(l) \longrightarrow C_2H_5OH$$

is the same as the meaning of

$$C_2H_5OH(l) \longrightarrow C_2H_5OH(aq)$$

but the first expression conveys the meaning in fewer symbols. The expression describes alcohol dissolving in water. In the same way, the equation

$$C_6H_{12}O_6 \longrightarrow C_6H_{12}O_6(s)$$

describes the crystallization of glucose from aqueous solution.

We mentioned earlier that complete dissociation to ions occurs when salts, strong acids, or strong bases dissolve in water. The formulas of these substances reflect that fact. When $NaCl(s)$ dissolves in water, it is no longer $NaCl(s)$. And it is not $NaCl$ or $NaCl(aq)$. It is $Na^+ + Cl^-$. An equation showing the reaction of sodium chloride in solution may include Na^+ or Cl^-; it will not include $NaCl$ or $NaCl(s)$. We can generalize and say that the formula for any salt that is not in solution is followed by a letter subscript, and that the formula for a salt in solution must reflect the fact that it has dissociated into ions. Thus, the formation of a solution of silver chloride, a salt, is described by

$$AgCl(s) \longrightarrow Ag^+ + Cl^-$$

The precipitation of calcium fluoride, another salt, from solution is described by

$$Ca^{2+} + 2F^- \longrightarrow CaF_2(s)$$

The formulas for strong acids and strong bases that dissociate into ions when in solution are written in the same way. Hydrogen chloride is $HCl(g)$ when pure and $H^+ + Cl^-$ when in solution. Sulfuric acid is $H_2SO_4(l)$ when pure and $H^+ + HSO_4^-$ (with a little SO_4^{2-}) in solution. Potassium hydroxide is $KOH(s)$ when pure and $K^+ + OH^-$ in solution. The formation of a solution of sulfuric acid is described by

$$H_2SO_4(l) \longrightarrow H^+ + HSO_4^-$$

and the precipitation of chromium(III) hydroxide is described by

$$Cr^{3+} + 3OH^- \longrightarrow Cr(OH)_3(s)$$

The treatment of aqueous solutions of weak acids and bases is simpler. Only a small fraction of each compound dissociates into ions when dissolved in water, so we indicate an aqueous solution by omitting the state symbol. The formation of a solution of ammonia is described by

$$NH_3(g) \longrightarrow NH_3$$

and the formation of a solution of acetic acid is described by

$$HC_2H_3O_2(l) \longrightarrow HC_2H_3O_2$$

When writing a net ionic equation, you must write the formula for each substance in the equation to indicate:

1. Whether the substance is in solution. This is done by using a state symbol if the substance is a liquid, solid, or gas, and by using no state symbol if the substance is in solution.
2. If the substance is in solution, whether it exists as ions. This is done by writing the formula for the substance either as an intact molecule or as ions.

One other detail is important in writing a correct net ionic equation. The equation must be **net**—that is, **it must include only substances that take part in the reaction.** The need to write net equations was mentioned earlier (Section 3.5). But it is not always easy to write a net equation.

Consider the metathesis reaction in which a solution of lead(II) nitrate is mixed with a solution of lithium iodide to yield a precipitate of lead(II) iodide and a solution of lithium nitrate. To write the net ionic equation for this process, we start by writing the formulas for the reactants and products in their proper states. A solution of lead(II) nitrate contains Pb^{2+} and NO_3^- ions. A solution of lithium iodide contains Li^+ and I^- ions. A solution of lithium nitrate contains Li^+ and NO_3^- ions. A precipitate of lead(II) iodide is $PbI_2(s)$. We can write

$$Pb^{2+} + 2NO_3^- + Li^+ + I^- \longrightarrow PbI_2(s) + Li^+ + 2NO_3^-$$

This expression is easily balanced:

$$Pb^{2+} + 2NO_3^- + 2Li^+ + 2I^- \longrightarrow PbI_2(s) + 2Li^+ + 2NO_3^-$$

But this equation is unnecessarily complicated because it is not *net*. As you can see, Li^+ and NO_3^- appear on both sides of the equation, but nothing has happened to them in the reaction. Therefore, they should be canceled from both sides of the equation. The correct net equation is

$$Pb^{2+} + 2I^- \longrightarrow PbI_2(s)$$

You might protest that this net equation omits a central feature of the reaction, the presence of lithium iodide. It is true that the final equation makes it impossible to tell whether sodium iodide, potassium iodide, or yet another iodide took part in the reaction. We can answer this objection by saying that, for the purposes we have in mind, most iodides are created equal. At the level of approximation we shall use, it makes no difference which iodide was the source of the I^- ions—or, indeed, which lead(II) salt was the source of the Pb^{2+} ions. The feature of interest of this reaction, the formation of a precipitate of lead iodide, will be the same for all these solutions of salts, at least until one is working on a more sophisticated level.

This commonsense approach simplifies the discussion of many chemical reactions. Take the reaction in which hydrochloric acid is mixed with a solution of sodium hydroxide. If all the substances are included, the equation is

$$H^+ + Cl^- + Na^+ + OH^- \longrightarrow H_2O + Na^+ + Cl^-$$

but the net equation is

$$H^+ + OH^- \longrightarrow H_2O$$

Or take the case in which a solution of nitric acid is mixed with a solution of barium hydroxide, forming water and barium nitrate as the products. If all the substances are included, the expression reads

$$H^+ + NO_3^- + Ba^{2+} + 2OH^- \longrightarrow H_2O + Ba^{2+} + NO_3^-$$

but the net equation is

$$H^+ + OH^- \longrightarrow H_2O$$

Indeed, this is the net equation for the reaction of any strong acid with any strong base.

The method of writing equations described in this chapter makes it possible to convey a great deal of information in a small amount of space. Consider these four net ionic reactions involving the same two substances:

(1) $NH_3(g) + HCN(l) \longrightarrow NH_4CN(s)$
(2) $NH_3(g) + HCN \longrightarrow NH_4^+ + CN^-$
(3) $NH_3 + HCN(l) \longrightarrow NH_4^+ + CN^-$
(4) $NH_3 + HCN \longrightarrow NH_4^+ + CN^-$

Each equation describes a different form of the reaction of ammonia with hydrogen cyanide to form the salt ammonium cyanide. In equation (1), the reaction occurs in the absence of water, and each substance has a state symbol.

The following three equations describe different reactions in aqueous solutions. In equation (2), ammonia gas is added to a solution of HCN, a weak acid, and the salt forms as dissociated ions in solution. In equation (3), liquid HCN is added to a solution of ammonia, a weak base. And in equation (4), both reactants are in solution. Many words are needed to convey differences that are described by simply making small changes in the notation.

EXERCISES

2.1 Write the chemical formulas of the following elements in the common form found under normal conditions: nitrogen, fluorine, sodium, phosphorus, sulfur, bromine, and tin.

2.2 Use the data given in Example 2.1 to find the mass of sodium contained in 32 g of sodium chloride.

2.3 Use the data given in Example 2.1 to find the mass of chlorine contained in 86 g of sodium chloride.

*2.4 A 7.83-g sample of HCN is found to contain 0.290 g of H and 4.06 g of N. Find the mass of carbon in a sample of HCN with a mass of 3.37 g.

2.5 The ratio of sulfur to oxygen by mass in SO_2 is 1.0:1.0.
(a) Find the ratio of sulfur to oxygen in SO_3.
(b) Find the ratio of sulfur to oxygen in S_2O.

2.6 The ratio of oxygen to carbon by mass in carbon monoxide is 1.33:1.00. Find the formula of an oxide of carbon in which the ratio of oxygen to carbon is 2.00:1.00.

2.7 Two other binary compounds of nitrogen and oxygen that are not listed in Table 2.2 are N_2O_4 and NO_3. Find the value of the relative mass of oxygen that should be entered in the last column of the table for these two substances.

2.8 Write the formulas of binary compounds of the following elements with hydrogen: (a) sodium, (b) magnesium, (c) aluminum, (d) silicon, (e) phosphorus, (f) sulfur, (g) chlorine.

2.9 Write the formulas of two binary compounds of each of the following elements with fluorine: (a) thallium, (b) lead, (c) tin.

2.10 Find the oxidation number of the metal in each of the following compounds: (a) MnO_2, (b) Mn_2O_7, (c) V_2O_5, (d) OsO_4, (e) AuF_3, (f) $TiCl_4$, (g) H_2CrO_4, (h) NH_4AuCl_4, (i) $H_2Pt(OH)_6$, (j) $Mo_3Br_4(OH)_2$.

2.11 The metal ruthenium (Ru) is un-usual in displaying all the oxidation states from +1 to +8. Write the formula of a binary compound of ruthenium and oxygen corresponding to each oxidation state of the metal.

2.12 Write the formulas of the following compounds: (a) barium sulfate, (b) calcium bicarbonate, (c) potassium hydrogen phosphate, (d) sodium thiosulfate, (e) aluminum nitrite, (f) ammonium dichromate, (g) magnesium hydrogen sulfate.

2.13 Name the following compounds: (a) CaC_2, (b) Li_3N, (c) Sr_3P_2, (d) Na_3As, (e) K_2O, (f) K_2O_2, (g) KO_2, (h) Al_2S_3, (i) BaF_2, (j) $MgBr_2$, (k) CsI.

2.14 Name the following compounds of metals that display more than one oxidation state: (a) Cu_2O, (b) CuO, (c) UCl_3, (d) UCl_4, (e) UCl_6, (f) PbS, (g) PbS_2.

2.15 Write the formulas of the following compounds: (a) vanadium(III) oxide, (b) tungsten(VI) bromide, (c) iron(III) sulfide, (d) cobalt(II) sulfate, (e) nickel(II) phosphate, (f) chromium(VI) cyanide.

2.16 Name the following compounds: (a) BrF, (b) BrF_3, (c) BrF_5, (d) N_2O_4, (e) P_4S_3, (f) OF_2.

2.17 Write the formulas of the following compounds: (a) silane, (b) germane, (c) phosphine, (d) arsine, (e) iodine heptafluoride, (f) phosphorus pentachloride, (g) tetraphosphorus decoxide.

2.18 Balance the following expressions of chemical change: (a) $As + O_2 \rightarrow As_4O_6$, (b) $N_2 + F_2 \rightarrow NF_3$, (c) $As + Cl_2 \rightarrow AsCl_5$, (d) $ClO_2 + O_3 \rightarrow Cl_2O_6 + O_2$.

2.19 Write balanced chemical expressions for the combustion (the rapid reaction with O_2) of the following organic substances. Assume that the only products of combustion are carbon dioxide and water. (a) CH_4, (b) C_2H_6, (c) C_6H_6, (d) C_8H_{16}.

*2.20 Fill in the missing substance to complete the following equations:
(a) $2NH_3 \rightarrow N_2 + 3$ _____

(b) $Ge + GeCl_4 \rightarrow 2$ _____
(c) $AgCl \rightarrow Ag^+ +$ _____
(d) $PbI_2 \rightarrow Pb^{2+} + 2$ _____
(e) $H_3PO_4 \rightarrow 2H^+ +$ _____
(f) $2N_2O_5 \rightarrow O_2 + 4$ _____
(g) $2N_2O_5 \rightarrow O_2 + 2$ _____

2.21 Classify the following substances as (1) salts, or (2) acids, or (3) bases, or (4) nonelectrolytes, or (5) oxides: (a) HCN, (b) CH_4, (c) NH_3, (d) NH_4I, (e) $HClO_4$, (f) SO_3, (g) $KHSO_4$, (h) Li_2O, (i) LiOH, (j) $Na_2Cr_2O_7$, (k) $SrMoO_4$, (l) $CHCl_3$.

2.22 Write balanced chemical expressions for the reactions of the following substances with a large excess of water: (a) K_2O, (b) CaO, (c) NaH, (d) SO_3, (e) CrO_3, (f) MgH_2.

2.23 Write equations to represent the following processes occurring at room temperature (from this point on equations should include the necessary state symbols):
(a) the evaporation of carbon tetrachloride
(b) the condensation of bromine vapor
(c) the condensation of iodine vapor
(d) the melting of benzene
(e) the formation of carbon dioxide by the reaction of carbon and oxygen
(f) the formation of hydrogen chloride from hydrogen and chlorine
(g) the combustion of butane

2.24 Write equations for the formation at 25°C of aqueous solutions of the following substances: (a) chlorine, (b) oxygen, (c) methyl alcohol (CH_3OH), (d) barium sulfate, (e) magnesium fluoride, (f) calcium phosphate, (g) ammonium chloride.

2.25 Classify the following substances as strong acids, weak acids, strong bases, or weak bases: (a) HCN, (b) H_3PO_4, (c) $Mg(OH)_2$, (d) HNO_3, (e) HI, (f) $Ba(OH)_2$, (g) NH_3, (h) LiOH.

2.26 Write equations for the formation of aqueous solutions of the substances listed in Exercise 2.25 at 25°C.

2.27 Write equations for the precipitations that take place at 25°C when the following solutions are mixed:

(a) a solution of silver nitrate with a solution of lithium bromide, to precipitate silver bromide

(b) a solution of sodium sulfate with a solution of barium chloride, to precipitate barium sulfate

(c) a solution of hydrogen sulfide with a solution of zinc chloride, to precipitate zinc sulfide

(d) a solution of sodium hydroxide with a solution of chromium(III) sulfate, to precipitate chromium(III) hydroxide

*2.28 Write equations for the following processes at 25°C:

(a) A solution of barium hydroxide is mixed with a solution of nitric acid.

(b) A solution of sodium hydroxide is mixed with a solution of hydrogen cyanide.

(c) A solution of ammonia is mixed with a solution of hydrogen cyanide.

(d) Pure hydrogen chloride is added to a solution of potassium hydroxide.

(e) Pure hydrogen chloride is added to a solution of ammonia.

(f) Pure ammonia is added to a solution of nitric acid.

3

Stoichiometry is the study of the quantities of reactants and products in chemical reactions. One of the important steps in the history of modern chemistry was the precise measurement of the relative proportions of elements and compounds that participate in chemical reactions. Indeed, modern chemistry would be impossible without such measurements. Answers to problems as diverse as the minimum human daily requirement of a vitamin, the potential yield of metal from a mine, the life of an electric battery, or the explosive power of a stick of dynamite are all, in a broad sense, obtained from stoichiometry.

In this chapter we shall explore the methods used for specifying quantities of materials and the techniques by which some important types of stoichiometric calculations are carried out.

3.1 THE MOLE AND AVOGADRO'S NUMBER

A common way of specifying a quantity of material is to give its **mass.** The mass of an object is a measure of its resistance to acceleration, that is, to having its speed or its position changed by the application of a force. Mass is defined by Newton's second law, the familiar expression:

$F = ma$

where m is mass, F is force, and a is acceleration. In everyday usage, we often erroneously say ''weight'' when we really mean ''mass.'' Mass and weight are not the same, although the two are closely related. The relationship between them is defined by the acceleration of gravity. The SI base unit for mass is the kilogram, abbreviated kg, which is defined as the mass of a specific piece of platinum-iridium alloy that is stored in Sèvres, France.

There is a second way to specify a quantity of material. If the material is known to consist of discrete units or entities, we can specify the **amount** of material as the number of those entities, or as a number that is proportional to the number of those entities.

We use both methods of specifying quantity almost every day. If you buy some apples, you can indicate either the mass of the apples or the amount of the apples: 2.3 kg of apples or 24 apples, for instance. By custom, one method is more widely used than the other in specific situations. We always buy eggs by amount, one dozen eggs rather than 1 kg of eggs, because eggs tend to break when weighed. But we buy caviar by mass, because it is impractical to count the very many fish eggs in a small sample of caviar.

It is important to understand that mass and amount, although they are related for a given substance, are not at all the same thing. A mass of caviar and the same mass of apples represent quite different amounts, because of the difference between their basic units. There is the same difference between mass and amount in chemistry, even though the basic units are much smaller.

When we deal with chemical substances whose composition is well defined, we can designate quantities in either units of mass or units of amount. Chemical substances consist of identifiable basic units, or entities—usually molecules, atoms, or ions. In the chemical laboratory, we most often deal with quantities whose mass is conveniently expressed in grams, g, rather than in kilograms. There are 1000 grams in one kilogram.

We can express the quantity of a chemical substance as an amount, by specifying the number of basic entities—for example, the number of molecules in a given quantity of a substance. But counting molecules is much more difficult than counting the number of fish eggs in a sample of caviar. There is an enormous number of molecules in any sample that is big enough to be visible. But there are compelling reasons for expressing the quantity of a chemical substance as an amount—a chemical equation is an expression of the number of entities of the substance that participate in a chemical reaction.

The equations that express chemical changes include chemical formulas. A chemical formula is a representation of the basic unit of structure of a substance. For example, the equation

$$2H_2 + O_2 \longrightarrow 2H_2O$$

can be read as the formation of two molecules of water from two molecules of hydrogen and one molecule of oxygen. We tend to interpret all chemical changes in terms of the individual molecules, atoms, or ions that take part in reactions. However, an overwhelmingly large number of entities take part in most reactions. If we are to express the quantity of chemical substances in terms of amount, we need a way to handle such large numbers. When we deal with apples, we talk about dozens. We can say we have bought one dozen apples, rather than 12 apples. When we deal with pencils, we are apt to use the gross as the unit of amount—one gross of pencils, rather than 144 pencils. As the number of apples or pencils increases, so does the convenience of using such units. Eleven dozen apples is a more manageable number than 132 apples, and six gross of pencils is a more manageable number than 864 pencils.

The **mole** (abbreviated mol) is the SI unit for amount. It is defined thus: "**The mole is the amount of substance of a system which contains as many elementary entities as there are atoms in exactly 12 g of ^{12}C.**" The symbol ^{12}C refers to the most common isotope of carbon; an isotope is one of two or more atoms of the same chemical element that have different atomic masses (Section 5.4). This definition of the mole has a long, involved history that we need not explore. It is sufficient to say that in 1961, to settle a long-standing dispute between chemists and physicists, the masses of all the elements were defined relative to the mass of ^{12}C. For convenience, the definition of the mole is also based on this isotope of carbon.

The mole contains a very large number of entities indeed. The best measurements now available show that one mole of a substance contains 602 209 430 000 000 000 000 000 entities of that substance. This number, which is more conveniently written in exponential form as 6.0221×10^{23}, is represented by the symbol N or N_A. It is called *Avogadro's number* in honor of a nineteenth-century chemist whose pioneering work on gases, largely ignored in his time, later proved to be valuable in determining accurate atomic weights.

We said that one mole of any substance contains Avogadro's number of entities. This means that one mole of Fe contains 6.0221×10^{23} atoms of Fe, while one mole of O_2 contains 6.0221×10^{23} molecules of O_2 and three moles of HCl contain $(3)(6.0221 \times 10^{23})$ molecules of HCl. The fact that we are dealing with very large numbers of very small entities may cause some confusion, but the principle is the same as when we talk about a dozen apples. The only difference between a dozen and a mole is the number of basic entities in each. A dozen has 12, a mole has 6.0221×10^{23}. Otherwise they are the same.

The relationship between the mass and amount of a quantity of a given substance depends on the nature of the substance. If the mass of one baseball is 100 g, the mass of one dozen baseballs is 1200 g. If the mass of one Ping-Pong ball is 5 g, then the mass of one dozen Ping-Pong balls is 60 g. In other words, the mass of a given amount of a substance depends on the mass of the basic unit of that substance. When we deal with baseballs or Ping-Pong balls, the numbers are relatively small and this relationship is clear. But when we deal with chemical substances, we deal with extremely large numbers, so we must take more care in defining the relationship between mass and amount.

We mentioned that the definition of the mole is based on ^{12}C, and that the masses of all atoms are defined relative to the mass of an atom of ^{12}C. The masses of atoms can be expressed in terms of the atomic mass unit (amu). One atomic mass unit is exactly one-twelfth the mass of a single atom of ^{12}C. We say that one atom of ^{12}C has a mass of exactly 12 amu. If another atom has exactly twice the mass of a ^{12}C atom, we can say that it has a mass of 24 amu. An atom with one-third the mass of ^{12}C has a mass of 4 amu.

Using these definitions, we can fix a relationship between the atomic mass unit, which is a unit of mass that is suitable for single atoms, and the gram, a unit suitable for macroscopic quantities of substances. We can then go on to find the relationship between the mass and the amount of any given substance, providing we can determine the mass of the basic unit of that substance.

The mass in grams of one mole of ^{12}C atoms is numerically equal to the mass in atomic mass units of a single atom of ^{12}C. That is, the mass of a single atom of ^{12}C is 12 amu, and the mass of one mole of ^{12}C is 12 g. The same relationship exists for all the other elements. When we find that one mole of ^{1}H, the lightest isotope of hydrogen, has a mass of 1.007825 g, we know that one atom of ^{1}H has a mass of 1.007825 amu. (We can also note that the assignment of the ^{12}C atom as the standard for the atomic mass unit was not accidental. The relative mass of the lightest atom known has a value very close to 1 because of this assignment.) When we find that one mole of ^{19}F atoms has a mass of 18.99840 g, we know that one atom of ^{19}F has a mass of 18.99840 amu.

In other words, both the mole and the atomic mass unit are defined relative to ^{12}C. The ratio of the mass of any given atom to the mass of an atom of ^{12}C is the same as the ratio of one mole of that atom to one mole of ^{12}C atoms.

In this respect, there is no difference between a mole, a dozen, and a gross. If we find that, for example, one baseball has a mass of 100 g while one Ping-Pong ball has a mass of 5 g, we have determined a ratio of 100:5, or 20:1, that also applies when we talk about dozens or grosses. If we compare three dozen baseballs to three dozen Ping-Pong balls, the ratio of their masses is 3600 g:180 g, or 20:1. If we had a mole of baseballs—that is, 6.0221×10^{23} baseballs—and a mole of Ping-Pong balls, the ratio of their masses would still be 20:1. Once it has been decided, as it was by international agreement, to express the masses of all atoms relative to that of a ^{12}C atom, and once one-twelfth of its mass is called an atomic mass unit, the relative masses of all atoms, ions, or molecules can be expressed in terms of the atomic mass unit.

There is, however, a complication. Most of the substances around us are not isotopically pure. Instead, they are mixtures of isotopes. A mole of hydrogen atoms does not consist solely of ^{1}H atoms. It also contains a quantity of ^{2}H atoms, the heavier isotope of hydrogen called deuterium. The fraction of ^{2}H atoms (or of any isotope found in a typical sample of any element) is called the *natural abundance* of the isotope.

A macroscopic sample of hydrogen, H, of natural origin contains about 0.015% ^{2}H atoms and 99.985% ^{1}H atoms. As we mentioned, a single ^{1}H atom has a mass of 1.007825 amu, and therefore a mole of ^{1}H atoms has a mass of 1.007825 g. Similarly, a single ^{2}H atom has a mass of 2.014102 amu and a mole of ^{2}H atoms has a mass of 2.014102 g. The molar mass, the mass of a mole of naturally occurring hydrogen, is found by calculating the weighted average of the molar masses of each isotope in a typical sample.

molar mass H = (fraction ^{1}H)(molar mass ^{1}H) + (fraction ^{2}H)(molar mass ^{2}H)

$$= \frac{99.985}{100}(1.0078 \text{ g}) + \frac{0.015}{100}(2.0141 \text{ g})$$

$$= 1.0079 \text{ g}$$

The presence of a small amount of ^{2}H in naturally occurring hydrogen causes the molar mass to be slightly higher than the molar mass of pure ^{1}H. This means that the mass of one mole of naturally occurring hydrogen, expressed in grams, does not correspond either to the mass of one atom of ^{1}H in atomic mass units or to the mass of one atom of ^2H in atomic mass units. Rather, it corresponds to a weighted average that reflects the natural abundances and atomic masses of the two isotopes. Traditionally, this mass is called the *atomic weight* of the element.

Chemists recognize that the word "weight" is used incorrectly in this phrase. But since the term "atomic weight" is embalmed in tradition, it remains in use with the understanding that it refers to masses of elements relative to ^{12}C, taking into account the existence of isotopes and their natural abundances. Table 3.1 lists the atomic weights of all the known elements. An entry in this table can be read as the mass of an "average" atom in atomic mass units (relative to the mass of ^{12}C as 12 amu) or, alternatively, as the mass of one mole of the element in grams. A table of atomic weights, such as Table 3.1, is of basic importance in chemistry. Since chemists usually work with naturally occurring elements, the information in a table of atomic weights is assumed to be part of the data given in any problem dealing with quantities of material.

It is worth stressing that the use of the mole as a unit of amount is not limited to atoms. We can have a mole of any entity—a mole of molecules, a mole of ions, a mole of electrons, and so on.

The concept of a mole helps us understand many things in chemistry. Start with the simple example of the formula for water, H_2O. We know that one mole of H_2O is composed of 6.0221×10^{23} molecules. Since each molecule has two hydrogen atoms and one oxygen atom, one mole of H_2O consists of two moles, $(2)(6.0221 \times 10^{23})$, of H atoms and one mole, 6.0221×10^{23} of O atoms, combined in H_2O molecules.

An analogous everyday situation might be: We have one dozen tricycles. Each one has two small rear wheels and one large front wheel. Therefore, there are one dozen, 1×12, large wheels and two dozen, $2 \times 12 = 24$, small wheels in the dozen tricycles.

The concept of a mole adds a new meaning to chemical formulas. A formula indicates not only the number of atoms in a molecule but also the number of moles of each atom in a mole of that substance. For example, we saw that one mole of H_2O consists of two moles of hydrogen and one mole of oxygen. Similarly, one mole of CF_4 consists of one mole of carbon atoms and four moles of fluorine atoms. And while NaCl does not consist of molecules, one mole of NaCl does consist of one mole of Na^+ ions and one mole of Cl^- ions.

The mass in grams of one mole of a molecular substance is called its *molecular weight,* although a more accurate term would be "relative molecular mass." The molecular weight of a molecule is the sum of all the atomic weights of all the atoms in the molecule. Molecular weight is readily computed by using the data in a table of atomic weights.

We can calculate the molecular weight of H_2O by noting that the atomic weight of H is 1.0079 and the atomic weight of O is 15.9994. The molecular weight of H_2O is the sum of the atomic weights of two H atoms and one O atom, or $2 \times 1.0079 + 15.9994 = 18.0152$. Therefore, the mass of the average H_2O molecule is 18.0152 amu, and the mass of one mole of H_2O is 18.0152 g. The word *average* should be stressed. Since these values are based on the natural abundances of the isotopes of hydrogen and oxygen, the actual measured values will vary slightly, depending on where the sample of water comes from.

These relationships between moles, formulas, masses, and Avogadro's number are at the heart of many calculations in chemistry. In doing these calculations, it is most important to remember that **a mole of any substance consists of 6.0221×10^{23} elementary entities and has a mass in grams that is numerically equal to the sum of the atomic weights of the atoms in its formula.** This can also be stated:

$$\text{the number of moles of a sample of a substance} = \frac{\text{the mass of the given substance in grams}}{\text{the "weight" of one mole of the entity (or formula) in grams}} \quad (3.1)$$

TABLE 3.1
Table of Atomic Weights[a]

Name	Symbol	Atomic Number	Atomic Weight	Name	Symbol	Atomic Number	Atomic Weight
actinium	Ac	89	(227)[b]	mercury	Hg	80	200.59
aluminum	Al	13	26.98154	molybdenum	Mo	42	95.94
americium	Am	95	(243)	neodymium	Nd	60	144.24
antimony	Sb	51	121.75	neon	Ne	10	20.179
argon	Ar	18	39.948	neptunium	Np	93	237.0482[c]
arsenic	As	33	74.9216	nickel	Ni	28	58.70
astatine	At	85	(210)	niobium	Nb	41	92.9064
barium	Ba	56	137.34	nitrogen	N	7	14.0067
berkelium	Bk	97	(247)	nobelium	No	102	(255)
beryllium	Be	4	9.01218	osmium	Os	76	190.2
bismuth	Bi	83	208.9804	oxygen	O	8	15.9994
boron	B	5	10.81	palladium	Pd	46	106.4
bromine	Br	35	79.904	phosphorus	P	15	30.97376
cadmium	Cd	48	112.40	platinum	Pt	78	195.09
calcium	Ca	20	40.08	plutonium	Pu	94	(244)
californium	Cf	98	(251)	polonium	Po	84	(209)
carbon	C	6	12.011	potassium	K	19	39.098
cerium	Ce	58	140.12	praseodymium	Pr	59	140.9077
cesium	Cs	55	132.9054	promethium	Pm	61	(145)
chlorine	Cl	17	35.453	protactinium	Pa	91	231.0359[c]
chromium	Cr	24	51.996	radium	Ra	88	226.0254[c]
cobalt	Co	27	58.9332	radon	Rn	86	(222)
copper	Cu	29	63.546	rhenium	Re	75	186.207
curium	Cm	96	(247)	rhodium	Rh	45	102.9055
dysprosium	Dy	66	162.50	rubidium	Rb	37	85.4678
einsteinium	Es	99	(254)	ruthenium	Ru	44	101.07
erbium	Er	68	167.26	samarium	Sm	62	150.4
europium	Eu	63	151.96	scandium	Sc	21	44.9559
fermium	Fm	100	(257)	selenium	Se	34	78.96
fluorine	F	9	18.99840	silicon	Si	14	28.086
francium	Fr	87	(223)	silver	Ag	47	107.868
gadolinium	Gd	64	157.25	sodium	Na	11	22.98977
gallium	Ga	31	69.72	strontium	Sr	38	87.62
germanium	Ge	32	72.59	sulfur	S	16	32.06
gold	Au	79	196.9665	tantalum	Ta	73	180.9479
hafnium	Hf	72	178.49	technetium	Tc	43	(97)
helium	He	2	4.00260	tellurium	Te	52	127.60
holmium	Ho	67	164.9304	terbium	Tb	65	158.9254
hydrogen	H	1	1.0079	thallium	Tl	81	204.37
indium	In	49	114.82	thorium	Th	90	232.0381[c]
iodine	I	53	126.9045	thulium	Tm	69	168.9342
iridium	Ir	77	192.22	tin	Sn	50	118.69
iron	Fe	26	55.847	titanium	Ti	22	47.90
krypton	Kr	36	83.80	tungsten	W	74	183.85
lanthanum	La	57	138.9055	uranium	U	92	238.029
lawrencium	Lr	103	(260)	vanadium	V	23	50.9414
lead	Pb	82	207.2	xenon	Xe	54	131.30
lithium	Li	3	6.941	ytterbium	Yb	70	173.04
lutetium	Lu	71	174.97	yttrium	Y	39	88.9059
magnesium	Mg	12	24.305	zinc	Zn	30	65.38
manganese	Mn	25	54.9380	zirconium	Zr	40	91.22
mendelevium	Md	101	(258)				

[a] The atomic weights of many elements are not invariant. They depend on the origin and treatment of the material. The values given here are from a 1973 report of the Commission on Atomic Weights of IUPAC. These values apply to elements as they exist on earth and to certain artificial elements. Values in parentheses are used for radioactive elements whose atomic weights cannot be quoted precisely without knowledge of origin.

[b] A value in parentheses is the mass number of the least unstable isotope of a radioactive element whose atomic weight cannot be given more precisely.

[c] Relative atomic mass of the most commonly available long-lived isotope.

THE REVISION OF ATOMIC WEIGHTS

The International Union of Pure and Applied Chemistry (IUPAC) makes changes in the table of atomic weights from time to time. Revisions are required because of improved techniques of analysis, changes in the isotopic abundances of elements due to mankind's activities, and the laboratory synthesis of heavy elements.

In 1969, for example, the IUPAC Commission on Atomic Weights quoted values for radium, neptunium, and protactinium for the first time. While integral values for the atomic mass of the most stable isotope of these three elements had been printed in many tables before then, the commission had not previously accepted any of these values as the atomic weight because of the rarity of the elements.

At the same time, IUPAC refined the previously listed values of 22 elements. For 20 elements, the values were extended by one significant figure. But IUPAC also reduced the number of significant figures in the atomic weights of boron, carbon, hydrogen, lead, samarium, and sulfur, in most cases because of the variability of the isotopic content of natural sources.

The commission on atomic weights also took note of a new factor: the effect of mankind's activities on the listed values of atomic weights. It was noted that the atomic weights of uranium, boron, lithium, and several other elements are susceptible to wide variations because the nuclear industry processes natural samples to separate out certain pure isotopes. The atomic weight values of some samples could be thrown into uncertainty if either the desired isotopes or spent material from which certain isotopes have been separated get out into the environment. The United States government, which conducts isotope separation activities on a large scale, suggested that chemists who had doubts about the isotopic composition of any compound should get a certificate of origin stating whether the material is from a natural source.

The effect of space exploration was also noted by the IUPAC commission. For the first time, the commission said, samples of material from other worlds are being brought to earth. It is possible that the isotopic distribution of the earth's elements could be altered by the presence of extraterrestrial material, IUPAC said. Thus far, material has been brought back only from the moon. However, there are now discussions of a possible mission to Mars in which an attempt will be made to return Martian soil samples to earth. While biologists have fears about contamination of other planets by terrestrial organisms—fears that resulted in strong efforts to sterilize the Viking spacecraft that went to Mars in 1976—the fear in chemistry is that the natural isotopic distributions on earth may be contaminated by extraterrestrial samples.

The IUPAC commission also said that future changes in listed values of atomic weights are certain, since more accurate determinations are needed for many elements. For example, values for osmium, palladium, and samarium are known to only one decimal place, and the values for many other elements are known to only two decimal places.

EXAMPLE 3.1

The formula of sulfuric acid, the most important strong acid used commercially, is H_2SO_4.
(a) What is the mass of one mole of H_2SO_4?
(b) What amount of H_2SO_4 is in 375 g of the pure acid?
(c) How many H_2SO_4 molecules are in 500 g of pure H_2SO_4?
(d) What amount of O is in 3.0 mol of H_2SO_4?
(e) How many H atoms are in 175 g of H_2SO_4?

SOLUTION

(a) The mass of one mole of H_2SO_4 is the numerical sum of the atomic weights of the constituent atoms in grams:

$$
\begin{aligned}
2 \times H &= 2 \times 1.0079 &&= 2.016 \\
1 \times S &= 1 \times 32.06 &&= 32.06 \\
4 \times O &= 4 \times 15.999 &&= \underline{63.996} \\
\text{molecular weight} && &= 98.07 \\
\text{mass of one mole (molar mass)} && &= 98.07 \text{ g}
\end{aligned}
$$

(b) This is probably the most common calculation in chemistry: the conversion of a given mass of a substance to an amount of substance. Direct substitution into Equation 3.1 gives:

$$
\text{amount } H_2SO_4 = \frac{375 \text{ g } H_2SO_4}{98.07 \text{ g } H_2SO_4 / \text{mol } H_2SO_4} = 3.82 \text{ mol } H_2SO_4
$$

Alternatively, the problem can be viewed as the conversion of a number of grams to a number of moles. For this approach, we need a conversion factor that gives the relationship between grams (mass) and moles (amount). This is probably the most important conversion factor in chemical calculations. We can get this conversion factor from the result of part (a), which states that there are

$$
\frac{98.07 \text{ g } H_2SO_4}{1 \text{ mol } H_2SO_4} \quad \text{or} \quad \frac{1 \text{ mol } H_2SO_4}{98.07 \text{ g } H_2SO_4}
$$

Since part (b) gives grams and asks us to find moles, the second factor is used:

$$
375 \text{ g } H_2SO_4 \times \frac{1 \text{ mol } H_2SO_4}{98.07 \text{ g } H_2SO_4} = 3.82 \text{ mol } H_2SO_4
$$

The units of grams appear in both the numerator and denominator and so cancel; only the units of moles remain. When the proper conversion factor is chosen for a problem such as this, the units in which the solution is expressed—in this case, moles—will be in the numerator only. If the wrong conversion factor is used, a check will reveal the mistake immediately. If we had used:

$$
375 \text{ g } H_2SO_4 \times \frac{98.07 \text{ g } H_2SO_4}{1 \text{ mol } H_2SO_4} = 3680 \frac{g^2 H_2SO_4}{\text{mol } H_2SO_4}
$$

the units would make no sense, and we would know that the result must be wrong. But it should be noted that any conversion factor can be written as its reciprocal and still be correct, if used in the proper situation.

(c) This problem gives a mass (the known) and asks for a number of molecules (the unknown). To solve the problem, we must have the relationship between mass and number of molecules. This is known, since the relationship between moles and molecules is defined for all substances and the conversion factor between mass and amount for this substance was calculated in part (a).

Using N as the symbol for Avogadro's number, we have as the relationship between moles and molecules the conversion factor

$$\frac{N \text{ molecules}}{1 \text{ mol}} \quad \text{or} \quad \frac{1 \text{ mol}}{N \text{ molecules}}$$

We set up the conversion factor for this problem to eliminate the units of grams and moles, leaving molecules:

molecules H_2SO_4

$$= 500 \text{ g } H_2SO_4 \times \frac{1 \text{ mol } H_2SO_4}{98.1 \text{ g } H_2SO_4} \times \frac{6.02 \times 10^{23} \text{ molecules } H_2SO_4}{1 \text{ mol } H_2SO_4}$$

$$= 3.07 \times 10^{24} \text{ molecules } H_2SO_4$$

In problems such as this, the conversion factor units must be included throughout, to be sure that the factors have been used correctly. There is a rule to follow: The units in the given (known) quantity and those in the conversion factor(s) must cancel so as to leave only the desired units of the answer.

(d) Simple reasoning shows that since 1 mol of H_2SO_4 contains 4 mol of O atoms, 3.0 mol of H_2SO_4 will contain $3.0 \times 4 = 12$ mol of O atoms. This statement reveals an important feature of the concept of the mole: Almost any statement that is correct for atoms and molecules will still be correct if the phrase "moles of" is placed before the words "atom" and "molecule" wherever they appear. In this case, the statement, "One H_2SO_4 molecule contains four O atoms" is correct when changed to read, "One mole of H_2SO_4 molecules contains four moles of O atoms."

More formally, this problem can be solved by using the appropriate conversion factors:

$$3.0 \text{ mol } H_2SO_4 \times \frac{4 \text{ mol O atoms}}{1 \text{ mol } H_2SO_4} = 12 \text{ mol O atoms}$$

(e) The known in this problem is a mass of H_2SO_4 in grams. The unknown is a number of H atoms. The problem can be solved by the following line of reasoning:

The number of moles in the given mass can be found by using the molar mass:

$$175 \text{ g } H_2SO_4 \times \frac{1 \text{ mol } H_2SO_4}{98.07 \text{ g } H_2SO_4} = 1.784 \text{ mol } H_2SO_4$$

Each mole of sulfuric acid has N molecules of H_2SO_4. The number of molecules in 1.784 mol of sulfuric acid is

$$1.784 \text{ mol } H_2SO_4 \times \frac{6.022 \times 10^{23} \text{ molecules } H_2SO_4}{1 \text{ mol } H_2SO_4}$$

$$= 1.074 \times 10^{24} \text{ molecules } H_2SO_4$$

Since H has the subscript 2 in the formula H_2SO_4, we know that there are two H atoms in each molecule of sulfuric acid. The number of H atoms in the sample therefore is

$$1.074 \times 10^{24} \text{ molecules } H_2SO_4 \times \frac{2 \text{ atoms H}}{1 \text{ molecule } H_2SO_4}$$

$$= 2.15 \times 10^{24} \text{ atoms H}$$

This calculation can also be performed in a single step by listing all the conversion factors in sequence:

$$175 \text{ g H}_2\text{SO}_4 \times \frac{1 \text{ mol H}_2\text{SO}_4}{98.07 \text{ g H}_2\text{SO}_4} \times \frac{6.022 \times 10^{23} \text{ molecules H}_2\text{SO}_4}{1 \text{ mol H}_2\text{SO}_4}$$

$$\times \frac{2 \text{ atoms H}}{1 \text{ molecule H}_2\text{SO}_4} = 2.15 \times 10^{24} \text{ atoms H}$$

3.2 MASS RELATIONSHIPS IN CHEMICAL FORMULAS

A table of atomic weights is assumed to be part of the information given in any stoichiometry problem. Therefore, a chemical formula conveys information not only on the number of atoms in a molecule of a substance but also on the relative masses of the components of the molecule. We can also turn this around and say that if we know the relative masses of the components of an unknown substance, we can find out something about the chemical formula of the substance.

EXAMPLE 3.2
The Comstock Lode in Nevada is a rich silver deposit containing silver sulfide, Ag_2S, the same black substance that is the tarnish on silver objects. What percentage of Ag_2S by mass is silver?

SOLUTION
The chemical formula shows that one mole of Ag_2S contains two moles of Ag atoms and one mole of S atoms. Two moles of Ag atoms have a mass of

$$2 \times (\text{atomic weight of Ag}) = 2 \text{ mol Ag} \times 107.87 \text{ g Ag/mol Ag}$$
$$= 215.74 \text{ g Ag}$$

One mole of S has a mass of 32.06 g. The mass of one mole of Ag_2S is 247.80 g. The percentage by mass of Ag in Ag_2S can be considered to be

The principles of stoichiometry can be used to determine the percentage of pure silver that can be extracted from the ore of a silver mine such as this one. (*Wide World*)

the number of grams of Ag in a 100-g sample of Ag_2S:

$$\text{mass Ag} = 100 \text{ g Ag}_2\text{S} \times \frac{215.74 \text{ g Ag}}{247.80 \text{ g Ag}_2\text{S}} = 87.06 \text{ g Ag}$$

or 87.06% Ag.

More simply, a percentage is calculated by dividing the quantity of interest by the total quantity and multiplying by 100.

$$\text{percentage Ag} = \left(\frac{\text{mass Ag}}{\text{mass compound}}\right)100 = \left(\frac{215.74 \text{ g Ag}}{247.80 \text{ g Ag}_2\text{S}}\right)100$$
$$= 87.06\% \text{ Ag}$$

Knowing the percentage by mass of Ag in Ag_2S, we also know the percentage of sulfur. The sum of the percentages of all the components is 100, so the mass percentage of sulfur is obtained by subtraction:

$$\text{percentage S} = 100\% - \text{percentage Ag} = 12.94\% \text{ S}$$

EXAMPLE 3.3

Calculate the percentage by mass of each element in the substance cryolite, Na_3AlF_6, which is used in glassmaking and in many metallurgical processes.

SOLUTION

As we saw in Example 3.2, the molecular weight of the substance must be calculated to solve such a problem. We can simplify the calculation by setting out the data in an orderly manner:

$$
\begin{array}{rcl}
3 \times Na = 3 \times 22.99 = & 68.97 \\
1 \times Al = 1 \times 26.98 = & 26.98 \\
6 \times F = 6 \times 19.00 = & \underline{114.00}
\end{array}
$$

molecular weight (molar mass) Na_3AlF_6 = 209.95

The percentage by mass of each element in the substance can be calculated directly from these data:

$$\text{percentage Na} = \frac{68.97}{209.95} \times 100 = 32.85\%$$

$$\text{percentage Al} = \frac{26.98}{209.95} \times 100 = 12.85\%$$

$$\text{percentage F} = \frac{114.00}{209.95} \times 100 = 54.30\%$$

A check shows that these three percentages sum to 100%.

We can find a formula for an unknown compound by doing the same sort of calculation as in Example 3.3, but in reverse—that is, by starting with an unknown substance and determining the percentage by mass of each element in the substance. When we know the relative mass of each element, we can find the relative number of atoms of each element in the substance. However, this procedure gives only an **empirical formula**, which indicates only the relative numbers of different atoms in a molecule of the substance.

EXAMPLE 3.4

One area of current interest in chemistry is the study of the carboranes, a class of compounds composed of carbon, boron, and hydrogen. One carborane is found to contain 32.77% C and 59.00% B by mass. Calculate its empirical formula.

SOLUTION

Since the empirical formula (or any other chemical formula) gives the relative number of moles of each constituent atom of a compound, the first step is to convert the data on mass into data on amount.

If a substance contains 32.77% C, then 100 g of the substance contains 32.77 g of carbon.

$$32.77 \text{ g C} \times \frac{1 \text{ mol C}}{12.01 \text{ g C}} = 2.729 \text{ mol C}$$

This 100-g sample also contains 59.00 g of B:

$$59.00 \text{ g B} \times \frac{1 \text{ mol B}}{10.81 \text{ g B}} = 5.458 \text{ mol B}$$

The mass of H in the 100-g sample is obtained by subtraction:

$$
\begin{aligned}
\text{mass H} &= 100 \text{ g} - (\text{mass C}) - (\text{mass B}) \\
&= 100 \text{ g} - 32.77 \text{ g} - 59.00 \text{ g} \\
&= 8.23 \text{ g H}
\end{aligned}
$$

Thus there are:

$$8.23 \text{ g H} \times \frac{1 \text{ mol H}}{1.008 \text{ g H}} = 8.165 \text{ mol H}$$

The ratio of moles of C to moles of B to moles of H is C:B:H = 2.729:5.458:8.165. The subscripts in the empirical formula for the compound are obtained by converting this ratio to an equivalent integer ratio. This can be done in two steps:

(i) Divide all the terms of the ratio by the smallest number in the ratio:

$$\frac{2.729}{2.729} : \frac{5.458}{2.729} : \frac{8.165}{2.729} = 1:2:3$$

(ii) In this case, division by the smallest number in the ratio gives a ratio containing only integers. If division by the smallest number produces a ratio that still includes decimals, multiply by the smallest factor that converts all the terms to integers. For example, the ratio 2:3.25:1.5 can be multiplied by 4 to give 8:13:6.

A check shows that the ratio 1:2:3 for C:B:H is reasonable for an empirical formula, since the numbers do not have any factors in common and the ratio is thus in its lowest terms. The subscripts of the empirical formula are 1, 2, and 3, and the empirical formula is written CB_2H_3.

An empirical formula may or may not be the same as a **molecular formula,** which gives the exact number of each kind of atom in a molecule. In the case of water, H_2O, the empirical formula and the molecular formula are identical. The empirical formula indicates that the ratio of hydrogen atoms to oxygen atoms is 2:1, and the molecular formula shows that there are exactly two atoms of hydrogen and one atom of oxygen in an individual molecule of water. On the other hand, hydrogen peroxide has the empirical formula HO, which indicates that the ratio of hydrogen atoms to oxygen atoms in the substance is 1:1. But the molecular formula of hydrogen peroxide is H_2O_2. The ratio of hydrogen atoms to oxygen atoms in a single molecule of hydrogen peroxide is 1:1, but each molecule actually has two atoms of hydrogen and two atoms of oxygen.

There are many cases in which a number of different substances have the same empirical formula but different molecular formulas. One such case is the cycloalkanes, a group of organic compounds that contain only carbon and hydrogen atoms and include one closed ring of carbon atoms. The empirical formula of all the cycloalkanes is CH_2, but the molecular formulas of individual cycloalkanes include C_3H_6, $C_{20}H_{40}$, and many others.

All the subscripts in a molecular formula are always integral multiples of the subscripts in the empirical formula. An empirical formula can be converted to a

THE MASS SPECTROMETER

The mass of an atom or a molecule can be measured with great accuracy by using an instrument called a mass spectrometer. The atom or molecule is first ionized to a cation by bombardment with electrons. The cation is accelerated by a high-voltage electric field, and then passes through a slit into a magnetic field. The cation is deflected by the magnetic field; the amount of deflection depends on the mass and the charge of the cation. If two cations have the same electric charge, the heavier one will be deflected less. By adjustment of the strength of the magnetic field, any cation can be made to pass through a second slit and arrive at a collector. Since the strengths of the electric and magnetic fields are known, the mass of the cation can be determined quite accurately. It is possible to link a mass spectrometer with other instruments that enable fast analysis of complex samples. For example, the Viking spacecraft that landed on Mars in 1976 carried mass spectrometers that were linked to instruments called gas chromatographs. A sample of Martian soil is collected by the Viking spacecraft's automated scoop. The gas chromatograph separates pure substances from the soil sample. The substances are then identified by the mass spectrometer, giving an accurate determination of the composition of Martian soil.

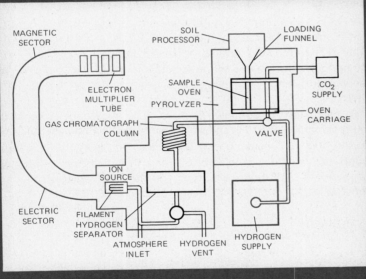

The gas chromatograph-mass spectrometer experiment on the Viking spacecraft.

molecular formula if one more piece of information is available—the molecular weight of the substance. The molecular weight can be measured experimentally.

EXAMPLE 3.5
The molecular weight of the carborane in Example 3.4 is found to be 72.5. Calculate the molecular formula of the compound.
SOLUTION
The subscripts in a molecular formula can be obtained by multiplying the

subscripts in an empirical formula by an appropriate integer. The problem is to find the correct integer.

First, calculate the "empirical formula weight" of the empirical formula as if it were an ordinary chemical formula:

$$1 \times C = 1 \times 12.0 = 12.0$$
$$2 \times B = 2 \times 10.8 = 21.6$$
$$3 \times H = 3 \times 1.0 \ = \underline{\ \ 3.0}$$
$$36.6$$

Divide this value into the experimentally determined molecular weight:

$$\frac{72.5}{36.6} = 1.98$$

and round off the result to the nearest integer, in this case 2. Then multiply each subscript in the empirical formula by this integer to obtain the molecular formula, in this case:

$$C_2B_4H_6$$

One technique used to find the composition of unknown substances in organic chemistry is combustion analysis. Compounds containing carbon and hydrogen burn to form carbon dioxide and water. The empirical formula of the substance that was burned can often be calculated by collecting and weighing the CO_2 and H_2O.

EXAMPLE 3.6

(a) A 1.037-g sample of a substance containing only C, H, and O is burned in a stream of excess $O_2(g)$ to yield 1.900 g of CO_2 and 0.521 g of H_2O. What is the empirical formula of the substance?

(b) The molecular weight of the substance is found to be 288. What is its molecular formula?

SOLUTION

(a) To calculate the empirical formula, it is necessary to know the relative number of moles of C, H, and O atoms in the compound. Since all the C atoms in the original substance now are in the CO_2 molecules produced by combustion, the number of moles of C atoms in the original compound is the same as the number of moles of C atoms in the CO_2, which is the same as the number of moles of CO_2. We have 1.900 g of CO_2, so:

$$1.900 \text{ g } CO_2 \times \frac{1 \text{ mol } CO_2}{44.010 \text{ g } CO_2} = 0.0432 \text{ mol } CO_2$$

$$0.0432 \text{ mol } CO_2 \times \frac{1 \text{ mol C}}{1 \text{ mol } CO_2} = 0.0432 \text{ mol C}$$

By the same line of reasoning, all the H atoms in the original compound are found in the 0.521 g of H_2O, so:

$$0.521 \text{ g } H_2O \times \frac{1 \text{ mol } H_2O}{18.015 \text{ g } H_2O} = 0.0289 \text{ mol } H_2O$$

$$0.0289 \text{ mol } H_2O \times \frac{2 \text{ mol H}}{1 \text{ mol } H_2O} = 0.0578 \text{ mol H}$$

The substance also contains oxygen. The amount of oxygen in the compound cannot be calculated from the amount of oxygen in the H_2O and CO_2 created by combustion, since excess O_2 was added during the burning. But since we know the total mass of the sample compound, and we can find the masses of C and H in the compound, the mass of O in the compound can be calculated by difference:

total mass = mass C + mass H + mass O

The masses of C and H are calculated from the number of moles of C and H in the sample compound:

$$\text{mass C} = 0.0432 \text{ mol C} \times \frac{12.011 \text{ g C}}{1 \text{ mol C}} = 0.519 \text{ g C}$$

$$\text{mass H} = 0.0578 \text{ mol H} \times \frac{1.008 \text{ g H}}{1 \text{ mol H}} = 0.0583 \text{ g H}$$

We can now find the mass of O in the sample by subtracting the masses of C and H from the known mass of the sample compound:

total mass = 1.037 g = mass C + mass H + mass O

1.037 g = 0.519 g + 0.0583 g + mass O

mass O = 0.460 g O

Now that we know the mass of O in the sample, we can calculate the number of moles of O in the sample:

$$\text{amount O} = 0.460 \text{ g O} \times \frac{1 \text{ mol O}}{15.999 \text{ g O}} = 0.0288 \text{ mol O}$$

We then can determine the molar ratios of the three elements, which are C:H:O = 0.0432:0.0583:0.0288. If we divide all the terms of the ratio by 0.0288, the smallest number in the ratio, we get 1.50:2.02:1. We can convert this ratio to integers by multiplying by 2 and rounding off. This gives C:H:O = 3:4:2. The empirical formula thus is:

$C_3H_4O_2$

(b) The formula weight of this empirical formula is 72.1. The molecular weight of the compound is 288. The integral factor relating the two is 288/72.1 = 3.99 or ~4. Multiplying every subscript in the empirical formula by 4 gives the molecular formula:

$C_{12}H_{16}O_8$

3.3 MASS RELATIONSHIPS IN CHEMICAL REACTIONS

In Section 2.5, page 33, we discussed some attributes of balanced chemical equations. If this information is combined with an understanding of the concept of the mole and with the data in a table of atomic weights, many useful calculations concerning the mass relationships among reactants and products in chemical processes are possible.

Mass relationships among the constituents of a chemical process depend in part on the amount of reactants used. In the general process:

$2A + B \longrightarrow C$

the equation can be taken to mean that two molecules of A combine with one molecule of B to form one molecule of C or that two moles of A combine with one mole of B to form one mole of C. If this reaction is carried out by bringing together quantities of A and B whose molar ratio is exactly 2:1, as it is in the equation, we say that we have *stoichiometric amounts* of A and B, and both of them can be completely consumed. Stoichiometric amounts of reactants are those amounts that can combine completely, leaving no excess of any reactant. In practice, we rarely work with stoichiometric amounts. One of the reactants usually is present in excess, so the quantity of the *other* reactant determines how much of the product is produced. For example, if two moles of A and two moles of B were mixed, B would be present in excess. It would be the quantity of A that would determine the amount of product produced. We would call A the *limiting quantity* or *limiting reagent*.

There is often a practical reason why an excess of one reagent is needed. In theory, the equation above says that one mole of C is obtained when two moles of A and one mole of B are mixed. When this occurs, the reaction is said to go to completion; that is, all molecules of A and B react to form C. In practice, many processes do not go to completion. Less than the "theoretical" amount of C is obtained because not all the molecules of the starting materials react. The presence of impurities in the reactants and the occurrence of unwanted processes called side reactions may also contribute to the production of less than the theoretical amount of product. These problems will be ignored at this point. Unless otherwise specified, the chemical equations associated with stoichiometric problems will be treated as going to completion.

EXAMPLE 3.7

A pure metal usually is quite shiny, but the metals that we observe tend to be dull and grayish. The loss of sheen is often due to the reaction between the metal surface and oxygen in the atmosphere that causes formation of an oxide. Lithium, the lightest metal, is no exception. It reacts with oxygen in the atmosphere to form lithium oxide.

(a) What amount of lithium oxide is formed from the reaction of 5.42 mol of lithium metal with oxygen?

(b) What mass of oxygen gas is required to form 2.96 mol of lithium oxide?

(c) What mass of lithium oxide is formed when 4.57 g of lithium undergoes this reaction?

(d) What mass of oxygen gas will be consumed in the reaction of part (c)?

SOLUTION

In a problem of this type, the essential first step is to write the chemical equation:

$$4Li(s) + O_2(g) \longrightarrow 2Li_2O(s)$$

The calculations that we shall perform are based on this equation, which states that 4 mol of lithium reacts with 1 mol of oxygen (O_2) to form 2 mol of lithium oxide.

(a) Since 4 mol of Li forms 2 mol of Li_2O, 1 mol of Li will form $\frac{2}{4} = \frac{1}{2}$ mol of Li_2O. Since 1 mol of Li forms $\frac{1}{2}$ mol of Li_2O, 5.42 mol of Li forms $5.42 \times \frac{1}{2} = 2.71$ mol of Li_2O. This reasoning can also be expressed by using a conversion factor:

$$5.42 \text{ mol Li} \times \frac{2 \text{ mol } Li_2O}{4 \text{ mol Li}} = 2.71 \text{ mol } Li_2O$$

(b) The data in the table of atomic weights, which is always assumed to be part of the data given in a chemical problem, allow us to find the relationship between amount (mol) and mass (g). In this problem, the equation gives the relationship between the moles of O_2 consumed and the moles of Li_2O formed. Two moles of Li_2O require 1 mol of O_2. Therefore, 1 mol of Li_2O requires $\frac{1}{2}$ mol of O_2. The formation of 2.96 mol of Li_2O requires $2.96 \times \frac{1}{2} = 1.48$ mol of O_2. Since the mass of one mole of O_2 is readily found to be 32.0 g, the number of grams of O_2 required is 1.48 mol of $O_2 \times 32.0$ g/mol of $O_2 = 47.4$ g of O_2.

This reasoning can be carried out in one step by using the appropriate conversion factors:

$$2.96 \text{ mol Li}_2O \times \frac{1 \text{ mol O}_2}{2 \text{ mol Li}_2O} \times \frac{32.0 \text{ g O}_2}{1 \text{ mol O}_2} = 47.4 \text{ g O}_2.$$

(c) The equation gives the relationship between the moles of Li and the moles of Li_2O. The table of atomic weights allows us to find the mass of one mole of each of these substances. Thus, 4.57 g of Li can be converted to moles by the calculation:

$$4.57 \text{ g Li} \times \frac{1 \text{ mol Li}}{6.94 \text{ g Li}} = 0.659 \text{ mol Li}$$

As we saw in part (a), we can find the number of moles of Li_2O formed from this amount of lithium:

$$0.659 \text{ mol Li} \times \frac{2 \text{ mol Li}_2O}{4 \text{ mol Li}} = 0.329 \text{ mol Li}_2O$$

The molecular weight of Li_2O can be calculated as 29.88. Therefore the mass of Li_2O formed here is:

$$0.329 \text{ mol Li}_2O \times \frac{29.88 \text{ g Li}_2O}{1 \text{ mol Li}_2O} = 9.83 \text{ g Li}_2O$$

The sequence of steps followed in this problem is an important one that is often used in chemical calculations. A mass (grams) is converted to an amount (moles) of substance, which is converted to an amount (moles) of another substance, using the molar relationships given by the chemical equation. The amount of the new substance is then converted to mass (grams). We can summarize the sequence as $\text{mass}_A \rightarrow \text{amount}_A \rightarrow \text{amount}_B \rightarrow \text{mass}_B$.

This entire calculation can also be performed in one step with the appropriate conversion factors:

$$4.57 \text{ g Li} \times \frac{1 \text{ mol Li}}{6.94 \text{ g Li}} \times \frac{2 \text{ mol Li}_2O}{4 \text{ mol Li}} \times \frac{29.88 \text{ g Li}_2O}{1 \text{ mol Li}_2O} = 9.83 \text{ g Li}_2O$$

(d) The solution to this problem can be found in the same way as the solution in part (c). The chemical equation defines not only the relationship between reactants and products, but also the relationships between the reactants themselves or the products themselves. The mass of O_2 consumed in the reaction of part (c) can be found by using the appropriate conversion factors:

$$4.57 \text{ g Li} \times \frac{1 \text{ mol Li}}{6.94 \text{ g Li}} \times \frac{1 \text{ mol O}_2}{4 \text{ mol Li}} \times \frac{32.0 \text{ g O}_2}{1 \text{ mol O}_2} = 5.27 \text{ g O}_2$$

EXAMPLE 3.8

When an unknown quantity of butane, C_4H_{10}, the fuel in some cigarette lighters, is burned in excess oxygen, 127 g of H_2O is collected. (a) What mass of butane is burned? (b) What mass of $CO_2(g)$ is produced at the same time?

SOLUTION

The equation is:

$$2C_4H_{10}(g) + 13O_2(g) \longrightarrow 8CO_2(g) + 10H_2O(l)$$

Although the numbers here are more unwieldy, this problem is solved by using the same method developed in Example 3.7. This equation states that 2 mol of butane and 13 mol of O_2 form 8 mol of carbon dioxide and 10 mol of water. Using the table of atomic weights, we find that the relevant molecular weights are: $C_4H_{10} = 58.1$; $H_2O = 18.0$; $CO_2 = 44.0$.

(a) We are given the mass of H_2O formed. We shall convert this mass of water to an amount of water. Then we shall convert this amount of water to the amount of butane that produced it, using the relationships in the chemical equation. Finally, we shall convert the amount of butane to a mass of butane.

The steps are:

$$127 \text{ g } H_2O \times \frac{1 \text{ mol } H_2O}{18.0 \text{ g } H_2O} = 7.06 \text{ mol } H_2O \quad \text{(mass} \longrightarrow \text{amount)}$$

$$7.06 \text{ mol } H_2O \times \frac{2 \text{ mol } C_4H_{10}}{10 \text{ mol } H_2O} = 1.41 \text{ mol } C_4H_{10} \quad \text{(amount} \longrightarrow \text{amount)}$$

$$1.41 \text{ mol } C_4H_{10} \times \frac{58.1 \text{ g } C_4H_{10}}{1 \text{ mol } C_4H_{10}} = 82.0 \text{ g } C_4H_{10} \quad \text{(amount} \longrightarrow \text{mass)}$$

(b) The reasoning here is the same as in part (a), since the chemical equation defines the molar relationship between the products. The calculation, in one step, is:

$$127 \text{ g } H_2O \times \frac{1 \text{ mol } H_2O}{18.0 \text{ g } H_2O} \times \frac{8 \text{ mol } CO_2}{10 \text{ mol } H_2O} \times \frac{44.0 \text{ g } CO_2}{1 \text{ mol } CO_2} = 248 \text{ g } CO_2$$

For the common situation in which reactants for a given process are not present in exactly the correct molar ratios called the stoichiometric amounts, the reaction continues until one of the reactants is exhausted. The reactant that is exhausted first is the *limiting reagent*. As an analogy, we can think of a dance attended by 60 men and 50 women. There can be no more than 50 male-female couples dancing, so the limiting "reagent" is the number of women.

When reactants are not present in stoichiometric amounts, an extra step is needed to determine the limiting reagent, the reactant that determines or limits the extent to which the reaction occurs. When nonstoichiometric quantities of reactants are present, there is an excess of all the reactants but one. The reactant or reagent that is not present in excess will be consumed before any of the others. When this happens, the reaction stops. Therefore, the one reactant that is not present in excess is the limiting reagent.

EXAMPLE 3.9

Potassium nitrate, KNO_3, sometimes called saltpeter, is an important fertilizer. One process for its manufacture includes the reaction, carried out at elevated temperatures:

$$3KCl(s) + 4HNO_3(l) \longrightarrow Cl_2(g) + NOCl(g) + 2H_2O(g) + 3KNO_3(s)$$

What mass of KNO_3 will be produced when 100-g quantities of each reactant are combined?

SOLUTION

Since the equation is defined in terms of moles, the given masses of the reagents must be converted to moles by dividing by the appropriate molar masses:

$$\text{amount KCl} = \frac{100 \text{ g KCl}}{74.6 \text{ g KCl/mol KCl}} = 1.34 \text{ mol KCl}$$

$$\text{amount HNO}_3 = \frac{100 \text{ g HNO}_3}{63.0 \text{ g HNO}_3/\text{mol HNO}_3} = 1.59 \text{ mol HNO}_3$$

Start with either of the reactants, say KCl, and calculate the number of moles of the other reactant that are needed to combine completely with the first reactant. If KCl is chosen:

$$\text{amount HNO}_3 \text{ needed} = 1.34 \text{ mol KCl} \times \frac{4 \text{ mol HNO}_3}{3 \text{ mol KCl}}$$

$$= 1.79 \text{ mol HNO}_3 \text{ needed}$$

Compare the calculated number of moles of the reactant with the number of moles that are actually present. If the number of moles present is less than the calculated number, this reactant is the limiting reagent; it will be consumed first.

In this case, 1.59 mol of HNO_3 is present, but 1.79 mol of HNO_3 is needed for conversion of all the KCl. There is insufficient HNO_3, so HNO_3 is the limiting reagent. If the amount of HNO_3 present had been greater than the calculated amount, then the other reactant, KCl, would have been limiting.

The remainder of this calculation is based on the quantity of the limiting reagent. The method of calculation is the same as that of Example 3.7. The mass of HNO_3 is converted to amount:

$$100 \text{ g HNO}_3 \times \frac{1 \text{ mol HNO}_3}{63.0 \text{ g HNO}_3} = 1.59 \text{ mol HNO}_3$$

The amount of KNO_3 that can be produced from this amount of HNO_3 is then calculated:

$$1.59 \text{ mol HNO}_3 \times \frac{3 \text{ mol KNO}_3}{4 \text{ mol HNO}_3} = 1.19 \text{ mol KNO}_3$$

The amount of KNO_3 is converted to a mass of KNO_3, using the value of the molar mass calculated with the aid of a table of atomic weights:

$$1.19 \text{ mol KNO}_3 \times \frac{101.1 \text{ g KNO}_3}{1 \text{ mol KNO}_3} = 120 \text{ g KNO}_3$$

Note that 120 g of KNO_3 is produced from 100 g of HNO_3, no matter whether 1000 g or 1 000 000 g of KCl is present. The amount of the limiting reagent, HNO_3, that is present determines the quantities of materials that take part in the reaction and the quantities of products formed.

EXAMPLE 3.10

How much KCl remains after the above reaction is completed?

SOLUTION

First we calculate the mass of KCl consumed. This calculation can be done in one step, as we have seen:

$$100 \text{ g HNO}_3 \times \frac{1 \text{ mol HNO}_3}{63.0 \text{ g HNO}_3} \times \frac{3 \text{ mol KCl}}{4 \text{ mol HNO}_3} \times \frac{74.6 \text{ g KCl}}{1 \text{ mol KCl}} = 88.8 \text{ g KCl}$$

The quantity of KCl remaining after the reaction goes to completion is the original quantity less the quantity consumed: 100 g − 88.8 g = 11 g KCl remaining. (Since KCl is present in excess, some must remain when the reaction is completed.)

The concept of a limiting reagent is familiar in everyday life, as the following example shows.

EXAMPLE 3.11

Humans consume food, which provides energy for life processes when the food is oxidized in the body. This oxidation can be represented, in greatly oversimplified fashion, as the oxidation of sugar, the reaction of glucose with oxygen:

$$C_6H_{12}O_6 + 6O_2 \longrightarrow 6CO_2 + 6H_2O + energy$$

In an average day, an individual consumes the equivalent of 500 g of glucose and inhales about 3000 g of oxygen. Calculate the amount of CO_2 formed during the individual's daily activity.

SOLUTION

First convert the given masses to moles:

$$\text{amount glucose} = \frac{500 \text{ g glucose}}{180 \text{ g glucose/mol glucose}} = 2.8 \text{ mol glucose}$$

$$\text{amount } O_2 = \frac{3000 \text{ g } O_2}{32 \text{ g } O_2/\text{mol } O_2} = 96 \text{ mol } O_2$$

The number of moles of O_2 needed to combine completely with 2.8 mol of glucose is:

$$2.8 \text{ mol glucose} \times \frac{6 \text{ mol } O_2}{1 \text{ mol glucose}} = 17 \text{ mol } O_2$$

Since only 17 mol of O_2 is needed and 96 mol is present, O_2 is in excess and glucose is the limiting reagent.

The number of moles of CO_2 produced is:

$$2.8 \text{ mol glucose} \times \frac{6 \text{ mol } CO_2}{1 \text{ mol glucose}} = 17 \text{ mol } CO_2$$

The calculation shows that the amount of energy that the human body produces to sustain itself, which is directly related to the amount of CO_2 produced, is limited by the quantity of food consumed, not by the quantity of atmospheric oxygen that is available normally.

We mentioned earlier that a reaction carried out in a laboratory will rarely produce the theoretical yield. Almost invariably, the amount of product actually obtained is less than the amount expected theoretically. The reaction does not go to completion because of side reactions or for other reasons, such as poor experimental techniques on the part of the chemist. The quantity of product that is actually obtained is expressed as a *percent yield,* which is defined by the expression:

$$\text{percent yield} = \frac{\text{quantity obtained}}{\text{theoretical quantity}} \times 100 \qquad (3.2)$$

EXAMPLE 3.12

Zinc oxide, ZnO, which is used as an antiseptic and as a white pigment in paint, can be prepared by burning zinc. A sample of zinc metal of mass 125 g is vaporized and is burned in excess oxygen. It is possible to isolate 131 g of ZnO after the reaction is completed. Calculate the percent yield of ZnO.

SOLUTION

The theoretical yield of ZnO is calculated from the equation:

$$2Zn(g) + O_2(g) \longrightarrow 2ZnO(s)$$

which indicates that 1 mol of Zn theoretically forms 1 mol of ZnO. The starting mass of Zn is 125 g:

$$125 \text{ g Zn} \times \frac{1 \text{ mol Zn}}{65.4 \text{ g Zn}} = 1.91 \text{ mol Zn}$$

and therefore 1.91 mol of ZnO theoretically could be formed. The theoretical mass of ZnO to be expected is therefore:

$$1.91 \text{ mol ZnO} \times \frac{81.4 \text{ g ZnO}}{1 \text{ mol ZnO}} = 155 \text{ g ZnO}$$

Equation 3.2 then gives:

$$\text{percent yield} = \frac{131 \text{ g ZnO obtained}}{155 \text{ g ZnO theoretical}} \times 100 = 84.3\%$$

Note that the theoretical amount is calculated from the limiting reagent, in this case Zn.

The relationships implied by chemical formulas and chemical equations can be used in many different ways. In the nineteenth century, the atomic weights of newly discovered elements were determined by measuring the quantities used in chemical conversions.

EXAMPLE 3.13

Imagine that it is the middle of the nineteenth century and that a new element has been discovered that seems to have chemical properties similar to those of sodium and potassium. Because of this, the element is assigned an oxidation state of $+1$ in its salts. Because the element emits deep red light when heated in a flame, it has been named rubidium. Its symbol is Rb. A sample of rubidium carbonate (which presumably has the formula Rb_2CO_3, by analogy with Na_2CO_3 and K_2CO_3) is treated with a strong acid. The following reaction occurs:

$$Rb_2CO_3(s) + 2H^+ \longrightarrow CO_2(g) + H_2O + 2Rb^+$$

It is found that a 0.1475-g sample of Rb_2CO_3 produces 0.0281 g of CO_2. Assuming that the reaction goes to completion, that the atomic weight of C is known to be 12.01, and that the atomic weight of O is known to be 16.00, calculate the atomic weight of Rb.

SOLUTION

From the chemical equation:

$$\text{moles } Rb_2CO_3 = \text{moles } CO_2$$

and from Equation 3.1 we can get:

$$\frac{\text{mass } Rb_2CO_3}{\text{MW } Rb_2CO_3} = \frac{\text{mass } CO_2}{\text{MW } CO_2}$$

where MW is the molecular weight.

All the quantities in the relationship but MW of Rb_2CO_3 are given. Substituting:

$$\frac{0.1475 \text{ g } Rb_2CO_3}{\text{MW } Rb_2CO_3} = \frac{0.0281 \text{ g } CO_2}{44.01}$$

$$\text{MW } Rb_2CO_3 = 231.0$$

If we let x be the atomic weight of Rb, then:

$$\text{MW } Rb_2CO_3 = 2x + 12.01 + (3)(16.00) = 231.0$$

$$x = 85.5 = \text{atomic weight of Rb}$$

3.4 THE STOICHIOMETRY OF SOLUTIONS

Many chemical reactions occur in solutions, particularly in water solution, so we need convenient ways to indicate the composition of solutions. We can indicate the composition of solutions in several ways. Although we shall not discuss solutions in detail until Chapter 12, we introduce them here because many of the experiments you will carry out in chemistry laboratory are reactions in aqueous solution.

Most aqueous solutions are formed of relatively small quantities of substances called *solutes* that are dissolved in water, which is called the *solvent*. A solution is homogeneous, having the same composition throughout, and so is defined by indicating the relative quantities of solutes and solvent, or the *concentration* of the solution.

The most direct way to describe concentration is to describe the mass percent of the solution, defined as:

$$\text{mass percent} = \frac{\text{mass of solute}}{\text{mass of solution}} \times 100$$

Thus, the expression "a 10% solution of sodium chloride" refers to a solution consisting of 10% NaCl, the solute, and 90% H_2O, the solvent. If we had 100 g of such a solution, it would contain 10 g of NaCl and 90 g of H_2O.

EXAMPLE 3.14
What mass of ammonium nitrate must be dissolved in 500 g of water to make a 25% solution?
SOLUTION
Let x = mass of NH_4NO_3, ammonium nitrate, in grams required to form a 25% solution. From the definition of mass percent:

$$25\% = \left(\frac{x}{500 + x} \right) \times 100$$

$$x = 167 \text{ g } NH_4NO_3$$

While nonchemists usually describe solutions in terms of mass percent, this expression is somewhat inconvenient in the chemical laboratory. It is usually easier to measure the volume of a solution than to measure the mass of a solution. The most common expression used by chemists to indicate the quantity of solute in a given quantity of solution is **molarity**. Molarity, symbolized by M, is an expression of concentration defined as:

$$M = \frac{\text{moles of solute}}{\text{volume of solution}} \qquad (3.3)$$

where the unit of volume is the cubic decimeter (dm^3), more commonly called the liter. That is, **molarity is the number of moles of solute in one liter of solution.**

Note that the denominator of the expression in Equation 3.3 refers to the entire solution, not just to the amount of solvent. Thus, the expression "a $2M$ solution of NaCl" means that one liter of this solution contains two moles of dissolved NaCl.

We can use other information to expand the meaning of such a statement about a solution. For example, a table of atomic weights can be used to determine the mass of the two moles of NaCl in a liter of solution. The number of particles in solution can be described, if we use the fact that NaCl in aqueous solution dissociates into Na^+ and Cl^- ions (Section 2.6, page 38), together with the fact that N ions of each kind are formed when a mole of NaCl dissolves. Thus, it can be stated that one liter of a $2M$ NaCl solution contains two moles of Na^+ ions and two moles of Cl^- ions, whose combined mass is 116.9 g (twice the molar mass of NaCl) and whose total number is $(2 + 2) (6.02 \times 10^{23}) = 2.41 \times 10^{24}$.

The relationship given as Equation 3.3 is often used when working with solutions. The equation has three terms: molarity, moles, and volume in liters. Given any two terms, the third is readily determined.

EXAMPLE 3.15
What mass of $CaCl_2$ is needed to prepare 0.32 liter of a 0.75M solution?
SOLUTION
Let x = number of moles of $CaCl_2$ required. Substitution into Equation 3.3 gives:

$$0.75M = \frac{x \text{ mol } CaCl_2}{0.32 \text{ liter solution}}$$

$$x = 0.24 \text{ mol } CaCl_2$$

$$\text{mass } CaCl_2 = 0.24 \text{ mol } CaCl_2 \times \frac{111 \text{ g } CaCl_2}{1 \text{ mol } CaCl_2}$$

$$= 27 \text{ g } CaCl_2$$

EXAMPLE 3.16
What volume of the above solution contains 38 g of $CaCl_2$?
SOLUTION
Convert 38 g of $CaCl_2$ to moles:

$$\frac{38 \text{ g } CaCl_2}{111 \text{ g } CaCl_2/\text{mol } CaCl_2} = 0.34 \text{ mol } CaCl_2$$

Let x = the number of liters of solution required. Substitute into Equation 3.3:

$$0.75M = \frac{0.34 \text{ mol } CaCl_2}{x \text{ liter solution}}$$

$$x = 0.46 \text{ liter required}$$

In most cases, we shall have information about the volume and molarity of a solution. These two pieces of information give the number of moles of the solute, through straightforward application of Equation 3.3. Once the number of moles of the solute is known, the mass of the solute can be found by the standard method.

EXAMPLE 3.17
When 0.100 liter of a 1.30M solution of $AgNO_3$ is mixed with an excess of an NaCl solution, an AgCl precipitate forms according to the reaction:

$$Ag^+ + Cl^- \longrightarrow AgCl(s)$$

What mass of AgCl(s) is formed if essentially all the $AgNO_3$ is converted to AgCl(s)?
SOLUTION
Since there is one Ag^+ ion in each unit of $AgNO_3$, the moles of Ag^+ ion in solution is equal to the moles of $AgNO_3$ dissolved. The equation shows that moles of AgCl formed is equal to moles of Ag^+ and therefore equal to moles of $AgNO_3$. The number of moles of $AgNO_3$ in the solution thus can be calculated:

$$1.30M = \frac{x \text{ mol } AgNO_3}{0.100 \text{ liter solution}}$$

$$x = 0.130 \text{ mol } AgNO_3$$

Therefore, 0.130 mol of AgCl is formed.

$$\text{mass AgCl} = 0.130 \text{ mol AgCl} \times \frac{(108 + 35.5) \text{ g AgCl}}{1 \text{ mol AgCl}}$$

$$= 18.7 \text{ g AgCl}$$

EXAMPLE 3.18

How many liters of 0.326M NaCl solution contains just the quantity of Cl⁻ necessary for the conversion in Example 3.17?

SOLUTION

From the equation, we see that 0.130 mol of NaCl is required, equal to the moles of $AgNO_3$. Let x = number of liters of solution. Then:

$$0.326M = \frac{0.130 \text{ mol NaCl}}{x \text{ liter solution}}$$

$$x = 0.400 \text{ liter solution}$$

We can formulate many variations of the previous examples. Solution problems of this type can be solved by using the definition of molarity and its relationship to the concept of the mole. As soon as you know the number of moles of the reactants, the problem is one of ordinary stoichiometric relationships.

EXAMPLE 3.19

One way to remove phosphates from sewage is by precipitation with calcium oxide, CaO. What mass of calcium oxide is needed to precipitate all the phosphate from 2000 liters of sewage that has a concentration of phosphate equal to 0.0022M, assuming that the precipitation is complete? The relevant reactions are:

$$CaO(s) + H_2O \longrightarrow Ca^{2+} + 2OH^-$$
$$3Ca^{2+} + 2PO_4^{3-} \longrightarrow Ca_3(PO_4)_2(s)$$

SOLUTION

Calculate the amount of phosphate in the quantity of solution:

$$0.022M = \frac{x \text{ mol } PO_4^{3-}}{2000 \text{ liters sewage}}$$

$$x = 4.4 \text{ mol } PO_4^{3-}$$

From the equations, we can write the conversion factors:

$$\frac{1 \text{ mol CaO}}{1 \text{ mol } Ca^{2+}} \quad \text{and} \quad \frac{3 \text{ mol } Ca^{2+}}{2 \text{ mol } PO_4^{3-}}$$

Therefore:

$$\text{amount CaO(s)} = 4.4 \text{ mol } PO_4^{3-} \times \frac{3 \text{ mol } Ca^{2+}}{2 \text{ mol } PO_4^{3-}} \times \frac{1 \text{ mol CaO}}{1 \text{ mol } Ca^{2+}}$$

$$= 6.6 \text{ mol CaO}$$

$$6.6 \text{ mol CaO} \times \frac{56 \text{ g CaO}}{1 \text{ mol CaO}} = 370 \text{ g CaO}$$

Relationships between the mass and the volume of a solution often are important in calculations concerning the stoichiometry of solutions. For example, we have already defined two units of concentration, mass percent and molarity, without defining the relationship between them. To find this relationship we must know the relationship between the mass and the volume of a given solution.

The relationship between mass and volume usually is expressed by the density (d) of the solution. Density can be understood to be the mass of a given volume of solution or of any homogeneous material:

$$d = \frac{\text{mass}}{\text{volume}}$$

For liquids and solids, mass is usually measured in grams and volume in

TABLE 3.2
Densities of Substances at Room Temperature

Substance	Density (g/cm^{-3})
Liquids	
butane, C_4H_{10}	0.58
octane, C_8H_{18}	0.70
ethanol, C_2H_5OH	0.79
water, H_2O	1.00
bromine, Br_2	3.12
mercury, Hg	13.6
Solids	
lithium, Li	0.53
magnesium, Mg	1.74
sodium chloride, NaCl	2.16
graphite, C	2.27
aluminum, Al	2.70
limestone, $CaCO_3$	2.71
lime, CaO	3.35
diamond, C	3.52
corundum, Al_2O_3	3.99
copper, Cu	8.93
lead, Pb	11.3
platinum, Pt	21.5
osmium, Os	22.5

cubic centimeters (cm^3). One cubic centimeter is one-thousandth of a liter.
Density has the units grams per cubic centimeter. The density of liquid water, for
example, has a value close to 1 g/cm^3. The density of pure solids, liquids, and
gases, as well as of solutions can be measured. Table 3.2 lists the densities of
some common substances.

EXAMPLE 3.20
A solution is prepared by dissolving 13.8 g of sucrose, $C_{12}H_{22}O_{11}$, in 38.4 g
of H_2O. The solution has a volume of 0.0470 liter. Find the density of the
solution.
SOLUTION
To express the volume of solution in cubic centimeters, we use the conversion factor 1000 cm^3/1 liter:

$$0.0470 \text{ liter solution} \times \frac{1000 \text{ cm}^3}{1 \text{ liter}} = 47.0 \text{ cm}^3 \text{ solution}$$

$$d = \frac{38.4 \text{ g H}_2\text{O} + 13.8 \text{ g sucrose}}{47.0 \text{ cm}^3 \text{ solution}}$$

$$= 1.11 \text{ g/cm}^3$$

EXAMPLE 3.21
A water solution of H_2SO_4 is 28% H_2SO_4 by mass and has a density of
1.20 g/cm^3. Calculate the molarity of the solution.
SOLUTION
We can use density as a conversion factor to get the mass of one liter of
solution.

$$1 \text{ liter solution} \times \frac{1000 \text{ cm}^3 \text{ solution}}{1 \text{ liter solution}} \times \frac{1.20 \text{ g solution}}{1 \text{ cm}^3 \text{ solution}} = 1200 \text{ g solution}$$

We are told that the solution is 28.0% H_2SO_4. Now that we know the mass of one liter of the solution, we can calculate that 1200 g of the solution contains:

$$1200 \text{ g solution} \times \frac{28 \text{ g } H_2SO_4}{100 \text{ g solution}} = 336 \text{ g } H_2SO_4$$

$$336 \text{ g } H_2SO_4 \times \frac{1 \text{ mol } H_2SO_4}{98.1 \text{ g } H_2SO_4} = 3.43 \text{ mol } H_2SO_4$$

Since 1200 g of solution has a volume of 1 liter and there is 3.43 mol of H_2SO_4 in this solution, the concentration of H_2SO_4 is 3.43M.

3.5 COMPOSITION OF MIXTURES

Stoichiometric reasoning can be used to calculate the composition of mixtures. We shall present a general method that can be used to calculate the composition of mixtures. Once the general method is mastered, it can be simplified in many cases.

To calculate the composition of a two-component mixture, we need two equations in two unknowns. One equation can be called the mass equation, since it usually depends on or expresses a relationship between the masses of the components and the total mass of the mixture. The second equation can be called the amount or molar relationship equation. This equation usually depends on or expresses a relationship between the number of moles of the components of the mixture and the number of moles of some product formed from them. To set up the amount equation, we usually need a chemical formula and a table of atomic weights. Information about a specific chemical reaction may also be needed.

EXAMPLE 3.22

A 2.00-g sample of a mixture of NaCl and NaBr is found to contain 0.75 g of Na. What is the fraction of NaCl in the mixture?

SOLUTION

Let x = mass of NaCl in grams and y = mass of NaBr in grams. The mass equation follows directly:

$$x + y = 2.00$$

The molar relationship equation can be set up by following this line of reasoning: The mass of Na in the mixture is given. Therefore, the number of moles of Na in the mixture can be found by using the atomic weight of Na. The relationship between the number of moles of Na and the number of moles of the other components is defined by the chemical formulas of the components of the mixture. There is one mole of Na in one mole of NaCl, and one mole of Na in one mole of NaBr. This can be presented as:

moles Na in NaCl + moles Na in NaBr = total moles Na

or

moles NaCl + moles NaBr = total moles Na

$$\text{moles NaCl} = \frac{\text{mass NaCl}}{\text{MW NaCl}} = \frac{x}{58.4}$$

$$\text{moles NaBr} = \frac{\text{mass NaBr}}{\text{MW NaBr}} = \frac{y}{103}$$

$$\text{moles Na} = \frac{\text{mass Na}}{\text{atomic weight Na}} = \frac{0.75}{23.0}$$

Substituting these molar expressions into the above word equation gives the molar relationship equation:

$$\frac{x}{58.4} + \frac{y}{103} = \frac{0.75}{23.0}$$

The mass equation can be written as:

$$y = 2 - x$$

Substituting this expression for y into the molar relationship equation gives:

$$x = 1.78 \text{ g NaCl} = \text{mass NaCl}$$
$$y = 0.22 \text{ g NaBr} = \text{mass NaBr}$$

$$\text{fraction NaCl} = \frac{\text{mass NaCl}}{\text{total mass}} = \frac{1.78 \text{ g NaCl}}{2.00 \text{ g mixture}} = 0.89$$

If the identities of the components of a mixture are known, the composition of the mixture can be analyzed by converting the components to a substance whose mass can then be measured.

EXAMPLE 3.23

A 3.20-g sample of a mixture of NaCl and $CaCl_2$ is dissolved in water and treated with excess silver nitrate solution. All the Cl in the mixture is converted to a precipitate of silver chloride, AgCl, according to the reaction:

$$Ag^+ + Cl^- \longrightarrow AgCl(s)$$

The precipitate is found to have a mass of 7.94 g. Calculate the mass of sodium chloride in the mixture.

SOLUTION

Let x = mass of NaCl in grams and y = mass of $CaCl_2$ in grams. The mass equation follows directly:

$$x + y = 3.20$$

The molar relationship can be set up by following this line of reasoning: The mass of the AgCl precipitate is given. Therefore, the number of moles of AgCl can be found by using the table of atomic weights:

$$7.94 \text{ g AgCl} \times \frac{1 \text{ mol AgCl}}{143.3 \text{ g AgCl}} = 0.0554 \text{ mol AgCl}$$

Each mole of NaCl forms one mole of AgCl, and each mole of $CaCl_2$ forms two moles of AgCl. Therefore, the number of moles of AgCl equals the number of moles of NaCl plus twice the number of moles of $CaCl_2$. That is:

$$\text{moles NaCl} = \frac{\text{mass NaCl}}{\text{MW NaCl}} = \frac{x}{58.44}$$

$$\text{moles CaCl}_2 = \frac{\text{mass CaCl}_2}{\text{MW CaCl}_2} = \frac{y}{111.0}$$

$$\text{moles NaCl} + (2)\text{moles CaCl}_2 = \text{moles AgCl}$$

or

$$\frac{x}{58.44} + \frac{2y}{111.0} = 0.0554$$

Solution of these two equations in x and y gives:

$$x = 2.49 \text{ g NaCl}$$

Calculating molar relationships and mass relationships in chemical processes is basic to the performance of chemistry. Such calculations are done constantly by practicing chemists. The beginning student of chemistry should take pains to master these calculations. One observation may be helpful: If there is a common denominator in these calculations, it is the conversion from mass (grams) to amount (moles), and from amount to mass, using the data given in the table of atomic weights.

EXERCISES

3.1 The mass of one pencil is 10 g. Find the amount of pencils that has a mass of 500 g. Express the answer in three different units of amount suitable for pencils.

3.2 The mass of two dozen eggs is 1200 g. Find the mass of an average egg.

3.3 A unit of measure sometimes used for blueberries is the cup. Let us define one cup of blueberries to be 2.0×10^2 blueberries. The mass of an average blueberry is 0.75 g. A bushel of blueberries contain 150 cups. Find the mass of blueberries in a bushel.

3.4 The mass of an average blueberry is 0.75 g. Find the mass of one mole of blueberries.

3.5 The mass of an automobile is 2.0×10^3 kg. Find the number of automobiles whose total mass is the same as the mass of the mole of blueberries in the previous exercise.

3.6 The mass of an average blueberry is 0.75 g. Find the mass of one blueberry expressed in atomic mass units.

3.7 The ratio of the mass of a gold atom to the mass of a ^{12}C atom is 50:3. Find the mass of one mole of gold atoms.

3.8 The ratio of the mass of an oxygen atom to an atom of ^{12}C is 4:3. The ratio of sulfur and oxygen by mass in SO_2 is 1:1. Find the mass of one mole of S atoms.

3.9 A sample of bromine of natural origin is found to consist of two isotopes: ^{79}Br, natural abundance 50.54%, and ^{81}Br, natural abundance 49.46%. The relative atomic masses of these isotopes are 78.918 and 80.916 respectively. Find the atomic weight of bromine.

3.10 A sample of strontium of natural origin is found to consist of three major isotopes: ^{86}Sr, natural abundance 9.86%, ^{87}Sr, natural abundance 7.02%, and ^{88}Sr, natural abundance 82.56%.

Their relative atomic masses are 85.909, 86.909, and 87.906 respectively. Find the atomic weight of strontium.

3.11 The atomic weight of chlorine is 35.453. It consists of only two isotopes. One of them is ^{35}Cl, natural abundance 75.53%, relative atomic mass 34.969. Find the relative atomic mass of the other isotope of chlorine.

3.12 Naturally occurring silver has an atomic weight of 107.87 and consists of only two isotopes: ^{107}Ag, relative atomic mass 106.905, and ^{109}Ag relative atomic mass 108.905. Find the natural abundance of these two isotopes.

3.13 Find the sum of the atomic weights of all elements that have four letters in their names.

3.14 Find the molecular weight of the following substances: (a) hydrogen cyanide, (b) arsenic trichloride, (c) dinitrogen pentoxide, (d) phosphoric acid.

3.15 Find the mass of one mole of the following substances: (a) lithium fluoride, (b) calcium nitrate, (c) ammonium phosphate, (d) aluminum sulfate.

3.16 The general formula for one class of hydrocarbons called the alkenes is C_nH_{2n}, where n is the number of carbon atoms in any given member of the class. Octene and undecene are the names of two alkenes. Undecene has three more carbon atoms than octene. Find the difference in the molecular weight of the two compounds.

3.17 Find the amounts of the following substances that have a mass of 125 g: (a) hydrogen cyanide, (b) arsenic trichloride, (c) glucose, $C_6H_{12}O_6$.

3.18 Find the mass of 0.722 mol of the following substances: (a) lithium fluoride, (b) calcium nitrate, (c) aluminum sulfate.

3.19 Elemental sulfur exists in a number of different forms at different temper-

atures. At room temperature it is primarily S_8, but as the temperature is raised it can be converted virtually completely to S_2 gas. At still higher temperatures the S_2 can be converted completely to gaseous sulfur atoms. Find the amount of each of these three different forms of sulfur that has a mass of 64.1 g.

3.20 The mass of 0.234 mol of a substance is 53.7 g. Find the molecular weight of the substance.

3.21 The mass of 0.0863 mol of an acid HXO_4 of the element X is 12.5 g. Find the atomic weight of the element X. Identify X.

3.22 The mass of an average molecule of a substance is 2.59×10^{-22} g. Find the molecular weight of the substance.

3.23 Find the number of molecules of H_3PO_4 in a 45.7-g sample of pure H_3PO_4.

3.24 Find the number of atoms in a 35.2-kg sample of uranium.

3.25 Find the total number of atoms in 12.9 g of $(NH_4)_2Cr_2O_7$.

3.26 Find the mass of 2.41×10^{30} molecules of water.

3.27 Find the amount of O contained in 0.85 mol of the following substances: (a) NO_2, (b) N_2O_4, (c) $Al_2(SO_4)_3$, (d) $Fe_2(CO)_9$.

3.28 Find the number of molecules of H_2 in 1.0 fg (10^{-15} g) of H_2.

3.29 Find the number of hydrogen atoms in a 31.7-g sample of acetic acid, CH_3CO_2H.

3.30 Find the amount of N in a 192-g sample of TNT, $C_7H_5N_3O_6$.

3.31 Find the amount of penicillin, $C_{16}H_{18}N_2O_4S$, that contains 0.10 mol of C.

3.32 Find the amount of SiO_2 that contains 1.0×10^{20} atoms of O.

3.33 Find the percentage by mass of oxygen in water.

3.34 Find the percentage by mass of each element in the insecticide DDT, $C_{14}H_9Cl_5$.

3.35 Find which of the oxides of vanadium, VO, V_2O_3, VO_2, or V_2O_5, has the highest percentage by mass of vanadium. Find the percentage.

3.36 Some of the important ores of zinc are ZnS, zinc blende, $ZnCO_3$, franklinite; and Zn_2SiO_4, willemite. Find the percentage by mass of zinc in the ore that is richest in zinc.

3.37 It is found that 50.7 g of S combines with 93.8 g of Sn to form one compound. Find the empirical formula of the compound.

3.38 A sample of mass 60.5 g of a compound containing only nickel, carbon, and oxygen is found to contain 20.8 g of Ni and 17.0 g of C. Find the empirical formula of the compound.

3.39 An unknown substance is found to contain 55.8% carbon, 11.6% hydrogen, and 32.6% nitrogen by mass. Find the empirical formula of the compound.

3.40 A compound is found to have the empirical formula CH and the molecular weight 103. Find the molecular formula of the compound.

*3.41 A binary compound of the unknown elements X and Y is $\frac{1}{3}$ X by mass. The atomic weight of element X is $\frac{3}{4}$ the atomic weight of element Y. Calculate the empirical formula of the compound.

3.42 Complete combustion of a sample of a compound containing only carbon and hydrogen produced 1.404 g of CO_2 and 0.764 g of H_2O. Find the empirical formula of the compound.

3.43 A 1.46-g sample of a compound containing only carbon, hydrogen, and oxygen is burned in excess O_2 to form 3.57 g of CO_2 and 1.45 g of H_2O. Find the empirical formula of the compound.

3.44 A 1.05-g sample of a compound containing only carbon, hydrogen, and nitrogen is burned in excess O_2 to form 2.92 g of CO_2 and 0.581 g of H_2O. The molecular weight of the compound is found to be 238. Find the molecular formula of the compound.

*3.45 The following data were obtained from experiments to find the molecular formula of benzocaine, a local anesthetic, which contains only carbon, hydrogen, nitrogen, and oxygen. Complete combustion of a 3.54-g sample of benzocaine with excess O_2 formed 8.49 g of CO_2 and 2.14 g of H_2O. Another sample of mass 2.35 g was found to contain 0.199 g of N. The molecular weight of benzocaine was found to be

165. Find the molecular formula of benzocaine.

3.46 At elevated temperatures the compound NF_3 decomposes to N_2 and F_2. Find the amounts of N_2 and F_2 formed by the decomposition of: (a) exactly 3 mol of NF_3, (b) 0.268 mol of NF_3.

3.47 Direct reaction of P with Cl_2 can form PCl_5 under suitable conditions. Find the mass of Cl_2 required to form 2.70 mol of PCl_5.

3.48 Find the mass of $AsCl_3$ formed by the reaction of 0.133 mol of Cl_2 with arsenic.

3.49 An important step in the manufacture of nitric acid is the reaction of ammonia with O_2 to form NO and water. Find the amount of O_2 required to react completely with 76 g of ammonia. Find the mass of water and of NO produced.

3.50 The Cl_2 formed by the decomposition of 1.3 mol of PCl_3 is used to convert carbon to CCl_4. Find the amount of CCl_4 formed.

*3.51 Hydrolysis of the compound B_5H_9 forms boric acid, $B(OH)_3$. Fusion of boric acid with Na_2O forms a borate salt, $Na_2B_4O_7$. Find the amount of B_5H_9 that forms 0.75 mol of the borate salt by this reaction sequence. (*Hint:* The solution can be obtained without writing complete equations.)

3.52 Carbon disulfide, CS_2, is a very flammable substance that reacts with O_2 to form carbon dioxide and sulfur dioxide. Find the mass of O_2 that is required to react with 9.34 g of CS_2. Find the mass of carbon dioxide and of sulfur dioxide formed in this reaction.

3.53 An important reaction that takes place in a blast furnace during the production of iron is the formation of iron metal and carbon dioxide from Fe_2O_3 and carbon monoxide. Find the mass of Fe_2O_3 required to form 910 kg of iron. Find the amount of carbon dioxide that forms in this process.

3.54 A certain oxide of lead is converted by H_2 to lead metal and water. One mole of the oxide reacts with 8.1 g of H_2 and forms 622 g of lead. Find the formula of the oxide.

3.55 Incomplete combustion of hydrocarbons is an important cause of air pollution because it produces carbon monoxide. Find the difference between the mass of O_2 required to convert one mole of octane, C_8H_{18}, completely to CO and the mass of O_2 required to convert one mole of octane completely to CO_2.

3.56 Quantities of 11.1 g of H_2 and 33.3 g of Cl_2 are mixed. Find the mass of hydrogen chloride that forms.

3.57 Reaction of tungsten with Cl_2 forms WCl_6. Find the mass of the unreacted starting material when 12.6 g of tungsten is treated with 13.6 g of Cl_2 and the reaction takes place. Find the mass of WCl_6 that forms.

3.58 Gold is generally a very unreactive metal, one reason it is so highly prized. However, treatment of gold with BrF_3 and KF leads to the formation of Br_2 and the salt $KAuF_4$. Find the limiting reagent when equal masses of the three reagents are mixed. Find the mass of the gold salt formed from 24 g of such a mixture.

*3.59 Metallic aluminum reacts with MnO_2 at elevated temperatures to form manganese metal and aluminum oxide. A mixture of the two reactants is 39% Al by mass. Find the limiting reagent in this mixture. Find the mass of manganese that forms from the reaction of 250 g of this mixture.

3.60 When HgO is heated it decomposes to mercury and O_2. Careful technique makes it possible to isolate 62.7 g of mercury from the decomposition of 75.8 g of the oxide. Find the percent yield of the reaction.

3.61 The reaction of carbon with calcium oxide produces carbon monoxide and calcium carbide, CaC_2. A total of 2.45 g of CaC_2 is isolated from the reaction of 5.00 g of calcium oxide with 2.50 g of carbon. Find the percent yield of the CaC_2.

3.62 Large quantities of formic acid, HCOOH, can be prepared in 76% yield from the reaction of carbon monoxide with sodium hydroxide. Find the mass of carbon monoxide needed to prepare 125 kg of formic acid.

3.63 The oxide of an unknown element is believed to have the formula XO_2. Heating a 25.2-g sample of the compound decomposes it completely to form 3.2 g of O_2. Find the atomic weight of X, the unknown element.

*3.64 The chloride of an unknown metal is believed to have the formula MCl_3. A 2.395-g sample of the chloride is dissolved in water and treated with excess silver nitrate solution. The mass of the AgCl precipitate formed is found to be 5.168 g. Find the atomic weight of M, the unknown metal.

3.65 Find the mass of sodium chloride and the mass of water needed to prepare 125 g of an 18.0% sodium chloride solution by mass.

3.66 Find the amount of hydrogen chloride that must be dissolved in 750 g of water to prepare a 25.0% hydrochloric acid solution by mass.

3.67 Find the mass of potassium sulfate required to prepare 0.800 liter of a $0.752M$ solution of this salt.

3.68 Find the molarity of a solution that contains 377 g of acetone, C_3H_6O, in a volume of 0.762 liter.

3.69 Find the volume of $0.227M$ nitric acid solution that contains 1.45 mol of nitric acid.

3.70 Find the mass of $PbCl_2$ that precipitates when excess lead nitrate solution is added to 0.250 liter of $1.08M$ NaCl solution.

3.71 Find the volume of $0.89M$ sodium iodide solution necessary to precipitate all the mercury from 0.051 liter of a $0.65M$ $Hg(NO_3)_2$ solution. The precipitate is HgI_2.

3.72 Find the volume of $0.110M$ hydrochloric acid necessary to react completely with 1.52 g of $Al(OH)_3$.

3.73 Find the molarity of Cl^- in a solution prepared by mixing 0.10 liter of $0.12M$ sodium chloride solution with 0.23 liter of $0.18M$ magnesium chloride solution.

*3.74 Find the mass of barium metal that must react with O_2 in order to produce enough barium oxide to prepare 1.0 liter of a $0.30M$ solution of OH^-.

3.75 Use the data in Table 3.2 to find the mass of 15 cm³ of the following: (a) octane, (b) bromine, (c) diamond, (d) lead, (e) osmium.

3.76 The density of a 20.0% by mass ethylene glycol ($C_2H_6O_2$) solution in water is 1.03 g/cm³. Find the molarity of the solution.

3.77 The density of a $10.0M$ solution of methanol, CH_3OH, in water is 0.947 g/cm³. Find the mass percent of methanol in the solution.

3.78 A mixture of carbon and sulfur has a mass of 9.0 g. Complete combustion with excess O_2 gives 23.3 g of a mixture of carbon dioxide and sulfur dioxide. Find the mass of sulfur in the original mixture.

3.79 A mixture of $CaCO_3$ and $(NH_4)_2CO_3$ is 61.9% by mass CO_3. Find the mass percent of $CaCO_3$ in the mixture.

*3.80 A mixture of 50.0 g of S and 100 g of Cl_2 reacts completely to form S_2Cl_2 and SCl_2 and no other products. Find the mass of S_2Cl_2 formed.

*3.81 A solution contains Cr^{3+} ion and Mg^{2+} ion. The addition of 1.00 liter of $1.51M$ NaF solution is just required to cause the complete precipitation of these ions as $CrF_3(s)$ and $MgF_2(s)$. The total mass of the precipitate is 49.6 g. Find the mass of the $CrF_3(s)$ precipitate and the mass of Cr^{3+} in the original solution.

*3.82 A mixture of C_3H_8 and C_2H_2 has a mass of 2.0 g. It is burned in excess O_2 to form a mixture of water and carbon dioxide that contains 1.5 times as many moles of CO_2 as of water. Find the mass of C_2H_2 in the original mixture.

4

4.1 STATES OF MATTER

Matter on earth exists in three different states: solid, liquid, and gas. The vast majority of the substances on earth are solids: the book you are reading, the table on which the book rests, the dwelling that contains the table all are composed of solid substances. A single liquid covers three-quarters of the earth's surface. And above the earth's surface is a layer composed of several substances—nitrogen, oxygen, argon, and others—that are gases.

These substances are in the solid, gaseous, or liquid state by an accident of temperature and pressure. The sun, which contains many of the same elements (although in much different proportions) is appreciably hotter and is totally gaseous. In the interior of the earth, where temperatures are higher than on the surface, substances that are solid on the surface are liquid.

Although all of us have been familiar from childhood with the differences between solids, liquids, and gases, we are not equally familiar with the idea that the differences are a function of temperature. We can see the macroscopic differences between solids, liquids, and gases with our eyes. But a knowledge of the microscopic structure of solids, liquids, and gases helps us to understand

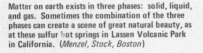

Matter on earth exists in three phases: solid, liquid, and gas. Sometimes the combination of the three phases can create a scene of great natural beauty, as at these sulfur hot springs in Lassen Volcanic Park in California. (*Menzel, Stock, Boston*)

THE LIQUID ROCK BENEATH US

While rock on the earth's surface appears to be strong and brittle, under extremely great pressures and high temperatures rock can flow slowly. Such conditions exist in the earth's interior. Near the surface, the temperature rises at a rate of about 1 K for each 30 m of depth. While the temperature gradient is not as steep deeper within the earth, it is estimated that the temperature is 1400 K 100 km beneath the surface and 2000 K at a depth of 1000 km. The pressure at a depth of 1000 km is estimated to be 400 000 atm. At the very center of the earth, the pressure is estimated to be 3.5 million atm.

By studying the transmission of earthquake shock waves through the earth, geologists have determined that the rigid outer layer of rock, which is called the crust, is only 70 km thick under the oceans and 150 km thick under the continents. Below the crust is a region called the athenosphere, in which increasing temperature and pressure cause the rock to become increasingly fluid with depth. In essence, the crust is floating on melted rock.

In the past decade, this picture of the earth's structure has been elaborated into the theory of plate tectonics, which is supported by a mass of evidence. The earth's surface is divided into a number of plates of rigid rock of unequal size. These plates sometimes rub against one another and sometimes collide head-on. Many of the phenomena studied in geology can be explained in terms of plate tectonics.

For example, the San Andreas fault, which runs through part of California, is a boundary where two plates are sliding past each other. The Pacific plate, the largest plate of the Pacific Ocean, is moving northward at a rate of about 6.5 cm a year. At the present rate of motion, Los Angeles will be at the same latitude as San Francisco in about ten million years. Along part of the fault, the slippage occurs smoothly. However, the two plates tend to stick together in many areas and then to slip suddenly to release the growing strain. An earthquake results.

The ridges that run down the center of some oceans, including the Atlantic, have been identified as regions where molten rock wells up from the interior, and the crustal plates are moving apart. There are also ocean trenches, where plates move together. In these trenches, the edge of one plate is forced under the other, creating a deep valley. The rock of the crustal plate melts deep in the interior. In many parts of earth, the transformation of rock between the liquid and solid phases goes on continuously.

better the differences in the behavior of the different states of matter. By ''microscopic structure'' we mean the arrangement of the elementary entities—atoms, molecules, or ions—of which a substance is composed.

Why do three different states of matter exist, and how do they differ from each other? The simplest explanation is that the state of a given substance at any given temperature is determined by a balance between two opposing tendencies, one acting to separate the structural entities, the other acting to keep them together.

The separating factor is thermal motion. Atoms, molecules, and ions are in constant motion. The movement of these entities increases with increasing temperature. But average velocity is also related to mass. Average velocity decreases with increasing mass. At a given temperature, the average velocity of light entities is greater than the average velocity of heavy entities.

There are attractive forces between entities that counteract this separating tendency. These attractive forces, which become stronger as the entities approach each other, are electrostatic in nature; that is, they are due to the attraction between positive and negative electric charge. To a large extent, the state of matter is determined by the balance between the separative tendency of thermal motion and the attractive tendency of electrostatic forces.

Figure 4.1(a) shows two Br_2 molecules that approach each other. Their thermal motion is great enough to overcome the attractive forces between them. Figure 4.1(b) shows the same situation, but at a lower temperature. In this case, the thermal motion of the two Br_2 molecules is not great enough to overcome their mutual attraction, and they "stick" together.

The most significant characteristic of a gas is that its molecules (or atoms, or whatever its elementary entities are) move freely in space. A substance is in the **gas** phase when the temperature is such that the attractive forces are not sufficient to overcome the tendency for its atoms or molecules to fly apart due to thermal motion. A gas is mostly empty space. The total volume that a gas occupies is large compared to the volume of the molecules of gas. Air at room temperature and atmospheric pressure is more than 99.9% empty space. A gas has no definite shape or volume; it assumes the shape and volume of its container. A gas that is transferred from a small container to a large container expands to fill the large volume; we say that it is expansible. If the container is squeezed to lessen its volume, the gas assumes the smaller volume (subject to certain limitations); the gas is said to be compressible.

We can think of thermal motion as being essentially random motion. The molecules in a gas move independently of one another, only occasionally approaching each other closely enough to interact—and even then, they quickly separate. As the temperature of the gas is lowered, the thermal motion of the molecules decreases. When the temperature is low enough, the attractive forces between molecules, which generally do not change with temperature, become relatively more important, and molecules begin "sticking" to each other. The result is a **condensed phase,** as shown in Figure 4.2. Thermal motion is still substantial in this phase, but the attractive forces are great enough to prevent a completely random arrangement of the molecules. The condensed phase is the **liquid** state of matter.

Unlike a gas, a liquid has a definite volume. While 1 g of gaseous Br_2 has the volume of the container it happens to occupy, 1 g of liquid $Br_2(l)$ occupies a volume of 0.32 cm^3. But while a liquid has a definite volume, it does not have a definite shape. Just as a gas takes its shape from its container, usually a liquid takes its shape from the portion of the container it occupies.

If the temperature of a liquid is lowered, a point is reached at which thermal motion becomes relatively unimportant and the molecules, atoms, or ions arrange themselves in a way that maximizes the attractive forces between them. At lower temperatures, thermal motion has very little influence on the arrangement of the molecules.* The complete randomness of the gas phase and the partial randomness of the liquid phase are replaced by a more ordered arrangement. In this

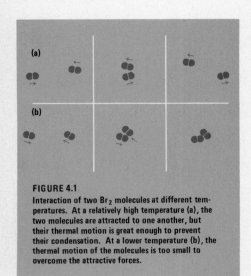

FIGURE 4.1
Interaction of two Br_2 molecules at different temperatures. At a relatively high temperature (a), the two molecules are attracted to one another, but their thermal motion is great enough to prevent their condensation. At a lower temperature (b), the thermal motion of the molecules is too small to overcome the attractive forces.

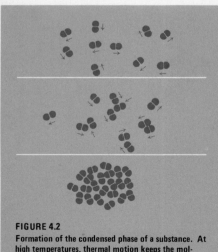

FIGURE 4.2
Formation of the condensed phase of a substance. At high temperatures, thermal motion keeps the molecules in the gas phase, moving randomly. As the temperature is lowered, the attractive forces between molecules become relatively more important, and some of them "stick" together. When the temperature is low enough, the molecules condense; the substance is in the liquid phase.

---

*For convenience, in the following discussion we shall use the word "molecule" to describe all species, including atoms or ions, that make up substances.

condensed phase, called the **solid** state, the molecules generally are closer together than in the liquid state. In such a case, we say that the solid is denser.

The molecules of the solid, as shown in Figure 4.3, are usually in an ordered, nonrandom arrangement that maximizes the attractions between them. In most cases, the ordered arrangement determines the shape of any macroscopic quantity of the substance. The major difference in the microscopic structure of gases, liquids, and solids involves randomness and order. A gas has essentially no order. A liquid has *short-range order;* the molecules in a small region of the liquid may have an orderly arrangement, but this arrangement is not repeated throughout the liquid. A solid has *long-range order.* The arrangement of the molecules in a small region is repeated throughout the solid. The molecules in a solid are ordered in a regular array, or crystal structure.

It is important to note that solids usually are crystalline. Their elementary units—molecules, atoms, or ions—are in an ordered arrangement that creates a regularly repeating three-dimensional pattern. On the macroscopic level, such an ordered arrangement appears as a crystal that is a polyhedron, a solid whose faces are planes. Some crystals are very large, even several meters across. Some are very small and can be seen only with the help of an electron microscope. For any given substance, the three-dimensional pattern of its elementary entities is always the same, no matter how large or how small the crystal is.

The study of the shape of crystals is called *crystallography.* We shall discuss it in detail in Chapter 10.

Not every substance can be classified neatly as a solid, liquid, or gas. Some substances, under appropriate conditions, seem to have intermediate properties. A glass is a material that has characteristics of both a liquid and a solid. While a glass seems to have the mechanical properties of a solid, it does not have a crystalline structure, and rather than melting at a given temperature to a liquid, it softens gradually when heated. Liquid crystals, a recent arrival on the scene, have both liquid and solid characteristics. As we shall see in Chapter 10, glasses and liquid crystals play important roles in technology and biology.

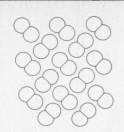

FIGURE 4.3
The solid phase. Molecules are in an ordered, nonrandom arrangement, rather than the relatively disordered arrangement of the gas phase. The arrangement of the molecules in the solid phase can influence the shape of the solid.

4.2 VAPOR PRESSURE

Temperature and pressure determine whether the elementary units of a substance come together to form a condensed phase, either solid or liquid. Most chemistry is performed at normal atmospheric pressure. Therefore, we can discuss the effect of temperature changes with the assumption that the pressure of the system being considered is close to atmospheric pressure.

Condensed Phases and Vapor

We have noted that when temperatures are high, thermal motion is great enough to overcome attractive forces, so a substance will exist as a gas. As temperature is lowered, a condensed phase—usually but not always a liquid—forms from the gas. The system then includes not only a condensed phase but also some gas.

For example, Br_2 is a dark reddish brown liquid at room temperature. A red vapor, gaseous Br_2, can be seen rising from its surface. Indeed, a sample of liquid Br_2 that is left in an open container soon evaporates completely. If the container is closed and is not too large, the red vapor fills the container, although most of the Br_2 remains liquid.

A similar observation can be made with I_2, which is a solid at room temperature but gives off a purple vapor. Crystals of solid I_2 will disappear if left in an open container.

Observations such as these led, long ago, to the conclusion that some vapor or gas is always associated with a condensed phase. Thermal motion causes the

A condensed phase of a substance is always associated with a vapor. Usually, the vapor is not visible, as in the case of water. This visible iodine vapor formed when crystals of iodine were dropped into the beaker. Solid iodine vaporizes to form violet gas at room temperature. (*Beckwith Studios*)

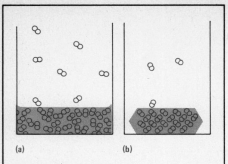

FIGURE 4.4
(a) Evaporation. Some molecules on the surface of a liquid have sufficient thermal motion to break free and enter the gas phase. (b) Sublimation. Some molecules on the surface of a solid have sufficient thermal motion to break free and enter the gas phase.

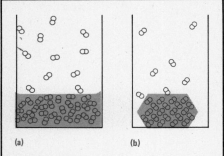

FIGURE 4.5
Condensation. Molecules in the gas phase come close to the surface of a liquid (a) or solid (b). Some of the molecules "stick" to the surface and become part of the condensed phase.

vapor to form. The molecules of a liquid are in thermal motion. Even the entities of a solid move (though not very far) with respect to one another. Although we have implied that the thermal motion of any given molecule is proportional to temperature, in fact it is the *average* thermal motion of a large number of molecules that is proportional to temperature. At any given temperature, there is a distribution of the molecular energies associated with thermal motion. Some molecules have less thermal motion than the average, and some have more. Every now and then, the thermal motion of one molecule or another on the surface of a liquid or a solid is great enough to overcome the attractive forces, and the molecule escapes into the gas phase (Figure 4.4). The process by which entities escape from the liquid phase to the gas phase is called **evaporation**. The same process for entities of a solid is called **sublimation**.

Knowledge of these processes raises many questions. Why don't all solids sublime? Why don't all liquids (including the oceans) evaporate? In small, closed systems, why do we see such substances as Br_2 and I_2 existing partly in the condensed phase and partly in the gas phase?

Most solids don't sublime because the attractive forces between their units of structure are so great that only a negligibly small fraction of entities leave the surface at room temperature. The oceans do evaporate, constantly. But the water vapor condenses into clouds and falls as rain, which eventually finds its way back to the oceans. A similar sort of balance exists in a small, closed system. In a closed container, the molecules of Br_2 or I_2 that are in the gas phase cannot get far from the condensed phase of the substance. Every now and then, a molecule will strike the surface of the condensed phase and will be held on that surface by attractive forces. This process is called **condensation** (Figure 4.5).

The Equilibrium State

When the condensed state of a substance is placed in a closed container, evaporation or sublimation occurs. The rate at which vapor forms is determined primarily by a balance between attractive forces, which depend on the nature of the substance, and thermal motion, which depends on the temperature and the mass of the elementary entities. The rate at which the substance condenses is determined by these two factors and by the rate at which molecules in the vapor collide with the condensed state.

When evaporation or sublimation begins, there are few molecules in the vapor, so the rate of condensation is small. As evaporation or sublimation continues, the number of molecules in the vapor increases, and so does the rate of condensation. Eventually, the rate of condensation equals the rate of evaporation or sublimation. If the number of molecules going into the gas phase equals the number of molecules leaving the gas phase, the system is static from a macroscopic point of view. But if the system is examined molecule by molecule, this apparently unchanging picture becomes dynamic and full of change. The net result of the changes on the molecular level is no overall macroscopic change. **When any system exists in a state that is macroscopically static, the system is said to be at equilibrium.** The equilibrium state is a dynamic state in which there is no net change in the total amounts of the components of the system. For every molecule that condenses, a molecule evaporates.

If any quantity of a condensed phase of a substance is placed in a closed container that is not too large, such a dynamic equilibrium will be reached eventually. The actual quantity of the substance that goes into the vapor phase depends on several factors. One factor is the volume of the container. At a given temperature, the amount of a substance in the vapor phase increases with the volume of the container.

There is an approach that makes it unnecessary to consider the actual volume of the container. Instead of attempting to deal with the *total* amount of the substance that is in the vapor phase, we can measure the amount in the vapor

phase per unit of volume. Measurement by units of volume—for example, the
number of gas molecules in 1 cm^3 of volume, or the number of moles of gas in
1 liter—means that the size of the container can be ignored.

Pressure

In practice, a direct measurement of the number of gas molecules in 1 cm^3 or the
number of moles of gas per liter is not easy. In theory, we could first weigh the
empty container and then get the weight of the gas by weighing the full container.
But this is impractical because the weight of a gas usually is negligibly small
compared to the weight of the container. Instead, it is customary to use another,
more easily measured quantity to describe the amount of substance in the gas
phase. This quantity is **pressure**.

*Pressure is the force exerted on a given unit of area of surface with which a
gas is in contact.* In a small, closed container, pressure is exerted on the walls of
the container. On earth, the gases of the atmosphere exert pressure on the
surface of the earth and everything exposed to the atmosphere.

Because we live in the atmosphere, the phenomenon of gas pressure is
familiar to all of us. A child sees what air pressure can do when a balloon explodes
because it cannot withstand the force exerted by the gas within. When the child
sucks on a straw that is dipped into a liquid, a partial vacuum is created within the
straw; atmospheric pressure then pushes the liquid up the straw. Evangelista
Torricelli (1608–1647), a student of Galileo, first recognized that the pressure
exerted by the atmosphere could be measured from such phenomena. Torricelli
used the height of a column of liquid in the simple device called the barometer
(Figure 4.6) as a measure of atmospheric pressure. A decrease in atmospheric
pressure causes a drop in the height of the column, while an increase in atmospheric pressure increases the height of the column of liquid.

The height of a column of liquid supported by atmospheric pressure depends
in part on the density of the liquid. At sea level, the atmosphere can support a
column of water more than 10 m high, or a column of mercury (a liquid element
that is much denser than water) 0.76 m high. For convenience, mercury is used
in most liquid-level barometers.

Gas pressure can also be measured by a "U" tube manometer (Figure 4.7),
which works on the same principle as a barometer. In a manometer, the pressure
of a gas is related to the difference between the heights of two columns of liquid.

The units of pressure in everyday use arose from measurements made with
the mercury barometer and the manometer. Chemists have most commonly
measured pressure in units of atmospheres (atm), millimeters of mercury (mmHg)
or torr. Torr and mmHg are virtually identical. There is a long history behind these
units. The unit atmosphere has a simple origin: Atmospheric pressure at sea level
averages about 1 atm. The other units are related to the atm; 1 atm =
760 mmHg = 760 Torr. Because so much chemistry is done at pressures close
to 1 atm, it is generally convenient for our purposes to use the atmosphere as the
unit for pressures. It would be more precise to describe units of pressure as so
much force per area, since pressure is a measure of force per area. The SI uses
such a unit. In the SI, the fundamental unit of force is called the newton (N);
$1 N = 1 kg\ m\ s^{-2}$. The unit of area is the square meter (m^2). The unit of pressure
(force per area) is therefore the newton per square meter (Nm^{-2}), which is called
the pascal (Pa). A pressure of 1 atm is defined to be exactly 101 325 Pa, or
101.325 kPa, while a pressure of 1 mmHg or 1 Torr is about 133 Pa.

Vapor Pressure

The pressure of the vapor of a substance in equilibrium with its condensed phase
is called the **vapor pressure** of the substance. Since pressure is, in a sense, a
measurement of the amount of gas per unit volume, **the value of the vapor**

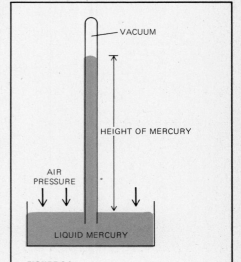

FIGURE 4.6
A barometer. The column of mercury is supported
by the pressure of air on the open pool of mercury.
As air pressure increases or decreases, the height
of the column of mercury increases or decreases.
Thus, air pressure can be measured by measuring
the height of the column of mercury.

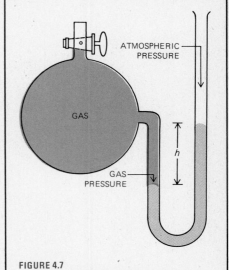

FIGURE 4.7
A manometer. Similar in principle to a barometer,
the manometer allows measurement of gas pressure.
Since the atmospheric pressure is known, the gas
pressure can be determined by measuring h, the
difference in levels of mercury in the two arms of
the tube.

pressure depends only on the nature of the substance in the system and the temperature of the system.

Innumerable experiments have shown that the vapor pressure of any given substance increases as the temperature increases. Vapor pressure also depends on the attractive forces between the units of structure in the condensed phase of the substance. The weaker the attractive forces in the condensed phase, the higher is the vapor pressure of a substance at a given temperature.

Thus, I_2 has a relatively high vapor pressure at room temperature because the attractive forces between the I_2 molecules are relatively weak. But NaCl, an ionic crystal with very strong attractive forces between ions in the crystal, has an extremely low vapor pressure at room temperature.

Boiling Point

If a liquid in an open system is heated, eventually it reaches a temperature at

which its vapor pressure equals atmospheric pressure. This temperature is called
the *boiling point*. The boiling point depends on the atmospheric pressure at the
place where the liquid is being heated. At sea level, where atmospheric pressure
is 1 atm, water boils at 100°C (373 K). On a mountaintop, where atmospheric
pressure is lower than at sea level, the boiling point is lower—sometimes substantially lower, depending on the height of the mountain. At 3000 m, water
boils at 363 K. Because of the lower boiling point at high altitudes, boiling an egg
on a mountain can be a problem.

    *The boiling point of a liquid is the highest temperature, at a given pressure, at
which there can be a liquid component in a system at equilibrium*. At the boiling
point, the liquid vaporizes, and bubbles form below the surface of the liquid. At
temperatures below the boiling point, the external pressure prevents such gas
bubbles from forming. At temperatures above the boiling point all of the substance is in the gas phase. Therefore, the boiling point can be regarded as the
temperature at which the system changes from liquid to gas if the temperature is
rising or from gas to liquid if the temperature is dropping. The temperature at
which the vapor pressure is equal to 1 atm is called the *normal boiling point*.

4.3 THE GASEOUS STATE

Gases have been the subject of a great deal of chemical study, both theoretical
and experimental. Both lines of work have been quite productive. A theory of
gases that is reasonably complete and highly useful has been developed, based
on the random and essentially independent behavior of the basic units of a gas.
And while the invisibility of most gases held back early experimental work somewhat, chemists were measuring such properties as gas pressure by the seventeenth century, and experimental skills have grown steadily since then. Theory
and experiment have been combined to give an excellent picture of the gaseous
state.

The Ideal Gas

Regardless of their chemical composition, all gases have some properties more or
less in common. To some degree, all gases follow certain "ideal" laws of
behavior. It has been found useful to imagine an **ideal gas** or **perfect gas** whose
behavior can be explained completely by these ideal laws.

    The behavior of any real gas only approximates that of an ideal gas. The
closeness of the approximation depends on the chemical composition of a real
gas and the conditions under which the gas is studied. The ideal laws of behavior
are a good approximation of the behavior of a real gas at low pressures and
reasonably high temperatures—that is, at atmospheric pressure or lower and at
temperatures somewhat higher than the temperature at which a gas condenses
to a liquid. As the pressure increases, or as temperature decreases, real gases
tend to deviate more and more from ideal behavior. The early observations of
gases, made at ordinary conditions, led to formulation of laws that were believed
to describe the behavior of real gases completely. In fact, these were ideal gas
laws. The methods of measurement then in use were not accurate enough to
detect the slight differences in behavior between real gases under those conditions.

Boyle's Law

Shortly after the invention of the barometer, Robert Boyle (1627–1691) devised
a simple apparatus for studying the relationship between the pressure and volume
of a gas (Figure 4.8). The device he used was a U-shaped tube, closed at one
end. The pressure on the gas in the closed end of the tube could be increased by

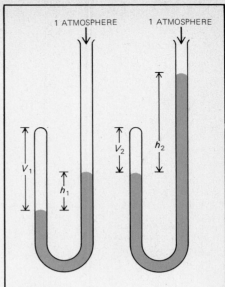

FIGURE 4.8
Boyle's device for studying the relationship between
gas pressure and volume. Pressure on the gas can be
increased by adding mercury to the open end of the
tube. The effect of the increased pressure can be
determined by measuring the change in the volume
of the gas in the closed end of the tube. Here
$P_1 = h_1 + P_{atm}$, $P_2 = h_2 + P_{atm}$, and $P_1V_1 = P_2V_2$.

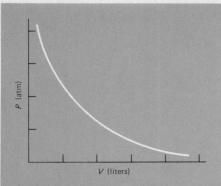

FIGURE 4.9
The relationship between gas pressure (P) and gas volume (V), shown graphically. As pressure increases, the volume of a gas decreases; as pressure decreases, the volume of a gas increases. This is Boyle's law.

adding a liquid, such as mercury, to the open end. The change in pressure was obtained by noting the height of the column of mercury, while the change in volume was obtained from a scale on the closed end of the tube.

In 1662, Boyle published the results of his experiments, which indicated that an increase in pressure is accompanied by a predictable decrease in the volume of the gas. This observation has come to be called Boyle's law. The modern statement of Boyle's law takes into account the fact that all of Boyle's observations were made at room temperature. The law is stated: **At constant temperature the volume of a sample of gas is inversely proportional to its pressure.** This can be written as:

$$PV = k \tag{4.1}$$

where P is pressure, V is volume, and k is a proportionality constant. When V increases, P decreases; when V decreases, P increases. Figure 4.9 shows a graphical representation of this relationship.

Real gases obey this ideal relationship fairly well when P is not too great. Equation 4.1 defines the relationship between the pressure and the volume of a given amount of gas at a given temperature: the product of pressure and volume is a constant. This relationship can be written in a form that is more suitable for calculations:

$$P_1V_1 = P_2V_2 \tag{4.2}$$

where P_1 and V_1 represent one set of pressure and volume conditions and P_2 and V_2 represent another set of conditions for the same amount of gas at the same temperature.

EXAMPLE 4.1

A sample of CO(g) occupies a volume of 2.2 liters and exerts a pressure of 0.95 atm. The pressure of the CO is increased to 1.1 atm. Assuming ideal gas behavior and no change in temperature, what is the new volume of the CO?
SOLUTION
Equation 4.2 has four terms, and we are given the value of three. Thus:

$P_1 = 0.95$ atm $V_1 = 2.2$ liters
$P_2 = 1.1$ atm $V_2 = ?$

Substituting gives:

$(0.95 \text{ atm})(2.2 \text{ liters}) = (1.1 \text{ atm})V_2$
$V_2 = 1.9$ liters

EXAMPLE 4.2

A sample of gas at room temperature is placed in an evacuated bulb of volume 0.51 liter and is found to exert a pressure of 0.24 atm. This bulb is connected to another evacuated bulb whose volume is 0.63 liter, and the gas is allowed to fill both bulbs, as shown in Figure 4.10. What is the new pressure of the gas at room temperature?
SOLUTION
List the information given:

$V_1 = 0.51$ liter $P_1 = 0.24$ atm
$V_2 = 0.51$ liter $+$ 0.63 liter $P_2 = ?$

Substitute into Equation 4.2:

$(0.24 \text{ atm})(0.51 \text{ liter}) = (1.14 \text{ liters})P_2$
$P_2 = 0.11$ atm

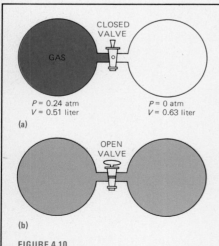

FIGURE 4.10
Example 4.2 illustrated. At the beginning of the experiment, the bulb at the left contains a sample of gas, while the bulb at the right is evacuated. When the valve between the two bulbs is opened, some gas goes from the left bulb to the right bulb.

The Law of Charles and Gay-Lussac

More than a century after Boyle's experiments, two French physicists, Jacques Alexandre Charles (1746–1823) and Joseph Louis Gay-Lussac (1778–1850) became interested in hot-air balloons. Their balloon work led them to study the relationship between the volume and the temperature of gases.

They knew from experience that a gas expands when heated and contracts when cooled. Charles measured the change in volume caused by a given change in temperature. He found that gases expand by the same fractional amount of their original volume for equal increases in temperature. Later, Gay-Lussac showed that for every 1 °C rise in temperature, any gas expands by 1/273 of its volume at 0°C. Thus, if a gas occupies a volume of 273 cm³ at 0°C, it will occupy a volume of 273 + 1 = 274 cm³ at 1°C, of 273 + 2 = 275 cm³ at 2°C, of 273 + 10 = 283 cm³ at 10°C, and so on, providing that the pressure remains constant (Figure 4.11).

These observations, based on the Celsius temperature scale, can be expressed as:

$$V = V_0 + V_0(t/273) = V_0\left(1 + \frac{t}{273}\right) \qquad (4.3)$$

where V is the volume of the gas at a temperature of $t°C$, V_0 is the volume of the gas at 0°C, and t, the temperature of the gas on the Celsius scale, is also the number of degrees away from 0°C.

Because the relationship is based on the Celsius temperature scale, the expression is more complicated than it need be. Volume and temperature appear to be proportional, since volume increases by the same amount for equal increases in temperature. However, Equation 4.3 is not a simple expression of a direct proportionality.

A simpler relationship between gas volume and temperature can be formulated with a different temperature scale. We can use a temperature scale in which the interval of temperature—the degree—is the same as that of the Celsius scale but the zero point is 273°C below the zero point on the Celsius scale. If this temperature is called T, then:

$$T = t + 273 \qquad (4.4)$$

where t is the Celsius temperature.

Equation 4.3 can be written as:

$$V = V_0\left(\frac{273 + t}{273}\right)$$

Combining this expression with Equation 4.4 gives:

$$V = V_0\left(\frac{(T)}{273}\right) \qquad (4.5)$$

Since V_0 is a constant, we can let $k = V_0/273$:

$$V = kT \qquad (4.6)$$

which is a more convenient expression of the law of Charles and Gay-Lussac.

The temperature scale used above was derived in a number of ways during the nineteenth century. It is called **the absolute or Kelvin temperature scale. It should be used in all calculations involving gases.** The interval of temperature on this scale is called the kelvin (K); 1 K = 1°C. The only difference between the Kelvin scale and the Celsius scale is in their zero points. The zero point of the Celsius scale is based on the freezing point of water. The zero point on the Kelvin temperature scale can be found from Equation 4.3. It is the temperature at which the volume (V) of an ideal gas is zero. Since the initial volume (V_0) of an ideal gas

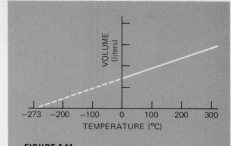

FIGURE 4.11
The relationship between temperature and gas volume. As temperature increases by 1°C, the volume of a gas increases by approximately 1/273 of its volume at 0°C. This is the law of Charles and Gay-Lussac.

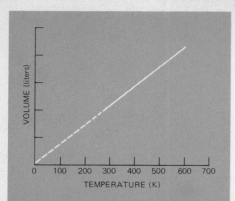

VOLUME (liters)

0   100 200 300 400 500 600 700
TEMPERATURE (K)

FIGURE 4.12
The relationship between temperature and gas volume extrapolated to zero. Theoretically, gas volume should be zero at 0 K (−273°C). Practically, zero gas volume and 0 K can never be achieved.

at 0°C is reduced by $V_0/273$ for every reduction of 1°C, it can be seen that 0 K = −273°C. Figure 4.12 shows this relationship graphically. Extremely accurate measurements give 0°C = 273.15 K.

In practice, the volume of a gas has never been reduced to zero. Any gas condenses to a liquid or a solid well before 0 K, the absolute zero, is reached. It has been shown theoretically that a temperature of 0 K can never be reached.

Equation 4.6 states that **the volume of a given amount of an ideal gas is directly proportional to its absolute temperature at constant pressure.** When V increases, T increases, and when V decreases, T decreases. Equation 4.7 expresses the same relationship in a form more suitable for calculations:

$$\frac{V_1}{T_1} = \frac{V_2}{T_2} \tag{4.7}$$

where V_1 and T_1 represent one set of volume and absolute temperature conditions for a quantity of gas and V_2 and T_2 another set for the same quantity of gas at the same pressure.

EXAMPLE 4.3
The gas inside a balloon at a temperature of 20°C is heated until the volume of the balloon is tripled. The pressure remains constant at 1 atm. What is the new temperature of the gas?
SOLUTION
First convert the temperature to the absolute scale:

$$T = t + 273 = 20 + 273 = 293 \text{ K}$$

If the initial volume is V_1 and the final volume is V_2, the problem states that V_2 is triple V_1; that is, $V_2 = 3V_1$. Substituting into Equation 4.7 gives:

$$\frac{V_1}{293 \text{ K}} = \frac{3V_1}{T_2}$$

$$T_2 = (3)(293 \text{ K}) = 879 \text{ K}$$

This problem could be solved with less formal calculation once it is understood that gas volume and absolute temperature are directly proportional. If the gas volume is tripled, the absolute temperature also triples.

EXAMPLE 4.4
The air in a 50-liter balloon is heated from 300 K to 450 K at a constant pressure of 1 atm. Calculate the new volume of the balloon.
SOLUTION
Equation 4.7 has four terms. Given the value of any three, the fourth can be found. In this case:

$V_1 = 50$ liters $V_2 = ?$

$T_1 = 300$ K $T_2 = 450$ K

$$\frac{50 \text{ liters}}{300 \text{ K}} = \frac{V_2}{450 \text{ K}}$$

$$V_2 = 75 \text{ liters}$$

The two laws for the behavior of ideal gases as expressed by Equations 4.2 and 4.7 can be combined into a single relationship. Suppose we have a gas with volume V_1, pressure P_1, and temperature T_1, and that these are changed to V_2, P_2, and T_2. Is there a simple relationship between the two sets of variables?

Let us suppose that the change is brought about by holding the temperature at T_1 while the pressure is changed to P_2. In this situation, Boyle's law applies. If the volume of the gas is called V_i, then:

$$P_1V_1 = P_2V_i \quad \text{or} \quad V_i = \frac{P_1V_1}{P_2}$$

Now the pressure is held constant at P_2 while the temperature is changed from T_1 to T_2. The volume also changes from V_i to V_2, and Charles's law applies:

$$\frac{V_i}{T_1} = \frac{V_2}{T_2}$$

We get the desired single relationship by substituting the value of V_i found from Boyle's law into the expression above:

$$\frac{P_1V_1}{T_1} = \frac{P_2V_2}{T_2} \tag{4.8}$$

This equation summarizes the relationships between P, V, and T for a given sample of an ideal gas. It is useful for a variety of calculations. Equation 4.8 connects two sets of pressure, volume, and temperature conditions for any given quantity of an ideal gas. It can also be read as: The quantity PV/T is a constant for a given amount of gas.

EXAMPLE 4.5

An automobile tire is inflated to a total pressure of 1.9 atm with air at 20°C. After five hours of driving, the volume of the tire has increased by 5.0% and the total pressure is 2.1 atm. Calculate the temperature of the air inside the tire, assuming ideal gas behavior.

SOLUTION

In such a problem, it is advisable to list all the data in suitable units:

$P_1 = 1.9$ atm $\qquad P_2 = 2.1$ atm

$T_1 = 20 + 273$ K $\qquad T_2 = ?$

$V_1 = V_1 \qquad\qquad V_2 = V_1 + \dfrac{5.0}{100}V_1$

Using Equation 4.8, we get:

$$\frac{(1.9 \text{ atm})(V_1)}{293 \text{ K}} = \frac{2.1 \text{ atm}(1.05\ V_1)}{T_2}$$

The V_1 can be canceled, giving:

$$T_2 = \frac{(2.1 \text{ atm})(293 \text{ K})(1.05)}{(1.9 \text{ atm})}$$

$$= 340 \text{ K}$$

The increase in temperature is consistent with the increase in pressure, since the pressure of an ideal gas is also directly proportional to its absolute temperature.

EXAMPLE 4.6

A balloon whose volume is 91 liters contains helium at a temperature of 295 K and a pressure of 1.0 atm. When the balloon reaches an altitude of 20 km, the pressure is 0.083 atm and the temperature is 225 K. Calculate the volume of the balloon at that altitude.

SOLUTION

List the data given:

$P_1 = 1.0$ atm $\qquad P_2 = 0.083$ atm

$V_1 = 91$ liters $\qquad V_2 = V_2$

$$T_1 = 295 \text{ K} \qquad T_2 = 225 \text{ K}$$

Substitution into Equation 4.8 gives:

$$\frac{(1.0 \text{ atm})(91 \text{ liters})}{295 \text{ K}} = \frac{(0.083 \text{ atm})V_2}{225 \text{ K}}$$

$$V_2 = 840 \text{ liters}$$

The volume of the helium has increased by almost a factor of ten. This substantial increase in volume must be considered in designing high-altitude balloons.

For convenience, a standard set of conditions is defined for an ideal gas: a pressure of 1 atm and a temperature of 0°C (273 K). These conditions are referred to as standard temperature and pressure, **STP.**

EXAMPLE 4.7

A sample of NO(g) occupies a volume of 3.2 liters at a temperature of 298 K and a pressure of 0.50 atm. What is the volume of the sample at STP?

SOLUTION

List the data given:

$$P_1 = 0.50 \text{ atm} \qquad P_2 = 1.0 \text{ atm}$$
$$V_1 = 3.2 \text{ liters} \qquad V_2 = V_2$$
$$T_1 = 298 \text{ K} \qquad T_2 = 273 \text{ K}$$

Substitution into Equation 4.8 gives:

$$\frac{(0.50 \text{ atm})(3.2 \text{ liters})}{(298 \text{ K})} = \frac{(1.0 \text{ atm})V_2}{(273 \text{ K})}$$

$$V_2 = 1.5 \text{ liters}$$

Avogadro's Law

So far we have discussed the relationships between the pressure, volume, and temperature of an ideal gas without considering a fourth factor: the amount of gas. Yet P, V, and T are related to the amount of the ideal gas in any sample under consideration.

In 1811, Amadeo Avogadro proposed an important and remarkable relationship. Avogadro said that at a given pressure and temperature, the number of molecules of gas in any sample depends only on the volume of the sample, not on the nature of the gas. This can be stated as: **Equal volumes of different gases under the same conditions of temperature and pressure contain equal numbers of molecules.**

At the time, Avogadro's law was largely ignored. After Avogadro's death, the law was found to be very useful for the determination of chemical formulas and molecular and atomic weights.

At constant T and P, the volume V of a gas is directly proportional to the number of molecules N of the gas:

$$V = kN \tag{4.9}$$

When we increase the number of molecules in a sample of gas at constant temperature and pressure, the volume increases in direct proportion. If, for example, the number of molecules is doubled, the volume that the sample occupies at the same total pressure doubles. The constant k in Equation 4.9 is the same for all gases, no matter what their nature.

Since the number of moles n is directly proportional to the number of molecules N, it is more convenient to write:

$$V = kn \tag{4.10}$$

or, in a form more suitable for calculations:

$$\frac{V_1}{n_1} = \frac{V_2}{n_2} \tag{4.11}$$

EXAMPLE 4.8
A balloon whose volume is 5.0 liters contains 6.1 g of N_2. What mass of N_2 must be added to the balloon to expand its volume to 11 liters at the same temperature and pressure?
SOLUTION
List the data given:

$V_1 = 5.0$ liters $\qquad\qquad\qquad V_2 = 11$ liters

$n_1 = \dfrac{6.1 \text{ g N}_2}{28 \text{ g N}_2/1 \text{ mol N}_2} = 0.22 \text{ mol N}_2 \qquad n_2 = n_2$

Using Equation 4.11, we get:

$$\frac{5.0 \text{ liters}}{0.22 \text{ mol N}_2} = \frac{11 \text{ liters}}{n_2}$$

and $n_2 = 0.48$ mol N_2, which is the total number of moles of gas required for the volume to be 11 liters.

Therefore

$$0.48 \text{ mol N}_2 - 0.22 \text{ mol N}_2 = 0.26 \text{ mol N}_2$$

which must be added, or

$$0.26 \text{ mol} \times \frac{28 \text{ g N}_2}{1 \text{ mol N}_2} = 7.3 \text{ g N}_2$$

must be added.

4.4 THE IDEAL GAS EQUATION

We have seen that PV/T is a constant for a given quantity of ideal gas. The numerical value of the constant is proportional to the quantity of gas. Therefore, we can write:

$$\frac{PV}{T} = Rn \tag{4.12}$$

where n is the number of moles of an ideal gas and R is the conventional symbol for the proportionality constant that is called the *universal gas constant*.

The numerical value of the universal gas constant can be determined. This value depends on the units of measurement that are chosen for V and P. While T is always expressed in kelvins and n is always expressed in moles, there is a choice of units for V and P. If we measure V in liters and P in atmospheres, the value of R is 0.082056 liter atm mol^{-1} K^{-1}.

Equation 4.12 usually is written in the form:

$$PV = nRT \tag{4.13}$$

and is called the **ideal gas equation**. The ideal gas equation combines in one relationship the laws of Boyle, Charles and Gay-Lussac, and Avogadro. It provides essentially a complete description of the state of an ideal gas, and is thus one of the most useful relationships in chemistry. Since there are four unknowns in the ideal gas equation, when the values of any three are specified the value of the fourth is fixed.

EXAMPLE 4.9

Calculate the volume of one mole of an ideal gas at STP.

SOLUTION

List the data given:

$P = 1$ atm $\qquad n = 1$ mol
$V = V \qquad\qquad T = 273$ K

Do the calculation, using Equation 4.13. Be sure that all units are consistent with the units of R:

$$(1 \text{ atm})V = (1 \text{ mol})\left(0.0821 \frac{\text{liter atm}}{\text{mol K}}\right)(273 \text{ K})$$

$$V = 22.4 \text{ liters}$$

The volume 22.4 liters is called the *standard molar volume* of an ideal gas.

A large number of problems can be solved by using the ideal gas equation, if the problems specify or imply the values of any three of the four variables in the equation.

EXAMPLE 4.10

A mouse is placed in a gas bulb that is filled with O_2 and contains some KOH to absorb CO_2 and H_2O. The volume of the bulb, after allowing for the mouse and the KOH, is 1.02 liters. The pressure of the O_2 is 0.953 atm; and the temperature of the gas is 37.0°C. After two hours, the pressure in the bulb drops to 0.774 atm. What mass of O_2 was consumed by the mouse?

SOLUTION

Assume that the consumption of $O_2(g)$ is the only reason for the drop in pressure. Then the pressure of $O_2(g)$ consumed is 0.953 atm − 0.774 atm = 0.179 atm. The other data given are $V = 1.02$ liters and $T = 37 + 273$ K. Substituting these data into the ideal gas equation gives:

$$(0.179 \text{ atm})(1.02 \text{ liters}) = n\left(0.0821 \frac{\text{liter atm}}{\text{mol K}}\right)(310 \text{ K})$$

$$n = 0.00717 \text{ mol } O_2$$

$$\text{mass } O_2 = 0.00717 \text{ mol } O_2 \times \frac{32.0 \text{ g } O_2}{1 \text{ mol } O_2}$$

$$= 0.230 \text{ g } O_2(g) \text{ consumed}$$

We can get more information about the behavior of ideal gases by using the ideal gas equation to derive different but equivalent relationships. We can obtain these new relationships by combining the ideal gas equation with expressions that define variables such as n. For example, we know that the number of moles of a sample of a substance is related to the substance's mass and molecular weight. The relationship can be expressed as:

$$n = \frac{g}{(MW)}$$

where g is the mass in grams and (MW) is the molecular weight or the mass of one mole.

If we substitute this expression for n into the ideal gas equation, we get:

$$PV = \frac{gRT}{(MW)} \qquad\qquad (4.14)$$

If a real gas is studied under conditions where it displays ideal gas behavior, the molecular weight of the gas can be determined by making the appropriate measurements and using Equation 4.14.

EXAMPLE 4.11

A sample of an unknown liquid whose mass is 1.35 g is introduced into an evacuated bulb of volume 502 cm³. All of the liquid is vaporized, and the resulting gas pressure is 0.477 atm at 333 K. Assuming ideal gas behavior, calculate the molecular weight of the liquid.

SOLUTION

List the given data in appropriate units:

$$V = 502 \text{ cm}^2 \times \frac{1 \text{ liter}}{1000 \text{ cm}^3} = 0.502 \text{ liter} \qquad T = 333 \text{ K}$$

$$P = 0.447 \text{ atm} \qquad\qquad\qquad \text{mass} = 1.35 \text{ g}$$

Using Equation 4.14, we can calculate:

$$(0.447 \text{ atm})(0.502 \text{ liter}) = \frac{1.35 \text{ g}}{(MW)}\left(0.0821 \frac{\text{liter atm}}{\text{mol K}}\right)(333 \text{ K})$$

$$MW = 154 \text{ g/mol}$$

Another version of the ideal gas equation can be derived by recalling that the density d of a gas is most conveniently expressed as the mass in grams per liter of the gas:

$$d = \frac{g}{V}$$

where g is the mass in grams and V is the volume in liters. Equation 4.14 can then be rewritten as:

$$\frac{g}{V} = \frac{P(MW)}{RT}$$

or

$$d = \frac{P(MW)}{RT} \qquad\qquad (4.15)$$

which expresses the relationship between the molecular weight and the density of a gas at a given temperature and pressure.

EXAMPLE 4.12

The density of an unknown gas at STP is 9.91 g/liter. Calculate the molecular weight of the gas.

SOLUTION

List the data given:

$d = 9.91$ g/liter
$P = 1$ atm
$T = 273$ K

Using Equation 4.15, we get:

$$9.91 \text{ g/liter} = \frac{(1 \text{ atm})(MW)}{(0.0821 \text{ liter atm/mol K})(273 \text{ K})}$$

$$MW = 222 \text{ g/mol}$$

The problem can be approached in another way. At STP, one mole of ideal gas occupies a volume of 22.4 liters. Therefore, the mass of 22.4 liters of the gas at STP is the molecular weight of the gas. We are given the density, and hence the mass, of one liter of the gas. Therefore, the molecular weight of the gas can be obtained by multiplying the density of one liter of the gas by 22.4; that is, $MW = (22.4)(d)$ at STP:

$$MW = 22.4 \text{ liters/mol} \times 9.91 \text{ g/liter} = 222 \text{ g/mol}$$

The ideal gas equation is often used to determine the amount of gas that is consumed or produced in chemical processes. If we assume that the gas displays ideal behavior, we can determine the quantity of the gas by measuring other variables, such as pressure, volume, and temperature.

EXAMPLE 4.13

The thermal decomposition of potassium chlorate is a standard (although potentially hazardous) laboratory procedure for producing oxygen by the reaction:

$$2KClO_3(s) \longrightarrow 2KCl(s) + 3O_2(g)$$

A quantity of $KClO_3$ is heated and partially reacts. The oxygen that forms is collected in a balloon at a pressure of 1.0 atm and a temperature of 300 K. The volume of the balloon is found to be 0.79 liter. What mass of $KClO_3$ has reacted?

SOLUTION

The amount of O_2 liberated can be calculated from the given data. The chemical equation gives the relationship between the moles of O_2 and the moles of $KClO_3$.

List the data:

$$P = 1.0 \text{ atm} \qquad T = 300 \text{ K} \qquad V = 0.79 \text{ liter}$$

The ideal gas equation gives the relationship between pressure, volume, temperature, and amount of gas:

$$(1.0 \text{ atm})(0.79 \text{ liter}) = n \left(0.0821 \frac{\text{liter atm}}{\text{mol K}} \right) (300 \text{ K})$$

$$n = 0.032 \text{ mol } O_2$$

Then, to find the mass of $KClO_3$ that reacted:

$$\text{mass } KClO_3 = 0.032 \text{ mol } O_2 \times \frac{2 \text{ mol } KClO_3}{3 \text{ mol } O_2} \times \frac{122.6 \text{ g } KClO_3}{1 \text{ mol } KClO_3}$$

$$= 2.6 \text{ g } KClO_3 \text{ reacted}$$

EXAMPLE 4.14

A 0.52-liter bulb is filled with $CO_2(g)$ at a pressure of 0.86 atm and a temperature of 298 K. A quantity of $CaO(s)$ of negligible volume is introduced into the bulb. After some time, the pressure of $CO_2(g)$ is found to fall to 0.39 atm due to its consumption by the process:

$$CaO(s) + CO_2(g) \longrightarrow CaCO_3(s)$$

What is the mass of the calcium carbonate that is formed?

SOLUTION

First use the ideal gas equation to calculate the number of moles of $CO_2(g)$ consumed. We can use the ideal gas equation because the drop in pressure, which is 0.86 atm $- 0.39$ atm $= 0.47$ atm, corresponds to the consumption of $CO_2(g)$. The ideal gas equation gives the relationship between pressure, temperature, volume, and amount of gas. Since we are given the pressure, temperature, and volume, we can calculate the amount of gas:

$$(0.47 \text{ atm})(0.52 \text{ liter}) = n \left(0.082 \frac{\text{liter atm}}{\text{mol K}} \right) (298 \text{ K})$$

$$n = 0.010 \text{ mol } CO_2 \text{ consumed}$$

The chemical equation indicates that the number of moles of $CaCO_3(s)$ formed will be the same as the number of moles of $CO_2(g)$ consumed. Thus:

$$\text{mass } CaCO_3(s) = 0.010 \text{ mol } CaCO_3 \times \frac{100 \text{ g } CaCO_3}{1 \text{ mol } CaCO_3}$$

$$= 1.0 \text{ g } CaCO_3(s) \text{ formed}$$

4.5 DALTON'S LAW OF PARTIAL PRESSURE

Early in the nineteenth century, John Dalton's experiments led him to recognize that **the pressure exerted by a gas in a mixture of a number of different gases is the same as if the gas were present alone.** We now know that this is ideal gas behavior. We noted that the behavior of ideal gases does not depend on their chemical composition. In a mixture of two or more gases, all displaying ideal behavior, each gas can be regarded as acting independently of all the others.

When a mixture of gases is present in a container, it is convenient to define a **partial pressure** of each gas—that is, the pressure exerted by that gas alone. The total pressure of all the gases in the mixture is the sum of these partial pressures. For example, if a container holds a quantity of He with partial pressure P_{He}, a quantity of Ne with partial pressure P_{Ne}, and a quantity of Ar with partial pressure P_{Ar}, the total pressure P_T is given by:

$$P_T = P_{He} + P_{Ne} + P_{Ar}$$

Usually, only P_T can be measured directly.

Dalton's law of partial pressures, which states that **the total pressure of a mixture of gases is the sum of the partial pressures of the gases,** has a broad range of applications. It is useful in the study of systems where substances in the gas phase are in contact with condensed phases of other volatile substances. Figure 4.13 shows such a system, which is frequently encountered: the collection of a gas by displacement of water. The gas is produced by a chemical reaction and is directed into a water-filled tube. As the gas collects in the container, it mixes with the water vapor that is already present. Therefore, the total gas pressure is the sum of the partial pressure of the gas that is being collected and the partial pressure of water vapor (more exactly, of the vapor pressure of water at the temperature of the experiment).

Table 4.1 lists the vapor pressure of water at a number of temperatures. To find the actual pressure of the gas being collected, the vapor pressure of water is subtracted from the total measured pressure.

EXAMPLE 4.15

One method of preparing $H_2(g)$ is by the reaction of an active metal with acid, for example:

$$Mg(s) + 2H^+ \longrightarrow Mg^{2+} + H_2(g)$$

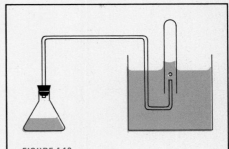

FIGURE 4.13
Dalton's law of partial pressures can be used to calculate the pressure of a gas collected by displacement of water. Some of the water vaporizes in the closed end of the tube. The pressure exerted by the gas that is being collected is determined by subtracting the vapor pressure of water from the total pressure in the tube.

TABLE 4.1
Vapor Pressure of Water

T (K)	P (atm)	T (K)	P (atm)
273	0.0061	298	0.0313
274	0.0065	299	0.0332
275	0.0070	300	0.0351
276	0.0075	301	0.0372
277	0.0080	302	0.0395
278	0.0086	303	0.0418
279	0.0092	304	0.0443
280	0.0099	305	0.0470
281	0.0105	306	0.0496
282	0.0113	307	0.0525
283	0.0121	308	0.0555
284	0.0129	313	0.0728
285	0.0138	318	0.0946
286	0.0147	323	0.1217
287	0.0158	328	0.1553
288	0.0168	333	0.1966
289	0.0179	338	0.2467
290	0.0191	343	0.3075
291	0.0204	348	0.3804
292	0.0217	353	0.4672
293	0.0230	358	0.5705
294	0.0246	363	0.6918
295	0.0261	368	0.8341
296	0.0278	373	1.0000
297	0.0295	378	1.1922

The hydrogen gas can be collected over water, as shown in Figure 4.17. The volume of the gases in the collecting bottle is found to be 0.350 liter and the temperature is 299 K. Because the water levels inside and outside the bottle are equal, the total pressure of gases in the bottle is equal to the pressure in the room, which is found by barometer measurement to be 1.041 atm. What mass of $H_2(g)$ is collected?

SOLUTION

Since the volume and temperature of the $H_2(g)$ are given and the mass is to be found, the ideal gas equation could be used if P_{H_2}, the pressure of H_2, were known. The P_{H_2} can be found by using Dalton's law and the data in Table 4.1.

At 299 K, the vapor pressure of H_2O is 0.0332 atm.

$$P_T = P_{H_2} + P_{H_2O}$$

Substituting:

$$1.041 \text{ atm} = P_{H_2} + 0.033 \text{ atm}$$
$$P_{H_2} = 1.008 \text{ atm}$$

Since we now have three of the variables in the ideal gas equation, we can find the fourth unknown. Using 2.016 as the molecular weight of H_2 gives:

$$(1.008 \text{ atm})(0.350 \text{ liter}) = \frac{g}{2.016}\left(0.0821 \frac{\text{liter atm}}{\text{mol K}}\right)(299 \text{ K})$$

$$\text{mass } H_2 = 0.0290 \text{ g } H_2 \text{ is collected}$$

We see from the ideal gas laws that the pressure of an ideal gas at constant volume and temperature is directly proportional to the number of moles of the gas:

$$P = kn$$

If there are two or more gases in the same container, they naturally have the same volume and temperature. We know from the law of partial pressures that the pressure exerted by each gas in such a mixture is not affected by the presence of the other gases. Therefore, the partial pressure of each gas is directly proportional to the amount of that gas in the container. When we have such a mixture, we can determine the amount of a gas by finding its partial pressure. Since it is relatively easy to measure the total pressure of a mixture of ideal gases, using partial pressures for stoichiometric calculations is often convenient.

EXAMPLE 4.16

A sample of air collected over Los Angeles during a smog alert occupies a volume of 1.0 liter and contains 0.00010 g of $CO(g)$. If the temperature is 308 K, what is the partial pressure of CO in the sample?

SOLUTION

The partial pressure of CO in the mixture of gases is the same as if CO were present alone. Since we know the values of four of the variables in Equation 4.14, we can find the value of the fifth:

$$P(1.0 \text{ liter}) = \frac{0.00010 \text{ g}}{28 \text{ g/mol}}\left(0.0821 \frac{\text{liter atm}}{\text{mol K}}\right)(308 \text{ K})$$

$$P_{CO} = 9.0 \times 10^{-5} \text{ atm}$$

$$= 9.0 \times 10^{-5} \text{ atm} \times 1.0 \times 10^5 \frac{\text{Pa}}{\text{atm}}$$

$$= 9.1 \text{ Pa}$$

EXAMPLE 4.17

A container with a volume of 3.00 liters holds $N_2(g)$ and $H_2O(l)$ at a temperature of 29°C. The pressure is found to be 1.00 atm. The water is then converted to hydrogen and oxygen by electricity according to the reaction:

$$H_2O(l) \longrightarrow H_2(g) + \tfrac{1}{2}O_2(g)$$

After the reaction is complete, the pressure is 1.86 atm. What mass of water was in the container?

SOLUTION

If the amount of $H_2(g)$ or $O_2(g)$ present after the reaction were known, the amount of $H_2O(l)$ could be determined. The amount of $H_2(g)$ or $O_2(g)$ can be found if the partial pressure of either gas is found, since their volume and temperature are given. The problem can be solved by using the total pressure (which is given) and the stoichiometric relationships to calculate P_{H_2}.

After the reaction is over:

$$P_T = P_{N_2} + P_{H_2} + P_{O_2} = 1.86 \text{ atm}$$

From the stoichiometry, 1 mol of H_2O produces 1 mol of H_2 and $\tfrac{1}{2}$ mol of O_2. Therefore, $n_{H_2} = 2n_{O_2}$, or:

$$P_{H_2} = 2P_{O_2} \quad \text{or} \quad P_{O_2} = \tfrac{1}{2}P_{H_2}$$

Thus:

$$P_T = P_{N_2} + P_{H_2} + \tfrac{1}{2}P_{H_2}$$

The P_{N_2} can be calculated from the original conditions. Initially, $P_T = P_{N_2} + P_{H_2O} = 1.00$ atm, where P_{H_2O} is the vapor pressure of water at $(29 + 273)$ K. Table 4.1 gives this vapor pressure as 0.0395 atm. Thus:

$$P_{N_2} + 0.040 \text{ atm} = 1.00 \text{ atm}$$
$$P_{N_2} = 0.96 \text{ atm}$$
$$0.96 \text{ atm} + \tfrac{3}{2}P_{H_2} = 1.86 \text{ atm}$$
$$P_{H_2} = 0.60 \text{ atm}$$

$$(0.60 \text{ atm})(3.00 \text{ liters}) = n\left(0.0821 \frac{\text{liter atm}}{\text{mol K}}\right)(29 + 273 \text{ K})$$

$$n_{H_2} = 0.073 \text{ mol} = n_{H_2O}$$

Therefore, the mass of H_2O is

$$(0.073 \text{ mol } H_2O)\left(\frac{18.01 \text{ g } H_2O}{1 \text{ mol } H_2O}\right) = 1.3 \text{ g } H_2O$$

EXAMPLE 4.18

One process related to the formation of smog is:

$$2NO(g) + O_2(g) \longrightarrow 2NO_2(g)$$

A convenient way to study this reaction in the laboratory is to monitor the total pressure of a mixture of the gases. Since the reaction converts every 3 mol of reactant gases to 2 mol of gaseous products, the total pressure of a mixture of NO and O_2 will drop as the reaction proceeds.

A bulb of volume 0.501 liter is filled with a mixture of $NO(g)$ and $O_2(g)$. The initial total pressure is 1.22 atm at 298 K. After six hours, the total pressure is 0.82 atm. What mass of $NO_2(g)$ has been formed?

SOLUTION

The chemical equation defines the relationship between the amounts of the gases. Therefore, it also defines the relationship between the changes in their partial pressures. In a problem of this kind, it is often convenient to tabulate the relationships under the formulas of the substances in the chemical equation. If we let x = the pressure of NO that undergoes reaction:

$$2NO(g) \quad + O_2(g) \quad \longrightarrow \quad 2NO_2(g)$$

before reaction	P_{NO}	P_{O_2}	0
after reaction	$(P_{NO} - x)$	$\left(P_{O_2} - \dfrac{x}{2}\right)$	x

The equation states that two moles of NO react with one mole of O_2 to form two moles of NO_2. From Dalton's law, the same statement can be made with units of pressure substituted for moles. If we begin with a quantity of NO represented as P_{NO}, and a quantity x of NO undergoes reaction, then $(P_{NO} - x)$ remains. Similarly, if we begin with P_{O_2} and a quantity $\frac{x}{2}$ undergoes reaction, then $(P_{O_2} - \frac{x}{2})$ remains. Further, when x of NO undergoes reaction, x of NO_2 is formed.

The total pressure of gases in the system after reaction is the sum of the partial pressures of NO, O_2, and NO_2. An expression for each of these partial pressures is tabulated under the formula of the gas in the chemical reaction.

$$P_T = (P_{NO} - x) + \left(P_{O_2} - \frac{x}{2}\right) + x = 0.82 \text{ atm}$$

or

$$P_T = P_{NO} + P_{O_2} - \frac{x}{2} = 0.82 \text{ atm}$$

From the initial conditions:

$$P_{NO} + P_{O_2} = 1.22 \text{ atm}$$

Therefore:

$$1.22 \text{ atm} - \frac{x}{2} = 0.82 \text{ atm}$$

$$x = 0.80 \text{ atm}$$

Since we now have the values of four of the variables in Equation 4.14, we can calculate the value of the fifth, the mass of NO_2:

$$(0.80 \text{ atm})(0.510 \text{ liter}) = \frac{g}{46.0 \text{ g } NO_2/1 \text{ mol } NO_2}\left(0.0821\frac{\text{liter atm}}{\text{mol K}}\right)(297.9 \text{ K})$$

$$\text{mass } NO_2 = 0.77 \text{ g}$$

A problem of this type can sometimes be solved by reasoning directly from the chemical equation. In the formation of NO_2 from O_2 and N_2, 3 atm of gas is required to form 2 atm of gas; therefore when the total pressure drops by 1 atm, 2 atm of the product has been produced. Once this relationship is known, we can calculate the amount of product that is formed if we know the amount of change in pressure. The pressure drop in this problem is given as $1.22 \text{ atm} - 0.82 \text{ atm} = 0.40 \text{ atm}$. Therefore, $(2)(0.40 \text{ atm}) = 0.80 \text{ atm}$ of the product, NO_2, has been formed. In any case where a simple relationship between change in total pressure and the pressure of product formed can be recognized, the relationship can be used for calculations.

4.6 THE KINETIC THEORY OF GASES

If we can account for some of the observable macroscopic properties of substances with a mental or mathematical picture based on the collective behavior of their component molecules, we can use that picture to predict and understand many other properties of the substances. We have such a picture for gases. The behavior of ideal gases can be predicted by starting with a simple picture of their structure on the molecular level, called a ''model.'' This model not only predicts those aspects of ideal gas behavior that we have already discussed but also helps us understand a number of other properties of gases. In addition, the model helps us understand the reasons why real gases do not behave like ideal gases.

A model for an ideal gas should be consistent with such observations as Boyle's law, Charles's law, and Avogadro's law. It should account for the observation that gases are compressible and expansible. The important features of a model that can account for these observations are:

1. An ideal gas consists of molecules (or atoms) in constant, random motion.
2. These molecules do not exert forces on one another except when they collide.
3. The diameter of the molecules is negligible compared to the distance between molecules and the space they occupy. An ideal gas consists primarily of empty space. Its molecules can be regarded as point masses that have no volume.
4. Because they are in motion, the molecules of an ideal gas will collide with the walls of their container. All of these collisions are elastic—that is, there is no loss in total kinetic energy due to the collisions.
5. When a large number of ideal gas molecules are in motion, there is an average speed of the molecules whose value does not change unless the gas is disturbed by outside influences.

It must be made clear that many of the features of this model cannot be correct for real gases. This is an *ideal* model. The deviations of real gas behavior from ideal gas behavior can be understood by comparing the ideal model with a real gas.

Some features of an ideal gas follow directly from the model. If an ideal gas is considered to be primarily empty space, the gas will be compressible and expansible. If there are no interactions between molecules, Dalton's law of partial pressures, which states essentially that gases that behave in an ideal way act independently of one another, must be correct.

Derivation of Boyle's Law

Boyle's law can be derived from the model by making some simplifying approximations about the random motion of the gas molecules. Consider an ideal gas in a cubical container of side *l*, as shown in Figure 4.14. There is a very large number, *N*, of gas molecules in the container, all moving randomly. To keep the mathematics simple, the random motion of the molecules can be described in an approximate way by dividing the molecules into three groups of equal number. One group consists of molecules that bounce back and forth between the left and right walls, moving perpendicular to each wall. This can be the *x* direction. The second group consists of molecules that bounce back and forth between the top and bottom walls of the cube. This can be the *y* direction. The third group consists of molecules bouncing between the front and back walls. This can be the *z* direction. Each group contains *N*/3 molecules.

The pressure on the walls of the cube is created by the collision of molecules with the walls. Consider the pressure on the right-hand wall. The pressure on this wall (or on any other wall) is the same as the total pressure of the gas, because the pressure exerted by a gas is exerted equally on all walls of its container. Pressure is defined as force per unit of area. The force imparted to the wall by the

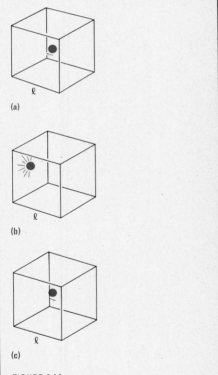

FIGURE 4.14

An approximation of ideal gas behavior, using three groups of molecules. One group of molecules bounces between the left and right walls of the container, a second group bounces between the top and bottom walls, and the third group bounces between the front and back walls. The pressure exerted by the gas is due to the collisions by gas molecules with walls of the container. Here we see one of the molecules bouncing back and forth between the left and right walls of the container.

collision of a single molecule is given by Newton's second law:

$$f = ma \qquad (4.16)$$

where f is the force, m is the mass of a molecule, and a is its acceleration. Since acceleration is the change in velocity in an elapsed time:

$$f = m\frac{\Delta v}{\Delta t} \qquad (4.17)$$

where Δv is the change in velocity and Δt is the elapsed time.

To obtain a useful expression for the force imparted to the wall by collisions with a molecule of mass m and velocity v_x in the x direction, we must answer two questions: How much does the velocity of the molecule change as a result of the collision? How often does a collision occur?

The molecule hits the wall perpendicularly, and the collision is elastic. Therefore, it changes direction by 180°, and its velocity changes from v_x to $-v_x$. Thus:

$$\Delta v_x = v_x - (-v_x) = 2v_x \qquad (4.18)$$

The time between collisions depends on two factors: the velocity of the molecule and the distance the molecule must travel before it collides with the same wall again. After a collision with the wall, the molecule moves a distance l from right to left. It collides with the left-hand wall, bounces back, travels a distance l again, and collides with the right-hand wall once more. The distance it travels between right-hand-wall collisions is $2l$.

The time between collisions is the distance traveled divided by the velocity. Thus:

$$\Delta t = \frac{2l}{v_x} \qquad (4.19)$$

Substitution of Equations 4.18 and 4.19 into Equation 4.17 gives:

$$f = \frac{m(2v_x)}{\dfrac{2l}{v_x}} = \frac{mv_x^2}{l} \qquad (4.20)$$

as the force imparted to the wall by one molecule. Since $N/3$ molecules are pictured as hitting the wall, the total force imparted is:

$$F = \left(\frac{N}{3}\right)\left(\frac{m\overline{v_x^2}}{l}\right) \qquad (4.21)$$

where $\overline{v_x^2}$ is the average value of v_x^2 of the $N/3$ molecules.

The pressure on one wall, and therefore the pressure of the gas, is obtained by dividing the force on the wall by the area of the wall, which is l^2. Thus:

$$P = \frac{\left(\dfrac{N}{3}\right)\left(\dfrac{m\overline{v_x^2}}{l}\right)}{l^2} = \left(\frac{N}{3}\right)\left(\frac{m\overline{v_x^2}}{l^3}\right) \qquad (4.22)$$

Since the volume of the cube (V) is l^3. Equation 4.22 can be written as:

$$P = \frac{N}{3}\frac{m\overline{v_x^2}}{V} \qquad (4.23)$$

where P is the pressure on one wall. Since we have approximated random motion by dividing the gas molecules into three groups moving in three perpendicular directions, $\overline{v_x^2}$ is equal to $\overline{v^2}$. Therefore:

$$PV = \frac{N}{3}m\overline{v^2} \qquad (4.24)$$

where $\overline{v^2}$ is the average value of the square of the velocity of all the molecules moving in all directions; that is, the average of $\overline{v_x^2}$, $\overline{v_y^2}$, and $\overline{v_z^2}$.

Equation 4.24 is Boyle's law; it states that PV depends on quantities that are constants for any given amount of gas. For any given sample, it is evident that N and m do not vary, and $\overline{v^2}$ does not vary when the temperature remains constant. In other words, Equation 4.24 is another way of writing Equation 4.1.

Kinetic Energy and Temperature

To point out some other characteristics of an ideal gas, Equation 4.24 can be rewritten as:

$$PV = \tfrac{2}{3}N(\tfrac{1}{2}m\overline{v^2}) \qquad (4.25)$$

The quantity $\tfrac{1}{2}m\overline{v^2}$ is called the mean kinetic energy per molecule of the gas. (A body in motion has energy by virtue of that motion. This energy is called kinetic energy and is equal to $\tfrac{1}{2}mv^2$, where m is the mass and v is the velocity of the moving body.)

Equation 4.25 can be modified to include some useful quantities. Thus:

$$N = nN_A \qquad (4.26)$$

The number of molecules N in a sample of a substance is the number of moles of the substance, n, multiplied by Avogadro's number, N_A. Substituting Equation 4.26 into Equation 4.25 gives:

$$PV = \tfrac{2}{3}nN_A \,(\tfrac{1}{2}m\overline{v^2}) \qquad (4.27)$$

From the ideal gas law, which is based on experimental data, we know that:

$$PV = nRT$$

Therefore:

$$nRT = \tfrac{2}{3}nN_A(\tfrac{1}{2}m\overline{v^2})$$

or, eliminating n and rearranging:

$$N_A(\tfrac{1}{2}m\overline{v^2}) = \tfrac{3}{2}RT \qquad (4.28)$$

This equation gives us important information about kinetic energy. Since N_A is the number of molecules in a mole, the left side of Equation 4.28 is the total kinetic energy of one mole of an ideal gas. This total kinetic energy is seen to be directly proportional to the absolute temperature of the gas. Thus, the kinetic energy of an ideal gas does not depend on its chemical composition, but only on the absolute temperature. Equation 4.28 may also be written as:

$$\tfrac{1}{2}m\overline{v^2} = \tfrac{3}{2}\frac{RT}{N_A} = \tfrac{3}{2}kT \qquad (4.29)$$

where the left side of the equation is the mean kinetic energy per molecule of the gas. The proportionality constant k is called *Boltzmann's constant*. It is the gas constant per molecule, R/N_A.

One might wonder why the kinetic energy of a given quantity of an ideal gas is proportional to the absolute temperature, since the absolute temperature scale is based on the volume of an ideal gas. The explanation is as follows:

The definition of the absolute temperature scale from Charles's law indicates that when a given quantity of gas is held at constant pressure and the temperature is lowered, the volume should become zero at 0 K. We mentioned earlier that the volume of a sample of gas at 273 K (0°C) changes by 1/273 for a change of 1°C; the temperature at which the gas has zero volume is thus put at −273°C, which is 0 K. Suppose we perform an imaginary experiment in which the volume of a gas is held constant, the temperature is lowered, and the pressure is allowed

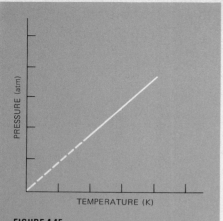

FIGURE 4.15
The relationship between pressure of a gas and temperature. As temperature decreases, the pressure of the gas decreases. According to the ideal gas law, gas pressure should become zero at 0 K.

to vary. As Figure 4.15 shows, P would decrease as T decreased. However, Boyle's law states that PV is a constant at a fixed temperature, so PV must be equal to zero at 0 K. We know that V, which is being held constant, is not zero. But PV equals zero. Therefore, P must equal zero.

Thus, for a given quantity of gas in a container of fixed volume, the pressure will be zero at 0 K. If the pressure is zero, no molecules are colliding with the walls of the container. Therefore, the molecules are no longer moving in the container; that is, $v = 0$ and so $\frac{1}{2}mv^2 = 0$. In other words, kinetic energy is zero at 0 K, and kinetic energy increases in direct proportion to increases in absolute temperature.

The relationship between the kinetic energy and the absolute temperature leads directly to Avogadro's law. Consider two different ideal gases, X and Y, under the same conditions of pressure and volume; $P_X = P_Y$ and $V_X = V_Y$. Then:

$$P_X V_X = P_Y V_Y$$

and, from Equation 4.25:

$$\tfrac{2}{3}N_X(\tfrac{1}{2}m_X\overline{v_X^2}) = \tfrac{2}{3}N_Y(\tfrac{1}{2}m_Y\overline{v_Y^2}) \tag{4.30}$$

If gas X and gas Y are at the same temperature, their kinetic energies are equal: $\frac{1}{2}m_X v_X^2 = \frac{1}{2}m_Y \overline{v_Y^2}$ and Equation 4.30 becomes:

$$N_X = N_Y$$

That is, under fixed conditions of T and P, equal volumes of two gases ($V_X = V_Y$) contain equal numbers of molecules ($N_X = N_Y$). This is Avogadro's law.

4.7 EFFUSION AND DIFFUSION

Expressions such as Equation 4.28 define the relationship between the average speed of ideal gas molecules and such measurable quantities as the molecular weight of the gas and the temperature. Since the molecular weight is the mass of one molecule multiplied by Avogadro's number ($MW = N_A m$), Equation 4.28 can be rewritten as:

$$\overline{v^2} = \frac{3RT}{MW} \tag{4.31}$$

Taking the square root of both sides of this equation gives:

$$\overline{v} = \sqrt{\frac{3RT}{MW}} \tag{4.32}$$

The quantity $\overline{v}$ is called the root-mean-square speed. It is the square root of $\overline{v^2}$, which is the average value of v^2. The root-mean-square speed is almost the same as the average speed of the gas molecules, and we shall use the terms "average speed" and "root-mean-square speed" interchangeably.

From Equation 4.32, it can be shown that if two different ideal gases are at the same temperature, the average speed of their molecules is inversely proportional to the square root of their molecular weights. For example, when H_2 ($MW = 2$) and O_2 ($MW = 32$) are at the same temperature, for H_2:

$$\overline{v} = \sqrt{\frac{3RT}{2}}$$

and for O_2:

$$\overline{v} = \sqrt{\frac{3RT}{32}}$$

Thus:

$$\frac{(\bar{v}H_2)}{(\bar{v}O_2)} = \frac{\sqrt{\dfrac{3RT}{2}}}{\sqrt{\dfrac{3RT}{32}}} = \sqrt{\frac{32}{2}} = 4$$

That is, the H_2 molecules, on the average, are moving four times faster than the O_2 molecules at the same temperature.

Some aspects of the behavior of gases are related to the speed of the gas molecules. One of the first of these to be studied is illustrated in Figure 4.16. If a small hole is made in the wall of a container holding a gas, and if the hole leads to a vacuum, the gas in a container will pass through the hole. This phenomenon is called **effusion**. It is reasonable to assume that the rate of effusion will be directly proportional to the speed of the molecules of a gas. Therefore, the rate of effusion should be inversely proportional to the square root of the molecular weight of the gas. The relative rates of effusion of two different gases, A and B, can be expressed as:

$$\frac{\bar{v}_A}{\bar{v}_B} = \frac{r_A}{r_B} = \sqrt{\frac{MW_B}{MW_A}} \tag{4.33}$$

where r_A and r_B are the rates of effusion of gas A and gas B, MW_A and MW_B are the molecular weights of the gases, and $\bar{v}_A$ and $\bar{v}_B$ are their root-mean-square speeds.

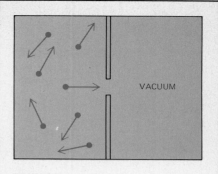

FIGURE 4.16
Effusion. If a small hole is made in a wall that separates a gas and a vacuum, the gas will pass through the hole, by a process called effusion. The rate of effusion is proportional to the speed of the gas molecules.

EXAMPLE 4.19

A sample of N_2 is placed in an effusion apparatus of the type shown in Figure 4.16. After 15 minutes, it is found that the pressure in the evacuated portion of the apparatus has risen to 0.21 atm. The N_2 is then removed from the apparatus and the experiment is repeated with a quantity of $Br_2(g)$ at the same temperature. What will be the pressure of Br_2 in the evacuated portion of the apparatus after 15 minutes?

SOLUTION

The rate of increase in pressure depends on the rate of effusion. The rate of effusion depends on the molecular weights of the gases, as indicated by Equation 4.33. The molecular weight of Br_2 is 160 and of N_2 is 28. Equation 4.33 gives the relationship between molecular weight and relative rate of effusion:

$$\frac{r_{N_2}}{r_{Br_2}} = \sqrt{\frac{160}{28}} = 2.4$$

The N_2 effuses 2.4 times faster than the Br_2. If the pressure of N_2 after 15 minutes is 0.21 atm, then the pressure of Br_2 after the same time period is $(0.21)(1/2.4) = 0.088$ atm.

The relationship between the rate of effusion and the chemical nature of the gas was first observed early in the nineteenth century by Thomas Graham (1805–1869). He found that the relative rates of effusion of two gases under the same conditions of temperature and pressure are inversely proportional to the square root of their densities:

$$\frac{r_A}{r_B} = \sqrt{\frac{d_B}{d_A}} \tag{4.34}$$

This relationship is called **Graham's law**. It is consistent with Equation 4.15, which indicates that under the same conditions of temperature and pressure, the densities of gases are proportional to their molecular weights.

Another closely related aspect of gas behavior is the mixing of gases. If a gas is introduced into a container that already holds another gas—for example, by opening a stopcock as shown in Figure 4.17—the newly introduced gas eventually spreads throughout the container. The process by which a gas spreads throughout a space is called **diffusion**. Again, it is reasonable to assume that the rate of diffusion of a gas will be proportional to the speed of its molecules. Therefore, Equations 4.33 and 4.34 also apply to the relative rates of diffusion of two gases.

The type of apparatus shown in Figure 4.18 is often used to study what is commonly called the diffusion of gases but is, in reality, relative effusion. A mixture of gases is allowed to pass through a porous wall. The relative rates of diffusion determine the relative amounts of the gases that pass through the barrier in a given time. An experiment of this kind can be used to find the molecular weight of an unknown gas.

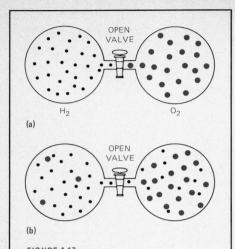

FIGURE 4.17
Diffusion. If a gas is introduced into a container that already holds a different gas by opening a stopcock, the newly introduced gas will spread—diffuse—throughout the container. The rate of diffusion is proportional to the speed of the molecules of the newly introduced gas.

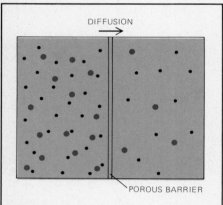

FIGURE 4.18
An apparatus for studying the diffusion of gases. A mixture of two or more gases is placed in the left side of the container and is allowed to pass through the porous barrier into the right side of the container. The relative amount of each gas that passes through the barrier in a given period of time indicates the rate of diffusion of the gas.

EXAMPLE 4.20

A mixture of carbon dioxide and an unknown gas, each with the same partial pressure, is placed in the left side of the apparatus shown in Figure 4.18. The right side of the apparatus is evacuated. After a short period, the pressure in the right side of the apparatus is found to be 0.041 atm. The CO_2 is then removed by absorption with a solid, and the pressure on the right side drops to 0.016 atm. Calculate the molecular weight of the unknown gas.

SOLUTION

In this experiment, pressure measurements are used to determine relative amounts of the gases. The pressure of CO_2 is indicated by the drop in the total pressure when the CO_2 is removed:

$$P_{CO_2} = 0.041 \text{ atm} - 0.016 \text{ atm} = 0.025 \text{ atm}$$

The pressure of the unknown gas is 0.016 atm, the remainder. Therefore, the relative rates of diffusion of CO_2 and the unknown gas are, from Equation 4.33:

$$\frac{r_{CO_2}}{r_{unknown}} = \frac{0.025}{0.016} = \sqrt{\frac{MW_{unknown}}{MW_{CO_2}}}$$

The MW_{CO_2} is known to be 44. Therefore, we have the values of three of the four unknowns in the equation. Solving for the fourth gives $MW_{unknown} = 107$.

Suppose we carry out the experiment described in Example 4.20 and then measure the relative concentrations of the gases on both sides of the barrier. We shall find that the relative concentration of the lighter gas—in this case CO_2—is higher on the right side of the apparatus than it was in the original mixture. The lighter gas passes through the barrier more readily than the heavier gas. Thus, we can use an apparatus of this kind to separate gases of different molecular weights. The most important application of this method occurred during World War II.

To make an atomic bomb, relatively pure $^{235}_{92}U$ was needed. Naturally occurring uranium consists almost entirely of $^{238}_{92}U$, which is not suitable bomb material; $^{235}_{92}U$ is a minor isotope. Chemical methods could not be used to separate the two uranium isotopes, since the chemical behavior of different isotopes of the same element is essentially the same. The problem was solved by using uranium hexafluoride, UF_6, a compound that has a high vapor pressure at room temperature.

Vapors of UF_6 consist of $^{235}UF_6$, $MW = 235 + 6(19) = 349$, and $^{238}UF_6$,

THE USES OF DIFFUSION

Diffusion, the process by which molecules or other particles mix because of their random thermal motion, occurs not only in gases but also in liquids, solutions, and even solids. Diffusion is seen in many chemical processes, in both natural and man-made systems. The rates of chemical transformations depend on the concentration of the reactants. As the initial concentration is depleted, it is nearly always replenished by some diffusion process. The rate of diffusion thus can determine the rate of the chemical transformation.

One commercial system that puts diffusion to practical use is called the iodine lamp. The hot tungsten filament of the lamp produces a relatively high vapor pressure of tungsten, whose atoms diffuse to the glass wall of the lamp. As tungsten accumulates on the glass, it lessens light output and shortens lamp life. If a trace of iodine vapor is added to the bulb, tungsten iodide tends to form in the relatively cool region near the glass. The tungsten iodide diffuses toward the filament, where it decomposes to release the tungsten. Thus, the rate at which tungsten accumulates on the glass is lessened.

Diffusion also plays an important role in microbes. The cell of a microbe can be regarded as a unit in which a number of different chemical reactions occur—the reactions by which a microbe maintains itself. For microbes, the diffusion of oxygen through a liquid medium into the cell is of major importance. It has been shown that the rate of oxygen diffusion into yeast cells can be the controlling factor in the activity of the cells under some conditions. These studies have been used to help develop the most efficient methods for a new food technology of great promise. Certain strains of yeast can ferment petroleum and produce edible proteins. The rate at which oxygen is supplied is an important part of the technology of such systems.

Diffusion also plays an interesting role in the process by which a spermatozoon, or sperm cell, manages to fertilize an egg. A spermatozoon "swims" toward the egg by wiggling its tail. The energy for the wiggling is derived from a molecule called adenosine triphosphate (ATP). However, ATP is produced in the body of the sperm cell, at the front of the tail. The spermatozoon is able to wiggle forward successfully because ATP diffuses all through the tail. One study of diffusion in the spermatozoon resulted in the calculation that diffusion could supply an adequate amount of ATP only over a length of 5×10^{-3} cm to 9×10^{-3} cm. In fact, the length of the tail of a sperm cell is about 5×10^{-3} cm.

$MW = 238 + 6(19) = 352$. When these vapors diffuse through a porous barrier, the lighter gas comes through slightly faster:

$$\frac{r_{(^{235}UF_6)}}{r_{(^{238}UF_6)}} = \sqrt{\frac{352}{349}} = 1.004$$

Therefore, the fraction of the desired $^{235}UF_6$ will be very slightly higher in the vapor that has passed through the porous barrier. If this diffusion process is repeated thousands of times, eventually ^{235}U of the desired purity can be obtained.

4.8 MOLECULAR SPEEDS

The theoretical and experimental work done on the speed of molecules make it possible to examine some aspects of this phenomenon more closely. Equation 4.33 gives a way to calculate the average speed of the molecules of a gas of known molecular weight at any temperature. To make this calculation, we must express R in units that are also appropriate for measuring speed, which is measured in units of length per time. We can do so by this line of reasoning: Pressure is force per unit of area, which is equivalent to force per unit of length squared. Volume is length cubed. Therefore, pressure times volume is equivalent to (force/length^2)(length^3), or force $\times$ length.

The SI unit of force is the newton; 1 N = 1 kg m s^{-2}. The SI unit of length is the meter, m. Pressure $\times$ volume in the SI has units called joules, J; 1 J = 1 kg m^2 s^{-2}. The joule is the SI unit of energy, since it is force $\times$ length. (It should be noted that the product of liters $\times$ atmospheres is also an expression of energy, as will be seen in Chapter 13.) Thus, in SI units, R = 8.3146 J mol^{-1} K^{-1} or 8.3146 kg m^2 s^{-2}mol^{-1} K^{-1}. This value of R is expressed in terms of length (m) and time (s), so it can be used to calculate speed (m/s).

EXAMPLE 4.21

Calculate the average speed of N_2 molecules in the atmosphere at a temperature of 300 K.

SOLUTION

Speed is expressed in units of m s^{-1}, so we must take care that all the units cancel except m s^{-1}. Since R has units of kg m^2 s^{-2}, the molecular weight or mass of one mole of N_2 must be expressed as kg mol^{-1}. Equation 4.32 gives the expression for the average speed:

$$\bar{v} = \sqrt{\frac{3RT}{MW}}$$

$$\bar{v} = \sqrt{\frac{(3)(300 \text{ K})(8.31 \text{ kg m}^2 \text{ s}^{-2} \text{ mol}^{-1} \text{ K}^{-1})}{0.028 \text{ kg mol}^{-1}}}$$

$$= \sqrt{267\,000 \text{ m}^2 \text{ s}^{-2}}$$

$$\bar{v} = 517 \text{ m s}^{-1}$$

The average speed of N_2 molecules calculated in Example 4.21 is quite large. For comparison, the speed of sound in air is 330 m s^{-1}.

The average speed, or root-mean-square speed, is a convenient way to describe molecular motion in gases. But any macroscopic sample of a gas contains an enormous number of molecules, and these molecules are actually moving with many different speeds. There is a distribution of molecular speeds. Some molecules move slower than the average and some move faster than the average. Because of the large number of molecules in such a gas sample, it is impossible to specify all their speeds individually. However, in the middle of the nineteenth century, James Clerk Maxwell (1831–1879) and Ludwig Boltzmann (1844–1906) found a statistical relationship between the temperature of a gas and the distribution of molecular speeds.

Maxwell and Boltzmann derived a complex equation that can be used to calculate the fraction of molecules that have a given range of speeds at a given temperature. A detailed discussion of the Maxwell-Boltzmann equation is beyond our scope. We shall say only that the equation can be used to calculate the distribution of molecular speeds at various temperatures, and that the results can be plotted. Figure 4.19 shows some of these curves, which can be regarded as probability curves. The maximum of the curve gives the most probable speed for a molecule; that is, the average molecular speed is close to the maximum of the

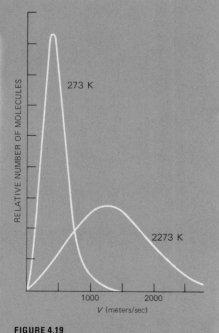

FIGURE 4.19
The distribution of molecular speeds at different temperatures, calculated from the Maxwell-Boltzmann equation. The maximum of each curve gives the most probable speed of a gas molecule at a given temperature, while the curve as a whole shows the distribution of molecular speeds for a given temperature.

curve. The theoretical predictions of the distribution of molecular speeds made by the Maxwell-Boltzmann equation have been verified experimentally.

4.9 REAL GASES

The ideal gas model that was proposed in Section 4.6 leads to the derivation of an equation of state that describes the properties of an ideal gas. Real gases sometimes behave very much like ideal gases, but they sometimes behave quite differently. Experimentally, it has been observed that a real gas behaves most like an ideal gas when the temperature is relatively high in relation to the boiling point of the gas and when the pressure is relatively low—not much higher than atmospheric pressure.

A convenient way to evaluate the extent to which a real gas deviates from ideal gas behavior is to measure the four variables of interest: pressure, volume, temperature, and amount. When these measured values are substituted into the expression:

$$\frac{PV}{nT}$$

the value obtained for an ideal gas is 0.082 liter atm mol^{-1} K^{-1}. A different value is obtained for a real gas that deviates from ideal gas behavior. If the value is greater than 0.082, the real gas is said to display a *positive deviation*. If the value is less than 0.082, the real gas is said to display a *negative deviation*. The behavior of some real gases is shown in Figure 4.20, in which experimentally observed values of PV/nT are plotted for some real gases at various pressures.

To understand how and why real gases deviate from the ideal gas model, we can analyze the approximations in the model. The molecules of the ideal gas are assumed to have no volume, so the calculated volume of an ideal gas is only the volume of the empty space. The molecules of a real gas have volume, so the measured volume of a real gas is larger; it is the volume of the empty space plus the volume of the molecules. At high pressures, the molecules of the real gas are closer to each other than at low pressures. Thus, a larger fraction of the gas volume is occupied by the molecules at higher pressures. Therefore, when the pressure is high a real gas shows a positive deviation, as can be seen in Figure 4.20. The measured volume is greater than the ideal gas volume at the same pressure, so $PV/nT > 0.082$.

The volume of the molecules of a real gas also affects the gas pressure. The measured pressure of a real gas is greater than the pressure of the same volume of ideal gas. This effect can be understood by referring back to the derivation of Boyle's law. The pressure on the walls of the container is due to collisions by molecules; more frequent collisions mean higher pressure. For an ideal gas, the frequency of collisions of a molecule with a wall is calculated on the assumption that the entire volume of a container is empty space. But since the molecules of a real gas have volume, not all of the container is empty space. We must exclude the volume occupied by the molecules when we calculate the distance that the molecules travel between collisions with the walls. A molecule of a real gas travels less distance than a molecule of an ideal gas before colliding with the wall. Collisions are more frequent, so the actual pressure of a real gas is higher than the calculated pressure of the same volume of an ideal gas.

Figure 4.21 illustrates the effect of the volume of gas molecules on gas pressure. A molecule leaves the right-hand wall of the container and is due to hit the left-hand wall at a time that is calculated on the basis of the velocity of the molecule and the distance between walls. At the same time, another molecule leaves the left-hand wall headed for the right-hand wall. The two molecules collide in the center of the container and rebound toward the walls they left originally.

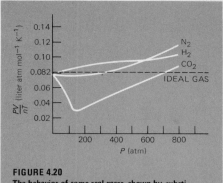

FIGURE 4.20
The behavior of some real gases, shown by substituting measured values into the expression PV/nT. For an ideal gas, the value of this ratio is 0.082 liter atm mol^{-1}K^{-1}. Hydrogen always shows a positive deviation. Other gases show a negative deviation at low pressure but a positive deviation at high pressure.

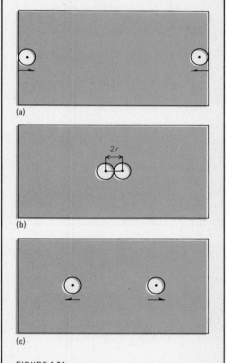

FIGURE 4.21
Why real gases can have higher pressures than are calculated for an ideal gas. The ideal gas approximation assumes that molecules of the gas do not occupy any volume. In a real gas, when molecules collide with one another, the distance that the molecules must travel between collisions with the container wall is less by $2r$, where r is the radius of a gas molecule. This results in an increased frequency of collisions with the walls and therefore a higher pressure.

Table 4.2
Van der Waals Constants

Substance	a (atm liter^2 mol^{-2})	b (liter mol^{-1})	Substance	a (atm liter^2 mol^{-2})	b (liter mol^{-1})
He	0.034	0.0237	CO_2	3.592	0.0427
Ne	0.211	0.0171	HCl	3.667	0.0408
H_2	0.244	0.0266	NH_3	4.170	0.0371
NO	1.340	0.0279	C_2H_2	4.390	0.0514
Ar	1.345	0.0322	H_2S	4.431	0.0429
O_2	1.360	0.0318	H_2O	5.464	0.0305
N_2	1.390	0.0391	Cl_2	6.493	0.0562
CO	1.485	0.0399	Hg	8.093	0.0170
CH_4	2.253	0.0428	C_2H_5OH	12.02	0.0841
Kr	2.318	0.0398	(ethanol)		

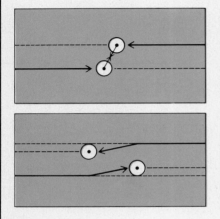

FIGURE 4.22
Molecular attractions in real gases. Two molecules attract one another, increasing the distance they must travel before hitting the walls of the container and thus lowering the measured pressure. The ideal gas law assumes that no such attractions take place.

Because the molecules have volume, the distance each one travels before colliding with a wall is less by $2r$, r being the radius of one molecule. Therefore, the frequency of collisions is higher. The volume of the gas molecules becomes more significant at high pressures, where real gases display positive deviation.

A second approximation of the ideal gas model is the assumption that gas molecules do not attract each other. In fact, there are rather substantial attractive forces between molecules in the gas phase. To a great degree, the magnitude of the attractive forces is related to the size of the gas molecules; larger molecules have greater attractive forces.

The major effect of these attractive forces is to reduce pressure. Figure 4.22 shows two molecules that pass near one another on the way to opposite walls of a container. Their mutual attraction deflects them from their straight paths, so they must travel a longer distance between collisions with the walls. If longer distances are traveled, the frequency of collisions with the walls is reduced, so gas pressure is reduced. Thus, attractive forces cause negative deviations of real gases from ideal gas behavior.

Most gases show negative deviations when pressure is not high enough to make molecular volume an important factor, as Figure 4.20 shows. The extent of the negative deviation at a given temperature depends on the magnitude of the attraction between molecules, which depends primarily on the size of the molecules. In gases that have large molecules and are not at high pressures, the intermolecular forces are more important than the volume of the molecules. Only gases made up of very small molecules, such as H_2 and He, always show positive deviations. In such gases, the force of intermolecular attraction is quite small.

Attempts to improve the ideal gas equation so that it provides a better description for real gases have been based on an empirical approach. Many measurements are made of real gases, and an attempt is made to fit their behavior to corrected versions of the ideal gas equation. One of the first and most successful of these approaches was made by Johannes van der Waals (1837–1923). He corrected the measured pressure of a gas for attractive forces and the measured volume of the gas for the volume of the molecules. The magnitude of the changes varies for each gas, but the changes can be determined by experiment and expressed as a constant for each gas. Table 4.2 lists the constants obtained in this manner. The constant a corrects P, and the constant b corrects V. The new equation of state, called the *van der Waals equation*, is written as:

$$\left(P + \frac{n^2a}{V^2}\right)(V - nb) = nRT \tag{4.35}$$

EXAMPLE 4.22
A heavy-walled evacuated container with a volume of 1.98 liters is filled with 215 g of dry ice, CO_2(s). After a time, all the dry ice sublimes to CO_2(g), and the temperature of the gas is found to be 299.2 K. Using the data of

Table 4.2, calculate the pressure of the $CO_2(g)$ and compare it to the ideal gas pressure.

SOLUTION

The van der Waals equation, like the ideal gas equation (Equation 4.35), has four variables, P, V, n, and T. The values of three of the variables are given in the problem. The van der Waals equation has three constants, R, a, and b, whose values are known. Therefore, the value of the unknown variable, P, can be calculated. First list the data:

$V = 1.98$ liters $\qquad\qquad R = 0.0821$ liter atm mol^{-1} K^{-1}

$T = 299.2$ K $\qquad\qquad a = 3.59$ liter^2 atm mol^{-2}

$n = \dfrac{215 \text{ g } CO_2}{44 \text{ g/mol } CO_2} = 4.89$ mol $\qquad b = 0.0427$ liter mol^{-1}

Then carry out the calculations:

$$\left[P + \frac{(4.89 \text{ mol})^2 (3.59 \text{ liter}^2 \text{ atm mol}^{-2})}{(1.98 \text{ liters})^2} \right] [1.98 \text{ liters} - (4.89 \text{ mol})(0.0427 \text{ liter/mol})]$$

$$= (4.89 \text{ mol}) \left(0.0821 \frac{\text{liter atm}}{\text{mol K}} \right) (299.2 \text{ K})$$

$P = 45.9$ atm

If we use the same values of n, T, and V in the ideal gas equation, we obtain a different value:

$$P(1.98 \text{ liters}) = (4.89 \text{ mol}) \left(0.0821 \frac{\text{liter atm}}{\text{mol K}} \right) (299.2 \text{ K})$$

$$P = 60.7 \text{ atm}$$

The measured pressure of $CO_2(g)$ under these conditions is found to be 44.8 atm; the van der Waals equation gives an answer closer to the measured pressure than does the ideal gas equation. Since the pressure calculated from the van der Waals equation is substantially less than the ideal pressure, $CO_2(g)$ shows a negative deviation from ideality. The negative deviation is consistent with the relatively strong intermolecular attraction expected at moderately high pressure for a gas with molecules the size of those of CO_2.

4.10 CRITICAL PHENOMENA

Suppose we are trying to liquefy a sample of a real gas. We could lower the temperature until it is at or below the condensation point, so that the gas condenses spontaneously. Theoretically, we could also liquefy the gas by raising the external pressure without lowering the temperature. As the pressure increases, the molecules come closer together, until the attractive forces between them become strong enough to cause condensation.

However, some gases will not liquefy at room temperature, no matter how much the pressure is increased. Indeed, this is true of most of the common gases, such as N_2, O_2, H_2, and CH_4. We can liquefy these gases by raising the external pressure, but only if the temperature is lowered. Experiments show just how low the temperature must be for each gas to liquefy. While N_2 can be liquefied at 154 K if the pressure is increased sufficiently, H_2 cannot be liquefied with high pressure until the temperature goes down to 33 K.

All substances that can exist as gases have a characteristic minimum temperature, called the **critical temperature**, above which liquefaction is impossible no matter how high the pressure. The minimum pressure required to liquefy the gas at the critical temperature is called the **critical pressure**. Table 4.3 lists

TABLE 4.3
Critical Temperatures and Pressures

Substance	Critical T (K)	Critical P (atm)
He	5.3	2.26
H_2	33.3	12.8
Ne	44.5	26.9
N_2	126	33.5
CO	133	34.5
F_2	144	55
Ar	151	48
O_2	155	50.1
NO	180	64
CH_4	191	45.8
Kr	209	54.3
CO_2	304	72.9
C_2H_6	305	48.2
C_2H_2	309	61.6
HCl	325	82.1
H_2S	374	88.9
NH_3	406	112.5
Cl_2	417	76.1
C_2H_5OH	515	63
H_2O	647	218.3
Hg	1823	200

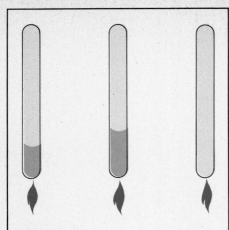

FIGURE 4.23

Critical temperature. As the temperature of a liquid in a closed container is raised, the density of the liquid decreases and the density of the gas increases. When the temperature is high enough, the density of the liquid and the density of the gas phase are equal and the two phases become one. This occurs at a temperature just above the critical temperature of the substance.

values of the critical temperatures and pressures of some common substances.

The existence of critical temperatures and pressures can be explained in several ways. Above the critical temperature, thermal motion is so great that no amount of pressure can push the molecules close enough so that the attractive forces overcome thermal motion. Table 4.3 shows that gases with small molecules (and hence weak attractive forces), such as H_2 and He, have very low critical temperatures. Gases with larger molecules (and stronger attractive forces), such as CO_2, H_2O, or NH_3, have higher critical temperatures. In fact, the critical temperature can be regarded as an indicator of the relative attractive forces for any gas, and can be shown to be related to the van der Waals constant a.

The existence of a critical temperature can also be observed and explained macroscopically. Suppose a liquid is placed in a sealed container, as shown in Figure 4.23. The liquid stays on the bottom of the container, while the vapor phase of the substance fills the rest of the container. There is a division between the two phases, gas and liquid, that is called a *meniscus*.

If the temperature of the sealed system is raised, the vapor pressure increases. The quantity of the substance in the gas phase increases, while its volume does not change appreciably. The density of the gas, which is mass divided by volume, thus increases as the temperature rises. The liquid, meanwhile, expands somewhat as the temperature rises. The increase in volume means that the density of the liquid decreases as the temperature increases.

When the temperature is high enough, the density of the gas will equal the density of the liquid. The meniscus disappears, and the liquid and the gas are indistinguishable. The temperature at which this happens is just above the critical temperature. At this or any higher temperature, there is no difference between the liquid state and the gas state, and we can say that it is impossible to liquefy a gas.

We can summarize our discussion of gases by saying that the ideal gas law and the model of gas behavior on which it is based provide the framework for understanding the behavior of real gases. However, the deviations of real gases from ideal gas behavior can be explained only by detailed experimental study of the nature of real gases and of the interactions between the molecules and atoms in the gaseous state.

EXERCISES

4.1   Suggest an example and a counterexample of the following generalizations in Section 4.1:
(a) When the temperature of a gas is lowered it condenses to a liquid.
(b) A liquid takes its shape from the portion of the container it occupies.
(c) A solid is denser than a liquid.
(d) Solids are crystalline.
4.2   A system at equilibrium is sometimes called a dynamic equilibrium. Explain why.
4.3   A liquid is in a sealed container at equilibrium with its vapor at 300 K. Decide whether each of the following statements is true or false and justify your answer.

(a) Raising the temperature of the system will increase the number of molecules in the vapor phase.
(b) Introduction of some more liquid into the container at 300 K will decrease the number of molecules in the vapor.
(c) Increasing the size of the container by a relatively small amount while keeping the temperature at 300 K will eventually cause the pressure of the vapor to decrease.
(d) Opening the container will cause the temperature of the liquid to fall.
(e) Lowering the temperature until the liquid freezes will result in the removal of all the molecules from the vapor phase.
(f) Squeezing the container at 300 K so

as to decrease its volume will result in an increase in the quantity of liquid.
(g) As the temperature is lowered the rate of collisions between molecules in the vapor increases.
(h) As the temperature is lowered the rate of collisions of molecules in the vapor with the surface of the liquid decreases.
4.4   The unit of pressure called the torr is defined as exactly 1/760 of 1 atm. The unit called the millimeter of mercury is defined as the pressure exerted by a liquid column exactly 1 mm high when the density of the liquid is 13 595.1 kg/m^3 and the acceleration of gravity is 9.806 65 m/s^2 at the location

of the column of liquid. Use the definition of the pascal and of the atmosphere to find the difference, in pascals, between 1 Torr and 1 mmHg. (*Hint:* pressure is density × acceleration of gravity × height of liquid.)

4.5 Molecules of a given vapor will generally condense more rapidly on a cold surface than on a warm one. Many solids with high vapor pressures can be purified by sublimation. Design a simple apparatus to separate an intimate mixture of sodium chloride and naphthalene (mothballs) by sublimation. Describe how the apparatus would bring about the desired separation.

4.6 If the normal boiling points of many organic compounds could be measured, they would be found to be substantially higher than the temperatures at which the compounds undergo decomposition. Yet many of these compounds can be purified by a process called distillation, in which they are boiled and recondensed many times. Suggest how such a process might be carried out.

4.7 A cylinder of compressed gas has a volume of 0.35 liter and the gas it contains is at a pressure of 48 atm. All the gas in the cylinder is then used to fill a balloon. The final pressure of gas in the balloon is 0.98 atm. Find the volume of the balloon.

4.8 A cylinder of an automobile engine has a volume of 0.83 liter at the point in its operating cycle when it fills with the mixture of gasoline vapor and air. The pressure of the mixture of gases is 0.62 atm. The piston then moves into the cylinder compressing the mixture of gases to a volume of 0.12 liter before they are ignited by the spark plug. Assume ideal gas behavior and find the pressure of the gases at this volume.

4.9 The gas inside a balloon of volume 12.3 m^3 is heated from 25°C to 125°C. Find the volume of the balloon.

4.10 A natural gas storage tank is a cylinder with a movable top whose volume can change only as its height changes. Its radius remains fixed. The height of the cylinder is 22.6 m on a day when the temperature is 22°C. The next day the height of the cylinder increases to 23.8 m as the gas expands because of a heat wave. Find the temperature, assuming that the pressure and amount of gas in the storage tank have not changed.

4.11 A sample of a gas at a pressure

of 1.0 atm and a temperature of 301 K formed in a chemical reaction is transferred from a container of volume 0.98 liter to one of volume 24 cm^3. The temperature of the gas is 195 K. Find the new pressure of the gas.

4.12 Predict whether the volume of a gas increases, decreases, or stays the same when the following changes are made:
(a) The pressure is increased from 2 atm to 3 atm; the temperature is increased from 200°C to 300°C.
(b) The pressure is increased from 2 atm to 3 atm; the temperature is decreased from 300 K to 200 K.
(c) The pressure is decreased from 4 atm to 2 atm; the temperature is decreased from 237°C to −23°C.
(d) The pressure is decreased from 4 atm to 2 atm; the temperature is increased from 200 K to 400 K.
(e) The pressure is decreased from 4 atm to 2 atm; the temperature is decreased from 300 K to 200 K.

4.13 A gas occupies a volume of 24.5 liters at 298 K and 2 atm. Find its volume at STP.

4.14 A gas is in a sealed container whose volume cannot be changed; the temperature is 297 K. Propose a method of tripling the pressure of the gas.

4.15 A 3.2-liter bulb contains O_2 and a 6.4-liter bulb contains N_2. Both bulbs are at the same temperature and pressure. Find the ratio of the amount of O_2 to the amount of N_2. Find the ratio by mass of O_2 to N_2.

4.16 A compressed gas cylinder contains 106 g of acetylene, C_2H_2, at a pressure of 51 atm. After some use of the acetylene by a welder the pressure in the cylinder falls to 42 atm. Find the mass of acetylene consumed, assuming the volume and the temperature of the cylinder do not change.

4.17 Some water is placed in a can and boiled. While the water is boiling the can is sealed. When the can is cooled it collapses. Explain this observation using only Avogadro's law and Boyle's law.

4.18 A steel container of volume 0.35 liter can withstand pressures up to 88 atm before exploding. Find the mass of helium that can be stored in this container at 299 K.

4.19 Find the mass of carbon contained in 2.34 liters of carbon dioxide at STP.

4.20 Zinc metal reacts with solutions

of strong acids to form $H_2(g)$ and Zn^{2+} ion. Find the mass of a sample of zinc that liberates 0.432 liter of H_2 at 295 K and 1.02 atm.

4.21 A 1.00-g sample of H_2O is placed in a sealed evacuated container of volume 1.00 liter and heated to 500 K. Assume all the water has evaporated and find the pressure in the container.

4.22 A person exhales about 800 g of CO_2 in a day. Find the pressure of this mass of CO_2 at 298 K in a room 3 m × 3 m × 2.5 m. (*Hint:* 1 m^3 = 1000 liters.)

4.23 A neon sign is made from 5.2 m of glass tubing of diameter 2.0 cm. The pressure of neon required for the proper operation of the sign is 1.5 mmHg and the sign is to be operated at 310 K. Find the mass of neon required to fill the sign.

4.24 A 1.07-g sample of liquid is evaporated completely at 341 K so that its vapors fill a 0.252-liter container at a pressure of 1.00 atm. Find the molecular weight.

4.25 The density of CO is 1.4 g/liter at certain conditions of temperature and pressure. Find the density of CO_2 at the same conditions.

4.26 Find the density of Xe at STP.

4.27 A quantity of N_2 occupies a volume of 1.0 liter at 300 K and 1.0 atm. The gas expands to a volume of 3.0 liters as the result of a change in both temperature and pressure. Find the density of the gas at these new conditions.

4.28 Find the volume of O_2 measured at STP required for the complete combustion of 1.0 mol of C_6H_6 to carbon dioxide and water. Find the volume of the carbon dioxide that forms at STP.

4.29 A mixture of 3.5 × 10^{23} molecules of O_2 and 6.5 × 10^{23} atoms of Ar has a total pressure of 0.82 atm. Find the partial pressure of Ar.

4.30 Binary compounds of alkali metals and hydrogen react with water to liberate $H_2(g)$. The H_2 from the reaction of a sample of NaH with an excess of water fills a volume of 0.49 liter above the water. The temperature of the gas is 305 K and the total pressure is 1.0 atm. Find the amount of H_2 liberated. Find the mass of NaH that reacted.

4.31 A mixture of 8.0 g CH_4 and 8.0 g Xe is placed in a container and the total pressure is found to be 0.44 atm. Find the partial pressure of CH_4.

4.32 A sample of $N_2O_3(g)$ has a pressure of 0.034 atm. It then undergoes complete decomposition to $NO_2(g)$ and $NO(g)$. Find the total pressure of the mixture of gases, assuming the volume and temperature do not change.

*4.33 A mixture of $CO(g)$ and $O_2(g)$ in a 1.0-liter container at 1000 K has a total pressure of 2.2 atm. After some time the total pressure falls to 1.9 atm as the result of the formation of CO_2. Find the amount of CO_2 that forms.

4.34 A sample of $C_2H_2(g)$ has a pressure of 7.8 kPa. After some time a portion of it reacts to form $C_6H_6(g)$. The total pressure of the mixture of gases is then 3.9 kPa. Assume the volume and the temperature do not change. Find the fraction of C_2H_2 that has undergone reaction.

4.35 A 10-liter container is filled with 0.10 mol of $H_2(g)$ and heated to 3000 K. The pressure is found to be 3.0 atm. Find the partial pressure of the $H(g)$ that forms from H_2 at this temperature.

*4.36 A mixture of $CH_4(g)$ and $C_2H_6(g)$ has a total pressure of 0.53 atm. Just enough $O_2(g)$ is added to the mixture to bring about its complete combustion to $CO_2(g)$ and $H_2O(g)$. The total pressure of these two gases is found to be 2.2 atm. Assuming constant volume and temperature, find the fraction of CH_4 in the mixture.

*4.37 A mixture of $NH_3(g)$ and $N_2H_4(g)$ is placed in a sealed container at 300 K. The total pressure is 0.50 atm. The container is heated to 1200 K, at which time both substances decompose completely according to the equations:

$$2NH_3(g) \longrightarrow N_2(g) + 3H_2(g)$$

$$N_2H_4(g) \longrightarrow N_2(g) + 2H_2(g)$$

After decomposition is complete the total pressure at 1200 K is found to be 4.5 atm. Find the percent of $N_2H_4(g)$ in the original mixture.

4.38 One mole of $H_2(g)$ in a container of volume 1.0 m³ exerts a pressure of 1.0×10^5 Pa. Find the average of the square of the velocity. (*Hint:* the mass should be in kilograms.)

4.39 Find the total kinetic energy of one mole of Kr at 300 K. Find the root-mean-square speed.

4.40 Find the ratio of the average speed of N_2 at 300 K to that of Ar at 400 K.

4.41 Samples of $I_2(g)$ and $Br_2(g)$ are introduced at the ends of glass tubes of identical length at the same temperature, and the movement of the colored vapors is observed. Predict which gas will reach the other end of the tube first. If it takes this gas 5.0 s to reach the end of the tube find how long it will take for the other gas to reach the end of the tube.

4.42 A mixture of N_2 and an unknown gas is placed in an apparatus of the type shown in Figure 4.22. The partial pressure of the two gases are equal. After 10 minutes it is found that the partial pressure of N_2 in the mixture that has effused through the barrier is 2.5 times that of the unknown. Find the molecular weight of the unknown gas.

4.43 Show how Graham's law can be derived from Equation 4.32 and the ideal gas equation.

4.44 One mole of a real gas in a volume of 0.13 liter at 273 K is found to exert a pressure of 2.0×10^2 atm. Find whether this gas is displaying a positive

or a negative deviation from ideality at these conditions.

*4.45 The radius of an Xe atom is 1.3×10^{-8} cm. A 100-cm³ container is filled with Xe at a pressure of 1.0 atm and a temperature of 273 K. Calculate the fraction of the volume that is occupied by Xe atoms. (*Hint:* the volume of a sphere is $\frac{4}{3}\pi r^3$)

4.46 Repeat the calculations of Exercise 4.45 for Xe at 273 K in a 100-cm³ container at a pressure of 200 atm.

4.47 Predict which one of the following systems behaves closest to an ideal gas: (a) CO_2 at STP, (b) Ar at STP, (c) Ar at 100 K and 1 atm, (d) Ar at 300 K and 1 atm, (e) CO_2 at 200 K and 1 atm.

4.48 Use the data in Table 4.2 to calculate the pressure of 101 mol of Ne in a 22.6-liter container at 273 K. Compare this value to the value calculated from the ideal gas equation. Account for the difference.

4.49 Consider 2.00 mol of Cl_2 at STP. Use the data in Table 4.2 to find the correction to the ideal gas volume that is to be used in the van der Waals equation. Correct the ideal gas volume by this amount to find the correction to the ideal gas pressure to be used in the van der Waals equation.

4.50 A sample of air is compressed to 60 atm and then its temperature is gradually lowered. Describe in sequence the changes in phase that occur, using the data in Table 4.3.

4.51 It can be shown that the value of the critical temperature of a gas is directly proportional to the value of its van der Waals constant, *a*. Suggest a reason for this relationship.

5

THE STRUCTURE OF ATOMS

"If in some cataclysm, all of scientific knowledge were to be destroyed, and only one sentence passed on to the next generation of creatures, what statement would contain the most information in the fewest words? I believe it is the atomic hypothesis (*or the atomic fact, or whatever you wish to call it*) *that* all things are made of atoms—little particles that move around in perpetual motion, attracting each other when they are a little distance apart, but repelling upon being squeezed into one another. *In that one sentence you will see there is an* enormous *amount of information about the world, if just a little imagination and thinking are applied."*

Richard P. Feynman, Robert B. Leighton, and Matthew Sands, The Feynman Lectures on Physics

5.1 THE DEVELOPMENT OF THE ATOMIC THEORY

We can describe chemistry as the study of how atoms combine to form molecules and how molecules are transformed into other molecules. To study how atoms combine, it is necessary to know what atoms are and what the structure of an atom is. Indeed, it is necessary to know that such things as atoms exist. Today, the existence of atoms is taken for granted. The existence of atoms was doubted by many scientists until the beginning of this century, and the structure of the atom has been known in detail for only a few decades.

In Western culture, the concept of the atom—the idea that matter is composed of extremely small, ultimately indivisible units—goes back to the fifth century before Christ. The Greek natural philosopher Leucippus and his student Democritus speculated that the universe consisted of the void and atoms, that atoms of different substances were fundamentally the same although different in detail, and that atoms were in constant motion.

These ideas were debated by other philosophers of the Hellenic world. Lucretius accepted them; Aristotle rejected them. But the debate was purely philosophical. None of the arguments, pro or con, was based on experiment. None of the philosophers thought of testing the theory by trial in the real world. The ancient Greeks did not think that way; they were not experimental scientists. Democritus had made a lucky guess with the theory of atoms. The failure to test the theory by experiment made it just one of many competing theories about the nature of matter. Some twenty centuries had to pass before the atomic theory was supported by experimental data.

It was John Dalton, in the early years of the nineteenth century, who first enunciated the atomic theory in what could be called modern terms. Dalton's picture of the atom is not the one we have today, but it did sum up some valid basic principles about the composition and behavior of chemical substances. These principles were consistent with the data available to Dalton and other scientists of the time. The major assumptions of Dalton's atomic theory were:

1. Elements are composed of small indivisible particles called atoms.
2. All the atoms of a given element are identical; they have the same mass, size, and properties. Atoms of one element have properties different from those of all other elements.
3. Atoms of two or more elements can combine to form new substances in which the atoms are held firmly together. The relative numbers of atoms in a given pure compound are constant.

114

These assumptions opened fruitful lines of investigation. For example, the assumption that all atoms of a given element have the same weight made it meaningful to calculate the weights of atoms—but only relative weights, since it was clear that atoms are so small that an individual atom could not be weighed.

Using the assumption that the relative numbers of atoms in any pure compound are constant, Dalton set out to assign relative weights to atoms. He gave the hydrogen atom a relative weight of 1 and measured all other atoms by that standard. For example, experiments showed that one gram of hydrogen would combine with eight grams of oxygen to form water. Using Occam's razor—which says essentially that the most economical explanation should be adopted whenever possible—Dalton proposed that each molecule of water had one atom of oxygen and one atom of hydrogen. He therefore assigned the oxygen atom a weight of 8 relative to the hydrogen atom. He also noted that three grams of carbon combined with one gram of hydrogen to form methane. Assuming methane to be CH (not CH_4 as we now know it to be), he assigned carbon the relative weight of 3, meaning that one carbon atom was three times heavier than one hydrogen atom.

5.2 THE ELECTRON

Dalton thought of atoms as tiny, featureless spheres, rather like infinitesimally small, sticky billiard balls, which combined in some mysterious way to form molecules. Today we know that the combination of atoms to form molecules is governed by a number of factors, the most important of which is the distribution of electrons in atoms. Much of the work of physicists in the nineteenth and twentieth centuries has been devoted to two closely related subjects—the structure of the atom and the nature of the forces in and between atoms.

The Nature of Electricity

The recognition that forces exist between atoms came to mankind long before science did. Science has made a difference in the way these forces are perceived. The most primitive man knows that when an object is thrown into the air, it falls to the ground. Newtonian physics says that the fall is due to the gravitational attraction between the object and the earth. Physicists have measured and defined gravitational force. It is always positive; objects are attracted but never repelled. It is inversely proportional to the square of the distance between objects. And it is relatively small. Gravitational attraction is important for relatively large objects, such as apples and planets. It is inconsequentially small for very small objects, such as individual atoms and molecules.

Gravity is not the only force that acts on objects. If a child rubs a balloon vigorously against a wall, the balloon will cling to the wall without falling to the floor. There is some attractive force between the balloon and the wall that overcomes the gravitational attraction that would otherwise make the balloon fall.

The ancient Greeks demonstrated the existence of this second attractive force by rubbing a piece of amber (the petrified sap of a tree) with wool or fur, an experiment that can be performed today (Figure 5.1). Once rubbed, the amber attracts light objects such as feathers. But if a second piece of amber is also rubbed and is held near the first, the two bits of amber repel each other. And if a piece of glass is rubbed with silk and held near the amber, the glass and the amber will attract each other.

We can explain this behavior by saying that there are *electrical forces* between the pieces of amber and glass, and that the rubbing causes the amber and the glass to be electrically charged. These are classical terms that are

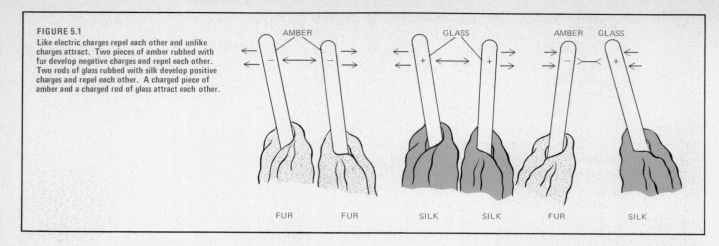

FIGURE 5.1
Like electric charges repel each other and unlike charges attract. Two pieces of amber rubbed with fur develop negative charges and repel each other. Two rods of glass rubbed with silk develop positive charges and repel each other. A charged piece of amber and a charged rod of glass attract each other.

outdated, but they at least provide a name for the phenomenon. As there is a gravitational force that always attracts, there is an electrical force that is more powerful than gravitational force and that may either attract or repel.

By convention, electricity was described as positive (+) or negative (−). Rubbed amber was described as having a negative electric charge, while rubbed glass had a positive charge. Like electrical charges repel each other, so anything that repelled rubbed glass had a positive charge and anything that repelled amber had a negative charge. The magnitude of the charge is measured by determining the force of attraction or repulsion between two charged objects.

The unit for the magnitude of charge is the coulomb, symbol C, named after the French scientist Charles de Coulomb (1736–1806). Experiments by eighteenth-century scientists, notably Coulomb, described how the force between two charges varies with the magnitude of the charges and the distance between them. The relationship known as **Coulomb's law**, is:

$$\text{force} = \frac{kq_1q_2}{r^2} \tag{5.1}$$

In Equation 5.1, the magnitude of one charge is represented by q_1 and the magnitude of the other charge by q_2. The distance between the charges is r. (The charges are considered to be concentrated at points in space.) The symbol k is a constant whose value depends on the choice of units for q and r and also depends on the nature of the medium between the two charges. Coulomb's law is often stated: *The force of attraction between two opposite charges is inversely proportional to the square of the distance between them and directly proportional to their magnitudes.*

The resemblance between Coulomb's law and the law of gravity should be noted. Gravitational attraction is directly proportional to the masses of two objects and inversely proportional to the square of the distance between them.

The Discovery of the Electron

Nineteenth-century experiments on electricity showed that the atom is not the indivisible particle pictured by Dalton. It was found that the atom has a structure, and that it contains subunits—particles smaller than atoms.

The British physicist Michael Faraday (1791–1867) played a major role in this work. In his experiments, Faraday studied the relative weights assigned by Dalton and later scientists to atoms of different elements. He found a relationship between these relative weights and the quantity of electricity needed to free a given amount of an element from a compound. A basic quantity of electricity in such electrochemical reactions is called a *faraday,* which is defined as 96 490 C.

A. PICCARD　　E. HENRIOT　P. EHRENFEST　Ed. HERZEN　Th. DE DONDER　　E. SCHRÖDINGER　E. VERSCHAFFELT　W. PAULI　W. HEISENBERG　R.H. FOWLER　L. BRILLOUIN

P. DEBYE　　M. KNUDSEN　　W.L. BRAGG　　H.A. KRAMERS　　P.A.M. DIRAC　　A.H. COMPTON　L. de BROGLIE　　M. BORN　　　N. BOHR

I. LANGMUIR　　M. PLANCK　　Mme CURIE　　H.A. LORENTZ　　A. EINSTEIN　P. LANGEVIN　Ch.E. GUYE　　C.T.R. WILSON　O W. RICHARDSON

Absents : Sir W.H. BRAGG, H. DESLANDRES et E. VAN AUBEL

One faraday, for example, liberates 1 g of hydrogen from water. We will discuss electrochemical reactions in detail in Chapter 16.

On the basis of Faraday's experiments, G. J. Stoney (1826–1911) suggested in 1874 that electricity exists in units associated with atoms, and that atoms are neutral because they contain not only such units of electricity but also units of opposite charge. Stoney proposed in 1891 that the unit of electricity be called an **electron.**

The name was adopted, but not without controversy. Many scientists of the time doubted not only the existence of units within the atom but also the existence of the atom itself. Even in the twentieth century, many distinguished scientists maintained that the atom was no more than an intellectual concept that was useful in making calculations. These scientists believed that the existence of the atom could never be proved, because it was too small to be measured.

The existence of atoms and electrons was established beyond doubt by a long series of experiments in many laboratories. Much of this work involved the measurement of electricity flowing through gas-filled glass tubes of a kind that have become quite familiar in recent years.

When an electric potential of about 10 000 V is applied across two electrodes to a tube filled with gas at low pressure, the gas emits light. However, if the pressure of the gas is reduced even further, a glow can be seen from the part of the glass tube that is directly opposite to the negative electrode, called the cathode. The first phenomenon is used in neon lights. The second is used in television tubes.

The forerunner of the television tube was invented by the English physicist William Crookes (1832–1919). In 1879, Crookes showed that the glow of the glass is due to rays emitted by the cathode. In the experiment shown in Figure 5.2, Crookes demonstrated that these cathode rays travel in straight lines. In the experiment, a piece of metal in the shape of a Maltese cross is placed between the cathode and the wall of the tube. When the electric current flows, the glass around the cross glows from the rays emitted by the cathode, while a shadow is cast on the glass behind the metal cross.

One aspect of the Crookes tube experiment could be interpreted equally well

The participants in the 1927 Solvay conference, held in Brussels, pose for a group portrait. The conferences were sponsored by the Solvay Institute, established by Ernest Solvay, who became wealthy by developing a process for manufacturing sodium carbonate. Only the greatest chemists and physicists of the era were invited to the conferences, and the scientists in the picture represent an almost unmatchable gathering of genius. (*Institute International De Physique Solvay*)

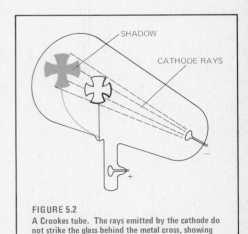

FIGURE 5.2
A Crookes tube. The rays emitted by the cathode do not strike the glass behind the metal cross, showing that cathode rays travel in straight lines.

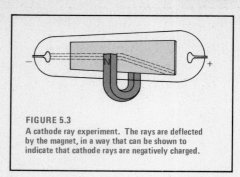

FIGURE 5.3
A cathode ray experiment. The rays are deflected
by the magnet, in a way that can be shown to
indicate that cathode rays are negatively charged.

in two different ways. The cathode rays could be either waves or particles; the experiment did not settle the question. An answer came from a series of experiments which showed that cathode rays could be deflected by magnetic and electric fields, as particles would be. In one experiment, a fluorescent screen was placed in the tube, so that the path of a beam of cathode rays could be seen (Figure 5.3). When a magnet was placed near the tube, the beam was bent in a way consistent with the rays being composed of negatively charged particles.

The experiment that did the most to confirm the existence of such negatively charged particles—electrons—was performed in 1897 by J. J. Thomson (1856–1940). Although Stoney gave the electron its name, Thomson is given the credit for discovering the particle, because he was the first to actually measure a property of the electron. The property measured by Thomson was the ratio of the charge of the electron to its mass, represented as e/m.

To measure this ratio, Thomson built an apparatus that could determine the angle of deflection and the velocity of a beam of cathode rays as it passed through an electric or magnetic field of known strength. Figure 5.4 shows the apparatus.

Thomson found that the cathode rays were moving at about one-fifth the speed of light. Given this speed and the measured angle of deflection, he was able to calculate a value for e/m. He could then compare the value of e/m obtained for the electron with the values of e/m that had been measured for several ions. The largest value for e/m that had been measured until then was for the hydrogen nucleus. It had been observed that 96 490 C produced 1 g of hydrogen, so that $e/m = $ 96 490 C/1 g. The value of e/m for the electron was found to be almost 2000 times larger than for hydrogen—to be exact, 1837 times larger. The magnitude of the charge e is the same for the hydrogen nucleus and the electron. Therefore, the mass of the electron is 1837 times smaller than that of hydrogen.

Thomson could determine only the relative mass of the electron. To measure the actual mass, it was necessary to get an exact measurement for either e or m; if one was known, the value of the other could be calculated. There were a number of attempts to make such measurements. Robert A. Millikan (1868–1953) of the University of Chicago succeeded with the apparatus shown in Figure 5.5.

The Millikan Oil Drop Experiment

Millikan's experiment started with a spray of small drops of oil into a chamber. Some of the drops acquired an electrical charge by collision with ions in the air. The rate at which the drops fell under the influence of gravity was measured. Then the drops were allowed to fall between two oppositely charged electrical plates. Uncharged drops will all fall at the same rate. If a drop acquires a change, it will be attracted to the oppositely charged plate, and its velocity will thus change. By turning the charge on the plates off and on, Millikan could make the drops rise and fall while measurements of their velocities were made. From the relative velocities of the drops with the electricity on and off, and from the known charge on the plates, Millikan could calculate the total charge on the drops.

He found that the smallest value for the charge on a drop was 1.6×10^{-19} C, and that the charge on any drop was either this value or a simple multiple of this value. Some drops had a charge of 3.2×10^{-19} C, which is twice 1.6×10^{-19} C; some had a charge of 4.8×10^{-19} C, which is three times the smallest value; and so on. But there was no drop whose charge was a fractional value of 1.6×10^{-19} C. On the basis of these data, Millikan proposed that the value of e, the charge on the electron, is this smallest value, 1.6×10^{-19} C. Considering that the work was done in 1909 with relatively unsophisticated equipment, Millikan's result was remarkably accurate. The latest figure for the charge on the electron, made with the most modern equipment, is $e = 1.602191 \times 10^{-19}$ C.

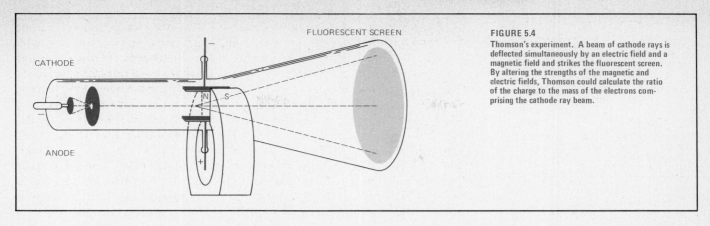

FIGURE 5.4
Thomson's experiment. A beam of cathode rays is deflected simultaneously by an electric field and a magnetic field and strikes the fluorescent screen. By altering the strengths of the magnetic and electric fields, Thomson could calculate the ratio of the charge to the mass of the electrons comprising the cathode ray beam.

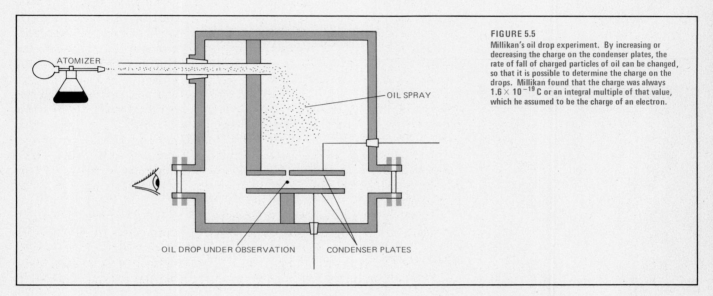

ATOMIZER

OIL SPRAY

OIL DROP UNDER OBSERVATION   CONDENSER PLATES

FIGURE 5.5
Millikan's oil drop experiment. By increasing or decreasing the charge on the condenser plates, the rate of fall of charged particles of oil can be changed, so that it is possible to determine the charge on the drops. Millikan found that the charge was always 1.6×10^{-19} C or an integral multiple of that value, which he assumed to be the charge of an electron.

With e known, the mass of the electron could be calculated. The mass is $m = 9.10956 \times 10^{-31}$ kg, which is $1/1837$ the mass of a hydrogen atom. Since the hydrogen atom is the smallest known atom, the electron has to be a subunit, or a subatomic particle.

5.3   X RAYS AND RADIOACTIVITY

While cathode rays were being studied, scientists were discovering and investigating a number of other rays that proved to be powerful tools for examining the details of atomic structure.

In 1895, Wilhelm Roentgen (1845–1923) discovered that a fluorescent screen that was placed near a cathode ray tube would glow when the tube was in operation. The glow was due to rays emitted by the walls of the cathode ray tube. Since the nature of the rays was unknown, Roentgen called them *X rays,* a name that has stuck.

Roentgen found that X rays had great penetrating power. They could cause a photographic plate to darken even when relatively thick objects were placed between the source of the X rays and the plate—a discovery that was quickly put to medical use. Roentgen showed that X rays were produced when fast-moving electrons hit a substance. X rays were found to travel in straight lines. Unlike

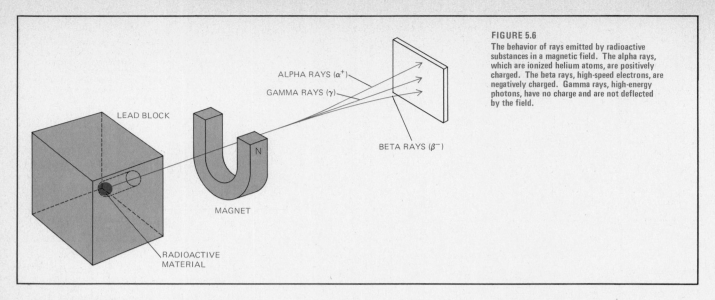

FIGURE 5.6
The behavior of rays emitted by radioactive substances in a magnetic field. The alpha rays, which are ionized helium atoms, are positively charged. The beta rays, high-speed electrons, are negatively charged. Gamma rays, high-energy photons, have no charge and are not deflected by the field.

ALPHA RAYS (α^+)

GAMMA RAYS (γ)

BETA RAYS (β^-)

LEAD BLOCK

N

MAGNET

RADIOACTIVE
MATERIAL

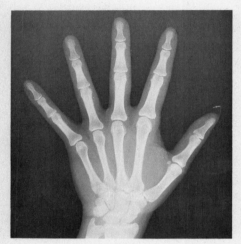

X rays were being used in medicine within months of their discovery by Wilhelm Roentgen in 1895. Modern medical practice would be impossible without X-ray images such as this one (*Cover to Cover*)

cathode rays, they are not deflected by electric or magnetic fields. Eventually, X rays were shown to be radiation of the same sort as ordinary light, but of much higher energy. In addition to their medical value, X rays are also extremely useful in determining the structure of molecules (Chapter 10).

Soon after Roentgen discovered X rays, Henri Becquerel (1852–1908) observed that a sample of a compound of uranium, left in a drawer with a photographic plate, produced rays that darkened the plate. These rays were studied by a number of investigators, including Marie Sklodowska Curie (1867–1934) and her husband Pierre Curie (1859–1906). It was found that other compounds and pure elements give off similar rays, a property that was named *radioactivity*. Eventually, it was shown that radioactive elements can emit three different kinds of rays, which were called alpha, beta, and gamma rays.

These three rays were distinguished by their behavior in a magnetic field, as shown in Figure 5.6. Gamma rays, which are the same sort of radiation as X rays but with even higher energies, do not change direction in the field. Beta rays, which are high-speed electrons, are deflected in a way consistent with their negative charge. The third type, alpha rays, are deflected in a way indicating that they are positively charged. Alpha rays are completely ionized helium atoms, that is, helium atoms with no electrons. They are also called alpha particles.

5.4   THE NUCLEAR ATOM

By the beginning of the twentieth century, scientists accepted the notion that atoms exist. The idea that some atoms, those of the radioactive elements, emit particles was also accepted. At least one subatomic particle, the electron, was known to exist. Therefore, it was evident that the atom has a structure. The effort to determine that structure began.

Atoms were known to be electrically neutral. They were also known to contain electrons, which are negatively charged. It was obvious that atoms had to contain subatomic particles with positive charges that cancel out the electrons' negative charges. But how were these negative and positive charges distributed within the atom?

Several models were proposed. One model had the atom resembling the planet Saturn, with a large, positively charged center sphere surrounded by a ''halo'' of electrons. Another model, proposed by Thomson, compared the atom to a plum pudding, with the electrons scattered through a positively charged mass in the way that raisins are embedded in a pudding.

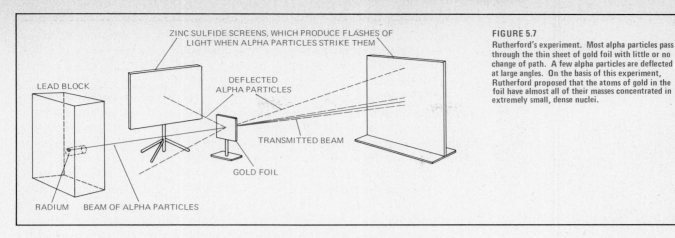

ZINC SULFIDE SCREENS, WHICH PRODUCE FLASHES OF
LIGHT WHEN ALPHA PARTICLES STRIKE THEM

DEFLECTED
ALPHA PARTICLES

LEAD BLOCK

TRANSMITTED BEAM

GOLD FOIL

RADIUM   BEAM OF ALPHA PARTICLES

FIGURE 5.7
Rutherford's experiment. Most alpha particles pass
through the thin sheet of gold foil with little or no
change of path. A few alpha particles are deflected
at large angles. On the basis of this experiment,
Rutherford proposed that the atoms of gold in the
foil have almost all of their masses concentrated in
extremely small, dense nuclei.

The Rutherford Model

Thomson's ''plum pudding'' atom seemed to fit the known facts, but experiments were needed to test the theory. The great physicist Ernest Rutherford (1871– 1937) conducted such an experiment, using alpha particles to probe into the atom. Rutherford's experiment, in which alpha particles were sent through a thin metal foil, as shown in Figure 5.7, is probably the single most important one ever done in the development of modern ideas of atomic structure.

The experiment was simple in principle. The results were a staggering surprise. A beam of alpha particles was directed at a thin sheet of gold foil, with a fluorescent screen behind the foil. Most of the alpha particles went straight through the foil, with little or no deflection. A few of the alpha particles were deflected sharply, some of them ricocheting off at angles of 90° or more. As Rutherford described the results: ''. . . as if you fired a fifteen-inch shell at a piece of tissue paper and it came back and hit you.''

Obviously, the ''tissue paper''—the metal foil—had an unusual structure on the atomic level. Only one conclusion was possible. The atoms that made up the foil had almost all of their mass concentrated in a very small positively charged region, which Rutherford called the **nucleus.** The tiny nucleus was surrounded by negatively charged electrons, whose distance from the nucleus was great on the subatomic scale. To explain why the negatively charged electrons were not pulled into the positively charged nucleus, Rutherford proposed that the electrons orbit the nucleus much as the planets of the solar system orbit the sun.

Rutherford's picture explained why most of the alpha particles passed through the foil undeflected, while a few were deflected at large angles. Most of the atom was empty space, through which the alpha particles passed unaffected. A few particles collided with the atomic nuclei and bounced off at large angles (Figure 5.8). By counting the percentage of alpha particles that were deflected, Rutherford was able to calculate the relative diameter of the atomic nucleus, which he found to be only 1/10 000 of the atom's total diameter.

The Rutherford atomic model provided the framework for the modern description of the atom. It is also the model of the atom that exists in the minds of most people. However, it later underwent some major modifications.

The Atomic Nucleus

The discovery that the atom has a concentrated nucleus that carries a positive charge raised a number of new questions. To begin with, what is the magnitude of the positive charge of the nucleus?

That question was answered by H. G. J. Moseley (1887–1915), working at the University of Manchester, England. Moseley studied the X rays that are

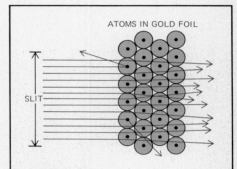

ATOMS IN GOLD FOIL

SLIT

FIGURE 5.8
The Rutherford experiment on the atomic level. The majority of alpha particles pass through the layers of gold atoms without hitting the nucleus of an atom and are deflected only slightly or not at all. An occasional alpha particle comes close to an atomic nucleus and is deflected sharply.

emitted when an element is bombarded with high-energy cathode rays. He found that each element emits a characteristic pattern of X rays, and that this characteristic pattern of X rays is related to the magnitude of the positive charge of the nucleus. Moseley then calculated the value of the positive charge of the nuclei of many elements.

The positive charge of the nucleus was attributed to positively charged particles in the nucleus. These particles are called **protons.** The number of protons in the nucleus is equal to the number of negatively charged electrons in the neutral atom. The number of protons in the nucleus of a given element is fixed and is called the **atomic number** of the element.

But this information is not enough to determine the mass of the atomic nucleus. From detailed studies of the effect of alpha particles on various elements, Rutherford concluded that the nucleus of a hydrogen atom is nothing more than a single proton. He also concluded that the nucleus of any atom heavier than hydrogen consists of a number of protons held together in some unknown manner. However, the mass of the atomic nucleus is substantially greater than the sum of the masses of the protons in the nucleus. The relative atomic mass, the total mass of the nucleus, is substantially greater than the atomic number.

It was not until 1932, when James Chadwick (1891–1974) discovered a new particle in the nucleus, that the difference between the atomic number and the atomic mass of the nucleus was accounted for. The new particle, called the **neutron,** has no electric charge and has a mass that is slightly larger than that of a proton.

The nucleus consists of both protons and neutrons. **The number of protons, and therefore the positive charge of the nucleus, is the atomic number of the element.** If an atom gains or loses protons, it becomes an atom of a different element. The number of protons in the nucleus equals the number of electrons in the electrically neutral atom. **The total number of protons and neutrons in the nucleus is called the mass number of the nucleus.** The relative atomic mass of the atom is very close to the mass number. (The electron, whose mass is only 1/1837 that of a proton, can be ignored in this reckoning.) Thus, since an atom of fluorine has an atomic number of 9 and a relative atomic mass that is very close to 19, we know that the nucleus of a fluorine atom has nine protons and $19 - 9 = 10$ neutrons.

The details of nuclear structure still are not understood completely. But for most purposes, the nucleus can be pictured as consisting of particles called **nucleons,** held together by extremely strong forces (Section 20.1). Some of the nuclear particles are protons, which have both mass and positive electric charge. The others are neutrons, with mass but no charge. Figure 5.9 shows this picture for several different nuclei. The number of nucleons in each nucleus approximates the numerical value of the relative atomic mass of the atom.

Isotopes

In general, the atomic weight* of a sample of an element that is found in nature is not an integer—that is, the atomic weight is not exactly equal to the mass number. The difference is too great to be accounted for by the mass of the electrons in the nucleus, or by the nonintegral relative masses of the nucleons. For example, the atomic weight of boron is found to be 10.8. The element boron has an atomic number of 5. Since it is the number of protons in the nucleus of

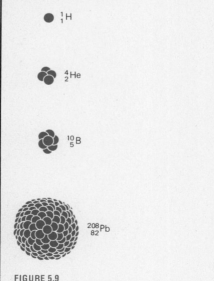

FIGURE 5.9
Relative sizes of nuclei. The hydrogen nucleus consists of a single proton. The helium nucleus contains two protons and two neutrons. The boron nucleus has five protons and five neutrons. The lead nucleus contains 82 protons and 126 neutrons.

---

*Chemists commonly use the term ''atomic weight'' interchangeably with the term ''relative atomic mass'' when referring to samples of elements obtained from natural sources.

an atom that defines the element and gives it its atomic number, the nucleus of a boron atom must contain five protons. There cannot be a fraction of a proton or a fraction of a neutron in an atomic nucleus, so how can there be fractional atomic weights? The answer, it developed, is that different atoms of the same element can have different numbers of neutrons in their nucleus.

For example, boron consists of atoms with two different masses. There are boron atoms whose nuclei have five protons and five neutrons, and boron atoms whose nuclei have five protons and six neutrons. These two atoms are present in such proportions that the measured atomic weight of boron is 10.8. Any sample of boron from natural sources is found to contain about 20% of atoms with five protons and five neutrons (relative atomic mass 10.0) and 80% of atoms with five protons and six neutrons (relative atomic mass 11.0). The atomic weight is the weighted average of the relative atomic masses of these two forms of boron:

$$(10.0)\left(\frac{20}{100}\right) + (11.0)\left(\frac{80}{100}\right) = 10.8$$

These two types of boron are called **isotopes.** They differ in atomic mass but not in chemical properties, since chemical properties are determined by the number of protons and the number of electrons in an atom. Every known element has at least two isotopes. Some elements have only one stable isotope; the other isotopes decay by radioactive processes. Among these elements are Be, F, Al, P, and As. Some elements have no known stable isotopes. These include technetium (atomic number 43), promethium (atomic number 61), and all elements of atomic number greater than 83.

For practical purposes, the atomic mass of an atom is less important than the atomic number in most processes of interest to chemists. All isotopes of a given element have very similar chemical behavior. The number and energy of the electrons and the nuclear charge of an element are mainly responsible for its chemical behavior. Since all isotopes of a given element have the same number of protons and the same number of electrons in the neutral atom, they have the same chemical behavior.

A system of notation has been devised to specify both the atomic number and the mass number of any atom of any element. We start with the symbol for the element. Two numbers precede the symbol for the element. There is a superscript, the uppermost number, which indicates the mass number of the isotope. And there is a subscript, written below the superscript, which indicates the atomic number of the isotope. This notation can be represented as:

$$_Z^A X$$

where X is the symbol for the element, A is the symbol for mass number, and Z is the symbol for atomic number (or nuclear charge, or the number of protons in the nucleus). Thus, the two isotopes of boron are designated $_5^{10}B$ and $_5^{11}B$. To calculate the number of neutrons in each of these isotopes, subtract the atomic number from the mass number, since:

$$A = Z + N \tag{5.2}$$

where N is the number of neutrons.

Because elements as found in nature usually are mixtures of isotopes, and because the combination of isotopes in the mixture may vary slightly depending on the source of the sample, there is some difficulty in establishing a very precise scale of relative atomic weights. Furthermore, the slight differences in mass between a proton and a neutron, as well as the small mass of the electrons, also are noticeable when atomic masses of pure isotopes are measured to more than three decimal places. Fractional atomic weights are the rule. An exception is the isotope $_6^{12}C$ which, as was mentioned earlier, has been assigned an atomic mass of exactly 12. All other atomic weights are calculated relative to this mass.

EXAMPLE 5.1

A sample of naturally occurring copper contains two stable isotopes, $^{63}_{29}Cu$ and $^{65}_{29}Cu$, whose abundances are 69.09% and 30.91% respectively, and whose atomic masses relative to $^{12}_{6}C$ are 62.93 and 64.93 respectively. Calculate the observed atomic weight of copper.

SOLUTION

The atomic weight of natural copper is found by calculating a weighted average. If we know the abundance of each isotope and the atomic mass of that isotope, we can find the atomic weight of natural copper.

We calculate a weighted average by multiplying the mass of each isotope by its abundance, then adding the products. Thus:

$$62.93 \times 0.6909 = 43.48$$
$$64.93 \times 0.3091 = 20.06$$

The atomic weight of natural copper is the sum of these products:

$$43.48 + 20.06 = 63.54.$$

EXAMPLE 5.2

Natural chlorine has an atomic weight of 35.45 and contains only two isotopes, $^{35}_{17}Cl$, of mass 34.97, and $^{37}_{17}Cl$, of mass 36.96. Calculate the relative abundance of these isotopes.

SOLUTION

Let x equal the fraction of $^{35}_{17}Cl$ in natural chlorine. Then $1 - x$ will equal the fraction of $^{37}_{17}Cl$ present in natural chlorine.

Using the method of Example 5.1:

$$(x)(34.97) + (1 - x)(36.96) = 35.45$$
$$36.96 - 35.45 = x(36.96 - 34.97)$$
$$1.51 = x(1.99)$$
$$x = \frac{1.51}{1.99} = 0.759, \text{ or } 75.9\% \text{ (abundance } ^{35}_{17}Cl)$$
$$1 - x = 0.241, \text{ or } 24.1\% \text{ (abundance } ^{37}_{17}Cl)$$

5.5 THE DEVELOPMENT OF THE QUANTUM THEORY

The atomic model in which most of the atom's mass was concentrated in a small, positively charged nucleus surrounded by orbiting electrons fit the results of Rutherford's experiments on alpha particle scattering. But it was inconsistent with some other observed properties of substances, and with the well-established laws of electricity and magnetism.

At the beginning of the twentieth century, scientists were faced with many such inconsistencies, experimental results that seemed to violate inviolable laws or did not agree with other equally valid experimental results. These dilemmas eventually were solved, but only by rewriting many of the old laws and by discarding many seemingly unshakable concepts—in short, by arriving at a new understanding of the nature of matter and energy.

Electromagnetic Radiation and Waves

The modern view is that the universe is composed of matter and radiant energy, and that there is an equivalence between the two. Radiant energy travels with the speed of light, while matter moves more slowly.

The road that led to this view began with the first serious investigations of light, one form of radiant energy, in the seventeenth century. Out of these investigations came two contradictory theories of the nature of light. One school of thought, led by Sir Isaac Newton (1642–1727), held that light consists of small particles, or corpuscles. The second theory, held by Christian Huygens (1629–1695), was that light consists of waves. Studies on the nature and behavior of light, especially on the diffraction of light continued until, by the middle of the nineteenth century, the wave theory of light was generally accepted. Later, when it was found that other forms of radiant energy, such as X rays, gamma rays, and radio waves, are essentially similar to visible light, these forms of radiant energy were also assumed to consist of waves. While the wave theory of light eventually had to be modified, it was a powerful tool for understanding the nature of radiant energy.

The Nature of Waves

A wave can be thought of as a disturbance that travels through a medium in a given direction. The medium itself is not carried along in this direction. For example, consider a swimmer floating in the ocean far from shore. The swimmer bobs up and down with each passing crest and trough of a wave, but is not carried toward the shore. If molecules of water in the ocean could be followed, it would be seen that they do not travel toward shore, although the peaks of the waves do. (Strictly speaking, objects are washed up on a beach by currents or turbulence, not by waves.)

Similarly, sound waves are disturbances that travel through the air but do not produce any substantial net movement forward of the gas molecules that make up air. This can be shown by floating a balloon in front of a rock band at a concert. The balloon will not move toward the back of the hall, no matter how overwhelming the volume of the sound.

Because a wave is a moving disturbance, its location cannot be specified exactly. The disturbance is a series of peaks and valleys, crests and troughs, whose position changes constantly as the wave travels. Because a wave is spread out over a region of space, it can be described only by specifying a number of its features.

Figure 5.10 shows a representation of an idealized wave. One feature of a wave is the distance between two adjacent peaks. This is called the **wavelength**, and it is commonly represented by λ, the Greek letter lambda. Another characteristic of a wave is the distance from a horizontal midline to either the peak or the trough. The distance from the midline to the peak or the trough is called the **wave amplitude**.

These two measurements, wavelength and wave amplitude, are not enough to describe a wave. Information is also needed on the velocity, c, of the wave, which can be described as the rate of motion of the peak (or any other point) of the wave in the direction of propagation. The value of c for light waves and all other electromagnetic radiation is approximately 3×10^8 m/s.

The **frequency** of the wave, which is represented by ν, the Greek letter nu, is directly related to the wavelength and velocity. Suppose there is an observer looking at a fixed point in space as a wave goes by (Figure 5.11). If the observer counts the number of peaks that pass the fixed point in a given time period, he can specify the frequency of the wave as so many units per second. If there is a long distance between peaks—that is, if the wavelength is long—fewer peaks will pass in a given time period. In other words, *the frequency of a wave is inversely proportional to the wavelength*. The number of peaks that pass by in a given time period also depends on the velocity of the wave. We can sum it up by saying that the frequency is a function of two characteristics: the velocity c of the wave and the wavelength λ. The formula that describes this relationship is:

$$c = \lambda\nu \qquad (5.3)$$

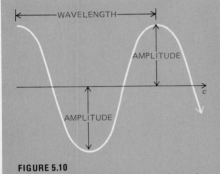

FIGURE 5.10
An idealized wave. The distance from peak to peak is the wavelength. The distance from the horizontal midline to the peak or the trough is the wave amplitude. The wave is traveling from left to right across the page.

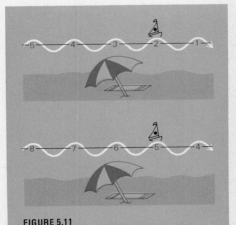

FIGURE 5.11
Wave frequency. An observer sitting under the beach umbrella and looking at the boat riding at anchor counts the number of times the boat bobs up and down in a given time period to calculate the wave frequency.

Since the velocity c of light is known, the wavelength of light can be calculated if the frequency is known, and vice versa.

EXAMPLE 5.3

A man on the deck of a ship anchored in the ocean observes that the crests of passing waves are 10 m apart and that a crest hits the bow of the ship every three seconds. Calculate the velocity of the waves.

SOLUTION

Since the values of λ (wavelength) and ν (frequency) of the waves are known, we can calculate the velocity v:

$$v = \lambda\nu = (10 \text{ m})\left(\frac{1}{3 \text{ s}}\right) = 3.3 \text{ m/s}$$

EXAMPLE 5.4

Calculate the frequency of light whose wavelength is 5×10^{-7} m.

SOLUTION

The value of c is known to be 3×10^8 m/s. The value of λ is given. Equation 5.3 gives the relationship between speed, wavelength, and frequency:

$$c = \lambda\nu$$
$$3 \times 10^8 \text{ m/s} = (5 \times 10^{-7} \text{ m})\nu$$
$$\nu = 6 \times 10^{14} \text{ s}^{-1}$$

The three parameters, c, λ, and ν, are related. If we are given any two for a wave, the value of the third can be calculated. For light waves, or any other waves traveling at fixed velocity, wavelength and frequency are inversely proportional. As wavelength increases, frequency decreases, and vice versa.

If light is a wave, an important question about the physical universe is raised. Since a wave is a disturbance in a medium, what is the medium through which light travels? A great deal of effort was devoted to answering that question in the late nineteenth century and early years of the twentieth century, and the answer proved surprising.

Some major steps that led to this work were experiments by James Clerk Maxwell (1831–1897) and Heinrich Hertz (1857–1894) which showed that visible light is one of many different waves, all of which are electromagnetic disturbances. One crucial experiment that supported this view showed that an oscillating electric charge produces electromagnetic waves that are not visible but have all the other characteristics of visible light.

We now know that all radiant energy is electromagnetic waves. Light waves, radio waves, gamma rays, and all other electromagnetic waves travel at a velocity of about $c = 3 \times 10^8$ m/s. These electromagnetic waves differ in wavelength and frequency, but their other characteristics are the same.

An electromagnetic wave can be regarded as having two components, an oscillating electric field and an oscillating magnetic field. The two fields oscillate at right angles to each other and at a right angle to the direction of travel (Figure 5.12). For several decades, it was believed that these waves were a disturbance in something called the "ether," which filled all of space but was undetectable. One of Albert Einstein's major achievements was to show that the ether does not exist. Ocean waves cannot exist without water, sound waves cannot exist without air, but no medium is necessary for electromagnetic waves.

Electromagnetic radiation is classified according to wavelength or frequency in an orderly arrangement called the *electromagnetic spectrum*. For convenience, the different wavelength (or frequency) regions of the electromagnetic spectrum often are given names, such as visible light, X rays, gamma rays, ultraviolet rays,

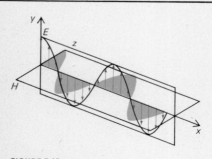

FIGURE 5.12
Electromagnetic radiation has two components, an electric field and a magnetic field, oscillating at right angles to each other and to the direction in which the radiation is traveling.

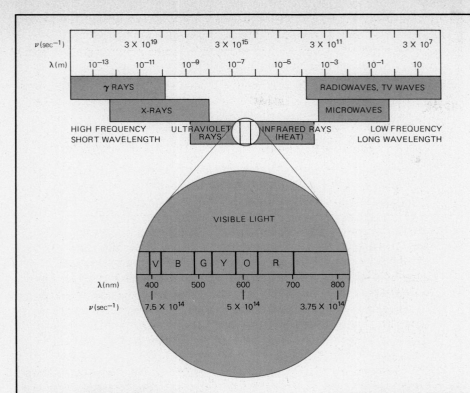

FIGURE 5.13
The electromagnetic spectrum. Visible light occupies a small section toward the center of the spectrum.

infrared rays. Figure 5.13 shows a part of the electromagnetic spectrum. Among other things, Figure 5.13 shows that visible light, the only part of the spectrum that we humans can detect with our eyes, is a very small sliver of the entire range of electromagnetic radiation.

We must understand radiant energy to understand the modern model of the atom, since much of the information on which this model is based comes from studies of the interaction between matter and radiant energy.

Atomic Spectrum of Hydrogen

Atoms of a pure element in the gas phase can emit light when heated. When studies of this phenomenon began in the mid-nineteenth century, it was expected that the results would be similar to those obtained in a somewhat comparable experiment done two centuries earlier by Isaac Newton.

Newton had passed ordinary sunlight through a prism. He found that a prism breaks the sunlight into a continuous spectrum that contains the colors of visible light. But the same sort of continuous spectrum is not obtained when light is emitted by gaseous atoms of a pure element. Instead, it was found that the emitted light consists of a relatively small number of narrow bands (or lines) of color, with large dark spaces between them (see the color plate facing page 142).

These lines are called *spectral lines*. Each element has a characteristic pattern of spectral lines, different from the patterns of all other elements. The pattern of spectral lines is called an **atomic spectrum** and serves as a "fingerprint" to identify the element (Figure 5.14). Spectral lines are emitted not only in the visible region of the electromagnetic spectrum, but also in the ultraviolet region, at wavelengths shorter than those of visible light, and in the infrared region, at wavelengths longer than those of visible light.

A considerable library of atomic spectra for different elements was accumulated during the nineteenth century. But nineteenth-century physics could not explain why different elements should have different spectral patterns. The first

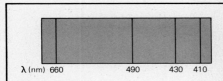

λ (nm) 660   490   430 410

FIGURE 5.14
The visible portion of the atomic spectrum of hydrogen. Each element has a unique atomic spectrum and can be identified by the pattern of emission lines.

SEPARATING ISOTOPES WITH LASER LIGHT

Isotopes of the same element have identical chemical properties; they cannot be separated chemically. Isotope separation has been a complicated and expensive undertaking, most notably in the World War II program to make an atomic bomb by separating fissionable uranium 235 from nonfissionable uranium 238 (Section 4.7).

But, the different isotopes of an element have very slightly different patterns of spectral lines, which means that they emit and absorb slightly different wavelengths of light. The possibility of exploiting the slight differences in absorption spectra to separate the two isotopes of uranium was examined and rejected, because the available light sources were unsuitable. In recent years, however, the development of lasers has provided a source of light whose wavelength may be controlled precisely enough so that one isotope will absorb the energy while another isotope will not. In addition, laser light is intense enough to be suitable for efficient isotope separation.

Laser light differs from ordinary light because it is "coherent"; that is, its waves all have the same wavelength, frequency, and orientation. Laser light is produced by inducing a large number of molecules to emit radiation of the same wavelength simultaneously. If the light is of the appropriate wavelength, it can be used to add energy to only one isotope of an element but not other isotopes. The more energetic isotopes could then be separated from the others.

Potentially, laser isotope separation of uranium is 1000 times more efficient than gaseous diffusion separation. It has been estimated that the use of laser light instead of gaseous diffusion to enrich uranium for nuclear generating plants could save $100 billion by the year 2000. However, there are a number of practical problems in achieving useful laser separation of isotopes.

One problem is the difficulty of developing lasers that emit the desired wavelengths of light. Another problem is that the thermal motion of atoms in a gas lessens the spectral distinctions between isotopes. Because thermal motion affects the absorption and emission of radiation, the spectral lines of the atoms may be blurred, making it difficult to excite only the desired isotope. There is also the problem of separating the excited isotope from the other isotopes.

Several approaches are being tried to solve these problems. Some laboratories are using "scavengers" that absorb all but a very narrow wavelength of light, the wavelength that will be absorbed by the desired isotope. Thermal motion can be lessened by cooling the atoms to a temperature approaching absolute zero. The energy absorbed by the isotope can be used to promote a chemical reaction that will make separation easier to accomplish.

Several laboratories have already reported successful separation of several isotopes, including those of chlorine and sulfur. An intensive effort is being made in several countries to achieve large-scale laser separation of uranium isotopes.

step toward an understanding of atomic spectra was made by an obscure Swiss schoolteacher, J. J. Balmer (1825–1898). In 1885, Balmer developed a formula that relates the wavelengths of the visible lines in the atomic spectrum of hydrogen:

$$\frac{1}{\lambda} = R_{\mathrm{H}} \left(\frac{1}{2^2} - \frac{1}{n^2} \right) \tag{5.4}$$

where λ is the wavelength of a given spectral line, n is any small integer greater than 2, and R_{H} is a proportionality constant, now called the *Rydberg constant*, whose value is known to be 1.09677578×10^7 m^{-1}, one of the most accurately measured quantities known.

By substituting different integers for n in this formula, one can calculate wavelengths that correspond to those in the observed spectrum of hydrogen. The formula even had a predictive value. It gave not only the wavelengths of the known spectral lines in the visible region but also the wavelengths of several ultraviolet lines of hydrogen, which had not been detected at the time. The prediction led to the discovery of the lines.

Later, Balmer's original formula was generalized to account for new sets of lines in the atomic spectrum of hydrogen that were discovered in the ultraviolet and infrared regions. The more generalized formula reads:

$$\frac{1}{\lambda} = R_{\mathrm{H}} \left(\frac{1}{n_f^2} - \frac{1}{n_i^2} \right) \tag{5.5}$$

in which the original term $1/2^2$ has been replaced by $1/n_f^2$ and $1/n^2$ has been replaced by the term $1/n_i^2$. In this formula, n_f and n_i may be any integers, as long as n_i is greater than n_f. If an integer value is set for n_f, a series of spectral lines can be found for the different allowed values of n_i that are obtained by letting $n_i = n_f + 1$; $n_i = n_f + 2$; $n_i = n_f + 3$; and so on.

Balmer's formula had one major weakness. It had no theoretical basis at all. The formula worked. It led to the discovery of other lines in the atomic spectrum of hydrogen. But no one had any idea of why it worked. Even though the Balmer formula could be used to predict the wavelengths of the spectral lines of hydrogen, classical physics could not explain why atomic spectra consisted of narrow spectral lines, rather than a continuous spectrum.

EXAMPLE 5.5

Calculate the wavelengths of the two spectral lines with the longest wavelengths (called the first two lines) in the visible region of the atomic spectrum of hydrogen.

SOLUTION

Equation 5.4, the Balmer formula, can be used to calculate the wavelengths of lines in the atomic spectrum of hydrogen. To calculate the wavelengths of the first two lines, we use the two smallest allowed integers, $n = 3$ and $n = 4$, in the Balmer formula:

$$\frac{1}{\lambda} = R_{\mathrm{H}} \left(\frac{1}{2^2} - \frac{1}{n^2} \right) = 1.1 \times 10^7 \text{ m}^{-1} \left(\frac{1}{4} - \frac{1}{9} \right) = 1.1 \times 10^7 \text{ m}^{-1}(0.14)$$

$$= \frac{1}{1.1 \times 10^7 \text{ m}^{-1}(0.14)}$$

$$\lambda = 6.5 \times 10^{-7} \text{ m}$$

$$\frac{1}{\lambda} = R_{\mathrm{H}} \left(\frac{1}{2^2} - \frac{1}{n^2} \right) = 1.1 \times 10^7 \text{ m}^{-1} \left(\frac{1}{4} - \frac{1}{16} \right)$$

$$\lambda = 4.8 \times 10^{-7} \text{ m}$$

EXAMPLE 5.6

In the original Balmer formula, the wavelengths of lines in the visible region of the atomic spectrum of hydrogen were calculated by letting $n_f = 2$. The ultraviolet region of the atomic spectrum of hydrogen has lines of shorter wavelengths than the lines in the visible region. Find the value of n_f that defines a series of lines in the ultraviolet portion, and then assign values to n_i to predict the positions of the two lines of longest wavelength in the ultraviolet region.

SOLUTION

Since the left side of Equation 5.5 is a fraction that has λ in the denominator, the value of λ, the wavelength, decreases as the value of the term on the right side of the equation increases. Since the ultraviolet region is of shorter wavelength than the visible region, the solution requires values for n_f and n_i that increase the value of the term

$$R_H \left(\frac{1}{n_f^2} - \frac{1}{n_i^2} \right)$$

Since n_f is in the denominator of a fraction, a smaller value for n_f means a larger value for the entire term. The smallest value that we can give to n_f is 1. Therefore, the smallest values we can give to n_i are $n_i = 2$ and $n_i = 3$. If we use these values to calculate values for λ, the results are:

$$\frac{1}{\lambda} = R_H \left(\frac{1}{1^2} - \frac{1}{2^2} \right)$$

$$\lambda = 1.2 \times 10^{-7} \text{ m}$$

$$\frac{1}{\lambda} = R_H \left(\frac{1}{1^2} - \frac{1}{3^2} \right)$$

$$\lambda = 1.0 \times 10^{-7} \text{ m}$$

Blackbody Radiation

If a bar of iron is heated, it first glows dull red, then emits a more yellowish glow, and eventually glows white-hot. If the radiation emitted by the bar of iron, or any other solid, is studied, it quickly becomes evident that this radiation is different from the radiation emitted by gaseous atoms. The radiation emitted by a solid does not have individual spectral lines separated by dark spaces. Instead, the radiation is emitted in a continuous spectrum through the ultraviolet, visible, and infrared regions.

There is a practical problem to overcome in studying the radiation from solids: the problem of reflection. Most solids reflect a considerable proportion of the light that falls on them. This is especially true of metals. When we study the radiation from a hot solid, we may thus be studying a mixture of emitted light and reflected light, which confuses the issue. To get around this difficulty, we need something that reflects none of the light falling on it. Because black objects reflect the least light, this imaginary object is called a *blackbody*. An ideal blackbody does not exist, since every object in nature does reflect some proportion of light. But a close approximation of the hypothetical blackbody can be created by using a heated cavity inside a block of metal, with a tiny hole that allows radiation to escape for analysis.

Early studies of blackbody radiation concentrated on the way in which the emitted energy is distributed over the range of wavelengths. Curves such as those shown in Figure 5.15 were drawn to show how the wavelength distribution

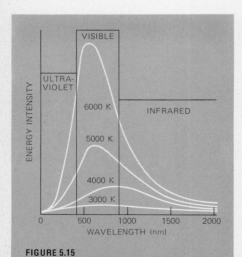

FIGURE 5.15

Curves showing the change in the emission of radiation as a blackbody is heated. As the temperature goes up, the total amount of radiation emitted by the blackbody increases. Both the amount and fraction of the total radiation that is in the visible part of the spectrum increase as temperature rises.

changed with the temperature of the blackbody. The curves show that as temperature rises, the region of maximum emission of energy moves from the infrared region to the visible. These curves are consistent with the everyday observation that the light emitted by, say, a bar of metal, changes from red to yellow—that is, from longer to shorter wavelengths—as the bar is heated.

These results created a major theoretical difficulty for classical physics. The classical theories could not explain the wavelength distribution of blackbody radiation. There were a number of attempts to develop theories explaining blackbody radiation, but they were only partially successful. One theory explained the short wavelength distribution but failed for the longer wavelengths. Another theory successfully explained the longer wavelengths but failed spectacularly with the short wavelengths. The second theory, which seemed to be consistent with all the known laws of physics, predicted that a blackbody would emit an infinite amount of energy in the ultraviolet region, which is a physical impossibility. This prediction was so disturbing to physicists that it was called the "ultraviolet catastrophe."

The dilemma was resolved by Max Planck (1858–1947), who introduced a principle that required a basic revision of classical theory. In classical theory, it was assumed that there was no basic unit of energy that a blackbody emitted. In other words, it was assumed that the radiant energy emitted by a blackbody could have any value within a continuous range.

Planck proposed that radiant energy could not be emitted with any value within a continuous range. Rather, he said, radiant energy can be emitted only in certain fixed quantities. The word that Planck used to describe the smallest such quantity was **quantum** (plural *quanta*), from the Latin word for "how much." The energy emitted by a blackbody is always an integral multiple of the quantum, and is never less than a quantum.

Planck said that the quantum of energy is given by:

$$E = h\nu \tag{5.6}$$

where E is the energy of the quantum, ν is the frequency of the emitted radiation, and h is a constant with the value 6.6262×10^{-34} J s.

By substituting from Equation 5.3, Equation 5.6 can be written as:

$$E = \frac{hc}{\lambda} \tag{5.7}$$

Equations 5.6 and 5.7 tell us that **the magnitude of a quantum of energy is directly proportional to the frequency of the radiant energy,** and that **the magnitude of the quantum is inversely proportional to the wavelength λ of the radiant energy.** Planck's hypothesis explained the distribution of frequencies in blackbody radiation that had been inexplicable by classical theory. We shall not explore the details of this explanation, since it is beyond the scope of our discussion.

Planck also proposed that the total radiant energy of any given frequency ν had to be $E = nh\nu$, where n could have only integer values: 1, 2, 3, 4, ... A blackbody could emit $2h\nu$, $3h\nu$, $4h\nu$, or any other integral multiple of $h\nu$, but it could not emit $3.27h\nu$, $5.9h\nu$, or any other fractional value of $h\nu$.

The concept of the quantum—the idea that energy is not infinitely divisible but rather is emitted in discrete units—is fundamental to modern science. On a different scale, quantum phenomena are common in everyday life. The production of eggs by chickens is a quantized phenomenon. In this case, the quantum is a single egg. A flock of chickens may lay any integral number of eggs in a day, but a chicken cannot lay a fraction of an egg.

The idea that radiant energy is not infinitely divisible was revolutionary—so much so that Planck's quantum theory was intellectually disturbing to many scientists of the day. Planck himself found the quantum hypothesis difficult to accept. He worked unsuccessfully for years to find an alternative explanation.

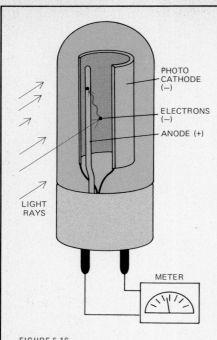

PHOTO CATHODE (−)

ELECTRONS (−)

ANODE (+)

LIGHT RAYS

METER

FIGURE 5.16
The photoelectric effect. If the frequency of radiation striking the cathode is low, no electrons are emitted. If the frequency of the radiation increases above a threshold, electrons are emitted. This cell can be used to measure the number of electrons that are emitted and their kinetic energy.

Today, the idea of a quantum is a cornerstone of science, because it permits explanations of phenomena that otherwise would be incomprehensible. In particular, the idea of the quantum is behind today's model of atomic structure.

5.6 THE PHOTOELECTRIC EFFECT

In 1905, Albert Einstein (1879–1955), a poorly paid clerk in the Swiss patent office, published three scientific papers that revolutionized physics. One paper outlined the special theory of relativity. The second paper explained Brownian motion, the random movement of small objects in liquid. The third paper was a major step forward in the understanding of the nature of light. The publication of these three papers was a spectacular achievement; historians speak of 1905 as Einstein's "wonder year." While the theory of relativity had the greatest repercussions, it was the paper on the nature of light that was cited when Einstein was awarded the Nobel Prize.

In this paper Einstein applied the quantam theory to a phenomenon called the *photoelectric effect,* which had been unexplained until then. The photoelectric effect is the emission of electrons by the polished surface of a metal that is struck by light or other electromagnetic radiation. An instrument such as the one shown in Figure 5.16 is used to study the details of the photoelectric effect.

Using such an instrument, it can be found that the emission of electrons depends on the frequency—that is, the energy—of the radiation. If the electromagnetic radiation that strikes the surface of the metal is of low frequency (which means that each of its quanta has only a relatively small amount of energy), no electrons are emitted by the metal. If energy of progressively higher frequency is used (which means that the energy of each quantum is increased), one suddenly arrives at a "threshold"—a frequency at which electrons are first emitted by the metal. No electrons are emitted at frequencies lower than the threshold frequency, no matter how much radiation of that frequency strikes the metal. In other words, increasing the intensity—that is, the amount—of low-frequency radiation without changing the frequency will not cause the emission of electrons.

Once the threshold frequency is reached, the number of electrons given off by the metal is not changed if the frequency is increased. What does change as the frequency increases above the threshold value is the energy of the electrons that are emitted. As the frequency of the incident radiation increases, the energy of the emitted electrons increases.

These results were a puzzle in 1905 because they could not be reconciled with the wave theory of light. Einstein solved the puzzle by proposing that radiant energy exists as quanta called **photons.** In doing this, Einstein made the first practical application of Planck's quantum theory. Einstein not only used the same term, quantum, to describe a "bundle" of light, but he also used Planck's constant, *h,* in the formula for the energy of a quantum of light. The energy of a quantum of light, Einstein said, is $E = h\nu$; that is, the energy is the frequency times Planck's constant.

The idea that light is quantized was revolutionary to physicists. The quantum description of light conveys the idea that light consists of particles or corpuscles, an idea that had long been discarded in favor of the theory that light consists of waves. Nevertheless, the wave theory could not explain the photoelectric effect, but the quantum theory could.

To say that light consists of small bundles—quanta—of energy is the same as saying that light consists of particles, the photons. One can visualize electrons being knocked loose from a metal by a beam of incident radiation that consists of photons. We can picture an electron as being held by the nucleus of an atom with a certain amount of energy, called the *binding energy,* ε. Only a photon with sufficient energy to overcome the binding energy can knock the electron loose. In this visualization, if the photon that hits the electron does not have enough energy, it will bounce away and the electron will remain in the atom.

The fact that the energy of a photon is proportional to its frequency ($E = h\nu$) explains why low-frequency radiation does not produce the photoelectric effect. The photons of low-frequency radiation do not have enough energy to overcome the binding energy of the electron. At the threshold frequency, the photon has just enough energy to knock the electron loose. If the photon has more energy than is required to overcome the binding energy, then the excess energy simply gives the departing electron more energy of motion.

If the threshold frequency is known, the binding energy of the electron can be measured. It is:

$$h\nu_0 = \varepsilon \tag{5.8}$$

where ν_0 is the threshold frequency and ε is the binding energy. The energy of motion that is imparted to the electron, called kinetic energy (Section 4.6), is:

$$\text{K.E.} = \tfrac{1}{2}mv^2 \tag{5.9}$$

where m is the mass and v is the velocity of the moving object.

If the radiation hitting the metal has a frequency greater than the threshold frequency, the energy E of a photon goes in part to overcome the binding energy ε and in part to give extra kinetic energy to the emitted electron:

$$E = \varepsilon + \tfrac{1}{2}mv^2 \tag{5.10}$$

or, substituting from Equations 5.6 and 5.8:

$$h\nu = h\nu_0 + \tfrac{1}{2}mv^2 \tag{5.11}$$

where ν is the frequency of the incident radiation and ν_0 is the threshold frequency.

A large-scale model of the photoelectric effect is shown in Figure 5.17. A carnival game requires the player to knock a doll off a shelf by shooting at it. The bottom of the doll is glued to the shelf. If corks are used for ammunition, they bounce off the doll because they do not have enough energy to overcome the binding energy of the glue. The corks are like photons whose frequency is below the threshold frequency. If light metal shot is used for ammunition, a hit just knocks the doll off the shelf. The light shot has just enough energy to overcome the binding energy of the glue, as do photons at the threshold frequency. If heavy shot is used as ammunition, and we assume it does not penetrate the doll, a hit not only knocks the doll off the shelf but sends it flying, just as a photon above the threshold frequency sends the electron off with kinetic energy.

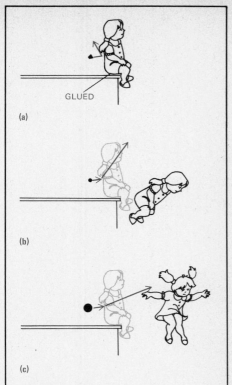

FIGURE 5.17
The carnival game of knock-the-doll-off-the-shelf can be used as a model of the photoelectric effect. Low-energy corks cannot overcome the binding energy of the glue holding the doll to the shelf (a). Light shot has just enough energy to overcome the binding of the glue and knock the doll off the shelf (b). Ammunition of higher energy not only knocks the doll loose but sends it flying off with kinetic energy (c).

EXAMPLE 5.7
If the threshold frequency, ν_0, of a metal is 6.7×10^{14} s^{-1}, calculate the kinetic energy of a single electron that is emitted when radiation of frequency $\nu = 1.0 \times 10^{15}$ s^{-1} strikes the metal.
SOLUTION
The relationship between ν_0, ν, and kinetic energy is given in Equation 5.11. Solving the equation for kinetic energy gives:

$$
\begin{aligned}
\text{kinetic energy} = \tfrac{1}{2}mv^2 &= h(\nu - \nu_0) \\
&= (6.6 \times 10^{-34} \text{ J s})(1.0 \times 10^{15} \text{ s}^{-1} - 6.7 \times 10^{14} \text{ s}^{-1}) \\
&= (6.6 \times 10^{-34} \text{ J s})(3.3 \times 10^{14} \text{ s}^{-1}) \\
&= 2.2 \times 10^{-19} \text{ J}
\end{aligned}
$$

By using Planck's quantum theory to explain the photoelectric effect, Einstein posed as many questions as he answered. Is the theory that light and all other electromagnetic radiation consist of waves incorrect? How can all the experiments that supported the wave theory of light be explained away? Can light be both a wave and a particle? Some answers came from experiments and

THE PHOTOELECTRIC EFFECT IN EVERYDAY LIFE

If you sit down to watch television tonight, you will literally be seeing a practical application of the photoelectric effect. The camera tube used in television systems, called an image orthicon, is made possible by the photoelectric effect.

When light enters the image orthicon, it falls on a screen made of a material that emits electrons. The screen is designed so that the electrons are released in proportion to the brightness of the light that reaches it. The emitted electrons are then focused onto a target surface by a magnetic field. The target surface releases what are called secondary electrons. The image created in this way is scanned by a beam of electrons from a cathode emitter at the other end of the tube. The electrons in the beam are reflected from the target surface and return to a collector that generates the signal that is broadcast.

The same basic principle is used in image intensifiers such as the "night scopes" used by the military. Faint night light strikes a screen that emits electrons. These electrons are focused onto a luminescent surface that emits photons. The result is an image similar to the original image, but much brighter.

One scientific field in which the practical application of the photoelectric effect has had an enormous impact is astronomy. "For a while it seemed that electronic image intensification would make the photographic plate obsolete, to be replaced by vidicon tubes," Bart J. Bok, professor emeritus of astronomy at the University of Arizona, said in 1977. "One might say that the techniques of photography were being replaced by television-related techniques. To some degree this has happened. The observing astronomer can now do much of his telescope pointing and following from a comfortably heated room. He can precision-guide his telescope with ease, centering upon guide stars several magnitudes fainter than the sorts of stars he used in the past for direct work at the telescope."

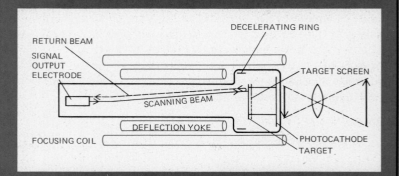

Diagram of an image orthicon.

theories that extended the findings of both Einstein and Planck and that developed a new model of the atom.

5.7 THE BOHR MODEL OF THE HYDROGEN ATOM

A new model of the atom was needed because the Rutherford model was far from being completely satisfactory. In fact, given the classical laws of physics, such an atom could not exist. The model described negatively charged electrons moving in a closed orbit around a positively charged nucleus. A body moving in a closed orbit must have acceleration to keep it going, and according to classical theory, a negatively charged body such as the electron must radiate energy under such conditions of acceleration. As the electron lost energy, it would spiral down toward the nucleus. The atom would radiate a continuous spectrum of radiant energy and it would collapse rapidly as the electron hit the nucleus. Observation of atoms gives a different picture. Atoms emit a noncontinuous spectrum of radiant energy, they do so only at high temperatures, and they stubbornly continue to exist. It was the great Danish physicist Niels Bohr (1885–1962) who developed a new model of the hydrogen atom whose characteristics are closer to those observed. Bohr made the crucial assumption that not all of the classical laws of electrodynamics apply to phenomena on the atomic scale.

Bohr's model is based on several postulates:

1. The energy of a hydrogen atom is not a continuous function that can have any value within a range. Rather, the hydrogen atom **can exist only in a limited number of energy states.** In other words, **the energy of the hydrogen atom is quantized,** a concept that is simply an extension of the Planck-Einstein quantum hypothesis. Each allowed energy state of the hydrogen atom is called a **stationary state.** There is a stationary state of minimum energy, called the **ground state.** All higher states are called **excited states.**

2. The atom does not radiate or absorb energy as long as it remains in a stationary state. When the atom goes from a higher energy state to a lower energy state, it emits radiation. When the atom absorbs radiation, it goes from a lower energy state to a higher energy state. The different energy states correspond to different orbits of the electron (only one electron, since this is a model of the hydrogen atom). The orbits of the electron are circular. A change in energy state corresponds to a jump of the electron from one circular orbit to another. When energy is absorbed, the electron jumps to a higher orbit. When energy is emitted, the electron jumps to a lower orbit. The discontinuous jumping of the electron cannot be visualized or explained by the classical laws of physics. It is a quantum phenomenon that occurs only on the atomic level.

3. Because the atom emits and absorbs energy only when the electron moves from one orbit to another, the energy emitted and absorbed by the atom is quantized. The energy of a quantum of such radiation is given by the formula $E = h\nu$. This quantum of energy is the difference between the lower and higher energy states of the atom. This relationship, called the *Bohr frequency rule,* is:

$$E_{\text{high}} - E_{\text{low}} = \Delta E = h\nu \qquad (5.12)$$

where ΔE is the difference in energy between the high and low states, h is Planck's constant, and ν is the frequency of a quantum of emitted or absorbed energy.

On the basis of these three postulates, Bohr dealt with much of the observed behavior of the hydrogen atom and with some of the problems of the Rutherford model. Using the observed atomic spectrum of hydrogen and the Bohr model, it was even possible to calculate the differences in energy between some stationary states.

EXAMPLE 5.8

The wavelength of one of the lines in the visible region of the atomic spectrum of hydrogen is 6.5×10^{-7} m. This radiation is emitted when a hydrogen atom goes from a high energy state to a lower energy state. Calculate the difference in energy between the two states.

SOLUTION

The difference in energy is related to the frequency of the radiation. The frequency is related to the wavelength.

$$\Delta E = h\nu \quad \text{and} \quad \nu = c/\lambda$$

Thus

$$\Delta E = \frac{hc}{\lambda} = \frac{(6.6 \times 10^{-34} \text{ J s})(3.0 \times 10^8 \text{ m s}^{-1})}{(6.5 \times 10^{-7} \text{ m})}$$

$$= 3.0 \times 10^{-19} \text{ J}$$

However, Bohr was able to go much further than this relatively simple description of the absorption and emission of energy by the hydrogen atom. His model had one more postulate.

4. The allowed orbits are defined most simply in terms of the *angular momentum* of an electron in the orbit. (Momentum is a measure of the tendency of a moving body to keep moving; angular momentum is a measure of the tendency of a body in rotational motion to keep rotating.) Since only certain orbits are allowed, only certain values of the angular momentum of the electron are allowed. Bohr postulated that the allowed values of the angular momentum are integral multiples of $h/2\pi$. In the ground state, the state of lowest energy, the angular momentum of the electron is $h/2\pi$. The angular momentum of an electron in any higher orbit is an integral multiple of $h/2\pi$. The angular momentum of the electron is $nh/2\pi$, where n is an integer (1, 2, 3, 4, . . .) corresponding to a given orbit. For the ground state, $n = 1$. For the next higher state, $n = 2$, and the angular momentum of an electron in this orbit is $2h/2\pi$. For the next higher state, $n = 3$, and so on. There can be no fractional value of $h/2\pi$. Thus, the integer n can be used to designate an orbit and a corresponding energy state; n is called the atom's *principal quantum number*.

The postulate that the angular momentum of the electron in a hydrogen atom is an integral multiple of $h/2\pi$ is central to the development of the Bohr model of the hydrogen atom. There was no deep theoretical reason for the use of the value $h/2\pi$. Bohr chose it because it gave the right answers.

Using these postulates and some classical laws of physics, Bohr could calculate the radius of each orbit of the hydrogen atom, the energy of the state associated with each orbit, and the wavelength of the radiation emitted in transitions between orbits. The wavelengths calculated by this method agreed with those in the actual atomic spectrum of hydrogen, which was a success for the Bohr model.

To calculate the size of the orbits, Bohr started with the knowledge that the coulombic attraction between the electron and the nucleus must be balanced exactly by the force of acceleration that keeps the electron in orbit.

From Coulomb's law, the coulombic attraction is e^2/r^2, where e is the magnitude of the negative charge on the electron and the positive charge on the hydrogen nucleus and r is the distance of the electron from the nucleus, that is, the radius of the circular orbit. The electron, like any body moving in a closed orbit, has an acceleration. From Newton's law of mechanics, the force due to acceleration is mv^2/r, where m is the mass and v is the velocity of the electron and r is the radius of the orbit. These two forces must balance each other exactly:

$$\frac{e^2}{r^2} = \frac{mv^2}{r} \tag{5.13}$$

or, multiplying both sides by r:

$$\frac{e^2}{r} = mv^2 \qquad (5.14)$$

Both e, the charge on the electron, and m, the mass of the electron, had been determined by experiment. If v, the velocity of the electron were known, then r, the radius of the orbit could be calculated.

The velocity of the electron is known to be related to the angular momentum. The angular momentum of a body moving in a circular orbit is mvr, from Newton's laws of motion. From one of Bohr's postulates, the angular momentum of the electron is $nh/2\pi$. Setting these as equal gives:

$$mvr = \frac{nh}{2\pi} \qquad (5.15)$$

or, after algebraic manipulation:

$$v = \frac{nh}{2\pi mr}$$

Substituting this expression for v in Equation 5.14 gives:

$$\frac{e^2}{r} = m\left(\frac{nh}{2\pi mr}\right)^2 \qquad (5.16)$$

and solving for r gives:

$$r = \frac{n^2 h^2}{4\pi^2 m e^2} \qquad (5.17)$$

Since values for all the quantities on the right-hand side of Equation 5.17 are known, values for r, the radii of the allowed orbits of the electron, can be calculated. Putting in the known values of h, m, and e gives a simple relationship between the size of the radius and the principal quantum number of the orbit:

$$r = 5.30 \times 10^{-11}\, n^2 \text{ m} \qquad (5.18)$$

Substituting $n = 1$ gives 5.30×10^{-11} m as the radius of the hydrogen orbit in the ground state, a value that is reasonably consistent with other information on the size of atoms. Substituting $n = 2$ gives $(2^2)(5.30 \times 10^{-11})$ m as the size of the atom in the first excited state orbit. It is possible to diagram the allowed orbits of the hydrogen atom as a series of concentric circles of known size (Figure 5.18), with the nucleus at the center and the radii increasing by factors of 2^2, 3^2, 4^2, and so on.

It can be shown from classical physics that the total energy of the electron in an orbit of the Bohr atom is

$$E = \frac{-e^2}{2r} \qquad (5.19)$$

The negative sign in Equation 5.19 means that the energy of the electron is negative. That is, the energy of the electron in the atom is lower than the energy of a free electron. As r increases, the numerical value of $-e^2/2r$ decreases; the energy becomes less negative, which means that the energy increases. In other words, the energy of the electron increases as its distance from the nucleus increases.

Substituting the expression for r from Equation 5.17 into Equation 5.19 gives the energy as:

$$E = \frac{-2\pi^2 m e^4}{n^2 h^2} \qquad (5.20)$$

Equation 5.20 allows us to calculate the energy for each orbit, since the

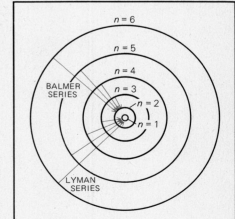

FIGURE 5.18
The hydrogen spectrum diagramed. The Balmer lines are caused by the emission of specific amounts of energy as the electron drops from higher orbits to the orbit $n = 2$. The Lyman series of lines is generated by the energy emitted as the electron drops from higher energy levels to the orbit $n = 1$. Each individual transition produces one photon with a wavelength and frequency characteristic of the energy change of the atom.

values of m, e, and h are known, while the appropriate integer can be assigned for n, the principal quantum number.

The most impressive result of the Bohr model came when the expression for E given in Equation 5.20 was substituted into Equation 5.12, the Bohr frequency rule, and c/λ was substituted for ν. If n_f is the principal quantum number of the final, or lower energy state and n_i is the principal quantum number of the initial, or higher energy state of emission, the resulting equation is:

$$\frac{1}{\lambda} = \frac{2\pi^2 me^4}{ch^3}\left(\frac{1}{n_f{}^2} - \frac{1}{n_i{}^2}\right) \tag{5.21}$$

an expression that appears to be the same as Equation 5.5, the Balmer expression for the atomic spectrum of hydrogen, if the term R_H in that expression equals $2\pi^2 me^4/ch^3$. Indeed, substituting experimentally determined values of m, e, c, and h gives a number that is in excellent agreement with the measured value of R_H.

In the Bohr model, each series of lines in the atomic spectrum of hydrogen is the result of radiant energy emitted in transitions of the atom's lone electron from different high-energy orbits to a single low-energy orbit. As Figure 5.18 shows, the Balmer series, those lines in the visible region of the atomic spectrum of hydrogen predicted by Balmer's formula, are due to transitions to the orbit $n = 2$. Using his model, Bohr was able to predict the existence of another series of lines, those produced by the electron's transitions to the orbit defined by $n = 1$. Those lines were later discovered. They are the Lyman series of lines in the ultraviolet region of the hydrogen spectrum. Balmer's formula could and did predict new lines. The Bohr theory was remarkable because it not only predicted new lines but also explained why the predictions were correct.

Qualitatively, Bohr's picture of quantized energy states and the emission and absorption of radiation works well for all atoms. But while the Bohr model's quantitative predictions were quite successful for the hydrogen atom, they were spectacularly unsuccessful for every other atom. The Bohr model could predict the atomic spectrum of an atom with only one electron but not the spectra of atoms with two or more electrons. The reason the Bohr model worked well for the hydrogen atom but not for any other atom was found by answering the question posed in Section 5.6: How can something appear to be both a particle and a wave?

5.8 WAVES AND PARTICLES

The suggestion that a particle and a wave can be confused with one another seems absurd. At the beginning of the twentieth century, the definitions of each were fixed precisely, with no danger of overlap. A wave was seen as a disturbance spread out over a region in space. A particle was seen as something localized in space, with definite boundaries.

The fact that a wave is in motion causes some difficulty in describing it precisely, but this is not an insuperable difficulty. The velocity of a particle can be measured by following the particle in motion, in the way that a baseball player measures the velocity of a pitch—by timing the trip from the pitcher's hand to the catcher's glove. The velocity of a wave can be measured by picking a point on the wave—a particular peak, for instance—and following that point for a given time.

These rules work well enough on a macroscopic scale, for particles as big as baseballs and waves of the ocean. But the rules do not work well for particles as small as atoms and molecules. When one tries to describe objects on the atomic scale, so small as to be invisible even with the most powerful magnifying devices, the rules of the game change. It took a long time for that point to be understood. But the fact that the principles that apply to macroscopic objects may not apply to objects on the atomic scale was the key to an understanding of the atom. A whole new set of rules is needed for the atomic game. Once this idea is accepted and we stop trying to understand the atom by analogy with baseballs and ocean waves, progress can be made.

But this idea leads to an apparent paradox. The idea seems to be that things on the atomic scale are fundamentally different from the large objects of everyday life, that atoms are fundamentally different from baseballs and ocean waves are completely different from light waves. But baseballs and ocean waves consist of atoms. Logically, there should be a fundamental similarity, not a fundamental difference.

In fact, it has been found that there is a basic similarity between subatomic phenomena and everyday objects. Baseballs, atoms, ocean waves, and light waves all have aspects in common. If the world is viewed properly, there are no waves and no particles. Instead, the world is made of wave particles, one phenomenon with two complementary aspects. On the everyday scale, one aspect or the other is dominant; the wave aspect of baseballs is undetectable and the particle aspect of ocean waves is also undetectable. But on the atomic scale, both aspects are detectable and important. Electrons, which were first thought of as particles, have an equally important wave aspect. Electromagnetic radiation, which was first thought of as waves, has an equally important particle aspect, as the existence of photons and the quantization of energy shows. The conception of a given object as either wave or particle depends on the point of view of the observer.

The de Broglie Hypothesis

By the 1920s, experiments had established the dual wave-particle nature of electromagnetic radiation. Even though momentum (the product of mass and velocity, mc) classically was associated only with particles and not with waves, it was even possible to calculate the momentum of a photon, a particle of light with wave characteristics. Combining Einstein's famous $E = mc^2$ with the relationship $E = h\nu$ gives:

$$h\nu = mc^2 \qquad\qquad (5.22)$$

which can be rewritten as:

$$mc = \frac{h\nu}{c}$$

and, since $\nu/c = 1/\lambda$, we can write:

$$mc = \frac{h}{\lambda} \qquad\qquad (5.23)$$

In 1924, Louis de Broglie (1892–   ) made a suggestion that was simple but startling: If light waves have particle characteristics, then particles should have wave characteristics. De Broglie noted that if a value for λ is substituted in Equation 5.23, it is possible to calculate a value for m, the mass of a photon of this wavelength λ. De Broglie suggested a different calculation: Replace m with the known mass of an electron, and replace c, the velocity of light, with v, the velocity of an electron. If these substitutions are made, a value for λ can be calculated. This value is the wavelength of an electron, an entity which, until that time, had been regarded solely as a particle.

The calculation suggested by de Broglie was performed. Experiments produced proof that a moving electron has an associated wavelength—a wavelength whose value was found to be the value calculated from Equation 5.23 by de Broglie's method.

The de Broglie relationship can be made more general. A wavelength can be calculated for any particle moving with a velocity v. The formula is:

$$\lambda = \frac{h}{mv} \tag{5.24}$$

Equation 5.24 means that any moving particle, even a baseball, has a wavelength associated with it. The formula also shows why the wavelengths associated with baseballs and other large-scale objects are undetectable under everyday conditions.

The value of h is small: 6.6×10^{-34} J s, or 6.6×10^{-34} kg m^2 s^{-1}. The wavelength of a moving particle will also be small unless the mv term in Equation 5.24 is of the same order of magnitude as h. The mass m of any large object, such as a baseball, is always large compared to the mass of a particle like the electron—so large that the momentum, mv, is enormous compared to h.

For example, a baseball thrown by a good pitcher has a momentum of about 5 kg m s^{-1} and a wavelength, $\lambda = (6.6 \times 10^{-34}$ kg m s$^{-1})/(5$ kg m s$^{-1}) = 1.3 \times 10^{-34}$ m. This means that the peaks of the waves associated with the baseball are separated from each other by a distance that is 10^{24} times smaller than the size of an atom. The waves associated with a heavy object are undetectable. Only on the atomic scale, where values of m are very small, does λ become significant.

The fact that the electron has a wave aspect explains Bohr's postulate that the angular momentum of an electron is quantized. If a wave goes around in a circular path, the circumference of the circle must be an integral multiple of the wavelength:

$$2\pi r = \lambda n \tag{5.25}$$

A circle whose circumference is a fractional part of the wavelength is not an allowed orbit. The peaks and troughs of the wave would overlap and the wave would cancel itself out, as Figure 5.19 shows. If we were to substitute into Equation 5.25 the value of the electron's wavelength λ given in Equation 5.24, we would get:

$$2\pi r = \frac{nh}{mv} \tag{5.26}$$

or, rearranging the terms:

$$mvr = \frac{nh}{2\pi}$$

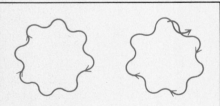

FIGURE 5.19
The electron as a wave. Only orbits whose circumference is an integral multiple of the electron's wavelength are possible. In other orbits the overlapping waves would interfere and cancel each other.

which is identical with Equation 5.15, part of the Bohr derivation.

Looking back at the long history that led to the discovery of the wave-particle nature of energy and matter, it seems rather a historical accident that light was first regarded as waves and electrons were first regarded as particles. But though the road to the present view of energy and matter has been long, it has reached the point where light, electrons, and all other "waves" and "particles" can be viewed as having the same essential nature: They are wave-particles moving with different velocities.

The Heisenberg Uncertainty Principle

Given this *wave-particle duality*, a new way was needed to describe the structure of the atom. The description must include not only electron particles but also electron waves.

The wave pictured in Figure 5.10 is a pure harmonic wave with a regular pattern extending to infinity. Such pure waves probably do not exist in the real world. A real wave starts from a point and ends at another point. Real waves tend to pile up on each other to produce what are called *wave packets*. Figure 5.20

FIGURE 5.20
A wave packet, with varying wavelengths and amplitudes. Compare with the idealized wave in Figure 5.10. Such wave packets moving through space are used to describe the motion of photons, whose velocity is c, 3×10^8 m/s and of matter, whose velocity is always less than c.

shows such a wave packet, whose wavelength and amplitude, unlike those of a pure wave, cannot be defined precisely. Because the packet extends through a region of space, its position also cannot be defined precisely.

The wave associated with a particle is such a wave packet, since a moving particle starts at one point and finishes at another point. Thus, the wavelength associated with a moving particle cannot be defined precisely. There is a narrow range of wavelengths associated with the particle. To put it another way, we can say that there is an uncertainty of wavelength within that range. That is not the only uncertainty. From Equation 5.24, it can be seen that the momentum of a particle is h, a constant, divided by λ, the wavelength. If there is uncertainty in the wavelength, there must also be uncertainty in the momentum. And since a particle such as an electron is actually moving as a wave packet, there is uncertainty about its position. All we know is that the particle is somewhere in the wave packet.

This thought—that there is a fundamental uncertainty about the wavelength, momentum, and position of wave-particles—is one of the more unsettling discoveries of twentieth-century science. Uncertainty now has been built into the accepted picture of natural phenomena, from subatomic particles on up. Werner Heisenberg (1901–1976), in his uncertainty principle, set forth the minimum amount of uncertainty to be expected in any description of the universe. The uncertainty principle can be stated in several ways. One statement is that the uncertainty in the momentum of any particle multiplied by the uncertainty of position can never be less than h, Planck's constant; that is:

$$\Delta(mv_x) \times \Delta x \geq h \qquad (5.27)$$

where mv_x is the momentum in the x-direction of the particle, x is the position, and Δ is the uncertainty or error in each.

The uncertainty principle has many consequences. Equation 5.27 can be interpreted to mean that it is impossible to know exactly both the position and the momentum of a particle at any moment. Any attempt to make an exact measurement of position and momentum must fail, because there is no way of making the measurement without changing one of the characteristics that is being measured.

This concept can be illustrated by inventing a party game, "Find the Balloon." In this game, a person is in a dark room and must find a floating balloon without turning on the lights. One method is to throw a dart, listen for a pop, and thus learn where the balloon was—except that the balloon is no longer there (and no longer a balloon). Another possible method is to line up some friends in the room, throw peanuts, and calculate the location by the interaction between peanuts, balloon, and friends. But if you hit the balloon, both the balloon and the peanut are deflected, and the location is still uncertain.

If the player decides to cheat by turning on the light for a quick peek, the results are better. The balloon can be seen and located instantly. It is true that the location of the balloon is changed because photons bounce off it, but the change is negligible for the purpose of the game.

But suppose the game is "Find the Electron"? In this case, turning on the light—that is, using photons in the search—causes problems. Care must be taken in choosing the photons. If a photon with the same momentum as the electron is used, the collision would deflect the electron measurably. If a photon of long wavelength is used, the accuracy of the measurement is decreased because the position of the electron can be determined only to an accuracy of a wavelength. Therefore, one wants a photon with low momentum and short wavelength—which is an impossibility. According to the de Broglie relationship, momentum and wavelength are inversely proportional; one increases as the other decreases. So this is a good game to avoid, because the nature of the wave-particles in the game sets an unconquerable limit on the accuracy of the measurement.

EINSTEIN AND UNCERTAINTY

"God does not play dice with the universe," Albert Einstein said. It was his brief way of rejecting the view of the universe that accompanied acceptance of the Heisenberg uncertainty principle. The implication of the uncertainty principle is that the universe is indeterminate on the most basic level. Since there is uncertainty in the most elementary events, no exact cause-and-effect relationship can be established. Instead, in the words of de Broglie, quantum physics appear to be "governed by statistical laws, and not by any causal mechanisms, hidden or otherwise."

Einstein could never accept this implication of quantum theory, a point of view that put him increasingly at odds with other physicists as the years went by. In 1944, in a letter to Max Born, Einstein expressed his beliefs directly:

"You believe in the God who plays dice, and I in complete law and order in a world which objectively exists, and which I, in a wildly speculative way, am trying to capture. I firmly *believe*, but I hope that someone will discover a more realistic way, or rather a more tangible basis than it has been my lot to do. Even the great initial success of the quantum theory does not make me believe in the fundamental dice game, although I am well aware that our younger colleagues interpret this as a consequence of senility."

Einstein made a number of attempts to describe "thought experiments" that would show that cause-and-effect exists on the subatomic level. A thought experiment is an experiment that can never be performed, only imagined, but that can nevertheless test a theory. But in every case, other physicists were able to show flaws in his thought experiments, and thus to uphold the uncertainty principle.

Einstein's viewpoint was summed up by another statement he made more or less casually when he heard of an experimental result that purported to disprove the theory of relativity. "God is subtle, but he is not malicious," Einstein said.

However, quantum mechanics and the uncertainty principle remain cornerstones of modern physics. If Einstein's world view is correct, it is on a deeper level than we have reached as yet.

5.9 THE QUANTUM MECHANICAL DESCRIPTION OF THE HYDROGEN ATOM

This new information showed why the Bohr model of the atom was only partially successful. First, the model did not take into account the wave aspect of the electron. Second, the model violated the uncertainty principle by defining the motion and position of the electron too precisely.

The impossibility of any attempt to define the electron's position precisely was recognized. The best that could be done was to define a *probability* for the electron being in a given location. It is not possible to say that at a given moment,

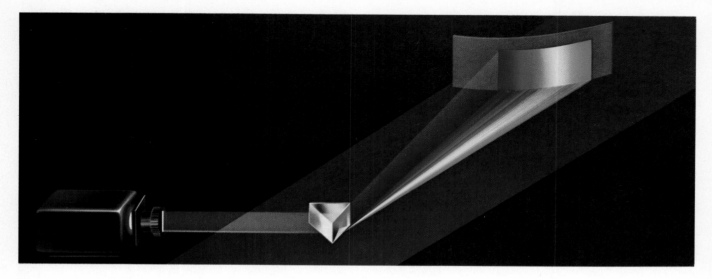

A beam of white light that passes through a prism is spread out as a spectrum. The spectrum is continuous because white light is made up of all visible wavelengths, from red to blue (above).

(1) A spectrum from an incandescent solid is also continuous; (2) The sun's spectrum shows several dark lines due to absorption by the indicated elements. The discontinuous emission spectra of sodium, hydrogen, calcium, mercury, and neon are shown (below).

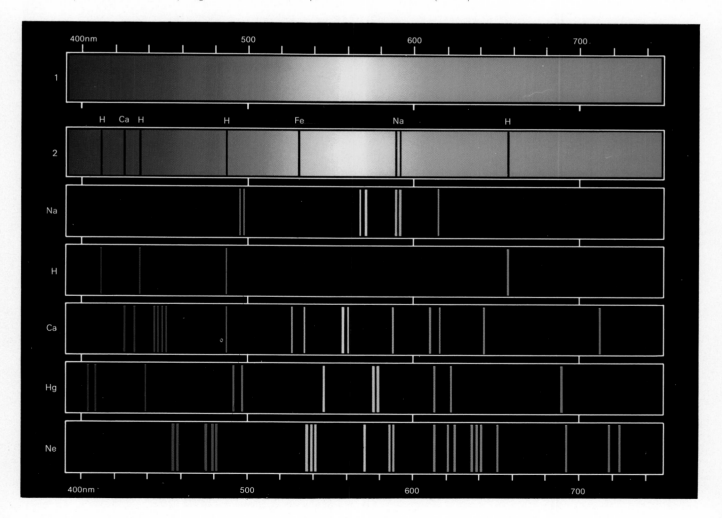

the electron is at point A rather than at point B. But it is possible to say that at some moment the electron is ten times as likely to be at point A as at point B.

Max Born (1882–1970) proposed that these probabilities could be obtained by studying the wave packet associated with the moving electron. In the quantum mechanical description of the atom that is used today, the idea of an electron particle is replaced by the concept of the electron wave. A mathematical function called the *wave function*, represented by ψ, the Greek letter psi, is used to describe the wave packet. The value of the wave function ψ at any point in space is related to the amplitude of the wave, its intensity of vibration, at that point. The probability that an electron is at a given point in space is proportional to the *square* of the wave function, $|\psi|^2$, at that point.

To find wave functions for electrons, we must describe how the forces that govern the behavior of electrons in atoms affect the behavior of waves. The simplest equation that can be used to calculate the electron wave function ψ was developed in 1926 by Erwin Schrödinger (1887–1961). The Schrödinger wave equation for the motion of a particle in any one direction is:

$$\frac{h^2}{8\pi^2 m}\frac{d^2\psi}{dx^2} + V\psi = E\psi \qquad (5.28)$$

Equation 5.28 seems complex, but analysis shows it to be a single equation in two unknowns, ψ and E. All the other quantities, such as m, the mass of the particle, and V, its potential energy, are known. Thus, the Schrödinger equation gives the relationship between the total energy E of the atom and the wave function ψ of its electron.

When the Schrödinger equation is applied to a real system such as a hydrogen atom, it is found that solutions are obtained only at certain values of the total energy E. These values of E are related to each other by small integers—the quantum numbers. Thus, quantum numbers and quantized energy of the atom are automatic consequences of the mathematics of the Schrödinger equation. More fundamentally, quantum numbers and quantized energy can be shown to be automatic consequences of the fact that electron waves in atoms have restrictions imposed on their motion.

The solutions of the Schrödinger equation give a set of allowed energies of the atom, with one or more wave functions, ψ corresponding to each energy. How can we picture such an atom?

The simplest and most widely used picture replaces an orbit, which is a path, with an **orbital.** An orbital can be thought of as a picture of the probability of finding the electron at every point in space. To help visualize where the electron is most likely to be found, a simple way of representing an orbital is needed.

We can develop a graphical representation of an orbital in a number of different ways. One simple way is to make a three-dimensional plot of all the points in space where the probability of finding the electron has an arbitrarily assigned constant value. Because of the properties of wave functions, such a plot is a continuous closed surface, called a *surface of constant probability*, which encloses a region of space.

Since most locations are highly improbable, the efficient course of action is to concentrate on the region where the electron is likely to be found most of the time—say 90% of the time. The plot is made in such a way that it encloses the region of space where the electron is found at least 90% of the time.

Other graphical representations of an orbital can be used, but the surface of constant probability is the most common one. When we use the term ''orbital,'' we shall also take it to mean a graphical representation of the orbital.

Because the value of the wave function ψ can be found by solving the Schrödinger equation, the exact size and shape of an orbital can be determined for the hydrogen atom. Having this orbital, we can represent it by a surface of constant probability enclosing a region in which there is a high probability of finding the electron. A "charge cloud" picture of the electron in an atom can thus be visualized. Instead of being a point in space, the electron is pictured as a cloud that takes its shape from the orbital. The density of the cloud varies as the probability varies. The cloud is thickest where the probability of finding the electron is highest, and the cloud thins out where the probability is low.

The quantum mechanical description of the atom has extremely important implications for chemistry. A detailed picture of the electronic structure of atoms has been developed, and has been used to explain not only the observed characteristics of atoms but also the way in which atoms join to form molecules. The way in which knowledge of the electronic structure of atoms can be used to explain chemical behavior will be examined in Chapter 6.

EXERCISES

5.1 Suggest an experiment that Democritus could have performed to test his hypothesis that atoms are in constant motion.

5.2 The value of k in Coulomb's law (Equation 5.1) can be given as 9.0×10^9 N m^2 C^{-2}. The value of the corresponding constant G in the law of universal gravitation ($F = Gm_1m_2/r^2$, where F is the force of gravitational attraction between two bodies, m_1 and m_2 are the masses, and r the distance between the two bodies) is 6.7×10^{-11} N m^2 kg^{-2}. Find the ratio of the coulombic attraction between an electron and a proton and the gravitational attraction between an electron and a proton at any given distance. The necessary values of physical constants can be found in Appendix I.

5.3 Use the value of the charge of the electron to find the number of electrons with a total charge of 96 490 C.

*5.4 An alpha particle, ^{4}He^{2+}, has a relative atomic mass of 4.00151. Find the value of e/m in coulombs per kilogram (C/kg) for the alpha particle.

5.5 The monopositive cation (X$^+$) of an isotope of an unknown element is found to have $e/m = 379.73$ C/g. Find the relative atomic mass of the isotope.

5.6 Find the number of neutrons and electrons in the following atoms or ions: (a) $^{15}_7$N, (b) $^{55}_{25}$Mn^{3+}, (c) $^{32}_{16}$S^{2-}, (d) ^{239}Pu, (e) ^{211}At$^-$.

5.7 Nuclei with the same mass number but with different numbers of protons are called isobars. Nuclei of mass number 40 are known for all the elements from chlorine to scandium. Use the system of notation to write symbols for these isobars.

5.8 Species that are isoelectronic have the same number of electrons. Write formulas for the ions of the isobars in Example 5.7 that have the same number of electrons as argon does.

5.9 Nuclei with the same number of neutrons but different mass numbers are called isotones. Write the symbols of four isotones of ^{238}U.

5.10 Find the elements that have the following nuclear composition: (a) $A = 51$, $N = 28$, (b) $A = 56$, $N = 30$; (c) $A = 59$, $N = 32$.

5.11 A sample of magnesium of natural origin is found to consist of three stable isotopes: ^{24}Mg, natural abundance 78.70%; ^{25}Mg, natural abundance 10.13%; and ^{26}Mg, natural abundance 11.17%. Their relative atomic masses are 23.985, 24.986, and 25.983 respectively. Find the atomic weight of magnesium.

5.12 The atomic weight of antimony of natural origin is 121.75. It consists of only two isotopes, ^{121}Sb, relative atomic mass 120.90, and ^{123}Sb, relative atomic mass 122.90. Find the natural abundances of the two isotopes.

5.13 Naturally occurring iodine has an atomic weight of 126.9045. A 12.3849-g sample of iodine is accidentally contaminated with 1.00070 g of ^{129}I, a synthetic radioactive isotope of iodine used in treatment of certain diseases of the thyroid gland. The relative mass of ^{129}I is 128.9050. Find the apparent "atomic weight" of the contaminated iodine.

5.14 The range of wavelengths of the vhf band used for television broadcasting is about 10 m–1 m. Find the frequency range of the electromagnetic radiation that is used in television broadcasting.

5.15 Find the wavelength of cosmic rays whose frequency is 1.0×10^{32} s^{-1}.

5.16 The speed of sound in air is 344 m/s at 20°C. The lowest frequency of a large organ pipe is 30 s^{-1} and the highest frequency of a piccolo is 1.5×10^4 s^{-1}. Find the difference between the wavelengths of these two sounds.

5.17 The distance from the earth to the sun is 1.5×10^8 km. Find the number of crests in a light wave of frequency 1.0×10^{14} s^{-1} traveling from the sun to the earth.

5.18 Find the smallest value of n_f that defines a series of hydrogen emission lines in the infrared region. Assign values to n_i and find the positions of the two lines of longest wavelength in this series.

5.19 Find the frequency of the shortest-wavelength radiation that can be emitted by hydrogen.

5.20 Evidence has been obtained that demonstrates the existence of elements such as technetium in the stars. Suggest what this evidence might be.

5.21 Find the energy of a quantum of cosmic rays whose frequency is 1.0×10^{32} s^{-1}.

5.22 Find the energy of a quantum of radio waves of wavelength 2.45 m.

5.23 The energy required to melt 1.00 g of ice is 333 J. Find the number of quanta of infrared radiation of frequency 4.67×10^{13} s^{-1} that must be absorbed in order to melt 1.00 g of ice.

5.24 A metal has a threshold frequency of 6×10^{14} s^{-1} for the photoelectric effect. Predict whether the number of electrons emitted and the kinetic energy of each electron increases, decreases, or stays the same as the result of the following changes:

(a) The frequency of radiation striking the metal is raised from 3×10^{14} s^{-1} to 4×10^{14} s^{-1}.

(b) The frequency of radiation striking the metal is raised from 6×10^{14} s^{-1} to 7×10^{14} s^{-1}.

(c) The distance of the metal from a source of radiation of 7×10^{14} s^{-1} is increased from 1 m to 3 m.

5.25   The threshold frequency of a metal is 5.25×10^{14} s^{-1}. Find the binding energy of an electron in the metal.

*5.26   The binding energy of electrons in a metal is 193 kJ/mol. Find the threshold frequency of the metal.

5.27   A metal emits electrons of kinetic energy 3.1×10^{-19} J/electron when exposed to radiation of frequency 2.4×10^{15} s^{-1}. Find the threshold fre-

quency of the metal.

*5.28   Use the mass of the electron to find the velocity of an electron emitted by a metal whose threshold frequency is 2.25×10^{14} s^{-1} when exposed to visible light of wavelength 5.00×10^{-7} m.

5.29   Find the velocity of an electron in the ground state orbit of the Bohr hydrogen atom.

5.30   Scientists living on a planet whose atmosphere is very rich in neon might have trouble determining whether or not there is neon on their sun. Suggest a reason for their difficulty.

5.31   Find the angular momentum of an electron in the orbit of the Bohr hydrogen atom that corresponds to the second excited state of hydrogen. Find the radius of this orbit.

5.32   Find the wavelength of the radiation required to remove an electron in the ground state orbit of the Bohr hydro-

gen atom completely and form $H^+ + e^-$.

5.33   In order for a thermonuclear fusion reaction of two deuterons ($_1^2H^+$) to take place, the deuterons must collide each with a velocity of about 1×10^6 m/s. Find the wavelength of such a deuteron.

5.34   Find the velocity of an electron whose wavelength is 2.2×10^{-10} m.

5.35   Find the longest wavelength of a wave that can travel around in a circular orbit of radius 1.8 m.

*5.36   The statement of the Heisenberg uncertainty principle in Equation 5.27 can also be formulated in terms of the variables energy and time. In the Bohr treatment of the hydrogen atom, the energy of each orbit and therefore the energy of any given electronic transition is known very accurately. Suggest the difficulty created by this accuracy.

ATOMS, ELECTRONS, THE PERIODIC TABLE

6

Chapter 5 briefly described a quantum mechanical picture of the electron in the hydrogen atom. We saw that this model defines regions in space where the probability of finding the electron is high. Such a region is defined by enclosing it in a surface. This method provides a convenient description of an orbital, which is actually a total picture of the probability of finding the electron at every point in space.

In this chapter, we shall learn why the electronic structure of atoms is of central importance in modern chemistry. We shall see why and how the electronic structure of atoms determines their chemical behavior. We shall also see how a knowledge of the electronic structure of atoms is used to explain an important unifying principle in chemistry, a principle that is embodied in the periodic table of the elements.

6.1 THE QUANTUM MECHANICAL DESCRIPTION OF THE HYDROGEN ATOM

The Schrödinger wave equation has been solved completely for all the allowed energy states of the hydrogen atom. These solutions agree with the experimental data. They are quite important, because they can be used as a basis for describing more complex atoms.

Quantum Numbers

The Schrödinger equation gives quantum numbers corresponding to allowed orbitals for the electron in a hydrogen atom. These quantum numbers can be used to define the allowed orbitals and to describe the behavior of an electron in an orbital. Each orbital corresponds to an allowed energy state. An electron of a hydrogen atom is described by assigning values to four quantum numbers, whose values are interrelated. The quantum numbers are:

1. **The Principal Quantum Number, n.** The principal quantum number can have any integral value greater than zero: $n = 1, 2, 3, 4, \ldots$. We can specify the energy of the hydrogen atom (but not of other atoms) by assigning it a value of n. The principal quantum number is the same as Bohr's quantum number. Bohr proposed his quantum number to explain the observed behavior of the hydrogen atom. The principal quantum number n is part of the solution of the Schrödinger wave equation, a much more general and successful description of the behavior of atoms. If we solve the Schrödinger equation for the hydrogen atom, we get an expression for the atom's total energy:

$$E = \frac{-2\pi^2 me^4}{n^2 h^2} \tag{6.1}$$

in terms of n, the principal quantum number of its electron. The expression is identical with Equation 5.20, which was the result obtained by Bohr. Once a value is assigned to n, the corresponding value of the energy of the hydrogen atom can be calculated.

Again, you should note that the total energy of the electron in an atom is negative and is lower than the energy of a free electron. It is important to remember that a decrease in the numerical value of E means that the energy has become less negative—that is, the energy has increased. If you look at Equation 6.1, you will see that n^2 is in the denominator of the fraction that determines the numerical value of E. Therefore, as the value of n increases, the numerical value of E, the energy, decreases. Since E is negative, a decrease in its numerical value means an increase in the total energy. The important thing to remember is that as n, the principal quantum number, increases, the energy E of the electron increases.

2. The Angular Momentum Quantum Number, ℓ. The values of ℓ depend on the value of the principal quantum number n. The quantum number ℓ has integral values from 0 to $n - 1$. If $n = 1$, then there is only one possible value of ℓ, $\ell = n - 1 = 0$. If $n = 2$, there are two possible values of ℓ, $\ell = 0$ or 1. If $n = 3$, there are three possible values of ℓ, $\ell = 0$, 1, or 2. And so on.

n	ℓ
1	0
2	0, 1
3	0, 1, 2
4	0, 1, 2, 3
⋮	⋮

The value of ℓ determines the shape of an orbital and the angular momentum of an electron occupying that orbital. In a hydrogen atom, the value of ℓ does not affect the total energy, which is determined completely by the value of n. But in atoms with more than one electron, the energy of an electron in an orbital, while described primarily by n, also depends to some extent on the value of ℓ. In multielectron atoms, the energy of electrons is found to increase with an increase in the value of ℓ. Given electrons with the same quantum number n, the energy of the electron with the quantum number $\ell = 2$ will be higher than that of the electron with $\ell = 1$; the electron with $\ell = 3$ will have a higher energy level than the electron with $\ell = 2$.

3. The Magnetic Quantum Number, m_ℓ. The magnetic quantum number can have a range of values, which is restricted by the value of ℓ. The magnetic quantum number is allowed all integral values, including 0, from $-\ell$ to $+\ell$. That is, if $\ell = 0$, then $m_\ell = 0$. If $\ell = 1$, the m_ℓ has three possible values, -1, 0, and $+1$. If $\ell = 2$, then m_ℓ has five possible values, -2, -1, 0, $+1$, and $+2$.

ℓ	m_ℓ
0	0
1	-1, 0, $+1$
2	-2, -1, 0, $+1$, $+2$
⋮	⋮

The value of the quantum number m_ℓ does not affect the energy of an electron, which is specified by the quantum numbers n and ℓ. Rather, each value of the quantum number m_ℓ corresponds to a different allowed orientation of a designated orbital. Once n and ℓ are specified, each value of m_ℓ corresponds to an orbital of the same shape and orientation. However, the orbitals described by the different values of m_ℓ have different orientations. Thus, m_ℓ is most simply described as a directional quantum number. It distinguishes between orbitals that have the same shape and energy but point in different directions. Therefore, the numerical value of m_ℓ is not important. What is important is the number of different values of m_ℓ that correspond to a given value of ℓ. When ℓ is 0, m_ℓ has only one value. When ℓ is 1, m_ℓ has three allowed values. When ℓ is 2, m_ℓ has five allowed values.

The three quantum numbers n, ℓ, and m_ℓ are sufficient to describe the orbitals in the hydrogen atom. However, there is another aspect of the behavior of the electron that must be considered. The electron behaves as if it were spinning. This property of the electron, called its **spin**, was discovered in 1925.

Spin is best visualized as a manifestation of the particle aspect of the electron. While it is difficult to visualize a spin associated with a wave packet, we can easily picture a particle spinning around an axis like a top. The electron can spin in one of two ways, clockwise or counterclockwise. The spin of the electron is specified by the *spin quantum number, m_s*. Since the electron has only two allowed directions of spin, m_s has only two values, which by convention are $+\frac{1}{2}$ and $-\frac{1}{2}$. There is no relationship between the spin quantum number and the values of n, ℓ, and m_ℓ.

We should not make the mistake of associating a particular value of m_s with a particular direction of spin. As we shall see, m_s describes relative, not absolute, direction of spin.

Quantum Numbers

$n = 1, 2, 3, \ldots$
$\ell = 0, 1, 2, \ldots \quad (n-1)$
$m_\ell = 0, \pm 1, \ldots \quad (\pm \ell)$
$m_s = +\frac{1}{2} \text{ or } -\frac{1}{2}$

The Set of Allowed Orbitals

The customary way of identifying orbitals is first to write the principal quantum number n, and then to use a letter instead of a numeral to identify the value of ℓ. The letters s, p, d, and f are used to designate different values of ℓ. These letters are used for historical reasons; they once were used to describe features of atomic spectra, and have never been abandoned. When $\ell = 0$, the letter s is used. When $\ell = 1$, the letter p is used. When $\ell = 2$, the letter d is used. When $\ell = 3$, the letter f is used. For values of ℓ greater than 3, the lettering proceeds alphabetically from f. In tabular form:

ℓ	type
0	s
1	p
2	d
3	f
4	g
5	h
⋮	⋮

Table 6.1 shows how the value of n and the letter corresponding to the value of ℓ are used to name orbitals. The three p orbitals, which possess the same values of n and ℓ, can be distinguished by using the subscripts x, y, and z to designate the three values of m_ℓ. Since these three orbitals differ only in their spatial orientation, the three subscripts can be considered to describe the relative orientation of the orbitals along three axes at right angles in space. For d orbitals, more complicated subscripts are used, as Table 6.1 indicates. It should be noted

TABLE 6.1
A Portion of the Set of Allowed Orbitals

n	ℓ	m_ℓ	Orbital Designation
1	0	0	$1s$
2	0	0	$2s$
	1	$-1, 0, +1$	$2p_x$, $2p_y$, $2p_z$
3	0	0	$3s$
	1	$-1, 0, +1$	$3p_x$, $3p_y$, $3p_z$
	2	$-2, -1, 0, +1, +2$	$3d_{z^2}$, $3d_{x^2-y^2}$, $3d_{xy}$, $3d_{xz}$, $3d_{yz}$

that there is no correspondence between a given subscript and a given value of m_l. Only the number of values of m_l is important.

Shape and Location of the Orbitals

Since the wave functions of the hydrogen atom are known, it is possible to find the shapes, sizes, orientations, and extents of each orbital corresponding to a given set of quantum numbers.

A change in the value of n, the principal quantum number, has a direct effect on the orbital description. As the value of n increases, we must go further from the nucleus to enclose a region of space that includes more than 90% of the electron density. In other words, a higher value of n means that the region of highest probability for the electron is further from the nucleus. Our description of a $3s$ orbital is larger than a $2s$ orbital, which is larger than a $1s$ orbital.

The orbital shapes defined by low values of l are easily described. For an s orbital ($l = 0$), the surface that represents the orbital is a sphere whose radius depends on the value of n (Figure 6.1). Note than when $l = 1$, m_l has only one value; spheres are not directional.

For p orbitals ($l = 1$), the geometry is more complicated. We can think of these orbitals as having two spherical lobes, as shown in Figure 6.2. The three p orbitals are identical in shape and size but differ in orientation. The three possible orientations of p orbitals are usually called p_x, p_y, and p_z. The surfaces drawn to represent the three p orbitals enclose different regions of space, because of their different orientations. If these three regions are superimposed, the result is a spherical space that can be enclosed by a spherical surface.

There are five d orbitals, and Figure 6.3 shows a convenient way to represent their shapes. Figure 6.3 appears to show two different shapes for a d orbital. One shape is four-lobed, the other two-lobed with an associated torus (the dough-nut-shaped figure). But all d orbitals have the same shape. They only seem to be different because of the methods used to draw orbitals. Figure 6.4 shows that the two ''different'' d orbital shapes are not really different. If two four-lobed d orbitals are superimposed, the result is the alternate shape, a two-lobed orbital with an associated torus. If all the d orbitals are superimposed, they enclose a spherical region of space, just as the p orbitals do.

The d orbitals are named on the basis of the direction of orientation of their maximum electron density. Three of the orbitals d_{xy}, d_{xz}, and d_{yz}, have maximum electron density in the plane indicated by their subscripts. These regions of maximum electron density of these orbitals are centered between the coordinate axes. The regions of maximum electron density of the other two d orbitals, $d_{x^2-y^2}$ and d_{z^2}, are centered on the coordinate axes, the $d_{x^2-y^2}$ orbital on the x and y axes, and the d_{z^2} orbital on the z axis.

6.2 ELECTRONIC CONFIGURATION OF MULTIELECTRON ATOMS

The description of the single electron in the hydrogen atom is relatively simple, since we must consider only the interaction between the negatively charged electron and the positively charged nucleus. An exact quantum mechanical description of any atom with more than one electron is a formidable mathematical challenge. In atoms with more than one electron, interactions between the electrons must also be considered. The effect of these interactions on the electronic structure of atoms cannot yet be completely included in the description of multielectron atoms. But if the description is not exact, it is close enough to serve most purposes, as experimental results show.

The set of allowed orbitals in a multielectron atom can be approximated closely by using the set of allowed orbitals in the hydrogen atom. Hydrogenlike

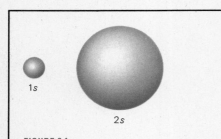

FIGURE 6.1
Two s orbitals, with different principal quantum number n. The orbital on the left has $n = 1$, the orbital on the right has $n = 2$. All s orbitals are represented as spheres whose radii are determined by the value of n. The $2s$ orbital is larger than the $1s$ orbital. A $3s$ orbital would be even larger than the $2s$ orbital.

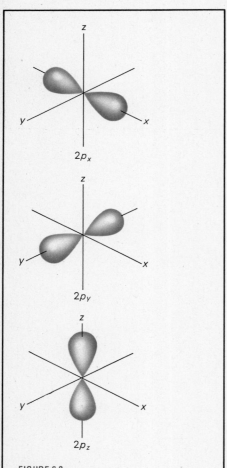

FIGURE 6.2
The three $2p$ orbitals of the hydrogen atom, each represented as two spheroidal lobes, but with different orientations for each orbital.

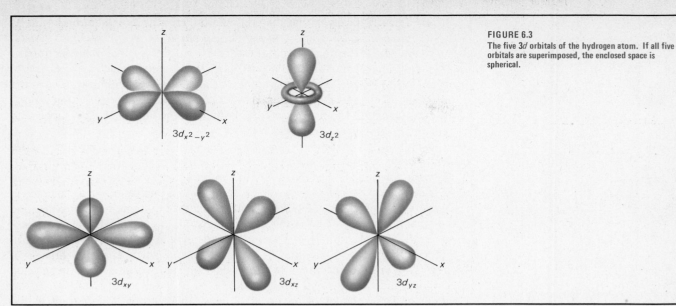

FIGURE 6.3
The five 3*d* orbitals of the hydrogen atom. If all five orbitals are superimposed, the enclosed space is spherical.

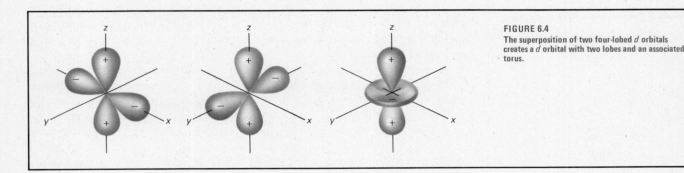

FIGURE 6.4
The superposition of two four-lobed *d* orbitals creates a *d* orbital with two lobes and an associated torus.

orbitals are used in the description of all atoms. This "hydrogenlike approximation" is used to build a picture of the electronic configuration of any atom. In other words, we use this approximation to determine the sets of quantum numbers that describe, as completely as possible, the orbitals occupied by an atom's electrons. We shall see that a description of an atom's electronic configuration is helpful in correlating and understanding the chemical properties of the atom. Generally, we shall be interested in the electronic configuration that corresponds to the lowest electronic energy, or *ground state*, of the atom.

To work out the electronic configuration of atoms, we use what is called the **Aufbau principle,** from the German word for "building up." The Aufbau principle can be stated simply: The electronic configuration of lowest energy is the one in which each electron is placed in the lowest-energy hydrogenlike orbital that is available. In some cases, placing an electron is not as simple as it might seem, but the electrons can be placed through a combination of theory and observation.

Let us start with the hydrogen atom. The lone electron has a principal quantum number $n = 1$, so the quantum number ℓ must have a value of 0, and the quantum number $m_\ell = 0$. The electron can have a spin quantum number m_s of either $+\frac{1}{2}$ or $-\frac{1}{2}$. With quantum numbers $n = 1$ and $\ell = 0$, the electron is in the 1*s* orbital.

Next consider the simplest multielectron atom, helium, which has two electrons. The lowest-energy orbital is the 1*s* orbital, and one electron goes there. Does the second electron go into the 1*s* orbital, or does it go into a higher-energy orbital with principal quantum number $n = 2$?

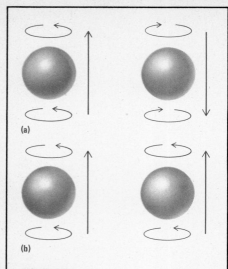

(a)

(b)

FIGURE 6.5

Two electrons may occupy the same orbital if they have different spins. In (a), the electron on the right has spin $+\frac{1}{2}$ and the electron on the left has spin $-\frac{1}{2}$. In (b), the electron on the right has spin $-\frac{1}{2}$ and the electron on the left also has spin $-\frac{1}{2}$. The signs are arbitrary and do not indicate actual directions of spin.

One possible view is that there is no room for a second electron in the $1s$ orbital, since there already is one electron spread out over this region of space. But experimental evidence indicates that this is not so, and that both electrons of helium in the ground state are in the $1s$ orbital. The electronic configuration of helium is written $1s^2$, the superscript 2 indicating the number of electrons in the $1s$ orbital. (The superscript is pronounced ''two,'' not ''squared,'' even though the notation is the same as for an exponent in mathematics. This rule applies to all superscripts describing the number of electrons in a set of orbitals.)

When we say that the two electrons of helium are in the same orbital, we are specifying three quantum numbers for each electron: $n = 1$, $\ell = 0$, and $m_\ell = 0$. We must also specify the spin quantum number. The most important point about m_s is that it can describe *relative* spins. If we picture two electrons as spinning in opposite directions, as in Figure 6.5(a), we can say that one has $m_s = +\frac{1}{2}$ and the other has $m_s = -\frac{1}{2}$. In such a case, we say that the spins of the electrons are *paired*. If we picture the two electrons as spinning in the same direction, as in Figure 6.5(b), we can say that they have the same value of m_s. Whether that value is $-\frac{1}{2}$ or $+\frac{1}{2}$ does not matter, as long as both electrons have the same value of m_s. In such a case, we say that the spins of the electrons are *parallel*, or *unpaired*. The concept of paired and unpaired electrons is important in the description of many atoms.

The two electrons in the $1s$ orbital of helium have paired spins. One electron has $m_s = +\frac{1}{2}$, and the other has $m_s = -\frac{1}{2}$. The helium atom is just one example of a general phenomenon: **When two electrons reside in the same orbital, their spins must be paired.**

One of the basic rules of atomic structure, formulated by Wolfgang Pauli (1900–1958), states that **in a given atom, no two electrons can have the same four quantum numbers.** This is the **Pauli exclusion principle,** first published in 1925. In its most basic form, it says that two things cannot be in the same place at the same time. Applied to electrons in atoms, the principle means that an orbital that is described by three quantum numbers, n, ℓ, and m_ℓ, can hold a maximum of two electrons, which must have paired spins, that is, different values of m_s.

There is room in the $1s$ orbital for a second electron because there are two allowed values of m_s. Both electrons in the ground state of helium have the quantum numbers $n = 1$, $\ell = 0$, $m_\ell = 0$. They differ in the fourth quantum number. One electron has $m_s = +\frac{1}{2}$, the other electron has $m_s = -\frac{1}{2}$. The spins are paired as required by the Pauli exclusion principle.

Now consider the lithium atom, atomic number 3. In accordance with the Pauli exclusion principle, the third electron in a lithium atom cannot also be in the $1s$ orbital, but will be in a higher-energy orbital, where $n = 2$. We ask: Which is the lowest-energy orbital available for this electron?

In the hydrogen atom, the $2s$ ($n = 1$, $\ell = 0$) orbital and the $2p$ ($n = 2$, $\ell = 1$) orbital have the same energy, since they have the same principal quantum number n. This is not true in multielectron atoms, where we must consider electron-electron interactions and the size of the nuclear charge. In a multielectron atom, the relative energy that corresponds to an electron in a given orbital depends on the value of two quantum numbers, ℓ as well as n. The $2s$ orbital ($n = 2$, $\ell = 0$) is of lower energy than the $2p$ orbitals ($n = 2$, $\ell = 1$), so lithium's third electron goes into the $2s$ orbital. The electronic configuration of lithium is $1s^2 2s$.

In beryllium, atomic number 4, it is found that the fourth electron is also in the $2s$ orbital. In accordance with the Pauli exclusion principle, the two $2s$ electrons of beryllium have paired spins. With the addition of the fourth electron, the $2s$ orbital is filled. The next orbital to fill is a $2p$ orbital. Thus, in boron, atomic number 5, one electron is in the $2p$ orbital. We can now summarize the electronic configurations of the ground states of the first five elements:

H : $1s$
He: $1s^2$
Li : $1s^2 2s$
Be: $1s^2 2s^2$
B : $1s^2 2s^2 2p$

Note that the superscript 1 is not essential when there is only one electron in an orbital. Also note that in the electronic configuration of boron, it is not necessary to specify whether the $2p$ electron is in the $2p_x$, $2p_y$, or $2p_z$ orbital. Since all three orbitals correspond to the same energy, they are indistinguishable.

By following Table 6.2 and a listing of the elements by atomic number, you can draw up your own list showing how orbitals are filled as the number of electrons in an atom increases. The number of electrons that can have any given principal quantum number n is limited. As Table 6.2 shows, the maximum number of electrons of any given quantum number n is $2n^2$. For example, there can be as many as $2(3)^2 = 18$ electrons of $n = 3$. We expect that there can be as many as $2(4)^2 = 32$ electrons of principal quantum number $n = 4$ in a given atom.

TABLE 6.2
The Electronic Distribution in Allowed Orbitals from $1s$ to $3d$

Principal Quantum Number n	ℓ	m_ℓ	m_s	Orbital Designations	Total Electrons
1	0	0	$+\frac{1}{2}, -\frac{1}{2}$	$1s$	2
2	0	0	$+\frac{1}{2}, -\frac{1}{2}$	$2s$	
	1	-1	$+\frac{1}{2}, -\frac{1}{2}$	$2p_x$	8
		0	$+\frac{1}{2}, -\frac{1}{2}$	$2p_y$	
		$+1$	$+\frac{1}{2}, -\frac{1}{2}$	$2p_z$	
3	0	0	$+\frac{1}{2}, -\frac{1}{2}$	$3s$	
	1	-1	$+\frac{1}{2}, -\frac{1}{2}$	$3p_x$	
		0	$+\frac{1}{2}, -\frac{1}{2}$	$3p_y$	
		$+1$	$+\frac{1}{2}, -\frac{1}{2}$	$3p_z$	
	2	-2	$+\frac{1}{2}, -\frac{1}{2}$	$3d_{xy}$	18
		-1	$+\frac{1}{2}, -\frac{1}{2}$	$3d_{xz}$	
		0	$+\frac{1}{2}, -\frac{1}{2}$	$3d_{yz}$	
		$+1$	$+\frac{1}{2}, -\frac{1}{2}$	$3d_{z^2}$	
		$+2$	$+\frac{1}{2}, -\frac{1}{2}$	$3d_{x^2-y^2}$	

The order in which orbitals fill can be predicted accurately for elements of relatively low atomic number. For the lower-energy orbitals, $1s$, $2s$, $2p$, $3s$, and $3p$, the relative energies of orbitals (and therefore the order in which they fill) parallel those predicted from the hydrogen atom, with the added consideration that for any given n, a higher value of ℓ leads to a slightly higher energy. It should be noted that the difference in energy between a $1s$ orbital and a $2s$ orbital, which have different principal quantum numbers, is much greater than the difference between a $2s$ and a $2p$ orbital, which have the same principal quantum number, as shown in Figure 6.6.

There are complications once we get to element 19, potassium, the first of the elements with more electrons than can be accommodated in the $1s$ through $3p$ orbitals. The relative energies of the higher orbitals depend on both the number of electrons in the atom and the atom's nuclear charge. We cannot

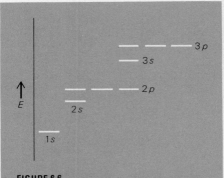

FIGURE 6.6
Qualitative energy differences between orbitals for a multielectron atom. There is a greater gap between the energy level of a $1s$ orbital and a $2s$ orbital than between the $2s$ and $2p$ orbitals or the $3s$ and $3p$ orbitals. In this group energy differences are greatest between orbitals with different principal quantum numbers.

predict the electronic configuration of the heavier elements on the basis of simple theory alone. Instead, electronic configurations must be assigned in a way that is consistent with experimental data, including atomic spectra and X-ray spectra, as well as the magnetic and chemical properties of the element.

To help work out electronic configurations, it is convenient to group orbitals into *shells* and *subshells*. There are two ways to describe the division of orbitals into shells. One way is favored by physicists. The other is useful to chemists. On the basis of X-ray spectra, the physicist uses the letter designation of shells shown in Table 6.3. Each value of the principal quantum number n is associated with a shell designation. For $n = 1$, the shell is designated K. For $n = 2$, the shell is designated L. For $n = 3$, the shell is designated M, and so on. The K shell contains only the $1s$ orbital, and helium ($1s^2$) is said to have a completed K shell. The L shell contains the $2s$ and the three $2p$ orbitals. Neon, element 10, has the electronic configuration $1s^22s^22p^6$ and thus is said to have a completed L shell. (Note that for convenience we group the three filled $2p$ orbitals as $2p^6$, rather than listing each orbital separately.)

Table 6.3 shows that each shell after K contains one more type of orbital. The idea that each shell consists of subshells was used in Table 6.2. Each subshell includes all the orbitals that have the same value of n and ℓ. The L shell has two subshells: $2s$, which can hold 2 electrons, and $2p$, which can hold 6 electrons. The M shell adds the $3d$ subshell, which can hold up to 10 electrons. The N shell adds the $4f$ subshell, which can hold up to 14 electrons, and so on. As we shall see, subshells are useful in working out the electronic configuration of the heavier atoms. In general, each lower subshell is filled before electrons begin filling the orbitals in the next higher subshell. We already have examples of this rule. In the beryllium and boron atoms, the $2s$ subshell must be filled before electrons are assigned to the $2p$ subshell.

The chemist's description of electronic configuration is more closely related to the chemical properties of the elements. Table 6.4 shows the chemist's description. Each shell in this description is named for a noble gas, and each shell contains the orbitals that are filled in the designated noble gas atom. The helium shell of the chemist corresponds to the K shell of the physicist; each contains the $1s$ orbital. The neon shell of the chemist corresponds to the L shell of the physicist; each contains the $2s$ and $2p$ orbitals that are filled in neon ($1s^22s^22p^6$). However, the argon shell of the chemist is not the same as the M shell of the physicist. The argon shell includes only the $3s$ and the three $3p$ orbitals, which are filled in the argon atom ($1s^22s^22p^63s^23p^6$). The $3d$ orbitals that are in the M shell of the physicist are found in the krypton shell of the chemist.

The chemist's description is related more closely to the arrangement of elements in the periodic table. It is also closer to the order in which orbitals are actually filled with increasing atomic number. One noble gas shell is filled before electrons start filling the next noble gas shell.

The idea of noble gas shells is helpful in working out the electronic configu-

TABLE 6.3
The Physicist's Shells

Shell	Orbitals or Subshells
K	$1s$
L	$2s$, $2p$
M	$3s$, $3p$, $3d$
N	$4s$, $4p$, $4d$, $4f$
O	$5s$, $5p$, $5d$, $5f$, $5g$
P	$6s$, $6p$, $6d$, $6f$, $6g$, $6h$

TABLE 6.4
Noble Gas Shells

Chemist's Name	Subshells	Maximum Number of Electrons in Shell
helium	$1s$	2
neon	$2s$, $2p$	8
argon	$3s$, $3p$	8
krypton	$3d$, $4s$, $4p$	18
xenon	$4d$, $5s$, $5p$	18
radon	$4f$, $5d$, $6s$, $6p$	32
eka-radon	$5f$, $6d$, $7s$, $7p$	32

ration of some of the elements, and in understanding some aspects of their chemical and physical behavior. However, to formulate the electronic configuration of the ground state of elements whose atomic number is greater than that of argon, element 18, we must know the order in which the subshells within a given noble gas shell are filled.

As we mentioned, we can use the hydrogen atom as a model to predict the order in which subshells are filled until the $3p$ subshell is completed. Thus, we begin filling the $2p$ subshell at boron, after the $2s$ subshell is completed. Each element from boron, atomic number 5, to neon, atomic number 10, adds another electron to the $2p$ subshell, which is filled at neon, whose configuration is $1s^2 2s^2 2p^6$. The $3s$ orbital begins filling with sodium, atomic number 11, whose electronic configuration is $1s^2 2s^2 2p^6 3s$.

There is an easier way to symbolize the electronic configuration of sodium. The method is based on the fact that the electronic configuration of sodium is simply the electronic configuration of neon with one additional electron, which is in the next noble gas shell. We say that the electronic configuration of sodium consists of a neon core, represented as [Ne], with one $3s$ electron, so we can write it as [Ne]$3s$. Similarly, the electronic configuration of magnesium, atomic number 12, is [Ne]$3s^2$.

When the $3s$ orbital is filled, the $3p$ subshell begins filling at aluminum, atomic number 13, whose electronic configuration is [Ne]$3s^2 3p$. The $3p$ subshell continues filling until the noble gas shell is completed at argon, element 18, which has the noble gas configuration [Ne]$3s^2 3p^6$. This electronic configuration is called the argon core, and is represented as [Ar].

When the argon shell is completed, the krypton shell begins to fill. In potassium, atomic number 19, the ground state electronic configuration is [Ar]$4s$. For calcium, atomic number 20, the ground state electronic configuration is found to be [Ar]$4s^2$. In these elements, the $4s$ subshell fills before the $3d$ subshell, something that is not predicted by the hydrogenlike approximation.

Table 6.5 relates the physicist's shells to the chemist's shells, and gives the usual order in which subshells are filled. The order of filling is indicated by the diagonal lines. But the rules are not always followed. Experimental data are needed to be sure of the exact electronic configuration of the heavier elements.

A problem arises when we attempt to write the correct electronic configuration of carbon, atomic number 6. The first five electrons of the carbon atom are readily assigned to the $1s$, $2s$, and one of the $2p$ orbitals, say the $2p_x$ orbital. Where should the sixth electron be placed? We know that it must go into a $2p$ orbital, since the neon shell must be completed before the $3s$ orbital in the argon shell begins to fill. But should the electron go into the $2p_x$ orbital, pairing with the

TABLE 6.5
Filling Order of Subshells[a]

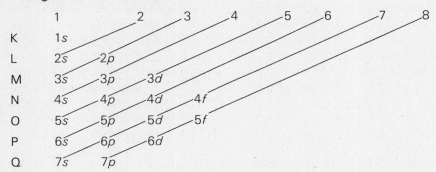

[a] The numerals at the top indicate the sequence that is followed in filling subshells. Subshells on diagonal 2 are filled before those on diagonal 3, which are filled before those on diagonal 4, and so on.

electron that is already there? Or should it go into an unoccupied $2p$ orbital, say $2p_y$, giving two half-filled $2p$ orbitals?

The latter choice is correct. There is a rule, called *Hund's rule*, that tells us that electrons go into unoccupied orbitals of a subshell whenever possible, rather than pairing with electrons in a partially occupied orbital. In more amplified form, Hund's rule says:

1. When electrons are filling a subshell, the lowest energy state is obtained if the maximum possible number of different orbitals in the subshell are used. By placing electrons in different orbitals, the unfavorable interactions between the like-charged electrons are minimized. In a carbon atom, this rule means that in the ground state there are two half-filled $2p$ orbitals, rather than one completely filled $2p$ orbital.

2. When there are two or more electrons in half-filled orbitals, the most stable state, the state of lowest energy, is the one in which the spins of all the electrons are parallel.

Relative spins of electrons can be designated by arrows. Arrows pointing in the same direction indicate parallel spins. Arrows pointing in opposite directions indicate paired spins. The ground state electronic configuration of carbon is represented as $1s^22s^22p_x{\uparrow}2p_y{\uparrow}$. This notation helps distinguish the ground state of carbon from a higher electronic energy state, in which the electron spins are paired, such as $1s^22s^22p_x{\uparrow}2p_y{\downarrow}$.

We can write the electronic configuration of the ground state of nitrogen, atomic number 7, to indicate the spins of the $2p$ electrons: $1s^22s^22p_x{\uparrow}2p_y{\uparrow}2p_z{\uparrow}$. Since Hund's rule is understood to apply, this usually is simplified: $1s^22s^22p^3$. Similarly, the ground state configuration of oxygen, atomic number 8, is $1s^22s^22p_x{}^22p_y{\uparrow}2p_z{\uparrow}$, or $1s^22s^22p^4$.

Figure 6.7 diagrams the ground state electronic configuration of these elements, showing electronic spins.

Certain electronic configurations tend to be especially stable. In one such stable configuration, a subshell is half-filled. Each orbital of the subshell holds one electron. All the electrons in the subshell have parallel spins. Another especially stable electronic configuration is the one in which a subshell is completely filled.

Table 6.6 lists the ground state electronic configurations of all the elements. Generally, the electronic configurations are those predicted by simple theory. But there are exceptions, some of which reflect the stability associated with half-filled and filled subshells.

For example, the electronic configuration of vanadium, element 23, is, as expected $1s^22s^22p^63s^23p^64s^23d^3$ or $[Ar]4s^23d^3$, with the $4s$ subshell filling before the $3d$ subshell. But in chromium, element 24, the configuration is $[Ar]4s3d^5$. The $4s$ and the $3d$ subshells are both half-filled, showing the extra stability associated with a half-filled subshell. The ground state configuration of copper, element 29, is $[Ar]4s3d^{10}$, another illustration of the same effect. One filled subshell and a half-filled subshell is a relatively stable arrangement.

In other words, the set of orbitals derived from the model of the hydrogen atom should not be considered a set of pigeonholes that are filled in precise sequence according to simple rules. The relative energy of orbitals, and therefore the order in which they fill, is altered by the magnitude of the nuclear charge and by the presence of the other electrons in the atom. Their distribution can change the relative energy of the orbitals. Therefore, the assignment of electrons to orbitals in some atoms is far from simple. It should also be noted that our discussion applies only to unbound atoms. For atoms in molecules, the situation is different.

The difficulty of assigning electrons to orbitals is shown not only by the irregularities that are evident in Table 6.6 but also in the study of ions, electrically charged species in which an atom has lost or gained one or more electrons, so that the number of electrons does not equal the nuclear charge. Table 6.6 shows

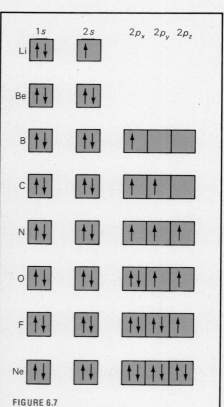

FIGURE 6.7
Ground state electronic configuration of the elements from lithium to neon showing electronic spins.

TABLE 6.6
Electronic Configuration of the Elements[a]

Atomic No.	Element	1s	2s	2p	3s	3p	3d	4s	4p	4d	4f	5s	5p	5d	5f	6s	6p	6d	7s
1	H	1																	
2	He	2																	
3	Li	2	1																
4	Be	2	2																
5	B	2	2	1															
6	C	2	2	2															
7	N	2	2	3															
8	O	2	2	4															
9	F	2	2	5															
10	Ne	2	2	6															
11	Na	2	2	6	1														
12	Mg	2	2	6	2														
13	Al	2	2	6	2	1													
14	Si	2	2	6	2	2													
15	P	2	2	6	2	3													
16	S	2	2	6	2	4													
17	Cl	2	2	6	2	5													
18	Ar	2	2	6	2	6													
19	K	2	2	6	2	6		1											
20	Ca	2	2	6	2	6		2											
21	Sc	2	2	6	2	6	1	2											
22	Ti	2	2	6	2	6	2	2											
23	V	2	2	6	2	6	3	2											
24	Cr	2	2	6	2	6	5[b]	1											
25	Mn	2	2	6	2	6	5	2											
26	Fe	2	2	6	2	6	6	2											
27	Co	2	2	6	2	6	7	2											
28	Ni	2	2	6	2	6	8	2											
29	Cu	2	2	6	2	6	10[b]	1											
30	Zn	2	2	6	2	6	10	2											
31	Ga	2	2	6	2	6	10	2	1										
32	Ge	2	2	6	2	6	10	2	2										
33	As	2	2	6	2	6	10	2	3										
34	Se	2	2	6	2	6	10	2	4										
35	Br	2	2	6	2	6	10	2	5										
36	Kr	2	2	6	2	6	10	2	6										
37	Rb	2	2	6	2	6	10	2	6			1							
38	Sr	2	2	6	2	6	10	2	6			2							
39	Y	2	2	6	2	6	10	2	6	1		2							
40	Zr	2	2	6	2	6	10	2	6	2		2							
41	Nb	2	2	6	2	6	10	2	6	4[b]		1							
42	Mo	2	2	6	2	6	10	2	6	5		1							
43	Tc	2	2	6	2	6	10	2	6	6		1							
44	Ru	2	2	6	2	6	10	2	6	7		1							
45	Rh	2	2	6	2	6	10	2	6	8		1							
46	Pd	2	2	6	2	6	10	2	6	10[b]									
47	Ag	2	2	6	2	6	10	2	6	10		1							
48	Cd	2	2	6	2	6	10	2	6	10		2							
49	In	2	2	6	2	6	10	2	6	10		2	1						
50	Sn	2	2	6	2	6	10	2	6	10		2	2						
51	Sb	2	2	6	2	6	10	2	6	10		2	3						
52	Te	2	2	6	2	6	10	2	6	10		2	4						
53	I	2	2	6	2	6	10	2	6	10		2	5						
54	Xe	2	2	6	2	6	10	2	6	10		2	6						
55	Cs	2	2	6	2	6	10	2	6	10		2	6			1			
56	Ba	2	2	6	2	6	10	2	6	10		2	6			2			
57	La	2	2	6	2	6	10	2	6	10		2	6	1		2			
58	Ce	2	2	6	2	6	10	2	6	10	2[b]	2	6			2			
59	Pr	2	2	6	2	6	10	2	6	10	3	2	6			2			
60	Nd	2	2	6	2	6	10	2	6	10	4	2	6			2			
61	Pm	2	2	6	2	6	10	2	6	10	5	2	6			2			
62	Sm	2	2	6	2	6	10	2	6	10	6	2	6			2			
63	Eu	2	2	6	2	6	10	2	6	10	7	2	6			2			
64	Gd	2	2	6	2	6	10	2	6	10	7[b]	2	6	1		2			
65	Tb	2	2	6	2	6	10	2	6	10	9	2	6			2			
66	Dy	2	2	6	2	6	10	2	6	10	10	2	6			2			
67	Ho	2	2	6	2	6	10	2	6	10	11	2	6			2			
68	Er	2	2	6	2	6	10	2	6	10	12	2	6			2			
69	Tm	2	2	6	2	6	10	2	6	10	13	2	6			2			
70	Yb	2	2	6	2	6	10	2	6	10	14	2	6			2			
71	Lu	2	2	6	2	6	10	2	6	10	14	2	6	1		2			
72	Hf	2	2	6	2	6	10	2	6	10	14	2	6	2		2			
73	Ta	2	2	6	2	6	10	2	6	10	14	2	6	3		2			
74	W	2	2	6	2	6	10	2	6	10	14	2	6	4		2			
75	Re	2	2	6	2	6	10	2	6	10	14	2	6	5		2			
76	Os	2	2	6	2	6	10	2	6	10	14	2	6	6		2			
77	Ir	2	2	6	2	6	10	2	6	10	14	2	6	7		2			
78	Pt	2	2	6	2	6	10	2	6	10	14	2	6	9		1			
79	Au	2	2	6	2	6	10	2	6	10	14	2	6	10		1			
80	Hg	2	2	6	2	6	10	2	6	10	14	2	6	10		2			
81	Tl	2	2	6	2	6	10	2	6	10	14	2	6	10		2	1		
82	Pb	2	2	6	2	6	10	2	6	10	14	2	6	10		2	2		
83	Bi	2	2	6	2	6	10	2	6	10	14	2	6	10		2	3		
84	Po	2	2	6	2	6	10	2	6	10	14	2	6	10		2	4		
85	At	2	2	6	2	6	10	2	6	10	14	2	6	10		2	5		
86	Rn	2	2	6	2	6	10	2	6	10	14	2	6	10		2	6		
87	Fr	2	2	6	2	6	10	2	6	10	14	2	6	10		2	6		1
88	Ra	2	2	6	2	6	10	2	6	10	14	2	6	10		2	6		2
89	Ac	2	2	6	2	6	10	2	6	10	14	2	6	10		2	6	1	2
90	Th	2	2	6	2	6	10	2	6	10	14	2	6	10		2	6	2	2
91	Pa	2	2	6	2	6	10	2	6	10	14	2	6	10	2[b]	2	6	1	2
92	U	2	2	6	2	6	10	2	6	10	14	2	6	10	3	2	6	1	2
93	Np	2	2	6	2	6	10	2	6	10	14	2	6	10	4	2	6	1	2
94	Pu	2	2	6	2	6	10	2	6	10	14	2	6	10	6[b]	2	6		2
95	Am	2	2	6	2	6	10	2	6	10	14	2	6	10	7	2	6		2
96	Cm	2	2	6	2	6	10	2	6	10	14	2	6	10	7	2	6	1	2
97	Bk	2	2	6	2	6	10	2	6	10	14	2	6	10	9[b]	2	6		2
98	Cf	2	2	6	2	6	10	2	6	10	14	2	6	10	10	2	6		2
99	Es	2	2	6	2	6	10	2	6	10	14	2	6	10	11	2	6		2
100	Fm	2	2	6	2	6	10	2	6	10	14	2	6	10	12	2	6		2
101	Md	2	2	6	2	6	10	2	6	10	14	2	6	10	13	2	6		2
102	No	2	2	6	2	6	10	2	6	10	14	2	6	10	14	2	6		2
103	Lr	2	2	6	2	6	10	2	6	10	14	2	6	10	14	2	6	1	2
104	Rf	2	2	6	2	6	10	2	6	10	14	2	6	10	14	2	6	2	2

[a] Electrons outside a completed noble gas shell are shown in color.
[b] Does not follow the order predicted by Table 6.5.

that the $4s$ subshell begins filling before the $3d$ subshell. In the *first transition series,* the elements from scandium (atomic number 21) to copper (atomic number 29), the $3d$ subshell is being filled. There is considerable evidence which indicates that the energy of the $3d$ subshell relative to the $4s$ subshell becomes progressively lower with increasing atomic number in this series of elements. Because of this phenomenon, which is caused by the increased charge on the nucleus of each succeeding element, a $4s$ electron can be removed more easily from these atoms than a $3d$ electron. In these elements, cations—species that have fewer electrons than the neutral atom—have ground state electronic configurations in which the $3d$ subshell fills before the $4s$ subshell. In general, when an ion is formed, electrons with higher principal quantum number n are removed first.

For example, if vanadium, $[Ar]4s^23d^3$, loses two electrons, the dipositive cation V^{2+} has the configuration $[Ar]3d^3$. The electrons have been lost from the $4s$ subshell. In the zinc atom, $[Ar]4s^23d^{10}$, further along in the series, the $3d$ subshell is much lower in energy than the $4s$ subshell. The cation Zn^{2+}, which has a completed $3d$ subshell and the configuration $[Ar]3d^{10}$, forms rather easily. But it is exceedingly difficult to remove more than two electrons from the Zn atom. Thus, it is not surprising that the $+2$ oxidation state of zinc is most important chemically. More evidence for the lower relative energy of the $3d$ subshell in this series of elements can be gained by examining the V^+ cation, $[Ar]3d^4$. When the V^+ cation is formed, one $4s$ electron is removed from the atom. The other $4s$ electron no longer is in the $4s$ subshell. It is in the $3d$ subshell in V^+.

6.3 ELECTRONS IN MULTIELECTRON ATOMS

A formulation of the ground state electronic configurations of the elements can be a useful tool for correlating, interpreting, and rationalizing many of their observed chemical and physical properties. Before we try to relate electronic configurations to these properties, it will be helpful to discuss in some detail the way in which electrons in hydrogenlike orbitals are influenced by the presence of other electrons and by the nuclear charge in multielectron atoms.

So far, we have used a graphical representation of an orbital, in which a surface of constant probability is plotted. Arbitrarily, the surface is drawn to enclose a region of space in which the electron is found at least 90% of the time. This way of representing an orbital gives us useful information about the size and shape of this region of space.

If we are interested in the details of the probability distribution of the electron in an orbital, we can use a different kind of representation. The probability of finding an electron that is in a given orbital at a given distance from the nucleus can be found from the wave function associated with the orbital. The results of such a calculation can be expressed by plotting what is called the *radial probability function* against the distance from the nucleus. Each point on the plot gives the probability of finding the electron in a very thin spherical shell at a distance r from the nucleus, which is at the center of the spherical shell.

Figure 6.8 shows the curves for the radial probability functions of the $1s$, $2s$, $2p$, $3s$, $3p$, and $3d$ orbitals. These curves reveal two important aspects of probability distributions within different orbitals.

First, the region of maximum probability moves away from the nucleus as the principal quantum number increases. Since the maximum of the curve is the most probable distance of the electron from the nucleus, the distance of the electron from the nucleus increases as the principal quantum number increases. We have already noted (Section 5.7, page 137) that the energy of the electron increases with its increasing distance from the nucleus and with an increase in principal quantum number.

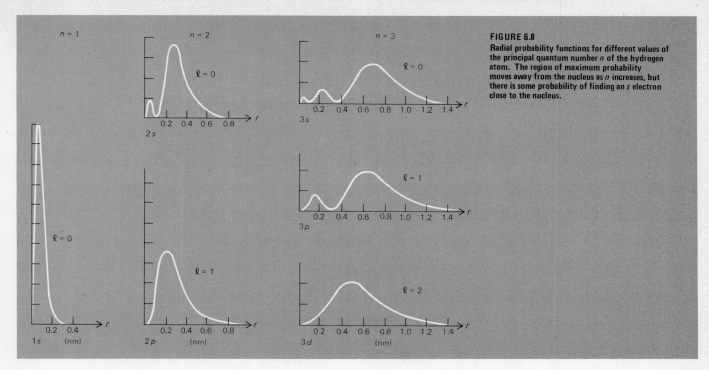

FIGURE 6.8
Radial probability functions for different values of
the principal quantum number *n* of the hydrogen
atom. The region of maximum probability
moves away from the nucleus as *n* increases, but
there is some probability of finding an *s* electron
close to the nucleus.

Second, Figure 6.8 also shows that there is a significant probability of finding an *s* electron in a region close to the nucleus, even if the *s* electron has a high principal quantum number. There is a lesser (but definite) probability of finding a *p* electron close to the nucleus, as Figure 6.8 shows. But it is much less probable that electrons in orbitals with higher values of ℓ, such as those in *d* orbitals ($\ell = 2$) or *f* orbitals ($\ell = 3$), are close to the nucleus. Thus, *s* orbitals and, to a lesser extent, *p* orbitals can be designated as "penetrating orbitals," while *d* and *f* orbitals can be designated as "nonpenetrating orbitals." This distinction is important in accounting for some properties of multielectron atoms.

Therefore, there are two important characteristics of orbitals that are revealed by the radial probability functions. A higher principal quantum number means a greater average distance of the electron from the nucleus. And a higher angular momentum quantum number means a lower probability of the electron ever being close to the nucleus. These two characteristics can help us understand how nuclear charge and the presence of other electrons affect relative electronic energies in multielectron atoms.

The total energy of the electron in a hydrogen atom is lower (more negative) than the energy of a free electron because of the coulombic attraction between the negatively charged electron and the positively charged nucleus. To study the effect of an increased nuclear charge on electronic energy, we can start by considering an atomic species that has only one electron and has a nuclear charge (more simply, an atomic number) designated as *Z*. At any given distance from the nucleus, the total energy of the single electron will be more negative than the energy of the electron in a hydrogen atom. An exact relationship for the energy of an electron in such an atomic species can be obtained by a suitable modification of Equation 6.1:

$$E = \frac{-2\pi^2 m e^4 Z^2}{n^2 h^2} \tag{6.2}$$

where *Z* is the nuclear charge (or the atomic number) of the one-electron species.

It is logical that the value of *E* depends on the value of *Z*. A larger nuclear charge *Z* means a greater coulombic attraction between the nucleus and the electron, and thus a lower value of E. While Equation 6.2 appears complicated, most of its terms are physical constants. Its most important feature is simply that the numerical value of the energy is directly proportional to the square of the nuclear charge and inversely proportional to the square of the principal quantum number:

$$E \propto \frac{-Z^2}{n^2} \qquad (6.3)$$

Suppose we want to extend this relationship to a multielectron atom. Generally, the atomic number Z is much higher than the highest principal quantum number n of any electron in the ground state of the atom. In cesium, for example, $Z = 55$ and $n = 6$. It would seem that every electron in the cesium atom would be of lower energy than the electron in the hydrogen atom. But this is not so. To understand why, we must consider the effect of the other electrons on the interaction between any given electron and the nucleus in a multielectron atom.

The radial probability functions in Figure 6.8 indicate that electrons with higher principal quantum numbers are further from the nucleus on the average than those with lower principal quantum numbers. In the case of a three-electron atom, the $1s$ electrons tend to lie between the nucleus and the $2s$ electron. Therefore, the negatively charged $1s$ electrons are said to screen, or shield, the $2s$ electrons from the coulombic attraction of the nucleus.

Figure 6.9 shows this screening effect schematically for a lithium atom. The nucleus, with a charge of $+3$, is surrounded by a sphere of two electrons, with a total charge of -2. The $2s$ electron is outside this sphere, and thus feels the attraction of an effective nuclear charge just slightly larger than $+1$. Therefore, the total energy of the $2s$ electron is not nearly as negative as is expected from the value of $Z^2/n^2 (3^2/2^2)$.

You might expect that the effective nuclear charge felt by the $2s$ electron would be exactly $+1$, the $+3$ nuclear charge minus the -2 charge of the $1s$ electrons. The radial probability distributions of Figure 6.8 show why the effective nuclear charge is not exactly $+1$. The $2s$ orbital is a penetrating orbital. For some part of the time, the electron in the $2s$ orbital will be close to the nucleus, penetrating the shield set up by the $1s$ electrons (Figure 6.10). The electron in the $2s$ orbital is not fully screened by the $1s$ electrons and will feel an effective nuclear charge that is somewhat larger than $+1$.

The phenomenon of screening is even more noticeable for electrons in p orbitals, since these orbitals are less penetrating than the s orbitals. The penetration of d orbitals and f orbitals is negligible, so the screening of them by electrons in orbitals nearer the nucleus is almost completely effective. The relative energy of an electron of a given principal quantum number becomes less negative as the effectiveness with which the other electrons shield it from the nuclear charge increases. It has also been found that electrons in penetrating orbitals are better at screening the nucleus than electrons in nonpenetrating orbitals. For a given value of n, electrons in s orbitals screen best. Electrons in p orbitals are less effective and electrons in d orbitals are still less effective.

Differences in penetration explain why the energy of electrons depends on both the value of ℓ and the value of n in multielectron atoms. We said that the energy of the electron in the hydrogen atom depends only on the value of n. The hydrogen atom has only one electron, so there is no screening. However, in other atoms, screening and penetration must be considered, so electronic energy depends on both n and ℓ.

The chemical behavior of the elements depends primarily on the behavior of the outermost, or **valence electrons.** *The valence electrons generally are those which lie outside the highest completed noble gas shell in the ground state electronic configuration of the atom.* Any description of the behavior of valence

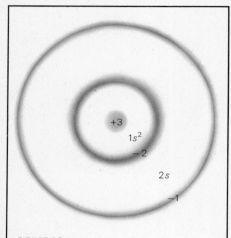

FIGURE 6.9
Electron shielding shown schematically for the lithium atom. The outermost $2s$ electron is shielded from the attractive force of the nucleus by the two $1s$ electrons, making the effective nuclear charge only slightly larger than $+1$.

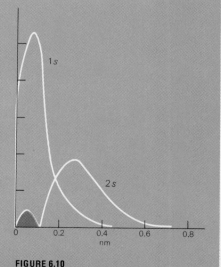

FIGURE 6.10
The effect of penetration on the effective nuclear charge of the lithium atom. Because the $2s$ is a penetrating orbital, as shown by the radial probability distribution curve, the outermost electron is not fully shielded by the two $1s$ electrons, making the effective nuclear charge larger than $+1$.

electrons must take into account not only the actual nuclear charge but also the effective nuclear charge that the valence electrons feel through the screen of the inner electrons.

6.4 IONIZATION ENERGY

One of the most fundamental properties of an atom is its **ionization energy**, sometimes called ionization potential. *Ionization energy is the energy required to remove the electron with the highest total energy from the ground state of an atom in the gas phase*. We have seen that electrons are held in an atom by coulombic attraction with the nucleus. The ionization energy is a positive quantity, and it provides a direct measure of the energy of the electron. The greater the ionization energy (the energy required to remove the electron from the atom), the greater is the coulombic attraction between the electron and the nucleus, and the lower the energy of the electron.

Ionization energies can be expressed in electron volts (eV), which can be converted to kilojoules per mole (kJ/mol): 1 eV = 96.5 kJ/mol.

Table 6.7 shows ionization energies for some elements. Note that data are available on the energy required to remove a second electron from an atom which has lost one electron, or a third electron from an atom which has lost two electrons, and so on. These energies are called the second ionization energy, the third ionization energy, etc. The first ionization energy always refers to the electron that is most easily removed from the intact atom. The second ionization energy refers to the electron most easily removed from the monopositive cation. The third ionization energy refers to the electron most easily removed from the dipositive cation:

$$A(g) + energy \longrightarrow A^+(g) + e^- \qquad \text{(first ionization energy)}$$
$$A^+(g) + energy \longrightarrow A^{2+}(g) + e^- \qquad \text{(second ionization energy)}$$
$$A^{2+}(g) + energy \longrightarrow A^{3+}(g) + e^- \qquad \text{(third ionization energy)}$$

The measurement of ionization energy is a valuable tool for determining the electronic structure of atoms. An electron can be removed from an atom by bombarding the atom with a beam of photons or free electrons. The energy of these photons or free electrons can be measured with accuracy. A common technique for measuring ionization energies is to find the minimum energy of a photon or free electron required to change a neutral atom into a cation. This minimum energy is the ionization energy. Ionization energies are always measured on single atoms in the gas phase.

Table 6.7 shows that the first ionization energy varies in a predictable way as we move down the list of elements by atomic number. First, there is a steady rise, then a sudden drop followed by another steady rise and a sudden drop, again and again. Ionization energy is just one example of the many periodic properties of the elements. The way in which chemical properties vary periodically with atomic number is made evident by the periodic table of the elements, which provides an invaluable aid to the study of chemistry.

Figure 6.11 shows the periodicity of ionization energies graphically. If first ionization energies are plotted as a function of atomic number, we get a pattern of fairly regular rises and falls. The first ionization energy is lowest for the group IA metals, which begin each horizontal row of the periodic table. The energy generally increases from left to right across a row, reaching a maximum for the noble gases at the far right of the row. As we go from the noble gas at the right of one row to the group IA metal at the left of the next row down, there is a noticeable drop in first ionization energy.

In addition to this periodicity, there is a general tendency for first ionization energy to decrease with increasing atomic number in any given column of the

TABLE 6.7
Ionization Energies (eV)^a

Z	Element	I	II	III	IV	V	VI	VII	VIII	IX	X
1	H	13.598									
2	He	24.587	54.416								
3	Li	5.392	75.638	122.451							
4	Be	9.322	18.211	153.893	217.713						
5	B	8.298	25.154	37.930	259.368	340.217					
6	C	11.260	24.383	47.887	64.492	392.077	489.981				
7	N	14.534	29.601	47.448	77.472	97.888	552.057	667.029			
8	O	13.618	35.116	54.934	77.412	113.896	138.116	739.315	871.387		
9	F	17.422	34.970	62.707	87.138	114.240	157.161	185.182	953.886	1103.089	
10	Ne	21.564	40.962	63.45	97.11	126.21	157.93	207.27	239.09	1195.797	1362.164
11	Na	5.139	47.286	71.64	98.91						
12	Mg	7.646	15.035	80.143	109.24						
13	Al	5.986	18.828	28.447	119.99						
14	Si	8.151	16.345	33.492	45.141						
15	P	10.486	19.725	30.18	51.37						
16	S	10.360	23.33	34.83	47.30						
17	Cl	12.967	23.81	39.61	53.46						
18	Ar	15.759	27.629	40.74	59.81						
19	K	4.341	31.625	45.72	60.91						
20	Ca	6.113	11.871	50.908	67.10						
21	Sc	6.54	12.80	24.76	73.47						
22	Ti	6.82	13.58	27.491	43.266						
23	V	6.74	14.65	29.310	46.707						
24	Cr	6.766	15.50	30.96	49.1						
25	Mn	7.435	15.640	33.667	51.2						
26	Fe	7.870	16.18	30.651	54.8						
27	Co	7.86	17.06	33.50	51.3						
28	Ni	7.635	18.168	35.17	54.9						
29	Cu	7.726	20.292	36.83	55.2						
30	Zn	9.394	17.964	39.722	59.4						
31	Ga	5.999	20.51	30.71	64						
32	Ge	7.899	15.934	34.22	45.71						
33	As	9.81	18.633	28.351	50.13						
34	Se	9.752	21.19	30.820	42.944						
35	Br	11.814	21.8	36	47.3						
36	Kr	13.999	24.359	36.95	52.5						
37	Rb	4.177	27.28	40	52.6						
38	Sr	5.695	11.030	43.6	57						
39	Y	6.38	12.24	20.52	61.8						
40	Zr	6.84	13.13	22.99	34.34						
41	Nb	6.88	14.32	25.04	38.3						
42	Mo	7.099	16.15	27.16	46.4						
43	Tc	7.28	15.26	29.54							
44	Ru	7.37	16.76	28.47							
45	Rh	7.46	18.08	31.06							
46	Pd	8.34	19.43	32.93							
47	Ag	7.576	21.49	34.83							
48	Cd	8.993	16.908	37.48							
49	In	5.786	18.869	28.03	54						
50	Sn	7.344	14.632	30.502	40.734						
51	Sb	8.641	16.53	25.3	44.2						
52	Te	9.009	18.6	27.96	37.41						
53	I	10.451	19.131	33							
54	Xe	12.130	21.21	32.1							
55	Cs	3.894	23.1								
56	Ba	5.212	10.004								
57	La	5.577	11.06	19.175							
58	Ce	5.47	10.85	20.20	36.72						
59	Pr	5.42	10.55	21.62	38.95						
60	Nd	5.49	10.72								
61	Pm	5.55	10.90								
62	Sm	5.63	11.07								
63	Eu	5.67	11.25								
64	Gd	6.13	12.1								
65	Tb	5.85	11.52								
66	Dy	5.93	11.67								
67	Ho	6.02	11.80								
68	Er	6.10	11.93								
69	Tm	6.18	12.05	23.71							
70	Yb	6.254	12.17	25.2							
71	Lu	5.426	13.9								
72	Hf	7.0	14.9	23.3	33.3						
73	Ta	7.89									
74	W	7.98									
75	Re	7.88									
76	Os	8.7									
77	Ir	9.1									
78	Pt	9.0	18.563								
79	Au	9.225	20.5								

TABLE 6.7 (continued)

Z	Element	I	II	III	IV	V	VI	VII	VIII	IX	X
80	Hg	10.437	18.756	34.2							
81	Tl	6.108	20.428	29.83							
82	Pb	7.416	15.032	31.937	42.32						
83	Bi	7.289	16.69	25.56	45.3						
84	Po	8.42									
85	At										
86	Rn	10.748									
87	Fr										
88	Ra	5.279	10.147								
89	Ac	6.9	12.1								
90	Th		11.5	20.0	28.8						
91	Pa										
92	U										
93	Np										
94	Pu	5.8									
95	Am	6.0									

[a] From C. E. Moore, "Ionization Potentials and Ionization Limits Derived from the Analyses of Optical Spectra," NSRDS-NBS 34, National Bureau of Standards, Washington, D.C., 1970.

periodic table. As atomic number increases in a given column, first ionization energy tends to decrease, although there are some exceptions to this general trend.

The seeming complexity of the ionization energy data can be clarified when we see how it is related to electronic configuration. The ionization energy of a hydrogen atom is 13.6 eV. Recalling that $E \propto Z^2/n^2$, we would predict that the first ionization energy of the helium atom ($Z = 2$) will be $2^2 = 4$ times greater than that of the hydrogen atom. But the first ionization energy of the helium atom is found to be 24.6 eV, about twice as large as that of the hydrogen atom. In the helium atom, the repulsion between the two electrons causes the first ionization energy to be lower than theory predicts. But the second ionization energy of helium is 54.4 eV, four times larger than the ionization energy of hydrogen, as the theory predicts. When one of the electrons of the helium atom is removed, the repulsion no longer exists and the ionization energy increases considerably.

The first ionization energy of lithium, atomic number 3, is lower than that of hydrogen, due to the screening effect of the 1s electrons. The effective nuclear charge felt by the 2s electron is not +3, but is closer to +1, and the 2s electron, which is further from the nucleus than the 1s electron of hydrogen, is removed

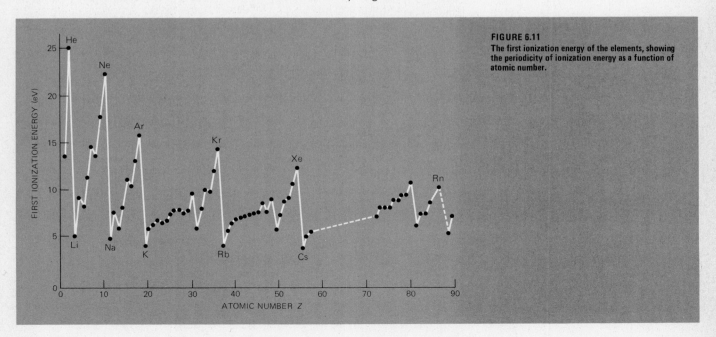

FIGURE 6.11
The first ionization energy of the elements, showing the periodicity of ionization energy as a function of atomic number.

relatively easily. But the second and third ionization energies of lithium are very large. In both cases, a $1s$ electron must be removed from an atom with a nuclear charge of $+3$, so we expect a large ionization energy.

The fact that the first ionization energy of beryllium is greater than that of lithium is consistent with the greater nuclear charge of Be (atomic number 4). The existence of electron-electron repulsions in the Be atom keeps the first ionization energy from being even larger. The second ionization energy of Be is much less than the second ionization energy of Li, because in Be the $2s$ electron is lost. In Li, it is the $1s$ electron that is lost.

For boron (atomic number 5), the first ionization energy (8.3 eV) is lower than that of Be (9.3 eV). The reason for this apparent anomaly is that a $2p$ electron is removed from the B atom, while a $2s$ electron is removed from the Be atom. The $2p$ electron is screened more effectively from the nuclear charge, because the $2p$ orbital is less penetrating than the $2s$ orbital, so the $2p$ electron is removed more readily.

The first ionization energies of the next two elements, carbon ($Z = 6$) and nitrogen ($Z = 7$) go up slightly, but the first ionization energy of oxygen ($Z = 8$) is lower than that of nitrogen. In nitrogen, the three $2p$ electrons are in separate orbitals, so electron-electron repulsions are minimized. There are four $2p$ electrons in the oxygen atom, and two of them must share an orbital. The increase in electron-electron repulsion due to the presence of two electrons in the same orbital offsets the effect of the larger nuclear charge of the oxygen atom.

We have already noted the increased stability associated with half-filled electronic subshells. In nitrogen, the $2p$ subshell is exactly half full, and thus the first ionization energy of nitrogen is relatively high. The second ionization energy of oxygen is also relatively high. A singly ionized oxygen atom, with seven electrons remaining, has the same half-filled subshell as a nitrogen atom with seven electrons. We can say that N and O^+ are *isoelectronic*. Two species with the same number of electrons are isoelectronic.

The increase in first ionization energy continues with fluorine ($Z = 9$) and reaches a maximum with neon ($Z = 10$). The large first ionization energy of neon represents the stability associated with a completely filled electron shell.

A major reason for paying such close attention to ionization energy is the close relationship between electronic configuration and chemical behavior. The chemical properties of any atom are related to the configuration of its outermost electrons, and this configuration is reflected in the atom's ionization energies. In going across the first row of the periodic table, from Li on the left to F on the right, there is a trend toward higher first ionization energies—that is, toward increasing difficulty in removing an electron. We find this trend reflected in the chemical properties of the elements. Cations formed by electron removal play an important role in the chemistry of the elements in columns IA and IIA of the periodic table. But cations of oxygen (column VIA) and fluorine (column VIIA) are almost never found under ordinary chemical conditions, a reflection of the high first ionization energies of these elements. The relationship between electronic configuration and the formation of chemical bonds will be examined in detail in Chapter 7.

The trends in ionization energy of elements in the second row of the periodic table parallel those of the first row. But the actual ionization energies are lower for each corresponding second-row element. The atoms of the second-row elements are larger than those of the first-row elements, so the outermost electrons are further away from the nuclei. Therefore, relatively less energy is required to remove these electrons. The stability of half-filled shells and subshells is demonstrated by two interruptions in the steady increase of ionization energies in the second row: at Al, where the $3p$ subshell starts filling, and at S, where the second half of the $3p$ subshell starts filling.

As we move along the list, we find that a substantial block of elements from

Sc ($Z = 21$) to Cu ($Z = 29$) are very close in first ionization energy. All these elements are metals, are close in atomic number to one another, and have many similarities in behavior. The small, steady increase in ionization energies in the group is consistent with increases in nuclear charge, but only if the electron is being removed from the same subshell in each element. By using ionization energy data and information from spectroscopic studies, we can form a picture of the electronic configuration of these elements and of the cations derived from them.

We mentioned in Section 6.2 that the $4s$ orbital begins filling before the $3d$ orbital does. But in elements in which the $3d$ subshell is being filled, the $4s$ orbital empties before the $3d$ orbital does. Thus, in the series of elements from Sc through Cu, each first ionization energy refers to the removal of a $4s$ electron. It might seem illogical that a $4s$ electron is removed first. Since the $4s$ orbital begins filling before the $3d$ orbital, it must be of lower energy than the $3d$ orbital, and thus the electrons in the $4s$ orbital should be held more tightly. How can an electron in a $4s$ orbital be removed more easily than an electron in the $3d$ orbital in these elements?

The answer is related to the differences in radial probability functions (Figure 6.8) between the $4s$ and $3d$ orbitals and the effect of an increase in Z on the relative energy of electrons in these orbitals. The electron in the $4s$ orbital is at a greater average distance from the nucleus than the electron in the $3d$ orbital. But the $4s$ electron penetrates closer to the nucleus, and thus is not screened from the attractive force of the nuclear charge as much as the $3d$ electron. In elements 19 and 20, shielding by the electrons in the filled Ar shell makes the effective nuclear charge relatively small for the outermost electrons. The penetration of the $4s$ orbital thus is important; it causes the $4s$ orbital to fill first. But for element 21 and succeeding elements, the effective nuclear charge outside the screen of the Ar core increases substantially. Penetration of the $4s$ orbital is then not as important a factor in determining electronic energy. The electrons in the $3d$ orbitals, whose average distance from the nucleus is not as great, are of lower energy than the $4s$ electrons. Thus, the $4s$ electrons are easier to remove than $3d$ electrons.

There is a noticeable increase in ionization energy at Zn ($Z = 30$), reflecting the stability of completed subshells. The electronic configuration of Zn is [Ar]$4s^2 3d^{10}$. With Ga ($Z = 31$), a new electronic subshell, the $4p$, begins to fill, and there is the predictable decrease in first ionization energy. There is one more break in the steady increase of ionization energies from Ga to Kr ($Z = 36$), which has a completed electron shell. The break comes at Se ($Z = 34$), whose first ionization energy is slightly lower than that of As ($Z = 33$). The lower ionization energy of Se is due to the presence of a half-filled subshell, $4p^3$, in As.

Another apparent anomaly in ionization energy trends is found in the transition elements in the fifth row of the periodic table. The ionization energies of these metals are unusually high. For example, the first electron to be removed from both Ba ($Z = 56$) and W ($Z = 74$) is a $6s$ electron. Yet the ionization energy of W is almost 60% higher than that of Ba. The high ionization energies of W and other transition elements is due to their relatively high effective nuclear charges. There is a difference of 18 in the atomic number, and therefore the nuclear charge, of comparable elements in the third and fourth rows of the periodic table. The difference between comparable elements in the fourth and fifth rows is much greater. Beginning with element 72, hafnium, there is a difference of 32 in the atomic number of comparable elements in the fourth and fifth rows. This larger difference in atomic number is caused by the presence of the lanthanoid series of elements, atomic numbers 57–71, in which the $4f$ electronic subshell is being filled. The 14 electrons in the $4f$ subshell are not very effective at screening the nucleus from the outermost electrons. Thus, the effective nuclear charge beginning at element 71 is relatively high, so the ionization energy is relatively high.

6.5 ELECTRON AFFINITY

Another property of atoms that gives us important information about electronic configuration is the **electron affinity**. Electron affinity is most simply defined as the energy required to remove an electron that has been added to the atom in the gas phase. Electron affinity is thus the energy required for the process:

$$A^-(g) + energy \longrightarrow A(g) + e^-$$

Electron affinity is important because it can be used to predict and explain the chemical behavior of elements.

Unlike ionization energy values, electron affinities can have negative values in some cases. A negative value means that the anion is less stable than the neutral atom and the free electron. In such a case, the electron-electron repulsion with the extra electron is greater than the attraction between the nucleus and the extra electron. However, electron affinities are positive for many atoms, meaning that $A^-(g)$ is more stable than $A(g) + e^-$; the acceptance of an electron by the neutral atom is a favorable process.

There are periodic trends for electron affinity, just as there are for ionization energy. Table 6.8, which gives the electron affinities of the first 18 elements, shows this periodicity.

TABLE 6.8
Electron Affinities (eV)[a]

H 0.75							He < 0
Li 0.62	Be < 0	B 0.24	C 1.27	N 0.0	O 1.47	F 3.34	Ne < 0
Na 0.55	Mg < 0	Al 0.46	Si 1.24	P 0.77	S 2.08	Cl 3.61	Ar < 0

[a]From E. C. M. Chen and W. E. Wentworth *J. Chem. Ed.* **52:**489 (1975).

In general, electron affinity values increase as one goes from left to right across any row of the periodic table, as does ionization energy (Figure 6.12). But the maximum value of electron affinity is not found in column 0, the furthest column on the right, but in column VIIA. This characteristic of electron affinity is to be expected. We know that all the elements in column 0 have the stability that accompanies completely filled noble gas electron shells. Each element in column VIIA is just one electron short of having such a completely filled shell. Adding an electron to create an anion of a column VIIA element makes that element isoelectronic with the corresponding column 0 element. For example, the anion F^- has the electronic configuration $[He]2s^22p^6$, identical with that of Ne.

As with ionization energy, there are irregularities in the periodicity of electron affinity that are primarily associated with filled or half-filled subshells. One difference worth noting is that, unlike ionization energies, electron affinities of elements in the second row of the periodic table are higher than those of the corresponding first-row elements. For example, the ionization energy of oxygen (13.6 eV) is higher than the ionization energy of sulfur (10.4 eV), the element below it in the periodic table. But the electron affinity of oxygen (1.47 eV) is lower than that of sulfur (2.12 eV).

The smaller size of the oxygen atom explains its lower electron affinity. Adding an electron to the oxygen atom creates greater electron-electron repulsions than adding an electron to the larger sulfur atom. The effect of electron-electron repulsion on the value of electron affinity is especially important in dianions. For example, the electron affinity of O^- has a negative value. The process:

$$O^{2-}(g) \longrightarrow O^-(g) + e^-$$

is favorable, which means that the O^{2-} anion is unstable relative to the O^- anion

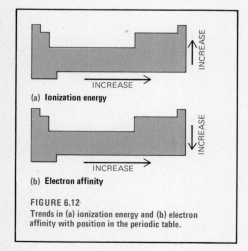

(a) Ionization energy

(b) Electron affinity

FIGURE 6.12
Trends in (a) ionization energy and (b) electron affinity with position in the periodic table.

and an electron and cannot exist by itself in the gas phase. The same is true for the S^{2-} and N^{3-} anions. Such ions can exist only if they are stabilized by the surroundings, which can occur in solution or in a crystalline lattice.

6.6 THE PERIODIC TABLE OF THE ELEMENTS

All the properties of atoms that have been discussed in this chapter have been related to their electronic configurations. The correlation of the physical and chemical properties of substances with the electronic structure of their atoms is one of the most impressive achievements of modern chemistry.

So far, we have discussed the properties of atoms entirely from the point of view of twentieth-century chemistry. The discoveries made in this century make it possible to state that the properties of elements that are of interest to chemists are determined primarily by the behavior of valence electrons, the electrons in the outermost shell of an atom. We have seen that our knowledge of quantum numbers allows us to arrange orbitals into shells and subshells. As orbitals are filled according to the Aufbau principle, the electronic configuration of valence shells tends to repeat. The chemical and physical properties of elements with the same valence electronic configurations tend to be similar. For example, the elements:

Li: [He]2s
Na: [Ne]3s
K: [Ar]4s
Rb: [Kr]5s
Cs: [Xe]6s

all have the same valence electronic configuration. Each of these elements has one s electron in the valence shell. Therefore, all these elements should have similar properties. We have noted the periodicity of such properties as ionization energy and electron affinity. We find the same sort of periodicity in studying a number of important chemical properties. When the elements are listed in order of increasing atomic number, elements with similar properties recur at predictable intervals. The recurrence of elements with similar properties reflects the recurrence of similar valence electronic configurations. This recurrence is made evident by arranging the elements in a periodic table.

There is a tendency for today's students to think that the idea of periodicity of the elements derives from knowledge about electrons and orbitals. In fact, the descriptions of orbitals and electrons used thus far are relatively recent concepts that provide a theoretical framework on which to hang all of the observations and correlations made by generations of chemists who had no knowledge of electronic configurations. Indeed, these observations have helped to create the current picture of electronic configuration. The history of the development of the modern version of the periodic table is instructive.

Early Periodic Tables

Early in the nineteenth century, chemists began to recognize that there are similarities in the chemical behavior of the elements, and to suggest systematic groupings of certain elements based on their properties. The earliest such arrangement was made by J. W. Döbereiner (1780–1849). In 1829, Döbereiner set forth a scheme of triads, groups of three elements of similar properties, with the middle element having an atomic weight that was the average of the atomic weights of the first and third elements. Among the triads proposed by Döbereiner were:

I	II	III
Lithium	Calcium	Chlorine
Sodium	Strontium	Bromine
Potassium	Barium	Iodine

Dmitri Ivanovich Mendeleev (1834–1907) was born in Siberia, went to college in France and Germany, and returned to Russia to become a professor of chemistry in St. Petersburg. In addition to developing the periodic table of the elements, Mendeleev wrote an outstanding chemistry textbook and did much to improve chemistry education in Russia. In 1906, he failed by one vote to receive the Nobel Prize in chemistry. (*Culver*)

But in general, attempts to find numerical relationships based on nineteenth-century atomic weights were not successful, partly because there were errors and uncertainties in the values of atomic weights and partly because a substantial number of elements were then undiscovered.

It was a Russian chemist, Dmitri I. Mendeleev (1834–1907), who in 1869 developed the systematic classification of the elements that incorporated most of the essential principles of the periodic table we use today. As often happens, another investigator, the German chemist Lothar Meyer (1830–1895) made the same discovery at almost the same time. But it is Mendeleev who is recognized today as the discoverer of the periodic table of the elements.

The essence of Mendeleev's achievement can be stated easily. He found that when the elements are arranged in order of increasing atomic weight, there is a repetition of properties at approximately regular intervals. This periodicity of properties was most apparent when the elements were grouped in tabular form. But Mendeleev did more than list elements by column and row. His genius enabled him to recognize underlying principles that make the periodic table a powerful tool of modern chemistry.

Table 6.9 shows a portion of Mendeleev's periodic table of 1871, with the atomic weights he assigned to the elements. To start with, Mendeleev separated hydrogen from the rest of the elements. He also left blank spaces in places where he believed that elements remained to be discovered. Mendeleev could not list the elements in order of increasing atomic number, as today's periodic table does, since atomic numbers were not known at that time. But he did recognize that a listing of elements by atomic weight was not satisfactory. Mendeleev had the courage and intelligence to assume that the *observed chemical properties* of the elements were more important than the atomic weights assigned to the elements—a good assumption, considering the suspect nature of many atomic weights of the time.

For example, the atomic weight of the element indium, which had been discovered eight years earlier in 1863, was put at 76.6. This atomic weight would have placed indium in the periodic table between As (75) and Se (78). But on the basis of their chemical properties, As had to be in group V and Se had to be in group VI, leaving no room between them for indium. Mendeleev suggested that the reported atomic weight of indium was incorrect, because it was calculated on the basis of a valence of 2. If the valence of In was in fact 3, its atomic

TABLE 6.9
A Portion of Mendeleev's Periodic Table

Row	Group I	Group II	Group III	Group IV	Group V	Group VI	Group VII	Group VIII
1	H = 1							
2	Li = 7	Be = 9.4	B = 11	C = 12	N = 14	O = 16	F = 19	
3	Na = 23	Mg = 24	Al = 27.3	Si = 28	P = 31	S = 32	Cl = 35.5	
4	K = 39	Ca = 40	— = 44	Ti = 48	V = 51	Cr = 52	Mn = 55	Fe = 59, Co = 59, Ni = 59
5	Cu = 63	Zn = 65	— = 68	— = 72	As = 75	Se = 78	Br = 80	
6	Rb = 85	Sr = 87	Yt = 88	Zr = 90	Nb = 94	Mo = 96	— = 100	Ru = 104, Rh = 104, Pd = 106
7	Ag = 108	Cd = 112	In = 115	Sn = 118	Sb = 122	Te = 125	I = 127	
8	Cs = 133	Ba = 137	Di = 138?	Ce = 140?	—	—	—	————
9	—	—			—		—	
10	—	—	Er = 178?	La = 180?	Ta = 182	W = 184	—	Os = 195, Ir = 197, Pt = 198
11	Au = 199	Hg = 200	Tl = 204	Pb = 207	Bi = 208			

weight would be 115, putting In in a space between cadmium and tin. Mendeleev showed that compounds of indium resembled compounds of aluminum and thallium, which are both in Group III of the periodic table. This observation was further evidence that In belonged in the place assigned for it, the seventh row of Group III. He made similarly accurate placements for other elements, including uranium (92) and beryllium (4), which had been misplaced in other efforts to develop periodic listings of the elements.

In other cases, Mendeleev observed that the accepted values of atomic weights for some elements did not correspond with the properties of those elements. Mendeleev stated that the atomic weights assigned to these elements were incorrect. He placed gold, whose atomic weight was believed to be 196.2, after osmium, iridium, and platinum, whose atomic weights were believed to be 198.6, 196.7, and 196.7 respectively, contending that these values were wrong. Mendeleev was right. The modern values for the atomic weights of these elements are Os, 190.2; Ir, 192.2; Pt, 195.1; Au, 197.0. In some other instances, Mendeleev was wrong in assuming that the atomic weights of elements were incorrect. But most of his judgments based on chemical properties were correct.

While Mendeleev's correction of atomic weights on the basis of his observations of chemical properties was impressive, his predictions of the existence, and even the properties, of undiscovered elements were spectacular. In at least three instances, Mendeleev said that a new element would be found, and he described its properties with such accuracy that he could even suggest where and how the element could be found.

In Table 6.9, space is left for an element in group III intermediate in properties between Al and In. Mendeleev called this unknown element *eka-aluminum* (*eka* means "first" in Sanskrit), and he formulated an elaborate description of the undiscovered element. The element eventually was discovered (it was named gallium, because the discovery was made in France and Gallic patriotism demanded a name that honored French science), and every prediction made by Mendeleev was found to be correct:

	Predicted Properties	*Observed Properties*
atomic weight	68	69.72
density (relative to water)	5.9	5.91
melting point	low	29.78°C
boiling point	high	1983°C
product with oxygen	X_2O_3	Ga_2O_3
hydroxide	$X(OH)_3$	$Ga(OH)_3$
product with chlorine	XCl_3	$GaCl_3$

Mendeleev made equally accurate predictions about the properties of the unknown elements he named *eka-boron,* with a predicted atomic weight of 44, and *eka-silicon,* atomic weight 72. Eka-boron was discovered in 1879 in Scandinavia, and was named scandium; eka-silicon was discovered in 1886 in Germany, and was named germanium.

As more elements were discovered and better experimental data became available, the periodic table was improved and expanded. It was clear at the time that some unknown fundamental relationship was behind the periodicities in behavior. The nature of that relationship was not discovered until the quantum mechanical description of electronic configuration was developed.

The Modern Periodic Table

The periodic table systematizes and rationalizes so much chemical knowledge that a familiarity with the important features of the modern periodic table is

THE SUPERHEAVY ELEMENTS
AND THEIR CHEMISTRY

The heavier an atomic nucleus is, the more likely it is to be unstable. No element of greater atomic number than uranium, element 92, is known to occur naturally on earth. However, nuclear theory indicates that there should be an "island of stability"—or at least relative stability—in which there are some superheavy elements of atomic numbers around 110, 116, 126, and beyond, up to element 184.

In 1976, a group of physicists working at Florida State University reported evidence for the existence of elements 116, 124, and 126, and perhaps three other superheavy elements. The report was made on the basis of work with some samples of monazite, or mica, a mineral form of calcium phosphate, from Madagascar. The monazite contains what are called "halos," rings that form around radioactive inclusions in the mineral. About one in 1000 of the halos are too big to be explained by any known nuclear decay. When these halos were bombarded with low-energy protons, they emitted X rays whose spectra indicated that the emission came from superheavy elements.

If it had been confirmed, the work would have marked the first discovery of a new element in nature since 1925, and would indicate that other superheavy elements might be present on earth. However, a number of other laboratories that attempted to reproduce the findings got negative results, and thus doubt was cast on the report.

Nevertheless, a number of efforts are being made to predict the chemical properties of the superheavy elements, in the same way that Mendeleev predicted the properties of gallium, scandium, and germanium. Kenneth S. Pitzer of the Lawrence Berkeley Laboratory presented what he called "the striking conclusion" that elements 112, 114, and 118, if they exist, are relatively inert gases. On the basis of his calculations, Pitzer concluded that the three superheavy elements would have closed electron shells and thus would be relatively inert chemically. Their positions in the periodic table indicated to him that the elements would be either gases or volatile liquids.

He added that the chemistry of elements 112 and 114 would differ from that of mercury and lead, their congeners in the periodic table. Almost all the compounds formed by these two superheavy elements would be unstable. "These properties of great volatility and ease of reduction to the element would appear to provide better separation methods than procedures based on uncertain similarities in solution chemistry of 112 to mercury and 114 to lead," he said.

essential. Figure 6.13 shows a frequently used version of the periodic table that is designed to emphasize the grouping of similar elements. Some important features of this arrangement are:

1. The arrangement of elements in order of increasing atomic number from left to right and from top to bottom of the table is interrupted to remove two

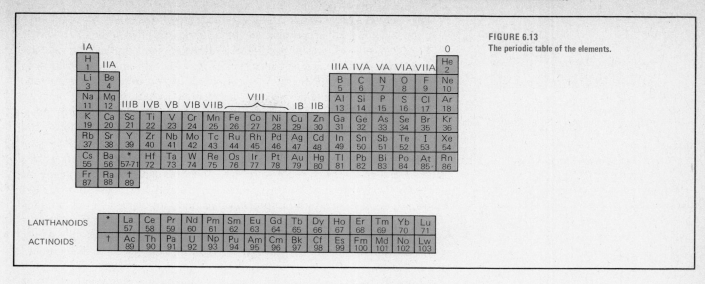

FIGURE 6.13
The periodic table of the elements.

horizontal rows, or periods, of elements (57–71 and 89–103) from the main body of the table. One reason for listing these elements separately is practical: If this were not done, the table would be too wide to fit on most pages. The separate listing also emphasizes the great similarity in the properties of elements in each of the two groups. The elements from 57 through 71 are called the *lanthanoids,* or *rare-earth metals.* The elements from 89 through 103 are called the *actinoids.* The elements in each group behave alike chemically. This similarity in chemical behavior can be explained by similarities in electronic configuration.

The lanthanoids have a filled $6s$ subshell, a filled $5p$ subshell, and $5d$ subshells that either are empty or hold only one electron. The differences in electronic configuration occur in the $4f$ subshell, which is filling in this series. But the $4f$ subshell, with its relatively low principal quantum number, is essentially buried in the large electron clouds of these elements and thus the $4f$ electrons have little effect on chemical behavior.

2. Each column of the periodic table is designated by a number, usually with a letter attached. Eight columns begin with elements in the first horizontal row, or period, of the table (we shall not include H and He in counting the rows because of their unique properties.) The number of the column gives the number of valence electrons of each element in the column, as can be seen by comparing Table 6.6 and Figure 6.13. The letter A is added to the designation of seven of these columns to indicate that each column starts from the first period of the table. The eighth column is column 0. It includes the elements called the noble gases, which have completely filled electronic shells.

The elements in each vertical column can be thought of as a family, or group. The elements in a column have similar chemical properties, although there are regular variations in those properties as the atomic number increases. Two elements in the same family are called *congeners.* The elements of family IA are called the *alkali metals.* Those of family IIA are the *alkaline-earth metals.* Those of family VIIA are the *halogens.*

3. Eight more families, or groups, begin with the third horizontal row of the periodic table. These families are designated by numerals, followed by the letter B when it is necessary to distinguish them from the A groups. The elements in column IB and columns IIIB–VIII are metals called *transition elements.* Each of these elements either has an incomplete d subshell or can readily give rise to cations that have an incomplete d subshell. The elements of column IIB are similar to the transition elements but do not conform to this definition because they have a completed d subshell and do not readily form cations with an incomplete d subshell. They are called *post-transition metals.*

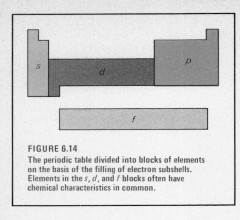

FIGURE 6.14
The periodic table divided into blocks of elements
on the basis of the filling of electron subshells.
Elements in the *s*, *d*, and *f* blocks often have
chemical characteristics in common.

Column VIII is unusual because it is three elements wide. All the members of group VIII have similar properties, so they are grouped together.

4. All the elements in a given family resemble one another. For example, group IB, the coinage metals, includes copper, silver, and gold, which are elements with similar properties. There is also a certain degree of resemblance between elements in columns with the same numeral but different letter designations. For example, strontium, in group IIA, and cadmium, in group IIB, have a number of similarities, including the tendency to form dipositive cations. Such similarities do not always exist. Cesium, in group IA, and gold, in group IB, are soft metals that form monopositive cations. But the resemblance stops there. Cesium is one of the most reactive metals, reacting explosively with water and nonmetals. Gold is one of the least reactive metals. It can be dissolved only by aqua regia, a potent mixture of concentrated nitric acid and hydrochloric acid.

As we observed in Chapter 2, elements to the right and near the top of the periodic table tend to be nonmetals, while the majority of elements, those to the left and near the bottom of the table, are metals. A diagonal band of semimetals starts with boron and runs to tellurium.

Figure 6.14 shows another way of dividing the elements into broad blocks—on the basis of the filling of electronic subshells. The elements in the *s* block to the left or the *p* block to the right are called the *representative* or *main group elements*. The *d* block contains transition elements, while the *f* block contains inner transition elements. The division is useful to chemists because elements in a given block often have some chemical characteristics in common.

However, the main emphasis in the periodic table is on vertical relationships—the similarities in the properties of the elements that are members of the same group. The tendency of the properties of elements in the same group to vary in a regular way is called a *group trend*.

There are also horizontal relationships, the regular changes in properties that are found in going from left to right across the table. These regular changes are called *horizontal trends*. There are also diagonal trends among some elements. Two elements often show similarities in behavior when the heavier element is one column to the right and one row below the lighter element. Lithium and magnesium have such a diagonal relationship, as do beryllium and aluminum.

6.7 PERIODIC TRENDS IN ATOMIC SIZES

One of the first periodic variations in the properties of the elements to be recognized was that of atomic size. As early as 1870, Lothar Meyer, the codiscoverer of the periodic table, published a graph in which a quantity called "atomic volume" was plotted against atomic number. Figure 6.15 shows an updated version of this graph. The atomic volume is calculated by dividing the atomic weight of an element by its density:

$$\text{atomic volume (cm}^3/\text{mol)} = \frac{\text{atomic weight (g/mol)}}{\text{density (g/cm}^3)}$$

which is taken to be the relative volume of the element. However, atomic volume has only an approximate relationship to the relative size of atoms, because the density used to calculate atomic volume depends on temperature and on the crystal structure of the element, as well as on the size of individual atoms. Nevertheless, the graph in Figure 6.15 is a striking indication of the periodic variation in atomic size.

The calculation or measurement of actual atomic size is not easy, because atoms do not have sharply defined boundaries. To determine atomic size, we must decide arbitrarily where the atom ends. This judgment is based primarily on the way the atom interacts with what is around it. Therefore, there are several ways in

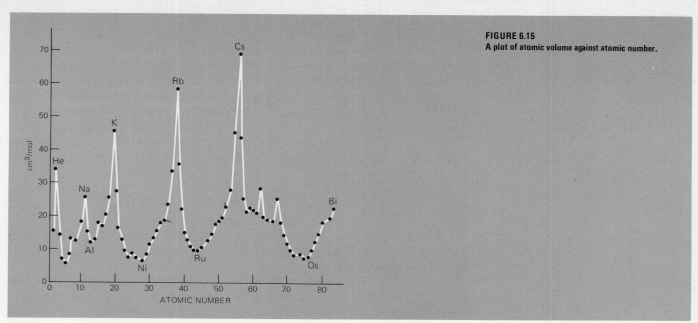

FIGURE 6.15
A plot of atomic volume against atomic number.

which to express the size of atoms. While these various methods may produce different values, they all show periodic trends that are consistent with what we know about quantum numbers, electronic shells, and nuclear charge.

For example, we have seen (Section 6.4, page 163) that the increase in effective nuclear charge is less important than the increase in the principal quantum number n when we compare elements in a family. Ionization energies tend to decrease in going from the top to the bottom of a column of the periodic table. The greater importance of the principal quantum number n produces an opposite trend in atomic sizes: The size of atoms in a family tends to increase as the atomic numbers of the elements increase. For example, one measure of atomic size is the atomic radius. In group IIA, the atomic radii in nanometers (nm) are: Be, 0.105; Mg, 0.15; Ca, 0.18; Sr, 0.20; and Ba, 0.22. The regular increase reflects the steady increase in the value of n for the valence electrons from Be, whose electronic configuration is $[He]2s^2$, to Ba, whose configuration is $[Xe]6s^2$.

In general, as we go from left to right across a row of the periodic table, atomic size decreases as atomic number increases. The principal quantum number of the valence electrons of the elements in one row of the table is the same; the increase in effective nuclear charge in successive elements tends to pull the electron cloud in a little tighter. This phenomenon is reminiscent of the increase in ionization energy that is observed in going from left to right in a row of the periodic table. The tendency for atomic size to decrease can be seen in the data for atomic radii, in nanometers (nm), of the first-row elements: Li, 0.15; Be, 0.11; B, 0.085; C, 0.070; N, 0.065; O, 0.060; F, 0.050.

The effect of an increase in nuclear charge on size is even more evident in the case of ions. The radius of an ion, another measure of atomic size, is of special interest because it is often related to the chemical properties of ionic compounds. We can measure the radii of a series of ions that are isoelectronic—that is, all the ions have the same number of electrons but different nuclear charges.

The ions O^{2-}, F^-, Na^+, and Mg^{2+} have the same electronic configuration, $1s^2 2s^2 2p^6$, the filled neon shell. The ionic radii of these species are: O^{2-} ($Z = 8$), 0.140 nm; F^- ($Z = 9$), 0.136 nm; Na^+ ($Z = 11$), 0.095 nm; and Mg^{2+} ($Z = 12$), 0.065 nm. The steady decrease in ionic radius reflects a steady increase in nuclear charge that is not offset by electronic screening.

The data for the ions of a group show another interesting trend. For example, the radii of the halogens in nanometers (nm) are: F^-, 0.136; Cl^-, 0.181; Br^-, 0.195; I^-, 0.216. The increase in radius is greater from F^- to Cl^- than from Cl^- to Br^- or from Br^- to I^-. We find the same phenomenon in a number of other families: The increase in radius is much greater between elements in the first and second row than between elements in the second and third row. This trend reflects the fact that nuclear charge increases by 8 between elements in the first and second rows of groups IIIA–VIIA, while the increase is 18 between elements in the second and third rows. The extra nuclear charge tends to contract the larger ions, reducing the expected increase in size due to the increase in n.

6.8 A GROUP TREND EXEMPLIFIED: THE ALKALI METALS

The physical and chemical properties of the group IA elements, the alkali metals, exemplify the vertical group trends of the periodic table more than those of any other family of elements.

However, we cannot predict all the physical and chemical properties of a family of elements, even the alkali metals, from the behavior of isolated atoms in the gas phase. In looking at isolated atoms, we are studying primarily how increasing atomic size and increasing principal quantum number affect the properties of the elements as we move down a column of the periodic table. Usually, however, we do not look at isolated atoms. We look at solids or liquids in which atoms interact with each other, or we look at substances in which different atoms are combined. These interactions have important effects on physical and chemical properties.

The alkali metals are lithium, sodium, potassium, rubidium, and cesium. They are all soft metals, of great chemical reactivity, with low boiling points. They are silvery white, are good conductors of electricity, and tarnish rapidly when exposed to air.

The great chemical reactivity of the alkali metals is a reflection of their relatively low first ionization energies. Predictably, their chemistry involves primarily the monopositive cations Li^+, Na^+, K^+, Rb^+, and Cs^+, formed by removal of one electron from the neutral atom. We mentioned that ionization energy decreases going down the column from Li to Cs. Reactivity also increases as we go from Li to Cs. All the alkali metals react violently with water. The reaction of lithium and water is vigorous. The reaction of sodium and water is noisy and is accompanied by fire. The reaction of cesium and water is explosive. The general equation for the reaction of an alkali metal with water is:

$$2M(s) + 2H_2O \longrightarrow 2M^+ + H_2(g) + 2OH^-$$

Such a reaction is accompanied by the release of a great deal of heat. With the exception of lithium, the reaction of an alkali metal with water releases enough heat to ignite the liberated $H_2(g)$, which reacts with the $O_2(g)$ in air.

Because alkali metals are so highly reactive, they are never found uncombined in nature. To obtain a pure alkali metal, it must be liberated from a salt. Lithium and sodium usually are prepared by passing an electric current through a molten salt, most commonly LiCl or NaCl. The monopositive cation of the alkali metal gains an electron to form the metal:

$$Na^+ + e^- \longrightarrow Na$$

while the anion of the salt (in this case chloride) gives up an electron:

$$Cl^- \longrightarrow \tfrac{1}{2}Cl_2 + e^-$$

Chemical transformations brought about by an electric current are called electrolysis. We shall discuss electrolysis in detail in Chapter 16.

TABLE 6.10
Properties of the Alkali Metals

Element	Valence Shell Electronic Configuration	M.P. (K)	B.P. (K)	Atomic Volume (cm³/mol)
Li	$2s$	453.7	1638	13.1
Na	$3s$	371.0	1156	23.7
K	$4s$	336.4	1030	45.3
Rb	$5s$	312.2	961	55.9
Cs	$6s$	301.7	954	70.0

The heavier alkali metals can also be prepared by electrolysis. However, the process usually is impractical on an industrial scale, because the greater reactivity of the heavier alkali metals makes it more difficult to collect them. Other methods in which the metal can be removed as a vapor are used. For example, cesium is prepared by a displacement reaction at temperatures of 1000 K–1200 K:

$$Ca(s) + 2CsCl(l) \longrightarrow CaCl_2(s) + 2Cs(g)$$

At these temperatures, cesium is a gas that is easily separated from the other components of the reaction, which are solids or liquids. A high conversion is achieved by using excess Ca and removing Cs as it forms.

We have seen that there is a fairly regular trend in such properties as ionization energy, electron affinity, atomic radius, and ionic radius in the family of alkali metals as atomic number increases. Table 6.10 lists other physical properties of the pure alkali metals. The data in the table show the effect of increases in atomic size on physical properties. Often, the attractive forces between neighboring atoms are greatest when the atoms are smallest. Since both the melting point and the boiling point of an alkali metal are determined by the magnitude of the attractive forces between atoms, both melting points and boiling points decrease as the size of the alkali metal atoms increases.

For example, the melting point of Rb is lower than that of K. The Rb atom is larger than the K atom, so the forces holding the Rb atoms together in the solid are not as great as the forces holding the K atoms together. Less thermal energy is needed to separate the Rb atoms, so they separate—melt—at a lower temperature.

We mentioned earlier that the change in a property is greater between first-row and second-row elements in a family than between elements in other rows. Table 6.10 shows that the alkali metals demonstrate this trend. The decrease in melting point from Li (first row) to Na (second row) is more than 80 K. The decrease from Na to K, from K to Rb, and from Rb to Cs totals about 70 K. This difference between the first-row element and the rest of the family reflects the fact that there is a greater difference in relative size between first-row and second-row elements of a family than between elements in other rows. The difference also shows up in other properties of the alkali metals, such as softness and electrical conductivity, (Chapter 19), which increase with increasing atomic number, because the attractive forces between atoms decrease as atomic size increases.

Although the high reactivity of the alkali metals makes them useless as structural metals, they are important in technology. The uses of the alkali metals depend on the properties of each metal and the cost of producing it.

Sodium is the least expensive alkali metal. In fact, its cost of $0.40/kg and its low density make sodium the cheapest of all metals by volume. Sodium chloride is abundant in nature, and producing sodium is much less costly than producing the heavier alkali metals. Thus, whenever an alkali metal is needed in industry, sodium is the first choice. Most uses of metallic sodium are based on its low first ionization energy, which makes sodium valuable as a supplier of electrons in chemical processes. One major use of sodium is in the manufacture of gasoline antiknock compounds.

Sodium is used to produce titanium metal by the high-termperature reaction:

$$4Na(l) + TiCl_4(g) \longrightarrow Ti(s) + 4NaCl(s)$$

Sodium metal is also used to manufacture some sodium compounds that cannot be obtained readily by other methods. One process burns sodium in air:

$$2Na(s) + O_2(g) \longrightarrow Na_2O_2(s)$$

to form the bleaching agent sodium peroxide.

The relatively high cost of the other alkali metals limits their use. Lithium, at $17/kg, is a starting material in many chemical processes. Compounds of lithium are used in the manufacture of high-strength glass and glass-ceramics for cookware. Lithium is also being used in newer high-power-density batteries. Some alloys of lithium with aluminum or magnesium are lightweight and maintain strength at high temperatures, which makes them valuable in aerospace applications. One such alloy, LA 141, contains 14% Li, 1% Al, and 85% Mg.

Potassium, which costs about $4/kg, is used primarily in processes where sodium is unsatisfactory. Sodium-potassium alloys that contain 40%–90% potassium by weight have an application with important implications for future energy supply. While both sodium and potassium are solids at room temperature, these NaK (called ''nack'' in the trade) alloys are liquid at room temperature. Alkali metals are among the best available conductors of heat, and NaK alloys are used as heat exchange liquids in fast-breeder nuclear reactors.

One combustion product of potassium has a major practical application. Unlike sodium, which burns to give the peroxide, potassium burns to give the superoxide:

$$K(s) + O_2(g) \longrightarrow KO_2(s)$$

High-efficiency sodium vapor lamps that were introduced in the 1960s now are widely used for outdoor lighting. The lamps contain gaseous sodium at high pressure within a corrosion-resistant ceramic globe and provide two-to-three times more light per watt than mercury lamps. (*General Electric*)

The most useful property of potassium superoxide is summed up by the equation:

$$4KO_2(s) + 2H_2O(l) + 4CO_2(g) \longrightarrow 4KHCO_3(s) + 3O_2(g)$$

This process consumes water and carbon dioxide, the products of human respiration, and produces oxygen, which makes this compound invaluable for breathing masks. Potassium superoxide is a store of an emergency supply of oxygen, which is released as the person wearing the mask needs it.

The uses of rubidium and cesium are based primarily on the ease with which these elements are ionized. Both are used in small amounts in radio tubes and photocells. Because of their low ionization energies, they have a low threshold frequency for the photoelectric effect. Therefore, they emit electrons even when exposed to low-energy visible light. Both cesium and rubidium hold promise for use in magnetohydrodynamic generators and thermionic generators. Magneto-hydrodynamic generators produce electricity by passing ions through a magnetic field, and thermionic generators produce an electric current by "boiling" electrons off suitable substances. Cesium is also being investigated as a possible fuel for ion propulsion engines, which could propel spacecraft in outer space. The current high prices for Cs ($250/kg) and Rb ($600/kg) restrict their use.

Generally, it is the compounds of alkali metals and not the pure elements that are of greatest importance. Compounds of both sodium and potassium are essential human nutrients. Potassium compounds are an essential constituent of fertilizers. The industrial uses of sodium and potassium compounds are legion. Lithium compounds also have a wide range of uses. One of the most striking, developed recently, is the medical use of lithium salts to treat manic depression. The uses of rubidium and cesium salts are more limited.

6.9 PERIODIC TRENDS IN THE OXIDES, HYDRIDES, AND HALIDES OF ELEMENTS

Since the properties of the elements affect the properties of their compounds, we find a number of correlations between the properties of compounds and the position of their constituent elements in the periodic table. Such correlations are especially clear in the oxides, binary compounds of oxygen, and in the hydrides, binary compounds of hydrogen. Correlations are also seen in the halides, binary compounds of the halogens.

Oxides

With the exception of some of the lighter noble gases, every element forms at least one binary compound with oxygen. Some elements form a number of different oxides. Vanadium, a transition metal, forms VO, V_2O_3, VO_2, and V_2O_5. Chlorine, a nonmetal, forms Cl_2O, Cl_2O_3, ClO_2, Cl_2O_6, and Cl_2O_7. The nature of the oxide or oxides of an element is determined primarily by the position of that element in the periodic table.

A number of periodic trends can be discerned in the behavior of oxides. Oxides of elements on the left side of the periodic table, such as those of the alkali metals and the heavier members of the alkaline earth metals of group IIA, generally are ionic solids. Oxides of the metallic elements toward the middle of the table and of the semimetals also are solids, but often are not ionic. Oxides of nonmetals are discrete, separate molecules, generally existing as liquids or gases at room temperature.

The oxides of the metals on the left side of the periodic table generally are easily soluble in water and give alkaline solutions because of the formation of OH^- ions. For example, lithium oxide dissolves readily in water:

$$Li_2O(s) + H_2O \longrightarrow 2Li^+ + 2OH^-$$

MAGNETOHYDRODYNAMICS: ION POWER

In today's conventional generators, electric current is produced by moving a conductor (a wire loop called an armature) through a magnetic field. An alternative method proposed for future generating plants would use a hot ionized gas, "seeded" with a metal such as cesium, as the conductor. Such a generator would have no moving parts and would generate 50% more electricity for a given amount of fuel than today's generating plants do, proponents say.

However, this magnetohydrodynamic or MHD generator, as it is called, is difficult to bring into creation because it places unusual demands on parts and materials. To generate electricity the MHD way, a gas would be heated to about 2700 K, compared to the 750 K operating temperature of conventional plants. Cesium or another metal that is easily ionized would be added to increase the conductivity of the gas, which would be completely ionized at these high temperatures. The seeded gas would be expanded through a nozzle and would be sent at the speed of sound down a channel lined with electrodes. The combination of great heat, erosion caused by the speed of the gas, and corrosion caused by the metal particles means that the channel would have to be built of ceramics that are far more durable than the best available today.

The total operating experience with MHD generators amounts to a few hundred hours. The only MHD generator that is tied into a working electric power grid was built several years ago in the Soviet Union. The generator, located near Moscow, has had only brief periods of operation and has never run at its design capacity. However, the lure of much greater fuel economy has kept MHD research projects alive in several countries, including the United States, although spending is at a relatively low level.

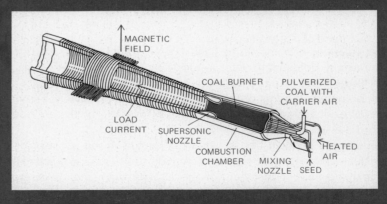

A diagram of a magnetohydrodynamic generator.

The oxides of the nonmetals on the right side of the periodic table generally dissolve in water to form acidic solutions (Section 2.6, page 40), as in the case of dichlorine heptoxide, which dissolves in water:

$$Cl_2O_7(l) + H_2O \longrightarrow 2H^+ + 2ClO_4^-$$

to form a solution of perchloric acid. Sulfur trioxide dissolves in water:

$$SO_3(g) + H_2O \longrightarrow H^+ + HSO_4^-$$

to form a sulfuric acid solution.

The behavior of the oxides of the representative, or main group, elements in the middle of the periodic table is less easily summarized, but there is a trend of increasing acidity and decreasing basicity in going from left to right across the table. Thus, some of the intermediate oxides formed by elements close to the left side of the periodic table are insoluble in water but dissolve in acidic solutions. Such oxides are classed as basic oxides. As an example, magnesium oxide (which in its hydrated form is the familiar milk of magnesia) dissolves only slightly in water but dissolves completely in acid:

$$MgO(s) + 2H^+ \longrightarrow Mg^{2+} + H_2O$$

Elements further to the right but still in the middle of the periodic table form some oxides that do not dissolve in water to any appreciable extent but do dissolve in both acidic and basic solutions. Such oxides are said to be *amphoteric,* meaning that they behave both as acids and bases. Aluminum oxide is amphoteric:

$$Al_2O_3(s) + 6H^+ \longrightarrow 2Al^{3+} + 3H_2O$$

and

$$Al_2O_3(s) + 2OH^- + 3H_2O \longrightarrow 2Al(OH)_4^-$$

It dissolves in acid to produce the equivalent of the Al^{3+} ion in solution and it dissolves in alkaline solution to produce the complex ion $Al(OH)_4^-$ in solution.

Further to the right in the periodic table are elements whose oxides do not dissolve in water but are considered to be acidic because they dissolve in alkaline solution. Silica, SiO_2, dissolves slowly in a strong alkaline solution:

$$SiO_2(s) + 2OH^- \longrightarrow SiO_3^{2-} + H_2O$$

Since glass is made primarily of SiO_2, this reaction explains why strongly alkaline solutions are stored in polyethylene containers rather than in glass bottles.

The trend of increasing acidity and decreasing basicity of oxides as we move from left to right across the periodic table is clear. Superimposed on this trend is a secondary trend of decreasing acidity and increasing basicity in moving down a given column of the periodic table. In group IIA, the oxide of magnesium, MgO, dissolves in acid solution but not to any appreciable extent in water. The oxide of barium, BaO, three rows lower in the same column, dissolves readily in water to produce an alkaline solution:

$$BaO(s) + H_2O \longrightarrow Ba^{2+} + 2OH^-$$

On the other side of the periodic table, such an increase in basicity can be seen in the oxides formed by the elements of group VA when they have the oxidation number $+5$. The oxide of nitrogen, N_2O_5, dissolves in water to produce a strongly acidic solution:

$$N_2O_5(s) + H_2O \longrightarrow 2H^+ + 2NO_3^-$$

The oxide of phosphorus, P_4O_{10}, dissolves in water to form phosphoric acid, an acidic solution of moderate strength:

$$P_4O_{10}(s) + 6H_2O \longrightarrow 4H_3PO_4$$

The corresponding oxide of arsenic, As_2O_5, behaves much the same as P_4O_{10}. But the oxides of antimony and bismuth, Sb_2O_5 and Bi_2O_5, are rather unstable and do not dissolve readily in water. They are neither markedly acid nor markedly basic, although there is some evidence suggesting that the oxide of bismuth is basic.

By classifying an oxide as acidic, basic, or amphoteric on the basis of its position in the periodic table, we can not only predict its behavior with water but we can also get a better understanding of other aspects of its chemical behavior. For example, acidic oxides and basic oxides react to form complex oxides that are actually salts:

$$Na_2O(s) + SO_3(g) \longrightarrow Na_2SO_4(s)$$

and

$$CaO(s) + SiO_2(s) \longrightarrow CaSiO_3(s)$$

The periodic correlation of oxides is complicated by the tendency of some elements to form a number of different oxides. Many of the transition metals and the nonmetals form several oxides. For nonmetals, the correlation of the properties of an oxide is best when the nonmetal has its highest oxidation number (Section 2.4, page 30). The properties of oxides such as CO_2, N_2O_5, SO_3, and Cl_2O_7 are closest to the periodic trends. Furthermore, there is a trend in a series of oxides of the same element: As the oxidation number of the element increases, the acidity of the oxide increases. Thus, while both oxides of sulfur, a nonmetal, are acidic, SO_3 (S = +6) is much more acidic than SO_2 (S = +4).

The relationship between the properties of an oxide and the oxidation number of the element forming the oxide is especially important among the transition elements, which tend to form many different oxides. The oxides of vanadium demonstrate this relationship:

Oxide	Oxidation Number of V	Properties
VO	+2	basic; dissolves only in strongly acidic solutions
V_2O_3	+3	basic; dissolves only in strongly acidic solutions
VO_2	+4	amphoteric; dissolves in both acidic and basic solutions but not in water
V_2O_5	+5	acidic; dissolves slightly in water to produce an acidic solution; but is much more soluble in acid solution than in water, so it resembles an amphoteric oxide.

Hydrides

Hydrides are binary compounds of hydrogen. There is a correspondence between the position of an element in the periodic table and the properties of its hydrides. The correspondence is not as strong as it is for oxides.

The hydrides of the alkali metals and the alkaline earth metals are saltlike ionic compounds that react with water to liberate hydrogen gas and form hydroxide ion, producing alkaline solutions. Sodium hydride reacts with water:

$$NaH(s) + H_2O \longrightarrow H_2(g) + OH^- + Na^+$$

The reaction of lithium hydride with water:

$$LiH(s) + H_2O \longrightarrow H_2(g) + OH^- + Li^+$$

was put to life-saving use in World War II, when it provided the hydrogen gas for inflating life rafts.

The hydrides of the main group elements toward the middle of the periodic table have complicated molecular structures, are often difficult to prepare, and are thermally unstable; that is, they decompose on heating. The hydrides of the transition metals usually are hard compounds with high melting points. They tend to have indefinite compositions, with varying numbers of hydrogen atoms fitted into spaces between the metal atoms in the solid. Hydrides of the semimetals usually are gases or liquids with low boiling points, usually not very stable thermally. Hydrides of the nonmetals also tend to be gases or liquids with low boiling points. They begin displaying some acid properties at group VA and are clearly acids at groups VIA and VIIA.

In addition to the general trend of increasing acidity of hydrides as we move from left to right across the periodic table, there are rough periodic trends in the stability of nonionic hydrides of the main group elements. If we exclude the hydrides of the alkali metals and the heavy alkaline earth metals, which are ionic, we find a tendency for increasing stability as we move from left to right across the periodic table, with an offsetting tendency toward decreasing stability as we move down a given column of the table. The hydrides of the heavier elements in groups IIIA–VIA are not stable, and often are virtually impossible to prepare.

The relatively stable hydrides, those which are most important to chemists, include many of the hydrides of the nonmetals and semimetals. Aqueous solutions of the hydrogen halides, HF, HCl, HBr, and HI, are important acids. Other familiar hydrides are methane, CH_4, the major component of natural gas; ammonia, NH_3, which has many practical uses; hydrogen sulfide, H_2S, whose rotten-egg smell is unmistakable; diborane, B_2H_6, an important rocket fuel; and H_2O, the most important hydride on earth.

Halides

The halides are binary compounds of an element and a halogen, one of the members of group VIIA. The common halides are the fluorides, the chlorides, the bromides, and the iodides. With the exception of the light noble gases, all the elements form compounds with halogens.

The halogen component of these compounds always has a negative oxidation number, except when oxygen is the other component. Fluorine alone has a negative oxidation number with oxygen. The other halogens all have positive oxidation numbers when combining with oxygen, and these compounds are more practically thought of as oxides.

Going from left to right across the periodic table there is a general trend in the properties of halogen compounds. Elements on the left side of the table, such as the alkali metals and the heavy alkaline earth metals tend to form stable ionic salts with high melting points. Elements on the right side of the table tend to form unstable nonionic halides that are gases or low-boiling liquids.

Halide salts do not react with water, but they usually are soluble in water. However, the nonionic halides of the transition elements and of many main group elements in the middle of the periodic table usually react with water, undergoing vigorous hydrolysis reactions (Section 2.7, page 41):

$$TiCl_4(l) + 2H_2O(l) \longrightarrow TiO_2(s) + 4HCl(g)$$
$$AlCl_3(s) + 3H_2O(l) \longrightarrow Al(OH)_3(s) + 3HCl(g)$$
$$PCl_5(s) + 4H_2O(l) \longrightarrow H_3PO_4(l) + 5HCl(g)$$

On the other hand, CCl_4 is a common solvent that does not undergo hydrolysis, and some other halides, such as NF_3 and SF_6, undergo hydrolysis very slowly. Often, it is difficult to discern clear periodic trends in the hydrolysis or general reactivity of the halides because of the interplay of a number of different factors.

Of the halides formed by elements in the first row of the periodic table, ionic lithium chloride does not undergo a hydrolysis reaction; both $BeCl_2$ and BCl_3 hydrolyze very readily; CCl_4 is almost completely unreactive toward water; NCl_3 and ClO_2 hydrolyze as readily as BCl_3; and ClF does not hydrolyze readily. On the basis of the position of these elements in the periodic table, the only pattern is one of unpredictability.

The same kind of irregularity is sometimes seen in the behavior of halides formed by elements in a given column of the periodic table. Of the halides formed by elements in group IVA, CCl_4, as we have seen, is unreactive toward water; $SiCl_4$ is very reactive; $GeCl_4$ undergoes only a mild hydrolysis reaction; and $SnCl_4$ is not hydrolyzed by pure water but is hydrolyzed by alkaline solution.

To explain this apparently irregular pattern of behavior of the halides, we must look not only at the position in the periodic table of the element that forms the halide but at the other factors that affect chemical behavior. The complex pattern of behavior of the halides illustrates the danger of looking for simple relationships between the properties of compounds and the position of the constituent elements in the periodic table. Nevertheless, the periodic table is by far the best single tool available for correlating patterns of chemical behavior.

We have seen that the quantum mechanical description of a hydrogen atom can be used as a basis for describing the electronic configurations of multielectron atoms. We saw how the orbitals in which electrons are placed can be described by sets of quantum numbers, and we learned how orbitals are grouped in shells and subshells on the basis of their energy. We saw that there is stability associated with half-filled and filled subshells, and we also saw how this stability helps explain the observed periodic variations in such properties as ionization energy and electron affinity. Finally, we saw how the periodic table of the elements presents much of this information in systematic, tabular form, and how many chemical properties of the elements and their compounds can be predicted on the basis of the position of the elements in the periodic table.

EXERCISES

6.1 Find the number that correctly completes each of the following statements:
(a) The number of possible values of l when $n = 4$ is _____ .
(b) The number of orbitals with the quantum numbers $n = 3$, $l = 2$ is _____ .
(c) The number of orbitals with the quantum numbers $n = 4$, $l = 4$ is _____ .
(d) The number of orbitals that have energy lower than the orbital with $n = 3$, $l = 0$ is _____ .
(e) The number of orbitals with quantum number $n = 5$ is _____ .
(f) The number of orbitals with $n = 4$, $l = 3$, $m_l = 2$ is _____ .
6.2 Indicate which of the following orbital designations is incorrect: (a) $3p_x$, (b) $9s$, (c) $3f$, (d) $4d_y$, (e) $5g$, (f) $2p$.
6.3 Arrange the following set of orbitals of a multielectron atom in order of increasing energy; group orbitals of the same energy together: $4d_{xy}$, $3d_{yz}$, $3d_{xy}$, $4p_z$, $3p_z$, $3p_y$, $2p_x$, $3s$, $2s$, $1s$.

6.4 Without consulting a periodic table write the electronic configuration of elements with the following atomic numbers: (a) 11, (b) 20, (c) 39, (d) 15, (e) 34, (f) 53.
6.5 Without consulting a periodic table find the atomic numbers of the noble gases, the elements at which a noble gas electronic shell is filled.
6.6 The alkaline-earth metals all have two more electrons than the noble gas nearest them in atomic number.
(a) Without consulting a periodic table, find the atomic numbers of the alkaline-earth metals. The last member of the group is radium, atomic number 88.
(b) Predict the atomic number of the next member of the group that may some day be discovered. (c) Predict the highest value of n for an electron in this element.
6.7 Write the electronic configurations of the following elements, showing the spins of electrons and the directional subscripts of orbitals in incompletely filled subshells: (a) Si, (b) P, (c) S, (d) Cr, (e) Fe, (f) Co.

6.8 List each element that has a partially filled krypton shell and only two electrons with unpaired spins.
6.9 Predict the electronic configuration of the element whose position in the periodic table is just below each of the following elements: (a) 5, (b) 15, (c) 25, (d) 50.
6.10 (a) Predict the electronic configuration of the element below Ni in the periodic table by analogy with the electronic configuration of Ni.
(b) Experiments reveal that this element does not have the electronic configuration you predicted. Suggest the most likely alternative for the electronic configuration.
*6.11 There is currently a dispute about the name of the recently discovered element 106. There is also some question about the expected electronic configuration for this element. Use the periodic table and the data in Table 6.6 to predict the two most likely possibilities.
6.12 Tabulate the quantum numbers of every electron in the O atom.

6.13 Predict the number of unpaired electrons that are found in the following ions: (a) Sc^+, (b) Ti^{2+}, (c) Mg^+, (d) Cl^-, (e) Si^-.

6.14 As part of a more positive approach to chemistry we wish to propose a theory which states that negative numbers are not permitted as values of the m_l quantum number. The Pauli exclusion principle and all other aspects of the Aufbau method are still valid. The elements still have the same names and atomic numbers, only their electronic configurations are changed. Name the elements described below:
(a) the first noble gas after helium
(b) the first element to have a d electron
(c) the first element in which the $3d$ subshell is filled
(d) the element below neon in the new periodic table

6.15 Find the ratio of the total energy of the electron in the ground state of hydrogen to:
(a) the electron in the ground state of He^+
(b) the electron in the ground state of Li^{2+}
(c) the electron in the first excited state of Be^{3+}
(d) the electron in the ground state of B^{4+}

6.16 Assuming that screening of the outermost electron by the inner electrons is complete find the ratio of the total energy of the electron in the ground state of hydrogen to:
(a) the $3p$ electron in Al
(b) the $3s$ electron in Na
(c) the $3d$ electron in Sc^{2+}

6.17 Predict which of the three results found in Exercise 6.16 will be most different from the experimentally determined result. Justify your prediction.

6.18 Find the relationships between the number of valence electrons of an atom and the group number of the atom in the periodic table.

6.19 The smooth decrease in ionization energy with increase in atomic number found in groups of the periodic table with the letter designation A is not found in group IIIA. Propose explanations for the trend in the ionization energies listed in Table 6.7 for the elements of this group.

6.20 Select one of the statements listed in each group below as the one that was probably made first by chemists. Justify your selection.
(a) (i) Because noble gases have only completed subshells they will be unreactive; or (ii) because noble gases are unreactive they have only completed subshells.
(b) (i) Chromium will have a half-filled $4s$ subshell because of the stability associated with half-filled shells; or (ii) the half-filled $4s$ subshell of chromium reflects the stability associated with half-filled shells.
(c) (i) Nitrogen has a relatively high first ionization energy because of the stability associated with a half-filled electronic subshell; or (ii) because of the stability associated with a half-filled electronic subshell nitrogen will have a relatively high first ionization energy.

6.21 Account for the observation that the electron affinity of chlorine is lower than the ionization energy of argon.

6.22 Without consulting a periodic table write the electronic configurations of: (a) the third alkali metal, (b) the fourth halogen, (c) the fifth element in group IVA, (d) the third element in group IIB.

6.23 Arrange the three elements Se, Br, and I in order of increasing: (a) ionization energy, (b) electron affinity, (c) atomic radius, (d) atomic volume, (e) electrical conductivity.

6.24 Arrange the three elements Li, Be, and Na in order of increasing:
(a) ionization energy, (b) atomic radius, (c) reactivity toward water.

6.25 The electron affinity of sodium is lower than that of lithium, while the electron affinity of chlorine is higher than that of fluorine. Suggest an explanation for this observation.

6.26 Arrange the isoelectronic species S^{2-}, Cl^-, Ar, K^+, Ca^{2+}, and Sc^{3+} in order of increasing: (a) ionization energy, (b) ionic radius, (c) electron affinity.

6.27 Table 6.10 does not include the element francium (Fr), the heaviest alkali metal, because all of its isotopes are unstable. No one has ever prepared enough Fr to weigh, but some of its properties have been measured by special techniques. Predict the values of the required entries for Fr in Table 6.10. Predict the nature and products of the reaction of Fr with water, chlorine, and oxygen.

6.28 Three of the oxides of the group VA elements in the $+3$ oxidation state are acidic, one is amphoteric, and one is basic. Write the empirical formula of each oxide and identify each as acidic, basic, or amphoteric.

6.29 Chromium forms oxides of the $+3$, $+4$, and $+6$ oxidation states. The $+3$ oxide is amphoteric. Write the formulas and predict the nature of the other two oxides.

6.30 Make a list of the formulas of the following oxides in order of decreasing basicity with the most basic oxide first and the most acidic one last: (a) sodium oxide, (b) potassium oxide, (c) calcium oxide, (d) thallium(I) oxide, (e) thallium(III) oxide, (f) silicon dioxide, (g) tetraarsenic decoxide, (h) selenium trioxide, (i) dibromine heptoxide, (j) diiodine heptoxide.

6.31 Write equations for the reaction of the hydrides of the following elements with water at 298 K. If no appreciable reaction takes place do not write an equation: (a) sodium, (b) calcium, (c) aluminum, (d) carbon, (e) chlorine.

6.32 Write equations for the reaction of the chlorides of the following elements with water: (a) beryllium, (b) boron, (c) nitrogen, (d) silicon, (e) zirconium.

6.33 You believe that you have cracked a code that uses elemental symbols to spell words. The code uses numbers to designate the elemental symbols. Each number is the sum of the atomic number and the principal quantum number of the highest occupied orbital of the element whose symbol is to be used. You have determined that the message may be written either forward or backward. Use the periodic table and Table 6.6 to decipher the following messages:
(a) 10, 12, 58, 11, 7, 44, 63, 66
(b) 9, 99, 30, 95, 19, 47, 79

THE CHEMICAL BOND

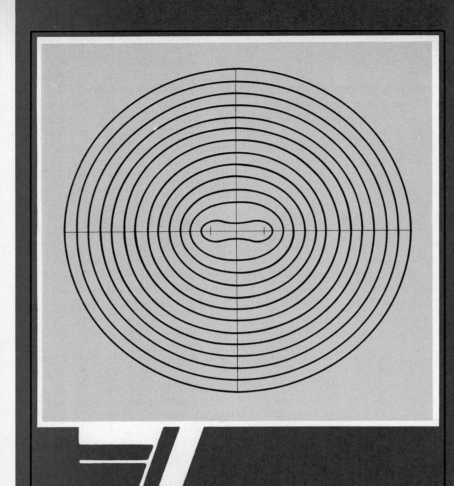

7

Chemistry usually does not deal with isolated atoms. Most of the time, chemistry studies atoms that are held together by attractive forces. The nature of these forces and the way in which they act to hold atoms together are of central importance in chemistry.

The attractive force between two atoms can be regarded in the simplest terms as the coulombic attraction between electric charges of opposite sign. The potential energy of a system is lowered as the charges come closer to each other and the coulombic attraction between charges of opposite sign increases. We have already discussed the effect for the nucleus and the electrons of a single atom. It should be kept in mind that lower potential energy is synonymous with greater stability. *The lower the potential energy of a given system, the more stable is the system*. The attractive forces between atoms can be classified by the extent to which they cause the potential energy of the system to be lowered compared to the potential energy of the isolated atoms. When the potential energy is lowered by at least 40 kJ/mol, we say that the atoms are held together by a chemical bond.

Two atoms will form a chemical bond when the net attractive forces make it more favorable for the atoms to be close to each other than to be apart. A more complete understanding of the chemical bond requires an understanding of the valence electronic configurations of the atoms that are joined together.

While the theory of chemical bonding can be quite complex, there is a useful simplication based on two idealized types of bonding that can serve as models for the description of most chemical bonds. The two types of bonds, which are opposite extremes, are:

1. Ionic bonds, in which there is a complete transfer of the electrons of the bond between the atoms that are bonded.
2. Covalent bonds, in which two atoms share the electrons of the bond equally.

Most chemical bonds are somewhere between the extremes of ionic and covalent bonding, closer to one than the other but having characteristics of each. In practice, a bond usually is described as being either ionic or covalent, depending on which extreme the bond comes closest to matching.

7.1 IONIC BONDS

In an ionic bond, the entities that are bonded together can be described as ions of opposite charge. An ionic bond is the result of the electrostatic attraction between positively charged cations and negatively charged anions. According to Coulomb's law, the force of this attraction is:

$$F \propto \frac{e_1 e_2}{r^2}$$

and the potential energy due to this attraction is:

$$E \propto \frac{e_1 e_2}{r}$$

where e_1 and e_2 are the magnitudes of the anionic and cationic charge and r is the distance between the centers of the ions. The energy E has a negative value, since e_1 and e_2 are of opposite sign, and the stability of the ionic bond increases as E becomes more negative.

In order for two atoms to form an ionic bond, they must transfer electrons.

185

One atom loses one or more electrons to become a cation; the other atom gains one or more electrons to become an anion. To understand the formation of ionic bonds, let us study sodium chloride, which is best described as an ionic compound consisting of Na^+ cations and Cl^- anions. An observation of the chemical reaction that forms sodium chloride tells us that it is a stable compound. When sodium metal and chlorine gas are mixed, there is a rapid, even violent reaction. Much heat is liberated and sodium chloride is formed:

$$Na(s) + \tfrac{1}{2}Cl_2(g) \longrightarrow NaCl(s) + heat$$

Sodium chloride can be converted back to elemental sodium and elemental chlorine, but a good deal of energy in the form of heat or electricity is needed to decompose the compound. The fact that a great deal of energy is liberated when sodium chloride is formed and that a great deal of energy is needed to decompose the compound tells us that sodium chloride is stable.

Formation of Ions and Ion Pairs

To understand the stability of a salt such as sodium chloride, we can analyze its formation as the result of several different processes. Some of these processes require the input of energy, and thus are unfavorable for the formation of sodium chloride. Other of these processes liberate energy, and thus are favorable for its formation. We can study the overall process in the way that an accountant goes over a balance sheet, totaling the energy ''assets'' and ''debits.'' When we do so we find that the balance sheet is favorable for sodium chloride.

Let us consider the transformations that $Na(s)$ and $Cl_2(g)$ might undergo to become sodium chloride. First we can imagine that $Na(s)$ becomes $Na(g)$ and the $Cl_2(g)$ becomes $Cl(g)$. Then we can imagine that there are three major transformations: the formation of $Na^+(g)$ ions from $Na(g)$; the formation of $Cl^-(g)$ ions from $Cl(g)$; and the formation of the ionic bond between the two ions of opposite charge. The formation of $Na(g)$ from $Na(s)$ and of $Cl(g)$ from $Cl_2(g)$, which must also occur, can be neglected at this stage.

The energy associated with the formation of Na^+ ions from Na in the gas phase is simply the ionization energy of Na (Section 6.4, page 162). The ionization energy of Na is 5.14 eV. We can write:

$$5.14 \text{ eV} + Na(g) \longrightarrow Na^+(g) + e^-$$

The ionization energy of sodium is a positive quantity. An input of energy is needed to remove an electron from a sodium atom. The energy input is 5.14 eV or 496 kJ/mol.

From the values of the electron affinities found in Table 6.8, we can see that adding an electron to a Cl atom:

$$Cl(g) + e^- \longrightarrow Cl^-(g) + 3.61 \text{ eV}$$

actually releases energy. We can say that -3.61 eV, or -347 kJ/mol is required. A negative sign for the energy of a given process means that the process liberates that quantity of energy. A positive sign tells us that a process absorbs energy. A negative sign tells us that a process leads to the formation of products with lower potential energy. A positive sign tells us that a process leads to the formation of products with higher potential energy. By extending this line of reasoning, we can see that the sign of the energy required for a given process is reversed for the reverse process.

The two processes by which sodium and chlorine form ions can be thought of as occurring together: The electron that is removed from the sodium atom is added to the chlorine atom. The stability of sodium chloride often is ascribed to the fact that both atomic species in NaCl have achieved the stable closed-shell electronic configuration of a noble gas; Na^+ has the neon configuration, $[He]2s^22p^6$ and Cl^- has the argon configuration, $[Ne]2s^22p^6$.

The energy data cited above show that the completion of noble gas shells cannot account for the stability of sodium chloride. Adding the two reaction equations and the associated energies, we find that energy must actually be added to bring about the electron transfer process:

Reaction	Energy Required
$Na(g) \longrightarrow Na^+(g) + e^-$	$+496$ kJ/mol
$Cl(g) + e^- \longrightarrow Cl^-(g)$	-347 kJ/mol
$Na(g) + Cl(g) \longrightarrow Na^+(g) + Cl^-(g)$	$+149$ kJ/mol

The overall process in which an electron is transferred from a gaseous Na atom to a gaseous Cl atom requires an input of 149 kJ/mol. Even though it is advantageous in energy for a Cl atom to gain an electron, the energy gained by adding an electron to a Cl atom does not compensate for the energy cost of removing an electron from an Na atom. Such an unfavorable energy balance is not unique to NaCl. In fact, an unfavorable energy balance must always be expected in the formation of a pair of ions. The data in Tables 6.7 and 6.8 show that the lowest ionization energy (3.89 eV for Cs) is greater than the highest electron affinity (3.6 eV for Cl).

Thus, we can calculate that an input of energy is always needed to form an anion and a cation from a pair of elements. But in the laboratory, we can see that the formation of most ionic compounds is accompanied by the release of energy. We must add to or refine our theory.

For example, we can include in our calculations the favorable effect of the electrostatic attractions between the anions and the cations of an ionic compound. We have calculated that the formation of the Na^+ cation and the Cl^- anion in NaCl requires an input of 149 kJ/mol of energy. But the process in which the two ions come together:

$$Na^+(g) + Cl^-(g) \longrightarrow NaCl(g)$$

releases 585 kJ/mol. That is, the energy required for the process is -585 kJ/mol. This release of energy is due to the coulombic attraction of the oppositely charged ions when the distance between their nuclei is reduced to that observed in NaCl(g), about 0.24 nm. The Na^+ cation and the Cl^- anion do not approach any closer because of coulombic repulsions between the inner-shell electrons.

There are even substances in which the release of energy due to the close approach of the oppositely charged ions is not enough to offset the energy required to form these ions. Yet these ionic substances form readily from their elements, and energy is released in their formation.

Ion Crystal Lattices

We can solve the problem if we realize that our calculations have assumed that ionic compounds exist as ion pairs in the gas phase. In fact, ionic compounds do not usually exist in this state. Under normal conditions, ionic compounds are solids. The ions are in a regular, repeating three-dimensional arrangement called a *lattice*.

The stability of ionic compounds is due primarily to favorable coulombic attractions. These attractions are much greater when ions are brought together in the lattice of a crystalline solid than when the same number of ions exist in the gas phase. In the solid phase, each ion interacts with a large number of ions of opposite charge. These interactions between many ions of opposite charge are responsible for the stability of ionic compounds.

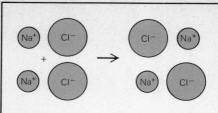

FIGURE 7.1
Formation of a sodium chloride crystal lattice from ion pairs. In the lattice, the distance between like ions generally is greater than the distance between unlike ions, maximizing the favorable coulombic forces.

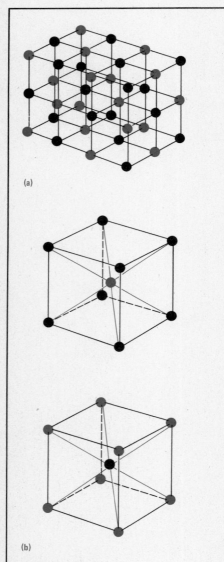

(a)

(b)

FIGURE 7.2
Two common types of ionic crystal lattices. In the sodium chloride lattice (a), each ion is surrounded by six ions of opposite charge. In the cesium chloride lattice (b), each ion is surrounded by eight ions of opposite charge.

Consider the simplest case, two sodium chloride ion pairs that come together as shown in Figure 7.1. Each Na^+ ion attracts two Cl^- ions, while each Cl^- ion attracts two Na^+ ions. Instead of the two favorable coulombic attractions that exist for two separate ion pairs in the gas phase, there are a total of four attractions. These attractions are offset somewhat by the repulsions that result when two Na^+ ions are brought close together and two Cl^- ions are brought close together. But the geometry of this NaCl arrangement shows that the gain of stability due to the attractive forces is much greater than the loss of stability due to repulsive forces.

If we regard the four ions in this arrangement as being at the corners of a square, the distance between the centers of like ions (Na^+ and Na^+ or Cl^- and Cl^-) is greater than the distance between the centers of the unlike ions. We have seen that the energy due to coulombic attraction or repulsion is proportional to $1/r$. Therefore, the loss in stability due to repulsion is not as great as the gain from the attraction.

A three-dimensional array can be built up by assembling ion pairs. The energy advantage increases every time an ion pair is added to the crystal lattice, but only if the arrangement of the ions results in greater attraction than repulsion. The geometry of the ions in the lattice plays an important role in determining the extent to which an arrangement of ions is more stable than isolated ion pairs. In order to form, a lattice must be more stable than the same number of ion pairs.

Figure 7.2 shows two common types of crystal lattices. In the sodium chloride lattice, each Na^+ ion is surrounded by six Cl^- ions, which are its nearest neighbors, and each Cl^- ion is surrounded by six Na^+ ions. For such a lattice, we say that each ion has a *coordination number* of 6. In the cesium chloride lattice, each ion has a coordination number of 8, because each Cs^+ ion is surrounded by eight Cl^- ions and each Cl^- ion is surrounded by eight Cs^+ ions.

The formation of crystal lattices, three-dimensional arrays of ions arranged for the maximum coulombic attraction between ions, is the major reason for the stability of ionic compounds.

We mentioned that the formation of an ionic lattice from isolated ions in the gas phase is accompanied by the release of energy. For the process:

$$Na^+(g) + Cl^-(g) \longrightarrow NaCl(s)$$

the energy is -765 kJ/mol; that is, the process releases 765 kJ/mol. The energy released by such a process is called the **lattice energy**; we say that the lattice energy of sodium chloride is 765 kJ/mol.

Lattice energy can be taken as a measure of the strength of an ionic bond. The greater the lattice energy, the stronger the bond. As expected, there is a greater gain in stability from the formation of the solid ionic lattice of sodium chloride (765 kJ/mol) than from the formation of ion pairs of sodium chloride in the gas phase (585 kJ/mol).

We can get the overall process by which an ionic solid is formed from gaseous atoms of the constituent elements by combining two processes. The first is the formation of ions in the gas phase from atoms of their elements. The second is the formation of the lattice from ions in the gas phase. The associated energies can also be combined:

Reaction	Energy Required
$Na(g) + Cl(g) \longrightarrow Na^+(g) + Cl^-(g)$	$+149$ kJ/mol
$Na^+(g) + Cl^-(g) \longrightarrow NaCl(s)$	-765 kJ/mol
$Na(g) + Cl(g) \longrightarrow NaCl(s)$	-616 kJ/mol

There is a difference of about 200 kJ/mol between the energy liberated by this two-step process and the energy actually liberated by the reaction between sodium and chlorine. The difference is accounted for by the energy needed to form Na(g) from Na(s) and to form Cl(g) from $Cl_2(g)$. We shall take this point up in more detail in Section 8.3.

Geometry and Lattice Energy

The geometry of the crystal lattice is an important influence on the magnitude of the lattice energy. We can define a geometric factor that is independent of the ionic charge and the distance between ions. This factor relates the geometry of the lattice to the difference in energy between ions in the lattice and ion pairs in the gas phase. This factor is called the *Madelung constant*.

The Madelung constant of every crystal lattice is greater than 1. In a sodium chloride crystal, for example, we can calculate the attractions between an ion and its six nearest neighbors, then add the repulsions between the same ion and its twelve next nearest neighbors, then include the more complex interactions with more distant ions. Eventually, we can calculate that the coulombic attraction per ion pair is 1.75 times greater in the crystal lattice than for a single NaCl ion pair. In other words, the Madelung constant for the sodium chloride crystal lattice is 1.75 and E, the potential energy of attraction is $(1.75)\ e_1e_2/r$.

All ionic solids with the same geometry have the same Madelung constant. For example, any ionic crystal with the same lattice geometry as sodium chloride has the Madelung constant 1.75. The Madelung constant for the cesium chloride lattice and all lattices with the same geometry is 1.76. Some other crystal lattice configurations have larger Madelung constants.

When we talk about the sodium chloride ionic bond, we are talking not about the attraction in an Na^+Cl^- pair but about the attractions between many Na^+ and Cl^- ions in the lattice. To start with, there are the attractions between one Na^+ ion and its six nearest-neighbor Cl^- ions in the crystal lattice. Each of these Cl^- ions, in turn, has six nearest-neighbor Na^+ ions. We can simplify this picture by assuming that each Na^+ ion "sees" only one-sixth of each nearest-neighbor Cl^- ion, and vice versa. Thus, the sum of the six one-sixth attractions is six sixths, or one ionic bond.

This simple accounting illustrates the characteristic feature of an ionic crystal—the fact that there really is no such phenomenon as a single anion-cation association in the lattice. Rather, the sodium chloride crystal can be visualized as a giant unit that includes all the Na^+ and Cl^- ions. It is impractical to write the formula for such a giant collection of ions. Instead, we write the formula NaCl, indicating the relative numbers of Na^+ and Cl^- ions, whose ratio in the ionic crystal is 1:1. It is rather misleading that the same formula can be used to refer to a specific NaCl(g) molecule in the gas phase, a species that is rarely encountered.

Ionic Radius

A second influence on lattice energy is the distance between ions in the crystal lattice. From Coulomb's law, we know that the strength of the attraction increases and the energy of the system decreases as the distance between ions of opposite charge decreases. But as the ions get closer together, the force of the repulsion between their inner electrons begins to become important (Figure 7.3). Eventually, the force of the repulsion between the inner electrons increases faster with decreasing distance than the coulombic attraction does. There is a point, r_0 (in Figure 7.3), which is the most favorable distance of separation for the ions. The distance that represents the minimum on the energy curve is the most likely distance between the centers of two ions in the gas phase.

If certain assumptions are made, each ion can be assigned an ionic radius that is a measure of its size. In most cases, the distance between the centers of any two adjacent ions in a crystal lattice is close to the sum of their ionic radii, as

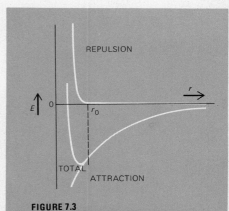

FIGURE 7.3
Variation of the potential energy of oppositely charged ions with distance. The point at which any further increase in the coulombic attraction of the ions with a decrease in distance is offset by the repulsion of the ions' inner electrons is the most favorable distance of separation for the ions in a lattice, and is the minimum of the curve.

TABLE 7.1
Ionic Radii (nm)

Li⁺	Be²⁺	B³⁺	N³⁻	O²⁻	F⁻
Li^+ 0.060	Be^{2+} 0.031	B^{3+} 0.020	N^{3-} 0.171	O^{2-} 0.140	F^- 0.136
Na^+ 0.095	Mg^{2+} 0.065	Al^{3+} 0.050	P^{3-} 0.212	S^{2-} 0.184	Cl^- 0.181
K^+ 0.133	Ca^{2+} 0.099			Se^{2-} 0.198	Br^- 0.195
Rb^+ 0.148	Sr^{2+} 0.113			Te^{2-} 0.221	I^- 0.216
Cs^+ 0.169	Ba^{2+} 0.135				

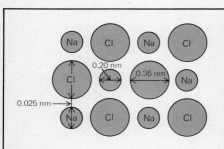

FIGURE 7.4
A section of an NaCl crystal lattice, showing ionic radii and distance between ions. In this and most other crystal lattices, the distance between the centers of two adjacent ions is roughly the sum of their ionic radii.

Figure 7.4 shows. In Section 6.7, we discussed trends in ionic radius as a function of the position of the elements in the periodic table. In a given column, ionic radius usually increases with increasing atomic number. In a given row, ionic radius tends to decrease with increasing atomic number. In any group of isoelectronic ions—ions with the same number of electrons—ionic radius decreases with increasing atomic number, because of the increase in nuclear charge. Table 7.1 shows some of these trends.

Figure 7.5 shows the relative sizes of a number of ions. Note that anions generally are much larger than cations.

The radii of ions in a crystal lattice play an important role in determining lattice energy. All other things being equal, lattice energy is highest when the ions forming the lattice are smallest and therefore closest together. Since $E \propto e_1 e_2 / r$, a smaller value of r increases the lattice energy.

Table 7.2 gives the lattice energy of alkali halides, that is, the value of the energy released in the reaction:

$$M^+(g) + X^-(g) \longrightarrow MX(s)$$

where M is an alkali metal and X is a halogen. The data show that for a given negative ion, the lattice energy decreases as the atomic number of the alkali metal increases. For a given positive ion, the lattice energy decreases as the atomic number of the halogen increases. The decrease in lattice energy is due to the larger ionic radii of elements of higher atomic number. Some of these crystal lattices have the sodium chloride geometry. Others have the cesium chloride geometry. There is no noticeable difference in lattice energy associated with the

TABLE 7.2
Lattice Energies of Alkali Halides (kJ/mol)[a]

LiF 1030	LiCl 840	LiBr 781	LiI 718
NaF 914	NaCl 770	NaBr 728	NaI 681
KF 812	KCl 701	KBr 671	KI 632
RbF 780	RbCl 682	RbBr 654	RbI 617
CsF 744	CsCl 630	CsBr 613	CsI 585

[a] Expressed as positive numbers. Energy actually is released when the lattice forms from gas phase ions.

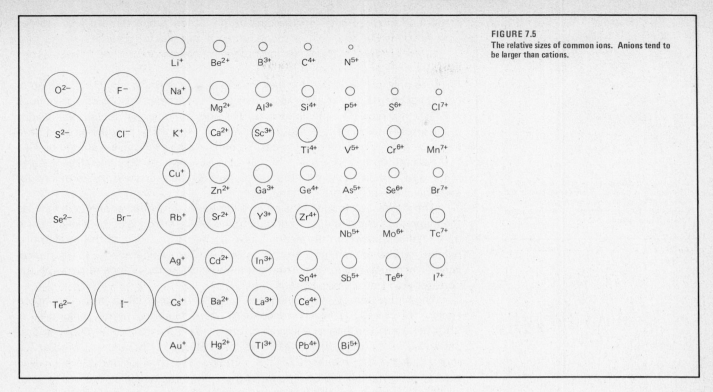

FIGURE 7.5
The relative sizes of common ions. Anions tend to
be larger than cations.

difference in geometry, because the difference in Madelung constants between these two geometries is negligible.

A good rule to remember is this: The stability due to coulombic interactions between ions in a crystal lattice is greater for small ions. The relative size of the ions in a crystal lattice can also have an important influence on the details of lattice geometry. For example, the radius of an ion and of its oppositely charged neighbor affects the coordination number of that ion—the number of its nearest neighbors of opposite charge.

Figure 7.6 illustrates the influence of ionic radii on coordination number. If there is a great difference in the size of two ions of opposite charge, it is not possible to place many of the larger ions around the smaller ion. In most of the alkali halides, where the anions are larger than the cations, we find the sodium chloride lattice geometry, with a coordination number of 6. Where the difference in the size of the ions is greater, the coordination number is smaller. In ZnS, the great difference in size between the Zn^{2+} cation and the S^{2-} anion leads to a coordination number of 4. Each Zn^{2+} ion has four S^{2-} ions as nearest neighbors, and each S^{2-} ion has four Zn^{2+} ions as nearest neighbors. When the anion and the cation are nearly the same size, as in CsCl, CsBr, or CsI, the ions form the cesium chloride lattice, with its coordination number of 8.

When there is an even more pronounced difference in size between ions, the simple sum of the ionic radii may not be enough to indicate the distance between the cations and the anions in a lattice. As Figure 7.7 shows, if the anion is much larger than the cation, the anions will "touch" before the anion and the cation touch. In such a lattice, the radius of the larger ion determines the distance between anion and cation.

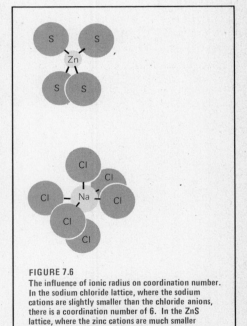

FIGURE 7.6
The influence of ionic radius on coordination number. In the sodium chloride lattice, where the sodium cations are slightly smaller than the chloride anions, there is a coordination number of 6. In the ZnS lattice, where the zinc cations are much smaller than the sulfur anions, the coordination number is 4.

Ionic Charge and the Ionic Bond

We have seen that lattice energy is affected by the geometry of the lattice as well as by the distance between ions, and that the strength of the ionic bond is directly related to the lattice energy. But the major determinant of lattice energy is the

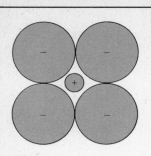

FIGURE 7.7
The effect of differences in size on ionic lattice formation. If the positive ion is much smaller than the negative ion, the anions will touch and the radius of the anions will determine the internuclear distances in the lattice.

magnitude of the charge on the ions in the lattice, since the attractive forces between ions depend on the magnitudes of the electric charges of the ions (Coulomb's law). It follows that lattice energy will be greater in lattices of ions with greater charges.

The only ionic solid we have examined closely is NaCl, which is made up of cations of charge $1+$ and anions of charge $1-$. There are also ionic compounds in which the anions and the cations can have greater charges, such as the halides of the alkaline-earth metals ($MgCl_2$, BaF_2, $CaBr_2$, etc.). The halide ions each have a charge of $1-$; but the alkaline-earth metal cations each have a charge of $2+$. The general formula for such compounds is MX_2, where M is any alkaline-earth metal and X is any halide. In the crystal lattice, there are two halide anions for each alkaline-earth metal cation.

The increased charge of the cations increases the lattice energy. We can see the magnitude of the effect by comparing the lattice energies of the alkaline-earth metal halides, in which the cations have a charge of $2+$, and the alkali metal halides, whose cations have a charge of $1+$. The lattice energy of each alkaline-earth metal halide is substantially larger than the lattice energy of the corresponding alkali metal halide.

The lattice energy of NaCl is listed in Table 7.2 as 770 kJ/mol. The lattice energy of $MgCl_2$, the corresponding alkaline-earth metal chloride, is 2500 kJ/mol. The lattice energies of other alkaline-earth metal chlorides are: $CaCl_2$, 2260 kJ/mol; $SrCl_2$, 2130 kJ/mol; $BaCl_2$, 2050 kJ/mol. Table 7.2 shows that the lattice energies of the corresponding alkali chlorides are lower in every instance. (It should be noted that there are three moles of ions in a mole of an alkaline-earth metal chloride and two moles of ions in a mole of an alkali metal chloride. Since lattice energy is measured per mole of formula, not per mole of total ions, the difference contributes to the relatively higher lattice energy of the alkaline-earth metal chlorides.) While such factors as lattice geometry play a role, the major reason for the higher lattice energies is the larger charge on the alkaline-earth metal cations. An even more striking example of the effect of ionic charge on the magnitude of the lattice energy can be seen in a compound such as calcium oxide, CaO. Here the anion has a charge of $2-$ and the cation has a charge of $2+$. The lattice energy of CaO is 3600 kJ/mol, much larger than the lattice energy of the halides.

Noble Gas Electronic Configurations

We have shown that the stability of ionic compounds is not due to the noble gas electronic configurations of the ions making up such compounds. The formation of these ions is actually an unfavorable process that requires an energy input. The stability of ionic solids is due to the lattice energy that is released when ions come together in a crystal array. But it is notable that the ionic compounds of main group elements almost always are made up of ions with noble gas electronic configurations. Why should this be so?

The best way to answer this question is to explain why some compounds are *not* formed. Why, for example, are there no salts in which an alkali metal forms a cation of charge $2+$, such as $NaCl_2$, when we would expect the dipositive Na^{2+} cation to produce a much larger lattice energy than is found in NaCl?

It is true that $NaCl_2$ would have a larger lattice energy than NaCl. But this predicted increase in lattice energy is not great enough to offset the very substantial amount of energy needed to form the Na^{2+} ion. Table 6.7 shows that the second ionization energy of sodium is enormous—47.286 eV, or approximately 4600 kJ/mol. This second ionization energy is much greater than the anticipated increase in lattice energy of $NaCl_2$.

Table 6.7 shows that an extremely large amount of energy is required to remove an electron from any species with a noble gas electronic configuration. The table lists very large values for the second ionization energies of the group IA metals, the third ionization energies of the group IIA metals, the fourth ionization energies of the group IIIA metals, and so on.

These high ionization energies can be explained by the relatively low principal quantum number of the electron that must be removed from a completed electron shell. For example, the first ionization of sodium removes the $3s$ electron, while the second ionization removes a $2p$ electron. The second ionization of Ba removes a $6s$ electron, while the third ionization removes a $5p$ electron. The ionization energies are so high that species such as Ba^{3+} are rarely formed. Therefore, we can expect that salts of metals will be composed of metal cations which have the noble gas electronic configuration, and that the maximum positive charge on the metal cation will be the same as the number of that metal's group in the periodic table.

The same reasoning can be applied to anions: An excessive amount of energy is needed to add electrons to an anion with a noble gas electronic configuration. Therefore, compounds such as Na_2Cl, which include a Cl^{2-} anion, do not exist. Energy is even needed to form multinegative anions such as O^{2-} and N^{3-}, which do have noble gas electronic configurations. The energy needed to add an electron to a species such as Cl^- is still greater. The lattice energy increase to be expected from a hypothetical Cl^{2-} ion is negligible by comparison. A great deal of energy is needed to form the Cl^{2-} anion because the electron that is added to the Cl^- anion is in a $4s$ orbital. The relatively small coulombic attraction between this electron and the Cl nucleus is not nearly as large as the electron-electron repulsions introduced by the addition of the extra electron to the Cl^- anion.

The same electron-electron repulsions work against the addition of electrons to other anions with noble gas electronic configurations. Therefore, anions found in salts do not have more electrons than can be accommodated by a noble gas electron shell. The maximum negative charge on an anion in an ionic compound can be determined by subtracting the group number of the element from 8. Group VIIA anions have a charge of $1-$, group VIA anions have a charge of $2-$, and so on.

But ionization energies do not explain why we do not observe ionic solids made up of ions that have electronic configurations on the "other side" of noble gas electronic configurations—for example, an Mg^+ cation, which has one more electron than a noble gas configuration, or an O^- anion, which has one electron less than a noble gas configuration. It can be shown that a compound such as MgCl(s), which contains the Mg^+ cation, is quite stable. The first ionization energy of Mg is not very large. It is readily offset by the electron affinity of Cl and the lattice energy of MgCl(s). Yet MgCl(s) is not encountered, because the compound can quickly react with itself to produce $MgCl_2(s)$, which has substantially greater lattice energy:

$$MgCl(s) + MgCl(s) \longrightarrow Mg(s) + MgCl_2(s) + energy$$

In the solid phase, MgCl(s) is so reactive that it does not exist. But in the gas phase, the chance of two MgCl molecules reacting with each other is much lower, and small amounts of MgCl(g) have been observed.

Compounds such as MgCl, in which one ion (in this case the Mg^+ cation) is on the "wrong side" of a noble gas electronic configuration, do not exist because there is a substantial gain in lattice energy when they are transformed into compounds whose ions do have the noble gas electronic configuration. They may be *stable*—lower in energy than their constituent elements—but they are excessively *reactive* because of a still more stable possibility. Such compounds are normally short-lived.

Electronegativity and Ionic Bonding

The ionic bond, in which electrons are transferred from one atom to another, is an idealized model for the description of some chemical bonds. We may now ask: Which bonds?

The answer depends on the elements that take part in the bonding. The basic premise of the ionic bond is that one or more electrons are transferred from atoms that become cations to atoms that become anions. Therefore, it is logical to assume that a bond between two atoms will be ionic if one atom has a much greater attraction for electrons than the other atom.

Ionization energy and electron affinity are measures of the facility with which an isolated atom in the gas phase loses or gains electrons. It would be desirable to have a similar measurement for the more complicated situation in which an atom that is combined with other atoms gains or loses electrons. Unfortunately, no such simple measurement can be made. To fill the need, an imaginary property of atoms called **electronegativity** was created.

The electronegativity of an atom is a measure of the power of that atom to attract electrons to itself when combined with other atoms. An atom that has a high electronegativity value has a relatively great attraction for electrons when it is combined with other atoms. A number of measured properties can be used to assign electronegativity values to atoms. One method is to assign electronegativity values on the basis of ionization energy and electron affinity values. Trends in electronegativity parallel the trends in ionization energy and electron affinity. Electronegativity increases from left to right across the periodic table. It also increases from the bottom to the top of the periodic table. In general, the increase is more pronounced from left to right than from bottom to top.

Since electronegativity is defined as the measure of the power of a combined atom to attract electrons, it gives us a measure of the ionic character of a bond. If there is a great difference between the electronegativity values of two atoms, those atoms are likely to form ionic bonds. Table 7.3 lists the electronegativity values calculated for most of the elements by Linus Pauling (1901–). There is an arbitrary but useful rule of thumb based on the values listed in Table 7.3: If the difference in electronegativity between two elements is greater than about 1.7, their bonding can be described as predominantly ionic. Therefore any compound that has a bond between atoms whose electronegativities differ by 1.7 or more is an ionic compound.

However, when we say that a binary compound has ionic bonding, we do not mean that there is a complete transfer of one or more electrons from one atom to the other, with the complete formation of ions. Rather, we mean that the electron

TABLE 7.3
Electronegativities of the Elements

H 2.2																
Li 1.0	Be 1.6											B 2.0	C 2.6	N 3.0	O 3.4	F 4.0
Na 0.9	Mg 1.3											Al 1.6	Si 1.9	P 2.2	S 2.6	Cl 3.2
K 0.8	Ca 1.0	Sc 1.4	Ti 1.5	V 1.6	Cr 1.7	Mn 1.6	Fe 1.8	Co 1.9	Ni 1.9	Cu 1.9	Zn 1.7	Ga 1.8	Ge 2.0	As 2.2	Se 2.6	Br 3.0
Rb 0.8	Sr 1.0	Y 1.2	Zr 1.3	Nb 1.6	Mo 2.2	Tc 1.9	Ru 2.2	Rh 2.3	Pd 2.2	Ag 1.9	Cd 1.7	In 1.8	Sn 1.8	Sb 2.1	Te 2.1	I 2.7
Cs 0.7	Ba 0.9	La 1.0	Hf 1.3	Ta 1.5	W 2.4	Re 1.9	Os 2.2	Ir 2.2	Pt 2.3	Au 2.5	Hg 2.0	Tl 1.6	Pb 1.9	Bi 2.0		

transfer is sufficient to make the ionic model the best description of the bond. There are degrees of ionic bonding. A greater degree of ionic character in a bond means a more complete degree of electron transfer. A greater difference in electronegativity between atoms means that the bond between them will have more ionic character. But it should be noted that no compound is completely ionic. The most ionic binary compound CsF, made up of the most electronegative and the least electronegative elements, is still not completely ionic; the electron transfer from the Cs atom to the F atom is not complete.

Using the values in Table 7.3, we can see that the ionic description will be best for binary compounds in which elements with the lowest electronegativity values—the alkali metals or the heavier alkaline-earth metals—combine with the elements that have the highest electronegativity values, the halogens and oxygen. Other binary compounds between metals and nonmetals have ionic characteristics, but the ionic description is not always the best one for such compounds. In particular, many binary compounds of the transition metals with the nonmetals can be described as nonionic.

In addition to binary ionic compounds, there are also ionic solids formed by polyatomic groups that behave as ions (Table 2.4). Most of these polyatomic groups are anions that form salts with metal cations. But there are also polyatomic cations, complex ions of nonmetals. The most common of these is NH_4^+. Others are nitronium, NO_2^+, and phosphonium, PH_4^+, whose salts tend to be extremely reactive, especially toward hydrolysis. The bonding in ionic solids that include polyatomic anions or cations is sometimes difficult to analyze because these polyatomic groups complicate the geometry of the solid.

7.2 THE COVALENT BOND

The ionic bond is one extreme model of bonding. The two species that form the bond are ions of opposite charge, held together in an ionic crystal in which each ion is associated equally with many neighboring ions of opposite charge.

The covalent bond is at the opposite extreme. The covalent bond model emphasizes the sharing of electrons between two neutral atoms, rather than the transfer of electrons to form ions. Neither of the two atoms in a covalent bond has either an excess or a deficiency of electrical charge as a result of bonding. We can say that the covalent bond is characterized by a symmetrical distribution of electrical charge. In the ionic bond, we study a large crystal lattice of many atoms. In the covalent bond, we focus our attention only on the two atoms that are bonded to each other.

A completely ionic bond, characterized by a complete transfer of one or more electrons, does not exist. But we can describe the bonds in such homonuclear diatomic molecules as H_2, Cl_2, and even Na_2 as completely covalent. It is reasonable to assume that in a molecule such as H_2, in which both atoms are identical, neither atom is relatively positive or relatively negative with respect to the other. Therefore, the two identical atoms will share electrons and will form a completely covalent bond.

The sharing of electrons by two atoms in a covalent bond is energetically favorable. It is advantageous for these two atoms to be close to one another rather than to be separated. Energy is liberated in the process by which two isolated atoms come together to form a covalent bond:

$$2H(g) \longrightarrow H_2(g) + energy$$

Unlike the ionic bond, the covalent bond cannot be understood fully on the basis of simple electrostatic theory. Most aspects of ionic bonding can be explained without referring to the quantum theory or to wave functions. A full explanation of covalent bonding requires some aspects of quantum mechanics

HARDNESS AND THE CHEMICAL BOND

Hardness is not easily defined. A simple way to test hardness is to scratch one substance with another. The harder substance scratches the softer one. However, tests based on indentation, grinding, boring, or abrasion are also used. The traditional method of judging the hardness of materials is based on the Mohs scale, which was introduced about 1812 by a German mineralogist. The Mohs scale is based on the scratch test. Ten minerals are arranged in order of increasing hardness, and each is assigned a hardness value from 1 to 10:

Mineral	Formula	Mohs Value
talc	$Mg_3Si_4O_{10}(OH)_2$	1
rock salt	$NaCl$	2
calcite	$CaCO_3$	3
fluorite	CaF_2	4
apatite	$Ca_5(PO_4)_3F$	5
feldspar	$KAlSi_3O_8$	6
quartz	SiO_2	7
topaz	Al_2SiO_4	8
corundum	Al_2O_3	9
diamond	C	10

Other materials are assigned values by comparison with the ten standard minerals. For example, on the Mohs scale, the hardness of a steel file is about 6 or 7. Window glass is about 5 Mohs, a knife blade is about 5 Mohs, a copper penny is about 3 Mohs, and a human fingernail is about 2 Mohs.

However, the Mohs scale is unsatisfactory because the differences in hardness between the ten standard minerals are not equal. A new scale has been developed in which hardness is interpreted in terms of binding energy per unit volume. Substances that bond ionically, such as sodium chloride, are relatively soft. All very hard substances, such as diamond and corundum, consist of crystals whose atoms are bonded covalently. In addition, if a substance is very hard, it cannot have weak bonds between any of its units of structure.

The second-hardest substance known is a synthetic material named cubic boron nitride. The atoms in a crystal of cubic boron nitride are covalently bonded in virtually the same way as the atoms in diamond. Cubic boron nitride was first made in 1956 by subjecting ordinary, relatively soft boron nitride to great pressures at high temperatures. Under these extreme conditions, the extra covalent bonds needed to transform boron nitride from a soft to a hard material are formed.

On the old Mohs scale, cubic boron nitride has a hardness that can be given as 9 + . But when we deal with materials as hard as cubic boron nitride or diamond, the Mohs scale does not give an adequate indication of relative hardness. On a more modern version of the Mohs scale, corundum retains its old value of 9 and harder substances are given proportionally larger values. On this scale, cubic boron nitride has a hardness value of 19 and diamond has a hardness value of 42 +. These values can be correlated with binding energy per unit volume.

In theory, it is possible to manufacture a harder material than diamond by finding a substance whose binding energy per unit volume is greater than the binding energy of diamond. However, no such substance is known to exist. Conceivably, diamond is the hardest material that can exist, because of the nature and strength of its chemical bonds.

that are beyond our scope. However, we can still examine many aspects of covalent bonding.

The first question to ask about covalent bonding is: Why is it energetically advantageous to share electrons between atoms? A partial answer is provided by classic electrostatics.

Consider the case in which two hydrogen atoms come together to form an H_2 molecule, with the two electrons of the molecule between the nuclei, as shown in Figure 7.8(b). What are the advantages and disadvantages compared to the isolated atoms of Figure 7.8(a)?

The main advantage is the increase in the number of favorable coulombic attractions in the H_2 molecule. In the isolated atom, there is a single electron-nucleus attraction. In the H_2 molecule, there are two negatively charged electrons and two positively charged nuclei. Therefore, there are four favorable interactions between oppositely charged species in the H_2 molecule, as compared to only two such interactions in the two isolated atoms.

There are several disadvantages that offset this advantage, at least partially. The two nuclei in the H_2 molecule are both positively charged, so they repel each other. However, this repulsion is reduced by the screening effect of the electrons, unless the nuclei are quite close together. The two negatively charged electrons also repel each other, but this repulsion is reduced because the electrons have opposite spins. Both of these unfavorable forces—nucleus-nucleus repulsion and electron-electron repulsion—are more than made up for by the greater number of favorable attractions between the electrons and the nuclei.

The electrostatic picture of the covalent bond shown in Figure 7.8 is an oversimplification. But it does allow some predictions about the major features of the covalent bond:

1. The electrons tend to lie between the two bonded nuclei. In this position, the attraction of the two electrons by both nuclei is maximized, while the screening effect of the electrons on nucleus-nucleus repulsion is also maximized.

2. We know that two electrons occupying the same orbital in an atom must have opposite spin, and that no more than two electrons can occupy the same orbital. By analogy, when the two electrons are in a covalent bond, in the region of space between two atoms, we can expect that the electrons will have opposite spin.

3. We can expect that a covalent bond will have two electrons. If there is only one electron between the two nuclei, the nuclei are not effectively screened from each other. Furthermore, there are only two favorable coulombic attractions in a one-electron bond. There are a few compounds with one-electron bonds, but the vast majority of covalent bonds have electron pairs—two electrons for each covalent bond.

4. Two atoms that form a covalent bond keep approaching each other until

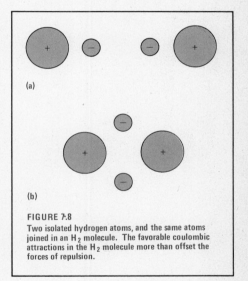

(a)

(b)

FIGURE 7.8
Two isolated hydrogen atoms, and the same atoms joined in an H_2 molecule. The favorable coulombic attractions in the H_2 molecule more than offset the forces of repulsion.

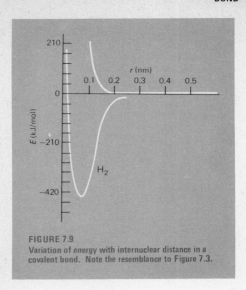

FIGURE 7.9
Variation of energy with internuclear distance in a
covalent bond. Note the resemblance to Figure 7.3.

the energy of the system reaches a minimum. Figure 7.9 plots energy against internuclear distance in a covalent bond. Note that it resembles Figure 7.3, a similar plot for ions of opposite charge. If the atoms get too close, the forces of repulsion between the two nuclei and between the inner electron shells (for all atoms but H) become significant compared to the force of attraction, and the energy of the system begins to increase. The point at which energy is at a minimum is the point of most favorable internuclear separation.

Let us examine some aspects of covalent bonding in detail.

7.3 LEWIS STRUCTURES

Until 1916, chemists drew diagrams of molecules in which the bonds between atoms were indicated by lines, called valence bonds, whose actual significance was not understood. In these diagrams, H_2 was written H—H; H_2O was written H—O—H; CH_4 was written

```
      H
      |
  H — C — H
      |
      H
```

But there was no feeling that the lines represented physical reality.

In 1916, G. N. Lewis (1875–1946), an American chemist, suggested that the dashes represented pairs of shared electrons. The suggestion was revolutionary, and it led to a new explanation of chemical bonding.

Lewis had a great deal to say about the nature of covalent bonds. He emphasized the importance of all the electrons in the valence shell, the atom's highest-energy electron shell, in the electronic structure of covalent species. He developed a method of diagraming the details of electronic structure in compounds with covalent bonds, using dots to represent electrons. These diagrams, only slightly modified, still are used to show the structure of covalent molecules. They are called **Lewis structures,** or **electron dot structures.**

A mastery of the technique of constructing Lewis structures is important for the study of chemistry. A correctly drawn Lewis structure indicates how all the atoms of a molecule are attached to one another and accounts for all the valence electrons of all the atoms in the molecule. Lewis structures can be constructed by following these rules:

A. Only valence (outer-shell) electrons are shown. For the main group elements in columns IA–VIIA of the periodic table, the valence electrons are the electrons of highest principal quantum number. *The number of valence electrons of a main group element is the same as the column number of the element in the periodic table.* For example, as we go across the first row of the periodic table, the number of valence electrons of each element is:

Li	1	Li·
Be	2	·Be·
B	3	·B·
C	4	·C·
N	5	·N·
O	6	·O·
F	7	:F·

where each dot next to the symbol of the element represents a valence electron.

The post-transition metals in group IIB generally have two valence electrons.

The number of valence electrons in a transition metal is not always easy to determine.

B. A shared electron pair can be represented in a Lewis structure either by two dots or by a line between the two atoms sharing the electrons. Although G. N. Lewis used only dots in the structures he drew, modern structures generally are written with lines to represent shared electron pairs. We shall call such structures Lewis structures, even though they are actually modified Lewis structures. When there is one line between two atoms, the atoms are said to be joined by a single bond. There are many compounds in which atoms share two or even three pairs of electrons; this is called **multiple bonding.** A double bond, in which two pairs of electrons are shared, is shown by two lines between the atoms. A triple bond is represented by three lines between the atoms. Double and triple bonds, the sharing of four or six electrons between two atoms, might seem to create prohibitively large electron-electron repulsions. But such bonds do exist. We shall discuss them in greater detail in Section 7.5.

C. The valence electrons that are not included in the covalent bonds between atoms are called *nonbonding electrons*. In a Lewis structure, nonbonding electrons are assigned to specific atoms and are represented by dots drawn next to the symbols for these atoms. Nonbonding electrons, like bonding electrons, almost always come in pairs.

Figure 7.10 shows some examples of Lewis structures. You will note that in Lewis structures, nonbonding electrons are grouped into pairs, with each pair placed on one side of the symbol for the element to which it belongs.

D. The Lewis structure that is drawn for a molecule must account for all the valence electrons of every atom in the molecule, using either lines to indicate bonds or dots to indicate nonbonding electrons.

E. Most of the Lewis structures that we shall study satisfy the **octet rule,** meaning that *each atom in the molecule achieves a noble gas electronic configuration by covalent bonding*. The octet rule gets its name from the fact that the noble gases Ne, Ar, Kr, and Xe all have the valence shell electronic configuration ns^2np^6, an octet of eight outer-shell electrons. (Since helium is $1s^2$, hydrogen achieves the noble gas configuration when it is surrounded by only two electrons.)

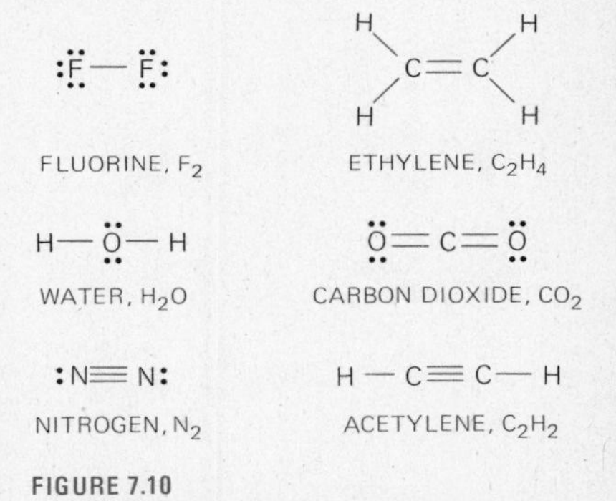

The octet rule is quite helpful in writing Lewis structures—so much so that the stability of the covalent bond often is incorrectly ascribed to the stability of the noble gas configuration. Figures 7.10 and 7.11 give examples of covalent bonds that can be explained in terms of noble gas electronic configurations.

When two hydrogen atoms come together, each with one electron, each nucleus becomes associated with two electrons and can be regarded as having the electronic configuration of He, $1s^2$. When two fluorine atoms come together, each with seven valence electrons ($2s^22p^5$), each fluorine atom gives one of its valence electrons to the bond. Thus, each fluorine atom is now associated with eight valence electrons: two bonding electrons, which are shared, and six nonbonding electrons, which are unshared. Each fluorine atom in such a bond can be regarded as having the electronic configuration of neon, $[He]2s^22p^6$. The octet rule is satisfied.

The same reasoning can be extended to multiple covalent bonds. Each isolated nitrogen atom ($[He]2s^22p_x2p_y2p_z$) has five valence electrons, two paired and three unpaired. When two nitrogen atoms come together, each atom can be associated with eight electrons if their unpaired electrons are shared. Thus, each atom in the N_2 molecule is associated with two nonbonding and six bonding electrons, and the octet rule is satisfied.

One helpful but not completely accurate way to explain covalent bonding is to say that there are not enough electrons for all the separate atoms to have complete octets. The insufficiency is overcome by the sharing of electrons in covalent bonds, so that the separate atoms complete their octets and obey the octet rule.

FLUORINE, F_2

ETHYLENE, C_2H_4

WATER, H_2O

CARBON DIOXIDE, CO_2

NITROGEN, N_2

ACETYLENE, C_2H_2

FIGURE 7.10
Lewis structures of some simple molecules. The lines represent bonds, and the dots represent nonbonding valence electrons.

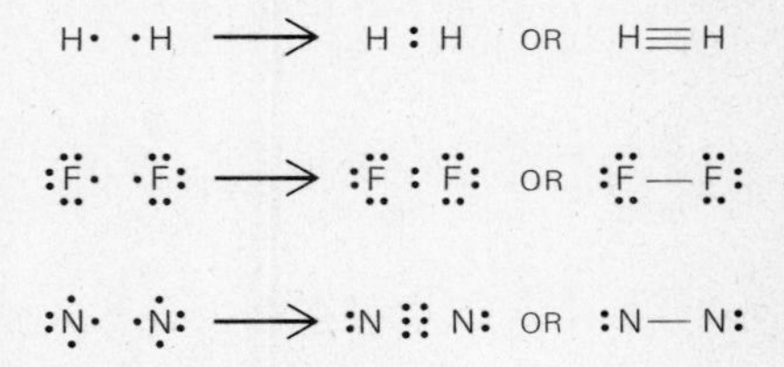

FIGURE 7.11
Electron sharing allows each atom to achieve a noble gas electronic configuration.

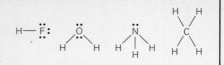

FIGURE 7.12
Hydrides of some elements in the first row of the
periodic table. The octet rule is obeyed.

We can say that each atom of a single bond has an unpaired electron to contribute to the formation of the bond. In a double bond between two atoms, each atom contributes two electrons. In a triple bond, each atom contributes three electrons. This concept is helpful but not essential in describing covalent bonds.

The octet rule can even help to predict the composition of molecules. Figure 7.12 shows the Lewis structures of the hydrides of some elements in the first row of the periodic table. We see that HF has one bond and six nonbonding electrons grouped in three pairs, that H_2O has two bonds and four nonbonding electrons grouped in two pairs, that NH_3 has three bonds and one pair of nonbonding electrons, and that CH_4 has four bonds. The element forming each hydride has achieved a noble gas structure by sharing electrons in one or more covalent bonds with one or more hydrogen atoms. The number of H atoms in each hydride is equal to the number of electrons that the other atom in the molecule needs to complete its octet.

The task of drawing Lewis structures is simplified by the fact that there are many compounds in which the atoms obey the octet rule. There are rules that help us to formulate Lewis structures in many common situations:

1. A hydrogen atom can never be associated with more than the two electrons needed to complete the helium shell. Thus, a hydrogen atom only forms one bond with one other atom.
2. The elements of the first row of the periodic table (Li–Ne) are never associated with more than eight valence electrons. An atom that is associated with more than eight electrons in a covalent compound is said to have an *expanded octet. Atoms of elements in the first row never have expanded octets*. In other words, atoms of these elements never have more than the eight valence electrons needed to complete the neon shell, $2s^2 2p^6$.
3. Elements of group IIIA often disobey the octet rule by being associated with only six electrons in covalent compounds. They are said to have *incomplete octets*.
4. Elements whose atomic number is 14 or greater often may be surrounded by more than eight electrons in covalent compounds. We say that these atoms have expanded octets.

We can explain why expanded octets exist in these elements but do not exist in elements of atomic number 13 or less. The $2s$ and $2p$ orbitals of the valence shell of a first-row element are filled by an octet of electrons. To accommodate more than eight electrons, such an element must use $3s$ and $3p$ orbitals. There is a large energy difference between the $2s$ and $2p$ orbitals and the $3s$ and $3p$ orbitals, which have different principal quantum numbers, and the use of the higher orbitals is energetically disadvantageous.

In the second-row elements, whose valence shell includes the $3s$ and $3p$ orbitals, the octet can be expanded if the electrons go into the next available orbitals. The next available orbitals are the $3d$ orbitals, which have the same principal quantum number as the other orbitals in the valence shell. The energy required to place electrons in these orbitals is often relatively small. The formation of extra bonds as a result of octet expansion releases much more energy. Since many heavy elements have empty orbitals whose energy is close to that of the orbitals occupied by the valence electrons, expanded octets are common in these elements.

There is a limit to the number of electrons in an expanded octet. Because of electron-electron repulsion, it is not energetically advantageous to have more than a certain number of electrons around a single atom. An expanded octet usually includes no more than 12 electrons; there usually will be no more than six bonds around any atom.

5. A compound that has an odd number of valence electrons must have at least one atom that does not satisfy the octet rule. Such a compound also has at least one unpaired electron.

To illustrate how these rules are applied, let us draw the correct Lewis structure for nitric acid, HNO_3.

a. The Lewis structure must account for all valence electrons, so the first step is to find the number of valence electrons in the molecule by totaling the number of valence electrons of the atoms. In HNO_3, H has one valence electron; N (group VA) has five valence electrons; each O (group VIA) has six valence electrons. The total for the molecule is $1 + 5 + (3 \times 6) = 24$ valence electrons.

b. Now construct the skeleton of the molecule by connecting the symbols for the atoms by single lines, indicating single bonds. This initial step may require a combination of experimental data, an understanding of chemical bonding, and even chemical intuition. With some experience, we can construct the skeleton for most molecules without too much trouble. There are rules, some of which we have already mentioned, that help in this construction:

i.   A hydrogen atom can accommodate only one electron pair, so it is always bonded by a single bond to one of the other atoms in the compound.

ii.   First-row elements do not have expanded octets, so they never bond to more than four atoms. (They may bond to fewer than four atoms.)

iii. In Section 2.4, page 30, we discussed the oxidation numbers commonly found for some of the elements. The oxidation number is often associated with the number of bonds formed by the element in covalent compounds. Whenever possible, it is preferable to write a structure in which the number of bonds to a given atom is the same as the value of the atom's oxidation number. Oxygen and nitrogen provide the most common examples of this rule. When oxygen is in the -2 oxidation state, it is preferable to write a structure in which oxygen has two bonds. When nitrogen is in the -3 oxidation state, it is preferable to write a structure in which it has three bonds.

iv.   In compounds with more than one oxygen atom, there usually are no bonds between the oxygen atoms. If there is an oxygen-oxygen bond, the name of the compound usually includes the word "peroxide" or "superoxide." As a rule, oxygen atoms are not connected to one another in a Lewis structure unless there is specific information that such a bond exists.

v.   Rings of atoms are not written unless there is specific information that they exist in a molecule.

vi.   Most oxyacids, acids containing oxygen and hydrogen, have the hydrogen atoms bound to the oxygen atoms.

vii. A halogen atom forms only one single bond, unless it is bonded to an oxygen atom or to the atom of a halogen of lower atomic number.

We now have enough information to assign a skeleton structure to nitric acid. We know that the molecule will have an O—H bond, no O—O bonds, and no rings of atoms, so only one structure is possible:

c. As the next step, count the electrons that have been used to construct the skeleton structure. Each dash is a single bond, representing two electrons. This number should be subtracted from the total number of valence electrons. In the HNO_3 skeleton structure, there are four dashes, representing eight electrons. Since 24 valence electrons were available, $24 - 8 = 16$ electrons remain to be accounted for.

d. Assume that all the atoms obey the octet rule and determine the number of electrons needed to complete all the octets in the molecule.

The hydrogen atoms in a skeleton structure can never accommodate more than two electrons (rule 1). The nitrogen atom has three bonds, representing six electrons, so two more electrons will complete its octet. The oxygen atom that is bound to the H atom and the N atom has two bonds; four more electrons will complete its octet. Each of the other oxygen atoms has one bond; each requires six electrons to complete its octet. Adding, we find that $2 + 4 + (2 \times 6) = 18$ electrons are needed to complete all the octets in HNO_3.

e. Compare the number of valence electrons that are available (step c) with the number needed to complete the octets (step d). If the number from step c is the same as the number from step d, the Lewis structure is completed by drawing in enough dots to give each atom a complete octet.

For HNO_3, 18 electrons are needed to complete all the octets but only 16 valence electrons remain after the single-bond structure is drawn. Therefore, for all the octets in HNO_3 to be complete, some of the remaining valence electrons must be shared. (As Figure 7.11 shows, sharing allows all the atoms in a molecule to have complete octets even though there are fewer than eight electrons for each atom.) To put it another way, one additional bond must be drawn in the Lewis structure for each difference of two electrons calculated in steps c and d.

In HNO_3, there is a two-electron difference, so one more bond is needed. This bond *cannot* be placed between the H atom and the O atom because H never has more than one bond. It is best *not* placed between the N atom and the O atom that is bonded to the H atom to give:

$$
\begin{array}{c}
O \\
\diagdown \\
\quad N{=}O{-}H \\
\diagup \\
O
\end{array}
$$

because the oxidation number of the O atom is -2, and it is best for the O to have only two bonds, rather than the three it would have in this structure. In general, we avoid drawing structures with three bonds to O in neutral molecules. By elimination, the double bond is placed to form the structure:

$$
\begin{array}{c}
O \\
\diagdown \\
\quad N{-}O{-}H \\
\diagup\diagup \\
O
\end{array}
$$

This drawing places the double bond between the N atom and the lower O atom. The double bond could just as well be placed between the N atom and the upper O atom, since these two O atoms are equivalent.

With the placement of the double bond established, there are $16 - 2 = 14$ valence electrons in the HNO_3 molecule remaining to be accounted for. The number of electrons needed to complete all octets is $18 - 4 = 14$. It remains only to place the dots representing these electrons around the appropriate symbols in the Lewis structure, giving:

$$
\begin{array}{c}
:\ddot{O}: \\
\diagdown \\
\quad N{-}\ddot{O}{-}H \\
\diagup\diagup \\
:\ddot{O}:
\end{array}
$$

as the final Lewis structure of HNO_3. To see whether you have written a correct structure, you can count the total number of valence electrons and also note whether the octet rule is obeyed for each atom.

EXAMPLE 7.1

Draw a Lewis structure for nitrogen trichloride, NCl_3.

SOLUTION

Step 1:   First calculate the number of valence electrons. The N atom has five and each Cl atom has three, giving $5 + (3 \times 7) = 26$ valence electrons.

Step 2:   Determine the skeleton structure. Since Cl forms only one bond unless it is bonded to O or F, the skeleton structure must be:

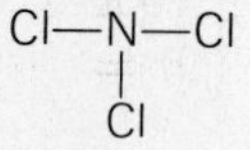

Cl—N—Cl
    |
   Cl

Step 3:   The number of electrons used to form the three bonds in the skeleton structure is 6, leaving $26 - 6 = 20$ electrons to complete the octets in the molecule.

Step 4:   The number of electrons needed to complete all octets is 2 (for the N atom) $+ (3 \times 6)$ (for the Cl atoms) $= 20$ electrons.

Step 5:   Since the number of electrons needed to complete all octets (Step 4) is the same as the number of available electrons (Step 3), the Lewis structure is finished by completing the octets, writing in the nonbonding valence electrons as pairs of dots:

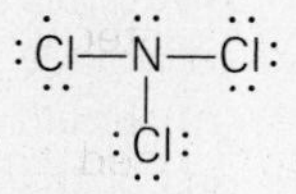

:Cl—N—Cl:
     |
    :Cl:

Step 6:   As a final step, check to be sure that the Lewis structure has the correct number of electrons. The structure we have written for NCl_3 has 26 electrons, 6 bonding and 20 nonbonding, which is the number of valence electrons.

EXAMPLE 7.2

Draw the Lewis structure of nitrous acid, HNO_2.

SOLUTION

Step 1:   There are $1 + 5 + (2 \times 6) = 18$ valence electrons.

Step 2:   The most likely skeleton structure is O—N—O—H, which avoids O—O bonds and includes the O—H bond expected for an oxyacid.

Step 3:   The number of electrons remaining to complete octets is $18 - 6 = 12$.

Step 4:   The number of electrons needed to complete octets is 14 (from left to right, 6 for the O atom, 4 for the N atom, 4 for the O atom, 0 for the H atom).

Step 5:   Since the number of remaining electrons is two less than the number of electrons needed to complete all octets, one more bond is required. Placing the bond in a way that avoids a ring structure or a triple bond to an O atom gives:

O=N—O—H

Completing the octets gives the Lewis structure:

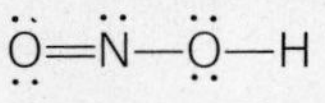

Ö=N̈—Ö—H

Step 6:   The number of electrons in the Lewis structure, 18, equals the number of valence electrons.

EXAMPLE 7.3

Draw the Lewis structure for formic acid, CH_2O_2.

SOLUTION

Step 1:   The number of valence electrons is $4 + 2 + (2 \times 6) = 18$.

Step 2:   There are several possible skeletons. But since we know that

CH_2O_2 is an acid, at least one O—H bond is necessary. Therefore, the only two possibilities are:

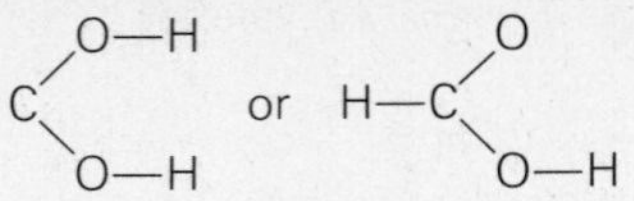

Only 10 electrons remain, while 12 are needed to complete all the octets. An additional bond is needed so that all the octets can be completed. The addition of a bond in the first structure will result in an O atom that has three bonds. The second structure therefore seems best. This choice is verified by experimental evidence. The structure with the required double bond and all the completed octets is:

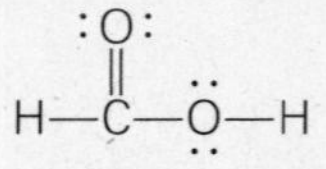

The formula of formic acid is often written HCOOH to reflect the bonding in this structure.

The same procedure can be used to write Lewis structures for polyatomic ions. For such species, the initial count of valence electrons is modified to include the charge on the ion. For a mononegative ion such as HCO_3^-, the count should include one more valence electron than the sum of the valence electrons of the neutral atoms. For HCO_3^-, we calculate four valence electrons for the C atom, one for the H atom, $3 \times 6 = 18$ for the three O atoms, and one for the negative charge on the ion, giving $4 + 1 + (3 \times 6) + 1 = 24$ valence electrons in all. A dinegative anion such as SO_4^{2-} has two more valence electrons than are present in the neutral atoms. The count is $6 + (4 \times 6) + 2 = 32$ valence electrons. For cations, the number of valence electrons is less than the number present in the atoms. Again, the difference is the charge on the ion. Thus, NH_4^+ has $5 + (1 \times 4) - 1 = 8$ valence electrons, while CH_3^+, the carbonium ion, has $4 + (1 \times 3) - 1 = 6$ valence electrons.

EXAMPLE 7.4
Draw a Lewis structure for the bicarbonate ion, HCO_3^-.
SOLUTION
Step 1:   The number of valence electrons of the atoms is 23, plus one for the negative charge, giving a total of 24.
Step 2:   The best skeleton structure is:

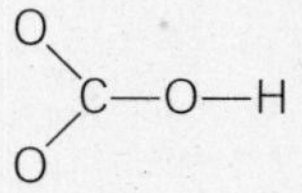

All three O atoms must be attached to the C atom to avoid O—O bonds. The H must also be attached to an O atom to allow the introduction of an additional bond that we shall require for the correct structure. If the H atom were attached to the C atom, no further bonds would be possible, since a C atom can have no more than four bonds.
Step 3:   The number of electrons remaining is $24 - 8 = 16$.
Step 4:   The number of electrons needed to complete all octets is $(2 \times 6) + 2 + 4 = 18$.
Step 5:   One additional bond is needed. It is best placed between the C atom and either of the O atoms not bonded to H:

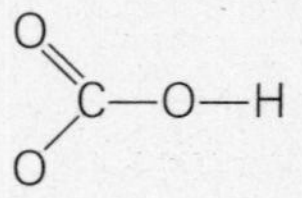

Step 6:   Completing the octets gives the structure:

which accounts for the required 24 electrons and indicates the negative charge of the species.

Incomplete Octets

In writing some Lewis structures, we may find that there are not enough valence electrons to complete octets for all the atoms in the skeleton structure, but that none of the atoms can accommodate multiple bonds. In such cases, some octets must be left incomplete.

In writing the Lewis structure for BF_3, the skeleton structure is:

since F can have only one bond. The total number of valence electrons is $3 + (3 \times 7) = 24$, of which six are accounted for in the skeleton structure. That leaves 18 valence electrons to complete octets for all the atoms in the molecule. However, a count shows that $2 + (3 \times 6) = 20$ valence electrons are needed to complete all octets. The further sharing of electrons is ruled out, since F cannot form double bonds. The only possible conclusion is that one of the atoms in the BF_3 molecule will be surrounded by only six electrons, rather than by a complete octet. The B atom is the atom with the incomplete octet, since it has lower electronegativity than the F atoms. The correct structure is:

$$:\!\ddot{F}\!:$$
$$|$$
$$B$$
$$:\!\ddot{F}\!:\quad:\!\ddot{F}\!:$$

EXAMPLE 7.5

Draw the Lewis structure of $HgCl_2$.

SOLUTION

Step 1:   The number of valence electrons is 16; $(2 \times 7) = 14$ from Cl and 2 from Hg (which is in group IIB).

Step 2:   Since Cl can form no more than one bond with Hg, the only possible structure is Cl—Hg—Cl.

Step 3:   The skeleton structure leaves $16 - 4 = 12$ electrons to be accounted for.

Step 4:   A total of $6 + 4 + 6 = 16$ electrons is needed to complete all octets. Only 12 electrons are available. Ordinarily, two multiple bonds would be added because of the shortage of four electrons. But since Cl can form only one bond, this solution is not possible. The $HgCl_2$ molecule must therefore have one incomplete octet. The Hg atom, which is less electronegative than Cl, has the incomplete octet, giving the Lewis structure:

$$:\!Cl\!-\!Hg\!-\!Cl\!:$$

Octet Expansion

We mentioned earlier that many compounds that include atoms of elements whose atomic number is 14 or greater (element 14 is silicon) can have expanded octets—that is, more than eight valence electrons associated with a single atom.

The presence of expanded octets in a compound can be recognized either from the rules of valence or, after a skeleton structure is written, by the presence of more valence electrons than are needed to complete octets for all the atoms in a molecule.

In the PCl_5 molecule, for example, the P atom must have an expanded octet. A Cl atom can form only one bond, so all five Cl atoms must be attached to the P atom. Any single atom that is covalently bonded to more than four other atoms must have an expanded octet. Therefore, the structure of PCl_5 is:

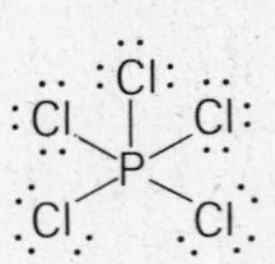

with the P atom accommodating 10 valence electrons.

In formulating a Lewis structure for BrF_3, the skeleton structure:

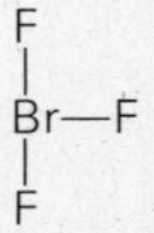

requires $(3 \times 6) + 2 = 20$ valence electrons to complete all octets. But there are $28 - 6 = 22$ valence electrons remaining. One of the atoms must have an expanded octet. Usually, it is the atom of lowest electronegativity, which is Br in BrF_3. The complete structure is:

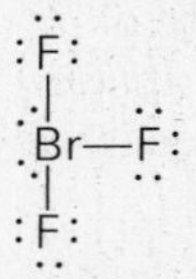

EXAMPLE 7.6
Draw the Lewis structure for SF_6.
SOLUTION
Since F can have only one bond, the skeleton must be:

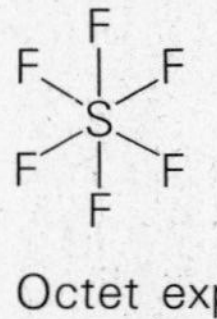

Octet expansion is necessary. The number of valence electrons is $6 + (6 \times 7) = 48$. The number remaining to complete octets is $48 - 12 = 36$, which is also the number required to complete the octets of all the F atoms. The complete structure will include three pairs of electrons around each F:

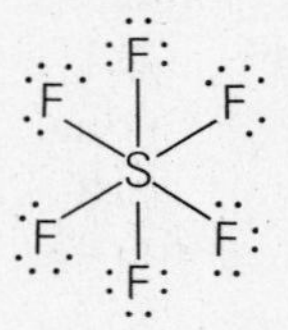

EXAMPLE 7.7
Draw the Lewis structure of IF_5.

SOLUTION

Step 1: The number of valence electrons is $7 + (5 \times 7) = 42$.

Step 2: Since only I can form more than one bond (it is bonded to another halogen of lower atomic number), the skeleton must be:

Step 3: The number of electrons remaining to complete octets is $42 - 10 = 32$.

Step 4: The number of electrons needed to complete octets is $5 \times 6 = 30$.

Step 5: Since there are two more electrons than are necessary to complete octets, there will be even more octet expansion than is found in the skeleton structure. The less electronegative atom is I, which is large enough to accommodate the two extra electrons even though it is already surrounded by 10 electrons. The structure is:

The Electroneutrality Principle and Formal Charge

We noted earlier that in covalent molecules that include nonmetals with complete octets, atoms of any given element tend to form a specific number of bonds. Usually, the number of bonds formed by an atom will be the same as its oxidation number when it is bonded to less electronegative atoms (Section 2.4, page 30). Elements of group IVA usually form four bonds. Elements of group VA form three bonds. Elements of group VIA form two bonds. Elements of group VIIA form one bond when they are covalently bonded to atoms of lower electronegativity.

The number of bonds and the oxidation number are the same as the number of valence electrons of each atom that are available for bonding. For example, the ground state electronic configuration of nitrogen, which forms three bonds, is $[\text{He}]2s^2 2p_x 2p_y 2p_z$; that of oxygen, which forms two bonds, is $[\text{He}]2s^2 2p_x^2 2p_y 2p_z$.

However, nonmetals can form compounds in which they do not have the number of bonds expected from their oxidation numbers. Nitric acid, HNO_3, is such a compound. In the Lewis structure of HNO_3, the N atom has four bonds and one of the O atoms has one bond. The N atom has five valence electrons. Three of these electrons are used to form three normal bonds, in which the atom bonded to N contributes an electron to the shared pair, as does the N atom. Two electrons are left for the fourth bond formed by the N atom.

By this method of electron bookkeeping, the fourth bond formed by the N atom seems to differ from the other three. Rather than being formed by one electron from each atom of the bond, it is formed by two electrons from one atom. This fourth bond is between a nitrogen atom and an oxygen atom. The oxygen atom has only one bond, rather than the usual two. It is surrounded by six nonbonding electrons, giving it a complete octet. Since the oxygen atom started with six valence electrons, it appears that both bonding electrons are contributed by the nitrogen atom.

A bond in which one atom seems to contribute both electrons is called a *dative* or *coordinate-covalent bond*. Such a bond does not differ in physical properties from a normal bond. We create this difference by the way we do our

TABLE 7.4
Number of Bonds and Formal Charge

Element	Oxidation State[a]	Number of Bonds	Formal Charge
C	−4	3	−1
N	−3	4	+1
		2	−1
O	−2	3	+1
		1	−1

[a]When combined with less electronegative elements.

electron bookkeeping. The existence of a dative bond in a compound is noted by the use of **formal charges,** pluses or minuses written next to the symbols. "Formal" is meant literally. *The formal charges do not indicate actual charge distribution* within the molecule. They are devices that are used to help us in electron bookkeeping.

In this bookkeeping, an atom (such as N in HNO_3) that contributes two electrons that are shared in a bond and receives no electrons has "lost" two "half-electrons." The atom bears a formal charge of +1. An atom (such as O in HNO_3) that does not contribute an electron to the covalent bond "gains" the two "half-electrons." It has a formal charge of −1. Note that this sort of electron bookkeeping, which splits electrons, does not correspond to any real physical process. It simply helps us to account for bonding.

Formal charges are easily assigned to atoms that have completed octets. An atom with one less bond than normal is given a formal charge of −1; an atom with one more bond than normal is given a formal charge of +1. Table 7.4 summarizes this rule for three important elements.

The structure of HNO_3 thus can be written:

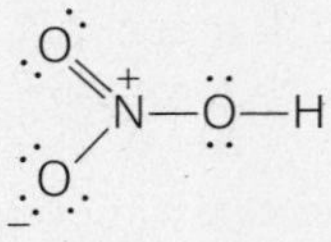

with the −1 charge assigned to the O atom that has only one bond and the +1 charge assigned to the N atom, which has four bonds. This structure can also be written:

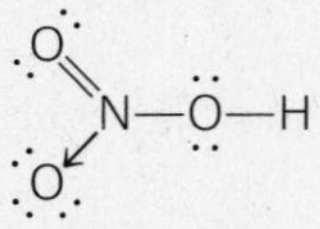

The arrow between the O and the N atoms represents a covalent bond and indicates that both electrons in the bond have been contributed by the N atom to the O atom.

Similarly, the structure of ozone, O_3, can be written:

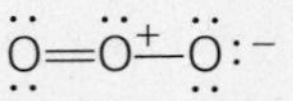

The central O atom, with three bonds, is assigned a formal charge of +1. (A formal charge of +1 is not assigned to oxygen unless it is unavoidable.) The O atom with only one bond has a formal charge of −1.

Formal charges are useful primarily as a guide in writing correct Lewis structures. There is a principle of **electroneutrality,** which states that Lewis structures with formal charges are to be avoided whenever possible, particularly structures with formal charges greater than +1 and −1 on any atom. This principle often helps us make a choice between two possible structures for a molecule.

For example, two skeleton structures are possible in drawing the Lewis structure of HCN: H—C—N or H—N—C. These two skeleton structures lead to final structures of H—C≡N: and H—N≡C:. The second structure has three bonds to the C atom, rather than the usual four, and four bonds to the N atom, rather than the usual three, so formal charges are necessary. The principle of electroneutrality says that the first structure which has no formal charges should be chosen. This choice is borne out by experimental evidence, which shows that HCN does have the first structure.

For H_2SO_4, the Lewis structure could be:

$$\text{H}-\overset{..}{\underset{..}{\text{O}}}-\overset{:\overset{..}{\text{O}}:^-}{\underset{:\overset{..}{\text{O}}:_-}{\overset{|}{\underset{|}{\text{S}^{2+}}}}}-\overset{..}{\underset{..}{\text{O}}}-\text{H}$$

This structure gives each atom a complete octet. But it requires formal charges of −1 for two O atoms, which have only one bond each. Since there is a total formal charge of −2 on the O atoms, there must be a formal charge of +2 on the S atom, because the *sum of the formal charges of a neutral atom is* 0. One possible way to avoid these formal charges is to expand the octet of the S atom as in the structure:

$$\text{H}-\overset{..}{\underset{..}{\text{O}}}-\overset{:\text{O}:}{\underset{:\text{O}:}{\overset{\|}{\underset{\|}{\text{S}}}}}-\overset{..}{\underset{..}{\text{O}}}-\text{H}$$

Here the choice is not between different skeleton structures but between different representations. You should note that both representations have exactly the same number of electrons; only the location of two pairs of electrons is different. Either structure can be regarded as correct.

EXAMPLE 7.8

Draw the most likely Lewis structure for formaldehyde, CH_2O.

SOLUTION

We must decide between two possible skeletons. The possibilities are:

$$\overset{\text{H}}{\underset{\text{H}}{\diagdown}}\text{C}-\text{O} \quad \text{and} \quad \text{H}-\text{C}-\text{O}-\text{H}$$

A count of electrons shows that one more bond is needed. When octets are completed, the two structures are:

$$\overset{\text{H}}{\underset{\text{H}}{\diagdown}}\text{C}=\overset{..}{\underset{..}{\text{O}}} \quad \text{and} \quad \text{H}-\overset{-}{\text{C}}=\overset{+}{\text{O}}-\text{H}$$

The first structure is preferred because it does not have formal charges. This choice is verified by experiment.

EXAMPLE 7.9

Draw the Lewis structure of chloric acid, $HClO_3$.

SOLUTION

Since this molecule is an acid, the skeleton will include an O—H bond. It will also include more than one bond to Cl, which is permissible when Cl is bonded to O. The best structure is:

$$\overset{\text{O}}{\underset{\text{O}}{\diagdown}}\text{Cl}-\text{O}-\text{H}$$

Electron bookkeeping gives the structure:

$$\overset{\displaystyle \overset{\cdot\cdot}{\underset{\cdot\cdot}{O}}}{\underset{\displaystyle \underset{\cdot\cdot}{\underset{\cdot\cdot}{O}}\cdot^-}{\overset{+}{C}l}}\!-\!\overset{\cdot\cdot}{\underset{\cdot\cdot}{O}}\!-\!H$$

which has formal charges. Since expansion of the octet of Cl is permissible, we may also write a Lewis structure with no formal charges:

$$\overset{\displaystyle \overset{\cdot\cdot}{\underset{\cdot\cdot}{O}}}{\underset{\displaystyle \underset{\cdot\cdot}{\underset{\cdot\cdot}{O}}}{Cl}}\!-\!\overset{\cdot\cdot}{\underset{\cdot\cdot}{O}}\!-\!H$$

7.4 RESONANCE

In writing Lewis structures, you will often find that there is more than one place to put a mulitiple bond to complete octets. In the case of HNO_3, the structure can be drawn in two ways:

$$\underset{O}{\overset{O}{\diagdown\!\!\!\diagup}}N\!-\!O\!-\!H \qquad \underset{O}{\overset{O}{\diagdown\!\!\!\diagup}}N\!-\!O\!-\!H$$

A similar choice arises in drawing the Lewis structure of O_3. The skeleton structure:

$$O\!-\!O\!-\!O$$

uses four electrons, leaving 14 of the original 18 valence electrons available for octet completion. A total of 16 electrons is needed to complete all octets, so one more pair of electrons must be shared. There are two possible structures:

$$O\!=\!O\!-\!O \quad \text{and} \quad O\!-\!O\!=\!O$$

Which of these structures is correct? The answer is neither, if the criterion is an accurate representation of the actual molecule. In a compound where there is more than one reasonable location for a multiple bond, no single Lewis structure can describe the molecule. More than one structure must be written for an accurate description. In such cases, some principles must be added to the ordinary Lewis structure method. These principles are summed up in the *theory of resonance,* which gives a method for representing such molecules:

1. If two or more Lewis structures that differ only in the distribution of multiple bonds can arise from the same skeleton structure, **resonance** is said to exist.
2. Two such Lewis structures (those for the HNO_3 molecule are examples) are called **contributing structures.** The molecule is said to be a **resonance hybrid** of its contributing structures. A contributing structure does not represent a real molecule. Rather, it only serves to help us draw the structure of a molecule that is a resonance hybrid.
3. The structure of a molecule that is a resonance hybrid is described as a blend of its contributing structures. The structure of a resonance hybrid is represented by the contributing structures, separated by the symbol ↔. This symbol does not mean that the molecule flips back and forth between the two contributing structures. Rather, it means that the real structure is a blend of the contributing structures, which are imaginary.

The structure of the resonance hybrid HNO_3 thus is represented by drawing two structures:

$$\overset{..}{\underset{..}{O}} = \overset{+}{N} - \overset{..}{\underset{..}{O}} - H \quad \longleftrightarrow \quad \overset{-\overset{..}{O}}{\underset{..}{O}} = \overset{+}{N} - \overset{..}{\underset{..}{O}} - H$$

Neither of these two structures can ever exist. We draw them to help us represent the actual structure of HNO_3. Since both contributing structures are identical, it is reasonable to assume that the HNO_3 molecule is an equal blend of the two structures. This statement can be interpreted to mean that two of the O atoms in the HNO_3 molecule are bound to the N atom by a bond that is intermediate between a single and a double bond.

EXAMPLE 7.10

Write the important contributing structures for N_2O_4. Experimental data reveal that N_2O_4 has an N—N bond.

SOLUTION

The skeleton structure is:

$$\begin{array}{ccc} O & & O \\ & \diagdown & \diagup \\ & N-N & \\ & \diagup & \diagdown \\ O & & O \end{array}$$

An electron count indicates that two additional bonds are required. Writing the possible combinations with the completed octets gives:

The correct structure of N_2O_4 is an equal blend of these four contributing structures.

EXAMPLE 7.11

Write the important contributing structures for NO_2.

SOLUTION

Since NO_2 has $5 + (2 \times 6) = 17$ valence electrons, an odd number, not all of its atoms can have complete octets. Without considering nonbonding electrons, we can write two structures:

$$O = N - O \quad \longleftrightarrow \quad O - N = O$$

In distributing the nonbonding electrons, we can write structures in which the N atom has an incomplete octet and has the unpaired or odd electron:

$$\overset{..}{O} = \overset{.}{N} - \overset{..}{\underset{..}{O}}: \quad \longleftrightarrow \quad :\overset{..}{\underset{..}{O}} - \overset{.}{N} = \overset{..}{O}$$

or structures in which the O has the incomplete octet, such as:

$$\overset{..}{O} = \overset{..}{N} - \overset{..}{O}\cdot$$

The first two structures, in which the N atom has an incomplete octet, make the most important contribution. Structures such as the last one, in which the O atom has the incomplete octet, are less important, since the more electronegative element usually has the completed octet.

The important point to remember is that resonance hybrids, when they are examined experimentally, are found to have bonds that are truly intermediate between those of the contributing structures, even though the contributing structures are imaginary constructs drawn for the sake of convenience. In HNO_3,

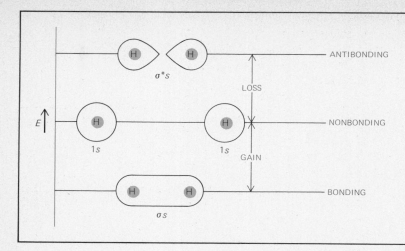

FIGURE 7.13
A covalent bond pictured as the combination of orbitals from two atoms. Each H atom contributes a 1s electron to the bond, and the resultant orbital is formed by the overlap of the two 1s orbitals. The new orbital is lower in energy than either of the original isolated nonbonding orbitals. Along with the bonding orbital, an antibonding orbital of higher energy than the atomic orbitals also forms.

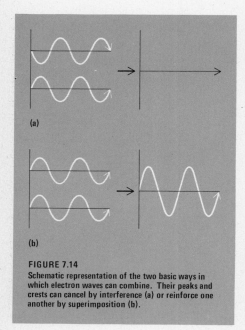

FIGURE 7.14
Schematic representation of the two basic ways in which electron waves can combine. Their peaks and crests can cancel by interference (a) or reinforce one another by superimposition (b).

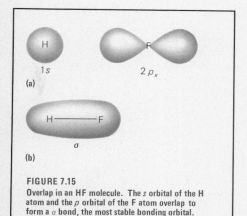

FIGURE 7.15
Overlap in an HF molecule. The s orbital of the H atom and the p orbital of the F atom overlap to form a σ bond, the most stable bonding orbital. Most single bonds between two atoms are σ bonds.

both N—O bonds are found by experiment to be midway in characteristics between a single and double bond, and both O atoms in these bonds are found to be identical. In NO_3^-, the nitrate ion, a resonance hybrid with three contributing structures:

all three O atoms are found to be identical, as are all three N—O bonds.

7.5 OVERLAP OF ATOMIC ORBITALS

As we saw in Chapters 5 and 6, the modern description of electrons in atoms is based on the concept of orbitals. The properties of covalent bonds can be understood better if the description of electrons in these bonds is also based on the concept of orbitals.

We describe a covalent bond as the sharing by two atoms of two electrons, one from each atom. This description suggests that the electrons will be found in an orbital that is formed by the combination of two orbitals, one from each atom. In the H_2 molecule, for example, each H atom contributes a 1s electron to the bond. Figure 7.13 shows that the resultant orbital containing the two electrons can be pictured most simply as being formed by the overlap or combination of the two 1s orbitals.

This picture is consistent with the known features of covalent bonds. The new orbital that forms in a covalent bond is located primarily between the two nuclei. It can hold a maximum of two electrons of opposite spin (obeying the Pauli exclusion principle) and corresponds to a lower energy state than either of the isolated 1s orbitals.

Such an orbital is called a **bonding molecular orbital**. The bonding orbital is spread out in space, but there is a high probability of finding the two electrons between the two nuclei. We can say that there is an **overlap** of the two atomic orbitals.

When two atomic orbitals overlap in a covalent bond, there is another allowed orbital for the electron. It is called the **antibonding molecular orbital**. The bonding molecular orbital is of lower energy and greater stability than either of the atomic orbitals. The antibonding orbital is of higher energy and lower stability than either of the atomic orbitals. If an electron occupies an antibonding orbital, its energy is higher than it would be if the electron were in an isolated atom.

When two or more atomic orbitals combine or overlap to form molecular orbitals, *the number of new orbitals created always equals the number of orbitals that have combined*. When two orbitals combine, they form one bonding orbital and one antibonding orbital. The gain of stability and lowering of energy in the bonding orbital is equal to the loss of stability and raising of energy in the antibonding orbital.

A highly simplified explanation for this rule can be based on the wave nature of electrons. As Figure 7.14 shows, there are two basic ways in which waves can overlap, or combine. The waves can be superimposed in a way that heightens the crests and troughs. Or they can be superimposed in a way that cancels the crests and troughs by interference. Since electrons in orbitals have wave characteristics, the orbitals can be combined in the same two ways. The bonding orbital corresponds to the superimposition that heightens the wave crests. The antibonding orbital corresponds to the superimposition that cancels the wave crests by interference.

The overlap between the two $1s$ orbitals in the H_2 molecule is one way in which two atomic orbitals can overlap. The bonding orbital in the H_2 molecule has cylindrical symmetry with respect to an imaginary line joining the nuclei of the two H atoms. This kind of symmetrical overlap is called sigma (σ) overlap and the orbital is called a σ orbital. A σ orbital is the most stable kind of bonding orbital, and σ overlap is the only kind of overlap possible between s orbitals. The antibonding orbital that corresponds to the σ bonding orbital is called a "sigma star" (σ^*) orbital. Even when an s orbital and a p orbital overlap to form a covalent bond, as in HF, it must be a σ bond (Figure 7.15). If two p orbitals are pointed at each other, σ overlap takes place. Indeed, there are very few exceptions to the rule that any single bond between two atoms is a σ bond.

But *only one σ bond is possible between two atoms*. A second kind of orbital overlap can occur in covalent bonds formed by p orbitals. When two p orbitals are parallel to each other, there can be a kind of orbital overlap, called pi (π) overlap. The bond formed in this way is called a π bond, as shown in Figure 7.16. The most probable region for the electrons of a π bond is outside the region of the σ bonding orbital. It should be noted that two p orbitals can overlap to form either a σ or a π bond, but only if the p orbitals have the same directional subscript can this be accomplished.

When a π bonding orbital forms, so does a π^* antibonding orbital. Since σ overlap usually is more favorable than π overlap, a π bonding orbital usually is not as low in energy as the corresponding σ orbital, and a π^* orbital is not as high in energy as the σ^* orbital.

The σ-π description can be used to explain the formation of double and triple bonds, as in the case of the $:N{\equiv}N:$ molecule (Figure 7.17). Each N atom contributes three unpaired electrons, which are in the $2p_x$, $2p_y$, and $2p_z$ orbitals. A σ bond can be formed by the overlap of the two $2p_x$ orbitals. No more σ bonds are possible. The other two electron pairs that are shared in the triple bond exist in two π bonding orbitals, one formed between the two $2p_y$ orbitals and the other formed between the two $2p_z$ orbitals. In both these π bonding orbitals, the most probable region for the electrons is outside the region of the σ bond.

The triple bond can thus be described as one σ bond in one direction and two π bonds at right angles to each other and to the direction of the σ bond. The π bonds are not as strong as the σ bonds, because π overlap is not as effective as σ overlap. Therefore, a triple bond generally is not three times more stable than a single bond. In the same way, a double bond that consists of one σ bond and one π bond usually is not twice as stable as a single bond.

By adding the orbital description of the covalent bond to the Lewis structure description, we get a better picture of bonding in molecules.

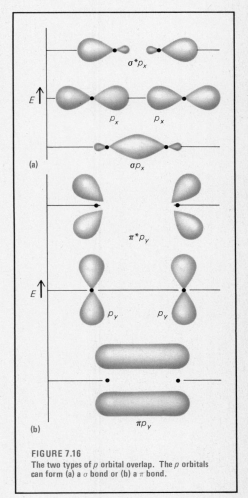

FIGURE 7.16
The two types of p orbital overlap. The p orbitals can form (a) a σ bond or (b) a π bond.

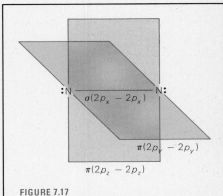

FIGURE 7.17
Bond formation in the N_2 ($:N{\equiv}N:$) molecule. The six unpaired electrons in the two atoms form one σ bond and two π bonds, with the π bonds at right angles to each other and to the σ bond.

7.6 THE MOLECULAR ORBITAL METHOD

Everything said thus far about bonding has been based on the assumption that the electrons in atoms can be divided into two groups, bonding and nonbonding, and that the two groups can then be considered separately. For example, in the Lewis structure $:F-F:$ and even in a simple orbital description of the F_2 covalent bond, the two shared electrons are described as being in a bonding orbital formed from the overlap of the two $2p_x$ orbitals. The interactions of the six nonbonding electrons of each F atom with the bonding electrons and with each other are not considered. The nonbonding electrons are simply placed around the atom.

This approach, is called the **valence bond (VB) method.** Lewis structures are at the heart of the VB method. There is a second approach, called the **molecular orbital (MO) method,** which also gives a good description of most molecules. We shall use the VB method in most cases, because it is easier. However, it is worth knowing how the MO method works and when it is more successful than the VB method.

The basic approach of the molecular orbital method resembles that of the Aufbau method, which is used to determine the electronic configuration of atoms. The MO approach assumes that there is a set of allowed orbitals for the molecule, rather than different sets of orbitals for each of the atoms in the molecule. The allowed orbitals are called molecular orbitals, and they give the method its name. Molecular orbitals (MOs) are not restricted to one or two atoms; they are associated with the entire molecule. Once the set of allowed orbitals is known, all the electrons of the molecule are placed in the orbitals, using the Aufbau method—that is, the orbitals are filled in order of increasing energy. As the Aufbau method gives the ground state electronic configuration of an atom, the MO method gives the ground state electronic configuration of a molecule. Just as an atomic orbital gives a probability distribution for an electron in the atom, a molecular orbital gives a probability distribution for an electron in the molecule.

However, there is a certain lack of precision in the MO method. Just as an exact set of orbitals cannot be constructed for a multielectron atom, an exact set of orbitals cannot be constructed for a molecule using the MO method. Only an approximate set of orbitals can be constructed for any molecule. Constructing even these approximate sets of orbitals, and arranging the orbitals in order of relative energy, can be a complex task.

Consider the MO description of the simplest molecules, the homonuclear diatomic molecules of the first two elements. The set of molecular orbitals associated with two hydrogen nuclei has been discussed, and is shown in Figure 7.13. Since each atom has only one valence orbital, there will be only two molecular orbitals, the $\sigma 1s$ orbital of relatively low energy and the $\sigma^* 1s$ orbital of relatively high energy. The same set of two MOs will be associated with any diatomic combination of H or He atoms.

Now consider the electronic configuration of diatomic species of H or He. The H_2^+ ion has one electron. As Figure 7.18 shows, the electron is in the $\sigma 1s$ bonding orbital. Since a covalent bond consists of two electrons in a bonding orbital, H_2^+ is said to have half of one bond. The observation that H_2^+ exists is consistent with this description.

The H_2 molecule has two electrons, both of which can be placed in the $\sigma 1s$ orbital. Two electrons equal one full bond, and the observed stability of the H_2 molecule corresponds to the existence of one full bond.

The He_2^+ molecule has three electrons. There is room for only two electrons in the $\sigma 1s$ bonding orbital. The third electron must go into the orbital of next lowest energy, the σ^* orbital. Thus, the electronic configuration of He_2^+ is $(\sigma 1s)^2 (\sigma^* 1s)^1$. But as was mentioned earlier, an antibonding orbital is destabilizing. The placement of an electron in an antibonding orbital can be regarded as canceling the favorable effect of placing an electron in a bonding orbital. For

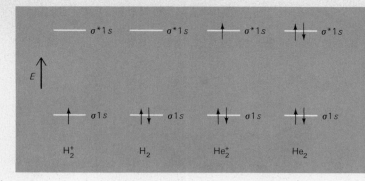

FIGURE 7.18
Electronic configuration of diatomic molecules of H and He. The H_2^+ molecule has one electron in the $\sigma 1s$ bonding orbital. The H_2 molecule has two electrons, both in the $\sigma 1s$ orbital. The He_2^+ molecule has three electrons, two in the $\sigma 1s$ bonding orbital and one in the $\sigma^* 1s$ antibonding orbital. The He_2 molecule would have two antibonding electrons, giving it no net bonds; this molecule does not exist.

bookkeeping purposes, placing an electron in an antibonding orbital cancels half of one bond. Thus, the number of bonds in a molecule is given by:

$\frac{1}{2}$(bonding electrons — antibonding electrons)

In $He_2{}^+$, there are two electrons in bonding orbitals and one in an antibonding orbital, giving $\frac{1}{2}(2 - 1) = \frac{1}{2}$ bond. This calculation is confirmed by observation. The stability of $He_2{}^+$ is found to be comparable to the stability of $H_2{}^+$, which also has half of one bond.

The power of the MO method is demonstrated by its correct prediction for He_2. This hypothetical species has four electrons and the electronic configuration $(\sigma 1s)^2 (\sigma^* 1s)^2$, in which there are two bonding and two antibonding electrons. There are $\frac{1}{2}(2 - 2) = 0$ bonds in this electronic configuration. A molecule with no bonds does not exist. In fact, no one has ever found evidence for the existence of the He_2 molecule.

First-Row Diatomic Molecules

The MO treatment of the first-row homonuclear diatomic molecules of the elements from Li to F is more complicated, even though the $1s$ orbitals and electrons can be omitted from our calculations on the grounds that only valence orbitals need be considered.

Each atom of an element in the first row has four valence orbitals: $2s$, $2p_x$, $2p_y$, and $2p_z$. Therefore, a total of eight MOs, four from each atom, are considered in a diatomic molecule.

One bonding orbital and one antibonding orbital correspond to each of these four valence orbitals. The types of orbitals to be expected for the diatomic molecules of these atoms are $\sigma 2s$ and $\sigma^* 2s$; $\sigma 2p$ and $\sigma^* 2p$; and $\pi 2p$ and $\pi^* 2p$. One pair of $2p$ orbitals overlaps in these diatomic molecules to form one $\sigma 2p$ orbital and one $\sigma^* 2p$ orbital. The remaining two pairs of $2p$ orbitals overlap to form two $\pi 2p$ orbitals of equal energy and two $\pi^* 2p$ orbitals of equal energy. Orbitals of equal energy are said to be *degenerate*.

The eight MOs must be arranged in order of increasing energy. While the finer points of this arrangement are beyond the scope of this book, some main points can be described. All s orbitals are considerably lower in energy than p orbitals, so the MO of lowest energy is the $\sigma 2s$ orbital. The next lowest is the $\sigma^* 2s$ orbital. The highest-energy orbital is the $\sigma^* 2p$ MO. Next highest are the two degenerate $\pi^* 2p$ MOs. The ranking by energy of the $\sigma 2p$ orbital and the two $\pi 2p$ orbitals presents some difficulty. Two arrangements of these orbitals are possible, as shown in Figures 7.19 and 7.20.

Figure 7.19 shows an energy level diagram that applies to diatomic molecules of Li, Be, B, C, and N. In these molecules, the $\sigma 2p$ MO is believed to be of higher energy than the two degenerate $\pi 2p$ orbitals. Figure 7.20 is an energy level diagram for the MOs in diatomic molecules of O and F, in which the $\sigma 2p$ MO is of lower energy than the two degenerate $\pi 2p$ MOs.

FIGURE 7.19
An energy level diagram for diatomic molecules of
Li, Be, B, C, and N in which the $\sigma 2p$ orbital is of
higher energy than the two $\pi 2p$ orbitals.

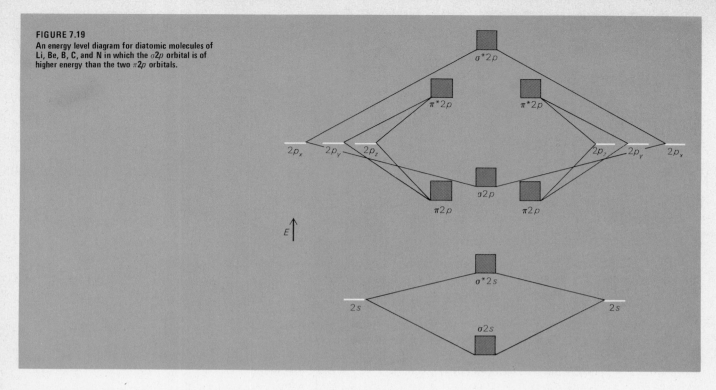

FIGURE 7.20
An energy level diagram for diatomic molecules of O
and F, in which the $\sigma 2p$ orbital is of lower energy
than the two $\pi 2p$ orbitals.

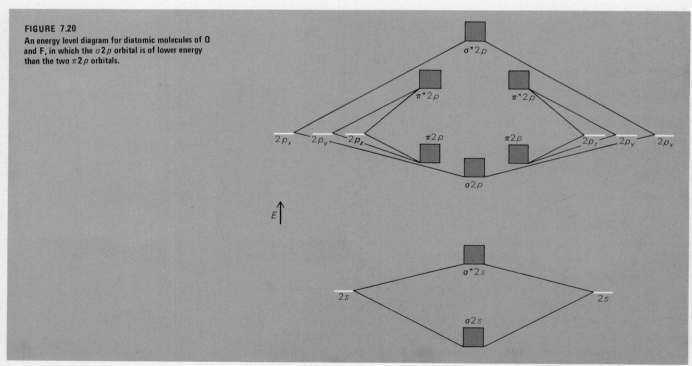

If the Aufbau method is applied to the set of allowed orbitals of the homonuclear diatomics of the first-row elements, the results are as shown in Table 7.5.

The Li_2 molecule (Figure 7.21) has two valence electrons, one $2s$ electron
from each Li atom. These two electrons can be placed in the $\sigma 2s$ bonding orbital.
Therefore, Li has $\frac{1}{2}(2 - 0) = 1$ bond, and thus should be stable. In fact, the Li_2
molecule has been found to exist in the gas phase and to be moderately stable.

TABLE 7.5
Electronic Configuration of First-Row Homonuclear Diatomics

Molecule	Lewis Structure	Number of Electrons								Bond	Unpaired Electrons
		$\sigma 2s$	$\sigma^* 2s$	$\pi 2p$	$\pi 2p$	$\sigma 2p$					
$Li_2{}^a$	Li–Li	2								1	0
$Be_2{}^b$		2	2							0	0
$B_2{}^a$	$(:B{-}B:)^c$	2	2	1	1					1	2
$C_2{}^a$	$:C{=}C:$	2	2	2	2					2	0
		$\sigma 2s$	$\sigma^* 2s$	$\sigma 2p$	$\pi 2p$	$\pi 2p$	$\pi^* 2p$	$\pi^* 2p$	$\sigma^* 2p$		
N_2	$:N{\equiv}N:$	2	2	2	2	2				3	0
O_2	$(:O{=}O:)^c$	2	2	2	2	2	1	1		2	2
F_2	$:F{-}F:$	2	2	2	2	2	2	2		1	0
$Ne_2{}^b$		2	2	2	2	2	2	2	2	0	0

^{a} Rarely encountered, very reactive.
^{b} Nonexistent molecule.
^{c} Incorrect structures.

The Be_2 molecule has four valence electrons, two $2s$ electrons from each Be atom. Two of these electrons can be placed in the $\sigma 2s$ bonding orbital; the other two are placed in the $\sigma^* 2s$ antibonding orbital. Therefore, Be_2 has $\frac{1}{2}(2 - 2) = 0$ bonds. A molecule with no bonds does not exist, and no evidence has been found for the existence of the Be_2 molecule.

The MO method predicts an unusual feature for the B_2 molecule: two unpaired electrons, as shown in Figure 7.22. Each B atom has three valence electrons, so there are six electrons that must be placed in the molecular orbitals of the B_2 molecule. Two electrons fill the $\sigma 2s$ molecular orbital. Two more fill the $\sigma^* 2s$ MO. The remaining two electrons fill the two degenerate (equal-energy) $\pi 2p$ orbitals. Each electron half fills one of the orbitals, consistent with Hund's rule.

There is a major difference between the B_2 electronic configuration of the MO method and that of the VB (Lewis structure) method. Several different Lewis structures can be written for the B_2 molecule. All have incomplete octets; none of them need have unpaired electrons. But the MO method indicates that the B_2 molecule has two unpaired electrons. There are four electrons in bonding orbitals and only two electrons in the $\sigma^* 2s$ antibonding orbital, giving $\frac{1}{2}(4 - 2) = 1$ bond, enough to hold the molecule together.

Which method gives the more accurate picture? The question can be answered most simply by examining the magnetic properties of the B_2 molecule, by placing the substance in a magnetic field (Figure 7.23). A substance that is slightly repelled by a magnetic field is said to be **diamagnetic.** The vast majority of substances are diamagnetic. All the electrons in a diamagnetic substance are paired. A relatively small number of substances are attracted by a magnetic field. They are classified as **paramagnetic.** A paramagnetic substance has one or more unpaired electrons. An experiment of the kind shown in Figure 7.23 demonstrates that B_2 is paramagnetic. Therefore, B_2 has unpaired electrons, and the MO picture of its electronic configuration is a useful and consistent model.

The B_2 molecule is rare. So is the C_2 molecule, whose electronic configuration, including eight electrons, is shown in Figure 7.22. All of the electrons are paired, so C_2 should be diamagnetic. Experiments have found that C_2 is diamagnetic.

EXAMPLE 7.12

Show how the fact that B_2 is paramagnetic and C_2 is diamagnetic confirms that the two degenerate $\pi 2p$ MOs are of lower energy than the $\sigma 2p$ MO and fill first.

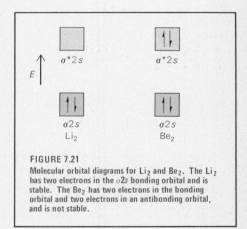

FIGURE 7.21
Molecular orbital diagrams for Li_2 and Be_2. The Li_2 has two electrons in the $\sigma 2s$ bonding orbital and is stable. The Be_2 has two electrons in the bonding orbital and two electrons in an antibonding orbital, and is not stable.

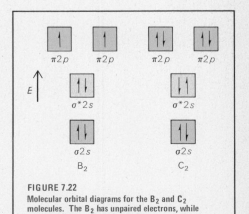

FIGURE 7.22
Molecular orbital diagrams for the B_2 and C_2 molecules. The B_2 has unpaired electrons, while the C_2 has only paired electrons.

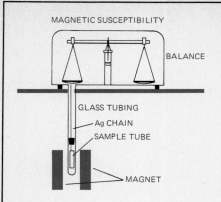

FIGURE 7.23
A device for determining the magnetic properties of substances. The substance is suspended in the sample tube by a slender copper wire attached to a silver chain, and is placed in a magnetic field. A substance that is repelled by the magnetic field is diamagnetic; one that is attracted by a magnetic field is paramagnetic.

SOLUTION

Figure 7.22 shows the electronic configurations that result when the $\pi 2p$ MOs have lower energy than the $\sigma 2p$ MO, as experimental results suggest. If the $\sigma 2p$ orbital filled first, B_2 would have the electronic configuration $(\sigma 2s)^2\ (\sigma^*2s)^2(\sigma 2p)^2$, with no unpaired electrons, so B_2 would be diamagnetic. If the $\sigma 2p$ orbital filled first, C_2 would have the electronic configuration $(\sigma 2s)^2(\sigma^*2s)^2(\sigma 2p)^2(\pi 2p)^1(\pi 2p)^1$, with two unpaired electrons, one in each of the degenerate $\pi 2p$ MOs, so C_2 would be paramagnetic. Experimental results indicate that these are not the ground state configurations of B_2 and C_2.

The MO picture of the N_2 molecule closely resembles the Lewis structure, as shown in Figure 7.24. The N_2 molecule has a total of six bonding electrons in the $2p$ molecular orbitals, and so has a triple bond. This picture is consistent with the Lewis structure $:N\equiv N:$ and with the observation that N_2 is diamagnetic.

The MO picture of the electronic configuration of the O_2 molecule, however, differs markedly from the VB picture. Historically, one of the first major successes of MO theory was the correct prediction of the properties of the O_2 molecule, which VB theory could not predict in any simple way.

A simple Lewis structure with completed octets is readily written for O_2:

$$\ddot{O}=\ddot{O}$$

But this structure, in which all the valence electrons are paired, is inconsistent with experimental evidence, which indicates that O_2 is paramagnetic and has two unpaired electrons. By contrast, the MO method predicts that two of the 12 valence electrons in the O_2 molecule will be unpaired, as shown in Figure 7.24.

The O_2 molecule has two more electrons than the N_2 molecule. These two electrons are placed in the two degenerate π^*2p orbitals, one in each orbital according to Hund's rule. The number of bonds in the O_2 molecule therefore is $\frac{1}{2}(8 - 4) = 2$, a double bond. The O_2 molecule thus has one less bond than the triple bond of N_2 because the two additional electrons in O_2 are in antibonding orbitals and cancel the effect of two electrons in a bonding orbital.

For the F_2 molecule, both the valence bond and molecular orbital methods

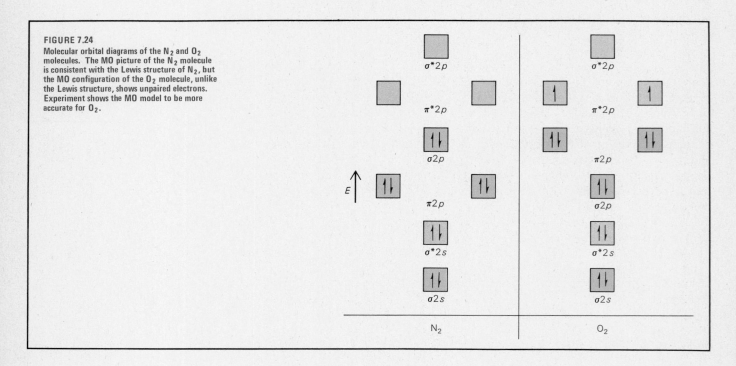

FIGURE 7.24
Molecular orbital diagrams of the N_2 and O_2 molecules. The MO picture of the N_2 molecule is consistent with the Lewis structure of N_2, but the MO configuration of the O_2 molecule, unlike the Lewis structure, shows unpaired electrons. Experiment shows the MO model to be more accurate for O_2.

CLEANING UP WITH PARAMAGNETISM

A mixture of iron filings and charcoal powder is readily separated by passing a magnet over the mixture. The iron is attracted to the magnet; the charcoal is not. Until recently, this technique of magnetic separation was possible on a large scale only for mixtures containing three elements: iron, nickel, or cobalt, the three ferromagnetic elements. These elements and some of their compounds are strongly magnetized by a weak magnetic field. Now, a new technique of magnetic separation has been developed that is potentially usable with the many elements and compounds that are paramagnetic. The technique could greatly increase the earth's iron resources, help purify water supplies, and even be used to remove sulfur from coal.

The successful magnetic attraction of paramagnetic substances requires not only a strong magnetic field but also one with a very steep gradient—that is, a rapid increase in the intensity of the field over a very short distance. The magnet must also have a very large surface for collecting the material. The development of strong magnets with steep gradients and large surfaces in recent years has made possible the separation of mixtures containing paramagnetic substances.

One practical application of this method is the separation of impurities out of kaolin, a white clay that is used as a coating for papers. These impurities, the most important of which is titanium dioxide, are extremely small paramagnetic particles. A single large magnetic separator now can treat thousands of kilograms of kaolin in an hour.

The use of the technique to separate iron from low-grade taconite ore is being explored. Most of the iron ore that has been mined until now contains magnetite, a compound of iron that is ferromagnetic. Magnetite is separated from iron ore with conventional magnetic separation techniques. But supplies of magnetite-containing iron ore are running low. Taconite ore contains hematite, a paramagnetic compound of iron. If hematite can be obtained magnetically from the ore using the new technique, enormous savings in energy and a large increase in usable iron reserves will result.

One potentially important application of high-gradient magnetic separation is for pollution control. Much of the available coal in the United States cannot be used easily because it has a relatively high sulfur content. The sulfur compounds that are formed when coal is burned can be harmful to human health. However, much of the sulfur in coal is in the form of pyrite, FeS_2, which is a paramagnetic substance. One technique that could possibly be used to remove sulfur from coal is magnetic separation. The coal could first be made into a liquefied form called slurry, and could then be passed through the magnetic field.

The same sort of process could be used for water purification. Many water pollutants are paramagnetic. Others can be made predict the same properties, a stable diamagnetic molecule with a single bond. Both methods also give the same result for the Ne_2 molecule: They say it does not exist. The VB method shows that a bond cannot be formed between two Ne atoms unless there are expanded octets. In the MO method (Figure 7.25), all the orbitals of the Ne_2 molecule are filled, and the number of antibonding electrons equals the number of bonding electrons. The number of bonds is $\frac{1}{2}(8 - 8) = 0$, and the molecule cannot exist. There is no experimental evidence for the existence of the Ne_2 molecule.

The simple molecular orbital approach is more successful than the simple valence bond approach for some molecules, such as B_2 and O_2. However, both methods usually work equally well in predicting the electronic configurations of most molecules. In fact, the VB method sometimes is more effective than the MO method. Molecular orbital theory does not readily account for the high reactivity and relative instability of such species as Li_2, B_2, and C_2. The Lewis structures written by the VB method have incomplete octets, which indicate high reactivity and relative instability.

Polyatomic Molecules

The MO method can become quite complex when applied to molecules of more than two atoms. The method must account for a larger number of orbitals that

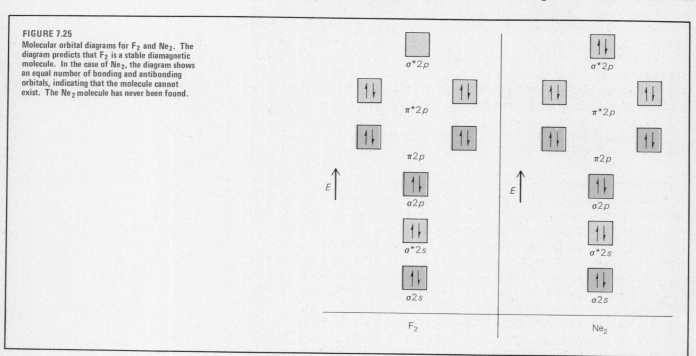

FIGURE 7.25
Molecular orbital diagrams for F_2 and Ne_2. The diagram predicts that F_2 is a stable diamagnetic molecule. In the case of Ne_2, the diagram shows an equal number of bonding and antibonding orbitals, indicating that the molecule cannot exist. The Ne_2 molecule has never been found.

have larger regions of high electron probability. But the molecular orbital method can be useful for many molecules that the valence bond method describes as resonance hybrids. A useful MO picture that can be combined with the VB picture to supplement it can be formulated for these molecules by making the simplifying assumption that the only molecular orbitals of interest are those of the π bonds, since contributing structures differ only in the arrangement of the π bonds.

Using this combined approach, we first put in all the σ bonds to form the skeleton structure. All the nonbonding electron pairs are then sketched in, leaving the other shared pairs of electrons to be accounted for. As a final step, we construct molecular orbitals with π overlap between one p orbital of each atom that takes part in multiple bonding in a contributing structure.

A simplified picture of the HNO_3 molecule can be obtained in this way (Figure 7.26). We recall from Section 7.4 that the N atom and two O atoms can take part in multiple bonding. Each of these three atoms has a p orbital available for π overlap. All three of these p orbitals have the same directional subscript and so must be parallel. These three p orbitals overlap to form three molecular orbitals that encompass all three atoms. The extra shared electron pair of the three atoms that is shown in the Lewis structure is placed into the molecular orbital of lowest energy.

In the NO_3^- anion (Figure 7.27), all four atoms can participate in multiple bonding. Molecular orbitals encompassing all four atoms can be constructed by using a p orbital from each atom. The extra shared electron pair is placed in the lowest-energy orbital, and thus is associated equally with all three O atoms. In HNO_3, one nonbonding valence electron pair is also placed in one of the three molecular orbitals; in NO_3^-, two nonbonding valence electron pairs are placed in two of the four molecular orbitals.

Molecules encountered in organic chemistry may have many such molecular orbitals, because of the large number of atoms that can take part in multiple bonding. Such molecular orbitals sometimes are spread over many atoms. For example, the benzene molecule, C_6H_6 (Figure 7.28), forms six molecular orbitals, three bonding and three antibonding. The benzene molecule has three pairs of π bonding electrons that can be accommodated by the three bonding molecular orbitals. Benzene is an unusually stable molecule, which is consistent with both this MO picture and the resonance hybrid VB picture.

When molecular orbitals can be formed from more than two p orbitals on more than two adjacent atoms, the system is said to be *conjugated*. If a large number of atoms are involved, the system is said to be one of multiple conjugation. Such molecules often absorb visible light and are intensely colored. Most organic dyes have this type of extended conjugation. Vitamin A is conjugated, as are the pigments that give carrots and tomatoes their distinctive colors.

7.7 POLAR BONDS

Both the ionic bond and the covalent bond that we have discussed are simplified models. No substance has a completely ionic bond, and only homonuclear diatomic molecules can be regarded as having completely covalent bonds. Most real bonds lie somewhere between these two extremes.

An idealized ionic bond includes the complete transfer of one or more electrons from one bonded atom that forms a cation to the other bonded atom that forms an anion. An electron pair thus is associated completely with the anion. The ideal covalent bond has equal sharing of the bonding pair, an equal association of the two electrons with both atoms.

In a covalent bond between two different atoms, the bond is partially ionic in character and the bonding electron pair is more closely associated with one of the atoms. As a result, the two atoms that are bonded have partial electrical charges.

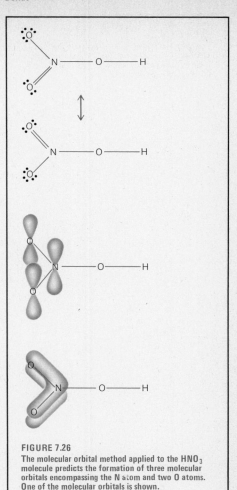

FIGURE 7.26
The molecular orbital method applied to the HNO_3 molecule predicts the formation of three molecular orbitals encompassing the N atom and two O atoms. One of the molecular orbitals is shown.

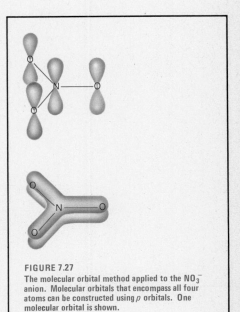

FIGURE 7.27
The molecular orbital method applied to the NO_3^- anion. Molecular orbitals that encompass all four atoms can be constructed using p orbitals. One molecular orbital is shown.

FIGURE 7.28
The molecular orbital method applied to the benzene molecule, C_6H_6. The molecule forms six molecular orbitals, three bonding and three antibonding. One bonding molecular orbital is shown.

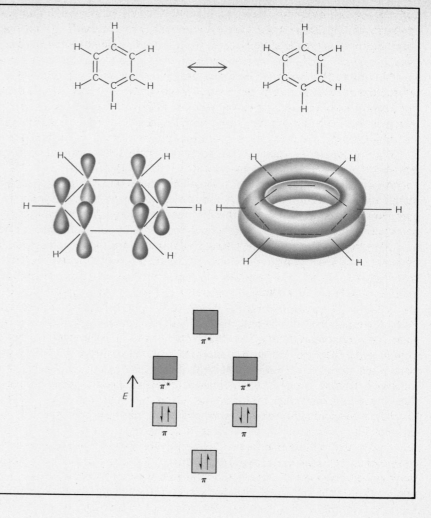

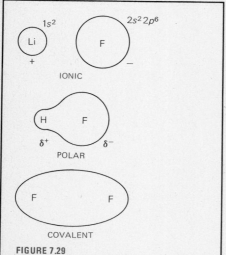

FIGURE 7.29
Three types of bonds of the fluorine atom. The LiF bond is largely ionic, but with some electron sharing. The HF bond is polar. The F_2 bond is completely covalent.

We call this pair of charges a **dipole**. The atom with the greater share of the bonding electron pair has a relative excess of negative charge, while the atom with the lesser share has a relative excess of positive charge. Because there are two equal but opposite charges, there is an electrical polarity associated with the bond. Such a bond is called a **polar bond**.

The extent to which the bonding electron pair is more closely associated with one of the two atoms of a bond depends largely on the difference in electronegativity values between the two atoms (Section 7.1, page 194). If the difference in electronegativity values is great, the bond is more ionic; if the difference is small, the bond is more covalent. Homonuclear diatomic molecules have the only completely covalent bonds because the atoms have the same electronegativity.

Figure 7.29 shows three types of bonds of fluorine. The F_2 bond is completely covalent. The HF bond is a polar bond. The LiF(g) bond is described as an ionic bond, but measurements show that there is some electron sharing by the atoms. In CsF(g), the binary compound whose atoms have the greatest difference in electronegativity, there is a small amount (about 8%) of electron sharing.

There is an arbitrary rule of thumb to help us make the distinction between bond types. In a nonpolar bond, the two bonded atoms have an electronegativity difference of 0.5 or less. In polar bonds, the electronegativity difference is between 0.5 and 1.7. In ionic bonds, the difference is greater than 1.7.

It is convenient to classify substances by bond type, using standard criteria. At one extreme, a substance is classified as ionic if it contains at least one ionic bond, which will be between two atoms that differ by more than 1.7 in electronegativity. At the other extreme is a substance that is nonpolar and contains only bonds between identical atoms or atoms that are relatively close in electronegativity.

Molecules that have polar bonds tend to have an overall polarity. Such molecules with overall polarity are called *polar molecules*. However, the polarity of the individual bonds sometimes cancels, and there is no overall separation of charge in the molecule. Among such molecules are CO_2, BF_3, and CCl_4. We shall see why they are nonpolar when we discuss their geometry in Chapter 8.

We discussed the types of ionic compounds in Section 7.1. Nonpolar compounds include homonuclear diatomic molecules as well as binary compounds of nonmetals and most compounds of hydrogen and carbon. Polar compounds include covalent compounds of metals and nonmetals, compounds of hydrogen with nonmetals, compounds of oxygen with nonmetals, and some compounds of halogens with nonmetals.

We can describe the properties of ionic, polar, and nonpolar compounds in general terms. Ionic compounds, such as NaCl and CaF_2, are hard, brittle solids with high melting points and high boiling points. They conduct electricity when they become liquid. They dissolve in polar solvents such as H_2O or liquid NH_3 but are insoluble in nonpolar solvents. Polar compounds, such as H_2O, SO_3, and P_4O_{10}, usually are solid or liquid. If solid, they usually are soft at room temperature. Their boiling points and melting points are much lower than those of ionic compounds. They may conduct electricity to a limited extent. They dissolve in polar solvents and, to a lesser degree, in nonpolar solvents. Nonpolar compounds, such as Cl_2 or CH_4, can be gases, liquids, or solids at room temperature. The solids have low melting points, generally are soft and waxy, and do not conduct electricity. They dissolve in nonpolar organic solvents, such as gasoline or kerosene.

The solubility of substances is described by a simple rule: Like dissolves like. Nonpolar substances dissolve best in nonpolar solvents; ionic or polar substances dissolve best in polar solvents. When we make salad dressing out of oil (nonpolar) and vinegar (polar), we see the effect of this rule.

7.8 OXIDATION NUMBERS

Every atom in every molecule can be assigned an *oxidation number,* a positive or negative integer that is related to the electronic structure of the molecule. We first introduced this concept in Section 2.4, page 29.

An oxidation number can be assigned to an atom in a molecule by imagining that every electron transfer is complete and calculating the resulting charge on each atom. The charge is the oxidation number. In any diatomic molecule, the more electronegative atom will have a negative oxidation number and the less electronegative atom will have a positive oxidation number.

For example, in the formation of LiF an electron is transferred almost completely from Li to F, so we think in terms of Li^+F^-. Thus, Li has an oxidation number of $+1$ and F has an oxidation number of -1. In HF, an electron is partially transferred from the H atom to the F atom, whose electronegativity is higher; H has an oxidation number of $+1$ and F has an oxidation number of -1. In the $N{=}O$ molecule, the oxidation numbers are $+2$ for N and -2 for O. In F_2, each F atom has an oxidation number of 0.

═══════════════════════════════════════

EXAMPLE 7.13
Calculate the oxidation numbers of the atoms in (a) PCl_5, (b) BrF_3, (c) NO_2.
SOLUTION
(a) The Lewis structure of PCl_5 is:

$$:\overset{..}{\underset{..}{Cl}}: \\ :\overset{..}{\underset{..}{Cl}} \overset{\overset{..}{\underset{..}{Cl}}:}{\underset{}{\overset{|}{\underset{|}{P}}}} \overset{..}{\underset{..}{Cl}}: \\ :\overset{..}{\underset{..}{Cl}} \quad \overset{..}{\underset{..}{Cl}}:$$

The Cl atom is more electronegative than the P atom. Let us assume that the pair of electrons in each P—Cl bond is transferred totally to the Cl atom. Its configuration will then be $:\overset{..}{\underset{..}{Cl}}:$. A neutral chlorine atom has seven electrons, while $:\overset{..}{\underset{..}{Cl}}:$ has eight. Therefore, $:\overset{..}{\underset{..}{Cl}}:$ has a charge of -1. Thus, the oxidation number of each Cl atom in PCl_5 is -1. If the transfer is total, the P atom will be left with no valence electrons. Since the atom originally had five valence electrons, its charge, and thus its oxidation number, is now $+5$.

(b) The Lewis structure of BrF_3 is:

$$:\overset{..}{\underset{..}{F}} - \overset{|}{\underset{|}{Br}} - \overset{..}{\underset{..}{F}}: \\ \quad :\overset{..}{\underset{..}{F}}:$$

The F atom is more electronegative than the Br atom. If the bonding electron pairs are transferred completely to the F atom, its configuration is $:\overset{..}{\underset{..}{F}}:$, which corresponds to an oxidation number of -1. The Br atom loses three of its seven electrons. Therefore, its oxidation number is $+3$.

(c) An important contributing structure for NO_2 is:

$$\overset{..}{\underset{..}{O}} \qquad \overset{..}{\underset{.}{O}} \\ \underset{\overset{|}{\underset{.}{N}}}{}$$

A complete transfer of the bonding electrons to the more electronegative O atoms gives each O atom eight electrons. Since the neutral O atom has only six electrons, the oxidation number of O is -2. The N atom is left with one electron. Since the neutral N atom has five electrons, the oxidation number of N in NO_2 is $+4$.

The oxidation number is merely an idealized concept, since there is no full transfer of electrons in molecules. But oxidation numbers are useful because they give information about both the combining power of elements and the bonding in molecules.

A positive oxidation number means that the atom, on balance, loses electrons in the molecule. A negative oxidation number indicates that the atom gains electrons. The magnitude of the oxidation number gives the number of electrons that the atom gains or loses, either partially or wholly. An oxidation number of $+2$ means that the atom contributes two electrons, or a share of two electrons, to bonds. A -2 oxidation number means that the atom gains two electrons, or a share of two electrons, from bonds.

Oxidation numbers also provide a guide to the chemical behavior of given elements in given molecules. They provide a convenient framework for systematizing a great deal of information.

It is not always easy to calculate oxidation numbers for all atoms in a polyatomic molecule. Sometimes it is tedious to calculate the hypothetical charge on every atom, and we can avoid this procedure. Instead, we can use some simple rules to assign oxidation numbers:

1. The sum of all the oxidation numbers of the atoms in an electrically neutral chemical substance must be zero. In H_2S, for example, each H atom has an oxidation number of $+1$, and the oxidation number of the S atom is -2. The sum is $2(+1) + (-2) = 0$.
2. In polyatomic ions, the sum of oxidation numbers must equal the charge on the ion. For example, in NH_4^+, each H atom has an oxidation number of $+1$,

and the oxidation number of the N atom is -3, making the total $4(+1) + (-3) = +1$, which is the charge on the ion. In SO_4^{2-}, each O atom has an oxidation number of -2. The sum of oxidation numbers for the ion must be -2, which is the charge on the ion. Since the sum of the oxidation numbers of the four O atoms is -8, the oxidation number of the S atom must be $+6$.

3. The oxidation number of an atom in a monoatomic ion is its charge. In Na^+, sodium has an oxidation number of $+1$. In S^{2-}, sulfur has an oxidation number of -2.

4. The oxidation number of an atom in a single-element substance is 0. Thus, the oxidation number of every sulfur atom in S_2, S_6, and S_8 is 0; the oxidation number of chlorine in Cl_2 is 0; the oxidation number of oxygen in O_2 or O_3 is 0.

5. Some elements have the same oxidation number in all or nearly all of their compounds. When F combines with other elements, its oxidation number is always -1. The halogens Cl, Br, and I have the oxidation number -1 except when they combine with oxygen or another halogen of lower atomic number. Oxygen usually has the oxidation number -2, except when it combines with F or with itself in such compounds as the peroxides or superoxides. Hydrogen always has the oxidation number $+1$ when combined with nonmetals and -1 when combined with metals. A metal in group IA of the periodic table always has an oxidation number of $+1$. A metal in group IIA always has an oxidation number of $+2$.

6. Metals almost always have positive oxidation numbers.

7. A bond between identical atoms in a molecule makes no contribution to the oxidation number of that atom because the electron pair of the bond is divided equally. In hydrogen peroxide, H_2O_2, for example, the two O atoms are bonded to one another. The oxidation number of O can be calculated by determining the contribution of the two H atoms, each of which has an oxidation number of $+1$. Since the sum of the oxidation numbers of the H atoms is $+2$ and the molecule is neutral, the sum of the oxidation numbers of the two O atoms is -2, giving each an oxidation number of -1.

EXAMPLE 7.14

Calculate the oxidation numbers of all the atoms in (a) $KMnO_4$, (b) H_5IO_6, (c) HPO_4^{2-}.

SOLUTION

(a) Since $KMnO_4$ is a neutral molecule, the sum of the oxidation numbers is zero. The oxidation number of oxygen is -2, making its contribution $4(-2) = -8$. The oxidation number of K (group IA) is $+1$. If we let $x =$ the oxidation number of Mn, then $(-8 + 1) + x = 0$; $x = +7 =$ oxidation number of Mn.

(b) The sum of the oxidation numbers in the neutral molecule H_5IO_6 is zero. The oxidation number of H is $+1$. The oxidation number of O is -2. The oxidation number of I can be obtained by letting $x =$ the oxidation number of I and solving $(5 \times +1) + (6 \times -2) + x = 0$; $x = +7$.

(c) The sum of the oxidation numbers for HPO_4^{2-} is -2, the charge on the ion. If $x =$ the oxidation number of P, solving for $(+1) + (4 \times -2) + x = -2$ gives $x = +5$.

EXAMPLE 7.15

Calculate the oxidation numbers of all the atoms in (a) $Cr_2O_7^{2-}$, dichromate ion, and (b) Fe_3O_4, magnetite.

SOLUTION

(a) Applying the method of Example 7.14, $2Cr + 7(-2) = -2$; $2Cr = +12$, and each Cr atom has an oxidation number of $+6$.

(b) The method of Example 7.14 does not give a simple answer for Fe_3O_4, because Fe seems to have a fractional oxidation number of $+8/3$. A better approach is to assume that there are two types of Fe atoms in this molecule, an assumption that is consistent with the experimental evidence. If Fe_3O_4 is considered to be $FeO + Fe_2O_3$, then one Fe atom has the oxidation number $+2$ and two Fe atoms have the oxidation number $+3$. We shall encounter other examples where either a fractional oxidation number or two different oxidation numbers are found for an element. Often, experimental evidence is needed to make correct assignments of oxidation numbers in such situations.

Oxidation numbers define **oxidation states**. The two terms often are used interchangeably. Oxidation states help us to understand chemical processes in which there is oxidation and reduction of elements.

The terms oxidation and reduction are defined as changes in oxidation state. If a chemical reaction changes the oxidation states of some atoms in the reactants, the process is called an **oxidation-reduction** or **redox** reaction. The substance in which an atom has an *increase* in oxidation state is said to be **oxidized**. The substance in which an atom has a *decrease* in oxidation state is said to be **reduced**. Oxidation and reduction always occur together. There can be no oxidation without an accompanying reduction.

In Section 6.8, we discussed a redox process that is brought about by an electric current; the electrolysis of molten sodium chloride:

$$NaCl \longrightarrow Na + \tfrac{1}{2}Cl_2$$

In this reaction, the oxidation number of Na goes down from $+1$ to 0 and the oxidation number of Cl goes up from -1 to 0. Sodium is reduced; chlorine is oxidized. We shall discuss redox processes in detail in Chapter 16.

As Figure 7.30 shows, the characteristic oxidation states of the main group elements have a rather direct relationship to the position of the elements in the periodic table. This relationship is to be expected, since the oxidation number of any atom is related to its electronic configuration. In many cases, there is a tendency for the oxidation number of a main group element to be the same as the number of electrons the element must gain or lose to achieve a noble gas electronic configuration. However, some atoms can lose p electrons but retain a complete s subshell. This tendency is found in the heavier elements of groups IIIA and IVA, which can have oxidation numbers of $+1$ and $+2$, respectively, in addition to the expected values of $+3$ and $+4$. For example, Tl can form either TlCl or $TlCl_3$, while Pb can form either PbO or PbO_2.

No element can have an oxidation number that is higher than the number of its group in the periodic table. The highest oxidation number of P is $+5$; the highest oxidation number of S is $+6$; and the highest oxidation number of Cl is $+7$. In general, positive oxidation numbers are less probable in elements with

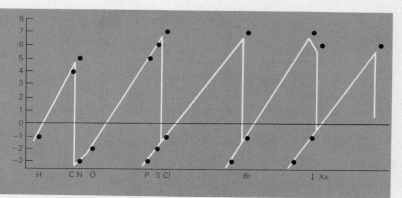

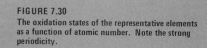

FIGURE 7.30
The oxidation states of the representative elements as a function of atomic number. Note the strong periodicity.

higher electronegativities. Likewise, no element can have an oxidation number more negative than the number of electrons needed to complete its octet. The lowest oxidation number of P is −3; the lowest oxidation number of C is −4 and the lowest oxidation number of S is −2.

7.9 WEAK INTERACTIONS

In addition to the strong forces that hold atoms together in chemical bonds, there are a number of weaker attractive forces between atoms that are not bonded to each other. These weak interactions can have a profound effect on the physical and chemical properties of substances. Usually, weak interactions are intermolecular—that is, they are attractions between portions of different molecules. But there can also be weak intramolecular attractive forces between different portions of the same molecule.

The simplest case of a weak interaction is coulombic attraction between the electrical dipoles of two polar diatomic molecules, as shown in Figure 7.31. The relatively high melting and boiling points of polar compounds can be attributed to such *dipole-dipole* interactions.

There can also be an attractive force between a substance with an electric dipole and a substance that ordinarily does not have a dipole. Such an attraction is called a *dipole-induced dipole* interaction. It results when a molecule with a dipole comes close enough to one without a dipole to induce a dipole in the latter. The molecule without the dipole becomes *polarized* by the other molecule. Figure 7.32 shows a dipole-induced dipole interaction. Usually, a dipole-induced dipole attraction is not as strong as a dipole-dipole attraction.

There can also be attractive forces even between completely nonpolar substances. The existence of such forces can be deduced from the observation that nonpolar substances such as I_2 and Br_2 are liquid or solid at room temperature. There must be some relatively strong intermolecular forces that hold the molecules of these substances together.

These intermolecular forces are called *London forces*. They are believed to arise from momentarily unequal distributions of electron density in molecules, even nonpolar molecules. At any instant, there may be an excess of electron density in one region of the molecule and a corresponding deficiency of electron density elsewhere in the molecule. The uneven distribution of electron density can create an "instantaneous dipole."

Suppose there is another nonpolar molecule nearby. As Figure 7.33 shows, a dipole will be induced in this second molecule by the instantaneous dipole of the first molecule. An attractive force results. This attraction between molecules can be maintained even though the instantaneous dipole disappears. A shift in the electron density of the first molecule can create another instantaneous dipole with a different orientation. If the induced dipole shifts in synchronization with the instantaneous dipole, the attraction between the molecules is maintained. This model is used to account for the attractive forces between nonpolar molecules.

The extent to which a dipole can be induced in a nonpolar molecule depends on two factors: the force that induces the dipole and the ease with which the electron distribution in the molecule can be distorted. The responsiveness of the electron distribution of a molecule to a distorting influence is called the *polarizability*. Several factors influence polarizability. An increase in the volume of a molecule or an increase in the number of electrons in the molecule results in an increase in the polarizability of the molecule. Since high polarizability indicates stronger attractive forces between molecules, molecules of the larger members of any family of elements will attract each other more strongly.

The physical properties of the halogens illustrate the increase in London forces in the larger members of a family of elements. At room temperature, Cl_2 is

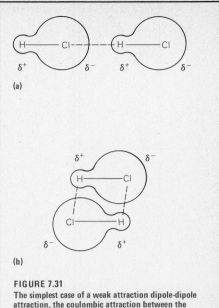

FIGURE 7.31
The simplest case of a weak attraction dipole-dipole attraction, the coulombic attraction between the electrical dipoles of two polar diatomic molecules.

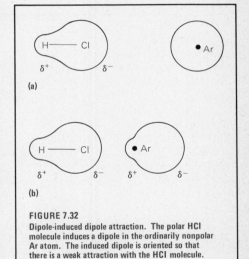

FIGURE 7.32
Dipole-induced dipole attraction. The polar HCl molecule induces a dipole in the ordinarily nonpolar Ar atom. The induced dipole is oriented so that there is a weak attraction with the HCl molecule.

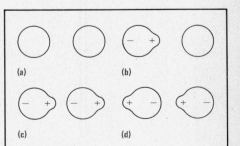

FIGURE 7.33
Instantaneous dipole-induced dipole attraction. A momentary displacement of electron density creates a dipole in a nonpolar molecule. This "instantaneous dipole" creates a dipole in a second nonpolar molecule, causing an attraction between the molecules.

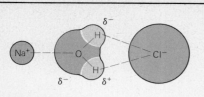

FIGURE 7.34
Ion-dipole interaction. The polar water molecule attracts both anions and cations in a solution of sodium chloride.

a gas, Br_2 is a liquid, and I_2 is a solid. The Cl_2 molecule is the smallest of the three and I_2 is the largest. The intermolecular forces between the Cl_2 molecules in the liquid or solid are weaker than the forces between the I_2 molecules in the liquid or solid.

Any of these weak interactions can occur between molecules, alone or in combination. Collectively, these weak interactions are called **van der Waals forces.** Usually, van der Waals forces are significant only at very short distances. They are negligible if the two interacting molecules are not close together.

There is a fourth type of interaction, called *ion-dipole interaction,* that is stronger than the other three weak forces. Ion-dipole interaction can exist in a system containing dissociated ions and a polar compound, such as a solution of an ionic substance in water. Figure 7.34 shows ion-dipole interactions in a solution of sodium chloride in water.

A special kind of weak interaction is the hydrogen bond. Hydrogen bonds occur primarily in compounds that contain O—H bonds or N—H bonds, although HF also has a hydrogen bond. Generally, compounds with hydrogen bonds display properties that are consistent with strong intermolecular attractions. Ion-dipole interactions and hydrogen bonding will be discussed at greater length in Chapter 11.

EXERCISES

7.1 Write the formulas of the following ionic solids: (a) barium fluoride, (b) ammonium chloride, (c) sodium sulfate, (d) gallium oxide, (e) magnesium phosphate.

7.2 List the following cations in order of the ease with which they form from atoms of their elements in the gas phase. Start with the cation that is most easily formed. Li^+, Na^+, K^+, Cs^+, Mg^+, Ca^+, Mg^{2+}, Ca^{2+}, Al^{3+}, H^+. (*Hint:* Table 6.7.)

List the following anions in order of the ease with which they form from atoms of their elements in the gas phase. Start with the anion that is most easily formed. F^-, Cl^-, I^-, O^-, O^{2-}, S^-, S^{2-}, N^{3-}. (*Hint:* Table 6.8.)

7.4 In each of the following groups select the pair of ions that requires the least amount of energy to form from atoms of its elements in the gas phase: (a) Li^+ and F^-, Na^+ and F^-, K^+ and F^-; (b) Ba^{2+} and O^{2-}, Ba^{2+} and S^{2-}, Ba^{2+} and C^{2-}; (c) Na^+ and F^-, Mg^+ and F^-, Na^+ and O^-, Mg^+ and O^-.

* 7.5 List the following gas phase ion pairs in order of the amount of energy released when they form from separated gas phase ions. Start with the pair that releases the least amount of energy. Na^+F^-, $Mg^{2+}F^-$, Na^+O^-, $Mg^{2+}O^{2-}$, $Al^{3+}O^{2-}$.

7.6 List the following gas phase ion

pairs in order of the amount of energy released when they form from separated gas phase ions. Start with the pair that releases the least amount of energy: $Mg^{2+}O^{2-}$, $Ca^{2+}O^{2-}$, $Ca^{2+}S^{2-}$, $Sr^{2+}S^{2-}$, $Ba^{2+}S^{2-}$.

* 7.7 The force of attraction between the two ions of an ion pair is found to be 2.6×10^{-9} N. Find the difference between the force of attraction of two of these isolated ion pairs and two of these ion pairs arranged at the corners of a square as shown in Figure 7.1.

7.8 Predict which halide of the alkaline-earth metals will have the largest lattice energy and which halide will have the smallest lattice energy.

7.9 Table 7.2 shows that the difference in lattice energies between the fluorides of the alkali metals and their corresponding chlorides is greater in every case than the difference in lattice energy between the chlorides and the corresponding bromides. Suggest an explanation for this observation.

7.10 Suggest an explanation for the observation that the ionic radii of anions are usually larger than the ionic radii of cations.

7.11 The three main influences on lattice energy may be listed succinctly as: (i) geometry, (ii) ionic radius, and (iii) ionic charge. Designate which influence is probably most important in caus-

ing each of the following observed effects:
(a) The lattice energy of MgO is greater than the lattice energy of NaF.
(b) The lattice energy of MgO is greater than the lattice energy of BaO.
(c) The lattice energy of MgS is greater than the lattice energy of BeS.

7.12 Although energy is required to form O^{2-} from O^-, ionic solids with the O^- anion are never observed, while solids with the O^{2-} anion are very common. Suggest an explanation for this observation. Suppose you were able to prepare the solid NaO under extraordinary conditions. Predict the chemical behavior of this solid under ordinary conditions.

7.13 We have suggested an explanation for the nonexistence of salts of the Cl^{2-} anion based on electron affinity. Another explanation, based on ionic size, is also possible. Formulate such an explanation.

7.14 List the following compounds in order of increasing ionic bond character: (a) LiF, (b) CsCl, (c) $BaCl_2$, (d) AlF_3, (e) $FeCl_2$, (f) ZnI_2.

7.15 List the sulfides of the following elements in order of increasing ionic bond character: (a) potassium, (b) calcium, (c) manganese(II), (d) copper(II), (e) zinc, (f) gallium, (g) carbon.

7.16 Suggest an explanation for the

smooth increase in electronegativity of the elements from lithium to fluorine in the first row of the periodic table.

7.17 List the following substances in order of decreasing ionic bond character: (a) HCl, (b) Cl_2, (c) NaCl, (d) CCl_4, (e) NCl_3, (f) ClF.

7.18 Find the relationship between the valence electronic configuration and the usual number of bonds formed by the elements from lithium, atomic number 3, to fluorine, atomic number 9.

7.19 Find the relationship between the electronic configuration of the p subshell of the valence shell of the nonmetals and the usual number of bonds formed by the nonmetals.

7.20 Indicate which of the following molecules cannot have an expanded octet: (a) H_2SO_4, (b) HNO_3, (c) $HClO_4$, (d) $HClO_3$, (e) C_2H_4, (f) AlF_3.

7.21 Indicate which of the following compounds cannot completely satisfy the octet rule: (a) N_2O_5, (b) N_2O_4, (c) NO_2, (d) NO, (e) ClO, (f) Cl_2O, (g) ClO_2.

7.22 Indicate which of the following elements may have expanded octets: (a) Al, (b) Cl, (c) K, (d) O, (e) Ba, (f) Sb.

7.23 Draw Lewis structures for the following compounds: (a) CF_4, (b) NF_3, (c) OF_2, (d) ClF.

7.24 Draw Lewis structures for the following compounds: (a) HNO, (b) H_2O_2 (hydrogen peroxide), (c) H_2CO_3, (d) H_2Se, (e) HClO.

7.25 Draw Lewis structures for the following compounds: (a) CO_2, (b) N_2O_3 (N—N bond), (c) N_2, (d) S_8 (cyclic).

7.26 Draw Lewis structures for the following ions: (a) CO_3^{2-}, (b) CN^-, (c) NH_4^+, (d) NO_2^+, (e) HSO_3^-.

7.27 Draw Lewis structures for the following species: (a) BCl_3, (b) AlF_3, (c) CH_3^+, (d) NO^+, (e) ZnI_2.

7.28 Draw Lewis structures for the following molecules: (a) AsF_5, (b) BrF_5, (c) SeF_4, (d) $SbCl_3$.

* *7.29* Draw Lewis structures for the following compounds: (a) $HBrO_4$, (b) $Te(OH)_6$, (c) H_3AsO_4, (d) Cl_2O_7 (no Cl—Cl bond), (e) H_3PO_3 (two O—H bonds).

7.30 Draw Lewis structures for the following species: (a) NO, (b) NO_2, (c) ClO_2, (d) CH_3, (e) BH_4^-.

7.31 Draw Lewis structures for the following ions: (a) HPO_4^{2-}, (b) I_3^-, (c) SiF_6^{2-}, (d) BrO_3^-.

* *7.32* A molecule has the formula C_8H_8 and does not contain any double

or triple bonds. Furthermore, all the carbon atoms are chemically identical and all the hydrogen atoms are chemically identical. Draw a Lewis structure for this compound.

7.33 Draw Lewis structures for the following molecules indicating the formal charges, if any, in each structure: (a) N_2O_4, (b) CH_3NO_2, (3 C—H bonds), (c) F_3BNH_3, (d) CO.

7.34 The compound Cl_2O_7 does not have a Cl—Cl bond. Draw two Lewis structures for this compound, one with the octets of the chlorine atoms expanded and one without octet expansion and with the formal charges indicated.

7.35 The compound SO_2Cl_2 is called sulfuryl chloride. Write two Lewis structures for sulfuryl chloride, one with formal charges and one with appropriate octet expansion.

7.36 The compound cyanogen chloride has the composition CNCl, and has no formal charges. Draw its Lewis structure.

7.37 Draw two Lewis structures for a molecule that has the composition NH_3O. Use the principle of electroneutrality to predict which of the two structures is more likely to be the correct one.

7.38 Phosgene, the poison gas used during World War I, has the composition $COCl_2$. Draw the most likely Lewis structure for phosgene. Draw another Lewis structure based on a different skeleton structure. Suggest how you could differentiate between the two structures experimentally.

7.39 Write the important contributing structures for the following molecules: (a) NO_2Cl, (b) O_3, (c) N_2O (NN bond), (d) N_2O_5.

7.40 Write the important contributing structures for the following ions: (a) CO_3^{2-}, (b) PO_4^{3-} include one P=O bond, (c) NO_2^-.

* *7.41* The azide ion, N_3^-, is a linear and symmetrical ion. All of its contributing structures have formal charges. Draw three important contributing structures for this ion.

7.42 Benzene has its six carbon atoms arranged in a cyclic structure. All the carbons are chemically identical as are all the hydrogens; all the CC bonds are identical and are intermediate in their properties between ordinary single and double bonds. Draw two contributing structures for benzene, C_6H_6.

7.43 Identify the atomic orbitals that may overlap to form the σ bonds in each

of the following molecules: (a) HCl, (b) H_2O, (c) NH_3, (d) F_2, (e) Na_2.

7.44 Identify each bond in the following molecules as a σ bond or as a π bond. Identify the atomic orbitals that may overlap to form each of these bonds: (a) NO, (b) CO, (c) O_3. (Consider just one contributing structure.)

7.45 The molecule N_2H_2 is called diimide and is very reactive. Draw a bonding diagram like that of Figure 7.17 for diimide.

7.46 The cyanide ion, CN^-, is a common anion. Draw a bonding diagram like that of Figure 7.17 for the cyanide anion.

7.47 Use the molecular orbital method to find the number of bonds in each of the following ions: (a) Li_2^+, (b) Be_2^+, (c) Be_2^{2+}, (d) B_2^{2+}.

7.48 Use the molecular orbital method to find the number of bonds and the number of unpaired electrons in each of the following ions: (a) O_2^+, (b) O_2^-, (c) O_2^{2-}, (d) O_2^{3-}. One of these ions is not known to exist. Suggest which one it is.

7.49 The molecular orbitals that we used for homonuclear diatomic molecules can also be used for heteronuclear diatomic molecules if the two atoms of the molecule are close in atomic number. Write electronic configurations of the type shown in Table 7.5 for the following molecules: (a) NO, (b) CO, (c) CN, (d) OF.

7.50 Indicate which of the following species are paramagnetic: (a) H, (b) He, (c) Li, (d) Li^+, (e) Cr, (f) Zn, (g) Cl^-.

7.51 Predict the electronic configuration and the magnetic properties of: (a) N_2^{2-}, (b) F_2^{2+}, (c) C_2^{2+}, (d) C_2^{2-}.

7.52 Suggest an explanation in VB terms and in MO terms for the unusual stability observed for the allyl cation:

7.53 Classify each of the following substances as ionic, polar, or nonpolar: (a) LiI, (b) NCl_3, (c) CH_4, (d) H_2O, (e) HBr, (f) CH_2Cl_2, (g) I_2.

7.54 Arrange the following substances in order of increasing melting point: (a) NaCl, (b) Cl_2, (c) PCl_3, (d) $POCl_3$.

7.55 Calculate the oxidation numbers of all the atoms in the following binary compounds: (a) N_2O_4, (b) Cl_2O_7, (c) Al_2S_3, (d) SF_6.

7.56 Calculate the oxidation numbers of all the atoms in the following molecules: (a) $COCl_2$, (b) NO_2Cl, (c) SO_2Cl_2, (d) $SOCl_2$, (e) NH_2OH.

7.57 Calculate the oxidation numbers of all the atoms in the following ions: (a) SO_4^{2-}, (b) PO_4^{3-}, (c) ClO_4^-, (d) NO_2^+, (e) O_2^{2-}.

7.58 Calculate the oxidation numbers of the transition metals in the following species: (a) OsO_4, (b) $[CuCl_4]^{2-}$, (c) $[Cr(NH_3)_3]^{3+}$, (d) $Pt(NH_3)_2Cl_2$, (e) $FeO(OH)$.

*7.59 Calculate the oxidation number of the sulfurs in the compound H_2S_4, which has a linear arrangement of its atoms.

7.60 Indicate whether each of the following conversions represents an oxidation, a reduction, or neither one: (a) sodium metal to the sodium cation, (b) chlorine to the chloride anion, (c) $HClO_3$ to $HClO_4$, (d) H_2O_2 to H_2O, (e) NO_2 to N_2O_4, (f) NH_4^+ to NH_3.

7.61 Explain in detail the observed increase in the boiling points of the noble gases with an increase in atomic number.

7.62 Explain the observation that ClF, which is more polar than Cl_2, has the lower boiling point.

7.63 The molecule CBr_4 is nonpolar because of its geometry, yet it is a solid at room temperature. Suggest two factors that may be responsible for its high melting point. The related molecule CF_4 is a gas at room temperature. Based on this observation select the more important of the two factors you proposed.

8

MOLECULAR STRUCTURE AND STABILITY

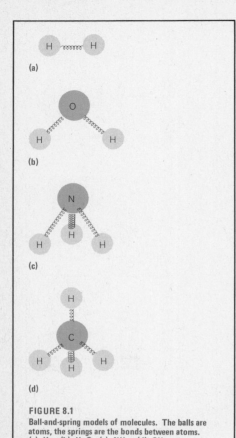

FIGURE 8.1
Ball-and-spring models of molecules. The balls are atoms, the springs are the bonds between atoms. (a) H_2. (b) H_2O. (c) NH_3. (d) CH_4.

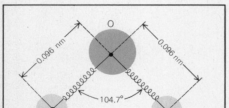

FIGURE 8.2
Bond angle and bond length illustrated for the H_2O molecule. Bond length is the distance between the centers of the nuclei of the O and H atoms, 0.096 nm in this molecule. Bond angle is the angle formed by imaginary lines connecting the O atom with the two H atoms, 104.7°.

The development of methods for representing molecular structures led to the growth of a major branch of chemistry called structural chemistry, which deals with the shapes of molecules and the nature of the bonds that hold them together.

When atoms bond together to form a molecule, the relative positions of the atoms in the molecule tend to be fixed, with little margin for change. Molecules are fairly rigid. One common way to picture a molecule is as a collection of balls, the atoms, connected by springs, the bonds. Figure 8.1 shows some molecules represented in this manner. But it should be remembered that the "springs" are not very flexible. The bonds between atoms in a molecule allow only limited movement of the atoms relative to one another.

In this chapter we shall discuss some details of the structure of molecules, and some properties of bonds that give us information about molecular structure. We shall also present some of the theory that explains the experimental data on molecular structure.

Molecular structure can be defined primarily by two parameters: **bond length** and **bond angle**. *Bond length is the distance between the centers of the nuclei of two bonded atoms,* as shown in Figure 8.2. A definition of bond angle requires a molecule with at least three atoms. *Bond angle is the angle formed by the two imaginary lines that connect the central atom with the two atoms on either side of it.* In the water molecule H—O—H, for example, the bond angle is the angle formed by the lines representing the two O—H bonds. The geometry of the water molecule can be described by saying that the H—O—H bond angle is 104.7° and that the H—O bond length is 0.096 nm.

Less precisely, we can just say that H_2O is a bent molecule. A molecule such as CO_2, whose O—C—O bond angle is 180°, is characterized as linear, rather than bent. Both CO_2 and H_2O are planar, meaning that the centers of the three atoms lie in an imaginary plane. All three-atom molecules are planar.

In molecules with more than three atoms, it is often necessary to indicate the relative orientation of atoms that are not bonded directly to one another. Many three-dimensional representations are used to describe such molecules. The branch of chemistry that is concerned with the three-dimensional structure, bond angles, and bond lengths of molecules is called stereochemistry.

Bond lengths and bond angles sometimes give us only an incomplete picture of molecular structure. To get a complete picture, we must study other properties of the molecule. One of these properties was discussed in Chapter 7: the polarity due to the separation of positive and negative charge in polar bonds. The *dipole moment* is a measure of the magnitude of the charges that are separated and of the distance separating them. Dipole moment measurements are often helpful in the determination of molecular structure.

Atoms that are connected by bonds form a more stable system than separate atoms. The relative stability of a molecule can often give us valuable information about its structure. The energy change that occurs in the formation of a molecule is a measure of its stability. When bonds form, energy is liberated, usually in the form of thermal energy (heat). To break bonds, energy must be supplied, usually in the form of thermal energy. The branch of chemistry that measures the heat effects associated with chemical processes is called **thermochemistry**. Thermochemistry gives us information about the relative stability of molecules and the strengths of the bonds holding the atoms together.

8.1 EXPERIMENTAL METHODS

Most of the experimental methods used to gather data that help us to understand the structure of molecules are based on the interaction of matter and radiant

A typical example of the sophisticated equipment used in today's chemistry laboratory to investigate the structure and properties of molecules. (*Robert Boikess*)

energy. These methods usually require complicated and expensive instruments that have become available to chemists only in the past few decades. In contrast to the relatively simple equipment found in the chemistry laboratory of the fairly recent past, today's well-equipped laboratory has instruments that cost tens of thousands, even hundreds of thousands of dollars. The modern chemist must know how to operate many of these instruments. The enormous increase in chemical knowledge in recent years is due in large part to the development of sophisticated chemical instrumentation, including the ever-present computer. Today, a team of chemists working on a large research project usually includes several members whose specialty is the "care and feeding" of instruments.

One powerful group of methods for structure determination is based on the interaction of radiation or subatomic particles with large assemblages of molecules. When radiation such as X rays, or particles such as electrons or neutrons, are directed at a substance, the interaction can create a diffraction pattern. By studying this pattern, experts can determine the molecular structure responsible for the deflection or refraction of the radiation or particles. Within limits, diffraction experiments give the location of atomic nuclei in a molecule quite accurately. One of the more memorable uses of diffraction patterns in recent decades was the determination of the structure of the DNA (deoxyribonucleic acid) molecule, the core material of the gene, an intellectual adventure described vividly in James Watson's book, *The Double Helix*.

Diffraction studies are the closest available approach to the direct observation of molecular structure. There are similar, highly specialized methods available for the measurement of dipole moments and thermochemical quantities.

Spectroscopy

Spectroscopy, which studies the interaction of matter with electromagnetic radiation, is the technique most often used to explore molecular structure. In Chapter 5, electromagnetic radiation was described as consisting of an oscillating electric-magnetic field. Ordinary spectroscopy studies the interaction of the atom's electrical charges with the electric component of electromagnetic radiation. The branch of spectroscopy called magnetic resonance studies the interaction of the magnetic component of electromagnetic radiation with the magnetic dipoles within molecules.

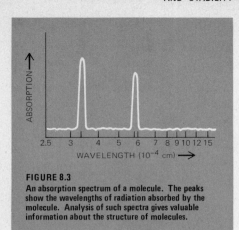

FIGURE 8.3
An absorption spectrum of a molecule. The peaks
show the wavelengths of radiation absorbed by the
molecule. Analysis of such spectra gives valuable
information about the structure of molecules.

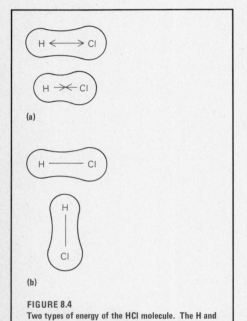

FIGURE 8.4
Two types of energy of the HCl molecule. The H and
Cl atoms move along the axis of their bond, giving the
molecule vibrational energy (a). The molecule as a
whole tumbles in space, giving it rotational energy (b).

Usually, a relatively narrow wavelength range of electromagnetic radiation is used for a specific spectroscopic study. We speak of infrared spectroscopy or ultraviolet-visible spectroscopy or microwave spectroscopy, depending on the wavelengths of the radiation used for the study.

In a typical experiment, a beam of electromagnetic radiation is directed at a sample of the substance to be studied. The wavelength of the beam of radiation is varied continuously within limits, and the extent to which radiation of a given wavelength is absorbed by the substance is measured. The results can be recorded as a graph of the extent of absorption for a given wavelength. This graph is called an *absorption spectrum* (Figure 8.3).

The determination of a spectrum once was a tedious job. Today, spectra are determined by instruments called *spectrometers,* which even draw the final graph. An infrared spectrum of the type in Figure 8.3 can be obtained in 10 minutes on a spectrometer that can be operated by a competent chemistry student after 30 minutes of training.

Spectroscopy is a valuable method of examining molecules because any given molecule absorbs only selected wavelengths of electromagnetic radiation. Parts of the absorption spectrum are unique for each compound, as a fingerprint is unique for a human. Analysis of the absorption spectrum pattern can yield a great deal of information about the molecule.

We mentioned the selective absorption of radiation by atoms in our discussion of the Bohr model of the hydrogen atom (Section 5.7, page 135). It is a general phenomenon. The absorption of radiation by molecules is quantized, so a molecule will absorb only those wavelengths of radiation that satisfy the Bohr frequency rule:

$$\Delta E = h\nu = \frac{hc}{\lambda}$$

Since ΔE of the Bohr frequency rule is the difference in energy between two allowed energy states of a system, the absorption spectrum of a molecule gives information about the allowed energy states of that molecule.

The energy states of a molecule are governed by more complex factors than are those of an atom. In our discussion of atoms, we were concerned only with the energy states corresponding to the allowed energy levels of the atom's electrons. With molecules, we must deal with different kinds of energy states.

Molecules possess *vibrational energy* and *rotational energy*. Consider the two atoms of the HCl molecule (Figure 8.4). The molecule as a unit tumbles through space, giving it rotational energy. At the same time, the two atoms move with respect to each other along the bond axis, giving the molecule vibrational energy.

Both vibrational energy and rotational energy are quantized. Only certain vibrational and rotational energy states are allowed in a given molecule. Compared to the energy levels of electrons in atoms or molecules, vibrational energy levels are more closely spaced. Rotational energy levels are even more closely spaced (Figure 8.5). The close spacing means that the magnitude of ΔE for transitions of a molecule between two vibrational energy levels is much smaller than ΔE for transitions between electronic energy levels.

The radiation emitted and absorbed by transitions between electronic energy levels is in the ultraviolet-visible part of the spectrum, with wavelengths of 200–700 nm. The smaller values of ΔE between vibrational energy levels correspond to the emission and absorption of electromagnetic radiation of lower frequency (longer wavelength). This radiation is in the infrared region (1–300 μm) for most molecules. The magnitude of ΔE between rotational energy levels corresponds to emission and absorption of radiation of still longer wavelength (>1 mm), in the microwave region.

These differences explain the use of one type of radiation or another in different spectroscopic studies. Infrared spectroscopy gives information about vibrational energy levels of a molecule, while microwave spectroscopy gives information about rotational energy levels. This information about energy levels helps us to determine the structure of molecules.

8.2 CHEMICAL BONDING AND MOLECULAR GEOMETRY

In Section 7.5, page 212, we discussed the advantage of visualizing the formation of a bond in terms of the formation of a bonding orbital and an antibonding orbital by the overlap of atomic orbitals from each of the two atoms in the bond. Two types of overlap are possible: σ overlap, in which the overlapping orbitals are directed toward each other along the internuclear axis (Figure 7.13), and π overlap, in which the overlapping orbitals are along parallel axes (Figure 7.16).

Tetrahedral Structures

But the concept of overlapping orbitals gives an oversimplified picture. The vast body of data about the structure of real molecules makes it clear that the picture of bond formation by overlap between s, p, and d orbitals must be modified to be consistent with experimental evidence.

The need for modifying the simplified picture is evident even in the study of such simple molecules as methane, CH_4. The Lewis structure:

$$H-\underset{\underset{H}{|}}{\overset{\overset{H}{|}}{C}}-H$$

and what we know about electronic configuration can be interpreted as showing that the carbon atom forms four bonds by σ overlap of one of its valence orbitals with the $1s$ orbital of each hydrogen atom. The carbon atom has one $2s$ and three $2p$ valence orbitals, so one might suggest that there should be two types of C—H bonds. In three of the bonds in the CH_4 molecule, the three $2p$ orbitals of the carbon atom should overlap with the $1s$ orbitals of three H atoms. The fourth bond should be formed by overlap of the $2s$ orbital of the C atom and the $1s$ orbital of the fourth H atom. Since the three $2p$ orbitals of the C atom are perpendicular to each other, we expect these three C—H bonds to form H—C—H angles of 90° relative to one another. The H atom that is bound by overlap to the $2s$ orbital of the C atom should be in some unspecified location.

Every experimental result obtained for the structure of CH_4 is at variance with this hypothetical structure. Experimentally, all four C—H bonds are found to be identical, all the H—C—H bond angles are found to be 109°28′, and all the C—H bonds are found to have the same length, 0.109 nm (Figure 8.6). If we draw imaginary lines connecting the H atoms, we find that the H atoms lie at the corners of a regular *tetrahedron,* a solid figure with four identical equilateral triangles as its faces (Figure 8.7).

The simple orbital picture is therefore wrong. We must seek other models to account for observations about the actual geometry of CH_4 and other molecules. A number of models are used currently. We shall discuss the two simplest models.

In one model, we do not concern ourselves with orbital types. Rather, we try to accommodate the observed geometry of molecules by an electrostatic picture based primarily on the Lewis structure. This model is called the *valence-shell electron pair repulsion* (**VSEPR**) *model.* The second model modifies the description of atomic orbitals when they overlap to form bonds, and uses this new description to rationalize the observed geometry. It is called the **hybridization** or *directed valence model.*

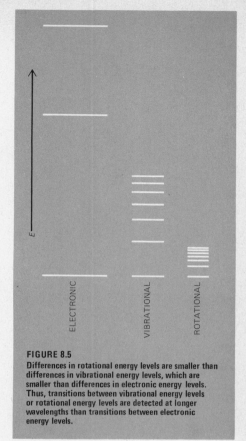

FIGURE 8.5
Differences in rotational energy levels are smaller than differences in vibrational energy levels, which are smaller than differences in electronic energy levels. Thus, transitions between vibrational energy levels or rotational energy levels are detected at longer wavelengths than transitions between electronic energy levels.

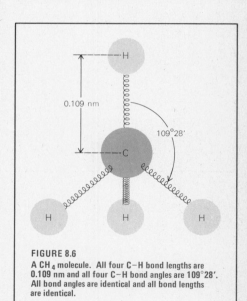

FIGURE 8.6
A CH_4 molecule. All four C—H bond lengths are 0.109 nm and all four C—H bond angles are 109°28′. All bond angles are identical and all bond lengths are identical.

MAKING MODELS OF MOLECULES

Most college bookstores sell relatively inexpensive collections of balls and sticks, reminiscent of the Tinker Toys of our younger days, called molecular model kits. With a molecular model kit you can build a three-dimensional model of a molecule that can help you visualize spatial relationships between atoms not directly bonded to each other. In these "ball-and-stick" models, the balls represent atoms and the sticks represent bonds. The holes in the balls hold the sticks at angles similar to the actual bond angles in the molecule.

Ball-and-stick models give us a somewhat approximate picture of molecules. Very precise scale models of molecules can be constructed by using other types of molecular model kits, which are usually quite expensive. One kit consists of stainless steel rods that snap into stainless steel sleeves to represent bonds. The bond lengths and bond angles are made accurately to scale in these kinds of models, but the atoms are not shown at all. They are understood to be at the intersections of the bonds. There are also space-filling models, which take the opposite approach. In space-filling models, the atoms are represented by truncated balls that are held together by snap fasteners and the bonds are not shown. The atomic radii are made accurately to scale in these models. Each type of model emphasizes different aspects of molecular geometry as you can see in the three models of ethane, C_2H_6, shown below.

BALL-AND-STICK   FRAMEWORK   SPACE-FILLING

Although these models are extremely useful, they have drawbacks. A model of a large complicated molecule is tedious and expensive to construct and is fragile. It must often be supported by rods that distract from the clarity of the model. In recent years, attempts have been made to obtain better models of molecules by developing computer programs whose outputs are visual displays.

There are a number of such computer programs, using different sets of rules. A typical program will include a memory store containing the basic bonding characteristics of atoms of many elements, as well as the characteristic bond angles of the atoms. To construct a molecular model, the user specifies the chemical composition of the molecule. Care must be taken to describe the exact kind of bond formed by each atom in a given molecule. For example, an oxygen atom may form one double bond or two single bonds, while several different kinds of bonding are possible for a carbon atom. If the computer is given the proper information, the end result is a three-dimensional display of molecular structure in which the bonds are shown as lines and the atoms are indicated as the points at which the lines intersect, much like a framework model. With

In the VSEPR model, the bond angles around a central atom are determined by the number and nature of the valence electrons associated with that atom; the inner-shell electrons are neglected. It is assumed that each electron pair occupies a well-defined region of space (or orbital) and that the electron pairs repel each other. The first and most important rule of the VSEPR model is that *the bond angles are those in which the total repulsion between the electron pairs in the valence shell of the central atom is a minimum*. The electron pairs will be as far from each other as is possible.

Let us apply this model to CH_4. The central atom of CH_4 is carbon. Its coordination number is 4; it is bonded to four other atoms. It has four pairs of electrons in its valence shell, one pair for each bonded hydrogen atom. A simple calculation shows that *the repulsive forces between four electron pairs are smallest when their orbitals are directed toward the corners of a regular tetrahedron*. According to the VSEPR model, the observed tetrahedral geometry of CH_4 is due simply to the fact that this geometry minimizes the repulsions associated with the four electron pairs in the bonds around the central carbon atom.

The VSEPR model also predicts that any time four atoms are bonded to one central atom—that is, when the central atom has a coordination number of 4—and the central atom has no additional nonbonding valence electron pairs, bond angles of 109°28' will be observed. Experiments indicate that these bond angles are found only when all four bonded atoms are identical, as in $SiCl_4$, BF_4^-, NH_4^+, and SO_4^{2-}. When the atoms are not identical, there will be some deviations from the ideal tetrahedral geometry, as we shall see later in this chapter.

The hybridization model invokes the concept of **hybrid orbitals.** A hybrid orbital is just what its name implies, an intermediate form obtained by mixing simple atomic orbitals. In CH_4, we can account for the four bonds that are actually formed by proposing that the four valence orbitals of the carbon atom "mix" to form four identical hybrid orbitals. The four hybrid orbitals in the CH_4 molecule are formed by mixing one 2s orbital and three 2p orbitals. Therefore, each hybrid orbital is composed of three parts of p orbital to one part of s orbital. In notation, the contribution of each orbital is indicated by a superscript, with the superscript 1 regarded as understood and thus omitted. The four hybrid orbitals of the CH_4 molecule are indicated by the notation $2sp^3$. The superscripts 1 (which is omitted) and 3 indicate the relative proportions of s and p contributions to the hybrid orbitals, while the coefficient 2 can be used to indicate the principal quantum number.

When hybrid orbitals are formed from atomic orbitals, the number of orbitals does not change. As Figure 8.8 indicates, hybridization may be considered to be simply the mixing of various "pure" orbitals to form the same number of hybrid orbitals.

The use of four equivalent sp^3 orbitals for the four σ bonds of CH_4 is consistent with the equivalency of the four bonds. Furthermore, it can be shown mathematically that the use of sp^3 orbitals for bonding results in bond angles of 109°28' and tetrahedral geometry.

Hybrid orbitals normally are not found in isolated atoms because unhybridized atomic orbitals are of lower energy than hybrid orbitals. But hybrid orbitals in molecules allow the formation of stronger bonds than do unhybridized orbitals.

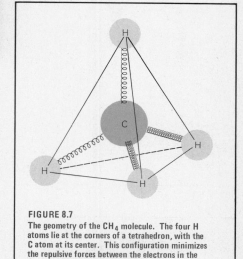

FIGURE 8.7
The geometry of the CH_4 molecule. The four H atoms lie at the corners of a tetrahedron, with the C atom at its center. This configuration minimizes the repulsive forces between the electrons in the bonds of the molecule.

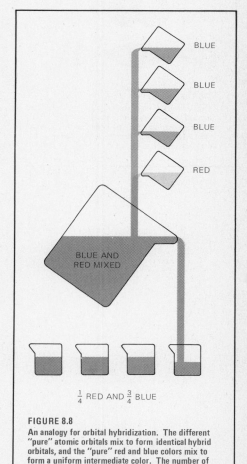

BLUE

BLUE

BLUE

RED

BLUE AND RED MIXED

$\frac{1}{4}$ RED AND $\frac{3}{4}$ BLUE

FIGURE 8.8
An analogy for orbital hybridization. The different "pure" atomic orbitals mix to form identical hybrid orbitals, and the "pure" red and blue colors mix to form a uniform intermediate color. The number of orbitals is the same before and after hybridization.

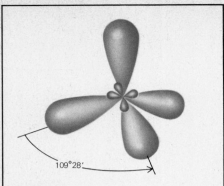

FIGURE 8.9
Hybrid orbitals are directional; that is, they point along the internuclear line. The hybrid orbitals of CH_4 are directed so that the bond angles are 109°28'.

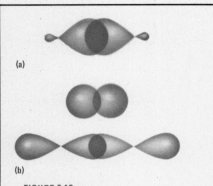

FIGURE 8.10
Hybrid orbitals (a) are asymmetrical, concentrating most of the electron density along the internuclear line, thus creating a stronger bond than a symmetrical orbital (b) because of greater overlap. The shaded area is the region of overlap.

These bonds are stronger because hybridized orbitals are less symmetrical than unhybridized ones—they "point" in various directions, as shown in Figure 8.9. When a hybrid orbital forms a σ bond, its region of high electron probability is directed toward the bonded atom. We can say that the hybrid orbital points at the bonded atom, so that there is a high probability that the electrons in the orbital lie between the two bonded atoms. The result is better orbital overlap and the formation of stronger bonds, as shown in Figure 8.10.

The concept of hybrid orbitals provides simple explanations of many observed features of molecular structures. A number of different hybrid orbital descriptions of bonding are possible in many molecules. The hybrid orbitals that we shall discuss are convenient because they are simple and give adequate explanations of observed properties of molecules.

The description of the CH_4 molecule in terms of four hybrid sp^3 orbitals also works well for other carbon compounds that have four atoms bonded to the carbon atom. Ethane, C_2H_6, has the Lewis structure:

$$\begin{array}{cc} H & H \\ | & | \\ H-C-C-H \\ | & | \\ H & H \end{array}$$

All bonds to the carbon atoms are σ bonds. For the C—H bonds, the sp^3 orbital of the carbon atom overlaps with the $1s$ orbital of hydrogen. In the C—C bonds, two sp^3 orbitals overlap. The bond angles in ethane are all close to 109°28'.

We can generalize even further and say that in many cases the bonding of any four atoms to any single central atom can be described in terms of four sp^3 orbitals of the central atom. The logic of this description is clear: Four orbitals are needed to form four bonds. It is natural to extend this description from a molecule with a central carbon atom to a molecule with a different central atom, such as nitrogen. A nitrogen atom is the central atom in the ammonium ion, which is isoelectronic with CH_4 and has the Lewis structure:

$$\left[\begin{array}{c} H \\ | \\ H-N-H \\ | \\ H \end{array}\right]^+$$

The bonding in the ammonium ion or in any other compound in which N is bonded to four other atoms is conveniently described by using four sp^3 orbitals. Indeed, the sp^3 explanation usually is satisfactory for any atom of a nontransition element that is bonded to four other atoms and has no nonbonding electrons in its valence shell.

EXAMPLE 8.1

Indicate which orbitals overlap to form each of the bonds in $CH_3\overset{+}{N}H_3$.

SOLUTION

Step 1: Draw the Lewis structure:

$$\begin{array}{cc} H & H \\ | & | \\ H-C-\overset{+}{N}-H \\ | & | \\ H & H \end{array}$$

Step 2: All the bonds of the H atoms are formed by overlap of the $1s$ orbitals of the H atoms and suitable orbitals of the other atoms. Note how many σ bonds each of the other atoms forms. Both the C atom and the

N atom form four σ bonds. Both are sp^3 hybridized, and hence each bond to an H atom is formed by the overlap of an sp^3 orbital with a $1s$ orbital. The C—N bond is formed by the overlap of two sp^3 orbitals.

Planar and Linear Structures

The VSEPR model can be applied to explain the observed geometry of molecules in which a central atom is bonded to less than four other atoms, that is, when the coordination number of the central atom is 2 or 3. For the moment, we shall restrict our discussion to atoms that have no nonbonding valence electrons, and that therefore must have either multiple bonds or incomplete octets.

In BF_3, for example, the central boron atom has a coordination number of 3 and no nonbonding electrons. Therefore, it has an incomplete octet. In C_2H_4, each carbon atom has a coordination number of 3. Its octet is complete because of the double bond:

$$
\begin{array}{c}
\text{H} \qquad\qquad \text{H} \\
\diagdown \qquad \diagup \\
\text{C}\!=\!\text{C} \\
\diagup \qquad \diagdown \\
\text{H} \qquad\qquad \text{H}
\end{array}
$$

Similarly, the coordination number of each C atom in C_2H_2 is 2, but each has a completed octet because of the triple bond:

$$\text{H—C}\equiv\text{C—H}$$

In $BeCl_2$, the Be atom has a coordination number of 2, but it has an incomplete octet because of the absence of nonbonding valence electrons.

The VSEPR model can be applied directly to compounds which have no nonbonding valence electrons, such as BF_3 and $BeCl_2$. *In three-coordinate molecules with no nonbonding valence electrons, such as BF_3, valence-shell electron pair repulsion is minimized when the three orbitals are directed toward the three corners of an equilateral triangle.* This geometry results in bond angles of $120°$. The centers of all three bonded atoms and of the central atom lie in the same plane, and we describe the geometry as *planar* or *trigonal planar*.

In two-coordinate molecules with no nonbonding valence electrons, such as $BeCl_2$, *valence-shell electron pair repulsion is minimized when the two orbitals lie along a straight line.* The resulting bond angle is $180°$, and the geometry is described as *linear*.

The VSEPR model can also be applied to compounds with multiple bonds. To a first approximation, the model ignores the extra electrons of the double and triple bond and treats only the electron pairs in the σ bonds. As a result, the model predicts that C_2H_4, in which the C atom has a coordination number of 3, and other related compounds with double bonds will have planar geometry and bond angles of $120°$, the same as in BF_3. The prediction is not completely accurate, because the extra electron pair of the double bond in C_2H_4 causes the bond angles to deviate slightly from the ideal values. Experimentally, we find that the H—C—H and H—C—C angles in C_2H_4 are close to $120°$. For C_2H_2 and other compounds with triple bonds, the VSEPR model predicts linear geometry with bond angles of $180°$.

The hybridization model can also be used to describe the bonding in these kinds of molecules. The shape and directional character of sp^3 orbitals are well suited for σ-type overlap. But they are not suited for π overlap, where the overlapping orbitals must be perpendicular to the internuclear line (Figure 7.16). If we picture a double bond as one σ bond and one π bond, a different kind of hybrid orbital is required. For example, in ethylene, C_2H_4, which has a double bond, a description based on sp^3 hybridization is awkward. Each carbon atom in

CHEMICAL STRUCTURE AND THE COMPUTER

In the past few years, the development of faster computers and of improved computer programs for molecular orbital calculations have enabled chemists to predict that some unusual carbon compounds may have structures that are quite different from those accepted as the norm. This sort of computer analysis has led to the prediction that a number of carbon compounds have unexpected structures.

The technique consists of selecting two or more likely geometries for a given molecule and then calculating their relative energies. If all the appropriate structures are included in the calculations, the most stable geometry is the one with the lowest energy.

This method can be validated by its application to molecules of known structure. For example, it can be calculated that a structure for methane, CH_4, in which all five atoms lie on the same plane, would have an energy that is about 600 kJ/mol higher than the energy of a methane molecule in which the four hydrogen atoms are arranged tetrahedrally around the carbon atom. Planar methane has never been observed.

Until recently, such relative energy calculations could be done only for simple molecules. The calculations can now be done for much more complicated molecules. The resulting predictions have opened some promising avenues of research, and have led to predictions that contradict some long-held chemical beliefs.

Until now, it has been assumed that any carbon atom with four substituents bonded to it prefers a tetrahedral geometry. The new calculations predict that some substituted methanes may have completely different structures. If two lithium atoms are substituted for two of the hydrogen atoms of methane, the resulting compound is dilithiomethane. Energy calculations indicate that instead of being tetrahedral, dilithiomethane may be planar, with all five atoms lying in the same plane. Similarly, a compound in which a carbon atom is double-bonded, such as ethylene, C_2H_4, has been assumed to have planar geometry. The new energy calculations predict that a compound called 1,1-dilithioethylene, in which two lithium atoms are substituted for two hydrogen atoms of the ethylene, will not have planar geometry. Instead, the substituents attached to one carbon atom are predicted to lie in a plane that is perpendicular to the substituents of the other carbon atom.

As this is written, the predictions made by the new energy calculations have not been verified by experiment. One reason for the lack of verification is that many of the compounds are difficult to study. Another reason is that such studies have not seemed worthwhile, since it has been assumed that the structures of the compounds would not be out of the ordinary. The results of the calculations suggest that the study of these compounds may be rewarding.

It is not known yet whether the predictions of new structures for carbon compounds will have any practical applications. But they do promise new insights into many chemical problems. The new computer techniques also offer a relatively easy and inexpensive way of exploring areas of chemistry that are inaccessible experimentally.

C_2H_4 forms only three σ bonds, and therefore only three hybrid orbitals are required for each.

To form three hybrid orbitals, three pure atomic orbitals are necessary: the $2s$ orbital and two of the carbon atom's $2p$ orbitals. Mixing these three orbitals creates three hybrid orbitals that are each two parts p and one part s; in notation, sp^2. It can be shown mathematically that sp^2 orbitals form bond angles of $120°$. These three sp^2 orbitals are used to form the three σ bonds of the carbon atom.

But only three of the four valence orbitals of carbon have been used to form the three hybrid sp^2 orbitals. The fourth valence orbital is not hybridized. It is a $2p$ orbital that is perpendicular to the plane of the sp^2 hybrid orbitals, as shown in Figure 8.11. This unhybridized $2p$ orbital is used for the π bond. There is π overlap between the two parallel unhybridized p orbitals in C_2H_4. The complete orbital picture of C_2H_4 is shown in Figure 8.12.

The observed geometry of many species in which a nontransition element is bonded to three other atoms and has no nonbonding valence electron pairs is consistent with sp^2 hybridization of the element. Compounds, such as BF_3 or $AlCl_3(g)$, or ions, such as CH_3^+, in which all three bonded atoms are identical have bond angles of exactly $120°$. Other substances of this type in which the three bonded atoms are not identical have bond angles close to the ideal value. When the sp^2 hybridized atom does not participate in a double bond, the unhybridized p orbital is simply empty; it does not participate in the formation of bonds.

The triple bond between the two carbon atoms in acetylene, C_2H_2, pictured as consisting of one σ bond and two π bonds, cannot be simply described in terms of sp^2 hybrids. Two σ bonds are necessary for each C atom, so it is plausible to form only two hybrid orbitals, using the $2s$ orbital and one of the $2p$ orbitals. Each of these two hybrid orbitals is one part s and one part p; in notation, sp.

The sp hybridization works well when an atom forms two π bonds, as it does in a triple bond or in two double bonds. Thus, the carbon atom in $O{=}C{=}O$ is sp hybridized. It can be shown mathematically that sp hybrid orbitals result in bond angles of $180°$, that is, linear geometry. However, in C_2H_2 there are two unhybridized $2p$ orbitals on each carbon atom, perpendicular to each other and to the line of the sp orbitals (Figure 8.13). The two π bonds of the $C{=}C$ bond are constructed from pairs of parallel, unhybridized $2p$ orbitals, one from each C atom (Figure 8.14).

The sp hybridization description also works well for atoms that are bonded to two other atoms and do not have nonbonding valence electron pairs. The observed linear geometry of compounds such as $BeCl_2$ and $HgCl_2$ can be accommodated by sp hybridization.

Nonideal Geometry

The VSEPR model makes exact predictions of the bond angles in substances such as CH_4 and BF_3, where all the substituent atoms are identical. But if these atoms are not all the same, or if there is a multiple bond, the simple VSEPR prediction only comes close to the correct bond angles.

Similarly, each of the three hybrid orbitals mentioned thus far is associated with an ideal bond angle: sp^3 hybrid orbitals with a bond angle of $109°28'$; sp^2 hybrid orbitals with a bond angle of $120°$; sp hybrid orbitals with a bond angle of $180°$.

A rule that helps us relate hybridization to Lewis structures is this: The hybridization of an atom *without expanded octets* and with *no nonbonding electron pairs* generally is related to the number of σ bonds that the atom forms. Four σ bonds indicate sp^3 hybridization. Three σ bonds indicate sp^2 hybridization. Two σ bonds indicate sp hybridization.

A hybridization model for the atoms in a molecule is best formulated from the measured bond angles. If the angles that are measured are closest to $109°28'$, sp^3 hybridization of the central atom is indicated. If the angles are closest to

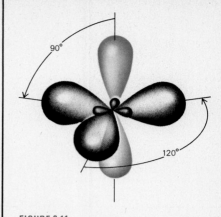

FIGURE 8.11
Each carbon in the C_2H_4 molecule has three hybrid sp^2 orbitals, with bond angles of $120°$, and one unhybridized $2p$ orbital, perpendicular to the plane of the hybrid orbitals.

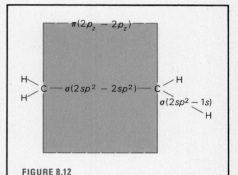

FIGURE 8.12
The complete orbital picture of C_2H_4, with three σ bonds and one π bond associated with each carbon atom. The designation of the directional subscripts of the p orbitals is arbitrary.

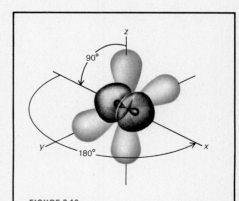

FIGURE 8.13
Each carbon of the C_2H_2 (acetylene) molecule has two sp hybrid orbitals, with $180°$ bond angles, and two $2p$ orbitals, at right angles to each other and to the hybrid orbitals.

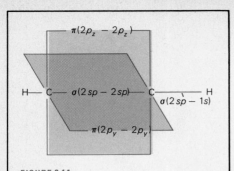

FIGURE 8.14
Orbital diagram of the C_2H_2 molecule, showing two σ bonds and two π bonds associated with each carbon atom.

120°, sp^2 hybridization of the central atom is indicated. If the angles are closest to 180°, sp hybridization of the central atom is indicated. In the molecule:

$$H-\overset{\overset{\displaystyle H}{|}}{\underset{\underset{\displaystyle H}{|}}{C}}-Cl$$

the H—C—H bond angles are found to be 111° and the H—C—Cl angles are 108°. The carbon atom is said to be sp^3 hybridized because its bond angles are closer to 109°28′ than to 120° or 180°. The deviation from the ideal value is explained as an effect of the different distribution of the electrons in C—H and C—Cl bonds. The bond angles in the molecule are those that minimize the forces of repulsion between the electron pairs.

Geometry Around Atoms with Nonbonding Valence Electrons

The observed geometry and bond angles of molecules that have atoms with nonbonding valence electron pairs can be explained by either the VSEPR model or the hybridization model. The simplest approach is to count each nonbonding valence electron pair in the same way as a pair of electrons in a σ bond. Thus, for the oxygen atom in the water molecule, H—$\overset{..}{O}$—H, we count four valence electron pairs, two in the σ bonds to H and two nonbonding valence electron pairs. For the nitrogen atom in $:NH_3$, we count four valence electron pairs, three in σ bonds to the hydrogen atoms and one nonbonding valence electron pair.

The hybridization description for atoms with nonbonding valence electron pairs is formulated in essentially the same way as for atoms without them. In H_2O, for example, we could say that the two σ bonds result from overlap between two of the four sp^3 hybrid orbitals of the O atom and the $1s$ orbital of the H atoms, and that the two nonbonding valence electron pairs of the O atom are in the remaining two sp^3 orbitals. In NH_3, three sp^3 hybrid orbitals of the N atom can be used for the three σ bonds, and the remaining sp^3 orbital for the nonbonding electron pair. We see that in this description all hybrid orbitals do not have to form bonds.

In both models, we associate bond angles of 109°28′ with the presence of four valence electron pairs. In the VSEPR model, these bond angles are said to minimize repulsions between the four valence electron pairs. In the hybridization model, the four sp^3 hybrid orbitals required for the four pairs of valence electrons are directed at angles of 109°28′.

Both models give the same result for the geometry of H_2O and NH_3. The sp^3 hybridization of the O atom of H_2O and of the N atom of NH_3 suggest bond angles close to 109°28′. The VSEPR model suggests the same bond angles because of the four pairs of valence electrons of each atom.

In general, the hybridization of nontransition elements that have both nonbonding electron pairs and bonds in molecules can be predicted by totaling the σ bonds *and* the nonbonding valence electron pairs. If the total is four, the hybridization is sp^3. If the total is three, the hybridization is sp^2. If the total is two, the hybridization is sp.

Thus, while the N atom in NH_3 is sp^3, the N atom in the molecule:

$$\overset{\displaystyle H}{\underset{\displaystyle H}{\diagup}}C=\overset{..}{N}-H$$

is better described as sp^2: It has only two σ bonds and one nonbonding valence electron pair, a total of three (the π bond is not counted). The sp^2 hybridization of the nitrogen atom suggests a bond angle close to 120° for the C—N—H bond. The VSEPR model gives the same result. We count three valence electron pairs around the nitrogen atom, two pairs in the σ bonds, and one nonbonding pair; the

second pair of the double bond is not counted. Bond angles close to 120°
minimize the repulsions between these three pairs of valence electrons.

In the H—C≡N: molecule, we can say that the N atom is *sp* hybridized,
since it has one **σ** bond and one nonbonding valence electron pair. Hybridization
can be assigned to an atom with one **σ** bond. However, there is no bond angle
that can be used to check the assignment, since two **σ** bonds are needed to
define a bond angle.

EXAMPLE 8.2

For the molecule HNO_2, estimate the bond angles using the VSEPR model
and indicate the hybrid orbitals which overlap to form each bond.

SOLUTION

Step 1:   Draw the Lewis structure:

H—Ö—N̈=Ö

Step 2:   Taking one atom at a time, total the number of **σ** bonds and
nonbonding valence electrons for each. From left to right:

The first O atom has two **σ** bonds and two nonbonding valence electron
pairs. The total of four suggests an H—O—N bond angle of about 109°28′.
(Note that the bonding of the central atom sets the value of the bond angle.)
This number of valence electron pairs is also consistent with sp^3 hybridization
of the O atom. Thus, the O—H bond forms from overlap of the $1s$ orbital of
the H atom and the sp^3 orbital of the O atom. The two nonbonding valence
electron pairs of the O atom are in sp^3 orbitals.

The N atom forms two **σ** bonds (and one **π** bond that is not counted)
and has one nonbonding valence electron pair. The total of three suggests
an O—N—O bond angle close to 120°. It also indicates sp^2 hybridization.
Thus, the single O—N bond is formed by sp^3-sp^2 overlap.

The remaining O atom has only one **σ** bond, so hybridization is not abso-
lutely necessary for its description. However, for convenience we could say
that the O atom is sp^2 hybridized, since it also has one **σ** bond and two nonbonding valence electron pairs, a total of three. The **σ** bond between the
N atom and this O atom is formed by sp^2-sp^2 overlap. The remaining nonbonding valence electron pairs of the N atom and of this O atom are in hybrid sp^2 orbitals. The **π** bond forms by overlap of two parallel unhybridized $2p$
orbitals, one on the N atom and one on the O atom. These $2p$ orbitals must
have the same directional subscript, for example, $2p_z$-$2p_z$.

The bond angles proposed by the simple application of either model to atoms
with nonbonding valence electrons generally are close to, but somewhat higher
than, the measured bond angles. For example, the H—O—H bond angle in water
is found to be 104.7°. The H—N—H bond angle in ammonia is found to be
106.8°. Both models can account for this reduced bond angle qualitatively by
proposing an additional rule: Nonbonding valence electron pairs take up more
space than bonding electron pairs. This is a reasonable rule, since the electrons in
a bond tend to be concentrated between the positive nuclei of the two atoms of
the bond. The bonding electrons thus tend to take up less space than the
nonbonding electrons, which are only attracted by one nucleus. Accordingly,
nonbonding valence electron pairs can be described as being relatively larger than
bonding pairs, so repulsions with nonbonding electron pairs are greater. This
difference affects bond angle. As Figure 8.15 shows, the angle between bonding
electron pairs in H—O—H narrows to 104.5° from the expected value of
109°28′ to make room for the two larger nonbonding pairs. The narrowing is less
in NH_3, which has only one nonbonding pair; the H—N—H angle is 107.3°.

As we have seen, the geometry of a molecule often can be described in one
of two ways. Either the bond angles can be given, or the overall shape of the

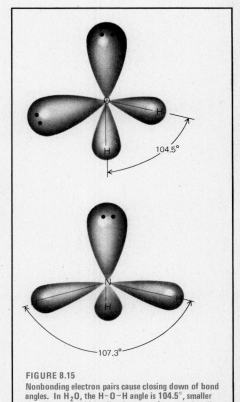

FIGURE 8.15
Nonbonding electron pairs cause closing down of bond
angles. In H_2O, the H–O–H angle is **104.5°**, smaller
than theory would predict, because of two nonbonding
electron pairs. In NH_3, the H–N–H angle is **107.3°**,
and the narrowing influence is smaller because there
is only one nonbonding electron pair in the molecule.

TABLE 8.1

Geometry Around Atoms with Four, Three, and Two Valence Electron Pairs

Total Number of Valence Electron Pairs[a]	Number of Nonbonding Pairs	Bond Angle	Geometry	Example[b]
4	0	109°28′	tetrahedral	$\underline{C}H_4$, $\underline{Si}F_4$, $\underline{N}H_4^+$
	1	<109°	pyramidal	$\underline{N}H_3$, $\underline{P}Cl_3$, $H_3\underline{O}^+$
	2	<109°	bent	$H_2\underline{O}$, $H_2\underline{S}$, $\underline{N}H_2^-$
3 (no π bonds)	0	120°	trigonal	$\underline{B}F_3$, $\underline{Al}Cl_3(g)$, $\underline{C}H_3^+$
(one π bond)	0	~120°	planar	$\underline{C}_2H_4$, $\underline{C}H_2O$, $\underline{N}O_3$
(one π bond)	1	<120°	bent	$\underline{N}OCl$, $\underline{N}O_2^-$
2	0	180°	linear	$\underline{C}O_2$, $\underline{C}_2H_2$, $\underline{N}O_2^+$

[a] Electrons in π bonds are not counted.
[b] The underlined atom is the central atom of the bond angle.

molecule can be given. For example, we describe CH_4 as tetrahedral, since the four H atoms are at the corners of a tetrahedron, with the C atom at its center. We can describe BF_3 as planar or trigonal planar, because all of its atoms lie in the same plane, and the three F atoms are at the corners of an equilateral triangle with the B atom at its center. Similarly, C_2H_4 is planar; all its atoms lie in the same plane. But C_2H_2 is linear, because all its atoms lie on the same line.

Alternatively, the overall shape of a molecule with nonbonding valence electron pairs can be given. However, only the atoms, and not the nonbonding electron pairs, are considered in describing the shape of the molecule. Thus, the geometry of NH_3 is described as pyramidal; the four atoms of the molecule lie at the corners of a pyramid. The water molecule has a planar or bent geometry. Its three atoms lie in the same plane, but not on the same line.

Table 8.1 summarizes our discussion of molecules with four, three, and two valence electron pairs. It should be kept in mind that only electron pairs in σ bonds or nonbonding valence electron pairs are counted. Electrons in π bonds are not counted. The VSEPR model neglects π bonding.

Geometry in Molecules with Expanded Octets

Any atom that satisfies the octet rule can be assigned one of the three idealized hybridizations, sp^3, sp^2, or sp. The maximum number of hybrid orbitals required for such atoms is four, as in the sp^3 hybridization, since only s and p orbitals are involved. However, for elements of atomic number 14 and higher that form expanded octets, the possibility of hybrid orbitals that include d orbitals must be considered.

Generally, either one or two d orbitals can mix with s and p orbitals to form hybrid orbitals. If one d orbital is mixed with one s orbital and three p orbitals, five hybrid orbitals are formed, each of which is called sp^3d. Such an orbital is 20% d, 20% s, and 60% p in character. This hybridization can be used to describe the geometry about an atom in which the sum of σ bonds and nonbonding valence electrons is five. An example is the PCl_5 molecule, whose Lewis structure:

```
       Cl
        |
 Cl — P — Cl
      /   \
    Cl     Cl
```

with five σ bonds to the P atom suggests sp^3d hybridization for phosphorus. In BrF_3, with the Lewis structure:

```
F —:Br:— F
     |
     F
```

the sum of σ bonds and nonbonding electron pairs is five. The two valence electron pairs and three bonds are accommodated conveniently by five sp^3d hybrid orbitals of the Br atom.

The ideal geometry associated with five sp^3d hybrid orbitals is the same geometry that the VSEPR model describes as minimizing repulsions between five valence electron pairs. The five Cl atoms of PCl_5 lie at the five corners of a geometric figure called a trigonal bipyramid, shown in Figure 8.16. Unlike the bonds found in the tetrahedral, planar, and linear hybrid orbitals discussed previously, the five positions—and hence the five bonds—of a trigonal bipyramid are not all equivalent. Two of the positions are called axial. In the PCl_5 molecule they form a linear system, Cl—P—Cl. The other three positions are called equatorial. They lie in a plane at the corners of an equilateral triangle, making the system

similar to that of an sp^2 hybrid such as BCl_3. The bond angle between the two axial Cl atoms is 180°, the bond angle between the equatorial Cl atoms is 120°, and the bond angle between an axial Cl atom and an equatorial Cl atom is 90°. The axial bonds are somewhat longer than the equatorial bonds, in this and similar molecules. The two axial Cl atoms and three equatorial Cl atoms in PCl_5 can be distinguished experimentally because of the different bond angles and bond lengths.

EXAMPLE 8.3
Predict the hybridization and geometry of the anion I_3^-.
SOLUTION

Step 1: Draw the Lewis structure:

$$\left[\ddot{\underset{..}{I}} - \ddot{\underset{..}{I}} - \ddot{\underset{..}{I}} \right]^-$$

Step 2: The sum of the σ bonds and the nonbonding valence electron pairs on the central I atom is five. The hybridization of the atom is thus sp^3d, and the ideal geometry is a trigonal bipyramid, with the central I of I_3^- at its center.

Step 3: The two iodine atoms attached to the central one and the three nonbonding valence electron pairs must be distributed among the five possible positions of the trigonal bipyramid. Experimental evidence indicates that it is more favorable for the nonbonding valence electron pairs to be in the equatorial positions. Therefore, the two end I atoms lie in the axial positions. That is, the three-atom I—I—I system is linear, (Figure 8.17), as confirmed by experiment.

If two d orbitals are mixed with the s and p orbitals, the result is six hybrid orbitals, designated sp^3d^2. In the compound SF_6, the Lewis structure:

indicates that the S atom, with six σ bonds, will have the hybridization sp^3d^2. The ideal geometry associated with this hybridization is the geometry that, according to the VSEPR model, minimizes repulsions between six valence electron pairs. It is called octahedral geometry. The six F atoms in SF_6 lie at the corners of an octahedron (Figure 8.18).

An octahedron is a regular solid figure whose corners are all identical.

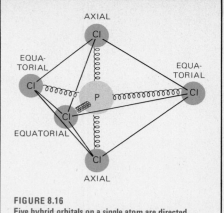

FIGURE 8.16
Five hybrid orbitals on a single atom are directed toward the corners of a trigonal bipyramid. In the PCl_5 molecule, the five Cl atoms lie at the five corners of the trigonal bipyramid.

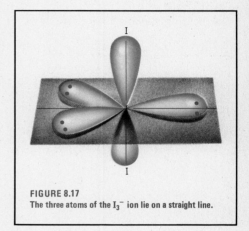

FIGURE 8.17
The three atoms of the I_3^- ion lie on a straight line.

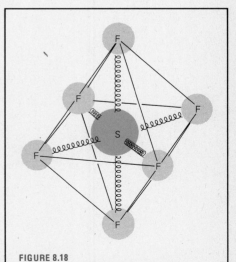

FIGURE 8.18
The ideal geometry of a molecule with six hybrid orbitals on a central atom is an octahedron. In the case of the SF_6 molecule, the six F atoms lie at the corners of the octahedron.

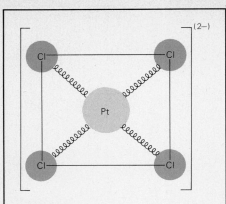

FIGURE 8.19
The ion $PtCl_4^{2-}$ has all five atoms lying in a plane, with the four Cl atoms at the corners of a square. This square planar geometry is found in molecules containing certain transition metals.

TABLE 8.2
Bond Lengths in the Halogens

Bond	Length (nm)
F—F	0.142
Cl—Cl	0.199
Br—Br	0.229
I—I	0.266

Therefore, in a molecule such as SF_6, all the bonds and all the F atoms are equivalent, and the bond angles are 90°.

Another important type of geometry is found in elements, such as some transition metals, that can use inner-shell d orbitals for bonding. Some compounds of these elements have four bonds to the transition metal atom, but not the predicted tetrahedral geometry. Rather, the four atoms that are bonded to the transition metal atom are in the same plane at the corners of a square, as shown in Figure 8.19. The VSEPR model cannot account for this geometry. One of the limitations of the VSEPR model is that it does not work well for the transition metals.

The hybridization that corresponds to this square planar geometry is believed to be dsp^2. It is most commonly found in molecules that include such metals as Ni, Pt, Pd, and Au. The chemistry of some of these substances will be discussed in Chapter 18.

Bond Lengths

A number of factors influence the length of bonds in covalent molecules. One important determinant is the size of the two atoms of the bond. In general, we can say that *larger atoms form longer bonds*. We can see the effect of atomic size in the lengths of the bonds in the homonuclear diatomic halogen molecules listed in Table 8.2.

The length of the bond between two atoms of similar size is affected in an important way by the *bond order*. For most of the bonds we shall discuss, *bond order is the number of electron pairs shared between the two atoms of the bond*. A small bond order means a longer bond. A single bond is longer than a double bond, and a double bond is longer than a triple bond. For example, the C—C bond length in C_2H_6 is 0.154 nm, while the C=C bond length in C_2H_4 is 0.134 nm and the C≡C bond length in C_2H_2 is 0.120 nm.

There is also a relationship between hybridization and the length of a bond of a given bond order. The bond length depends on the fraction of s character in the hybrid orbital forming the bond. A larger fraction of s character means a shorter bond. Single bonds of sp^3 orbitals, which are one-quarter s in character, are longer than single bonds of sp^2 orbitals, which are one-third s in character. These single sp^2 bonds, in turn, are longer than single bonds of sp orbitals, which are one-half s in character.

This hybridization effect can be seen in a series of C—H bonds. The C—H bonds in C_2H_6, where the C atom is sp^3, are 0.110 nm long. The C—H bonds in C_2H_4, where the C atom is sp^2, are 0.107 nm long. Those in C_2H_2, where the C atom is sp, are 0.106 nm long.

8.3 THERMOCHEMISTRY

Almost every chemical and physical change is accompanied by the evolution or absorption of heat, or thermal energy. The branch of chemistry that studies these thermal effects is called thermochemistry.

The amount of heat absorbed or released during a chemical change is of interest because it can tell us something about the structure and relative stability of the starting materials and the products. The quantity of thermal energy evolved or absorbed in a chemical or physical change is measured by application of calorimetric techniques, which we shall discuss in Chapter 13.

Processes that liberate heat usually cause a rise in the temperature of the surroundings and are said to be **exothermic**. The combustion of gasoline is an exothermic process. The temperature rise in the region surrounding the reaction can be used productively, as it is in an automobile engine.

Processes that absorb heat are said to be **endothermic.** Endothermic processes usually result in a drop in the temperature of the surroundings. The coolness felt on the skin during the evaporation of a liquid with a low boiling point, such as alcohol, is the result of an endothermic process.

A useful but by no means rigorous generalization is that most exothermic processes are favorable at room temperature (about 25°C) and most endothermic processes are unfavorable at room temperature. That is, if the process A → B liberates heat, compound B probably is more stable than compound A. If the process C → D absorbs heat, the reverse is true: C probably is more stable than D. Since the relative stability of any compound depends on the nature of its bonds and other details of molecular structure, a knowledge of heat effects can give information about molecular structure.

If a process is endothermic, the surroundings must supply thermal energy for the process to proceed, and the process will proceed only if sufficient thermal energy is available. An exothermic process will proceed by itself, with no help from the surroundings, and is said to be *spontaneous*. These statements are slight oversimplifications. Heat changes alone do not determine the relative stability of compounds and the spontaneity of processes, as we shall see in Chapter 13. But at room temperature, this simplification usually is correct.

The amount of heat liberated or absorbed in a given process depends on the quantities of materials and the conditions under which the process is carried out. We shall deal primarily with processes that occur in containers open to the atmosphere, which means that the processes occur at a *constant* pressure of 1 atm. In describing heat changes, it is convenient to imagine that every substance has a *heat content,* called its **enthalpy,** which is represented by the symbol H. For any process A → B carried out at constant pressure, the heat change is the difference between H_B, the enthalpy of the products, and H_A, the enthalpy of the reactants. The Greek letter Δ, *delta*, is used to indicate any difference between a final and an initial state. Thus:

$$\Delta H = H_B - H_A \tag{8.1}$$

The quantity ΔH, called the *enthalpy change,* is the heat absorbed or released by the process when it occurs at constant pressure. By the definitions cited above, an exothermic process will have a negative ΔH and an endothermic process will have a positive ΔH (Figure 8.20).

Some features of H and ΔH should be noted:

1. The actual value of the enthalpy, H, of a single substance cannot be measured. Only the difference in the enthalpy, ΔH, between the reactants and products of a chemical reaction or between two states of a substance can be measured.
2. Since the value of ΔH depends on the quantity of material, ΔH usually is expressed in units that indicate the amount of material as well as the quantity of energy. The SI units for ΔH are joules per mole (J/mol) or kilojoules per mole (kJ/mol), depending on the magnitude of the heat change. The most common non-SI units for ΔH are calories per mole or kilocalories per mole: 1 cal = 4.184 J. The quantities of material sometimes are implied by the equation for a process, as in 2B → D, which indicates that two moles of B undergo reaction.
3. If the magnitude and sign of ΔH are known for a process, ΔH of the reverse process is known to have the same magnitude but the opposite sign. In the evaporation of water:

$$H_2O(l) \longrightarrow H_2O(g)$$

ΔH is +44 kJ/mol at 298 K. The process is endothermic. Therefore, the reverse process, the condensation of steam:

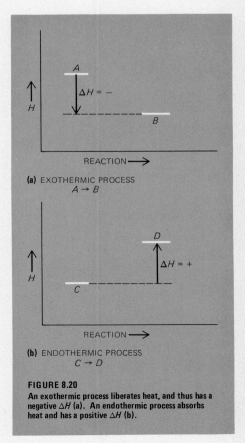

FIGURE 8.20
An exothermic process liberates heat, and thus has a negative ΔH (a). An endothermic process absorbs heat and has a positive ΔH (b).

$$H_2O(g) \longrightarrow H_2O(l)$$

is exothermic and ΔH is -44 kJ/mol at 298 K, if all conditions are the same.

4. The value of ΔH for a process depends on the physical state of each component, as is evident from the two processes discussed above. Therefore, it is essential to indicate the state of every component in a process by an appropriate state symbol. For components in aqueous solution, no subscripts are used (Section 2.8, page 43).

Heat of Formation

The heat content H of single substances cannot be measured. But it is possible to use standard, well-defined processes to give values of ΔH that can serve as well as values of H for tabulating the relative stability of substances. The formation of one mole of a substance from its constituent elements is used to provide values of ΔH that can help us to assess the relative stability of the substance, and to calculate the values of ΔH for other processes in which the substance takes part.

The exact conditions of the process by which one mole of a substance forms from its constituent elements must be carefully specified to make comparisons meaningful. For convenience, a specific set of conditions called a **standard state** is defined for this process.

The standard state value for pressure is 1 atm (101.325 kPa). This pressure is close to the atmospheric pressure in most laboratories. The physical and chemical state of an element that is most stable at this pressure and a specified temperature is defined as the standard state of the element. We shall specify the temperature as 298.15 K (25°C) for the purposes of this discussion of thermochemistry. Some elements in their standard states are: $H_2(g)$, $O_2(g)$, $Br_2(l)$, $I_2(s)$, $Na(s)$, $mg(s)$, $Fe(s)$, $Hg(l)$, $C(s)_{graphite}$, $P_4(s)$, and $S_8(s)$.

The value of ΔH for the reaction, carried out under standard conditions, in which one mole of a substance is formed from its constituent elements in their standard states is called the **standard heat of formation** of the substance. This value is written ΔH_f°, in which the superscript refers to the standard state and the subscript refers to *formation*. A tabulation of values of ΔH_f° for various substances is found in Appendix II.

The units of ΔH_f° usually are kilojoules per mole (kJ/mol). For example, ΔH_f° of water is the ΔH of the process:

$$H_2(g) + \tfrac{1}{2}O_2(g) \longrightarrow H_2O(l)$$

and has the value -286 kJ/mol. (The fractional coefficient is used for O_2 so that only one mole of $H_2O(l)$ is formed.)

By definition, ΔH_f°, the standard heat of formation of any element in its standard state, must be zero. Chemical forms other than the standard state of the element will have positive values of ΔH_f°. For example, ΔH_f° for $O(g)$ is the ΔH for the process:

$$\tfrac{1}{2}O_2(g) \longrightarrow O(g)$$

carried out under standard conditions. It has the value $+249$ kJ/mol. For ozone, O_3, ΔH_f° is defined by the process:

$$\tfrac{3}{2}O_2(g) \longrightarrow O_3(g)$$

and has the value $+142$ kJ/mol.

The standard heats of formation for both $O(g)$ and $O_3(g)$ are positive, which means that both are formed by endothermic processes. Since $O_2(g)$ is the most stable state of oxygen at standard conditions, any other form of oxygen is relatively unstable. Thermal energy must be provided to form such relatively unstable species (Figure 8.20).

Nonstandard states such as $O_3(g)$ and $O(g)$ are said to be *unstable relative to their elements*. Other common substances that are unstable relative to their elements are gaseous binary compounds of nitrogen and oxygen, such as the compound formed in the process:

$$\tfrac{1}{2}N_2(g) + \tfrac{1}{2}O_2(g) \longrightarrow NO(g)$$

The ΔH_f° of $NO(g)$ is $+90$ kJ/mol. Nitric oxide, NO, is unstable relative to its elements, as the positive value of its ΔH_f° shows. Any substance with a positive ΔH_f° is unstable relative to its elements. While substances with a positive ΔH_f° should revert back to their elements spontaneously, many of them can exist because they revert very slowly. It is important to note that thermally unstable substances can exist because a spontaneous or "thermally favorable" process can be exceedingly slow. When we say that a process is spontaneous, we mean only that it occurs without help from the surroundings. We do not say anything about the time needed for the process to occur. We shall discuss these points in greater detail in Chapter 13.

Water is a more typical molecule than NO. Its standard heat of formation is negative, which means that water is stable with respect to its elements and the process by which it forms is exothermic. We can use the ΔH_f° of water listed in Appendix II as a guide to its stability compared to other substances. In general, the stability of a substance is high if its ΔH_f° has a large negative value and low if ΔH_f° has a positive value.

Liquid water, $H_2O(l)$, has a ΔH_f° of -286 kJ/mol. The ΔH_f° of hydrogen peroxide, $H_2O_2(l)$, is -188 kJ/mol, which is less negative than the ΔH_f° of $H_2O(l)$. We can predict that $H_2O_2(l)$ will decompose spontaneously to $H_2O(l)$ and $O_2(g)$ at 298 K, while the reverse process will not occur spontaneously. The prediction is borne out by observation. Similarly, we note that for $H_2O(g)$, the ΔH_f° is -242 kJ/mol. The difference between this value and the ΔH_f° for $H_2O(l)$ is the amount of heat required for one mole of H_2O to evaporate at standard conditions.

We can use ΔH_f° to calculate the standard heat of reaction, a quantity whose symbol is ΔH°. The standard heat of reaction is the difference in enthalpy between the products and the reactants in any process carried out at 1 atm. We shall specify the temperature as 298 K. If ΔH_f° is known for all the components in a process, then:

$$\Delta H^\circ = \Sigma\, \Delta H_f^\circ \text{ (products)} - \Sigma\, \Delta H_f^\circ \text{ (reactants)} \tag{8.2}$$

(The symbol Σ, the Greek letter sigma, means "the sum of.") Equation 8.2 is equivalent to Equation 8.1.

EXAMPLE 8.4
Calculate ΔH° for the reaction:

$$2H_2O(l) \longrightarrow H_2O_2(l) + H_2(g)$$

SOLUTION
Step 1: Calculate the $\Sigma\, \Delta H_f^\circ$ of the products. For $H_2O_2(l)$, ΔH_f° is -188 kJ/mol. For $H_2(g)$, ΔH_f° is zero, since $H_2(g)$ is an element in its standard state. Thus, $\Sigma\, \Delta H_f^\circ$ of the products is -188 kJ/mol.

Step 2: Calculate the $\Sigma\, \Delta H_f^\circ$ of the reactants. The ΔH_f° of $H_2O(l)$ is -286 kJ/mol. Since there are two moles of $H_2O(l)$ as reactants, $\Sigma\, \Delta H_f^\circ$ (reactants) = (2 mol $H_2O(l)$)(-286 kJ/mol $H_2O(l)$) = -572 kJ. (In calculations of this type, every ΔH_f° is expressed in units of kilojoules per mole (kJ/mol) and must be multiplied by the coefficient of the substance, since the coefficient indicates the number of moles of the substance in the reaction.)

Step 3: From steps 1 and 2, $\Delta H^\circ = -188$ kJ $- (-572$ kJ$) =$

$+384$ kJ. The positive value of $\Delta H°$ indicates that this process probably does not occur spontaneously at room temperature.

Hess's Law

Hess's law states that *the heat change (ΔH) that accompanies a chemical reaction is the same whether the reaction occurs in one step or in a number of steps*. That is, the enthalpy difference (ΔH) between A and B depends only on the nature of A and B and not on the path followed to change one to the other. One of the consequences of this law is that chemical equations can be manipulated in the same way as algebraic equations. They can be added to or subtracted from one another or multiplied by a common factor to obtain new equations. If the ΔH is known for each of a starting set of equations and the same arithmetic operations performed with these equations are performed with their ΔH's, then we will obtain the ΔH for the new equation.

A simple example is a two-step process:

(1) $A \longrightarrow B \qquad \Delta H_1$
(2) $B \longrightarrow C \qquad \Delta H_2$

We can perform a simple addition of equations to obtain:

$$A + B \longrightarrow B + C$$

so,

$$A \longrightarrow C \quad \text{and} \quad \Delta H_3 = \Delta H_1 + \Delta H_2$$

In this example, there are two steps between A and C. No matter how many steps there are, ΔH for the overall reaction $A \rightarrow C$ is always the sum of the enthalpy changes for all the individual steps. This is shown in Figure 8.21. The value of ΔH, the difference in H between A and C does not depend on the path taken to get from A to C but only on the values of H_A and H_C.

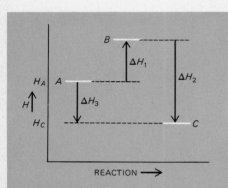

FIGURE 8.21
The heat change for any overall reaction, such as $A \rightarrow C$, is always the same, no matter how many steps occur in the reaction.

EXAMPLE 8.5

Given that 44 kJ of heat is needed to evaporate one mole of water at 298 K, calculate $\Delta H_f°$ of $H_2O(g)$, using the data already given.
SOLUTION
The $\Delta H_f°$ of $H_2O(g)$ refers to the process:

$$H_2(g) + \tfrac{1}{2}O_2(g) \longrightarrow H_2O(g)$$

We know that ΔH for the process

(1) $\quad H_2(g) + \tfrac{1}{2}O_2(g) \longrightarrow H_2O(l)$

is -286 kJ/mol, and that ΔH for the process

(2) $\quad H_2O(l) \longrightarrow H_2O(g)$

is 44 kJ/mol.

We can sum these two equations to get:

$$H_2(g) + \tfrac{1}{2}O_2(g) + H_2O(l) \longrightarrow H_2O(l) + H_2O(g)$$

or, simplifying:

(3) $\quad H_2(g) + \tfrac{1}{2}O_2(g) \longrightarrow H_2O(g)$

Summing ΔH for Equations 1 and 2 gives:

$$\Delta H_1 + \Delta H_2 = -286 \text{ kJ/mol} + 44 \text{ kJ/mol} = -242 \text{ kJ/mol} = \Delta H_3$$

In Example 8.5, Equation 2 represents a change in state or phase, rather than a chemical change. There are a number of such phase changes, and there is a specific name for the heat change associated with each phase change.

The process in which a liquid becomes a gas is called vaporization. The heat change associated with it is called the **heat of vaporization.** (ΔH_{vap}) *The heat of vaporization always has a positive value.*

The reverse process, in which a gas becomes a liquid, is called condensation. Under the same conditions, the **heat of condensation** is equal in magnitude to the heat of vaporization and has the opposite sign. As the heat of vaporization always has a positive value, *the heat of condensation always has a negative value.*

The heat change associated with the change from solid to gas is the **heat of sublimation** (ΔH_{sub}). *The heat change associated with the change from solid to liquid is the* **heat of fusion** (ΔH_{fus}). *Often a heat change accompanies the formation of a solution* and is called the **heat of solution** (ΔH_{sol}).

EXAMPLE 8.6
Given the data:

(1) $N_2O_4(g) \longrightarrow 2NO_2(g)$ $\qquad\qquad \Delta H_1 = 24$ kJ

(2) $NO(g) + \frac{1}{2}O_2(g) \longrightarrow NO_2(g)$ $\qquad \Delta H_2 = -56$ kJ

calculate ΔH for the process:

(3) $2NO(g) + O_2(g) \longrightarrow N_2O_4(g)$

SOLUTION
All the substances in Equation 3 appear in Equations 1 and 2. By appropriate manipulations of Equations 1 and 2, Equation 3 can be obtained. The same manipulations can then be repeated for the ΔH value of each equation to obtain ΔH_3.

Step 1: The first substance on the left side of Equation 3 is 2NO(g). This substance appears in Equation 2 as NO(g). Multiplying Equation 2 by two, we get:

$2NO(g) + O_2 \longrightarrow 2NO_2(g)$ $\qquad 2\Delta H_2 = 2(-56$ kJ$) = -112$ kJ

Step 2: The next substance on the left side of Equation 3 is $O_2(g)$. This substance is already included in Step 1.

Step 3: The product on the right side of Equation 3 is $N_2O_4(g)$, which appears on the left side of Equation 1. We can reverse Equation 1, which means that we must reverse the sign of its ΔH. This procedure gives us $-$(Equation 1):

$2NO_2(g) \longrightarrow N_2O_4(g)$ $\qquad -\Delta H_1 = -24$ kJ

Step 4: Combining the equations in steps 1 and 3 gives:

[2 × (Equation 2)] + [−(Equation 1)]

or

$2NO(g) + O_2(g) + 2NO_2(g) \longrightarrow N_2O_4(g) + 2NO_2(g)$

or, simplifying:

(3) $2NO(g) + O_2(g) \longrightarrow N_2O_4(g)$

Since Equation 3 = [2 × (Equation 2)] + [−(Equation 1)]

$\Delta H_3 = [2 \times (\Delta H_2)] + (-\Delta H_1)$

$\qquad = 2(-56$ kJ$) - (24$ kJ$)$

$\qquad = -136$ kJ

FUELS FOR THE FUTURE

As the earth's supply of petroleum dwindles, the search has begun for an alternate fuel that can be made from renewable resources to replace gasoline, which comes from a nonrenewable resource. Major attention has focused on alcohol—either methanol, CH_3OH, or ethanol, C_2H_5OH—and on hydrogen gas. The studies done thus far indicate that any of these substances could eventually replace gasoline and other fuels from petroleum. But the studies also indicate why gasoline was chosen originally as the fuel for most internal combustion engines: It has a combination of good properties that is not easy to duplicate.

The heat of combustion of gasoline—the amount of energy released when it burns—compares favorably with that of the other fuels. The heat of combustion of typical unleaded gasoline is 43 megajoules per kilogram (MJ/kg). For methanol, the heat of combustion is 20 MJ/kg; for ethanol, it is about 39 MJ/kg; and for hydrogen, it is about 115 MJ/kg.

Hydrogen seems to offer the most advantages. In addition to its high heat of combustion, hydrogen is also a literally inexhaustible fuel. It can be extracted from water and it burns to produce water, from which it can again be extracted. However, it is the most distant possibility. The cost of obtaining hydrogen by the electrolysis of water is prohibitively high. Almost all H_2 is obtained from natural gas, itself a fuel in short supply. Efforts are being made to develop low-cost methods of decomposing water. The future of the proposed ''hydrogen economy'' based on the use of H_2 gas as fuel, depends on the success of these efforts.

Other apparent drawbacks, such as the low density of hydrogen and the danger of explosions, are more easily overcome. Studies have shown that the danger of hydrogen explosions is no greater than the danger of natural gas explosions. The fact that hydrogen cannot be liquefied above 33.3 K has created interest in special systems for storage. One such system is based on the ready absorption of hydrogen by such metals as iron and titanium. The hydrogen is easily released by a small amount of heat as needed.

The use of alcohol is a more immediate possibility. Automotive experts agree that existing automobile engines could operate well on a mixture of 90% gasoline and 10% alcohol. In the United States, research is concentrating on methanol, which can be made from coal or wood. In other nations, studies are being made of ethanol, which is made from plants (the ethanol in alcoholic beverages is made from grains such as wheat and corn).

Brazil has begun a full-scale program designed to eliminate oil imports by a complete conversion to ethanol-fueled automobiles. This effort requires a complete redesign of existing auto engines and a massive effort to grow enough crops to provide ethanol. The Brazilians propose to use manioc, a starchy plant whose root now is grown for food. In addition to expanding manioc production manyfold, they must also develop commercial methods of distilling large quantities of ethanol from manioc.

Such methods already exist for methanol, but there are other problems standing in the way of quick use of a gasoline-methanol mixture in U.S. automobiles. The limited supply of methanol is one problem. The annual production of 4.5 billion liters is the energy equivalent of only a two days' supply of gasoline. Some mixtures of methanol and gasoline have been found to separate during transportation, causing problems of distribution. Nevertheless, these problems seem surmountable in light of the ultimate necessity for developing a supply of fuel for the future.

Average Bond Energy

The way in which thermochemical observations give us information about molecular structure can be illustrated by examining the reaction:

$$H_2(g) + Cl_2(g) \longrightarrow 2HCl(g)$$

This process is exothermic at 298 K, so the products are more stable than the reactants. The increased stability of two HCl molecules as compared to separate H_2 and Cl_2 molecules is related to the molecular structure of these species. More specifically, the difference in stability is related to the differences in bond strengths of the molecules.

Examining Lewis structures, we see that H_2 has a single H—H bond, Cl_2 has a single Cl—Cl bond, and HCl has a single H—Cl bond. Since the process by which 2HCl is formed is exothermic, two HCl bonds must be stronger than the sum of one H—H bond and one Cl—Cl bond.

We can use this kind of analysis on many reactions, such as the reaction of Example 8.4:

$$2H_2O(l) \longrightarrow H_2O_2(l) + H_2(g)$$

In this reaction, H_2O has the structure H—O—H, which has two O—H bonds. The product H_2O_2 has the structure H—O—O—H, which has two O—H bonds and one O—O bond, while H_2 has one H—H bond.

The conversion shown by this reaction can be described in this way: Four O—H bonds (from $2H_2O$) are broken. Two O—H bonds, one O—O bond, and one H—H bond form. The net result is that two O—H bonds break and one O—O bond and one H—H bond form. Since the process is endothermic, the bonds on the left side of the equation are stronger, in sum, than the bonds on the right side of the equation. It is important to understand that this does not necessarily mean that the compounds on the right side of the equation are unstable. In this case, they are stable relative to their elements, but the compounds on the left side of the equation have stronger bonds and so are even more stable. It is an essential principle of chemistry that **the formation of chemical bonds between two atoms is always an exothermic process.** Otherwise no bonds would be formed.

We can analyze reactions in terms of bond formation and bond breaking with some accuracy by defining a thermochemical quantity called the **average bond energy.** *Average bond energy is defined as the quantity of energy required to break a bond between atoms in a molecule that is in the gas phase.* It is a positive quantity. The concept of average bond energy is based on the idea that a given type of bond, such as N—H or C=O or C≡C, has about the same strength in any compound.

The bond energy of the H—Cl bond is defined by ΔH for the process:

$$HCl(g) \longrightarrow H(g) + Cl(g) \qquad \Delta H° = 432 \text{ kJ/mol}$$

while the bond energy of the I—I bond is defined by ΔH for the process:

$$I_2(g) \longrightarrow 2I(g) \qquad \Delta H° = 151 \text{ kJ/mol}$$

253

Note that standard states are not necessarily involved in defining bond energies. The bond energies shown here do not have the same values as the heats of formation (ΔH_f°) of the substances. All substances must be in the gas phase to define standard bond energies, and the gas phase is not the standard state of many substances. It is relatively simple to find the bond energy for a diatomic molecule, because it has only one bond. But it is more difficult to assign bond energies to bonds that do not occur in diatomic molecules. We must first find the energy required to break all the bonds in a polyatomic molecule. Then we must divide the total energy by the number of bonds to find the average bond energy. The process:

$$H\text{---}O\text{---}H(g) \longrightarrow 2H(g) + O(g) \qquad \Delta H = 926 \text{ kJ/mol } H_2O$$

can define the average bond energy of the O—H bond. Since there are two O—H bonds, the average bond energy is $\Delta H/2 = 463$ kJ/mol.

The process:

$$CH_4(g) \longrightarrow C(g) + 4H(g) \qquad \Delta H = 1660 \text{ kJ/mol } CH_4$$

can define the average bond energy of the C—H bond. Since there are four C—H bonds, the average bond energy is $\Delta H/4 = 415$ kJ/mol. The average bond energy is not based on ΔH for the process $CH_4(g) \rightarrow CH_3(g) + H(g)$, because this process does not break all the bonds in the CH_4 molecule. Only those processes that break all the bonds in a polyatomic molecule can be used to calculate *average* bond energies. The ΔH for the process $CH_4(g) \rightarrow CH_3(g) + H(g)$, which is 423 kJ/mol, defines another quantity called the *bond dissociation energy*.

A table of average bond energies is found in Appendix III. Such a table is useful because it describes the relative strengths of different types of bonds in a straightforward way: The greater the bond energy, the stronger the bond. The fact that the bond energy of the N≡N bond in $N_2(g)$ is 946 kJ/mol, a high value, is consistent with the observation that N_2 in air is highly unreactive.

Average bond energies can be used to estimate approximate values of the ΔH of reactions in a way similar to that in which ΔH_f° is used, although calculations using ΔH_f° give more accurate results. The two methods often are complementary. If a process is analyzed in terms of bonds broken and bonds made, then:

$$\Delta H = \Sigma \text{ bond energy of bonds broken} - \Sigma \text{ bond energy of bonds made} \quad (8.3)$$

EXAMPLE 8.7

Using the average bond energy values in Appendix III, estimate ΔH for the reaction:

$$CH_4(g) + Cl_2(g) \longrightarrow CH_3Cl(g) + HCl(g)$$

SOLUTION

If we analyze changes in bonds using Lewis structures, we find that in this reaction, one C—H bond and one Cl—Cl bond are broken, and one H—Cl bond and one C—Cl bond are formed. Since all molecules in the reaction are in the gas phase, average bond energies can be used directly in the calculation. Equation 8.3 defines ΔH in terms of bonds broken and bonds made:

$$\Delta H = [(415 \text{ kJ/mol}) + (243 \text{ kJ/mol})] - [(328 \text{ kJ/mol}) + (423 \text{ kJ/mol})]$$
$$= -102 \text{ kJ}$$

This type of calculation can be combined with ΔH_f° data to determine other average bond energies.

EXAMPLE 8.8

Calculate the average bond energy of an N—F bond in NF_3 using only the following data: ΔH_f° of $NF_3(g) = -114$ kJ/mol; bond energy of N≡N = 946 kJ/mol; bond energy of F—F = 158 kJ/mol.

SOLUTION

The ΔH_f° is the ΔH of the reaction:

$$\tfrac{1}{2}N_2(g) + \tfrac{3}{2}F_2(g) \longrightarrow NF_3(g)$$

The ΔH of this reaction can also be defined in terms of bonds made and bonds broken. In the reaction as written, $\tfrac{1}{2}$ mol of N≡N bonds and $\tfrac{3}{2}$ mol of F—F bonds are broken, while 3 mol of N—F bonds is formed. (Each NF_3 has three N—F bonds.) The average bond energies of the N≡N bonds and the F—F bonds are known, while the average bond energy of the N—F bond is not known. If we designate the N—F bond energy as x and use Equation 8.3 to calculate ΔH for the reaction, we get:

$$\Delta H = [\tfrac{1}{2}(946 \text{ kJ/mol}) + \tfrac{3}{2}(158 \text{ kJ/mol})] - [3x] = -114 \text{ kJ/mol}$$
$$x = 275 \text{ kJ/mol}$$

Although average bond energies are defined for the gas phase, they can also be used in calculations involving other phases, provided that the data on the ΔH of the relevant phase change are available.

EXAMPLE 8.9

Calculate the average bond energy of the O—H bond using only the following data: ΔH_{vap} of H_2O at 298 K = 44 kJ/mol; ΔH_f° of $H_2O(l)$ = −286 kJ/mol; bond energy of H—H bond = 436 kJ/mol; bond energy of O═O bond = 498 kJ/mol.

SOLUTION

The average bond energy can be calculated from ΔH for the reaction:

$$H_2(g) + \tfrac{1}{2}O_2(g) \longrightarrow H_2O(g)$$

in which two O—H bonds are made, while one H—H and one O═O bond are broken, all in the gas phase. In Example 8.5, the ΔH of this reaction was calculated from ΔH_f° and ΔH_{vap} to be −242 kJ. If we designate the O—H bond energy as x and use Equation 8.3 to calculate ΔH, we get:

$$\Delta H = [(436 \text{ kJ/mol}) + \tfrac{1}{2}(498 \text{ kJ/mol})] - [2x] = -242 \text{ kJ/mol}$$
$$x = 463 \text{ kJ/mol}$$

Resonance Energy

A number of factors other than bond strength affect the stability of molecules. For example, it was mentioned in Section 7.4 that resonance hybrids have greater stability than would be predicted on the basis of bond energies alone. The extra stability, called the resonance energy, can be given a numerical value. By convention, resonance energy is given a positive value to simplify calculations with bond energies, which also have positive values.

To find the value of the resonance energy of a molecule, we must find the difference between the ΔH that is calculated from the bond energies of an important contributing structure of the resonance hybrid and the ΔH that is determined experimentally.

The determination of resonance energy starts with the calculation of ΔH for a given reaction of the molecule on the basis of bond energies alone. If a molecule has resonance energy, the calculation of ΔH on the basis of bond energies alone will be incorrect by an amount equal to the resonance energy. If the correct value of ΔH is determined experimentally, the difference between the observed ΔH (ΔH_{obs}) and the calculated value of ΔH (ΔH_{calc}) is the resonance energy. Figure 8.22 shows this process graphically.

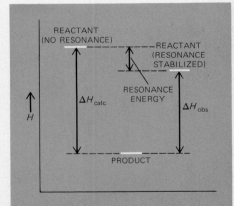

FIGURE 8.22
Calculation of resonance energy for a resonance hybrid starts with the calculation of the ΔH of a reaction based on an important contributing structure of the molecule. This calculated value is then compared with the measured value of ΔH of the reaction. The difference between the two values of ΔH is the resonance energy of the molecule.

EXAMPLE 8.10

Calculate the resonance energy of N_2O from the following data: ΔH_f° of N_2O = 82 kJ/mol; bond energy of $N\equiv N$ = 946 kJ/mol; bond energy of $O=O$ (in O_2) = 498 kJ/mol; bond energy of $N=O$ = 607 kJ/mol; bond energy of $N=N$ = 418 kJ/mol.

SOLUTION

The value of ΔH_f° is measured and it refers to the reaction:

$$N_2(g) + \tfrac{1}{2}O_2(g) \longrightarrow N_2O(g)$$

The value of ΔH for this reaction can be calculated from the bond energies. Draw an important contributing structure for N_2O:

$$\ddot{N}=N=\ddot{O}$$

On the basis of this contributing structure, we can say that in this reaction, an $N=N$ bond and an $N=O$ bond are made and that an $N\equiv N$ bond and one-half of an $O=O$ bond are broken. We can use the values of the bond energies and Equation 8.3 to calculate ΔH:

$$\Delta H_{calc} = [(946 \text{ kJ/mol}) + \tfrac{1}{2}(498 \text{ kJ/mol})] - [(607 \text{ kJ/mol}) + (418 \text{ kJ/mol})]$$
$$= 170 \text{ kJ/mol}$$

ΔH_{obs} = 82 kJ/mol, the ΔH_f°

resonance energy = $\Delta H_{calc} - \Delta H_{obs}$
$$= 170 \text{ kJ/mol} - 82 \text{ kJ/mol} = 88 \text{ kJ/mol}$$

Molecular stability can be influenced by factors other than bond strength and resonance energy—for example, interactions between portions of a molecule that are not directly attached to one another. The calculation of such nonbonded interactions is beyond the scope of this book. But it should be noted that these interactions are important in large, complex organic molecules such as proteins and nucleic acids.

The Born-Haber Cycle

One important application of the principles of thermochemistry, and of Hess's law in particular, is the analysis of a chemical process as a series of thermochemical steps. One such analysis is called the Born-Haber cycle.

The Born-Haber cycle is often used to analyze the formation of an ionic solid. The Born-Haber cycle is useful when ΔH can be measured for an overall process and for all but one of the steps in the process; the missing ΔH can then be calculated. The factors that affect the stability of a substance can be understood with the help of the Born-Haber cycle.

To show how a Born-Haber cycle can be used, let us construct one for sodium chloride. We can measure ΔH_f° for NaCl:

$$Na(s) + \tfrac{1}{2}Cl_2(g) \longrightarrow NaCl(s) \qquad \Delta H_f^\circ = -410 \text{ kJ/mol} \qquad (8.4)$$

Clearly, NaCl(s) is a stable species. What is the source of its stability?

The question was discussed semiquantitatively in Section 7.1. The Born-Haber cycle gives a more precise answer. We know that the stability of NaCl(s) is due primarily to lattice energy, which is the ΔH for the process:

$$Na^+(g) + Cl^-(g) \longrightarrow NaCl(s) \qquad (8.5)$$

the formation of the NaCl lattice from gaseous ions. Lattice energy cannot be measured directly, but we can find ΔH for the processes that lead to formation of $Na^+(g)$ from Na(s) and of $Cl^-(g)$ from $\tfrac{1}{2}Cl_2(g)$. Using Hess's law and the measured ΔH_f°, we can then calculate the lattice energy for NaCl.

Two steps, each with measurable ΔH, lead from Na(s) to $Na^+(g)$:

(1) $Na(s) \longrightarrow Na(g)$ ΔH_{sub} (heat of sublimation) = 109 kJ

(2) $Na(g) \longrightarrow Na^+(g)$ ΔH_{ion} (ionization energy) = 495 kJ

Two steps, each with measurable ΔH, lead from $\frac{1}{2}Cl_2(g)$ to $Cl^-(g)$:

(3) $\frac{1}{2}Cl_2(g) \longrightarrow Cl(g)$ $\frac{1}{2}\Delta H_{diss}$ ($\frac{1}{2}$ Cl—Cl bond energy) = 122 kJ

(4) $Cl(g) + e^- \longrightarrow Cl^-(g)$ $- \Delta H_{e.a.}$ (electron affinity) = −347 kJ

According to Hess's law, ΔH for the process that is the sum of steps (1)–(4) is the sum of the ΔH values of these four steps:

$$Na(s) + \tfrac{1}{2}Cl_2(g) \longrightarrow Na^+(g) + Cl^-(g) \qquad \Delta H = 379 \text{ kJ} \qquad (8.6)$$

The sum of Equation 8.6 and Equation 8.5, the process in which the NaCl lattice is formed from gaseous ions, is:

$$Na(s) + \tfrac{1}{2}Cl_2(g) \longrightarrow NaCl(s)$$

which is identical with Equation 8.4. Since the sum of Equations 8.5 and 8.6 is Equation 8.4, by Hess's law:

$$\Delta H(8.5) + \Delta H(8.6) = \Delta H(8.4)$$

or

$$\Delta H(8.5) + 379 \text{ kJ} = -410 \text{ kJ}$$

$$\Delta H(8.5) = -789 \text{ kJ} = \text{lattice energy}$$

Figure 8.23 shows the cyclical nature of this thermochemical construction.

The Born-Haber cycle shows that the major reason for the stability of NaCl(s) is its lattice energy, which has a large negative value. All the other factors, except the electron affinity of Cl, have positive ΔH values, and therefore are actually destabilizing influences.

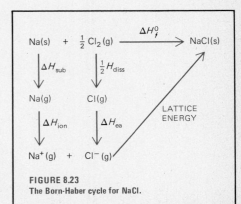

FIGURE 8.23
The Born-Haber cycle for NaCl.

EXAMPLE 8.11

Acetylene, $C_2H_2(g)$, is unstable. Identify reasons for this instability with the aid of a thermochemical cycle, using data in Appendix II and Appendix III.

SOLUTION

The reported value of ΔH_f° for C_2H_2 is +227 kJ/mol. The large positive value means that acetylene is unstable relative to its constituent elements, carbon and hydrogen, in their standard states. Acetylene exists because its rate of decomposition at room temperature is very slow. The ΔH_f° is defined by the reaction:

$$2C(s)_{graphite} + H_2(g) \longrightarrow C_2H_2(g)$$

We can identify the reason for the instability of C_2H_2 by breaking this reaction down into a series of steps and constructing a thermochemical cycle.

The atoms of the C_2H_2 molecule are held together by strong bonds. Indeed, the process in which C_2H_2 forms from gaseous atoms is exothermic:

(1) $2C(g) + 2H(g) \longrightarrow C_2H_2(g)$ $\Delta H = -\Sigma$ bonds formed

The Lewis structure of C_2H_2 is H—C≡C—H. In step 1, two C—H bonds and one C≡C bond are formed. Thus: $\Delta H = -[2(415 \text{ kJ})] + (812 \text{ kJ}) = -1642$ kJ. This is a highly favorable process. The reason for the relative instability must lie elsewhere.

The steps required to produce the gaseous atoms on the left side of the reaction in step 1 are:

(2) $H_2(g) \longrightarrow 2H$ ΔH_{diss} = H—H bond energy = 435 kJ

(3) $2C(s)_{graphite} \longrightarrow 2C(g)$ ΔH_{sub} = 2(718 kJ) = 1436 kJ

The large positive value of the ΔH of step 3 gives the major reason why

C_2H_2 is unstable relative to its elements. In graphite, the standard state of carbon, the C atoms are held together tightly. A large amount of energy is needed to separate them before compounds of carbon can be formed.

The sum of steps 1–3 defines ΔH_f° for C_2H_2. The sum of the values of ΔH for each of these steps is 230 kJ/mol, which is close to the experimentally determined value, 227 kJ/mol. A similar line of reasoning was behind the statement that reactions of N_2 are difficult to bring about because of the strength of the nitrogen-nitrogen triple bond.

In this chapter, we have examined the size and shape of molecules and the nature and strength of the bonds that hold atoms together in molecules. We have seen that the geometry of molecules can be explained by two models, the valence shell electron pair (VSEPR) model and the hybridization model. Each model offers a different rationalization for the bond angles that are found experimentally. The VSEPR model predicts bond angles that will minimize repulsions between valence electron pairs, while the hybridization model describes the formation of hybrid bonding orbitals from unhybridized atomic orbitals.

The stability of molecules can be described in an approximate way in terms of the amount of heat that is absorbed or released as the result of chemical changes in which they participate. By measuring heat changes that accompany chemical processes, we can describe the stability of the substances that are formed or react as a result of these processes. Using such measurements, we can calculate the strength of bonds in specific molecules, and can predict the relative stability of these molecules. A thorough understanding of heat changes will give us a deeper understanding of chemical processes and molecular stability.

EXERCISES

8.1 Indicate which of the following molecules have ideal tetrahedral geometry, which have close to ideal tetrahedral geometry, and which have some other geometry: (a) CCl_4, (b) CH_2Cl_2, (c) NO_2Cl, (d) CHI_3, (e) C_2H_6, (f) PCl_5.

8.2 Indicate which of the following ions have ideal tetrahedral geometry, which have close to ideal tetrahedral geometry, and which have some other geometry: (a) NH_4^+, (b) BF_4^-, (c) H_3O^+, (d) NH_2^-.

8.3 Indicate which orbitals overlap to form each of the bonds in: (a) CCl_4, (b) $CHCl_3$, (c) SiH_4.

8.4 Indicate which orbitals overlap to form each of the bonds in ethyl bromide, CH_3CH_2Br.

8.5 Indicate which orbitals overlap to form each of the bonds in the following ions: (a) PH_4^+, (b) BF_4^-, (c) AlH_4^-.

8.6 Indicate which of the following molecules have ideal trigonal planar geometry, which have close to ideal trigonal planar geometry, and which have some other geometry: (a) NH_3, (b) CH_2O, (c) N_2H_4, (d) C_2H_4, (e) BF_2Cl.

8.7 Indicate which of the following ions have ideal trigonal planar geometry, which have close to ideal trigonal planar geometry, and which have some other geometry: (a) CH_3^+, (b) CH_3^-, (c) CH_2Cl^+, (d) NH_2^-, (e) NO_3^-.

8.8 Indicate which of the following molecules are linear and which are not linear: (a) $BeBr_2$, (b) $HgCl_2$, (c) $HOCl$ (d) $PbCl_2$.

8.9 All of the following species have multiple bonds. Indicate which are linear and which are not: (a) CO_2, (b) $ClCN$, (c) N_3^-, (d) $NOCl$.

8.10 Indicate which orbitals overlap to form the σ bonds in: (a) BF_2Cl, (b) CH_2O, (c) CH_2Cl^+.

8.11 Indicate which orbitals overlap to form the σ bonds in: (a) $BeBr_2$, (b) $HgCl_2$, (c) ICN.

8.12 Draw a diagram of the type shown in Figures 8.12 and 8.14 for CH_2O.

8.13 Draw a diagram of the type shown in Figures 8.12 and 8.14 for HCN.

8.14 Draw a diagram of the type shown in Figures 8.12 and 8.14 for CO_2.

8.15 Draw a diagram of the type shown in Figures 8.12 and 8.14 for formyl chloride, HCOCl.

8.16 Indicate the most likely hybridization for the starred atoms in each of the following molecules: (a) N^*F_3, (b) $HClO^*$, (c) $H_2O_2^*$, (d) $N_2^*O_3$, (e) $N^*H_2O^*H$.

8.17 Indicate the most likely hybridization for the starred atoms in the following ions: (a) $N^*H_2^-$, (b) $N^*O_2^-$, (c) $C^*H_3^-$, (d) H_3O^{*+}, (e) $N^*H_2N^*H_3^+$.

8.18 Use the designations of geometry given in Table 8.2 to describe the geometry of the following molecules: (a) CH_2F_2, (b) ClCN, (c) $COCl_2$, (d) NCl_3.

8.19 Use the designations of geometry given in Table 8.2 to describe the geometry of the following ions: (a) CH_3^-, (b) CO_3^{2-}, (c) NCO^-.

8.20 Carbon suboxide, C_3O_2, is an unusual oxide of carbon that is found to have linear geometry. Write a reasonable Lewis structure for C_3O_2 consistent with this observed geometry.

*8.21 Diazomethane, CH_2N_2, is an explosive, highly poisonous gas, which is nonetheless commonly used as a reagent for organic synthesis. The CNN bond angle is found to be 180° and the HCH bond angle is close to 120°. Write a Lewis structure for CH_2N_2 consistent with these measurements. (Hint: formal charges cannot be avoided.)

8.22 Propose a reasonable hybridization for the central atom in each of the following molecules: (a) PF_3, (b) PF_5, (c) SeF_4, (d) SeF_6.

8.23 Propose a reasonable hybridization for the central atom in each of the following species: (a) SiF_4, (b) SiF_6^{2-}, (c) ClF_3, (d) ClF_5.

8.24 Predict the geometry of each of the following phosphorus-containing species: (a) PF_3, (b) F_3PO, (c) PF_5, (d) PF_6^-.

8.25 The sum of the σ bonds and the nonbonding valence electron pairs in each of the following molecules is five, yet a different geometry is observed for each. Account for this observation and predict each geometry: (a) AsF_5, (b) SeF_4, (c) BrF_3, (d) KrF_2.

8.26 The compound TeF_6 has ideal octahedral geometry. Write the formulas for the binary ions of fluorine with tin, antimony, and iodine that also have ideal octahedral geometry.

8.27 The sum of the σ bonds and the nonbonding valence electron pairs in each of the following molecules is six, yet a different geometry is observed for each. Account for this observation and predict each geometry: (a) SF_6, (b) BrF_5, (c) XeF_4.

*8.28 Neither the VSEPR model nor the hybridization model is able to account for the experimental observation that the FBaF bond angle in $BaF_2(g)$ is less than 180°. Suggest a possible explanation for this experimental observation.

8.29 Arrange the following molecules in order of increasing length of the bond of the central atom to chlorine: (a) NCl_3, (b) PCl_3, (c) $AsCl_3$, (d) $SbCl_3$.

8.30 Predict the relative lengths of the two bonds in N_2O from a consideration of the important contributing structures of this resonance hybrid.

8.31 Arrange the following molecules in order of increasing length of the CN bond: (a) CH_3NH_2, (b) CH_3NHCH_3, (c) CH_2NH, (d) HCN.

8.32 The ΔH_f° of $H_2O(l)$ is -286 kJ/mol. Calculate the heat liberated during the formation of 1.00 g of H_2O. Calculate the heat required to convert 100 g of H_2O to its constituent elements in their standard states.

8.33 Calculate^a ΔH° for the reaction: $HBr(g) + NH_3(g) \rightarrow NH_4Br(s)$.

8.34 Calculate^a ΔH° for the reaction: $MgCO_3(s) \rightarrow MgO(s) + CO_2(g)$.

8.35 Calculate^a ΔH° for the reaction: $CH_4(g) + 3Cl_2(g) \rightarrow CHCl_3(g) + 3HCl(g)$.

8.36 Calculate^a ΔH° for the reaction: $P_4O_{10}(s) + 6H_2O(l) \rightarrow 4H_3PO_4(s)$.

8.37 Calculate^a ΔH° for the reaction: $4NH_3(g) + 5O_2(g) \rightarrow 4NO(g) + 6H_2O(g)$.

8.38 The heat of combustion, ΔH_{comb}, designates the heat liberated when one mole of a substance undergoes complete reaction with oxygen at standard state conditions. The products of combustion of substances composed of carbon and hydrogen are assumed to be $CO_2(g)$ and $H_2O(l)$. Calculate^a ΔH_{comb} of $C_2H_6(g)$.

8.39 The ΔH_{comb} (see Exercise 8.38) of benzene, $C_6H_6(l)$, is 3270 kJ/mol. (Note that the ΔH_{comb} is a positive quantity by definition although heat is always liberated in a combustion reaction.) Calculate^a the ΔH_f° of $C_6H_6(l)$.

8.40 The combustion of 2.20 g of propane, $C_3H_8(g)$, liberates 111 kJ of thermal energy. Calculate^a ΔH_f° of propane.

8.41 The ΔH_f° of $HI(g)$ is 26 kJ/mol and the heat of sublimation of $I_2(s)$ is 62 kJ/mol. Calculate ΔH° for the reaction: $2HI(g) \rightarrow H_2(g) + I_2(g)$.

8.42 The reaction $N_2O_3(g) \rightarrow NO(g) + NO_2(g)$ absorbs 40.2 kJ of thermal energy and the reaction $2NO_2(g) \rightarrow N_2O_4(g)$ liberates 23.4 kJ of thermal energy. Using only this information, calculate ΔH° for the reaction $2NO(g) + N_2O_4(g) \rightarrow 2N_2O_3(g)$.

8.43 The reaction $CHCl_3(g) + Cl_2(g) \rightarrow HCl(g) + CCl_4(l)$ is found to have ΔH° of -131 kJ/mol. Calculate^a the heat of vaporization of $CCl_4(l)$.

8.44 Under nonstandard conditions, the heat of formation of $SO_3(g)$ from $O_2(g)$ and $S(g)$ is found to be -111 kJ/mol. The ΔH of the reaction $SO_2(g) + \frac{1}{2}O_2(g) \rightarrow SO_3(g)$ under these conditions is -23 kJ/mol. Calculate the ΔH_f of $SO_2(g)$ under these conditions.

8.45 Calculate^b ΔH° for the reaction $H_2(g) + Br_2(g) \rightarrow 2HBr(g)$. This calculated value is not twice the ΔH_f° for $HBr(g)$ listed in Appendix II. Account for the difference.

8.46 Use the result of Exercise 8.45 to find the heat of vaporization of $Br_2(l)$.

8.47 The *heat of atomization* is the heat required to convert a molecule in the gas phase into its constituent atoms in the gas phase. The heat of atomization is used to calculate average bond energies. Without using any tabulated bond energies, calculate the average CCl

bond energy from the following data: The heat of atomization of CH_4 is 1660 kJ/mol; the heat of atomization of CH_2Cl_2 is 1486 kJ/mol.

8.48 Calculate^b the heat of atomization of $CH_2{=}CHBr$.

8.49 Calculate^b ΔH° for the reactions: (a) $N_2(g) + 3H_2(g) \rightarrow 2NH_3(g)$ (b) $CH_4(g) + 4Cl_2(g) \rightarrow CCl_4(g) + 4HCl(g)$ (c) $C_2H_4(g) + HI(g) \rightarrow C_2H_5I(g)$

*8.50 The standard state of sulfur is $S_8(s)$. Use the values given in Appendix II for ΔH_f° of $S(g)$, $O(g)$, and $SO_2(g)$ to find the average bond energy of the SO bond in SO_2.

8.51 The ClCl and FF bond energies are 243 kJ/mol and 158 kJ/mol, respectively. The ΔH_f° of $ClF_3(g)$ is -159 kJ/mol. Calculate the average bond energy of the ClF bond in ClF_3.

8.52 Repeat the calculation of Exercise 8.51 on $ClF_5(g)$, $\Delta H_f^\circ = -240$ kJ/mol. Account for the difference in the average ClF bond energies in ClF_3 and in ClF_5.

8.53 Calculate^b the resonance energy of NO_2 ($\ddot{O}{=}\ddot{N}{-}\ddot{O}{:}$), using the measured ΔH_f° of 34 kJ/mol.

8.54 Calculate^b the ΔH_f° of $N_2O_5(g)$, given that its resonance energy is 296 kJ/mol.

*8.55 The measured ΔH° for the combustion of benzene, $C_6H_6(g)$, to $CO_2(g)$ and $H_2O(g)$ is -3170 kJ/mol. Calculate^b the ΔH° for this reaction using bond energies, and then calculate the resonance energy of benzene.

8.56 The ΔH_f° of $KF(s)$ is -563 kJ/mol, the ionization energy of $K(g)$ is 419 kJ/mol, and the heat of sublimation of potassium is 88 kJ/mol. The electron affinity of $F(g)$ is 322 kJ/mol and the FF bond energy is 158 kJ/mol. Calculate the lattice energy of $KF(s)$ using a Born-Haber cycle.

*8.57 The ΔH_f° of $CaBr_2$ is -675 kJ/mol. The first ionization energy of $Ca(g)$ is 590 kJ/mol, and its second ionization energy is 1145 kJ/mol. The heat of sublimation of calcium is 178 kJ/mol. The bond energy of Br_2 is 193 kJ/mol, the ΔH_{vap} of $Br_2(l)$ is 31 kJ/mol, and the electron affinity of $Br(g)$ is 325 kJ/mol. Calculate the lattice energy of $CaBr_2(s)$.

*8.58 The ΔH_f° of $PI_3(s)$ is -24.7 kJ/mol and the PI bond energy in $PI_3(g)$ is 184 kJ/mol. Construct a thermochemical cycle of the type illustrated in Example 8.11, using the necessary data in Appendix II and Appendix III. Calculate the heat of sublimation of $PI_3(s)$ with the aid of the cycle.

---

^{a} The necessary values of ΔH_f° can be found in Appendix II.

^{b} The necessary values of the bond energies can be found in Appendix III.

THE CHEMISTRY OF THE NONMETALS

9

The nonmetals are a relatively small but extremely important group of elements found on the right side and toward the top of the periodic table (Figure 9.1). The nonmetals and the compounds they form play a major role in chemistry. Most of the atoms in the universe are nonmetals, as are most of the atoms in the human body and the bodies of all other living organisms. If we wish to study the evolution of the universe or the evolution of life on earth, we must understand the chemistry of the nonmetals.

In their elemental form, the nonmetals usually are relatively small molecules, although they can also form quite large molecules. Some of the compounds the nonmetals form with each other can be classified as polar; others can be classified as nonpolar. In either case, the bonding is covalent.

In this chapter, we shall describe the behavior of the nonmetals and the compounds they form with each other. We shall focus our attention primarily on the most common nonmetals: hydrogen; carbon, which is in group IVA; nitrogen and phosphorus, which are in group VA; oxygen and sulfur, in group VIA; the halogens (fluorine, chlorine, bromine, and iodine) in group VIIA; and the noble gases (helium, neon, argon, krypton, xenon, and radon) in group 0.

9.1   HYDROGEN

Periodic Classification

Alone among the elements, hydrogen cannot clearly be classified as a member of any chemical family. Because the electronic configuration of hydrogen is $1s$, it has been suggested that hydrogen be classified as a member of group IA, the alkali metals, which have the valence electronic configuration ns. But many properties of the alkali metals are quite different from the properties of hydrogen. Because hydrogen is just one electron short of a noble gas configuration, some have classified it as a member of group VIIA, the halogens, whose electronic configuration, ns^2np^5, is also one electron short of a noble gas configuration. Again, the substantial differences between the properties of hydrogen and those of the halogens make this classification untenable. Finally, because hydrogen has a half-filled noble gas shell, some have tried to place it in group IVA, whose elements have the valence electronic configuration ns^2np^2, a half-filled ns^2np^6 shell. Hydrogen does have virtually the same electronegativity as carbon, a member of group IVA. But again, the differences between hydrogen and carbon are greater than the similarities. Hydrogen cannot really be assigned to any chemical family.

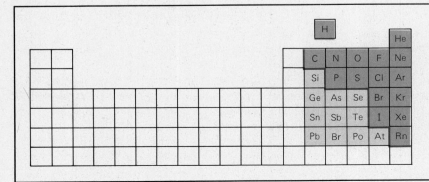

FIGURE 9.1
The nonmetals are grouped at the right side and the top of the periodic table, in the colored boxes. The semimetallic elements and the metallic elements which are found in the same columns of the periodic table are in the gray boxes.

A photograph of the sun showing solar flares (white swirls at center) and solar prominences (narrow black lines). The sun consists mostly of hydrogen, and the prominences are tongues of burning hydrogen shooting out from the surface of the sun. (*Wide World*)

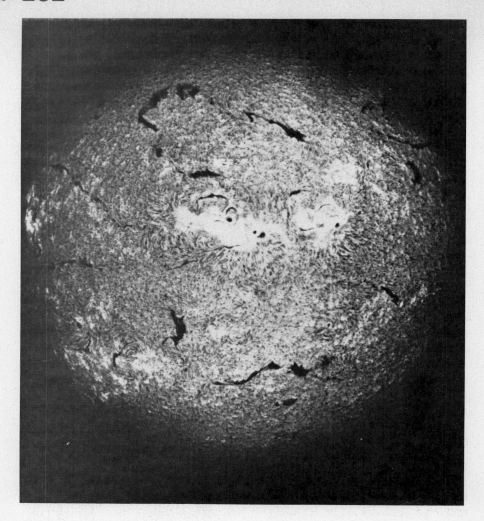

Discovery and Occurrence

While observations of "inflammable air" go back to the sixteenth century, hydrogen was first isolated in 1766 by Henry Cavendish (1731–1810). Hydrogen is by far the most abundant element in the universe. Many stars consist mainly of hydrogen. The fusion process by which hydrogen is converted to helium is the major source of energy for stars, and is the first step in the series of fusion reactions that leads to the formation of the heavier elements (Chapter 20). While hydrogen is less common on earth than it is in stars, it is found here in relative abundance. Fifteen percent of all the atoms in the atmosphere, the oceans, and the top kilometer of the earth's crust are hydrogen atoms. However, because hydrogen is the lightest element, it is only 1% of the oceans, the atmosphere, and the crust by mass. Most of the earth's hydrogen is combined in water and minerals. Because hydrogen gas, H_2, has such a low density, it escapes easily from the earth's gravitational pull and so is only a minor constituent of the atmosphere.

In its pure form, hydrogen exists over a wide range of conditions as H_2, a colorless, odorless, tasteless gas. Its critical temperature is 33 K. No amount of pressure will liquefy H_2 when it is above this temperature. At atmospheric pressure, H_2 becomes a liquid if the temperature is lowered to 20 K. It freezes at 14 K. The H—H bond has a large bond energy, 436 kJ/mol, so H_2 dissociates into H atoms only at extremely high temperatures. Even at 2000 K, less than 0.1% of H_2 is dissociated.

Preparation

Hydrogen is widely used commercially and is available in large quantities. For industrial purposes, a major source of H_2 is methane, CH_4, from natural gas. Another important source of hydrogen is the water-gas reaction, the reaction of red-hot coke with steam:

$$H_2O(g) + C(s) \longrightarrow CO(g) + H_2(g)$$

For laboratory needs, hydrogen can be bought in small quantities, or it can be made by several methods. The most common method is the reaction of a metal, such as zinc or iron, with a strong acid such as hydrochloric acid:

$$Zn(s) + 2H^+ \longrightarrow Zn^{2+} + H_2(g)$$

Most hydrogen is produced as a by-product of petroleum refining. Extra-pure hydrogen is prepared by the electrolysis of brine solutions:

$$H_2O(l) \longrightarrow H_2(g) + \tfrac{1}{2}O_2(g)$$

Most industrial hydrogen is used as a raw material for the synthesis of ammonia, which has many uses. Hydrogen itself is used to hydrogenate vegetable oils, a process that changes liquid oils to solid fats for use in margarine. Hydrogen is also used in the hydrogenation of gasoline, in the manufacture of hydrogen chloride, and as a fuel in oxyhydrogen torches and in jet engines. The fact that hydrogen must be handled with care because mixtures of hydrogen and air can explode in the presence of a spark is well known. The most famous of these explosions occurred at Lakehurst, N.J., on May 6, 1937, destroying the zeppelin *Hindenburg* and effectively ending the era of the lighter-than-air ship.

Chemical Properties

Because the H—H bond is so strong, H_2 is relatively unreactive at room temperature. It does combine vigorously with fluorine, even at low temperatures, but hydrogen does not react with other halogens, with oxygen, or with the other nonmetals under ordinary conditions. In general, H_2 must be activated to participate in chemical reactions. Hydrogen can be activated by heating, by being adsorbed on the surface of a solid (usually a metal), or by treatment with a reactive chemical species.

The reaction of H_2 and Cl_2 is typical. At room temperature, mixtures of these two gases are quite stable. When the temperature is raised to 650 K, a reaction takes place rapidly:

$$H_2(g) + Cl_2(g) \longrightarrow 2HCl(g) \qquad \Delta H^\circ = -184 \text{ kJ}$$

Mixtures of H_2 and Cl_2 will also react rapidly—indeed, explosively—at room temperature under powerful illumination. The energy from a bright light acts in much the same way as thermal energy. Both start the reaction by causing the dissociation of some Cl_2:

$$Cl_2(g) \rightleftharpoons 2Cl(g)$$

When the mixture of $H_2(g)$ and $Cl_2(g)$ is heated, the Cl_2 dissociates, rather than the H_2, because the Cl—Cl bond is weaker than the H—H bond. When the mixture is illuminated, the Cl_2 dissociates because it absorbs light and H_2 does not. We can tell that hydrogen does not absorb radiation in the visible part of the spectrum because H_2 is colorless. The greenish yellow color of Cl_2 shows that it absorbs some wavelengths of visible light. The Cl atoms that form from the dissociation of Cl_2 are very reactive, and the reaction:

$$Cl(g) + H_2(g) \longrightarrow HCl(g) + H(g) \qquad \Delta H^\circ = 4.6 \text{ kJ}$$

occurs rapidly. The H atoms formed in this step are also very reactive, and the

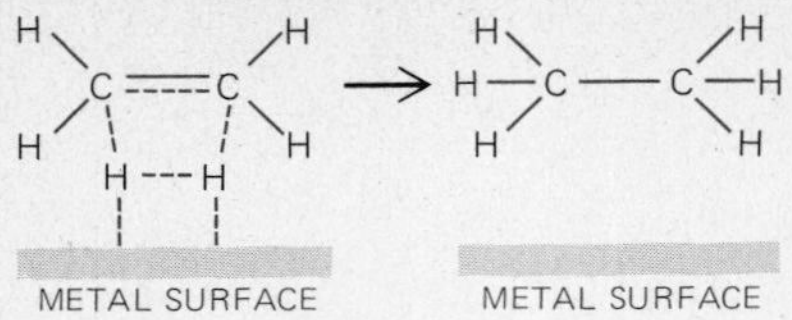

FIGURE 9.2
The hydrogenation of an unsaturated compound can occur under relatively mild conditions in the presence of a powdered metal, such as palladium or platinum.

reaction:

$$H(g) + Cl_2(g) \longrightarrow HCl(g) + Cl(g) \qquad \Delta H° = -188 \text{ kJ}$$

takes place. These two steps, taken together, are called a *chain reaction,* because the Cl atoms that are a product of the second step immediately participate in the first reaction, which forms more H atoms that take part in the second reaction, and so on. The first step is very slightly endothermic because the H—Cl bond is slightly weaker than the H—H bond. The second step is exothermic because the H—Cl bond is stronger than the Cl—Cl bond.

Hydrogenation, the addition of hydrogen atoms to a molecule, is an important reaction. In the manufacture of margarine, for example, vegetable oils, which are liquid at room temperature, are converted to fats, which are solid at room temperature, by hydrogenation. The simplest reaction of this sort is the hydrogenation of ethylene to ethane:

$$C_2H_4(g) + H_2(g) \longrightarrow C_2H_6(g) \qquad \Delta H° = -137 \text{ kJ}$$

This and other hydrogenation reactions can be carried out at relatively low temperatures and pressures in the presence of metals such as platinum or palladium. The H_2 molecule adheres to the surface of the metal, and the H—H bond is weakened as a result. The molecule that is to be hydrogenated approaches the surface of the metal to which the activated hydrogen adheres, and two or more hydrogen atoms are added to the molecule as shown in Figure 9.2. In hydrogenation, the hydrogen atoms are added to molecules in which multiple bonds exist. Any compound with a multiple bond is said to be *unsaturated*. A compound with no multiple bonds is said to be *saturated*. Hydrogenation produces saturated compounds from unsaturated compounds.

Hydrides

A binary compound of hydrogen and another element is called a hydride. In Section 6.9, page 180, we mentioned that almost all elements form hydrides, and we discussed some periodic trends in the behavior of hydrides. Hydrides include some of the most important compounds in nature. We shall devote an entire chapter to water, the most common and the most important of the hydrides (Chapter 11). Another chapter will discuss the hydrocarbons, hydrides of carbon and hydrogen, a most important class of organic compounds (Chapter 20). Indeed, virtually every organic compound includes both carbon and hydrogen. In Chapter 15, we shall discuss the many hydrides that are acids. In this section, we shall discuss some other hydrides.

One interesting group of hydrides is the boranes, hydrides of boron. More than 20 boranes have been identified. Many of them have unusual structures and properties. The simplest borane that exists as a stable compound is diborane, B_2H_6. The compound BH_3 cannot be isolated and seems to have only a brief existence, even under extraordinary conditions. More complex boranes include B_5H_9, $B_{10}H_{14}$, and $B_{20}H_{16}$.

Formulating the structure of diborane, shown in Figure 9.3, is not easy. The compound cannot be represented by using only ordinary covalent bonds. By conventional rules, the structure of B_2H_6 requires a minimum of seven bonds (six B—H bonds and one B—B bond) and therefore a minimum of 14 valence electrons. But the compound has only 12 valence electrons, so a different formulation is needed. An unusual kind of bond is needed to represent diborane. The ordinary covalent bond has two electrons shared by two atoms. Some of the bonds in diborane consist of two electrons that are shared by three atoms. These bonds are called *three-center bonds*. Diborane has two three-center bonds. In each of these bonds, an electron pair is shared by two boron atoms and one hydrogen atom, as shown in Figure 9.3. The hydrogen atom in this bond is called a bridge atom. The other hydrogen atoms in diborane are called terminal hydrogen atoms.

The bonding in diborane can be explained in this way: The boron atom is an sp^2 hybrid. Its bonds to the two terminal hydrogen atoms are the result of σ overlap of the $2sp^2$ orbital of boron with the $1s$ orbitals of the hydrogen atoms. This bonding is consistent with the fact that the terminal H—B—H angles are close to 120°. The third $2sp^2$ orbital of boron combines with the atom's remaining unhybridized $2p$ orbital to give two new equivalent hybrid orbitals, each of which overlaps with the $1s$ orbital of a bridge hydrogen atom. This further hybridization helps to account for the observed equivalence of the two bridge hydrogen atoms. In some other boranes, we find three-center bonds in which one electron pair is shared by three boron atoms.

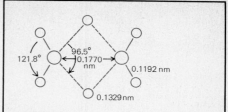

FIGURE 9.3
The structure of diborane. There are two three-center bonds, in each of which two boron atoms and one hydrogen atom share an electron pair.

The chemistry of the boranes has been studied extensively, and not only because of the unusual structure of these compounds. There are some important practical applications for boranes. They are valuable reagents for the synthesis of many organic compounds, which are otherwise difficult to prepare. Boranes also have high heats of combustion; they undergo combustion reactions that are similar to those of hydrocarbons but release more heat. The heat of combustion is the quantity of heat liberated when one mole of a substance undergoes the combustion reaction with oxygen. For ethane, C_2H_6, the heat of combustion is 970 kJ/mol. For diborane, the heat of combustion is 2020 kJ/mol. The reaction is:

$$B_2H_6(g) + 3O_2(g) \longrightarrow B_2O_3(s) + 3H_2O(l)$$

Because they release so much heat, boranes have been considered for use as high-energy rocket fuels. Unfortunately, they cost too much to be used as rocket propellants.

9.2 CARBON

Periodic Classification

Carbon is the lightest element in group IVA of the periodic table, and the only nonmetal in the group. Silicon and germanium, the next two elements in the group, are best classified as semimetals. Tin and lead, the two heaviest elements in the group, are metals.

The valence electronic configuration of carbon is $2s^2 2p^2$, which means that the carbon atom requires more electrons to achieve the noble gas electronic configuration than any other nonmetal. To achieve the electronic configuration of neon, a carbon atom has to gain four electrons by electron sharing. There are a small number of compounds, such as Be_2C and Al_4C_3, members of the group called carbides, in which carbon does seem to exist as the C^{4-} anion. In theory, a carbon atom could achieve the electronic configuration of helium by losing four electrons, but such a loss is never observed under ordinary conditions.

Carbon forms an unusually large number of compounds, in large measure because a carbon atom forms strong covalent bonds with other carbon atoms as well as with most of the common nonmetals. Carbon atoms easily form long chains, a process called *catenation*. Often, there are many hydrogen atoms attached to the carbon atoms of such a chain. This ability to form stable long chains of identical atoms even when it is combined with other elements is unique to carbon. It accounts for the importance of carbon in the chemistry of living things. The study of exobiology—life on other planets—generally starts with the assumption that life anywhere in the universe must be based on carbon.

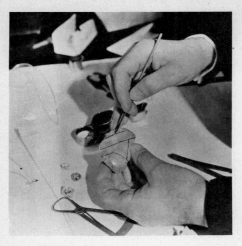

A large diamond is marked for cutting. After such a gem diamond is cleaved, diamond saws and discs coated with diamond dust will be used to create the facets that bring out its brilliance. (*American Museum of Natural History*)

Discovery and Occurrence

As charcoal or soot, carbon in relatively pure form was known to primitive man. But it was not until the late eighteenth century that carbon was recognized as an element. The discovery was made in France, where the word *carbone* was introduced as the name of the element, to distinguish it from *charbon,* the French word for charcoal.

Carbon is the third most abundant element in the universe. It plays an important part in the energy-producing fusion reactions of the stars. On earth, carbon is much less abundant. The earth's crust contains about 0.027% carbon by mass, mainly in the form of carbonates. The most common of these are $CaCO_3$ and $FeCO_3$. The former, calcium carbonate, occurs in many different forms, including limestone, chalk, marble, calcite, and seashells. A small fraction of the earth's atmosphere, about 0.013% by mass, consists of carbon, mainly in the form of carbon dioxide. It is this relatively miniscule percentage of carbon that provides most of the raw material for life on earth. Carbon is constantly being recycled through living organisms, which use it for the complex chemicals of life. It is worth mentioning here that the oil, coal, and natural gas that provide almost all the energy on which our modern technological society is based consist of the fossilized remains of ancient living organisms. These fossil fuels contain only a small fraction of the carbon in the earth's crust, but it is an important part to mankind.

Allotropes of Carbon

Different forms of the same element, especially in the same phase, are called **allotropes.** The two most prominent solid allotropes of carbon are graphite and diamond. Amorphous carbon, which includes such substances as coke, wood charcoal, and carbon black, consists of aggregates of microscopically small graphite particles and various surface impurities.

The differences between graphite and diamond could not be more striking. Graphite is black and so soft that it can be used as a lubricant. Pure diamonds are extremely hard, colorless crystals. Yet both consist only of carbon atoms. Both allotropes have been known to mankind since ancient times, but the reason for the startling difference in their properties is a modern discovery.

Graphite gets its name from the Greek word *graphein,* meaning ''to write.'' This Greek word indicates that the ancients used this soft, black, almost greasy substance in the same way we do today, as the ''lead'' in lead pencils. Graphite has other useful properties. It has great resistance to heat and thus is used in furnaces and in similar applications. It conducts electricity (unlike the other nonmetals) and thus is used in electrodes, dynamos, and microphones. Graphite is manufactured by heating an amorphous form of carbon, such as coke, in an electric furnace to temperatures of 2800 K, driving off impurities and leaving only pure graphite.

Diamonds were also known to the ancients (although the ''diamonds'' mentioned in the Old Testament probably were corundum, a hard, colorless form of aluminum oxide). The name is derived from the Greek word *adamas,* meaning ''invincible.'' Most diamonds are found in South Africa, where the largest diamond known thus far, the Cullinan diamond, weighing 570 g—2850 carats—was found in 1905. Aside from their value as gems, diamonds have important industrial uses based on their hardness.

A great deal of effort has gone into developing methods for making synthetic diamonds, and several techniques now are available. These methods generally use temperatures above 1200 K and pressures greater than 70 000 atm. Even under these conditions, diamonds of gem quality cannot be synthesized. Indeed, the way in which diamonds are created naturally remains a mystery.

Figure 9.4 shows the structure of diamond—actually a portion of a diamond

FIGURE 9.4
The structure of diamond, a giant molecule made up of carbon atoms in a regular tetrahedral array.

molecule. A diamond can be thought of as a single giant molecule made up of nothing but carbon atoms, each atom being bonded to four other carbon atoms that are at the corners of a regular tetrahedron. In this crystal structure, all the C—C bonds are identical. It is the strength and number of the C—C bonds that gives diamond its hardness. The strength of the C—C bond in diamond is comparable to the strength of a C—C bond in an ordinary stable molecule such as ethane. But the C—C bonds in diamond extend throughout the crystal lattice. The carbon atoms in the diamond crystal are held together by covalent bonds. These covalent bonds represent much stronger forces of attraction than those which hold the units of structure together in most other crystals. To deform a solid, it is necessary to overcome some of the forces holding its units of structure together. These forces are more difficult to overcome in diamond than in almost any other substance.

Despite the great strength of the forces holding carbon atoms together in its crystal lattice, diamond is the less stable allotrope of carbon under ordinary conditions. The standard state of carbon at 298 K is graphite. The standard heat of formation (ΔH_f°) $C(s)_{graphite} \rightarrow C(s)_{diamond}$ of diamond is a positive quantity, 1.9 kJ/mol. Diamond converts spontaneously to graphite. But under ordinary conditions, the conversion takes place at a rate that is so slow as to be negligible. The rate of conversion is greater at high temperatures. At 1300 K, diamond becomes graphite at an appreciable rate.

Diamond forms naturally because it is the more stable allotrope of carbon at pressures substantially higher than 1 atm. It might seem that diamonds could be synthesized by applying high pressure to graphite. But high pressure alone is not enough. High temperature is also needed so that the diamond will form from the graphite at a reasonable rate. However, as the temperature increases, the relative stability of diamond decreases. The decrease in stability of diamond relative to graphite as the temperature increases helps explain why it is so difficult to prepare synthetic diamonds. Extremely high pressure is needed so that diamond will be the more stable form of carbon at the elevated temperature required for the conversion to occur in a reasonable time.

A diamond crystal is a single large molecule. Graphite, by comparison, consists of a large number of flat molecules stacked rather loosely on one another. As Figure 9.5 shows, each carbon atom in a flat graphite molecule is bonded to three other carbon atoms that lie at the corners of an equilateral triangle. All the bonds are equivalent, and the bonding between the carbon atoms in a molecule is quite strong. If graphite absorbs water, the flat molecules themselves can be easily moved or separated. The adsorbed film enables the flat molecules of graphite to slide over each other, giving graphite its softness and making it useful as a lubricant. In the absence of water, graphite becomes abrasive. Graphite has a structure that is much more open than the structure of diamond, which means that graphite has a lower density than diamond. High pressure favors the diamond structure. As molecules are squeezed under high pressure, the atoms tend to assume the more compact, denser form.

The greater stability of graphite under ordinary conditions can be explained by thermochemical reasoning. All the bonds in graphite are found to be intermediate in length between a C—C single bond and a C=C double bond. This intermediate bond length suggests that graphite is a resonance hybrid of contributing structures with alternating single and double bonds. Figure 9.6 shows two contributing structures for a portion of the graphite molecule.

If we picture the process by which a single carbon atom becomes part of a graphite molecule, we can see that this atom forms two single C—C bonds and one C=C double bond. We know the bond energies of these bonds, so we can estimate that $2(346 \text{ kJ/mol}) + (610 \text{ kJ/mol}) = 1300 \text{ kJ/mol}$ is liberated by this process. If we picture the process by which a single carbon atom becomes part of a diamond molecule, we can see that this atom forms four single C—C

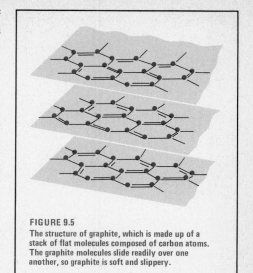

FIGURE 9.5
The structure of graphite, which is made up of a stack of flat molecules composed of carbon atoms. The graphite molecules slide readily over one another, so graphite is soft and slippery.

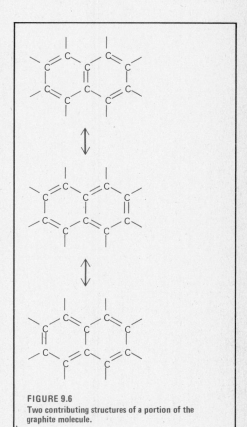

FIGURE 9.6
Two contributing structures of a portion of the graphite molecule.

TABLE 9.1
**Standard Heats of Formation of Compounds
of Carbon at 298 K**

Compound	Name	ΔH_f° (kJ/mol)
C(s)	graphite	0
C(s)	diamond	1.9
C(g)		718
C_2(g)		808
CO(g)	carbon monoxide	−110.5
CO_2(g)	carbon dioxide	−393.5
CH_4(g)	methane	−75
C_2H_2(g)	acetylene	227
C_2H_4(g)	ethylene	52
C_2H_6(g)	ethane	−85
CCl_4(g)	carbon tetrachloride	−107
$CHCl_3$(g)	chloroform	−100
CBr_4(g)	carbon tetrabromide	50
CS_2(g)	carbon disulfide	115
CH_2O(g)	formaldehyde	−116

bonds, liberating 4(346 kJ/mol) = 1380 kJ/mol of heat. This result is not what we expect. Since graphite is more stable than diamond, more heat should be given off in its formation. Arguing from bond energy alone, diamond appears to be more stable than graphite by $\frac{1}{2}$(1380 kJ/mol − 1300 kJ/mol) = 40 kJ/mol. (The difference in bond energy is divided by two because each bond in a complete molecule includes two atoms, rather than the single atom of our hypothetical case.) The reason graphite is found to be more stable than diamond by 1.9 kJ/mol is that graphite has resonance energy, which actually amounts to about 42.4 kJ/mol.

While the carbon atom forms strong covalent bonds with itself and with the other nonmetals, many carbon compounds still have relatively small negative heats of formation. As Table 9.1 shows, some carbon compounds even have positive heats of formation. The fact that graphite has such strong bonds explains why carbon compounds often are not much more stable than the elements from which they are formed. Example 8.11 (page 257) gave an illustration of this effect for a specific reaction, the formation of acetylene.

Carbon Dioxide and the Carbon Cycle

Life on earth is based on carbon. The major source of carbon for the compounds that make life on earth possible is the carbon dioxide in the atmosphere, and the group of reactions by which CO_2 is removed from and returned to the atmosphere is called the *carbon cycle*. Figure 9.7 shows some of the major features of the carbon cycle.

Photosynthesis in plants removes carbon dioxide from the atmosphere, in a process that converts solar energy to chemical energy. In photosynthesis, sunlight is used to produce glucose and other organic compounds from carbon dioxide and water. The plants that carry on photosynthesis and the animals that eat those plants use the chemical energy produced by photosynthesis, converting the organic substances back to carbon dioxide in the process. When organisms die, their carbon compounds are returned to the atmosphere in the form of CO_2 as part of the process called decay. Some dead plants are converted to fossil fuels. When we burn these fuels, CO_2 is added to the atmosphere.

Fossil fuel combustion has reached significant levels in recent years. The atmosphere contains a relatively small amount of carbon dioxide. Only 0.0325% of the atmosphere—325 parts per million by volume (ppm)—is CO_2. But in recent

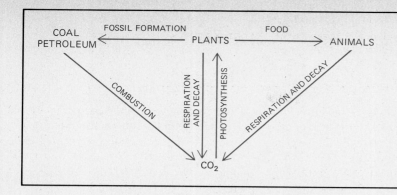

FIGURE 9.7
The carbon cycle. Plants obtain CO_2 from the atmosphere for photosynthesis. The carbon is returned to the atmosphere by respiration, decay, or combustion of organic material.

decades, the concentration of CO_2 in the atmosphere has been going up by about 1 ppm per year. The increase appears to be caused primarily by the burning of fossil fuels. The air above the cities where large amounts of fossil fuels are burned tends to have unusually high concentrations of CO_2. Concentrations as high as 1000 ppm have been measured above New York City, for example. Such concentrations are not toxic. But on a global scale, the rising CO_2 content of the atmosphere could have serious long-term consequences. If the century-long pattern of rising consumption of fossil fuels continues, the amount of CO_2 in the atmosphere could double by early in the twenty-first century. Climatologists calculate that such an increase could raise the global temperature by an average of between 0.5 K and 1 K. For the first time, human activity will be on a scale great enough to change the climate, with unknown consequences.

Chemically, carbon dioxide is quite stable, with $\Delta H_f^\circ = -393.51$ kJ/mol. It is formed by the oxidation of carbon or carbon-containing compounds in the presence of excess O_2:

$$C(s) + O_2(g) \longrightarrow CO_2(g) \qquad \Delta H^\circ = -393.5 \text{ kJ/mol}$$
$$CH_4(g) + 2O_2(g) \longrightarrow CO_2(g) + 2H_2O(l) \qquad \Delta H^\circ = -890.3 \text{ kJ/mol}$$

One reason why combustion reactions are exothermic is that the formation of a molecule as stable as carbon dioxide is accompanied by the release of a great deal of heat.

Carbon dioxide can also be formed by heating metal carbonates. A typical reaction is:

$$CaCO_3(s) \longrightarrow CO_2(g) + CaO(s)$$

Carbon dioxide dissolves in water to form a weakly acidic solution that contains carbonic acid (Section 15.6), and thus has a slightly sour taste. Solid carbon dioxide is familiar to us as dry ice, a common refrigerant.

The carbon dioxide molecule has a linear structure, conveniently represented by $\ddot{O}{=}C{=}\ddot{O}$. The $O{=}C{=}O$ bond angle is $180°$. However, the $C{=}O$ bond in CO_2 is both shorter and stronger than the $C{=}O$ bond in other compounds. A possible explanation is that CO_2 is really a resonance hybrid of the structure above and two other contributing structures:

$$^-:\ddot{O}{-}C{\equiv}O:^+ \longleftrightarrow ^+:O{\equiv}C{-}\ddot{O}:^-$$

Carbon Monoxide

Carbon monoxide is a colorless, odorless, tasteless gas that forms when carbon or carbon-containing compounds are burned with an insufficiency of oxygen:

$$C(s) + \tfrac{1}{2}O_2(g) \longrightarrow CO(g) \qquad \Delta H^\circ = -110.5 \text{ kJ/mol}$$

Carbon monoxide is unstable relative to carbon dioxide and is readily converted to

CO_2 in an exothermic reaction with O_2:

$$CO(g) + \tfrac{1}{2}O_2(g) \longrightarrow CO_2(g) \qquad \Delta H° = -283.0 \text{ kJ/mol}$$

It also can react with itself to form CO_2:

$$2CO(g) \longrightarrow CO_2(g) + C(s) \qquad \Delta H° = -172.5 \text{ kJ/mol}$$

While this reaction is favorable, it occurs rapidly only at high temperatures. The reaction becomes important above 750 K. Thus, even though carbon monoxide can convert to the very stable CO_2, it is quite a stable compound, with $\Delta H_f° = -110.5$ kJ/mol.

Carbon monoxide has the same number of electrons as N_2, but the carbon-oxygen bond in CO is even stronger than the nitrogen-nitrogen triple bond in N_2. In fact, the CO bond is the strongest bond known, with a bond energy of 1070 kJ/mol. The great strength of this bond is due in part to resonance between two important contributing structures: $:C{=}\ddot{O}: \longleftrightarrow {}^-:C{\equiv}O:^+$. This picture of CO is consistent with the experimental observation that there is very little net electric charge separation between the atoms of CO. Because the O atom is much more electronegative than the C atom, we expect to find a relative negative charge on the O atom and a relative positive charge on the C atom. This charge distribution is essentially canceled by the contribution of the structure with the formal charges. The CO molecule is thus relatively nonpolar.

Carbon monoxide is an important industrial raw material, prepared in large quantities by the water-gas reaction, the reaction between coke and steam. It reacts with many different substances. Reactions of CO with such substances as hydrogen, water, or simple organic compounds are used to manufacture many more complex organic compounds. Because CO has a nonbonding valence electron pair on the carbon atom, it can react with transition metals to form a useful class of compounds called metal carbonyls.

The carbon monoxide content in the atmosphere under natural conditions is quite small, from 0.1 to 0.2 ppm. About 70% of this CO is produced by human activities, primarily in the industrial areas of North America, Europe, and Japan. Most CO is produced in combustion processes where the ratio of fuel to oxygen is too high; the internal combustion engine is a major source of carbon monoxide. The CO content of the atmosphere in large cities with many automobiles often is as high as 10–20 ppm, concentrations that are a matter of medical concern. In warm-blooded animals, red blood cells transport oxygen in the body. The oxygen-carrying component of red blood cells is a molecule called hemoglobin, which forms a complex with O_2 and carries it to the cells. Carbon monoxide interferes with this process, since its affinity for hemoglobin is 300 times greater than that of oxygen. A CO molecule holds so tightly to a hemoglobin molecule that the hemoglobin molecule cannot perform its normal function of carrying O_2. Exposure to high concentrations of CO can be fatal. The effects of long-term exposure to lower levels of CO are not known, but they are believed to be damaging. For example, one reason why cigarette smoking is believed to increase the risk of heart disease is the high concentration of carbon monoxide in cigarette smoke. The reduction of carbon monoxide emissions from internal combustion engines is a leading goal of pollution control agencies.

9.3 NITROGEN

Periodic Classification

Nitrogen is the lightest member of group VA of the periodic table. Nitrogen and phosphorus are the only nonmetals in this group. Arsenic and antimony are semimetals, and bismuth, the heaviest member of the group, is a metal. The electronic configuration of nitrogen is $2s^2 2p^3$. A nitrogen atom thus needs three electrons to achieve the noble gas electronic configuration. Nitrogen, like carbon, forms primarily covalent bonds. An exception to this rule is a small group of compounds called nitrides, which seem to have the N^{3-} ion, but these are rare. There are no long chains of nitrogen atoms similar to the long chains of carbon atoms, because the N—N single bonds that would exist in such chains are relatively weak. There are only a few types of compounds in which two or more nitrogen atoms are bonded to each other.

Many compounds of nitrogen are known. Most of them are classified as organic compounds, because their nitrogen atoms are attached to a framework of carbon atoms. Inorganic nitrogen compounds are in the minority.

Discovery and Occurrence

Although nitrogen compounds were known to the ancients, nitrogen was not recognized to be an element until 1772. The identification of nitrogen as an element usually is credited to a physician, Dr. Daniel Rutherford (1749–1819), an uncle of Sir Walter Scott, the famous novelist. Nitrogen, which exists as N_2 under ordinary conditions, is a colorless, odorless, tasteless gas. The earth's atmosphere is 78.09% nitrogen by volume, and the atmosphere also contains trace amounts of various compounds of nitrogen. The earth's crust and the oceans contain many compounds of nitrogen, mainly nitrate salts such as KNO_3 and $NaNO_3$. Nitrogen is a component of many compounds that are essential to life, including proteins and nucleic acids.

The boiling point of N_2 is 77.36 K and the critical temperature of N_2 is 126.6 K. The Lewis structure of N_2 is $:N\equiv N:$. The bond energy of the triple bond is 945 kJ/mol. This triple bond is so strong that N_2 does not dissociate into nitrogen atoms except under the most extreme chemical conditions. Single nitrogen atoms usually play no role in the chemistry of nitrogen.

Nitrogen Fixation and the Nitrogen Cycle

Living organisms seeking nitrogen for proteins (which average 16% nitrogen), nucleic acids, and other organic compounds can be compared to a thirsty mariner adrift in a salty sea. There is an abundance of N_2 in the atmosphere, but most living organisms can make no use of N_2. To be biologically usable, N_2 must be converted into compounds of nitrogen. This conversion is called *nitrogen fixation*. A variety of microorganisms can carry out the essential first step in nitrogen fixation by converting N_2 to NH_3 or an equivalent substance. Soil bacteria, yeasts, and blue-green algae can perform this transformation by themselves. Another important source of biologically useful nitrogen compounds is the cooperative effort of plants and bacteria, a process called symbiotic nitrogen fixation. The bacteria that take part in this process are found in the root nodules of such plants as beans, peas, alfalfa, and clover, the so-called leguminous plants. Other plants rely on relatively simple inorganic compounds of nitrogen in the soil for the raw material out of which more complex nitrogen compounds are made. Without nitrogen fixation processes, however, the supply of nitrogen compounds in the soil would be depleted quickly and life would cease.

The growing demands of a hungry human race have made the natural replenishment of nitrogen in the soil inadequate for modern agriculture. Farmers add simple compounds of nitrogen to the soil in the form of fertilizers. The first fertilizers were animal wastes. Later, inorganic nitrates such as Chile saltpeter, $NaNO_3$, were mined for use as fertilizer. Now farmers rely on industrial processes for nitrogen fixation. The most important of these is the Haber process, developed originally to produce munitions in Germany during World War I by Fritz Haber (1868–1934). In the Haber process, N_2 and H_2 react at high temperature and

NITROGEN FIXATION: IMPROVING ON NATURE

The revolutionary new technique of genetic manipulation that has resulted from recombinant DNA research has raised the possibility that food production could be increased enormously by giving crops such as wheat and corn the ability to do their own nitrogen fixation. As this is written, the first tentative steps toward such a revolution have been taken, although great obstacles remain to be overcome.

The technique for producing recombinant DNA, developed in the early 1970s, makes it possible to transfer a gene for a specific trait from one organism to another. The potential hazards of this kind of genetic engineering have led to strict controls on recombinant DNA research. But the research is continuing because of the great potential benefits from application of genetic engineering.

Biological nitrogen fixation is still far more important than human fertilizer production. It is estimated that microorganisms fix some 160 billion kilograms of nitrogen each year, about four times more than the amount of nitrogen fertilizers produced annually. Because the energy crisis has caused the price of nitrogen fertilizer to increase more than sixfold since 1970, the search for more efficient methods of nitrogen fixation is being pursued actively.

One such method would be to take the *nif* genes that control nitrogen fixation out of microorganisms and put them into plants. Thus far, it has been possible to transfer *nif* genes out of one bacterium, *Klebsiella pneumoniae,* into another one, *Escherechia coli,* a familiar inhabitant of the human intestine and a favorite species for bacterial research. A few laboratories have managed to transfer the *nif* genes to other bacteria and algae.

However, the great leap to the cells of higher plants is yet to be accomplished. The research must do more than merely transplant the genes; it must also assure that the genes function effectively in the new environment. A plant cell is far different from the microorganisms in which the *nif* genes ordinarily function. For one thing, the enzyme called nitrogenase, which plays a central role in nitrogen fixation, is acutely sensitive to oxygen, and must be protected from destruction. Perhaps even more important, simply transferring a gene might accomplish nothing if the cell into which it is transferred lacks the components needed for the gene to function effectively.

Meanwhile, research is also being conducted on better nonbiological methods of nitrogen fixation, methods that would not require the high energy expenditure of the Haber process. Nitrogen fixation at, or near, room temperature would be a revolution almost as great as successful transfer of the *nif* genes. Laboratories in the United States, the Soviet Union, and Europe are trying a number of approaches. Some success has been reported with methods using complexes of transition metals and N_2. However, a practical low-energy system of nitrogen fixation has not yet been achieved. It is estimated that at least a decade of research may be needed to develop such a method.

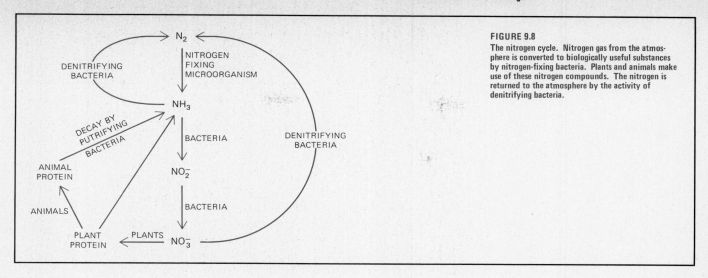

FIGURE 9.8
The nitrogen cycle. Nitrogen gas from the atmosphere is converted to biologically useful substances by nitrogen-fixing bacteria. Plants and animals make use of these nitrogen compounds. The nitrogen is returned to the atmosphere by the activity of denitrifying bacteria.

high pressure to form ammonia, NH_3, which can be used directly for fertilizer or can be converted to other compounds of nitrogen. Since the major source of H_2 for this process is natural gas, which is in increasingly short supply, and since the process uses a great deal of energy, the development of methods for converting nitrogen to ammonia at room temperature and atmospheric pressure is a topic of considerable current interest in chemical research.

The fixation of nitrogen by microbes is just one step in the processes by which nitrogen is made available to living organisms and is eventually returned to the atmosphere in the form of N_2. This group of processes is called the *nitrogen cycle,* and it is outlined in Figure 9.8. In the nitrogen cycle, ammonia usually is converted to nitrate, which moves more freely through the soil. Plants use the nitrate to produce protein. Animals eat the plants and are eaten by other animals in turn. Animal wastes are rich in nitrogen compounds, which are converted back into ammonia by soil bacteria. These bacteria also convert nitrogen compounds into ammonia during the decay of dead animals or plants. Other bacteria then convert ammonia or nitrate back to N_2, completing the cycle.

Compounds of Nitrogen

There are simple inorganic compounds that correspond to every oxidation state of nitrogen between -3 and $+5$. Table 9.2 lists some of these compounds and their standard heats of formation. You will note that ΔH_f° for many of these compounds is positive, suggesting that these compounds probably are unstable relative to the standard states of their constituent elements. These standard heats of formation generally reflect the great strength of the NN triple bond in N_2.

Positive oxidation states for nitrogen are possible only when it combines with oxygen and fluorine, the only two elements clearly more electronegative than nitrogen.

A discussion of the chemistry of inorganic compounds of nitrogen could—and does—fill many volumes. We shall survey some main features of a few of the more important of these compounds.

Ammonia

Ammonia, NH_3, perhaps the most important simple compound of nitrogen, is a colorless gas at room temperature. It has a distinctive, irritating odor. With a boiling point of 239.7 K and a critical temperature of 405.6 K, ammonia can be condensed rather easily. Liquid ammonia is made in large quantities and has

TABLE 9.2
Standard Heats of Formation of Compounds of Nitrogen at 298 K

Oxidation State	Compound	Name	ΔH_f° (kJ/mol)
−3	$NH_3(g)^a$	ammonia	−46
−2	$N_2H_4(l)$	hydrazine	50
−1	$NH_2OH(s)$	hydroxylamine	−107
0	$N_2(g)$	nitrogen	0
+1	$N_2O(g)$	nitrous oxide	82
+2	$NO(g)$	nitric oxide	90
	$N_2F_4(g)$	tetrafluorohydrazine	−7
+3	$N_2O_3(g)$	dinitrogen trioxide	84
	$HNO_2{}^b$	nitrous acid	−119
	$NF_3(g)$	nitrogen trifluoride	−114
+4	$NO_2(g)$	nitrogen dioxide	34
	$N_2O_4(g)$	dinitrogen tetroxide	10
+5	$N_2O_5(s)$	dinitrogen pentoxide	−42
	HNO_3	nitric acid	−207

[a] State symbol gives the state at 298 K.
[b] The absence of a state symbol denotes a $1M$ aqueous solution.

many uses. In the chemical laboratory, ammonia is used as a polar solvent for many reactions for which water is unsuitable. One property that makes ammonia valuable as a solvent is its ability to dissolve substantial quantities of some metals, including the alkali and alkaline-earth metals. Solutions of these metals in liquid ammonia are characterized by a beautiful deep blue color. These solutions have the highest electrical conductivity of any known solutions, apparently because dissolution in ammonia separates the metal and its valence electrons. For example, it is believed that when sodium dissolves in liquid ammonia, the $3s$ electron leaves the Na atom, which becomes the Na^+ ion. The independent, or "solvated," electron apparently becomes loosely associated with a number of NH_3 molecules. Many interesting reactions can be brought about by these solvated electrons. If O_2 is bubbled into the solution, for instance, an electron joins the O_2 molecule to form superoxide ion, O_2^-. The reaction is $O_2(g) + e^- \rightarrow O_2^-$, where e^- represents a solvated electron.

Ammonia is the initial product of both biological and industrial nitrogen fixation. Ammonia itself has a large number of applications, ranging from its use in the manufacture of synthetic fibers to service as a nutriment in fermentation processes to its everyday use in water solution as a household cleaner. Ammonia is used as a fertilizer and is also the starting material for the manufacture of many commercially important nitrogen compounds.

One of the most widely used of these compounds is nitric acid, HNO_3, which is produced by the oxidation of ammonia by what is called the Ostwald process. The first and key step in this process is

$$4NH_3(g) + 5O_2(g) \longrightarrow 4NO(g) + 6H_2O(g) \qquad \Delta H^\circ = -906 \text{ kJ}$$

In this reaction, the ammonia is oxidized. The oxidation state of its nitrogen atom changes from −3 to +2. The reaction is highly exothermic, with a ΔH° of −906 kJ for 4 mol of NH_3. The formation of NO as a product of this reaction is something of a surprise, since the data in Table 9.1 indicate that nitric oxide has the highest positive ΔH_f° of all the oxides of nitrogen, and is thus the least stable of these oxides. As the data in Table 9.1 show, there are several reactions with more favorable enthalpy changes: the oxidation of NH_3 to N_2O in the +1 state, to NO_2 in the +4 state, or, most strikingly, to N_2 in the 0 oxidation state. For this last reaction:

$$4NH_3(g) + 3O_2(g) \longrightarrow 2N_2(g) + 6H_2O(g)$$

$\Delta H^\circ = -1268$ kJ for 4 mol of NH_3.

Liquid ammonia is applied to a farmland before planting. Heavy application of this fertilizer helps American farmers get high yields from their land. (*Conklin, Monkmeyer*)

The oxidation of ammonia takes the desired pathway to NO only under special conditions. In the Ostwald process, a preheated mixture of ammonia and air is passed quickly over a platinum-rhodium gauze at temperatures above 925 K. Some of the platinum is lost because the metal has a high vapor pressure at these temperatures, so the process is rather expensive. The nitric oxide formed in this step is converted to nitric acid in subsequent steps of the Ostwald process.

Because ammonia has a nonbonding valence electron pair, NH_3 can readily accept a proton to form the ammonium cation:

$$NH_3 + H^+ \rightleftharpoons NH_4^+$$

One of the many ammonium salts is ammonium nitrate, NH_4NO_3. Ammonium nitrate ranks in the top ten products of the American chemical industry. It is used chiefly as a fertilizer, but it also is an explosive. High temperatures, or the shock wave from a detonator, can start a reaction in which ammonium nitrate is converted very rapidly to $N_2(g)$, with an enormous release of heat. It is the formation of the extremely stable triple bond in N_2 that causes the great release of heat and gives ammonium nitrate and other nitrogen compounds explosive properties. One of the worst disasters in American history occurred in 1947, when an explosion of NH_4NO_3 killed 576 people in Texas City, Texas.

The decomposition of ammonium nitrate can be controlled by careful heating. The product is nitrous oxide:

$$NH_4NO_3(s) \longrightarrow N_2O(g) + 2H_2O(g)$$

If ammonium nitrate is heated even more carefully, it can revert to ammonia and pure nitric acid:

$$NH_4NO_3(s) \longrightarrow NH_3(g) + HNO_3(g)$$

Nitrous Oxide

Nitrous oxide, N_2O, is best known as "laughing gas," from its ability to induce hysterical excitement and laughter when inhaled. Mixed with oxygen, nitrous oxide is used as an anesthetic in dentistry and obstetrics. Because it is highly soluble in cream, nitrous oxide is also used as a propellant in canned whipped cream imitations. As was mentioned earlier, it is manufactured by the controlled decomposition of ammonium nitrate.

Nitrous oxide is a resonance hybrid of two contributing Lewis structures, each of which has formal charges:

$$:N\equiv\overset{+}{N}-\overset{..}{\underset{..}{O}}:^- \longleftrightarrow {}^-:\overset{..}{N}=\overset{+}{N}=\overset{..}{\underset{..}{O}}$$

The molecule is linear. The length of the NO bond is intermediate between that of a single and a double bond, while the length of the NN bond is intermediate between that of a double and a triple bond. While nitrous oxide has a positive heat of formation and is relatively unstable, it is also relatively unreactive, and so is easily transported and is widely available.

Nitric Oxide

Nitric oxide, NO, is the simplest stable molecule known that has an odd number of electrons. The length of the NO bond is intermediate between the length of a double bond and that of a triple bond. As in CO, there is not much charge separation between the atoms in NO; the molecule is relatively nonpolar. This molecule, like O_2, is one for which the molecular orbital method (Section 7.6, page 218) is more successful than the valence bond description. It is possible to write contributing Lewis structures for NO, such as $\cdot\ddot{N}{=}\ddot{O} \longleftrightarrow :\ddot{N}{=}\ddot{O}\cdot$, but this approach does not explain the properties of NO very well.

Nitric oxide is a colorless gas at room temperature and a blue liquid or solid at low temperatures. It is manufactured by the oxidation of ammonia by the Ostwald process. Its most common reaction is further oxidation to NO_2:

$$2NO(g) + O_2(g) \longrightarrow 2NO_2(g) \qquad \Delta H° = -112 \text{ kJ}$$

Nitric oxide has one more electron than N_2 or CO, both of which are very stable molecules. It is not surprising, therefore, that NO readily gives up an electron to form the nitrosyl cation, NO^+. The nitrosyl cation forms bonds to metals similar to those formed by carbon monoxide. Using an orbital overlap description of nitric oxide, we would say that the odd electron is in an antibonding orbital (Section 7.5, page 212). Since NO^+ is formed by the loss of this electron, the theory predicts that the NO bond in the nitrosyl cation will be stronger than the NO bond in nitric oxide. This prediction is consistent with experimental observations.

At 298 K, NO does not form from N_2 and O_2. But at much higher temperatures, the reaction

$$N_2(g) + O_2(g) \longrightarrow 2NO(g) \qquad \Delta H° = 180 \text{ kJ}$$

proceeds rather easily. If the hot mixture of these three gases is cooled rapidly to room temperature, the nitric oxide remains. This process occurs in automobile engines: Nitric oxide forms from the hot air in the cylinders and the NO cools rapidly as it is exhausted. Automobiles are sources of the nitric oxide that is a major component of photochemical smog.

Nitrogen Dioxide

Nitrogen dioxide, NO_2, cannot be discussed without mentioning dinitrogen tetroxide, N_2O_4. Whenever NO_2 is present, we find some N_2O_4. The relative amount of N_2O_4 decreases as the temperature rises. The relevant reaction is

$$2NO_2(g) \rightleftharpoons N_2O_4(g) \qquad \Delta H° = -58 \text{ kJ}$$

NO_2 is reddish brown and N_2O_4 is colorless, so the composition of a mixture of the two gases can be determined by measurements of color intensity. At low temperatures, both oxides exist together as liquids, but N_2O_4 alone is found in the solid.

At 298 K, $\Delta H°$ for the formation of N_2O_4 from $2NO_2$ is -58 kJ/mol. The release of energy is due to the formation of the NN bond in N_2O_4. Like NO, NO_2 has an odd number of electrons, and the NN bond in N_2O_4 is formed by the unpaired electron of each NO_2 molecule:

Nitrogen dioxide is a toxic substance. Prolonged exposure to 5 ppm in air is harmful to humans, and 100 ppm in air can be fatal. When one realizes that the dirty brown color of the air over many large cities is due in part to NO_2, the ill effects of pollution are clear. Nitrogen dioxide is also an occupational health hazard, since NO_2 is a side product of high-temperature processes such as welding.

Nitric Acid

Nitric acid, HNO_3, was known to the alchemists, who used it to refine precious metals. It remains one of the most widely used strong acids. Nitric acid is produced by the Ostwald process; the final product of the oxidation of ammonia in this process is nitric acid. The nitric oxide formed in the first step of the Ostwald process:

$$4NH_3(g) + 5O_2(g) \longrightarrow 4NO(g) + 6H_2O(g)$$

is then oxidized to nitrogen dioxide:

$$NO(g) + \tfrac{1}{2}O_2(g) \longrightarrow NO_2(g)$$

The nitrogen dioxide then reacts with water in an absorption tower to form a solution of nitric acid and gaseous nitric oxide:

$$3NO_2(g) + H_2O(l) \longrightarrow 2H^+ + 2NO_3^- + NO(g)$$

The nitric oxide is then recycled. This process produces what is called concentrated nitric acid, which is about 55–60% HNO_3 by mass in water, and which is subsequently concentrated to 70% HNO_3. Pure nitric acid can be obtained by heating potassium nitrate with pure sulfuric acid:

$$2KNO_3(s) + H_2SO_4(l) \longrightarrow 2HNO_3(g) + K_2SO_4(s)$$

Cooling the vapor of HNO_3 yields solid nitric acid, which is a white crystalline material with a low melting point.

If nitric acid is treated with dehydrating agents—that is, reagents that remove water—such as P_2O_5, the product is dinitrogen pentoxide, the anhydride of nitric acid:

$$2HNO_3(s) + P_2O_5(s) \longrightarrow N_2O_5(s) + 2HPO_3(s)$$

If water is added to N_2O_5, which is a solid at room temperature, the product is nitric acid:

$$N_2O_5(s) + H_2O \longrightarrow 2H^+ + 2NO_3^- \qquad \Delta H^\circ = -86 \text{ kJ}$$

Interestingly, the structure of N_2O_5 alters with its physical state. In the gas phase, N_2O_5 has a covalent structure that can be represented by Lewis structures such as

But in the solid phase, N_2O_5 has a saltlike structure, composed of the NO_2^+ cation and the NO_3^- anion.

9.4 PHOSPHORUS

Periodic Classification

Phosphorus lies immediately below nitrogen in the periodic table. Its valence electron configuration is $3s^2 3p^3$. There are some similarities between phosphorus and nitrogen. Phosphorus, like nitrogen, rarely forms the 3− ion, although it does so in some rare compounds called phosphides. The chemistry of both phosphorus and nitrogen is that of the covalent bond. The composition of some compounds of phosphorus resembles the composition of the corresponding compounds of nitrogen. For example, phosphine, PH_3, the hydride of phosphorus, corresponds to ammonia, NH_3. However, the differences between phosphorus and nitrogen are more important than the similarities.

The differences start at the most basic level, the types of covalent bonds formed by each. The chemistry of nitrogen abounds with compounds in which there is a double or triple bond between two nitrogen atoms, or between nitrogen and another atom. These bonds often are quite strong. The triple bond in N_2 and the NO double bond in oxides and oxyacids are outstanding examples of such bonds. Phosphorus does not form this kind of multiple bond.

The multiple bonds in nitrogen compounds are due to π overlap between $2p$ orbitals. In the NO double bond, for example, we can picture π overlap occurring between a $2p$ orbital of the N atom and the corresponding $2p$ orbital of the O atom. But the $3p$ orbitals of phosphorus do not form bonds by π overlap except in unusual cases, and thus phosphorus does not form the kind of multiple bonds formed by nitrogen.

The $3d$ orbitals of phosphorus, which are close in energy to its occupied valence orbitals in compounds of phosphorus, are important in its chemistry. We often find multiple bonds of considerable strength in which a $3d$ orbital of P overlaps with a p orbital of another element. In addition, the $3d$ orbitals of phosphorus allow octet expansion. There are many compounds with five or even six groups bonded to a single P atom. By comparison, we do not find octet expansion in nitrogen, which has no valence-shell d orbitals.

The similarities between phosphorus and the next two elements in group VA, arsenic and antimony, both of which are classified as semimetals, are closer than those between nitrogen and phosphorus. Table 9.3 lists some important compounds of phosphorus.

Discovery and Occurrence

About 1669, a Dr. Brand, a physician and alchemist living in Hamburg, Germany, got the idea that human urine might contain a substance capable of converting

TABLE 9.3
Standard Heats of Formation of Compounds of Phosphorus at 298 K

Compound	Name	ΔH_f° $(kJ/mol)^a$
$P(g)$		334
$P_2(g)$		179
$P_4(s)$	white phosphorus	18
$PH_3(g)^b$	phosphine	23
$PCl_3(l)$	phosphorus trichloride	-300
$PCl_5(s)$	phosphorus pentachloride	-400^c
$POCl_3(l)$	phosphorus oxychloride	-578
$PBr_3(l)$	phosphorus tribromide	-167
$PBr_5(s)$	phosphorus pentabromide	-293
$P_4O_6(s)$	phosphorus trioxide	-1593
$P_4O_{10}(s)$	phosphorus pentoxide	-2940
$P_4S_3(s)$	tetraphosphorus trisulfide	-155
HH_2PO_2	hypophosphorous acid	-592
H_2HPO_3	phosphorous acid	-955
$H_3PO_4(s)$	phosphoric acid	-1260

^{a} Based on red phosphorus as the standard state.
^{b} State symbol gives the state at 298 K; absence of a state symbol denotes a $1M$ aqueous solution.
^{c} Estimated from ΔH_f° of $PCl_5(g)$.

silver to gold. He did not find such a substance but he did succeed in isolating from a large quantity of urine a small quantity of a waxy white substance that glowed in the dark and that burst into flame when exposed to air. The substance was later named *phosphorus* from the Greek words for "light I bear."

Phosphorus is estimated to be the twelfth most abundant element in the earth's crust. Large deposits of phosphorus, in the form of phosphates such as $Ca_3(PO_4)_2$ and $Ca_5(PO_4)_3F$, are found in many areas. Most of these deposits consist of the accumulated shells, bones, and tissues of marine organisms that collected on the bottom of ancient oceans, and that are now found on dry land because of geological changes.

Phosphorus and its compounds have been obtained from these deposits for more than a century. Earlier, phosphorus-containing substances were obtained from biological materials, such as guano—deposits of bird droppings—and fish. The bones of both animals and humans were used as a source of phosphorus during the early nineteenth century, when phosphorus was obtained from bones collected from the battlefields of Europe. These phosphorus-containing materials have been used for fertilizer since antiquity. Guano, bones, and similar biological materials contain phosphates that are necessary for plant growth.

Every method for producing phosphorus, from Dr. Brand's original process for isolating phosphorus from urine to the technique in use today for manufacturing phosphorus from phosphate rock, is essentially the same. The phosphorus is reduced from the $+5$ to the 0 oxidation state, using coke at a temperature of 1500 K, in the presence of SiO_2, ordinary sand. The overall reaction can be represented as

$$Ca_3(PO_4)_2 + 3SiO_2 + 5C \longrightarrow 3CaSiO_3 + 5CO(g) + P_2(g)$$

Phosphorus is produced in large quantities for industrial use. In addition to the many compounds of phosphorus used by industry, the element itself is used in the striking surface for safety matches, as an ingredient for incendiary devices, and in the caps for children's cap pistols.

Allotropes of Phosphorus

The $P_2(g)$ that is the product of the process for extracting phosphorus from phosphate rock exists only at very high temperatures. When P_2 is cooled, it is converted to P_4, a much more stable form of phosphorus. The P_2 molecule can be represented by the Lewis structure $:P\equiv P:$, which is analogous to the structure of

A drag line in operation at a phosphate rock strip mine in Florida. Large quantities of phosphorus-containing ores are mined for use in fertilizers that are essential to modern agriculture. (*Glinn, Magnum*)

N_2. But the energy of the PP bond is only 490 kJ/mol, which is much less than the bond energy of N_2. The relative weakness of the bond in P_2 is due to a poor π overlap between $3p$ orbitals.

The structure of P_4 is shown in Figure 9.9. The four phosphorus atoms lie at the corners of a regular tetrahedron. The bonds are identical. The same P_4 structure is found in liquid phosphorus. It is also found in white phosphorus, the most common solid form of the element, which is formed when $P_4(g)$ is condensed under water. White phosphorus is the most reactive form of the element. It bursts into flame spontaneously when exposed to air, so it must be stored under water.

When white phosphorus is heated for some time, it is converted into another solid allotrope called red phosphorus, an inexact term used to describe a number of different forms that are more or less red in color and that are much less reactive than white phosphorus. The structures of the various kinds of red phosphorus have not been well characterized. They seem to consist of three-dimensional networks of phosphorus atoms that are covalently bonded more or less randomly, to form giant molecules. It has been proposed that the change from white phosphorus to red phosphorus starts when one of the P—P bonds in the P_4 tetrahedron breaks and long-chain molecules of many P_4 units form. Figure 9.10 shows a section of such a chain.

The most stable form of phosphorus is a third allotrope, black phosphorus, which is quite stable in air and is not easily ignited by a flame. Black phosphorus is difficult to prepare from red or white phosphorus, and is therefore the least commonly encountered allotrope. It consists of P atoms, each bonded to three other P atoms in a giant double-layered network. These double layers are stacked one on top of the other in the crystal. The structure of black phosphorus has features in common with the structure of graphite.

Phosphorus Oxides

Figure 9.11 shows the structure of the oxide P_4O_6, tetraphosphorus hexoxide, which is prepared by the controlled reaction of phosphorus with oxygen. Its common name is phosphorus trioxide, from its empirical formula, P_2O_3. The structure of P_4O_6 is closely related to the structure of P_4. In P_4O_6, each of the four P atoms is at the corner of a regular tetrahedron, with an O atom interposed between each pair of P atoms. Thus, we can visualize a structure for the P_4O_6 molecule by starting with a P_4 molecule and inserting an O atom into each PP bond. Since there are six PP bonds in P_4, six O atoms are added.

A more important oxide of phosphorus is P_4O_{10}, tetraphosphorus decoxide, commonly called phosphorus pentoxide because its empirical formula is P_2O_5. The structure of P_4O_{10} can be derived from that of P_4O_6. Each of the four P atoms bears an additional O atom, giving the structure shown in Figure 9.12. The extra four O atoms are bonded to the P atoms by double bonds formed by π overlap between a $2p$ orbital of O and a $3d$ orbital of P.

Phosphorus pentoxide is prepared by burning phosphorus in an excess of oxygen. It has a great affinity for water, and is widely used in the chemistry laboratory as a drying and dehydrating agent. For example, it converts HNO_3 to N_2O_5 and H_2SO_4 to SO_3. Phosphorus pentoxide represents the +5 oxidation state of phosphorus. It is used as a starting material for the production of phosphates and phosphoric acids, which are compounds of the +5 oxidation state of phosphorus.

Oxyacids of Phosphorus

Figure 9.13 shows the three most important oxyacids of phosphorus, which correspond respectively to the +1, +3, and +5 oxidation states of phosphorus: hypophosphorous acid (+1), phosphorous acid (+3), and phosphoric acid (+5).

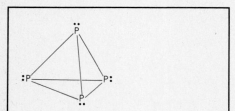

FIGURE 9.9
The structure of P_4. The four phosphorus atoms lie at the corners of a regular tetrahedron. All the P—P bonds are identical.

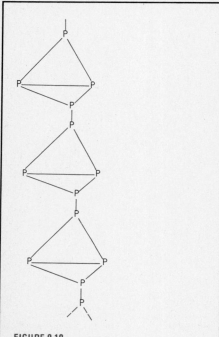

FIGURE 9.10
A section of a chain formed from many P_4 units. Red phosphorus can consist of such long-chain molecules.

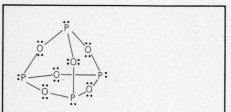

FIGURE 9.11
The structure of P_4O_6, phosphorus trioxide. We can visualize P_4O_6 as being formed by the insertion of an oxygen atom into each P—P bond of the P_4 molecule shown in Figure 9.9.

All of these compounds have tetrahedral structures. In each molecule, there is a central phosphorus atom surrounded by four groups that lie roughly at the corners of a regular tetrahedron. The hydrogens that are bonded to the P atom are not acidic. Only the hydrogens of the OH groups may ionize in polar solvents. This distinction can be indicated by separating the ionizable hydrogens from the others when writing the formulas of the oxyacids. Thus, the formula $H_2[HPO_3]$ indicates that phosphorous acid has two ionizable hydrogens, and the formula $H[H_2PO_2]$ indicates that hypophosphorous acid has one ionizable H atom. All three oxyacids have a double bond between P and O. The P=O double bond is much stronger than a P—O single bond. The P=O double bond is a characteristic structural feature of many phosphorus-oxygen compounds.

Phosphoric acid, H_3PO_4, the +5 oxyacid, and substances related to it, are by far the most important compounds of phosphorus. If we compare phosphoric acid to the corresponding nitrogen compound, nitric acid, we find major differences. Nitric acid, HNO_3, is a good oxidizing agent and is easily reduced to lower oxidation states of nitrogen. Phosphoric acid is not an oxidizing agent under ordinary conditions, and the +5 state of phosphorus is clearly the most stable state of phosphorus in its compounds with oxygen. This difference between P and N has been ascribed in part to electronegativity. There is a greater difference between the electronegativities of phosphorus and oxygen than between the electronegativities of nitrogen and oxygen. The greater electronegativity difference leads to greater stability, and a phosphorus atom thus bonds to more oxygen atoms than a nitrogen atom.

Phosphoric acid is prepared by burning phosphorus and treating the resulting oxide with water:

$$P_4O_{10}(s) + 6H_2O(l) \longrightarrow 4H_3PO_4(l)$$

or by treating phosphate rock with H_2SO_4. Both processes yield a syrupy liquid that is 85% phosphoric acid. Pure H_3PO_4 is a solid at room temperature and melts at 315 K.

Phosphoric acid is used extensively in foods. For example, it gives cola drinks and root beer their tart taste. It is mixed with molasses as an animal feed supplement. It is also used in acid baths to polish aluminum or stainless steel.

The phosphates are substances in which one or more of the hydrogen atoms of phosphoric acid are replaced by other chemical species. Inorganic phosphates in which metal atoms or other cations replace the hydrogen atoms are common and are of great practical importance. In the human body, several calcium phosphates, the most notable of which is hydroxyl apatite, $Ca_5(PO_4)_3OH$, help form bones and teeth. In agriculture, calcium phosphates, especially $Ca[H_2PO_4]_2$, are used as fertilizers. Sodium phosphates have many applications. The acidic properties of NaH_2PO_4 make it useful in effervescent laxative tablets; Na_2HPO_4 is used to pickle hams and to manufacture American cheese; Na_3PO_4 is added to scouring powders because its high alkalinity helps remove kitchen grease.

Phosphate esters are compounds in which one or more of the hydrogen atoms of phosphoric acid are replaced by organic groups. We can picture their formation as the joining together of an OH group of the phosphoric acid molecule and an OH group of the organic substance, with an H_2O molecule being formed in the process. Figure 9.14 shows the formation of methyl phosphate, a phosphate ester of methyl alcohol, in this way. As Figure 9.14 also shows, the process can be repeated to form dimethyl phosphate, a diester, or trimethyl phosphate, a triester. The formulas of these three esters are, respectively

$$CH_3OP(OH)_2 \qquad (CH_3O)_2POH \qquad (CH_3O)_3P{=}O$$
$$\quad\; \overset{\|}{O} \qquad\qquad\quad \overset{\|}{O}$$

Sugars have OH groups, and phosphate esters of sugars are of great biological importance. Two of these esters, DNA (deoxyribonucleic acid) and RNA (ribonu-

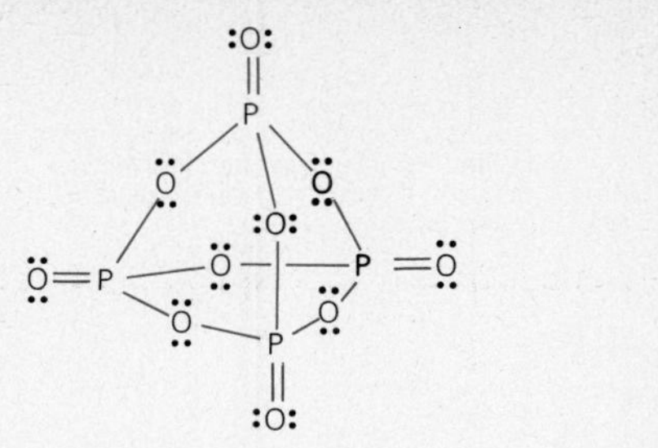

FIGURE 9.12

The structure of P_4O_{10}, phosphorus pentoxide. We can visualize P_4O_{10} as being formed by the addition of an oxygen atom to each of the phosphorus atoms of the P_4O_6 molecule shown in Figure 9.11.

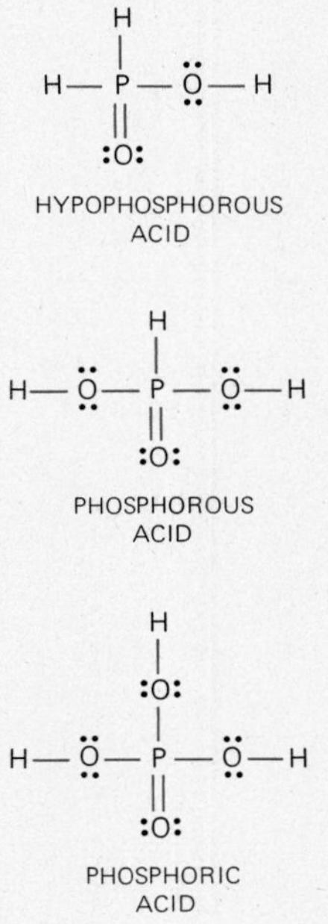

FIGURE 9.13

Three oxyacids of phosphorus. They represent respectively the +1 oxidation state of phosphorus (hypophosphorous acid), the +3 oxidation state (phosphorous acid), and the +5 oxidation state (phosphoric acid).

FIGURE 9.14
The formation of methyl phosphate. The OH group
of phosphoric acid joins to the OH group of methyl
alcohol. An H_2O molecule is eliminated as the two
molecules join. The process can be repeated twice
more with the same phosphoric acid molecule, to
form trimethyl phosphate.

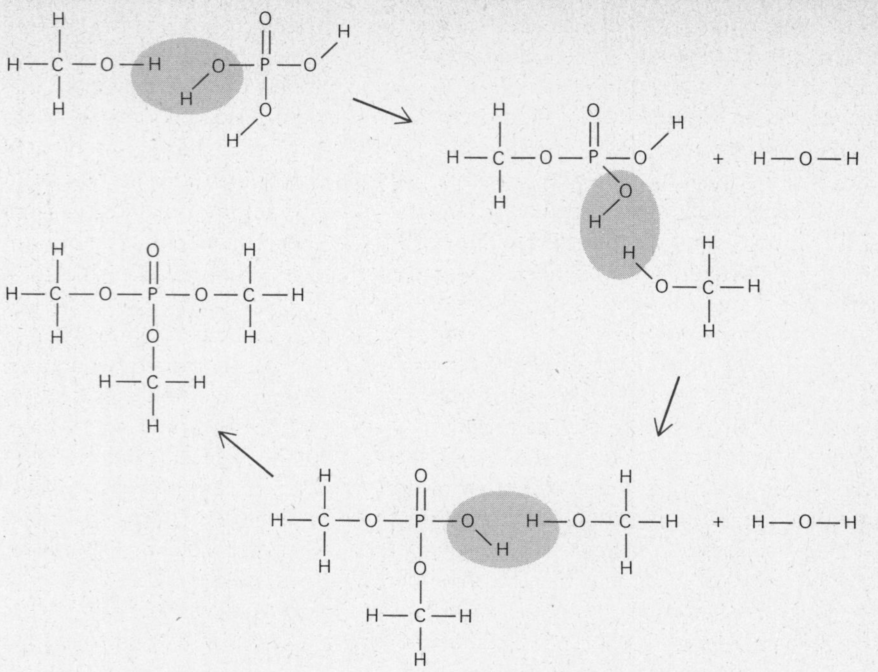

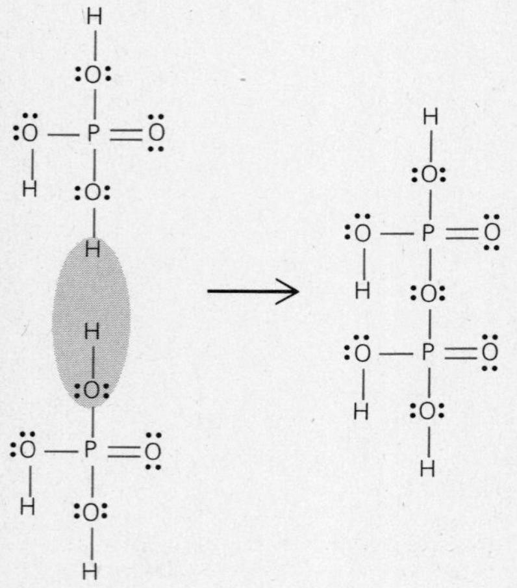

FIGURE 9.15
The formation of pyrophosphoric acid. The structure
of the molecule is best understood by visualizing it
as a combination of two phosphoric acid molecules,
with the elimination of a molecule of H_2O.

cleic acid), the genetic materials of living organisms, are long-chain molecules
whose backbones consist of phosphate esters of the sugars deoxyribose or ribose
(Chapter 21).

The elimination of H_2O can also occur between OH groups of two phosphoric
acid molecules. When phosphoric acid is heated, such a reaction occurs:

$$2H_3PO_4(l) \longrightarrow H_4P_2O_7(s) + H_2O(g)$$

The product is pyrophosphoric acid, with the prefix *pyro* meaning "heat." If we
visualize $H_4P_2O_7$ as two H_3PO_4 molecules that have been joined with the elimination of one H_2O, its structure is easily understood (Figure 9.15). Reactions in
which more than two molecules of H_3PO_4 condense in this way are common.
Esters of such condensed phosphoric acids get their names from the number of
phosphoric acid units in the molecule. Esters of pyrophosphoric acid are called
diphosphates, and esters in which three phosphoric acid molecules are joined are
called triphosphates.

A substance called adenosine can form either a diphosphate ester or a
triphosphate ester—adenosine diphosphate and adenosine triphosphate, usually
abbreviated ADP and ATP. These compounds play a pivotal role in the basic
energy processes in most living cells (Section 16.8). The reaction
$ATP + H_2O \longrightarrow ADP + H_2PO_4^-$ is crucial in biology. The structures of these
two molecules can be represented schematically:

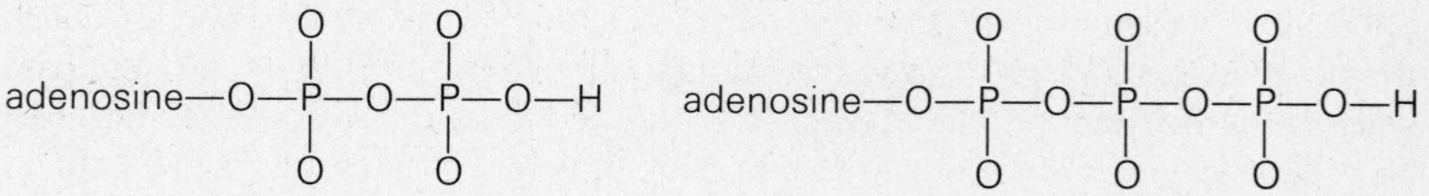

9.5   OXYGEN

Oxygen is only a minor constituent of the universe as a whole, but it is the most
abundant element on earth. Indeed, the great abundance of oxygen on earth is

Fed by oxygen in the air, a fire burns out of control. Fire-fighters extinguish such blazes by using a material that prevents oxygen from reaching the combustible substance. Firemen use water, because it can cut off the supply of oxygen, not because it is wet. (*Wide World*)

presumed to be one factor that makes this planet a hospitable place for life. The crust of the earth is 46.5% oxygen by mass. Most of the oxygen is found in oxides of metals or, combined with metals and silicon, in silicates. The water of the oceans is 88.8% oxygen by mass, and the atmosphere is 23.0% oxygen by mass. Most of the oxygen in the air exists in the elemental form, as $O_2(g)$.

Oxygen is the lightest element in group VIA of the periodic table. Its valence electronic configuration is $2s^2 2p^4$; it is two electrons short of the noble gas electronic configuration. Oxygen is second in electronegativity only to fluorine. Oxygen may occasionally display positive oxidation states, primarily in compounds in which it is combined with fluorine. In the vast majority of cases, however, oxygen displays negative oxidation states, either when in covalent compounds or as the O^{2-} ion, which is found in many metal oxides.

Oxygen forms binary compounds called oxides with all the other elements except argon, neon, and helium. We discussed trends in the behavior of oxides in Section 6.9, page 177, and we shall discuss many of these oxides further as we describe the chemistry of the elements. The most important oxide, H_2O, is described in detail in Chapter 11. In this section, we shall discuss the properties of some forms of elemental oxygen and some compounds in which oxygen is not in the −2 oxidation state. Table 9.4 lists some of these compounds.

Molecular Oxygen

The O_2 molecule, which is the stable form of oxygen at ordinary conditions, is usually, but imprecisely, called either oxygen or molecular oxygen. As work with other forms of oxygen has become more common, the systematic name dioxygen has been proposed for the O_2 molecule.

At room temperature, O_2 is a colorless, odorless gas. Its boiling point is 90.18 K and its critical temperature is 154.8 K. It condenses to a blue liquid. As gas, liquid, or solid, O_2 has the unusual property of being attracted fairly strongly by a magnetic field. In the terms introduced in Section 7.6, page 217, O_2 is one of the small group of substances of simple structure that are paramagnetic. Most substances are diamagnetic; they are repelled slightly by a magnetic field.

All the electrons of a diamagnetic substance have paired spins. A paramag-

TABLE 9.4
Unusual Oxidation States of Oxygen

State	Species	Name
+2	OF_2	oxygen difluoride
+1	O_2F_2	dioxygen difluoride
+$\frac{1}{2}$	O_2^+	dioxygenyl cation
0	O_3	ozone
−$\frac{1}{2}$	O_2^-	superoxide anion
−1	H_2O_2	hydrogen peroxide

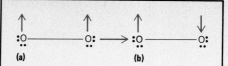

FIGURE 9.16
An O_2 molecule goes from (a) the triplet state, in which it has two unpaired electrons, to (b) the singlet state, in which these two electrons have paired spins, when one of the electrons changes its spin spontaneously.

netic substance has one or more electrons with unpaired spins. The O_2 molecule is paramagnetic because it has two unpaired electrons. This electronic configuration is predicted by the molecular orbital description of O_2 (Section 7.6, page 218), which places the two highest energy electrons of the molecule in two different molecular orbitals of the same energy. By Hund's rule, these electrons have parallel spins in the ground state of O_2. Any electronic state in which we find two unpaired electrons with parallel spins is called a *triplet* state; O_2 is quite unusual because it is a molecule that is a triplet in its ground state.

If one of the unpaired electrons in O_2 flips its spin, as represented in Figure 9.16, a higher energy state called a *singlet state* is formed, in which all the electron spins of the molecule are paired. The energy of visible light can bring about the formation of singlet O_2 from triplet O_2. Singlet O_2 is much more reactive than triplet O_2. It is believed to play a major role in oxidations with O_2 that are brought about by light, including some that occur in living organisms.

Ozone

Ozone, O_3, is an allotrope of oxygen. It is a blue gas at room temperature. It liquefies at 161 K to a dark blue liquid. When it freezes, at 81 K, ozone is a dark purple solid. Ozone is highly explosive and quite poisonous, so it is rarely handled in pure form. It has a distinctive, sharp odor, which is often noticeable after a thunderstorm, when lightning has formed some ozone from O_2 in the atmosphere.

Ozone is a resonance hybrid of several contributing structures. Two of the major contributing structures are:

$$\text{structures}$$

The OOO bond angle has been measured; it is 116.6°. Ozone, O_3, is thermally unstable with respect to $O_2(g)$, the standard state of oxygen. The standard heat of formation of $O_3(g)$ is $\Delta H_f^\circ = 142 \text{ kJ/mol}$.

For laboratory or industrial use, ozone usually is prepared by a device called an ozonizer, which produces ozone by passing an electric spark through a stream of O_2. Only a small fraction of the O_2 is converted to O_3 by this process. In the atmosphere most ozone is formed by the action of sunlight on O_2. Although ozone is a minor constituent of the atmosphere, it plays a significant role in the well-being of living organisms on earth. The concentration of ozone is highest in the upper atmosphere, reaching a maximum at some 24 km above the surface of the earth, a region called the ozone layer. At this altitude, ozone is formed in a two-step process in which short-wavelength ultraviolet light (~200 nm) acts on O_2:

$$O_2 + \text{light} \longrightarrow O + O$$
$$O + O_2 \longrightarrow O_3$$

The ozone layer effectively absorbs ultraviolet radiation in the 200–350 nm wavelengths, by the process

$$O_3 + \text{light} \longrightarrow O_2 + O + \text{heat}$$

which is doubly important. First, the ozone layer prevents this ultraviolet radiation, which is extremely damaging to living things, from reaching the earth's surface. Second, the heat released by the chemical change creates a warm layer in the upper atmosphere that helps to regulate the temperature of the lower atmosphere. In recent years, atmospheric chemists have warned about threats to the ozone layer from high-flying supersonic aircraft, the release of fluorocarbons into the atmosphere, nuclear explosions, and other human activities.

The industrial uses of ozone are based on its powerful oxidizing action. Ozone is being used in increasing quantities in water purification, in bleaching, and in

THE CHEMISTRY OF PHOTOCHEMICAL SMOG

There are at least two distinct types of smog. One smog has reducing properties and is called sulfurous smog because of its high concentration of oxides of sulfur. This kind of smog is common in Europe and the eastern United States. Another kind of smog has oxidizing properties and is called photochemical smog, because sunlight plays a crucial role in its formation. It is common in the western United States.

The word *smog* was coined near the beginning of the twentieth century to describe a combination of coal smoke and fog. Now it is used to refer to many kinds of atmospheric pollution. In the late 1940s and early 1950s, atmospheric pollution in Los Angeles was recognized as photochemical smog. Years of research were needed to demonstrate the role of the automobile in the formation of this kind of smog. Although many details of the chemical reactions in photochemical smog remain uncertain, the major reactions are reasonably well known.

One of the major pollutants in photochemical smog is the oxidant ozone, O_3. Ozone is formed naturally in relatively large amounts in the stratosphere, at altitudes of 25–40 km. Air currents bring some of this ozone down to the surface. While only relatively small traces of natural ozone are found in air, these small amounts seem to play an essential role in the chemistry of smog formation. A chain of reactions is begun when O_3 absorbs ultraviolet radiation in a process that leads to formation of an excited O atom (O^*):

$$O_3 + h\nu \longrightarrow O^* + O_2$$

In a small percentage of cases, the excited oxygen atom reacts with water vapor to yield the highly reactive hydroxyl radical:

$$O^* + H_2O \longrightarrow 2OH$$

The hydroxyl radicals can then combine with hydrocarbons, which can be denoted as HRH, where R contains at least one carbon atom:

$$OH + HRH \longrightarrow H_2O + RH$$

A complex series of steps follows. The RH reacts with O_2 to form products that react with nitric oxide from automobiles. One product of these reactions is nitrogen dioxide, NO_2, which is decomposed by the action of sunlight:

$$NO_2 + h\nu \longrightarrow NO + O$$

The reaction of these O atoms with O_2 is responsible for the formation of the O_3 that is a significant pollutant in photochemical smog.

This sequence of reactions is not the only pathway by which ozone is formed. The OH radicals can also react with carbon monoxide to begin a similarly complex series of steps. The overall reaction in this sequence in given by:

$$CO + 2O_2 \longrightarrow CO_2 + O_3$$

chemical manufacturing processes where a strong oxidizing agent is needed. But this oxidizing ability makes ozone an unwelcome constituent of the lower atmosphere. Ozone is an air pollutant, reacting with oxides of nitrogen and hydrocarbons in the formation of photochemical smog. As an air pollutant, ozone causes rubber to crack and is a hazard to health. Some vestiges of the old belief that "ozone" is synonymous with "fresh air" still survive, but the truth is that ozone is dangerous, and its concentration is higher in smog than in fresh air.

Hydrogen Peroxide

Water is not the only possible product of the reaction between H_2 and O_2. Hydrogen peroxide, H_2O_2, is another stable compound of hydrogen and oxygen that can form under suitable conditions. The standard heat of formation of aqueous hydrogen peroxide is $\Delta H_f^\circ = -191$ kJ/mol. For $H_2O_2(l)$, $\Delta H_f^\circ = -188$ kJ/mol. But H_2O_2 is rarely encountered in pure form because it is unstable with respect to the formation of water and oxygen:

$$H_2O_2(l) \longrightarrow H_2O(l) + \tfrac{1}{2}O_2(g) \qquad \Delta H^\circ = -98 \text{ kJ}$$

This reaction is quite exothermic. But it proceeds very slowly, so pure hydrogen peroxide or very concentrated solutions of it can be prepared at room temperature. However, even small amounts of impurities such as dust, grease, or metals can cause the decomposition of H_2O_2 to proceed with explosive speed. While 3% solutions of H_2O_2 are used as antiseptics and 30% solutions are commonly used in the chemistry laboratory, solutions more concentrated than 50% H_2O_2 must be handled with extreme caution.

The oxygen in H_2O_2 can be regarded as being in the -1 oxidation state, intermediate between the -2 state of oxygen in oxides and the 0 state in O_2. The structure of the hydrogen peroxide molecule is H—Ö—Ö—H. The HOO bond angle is $94.8°$.

Hydrogen peroxide forms salts with metals in which its hydrogen atoms are replaced by metal atoms. Two such salts are sodium peroxide, Na_2O_2, and barium peroxide, BaO_2. Hydrogen peroxide is used industrially as an oxidizing agent and as a bleach. It is produced commercially by the oxidation of water, using peroxydisulfate, $S_2O_8^{2-}$, as a reagent in a process that yields an aqueous solution that is 30–35% H_2O_2. Aqueous solutions of H_2O_2 can also be manufactured by the reduction of O_2 by H_2 under special conditions, in an overall reaction that can be represented as

$$H_2(g) + O_2(g) \longrightarrow H_2O_2$$

The biological role of hydrogen peroxide has attracted considerable attention in recent years. It is produced as an intermediate in the reduction of O_2 to H_2O, a reaction that occurs as a part of many energy-supplying processes in the cell. Because H_2O_2 is toxic, living cells have developed pathways that remove it efficiently.

Superoxide

The superoxide anion, O_2^-, is found combined with a number of cations. The superoxide anion represents an oxidation state of oxygen midway between that of O_2 and that of H_2O_2. Since oxygen is in the 0 oxidation state in O_2 and the -1 oxidation state in H_2O_2, it might appear difficult to assign an oxidation state to the oxygen atoms in superoxide. We can solve the problem either by assigning one oxygen atom in O_2^- the oxidation number -1 and the other the oxidation number 0, or by assigning each oxygen atom an oxidation number of $-\frac{1}{2}$. Since experimental evidence indicates that both oxygen atoms in the superoxide anion are equivalent, the use of fractional oxidation numbers is preferable. The existence of species in which fractional or mixed oxidation states occur is not uncommon. Other examples include N_3^-, the azide anion, and O_2^+, the dioxygenyl cation.

Lewis structures are readily drawn for the superoxide anion by adding one electron to the O_2 molecule:

$$\left[:\overset{..}{O}-\overset{..}{\underset{..}{O}}: \longleftrightarrow :\overset{..}{\underset{..}{O}}-\overset{..}{O}: \right]^-$$

The observation that the superoxide anion is paramagnetic is consistent with the fact that it has an odd number of electrons.

Superoxides react vigorously with water to form peroxides and oxygen:

$$2O_2^- + H_2O \longrightarrow O_2(g) + HO_2^- + OH^-$$

There is evidence that the superoxide anion also forms in living organisms as an intermediate in the biological reduction of O_2 to H_2O. The superoxide anion is believed to be highly toxic, and some have suggested that it is one of the causes of aging. If the superoxide anion does play such a role, it acts quickly. Its lifetime in cells is extremely short since it is converted to O_2 and H_2O_2 almost as quickly as it forms.

9.6 SULFUR

Sulfur is the second lightest element in group VIA of the periodic table. It is found in its uncombined state in many parts of the earth, and has been known since the earliest times. Under the name "brimstone," sulfur is mentioned frequently in the Bible. Its fiery tendencies led the Hebrews to associate it with torment and suffering. In our industrial age, sulfur has lost its hellish connotations and is mined in quantity for industrial use. The largest known deposits of elemental sulfur are in Texas and Louisiana. These are underground deposits, apparently formed by the reduction of sulfur compounds from living organisms by bacteria. To obtain sulfur from these deposits, superheated steam is forced into the ground. The sulfur melts and is forced to the surface by air pressure. The sulfur obtained by this method, the Frasch process, can be 99.5% pure.

Sulfur atoms have a tendency to form rings or chains. Sulfur is second only to carbon in its tendency to catenation, which is observed not only in elemental sulfur but also in many of its compounds. Oxygen, the lightest member of group VIA, displays no such tendency.

The stable form of elemental sulfur at room temperature is cyclo-octasulfur, a molecule whose formula is S_8 and whose eight sulfur atoms are arranged in a crown-shaped ring (Figure 9.17). Cyclo-octasulfur occurs in a number of different solid forms, depending on the method of preparation. There are also less stable molecular forms of sulfur consisting of rings containing from six to 12 sulfur atoms. All of these forms of sulfur are unstable at room temperature, but they can be stored and studied at lower temperatures. Another form can be obtained by

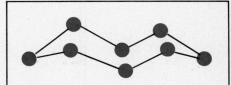

FIGURE 9.17
The structure of cyclo-octasulfur, S_8, the stable form of sulfur at room temperature. The eight sulfur atoms form a ring.

These two photographs of a statue, taken some 60 years apart, show the corrosive power of air pollutants such as sulfur oxides. When the top picture was taken in 1908, the statue at Harten Castle in Westphalia, Germany, had survived almost intact for more than two centuries. The bottom photograph was taken in 1969 and shows the damage done by an increasingly polluted atmosphere. (From E. M. Winkler, *Stone: Properties, Durability in Man's Environment,* Springer-Verlag, New York, 1973. Reproduced with permission.)

pouring molten sulfur into ice water. This "plastic sulfur" resembles rubber. It can be stretched into long, elastic fibers and consists of long-chain molecules of sulfur atoms. Left alone, it changes spontaneously to S_8.

Sulfur has a substantial vapor pressure, which reaches 1 atm at 717.8 K. The composition of the vapor depends on temperature and pressure. At very high temperatures (>2500 K) and very low pressures ($<1 \times 10^{-8}$ atm), sulfur vapor consists predominantly of S atoms. At lower temperatures, the predominant form is the S_2 molecule, which is paramagnetic and which gives sulfur vapors their blue color. At still lower temperatures, the vapor contains a mixture of different forms of sulfur containing from two to eight atoms. Below 717 K, sulfur forms liquid sulfur, whose unusual behavior and structure will be discussed in Section 10.1. Sulfur is a solid below 392.4 K.

One difference between sulfur and oxygen has already been mentioned; sulfur has a tendency to catenation and oxygen does not. There are other differences. Sulfur has the valence electronic configuration $3s^2 3p^4$ and, like oxygen, often displays the -2 oxidation state that results when it achieves the noble gas electronic configuration. But sulfur is less electronegative than oxygen, and compounds of sulfur in the -2 oxidation state usually are less ionic than the corresponding oxides. The $3d$ orbitals of the sulfur atom in many compounds are believed to be relatively close in energy to the $3s$ and $3p$ orbitals that hold the valence electrons. These $3d$ orbitals may often be involved in the chemistry of sulfur. There are no available d orbitals in oxygen that can be used for bonding. Because the $3d$ orbitals are available for bonding in sulfur, we find many compounds in which the coordination number of sulfur is as high as 6. By contrast, the coordination number of oxygen is usually only 2 or 3. In a few rare cases, the coordination number of oxygen may be as high as 4, but it is never higher.

There is some evidence that the $3d$ orbitals of a sulfur atom can also form π bonds, as the $3d$ orbitals of the phosphorus atom do. In addition, sulfur, unlike oxygen, often displays positive oxidation states in many of its compounds. The $+4$ and $+6$ states are the most common. The combination of the lower electronegativity of sulfur and the accessibility of the $3d$ electrons in sulfur are believed to be responsible for the existence of these positive oxidation states.

Table 9.5 lists standard heats of formation of compounds of sulfur. We shall discuss some of these compounds in detail. First, however, we should consider

TABLE 9.5
Standard Heats of Formation of Compounds of Sulfur at 298 K

Compound	Name	ΔH_f° $(kJ/mol)^a$
$S(g)$		277
$S_2(g)$		129
$H_2S(g)^b$	hydrogen sulfide	-20.4
$S_2Cl_2(l)$	disulfur dichloride	-60
$SF_4(g)$	sulfur tetrafluoride	-780
$SF_6(g)$	sulfur hexafluoride	-1220
$SOCl_2(g)$	thionyl chloride	-210
$SO(g)$	sulfur monoxide	6.9
$SO_2(g)$	sulfur dioxide	-297
$SO_3(g)$	sulfur trioxide	-396
H_2SO_3	sulfurous acid ($SO_2 + H_2O$)	-633
$H_2SO_4(l)$	sulfuric acid	-814
H_2SO_4	sulfuric acid	-908
$FSO_3H(l)$	fluorosulfuric acid	-800

^{a} Based on $S_8(s)$ as the standard state.
^{b} State symbol gives the state at 298 K; absence of state symbol means that the substance is in $1M$ aqueous solution.

the other elements of group VIA: selenium, tellurium, and polonium. Selenium conducts electricity when exposed to light and thus behaves somewhat like a semimetal, but it is best classified as a nonmetal. The most stable form of this element is gray selenium, which consists of long, spiral chains of Se atoms. The Se_8 molecule can also be prepared, but it is unstable. Tellurium is a semimetal that exists only as a long-chain molecule with a structure comparable to that of gray selenium. No one has succeeded in preparing Te_8. The last element in the group, polonium, is a radioactive metal that is quite rare. As we look over group VIA, we see a gradual increase in the metallic properties of its elements, moving from the lightest to the heaviest.

Hydrogen Sulfide

The rotten-egg smell of hydrogen sulfide is one of the most distinctive on earth. Hydrogen sulfide is extremely poisonous; more so than hydrogen cyanide, for example, and its odor serves a useful purpose in giving a warning signal before H_2S reaches lethal concentrations. In passing, we should note that many compounds of sulfur, selenium, and tellurium have unpleasant odors, especially compounds in which these elements are in the -2 oxidation state. Many of these odors are associated with decay and putrefaction. An organic derivative of sulfur, *n*-butyl mercaptan, gives a skunk its odor.

In aqueous solution, H_2S is a weak acid. Binary compounds called sulfides form between sulfur and the metals. With the exception of the sulfides of the group IA and group IIA metals, threse sulfides are very insoluble in water. In everyday life, hydrogen sulfide makes itself known by causing the formation of tarnish on silver objects. Even small amounts of hydrogen sulfide in the air can cause tarnishing, which is actually the formation of a black compound, Ag_2S, on silver surfaces.

Oxides of Sulfur

Sulfur dioxide, SO_2, and sulfur trioxide, SO_3, are both of great practical interest. Sulfur dioxide is a gas at room temperature and liquefies at 263 K. Its structure can be written as a resonance hybrid of two contributing structures with formal charges:

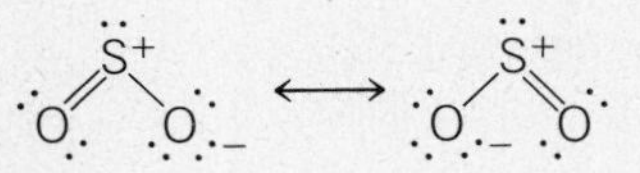

or as a single structure with no formal charges but with an expanded octet of sulfur:

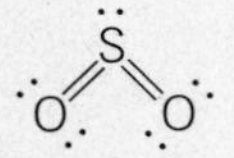

In this structure, one of the S=O double bonds can be visualized as being formed by π overlap between a $3d$ orbital of the S atom and the corresponding $2p$ orbital of the O atom, thus accounting for the expanded octet of the sulfur atom. The observed OSO bond angle of 119.5° is consistent with either representation.

Sulfur dioxide has many uses—as a bleach, a preservative, and a disinfectant, among others. It is prepared commercially by the combustion of sulfur:

$$\tfrac{1}{8}S_8(s) + O_2(g) \longrightarrow SO_2(g) \qquad \Delta H° = -297 \text{ kJ/mol}$$

Sulfur dioxide dissolves readily in water to form a solution that behaves as a weak acid. This solution usually is given the formula H_2SO_3 and the name sulfurous acid, although there is no evidence that a sulfurous acid molecule exists. Salts of sulfurous acid, such as sodium bisulfite, $NaHSO_3$, and sodium sulfite, Na_2SO_3, are well known.

The unintentional production of sulfur dioxide by the combustion of the sulfur compounds found in coal and oil is of major concern in most urban areas. As an air pollutant, sulfur dioxide is highly undesirable. It is irritating in even small concentrations and it is believed to be cumulatively toxic; as little as 1 ppm of SO_2 can damage plants. Many cities have mandated the use of low-sulfur fuels by industry and electric utilities as a central part of their clean-air efforts.

Sulfur trioxide, SO_3, is more difficult to prepare than SO_2 under ordinary conditions, even though it is more stable. The conversion of SO_2 to SO_3 is a key step in the manufacture of sulfuric acid, one of the most widely used industrial chemicals. The reaction that can be represented as:

$$SO_2(g) + \tfrac{1}{2}O_2(g) \longrightarrow SO_3(g) \qquad \Delta H° = -99 \text{ kJ}$$

and that can be brought about at high temperatures in the presence of a substance such as platinum or V_2O_5 is carried out commercially on a large scale.

Because sulfur trioxide has a low boiling point, it is often handled as a gas. In the gas phase, the SO_3 molecule is planar. All three oxygen atoms are equivalent, and the OSO bond angles thus are 120°. The structure of sulfur trioxide can be written as a resonance hybrid of three contributing structures that obey the octet rule and have formal charges:

or as a single structure with no formal charges but with an expanded octet of sulfur:

FIGURE 9.18
A fibrous form of SO_3. The sulfur trioxide molecules can link together to form a long chain.

Sulfur trioxide is an extremely reactive substance in the presence of reagents that can add to an S=O double bond to form two single bonds. Sulfur trioxide even reacts with itself. There is a form of solid SO_3 that has a fibrous asbestoslike structure. The fibers consist of long-chain molecules made up of SO_3 subunits, as shown in Figure 9.18. Each S atom has four bonds, and there is tetrahedral geometry about the S atom.

Sulfuric Acid

Sulfuric acid is formed when sulfur trioxide reacts with water. Pure sulfuric acid, H_2SO_4, is a colorless, oily liquid that has long had a variety of uses. A method for preparing H_2SO_4 was recorded as long ago as the thirteenth century, when the compound was called "oil of vitriol." Sulfuric acid today is an extremely important industrial chemical—not only the most prominent compound of sulfur but also the most widely used strong acid. More than 30 million tons of sulfuric acid are produced annually in the United States, and H_2SO_4 is so essential to industry that the level of sulfuric acid production can be used as an accurate economic barometer. Most sulfuric acid is manufactured by the contact process, in which sulfur is oxidized by air in two steps; first to SO_2 and then to SO_3. In the final step, the SO_3 is dissolved in sulfuric acid that contains some water:

$$SO_3(g) + H_2O(l) \longrightarrow H_2SO_4(l)$$

Since the raw materials for this process are sulfur, water, and air, sulfuric acid can be produced cheaply.

The uses of sulfuric acid are legion. A full 40% of sulfuric acid production goes to produce phosphate fertilizers. The phosphoric acid needed to produce fertilizer is prepared by treating phosphate rock with H_2SO_4. Any industrial process that requires a strong acid is likely to use sulfuric acid. Its uses include the manufacture of drugs and dyes, petroleum refining, the cleaning of steel, and the manufacture of other strong acids. The high boiling point of sulfuric acid makes it especially suitable for the preparation of lower-boiling compounds, such as nitric acid, which can be separated from the reaction mixture by boiling them out.

Many applications of H_2SO_4 are based on its well-known affinity for water. Gases can be dried by bubbling them through H_2SO_4. Sulfuric acid can also remove water from organic compounds that contain hydrogen and oxygen, and it can be used to promote reactions in which water is produced, such as

$$C_3H_5(OH)_3 + 3HNO_3 \longrightarrow C_3H_5(NO_3)_3 + 3H_2O$$

in which the explosive nitroglycerine is produced from glycerine. The nitric acid used in this process has some sulfuric acid added to it, and the affinity of H_2SO_4 for water enhances the formation of the products.

Sulfuric acid must be handled carefully. A good deal of heat, about 880 kJ/mol, is evolved when sulfuric acid is mixed with water. This heat can readily vaporize the water, so the mixing must be done cautiously to avoid dangerous splashing. Sulfuric acid is always added to water with stirring to disperse the heat that is evolved. For safety's sake, water should never be added to sulfuric acid.

In the H_2SO_4 molecule, the four oxygen atoms are at the corners of a tetrahedron around a central sulfur atom. Two of the SO bonds can be represented as double bonds. The measured bond lengths are consistent with some π overlap between the vacant $3d$ orbitals of the S atom and the $2p$ orbitals of the two O atoms. A Lewis structure of H_2SO_4 is thus:

```
        :O:
         ‖
H—O—S—O—H
         ‖
        :O:
```

A comparison with the Lewis structure of SO_3 suggests why H_2SO_4 is formed so readily by the reaction of SO_3 with water. One S=O double bond of SO_3 becomes two single S—O bonds in sulfuric acid. The same thing happens when SO_3 adds to H_2SO_4 to form disulfuric acid, $H_2S_2O_7$, whose Lewis structure is

```
      :O:    :O:
       ‖      ‖
H—O—S—O—S—O—H
       ‖      ‖
      :O:    :O:
```

and the process can be repeated to form trisulfuric acid, $H_2S_3O_{10}$, and higher sulfuric acids. Mixtures of sulfuric acid and sulfur trioxide are called oleum or fuming sulfuric acid, because they emit white fumes of SO_3.

In sulfuric acid, sulfur is in the $+6$ oxidation state, its highest possible state. Sulfuric acid is commonly regarded as an oxidizing agent, although a poor one. Nevertheless, sulfuric acid itself can be oxidized by powerful oxidizing agents. It is not the sulfur but the oxygen that is oxidized. In the oxidation reaction that produces peroxysulfuric acid, H_2SO_5, one of the O atoms in H_2SO_4 is oxidized from the -2 state to the -1 state. Peroxysulfuric acid has the O—O bond of peroxides:

```
        :O:
         ‖
H—O—S—O—O—H
         ‖
        :O:
```

Sulfur-Sulfur Bonds

The strong tendency of sulfur atoms to catenate—form bonds with other sulfur atoms—also appears in compounds of sulfur. There are, for example, the sulfanes, a family of compounds each of which consists of a straight chain of sulfur atoms with a hydrogen atom at each end of the chain. The general formula for a sulfane is H_2S_x, where x is an integer from 2 to 6. The Lewis structure of one sulfane, H_2S_4, is H—S̈—S̈—S̈—S̈—H. The sulfanes are unstable, decomposing to H_2S and elemental sulfur, but they exist long enough to be studied, if appropriate techniques are used. Preliminary studies suggest that the sulfanes may have some practical applications in addition to their theoretical interest.

Another family of sulfur compounds, the polysulfide salts, can be prepared by a number of methods, including the addition of sulfur to boiling aqueous solutions of metal sulfides, or the direct reaction of sulfur with a group IA or IIA metal. A polysulfide ion has the general formula S_x^{2-}, where x is an integer from 2 to 6. Polysulfide ions have the same straight-chain structure as the sulfanes.

An interesting ion, $S_2O_3^{2-}$, the thiosulfate ion, can be prepared by adding sulfur to a boiling solution of SO_3^{2-} ion. The reaction can be represented as:

$$SO_3^{2-} + S(s) \longrightarrow S_2O_3^{2-}$$

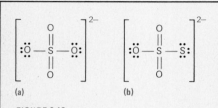

(a)   (b)

FIGURE 9.19
The thiosulfate ion (b) can be regarded as being formed by replacement of one of the oxygen atoms in a sulfate ion (a) by a sulfur atom.

The acid $H_2S_2O_3$, corresponding to the thiosulfate ion, is unstable under ordinary conditions, but there are several stable salts of the ion. The best known of these salts is sodium thiosulfate, $Na_2S_2O_3$, often called hypo, which is used to develop photographic film (Chapter 19). Sodium thiosulfate helps dissolve the unreacted silver bromide from film after the film is exposed. We can visualize the thiosulfate ion as being formed from sulfate ion, SO_4^{2-}, by formal replacement of one O atom by one S atom, as shown in Figure 9.19. The prefix *thio* is commonly used to indicate the replacement of one O atom of a compound by an S atom in a new compound. On this basis, the central S atom in thiosulfate can have a $+6$ oxidation number, just as it has in sulfate. If this is so, the other S atom has the same -2 oxidation number as the O atom it replaces. Thiosulfate thus is a compound in which one element has two oxidation states.

Sulfur is vital in the chemistry of life, and carbon-sulfur bonds are prominent in organic chemistry. We often find a disulfide group, an S—S bridge, linking two parts of a protein molecule. These sulfur bonds maintain the convoluted yet organized structure that enables protein molecules to perform their vital functions. There is a complex sequence of reactions, the sulfur cycle, in which sulfur passes through different oxidation states while being cycled from organic sulfur compounds in plants and animals through elemental sulfur and sulfate compounds and back to plants and animals. The sulfur cycle is carried on by bacteria in water and soil.

9.7   THE HALOGENS

The Elements of Group VIIA

The halogens are the elements of group VIIA of the periodic table: fluorine, chlorine, bromine, iodine, and astatine. The halogens all are nonmetals. While many of the properties of these elements change predictably with increasing atomic size (Section 6.7, page 172), there is a discontinuity between the behavior of fluorine and that of the other halogens. (This behavior is typical of a number of first-row members of other families of elements.) Astatine is a fast-decaying radioactive element. It has not been possible to prepare enough of this element or any of its compounds for detailed study, but it is believed that astatine has properties similar to those of the other halogens.

FLUORIDATION AND FALLOUT IN TEETH AND BONES

More than 100 million Americans drink water to which small amounts of fluoride compound have been added to reduce tooth decay. The addition of fluorides to water supplies resulted from observations that there is less tooth decay in communities whose water is naturally fluoridated. More recent studies have found that fluoridation achieves its effect by strengthening the crystal structure of the enamel, the tough outermost material of teeth.

An essential part of both bones and teeth is a hexagonal crystal called apatite, a complex phosphate of calcium. Several kinds of apatite are found in varying mixtures in bones and teeth. These include hydroxylapatite, $Ca_5(PO_4)_3OH$, which contains a hydroxyl group and fluorapatite, $Ca_5(PO_4)_3F$, which contains fluorine. Tooth enamel is about 99% apatite, while dentine, the material of the interior of teeth, is about 78% apatite.

Most of the apatite in teeth is hydroxyapatite. The percentage of fluorapatite is small, but apparently it is of great importance in protecting against tooth decay. Cavities are caused by acid that is produced by bacteria in plaque, a sticky substance that coats the teeth, and fluorapatite seems to be more resistant to acid attack than is hydroxylapatite. Apparently, the fluorapatite concentration is fixed during the early years of dental development. Water that contains about one part per million of fluoride compounds has been found to produce a reduction of tooth decay of about 75% in children who drink it during these years, by increasing the proportion of fluorapatite in the enamel. The protection against tooth decay provided by fluoridation appears to be lifelong.

The fact that apatite is an essential part of bone helps explain some of the concern expressed about radioactive fallout from nuclear testing. One element in fallout is ^{90}Sr, a radioactive isotope of strontium. Calcium and strontium are in the same group of the periodic table, and they form similar compounds. In fact, a little of the calcium in bone is always replaced by strontium, so there is some strontium apatite among the calcium apatite of bone. If radioactive ^{90}Sr is incorporated in bone instead of the stable isotope ^{88}Sr, radiation is emitted into the bone marrow, where blood cells are formed, as the ^{90}Sr decays. Since the bones of children are growing rapidly, they will incorporate more ^{90}Sr than the bones of adults. The realization that fallout is most damaging to the children of the world is one of the major factors in the attempt to eliminate all nuclear bomb tests.

The halogens, whose valence electron configuration is ns^2np^5, are each one electron short of the noble gas electronic configuration. Both ionic and covalent compounds of the -1 oxidation state, in which one electron is added to a halogen, are common. Salts of the halide anions X^- (where X symbolizes any halogen), as well as many compounds with other nonmetals, are well known. Fluorine is the most electronegative element and has the -1 oxidation state in all

TABLE 9.6
Properties of the Halogens

Property	F	Cl	Br	I
Atomic Properties				
atomic number	9	17	35	53
atomic weight	19.0	35.5	79.9	126.9
valence electronic configuration	$2s^2 2p^5$	$3s^2 3p^5$	$4s^2 4p^5$	$5s^2 5p^5$
ionization energy (kJ/mol)	1681	1251	1140	1008
electron affinity (kJ/mol)	322	348	324	295
electronegativity	3.98	3.16	2.96	2.66
ionic radius of X^- (nm)	0.133	0.182	0.198	0.220
ΔH_f° of X(g) (kJ/mol)	78.99	121.3	111.8	106.8
ΔH° of hydration of X^- (kJ/mol)	−510	−372	−339	−301
Properties of X_2				
melting point (K)	53.5	172	266	387
boiling point (K)	85.0	239	332	458
bond energy of X—X (kJ/mol)	158	243	193	151
bond length of X—X (nm)	0.142	0.198	0.228	0.266
ΔH° of vaporization (kJ/mol)			30.91	62.42

of its compounds. The other halogens can display positive oxidation states when they combine with more electronegative elements—oxygen or halogens of lower atomic number. The most common positive oxidation states encountered for chlorine, bromine, and iodine are +1, +3, +5, and +7. Note that the maximum oxidation state, +7, is equal to the group number.

At room temperature, the halogens are covalently bonded diatomic molecules, X_2, in which each X atom achieves the noble gas electronic configuration by electron sharing. Fluorine, F_2, is a yellow gas; chlorine, Cl_2, is a greenish yellow gas; bromine, Br_2, is a dark red-brown liquid; and iodine, I_2, is a shiny black solid at room temperature.

Table 9.6 gives some of the properties of the halogen atoms, X, and the halogen molecules, X_2. We find the expected periodic trends as well as some anomalies. The regular decrease in ionization energy and in electronegativity as atomic number increases is predictable. So is the increase in melting points and boiling points, due to the stronger interactions between molecules of larger atoms. We also observe the expected increase in ionic and covalent size with increasing atomic numbers. However, fluorine, as F^- or F_2, is much smaller than the other halogens.

A number of consequences stem from the fact that fluorine is very small. Most striking of these is the anomalous electron affinity of fluorine. Looking at the values for the rest of the halogens, we expect fluorine to have a larger electron affinity than chlorine. In fact, the electron affinity of fluorine is smaller than that of chlorine. We find a similar anomaly in bond strength. Based on the trends in the other halogens, we expect the bond energy of the F—F bond to be greater than that of the Cl—Cl bond. But the F—F bond is weaker than the Cl—Cl bond. The anomaly in bond strength appears to be related to the anomaly in electron affinity. Both seem to be a result of the unusually small size of fluorine.

In such a small atom, the repulsions between an added electron and the valence electrons are great enough to offset, at least to some extent, the favorable interaction between the added electron and the nucleus. The interactions between the added shared electron and the nonbonding valence electrons weaken the F—F bond. This weakening has important chemical consequences. Fluorine, F_2, is the most reactive element, combining explosively with a large number of substances. It is also one of the strongest oxidizing agents known. The great reactivity of fluorine is due at least in part to the weakness of the F—F bond.

Interhalogen Compounds

Table 9.7 lists the standard heats of formation of some halogen compounds. As we mentioned earlier, the most common of these compounds are the halides, in which the halogens have the -1 oxidation state. Periodic trends in the behavior of the halides were discussed in Section 6.9, page 181.

There is a large group of binary compounds of one halogen with another, called interhalogen compounds. In such a compound, the halogen of higher atomic number is in a positive oxidation state and the halogen of lower atomic number is in the -1 state. Diatomic interhalogen compounds include ClF, BrCl, and IBr. The more complex interhalogen compounds are fluorides, with the sole exception of ICl_3, they include ClF_3, ClF_5, BrF_3, BrF_5, IF_3, IF_5, and IF_7. Compounds with seven fluorine atoms around a central bromine or chlorine atom are theoretically possible, but none are known to exist, presumably because the bromine and chlorine atoms are too small to accommodate so many substituents.

The geometries of the complex interhalogen compounds are worth studying. We can describe the structure of these compounds quite well by the VSEPR method, which assumes that their geometry will be such as to minimize the repulsion between the bonding electrons and the relatively larger nonbonding

TABLE 9.7
Standard Heats of Formation, ΔH_f°, of Halogen Compounds at 298 K (kJ/mol)

Compound	Name[a]	X is: F	Cl	Br	I
HX(g)	hydrogen chloride	-273	-92	-36	26
KX(s)	potassium chloride	-529	-437	-394	-328
XF(g)	chlorine fluoride		-51	-59^b	-95^b
XF_3(g)	chlorine trifluoride		-159	-256	-485^b
XF_5(g)	chlorine pentafluoride		-240	-429	-840
XF_7(g)	iodine heptafluoride				-961
XO(g)	chlorine monoxide	109^b	101^b		
XO_2(g)	chlorine dioxide		103	c	
X_2O(g)	dichlorine monoxide	28	88	c	
X_2O_5(s)	iodine pentoxide			c	-158
X_2O_7(l)	dichlorine heptoxide		272		
HXO	hypochlorous acid		-121	-113	-138
XO^-	hypochlorite ion		-107	-94	-108
HXO_2	chlorous acid		-52		
XO_2^-	chlorite ion		-67	c	c
HXO_3	chloric acid		-98	-40	-230
XO_3^-	chlorate ion		-104	-67	-220
XO_4^-	perchlorate ion		-128	13	-145
KXO_4(s)	potassium perchlorate		-432	-287	-461
HXO_4(l)	perchloric acid		-41		
H_5XO_6(s)	periodic acid				-834

[a] The name given is that of the chlorine compound, if there is one. The name of the corresponding fluorine, bromine, or iodine compound is obtained by substituting the appropriate stem for *chlor*.
[b] Very reactive and not isolable under normal conditions.
[c] Substance is known but ΔH_f° has not been measured.

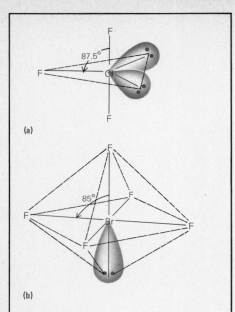

(a)

(b)

FIGURE 9.20
The structure of ClF_3 (a) and of BrF_5 (b). In ClF_3, the three fluorine atoms are at three of the five corners of a trigonal bipyramid, with the chlorine atom at the center. In BrF_5, the five fluorine atoms are at five of the six corners of a regular octahedron, with the chlorine atom at the center. In each molecule, the presence of nonbonding valence electrons affects the bond angles.

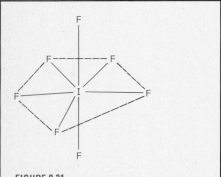

FIGURE 9.21
The structure of IF_7. The single iodine atom is in the center of an almost planar pentagon of fluorine atoms. The other two fluorine atoms lie directly above and below the iodine atom.

valence electron pairs (Section 8.2, page 237). The Lewis structure of ClF_3, for example, shows a central chlorine atom bearing three substituent atoms and two nonbonding valence electron pairs. As we mentioned earlier, five groups around a central atom lie at the corners of a trigonal bipyramid (Figure 8.16), and the geometry of ClF_3 can be explained on this basis. Repulsions are minimized in trigonal bipyramid structures when the nonbonding valence electron pairs of the central atom lie in the equatorial positions. The two nonbonding valence electron pairs of the Cl atom in ClF_3 do indeed seem to occupy equatorial positions. The three F atoms must therefore lie at the remaining three corners of the trigonal bipyramid. As Figure 9.20(a) shows, the ClF_3 molecule is observed to be roughly T-shaped. The FClF bonds do not have the ideal 90° angles of a T, since the relatively large nonbonding valence electron pairs push the atoms together.

In BrF_5, we find a Lewis structure in which the central bromine atom bears five fluorine atoms and one nonbonding valence electron pair. Six units around a central atom lie at the corners of a regular octahedron (Figure 8.18). The observed geometry of BrF_5 is one in which the five F atoms are at the corners of a slightly distorted pyramid with a square base. The Br atom is close to the center of the base. This is the geometry that results when atoms are placed at five of the six corners of an octahedron and the nonbonding valence electron pair is placed at the sixth corner. Again, the FBrF bonds have less than the ideal 90° angles because of the relatively large nonbonding electron pair on the Br atom.

Another interhalogen compound, IF_7, is interesting because its geometry cannot be predicted solely on the basis of minimizing valence electron repulsions. The IF_7 molecule is found to have the geometry of a slightly distorted pentagonal bipyramid. Five of the F atoms lie almost in a plane at the corners of a regular pentagon. The other two F atoms lie above and below this plane, on a perpendicular line through the I atom, which is at the center of the pentagon (Figure 9.21). Note the relationship between the trigonal bipyramid of ClF_3, the octahedron of BrF_5, and the pentagonal bipyramid of IF_7.

Halogen-Oxygen Compounds

There are two classes of binary compounds between halogens and oxygens. One is the oxyfluorides, in which oxygen is in a positive oxidation state and fluorine is in the −1 oxidation state. The second is the family of halogen oxides, compounds between oxygen and chlorine, bromine or iodine, in which oxygen generally is in the −2 oxidation state and the halogens are in positive oxidation states. About 20 halogen-oxygen compounds have been identified and characterized—a remarkably large number, since these compounds (with the exception of some oxides of iodine) have positive standard heats of formation and are thermally unstable. Most halogen-oxygen compounds are exceedingly reactive; many are explosive.

Many of the halogen oxides are of theoretical interest. One of them, the gas ClO_2, is of practical interest. This gas is widely used as a bleach and an oxidant; about 10 000 tons are consumed annually in the bleaching of paper pulp. The compound is quite explosive, and it must be handled with considerable care. Table 9.7 includes many halogen-oxygen compounds.

Chlorine, bromine, and iodine form a number of oxyacids in aqueous solution. There are four known oxyacids of chlorine: hypochlorous acid, $HClO$; chlorous acid, $HClO_2$; chloric acid, $HClO_3$; and perchloric acid, $HClO_4$. Perchloric acid has been isolated in pure form; the other three oxyacids are known only in solution. These four acids represent, respectively, the +1, +3, +5, and +7 oxidation states of chlorine. In theory, these oxidation states can be produced consecutively by a sequence of disproportionations, which are oxidation-reduction reactions in which a compound reacts with itself.

When Cl_2 is dissolved in water, an oxidation-reduction reaction occurs. The chlorine is oxidized to the $+1$ state and reduced to the -1 state:

$$Cl_2 + H_2O \rightleftharpoons H^+ + Cl^- + HClO$$

The reaction occurs much more readily in basic solution:

$$Cl_2 + 2OH^- \rightleftharpoons Cl^- + ClO^- + H_2O$$

The ClO^- can then react with itself, so that the chlorine goes up to the $+3$ state and down to the -1 state:

$$2ClO^- \longrightarrow Cl^- + ClO_2^-$$

In practice, this reaction occurs but does not stop here. Instead, if chlorous acid or chlorites, salts of ClO_2^-, are desired, they are prepared from $ClO_2(g)$. The isolable products of the reaction in which hypochlorite ion, ClO^-, disproportionates are the chlorate ion, in which chlorine is in the $+5$ state, and chloride ion, in which chlorine is in the -1 state:

$$3ClO^- \longrightarrow 2Cl^- + ClO_3^-$$

The $+7$ state can then be formed by disproportionation of the $+5$ state:

$$4ClO_3^- \longrightarrow Cl^- + 3ClO_4^-$$

However, this reaction occurs slowly, so perchlorates, salts of ClO_4^-, are prepared by other methods.

Perchloric acid is used commercially as an oxidizing agent of organic material—with care, because it reacts so powerfully that it is potentially explosive. Many perchlorates must also be handled with care for the same reason. Perchlorate salts often are used in the laboratory to study the behavior of metal cations in solution, since ClO_4^- and these cations hardly interact at all in aqueous solution. The structure of ClO_4^- is tetrahedral. It can be represented by a number of different Lewis structures in which there are varying numbers of formal charges and of double bonds that are due to π overlap between the $3d$ orbitals of the chlorine atom and $2p$ orbitals of oxygen. Since all the O atoms in the perchlorate anion are equivalent, the anion is a resonance hybrid of at least four contributing structures. One of the four contributing structures with no formal charges but an expanded octet of chlorine is:

$$\left[\begin{array}{c} :\overset{..}{O}: \\ \| \\ :\overset{..}{O}=\overset{..}{Cl}-\overset{..}{O}: \\ \| \\ :\overset{..}{O}: \end{array} \right]^-$$

It is often more difficult to prepare the oxygen acids of bromine and iodine, and the salts of these acids, than the corresponding compounds of chlorine. One such compound, perbromic acid, $HBrO_4$, provides a striking illustration of the rule that *negative experimental evidence must be interpreted with great care*. For years, all attempts to prepare perbromic acid and perbromate salts, in which bromine displays the $+7$ oxidation state, were complete failures. As a result, elaborate theories were devised to explain why these compounds were unstable and could not exist. In the late 1960s, however, perbromic acid and perbromate salts were prepared by several methods. They proved to be quite stable. For example, $KBrO_4$ must be heated to 550 K before it decomposes.

Periodic acid, the $+7$ oxyacid of iodine, differs somewhat from perchloric and perbromic acids. In aqueous solution, periodic acid is best represented as H_5IO_6 rather than by the formula HIO_4, which corresponds to the $HClO_4$ of perchloric acid and the $HBrO_4$ of perbromic acid. The iodine atom is large enough to accommodate six oxygen atoms, which lie at the corners of an octahedron (Figure 9.22). If H_5IO_6 loses two H_2O molecules, then HIO_4 remains. While HIO_4 is unstable with respect to H_5IO_6, salts of the IO_4^- anion are common.

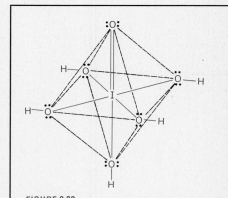

FIGURE 9.22
The structure of periodic acid. The formula H_5IO_6 best fits the molecule, which has six oxygen atoms at the corners of an octahedron, with the single iodine atom at the center. Hydrogen atoms are on five of the oxygen atoms.

9.8 THE NOBLE GASES

Until 1962, the elements of column 0 of the periodic table, the noble gases, were believed to be chemically inert. They were called "noble" in the belief that they would not join chemically with lesser elements. It is still true that the lighter noble gases—helium, neon, and argon—remain chemically inert. But, in one of the more notable achievements of recent chemical research, it has been found possible to prepare compounds of krypton, xenon, and radon.

All the elements of column 0 exist as monatomic gases that are present in the atmosphere and the earth's crust. Helium is a constituent of natural gas, from which it is separated in commercial quantities. Radon is one of the daughter products of the radioactive decay of radium. These two elements occur together. The other noble gases are produced commercially by separation from the atmosphere.

The existence of the noble gases was first suspected in the late eighteenth and early nineteenth centuries, when it was found that a small proportion of the atmosphere is chemically inert. The first noble gas to be identified was helium, which was detected, not on earth, but in the sun. Hence its name, from *helios*, the Greek word for "sun." Helium was first identified because of the presence of a strange spectral line in the sun. Later, the same spectral line was found in gases emitted from Mt. Vesuvius, confirming the presence of helium on earth. The other noble gases were discovered in the late nineteenth century as minor components of air.

Helium is second only to hydrogen in cosmic abundance. Between them, hydrogen (76%) and helium (23%) are believed to make up about 99% of the universe. Helium is relatively rare on earth, since it is so light that it can drift out of the earth's gravitational pull. Argon is more abundant in the atmosphere. Air is 0.94% argon.

Table 9.8 gives some of the physical properties of the noble gases. These properties change quite regularly with increasing atomic number.

The special characteristics of helium make it useful for a number of applications. Because of its low boiling point, helium is used as a coolant in low-temperature systems. Because of its low density and nonflammability, helium is used in balloons and other lighter-than-air craft. Because of its low solubility, helium is used instead of nitrogen in breathing mixtures for deep-sea divers to prevent the "bends," the formation of gas bubbles in the blood of divers who surface too rapidly. The other noble gases have one application that is quite visible. They are used in advertising signs, generically known as neon signs. When an electric discharge is passed through a tube containing neon, the neon atoms emit a red light that has delighted advertising agencies. Argon is used to provide a chemically inert atmosphere for laboratory and industrial purposes. The noble gases can be used as anesthetics, but other agents have been found to be more suitable.

Compounds of the Noble Gases

Before 1962, this section could not have existed; there were no compounds of noble gases. In that year, the compound $XePtF_6$, a yellow crystalline solid, was

TABLE 9.8
Physical Properties of the Noble Gases

Property	Helium	Neon	Argon	Krypton	Xenon	Radon
atomic number	2	10	18	36	54	86
boiling point (K)	4.23	27.1	87.3	119.8	165.0	211
critical temperature (K)	5.26	44.5	150.9	209.4	289.8	378
melting point (K)	0.96[a]	24.5	84.0	116	161	196

[a] At 26 atm. Helium does not solidify at 1 atm.

prepared. Immediately, a new field of chemical research came into existence. In early 1963, when most reference books on library shelves were still describing the noble gases as inert, a 400-page volume, *Noble Gas Chemistry,* was published on the known compounds of xenon, radon, and krypton.

These three noble gases all display positive oxidation states and expanded octets in their compounds. Powerful oxidizing agents are needed to make the noble gases react, so their compounds always include oxygen or fluorine or both. One of the simplest methods of preparing a noble gas compound is to react xenon and fluorine at high temperatures, obtaining a mixture of three compounds in which xenon has the $+2$, $+4$, and $+6$ oxidation states:

$$Xe(g) + F_2(g) \longrightarrow XeF_2(g) + XeF_4(g) + XeF_6(g)$$

The relative quantities of the three products can be altered by changes in the relative quantities of the reactants and changes in the conditions. All three of these fluorides are colorless solids at room temperature. The ΔH_f° of each is negative. In all three compounds, the Xe—F bond has an energy of about 134 kJ/mol. The compound XeF_8 has not yet been prepared, but the $+8$ oxidation state of xenon is found in perxenate salts such as sodium perxenate, Na_4XeO_6, and in XeO_4.

The structures of the noble gas compounds have been studied extensively. In most cases, we find structures consistent with minimizing valence electron pair repulsion. The Lewis structure of XeF_2 is

:F—Xe—F:

The two fluorine atoms and three nonbonding electron pairs of Xe lie at the corners of a trigonal bipyramid, with the two F atoms at the axial positions. The observed geometry is linear, as shown in Figure 9.23(a). The structure of XeF_2 is quite similar to the linear structure of I_3^- (Example 8.3), which has the same number of valence electrons. The Lewis structure of XeF_4 is:

:F—Xe—F:
 :F: :F:

The four fluorine atoms and the two nonbonded electron pairs lie at the corners of an octahedron. The two electron pairs lie on the same axis, and the four fluorine atoms lie at the corners of a square, as shown in Figure 9.23(b). In XeF_6, there are six fluorine atoms and one nonbonding electron pair around the Xe atom, resulting in a more complicated geometry. In XeO_4, in which Xe has no nonbonding valence electron pairs, we find the expected tetrahedral arrangement of four groups around a central atom.

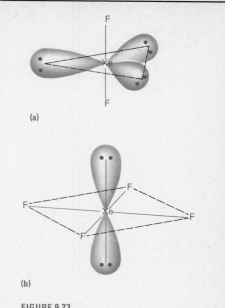

(a)

(b)

FIGURE 9.23
(a) The structure of XeF_2, with three nonbonding valence electron pairs at the corners of an equilateral triangle. The xenon atom is in the center, with the two fluorine atoms above and below the plane of the electron pairs. The geometry is linear. (b) The XeF_4 molecule. The Xe atom lies in the center of a square, with four fluorine atoms at the corners. The nonbonding electron pairs are above and below the plane of the square.

EXERCISES

9.1 List the nonmetals in order of increasing electronegativity. Do not include the noble gases. List any semimetals with greater electronegativity than any nonmetal.

9.2 The ionization energy of hydrogen is at least $2\frac{1}{2}$ times greater than the ionization energy of any alkali metal. Yet in all these elements, the electron that is ionized from the atom is the *s* electron of a half-filled *s* subshell. Account for

the unusually large ionization energy of hydrogen.

9.3 The electron affinities of the halogens are at least four times larger than the electron affinity of hydrogen. Yet in all these elements, addition of an electron completes a noble gas shell. Account for the unusually low electron affinity of hydrogen.

9.4 An important commercial method of preparing H_2 is the reaction of meth-

ane, from natural gas, with steam. Write an equation for this process. A method of preparing H_2 in the laboratory is the treatment of a metal with a solution of a strong acid. Write the equation for the reaction of magnesium metal with nitric acid solution.

9.5 The reaction between hydrogen and chlorine proceeds by the simultaneous occurrence of many chain reactions of the kind shown in Section 9.1. Any given chain reaction usually ends before a reactant is exhausted; we say the chain terminates. Write two possible reactions that result in the termination of a chain.

9.6 A table of food compositions lists a higher "calorie content" for vegetable oil than for margarine. Account for this difference in terms of the bond energies of the bonds made and broken in hydrogenation.

*9.7 The structure of B_4H_{10}, one of the simpler boranes, resembles that of diborane in a number of ways. It has four three-center bonds in each of which two boron atoms and one hydrogen atom share two electrons. Unlike B_2H_6, however, B_4H_{10} has one normal B—B bond, so two of its boron atoms are bonded to only one hydrogen each by normal B—H bonds. Each of the other two boron atoms is bonded to two hydrogen atoms by normal B—H bonds. Draw a structure for B_4H_{10}.

9.8 Calculate the mass percent of carbon in calcium carbonate.

*9.9 Find the number of moles of C—C bonds that must be broken when one mole of C(g) is formed from diamond. Find the heat of sublimation of diamond from the data in Table 9.1. Calculate the heat of sublimation of diamond from the C—C bond energy of 346 kJ/mol. Suggest a reason for the difference between these two values.

9.10 Calculate the $\Delta H°$ for the combustion of $C_8H_{18}(g)$, an octane, to $H_2O(g)$ and $CO_2(g)$ using the bond energies in Appendix III.

9.11 Draw a Lewis structure for carbonic acid, H_2CO_3. Indicate the most likely hybridization of the carbon atom in carbonic acid and in CO_2.

9.12 Calculate the mass of water gas that is produced from 2300 kg of coke, assuming that a 90% conversion is possible.

9.13 Breathing air that contains 0.13% CO(g) by volume for 30 minutes will cause death. Find the mass of CO present in air of this composition in a garage of volume 40 m^3 at STP.

9.14 Use the average bond energies listed in Appendix III to calculate the heat of formation of the nonexistent compound N_4H_6, assuming a structure with four nitrogen atoms attached in a straight chain.

9.15 Draw Lewis structures for the following compounds of nitrogen, all of which have negative values of $\Delta H_f°$:
(a) N_2F_4, (b) NH_2OH, (c) NF_3.

9.16 Draw Lewis structures for the following compounds of nitrogen, both of which have positive values of $\Delta H_f°$:
(a) N_2H_2, (b) HN_3.

9.17 Two compounds are known that have the molecular formula $H_2N_2O_2$. One of them, called hyponitrous acid, has two O—H bonds as part of its molecular structure. The other, called nitramide, has no O—H bonds. Draw Lewis structures for these two compounds.

9.18 Use the values of $\Delta H_f°$ in Appendix II to find $\Delta H°$ for the decomposition of $NH_4NO_3(s)$ to $O_2(g)$, $N_2(g)$, and $H_2O(g)$.

9.19 Draw two contributing structures for the NO^+ cation.

9.20 Write the overall chemical equation for the Ostwald process.

9.21 An anhydride of an oxyacid is the oxide that is formed by the removal of water from the oxyacid. Indicate the anhydride of each of the following oxyacids: (a) HNO_3, (b) HNO_2, (c) $H_2N_2O_2$.

9.22 Propose a hybridization picture and a likely geometry for: (a) NH_2OH, (b) NO_2^-.

9.23 The measured ONO bond angles are 180° in NO_2^+, 134° in NO_2, and 115° in NO_2^-. Account for this trend in bond angles.

9.24 Find the PPP bond angle in P_4. Suggest a reason for the great reactivity of P_4 related to this bond angle.

9.25 Write chemical equations for the following processes:
(a) the combustion of white phosphorus in excess O_2
(b) the hydrolysis of phosphorus trioxide
(c) the dehydration of sulfuric acid by phosphorus pentoxide
(d) the hydrolysis of phosphorus pentachloride

9.26 In addition to the common oxides of phosphorus discussed in Section 9.4, the oxide P_4O_8 has been prepared. Draw a structure for P_4O_8 by analogy with the structures of the common oxides of phosphorus.

9.27 Predict the hybridization and the geometry of the phosphonium cation, the phosphorus analog of the ammonium cation.

9.28 The adenosine portion of ATP has the molecular formula $C_{10}H_{12}N_5O_3$. Calculate the mass percent of phosphorus in ATP.

9.29 At 1100 K, $P_4(g)$ undergoes appreciable decomposition to $P_2(g)$. A sample of P_4 of mass 1.24 g is placed in a container of volume 0.250 liter and heated to 1100 K. The pressure is found to be 4.61 atm. Calculate the fraction of P_4 that has reacted to form P_2, assuming ideal gas behavior.

9.30 Suggest a sequence of reactions that can be used to prepare $POCl_3$ from its elements in the laboratory.

9.31 The reaction between nitric oxide and ozone to form nitrogen dioxide and O_2 is important in the formation of photochemical smog. In this reaction there is an increase in the oxidation number of nitrogen from +2 to +4. Find the corresponding decrease in oxidation number in this reaction.

9.32 The formation of ozone as a pollutant in photochemical smog occurs by a number of multistep pathways. The overall change in one of these pathways is: $CO(g) + 2O_2(g) \rightarrow CO_2(g) + O_3(g)$. Calculate $\Delta H°$ for this overall change from the values of $\Delta H_f°$ in Appendix II.

9.33 Write an equation for the oxidation of methane by concentrated hydrogen peroxide solution.

9.34 Before the development of the methods now used for the synthesis of hydrogen peroxide, dilute solutions of hydrogen peroxide were prepared by the oxidation of barium metal with $O_2(g)$ followed by the reaction of the product with a solution of dilute sulfuric acid. Write chemical equations for this process.

9.35 Treatment of potassium hydroxide with $O_3(g)$ leads to the formation of an ionic solid with the formula KO_3, called potassium ozonide. Draw a Lewis structure for the O_3^- anion and predict its magnetic properties.

9.36 Potassium superoxide is used in breathing masks because it converts both water and carbon dioxide, the products of human respiration, into O_2. The products of the reaction of potassium superoxide with carbon dioxide are O_2 and potassium carbonate. Find the mass of potassium superoxide necessary to react completely with 5.5 liters of CO_2 at a pressure of 0.031 atm and a temperature of 300 K.

9.37 Predict the atomic number, electronic configuration, and chemical properties of the as yet undiscovered element after polonium in group VIA of the periodic table.

9.38 Suggest a reason why the eight atoms of cyclo-octasulfur do not all lie in the same plane, but rather assume the crownlike arrangement shown in Figure 9.17.

9.39 The compound S_2O is unstable, but has been isolated at 77 K. It is believed to have the SSO skeleton. Draw two contributing structures for this oxide by analogy with SO_2. Predict the value of the SSO bond angle.

9.40 Thionyl chloride, $SOCl_2$, is a commonly used reagent in the organic chemistry laboratory. Draw its Lewis structure. It reacts vigorously with water to liberate two gases. Write the chemical equation for this reaction.

9.41 Draw the Lewis structure of trisulfuric acid, $H_2S_3O_{10}$.

9.42 The sulfate, phosphate, and perchlorate anions are isoelectronic and have the same geometry. Use the VSEPR model to predict this geometry.

9.43 Use the data in Table 9.5 and any necessary data from Appendix II to calculate the $\Delta H°$ for the formation of a $1M$ solution of SO_2 in water from $SO_2(g)$.

9.44 Use the data in Table 9.5 and any necessary data from Appendix II to calculate the $\Delta H°$ for the formation of a $1M$ solution of H_2SO_4 from $SO_3(g)$.

9.45 The data in Table 9.6 show that the electronegativity of fluorine is greater than that of chlorine. But the electron affinity of chlorine is greater than the electron affinity of fluorine. Account for this observation.

9.46 The $\Delta H°$ of hydration listed in Table 9.6 is the heat change that accompanies the formation of a $1M$ solution of the X^- anion. Account for the observed trend on heats of hydration listed for the halogens.

9.47 Predict which of the alkali metal halides will have the greatest lattice energy and which will have the least lattice energy.

9.48 The interhalogen compounds are very reactive. They react vigorously with water to form hydrogen halides and oxyacids. Write chemical equations for the reactions of the following compounds with water: (a) ClF, (b) BrF_3, (c) BrF_5, (d) IF_7.

9.49 In addition to the neutral interhalogen compounds, many interhalogen ions are known. Draw the Lewis structure and predict the geometry of BrF_4^-.

9.50 Chlorine dioxide is prepared commercially by the reaction of sodium chlorate, sulfur dioxide, and sulfuric acid. The other product of this reaction is sodium bisulfate. Write a chemical equation for this process. Indicate which atoms undergo changes in oxidation numbers and what these changes are.

9.51 Write the formulas and names of the anhydrides of the following oxyacids: (a) perchloric acid, (b) iodic acid, (c) chlorous acid, (d) hypobromous acid.

9.52 Explain the trend in the critical temperatures of the noble gases listed in Table 9.8.

9.53 Draw the Lewis structures and predict the geometry of the following oxygen-containing compounds of xenon: (a) XeO_3, (b) $XeOF_2$, (c) $XeOF_4$.

STATES OF MATTER: CONDENSED STATES

10

In Chapter 4, we discussed some of the similarities and differences between the three most important states of matter on earth, solid, liquid, and gas. We went on to discuss the characteristics of the gaseous state in detail. A model to account for the observed properties of gases can be developed relatively easily, because the molecules of an ideal gas are considered to be moving randomly and independently. Models of the solid and liquid states are more complex. In this chapter, we shall discuss some of the characteristics of the liquid and solid states in more detail, and we shall try to account for some of the characteristics of these states of matter.

10.1 THE LIQUID STATE

The structural units of a liquid can be atoms, as in Hg(l). They can be molecules, as in Br_2(l) or H_2O(l). They can even be ions, as in NaCl, which exists as a liquid above temperatures of 1100 K. For convenience, we shall use the word "molecule" to refer to all these possibilities.

As temperature falls, the thermal motion of molecules in the gas phase decreases. When the temperature falls low enough, the thermal motion of the molecules is not great enough to overcome the attractive forces between molecules, and the gas condenses to a liquid. However, the thermal motion in a liquid still is great enough to prevent the molecules from assuming the ordered state characteristic of a crystalline solid. There is a good deal of disorder in the structure of a liquid.

Figure 10.1 shows that a small group of molecules in one part of a liquid may have a regular arrangement. But this regularity is not repeated throughout the liquid because the attractions between the molecules of a liquid are not great enough to create a completely ordered arrangement.

It is important to remember that one fundamental difference between molecules in the gas phase and molecules in the liquid phase is a question of energy. Thermal energy is liberated when molecules in the gas phase form a liquid through their mutual attraction. Thermal energy must be added to overcome the attractions between molecules in the liquid phase to form the gaseous state.

Surface Tension

In Section 7.9, page 227, we mentioned the various kinds of attractive forces between molecules. As Figure 10.2 suggests, molecules at the surface of a liquid do not have as many attractive interactions with other molecules as do molecules that are below the surface. Attractions between molecules are favorable interactions that lower the relative energy of the molecules, in the same way that the formation of chemical bonds lowers the relative energy of atoms. Therefore, the energy of molecules at the surface of a liquid is higher than the energy of molecules away from the surface. A liquid tends to assume a shape in which the number of molecules at the surface is minimized. This shape maximizes the number of attractions between molecules and thus results in the lowest energy for the liquid. The lowest-energy shape assumed by a liquid is a sphere. Drops of a liquid or bubbles of a gas tend to be spherical because a sphere has the smallest surface area for a given volume (more briefly, the smallest surface-to-volume ratio) of all shapes.

The surface of a liquid can be visualized as being under tension, much as if it were a stretched membrane made of an elastic material such as rubber. **Surface tension** *is the force that pulls the surface inward and resists an increase in surface area*. Surface tension can be expressed in units of force per length. The SI

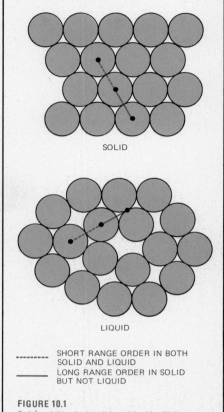

FIGURE 10.1
Order and disorder in solids and liquids. While some molecules in one part of a liquid may have a regular, ordered arrangement (bottom), this order is not maintained throughout the liquid. Molecules in a solid (top) are in an ordered, regular arrangement.

SOLID

LIQUID

-------- SHORT RANGE ORDER IN BOTH SOLID AND LIQUID
———— LONG RANGE ORDER IN SOLID BUT NOT LIQUID

FIGURE 10.2
Surface tension. Molecules in the interior of a liquid (right) are surrounded by other molecules and experience equal attractions from all directions. Molecules on the surface of the liquid (left) are subjected to unequal forces, being attracted only by liquid molecules below the surface.

303

TABLE 10.1
Surface Tension of Liquids

Liquid	Temperature (K)	Surface Tension (N/m)
H_2O	273	76.5×10^{-5}
H_2O	293	72.8×10^{-5}
H_2O	333	66.2×10^{-5}
H_2O	373	58.9×10^{-5}
$CHCl_3$ (chloroform)	293	27.1×10^{-5}
CCl_4 (carbon tetrachloride)	293	26.9×10^{-5}
C_2H_5OH (ethanol)	293	22.8×10^{-5}
C_8H_{18} (*n*-octane)	293	21.8×10^{-5}
Hg	293	485×10^{-5}
Ag	1273	900×10^{-5}
NaF	1283	260×10^{-5}
NaCl	1273	98×10^{-5}
NaBr	1273	88×10^{-5}

unit of force is the newton (N); $1\ N = 1\ kg\ m/sec^2 = 1\ J/m$. The SI unit of length is the meter (m). Thus, the units of surface tension are newtons per meter (N/m). Table 10.1 lists values for the surface tension of some common liquids.

The magnitude of the surface tension is related to the strength of the attractive forces between molecules. Liquids with large attractive forces have relatively large surface tensions. The surface tension of a given liquid decreases as the temperature of the liquid increases and the thermal motion of the molecules increases. Surface tension becomes zero when the temperature is close to the critical temperature (Section 4.10, page 109) of the substance.

Capillary Action

When a liquid is in contact with a solid surface such as glass, there are forces of attraction between the molecules of the liquid and the surface of the solid. When these forces are relatively strong compared to the attractions in the liquid itself, the liquid is said to *wet* the solid. If these forces are not too strong, the liquid does not wet the solid. The curvature of the liquid surface, the *meniscus*, can be used to determine whether the liquid wets the solid.

When water is in a glass tube, its surface is concave, as shown in Figure 10.3. The water wets the glass, and the shape of the meniscus tends to increase the contact between the liquid and the solid surface. Mercury does not wet the glass. Its meniscus is convex, which minimizes the contact between the liquid and the solid surface.

If a capillary tube, a glass tube of narrow bore, is placed in a vessel containing liquid, there is a readily apparent difference between the height of the liquid in the tube and the height of the liquid in the vessel. If the liquid wets the glass, it rises in the tube. A liquid that does not wet the glass falls in the capillary tube. Surface tension is one of the factors that determines the difference between the level of the liquid in the tube and the liquid in the vessel. Measuring this difference in height is one way to measure the value of the surface tension.

Phenomena in which solid surfaces are wet by liquids often are discussed in terms of capillarity and surface tension. Such phenomena are important in biology and are of great interest in many areas of technology.

Viscosity

Liquids flow. Viscosity is the resistance of liquids to flow. The viscosity of liquids is a property with great practical consequences. For example, a major consideration

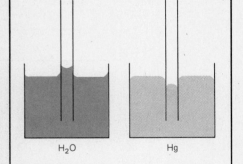

H_2O Hg

FIGURE 10.3
Capillarity. The forces between molecules of water and the walls of a glass tube (left) are relatively strong compared to the attractive forces between the water molecules. Water wets glass; the liquid surface becomes concave as the liquid rises in the tube. The forces between atoms of mercury and a glass tube are relatively weak compared to the forces between mercury atoms. Mercury doesn't wet glass; the surface of the liquid mercury becomes convex as the liquid is pulled down within the tube.

HOW TREES GET WATER

Atmospheric pressure can support a column of water less than 10 m high. Yet plants can move water many times higher; the sequoia tree can pump water to its very top, more than 100 m above the ground. Until the end of the nineteenth century, the movement of water in trees and other tall plants was a mystery. Some botanists hypothesized that the living cells of plants acted as pumps. But many experiments demonstrated that the stems of plants in which all the cells are killed can still move water to appreciable heights. Other explanations for the movement of water in plants have been based on root pressure, a push on the water from the roots at the bottom of the plant. But root pressure is not nearly great enough to push water to the tops of tall trees. Furthermore, the conifers, which are among the tallest trees, have unusually low root pressures.

If water is not pumped to the tops of tall trees, and if it is not pushed to the tops of tall trees, then we may ask: how does it get there? According to the currently accepted cohesion-tension theory, water is pulled there. The pull on a rising column of water in a plant results from the evaporation of water at the top of the plant. As water is lost from the surface of the leaves, a negative pressure, or tension, is created. The evaporated water is replaced by water moving from inside the plant in unbroken columns that extend from the top of a plant to its roots. The same forces that create surface tension in any sample of water are responsible for the maintenance of these unbroken columns of water. When water is confined to tubes of very small bore, the forces of cohesion; that is, the attraction between water molecules, are so great that the strength of a column of water compares with the strength of a steel wire of the same diameter. This cohesive strength permits columns of water to be pulled to great heights without being broken.

The theory has been tested both in the laboratory and in the forest. In laboratory experiments, the pull exerted by evaporating water has been used to draw a fine column of mercury 226 cm high in a fine glass tube, the equivalent of maintaining a column of water about 30 m high. And in 1960, a team of plant physiologists in the jungles of northeastern Australia carried out several experiments on rattan vines, which can climb up to 50 m high in trees. When the base of the vine was severed and then immersed in a basin of water, the vine continued to draw up water at a steady rate. It was found that a rattan vine 2 cm in diameter will take up about 12 cm^3 of water per minute indefinitely. The pull exerted by the rattan vine, the sequoia, and the 100-m eucalyptus tree of Australia can be equivalent to a pressure of 100 atm.

The same sort of phenomenon is responsible for much of the water content of soil. The water table ordinarily is rather far below the surface. In gravels and coarse soils, the water moves only a few centimeters, but in fine silts it can move several meters. As water evaporates from the surface, more is drawn up from below. The attractive forces between water molecules can also maintain unbroken columns of water in the soil.

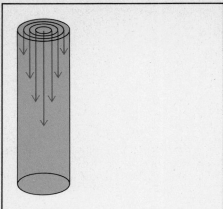

FIGURE 10.4
Flow. When layers of molecules of a liquid move in
a regular way with respect to one another, the liquid
is flowing in an ideal way.

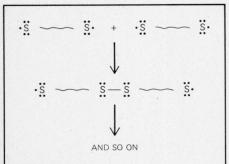

AND SO ON

FIGURE 10.5
The viscosity of sulfur increases above 160°C, as short-
chain molecules join to form long-chain molecules,
decreasing the ability of the liquid to flow. Above
200°C, the long chains begin to break, and the
viscosity of sulfur decreases.

in making motor oils is to provide the desired viscosity in a given temperature
range for specific applications.

Flow can be regarded as the movement of layers of molecules in a regular
way with respect to one another (Figure 10.4). A liquid that flows with ease is
said to be *mobile;* one that does not is said to be *viscous*. To some extent, we can
picture a mobile liquid as one whose molecules flow smoothly over each other,
with few tangles between layers of molecules. A viscous liquid is one in which
tangles between molecules in different layers interfere with the smooth flow of the
liquid.

Liquids whose molecules consist of long chains of atoms often are very
viscous, because tangles occur easily between layers of such long molecules. For
example, gasoline, lubricating oil, and tar all consist predominantly of chains of
carbon atoms with hydrogen atoms attached. Gasoline flows easily, lubricating oil
is less mobile, and tar is quite viscous. The length of the carbon chains in the
three substances parallels their viscosity. Gasoline has the shortest molecules, tar
the longest; the molecules of lubricating oil are of intermediate length.

The viscosity of a liquid usually decreases as the temperature rises. Molasses
flows more freely in the summer than in the winter. An interesting exception to
this rule is the behavior of sulfur. At room temperature, sulfur is a solid composed
of S_8 molecules in which the eight S atoms are joined in a ring:

When $S_8(s)$ is heated slowly, it melts to a mobile, straw-colored liquid at 119°C.
If this liquid is heated further, it becomes more mobile at first. But at about
160°C, the viscosity of liquid sulfur starts to increase as the temperature rises.
The viscosity reaches a maximum at about 200°C.

This behavior is due to a change in molecular structure in S_8 just above
160°C. At this temperature, the S_8 rings break, and the eight S atoms in each
ring form a chain. Each end of the chain has an S atom with only seven valence
electrons:

These ends are reactive. The eight-atom chains join together to form long-chain
molecules, causing the increase in viscosity, as shown in Figure 10.5. When the
temperature goes above 200°C, the long chains begin breaking, and the liquid
sulfur becomes less viscous again.

10.2 THE CRYSTALLINE STATE

In everyday language, we call a substance a solid if we cannot see the flow and
change in shape that we associate with liquids. But a rigorous definition of a solid
demands not only the apparent maintenance of shape but also long-range order
on the molecular scale. The existence of this long-range molecular order usually is
visible. Most true solids occur in characteristic shapes, called **crystals.** A crystal is
a three-dimensional geometric solid whose faces are planes. The existence of
crystals is a consequence of long-range order, or repetition of basic units, on the
atomic scale. There are materials that appear to have the mechanical properties
of solids but do not have the crystalline form and the accompanying long-range
atomic order. They are called *amorphous solids* or glasses.

(a)

(b)

(c)

(d)

A variety of crystals: (a) tourmaline, (b) stibnite,
(c) copper, (d) sulfur. (*American Museum of
Natural History*)

Classification of Crystals

The structural entities of a solid, such as atoms, molecules, or ions, are in
relatively fixed positions. These fixed positions maximize the attractive forces
between neighboring entities. We can classify solids on the basis of the nature of
the attractive forces between their entities.

We have already discussed the coulombic forces that exist in *ionic crystals*,
where the repeating units are ions of opposite charge. In other crystals, the
repeating units are molecules, and the attractive forces may be any of the weak
interactions. Such crystals are called *molecular crystals*. An especially important
type of molecular crystal is one in which the forces of attraction between the units
of structure are hydrogen bonds (Chapter 11). Ice is one such crystal. There is
another type of crystal in which the forces of attraction are chemical bonds
between the atoms that are the structural units of the solid. Such substances are
called *covalent network solids*. Well-known examples of covalent network solids
are diamond, graphite (Chapter 9), and quartz.

Still another important type of crystal is one in which a metal atom is the
repeating structural unit. The forces of attraction in a *metallic crystal* are somewhat different from those we have discussed. They will be described in detail in
Chapter 19.

Crystals range in size from those large enough for their crystalline structure to
be evident at a glance to those so small that they seem to be powders. But these
powders are, in fact, heaps of tiny crystals which require the use of a magnifying
device, perhaps one as powerful as an electron microscope, to detect their
crystalline structure. The way in which a crystal of a given substance grows
affects the size of the crystal. Size is in inverse proportion to the rate of growth. If
a crystal grows slowly from the liquid state or from solution, there will be greater
opportunity for the formation of large crystals.

Crystals exist in a variety of shapes. The way in which a crystal of a given
substance is formed may affect its shape. A sodium chloride crystal that is grown
suspended in a solution of NaCl, as shown in Figure 10.6(a), will be cubical,
because NaCl deposits at an equal rate on all sides of the crystal. But if the NaCl
crystal grows on the bottom of the container (Figure 10.6(b)), material is not
deposited on the face of the crystal that touches the bottom. Therefore, the
crystal grows faster horizontally than vertically, and the width of the crystal will be
at least twice the height.

The shape of a crystal of sodium chloride can be changed even further by
varying the conditions of growth. All the faces of both crystals shown in Figure
10.6 meet at 90° angles. Under certain conditions, we can form crystals of
sodium chloride that superficially do not resemble those of Figure 10.6. One
example is shown in Figure 10.7. These crystals, grown from certain solutions,

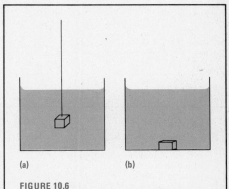
(a)   (b)

FIGURE 10.6
A crystal of sodium chloride grown by suspension in
the center of a solution will have a symmetrical, cubic
shape, as all faces of the crystal grow equally (a). If
the crystal is grown on the bottom of a container, it
will grow faster horizontally than vertically (b).

FIGURE 10.7
An octahedral crystal of sodium chloride. Although
this crystal appears anomalous for sodium chloride,
it is closely related to a cubic crystal.

are octahedral. At first glance, this shape does not resemble the cube or rectangle shown in Figure 10.6. But closer inspection shows that both the cube and the octahedron are regular solids that have many features in common, including the same symmetry.

10.3   UNIT CELLS AND LATTICES

We find the same angles between corresponding faces of all the crystals of a given substance because of the long-range order that is characteristic of the solid state. In fact, a crystal can be defined as an object whose basic units of structure are arranged in a regularly repeating three-dimensional order.

The idea of a regularly repeating pattern is familiar from everyday life. We can understand some characteristics of crystals by examining a tile floor (Figure 10.8), which represents a regularly repeating two-dimensional pattern. To lay a tile floor such as the one in Figure 10.8, we must repeat the basic unit of design, the four tiles that together form one complete pattern. We can describe the tile floor by describing the pattern made by the four-tile group, and then stating that this pattern is repeated to cover the entire area of the floor.

The shape of the area covered by the tiles depends on how the tiles are laid. A square floor is made by laying tiles equally on all four sides. A long, narrow floor is made by adding tiles to the two narrow ends of the floor. An irregularly shaped floor can be made by varying the way in which the tiles are laid. But no matter how irregular the floor might be, the angles between corresponding sides of the floor will always be 90°, because the basic repeating unit is the same square of four tiles (Figure 10.9).

There is another feature of tile floors that is worth noting. Only a limited number of patterns based on certain basic shapes will cover an area completely, with no gaps. An area can be covered entirely by square tiles. An area can also be covered entirely by tiles that are triangles, parallelograms, or regular hexagons.

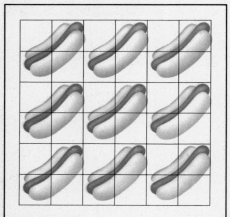

FIGURE 10.8
A tile floor consisting of a regularly repeating two-dimensional pattern, in this case a hot dog. No matter how large an area it covers, the pattern of this floor will always be a repetition of hot dogs.

FIGURE 10.9
Variations on the theme of a regular two-dimensional pattern. This tile floor consists of four tiles fitted together to depict a hot dog. No matter how the pattern is shaped—as a rectangle (a), a square (b), or any irregular shape (c) — the angles between corresponding sides of the floor are still the same.

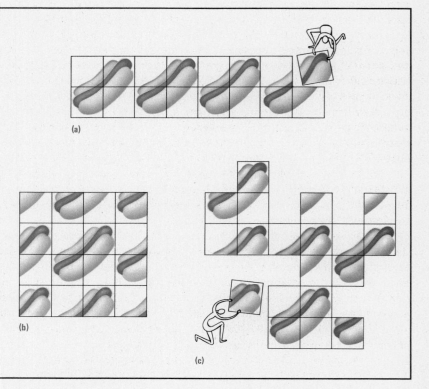

(a)

(b)

(c)

But an area cannot be covered entirely if we use tiles that are regular pentagons or heptagons, or tiles of irregular shape. In fact, there are only a relatively small number of polygons that will cover an area completely.

Tile floors are two-dimensional, but they have a number of features in common with three-dimensional crystals. *A crystal also has a characteristic pattern based on a relatively small number of basic structural units,* which can be atoms, ions, or molecules. This basic repeating three-dimensional pattern is called the **unit cell.** A single unit cell is somewhat analogous to the set of four tiles that make up the basic repeating pattern of the floor. For example, just as the shape of the tiles determines the angles between corresponding sides of the floor, the shape of the unit cell determines the angles between the corresponding faces of a crystal. And just as only tiles of certain shapes can cover a floor area completely, only unit cells of certain shapes can be used to fill three-dimensional space completely.

But the analogy between floor tiles and crystals breaks down if it is carried too far. While a floor tile is a clearly defined object with real sides, the unit cell of a crystal is an idealized shape that is generated by using imaginary lines to connect the points that form the regularly repeating pattern called the *crystal lattice* (Section 7.1, page 187).

An effective way to describe the arrangement of the basic units in a crystal is to describe this pattern of points. Suppose that a crystal is formed by repetition of the group of eight spherical atoms shown in Figure 10.10(a). The most convenient way to describe the repeating pattern of the crystal is to select a set of points corresponding to this pattern. For the group of atoms shown in Figure 10.10(a), we can conveniently place these points in the centers of the atoms, as shown in Figure 10.10(b). When these points are connected by lines, a cube is formed, as shown in Figure 10.10(c). When this pattern of points is continued in three dimensions, it forms a lattice whose unit cell is a cube, as shown in Figure 10.10(d). It should be noted that each point at the corner of a unit cell actually belongs to eight cubes in the lattice, as shown in Figure 10.11. It should also be noted that the points in a representation of a unit cell do not convey information about the size of the atoms relative to the size of the cell. In most crystals, there is very little empty space; the units of structure are almost touching.

For structural entities that are more complicated than spherical atoms—molecules, for instance—an appropriate set of points can be chosen to represent the unit cell. It might seem that a great variety of unit cell shapes is possible. But the variety of shapes is restricted sharply by the requirement that the unit cells must be chosen to fill space completely when repeated in a lattice.

All unit cells are parallelepipeds, solids that have six faces, all of which are parallelograms. In the nineteenth century, Auguste Bravais (1811–1863) devised a classification of unit cells and the crystal lattices that can result from them. The classification is based on the symmetry of the crystals and is designed to help one understand the arrangement of the points of the lattice. The Bravais lattices and the unit cells are most conveniently classified by specifying the relationship between three sides and three angles of the parallelepiped, as shown in Figure 10.12.

There are seven *simple* unit cells. These unit cells are shown in Figure 10.13, which also classifies the simple unit cells by the relationship between three sides and three angles of each. It can be seen that the unit cell is the highly symmetrical cube when all three sides are equal ($a = b = c$) and all three angles are 90° ($\alpha = \beta = \gamma$). The least symmetrical unit cell, called triclinic, has three sides of unequal length ($a \neq b \neq c$) and three angles that are unequal and do not equal 90° ($\alpha \neq \beta \neq \gamma$). The unit cells between these two extremes have various combinations of equal and unequal sides and angles.

In the seven simple unit cells, the lattice points lie only at the corners of the unit cells, and each lattice point belongs to eight unit cells. There is another type

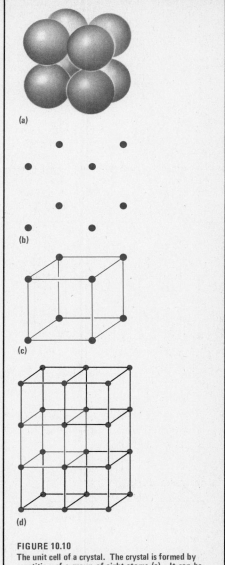

(a)

(b)

(c)

(d)

FIGURE 10.10
The unit cell of a crystal. The crystal is formed by repetition of a group of eight atoms (a). It can be represented by points at the centers of those atoms (b). The points can be connected by lines to form a cube (c), and that cubical pattern can be extended through space to form a lattice (d).

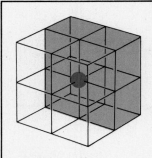

FIGURE 10.11
In a crystal made up of cubic unit cells, each point in the lattice is at a corner of eight adjoining cubes.

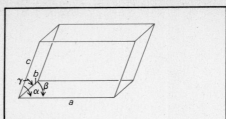

FIGURE 10.12
Unit cells and the crystal lattices that result from them can be classified by specifying the relationship between three sides of a unit cell (a, b, and c) and three angles of the unit cell (α, β, and γ).

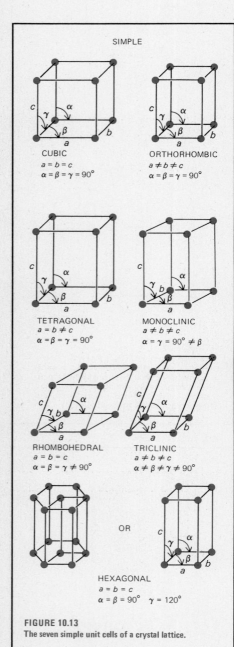

FIGURE 10.13
The seven simple unit cells of a crystal lattice.

of unit cell in which an additional lattice point is found in the center of the unit cell. Such a lattice is described as *body-centered*. There are three body-centered unit cells, which are shown in Figure 10.14. Note that the point in the center of a body-centered unit cell belongs only to that cell.

Still other unit cells have one or more lattice points centered in the faces, in addition to the points at the corners. These lattices are called *face-centered*. There are four face-centered unit cells, which are shown in Figure 10.15. A point on the face of a unit cell belongs to two unit cells, which share a face in the lattice (Figure 10.16).

The relationship between the type of unit cell and the number of atoms that it contains is important. In the seven simple unit cells, there is one lattice point per unit cell (eight lattice points, each of which is shared by eight cells). The three body-centered unit cells each have an additional atom in the center, and therefore a total of two atoms per unit cell. The two face-centered cells have six additional half-atoms on the faces, giving a total of four atoms per unit cell. There are two *end-centered* cells, which have two additional half-atoms, one in each end, giving them a total of two atoms per unit cell.

The crystal lattices observed in solids can be accounted for by fewer than 14 lattices. The Bravais lattice system was devised to help recognize the symmetry of crystals. Being aware of these 14 lattices can help us interpret crystal structures.

You should recognize that all lattice points in a crystal lattice are equivalent when the crystal has one basic unit of structure. Intuitively, it might seem that some points are more important than others. For example, the single point inside a body-centered unit cell, which belongs to that cell alone, might seem unique and different from the points at the cell's corners, which belong to other cells as well. But we can see that all the points are equivalent if we ignore the underlying unit cell and look only at the three-dimensional lattice. There is no difference between any of the points in such a lattice. It is only when we draw imaginary lines to outline unit cells that some lattice points appear to be different from others.

Since macroscopic quantities of solid substances are very large repeating arrays of unit cells, there is a relationship between the properties of the unit cell and properties of the solid. We saw earlier that Avogadro's number often defines a relationship between a measurement on the atomic scale and a measurement on the macroscopic scale. There is a method of determining Avogadro's number that is based on measurements of unit cells in a crystalline solid.

EXAMPLE 10.1
The unit cell in a crystal of metallic calcium is found to be a face-centered cube with a side of 0.556 nm. The density of calcium is 1.54 g/cm^3 and its atomic weight is 40.08. Calculate Avogadro's number from these data.
SOLUTION
The volume of a cube of side a is a^3. Therefore the volume of this cubic unit cell is:

$$(0.556)^3 \text{ nm}^3 = 0.172 \text{ nm}^3$$

The mass of this volume of calcium can be found from the density:

$$\text{mass unit cell} = 0.172 \text{ nm}^3 \times (10^{-7})^3 \frac{\text{cm}^3}{\text{nm}^3} \times 1.54 \frac{\text{g}}{\text{cm}^3}$$

$$= 2.649 \times 10^{-22} \text{ g}$$

Since a face-centered cubic unit cell has four atoms, the mass of a single atom is:

$$\text{mass atom} = 2.649 \times 10^{-22} \text{ g}/4 = 6.622 \times 10^{-23} \text{ g}$$

The ratio of the mass of one mole of a substance to the mass of one atom of a substance is Avogadro's number:

$$\frac{40.02 \text{ g/mol}}{6.622 \times 10^{-23} \text{ g/atom}} = 6.04 \times 10^{23} \text{ atoms/mol}$$

which is the value of Avogadro's number found from these measurements on calcium.

The relationship between the macroscopic properties of a solid and the properties of the unit cell is illustrated further in the following example.

EXAMPLE 10.2

Tungsten has a density of 19.35 g/cm³ and an atomic weight of 183.85. The unit cell in the most important crystalline form of tungsten is the body-centered cubic unit cell. Calculate the length of a side of the unit cell.
SOLUTION
Since there are two atoms per unit cell in a body-centered cubic lattice, the mass of the unit cell is the mass of two atoms of tungsten:

$$\text{mass unit cell} = 2 \frac{\text{atoms}}{\text{cell}} \times 183.85 \frac{\text{g}}{\text{mol}} \times \frac{1}{6.0221 \times 10^{23}} \frac{\text{mol}}{\text{atom}}$$

$$= 6.106 \times 10^{-22} \text{ g}$$

The volume of the unit cell can be found from the density:

$$6.106 \times 10^{-22} \text{ g} \times \frac{1 \text{ cm}^3}{19.35 \text{ g}} = 3.156 \times 10^{-23} \text{ cm}^3$$

The length of the side of a cube of known volume is found by taking the cube root of the volume:

$$(0.03156 \times 10^{-21} \text{ cm}^3)^{1/3} = 0.3160 \times 10^{-7} \text{ cm} = 0.3160 \text{ nm}$$

10.4 CLOSEST-PACKED STRUCTURES

The description of crystal structure by means of lattice points and unit cells is a general method that can be used for any crystal, no matter how complex its repeating unit may be. The method works for a crystal of a complicated organic molecule, such as a protein or a nucleic acid, as well as it does for a crystal of a simple monatomic substance, such as a metal.

There is another, alternative method of picturing the crystal structure of many substances that is simpler than the Bravais lattice approach. We can describe the crystal as an assembly of closely packed spheres. This alternative method works very well when the basic structural unit of a substance is an atom. Some molecular substances can also be described in this way, but only if the molecules are simple enough to be roughly spherical. Some ionic substances also lend themselves to a description of this sort.

The basic idea of the close-packing approach is to determine ways in which spheres of identical size can best be packed to fill a given volume as completely as possible. A variation of this approach often is encountered in everyday life, as when we consider how best to pack oranges or other spherical objects in a crate.

There are two ways to pack spheres most efficiently while preserving long-range order, "efficiently" meaning that the smallest possible fraction of the volume is empty space.

Figure 10.17 shows one two-dimensional layer of spheres packed as closely as possible. Most of the available volume is occupied by the spheres, but there

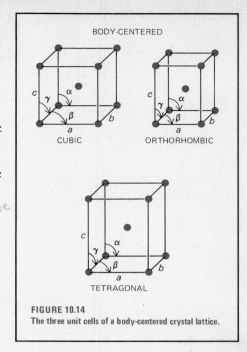

FIGURE 10.14
The three unit cells of a body-centered crystal lattice.

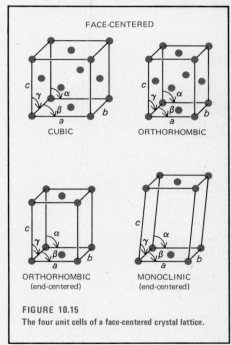

FIGURE 10.15
The four unit cells of a face-centered crystal lattice.

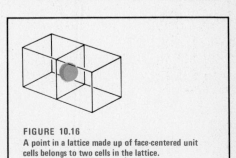

FIGURE 10.16
A point in a lattice made up of face-centered unit cells belongs to two cells in the lattice.

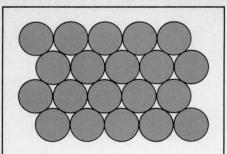

FIGURE 10.17
One layer of spheres packed as closely as possible.
Most of the volume is occupied by the spheres, but
an irreducible volume is represented by the holes,
or interstices, between the spheres.

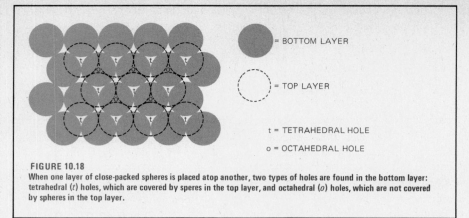

= BOTTOM LAYER

= TOP LAYER

t = TETRAHEDRAL HOLE

o = OCTAHEDRAL HOLE

FIGURE 10.18
When one layer of close-packed spheres is placed atop another, two types of holes are found in the bottom layer:
tetrahedral (t) holes, which are covered by speres in the top layer, and octahedral (o) holes, which are not covered
by spheres in the top layer.

are small empty spaces, or *holes,* between spheres. These holes, sometimes
called *interstices,* are almost as important as the spheres for an understanding of
close-packed structures.

In Figure 10.17, each sphere is surrounded symmetrically by six other
spheres, its nearest neighbors. In three dimensions, close-packed arrangements
are made by stacking layers of such spheres on one another. As Figure 10.18
shows, this stacking is done most efficiently by fitting the spheres of the top layer
into the holes of the bottom layer. Figure 10.18 also shows that only half of the
holes in the bottom layer are covered by spheres in the top layer. The other half of
the holes in the bottom layer lie directly below holes in the top layer.

We can say that there are two "types" of holes in the bottom layer once the
second layer is in place. There are holes that are covered by spheres in the second
layer; these are labeled *t* in Figure 10.18. And there are holes that lie below holes
in the second layer; these are labeled *o*.

When we place layers of spheres together, we create not only two types of
holes in each layer but also different types of space between the layers. These
spaces are called *sites*. In Figure 10.18, the spaces labeled *t* are called *tetrahedral sites*, because each is formed by four spheres whose centers lie at the
corners of a tetrahedron. Figure 10.19(a) shows that three of the spheres in a
tetrahedron are in one layer and the fourth is in the other layer. The spaces
labeled *o* are called *octahedral sites*, because they are formed by six spheres
whose centers lie at the corners of an octahedron, as shown in Figure 10.19(b).
Three of the spheres are in the bottom layer and the other three are in the top
layer.

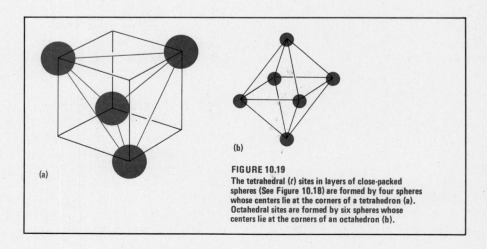

(a)

(b)

FIGURE 10.19
The tetrahedral (t) sites in layers of close-packed
spheres (See Figure 10.18) are formed by four spheres
whose centers lie at the corners of a tetrahedron (a).
Octahedral sites are formed by six spheres whose
centers lie at the corners of an octahedron (b).

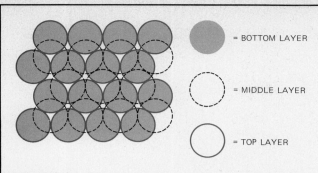

FIGURE 10.20
An *ababab* . . . pattern in three layers of close-packed spheres. Each sphere in the top layer lies directly over a sphere in the bottom layer. The pattern can be extended through an indefinite number of layers.

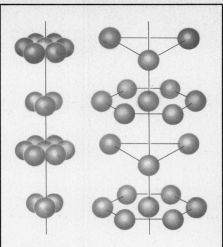

FIGURE 10.21
Another view of the layers of spheres shown in Figure 10.20, showing that the layers can be regarded as alternating units made up of hexagons whose corners are defined by spheres. This arrangement is sometimes called hexagonal closest packing.

If we want to place a third layer on top of the second, there are two ways to do it. One way is to have the third layer fit into the second layer so that each sphere in the third layer is directly above a sphere of the first layer, as shown in Figure 10.20. In this method, the layers of spheres have an *ababab* . . . pattern; that is, the spheres of the first, third, fifth, seventh, . . . layers lie directly above each other, and the spheres of the second, fourth, sixth, eighth, . . . layers also lie directly above each other. Figure 10.21 shows another view of this kind of packing, one that helps to explain why it is called *hexagonal closest packing*. This kind of packing does not correspond directly to one of the Bravais lattices.

Another way to fit the third layer of spheres into the second layer is to have the spheres of the third layer lie above the holes labeled *o*, the octahedral sites of the first layer, as shown in Figure 10.22. In this method of packing, the orientation of the spheres in the third layer differs from those of both the first and second layers. This is an *abcabc* . . . pattern, since it does not repeat until the fourth layer. Figure 10.23 shows why this arrangement is called *cubic closest packing*. Cubic closest packing corresponds to a face-centered cubic unit cell (see Figure 10.15).

A detailed study of these two arrangements shows that they have a number of features in common:

1. Both arrangements have the same fraction of empty space. In both cases, just over 74.0% of a given volume is occupied by spheres and just under 26.0% is empty space. It can be shown that this is the closest regular packing arrangement possible for spheres of equal size.
2. Each sphere has 12 nearest neighbors, six in its own horizontal layer and three in each adjacent horizontal layer.

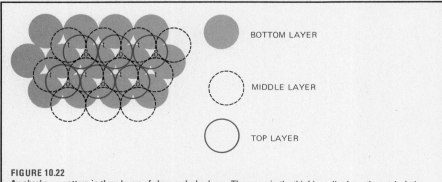

FIGURE 10.22
An *abcabc* . . . pattern in three layers of close-packed spheres. The speres in the third layer lie above the octahedral (o) holes in the first layer.

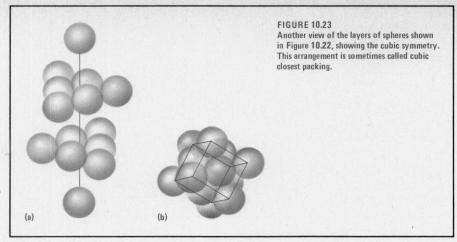

FIGURE 10.23
Another view of the layers of spheres shown in Figure 10.22, showing the cubic symmetry. This arrangement is sometimes called cubic closest packing.

(a)   (b)

hexagonal cp (hcp) – 12
cubic cp (ccp) – Face-centered –12
Body centered – 8
simple – 6

3. The number of octahedral sites is equal to the number of spheres. The number of tetrahedral sites is equal to twice the number of spheres.

Metals such as Be and Mg in group IIA, the metals in groups IIIB and IVB, and Zn and Cd in group IIB have hexagonal closest-packed crystal structures. Metals such as Ca and Sr in group IIA, Ni and Pt in group VIII, and Cu, Ag, and Au in group IB have cubic closest-packed crystal structures. At low temperatures, the noble gases (with the exception of helium) crystallize in a cubic closest-packed arrangement. Some small, more or less spherical, molecules such as H_2 may be found in closest-packed crystalline arrangements. The major advantage of a closest-packed arrangement is that the attractive forces are maximized, because each structural unit has a large number of nearest neighbors.

Some metals, such as those in groups IA, VB, and VIB, as well as Fe and Ba, are found to crystallize with a structure that is not packed quite as closely as the two structures mentioned thus far. These metals form a structure that has body-centered cubic unit cells, as was shown in Figure 10.14. It can be seen that each sphere in the body-centered cubic structure has only eight nearest neighbors (Figure 10.24), as compared to 12 in the two closest-packed arrangements. However, in the body-centered cubic arrangement there are six next-nearest neighbors, which lie relatively close to each sphere. In this arrangement, spheres occupy 68% of the available volume.

Once again, an understanding of closest-packed structures on the atomic level allows us to relate macroscopic properties and atomic properties.

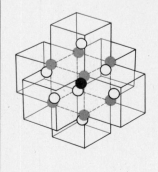

FIGURE 10.24
The body-centered cubic structure typical of some metals. Each sphere has eight nearest neighbors and six next-nearest neighbors.

CENTER ATOM (●) HAS
8 NEAREST NEIGHBORS (O)
6 NEXT-NEAREST NEIGHBORS (●)
- - - - - BODY-CENTERED UNIT CELL

EXAMPLE 10.3

Magnesium crystallizes with a hexagonal closest-packed structure that has a density of 1.74 g/cm^3. Calculate the volume and the radius of a magnesium atom.

SOLUTION

The volume occupied by one mole of magnesium is:

$$V_{molar} = \frac{1 \text{ cm}^3}{1.74 \text{ g Mg}} \times 24.3 \frac{\text{g Mg}}{\text{mol Mg}} = 14.0 \text{ cm}^3/\text{mol Mg}$$

This volume of Mg consists of 6.02×10^{23} (Avogadro's number) of spherical atoms of Mg and of empty space between the atoms. In a closest-packed arrangement, 26.0% is empty space and 74.0% is the actual volume of the Mg atoms:

$$14.0 \text{ cm}^3/\text{mol Mg} \times \frac{74\%}{100\%} = 10.3 \text{ cm}^3/\text{mol Mg}$$

The volume of an individual atom is:

$$10.3 \text{ cm}^3/\text{mol Mg} \times \frac{1 \text{ mol}}{6.02 \times 10^{23} \text{ atoms}} = 1.72 \times 10^{-23} \frac{\text{cm}^3}{\text{atom Mg}}$$

The individual atom is a sphere. To find the radius of a sphere from its volume, we employ the relationship:

$$V = \frac{4}{3}\pi r^3 \quad \text{or} \quad r = \left(\frac{3V}{4\pi}\right)^{1/3}$$

where V is the volume and r is the radius of the sphere. For this spherical atom, r can be found to be 1.60×10^{-8} cm = 0.160 nm.

The usefulness of picturing crystal structures as close-packed arrangements is not limited to monatomic or simple molecular substances. This picture can also help us to understand the structures of many ionic solids.

Monatomic ions can be regarded as spheres. In Chapter 7 (page 190), we mentioned that anions usually are considerably larger than cations. The structure of many salts made up of ions with a large difference in size can be pictured as a close-packed arrangement of the large anions, with the much smaller cations placed in the holes.

Sodium chloride has such a structure. The relatively large Cl^- anions are arranged in a cubic closest-packed structure. The relatively small Na^+ cations are located in a regular way throughout this close-packed lattice. As we know from the formula NaCl, there are an equal number of Na^+ cations and Cl^- anions.

We mentioned that the number of octahedral sites is equal to the number of spheres in a closest-packed lattice. In an NaCl crystal, all the Na^+ ions are found in the octahedral sites of the Cl^- lattice. Since an octahedral site is formed by six spheres, each Na^+ ion has six Cl^- ions as nearest neighbors, as shown in Figure 10.25. Examination of the structure shows that each Cl^- ion also has six Na^+ ions as nearest neighbors.

Cubic closest packing corresponds to a face-centered cubic cell. The Cl^- anions in NaCl make up such a unit cell. It can be shown that the Na^+ cations in the octahedral sites also define a face-centered cubic unit cell. Figure 10.26 shows that sodium chloride can be regarded as two interlocking lattices made up of face-centered cubic unit cells. However, it is easier for most people to visualize the Na^+ ions occupying the octahedral sites of a cubic closest-packed Cl^- lattice.

Many salts and oxides have the same crystal structure as NaCl. Since the number of octahedral sites is equal to the number of spheres, this type of structure appears in substances with equal numbers of cations and anions. A compound such as Na_2O cannot have the sodium chloride lattice because there

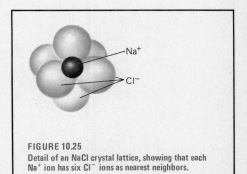

FIGURE 10.25
Detail of an NaCl crystal lattice, showing that each Na^+ ion has six Cl^- ions as nearest neighbors.

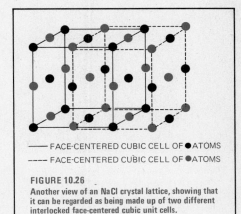

—— FACE-CENTERED CUBIC CELL OF ● ATOMS
---- FACE-CENTERED CUBIC CELL OF ● ATOMS

FIGURE 10.26
Another view of an NaCl crystal lattice, showing that it can be regarded as being made up of two different interlocked face-centered cubic unit cells.

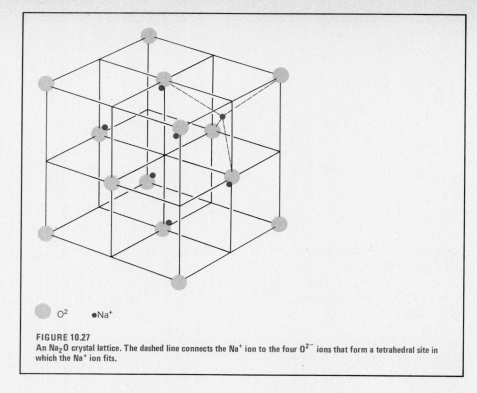

FIGURE 10.27
An Na_2O crystal lattice. The dashed line connects the Na^+ ion to the four O^{2-} ions that form a tetrahedral site in which the Na^+ ion fits.

○ O^2 •Na^+

are two cations for each anion, and there are not enough octahedral sites to hold all the cations. Substances of this kind have a structure in which the anion, O^{2-} in this case, forms a cubic closest-packed lattice, and the cations occupy the tetrahedral sites. This arrangement is possible because there are twice as many tetrahedral sites as spheres. Figure 10.27 shows that each Na^+ ion in Na_2O has only four nearest neighbors, while each O^{2-} ion can be seen to have eight Na^+ nearest neighbors if the lattice is extended.

10.5 X-RAY CRYSTALLOGRAPHY

One might ask how so much is known about the arrangement of units of structure in crystals. The information is obtained primarily by a method called X-ray crystallography, which is the study of the interaction between X rays and crystals. The details of this branch of chemistry can be quite complex. But a general understanding can be achieved by recalling certain characteristics of electromagnetic radiation.

Suppose beams of light of a given wavelength start out from two or more points of origin and arrive at the same point. If the light waves are in phase—that is, if the maxima and minima of each wave strike the point simultaneously—we will see a bright light. If the light waves are in phase, the waves add to, or reinforce each other, as was shown in Figure 7.14. It can be shown that the waves will reach a given point in phase if two conditions are met: (a) all the beams start at the same time and (b) the difference of the distances they travel is an integral multiple of the wavelength.

If the waves are out of phase when they strike the point, they will interfere, or cancel, to a greater or lesser extent. If the waves are out of phase, the intensity at the point of interest will be less than the intensity of the individual beams. In the extreme case where the beams are completely out of phase, so that a maximum of one wave and a minimum of another reach the point simultaneously, there will be a complete cancellation of the light, as was shown in Figure 7.14. A diffrac-

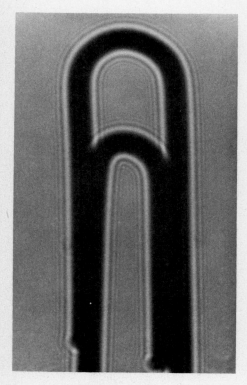

A diffraction pattern produced as light passes by the edges of a paper clip. The picture was taken without a lens in the camera, since the interference bands would be destroyed by focusing the image through a lens. (*Fundamental Photographs, Granger*)

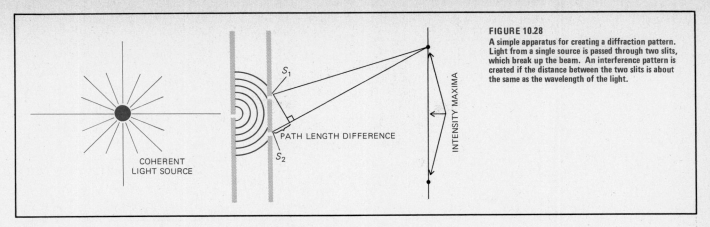

FIGURE 10.28
A simple apparatus for creating a diffraction pattern. Light from a single source is passed through two slits, which break up the beam. An interference pattern is created if the distance between the two slits is about the same as the wavelength of the light.

tion pattern is produced by wave interference of visible light. The bright bands are caused by wave reinforcement, the dark bands by wave cancellation.

Experimentally, the two conditions described above are met by using one light source, which is split into several beams by one method or another. Figure 10.28 shows a simple method that uses two slits as separate starting points for the light, creating an interference pattern. If there were a number of slits, we would create the more complex diffraction pattern.

In 1912, a group of physicists at the University of Munich showed that a diffraction pattern could be created by directing a beam of radiation at a crystalline substance. Their experiment is represented in Figure 10.29. The effect of directing the beam of radiation at the regularly spaced units of structure of the crystal is the same as sending radiation through regularly spaced slits. In each case, a diffraction pattern is created.

The physicists did not use visible light in their experiment with the crystal. To create a diffraction pattern, the "slits" through which the radiation travels must be separated by a distance that is roughly the same as the wavelength of the radiation. In the experiment, the "slits" were the spaces between units of structure of the crystal. The units are separated by about 10^{-1} nm, so X rays, which have wavelengths in this range, were used to produce diffraction patterns.

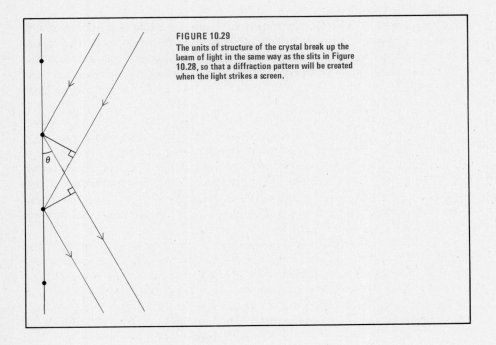

FIGURE 10.29
The units of structure of the crystal break up the beam of light in the same way as the slits in Figure 10.28, so that a diffraction pattern will be created when the light strikes a screen.

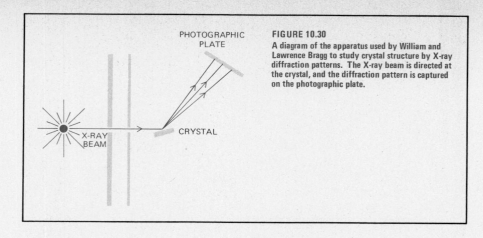

FIGURE 10.30
A diagram of the apparatus used by William and Lawrence Bragg to study crystal structure by X-ray diffraction patterns. The X-ray beam is directed at the crystal, and the diffraction pattern is captured on the photographic plate.

The characteristics of an X-ray diffraction pattern are related to the structure of the crystal, the wavelength of the X rays, and the angle at which the radiation strikes the crystal. Two English physicists, William Bragg (1862–1942) and his son Lawrence (1890–1971), were foremost in devising experimental methods for studying X-ray diffraction patterns and in developing the theory for interpreting the results. Figure 10.30 shows a schematic diagram of their apparatus.

The theory developed by the Braggs is based on the idea that if two waves arrive at the same point in phase—that is, if they produce maximum intensity at this point—then the difference between the distances they have traveled from their origins must be an integral multiple of their wavelength.

Figure 10.31 shows how one beam of X rays interacts with two units of structure in successive layers of a crystal. Using a picture of this type, the Braggs showed how to calculate the distance between these layers from X-ray diffraction data.

The distance is calculated by measuring the angle between the crystal and a spot of intense radiation in the diffraction pattern. This angle is labeled θ in the diagram. The intense spot is produced on the detector when waves of X rays that are "reflected" off different units of structure in the crystal arrive at one point on the detector in phase. For this to happen, the difference between the distances traveled by the waves after they are reflected by the crystal must be an integral multiple of the wavelength, $n\lambda$.

The geometric construction shown in Figure 10.31 allows us to find the relationship between the angle θ, the difference in path length, and the distance between layers of the crystal. If the distance between layers is called d, then the difference between the path lengths of the two waves is $2d \sin \theta$. We can thus write:

$$2d \sin \theta = n\lambda \qquad (10.1)$$

where n is any integer. This relationship is called the Bragg equation. Experimentally, θ and λ are measured, and d then can be calculated.

The study of the diffraction patterns made by X rays interacting with crystals can give much more information than just the spacing between planes of the lattice points in the crystal. If the crystal consists of molecules, the diffraction pattern can be used to map the electron density of the molecules. The structure of the molecule often can be determined from the electron density.

X-ray crystallography is one of the most powerful methods available for determining molecular structure. Computers can be used to analyze diffraction patterns, mapping the electron densities and thus the structure of the very large and complex molecules that play essential roles in living organisms. Proteins and nucleic acids have been mapped in this way. Among the major successes of X-ray crystallography are the determination of the structure of many complex protein molecules such as myoglobin, the oxygen-transporting protein of muscle cells,

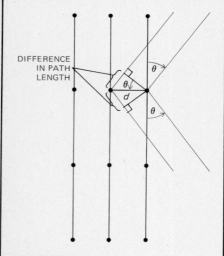

FIGURE 10.31
The interaction of a beam of X rays with units of structure in two layers of a crystal. By analyzing the diffraction pattern created by such an interaction, we can determine the distance between layers of the crystal. If the distance between layers is d, the difference between the path lengths of the two waves is $2d \sin \theta$.

X-RAY CRYSTALLOGRAPHY AND THE HEMOGLOBIN MOLECULE

The first protein whose structure was determined by X-ray crystallography was myoglobin. While the structure of myoglobin appears complex, it is relatively simple compared to that of hemoglobin, which was an even greater challenge for X-ray crystallographers. While myoglobin is a single chain of subunits with one iron-containing heme group, hemoglobin consists of four chains and four heme groups, containing a total of more than 10 000 atoms. The same set of techniques, developed in British laboratories in the 1950s, was used to determine the structure of myoglobin and hemoglobin.

Perhaps the most important part of the technique is the use of heavy atoms, such as those of mercury, to serve as reference points. The heavy atoms, attached at specific sites in the hemoglobin molecule, change the diffraction patterns. By attaching different atoms at different sites and comparing literally thousands of diffraction patterns, it is possible to build up a three-dimensional image of the molecule, layer by layer. Millions of arithmetic operations must be performed to obtain the image, and the task would be impossible without the use of computers to perform the calculations.

The structure of the myoglobin molecule was determined in 1957. The first accurate three-dimensional image of the hemoglobin molecule was obtained two years later. Somewhat to the surprise of the investigators, the basic structure of both myoglobin and the hemoglobin molecule proved to be essentially the same. Each of the four chains of the hemoglobin molecule has a shape closely resembling that of the myoglobin molecule. It has been found that this similarity in shape does not reflect a similarity of composition. The sequences of the amino acid subunits that make up the hemoglobin and myoglobin molecule are 90% different; that is, the same amino acid occurs in only 10% of the sites in the two chains. It is believed that the two molecules had a common ancestor, and that the differences are due to the different evolutionary paths that were followed. Thus the determination of the structure of hemoglobin and myoglobin helps us to understand how they function on the molecular level and also tells us something about their evolution.

and lysozyme, the antibacterial enzyme present in many body fluids. The double helix structure of DNA was proposed by Watson and Crick on the basis of X-ray diffraction data obtained by others. X-ray crystallography has had a profound influence on modern molecular biology.

10.6 PHASE TRANSITIONS AND PHASE EQUILIBRIA

Chapter 4 had a brief discussion of the way in which changes in temperature determine the most stable phase—solid, liquid, or gas—of a given substance. We know from experience that if the pressure remains constant and the temperature goes down, most substances change from gas to liquid to solid. There are exceptions to the rule. At atmospheric pressure, for example, CO_2 goes directly

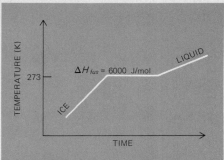

FIGURE 10.32
The warming curve of water. The temperature of solid water—ice—increases steadily until it reaches 273 K, then remains constant for a time as the heat melts the ice. When the ice is melted, the temperature of the liquid water increases as heat continues to be added. The length of the horizontal part of the warming curve depends on the size of the sample of water and the rate at which heat is added.

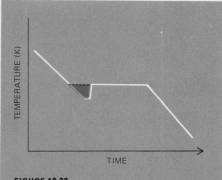

FIGURE 10.33
The cooling curve of water. The unexpected dip is caused by supercooling—failure of the liquid to freeze because its molecules cannot achieve the proper ordered arrangement.

from gas to solid. There are other details of phase changes and of the relationship between phases that are worth considering.

Figure 10.32 shows what is called a *warming curve*, which illustrates the way in which the temperature of a sample of water goes up as thermal energy is added at a constant rate. As expected, the temperature of solid water, ice, at first increases linearly with time as heat is added. But at 273 K, the melting point of ice, the addition of heat no longer brings an increase in temperature. The temperature remains the same because the heat melts the ice. The quantity of heat that must be added to the system to cause this change in phase is the heat of fusion. When all the ice has melted, the temperature of the water again rises as heat is added.

It is logical to assume that if we started with liquid water and removed heat at a constant rate, we would get a plot identical to the warming curve but running in the other direction. In fact, there are differences.

Figure 10.33 shows a plot, called a *cooling curve,* in which heat is removed at a constant rate from a sample of water. The unexpected dip in the cooling curve occurs because a liquid does not necessarily solidify when its temperature reaches the freezing point. Water often remains liquid when its temperature is well below 273 K. This phenomenon is called *supercooling*. It occurs because molecules must be arranged in a regular manner to form a crystalline solid. When the freezing point is reached, some time may pass before the necessary pattern of molecules forms in a portion of the liquid. Until this pattern forms, the substance remains liquid even though the temperature goes down.

Eventually, however, a small crystal forms in the liquid. This crystal acts as a template, or seed, for the rapid crystallization of part of the liquid. The release of heat that accompanies the crystallization causes the temperature of the liquid to rise back to the freezing point. The temperature then remains at the freezing point until the entire sample crystallizes. Only then does the temperature begin dropping again.

Thus, we see that during heating, a solid begins melting as soon as its melting point is reached, while there is a tendency for the liquid state to persist during cooling. For a liquid to solidify, the temperature must drop to the freezing point and an ordering of the molecules must occur. A regular arrangement of the molecules often does not occur instantly, and supercooling ensues.

Several different techniques are used to promote crystallization. A liquid can be stirred vigorously to create more motion and hasten the formation of the molecular arrangement needed for crystallization. Solid surfaces, such as dust particles or, ideally, small seed crystals, can be used to provide sites to promote the desired crystalline pattern. If water is kept both still and dust-free, it can remain liquid at 235 K, nearly 40 K below its freezing point.

Normally, crystallization is induced easily in supercooled liquids. But there are some important exceptions to this rule. The forces of attraction between the molecules or atoms of some liquids are strong, even though long-range order does not exist. When such liquids are supercooled and the thermal motion of their molecules is reduced, it may be virtually impossible for the molecules to arrange themselves into the regular pattern needed for crystallization. Such a liquid will remain liquid no matter how low the temperature goes.

We know that the viscosity of a liquid increases as the temperature goes down. The viscosity of a supercooled liquid may increase so much that the rate of flow becomes negligible. The flow is unobservable by any ordinary means, and the supercooled liquid has the appearance of a solid. But it is not a true solid. It lacks the long-range order of a solid, and its cooling curve and warming curve do not display the flat regions characteristic of the phase change from liquid to solid. Such supercooled liquids are called amorphous solids or *glasses,* the latter name coming from the best known of these materials.

Ordinary glass is a supercooled liquid prepared by cooling a mixture of molten silica, SiO_2, sodium carbonate, and calcium carbonate. The cooling curve for this liquid does not show the flat portion, or discontinuity, characteristic of crystallization. There are strong forces of attraction between the units of structure of glass, in the form of silicon-oxygen bonds. But these units of structure are not arranged in the regular repeating pattern expected in a crystal. Rather, they become parts of very large molecules that are not in a regular pattern. Glass does flow, although the rate of flow is so small as to be unnoticeable in most cases. But some centuries-old panes of glass can be found to be thicker at the bottom than at the top. The thickening is evidence that the glass in these panes is flowing.

The persistence of the liquid state that manifests itself as supercooling also can be found when we examine the behavior of liquids on warming. We expect the warming curve of a liquid to resemble that of a solid. There should be a flat portion where added heat does not raise the temperature but instead causes boiling, the change from liquid to gas. A substantial amount of heat, the heat of vaporization, may be needed to cause this phase change.

But boiling may not occur when the temperature of a liquid goes above the boiling point. This phenomenon is called *superheating*. When a liquid boils, bubbles of its vapor form below its surface. Bubbles can form only on some sort of site, such as a particle of dust or a sharp edge. If no such site exists, bubbles do not form and the liquid does not boil. Very clean water has been heated 150 K *above* its normal boiling point without the occurrence of boiling. When a super-heated liquid finally does boil, bubbles can form so violently that the liquid spatters out of its container. This potentially dangerous behavior is called bumping. In many laboratory operations that require boiling, inert solids with sharp edges, called boiling chips, are added to a liquid to prevent superheating and bumping.

Supercooled liquid and superheated liquid represent conditions different from those we have discussed until now. They are nonequilibrium states, and this is reflected in their unusual behavior—the sudden crystallization of supercooled liquid and the bumping of superheated liquid.

In Chapter 4, page 82, the equilibrium state of a substance was described as a function of temperature and pressure. There is a temperature at which the vapor pressure of a liquid is 1 atm. This temperature is called the *normal boiling point* of the liquid. For water, the normal boiling point is 373 K (100°C). When a liquid is in a system where the pressure is 1 atm, such as in an open container, it boils when the temperature reaches the normal boiling point.

At the normal boiling point, the liquid and its vapor are in equilibrium at a pressure of 1 atm. Below this temperature, the liquid and its vapor are in equilibrium at a pressure of less than 1 atm. Above this temperature, no liquid exists at equilibrium if the pressure is only 1 atm. Thus, superheated liquid is not at equilibrium.

Similarly, at a given pressure there is a temperature called the *freezing point* at which solid and liquid are at equilibrium. The freezing point of water is 273 K at 1 atm. Above this temperature, no solid is present at equilibrium. Below this temperature, no liquid is present at equilibrium. If liquid exists below the freezing point, the system is not at equilibrium. Indeed, while supercooled water can exist for some time in an isolated system, it cannot exist in contact with ice.

Phase Diagrams

We have seen that the equilibrium state of a system depends on two variables, temperature and pressure. The study of the variations of the equilibrium state of a system with changes in pressure and temperature is important in chemistry. Figure 10.34 shows an idealized apparatus for studying these variations. The temperature of the system can be varied by changing the temperature of the bath.

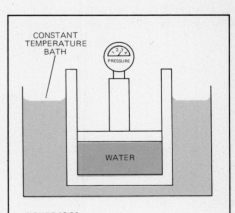

FIGURE 10.34
An idealized apparatus for studying the effect of variations in temperature and pressure on the equilibrium state of a system. Temperature can be controlled by heating or cooling the water in the bath, and pressure can be controlled by changing the force on the piston.

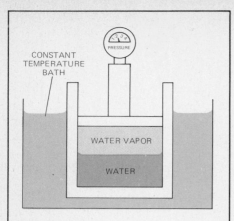

FIGURE 10.35
The measurement of the vapor pressure of water at a given temperature. The pressure on the piston is lowered to the point where it is equal to the vapor pressure and the piston therefore rises to create vapor space.

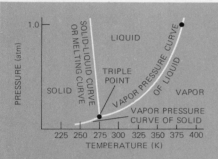

FIGURE 10.36
The phase diagram of water. All the values of P and T at which liquid and vapor are in equilibrium fall on the vapor pressure curve of liquid water. All the points at which vapor and solid are in equilibrium fall on the vapor pressure curve of solid water. All points at which solid and liquid are in equilibrium fall on the solid-liquid curve. The triple point is at the intersection of the three curves. It is the temperature and pressure at which all three phases are in equilibrium.

The pressure can be varied by changing the force on the piston, which is assumed to be weightless and frictionless.

Suppose the system in Figure 10.34 is at a temperature of 300 K and a pressure of 0.5 atm. Only liquid is present, because the vapor pressure of water at 300 K is only 0.035 atm (Table 4.1). The vapor pressure is not enough to push the piston off the surface of the liquid, so there is no vapor space. If the pressure on the piston is lowered to 0.035 atm, the vapor pushes the piston up, as shown in Figure 10.35. At a pressure of 0.035 atm and a temperature of 300 K, liquid and vapor are in equilibrium. But if the external pressure is kept constant at 0.5 atm and the temperature is raised, vapor appears at 355 K. At this temperature, the vapor pressure of water is 0.5 atm. Liquid and vapor are in equilibrium at $T = 355$ K and $P = 0.5$ atm.

The same sort of reasoning can be applied to the changes that occur as the temperature goes down. If supercooling does not occur, solid appears at the temperature at which solid and liquid are in equilibrium when $P = 0.5$ atm. For water, this temperature is very slightly above 273.15 K, the freezing point of water in air at a pressure of 1 atm.

Experiments of this type yield a great deal of data, which are most conveniently shown in graphical form. Figure 10.36 shows such a graph for water. All the points corresponding to values of P and T at which vapor and liquid are in equilibrium fall on a smooth curve, which is called the *vapor pressure curve* or *boiling curve* of liquid water. All the points corresponding to values of P and T at which vapor and solid are in equilibrium fall on another smooth curve, which is called the *vapor pressure curve* (Section 4.2, page 81) or *sublimation curve* of solid water, or ice. All the points corresponding to values of P and T at which solid and liquid are in equilibrium fall on a third smooth curve, which defines the melting behavior of water.

When all three of these curves are drawn using the same set of axes, we have what is called a **phase diagram**. Figure 10.36 is the phase diagram of water. The three curves intersect at only one point, the **triple point**. The triple point marks the temperature and pressure at which solid, liquid, and gas all are in equilibrium. For water, the triple point occurs at $T = 273.16$ K and $P = 0.006025$ atm, or 610.5 Pa. For any system composed of one chemical substance, there is only one point at which three phases can be in equilibrium. It should be noted that the triple point of water is the temperature at which the vapor pressure of solid and the vapor pressure of liquid are equal.

All P, T points that fall on curves of a phase diagram correspond to conditions where two phases are in equilibrium. But most pressure-temperature combinations do not fall on these curves. The regions between curves in phase diagrams correspond to combinations of pressure and temperature at which just one phase of a substance can exist at equilibrium.

For example, the region to the left of the melting curve and above the solid vapor pressure curve represents temperatures lower than the melting point and pressures higher than the vapor pressure. This region corresponds to a state that has only solid. The region between the melting curve and the liquid vapor pressure curve—that is, T higher than the melting point and P higher than the vapor pressure—corresponds to an equilibrium state in which only liquid exists. The region to the right of and below the vapor pressure curve, representing higher temperature than the boiling point and lower pressure than vapor pressure, corresponds to an equilibrium state in which only vapor exists. These regions are labeled accordingly in the phase diagram.

By studying phase diagrams, we can learn a great deal about the way substances respond to changes in pressure and temperature. In the phase diagram for water, we see that there is only one point, the triple point, at which all three phases are in equilibrium. If we start at the triple point and change either pressure or temperature, one or two of the phases will disappear.

The lines on the diagram represent pressure-temperature combinations at which two phases can exist simultaneously. If we start on one of the lines and change either pressure or temperature, one of the two phases disappears. The regions between the lines represent pressure-temperature combinations at which only one phase exists. In these regions, the values of pressure and temperature can be changed over a range without changing the phase of the system.

The ideas we use to discuss the phase diagram of water can be applied to other systems with only one component. Figure 10.37 shows the phase diagram for CO_2. The diagram has a general resemblance to the phase diagram for water, with three curves intersecting at a triple point. But there are a number of differences between the two phase diagrams. These differences are consistent with the differences between the properties of water and carbon dioxide.

The phase diagram for CO_2 is shifted to lower temperatures. At room temperature, water is a liquid but CO_2 is a gas. The pressure at the triple point of CO_2 is 5.1 atm. At pressures lower than 5.1 atm, it is possible to have solid CO_2 or CO_2 vapor or an equilibrium between the two, but it is not possible to have liquid CO_2. Liquid CO_2 is never observed at atmospheric pressure. "Dry ice," solid CO_2, does not melt as it warms at 1 atm. Rather, it goes directly to gas, as the phase diagram shows.

It is interesting to consider the shape of a warming curve for solid CO_2 at 1 atm pressure, because CO_2 does not melt. Figure 10.38 shows that the warming curve of CO_2 looks very much like the warming curve of ice. There is the same flat part of the curve where an input of thermal energy does not raise the temperature but changes the phase of the material. The phase change for CO_2 is from solid to vapor. The heat required for the change is the heat of sublimation, ΔH_{sub} at 1 atm. The phase change occurs at 195 K, the temperature at which the vapor pressure of the solid is 1 atm.

The vapor pressure curve of liquid CO_2 is shown as ending at the critical point (Chapter 4, page 109). There is no need to continue the line past the critical point, since there is no distinction between liquid and vapor above the critical point.

One difference between the phase diagrams of H_2O and CO_2 is the slopes of their melting point curves, the lines representing pressure-temperature conditions at which liquid and solid exist simultaneously. For water, the line has a negative slope; it goes to the left as pressure increases. That is, the melting point of ice decreases slightly as the pressure increases. For CO_2, the curve has a positive slope; it goes to the right as pressure rises. The melting point of CO_2 increases slightly as pressure increases. Most substances behave as CO_2 does, with their melting points increasing as pressure increases. Water is exceptional in this respect. The unusual melting behavior of ice will be discussed in more detail in Chapter 11.

10.7 LIQUID CRYSTALS

Some substances do not undergo the simple transition from the solid phase to the liquid phase that has been described for water. There can be phases whose structures lie between the complete long-range order of a crystal and the almost complete long-range disorder of a liquid. These phases can occur in any substances whose molecules are nonspherical, but are most commonly observed in substances with long-chain molecules.

Figure 10.39 shows how this occurs. In Figure 10.39(a), a crystal of long-chain molecules is ordered in three dimensions. In Figure 10.39(b), the molecules all have the same orientation and they are in equispaced planes, but the arrangement within the planes is irregular. The substance no longer is a true crystal, although it still has more order than a liquid. It is in what is called a

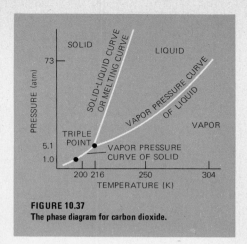

FIGURE 10.37
The phase diagram for carbon dioxide.

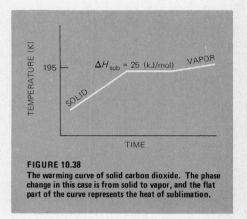

FIGURE 10.38
The warming curve of solid carbon dioxide. The phase change in this case is from solid to vapor, and the flat part of the curve represents the heat of sublimation.

FIGURE 10.39
Four different states of the same substance. (a)
Crystalline. (b) Smectic structure. (c) Nematic
structure. (d) Liquid structure.

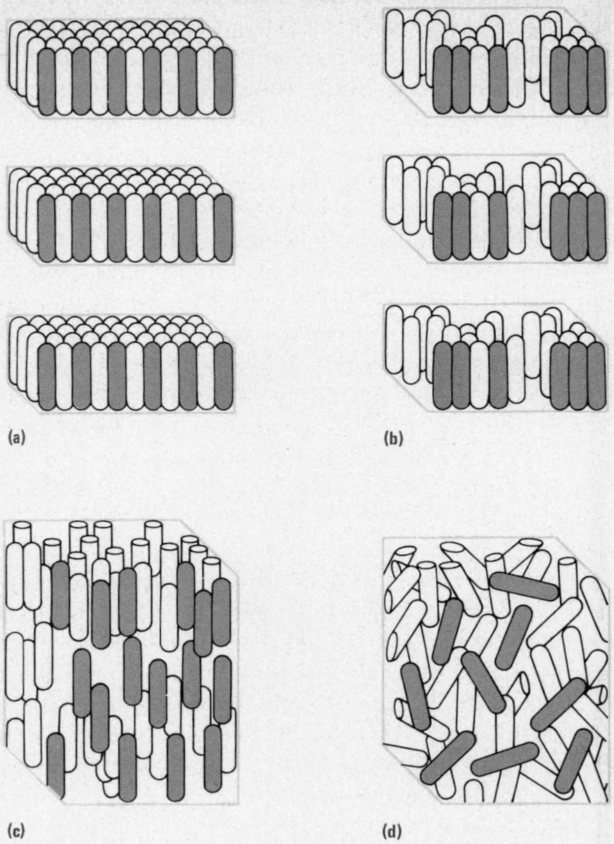

(a)

(b)

(c)

(d)

smectic state. When the molecules no longer are in equispaced planes (Figure 10.39(c)), more disorder is introduced, although some order is maintained if the molecules all have the same orientation; this is called the nematic state. When even this order is lost, the substance is in the liquid state (Figure 10.39(d)).

Substances that have the partially ordered arrangements are said to be in the *paracrystalline state*. They are called liquid crystals. A liquid crystal is a distinct phase, like a solid or a liquid. Substances that display the changes shown in Figure 10.42 can be seen to undergo sharp transitions from crystal to liquid crystal, from liquid crystal to liquid, or even from the smectic to the nematic state. These transitions are comparable to the melting or freezing of solids and liquids.

For some substances, a number of distinct paracrystalline phases between solid and liquid can be distinguished. The changes from one phase to another can occur abruptly as the result of a slight change in temperature. Often, the different paracrystalline states have different colors. A growing number of practical applications have been found for liquid crystals whose color changes in response to very slight temperature changes.

Liquid crystals can be used in thin, lightweight display devices that require very little energy. In recent years, liquid crystal display devices have been used increasingly in digital watches, pocket calculators, and similar instruments. A typical display device has a film of liquid crystal no more than a thousandth of an inch thick between two pieces of glass, one of which is coated on one side with a conductive material. The liquid crystal molecules are polar, so the application of a small electric current can align them to present a number or a letter. The image can be changed by altering the voltage.

The paracrystalline state has features of both liquids and solids. A substance in the paracrystalline state flows and mixes readily, but it also maintains an

TEMPERATURE IN LIVING COLOR

A physician paints a black liquid onto a patient's skin. In moments, the skin is alive with a pattern of vivid reds and blues that give a startlingly good picture of the blood vessels in the body.

A nurse in an intensive care unit for newborn babies checks the color of a small circle of material on an infant's stomach. That quick glance tells her whether the baby's body temperature is within normal limits.

A business executive who isn't feeling quite right holds a small square of plastic to his forehead. By looking in the mirror to see what appears on the plastic, the executive can tell whether he's running a fever.

All three of these devices use liquid crystals that change color in response to slight changes in temperature. The liquid painted on the patient's body shows warmer areas as blue, cooler areas as red, and intermediate temperatures as green or yellow. The dot on the baby's stomach is blue-green at normal temperature but turns bronze if the infant is too cold; if the baby is running a fever, the dot turns blue. The small piece of plastic used by the executive has been designed so that a warning letter will appear if the person using it is feverish.

The medical uses of temperature-sensitive liquid crystals are only a few of the applications of this new technology. In the Apollo missions to the moon, liquid crystal markers were placed on the scientific instruments that were left on the surface of the moon. The Apollo astronauts could tell at a glance whether the instruments were overheating. The "mood rings" that were a brief craze in 1976 consisted of liquid crystal patches. The changes in mood that the rings were supposed to mirror by color changes were actually temperature variations of the body. In a less emotional application, digital thermometers have been designed in which the temperature responses of liquid crystals are carefully orchestrated. As the temperature goes up or down, one number fades out and another fades in. The color of the number that is displayed gives a more precise reading of temperature: a blue 22 indicates that the temperature is closer to 23°C, while a white 22 indicates a reading closer to 21°C. Liquid crystal producers see another potential market among wine lovers, who monitor the temperature at which their wine is stored constantly to be sure that the wine is aging properly. A tiny liquid crystal dot on each bottle of wine could give assurance that the temperature is in the right range for proper aging.

In the future, liquid crystals might make possible a wall-size television screen no thicker than an ordinary painting. Attempts are being made to develop a system in which liquid crystals would change color in response to electronic signals, giving pictures in living color without the need for the cathode ray tubes used today.

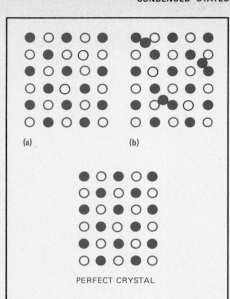

FIGURE 10.40
Point defects in a crystal. (a) A unit of structure can migrate to the surface of the crystal, leaving an empty space. In an ionic crystal, overall neutrality of electric charge is maintained despite such migrations. (b) A second type of point defect occurs when a unit of structure migrates to a hole, leaving a vacant site.

ordered arrangement of its units of structure. Such a combination of properties often is desirable for living systems. It is known that many portions of a living organism are in the paracrystalline state. Some recent work indicates that portions of cell membranes and organelle membranes are liquid crystals. The lipid, or fat, molecules in the membranes are in the long-chain shape favorable for liquid crystal formation.

10.8 DEFECT CRYSTALS

So far, we have described what might be called ideal or perfect crystal structures. Real crystals have defects or imperfections. To a much greater extent than is true for gases, the differences between the real and the ideal in crystal structure are important in determining the physical and chemical properties of a solid substance. Many properties of the solid state can be understood only in terms of the defects in real crystals. Moreover, many important practical applications of the properties of the solid state, especially in photography and in semiconductors, result from crystal defects.

There are two main categories of crystal defects: structural imperfections, which are flaws in the crystal lattice; and chemical imperfections, the presence of impurities in the crystal.

There are two types of structural imperfections, point defects and linear defects. Figure 10.40 illustrates the two main kinds of point defects. In one type of point defect, a unit of structure migrates to the surface of the crystal, leaving an empty space called a *vacant site* (Figure 10.40(b)). In an ionic crystal, vacant sites occur in such a way as to preserve electrical neutrality. Such defects account for some aspects of the conduction of electricity by ionic solids, and for diffusion in the solid state. Diffusion is important in such processes as the tarnishing of metals and in sintering, in which fine particles of solid bake into a strong, dense mass when heated.

The second type of point defect is more complicated. An example is shown in Figure 10.40(a), in which a unit of structure migrates to an interstitial site, or hole, in the lattice, leaving behind a vacant site. Such defects occur in salts such as AgBr, in which an Ag^+ ion migrates to the interstitial site. This phenomenon is put to practical use in photographic film, which consists of AgBr in a gelatin coating. The changes that occur in a small crystal of AgBr when it is struck by light and that can be used to create a photographic negative are believed to be due to the migration of Ag^+ ions to interstitial sites and the accompanying creation of vacant sites.

The second major kind of defect, the linear defect, is also called a *dislocation*. The two important types of linear defects are edge dislocation and screw dislocation. Both result from the imperfect orientation of planes with respect to one another in the crystal.

Figure 10.41(a) shows an edge dislocation, which can be described as the insertion of an extra plane of structural units partway into the crystal. A screw dislocation (Figure 10.41(b)) results from stresses that can occur while the crystal is forming. A screw dislocation resembles a spiral staircase within the crystal.

Recent evidence indicates that the structural strength of solids, especially metals, is closely related to the number and type of dislocations in the solids. Any dislocation reduces structural strength considerably. Even chemical reactions that occur on the surface of a crystal tend to occur at the site of dislocations. One approach to creating stronger structural materials is the effort to reduce or eliminate dislocations in metals such as steel.

Defects due to impurities in crystals play a vital role in electronics. Impurities often are added deliberately to make semiconductors, substances whose electrical conductivity increases with increasing temperature. Two important semicon-

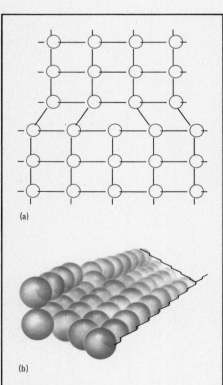

FIGURE 10.41
Linear defects, or dislocations, in a crystal. An edge dislocation can be pictured as the partial insertion of an extra plane of units in the crystal lattice (a). A screw dislocation resembles a spiral staircase jutting from the crystal lattice (b).

ductors are silicon and germanium. When either of these substances is extremely pure, its electrical conductivity is quite low. The deliberate addition of tiny amounts of impurities, such as arsenic or boron, a process called "doping," causes a substantial increase in the conductivity of Si and Ge, which then become semiconductors. The electronics industry today is based on semiconductors such as doped silicon and doped germanium. We shall discuss how doping is effective in creating semiconductors in Chapter 19.

EXERCISES

10.1   Explain why heat is always liberated when a gas condenses to a liquid and when a liquid freezes to a solid.

10.2   Certain insects can walk on the surface of water despite the fact that they are denser than water and sink to the bottom if placed below the surface. Explain why.

10.3   We have defined surface tension as a force acting on the surface of a liquid. It can also be defined as the energy of the surface of the liquid. Use the relationships between the SI units of force and energy to find the appropriate units for surface tension when it is defined as energy, and show the equivalence of these units to the units in Table 10.1.

10.4   Identify the attractive forces between the units of structure in crystals of: (a) I_2, (b) MgF_2, (c) SO_3, (d) Xe.

10.5   Examine, with the aid of a magnifying glass if necessary, crystals of table salt and of sugar. Describe their shapes.

10.6   Build models of the Bravais lattices using miniature marshmallows or gumdrops for the lattice points and toothpicks for the imaginary lines that connect the lattice points.

10.7   The unit cell in a crystal of metallic barium is found to be a body-centered cube with a side of 0.506 nm. The density of barium is 3.51 g/cm³. Calculate Avogadro's number from these data.

10.8   The unit cell in a crystal of metallic gold is found to be a face-centered cube with a side of 0.408 nm. Calculate the molar volume (the volume of one mole) of gold.

10.9   The unit cell in a crystal of diamond belongs to a cubic crystal system but does not correspond to one of the Bravais lattices. The volume of a unit cell of diamond is 0.0454 nm³ and the density of diamond is 3.52 g/cm³. Find the number of carbon atoms in a unit cell of diamond.

10.10   The unit cell of copper is a face-centered cube. The density of copper is 8.93 g/cm³. Find the length of the side of the face-centered cube.

10.11   The unit cell of nickel is a face-centered cube of volume 0.04376 nm³. The atom centered in the face of the unit cell is a sphere that just touches the other four atoms at the corners of its face. Find the atomic radius of nickel.

10.12   An unknown metal is found to have a density of 7.8748 g/cm³ and to crystallize in a body-centered cubic lattice. The edge of the unit cell is found to be 0.28664 nm. Calculate the atomic weight of the metal.

10.13   The atomic radius of silver is 0.1445 nm. The unit cell of silver is a face-centered cube. Calculate the density of silver.

10.14   Obtain a quantity of marbles and use them to study the regular closest-packed structures.

10.15   Cubic closest packing corresponds to a face-centered cubic lattice. Calculate the fraction of empty space in a face-centered cube to five significant figures.

10.16   When spheres of radius r are packed in a body-centered cubic arrangement, they occupy 68.0% of the available volume. Calculate the value of a, the length of the edge of the cube, in terms of r.

10.17   Zinc crystallizes with a hexagonal closest-packed structure that has a density of 7.134 g/cm³. Find the volume and the radius of a zinc atom.

10.18   The atomic radius of strontium is 0.215 nm and it crystallizes with a cubic closest-packed structure. Calculate the density of strontium.

10.19   The atomic radius of lead is 0.1750 nm and its density is 11.34 g/cm³. Find which of the close-packed structures is possible for lead.

*10.20   The atomic radius of zirconium is 0.160 nm and it crystallizes in a hexagonal closest-packed structure. Calculate the molar volume of zirconium.

10.21   The density of an unknown metal is 12.3 g/cm³ and its atomic radius is 0.134 nm. It has the hexagonal closest-packed structure. Find the atomic weight of the metal.

10.22   The ionic radius of Cl^- is 0.181 nm. Consider a closest-packed structure of Cl^- anions, in which all the anions are just touching. Find the radius of the cation that just fits into the octahedral holes of this lattice of anions.

10.23   The radius of a cation that just fits into a tetrahedral hole of a close-packed structure of Cl^- anions in which all the anions are just touching is 0.0407 nm. Compare this value to the result of Exercise 10.22. Suggest a relationship between the size of the cation and the site it will occupy based on the principle that ionic solids are more stable if the anions do not touch.

10.24   The smallest angle of reflection observed in the X-ray diffraction pattern of strontium is 14.7° when the wavelength of the incident X rays is 0.154 nm. Calculate the distance between the reflecting planes.

10.25   The distance between planes of atoms parallel to the face of a face-centered cubic unit cell is one-half the length of the edge of the cubic unit cell. Use the result of Exercise 10.24 to find the atomic radius of strontium, which crystallizes in a face-centered cubic lattice.

10.26   The distance between planes of ions parallel to the face of the unit cell of sodium chloride is 0.282 nm. The smallest angle of reflection observed in the X-ray diffraction pattern is 5.97°.

Find the wavelength of the incident
X rays.

10.27 Find the minimum angle of reflection in the X-ray diffraction pattern of a crystal with a distance of 0.190 nm between planes when it is exposed to X rays of wavelength 0.229 nm.

10.28 The maximum value of the sine of an angle is 1. Use the Bragg equation to find the minumum distance between planes of atoms that can give a diffraction pattern with X rays of wavelength λ.

10.29 The warming curve for water shown in Figure 10.32 ends at a temperature not much higher than 273 K. Draw a continuation of the curve to a temperature of about 400 K, but do not include any superheating.

10.30 Find the quantity of heat required to melt 100 g of ice. Find the quantity of heat released when 100 g of liquid water freezes at 273 K. Predict whether the quantity of heat released when supercooled water freezes at 263 K is greater than, less than, or the same as the quantity of heat released at 273 K.

10.31 Draw a warming curve for water from 350 K to 400 K that shows superheating before boiling. Draw a cooling curve for water in the same temperature range.

10.32 Suggest an explanation for the observation that the heat of fusion of a substance is always smaller than its heat of evaporation.

10.33 Draw a cooling curve for carbon dioxide from 225 K to 175 K at a pressure of 1 atm.

10.34 Describe the phase changes that occur when the pressure on water is increased from 0.001 atm to 1 atm at a constant temperature of (a) 270 K, (b) 273.15 K, (c) 273.155 K, (d) 273.16 K, (e) 275 K. Use Figure 10.36 and the data given for the triple point of water.

10.35 Describe the phase changes that occur when the temperature of water is increased from 260 K to 360 K at a constant pressure of (a) 0.001 atm, (b) 0.006025 atm, (c) 0.99 atm, (d) 1.1 atm.

10.36 Describe the phase changes that occur when the pressure on carbon dioxide is increased from 0.5 atm to 10 atm at a constant temperature of (a) 195 K, (b) 216 K, (c) 220 K, (d) 250 K, (e) 350 K. Use Figure 10.37 for this exercise.

10.37 Describe the phase changes that occur when the temperature of carbon dioxide is increased from 190 K to 350 K at a constant pressure of (a) 1 atm, (b) 5.1 atm, (c) 10 atm, (d) 100 atm.

10.38 The positive slope of the melting curve of carbon dioxide is characteristic of the melting of most substances. It results from the greater density of the solid compared to the liquid. Explain why an increase in pressure favors the solid.

10.39 Draw a phase diagram for krypton from the following data: Normal melting point, 117 K; normal boiling point, 121 K; triple point, 104 K, 0.18 atm; critical temperature, 209 K; critical pressure, 54.3 atm; vapor pressure at 101 K, 0.13 atm. For convenience you may compress part of the scale on the diagram.

10.40 Although long-chain molecules are best at forming liquid crystals, there are many long-chain molecules that show no tendency toward liquid crystal formation. Suggest a reason why a molecule such as $CH_3(CH_2)_nCH_3$, where n is 20, does not form liquid crystals, while a molecule with a much shorter chain but with groups such as N—O substituted on the chain does form liquid crystals.

10.41 Predict the effect of each of the two types of point defects shown in Figure 10.40 on the measured density of a crystalline solid.

"Blood is a strange juice," Goethe said. Water is nearly as strange. Because water is all around us and because it is intimately involved in so many chemical processes—including the processes of life—we tend to forget that it has very unusual properties indeed, including some that make it unique among liquids. These unusual properties are not just minor curiosities. If water behaved the way other liquids do, the earth would be a much different planet. The earth's inhabitants would also be different—if there were any. As you will learn from our discussion of water in this chapter, this is one case where a familiar substance is also quite extraordinary.

11.1 PROPERTIES OF WATER

To start with the obvious, water on earth usually exists as a liquid that appears transparent in small amounts but appears bluish green in large quantities. At a pressure of 1 atm, the boiling point of water is 100°C (373 K) and its melting point is 0°C (273 K). The Celsius temperature scale is based on the boiling point and melting point of water. Solid water is found in many places on earth. Since temperatures in the polar areas of the planet often are well below 273 K, these are regions of ice.

We are not surprised by the fact that water and ice exist simultaneously on earth. We should be. Water is the only common inorganic substance that can be found both as a liquid and a solid at temperatures that normally occur on the surface of the earth. This unique characteristic arises from a property of water that does not strike us as unusual only because it is so familiar to us. The boiling point and the melting point of water are much higher than those of similar compounds, notably other hydrides of nonmetals. Table 11.1 lists some of the data for these compounds. It can be seen, for example, that methane, CH_4 (MW = 16) boils at a temperature 264 K lower than water (MW = 18) and freezes at a temperature 182 K lower than water. Again, H_2S, the hydride of the element just below oxygen in group VIA of the periodic table, boils 161 K lower and freezes 86 K lower than H_2O. Indeed, none of the boiling points and melting points of the hydrides in Table 11.1 are close to those of water. The unusually high boiling point and melting point of H_2O are due to an unusually high degree of intermolecular attraction in water, which we shall discuss in Section 11.2.

It is not only the phase transition temperatures of water that are unusual. The general thermal behavior of water is also distinctive.

We know that when heat is added to a substance, its temperature will increase unless the substance is undergoing a phase transition. The temperature rise that occurs when a given quantity of heat is added to a substance depends on the nature of the substance and the mass of the sample that is being heated. If an unusually large amount of heat must be added to a substance to cause a given increase in temperature, we say that the substance has a high heat capacity.

Water has a very high heat capacity. It absorbs a great deal of heat for a given temperature increase and releases a great deal of heat for a given temperature decrease. In addition, water has a very high heat of vaporization compared to similar substances; a great deal of heat must be added to evaporate a given quantity of water. Even the heat of fusion of water is relatively high, so a substantial quantity of heat is needed to melt a given mass of ice. To some degree, the same intermolecular attractions that explain the high boiling point and melting point of water also explain its unusual thermal properties.

Life on earth is influenced profoundly by the unusual thermal properties of water. The high heat capacity and high heat of vaporization of water are of major

TABLE 11.1
Boiling Points and Melting Points of Hydrides of Nonmetals

Compound	Boiling Point (K)	Melting Point (K)
H_2O	373	273
CH_4	109	91
NH_3	240	195
HF	293	190
SiH_4	161	88
PH_3	185	140
H_2S	212	187
HCl	188	158
GeH_4	185	108
SbH_3	256	185
H_2Se	232	213
HBr	206	185

importance in creating favorable conditions for life on our planet. We can start with temperature changes. If the earth had no oceans, it would be subjected to extreme temperature fluctuations daily. During the day, the sun transfers large quantities of heat to the earth. A great deal of this heat is absorbed by the oceans, whose temperature does not increase appreciably because of the high heat capacity of water. At night, the water that covers more than 70% of our planet releases much of the heat that was absorbed during the day. As a result, temperature fluctuations are minimized. The moderating effect of the oceans is especially noticeable in seaside regions, which tend to be cooler in the summer and warmer in the winter than inland regions.

The effect of water on temperatures on earth should not be underestimated. On the moon, where there is no water, the temperature can vary more than 250 K from day to night. On earth, a variation of 25 K is regarded as large. The large heat of vaporization of water also plays a role in moderating temperature fluctuations. A good deal of the sun's heat is taken up by the evaporation of water, one of the processes that are responsible for the conditions called weather.

Evaporation of water is also important as a temperature regulator for the higher animals. The stability and properties of many compounds that are essential to life, such as proteins, are quite sensitive to temperature changes. While some organisms have developed special mechanisms that allow them to survive at external temperatures above the boiling point or below the freezing point of water, most living things are comfortable only within a much narrower range of temperatures. The major reason why humans and other mammals can withstand substantial external temperature fluctuations is a well-developed system for regulating body temperature that is based on the unusual thermal properties of water. The human body is cooled by the evaporation of water through pores in the skin. A high heat of vaporization means that only a relatively small amount of water must evaporate to cause considerable cooling. Therefore, humans can maintain body temperature within narrow limits without drinking inordinate amounts of water.

Some organisms do not have such water-based regulatory systems. For example, many marine organisms can function only in a very narrow range of external temperature. These organisms survive because they live in large bodies of water whose temperature normally varies by only a few degrees. The addition of heated water from industrial processes to natural bodies of water, called thermal discharges or thermal pollution, can cause serious problems for some organisms.

Another unusual property of water is the way that its density changes with temperature. Water is one of the very few exceptions to the rule that the solid phase of any substance is denser than the liquid phase. In almost every other case, a given volume of a solid has a greater mass than the same volume of the same substance in liquid form. In any mixture of the liquid and solid phases of such a substance, the solid sinks to the bottom.

Not so with water. When water freezes, it expands—a phenomenon that is familiar to anyone who has left a bottle of soda outside on a freezing night. Icebergs float on the surface of the ocean and ice cubes float on the surface of a drink. Ice—water in solid form—is *less* dense than liquid water.

The density behavior of water, that is, the fact that ice is less dense than liquid water, has important implications for living organisms. If the density behavior of water were like that of most liquids, ice would form on the bottom of rivers, lakes, and ponds in the winter. If water froze from the bottom up, a heavy coat of ice would interfere with the rich growth of organisms on the bottom and the diversity of life in these bodies of water would be reduced. Instead, the layer of ice that forms on the surface insulates the lower layers of water, helping fishes and other aquatic animals to survive cold weather.

The change in volume of H_2O with the addition of heat is also unusual. Most substances expand when heated. As the temperature of a given mass of a substance increases, the substance expands; its volume increases and its density decreases. Water follows this rule over much of the temperature range in which it is liquid. But as the temperature of liquid water increases from 273 K, the melting point, to 277 K, the density of liquid water *increases* as the temperature increases. In this temperature range, water contracts as its temperature goes up. Above 277 K, water follows the usual rule. Its density decreases as the temperature increases.

The list of the unusual properties of water also includes its surface tension, viscosity, and more. All of these anomalous properties are due to the structure and composition of the water molecule, the subject of the next section.

11.2 THE HYDROGEN BOND

When a hydrogen atom is bonded covalently to an atom of an element with high electronegativity, such as nitrogen, fluorine, or oxygen, the bond is highly polar. There is a partial positive charge on the H atom and a partial negative charge on the atom of the more electronegative element. We can represent these partial charges as δ^+ and δ^- and show this charge distribution, as for example $-\ddot{O}^{\delta-}-H^{\delta+}$

In Section 7.9, we mentioned that whenever such polar bonds exist, there will be dipole-dipole intermolecular attractions, particularly in condensed phases. For several reasons, the dipole-dipole attractions are unusually large if hydrogen is one of the atoms in a polar bond. The most important factor is the small size of the hydrogen atom. We know that coulombic attractions increase as the distance between charges of opposite sign decreases. Thus, the dipole-dipole attraction will be very large if the atom with the partial positive charge can get very close to the atom in the other molecule with the partial negative charge. The closeness of the approach is limited by the valence electrons and inner-shell electrons of the two positive atoms, since the negative charges of these electrons eventually will lead to repulsions between the atoms.

But a hydrogen atom in a polar bond has no inner-shell electrons. Therefore, a highly electronegative atom with a partial negative charge can get closer to a hydrogen atom than to any other atom. This closer approach means that the strength of dipole-dipole interactions involving hydrogen atoms will be appreciably larger than the dipole-dipole interactions of other atoms.

There is a second factor that enhances the strength of dipole-dipole interactions in polar molecules with hydrogen atoms. The electronegative atom in such a molecule usually has one or more nonbonding valence electron pairs. One of these electron pairs can be directed toward the positively charged hydrogen atom of another molecule, providing a strong localized negative charge that strengthens the dipole-dipole interaction.

Thus, there is the possibility of an unusually strong coulombic attraction whenever a hydrogen atom is bonded to an atom of great electronegativity. Such an attraction is called a *hydrogen bond*. A hydrogen bond is substantially weaker than a covalent or ionic bond, but it is substantially stronger than the weak interactions described in Section 7.9.

In discussing water, we shall be especially interested in hydrogen bonds between identical molecules. A hydrogen atom in one molecule can attract a highly electronegative atom of another identical molecule. In water, the hydrogen atom of one H_2O molecule attracts the oxygen atom in another H_2O molecule.

This kind of intermolecular interaction makes for strong cohesive forces between molecules of the substance. Figure 11.1 shows three common examples of this kind of hydrogen bond. The ordinary covalent bond is represented by a solid line, the hydrogen bond by a dashed line.

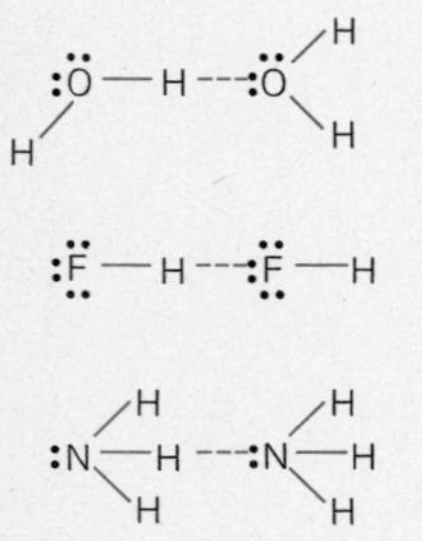

FIGURE 11.1
Three examples of hydrogen bonds between identical molecules. In each example, the hydrogen atom of one molecule forms a hydrogen bond (dashed line) with an electronegative atom of another molecule. All the hydrogen bonds are weaker than the covalent bonds (solid lines).

This kind of intermolecular hydrogen bond is important for many substances. Water is the most common of these substances. It is also the substance in which the effect of hydrogen bonding is most pronounced. Other substances in which hydrogen bonds are important are ammonia, NH_3, and hydrogen fluoride, HF. Because hydrogen bonds are present, we can predict that the phase transition temperatures of both substances will also be relatively high, since energy is required to break the hydrogen bonds. The data in Table 11.1 confirm this prediction. However, the melting point and boiling point of H_2O are still considerably higher than those of NH_3 and HF.

Before considering other types of hydrogen bonding, it is worth asking why hydrogen bonding seems especially strong in water. Why are the properties of water unusual compared to those of other substances that form hydrogen bonds? Why does water show a density maximum in the liquid phase, when substances such as ammonia and hydrogen fluoride do not?

The answer lies in the fact that an H_2O molecule is in some ways uniquely suited for the formation of intermolecular hydrogen bonds with other H_2O molecules. The partial positive charge on the H atom in H_2O is substantial because oxygen is highly electronegative; only fluorine is more electronegative. More important, each water molecule has two hydrogen atoms and two nonbonding valence electron pairs, so each water molecule can form four hydrogen bonds, as shown in Figure 11.2. The water molecule acts as both a donor and an acceptor. It supplies hydrogen atoms for two of the bonds and also accepts a hydrogen atom with each of its nonbonding electron pairs. The oxygen atom of a water molecule thus is surrounded by four hydrogen atoms—two of its own and two from two other H_2O molecules.

In a large group of water molecules, every H_2O molecule can form four hydrogen bonds. One mole of H_2O molecules forms a total of two moles of hydrogen bonds. By contrast, ammonia and hydrogen fluoride can form only one mole of hydrogen bonds per mole of substances. The hydrogen bonds in water are not necessarily stronger than the hydrogen bonds in HF and NH_3; there are just more of them.

In light of what was said in Section 8.2, page 236, it should come as no surprise to learn that the four hydrogen atoms surrounding the oxygen atom of a hydrogen-bonded water molecule lie at the corners of a tetrahedron. In a large array of water molecules, a regular, repeating arrangement of this kind can be achieved. Each oxygen atom is in the center of a tetrahedron whose four corners are defined by four hydrogen atoms. Two of the hydrogen atoms are covalently bonded to the O atom and two are hydrogen-bonded to the O atom. Figure 11.3 shows a portion of such an arrangement. This regular, repeating arrangement is the structure found in the crystals of ice, and the hexagonal pattern that is evident in this structure is reflected in the hexagonal patterns of snowflakes.

The crystalline structure of ice is relatively open, compared to the tightly packed crystals of many other substances. The "open" arrangement of the ice crystal maximizes favorable coulombic attractions by permitting the formation of a maximum number of hydrogen bonds. The relatively low density of ice is due to this open crystal structure. Ice has a low density because so much of it is open space.

This open structure is also responsible for the abnormal density behavior of water. When the temperature of ice reaches 273 K, the thermal motion of the H_2O molecules is great enough to start collapsing the open crystal structure by breaking some hydrogen bonds. As the crystal structure collapses, the density of water increases, so liquid water has a greater density than ice. But even when all the ice is melted, many of the hydrogen bonds remain. There still are some aggregates of H_2O molecules that have open, tetrahedral-type structures. These structures in the liquid continue to collapse as the temperature rises from 273 K to 277 K. In this temperature range, liquid water contracts and its density

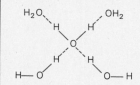

FIGURE 11.2
Why hydrogen bonding is important in water. The oxygen atom of the H_2O molecule has two nonbonding valence electron pairs and can form two hydrogen bonds. Each hydrogen atom can form one bond, making a total of four hydrogen bonds for each H_2O molecule.

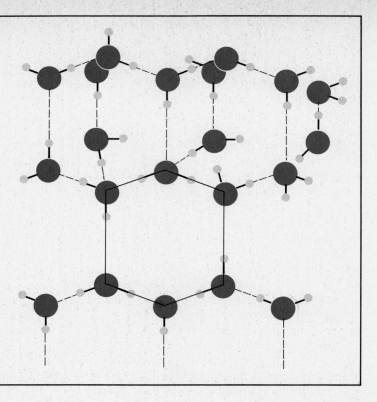

FIGURE 11.3
The hydrogen bonding of ice. Each oxygen atom, represented by a large sphere, is in the center of a tetrahedron whose corners are defined by hydrogen atoms, the small spheres. The overall pattern is hexagonal. Snowflakes are six-sided because their geometry is related to this structure.

TABLE 11.2
Density of Water

Temperature (K)	Density (g/cm^3)
273 (ice)	0.9168
273 (liquid)	0.999841
274	0.999900
275	0.999941
276	0.999965
277	0.999973
278	0.999965
279	0.999941
280	0.999902

FIGURE 11.4
A variety of hydrogen bonds. Hydrogen atoms that are bonded to oxygen or nitrogen atoms can form bonds (dashed lines) with other O and N atoms, in other molecules or in the same molecule.

increases as the temperature rises. Above 277 K, the normal thermal expansion of the liquid is more important than the collapse of the remaining tetrahedral structures, and the density of water follows the normal pattern of behavior: Density decreases as temperature rises. Table 11.2 shows how the density of water changes with temperature in this range.

What if the pressure is increased while the temperature remains below 273 K? The effect of increased pressure on the open tetrahedral structure of ice is to bend the hydrogen bonds. When the pressure is great enough, ordinary ice (sometimes called ice I) collapses. It is replaced by forms of ice whose tetrahedral structures are deformed because the hydrogen bonds are bent. The amount of bending depends on the amount of pressure. Seven other forms of ice have been identified as stable phases at pressures from about 2000 atm to 25 000 atm. The structures of these forms of ice have been studied by a number of methods, including X-ray crystallography.

One notable point is that these high-pressure forms of ice can be stable at temperatures well above 273 K. Ice VII (density = 1.65 g/cm^3) is stable at 373 K, the normal boiling point of water, but ice VII exists only at pressures above 24 000 atm. Fortunately, the imaginary substance called ice IX, which purports to be stable at a temperature well above 273 K at a pressure of only 1 atm, exists only within the pages of Kurt Vonnegut's novel *Cat's Cradle*, in which the world comes to a warm but frozen end.

Hydrogen bonds can form between nonidentical entities as well as between identical molecules. As we shall see in Chapter 22, the detailed structures of such biologically important molecules as proteins and nucleic acids are determined to a large extent by hydrogen bonding.

A hydrogen atom that is bonded to an oxygen or a nitrogen atom will form hydrogen bonds with other oxygen or nitrogen atoms. Such hydrogen bonding can occur between two different molecules. It can also be intramolecular, occurring between atoms that are in different parts of the same molecule. Figure 11.4 shows some of the possibilities.

THE POLYWATER AFFAIR

In the early 1970s, dozens of laboratories in the United States and Europe were conducting intensive studies of a strange new substance that was called variously water II, anomalous water, orthowater, superwater, and polywater. The story of polywater is an amusing chapter in recent chemical history. It gives us a fascinating insight into the way in which a scientific fad can waste scientists' time.

The existence of the substance was first reported in 1962 by Boris V. Derjaguin of the Institute of Physical Chemistry of the USSR Academy of Sciences. Derjaguin reported the preparation of minute quantities of a substance consisting of H_2O molecules, but with highly unusual properties. The substance was much denser than normal water, had the consistency of Vaseline, and displayed a number of other unexpected properties. Derjaguin called it anomalous water. Others called it polywater, on the assumption that the substance was formed by the aggregation of H_2O molecules into a giant unit, or polymer.

A number of prominent laboratories made more than 25 different measurements on the substance. No model could explain all the experimental results. One difficulty was that polywater could be prepared only in small quantities, usually of only a few micrograms. The standard method of preparing polywater was condensation of water into extremely small quartz tubes with diameters of 1–10 μm. Attempts to make larger quantities of the substance by other methods all failed.

It was suggested by some skeptics that the unusual properties of polywater could be explained on the assumption that it was not really pure water, but instead contained large quantities of impurities that were leached from the walls of the quartz tube. Derjaguin and others presented experimental data purporting to show that polywater consisted only of H_2O molecules, but the doubts persisted.

The doubts were ended in 1973, when Derjaguin presented a paper describing the use of a number of techniques to detect any possible impurities in polywater. One of those techniques was an electron probe method to determine the number of silicon, sodium, potassium, sulfur, and carbon atoms present relative to the number of oxygen atoms. The result was the evaporation of the belief in polywater.

The electron probe studies found that samples of polywater prepared under the most scrupulously clean conditions contained silicon atoms in quantities indicating that the water had partially dissolved the wall of the quartz tube. A gel or sol of a silicate acid was formed. "Consequently, these experiments do not support the hypothesis of anomalous or polymeric water," Derjaguin's paper said. "The formation of the anomalous water is probably attributable to an enhanced dissolving power of condensing vapors." Many scientists suddenly developed new research interests. The strange case of polywater was ended.

11.3   WATER AS A SOLVENT

Water is an excellent solvent. It can dissolve a wide range of materials that will not dissolve in other liquids. For example, salts do not dissolve in most common solvents, such as gasoline, kerosene, turpentine, and cleaning fluids. But many salts dissolve readily in water. So do many inorganic polar substances and even a variety of nonionic organic substances, such as sugars and alcohols of low molecular weight.

For a substance to dissolve in a solvent, some of the things that must happen are:

1. The attractive forces between the units of structure of the substance must be overcome, so that the structure can separate into units of atomic or molecular size that disperse evenly through the solution.
2. Some of the attractive forces between the molecules of the solvent must also be overcome, to allow dispersal of the dissolving substance.
3. The "cost" in energy of disrupting these attractive forces must be paid back by creation of new attractive forces between the solvent and the dissolved substance.

By using this general picture as a framework, we can understand why water is a good solvent for ionic salts. In these salts, the forces of attraction between the units of structure are the ionic bonds, or lattice energy, which we discussed in Section 7.1. The ionic bond is the attractive force between charges of opposite sign in a lattice. This force is proportional to the product of the magnitudes of the charges divided by the square of the distance between them. Coulomb's law, which expresses this relationship, can be written to include a proportionality constant:

$$F = \frac{1}{K}\frac{e_1 e_2}{r^2} \tag{11.1}$$

The proportionality constant is K, which is called the *dielectric constant*. The dielectric constant is a characteristic of the medium in which the charges are found. From Equation 11.1, it can be seen that an increase in the dielectric constant causes a decrease in the magnitude of the force between charges.

The dielectric constant of a substance is a measure of the ability of that substance to neutralize an applied electric field. An electric field can be neutralized by orienting the molecules of a substance so that their positively charged ends face the negative pole of the electric field (which means that their negatively charged ends face the positive pole of the field). The degree to which the electric field is neutralized is related directly to the polarity of the substance (Figure 11.5). Polar substances thus have relatively high dielectric constants. The dielectric constant of a vacuum is 1, and the dielectric constant of air is close to 1. The dielectric constant of water has a high value—about 80.

Water has a high dielectric constant not only because it is a polar substance but also because of its hydrogen bonds. In general, liquids whose molecules form hydrogen bonds have dielectric constants that are higher than might otherwise be expected.

In practical terms, the high dielectric constant of water (80) means that the attractive forces between the anions and cations in a salt are reduced by a factor of 80 when water is the medium between them. As shown in Figure 11.6, water molecules come between the anions and the cations. The water molecules line up as if in an electric field, with their negatively charged ends directed at the cations and their positively charged ends directed at the anions. The effect of interposing the water molecules in this way is to reduce the coulombic attractions between the anions and the cations of the salt.

Once the ions are separated, they interact with water molecules in a number

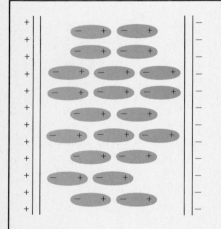

FIGURE 11.5
When a polar substance is placed in an electric field, its molecules are oriented so that their positive ends face the negative pole and their negative ends face the positive pole. This alignment partially neutralizes the field. The dielectric constant of a substance can be regarded as a measure of the extent of neutralization.

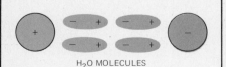

H₂O MOLECULES

FIGURE 11.6
One reason why water is a good solvent for ionic substances. The highly polar H₂O molecules reduce the attractions between cations and anions.

of ways. A cation is surrounded by water molecules whose oxygen atoms, which have a relative negative charge, are directed at the positive ion. Because of the favorable ion-dipole interactions in this arrangement, water molecules are held very tightly by cations, especially by bipositive and terpositive cations. When these ions hold water molecules, they are said to be *hydrated*. Usually, a given cation holds a specific number of water molecules. The number of water molecules is determined primarily by the size of the cation. A small ion, such as Be^{2+}, that is in solution is surrounded by four water molecules whose oxygen atoms are directed at the cation. Most cations are larger and can hold more water molecules. For example, Mg^{2+} can accommodate six water molecules, as Figure 11.7 shows.

The number of water molecules held by an ion is called the *ligancy* or coordination number of the ion. Thus, the ligancy of Be^{2+} is four and the ligancy of Mg^{2+} is six.

In many cases, cations hold water molecules so tightly that the water molecules accompany the cations when ionic solids crystallize from aqueous solution. This water is called the *water of crystallization*. Many crystalline salts, especially those of dipositive and terpositive cations, have stoichiometric quantities of water of crystallization. The formulas of such compounds are written to indicate the presence of water, whose formula is separated from the formula of the salt by a dot. Some examples are: $BeSO_4 \cdot 4H_2O$; $MgCl_2 \cdot 6H_2O$; and $Fe(NO_3)_2 \cdot 6H_2O$, three compounds in which each cation retains all its water of hydration when the salt crystallizes from aqueous solution; $FeSO_4 \cdot 7H_2O$, in which an additional water molecule is associated with the $SO_4{}^{2-}$ anion; and $KAl(SO_4)_2 \cdot 12H_2O$, or alum, in which each of the two cations accommodates six H_2O molecules. Salt crystals can also form in which the water of crystallization has been removed partly or completely, as with $MgSO_4 \cdot H_2O$ and $MgSO_4$.

Attraction between negative ions and water molecules is not nearly as important as attraction between cations and water molecules. Anions generally are much larger than cations, so their charge is more dispersed and the benefit of association with the water molecule is reduced. Nevertheless, anions in solution are surrounded by water molecules, whose positively charged ends point toward the anions. The water molecules may even form hydrogen bonds with the anions.

The solubility in water of nonionic substances such as sugars, alcohols, or amines is due primarily to the formation of hydrogen bonds. Molecules of these substances may contain one or more $-\ddot{O}-H$ or $-\ddot{N}-H$ groups. The hydrogen atoms in these groups can form hydrogen bonds with water molecules. In addition, the nonionic substances form hydrogen bonds with the hydrogen atoms in water molecules, as Figure 11.8 shows. Some other substances, such as formaldehyde, $H_2C{=}O$, are soluble in water because the water molecules can form hydrogen bonds with them, although the formaldehyde molecules cannot form hydrogen bonds with each other. The hydrogen bond between formaldehyde and water is represented by a dashed line:

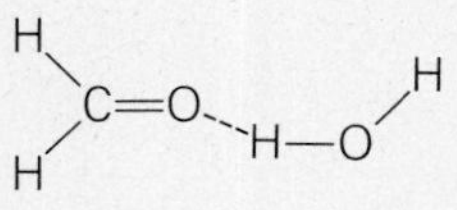

Although water is an excellent solvent, many substances do not dissolve in water to any great extent. A substance dissolves in water primarily because of the attractive forces that arise when the substance goes into solution. These attractions must be greater than the attractive forces that are lost between the units of structure of the pure substance and between water molecules. A substance will not dissolve in water if the attractions between its units of structure and water molecules are not strong enough. Such is the case with some cations that form salts with substantial lattice energies but do not form good hydrates. The ions in

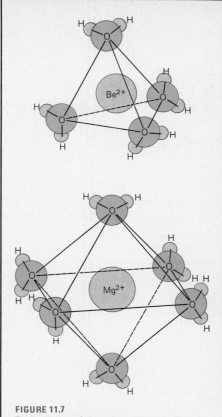

FIGURE 11.7
The interactions of two cations with water. The Be^{2+} ion is surrounded by four H_2O molecules, whose oxygen atoms are directed at the cation. The larger Mg^{2+} ion attracts six H_2O molecules. In general, larger cations attract a larger number of H_2O molecules.

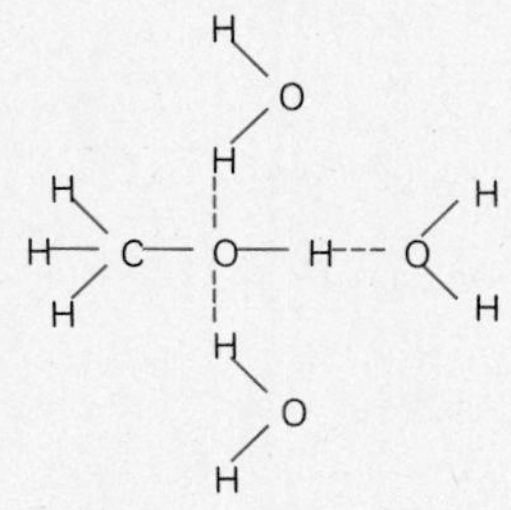

FIGURE 11.8
The solubility of nonionic substances can be explained by the formation of hydrogen bonds. A hydrogen atom of the CH_3OH molecule forms a hydrogen bond with the oxygen atom of a water molecule, while the oxygen atom of the CH_3OH molecule forms two hydrogen bonds with hydrogen atoms of water molecules. The strength of these attractions allows the substance to dissolve.

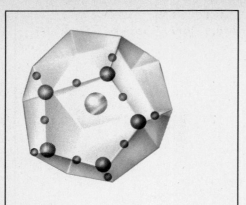

FIGURE 11.9
The structure of xenon hydrate, a clathrate crystal. The water molecules form a hydrogen-bonded three-dimensional network. The xenon atoms occupy the cavities in the structure.

these lattices do not separate in water, so these salts are not soluble in water. Many silver salts are in this class.

Clathrates

Nonpolar substances usually do not dissolve in water because there are no appreciable favorable interactions between the nonpolar molecules and the water molecules. Under special conditions, however, many nonpolar substances of simple structure can form crystalline hydrates with water, even though these substances do not dissolve in water. One such substance is CH_4, which does not dissolve appreciably in water but does form crystals of a hydrate whose composition is $CH_4 \cdot 5\frac{3}{4}H_2O$.

Many other substances, such as the noble gases, simple hydrocarbons, chloroform, chlorine, and bromine, form crystalline hydrates with complex structures. All of these hydrates have a hydrogen-bonded tetrahedral array of water molecules similar to that of ice but even more open. These crystal structures have so much open space that the lattice can be pictured as a three-dimensional array of cavities whose walls are made of water molecules. Each cavity holds an atom or a molecule of the substance that forms the hydrate. These crystals are called clathrate crystals—"clathrate" is from the Latin word for "cage"—because the molecule of the substance forming the hydrate seems to be enclosed in a cage of water molecules. Figure 11.9 shows the partial structure of one such hydrate.

Aside from their purely theoretical interest, clathrate hydrates are of practical importance because they may play a role in anesthesia. Some gases, such as the noble gases, are highly effective anesthetics even though they are chemically inert under physiological conditions. It has been suggested that these gases depress the functioning of the nervous system by forming clathrate hydrate crystals in the brain.

11.4 THE COMPOSITION OF NATURAL WATERS

While water is found almost everywhere on the surface of the earth, it is almost never found in a pure state because it is such a good solvent. Water dissolves a great variety of substances. It is also a medium for the suspension of many kinds of undissolved particles. The composition of any sample of water depends on the source of the sample, since the type and amount of substances that are dissolved or suspended in water are a function of the immediate environment.

The composition of a sample of water is of more than academic interest. Any organisms that live in water have a direct and vital interest in the composition of the water surrounding them. Humans do not live in water, but we do use a great deal of water for both personal and industrial purposes. It is important that the composition of water be suitable for these purposes. Failing that, the composition must be known so that the water can be treated to make it usable.

The substances found in water can be divided into two categories: those that are present as a result of natural biogeochemical cycles that have been functioning throughout geological time, such as the ions found in seawater; and those that have been introduced by human activity. Some substances, such as phosphate, fit in both categories. The second category includes both substances put into water for useful purposes and those that get into water as a by-product of industrial or municipal activity. Many of the latter substances can harm living things. They are pollutants.

Seawater

Most of the liquid water on earth is in the oceans. Seawater is a solution of many different substances, including salts, gases, and organic materials.

We all know that seawater contains dissolved salts. But in discussing the composition of seawater, we should remember that when salts dissolve in water, they exist as dissociated ions. Thus, a solution prepared from NaCl(s) and KBr(s) will contain Na^+ and K^+ cations and Cl^- and Br^- anions. A solution prepared from KCl(s) and NaBr(s) will have the same composition. The composition of a solution of salts thus is discussed only in terms of the ions in solution.

Seawater contains 11 major ionic species and a large number of minor constituents whose mass is negligible compared to that of the major ions. The amount of salts dissolved in seawater is measured by *salinity*, which is the total mass in grams of dissolved salts in 1 kg of seawater. Typically, the salinity of seawater is 35 g of salts per kilogram of seawater. Although the salinity of samples of seawater may vary, depending on the source of the samples, the *relative* quantities of the major dissolved ionic species in water from the open sea is remarkably invariant, and appears to have been invariant for the past 100 million years. Table 11.3 lists the relative quantities of the major ionic species to be found in a sample of seawater. As might be expected, the two species that are present in the largest quantities are Na^+ and Cl^-.

There is no similar constancy of composition for the minor ionic species, such as Ni^{2+}, Co^{2+}, I^-, and many others that are present in very small concentrations in seawater. There are two reasons for the large variations in the concentrations of these minor species in different samples of seawater: the way in which the ions reach the sea and their role in biological processes.

Ions are washed out of the land into the sea. If a land area happens to have minerals that are a rich source of one ion or another, the sea near this area will contain relatively high concentrations of these ions. Once in the sea, these ions can easily enter the food chain. Some of these ions contain elements such as nitrogen, phosphorus, silicon, and iron, which are important to the growth of the microscopic oceanic plants that are collectively called phytoplankton. Areas of the sea that are rich in these ions will support dense growths of phytoplankton. These microscopic plants then serve as the base of a food chain whose middle link is zooplankton, tiny oceanic animals, and whose final consumers are large fishes.

It is especially interesting to note that the relative ionic composition of seawater is similar to that of the blood and other body fluids of many animals. If you compare seawater and the body fluids of marine vertebrates and land animals, you will find the same relative proportions of the major dissolved ions in both, although seawater has a higher concentration of ions than the body fluids. If you go on to study the body fluids of marine invertebrates such as lobsters and clams, you will find a close resemblance to seawater, both in salinity and in the relative concentrations of different dissolved ions. The constancy of ionic concentrations in seawater and in body fluids is interpreted as evidence that all life originated in the sea and that all mammals, including humans, carry a memento of the distant past within them.

Freshwater bodies such as lakes, rivers, and ponds account for less than 0.03% of the earth's surface water. But they play an important role in many human activities.

Although they are called "freshwater" bodies, lakes and rivers do contain dissolved salts. However, the salinity of fresh water is usually much lower than that of the sea, and the major dissolved salts found in fresh water often are different from those found in seawater. In addition, the composition of individual freshwater bodies is much more variable than the composition of seawater.

Water is regarded as being fresh if it contains less than 0.01% dissolved solids (0.1 g of salts per 1 kg of water), but the usual United States standard for drinking water requires no more than 0.005% dissolved solids. These salts are not necessarily those which are found in seawater. The major dissolved ions in river water are not Na^+ and Cl^- but rather cations such as Mg^{2+} and Ca^{2+}, and anions such as HCO_3^- and SO_4^{2-}. In addition, silicon is found in fresh water in

TABLE 11.3
Major Ionic Species in Seawater

Ion	g/kg H_2O	Percent of Total Salt Content
Cl^-	18.98	55.04
Na^+	10.56	30.61
SO_4^{2-}	2.65	7.68
Mg^{2+}	1.27	3.70
Ca^{2+}	0.40	1.16
K^+	0.38	1.10
HCO_3^-	0.14	0.41
Br^-	0.065	0.19
Sr^{2+}	0.013	0.04
F^-	0.001	0.003

EL NIÑO AND FOOD FROM THE SEA

The cold water of the Peru Current that flows from the Antarctic northward along the western coast of South America wells to the surface just off Peru. The water carries with it a rich supply of phosphates, nitrates, and other nutrients, the remnants of dead marine plants and animals that have accumulated on the ocean bottom. Where these nutrients reach the surface, marine plants grow in great profusion. These plants support billions of animals, ranging in size from microscopic crustaceans to large fishes. The most numerous of these animals are anchovies, which abound in these waters. More than a fifth of the world's total fish catch is made off the coast of Peru in normal years.

But about one year in seven is not a normal year. In these years, the southeast winds that push the cold waters of the current northward shift to the west. Warmer water replaces the cold water of the deep, reducing the nutrient supply drastically. The food chain is broken, and the number of anchovies decreases, too. This catastrophic event is called El Niño, the Spanish name for the Christ child, because it often begins around Christmastime.

In the past, the anchovy fishery of Peru has recovered in a year or two, as soon as the normal flow of ocean currents is restored. Now, however, there are fears that human activities may be affecting the ability of the fishery to recover. There are indications that overfishing may reduce the anchovy population below the levels needed to sustain it. If this is so, other animal populations will suffer as well. The islands off the coast of Peru support vast numbers of birds, whose major food source is the anchovy. The droppings of these birds, called guano, is an excellent natural fertilizer; some islands have deposits of guano up to 50 m deep. If the anchovy population is overfished, the birds will not survive. The productive food chain of the waters off the Peruvian coast will be broken. A combination of the natural disturbance called El Niño and a disturbance added by mankind may result in the depletion of a vital natural resource.

various forms. All of these constituents are dissolved from soil in the riverbed.

You may have noted that while Ca^{2+} is a major ionic constituent of rivers, it is found in relatively low concentrations in the sea, even though river water flows into the sea. The activities of living organisms are responsible for the relatively low concentration of Ca^{2+} in seawater. The shells formed by marine invertebrates consist mainly of $CaCO_3$. These invertebrates collect Ca^{2+} from seawater to form their shells, thus reducing the concentration of Ca^{2+} in seawater.

The composition of the water in lakes is much more variable than that of either river water or seawater. Some lakes have negligibly small concentrations of dissolved salts. Others, such as the Great Salt Lake and the Dead Sea, are much more saline than the oceans. The composition of lake water depends not only on the type of soil surrounding the lake, which influences the amount of salts washed into it, but also on the age of the lake, the living organisms in the lake, and the rate of evaporation of lake water. Older lakes tend to have a higher concentration of dissolved substances.

Dissolved salts are not the only substances in seawater and fresh water. Natural waters also contain varying amounts of dissolved gases, some of which are of major importance to living organisms. The three most important dissolved gases are O_2, CO_2, and N_2. The first two are essential to marine life. Many of the problems of water pollution center on the concentrations of dissolved O_2 and CO_2 in bodies of water.

The concentration of O_2 in water is much lower than the concentration of O_2 in the air. Air is about 20% O_2, while water has no more than 8–10 ppm of O_2 by mass, the equivalent of about 7.5 cm^3 of O_2 at STP per liter of water. Marine animals that get their oxygen from water have evolved mechanisms for extracting a sufficient supply even though the concentration of oxygen is low. The most widely used mechanism is the gill, in which large amounts of water pass over a relatively thin membrane through which gas exchange takes place.

The solubility of gases in water depends on a number of factors, including the pressure of the gas in contact with the surface of the liquid and the temperature of the liquid. The solubility of any gas increases with pressure and decreases with temperature. The surface layer of any body of water generally is rich in oxygen, which dissolves from the atmosphere. In addition, water near the surface of the ocean is rich in oxygen because of the activity of phytoplankton, which carry out photosynthesis and release oxygen as a waste product. Typically, surface water will contain 5 cm^3 of oxygen per liter, while the oxygen content at a depth of 1 km may be 1 cm^3/liter.

This oxygen is consumed in part by living marine organisms. It is also consumed in the decay of dead organisms and of animal wastes. The microbes that decompose these substances consume oxygen. The dumping of raw sewage—human wastes—into rivers and lakes is undesirable because so much oxygen is consumed by the decay of these wastes that there is none left to support higher forms of life.

The consumption of oxygen by the processes of decay becomes more important with increasing depth, but the replenishment of the consumed oxygen by the absorption of oxygen from the atmosphere and the release of oxygen through photosynthesis decreases with depth. Thus, there is a general tendency for the concentration of oxygen in any natural body of water to decrease with depth. However, the oxygen content of deep ocean water is replenished by cold-water currents. The solubility of gases is greater in cold water, so the cold surface water at the poles of the earth are rich in oxygen. This dense cold water sinks to the bottom, carrying oxygen with it and replenishing the oxygen supply in the ocean depths.

11.5 WATER POLLUTION

In addition to the materials that occur naturally in the ocean and in freshwater bodies, increasing amounts of many substances are being introduced by mankind. These substances often are harmful to organisms in the water and to humans. Most of these substances are introduced because natural bodies of water are convenient carriers or dumping sites for waste products.

Mankind has many different wastes. One of them is domestic sewage, the waste material from our everyday activities. Domestic sewage consists of human excrement and its associated microorganisms, some of which cause disease; food wastes; soap and detergents; and a variety of miscellaneous materials that we wash down the sink, flush down the toilet, or dump down the drain.

Domestic sewage contains organic matter that is decomposed by microor-

ganisms in the water. These microorganisms are of two main types: aerobes, which require free O_2 to live, and anaerobes, which do not require free O_2 and may even be harmed by it. Aerobes decompose organic matter into CO_2, H_2O, and various ions that usually are not harmful. Anaerobes produce substances that are unpleasant or toxic, such as H_2S and CH_4.

Since the end products of anaerobic decomposition can kill fishes and other marine organisms, it is clearly desirable for the material in domestic sewage to undergo aerobic decomposition. But when this happens, substantial amounts of dissolved O_2 are consumed. The rate at which the microbial population consumes O_2 is called the *biochemical oxygen demand* (BOD). When massive amounts of organic wastes are dumped in a body of water, BOD becomes high and oxygen content drops because O_2 is consumed faster than it can be replenished. When oxygen content drops below about 4 ppm (3 cm^3/liter), all fish die. If the O_2 supply is exhausted, only anaerobes remain and the water becomes septic and noxious.

Agricultural wastes are also a source of water pollution. Substances such as fertilizers, pesticides, and animal wastes are carried off into bodies of water, such as rivers, lakes, and the ocean, by surface runoff.

Industrial waste is much more variable in composition and unusual in content. Most domestic sewage and agricultural waste are potentially dangerous because of their sheer volume, which can overwhelm the natural cleansing processes of bodies of water. Industrial waste often consists of materials that have been created by mankind and that can produce unexpected results when introduced into bodies of water. For example, the first detergents to be put on the market were found to pass through sewage treatment plants untouched and to create mountains of foam in rivers and streams. The detergents were marketed as a replacement for soap, which is made from animal fats. The microorganisms in sewage treatment plants can decompose natural fats, but they cannot decompose the molecules of the original synthetic detergents, which are derived from petroleum chemicals. The detergent industry had to convert to ''biodegradable'' products, which can be decomposed by the microorganisms in sewage treatment plants.

Some pesticides, such as the chlorinated hydrocarbons used as insecticides, are almost immune to decomposition by natural methods. Relatively small amounts of pesticides, introduced accidentally into rivers, have killed millions of fishes. The chlorinated hydrocarbons, including DDT, each have a chlorine-carbon bond that is rare in nature and that few microorganisms can break. The chlorinated hydrocarbons are relatively insoluble in water, but they are soluble in fats. Therefore, they tend to accumulate in fatty tissues of animals, where they can build to poisonous concentrations.

The manner in which pollutants affect the quality of water often is subtle and surprising. For example, it has been found that what seem to be small quantities of industrial waste may be harmful. An example is mercury-containing waste.

At one time, it was believed that the release of mercury and inorganic mercury products into water was relatively harmless. The mercury, being heavy and insoluble, was believed to sink harmlessly to the bottom of any body of water. It has now been found that inorganic mercury is converted by bacteria to organic mercury compounds, such as dimethyl mercury, $(CH_3)_2Hg$, which are extremely toxic. It has also been found that mercury compounds can be concentrated in body tissues as large animals feed on smaller organisms, until a toxic level of mercury is reached for animals at the top of the food chain.

Other heavy-metal ions that are released in industrial wastes, such as the ions of cadmium, lead, and nickel, can also be retained in body tissues. The concentration of these ions increases as they are passed up through the food chain. Smaller organisms, the primary consumers in the food chain, may not be damaged. But the large predators at the top of the food chain may be poisoned.

One effect of domestic sewage, agricultural fertilizers and some industrial wastes is to cause eutrophication, a serious problem for some lakes. Eutrophication is caused by an excess of phosphates, nitrates, and other nutrients. This excess leads to a number of objectionable developments, including an acceleration of the normal process by which a lake "ages."

When a lake is first formed by geological activity, it consists of relatively pure

The activities of a beaver are part of the natural cycle by which bodies of water age. Human activities can speed up the aging process for many bodies of water by adding excess nutrients that cause eutrophication. (*Wide World*)

water that can support very little life. As nutrients from the soil find their way into the lake, more and more organisms can be sustained in the water. As these organisms die, their decay is carried out by aerobic microbes, with the consumption of O_2. A solid residue is left, and this residue sinks to the bottom to form a sediment. Over many centuries, the sediment formed by dead organic material builds up, slowly filling in the lake until it becomes first a marsh and then finally dry land.

If nutrients such as phosphorus and nitrogen are added to the lake by human activities, the lake will be able to support more living organisms. Organic sediments will build up faster and the lake will grow ''old'' too quickly. The overgrowth of living organisms causes many other problems. The decomposition of dead organisms can increase the BOD level so much that the O_2 content of the water is decreased below the concentration needed to support aerobic organisms. Anaerobes take over, with all the unpleasant consequences they bring with them. The deeper levels of a lake may become completely depleted of oxygen, so that fish habitats are destroyed and desirable species are eliminated. Use of the lake for drinking water or human recreation may become impossible. The solution is to somehow limit the addition of phosphates from detergents, nitrates from agricultural fertilizers, and the varied nutrients from human sewage and other sources. Otherwise, too much of a ''good'' thing can destroy a lake.

11.6 WATER TREATMENT AND PURIFICATION

Water intended for human use must be treated to remove any objectionable or dangerous constituents. Water that has been used for domestic or industrial purposes requires treatment to remove the impurities that have been added. The techniques used in water purification and sewage treatment have many features in common, but there are differences due to the amount and quality of water that is being treated and the purpose of the treatment. For some purposes, virtually all impurities must be removed. In other cases, partial removal of impurities will do. We shall discuss some of the more widely used techniques of water treatment in the remainder of this chapter.

Drinking Water

Water for human use must not only be safe, it must also be free of odors, color, particulates, and other unpleasant impurities. Excessive ''hardness''—the presence of appreciable quantities of dissolved ions—is also undesirable. Water for domestic consumption goes through several treatment processes to meet modern standards.

The simplest way to remove particulate matter is by sedimentation. If water is allowed to stand, most solids settle to the bottom. Sedimentation occurs naturally in reservoirs or in settling tanks in water treatment plants. But some solid particles are so small that they do not settle out. They can be removed by chemical coagulation. If a substance such as aluminum sulfate, $Al_2(SO_4)_3$, is added to the water, Al^{3+} ions are formed. The ions react with water molecules to form a gelatinous precipitate of $Al(OH)_3(s)$, a substance to which small particulates and other substances become attached. These substances are removed with the precipitate, which is called floc. Water that is treated by this method, called flocculation, may also be passed through filter beds of sand to remove even finer solid particles and other impurities.

There are a number of ways of removing unpleasant tastes and odors from water. Aeration, mixing air with water, is a method that eliminates volatile bad-smelling substances such as H_2S. Many organic substances can be removed by passing water over powdered activated charcoal, which adsorbs the substances. Adsorption is the process by which a substance sticks to a solid surface. Activated charcoal consists of very fine particles of carbon that collectively have a very large surface area and so provide ample surface for adsorption.

Perhaps the best-known method of water treatment is chlorination. The application of chlorine or chlorine compounds to water kills microorganisms and oxidizes some undesirable organic compounds. Relatively uncontaminated water can be disinfected by the use of less than one part per million of chlorine. Several parts per million may be needed for contaminated water. Chlorine sometimes causes an objectionable smell or taste, which can be eliminated by activated charcoal treatment.

Although chlorination has been in widespread use for most of the twentieth century and is credited with eliminating epidemics of water-borne diseases such as typhoid, its safety has been questioned recently. Sophisticated instruments have found that chlorination is associated with the formation of a number of apparently undesirable compounds, such as chloroform and carbon tetrachloride, in drinking water. The concentrations of these compounds usually are no greater than a few hundred parts per billion, and their effect on human health is unknown. However, some communities have turned to an alternative method of disinfection called ozonization, in which ozone, O_3, is used instead of chlorine to kill microorganisms.

Water Softening

While fresh water is much less saline than seawater, it may still contain dissolved ionic material in concentrations that are too high for some uses. Water that contains appreciable quantities of such ions as Ca^{2+}, Mg^{2+}, and Fe^{2+} is called "hard" water. Hard water makes washing difficult because these ions form an unpleasant, scummy precipitate with soap. Hard water is also an industrial problem. Salts of calcium, such as $CaCO_3$ and $CaSO_4$, tend to precipitate and form scale in boilers and other water-holding tanks.

A common method of softening hard water is by treatment with lime and soda ash to precipitate the calcium and magnesium as carbonate and hydroxide, which can be removed by filtration. Another less common method uses substances called *zeolites*, which soften hard water by a process called *ion exchange*. A zeolite is a mineral that contains aluminum, silicon, oxygen, and a metal such as sodium or calcium. A typical empirical formula for a zeolite is $NaAlSi_2O_6$.

In general, a zeolite is a crystal whose aluminosilicate portions form a lattice structure with sizable vacancies. The metal cations, such as Na^+, are found in these corridorlike vacancies, and have considerable freedom of movement. When water flows over a collection of zeolite crystals, dipositive cations in the water will displace the Na^+ cations from the zeolite crystal. Unlike the dipositive cations that are responsible for the "hardness" of water, the Na^+ cations will not form a precipitate with soap and will not form scale in boilers. By replacing the dipositive cations with Na^+ cations, the water is "softened," in a process that can be represented by the reaction:

$$2Na^+Z^-(s) + Ca^{2+} \longrightarrow Ca^{2+}(Z^-)_2(s) + 2Na^+$$

where Z^- represents the zeolite framework.

When most of the sodium in the zeolite has been displaced, the zeolite can be regenerated by placing it in contact with concentrated sodium chloride solution, in the form of brine. When this is done, Na^+ ions reenter the zeolite, and the dipositive cations are washed away with the Cl^- anions.

Until recently, water softening was regarded as harmless. But epidemiological studies in Great Britain during the 1960s found a relatively high death rate

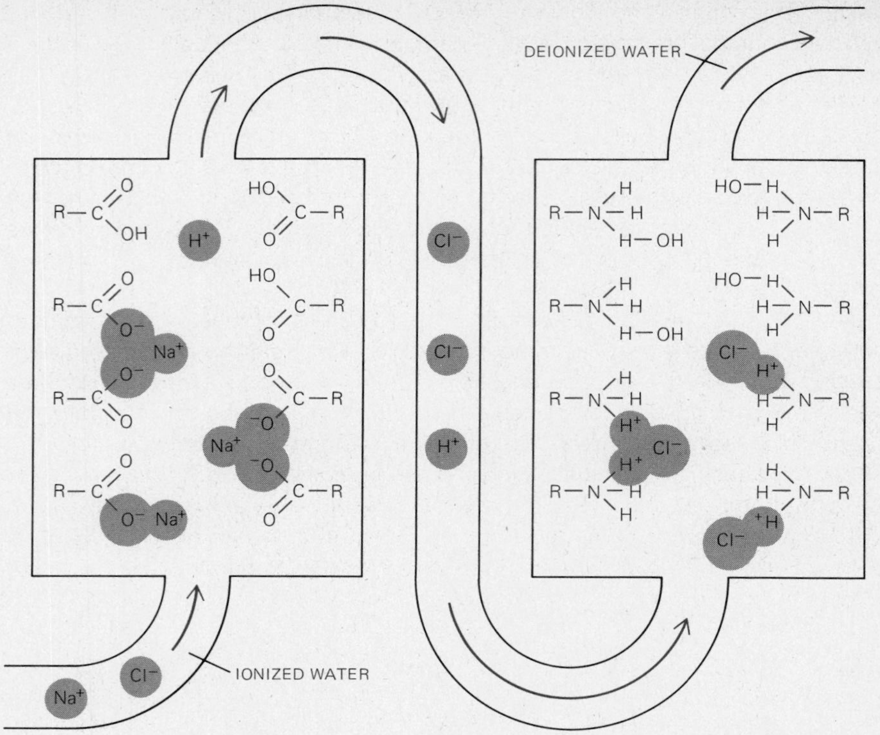

FIGURE 11.10
A water treatment system that uses two ion exchange resins in series. The first ion exchange resin removes Na$^+$ cations from the water. The second removes Cl$^-$ anions.

from heart disease in communities with water softened by this or other processes, compared to communities with hard water. It is not known whether the higher risk of heart disease is related to the removal of naturally occurring ions such as magnesium, or to the addition of sodium ions to artificially softened water or to another effect of the water softening process. While research goes on, British health officials have recommended that all communities with hard water discontinue the use of this water softening process.

Other methods of removing dissolved ionic material are available for processes that require water of exceptionally high purity. Since ionic impurities in water can interfere with or divert the course of chemical reactions, water that is pure enough to drink is sometimes further purified by distillation for laboratory use. By using appropriate materials for the still, it is possible to obtain water that is almost completely free of ionic impurities.

Another method uses an ion exchange resin that incorporates acid groups in its structure. All the cations in water can be exchanged with the H$^+$ ions of the acid. A general representation for one kind of organic acid used in these resins is RCOOH, where COOH represents the acid and R the organic portion. A typical reaction can be represented as:

$$RCOOH(s) + Na^+ \longrightarrow RCOO^-Na^+(s) + H^+$$

Water that is treated in this way can be passed over another resin that incorporates organic basic groups in its structure, to remove all anions and neutralize the H$^+$ ions. The process can be represented as:

$$(RNH_3)^+(OH)^-(s) + Cl^- + H^+ \longrightarrow (RNH_3)^+Cl^-(s) + H_2O$$

This ion removal process can be carried out by using columns containing the two kinds of resin in series, as shown in Figure 11.10. The process is expensive and is used only when water of unusual purity is needed, as in pharmaceutical manufacturing.

(1)

(2)

(3)

(4)

(5)

(1) In a typical municipal water treatment plant, chemicals such as chlorine are first added to soften the water, remove tastes and odors, and kill bacteria.

(2) Large paddles are used to insure that the added chemicals mix completely. Impurities in the water combine with the chemicals to form a substance called floc.

(3) The floc settles to the bottom. The basin must be drained periodically to remove the sediment.

(4) More impurities are removed by allowing the water to flow through a series of filtering materials.

(5) The treated water is stored in a reservoir, where it will be chlorinated again before use. (Photographs taken by Philip R. Pryde. From Lucy T. Pryde, *Environmental Chemistry: An Introduction,* Copyright © 1973 by Cummings Publishing Company, Inc., Menlo Park, California.)

Sewage Treatment

Sewage treatment varies greatly from community to community in the United States. Many suburban and rural areas have no central sewage treatment systems and rely on septic tanks. Some communities dump raw sewage into rivers or lakes. A few communities have sewage treatment systems capable of producing water that is pure enough for swimming or drinking. Most communities are somewhere in between.

Sewage treatment can be classified as primary, secondary, or complete. Primary treatment is the removal of suspended and floating solids by screening and sedimentation. Primary treatment collects about 60% of the total solids in sewage, in a form called primary sludge. The disposal of primary sludge can be a problem. Usually, sludge is decomposed by anaerobic bacteria. This process produces methane, CH_4, which can be burned to supply part of the energy needs of the sewage plant. However, the anaerobic decomposition is not complete. The remaining sludge may be used as fertilizer or landfill, or it may simply be dumped into the ocean.

Secondary treatment uses microorganisms to remove many of the impurities that remain after primary treatment. The sewage can be exposed to the microorganisms in sand beds, in tanks where compressed air is diffused through the sewage, or in oxidation ponds. The microorganisms metabolize the organic matter to such substances as CO_2, NH_4^+, NO_3^-, PO_4^{3-}, and SO_4^{2-}.

Treated sewage may also be chlorinated to kill the coliform group of bacteria, which can cause disease. Some communities also add another step, tertiary treatment, in which a high percentage of the solids and organic matter remaining after primary and secondary treatment are removed. Tertiary treatment uses a variety of chemical processes, each designed to remove specific impurities. In a few cases, tertiary treatment of sewage can produce water that is suitable for human use.

EXERCISES

11.1   Consider a planet that has roughly the same temperature range as the earth, but has a sea level air pressure of 3 atm. Predict how water would behave and function on such a planet. (*Hint:* Refer to Figure 10.36.)

11.2   Use the data in Table 11.1 to estimate the boiling points of H_2O, NH_3, and HF if hydrogen bonding did not exist.

11.3   The original choice of the unit of mass in the metric system was based on a property of water: One cubic centimeter of water at 4°C has a mass of approximately one gram. Suggest a reason for the selection of 4°C as the temperature in this definition.

11.4   Bromine and nitrogen have almost the same electronegativity values. Bromine does not form hydrogen bonds whereas nitrogen does. Suggest an explanation for this observation.

11.5   Indicate which of the following compounds can form intermolecular hydrogen bonds: (a) H_2SO_4, (b) H_3PO_4, (c) H_2O_2, (d) CH_3F, (e) NH_4I, (f) N_2H_4.

11.6   There are two known compounds with the molecular formula C_2H_6O. One boils at 352 K and the other boils at 250 K. Identify the two compounds and account for the dramatic difference in their boiling points.

11.7   A sample of HF of mass 0.033 g is vaporized at 315 K in a container of volume 0.10 liter. The pressure is measured and is found to be 0.14 atm. The difference between the measured pressure and the pressure calculated from the ideal gas equation can be attributed mainly to the formation of strongly hydrogen-bonded structures of composition H_6F_6 in the gas phase. Calculate the fraction of HF that forms H_6F_6 under these conditions.

11.8   In addition to the hydrogen bonds, ordinary van der Waals forces of attraction exist between the water molecules of ice. The strength of the van der Waals forces can be estimated as

11 kJ/mol. The heat of sublimation of ice is measured as 51 kJ/mol. Calculate the strength of the hydrogen bonds in ice.

11.9   Ammonium bifluoride is a salt with the composition NH_5F_2. Propose structures for the ions of this salt.

11.10   The crystal structure of NH_4F has been compared to that of ice, while the crystal structures of NH_4Cl, NH_4Br, and NH_4I are like that of NaCl. Suggest an explanation for this structural difference.

11.11   The bond energy of a hydrogen bond to fluorine is larger than the bond energy of a hydrogen bond to oxygen. Hydrogen fluoride melts at a much lower temperature than ice. Reconcile these two observations.

11.12   It has been stated that the HOH bond angle in water plays an important role in the hydrogen bonding of water. Explain the reason for this statement.

11.13   Draw a diagram of a portion of a solution of NaF showing the interactions between water molecules and the dissolved species.

11.14   Lithium fluoride is much less soluble in water than lithium chloride. Explain this observation.

11.15   Ammonia is very soluble in water; NCl_3 is insoluble in water. Explain this observation.

11.16   Cesium fluoride is more soluble in water than cesium chloride. Explain this observation, taking note of the observation of Exercise 11.14.

11.17   Ethanol, CH_3CH_2OH, is miscible with water, but octanol, $C_8H_{17}OH$, is insoluble in water. Explain this observation.

11.18   The oxides of many metals are much less soluble than their chlorides despite the fact that oxygen can form much stronger hydrogen bonds than chlorine. Explain this observation.

11.19   Describe the geometry of hydrated Be^{2+} and Ca^{2+} cations.

11.20   Describe the nature of the attractions between noble gas atoms and water molecules in a clathrate.

11.21   The volume of the Atlantic Ocean is 3.2×10^8 km³ and its density is 1.0 g/cm³. Calculate the mass of sodium chloride in the Atlantic Ocean.

11.22   The concentration of radium in the ocean is about 0.09 mg in 10^6 m³ of water. Calculate the volume of water that contains 1 g of radium.

11.23   Outline how each of the following pollutants harms bodies of water: (a) sodium nitrate (agricultural waste), (b) phosphates (from detergents), (c) chlorinated hydrocarbons (insecticides), (d) mercury (industrial waste), (e) domestic sewage.

11.24   Write a chemical equation for the formation of aluminum hydroxide as part of the flocculation method.

11.25   Write chemical equations for the precipitation of calcium and magnesium carbonates from hard water by the addition of soda.

11.26   Write the chemical equation for the regeneration of a zeolite that has been used to remove magnesium ions from water.

11.27   Suggest a reason for the necessity of keeping the two ion exchange resins shown in Figure 11.10 physically separated from each other.

11.28   One major problem in the maintenance of a marine aquarium is the prevention of the buildup of ammonia or nitrites that form from partial bacterial decomposition of metabolic waste products of fishes. When such an aquarium is first established the ammonia concentration may reach 7 ppm and the nitrite concentration may reach 10 ppm. With time, bacteria flourish that oxidize ammonia and nitrite to nitrate, which is not so harmful to the inhabitants of the aquarium. Calculate the volume of O_2 at STP required to oxidize the ammonia and nitrite to nitrate in a tank containing 200 liters of water, when each of these substances reaches its maximum concentration.

12

THE PROPERTIES OF SOLUTIONS

Even the most cursory examination of our surroundings shows that pure substances are rare. Most of the materials that are familiar to us are mixtures of two or more substances. One type of mixture that is of special interest is the **solution.** Solutions are everywhere. The air, the oceans, and the fluids in our bodies are solutions, to name just a few.

Most of the reactions that chemists study take place in solutions. Virtually all of the reactions that biologists study take place in solutions. In this chapter, we shall discuss the different sorts of solutions and their properties.

12.1 TERMINOLOGY

A solution is a homogeneous mixture of two or more substances. "Homogeneous" means that a sample taken from any part of a solution has the same relative composition as the entire solution, even if the sample is very small. You probably are accustomed to thinking of a solution as a liquid that contains dissolved substances. But a solution can be a solid, a liquid, or a gas. The air we breathe is a solution of gases (although air is not entirely homogeneous, since its composition can vary from place to place). Alloys such as brass, bronze, and sterling silver are solid solutions of metals.

In this chapter, we shall deal primarily with solutions that are liquids under the conditions of temperature and pressure at which they are studied. The liquid component, which is usually the major component of such a solution, is called the **solvent.** A substance that is dissolved in the solvent is called a **solute.** A single solution may contain a number of different solutes.

The terminology for a liquid solution is not always clear-cut, especially when all the components of a solution are liquids. For example, 100-proof vodka contains almost equal quantities of water and ethanol. We can ask whether vodka is a solution of water in alcohol or of alcohol in water. The best answer is to avoid the question by saying that vodka is a homogeneous system with two components.

In addition to describing a solution as a solvent and one or more solutes, we can use several terms to describe some aspects of the composition of a solution. The **solubility** of a substance is a measure of the maximum quantity of the substance that can dissolve in a given quantity of a solvent under a specified set of conditions. A solution that contains the maximum amount of solute is called a *saturated* solution.

For example, the solubility of NaCl in water at 273 K is 35.7 g of NaCl in 100 cm^3 of water. If more solid NaCl is added to such a saturated solution, the concentration of dissolved NaCl will not increase. If a solution of NaCl in water has less than 37.5 g of NaCl to 100 cm^3 of water, it is said to be *unsaturated*. If solid NaCl is added to an unsaturated solution, more NaCl will dissolve until the solution is saturated.

The concepts of solubility and saturation are useful for solid solutes. They are not always necessary for liquid-liquid solutions, because some liquids are soluble in one another in any proportions. Such liquids are said to be *miscible* in all proportions. For example, water and ethanol are miscible in all proportions and can form solutions of any desired composition.

12.2 EXPRESSIONS OF CONCENTRATION

In Section 3.4, page 68, we introduced some methods of expressing concentration that are used in quantitative studies of solutions. We shall now discuss these methods in detail.

350

An expression of concentration is a ratio, or fraction, that describes the quantity of solute in a given quantity of either solvent or of solution. The quantity of solute is the numerator in the fraction. Either the quantity of solvent or the total quantity of solution is the denominator. We can express these quantities in several ways: in units of mass, such as grams; in units of amount, such as moles; or in units of volume, such as liters. The most commonly used expressions of concentration can be classified by the type of units they employ.

Mass-Mass Expressions

Mass fraction and **mass percent** are expressions that describe the mass of solute in a given mass of *solution:*

$$\text{mass fraction} = \frac{\text{mass of solute}}{\text{mass of solution}} \tag{12.1}$$

$$\text{mass percent} = \frac{\text{mass of solute}}{\text{mass of solution}} \times 100\% \tag{12.2}$$

These expressions of concentration can be used conveniently for any kind of solution. Usually, the mass of solute and the mass of solution are expressed in the same units.

EXAMPLE 12.1

An alloy of copper and aluminum is prepared from 65.6 g of Cu and 423.1 g of Al. Calculate the mass percent of each component in this solid solution.

SOLUTION

The data tell us that the total mass of the solution is 65.6 g of Cu + 423.1 g of Al = 488.7 g of solution, and that there is 65.6 g of Cu in the solution. Therefore:

$$\text{mass percent Cu} = \frac{65.6 \text{ g Cu}}{488.7 \text{ g solution}} \times 100\% = 13.4\%$$

$$\text{mass percent Al} = 100\% - 13.4\% = 86.6\%$$

Note that the sum of the percentages of all the components must equal 100%.

Expressions of concentration such as mass fraction and mass percent are convenient for laboratory work. But they are not obviously related to many of the properties of solutions and so are rarely used in theoretical work.

Mole fraction and **mole percent** are defined in exactly the same way as mass fraction and mass percent, except that the unit used is the mole, the unit of amount. Just as the mass of a solution is the sum of the masses of all the components, so the number of moles in a solution is the sum of the number of moles of each component in the solution. For a two-component solution:

$$X_1 = \frac{n_1}{n_1 + n_2} \qquad X_2 = \frac{n_2}{n_1 + n_2} \tag{12.3}$$

and

$$X_1 + X_2 = 1$$

where X_1 is the mole fraction of one component, X_2 is the mole fraction of the other component, n_1 is the number of moles of component 1 and n_2 is the number of moles of component 2. These relationships can be extended for solutions with more than two components.

EXAMPLE 12.2

Calculate the mole fraction of sugar, $C_{12}H_{22}O_{11}$, in an aqueous solution that is 5.30% sugar by mass.

SOLUTION

To calculate the mole fraction, we must find the number of moles of sugar and of water in some fixed quantity of solution. From the given mass percent, we can say that there is 5.30 g of sugar in 100 g of solution, which is a convenient fixed quantity of solution. Since the solution consists of only sugar and water, 100 g of solution must contain 100 g − 5.30 g = 94.70 g of water. From a table of atomic weights, we can find that the molecular weight of water is 18.01 and the molecular weight of sugar is 342.3. Therefore, in the 100-g sample of solution there is:

$$\frac{5.30 \text{ g sugar}}{342.3 \text{ g sugar/mol sugar}} = 0.0155 \text{ mol sugar}$$

and

$$\frac{94.70 \text{ g } H_2O}{18.01 \text{ g } H_2O/\text{mol } H_2O} = 5.26 \text{ mol } H_2O$$

We can calculate the mole fraction of sugar in the solution by substituting these data into Equation 12.3, which gives the mole-fraction relationships for a two-component solution:

$$X_{C_{12}H_{22}O_{11}} = \frac{0.0155 \text{ mol sugar}}{0.0155 \text{ mol sugar} + 5.26 \text{ mol } H_2O} = 0.00294$$

$$X_{H_2O} = 1 - X_{C_{12}H_{22}O_{11}} = 1 - 0.00294 = 0.99706$$

It should be kept in mind that the mole is just the unit of amount, or of numbers of particles. Therefore, any property of a solution that is determined by the relative number of molecules of solute often can be calculated by using mole fractions.

Molality is an expression of concentration that differs from those we have discussed because its denominator is a quantity of *solvent*, rather than a quantity of solution. Molality is defined by:

$$m = \frac{n_{solute}}{g_{solvent}} \times 1000 \frac{g}{kg} \tag{12.4}$$

where m is the common abbreviation for molality and g is the mass in grams. Molality is the number of moles of solute that are dissolved in 1000 g (or 1 kg) of solvent. The units of molality are moles per kilogram (mol/kg).

EXAMPLE 12.3

Calculate the molality of sugar in the solution of Example 12.2.

SOLUTION

In Example 12.2 we determined that there is 0.0155 mol of sugar in a quantity of solution that also contains 94.7 g of water. According to the relationship expressed in Equation 12.4:

$$m = \frac{0.0155 \text{ mol sugar}}{94.7 \text{ g } H_2O} \times 1000 \frac{g}{kg} = 0.164 \text{ mol/kg}$$

The solution is said to be 0.164m in sugar.

Molality is commonly used as an expression of concentration in calculations of the boiling point and freezing point of solutions.

Molarity is the most commonly used expression of concentration. It is defined by:

$$M = \frac{n_{solute}}{V_{solution}} \qquad (12.5)$$

where M is used to express molarity and V is the volume of the *solution* in liters. The molarity of a solute is the number of moles of that solute dissolved in 1 liter of solution. Some calculations of molarity were carried out in Section 3.4, page 69.

To determine molarity, the volume of a liquid must be measured. One reason why molarity is used so often is that it is easier to measure the volume of a liquid than to measure the mass of a liquid with reasonable accuracy. (However, it is easier to measure the mass of a liquid than the volume of a liquid with extreme accuracy.) You should note that the value of the molarity of a solution changes as the temperature changes, since the volume of a liquid increases as the temperature rises and decreases as the temperature falls. This complication is not encountered when we use molality or other mass-mass expressions, since they do not include units of volume.

You may occasionally encounter the unit called formality (F). This unit exists because some chemists dislike talking about the molarity of a solution of a salt such as NaCl when there are no NaCl molecules in solid NaCl. Aside from the need to calculate the molecular weight of the formula (the formula weight) for formality, there is no difference between formality and molarity. The use of formality can be avoided by treating the formula NaCl as if it represented an actual molecule. When we say "a $1M$ NaCl solution," we mean that a mass of NaCl corresponding to the sum of the atomic weights of Na and Cl is dissolved in 1 liter of solution. We shall not use the term formality.

Another unit that closely resembles molarity is normality (N), which was devised to simplify stoichiometric calculations of reactions in solutions. We shall discuss the use of normality in calculations of acid-base reactions and reduction-oxidation reactions in later chapters.

Volume-Volume Units

Volume-volume units are convenient for liquid-liquid solutions.

Volume percent is a unit that is similar to mass percent but that refers to the volumes rather than the masses of the components. It is defined by:

$$\text{volume percent} = \frac{\text{volume of pure solute}}{\text{sum of the volumes of each component}} \times 100\% \quad (12.6)$$

For a two-component solution:

$$\text{volume percent}_1 = \frac{V_1}{V_1 + V_2} \times 100\% \qquad (12.7)$$

where V_1 is the volume of component 1 when it is pure and V_2 is the volume of component 2 when it is pure. In general, the volume of the solution is not the sum of the volumes of the individual components. For example, a mixture of 0.500 liter of water and 0.500 liter of ethanol has a volume of only 0.965 liter.

EXAMPLE 12.4
What volume of diethyl ether must be mixed with 0.125 liter of water to form a solution in which the volume percent of the diethyl ether is 3.2%?
SOLUTION
Let x equal the required volume of diethyl ether. Then the volume percent of diethyl ether is defined by Equation 12.7 as:

$$\frac{x_1}{x_1 + 0.125 \text{ liter}} \times 100\% = 3.2\%$$

$$x = 0.0041 \text{ liter diethyl ether}$$

You will not usually have to perform calculations to determine volume percent, since it is easier simply to measure the final volume of a solution. However, one volume percent measurement that does appear frequently in everyday life is *proof*, which is equal to twice the volume percent of ethanol in water. Thus, a wine that is 12% alcohol by volume is 24 proof. A rum that is 150 proof contains 75% alcohol by volume. A 3.2% beer is 6.4 proof.

You will often find it desirable to convert from one unit of concentration to another. If such a calculation requires you to express a given mass of liquid as a volume, you will have to know the relationship between a given volume of solution and its mass. This quantity, which we discussed in Chapter 3, is the density:

$$\text{density} = \frac{\text{mass}}{\text{volume}} \tag{12.8}$$

which is conveniently defined in units of grams per cubic centimeter (g/cm^3) for aqueous solutions.

EXAMPLE 12.5

The concentration of ethanol in the solution called 86-proof vodka is $6.5M$. The density of the solution is 0.95 g/cm^3. Calculate the molality and the mole fraction of ethanol in the vodka.

SOLUTION

The mass of 1 liter of the solution is found from the density:

mass solution $= 0.95 \text{ g/cm}^3 \times 1000 \text{ cm}^3 = 950 \text{ g}$

Since we are told that 1 liter of solution contains 6.5 mol of ethanol, and we know that the mass of 1 liter of solution is 950 g, we can see that 950 g of solution contains 6.5 mol of ethanol. The molecular weight of ethanol is 46, so the mass of 6.5 mol of ethanol is:

$$6.5 \text{ mol ethanol} \times \frac{46 \text{ g ethanol}}{\text{mol ethanol}} = 300 \text{ g ethanol}$$

The mass of the water in 950 g of solution can be determined by subtraction:

950 g solution $-$ 300 g ethanol $=$ 650 g H_2O

We now have the information necessary to calculate the molality, using Equation 12.4:

$$m = \frac{6.5 \text{ mol ethanol}}{650 \text{ g } H_2O} \times 1000 \frac{g}{kg} = 10 \text{ mol/kg}$$

This solution is $10m$ in ethanol.

To calculate the mole fraction of ethanol, we must know the number of moles of ethanol and the number of moles of water in a given quantity of solution. In 1 liter of this solution there is 6.5 mol of ethanol and

$$\frac{650 \text{ g } H_2O}{18 \text{ g } H_2O/\text{mol } H_2O} = 36 \text{ mol } H_2O$$

Thus,

$$X_{\text{ethanol}} = \frac{6.5 \text{ mol ethanol}}{6.5 \text{ mol ethanol} + 36 \text{ mol } H_2O} = 0.15$$

12.3 SOLUBILITY

As we mentioned in Chapter 11, several things must happen when a solute, especially a solid, dissolves in a liquid solvent (Figure 12.1). The attractions between the structural units of the solute must be overcome so that the units may be dispersed throughout the solution. Some interactions between solvent molecules must also be overcome to make room for the solute molecules between the solvent molecules. The cost in energy of overcoming these attractions is repaid by favorable interactions between solvent and solute molecules. To some extent, the cost is also repaid by the advantages that accompany the mixing of substances, as we shall discuss in Chapter 13.

"Like Dissolves Like"

Since we must consider at least these three interrelated factors in determining solubility, we can only make qualitative predictions about the solubility of one substance in another. The best rule for such predictions is that "like dissolves like," meaning that a solute tends to dissolve in solvents that are chemically similar to it. The most important characteristics in determining similarity when we apply this rule are the polarity and the electrical properties of substances. The formation of hydrogen bonds also has an important effect on solubility.

To see how these considerations apply, let us study the solubility of substances in three representative liquids. The first solvent we shall consider is octane, C_8H_{18}, a nonpolar hydrocarbon found in gasoline. The second is water, a highly polar hydrogen-bonding solvent. The third is acetone, an organic solvent with the structure:

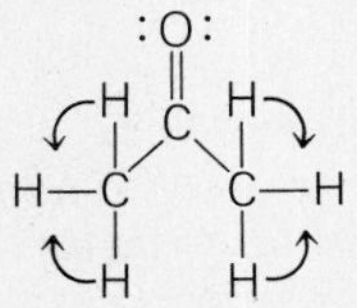

We first note that water and octane are essentially insoluble in one another. Because octane is nonpolar and water is polar, and because octane cannot form hydrogen bonds, there is no good way for the molecules of these two substances to interact strongly with each other. There are no favorable solute-solvent interactions that can compensate for the solute-solute attractions and solvent-solvent attractions that must be overcome if the solute is to disperse in the solvent. Therefore, a solution cannot form. Octane and water are not miscible.

But water and acetone are miscible. The acetone molecule has a relatively negative oxygen atom that can form a hydrogen bond with the hydrogen of a water molecule, as shown in Figure 12.2. Acetone and octane are also miscible. The interactions between molecules of octane and acetone are similar in magnitude to those between molecules of acetone alone or of octane alone. When acetone and octane are mixed, there is nothing to prevent the molecules of the two substances from mixing freely.

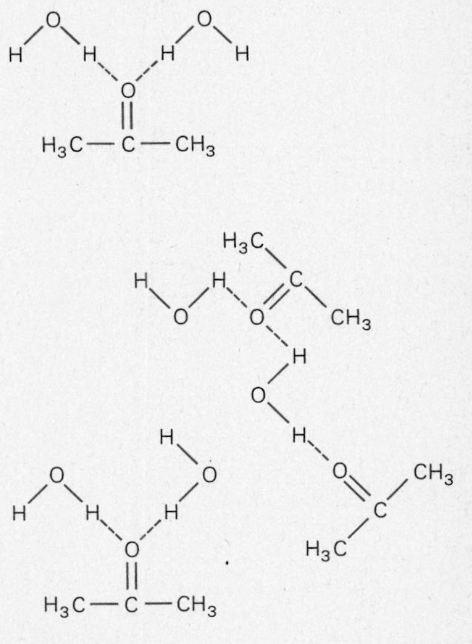

The rule that "like dissolves like" can be applied to predict the solubility of a salt such as LiCl in these three solvents. Lithium chloride, LiCl, is an ionic, highly polar salt that is most soluble in water, the polar solvent. There are several reasons for its solubility. The attractive forces between the anions and the cations of LiCl are reduced when they are dispersed in water because of the high dielectric constant of water (Section 11.3, page 336). In addition, ion-dipole attractions occur between the Li^+ cations and the relatively negative oxygen atoms of the water molecules. Hydrogen bonding between the Cl^- anions and the water molecules also occurs. These favorable solute-solvent interactions more than compensate for the loss of the attractions in pure water and pure LiCl.

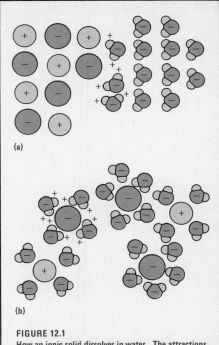

FIGURE 12.1
How an ionic solid dissolves in water. The attractions between the polar molecules and the ions making up the crystal lattice of the solid are greater than the attractive forces that hold the lattice together. The ions become dispersed among the water molecules. Note how the water molecules orient their positive ends toward the negative ions and their negative ends toward the positive ions.

FIGURE 12.2
The miscibility of acetone and water. Hydrogen bonds form between the water molecules and the acetone molecules.

Lithium chloride is less soluble in acetone. The acetone molecules do not interact as favorably with dissolved ions as do water molecules, and acetone has a lower dielectric constant than water. Finally, LiCl does not dissolve at all in octane, because there are no favorable interactions between the charged ions of LiCl and the nonpolar molecules of octane.

A nonpolar hydrocarbon solute such as naphthalene, $C_{10}H_8$, has a pattern of behavior that is opposite to that of an ionic salt. Naphthalene, a solid, dissolves in octane. The interactions between molecules of naphthalene and molecules of octane are similar in magnitude to those between molecules of naphthalene alone or of octane alone. Naphthalene is less soluble in acetone. There are some polar attractions between molecules of acetone, and naphthalene does not interact well enough with acetone to compensate for the loss of these attractions. Naphthalene is virtually insoluble in water, since there are no favorable attractions between water molecules and the nonpolar naphthalene molecules sufficient to compensate for the energy required to break the strong hydrogen bonds between the water molecules.

A polar organic solid such as glucose, $C_6H_{12}O_6$, dissolves in water because of the hydrogen bonds that form between the water molecules and the many OH groups of the glucose molecule. Glucose does not dissolve in octane because there are no favorable interactions between glucose molecules and octane molecules to compensate for the energy required to break the hydrogen bonds between glucose molecules in the solid.

While the ''like dissolves like'' rule is helpful in predicting relative solubilities of a given solute in different solvents, it should not be interpreted to mean that large quantities of a solute will dissolve in any solvent that is chemically similar. Many ionic substances are extremely insoluble in polar solvents, even water. Many nonpolar substances are extremely insoluble in nonpolar solvents such as octane. The rule tells us only *relative* solubilities. For example, a salt that has low solubility in water generally will be even less soluble in nonpolar solvents.

Soaps and detergents provide an interesting application of this concept. A molecule of a soap or of a detergent consists of a long nonpolar hydrocarbon chain with an ionic or hydrogen-bonding group at one end. The nonpolar portion of the molecule tends to dissolve in nonpolar substances, such as grease. The polar end of the molecule tends to dissolve in polar solvents, such as water. In a system containing both water and grease, the hydrocarbon portion of the soap or detergent molecule becomes associated with the grease, while the polar group is associated with the water molecules. The soap or detergent molecule is carried away along with the water, carrying grease and other nonpolar dirt with it, as shown in Figure 12.3.

The rule that ''like dissolves like'' also helps explain the orientation of large organic molecules in aqueous media—for example, of proteins in a living cell or of fats in the membrane surrounding the cell. Proteins and fats are long molecules that include both nonpolar hydrocarbon segments and polar subunits. A protein often folds itself in a way that buries the nonpolar units within the molecule and presents the polar subunits to the water that surrounds the protein, as Figure 12.4 shows. The cell membrane is believed to consist of long-chain phospholipid molecules whose polar ends point toward the aqueous medium outside the membrane and whose nonpolar portions are within the membrane (Figure 12.5).

Solubility of Ionic Compounds in Water

Much of the chemistry carried out in the laboratory occurs in aqueous solution. Most of the reactions that we shall study in later chapters in connection with chemical equilibrium are reactions of ionic substances in aqueous solutions. Accordingly, it is important to have a working knowledge of the solubilities of salts and related compounds in water.

For the sake of convenience, we shall divide solutes into three arbitrary

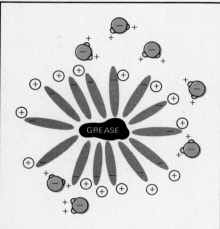

FIGURE 12.3

Soap and/or detergent molecules in action. Each molecule has a polar end and a nonpolar end. The polar end becomes associated with water molecules. The nonpolar end becomes associated with nonpolar substances, such as grease or other forms of dirt. The soap-water-dirt complex is washed away, leaving a clean fabric.

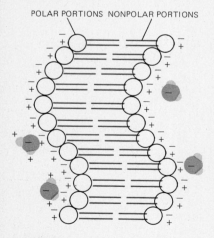

POLAR PORTIONS NONPOLAR PORTIONS

FIGURE 12.4

A large organic molecule such as a protein can have both polar and nonpolar portions. In aqueous solution, the conformation of the protein molecule is influenced by the surrounding polar water molecules. The protein molecule is folded so that its polar portions interact with the water molecules while its nonpolar portions are tucked away.

categories: soluble, slightly soluble, and insoluble. A substance will be regarded as *soluble* if more than 1 g dissolves in 100 cm³ of water. It will be regarded as *insoluble* if no more than 0.1 g dissolves in 100 cm³ of water. And it will be regarded as *slightly soluble* if the mass that dissolves in 100 cm³ of water is between 0.1 g and 1 g. The dividing line between "slightly soluble" and the other categories is understood to be arbitrary.

Some general rules for the solubility of the more common salts and hydroxides at room temperature are listed below. The listing is made easier by the fact that almost all the salts of some anions and cations tend to have the same solubility behavior.

The common salts of the following anions are soluble in water (exceptions are indicated):

1. All nitrates (NO_3^-) are soluble.
2. All chlorates (ClO_3^-) are soluble.
3. All acetates (CH_3COO^- or OAc^-) are soluble.
4. All sulfates (SO_4^{2-}) are soluble, except $BaSO_4$, $SrSO_4$, and $PbSO_4$, which are insoluble, and $CaSO_4$, Ag_2SO_4, and Hg_2SO_4, which are slightly soluble.
5. All chlorides (Cl^-), bromides (Br^-), and iodides (I^-) are soluble, except those of silver ($AgCl$, $AgBr$, AgI) and those of mercury(I) (Hg_2Cl_2, Hg_2Br_2, Hg_2I_2), which are insoluble, and those of lead ($PbCl_2$, $PbBr_2$, PbI_2), which are slightly soluble.

   The salts of some other groups of anions that are usually insoluble are listed below, with the exceptions to the rule:

6. All hydroxides (OH^-) are insoluble, except those of group IA metals (LiOH, NaOH, KOH, etc.) and $Ba(OH)_2$, which are soluble, and $Sr(OH)_2$ and $Ca(OH)_2$, which are slightly soluble.
7. All carbonates (CO_3^{2-}) and phosphates (PO_4^{3-}) are insoluble, except those of group IA metals and $(NH_4)_2CO_3$ and $(NH_4)_3PO_4$, which are all soluble.
8. All sulfides (S^{2-}) are insoluble, except those of the group IA and IIA metals and $(NH_4)_2S$, all of which are soluble.
9. As the above rules show, almost all the common salts of the cations NH_4^+, Na^+, and K^+ are soluble, a point that is worth remembering.

Solubility data are available for a large number of compounds. This information can be found in several standard compilations, such as *The Handbook of Chemistry and Physics,* which is published annually by the Chemical Rubber Company.

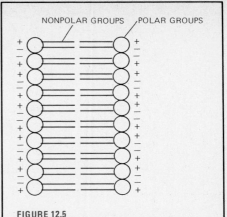

FIGURE 12.5
The schematic structure of a cell membrane. The structural units of the membrane are long-chain fatty acids. The polar ends of the fatty acid molecules are oriented toward the water molecules inside and outside the cell, while the nonpolar ends are in the middle of the membrane.

Solubility of Gases

Most gases dissolve at least partially in liquids. One reason for the solubility behavior of gases is the fact that there are almost no intermolecular attractions to be overcome in a gas.

The solubility of a given gas in a given solvent is determined primarily by the partial pressure of the gas in contact with the liquid and by the temperature of the system. In 1803, William Henry (1775–1836) found a simple relationship between the partial pressure of a gas above a liquid and the solubility of the gas in the liquid. There is a direct proportionality between these two measurements that can be expressed for a gas A as:

$$X_A = kP_A \tag{12.9}$$

where P_A is the partial pressure of gas A, X_A is the mole fraction of the gas in solution, and k is a proportionality constant, called the Henry's law constant.

Table 12.1 gives some values of Henry's law constants for gases in water. The higher the value of k, the greater is the solubility of the gas. Henry's law is an approximate, idealized description that works well only if the partial pressure of the gas is not too high and the gas is not extremely soluble.

Some practical consequences of the increase in the solubility of a gas with

TABLE 12.1
Henry's Law Constants for Gases in Water

Gas	T (K)	k (atm⁻¹)
O_2	273	4.5×10^{-5}
	298	2.3×10^{-5}
	333	1.6×10^{-5}
N_2	273	1.9×10^{-5}
	298	1.2×10^{-5}
Ar	298	2.7×10^{-5}
He	298	6.8×10^{-6}
CO_2	273	1.4×10^{-3}
	298	6.1×10^{-4}
	333	2.9×10^{-4}

an increase in pressure are well known. Carbonated soft drinks are bottled under a pressure of carbon dioxide slightly greater than 1 atm. As long as the cap is on, the pressure of CO_2 in the bottle remains greater than 1 atm. When the bottle is opened, the pressure of CO_2 above the soda suddenly becomes the CO_2 pressure of the atmosphere, about 0.0003 atm. The result is a decrease in the solubility of CO_2, which is evidenced by the appearance of bubbles as the gas comes out of solution.

A more dangerous manifestation of the same phenomenon is decompression sickness, commonly called "the bends," which is a threat to deep-sea divers. Decompression sickness occurs because large quantities of gases, especially nitrogen, dissolve in the blood of deep-sea divers, when they are subjected to high pressures deep under water. When a diver comes to the surface, the lower pressure decreases the solubility of the gases. If the pressure is lowered too quickly, bubbles of gas can form in body fluids, just as they do in soda when a bottle is opened. The results can be painful or even fatal.

Effect of Temperature and Pressure

Temperature usually has an important effect on the solubility of any substance. The nature of this effect depends on the nature of the solute and the solvent and on the specific temperature range.

The solubility of most salts in water increases with increasing temperature (Figure 12.6). However, there are many exceptions to this rule. For example, the solubility of sodium sulfate decreases as the temperature goes up. Even different hydrates of the same salt may behave differently. The stable hydrate of sodium carbonate at room temperature is $Na_2CO_3 \cdot 10H_2O$. The solubility of this hydrate increases as the temperature goes up to about 305 K. At this temperature, $Na_2CO_3 \cdot 10H_2O$ is no longer stable. The new stable hydrate is $Na_2CO_3 \cdot H_2O$, whose solubility decreases as the temperature continues to go up. If a quantity of $Na_2CO_3 \cdot 10H_2O$ is put into a beaker of water, you will see more and more of the solid dissolve as the beaker is heated. The solid will appear to stop dissolving at 305 K. If the solution is heated so that the temperature goes above 305 K, you will see a different solid starting to precipitate out of the solution.

Gases usually show a decrease in solubility with an increase in temperature. In Chapter 11, we discussed the replenishment of oxygen in deeper ocean waters by oxygen-rich cold water from the polar regions. You can see the effect of temperature on solubility in your own kitchen. Heat a pot of water on the stove. As the temperature of the water rises, bubbles form. Dissolved air is coming out of solution because the solubility of the air in water has been decreased by the increase in temperature. As the temperature continues to rise, more air continues to come out of solution. By the time the water boils, almost all the dissolved air has been expelled. If you turn off the heat, air will dissolve back into the water rather slowly after the water cools. You may notice a subtle sort of "flatness" if you drink some water that was boiled recently. We are used to having dissolved air in our drinking water, and we miss it if it is not there. If you put a fish into water that was boiled recently, the fish may suffocate.

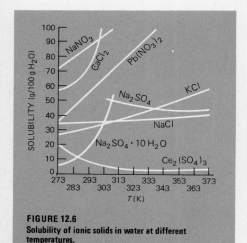

FIGURE 12.6
Solubility of ionic solids in water at different temperatures.

12.4   VAPOR PRESSURE OF SOLUTIONS

When we discussed the behavior of gases in Chapter 4, we used the concept of an ideal gas. This concept is a useful approach to a discussion of the properties of gases. The behavior of a real gas is often close to that of an ideal gas. Even when real gases deviate from ideal gas behavior, the ideal gas provides a standard against which real gas behavior can be judged.

Our approach to the description of the properties of solutions will resemble the ideal gas approach in many ways. We shall use the concept of an **ideal solution** as a model for the behavior of real solutions. The ideal solution is a model in which the strength of all the interactions between all the components of the solution is the same. That is, in an ideal solution of A and B, the interactions between A and A, between A and B, and between B and B are of the same magnitude.

We can come close to this uniformity of interactions in a real solution if the components of the solution are quite similar chemically. For example, benzene, C_6H_6, and toluene, C_7H_8, are very similar chemically. A solution of benzene and toluene will have properties that are close to the ideal. At the other extreme, the behavior of a solution of an ionic compound in water will deviate sharply from that of the ideal solution, because water and ionic compounds are quite different chemically.

In general, solutions of solids in liquids do not follow the ideal solution model. But if such solutions are dilute, their behavior comes fairly close to that of an ideal solution. In a dilute solution, we are dealing primarily with solvent-solvent interactions—that is, with interactions between identical molecules. There are only a relatively small number of solvent-solute interactions. These interactions do not have an appreciable effect on the properties of the solution, even when the solvent and the solute are very different chemically.

With these considerations in mind, let us discuss some properties of solutions. One of the most important properties of any solution is the vapor pressure of its volatile components. The relationship between the composition and the vapor pressure of an ideal solution can be described by starting with the simple picture of the vapor pressure of a pure liquid (Section 4.2, page 82). The magnitude of the vapor pressure of a liquid depends on the rate at which molecules leave and return to the surface of the liquid, in other words, on the rate of evaporation and the rate of condensation. Several factors determine these two rates. One factor is the number of molecules on the surface of the liquid.

Suppose we are studying the vapor pressure of the solvent in an ideal solution. In considering the factors that determine the vapor pressure of the solvent, we must take into account the fact that part of the surface is occupied by solute particles, as shown in Figure 12.7. Since this is an ideal solution, the interactions between the solvent and the solute are the same as those between molecules of the solvent. Therefore, the evaporation rate of the solvent will decrease in relation to the fraction of the surface that is occupied by the solute. Since the composition of a solution is uniform throughout, the portion of the surface that is occupied by solute molecules is identical to the proportion of solute molecules to solvent molecules in the entire solution. In other words, the evaporation rate of the solvent depends on the concentration of the solution, which means that it depends on the relative number of solvent molecules in the solution. Since the relative number of solvent molecules is expressed by the mole fraction of solvent, we can write:

$$r_e = k_e X_A \tag{12.10}$$

where r_e is the rate of evaporation, X_A is the mole fraction of solvent A, and k_e is a proportionality constant.

The rate of condensation of the solvent A is proportional to the partial pressure of A in the vapor above the solution. That is:

$$r_c = k_c P_A \tag{12.11}$$

where r_c is the rate of condensation, P_A is the partial pressure of A, and k_c is another proportionality constant. At equilibrium—that is, when P_A is the vapor pressure—the rate of evaporation is equal to the rate of condensation:

$$r_e = r_c$$

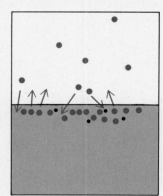

SOLUTION OF SOLUTE • AND SOLVENT ● MOLECULES

FIGURE 12.7
The vapor pressure of a solvent is related to the rate at which the solvent evaporates. The evaporation rate, in turn, depends on the number of solvent molecules on the surface of the solution. Solute molecules occupy part of the surface, reducing the evaporation rate of the solvent.

or, combining Equations 12.10 and 12.11:

$$k_e X_A = k_c P_A$$

or

$$P_A = \frac{k_e}{k_c} X_A \tag{12.12}$$

Equation 12.12 states that the vapor pressure of a solvent above a solution is directly proportional to the mole fraction of the solvent in the solution. We can evaluate the proportionality constant k_e/k_c if we know the vapor pressure that corresponds to a given value of the mole fraction of A. When $X_A = 1$, then P_A is the vapor pressure of the pure liquid, which is written as P_A^0. Substituting into Equation 12.12 for this case gives:

$$P_A = P_A^0 = \frac{k_e}{k_c} (1)$$

or

$$\frac{k_e}{k_c} = P_A^0 \tag{12.13}$$

Combining Equation 12.13 and Equation 12.12 gives:

$$P_A = P_A^0 X_A \tag{12.14}$$

which states that *at a given temperature, the vapor pressure of a volatile component of a solution is equal to its vapor pressure when pure at this temperature multiplied by its mole fraction in the solution.* This relationship was discovered in 1886 by François-Marie Raoult (1830–1901) and is called **Raoult's law.**

Raoult's law is exact only for an ideal solution. But it does produce results close to the data observed experimentally for dilute solutions of nonionic solids in liquids and for solutions of two or more chemically similar liquids. As we shall see in Section 12.7, with suitable modifications, Raoult's law also works well for dilute aqueous solutions of ionic solids.

EXAMPLE 12.6
The vapor pressure of water at 343 K is 0.308 atm. Calculate the vapor pressure of water above a solution prepared from 215 g of water and 25.0 g of glucose, $C_6H_{12}O_6$ (MW = 180) at 343 K.
SOLUTION
Calculate the amount of each component in the solution, so that the mole fraction of water in the solution can be found:

$$n_{H_2O} = \frac{215 \text{ g H}_2O}{18.0 \text{ g H}_2O/\text{mol H}_2O} = 11.9 \text{ mol H}_2O$$

$$n_{glucose} = \frac{25.0 \text{ g glucose}}{180 \text{ g glucose}/\text{mol glucose}} = 0.139 \text{ mol glucose}$$

$$X_{H_2O} = \frac{n_{H_2O}}{n_{H_2O} + n_{glucose}} = \frac{11.9 \text{ mol H}_2O}{11.9 \text{ mol H}_2O + 0.139 \text{ mol glucose}}$$

$$= 0.989$$

Since the vapor pressure of pure water is given and the mole fraction of water in the solution is known, Raoult's law can be used to find the vapor pressure of water above the solution:

$$P_{H_2O} = 0.308 \text{ atm} \times 0.989 = 0.305 \text{ atm}$$

DISTILLING COGNAC AND ARMAGNAC

The word "brandy" is said to derive from the Dutch term *brandewijn,* or burnt wine, which refers to the fire used to heat wine in the distillation process. Possibly the two greatest brandies in the world are Cognac and Armagnac, which come from regions of France that are only 130 km apart, and which are made by two distinctively different methods of distillation.

By law, Cognac must be distilled twice. The law requires use of a traditional and unique copper pot still. This still consists of a boiler in which the wine is placed, an onion-shaped head, and an elongated pipe called a *col de cygne,* or swan's neck, which leads to a curved copper coil that is surrounded by cold water.

In the first distillation, a fire is lit under the boiler. The volatile elements boil off first and are collected in the condenser, in the form of a milky liquid called the *broulis,* whose alcohol content is about 28%. The *broulis* is about one-third the quantity of the original wine. In the second distillation, the *broulis* is placed in the boiler. The fraction that distills first, called the head, is discarded. So is the fraction that distills last, the tails. Only the "heart" of the second distillation, a clear liquid of about 70% alcohol content, is retained. Several years of aging in oak barrels are required to tame the fiery liquid and to give it the distinctive color and aroma of fine Cognac, which comes from substances leached out of the barrel.

At one time, Armagnac was distilled in small, portable stills that were drawn from farmyard to farmyard by horse. Today, almost all Armagnac is distilled in what is called a continuous still, which was invented in about 1830 by an Irishman, Aeneas Coffey.

In a continuous still, wine passes from a storage tank into an adjoining unit, which consists of two or three boilers above a fire. The volatile fraction of the wine is vaporized by the heat of the fire and rises through several perforated plates, which stop some of the less volatile components. The vapor that condenses has an alcohol content of between 52% and 70%. Once again, several years of aging are required to give the distillate the smoothness and color that are typical of fine Armagnac.

The pot stills of the type used for Cognac once were the only stills available. Their antiquity is evidenced by their traditional name of *alembic,* which comes from the Arabic. Today, almost all brandy and other distilled spirits are prepared in continuous stills. These stills require much less labor than the pot stills, which must be watched round the clock. But the continuous still does not allow the distiller to separate out the best fraction of the distillate, as is possible with the pot still. The best malt whiskey also is prepared in pot stills, rather than in continuous stills.

EXAMPLE 12.7

Assuming ideal behavior, what mass of naphthalene, $C_{10}H_8$, would have to be dissolved in 200 g of octane, C_8H_{18}, to lower the vapor pressure of pure octane by 20%?

SOLUTION

If the vapor pressure of pure octane is P^0_{oct}, then the desired vapor pressure above the solution is 80% of P^0_{oct}, or $0.80\,P^0_{oct}$. According to Raoult's law:

$$0.80\,P^0_{oct} = P^0_{oct}X_{oct}$$
$$X_{oct} = 0.80$$
$$= \frac{n_{oct}}{n_{oct} + n_{C_{10}H_8}}$$

The molecular weight of octane is 114, and the molecular weight of naphthalene is 128. Thus:

$$0.80 = \frac{\dfrac{200 \text{ g octane}}{114 \text{ g octane/mol octane}}}{\dfrac{200 \text{ g octane}}{114 \text{ g octane/mol octane}} + \dfrac{\text{mass } C_{10}H_8}{128 \text{ g } C_{10}H_8/\text{mol } C_{10}H_8}}$$

$$\text{mass } C_{10}H_8 = 56 \text{ g}$$

Solutions consisting of a number of volatile components are interesting for several reasons. Let us compare a solution of ethanol and water, such as a glass of Scotch and water, with a solution of sugar and water. We find only water vapor above the solution of sugar and water. But the vapor above the Scotch and water consists of two gases, ethanol and water (as well as some minor components that give Scotch its smell and that we shall ignore). Given the proper information, we can find the composition of this mixture of gases.

EXAMPLE 12.8

Calculate the composition of the vapor above a solution containing 735 g of water and 245 g of ethanol, C_2H_5OH, at 323 K. At this temperature, the vapor pressure of pure ethanol is 0.292 atm and the vapor pressure of pure water is 0.122 atm.

SOLUTION

First calculate the mole fraction of each component:

$$X_{H_2O} = \frac{\dfrac{735 \text{ g } H_2O}{18.0 \text{ g } H_2O/\text{mol } H_2O}}{\dfrac{735 \text{ g } H_2O}{18.0 \text{ g } H_2O/\text{mol } H_2O} + \dfrac{245 \text{ g ethanol}}{46.0 \text{ g ethanol/mol ethanol}}} = 0.885$$

$$X_{ethanol} = 1 - X_{H_2O} = 1 - 0.885 = 0.115$$

According to Raoult's law:

$$P_{H_2O} = (0.122 \text{ atm})(0.885) = 0.108 \text{ atm}$$
$$P_{ethanol} = (0.292 \text{ atm})(0.115) = 0.0336 \text{ atm}$$

The vapor above the solution is a solution of gases. The mole fraction of each component in the vapor can be calculated directly if we recall that the partial pressure of a gas is directly proportional to the amount (moles) of the gas:

$$X_{H_2O(g)} = \frac{P_{H_2O(g)}}{P_{H_2O(g)} + P_{ethanol(g)}}$$

$$= \frac{0.108 \text{ atm}}{0.108 \text{ atm} + 0.0336 \text{ atm}} = 0.763$$

$$X_{\text{ethanol(g)}} = 1 - 0.763 = 0.237$$

The calculation in Example 12.8 illustrates an important point: We shall find a higher concentration of the more volatile component (the component with the higher vapor pressure) in the vapor than we find in the solution. The mole fraction of ethanol in Example 12.8 increases from 0.115 in solution to 0.240 in the vapor. This phenomenon is the basis of an important set of methods of separation and purification called *distillation*.

If the vapor above the solution described in Example 12.8 is removed and condensed, it gives us a new solution. This new solution has the same composition as the vapor; that is, $X_{\text{H}_2\text{O}} = 0.763$ and $X_{\text{ethanol}} = 0.273$. There will be vapor above this new solution, since part of the solution will evaporate. We can repeat the calculation of Example 12.8 to find the composition of this vapor. By Raoult's law (Equation 12.14):

$$P_{\text{H}_2\text{O}} = (0.122 \text{ atm})(0.763) = 0.0931 \text{ atm}$$

$$P_{\text{ethanol}} = (0.292 \text{ atm})(0.237) = 0.0692 \text{ atm}$$

The mole fraction of ethanol in this vapor is:

$$X_{\text{ethanol}} = \frac{0.0692 \text{ atm}}{0.0692 \text{ atm} + 0.0931 \text{ atm}} = 0.426$$

and the $X_{\text{H}_2\text{O}} = 1 - 0.426 = 0.574$.

If this vapor is again removed and condensed, we have a new solution, further enriched in the more volatile component, ethanol. By carrying the solution through many of these evaporation-condensation-evaporation steps, we can remove the more volatile component in extremely pure form, leaving behind the less volatile component.

Distillation is used in many ways and for many purposes. In the chemical laboratory a distillation apparatus can be used for the slow but highly efficient separation of small amounts of material. The use of a still to convert large quantities of wine (about 25 proof) to brandy, which is a more concentrated solution (about 85 proof) of ethanol in water, is also distillation.

12.5 BOILING POINT AND FREEZING POINT OF SOLUTIONS

The fact that the concentration of a solution is related to the vapor pressure of its components has an important consequence. If we assume ideal solution behavior, we know that the vapor pressure of the solvent is not influenced by the nature of the dissolved substance. We saw earlier that the vapor pressure of a solvent is related to the number of molecules of solvent that are displaced from the surface by solute particles. The properties of solutions that depend on the *number* of solute particles but not on their specific nature are called **colligative properties**. Vapor pressure is a colligative property.

The phase transition temperatures of a substance are closely related to its vapor pressure, and so we find that the boiling point and freezing point of liquid solutions are also colligative properties.

The boiling point is defined as the temperature at which the vapor pressure of the liquid is equal to the external pressure. The vapor pressure of water is 1 atm at 373 K, and this temperature is the normal boiling point of water. But an aqueous solution of a nonvolatile solute that is heated to 373 K at 1 atm does not boil, because the vapor pressure of water above the solution is always lower than the vapor pressure of pure water, as Figure 12.8 shows. At 373 K, the vapor pressure of water above a solution is less than 1 atm.

A typical laboratory distillation apparatus.
(*Robert Boikess*)

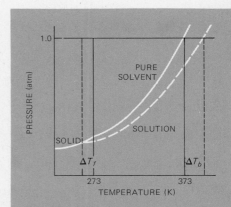

FIGURE 12.8
Pure water that is heated to 373 K at 1 atm will boil. An aqueous solution of a nonvolatile solute heated to the same temperature at the same pressure will not boil. The boiling point is the temperature at which the vapor pressure equals the external pressure, and the vapor pressure of a solution is lower than the vapor pressure of pure water at any given temperature. Pure water that is cooled to 273 K freezes, while an aqueous solution does not freeze until its temperature falls below 273 K.

TABLE 12.2
Molal Boiling Point and Freezing Point Constants

Substance	K_b	B. P. (K)	K_f	M. P. (K)
water	0.512	373.15	1.86	273.15
chloroform	3.63	334.9		
benzene	2.53	353.3	4.90	278.7
carbon tetrachloride	5.03	349.7		
acetic acid	3.07	391.1	3.90	298.8
ethanol	1.22	351.7		
n-octane	4.02	398.8		
naphthalene			6.8	353.7
camphor			40.0	453.0

If we continue to heat the solution, the vapor pressure of water above the solution will finally reach 1 atm and the solution will boil. The temperature at which a given aqueous solution boils depends on the vapor pressure of water above the solution, and the vapor pressure of water is determined by the concentration of solute in the solution. As the concentration of solute goes up, the vapor pressure of water at any given temperature goes down and the boiling point of the solution goes up. In other words, an increase in the concentration of solute means a boiling point *elevation*.

In an ideal solution, the magnitude of the boiling point elevation for the solvent depends on two factors: the nature of the solvent and the number of solute particles (that is, the concentration of the solute). This relationship usually is written in the approximate form:

$$\Delta T = K_b m \qquad (12.15)$$

where ΔT is the difference between the boiling point of the pure solvent and the boiling point of the solvent in solution, m is the molality of solute, and K_b is a proportionality constant whose value is determined by the nature of the solvent. The proportionality constant K_b is called the *molal boiling constant*. It is a measure of the way in which the boiling point of a solvent changes in response to the addition of different amounts of a solute. Values of K_b for various solvents can be found in Table 12.2.

Equation 12.15 works best for dilute solutions of nonvolatile, nonionic solutes in liquids. With some modifications, it also works for dilute aqueous solutions of ionic solids.

EXAMPLE 12.9

Using the data in Table 12.2, calculate the boiling point of a solution of 18.2 g of DDT ($C_{14}H_9Cl_5$), a nonvolatile nonionic substance, in 342 g of chloroform, $CHCl_3$.

SOLUTION

First calculate the molality of DDT in the solution. The molecular weight of DDT is 354.5. The molality is given by:

$$\frac{\dfrac{18.2 \text{ g DDT}}{354.5 \text{ g DDT/mol DDT}}}{342 \text{ g } CHCl_3} \times 1000 \, \frac{\text{g}}{\text{kg}} = 0.150m$$

From the molality calculated here and the value of K_b for chloroform in Table 12.2, we can use Equation 12.15 to calculate the boiling point elevation of chloroform that is caused by the addition of 18.2 g of DDT:

$$\Delta T = (3.63)(0.150) = 0.54 \text{ K}$$

The boiling point of pure chloroform, given as 334.9 K in Table 12.2, is raised by 0.54 K in this solution. The boiling point for the chloroform in the solution therefore is:

$$334.9 \text{ K} + 0.54 \text{ K} = 335.4 \text{ K}$$

EXAMPLE 12.10

Using the data in Table 12.2, calculate the mass of glucose, $C_6H_{12}O_6$, which must be dissolved in 100 g of water to raise the boiling point of the water to 375.2 K.

SOLUTION

One way to perform this calculation is to combine the definition of molality given in Equation 12.4 with the definition of boiling point elevation given in Equation 12.15:

$$\Delta T = K_b \frac{n_{C_6H_{12}O_6}}{g_{H_2O}} \times 1000 \frac{g}{kg}$$

Substituting the given data into this expression:

$$(375.2 - 373.15) \text{ K} = 0.512 \frac{n_{C_6H_{12}O_6}}{100 \text{ g H}_2\text{O}} \times 1000 \frac{g}{kg}$$

$$n_{C_6H_{12}O_6} = 0.400 \text{ mol } C_6H_{12}O_6$$

The molecular weight of glucose is 180, and the required mass of glucose therefore is:

$$\text{mass glucose} = 0.400 \text{ mol glucose} \times 180 \frac{g \text{ glucose}}{\text{mol glucose}}$$

$$= 72.0 \text{ g glucose}$$

The freezing behavior of liquid solutions is more complicated than the boiling behavior. It is well known that the freezing point of a solution of a nonvolatile solute is lower than the freezing point of the pure solvent. The lower freezing point of a solution is the reason why the salt water of the sea does not freeze as readily as freshwater lakes and rivers do. To illustrate the freezing behavior of a solution, let us examine how a solution of sodium chloride in water behaves as the temperature drops.

A solution of sodium chloride in water does not freeze when the temperature reaches 273 K, the freezing point of pure water. The temperature must go lower before the solution freezes. How much lower depends on the concentration of NaCl in the solution. When the appropriate temperature is reached, ice starts to form. This ice consists of pure H_2O. The remaining solution thus becomes more concentrated, since it contains the same amount of NaCl in a smaller amount of liquid H_2O.

Pure water maintains a constant temperature as it freezes. But the temperature of a solution of sodium chloride in water drops steadily as ice freezes out. As the ice freezes out, the solution becomes more concentrated, and the freezing point of the solution decreases as the concentration of the solute increases. Eventually, the solution becomes saturated in NaCl, and its temperature can go no lower. The solution becomes saturated at 252 K, which is called the *eutectic temperature* of the solution.

If the solution is maintained at the eutectic temperature, a solid with the composition $NaCl \cdot 2H_2O$ crystallizes out of the solution. Eventually, the entire solution freezes, leaving a solid called the eutectic solid. This solid consists of a mixture of ice and $NaCl \cdot 2H_2O$.

The lowered freezing point in a solution of salt and water is related to the lower vapor pressure of a solvent above a solution, as Figure 12.8 shows. The freezing point of a liquid at 1 atm of external pressure usually is close to the temperature at which both the liquid and the solid have the same vapor pressure; or, in other words, to the temperature at which the vapor pressure curves of the liquid and the solid cross. Because the vapor pressure of the solution is lower than that of the pure solid at any given temperature, the vapor pressure curve of the solution crosses that of the solid at a lower temperature than does the curve of the pure liquid, as Figure 12.8 shows. We speak of a freezing point *depression*, which is defined by a relationship almost identical to that for the boiling point elevation:

$$\Delta T = K_f m \tag{12.16}$$

which differs from Equation 12.15 only in the proportionality constant K_f. This proportionality constant is called the *molal freezing point constant*, and is characteristic of the solvent. Values of K_f are found in Table 12.2. It should be noted that ΔT is the difference between the freezing point of pure liquid and the freezing point of liquid in a solution, and it is a positive quantity.

Generally, the freezing point of a solvent is more sensitive than the boiling point to the addition of solute. For this reason, and because freezing points usually can be measured more accurately than boiling points, the freezing point is often used in studies of solutions.

EXAMPLE 12.11

The Rast method, one of the older techniques for finding the molecular weight of an unknown, measures the freezing point depression of solutions in camphor. Camphor is used as the solvent in this technique because its freezing point is very sensitive to added solute. A solution of 2.342 g of an unknown substance in 49.88 g of camphor freezes at 441.2 K. Using the data from Table 12.2, calculate the molecular weight of the unknown.

SOLUTION

The molality of the solution can be found from the freezing point of the solution, the freezing point of pure camphor, and the value of K_f given in Table 12.2 by using the relationship expressed in Equation 12.16:

$$(453.0 - 441.2) \text{ K} = (40.0)m$$
$$m = 0.295$$

From the molality, we can find the amount of solute in any mass of solvent. In this solution, the mass of solvent is 49.88 g. Therefore:

$$0.295m = \frac{n_{solute}}{49.88 \text{ g camphor}} \times 1000 \frac{g}{kg}$$
$$n_{solute} = 0.0147 \text{ mol}$$

Since 2.342 g of solute is 0.0147 mol, the molecular weight of the solute is:

$$\frac{2.342 \text{ g}}{0.0147 \text{ mol}} = 159$$

It is not unusual to find changes in the nature of a solute when it dissolves. In particular, one mole of undissolved solute does not necessarily form one mole of particles when it is in solution. The solute may dissociate partially or completely to form more than one mole of particles. On the other hand, some molecules of solute may come together, or associate, partially or completely to form less than one mole of particles. Information about the behavior of a solute sometimes can be obtained from the colligative properties of the solution. Often, this sort of information about solute behavior cannot be obtained easily by any other method.

The simplest way to determine the number of solute particles in a solution is to measure the molecular weight of the solute when it is in solution. This so-called ''apparent'' molecular weight often differs from the actual molecular weight. It is found by making appropriate measurements of a colligative property of a solution of known composition. If there is neither association nor dissociation, the apparent molecular weight of the solute in solution will be very close to the molecular weight that is calculated from the formula of the solute. If there is dissociation, the apparent molecular weight in solution will be lower than the molecular weight calculated from the formula. If there is association, the apparent molecular weight will be higher than the one calculated from the formula.

EXAMPLE 12.12

A solution of 1.43 g of acetic acid, CH_3COOH, in 12.3 g of benzene freezes at 273.9 K. Using the data in Table 12.2, determine the behavior of acetic acid dissolved in benzene.

SOLUTION

We can use the method of Example 12.11 to calculate the apparent molecular weight of acetic acid in benzene solution:

$$(278.7 - 273.9) \text{ K} = (4.90)m$$

$$m = 0.980$$

$$0.980 = \frac{n_{\text{solute}}}{12.3 \text{ g benzene}} \times 1000 \frac{\text{g}}{\text{kg}}$$

$$n_{\text{solute}} = 0.0121 \text{ mol}$$

$$\frac{1.43 \text{ g acetic acid}}{0.0121 \text{ mol acetic acid}} = 118$$

The apparent molecular weight of the acetic acid in benzene solution is 118. The molecular weight calculated from the formula of acetic acid, CH_3COOH, is 60. The discrepancy is due to the strong tendency of acetic acid molecules in solution to associate in pairs, called dimers, by hydrogen bonding. Each pair of molecules then acts as a single solute particle. The structure of the hydrogen-bonded dimer is:

$$CH_3 - C \begin{matrix} \ddot{O}:\text{---}H - \ddot{O} \\ \\ \ddot{O} - H\text{---}:\ddot{O} \end{matrix} C - CH_3$$

12.6 OSMOTIC PRESSURE

Another important colligative property is the **osmotic pressure** of a solution, which can be observed under special circumstances. The observation requires a *semipermeable membrane*, either natural or synthetic, which gets its name because it allows some substances but not others to pass through it.

The mechanisms by which semipermeable membranes work are not always well understood, but the practical uses of such membranes are of great interest. Consider a semipermeable membrane that separates an aqueous solution from pure water (Figure 12.9). If the membrane were not there, the solution and the water would mix. Pure water would pass into the tube containing the solution, and solution would pass out of the tube, until the concentration of the solute would be the same throughout the liquid phase. We would have another example of the natural tendency of substances to mix, which we shall discuss in detail in Chapter 13.

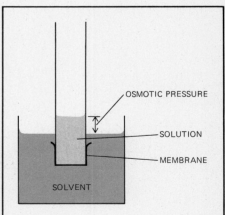

FIGURE 12.9
Osmosis. The membrane allows the passage of solvent but not solute. The level of the solution in the tube rises until the pressure on the membrane from within the tube is great enough to cause water to flow out of the tube at the same rate water flows into the tube. The height of the column of the solution in the tube is a measure of osmotic pressure.

The semipermeable membrane allows water to flow through it in either direction, but it does not allow solute to pass through it. Therefore, pure water flows into the tube containing the solution, but solute does not pass out. As a result, the level of the water in the tube rises. It might seem that the water in the tube would never stop rising, since the solution inside the tube can never be as dilute as the pure water outside. But as the level of water rises in the tube, water begins to flow more rapidly the other way, out of the tube. Eventually, the column of solution is high enough so that the pressure it exerts causes the rate of flow of water out of the tube to equal the rate of flow into the tube. The level of solution no longer rises. The pressure at which the rates of flow into and out of the solution are equal is called the **osmotic pressure** of the solution.

There are several ways to measure the osmotic pressure of a solution. We can measure the height of the column of liquid when it no longer rises. As we saw in Section 4.2, page 83, such a measurement of the height of a column of liquid is readily converted to a pressure. Alternatively, an external pressure just great enough to stop the net flow of water through a semipermeable membrane into the tube can be applied. The measurement of this applied pressure gives the osmotic pressure of the solution.

Studies have shown that the osmotic pressure of dilute solutions is a colligative property. That is, osmotic pressure depends on the amount of solute in the solution but not on the nature of the solute. The relationship between the osmotic pressure of a solution that is dilute enough to follow ideal behavior and the composition of the solution is given by the formula:

$$\pi V = nRT \tag{12.17}$$

where π is the osmotic pressure in atmospheres, V is the volume of the solution in liters, n is the number of moles of solute particles in this volume of solution, T is the absolute temperature, and R is 0.0821 liter atm mol^{-1} K^{-1}, the universal gas constant. The similarity between this relationship and the ideal gas equation is striking. Since the quantity n/V is an expression of concentration, Equation 12.17 can also be written:

$$\pi = cRT \tag{12.18}$$

where $c = n/V$.

Osmotic pressure is a colligative property of great interest because of the role it plays in the basic processes of life. Osmosis, one way in which a substance may move across a semipermeable membrane, takes place in all living organisms. The wall of a living cell is a semipermeable membrane; one of its roles is to keep some substances out of the cell while allowing others to enter.

In the laboratory, the measurement of osmotic pressure is often used to study the properties of substances with very high molecular weights, such as proteins and nucleic acids. Because of their high molecular weights, it is difficult to get substantial molar quantities of these substances into solution. But a solution of a relatively small quantity of such a dissolved solute develops a large osmotic pressure. For example, a solution that contains 0.1 mol of solute in 1 liter of solution (0.1M) has an osmotic pressure of 2.46 atm at 300 K. This osmotic pressure is equivalent to the pressure of a column of water more than 20 m high. Measurements of osmotic pressures of solutions as dilute as $10^{-4}M$ can be made with great accuracy.

One problem in conducting osmosis experiments is to find membranes strong enough to withstand extremely high osmotic pressures. New synthetic membranes have made it possible to measure pressures greater than 250 atm.

EXAMPLE 12.13
A solution of 4.68 g of hemoglobin, the oxygen-carrying protein in red blood cells, in 0.125 liter of water is found to have an osmotic pressure of 0.0135 atm at 300 K. Calculate the molecular weight of hemoglobin.

OSMOTIC POWER

One of the more unusual suggestions for meeting future energy needs is the possible exploitation of the salinity gradients that occur at the mouths of rivers to generate electricity. A 1976 report from the Scripps Institute of Oceanography estimated the "osmotic potential" in all the freshwater runoff to the earth's oceans at 2600 gigawatts, about the same as the amount of energy being generated by hydroelectric plants. The United States' six largest rivers were estimated to have a potential generating capacity of 70 gigawatts.

The concept of osmotic power is simple in principle. By using giant semipermeable membranes, the osmotic pressure of seawater could be used to raise a column of water in a suitably enclosed portion of the ocean. The elevated water could fall through turbines, generating electricity. For example, it is estimated that the pressure head between seawater and incoming fresh water at the mouth of the Columbia River is almost 25 atm, or about 240 m. This pressure is equal to the total height of all the dams on the Columbia. If half of the river's osmotic potential could be exploited at 30% efficiency, 2.3 gigawatts of electricity could be generated.

However, major questions remain to be answered before osmotic power could be generated on a large scale. One major question concerns the environmental disturbances that would result from any attempt to exploit the salinity gradient of a major river on a large scale. In addition, serious technological problems must be solved if osmotic power is to be generated economically. Several different schemes are being studied.

One Israeli proposal describes the use of the saline waters of the Dead Sea to generate electricity. In this scheme, brine from the Dead Sea would be pumped alongside tubes made of a semipermeable material filled with water of low salinity. Fresh water would permeate through the semipermeable material, thus providing the movement required to run a turbine generator.

The material that would be used in the proposed Israeli system is one developed for the desalination of water, in a technique called reverse osmosis. By applying sufficient pressure from an external source to saline water, water but not salt can be made to pass through the membrane. Reverse osmosis for water desalination is already economically practical and is already being used in some areas. The use of osmotic pressure to generate electricity is a promising concept that requires much more research and development to become feasible.

SOLUTION

All the data needed to calculate the number of moles of hemoglobin in the solution are given. Using the relationship expressed in Equation 12.17:

$$(0.0135 \text{ atm})(0.125 \text{ liter}) = n\left(0.0821 \frac{\text{liter atm}}{\text{mol K}}\right)(300 \text{ K})$$

$$n = 0.0000685 \text{ mol hemoglobin}$$

Since this amount of hemoglobin has a mass of 4.68 g, the molecular weight of hemoglobin is:

$$\frac{4.68 \text{ g hemoglobin}}{0.0000685 \text{ mol hemoglobin}} = 68\ 300$$

12.7 SOLUTIONS OF ELECTROLYTES

So far, our discussion of colligative properties has focused on solutions of substances that do not ionize. In general, the number of solute particles in such solutions is the same as the number of particles of solute that are dissolved. In some compounds, such as organic acids, association can occur, reducing the number of solute particles.

But the most important solutions with which we shall deal are aqueous solutions of substances that dissociate partially or completely into ions. The colligative properties of these solutions are affected by the increase in the number of solute particles that results from dissociation. These solutions are also found to deviate from ideal behavior at much lower concentrations than solutions that do not produce ions. In addition, the electrical properties of water are changed substantially by the addition of ionic solutes.

Solutes can be classified by the electrical conductivity of their aqueous solutions. Figure 12.10 shows a simple apparatus for measuring the electrical conductivity of a solution. Using such an apparatus, we find that water is a very poor conductor of electricity. Aqueous solutions of substances that do not dissociate into ions are also poor conductors of electricity. These substances, which include most organic compounds and gases such as O_2 and N_2, are called **nonelectrolytes.**

Substances that dissociate into ions in aqueous solution are classified as **electrolytes.** The electrical conductivity of an aqueous solution of an electrolyte usually will be much greater than that of water, because charged particles are present.

Electrolytes are divided into two groups: *strong electrolytes* and *weak electrolytes.* Strong electrolytes are completely dissociated in aqueous solution at concentrations around $1M$. Among the substances classified as strong electrolytes are all salts of group IA and group IIA metals, many salts of other metals, strong acids, and strong bases. Strong electrolytes usually are substantially ionic even when they are not in solution, but there are exceptions to this rule. One such exception, for example, is that HCl is a polar gas when pure, but a strong electrolyte when in solution.

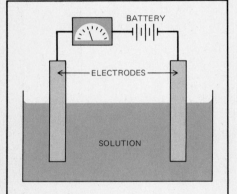

BATTERY

ELECTRODES

SOLUTION

FIGURE 12.10
An apparatus for demonstrating that a solution conducts electricity. The meter measures the current that passes between the electrodes through the solution.

EXAMPLE 12.14
Using the data in Table 12.2 and assuming ideal solution behavior, calculate the boiling point elevation of a solution prepared from 0.571 g of LiF, a strong electrolyte, and 111 g of H_2O.
SOLUTION
The boiling point elevation of the solution is determined by the molality of solute particles in the solution. Since LiF is a strong electrolyte, dissociating into ions in aqueous solution, the molality of solute particles is greater than the molality based on the original number of moles of undissolved solute. Each mole of LiF produces two moles of dissolved solute particles:

$$LiF(s) \longrightarrow Li^+ + F^-$$

Therefore, to calculate the molality of solute particles in a solution of LiF, we must multiply by a factor of 2 in the usual molality calculations:

$$m = 2 \times \frac{\dfrac{0.571 \text{ g LiF}}{25.9 \text{ g LiF/mol LiF}}}{111 \text{ g H}_2\text{O}} \times 1000 \frac{\text{g}}{\text{kg}}$$
$$= 0.397m$$

which is the total molality of solute particles.

We can now find the boiling point elevation in the usual way, using the relationship expressed in Equation 12.15:

$$\Delta T = (0.512)(0.397) = 0.203 \text{ K}$$

Weak electrolytes are those substances that are only partially dissociated into ions at concentrations around $1M$. While weak electrolytes increase the conductivity of water, the increase is not as great as with strong electrolytes. The weak electrolytes include some salts of heavy metals, such as mercury and cadmium; weak acids, such as acetic acid and hydrogen cyanide; and weak bases, such as ammonia.

The extent to which an electrolyte dissociates in solution can be determined by measuring any of the colligative properties of the solution, since these properties depend on the number of solute particles in solution. It was a careful study of the colligative properties and electrical conductivities of solutions that enabled the Swedish chemist Svante Arrhenius (1859–1927) to propose in 1887 that electrolytes dissociate into ions in solution.

EXAMPLE 12.15

A solution of 26.3 g of $CdSO_4$ in 1000 g of H_2O has a freezing point 0.285 K lower than pure water. Calculate the percentage of $CdSO_4$ that dissociates into ions in this solution.

SOLUTION

Using the value of K_f in Table 12.2 and Equation 12.16, the molality of solute particles can be found from the measured freezing point depression:

$$0.285 \text{ K} = (1.86)m$$
$$m = 0.153$$

The molality of the solution can also be calculated in the usual way from the mass and molecular weight of $CdSO_4$:

$$\frac{\dfrac{26.3 \text{ g CdSO}_4}{208.5 \text{ g CdSO}_4/\text{mol CdSO}_4}}{1000 \text{ g H}_2\text{O}} \times 1000 \frac{\text{g}}{\text{kg}} = 0.126 \ m$$

The molality found from the observed freezing point depression is larger than the molality calculated from the mass and molecular weight of solute. The larger molality is due to the dissociation of the solute in solution.

Let x equal the molality of $CdSO_4$ that undergoes dissociation. Both the molality of Cd^{2+} that is formed and the molality of SO_4^{2-} that is formed will be x. The molality of $CdSO_4$ remaining after the reduction of the molality by x will be $0.126 - x$. The molalities present before and after dissociation can be placed beneath the appropriate formulas in the chemical reaction:

	$CdSO_4$	$\longrightarrow Cd^{2+}$	$+ SO_4^{2-}$
molality before dissociation	0.126	0	0
molality after dissociation	$0.126 - x$	x	x

The total molality of all particles in solution after dissociation is

$$(0.126m - x) + x + x = 0.126m + x = 0.153m$$
$$x = 0.027m$$

The percentage of $CdSO_4$ that dissociates in solution thus is

$$\frac{0.027m}{0.126m} \times 100\% = 21.4\%$$

One of the factors that influences the extent to which a weak electrolyte dissociates is the concentration of the electrolyte in the solution. As the concentration of the electrolyte decreases, the *percentage* of dissociation increases. Thus, if more water is added to the solution of Example 12.15, the concentrations of all species in solution decreases, but the amount of ions increases relative to the amount of undissociated $CdSO_4$.

As we have mentioned, the ideal solution approximation does not work nearly as well for solutions of electrolytes as it does for solutions of nonelectrolytes. The deviation from ideality often is expressed by using a quantity i, the van't Hoff factor, which relates the observed value of a colligative property to the value calculated by one of the ideal relationships. For example, the relationship between observed and ideal value can be expressed for the freezing point depression by:

$$\Delta T = iK_f m \tag{12.19}$$

where ΔT is the observed freezing point depression and m is the value of the molality calculated on the basis of no dissociation of the solute. Assuming ideal behavior, a value of i for a strong electrolyte can be calculated from the formula of the substance. It is equal to the number of moles of ions formed in solution from complete dissociation of each mole of undissolved substance. For NaCl, i should have a value of 2, since each mole of NaCl produces two moles of ions or solute particles in solution. For $MgCl_2$, i should be 3, since each mole of $MgCl_2$ produces one mole of Mg^{2+} ions and two moles of Cl^- ions in solution.

However, i does not have these ideal values in real solutions. Even in very dilute solutions, the value of i deviates from the ideal value. The deviation increases as the concentration of the solution increases. For example, the values of i for solutions of NaCl and $MgSO_4$, each of which dissociates into two ions, are:

	Value of i	
	NaCl	$MgSO_4$
ideal	2	2
0.001m concentration	1.97	1.82
0.1m concentration	1.87	1.21

A number of theories have been developed to explain both the deviation of the value of i from ideality and some other characteristics of solutions of electrolytes. The problem is extremely complex, and it is far from solved for solutions of even moderate concentration, about 0.1m. Very dilute solutions are handled well by the Debye-Hückel theory, proposed in 1923 by Peter Debye (1884–1966) and Erich Hückel (1896–). Their theory is based on the idea that deviations from ideality are caused by electrical interactions between ions of opposite charge in solution. A positive ion tends to have negative ions in its vicinity. The effect of the resulting coulombic attractions is similar to the effect of incomplete dissociation. These attractions lead to the observed lowering of the value of i below the theoretical value. As expected, the effect is greater in solutions of ions with larger charges. The deviation from ideality is greater in a solution of 0.001m $MgSO_4$ than in a solution of 0.001m NaCl because the coulombic attraction between the ions of charge $+2$ and -2 that are found in the $MgSO_4$ solution is four times greater than the attraction between the ions of charge $+1$ and -1 found in the NaCl solution.

EXAMPLE 12.16

A 0.0103M solution of K_2SO_4 is found to have an osmotic pressure of 0.680 atm at 299 K. Calculate the value of i for the solution.

SOLUTION

Equation 12.18 can be rewritten to include i:

$$\pi = icRT$$

For this solution, the expression gives the value:

$$0.680 \text{ atm} = i(0.0103 \text{ mol/liter})\left(0.0821 \frac{\text{liter atm}}{\text{mol K}}\right)(299 \text{ K})$$

$$i = 2.69$$

The calculated value of i is substantially lower than the ideal value $i = 3$, even in this rather dilute solution.

It is important to remember that many of the solutions that are used in the laboratory and many that we shall discuss later are of relatively high concentration, and so are described poorly by the ideal solution approximation. More elaborate quantitative treatments of concentrated solutions have been developed, but they are beyond our scope. We shall use the ideal solution approximation throughout.

12.8 COLLOIDS

A solution and a suspension are two extreme ways in which two substances can mix. In a solution, the solute particles are so small that they cannot be seen. In a suspension of, say, a solid in a liquid, the particles of solid are large enough to be visible, by an ordinary microscope if not by the unaided eye. If a suspension is allowed to stand, the solid particles settle out after a while because of gravity.

Particle size makes the difference between a solution and a suspension. In a solution, the solute particles are smaller than about 1 nm in diameter. In a suspension, the solid particles are larger than 1000 nm in diameter. There can be mixtures of substances in which the size of particles is between the upper limit for solutions and the lower limit for suspensions. These mixtures often have properties of both suspensions and solutions. They are called **colloids**.

A colloid exists whenever particles smaller than those of a suspension and larger than those of a solution are dispersed in a medium. The particles do not settle out, because they are too small, and they are not directly visible. But particles of colloidal size scatter light, so a light beam that passes through a colloid is visible to an observer at right angles to the beam. An everyday example of this scattering effect can be seen when light beams pass through tobacco smoke, a colloid consisting of solid particles in air.

There are many different types of colloids. They are classified by the state of the components of the colloids when they are pure. As we define a solute and a solvent for solutions, so, by analogy, for colloids we define a *dispersed phase* (which corresponds to the solute) and a *dispersion medium* (which corresponds to the solvent).

There are a number of colloids in which the dispersion medium is a liquid. If the dispersed phase is a solid and the system appears to be liquid and flows, it is called a *sol*. If it has a solidlike structure that prevents it from flowing, it is called a *gel*. Milk of magnesia is a sol. Jell-O is a gel when it is cool but a sol when it is warm enough to flow. The names given to some colloids with various dispersed phases and dispersion media are listed in Table 12.3.

TABLE 12.3
Names of Colloids

Name	Dispersed Phase	Dispersion Medium	Example
emulsion	liquid	liquid	milk (fat in water)
foam	gas	liquid	suds (carbon dioxide in beer)
aerosol	solid	gas	smoke (particles in air)
	liquid	gas	fog (water in air)

Particles of colloidal size generally can be prepared either by condensing smaller particles or by dispersing larger particles. A variety of methods have been devised for carrying out both of these processes. Special methods sometimes are needed to form colloidal particles that will not coagulate and settle out. Often, the major reason for the stability of a colloid is the fact that all the colloidal particles have the same surface charge. This factor must be considered in preparing colloidal particles. It is also of importance in processes such as water treatment, in which coagulation of pollutants of colloidal size is achieved in part by neutralizing their surface charges. Emissions of smoke are reduced by a reverse procedure, called electrostatic precipitation. The particles in smoke are given electrical charges and are collected on high-voltage electrodes. Electrostatic precipitators commonly reduce smoke emissions by 90% or more.

EXERCISES

12.1 Calculate the mass fraction and the mass percent of bromine in a solution prepared from 22.7 g of bromine and 98.7 g of carbon tetrachloride.

12.2 Calculate the mass of sugar in 65.6 g of a solution that is 0.153 sugar by mass.

12.3 Calculate the mass of gold required to prepare 38.5 g of an alloy that is 18/24 parts gold by mass.

12.4 The density of ethyl alcohol at 293 K is 0.79 g/cm^3 and the density of chloroform at this temperature is 1.5 g/cm^3. Calculate the mass percent of chloroform in a solution prepared from equal volumes of ethyl alcohol and chloroform.

12.5 Calculate the mole fraction of copper in an alloy made from 124 g of copper and 124 g of aluminum.

12.6 Calculate the mole fraction of CCl_4 in a solution prepared from 22.7 g of Br_2 and 98.7 g of CCl_4.

12.7 Calculate the mole fraction of ethylene glycol, $C_2H_6O_2$, in a solution that is 74.5% by mass ethylene glycol in water.

12.8 The density of mercury is 13.6 g/cm^3 at 273 K. A solution of sil-

ver in mercury, called an amalgam, is prepared from 30.0 cm^3 of mercury and 2.34 g of silver. Calculate the mole fraction of silver in the silver amalgam.

12.9 Find the mass of glucose, $C_6H_{12}O_6$, that must be dissolved in 250 g of water to prepare a solution in which the mole fraction of glucose is 0.10.

12.10 Calculate the molality of Br_2 in a solution prepared from 22.7 g of bromine and 98.7 g of carbon tetrachloride.

12.11 Calculate the molality of glycerin, $C_3H_8O_3$, in a solution prepared from equimolar quantities (the same number of moles) of glycerin and water.

12.12 Calculate the molality of water in a solution that is 74.5% ethylene glycol by mass.

12.13 Calculate the quantities of water and acetone, C_3H_6O, necessary for the preparation of 120 g of a solution that is 2.2m in acetone.

12.14 Calculate the molarity of a solution that contains 1.22 g of hemoglobin (MW = 68 300) in 165 cm^3 of solution.

12.15 The density of a 22.0% solution of methanol in water by mass is

0.963 g/cm^3. Calculate the molarity of the solution.

12.16 A 0.790M solution of potassium iodide has a density of 1.093 g/cm^3. Calculate the molality of the solution.

12.17 The density of a solution of 18.0% lead nitrate by mass in water is 1.18 g/cm^3. Calculate the molarity of lead nitrate.

12.18 Suggest a procedure for preparing 100 cm^3 of a 0.001M solution of sugar from 100 cm^3 of a 1M solution of sugar, using a 100-cm^3 graduated cylinder.

* 12.19 A solution of 80.00% sulfuric acid by mass has a density of 1.727 g/cm^3. A volume of 50.00 cm^3 of this solution is diluted with water until the volume of the solution is 75.00 cm^3. Calculate the molarity of sulfuric acid in the solution.

12.20 Calculate the volume of H_2 at STP that is released when excess zinc is added to 36.2 cm^3 of a 0.895M solution of hydrochloric acid.

12.21 A solution of sugar is prepared by mixing 436 cm^3 of 2.05M sugar solution with 238 cm^3 of 3.34M sugar solution. Assume the volume of the resultant solution is the sum of the volumes of the two solutions. Calculate the molarity of sugar in the new solution.

*12.22 A solution is prepared by mixing 631 cm^3 of methanol with 501 cm^3 of water. The molarity of methanol in the resulting solution is 14.29M. The density of methanol at this temperature is 0.792 g/cm^3. Calculate the difference in volume between the solution and the total volume of the water and methanol that were mixed to prepare the solution.

12.23 Calculate the molality and mole fraction of ethanol in 25.4-proof wine at 293 K. The density of ethanol is 0.789 g/cm^3 and the density of water is 0.998 g/cm^3 at 293 K.

12.24 Formulate a picture of the solute-solvent interactions in the following solutions: (a) sodium fluoride in water, (b) lithium chloride in acetone, (c) methane in octane, (d) water in liquid ammonia.

12.25 The solubility of magnesium carbonate is 0.011 g/100 cm^3 H$_2$O. Calculate the molarity of magnesium carbonate in a saturated solution, assuming no volume change of the water on formation of the solution.

12.26 A saturated solution of KMnO$_4$ at 293 K is 0.40M. Calculate the solubility of KMnO$_4$ measured in grams per 100 cm^3 of H$_2$O assuming no volume change on formation of the solution.

12.27 Predict which of the following salts are soluble, which are insoluble, and which are slightly soluble in water: (a) barium carbonate, (b) lead iodide, (c) calcium hydroxide, (d) mercury(I) sulfate, (e) silver chlorate, (f) gold(III) acetate, (g) radium bromide, (h) tin sulfide.

12.28 The partial pressure of O$_2$ in air at sea level is 0.21 atm. Calculate the mole fraction of O$_2$ in a saturated solution in water at 298 K.

12.29 Calculate the volume of carbon dioxide released when a bottle containing a liter of soda water, bottled at a pressure of CO$_2$ of 1.1 atm, is opened. Assume that both the soda water and the surroundings are at STP.

12.30 Suggest a reason for the use of helium in the breathing mixtures used by deep-sea divers.

12.31 Calculate the mole fraction of dissolved N$_2$, O$_2$, and Ar in water at 298 K in contact with the atmosphere. The mole fractions of these three gases in the atmosphere are 0.7809, 0.2095, and 0.0093, respectively.

12.32 Calculate the vapor pressure of water above a solution prepared from 18.22 g of lactose, C$_{12}$H$_{22}$O$_{11}$, and 81.46 g of water at 338.0 K. The vapor pressure of pure water at this temperature is 0.2467 atm.

12.33 The density of a 3.742M solution of glycerol, C$_3$H$_8$O$_3$, in water at 298.0 K is 1.0770 g/cm^3. The vapor pressure of pure water at this temperature is 0.03126 atm. Calculate the vapor pressure of water above the solution.

12.34 The vapor pressure of pure benzene, C$_6$H$_6$, is 0.132 atm at 299 K. Calculate the mass of hexachlorobenzene, C$_6$Cl$_6$, that must be dissolved in 25 g of benzene to lower its vapor pressure to 0.126 atm.

12.35 A solution is prepared from 26.7 g of an unknown compound and 116.2 g of acetone, C$_3$H$_6$O, at 313 K. The vapor pressure of pure acetone at this temperature is 0.526 atm and the vapor pressure of acetone above the solution is 0.501 atm. Calculate the molecular weight of the unknown compound.

12.36 At 333 K the vapor pressure of C$_6$H$_6$ is 0.521 atm and the vapor pressure of toluene, C$_7$H$_8$, is 0.184 atm. A solution of equimolar amounts of these two liquids is prepared. Calculate the mole fraction of toluene in the vapor and the total vapor pressure above the solution.

*12.37 Two alcohols, isopropyl alcohol and propyl alcohol, have the same molecular formula, C$_3$H$_8$O. A solution of the two that is $\frac{2}{3}$ by mass isopropyl alcohol has a vapor pressure of 0.110 atm at 313 K. A solution that is $\frac{1}{3}$ by mass isopropyl alcohol has a vapor pressure of 0.089 atm at 313 K. Calculate the vapor pressure of each pure alcohol at this temperature.

12.38 The vapor pressure of CCl$_4$ is 0.354 atm and the vapor pressure of chloroform, CHCl$_3$, is 0.526 atm at 316 K. A solution is prepared from equal masses of these two compounds. Calculate the mole fraction of chloroform in the vapor above the solution. Calculate the mole fraction of chloroform in the vapor above the solution obtained by condensation of the vapor above the original solution.

*12.39 Use the result of Exercise 12.38 to calculate the mole fraction of chloroform in the vapor above a solution obtained by three successive separations and condensations of the vapors above the original solution of carbon tetrachloride and chloroform.

12.40 Calculate the boiling point of a solution of 4.39 g of naphthalene, C$_{10}$H$_8$, in 99.5 g of carbon tetrachloride. Use the data in Table 12.2.

12.41 Calculate the mass of sugar, C$_6$H$_{12}$O$_6$, that must be dissolved in 1000 g of ethanol to raise its boiling point 1.0 K.

12.42 A solution of 10.8 g of an unknown substance in 101.2 g of n-octane boils at 401.4 K. Using the data in Table 12.2, calculate the molecular weight of the unknown.

*12.43 A solution of a nonvolatile solute in water has a boiling point of 375.3 K. Calculate the vapor pressure of water above this solution at 338 K.

12.44 The solubility of two substances, A and B, in water is about the same at 273 K. However, the solubility of A increases, while the solubility of B decreases as the temperature decreases. Predict the relative eutectic temperatures of solutions of these two substances. Justify your prediction.

12.45 Calculate the freezing point of a 20.0% by mass ethylene glycol, C$_2$H$_6$O$_2$, solution in water.

12.46 A solution of 0.358 g of an unknown substance in 6.45 g of camphor freezes at 446.8 K. Calculate the molecular weight of the unknown.

12.47 Calculate the mass of glycerol, C$_3$H$_8$O$_3$, that must be dissolved in 1000 g of water to form a solution with a freezing point of 268.2 K.

12.48 Calculate the osmotic pressure at 298 K of 1.00% *by mass* solutions of nonionic solutes with the following molecular weights: (a) 1.00 × 10^2, (b) 1.00 × 10^4, (c) 1.00 × 10^6.

12.49 An aqueous solution of catalase, a liver enzyme, prepared from 0.115 g of catalase has a volume of 11.4 cm^3 at 300 K. The osmotic pressure of the solution is found to be 9.80 × 10^{-4} atm. Calculate the molecular weight of catalase.

12.50 The density of a 9.50% by mass solution of fructose, C$_6$H$_{12}$O$_6$, is 1.037 g/cm^3 at 293 K. Calculate the osmotic pressure of the solution.

12.51 Calculate the freezing point of a solution prepared from 5.46 g of potassium iodide and 134 g of water.

12.52 Calculate the boiling point of a solution prepared from 15.3 g of sodium sulfate and 98.3 g of water.

12.53 Calculate the vapor pressure above a solution prepared from 24.3 g of magnesium nitrate and 249 g of water at 308 K.

*12.54 The density of a 0.438M solution of potassium chromate at 298 K is 1.063 g/cm^3. Calculate the vapor pressure of water above this solution.

12.55 Calculate the molality of sodium chloride in water at the eutectic temperature of 252 K.

12.56 Calculate the mass of calcium phosphate that must be dissolved in water to form 125 cm^3 of a solution with an osmotic pressure of 0.0034 atm at 300 K.

12.57 Seawater is 3.5% by mass salt and has a density of 1.04 g/cm^3 at 293 K. Assume that all the dissolved salt is sodium chloride and calculate the osmotic pressure of seawater.

*12.58 A metal, M, of atomic weight 96 reacts with fluorine to form a salt that can be represented as MF$_x$. In order to determine x and therefore the formula of the salt, a boiling point elevation experiment is performed. A 9.18-g sample of the salt is dissolved in 100 g of water and the boiling point of the solution is found to be 374.38 K. Find the formula of the salt.

12.59 A solution of 2.25 g of a solute in 100 g of benzene freezes at 277.6 K, while a solution of 4.50 g of this substance in 100 g of water freezes at 270.7 K. Assuming that there is no association or dissociation of the substance in benzene, calculate the number of particles formed in water solution from each original solute particle.

12.60 A solution contains 3.22 g of HClO$_2$ in 47.0 g of water. The freezing point of the solution is 271.10 K. Calculate the fraction of HClO$_2$ that undergoes dissociation to H$^+$ and ClO$_2^-$.

12.61 Use the appropriate values of i to calculate the difference in freezing point between a 0.10m solution of sodium chloride in water and a 0.10m solution of magnesium sulfate in water.

*12.62 A solution is prepared from 4.5701 g of magnesium chloride and 43.238 g of water. The vapor pressure of water above this solution is found to be 0.3624 atm at 348.0 K. Calculate the value of i for magnesium chloride in this solution.

12.63 Sulfuric acid in water dissociates completely into H$^+$ and HSO$_4^-$ ions. The HSO$_4^-$ ion dissociates to a limited extent into H$^+$ and SO$_4^{2-}$. The freezing point of a 0.1000m solution of sulfuric acid in water is 272.76 K. Calculate the molality of SO$_4^{2-}$ in the solution, assuming ideal solution behavior.

12.64 Emulsions of oil and water often separate into two liquid layers quite rapidly. However, such emulsions can be stabilized by the addition of a detergent, a substance that lowers the surface tension of water. Suggest an explanation for this action of detergents on emulsions.

13

Thermodynamics is one of the major fields of the physical sciences, but the concepts of thermodynamics have such broad implications that they have been applied with striking success in fields ranging from the biological sciences to philosophy. The ideas of thermodynamics have been applied to subjects as diverse as the fate of the universe, the nature of change in social systems, the design of communications networks, and the use of fossil fuels in a modern society.

In the physical sciences, thermodynamics sprang from experimental observations. It is used to study relationships between those properties of matter which are directly measurable, such as volume, temperature, and pressure. Thermodynamics does not concern itself with the atomic structure of matter.

For the chemist, thermodynamics provides the answers to several important questions:

1. Can a physical or chemical change be expected to occur when one or more substances are in a given set of conditions?
2. How far can such a change proceed?
3. If a chemical or physical change occurs, what are the accompanying energy changes in the system and its surroundings?

In Chapter 8, we saw how we could obtain answers to such questions by studying the heat changes that accompany chemical processes. In this chapter, we shall use the same approach in a more general way, to study both physical and chemical changes.

One basic limitation of thermodynamics should be kept in mind from the outset: Thermodynamics does not concern itself with time. On the practical level, this limitation can be most important. For example, thermodynamics tells us that if 2 mol of $H_2(g)$ and 1 mol of $O_2(g)$ are mixed at room temperature, a chemical change will occur; the two gases will react to form 2 mol of H_2O, with the evolution of a considerable quantity of heat. But thermodynamics does not tell us that if the two gases are left alone, the reaction occurs so slowly that the conversion will not take place in a normal human lifetime. Nor does thermodynamics tell us that the presence of a spark in this mixture of gases makes the reaction proceed so quickly that there is an explosion as water is formed and heat is liberated. Thermodynamics ignores the time aspect of the reaction because the spark does not change the extent to which the reaction ultimately proceeds or the quantity of energy that is liberated; it just allows the process to go faster. The question of time in chemical reactions is the subject of the discipline called kinetics, which we shall discuss in Chapter 17.

The rate at which reactions occur is a subject of practical importance for the chemist. Many substances or mixtures of substances can be regarded as being thermodynamically unstable—that is, if left alone, they should change. But thermodynamics tells us only whether something *should* happen. Kinetics determines whether that something actually *does* happen at an appreciable rate. For example, dynamite is thermodynamically unstable. But it can be transported and stored because it is undergoing chemical change at an immeasurably slow rate. Only when dynamite is detonated does its instability become evident. Similarly, most of the gaseous oxides of nitrogen are thermodynamically unstable. But most of them are listed in the catalogs of chemical supply houses; they can be ordered, shipped, and stored for prolonged periods. All the gaseous oxides of nitrogen are decomposing to $N_2(g)$ and $O_2(g)$, but the rate of decomposition usually is immeasurably slow and of no practical importance.

Throughout our discussion, you should remember that thermodynamics tells us whether something should happen, but it does not tell us the length of time that it takes to happen.

13.1   DEFINITIONS

We have mentioned that thermodynamics deals entirely with the bulk properties of matter—volume, temperature, pressure—and not with the internal structure of matter. In thermodynamics, the behavior and structure of atoms and molecules can be ignored.

Thermodynamics starts by defining a **system,** a specific amount of one or more substances. The system is separated from the rest of the universe by a *boundary,* which can be a real or imaginary surface. The boundary confines the physical system to a specific location. The part of the universe outside the boundary is called the *surroundings* of the system.

Usually, a boundary prevents mass from entering or leaving the system. If the boundary also prevents energy from entering or leaving, so that there is no interaction of the system with the surroundings, we are dealing with an *isolated* system.

The physical characteristics of a system that can be perceived or measured, such as its volume, its mass, and its temperature, are the *properties* of the system. When these properties are defined, a **state** of the system is defined. An **equilibrium state** is a state in which the properties of a system do not change with time, unless the system is influenced by the surroundings.

Figure 13.1 shows a sealed container that holds one or more chemical substances. The system consists of the substances and the vapor space inside the container. The boundary is the walls of the container. Everything outside the container is the surroundings. The properties of the system include the quantities of the substances in the container, their temperature, their pressure, and their energy. All the properties, taken together, specify the state of the system. If the values of the properties do not change with time, the system is in an equilibrium state.

A system in an equilibrium state can be described by specifying the values of certain of its properties called **state functions.** The value of a state function is fixed when the system is in a given state and it is not influenced by the path that the system followed to reach the given state. The equilibrium state of a system can be described by specifying the values of a small number of state functions, two or three in most cases. We are already familiar with three state functions, pressure (P), temperature (T), and volume (V). In our discussion of thermodynamics, we shall be interested in four other state functions: internal energy (E), enthalpy (H), entropy (S), and free energy (G). (By convention, the symbol for a state function is a capital letter.) If a system is not in an equilibrium state, it is usually necessary to specify the values of more than two or three state functions to describe the state.

Thermodynamics deals with changes in state, that is, with processes in which a system in an initial state goes through a change that brings it to some final state. A convenient way to describe such a change in state is to give the values for the changes that occur in the state functions in going from the initial state to the final state. Changes in a function are represented by the symbol Δ, the Greek letter delta, placed before the symbol of the function. The symbol Δ denotes a difference that is found by taking the value of the state function in the final state and subtracting the value of the state function in the initial state. Thus:

$$\Delta T = T_{\text{final}} - T_{\text{initial}}$$

if temperature, T, is the state function of interest.

We said that the value of a state function depends only on the state of the system and not on how this state is reached. Therefore, the value of the change in a state function does not depend on the path followed by the system between its initial and final states. Hess's law and many of the thermochemical calculations we carried out in Section 8.3 are based on this property of state functions.

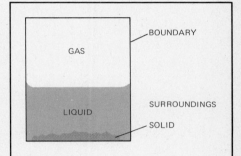

FIGURE 13.1
A system and its surroundings. The system is everything inside the container. The surroundings are everything outside the container. The walls of the container form the boundary between the system and the surroundings.

Not all thermodynamic functions are state functions. Many of the quantities associated with a change in state depend on the path that the system follows. As an analogy, consider a resident of New York who moves to Los Angeles. The change in latitude and the change in longitude that result from the trip are both changes in state functions, since neither depends on the path of the journey. But several other quantities called **path functions** depend on the path that is followed. For example, the distance traveled, the work performed by the person making the trip, and the time needed to make the trip depend on the route that is followed and the mode of transportation that is used. The traveler can fly from New York directly to Los Angeles, can ride a bike via Arizona or can drive via North Dakota. In each case, the path functions are different. In chemical thermodynamics, by convention, the symbols for path functions are lowercase letters.

Internal Energy

One important consideration in thermodynamics is the magnitude of the energy changes in a system and its surroundings that are associated with a change in state. We have used the word ''energy'' frequently in this text. We have often found it convenient to designate energy as kinetic energy or potential energy. Kinetic energy is the energy that an object has by virtue of its motion. Potential energy is the energy associated with position. In chemistry, we are especially interested in the potential energy that results from the attractions and repulsions between the electrons and nuclei of atoms and molecules. In thermodynamics, we simply define the **internal energy**, E, as the total of all the possible kinds of energy of a system. The internal energy of a system is a state function. It is usually not possible to know or measure the value of E of a system in a given state. But it is often possible to measure or find the value of ΔE for a change in state of a system.

The SI unit for internal energy is the joule (J). It is defined as $1\ J = 1\ kg\ m^2\ sec^{-2}$. Another unit of energy, which is not an SI unit, is called the calorie. It is now defined as $1\ cal = 4.184\ J$.

In thermodynamics, we approach ΔE, the change in the internal energy of a system that results from a change in state, by considering the ways in which the energy of a system can change. The energy of a system changes because energy is transferred into or out of a system across its boundary. Two methods of transferring energy across the boundary of a system are **heat** and **work**. These are familiar terms in everyday life, but we shall give them more technical meanings.

Work

We shall use the term ''work'' to refer to mechanical work, which is succinctly defined as the product of a force by a displacement. We can show that there is a relationship between the internal energy of a system and work by a few common examples.

The potential energy of a weight that is suspended two meters above the floor of a room is greater than its potential energy when the weight is one meter above the floor. When the weight falls, it can do work; it can do more work when it falls two meters than when it falls one meter. A mixture of carbon and hydrogen in the form of gasoline and of O_2 is of higher potential energy than a mixture of the same quantities of these three elements in the form of carbon dioxide and water. When it takes place in the cylinder of an automobile engine, the change in state from gasoline and O_2 to CO_2 and H_2O, called combustion, is used to do such work as moving the automobile (Figure 13.2).

The units of work are units of force (newtons in SI), multiplied by units of distance (meters in SI). Since $1\ J = 1\ N \times 1\ m$, the units of work are joules, the same as the units of energy. As we shall see, the quantity of work that

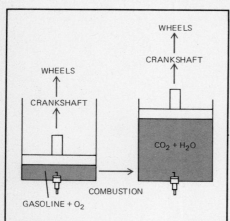

FIGURE 13.2
Both before and after combustion, this system contains hydrogen, carbon, and oxygen. Before combustion, the piston contains a mixture of gasoline and O_2. In the combustion reaction, the gasoline combines with the O_2 to produce CO_2 and H_2O and release energy. The piston moves and thus does work. The amount of work that can be done by a system is related to its internal energy.

accompanies a change in state depends on how the change in state is brought about. Work is a path function and is represented by the lowercase letter w.

In dealing with work, we designate not only the quantity of work in joules but also the direction of the work. A system can evolve energy by performing work on the surroundings. Energy is transferred from the system to the surroundings in the form of work, and the sign of w is negative. A system can absorb energy when the surroundings perform work on the system. For this type of energy transfer, the sign of w is positive. More simply, work done by the system on the surroundings has a negative sign $(-w)$, while work done on the system by the surroundings has a positive sign $(+w)$.

In our discussion of chemical thermodynamics, we shall limit ourselves to systems in which only work due to changes in volume occurs. For the sake of simplicity, we shall ignore all the other kinds of work, such as electrical work. Since we limit ourselves to considering only changes in volume, there are just three possibilities to discuss. If the volume of a system increases as the result of a change in state ($\Delta V > 0$), the system does work on the surroundings and w has a negative sign $(-w)$. If the volume of a system decreases because of a change in state ($\Delta V < 0$), the surroundings do work on the system and w has a positive sign $(+w)$. If there is no change in volume as a result of a change in state ($\Delta V = 0$), then $w = 0$ and no work is done.

Work associated with changes in volume is called *expansion work*. The assumption that only expansion work is possible allows considerable simplification in the thermodynamic description of chemical processes. For example, if a system has no mechanical link with the surroundings (as in Figure 13.1), no change of volume is possible and $w = 0$ for any change of state.

Furthermore, since the volume of a solid or a liquid to a first approximation does not change with a change in conditions, changes in state between liquids and solids are not accompanied by significant changes in volume. Thus, we can assume $w = 0$ for all systems that contain only solids or liquids.

The simplest type of expansion work occurs when the volume of a system changes at constant pressure, as in reaction vessels that are open to the atmosphere. If the pressure on the system is P, we can show that the work due to a volume change ΔV is given by:

$$w = -P\,\Delta V \tag{13.1}$$

In SI, the units of pressure, which is force per area, can be given as N/m^2, and the units of volume are m^3. The units of $P\,\Delta V$ are therefore $N \times m = J$. Since we are using atmospheres and liters for P and V, we use the relationship: 1 liter atm $= 101.3$ J to express expansion work in units of joules.

EXAMPLE 13.1

Calculate the work associated with the vaporization of 1.0 mol of water at 373 K and 1.0 atm. Assume ideal gas behavior.

SOLUTION

The value of ΔV is found by finding the difference in volume between the initial and final states of the system. The final volume is the volume of one mole of water vapor at the specified conditions. We find the volume by using the ideal gas equation:

$$V = \frac{nRT}{P} = \frac{(1.0 \text{ mol})(0.082 \text{ liter atm mol}^{-1}\text{ K}^{-1})(373 \text{ K})}{1.0 \text{ atm}}$$

$$= 31 \text{ liters}$$

The initial volume of the liquid water is much smaller, less than 0.02 liter. In general, the volumes of liquid or solid can be neglected in the calculation of ΔV because they are so much smaller than the volume of the same amount of gas. Thus, $\Delta V = V_{\text{final}} - V_{\text{initial}} = 31$ liters.

$$w = -P \Delta V = -(1.0 \text{ atm})(31 \text{ liters})(101 \text{ J/liter atm}) = -3100 \text{ J}$$

The negative sign of w indicates that the system does work on the surroundings, due to its increase in volume. The steam engine is a device that operates on this principle.

Heat

The second method by which energy is transferred across a boundary is harder to define. The transfer of energy by heat occurs as the result of processes such as thermal conduction or thermal radiation. Such processes are best explained by reference to atoms and molecules, whose existence is ignored by classical thermodynamics. Work is the result of measurable mechanical displacements of macroscopic objects. Heat is the result of microscopic effects. But both heat and work are methods for transferring energy across a boundary.

The symbol for heat is q, a lowercase letter, because heat is also a path function. It is convenient to picture heat as flowing in a direction. If heat flows from the surroundings into the system to raise the energy of the system, it is taken to be positive, $+q$. If heat flows from the system into the surroundings, lowering the energy of the system, it is taken to be negative, $-q$. For many years, the calorie was the unit used to express the magnitude of heat. The SI unit of heat is the joule, which is also used as the unit of work and the unit of energy. We often use the kilojoule, kJ, for large quantities of work, heat, or energy; 1 kJ = 1000 J.

We can summarize the sign conventions for q and w by noting that they are positive quantities when they result in an increase in the energy of a system and they are negative quantities when they result in a decrease in the energy of the system:

heat flows into the system, q is positive
work is done by the surroundings on the system, w is positive
heat flows out of the system, q is negative
work is done by the system on the surroundings, w is negative

13.2 THE FIRST LAW OF THERMODYNAMICS

There are many different ways of stating an important result derived from countless experiments and observations of the energy changes that accompany changes in state. These different but equivalent statements are known as the **first law of thermodynamics.**

The first law of thermodynamics is sometimes called the *law of conservation of energy.* Most generally, it states that the energy of the universe is constant. More specifically, it states that energy cannot be destroyed or created; it can only be transferred from place to place or transformed from one form to another. The first law gives us a method for keeping a balance sheet on the energy change of any system that undergoes a change in state. Since there are only two ways to transfer energy, heat and work, the first law may be stated:

$$\Delta E = q + w \tag{13.2}$$

Equation 13.2 states that the change in the energy of a system is equal to the sum of the heat that flows into the system and the work done on the system. The equation gives a complete accounting of the energy.

Another way to state the first law is to say that it is impossible to construct a machine that operates in a cycle and that yields more energy output than it receives as input. In other words, there cannot be a machine that does a greater quantity of work on the surroundings than the quantity of heat it receives from the surroundings. Nevertheless, untold effort has been expended for centuries in attempts to construct such a machine, which is called "a perpetual motion machine of the first kind."

Perpetual motion of this kind is impossible because a machine that operates in a cycle must return to its initial state at regular intervals. The value of ΔE for the cycle must be 0, because E is a state function and the initial and final states of a cycle are identical. Perpetual motion is possible only if the value of work is greater than the value of heat, so that $\Delta E < 0$. But we know that ΔE must equal 0. A perpetual motion machine cannot run continuously unless energy is somehow created. Since energy cannot be created, such a machine cannot be built.

This conclusion is the result of many years of futile effort by many diligent workers. There are still occasional claims that a perpetual motion machine of the first kind has been constructed. The first law of thermodynamics rests upon such a firm foundation that these claims are not taken seriously.

Stated loosely, the first law says, "You can't win," or, "You can't get something for nothing." In the form of Equation 13.2, the first law is a convenient method for reasoning both qualitatively and quantitatively about the changes in work, heat, and energy associated with a change in state.

As an example of this type of reasoning, let us consider a process for which $q = 0$, which is called an *adiabatic* process. For an adiabatic process, Equation 13.2 becomes:

$$\Delta E = w$$

Suppose we want to know what happens in an adiabatic volume change of an ideal gas. If the gas expands, the system does work on the surroundings and the sign of work is negative. Therefore, ΔE is negative; the energy of the system decreases. Since the energy of an ideal gas is directly proportional to its temperature, the temperature falls. The same reasoning allows us to state that an adiabatic compression of an ideal gas causes its temperature to increase.

EXAMPLE 13.2

The volume of a sample of an ideal gas contracts from 8.4 liters to 4.2 liters as the result of an applied pressure of 1.5 atm. The system is found to evolve 830 J of heat during this contraction. Find ΔE for this change in state.

SOLUTION

To find ΔE, we must know q and w. The value of q is given as -830 J. The value of w can be calculated:

$$w = -P_{ex}V$$
$$= -(1.5 \text{ atm})(4.2 \text{ liters} - 8.4 \text{ liters})\left(101 \frac{J}{\text{liter atm}}\right)$$
$$= 640 \text{ J}$$

The sign for w is positive because the surroundings do work on the system.

$$\Delta E = q + w$$
$$= -830 \text{ J} + 640 \text{ J}$$
$$= -190 \text{ J}$$

Since the temperature of an ideal gas is proportional to its energy, the temperature of the gas also decreases.

The energy change that accompanies a real chemical or physical transformation is of interest to chemists. The relative stability of products and reactants is important both theoretically and practically, and measuring the magnitude of ΔE

PERPETUAL MOTION: THE DURABLE FALLACY

It might not seem possible that anyone in today's sophisticated society would invest money in a perpetual motion machine. But the world still holds many inventors who, knowingly or otherwise, are trying to sell a modern version of perpetual motion. The key to these machines is their apparent complexity, and such machines still are capable of causing an occasional stir in Wall Street, where experts are more at home with price-earnings ratios and dividend rates than they are with the laws of thermodynamics. In the early 1970s, for example, the price of one stock went up sharply for a time (before the Securities and Exchange Commission intervened) because the company owned rights to a machine that was said to produce hydrogen from water in a self-sustaining reaction.

The machine contained two steel tanks, each holding granules of an unidentified metal. Supposedly, the metal reacted with steam to bind oxygen and release hydrogen, which could be used as fuel. Several short demonstrations were held, at which the machine did indeed produce hydrogen.

However, this machine cannot pass the ultimate test by being truly self-sustaining. Sooner or later (probably sooner), the reactant metal must be recycled by heating, driving off the oxygen, and regenerating the metal so that it reacts with steam again. The hydrogen-producing reaction in one steel tank must give off enough heat to drive the recycling reaction in the other tank. We can represent the two reactions as:

$$M + H_2O \longrightarrow H_2 + MO + \text{heat}$$
$$\text{heat} + MO \longrightarrow M + \tfrac{1}{2}O_2$$

The overall process is:

$$\text{heat} + H_2O \longrightarrow H_2 + \tfrac{1}{2}O_2 + \text{heat}$$

Since even the simplest measurement demonstrates that the decomposition of water into H_2 and O_2 is endothermic, the quantity of heat on the left side of the reaction is greater than the quantity on the right side. Heat is required for the overall process. Since energy cannot be created inside the machine, it must be supplied from an external source. Therefore, the machine cannot be self-sustaining, according to the first law.

Even if the inventor of the machine somehow thought of a way to circumvent the first law, the recycling process would require the conversion of a quantity of heat from the hydrogen-generating reaction into an equal amount of work to regenerate the metal. As we shall see, the second law of thermodynamics states that heat cannot be converted into work without some loss. Again, the machine requires some energy from the surroundings.

The marvelous hydrogen-making device was no more than a perpetual motion machine. The machine might be a useful source of hydrogen, if the energy input required to recycle the reactant is sufficiently small. But no machine could live up to the claims made for this one. A perpetual motion machine by any name will never work.

is an excellent way to obtain this information. The first law of thermodynamics tells us how this measurement can be made. Almost every process is accompanied by the flow of heat, and many processes are accompanied by work. The magnitude of ΔE can be found if the magnitudes of both q and w are known. But in practice, it is better to measure either q or w than to have to measure both. The easiest way to simplify the measurement is to carry out the transformation of interest at constant volume. If $\Delta V = 0$, $w = 0$, since only expansion work is possible. Equation 13.2 then becomes:

$$\Delta E = q_v \tag{13.3}$$

where the subscript v indicates constant volume. In other words, the ΔE of a process can be measured by carrying out the process in a sealed container and measuring the flow of heat into or out of the container.

At first glance, Equation 13.3 might seem to contain an inconsistency. We have stated that E is a state function, and that the value of ΔE depends only on the initial and final states of the system, not the path between those states. But q is a path function whose value for a given change in state depends on the path that is followed. Thus, Equation 13.3 defines a state function, E, in terms of a path function, q. This definition is not an inconsistency. In fact, such a definition is often encountered in thermodynamics. The procedure is legitimate because the path associated with the path function is also specified. By specifying the exact path, we give the value of the path function. In Equation 13.3, the subscript v, which says that the constant-volume path is followed, describes the path completely enough to fix the value of q.

Enthalpy

In practice, it is often difficult to carry out a change in state at constant volume. In processes where the amount of gas present changes, the resulting change in pressure may be great enough to require the use of thick-walled, sealed containers called "bombs." Most chemical reactions are not carried out in sealed containers but in containers that are open to the atmosphere. Therefore, most changes do not occur along constant-volume paths. They occur along constant-pressure paths, at the pressure of the atmosphere.

Therefore, we must define another thermodynamic state function, enthalpy, H, which we discussed briefly in Chapter 8. Enthalpy is related to internal energy by the definition:

$$H = E + PV \tag{13.4}$$

The change in enthalpy ΔH is given by:

$$\Delta H = \Delta E + \Delta(PV) \tag{13.5}$$

For a change in state that occurs at constant pressure:

$$\Delta H = \Delta E + P\,\Delta V \tag{13.6}$$

Since $\Delta E = q + w$ and $P\,\Delta V = -w$, we can write Equation 13.6 as:

$$\Delta H = (q + w) + (-w)$$

or $\Delta H = q$ when the change in state occurs at constant pressure. This relationship usually is written as:

$$\Delta H = q_p \tag{13.7}$$

where the subscript p means constant pressure. Once again a state function is defined in terms of a path function. This time, the path is specified as the constant-pressure path. The ΔH can be measured by measuring the heat of a process occurring at constant pressure.

It should be noted that the difference between ΔH and ΔE is a PV term related to expansion work. For processes that include only liquids and solids, changes in state do not cause significant volume changes; thus, $\Delta V \cong 0$ and $w \cong 0$. Consequently, for such processes, ΔH and ΔE are approximately the same.

The relationship between ΔH and ΔE is of interest for reactions in which there are changes in the amounts of gases. Assuming ideal behavior:

$$PV = nRT$$

and

$$\Delta(PV) = R\,\Delta(nT)$$

If the number of moles of an ideal gas changes in a process at constant T, then:

$$\Delta(PV) = RT\,\Delta n$$

and substitution into Equation 13.5 gives:

$$\Delta H = \Delta E + RT\,\Delta n \tag{13.8}$$

If both the temperature and the amount of ideal gas remain constant, $RT\,\Delta n = 0$ and $\Delta E = \Delta H$. Since the energy of a given amount of an ideal gas depends only on the temperature, ΔE and ΔH are 0.

EXAMPLE 13.3

The value of ΔH for the reaction:

$$2N_2(g) + O_2(g) \rightleftharpoons 2N_2O(g)$$

at 298 K is 164 kJ. Calculate ΔE.

SOLUTION

In this process, 3 mol of gas changes to 2 mol of gas at constant temperature. Assuming ideal gas behavior, Equation 13.8 can be used:

$$164 \text{ kJ} = \Delta E + R(298 \text{ K})(2 \text{ mol} - 3 \text{ mol})$$

To obtain a value for ΔE, R is expressed in units of J mol^{-1} K^{-1}:

$$R = 8.314 \text{ J mol}^{-1} \text{ K}^{-1}$$

You should use this value of R for all calculations of thermodynamic quantities.

Converting the value of ΔH from 164 kJ to 164 000 J gives:

$$164\ 000 \text{ J} = \Delta E + (8.31 \text{ J mol}^{-1} \text{ K}^{-1})(298 \text{ K})(-1 \text{ mol})$$
$$\Delta E = 166\ 000 \text{ J} = 166 \text{ kJ}$$

The difference between ΔE and ΔH is only 2000 J, a small fraction of either.

13.3 THE MEASUREMENT OF HEAT; CALORIMETRY

The branch of chemistry concerned with the measurement of heat is **calorimetry**. A device for carrying out such a measurement is called a **calorimeter**. There are several ways to measure the quantity of energy transferred in the form of heat; that is, the quantity of heat that is evolved or absorbed during a process. Many different designs have been used for calorimeters. All of them are based on the measurement of some property that changes in a known way in response to heat.

One of the first calorimeters was constructed in 1780 by Lavoisier and Laplace, who measured the heat liberated in various processes by measuring the quantity of ice melted by the heat. In addition to studying such reactions as the combustion of carbon, Lavoisier and Laplace also applied calorimetry to biology by placing a guinea pig in their ice calorimeter and measuring the quantity of ice melted by the heat liberated by the animal over a certain period of time.

Figure 13.3 shows a calorimeter that is often used to measure the heat evolved by exothermic reactions such as combustion. This device is called a bomb calorimeter. The heat liberated by a process carried out in the bomb is calculated by measuring the temperature change in the water surrounding the bomb.

In the operation of the bomb calorimeter, the heat transferred by a change of state occurring in the bomb causes a temperature change, ΔT, in the water that surrounds the bomb. But we must know the quantity of heat that causes a given temperature change in the calorimeter in order to use the value of ΔT to find the quantity of heat released by a change in state. The relationship between the heat released and the temperature change in the calorimeter is the **heat capacity** of the calorimeter, which usually is measured by finding the temperature change caused by a fixed amount of electrical energy. Knowledge of the heat capacity of the calorimeter allows the conversion of a measured temperature rise into a value for the quantity of heat liberated. The heat capacity of the calorimeter can be defined as:

$$\text{heat capacity} = \frac{\text{heat absorbed by the calorimeter}}{\Delta T} \tag{13.9}$$

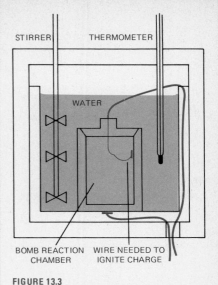

FIGURE 13.3
A bomb calorimeter. The heat produced by a change of state of the system in the bomb causes a change in the temperature of the water. The quantity of heat released by the process is found by measuring the temperature change of the water.

EXAMPLE 13.4

The reaction bomb of the calorimeter in Figure 13.3 is charged with 2.456 g of n-decane, $C_{10}H_{22}$, and excess O_2. It is then sealed and placed in the calorimeter. The temperature of the water in the calorimeter is 296.32 K. The decane is ignited by the resistance wire. After combustion is complete, the temperature of the water is 303.51 K. In a separate experiment, the heat capacity of the calorimeter is found to be 16.24 kJ/K. Calculate ΔE for the combustion of one mole of n-decane.

SOLUTION

A bomb calorimeter operates at constant volume. Therefore, the heat evolved is $q_v = \Delta E$. The combustion reaction is exothermic, since the temperature of the water in the calorimeter rises after the reaction occurs. Thus, ΔE and q are negative. We can use Equation 13.9, which defines the heat capacity, to find the heat absorbed by the water of the calorimeter:

$$16.24\,\frac{kJ}{K} = \frac{\text{heat}}{(303.51\ K - 296.32\ K)}$$

$$\text{heat} = 117\ kJ$$

For the system of decane and oxygen, $\Delta E = q_v = -117$ kJ. The negative sign indicates that the process is exothermic.

The heat absorbed by the calorimeter is the heat evolved by the combustion of 2.456 g of decane (MW = 142.3), which is:

$$\frac{2.456\ \text{g decane}}{142.3\ \text{g/mol decane}} = 0.0173\ \text{mol decane}$$

Therefore, $\Delta E = -117$ kJ/0.0173 mol decane $= -6760$ kJ/mol.

The ΔH of combustion at 296.3 K can also be calculated from these data by assuming ideal gas behavior of the gases in the system:

$$\Delta H = \Delta E + RT\,\Delta n$$

The equation for the combustion of one mole of n-decane is:

$$C_{10}H_{22}(l) + \frac{31}{2}O_2 \longrightarrow 10CO_2(g) + 11H_2O(l)$$

and

$$\Delta n = 10 \text{ mol} - \frac{31}{2} \text{ mol} = -5.5 \text{ mol}$$

$$\Delta H = -6760 \frac{\text{kJ}}{\text{mol}} \times 1000 \frac{\text{J}}{\text{kJ}} + \left(8.31 \frac{\text{J}}{\text{mol K}}\right)(296.3 \text{ K})(-5.5 \text{ mol})$$

$$= -6770 \text{ kJ/mol}$$

Heat Capacity

The relationship between heat and temperature change is not restricted to use with calorimeters. It is used extensively in thermodynamics. The heat capacity of a substance is defined as the heat (in joules) that changes the temperature of one mole of a substance by 1 K. Heat capacity can be regarded as the proportionality constant that relates heat and temperature change per mole of substance. It is defined by:

$$q = nC \, \Delta T \tag{13.10}$$

where C is the heat capacity and n is the number of moles of the substance whose temperature changes. A quantity that is less frequently used is the *specific heat* of a substance, which is defined as the heat that changes the temperature of one gram of a substance by 1 K.

The value of the heat capacity depends on the path taken by the system in changing from one temperature to another. We shall assume a constant-pressure path. Table 13.1 lists the heat capacities for some substances.

TABLE 13.1
Heat Capacities of Some Substances at Room Temperature

Substance	Heat Capacity ($J \, mol^{-1}K^{-1}$)
$C_{(graphite)}$	8.54
He	20.8
Al	24.3
Fe	24.8
Hg	27.9
H_2	28.8
NaCl	45.3
H_2O	75.2
ethanol	113

EXAMPLE 13.5

Calculate the heat required to raise the temperature of 1.0 kg of water from 298 K to 308 K. Calculate the heat required to bring about the same temperature change in the same mass of mercury.

SOLUTION

Equation 13.10 gives the relationship between the heat and the three quantities given: the quantity of material, the temperature change, and the heat capacity. Thus:

$$q = 1000 \text{ g} \times \frac{1 \text{ mol } H_2O}{18.0 \text{ g } H_2O} \times 75.2 \frac{\text{J}}{\text{mol K}} \times (308 \text{ K} - 298 \text{ K})$$

$$= 42\,000 \text{ J}$$

The same calculation can be performed for Hg, using its heat capacity and atomic weight:

$$q = 1000 \text{ g} \times \frac{1 \text{ mol Hg}}{200.6 \text{ g Hg}} \times 27.9 \frac{\text{J}}{\text{mol K}} \times (308 \text{ K} - 298 \text{ K})$$

$$= 1400 \text{ J}$$

The difference in the two quantities of heat reflects the high specific heat of water. A relatively large quantity of heat is required to raise the temperature of a given mass of water.

13.4 PATHS FOR CHANGES IN STATE

The distinction between state functions and path functions is important in thermodynamics. The value of the change in a state function that results from a change in state depends on the initial and final states of the system, not on the path followed by the system in changing from one state to another. The value of a path function depends directly on the way in which the system goes from its initial state to its final state.

The relationships between state functions and path functions are also very useful. A convenient way to determine the value of the ΔE for a change in state is to measure the heat absorbed or evolved when the change occurs by a constant-volume path. We use the relationship $\Delta E = q_v$. Similarly, the value of ΔH for a process is found by measuring the heat absorbed or evolved when the process occurs by a constant-pressure path. We use the relationship $\Delta H = q_p$. Constant-volume and constant-pressure paths are idealized ways of carrying out changes in state that are of special interest to chemists, since many real processes occur by pathways that closely approximate these ideal paths.

It will be useful to define one more idealized path for carrying out a change in state. It is called the **reversible path.** A system that undergoes a change in state by a reversible path is always in an equilibrium state. That is, every state of the system—the initial state, the final state, and all states in between—is an equilibrium state. A change of state by a reversible path is called a **reversible process.** A change in state in which there are one or more intermediate states that are not equilibrium states is an **irreversible process.**

A truly reversible process exists only in theory. But there are real processes that are good approximations of reversible processes. One such process occurs when ice melts at 273 K. Mixtures consisting of an increasing quantity of liquid and a decreasing quantity of solid are formed until all of the ice melts. At 273 K, any mixture of ice and liquid water is an equilibrium state, provided that there are no localized hot or cold regions in the system. The term "reversible" can be used to describe this process because the absorption of a small quantity of heat causes some ice to melt, while the withdrawal of the same quantity of heat reverses the process, causing the same amount of water to freeze.

Another real process that approximates an ideal reversible process is the evaporation of liquid water at 373 K against an opposing pressure of 1 atm (Figure 13.4). As the liquid water absorbs small quantities of heat, it forms vapor at a pressure of 1 atm (Figure 13.4b). As the amount of vapor increases, the piston rises. Eventually, all the water evaporates; only vapor at 1 atm and 373 K is present. At any stage along the way, the equilibrium state—liquid water and water vapor at a pressure of 1 atm—exists. At any stage along the way, the withdrawal of a small quantity of heat causes the condensation of a small amount of water vapor, with a consequent fall of the piston. (Note that this reversible process involves not only heat but also work.)

Both processes described above have the two important features of reversible processes: Every state during the process is an equilibrium state, and a very small change in conditions can reverse the process.

The work associated with a reversible path is of special interest. If a system does work on the surroundings by expanding, it does the maximum possible work for the given change in state if the expansion occurs at constant temperature by a reversible path. The opposite is also true. If the surroundings do work on the system, causing it to contract, the maximum work is required if the change in state occurs at constant temperature by a reversible path.

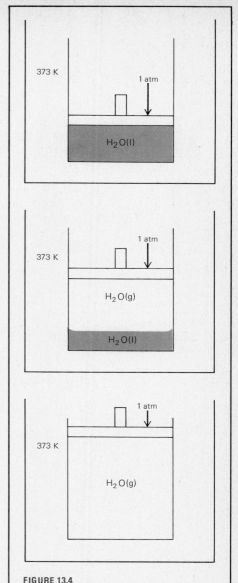

FIGURE 13.4
The evaporation of water against a pressure of 1 atm at 373 K is a real process that closely resembles an ideal process. The liquid water can absorb small amounts of heat, forming water vapor, which raises the piston. All the water can evaporate. The system is in an equilibrium state during the entire process.

13.5   THE SECOND LAW OF THERMODYNAMICS

Thermodynamics gives us a logical framework that allows us to unite a very large number of observations in a few basic rules. For example, the first law of thermodynamics tells us that no process in which energy is created or destroyed can occur. But the conservation of energy is not the only thermodynamic rule that has emerged from observations and experiments. Another rule, equally basic, is

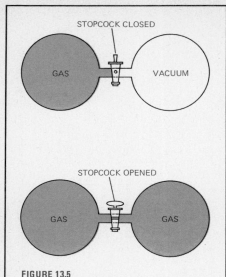

FIGURE 13.5
Gas flows from a filled bulb to an evacuated bulb until the pressures in both bulbs are equal. But when two bulbs containing gas at equal pressure are connected, a flow of gas that produces unequal pressures will not occur.

illustrated when two metal blocks, one hot and one cold, are placed in contact. Heat always flows from the hot block to the cold block until the two blocks are at the same temperature. We never see heat flow in the other direction, so that the cold block gets colder and the hot block gets hotter. Similarly, if we have two blocks at the same temperature, we never see heat flow so that one becomes hot and the other becomes cold. We can say that the two blocks at the same temperature are in an equilibrium state, a state that does not change with time. We will not see heat flow from one block to the other no matter how long we wait.

Now consider the system shown in Figure 13.5. When the stopcock is opened, gas flows into the evacuated bulb. The gas flows until the pressure is the same throughout the system. If the walls of the system are made of a material that does not allow the flow of heat, the process occurs adiabatically; $q = 0$. Since there is no mechanical link with the surroundings, $w = 0$, and therefore the first law tells us that $\Delta E = 0$.

We know from experience that the reverse process does not occur. If the system is left undisturbed, there will never be a net flow of gas from one bulb back into the other to create unequal pressures in the connected bulbs, even though the reverse flow also obeys the first law; $q = 0$, $w = 0$, and $\Delta E = 0$.

Both the processes we have described, the flow of heat in the metal blocks and the flow of gas, have a preferred direction. Heat flows from a hot body to a cold body and not the other way. Gas flows from a region of high pressure to a region of low pressure and not the other way. Chemical changes also have the same kind of preferred direction. A mixture of gasoline and oxygen in the presence of a spark forms H_2O and CO_2, but a mixture of CO_2 and H_2O does not form gasoline and oxygen.

These examples and countless others that illustrate a preferred direction are all *irreversible processes in which a system that is not in an equilibrium state changes spontaneously until it reaches an equilibrium state*. A system in an equilibrium state undergoes no further detectable change unless the system is disturbed by a change in the surroundings.

The meaning of the word "spontaneous" in thermodynamics is somewhat different from the ordinary connotation. The difference is that in thermodynamics, time is not a factor. A system is said to change spontaneously if the change occurs without the intervention of an external agent. But since this definition does not include time, a spontaneous change may occur very slowly in practice.

If the system is not in an equilibrium state, a spontaneous change is not merely possible; it is inevitable. All we have to do is wait and the system eventually will undergo the change in state that is necessary to reach equilibrium. In some cases, we may have to wait millions of years, but if we have patience (and longevity) we will see the system reach equilibrium. Once a system is in an equilibrium state, it does not undergo any further spontaneous change in state if it is left undisturbed. To take the system away from equilibrium, some external agent must act on the system. Work must be done by the surroundings, on the system.

The spontaneous direction of processes is the subject of the second law of thermodynamics. Like the first law, the second law is the result of countless observations and can be stated in many ways. In one statement of the second law, a new state function is used. This state function is **entropy**, represented by the symbol S. We shall defer briefly a discussion of the definition and significance of S and simply use it here to state the second law.

A change in state is accompanied by a change in entropy, ΔS. Stated in its most general sense, the second law refers to the change in the entropy of the entire universe that results from a change in state of a system. More specifically, we take the term "universe" to mean the system and its surroundings:

$$\Delta S_{\text{univ}} = \Delta S_{\text{sys}} + \Delta S_{\text{surr}}$$

The second law tells us how the entropy of the universe may change. When an irreversible spontaneous process occurs, the entropy of the universe increases, $\Delta S_{univ} > 0$. When a reversible process occurs, the entropy of the universe remains constant, $\Delta S_{univ} = 0$. At no time does the entropy of the universe ever decrease. Since the entire universe is undergoing spontaneous change, the second law can be most generally and concisely stated as: *The entropy of the universe is constantly increasing.*

We know what the other state functions, such as P, V, T, E, and H, measure. But what does S measure? We shall take up this point in detail later. For the moment, it can be said that classical thermodynamics does not require a physical explanation of the concept of entropy. All we need is an operational definition so that we can calculate the entropy change of the system and surroundings that accompanies a process.

A basic definition of the entropy change associated with a change in state of a system at constant temperature is:

$$\Delta S_{sys} = \frac{q_{rev}}{T} \tag{13.11}$$

where q_{rev} is the heat that accompanies the change in state if it takes place by a reversible path.

As we shall see, this definition can be applied to both the system and the surroundings. As is true for all state functions, the value of ΔS_{sys} is independent of the path followed for a change in state. Just as we did previously for ΔH and ΔE, we can define the change in a state function with a path function. Once again, the exact path must be specified. For ΔS we specify the reversible path. Equation 13.11 says that the entropy change accompanying a change in state at constant temperature can be determined by finding the value q would have if the change occurred reversibly, and dividing this value of q by the temperature.

For a phase change occurring at the normal transition temperature and constant pressure, q_{rev} is the ΔH of the phase change. For example, q_{rev} for the melting of ice or the freezing of water at 273 K is the ΔH of fusion of ice, 6.0 kJ/mol, or the ΔH of the freezing of water, -6.0 kJ/mol. Thus, ΔS_{sys} for 1 mol of $H_2O(s) \rightarrow 1$ mol of $H_2O(l)$ at 273 K and 1 atm is found by dividing ΔH of fusion by the melting point:

$$\frac{6000 \text{ J/mol}}{273 \text{ K}} = 22 \text{ J mol}^{-1} \text{ K}^{-1}$$

It can be seen from Equation 13.11 that when a solid melts or a liquid boils, both endothermic processes, the entropy of the system increases. For exothermic phase changes, such as the condensation of a gas or the freezing of a liquid, the entropy of the system decreases. The entropy changes that accompany phase changes provide simple examples of the relationship between entropy changes and spontaneity. Although ΔS_{sys} is positive for the melting of a solid, a solid does not always melt spontaneously. Below its melting point it remains solid. We must calculate the sum of ΔS_{sys} and ΔS_{surr} to find whether a process is spontaneous.

The calculation of ΔS_{surr} is simplified by assuming that the surroundings are much larger than the system, which means that the flow of heat into or out of the system because of a change in state does not change the temperature of the surroundings. It also means that the surroundings are in an equilibrium state during the entire change in state of the system. Therefore, the actual heat that flows into or out of the surroundings because of the change of state is q_{rev} with respect to the surroundings. Thus, we must know two things to calculate ΔS_{surr}: the temperature of the surroundings and the value of q for the change in state of the system. Then:

$$\Delta S_{surr} = \frac{-q}{T} \tag{13.12}$$

where T is the temperature of the surroundings.

WILL THE UNIVERSE DIE?

The German physicist Rudolf Clausius (1822–1888) formulated the laws of thermodynamics in two sentences:

1. The energy of the universe is constant.
2. The entropy of the universe tends toward a maximum.

The philosophers of nineteenth-century Europe dramatized this formulation into a picture of the "heat death" of the universe. Looking at the stars, they saw a continuing energy transfer process that would end only when all temperature differences were wiped out, the state of maximum entropy. Without a temperature difference, heat cannot be converted to mechanical work. Thus, all chemical, physical, and biological processes must eventually cease.

Recent cosmology has attempted to find an escape from this bleak forecast. One source of hope is the discovery that the universe is expanding. This expansion has given rise to at least two theories in which the universe will not die but will go on eternally.

One theory, formulated in the 1950s, sees a "steady-state" universe. According to this theory, the expansion of the universe is accompanied by the creation of new matter. Only one proton need be created per year in a billion liters of space to keep the mass of the universe constant, it has been calculated.

The alternative to the steady-state theory is the "big bang theory," which says that all the matter in the universe was gathered into one "superatom" some 15–20 billion years ago. The explosion of the superatom is believed to have started the expansion of the universe that is still going on. The big bang theory today is accepted by almost all cosmologists, because observations of the universe are consistent with almost all its predictions. In particular, background radiation that is believed to be the "ashes" of the primeval big bang has been detected.

The big bang theory opens another escape route from the heat death of the universe. It is possible that the expansion of the universe will not go on forever. Given enough mass, gravitational attraction will first slow, then reverse the expansion. Over a period of perhaps 50 billion years, all the matter in the universe will once again contract into a new superatom, and a new big bang will start the cycle again. If this picture is true, the universe will go on expanding, contracting, and exploding in an eternal cycle. However, if there is not enough matter in the universe to provide the needed gravitational attraction, expansion will be perpetual and the heat death is inevitable.

Observations made in the 1970s appear to support the latter pessimistic picture. Astronomers at a number of observatories, using a variety of techniques, have found that the mass of the matter in and between stars apparently falls far short of what is needed to reverse the expansion of the universe. However, both these observations and our knowledge of the universe are admittedly imperfect. There is still the possibility that the universe will not die the slow death predicted by our current understanding of the laws of thermodynamics.

Consider the melting of ice at 272 K, slightly below the melting point. The ΔH of fusion can be assumed to be virtually unchanged from the value, 6000 J, it has at 273 K. Thus:

$$\Delta S_{sys} + \Delta S_{surr} = \frac{6000 \text{ J}}{273 \text{ K}} + \frac{(-6000 \text{ J})}{272 \text{ K}} < 0$$

Since the total entropy change is negative, the process does not take place, according to the second law.

Above the melting point of ice, say at 274 K:

$$\Delta S_{sys} + \Delta S_{surr} = \frac{6000 \text{ J}}{273 \text{ K}} + \frac{(-6000 \text{ J})}{274 \text{ K}} > 0$$

The total entropy change is positive and the process is spontaneous.

At the melting point of ice we can write:

$$\Delta S_{sys} + \Delta S_{surr} = \frac{6000 \text{ J}}{273 \text{ K}} + \frac{(-6000 \text{ J})}{273 \text{ K}} = 0$$

The total entropy change is 0 and the process is reversible.

As is true of the first law, there are many different but equivalent statements of the second law of thermodynamics. Some of these statements concern what are called "perpetual motion machines of the second kind." One way to state the second law is to say that no machine operating in a cycle can convert a quantity of heat from the surroundings into an equal quantity of work on the surroundings. The first law does not forbid the existence of such a machine. The second law does.

Earlier, we gave, "You can't win," as a simplified version of the first law. A comparable version of the second law is, "You can't even break even." You can never get back as work all the heat that goes into a process.

13.6 THE MEANING OF ENTROPY

We said earlier that you do not have to know the meaning of entropy to use the second law to predict the spontaneous direction of a process. However, you will understand the behavior of systems better if you know what entropy actually measures. As an approach to the meaning of entropy, think of the microscopic composition of a macroscopic system.

Any macroscopic system consists of a very large number of small units; atoms, ions, or molecules. The various thermodynamic properties of the system, such as pressure, temperature, and energy, are ultimately determined by the state of all its microscopic units. If you will, there are specific microscopic states, specific arrangements of all the units of the system, corresponding to any macroscopic state.

If we were studying the thermodynamic property of temperature, we could in theory describe the state of an ideal gas by specifying the velocity of each molecule of the gas. In practice, there are so many molecules in even a small sample that we could not make the necessary measurements.

There is another fact to consider. On the microscopic scale, every system changes constantly. Even a system in an equilibrium state, which is macroscopically static, changes continually on the atomic level. We know that molecules are in constant motion, and that their individual positions, velocities, and energies change from moment to moment. Each change of each molecule creates a new microscopic state of the system, but not necessarily a new macroscopic state. In other words, any given macroscopic state of a system can have many different microscopic states.

The entropy of a system in a given state is a measure of the number of

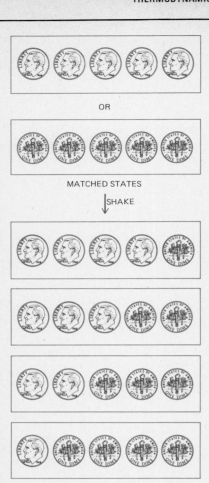

MATCHED STATES

SHAKE

SOME UNMATCHED STATES

FIGURE 13.6
Only two matched states are possible for a system of five coins, all heads or all tails. Thirty unmatched states, various combinations of heads and tails, are possible. If we shake a box containing the coins, the odds are that an unmatched state will result. As the number of coins is increased, the odds in favor of an unmatched, disorganized state increase sharply.

different microscopic states that correspond to a given macroscopic state. Entropy increases as the number of microscopic states increases.

There are important implications in this definition. It can help us understand why the direction of spontaneous change is always toward the equilibrium state, and why entropy is often described as *a measure of increasing disorder, or a measure of probability*.

We can illustrate the point with an analogy, shown in Figure 13.6. Suppose we have five coins. We can define two macroscopic states for the coins. They can be in an orderly arrangement, a "matched state," in which either five heads or five tails are showing. Or they can be in a disorderly, "unmatched" state, in which some of the coins show heads and some show tails. Each of these macroscopic states corresponds to a number of microscopic states. Only two microscopic states are possible for the matched state: all heads or all tails. A much larger number of microscopic states is possible for the unmatched state; we may have one head and four tails, two heads and three tails, and so on. With only five coins, 30 different microscopic states are possible for the unmatched macroscopic state.

Suppose we put the five coins in a box, shake the box, and then look at the coins. Since there are only two microscopic states that correspond to the matched macroscopic state and 30 microscopic states that correspond to the unmatched state, the odds are 15 to 1 that the coins are in the unmatched state. Whether we start with the matched state or the unmatched state, it is equally unlikely that the matched state will occur after we shake the box.

Even with five coins, we can say that there seems to be a direction associated with the process of shaking the box. The process "flows" toward the unmatched state. If we increase the number of coins, the direction of the process becomes more obvious. With 100 coins in the box, there are still only two microscopic states that produce a matched macroscopic state, but there are 1.3×10^{30} microscopic states that produce the unmatched state. For practical purposes, there is almost no chance of finding a matched state after shaking a box containing 100 coins.

To use the thermodynamic term, the unmatched state in this system is the equilibrium state. We may change the microscopic state when we shake the box, but we almost certainly will not move the system from the unmatched equilibrium state to the matched state. A human hand can put the system in the matched state by doing work, picking up and turning over the necessary number of coins. But again, a shake of the box will cause the system to proceed spontaneously to the unmatched equilibrium state.

Now consider a system whose units of structure are molecules, not coins. This page is such a system. The molecules of this page are in constant, random motion at equilibrium. There is always some small probability that the molecules making up the page will move spontaneously in an ordered way, out of their relatively disordered equilibrium state, so that the page will turn by itself. You are not advised to wait for this event to happen. It is extremely improbable that the page will turn of itself, even in the 12 billion or so years before the sun burns out. The probability of this nonequilibrium state forming is small enough to ignore.

Once we define entropy in this way, we can reason qualitatively and predict the sign of ΔS_{sys} for various changes in state. (It should be remembered that the total ΔS of system and surroundings, is always either positive or zero.)

When a substance undergoes an expansion at constant temperature, the entropy of the system increases, $\Delta S_{sys} > 0$. When a given quantity of material occupies a greater volume, it becomes more disordered. More microscopic states are possible because more positions are available for the molecules, and therefore the value of ΔS is greater. Conversely, $\Delta S_{sys} < 0$ when a substance contracts at constant temperature.

When the temperature of a substance increases at constant volume,

$\Delta S_{sys} > 0$. There is a greater distribution of molecular velocities at higher temperatures, and thus a greater number of microscopic states is possible.

Any process in which different substances are mixed without chemical change has $\Delta S_{sys} > 0$, because mixing means increased disorder. Gaseous solutions form by diffusion because of the increase in the entropy of the system due to mixing. As we hinted in Chapter 12, the solubility of one substance in another is often due largely to the increase in entropy that occurs when two materials mix.

Any process that increases the number of particles in the system has $\Delta S_{sys} > 0$. For a reaction like $H_2(g) \rightarrow 2H(g)$, which forms two particles from one, the entropy of the system increases. In the reverse process, $2H(g) \rightarrow H_2(g)$, in which the number of particles is reduced, the entropy of the system decreases, $\Delta S_{sys} < 0$.

The decrease or increase in the entropy of the system associated with a phase change is due to the relative order in the phases. A solid is more ordered than a liquid, which is more ordered than a gas. Therefore, an increase in the entropy of the system is associated with melting (the change from solid to liquid), boiling (the change from liquid to gas), and sublimation (the change from solid to gas). A decrease in the entropy of the system results from processes such as condensation (a change from gas to liquid or solid) and freezing (a change from liquid to solid).

The Third Law of Thermodynamics

We stated that it is impossible to find the absolute value of E or H of a given state. But we can find the ΔE or ΔH associated with a change in state. Therefore, we can choose an arbitrary standard state or zero point for these state functions. Elements in their standard states are assigned zero enthalpy of formation for the sake of convenience.

The choice of a state for the zero entropy point is not arbitrary. It is based on the fact that entropy measures the number of microscopic states that correspond to a given macroscopic state. When there is only one microscopic state, $S = 0$. The designation of this state is often called the third law of thermodynamics. The third law is stated: *The entropy of a perfect crystal of a pure substance at the absolute zero of temperature is zero.*

A perfect crystal is a completely ordered arrangement in which each unit of structure has a fixed position. At 0 K, each atom has only one possible energy state: the minimum state. Since the substance is perfectly pure, there is no disorder due to mixing. Therefore, a perfect crystal of a pure substance at the absolute zero of temperature is the most ordered system imaginable.

When $S = 0$, there can be only one microscopic state. The fact that $\log 1 = 0$ suggests that there is a logarithmic relationship between the number of microscopic states and S. The relationship can be expressed as:

$$S = k \ln \Omega \qquad (13.13)$$
$$= 2.303 \, k \log \Omega$$

where k is called *Boltzmann's constant* and is defined as R/N, the gas constant divided by Avogadro's number, and Ω is the number of microscopic states.

The third law makes it possible to find an actual value for the entropy of a substance at a given temperature. This value is called the absolute entropy, $S°$. Some absolute entropies are listed in Appendix II.

The value of $S°$ indicates the relationship between the molecular structure of a given substance in a given state and the entropy of the substance. Other things being equal, the absolute entropy generally increases as molecular size and complexity increase. For a substance of given molecular complexity, the gas has higher $S°$ than the liquid, and the liquid has higher $S°$ than the solid. The entropy of diamond, a rigid, highly organized form of carbon, is lower than that of graphite, a form of carbon whose molecular structure is not as organized. Hard substances such as diamond have rigid, symmetrical crystal structures with a minimum of disorder. Therefore, hard substances tend to have low $S°$.

The values of $S°$ can be used to find $\Delta S°$, the absolute entropy change of a system undergoing a chemical reaction. This calculation is performed in essentially the same way as the calculation of $\Delta H°$ outlined in Section 8.3, page 249:

$$\Delta S° = S°\text{(products)} - S°\text{(reactants)} \qquad (13.14)$$

EXAMPLE 13.6

Use the data in Appendix II to calculate $\Delta S°$ for the reaction:

$$N_2(g) + 3H_2(g) \rightleftharpoons 2NH_3(g)$$

at 298 K and 1 atm.

SOLUTION

The relevant values of $S°$ are: $NH_3(g)$, 193 J K^{-1} mol^{-1}; $N_2(g)$, 192 J K^{-1} mol^{-1}; $H_2(g)$, 131 J K^{-1} mol^{-1}. Using the relationship expressed in Equation 13.14:

$$\begin{aligned}
\Delta S° &= [(2\ \text{mol})(193\ \text{J K}^{-1}\ \text{mol}^{-1})] - [(192\ \text{J K}^{-1}\ \text{mol}^{-1}) \\
&\quad + (3\ \text{mol})(131\ \text{J K}^{-1}\ \text{mol}^{-1})] \\
&= -199\ \text{J/K}
\end{aligned}$$

In the reaction, the number of moles of gas decreases from four to two. The decrease in the entropy of the system is consistent with this reduction in the number of moles of gas. Since there is less disorder, entropy decreases.

13.7   FREE ENERGY

The use of the total entropy change (ΔS_{univ}) as a criterion for the spontaneity of reactions is somewhat awkward. It requires two calculations, one for the ΔS of the system, the other for the ΔS of the surroundings. It would be convenient to have a state function that is a property of the system alone and that could be used as a criterion for spontaneity.

We shall define such a state function that can be used to predict spontaneity for changes in state carried out at constant temperature and pressure, the conditions of most chemical reactions. This state function is the **Gibbs free energy**, G, which is defined in terms of other state functions of the system as:

$$G = H - TS \qquad (13.15)$$

For a change in state at constant temperature:

$$\Delta G = \Delta H - T\,\Delta S \qquad (13.16)$$

Equation 13.16 has many applications in chemistry.

The relationship between ΔG and the total entropy change (ΔS_{univ}) can be derived from Equation 13.16. For a process carried out at constant pressure, $\Delta H = q$. If the process is also carried out at constant temperature, substitution into Equation 13.12 gives a simple relationship between ΔH and ΔS_{surr}:

$$\Delta H = -T\,\Delta S_{surr}$$

The ΔS in Equation 13.16 is the ΔS_{sys}. Therefore, Equation 13.16 can be rewritten as:

$$\Delta G = -T\,\Delta S_{surr} - T\,\Delta S_{sys} = -T\,\Delta S_{univ} \qquad (13.17)$$

Therefore, $\Delta G = 0$ *for a reversible change at constant pressure and temperature*. Because of the negative sign in Equation 13.17, ΔG *is negative for a spontaneous change at constant pressure and temperature*. A change in state with a positive ΔG can occur only if work is done on the system; it is nonspontaneous.

It is important not to confuse the signs of ΔS_{univ} and ΔG for a change in state. For a reversible change, both these quantities are zero. But for a spontaneous change, ΔS_{univ} is positive while ΔG is negative.

Equation 13.16 tells us a great deal about changes of state, as you can see by looking at the example of water boiling at 373 K and 1 atm. The change of state is

$$H_2O(l) \longrightarrow H_2O(g) \ (P = 1 \text{ atm})$$

Since a gas is more disordered than a liquid, ΔS of the system for this change is positive. The value of ΔS_{vap} of H_2O has been found to be 108.8 J K^{-1} mol^{-1} at 373 K. The sign of ΔH for this change is also positive, since the system must absorb heat from the surroundings to make the liquid vaporize. The value of ΔH has been found to be 40 626 J/mol at 373 K.

Assume that the values of ΔH_{vap} and ΔS_{vap} do not vary appreciably as the temperature changes. Since $\Delta G = 0$ for the reversible phase change, the boiling point of water can be determined from the values of ΔH and ΔS, using Equation 13.16. The boiling point can be defined as the temperature at which any quantities of $H_2O(g)$ and $H_2O(l)$ are at equilibrium at 1 atm. Therefore, the change from $H_2O(l)$ to $H_2O(g)$ at 1 atm and the change from $H_2O(g)$ to $H_2O(l)$ at 1 atm are reversible changes, for which $\Delta G = 0$. Setting ΔG in Equation 13.16 equal to zero gives:

$$\Delta H - T \Delta S = 0$$

and

$$\Delta H = T \Delta S$$

or

$$T = \frac{\Delta H}{\Delta S} \tag{13.18}$$

Substituting the values of ΔH_{vap} and ΔS_{vap} for H_2O in Equation 13.18 gives:

$$T = \frac{40\ 626 \text{ J/mol}}{108.8 \text{ J/mol K}} = 373 \text{ K}$$

Another way to express the same idea is to say that at the boiling point of water, the favorable negative term, the product of temperature and entropy, just balances the unfavorable positive enthalpy term. At a temperature below 373 K, the product of T and ΔS is not great enough to cancel ΔH; that is, $\Delta H > T \Delta S$, and therefore $\Delta G > 0$. The positive value of ΔG indicates that water does not boil spontaneously below 373 K at 1 atm. When the temperature is greater than 373 K, then $T \Delta S > \Delta H$, and $\Delta G < 0$. The negative value of ΔG indicates that water will boil spontaneously if heated above 373 K at 1 atm.

For the reverse process:

$$H_2O(g)(P = 1 \text{ atm}) \longrightarrow H_2O(l)$$

both ΔS and ΔH are negative. When $T < 373$ K, the term ΔH is more important than the term $T \Delta S$. Since $\Delta H < 0$, then $\Delta G < 0$. This process, 1 atm of water vapor condensing to liquid at a temperature below 373 K, is spontaneous. At temperatures above 373 K, the term $T \Delta S$ is more important than the term ΔH. And since $\Delta S < 0$, then $-T \Delta S > 0$ and $\Delta G > 0$. The condensation of 1 atm of water vapor at temperatures above the boiling point of water is not a spontaneous process. But at exactly 373 K, $\Delta H = T \Delta S$, $\Delta G = 0$, and again the process is reversible.

TABLE 13.2
Temperature and Spontaneity

ΔH	ΔS	ΔG	
−	+	−	reaction is favored at all temperatures
+	−	+	reaction is never favored
−	−	±	decreasing temperature favors reaction
+	+	±	increasing temperature favors reaction

Since $\Delta G = 0$ for the process

$$H_2O(l) \rightleftharpoons H_2O(g)(P = 1 \text{ atm}, T = 373 \text{ K})$$

in either direction, we can say that the free energy of $H_2O(l)$ is equal to the free energy of $H_2O(g)$ at 373 K and 1 atm. Thus, a system containing these two components is in an equilibrium state.

This sort of analysis can be performed on many processes. The first step is to judge whether ΔH and ΔS are positive or negative for a given process. By considering the bonds made and broken (Section 8.3, page 252) and the phase changes that occur, it is often possible to determine whether ΔH is positive or negative. A similar judgment for ΔS is often possible from simple considerations of order and disorder, such as changes in the number of molecules. A process that is exothermic ($\Delta H < 0$) and causes increased disorder ($\Delta S > 0$) must result in a negative ΔG. Such a process is spontaneous. A process that is endothermic ($\Delta H > 0$) and causes increased order ($\Delta S < 0$) must result in a positive ΔG. Such a process is nonspontaneous. We need more information to make the analysis only when ΔH and ΔS have the same sign. Table 13.2 tabulates these possibilities.

Although both ΔH and ΔS do not change appreciably as the temperature changes, ΔG can vary considerably. According to Equation 13.16, the value of ΔG is determined by ΔH and by $T \Delta S$. Therefore, the contribution of ΔS to the value of ΔG becomes relatively more important as the temperature increases. Thus, an endothermic process that results in increasing disorder is more likely to proceed spontaneously as the temperature increases. At high enough temperatures, ΔS becomes the controlling factor. Conversely, an exothermic process that results in increasing order is more likely to proceed spontaneously as the temperature decreases. When T is small, $T \Delta S$ can be small compared to ΔH.

A good rule of thumb (which has numerous exceptions) is that ΔH is the controlling factor at temperatures close to those normally found on the surface of the earth. Exothermic processes generally are spontaneous. Endothermic processes generally are not. A simple change in state that illustrates these points is:

$$\tfrac{1}{2}H_2(g) \longrightarrow H(g)$$

The extent to which this change in state occurs depends on the temperature. The reaction does not proceed to any measurable extent at 300 K. It proceeds to an appreciable extent above 4000 K. Since the reaction breaks one bond and does not make any, it is endothermic, $\Delta H > 0$. Since it produces two particles from one, it increases disorder, $\Delta S > 0$. When T is not large, $T \Delta S$ is not large; ΔH is the controlling factor and ΔG has a positive sign. When T is large, $T \Delta S$ is large and ΔG has a negative value. Table 13.3 lists the values of ΔG, ΔH, and ΔS at different temperatures at 1 atm for this reaction.

This reaction is just one demonstration of the fact that all compounds will break down into their constitutent atoms at very high temperatures. We can look to the sun for confirmation. Temperatures inside the sun are so high that no amount of exothermic bond-making can overcome the bond-breaking associated with a large $T \Delta S$ term. There are no molecules in the sun. In fact, the sun is so

TABLE 13.3
Thermodynamic Functions for the Reaction $\frac{1}{2}H_2(g) \rightarrow H(g)$

Temperature (K)	$\Delta H°$ (kJ)	$\Delta S°$ (kJ/K)	$\Delta G°$ (kJ)
298	218	0.0494	203
500	219	0.0527	193
1000	222	0.0567	166
2000	227	0.0600	107
3000	230	0.0612	46
4000	231	0.0617	−15
5000	232	0.0618	−80

hot that it consists of plasma, a gas in which all the atoms have been completely ionized.

Standard Free Energy

In Section 8.3 (page 249), which you should review at this point, we used tabulated values of ΔH to calculate the enthalpy changes associated with physical and chemical changes. We can use an analogous approach for calculations with ΔG.

Once again, we start by defining a standard state. In this case, we define the standard state of a substance as its state at 1 atm pressure. The symbol for the free energy of a substance in the standard state is $G°$. To obtain the standard free energy change, $\Delta G°$, we subtract the standard free energy of the reactants from the standard free energy of the products at the temperature of the process.

$$\Delta G° = G°(\text{products}) - G°(\text{reactants}) \tag{13.19}$$

It is helpful to define a $\Delta G°$ for a specific process, the formation of one mole of a substance in its standard state from its constituent elements in their standard states. The $\Delta G°$ for this specific process is called the *standard free energy of formation,* and its symbol is $\Delta G_f°$. By definition, the $\Delta G_f°$ of an element in its standard state is zero. For example, the $\Delta G°$ of the reaction:

$$H_2(g) + \tfrac{1}{2}O_2(g) \longrightarrow H_2O(l)$$

is $\Delta G_f°$, the standard free energy of formation of water, when the pressures of all the substances in the reaction are 1 atm.

The relationship between the $\Delta G°$ of a change in state and the $\Delta G_f°$ of the substance involved in the change is:

$$\Delta G° = \Delta G_f°(\text{products}) - \Delta G_f°(\text{reactants}) \tag{13.20}$$

Note that this expression is essentially the same relationship as Equation 8.2, which we used earlier for $\Delta H°$.

There are extensive tables listing $\Delta G_f°$ values for many substances. Most of these lists give $\Delta G_f°$ values for a substance at 298 K. One such table is in Appendix II. You can use this information to calculate $\Delta G°$ for a change in state at 298 K in the same way that you would use the data for the standard enthalpy of formation.

EXAMPLE 13.7
Using the data in Appendix II, calculate $\Delta G°$ at 298 K for the combustion of acetylene:

$$C_2H_2(g) + \tfrac{5}{2}O_2(g) \longrightarrow 2CO_2(g) + H_2O(l)$$

SOLUTION
The values of $\Delta G_f°$ from Appendix II are: $C_2H_2(g)$, 209 kJ/mol; $O_2(g)$,

0 kJ/mol (an element in its standard state); $CO_2(g)$, -394 kJ/mol; $H_2O(l)$, -237 kJ/mol. We can calculate the value of $\Delta G°$ by using the relationship given in Equation 13.20:

$$\Delta G° = [(2 \text{ mol})(-394 \text{ kJ/mol}) + (-237 \text{ kJ/mol})] - (209 \text{ kJ/mol} + 0)$$
$$= -1230 \text{ kJ}$$

The standard free energy of formation of many compounds has a negative value at 298 K. These compounds form spontaneously from their constituent elements in their standard states. Compounds that have positive values of $\Delta G_f°$ usually are not formed directly from their constituent elements because they are present only in relatively small amounts at equilibrium. Indirect methods generally must be used to form them.

We can illustrate the meaning of $\Delta G°$ by using a general reaction:

$$A(g) \rightleftharpoons B(g)$$

Assume that both A and B are ideal gases in a bulb, each at a pressure of 1 atm; both A and B are in their standard states. If the $G°$ of A is the same as the $G°$ of B, then $\Delta G° = 0$ and the system is in an equilibrium state. No detectable change in composition occurs. But if the $G°$ of B is less than the $G°$ of A, then $\Delta G° < 0$. Some A will convert spontaneously to B. The free energy of A decreases as the amount of A decreases, and the free energy of B increases as the amount of B increases. Eventually, the two free energies are equal, $\Delta G = 0$, and the system is at equilibrium. The equilibrium state contains both A and B, but since $G_A° > G_B°$ and $\Delta G° < 0$, it contains more B than A.

Now assume the opposite case, that the $G°$ of A is less than the $G°$ of B. Now $\Delta G°$ for the reaction is positive, $\Delta G° > 0$. The reaction proceeds "backward," from right to left, spontaneously producing A from B until an equilibrium state is reached. In this equilibrium state, there is more A than B.

We have just described a qualitative relationship, in which only relative amounts of A and B were discussed. But the magnitude of $\Delta G°$ can be used to describe a quantitative relationship between reactants and products at equilibrium. As the numerical value of $\Delta G°$ increases, so do the relative quantities of the favored components.

Free Energy and Equilibrium

The relationship between $\Delta G°$ and the composition of equilibrium states is one of the most important in chemistry. We shall begin the discussion of this relationship by writing a general chemical reaction as:

$$aA(g) + bB(g) + cC(l) \rightleftharpoons dD(g) + eE(g) + fF(s)$$

The lowercase letters are the coefficients of the chemical equation, and the uppercase letters represent different substances whose states are indicated by state symbols. When each component of this reaction is in its standard state— that is, at a pressure of 1 atm—the free energy change of this change in state, $\Delta G = \Delta G°$. At equilibrium, however, the pressure of each component generally is not 1 atm. To find the relationship between $\Delta G°$ and the composition of the equilibrium state, we must first state how ΔG changes with changes in the pressure of each component. This relationship is:

$$\Delta G = \Delta G° + 2.303RT \log \frac{P_D^d\, P_E^e}{P_A^a\, P_B^b} \tag{13.21}$$

Equation 13.21 tells us that the relationship between the free energy change for a process in which all the components are at a pressure of 1 atm ($\Delta G°$) and the free energy change under any other set of pressure conditions (ΔG) depends on a logarithmic pressure term. The exact form of this term depends on the form of the chemical reaction. You can see that the numerator of the fraction includes all the gases on the right side of the general reaction. The denominator includes all the gases on the left side of the reaction. The coefficients of the gases in the reaction appear in the fraction as exponents of the terms for the pressures of the gases. None of the liquids or solids in the reaction appear in the pressure term, because the free energy of a liquid or solid does not change in any significant way with a change in pressure. The liquid and solid can always be considered to be in the standard state, a pressure of 1 atm.

EXAMPLE 13.8

Write the form of the pressure term in Equation 13.21 for each of the following reactions:

(a) $H_2(g) + Br_2(g) \rightleftharpoons 2HBr(g)$
(b) $H_2O(l) \rightleftharpoons H_2O(g)$
(c) $2C(s) + O_2(g) \rightleftharpoons 2CO(g)$
(d) $CO_2(g) + MgO(s) \rightleftharpoons MgCO_3(s)$

SOLUTION

(a) The numerator of the pressure term has the pressure of HBr raised to the power of 2. The denominator has the product of the two gases on the left. Therefore, the term is:

$$\frac{P^2_{HBr_2}}{P_{H_2}P_{Br_2}}$$

(b) Since liquids are not included in the pressure term, the term for this reaction is simply P_{H_2O}.

(c) Since solids are omitted from the term and coefficients become exponents, the term is P^2_{CO}/P_{O_2}.

(d) Since solids are not included, the term is $1/P_{CO_2}$.

EXAMPLE 13.9

Calculate ΔG for the conversion of $N_2O_4(g)$ at a pressure of 0.75 atm to $NO_2(g)$ at a pressure of 1.25 atm at 298 K. Use data from Appendix II.

SOLUTION

For the reaction:

$N_2O_4(g) \longrightarrow 2NO_2(g)$

$\Delta G° = \Delta G_f°(\text{products}) - \Delta G_f°(\text{reactants})$
$= [(2 \text{ mol})(51.24 \text{ kJ/mol})] - [(99.52 \text{ kJ/mol})]$
$= 2.96 \text{ kJ} = 2960 \text{ J}$

For this reaction, Equation 13.21 takes the form:

$$\Delta G = \Delta G° + 2.303RT \log \frac{P^2_{NO_2}}{P_{N_2O_4}}$$

$$\Delta G = 2960 \text{ J} + 2.303(8.31 \text{ J mol}^{-1} \text{ K}^{-1})(298 \text{ K}) \log \frac{(1.25 \text{ atm})^2}{(0.75 \text{ atm})}$$

$$= 4780 \text{ J}$$

Since ΔG has a positive value, this change is not spontaneous.

When the pressures of all the components of the system are such that it is in an equilibrium state, we know that $\Delta G = 0$. Therefore, when the system is at equilibrium, Equation 13.21 becomes:

$$\Delta G° = -2.303RT \log \frac{P^d_D P^e_E}{P^a_A P^b_B} \tag{13.22}$$

Here we have the desired relationship between $\Delta G°$ and the composition of the equilibrium state of the system.

Once we write a chemical equation that relates the components of a system, we fix the value of $\Delta G°$ at any given temperature. We can calculate the value of $\Delta G°$ when the temperature is close to 298 K by using the data in Appendix II.

Now look at each of the terms in Equation 13.22. At a given temperature, R, $\Delta G°$, and T are all constants. Therefore, the pressure term must also be a constant, which is a most interesting conclusion. It means that for any chemical reaction, we can write an expression that includes the pressures of the gases in the system and that has a constant value at a given temperature when the system is at equilibrium. This useful expression is called the **equilibrium constant.** It is usually represented by the symbol K. Equation 13.22 usually is written in a simpler form that uses the symbol for the equilibrium constant:

$$\Delta G° = -2.303RT \log K \tag{13.23}$$

The form of the equilibrium constant is a result of the way in which the free energy of the substance depends on changes in pressure. The free energy of pure liquids and solids does not change with pressure. Therefore, no information about liquids and solids appears in the expression for K.

The relationship between the free energy change of a gas and a change in its pressure is logarithmic. Therefore, the coefficients of the gases appear in the expression for K as exponents. You should also note that the expression for K includes products and exponents of only the gases in the reaction.

EXAMPLE 13.10

Calculate the value of K for the reaction:

$$N_2O_4(s) \rightleftharpoons 2NO_2(g)$$

at 298 K.

SOLUTION

In Example 13.9, we found $\Delta G°$ for this reaction to be 2960 J. Equation 13.23 therefore gives us:

$$2960 \text{ J} = -2.303(8.31 \text{ J mol}^{-1} \text{ K}^{-1})(298 \text{ K}) \log K$$
$$\log K = -0.519$$
$$K = 3.03 \times 10^{-1} \text{ atm}$$

We shall discuss the use and significance of the equilibrium constant in detail in Chapter 14. For the moment, let us review some of the features of K and ΔG:

1. *Any spontaneous change in state of a system is accompanied by a negative ΔG,* that is, a decrease in the free energy of the system. The equilibrium state is said to be the most stable state of the system; it is *the state of minimum free energy*. There is, in fact, a clear relationship between the free energy of a system and its stability. The lower the free energy, the greater the stability. A substance with a negative free energy of formation is stable with respect to decomposition to its elements. A substance with a positive free energy of formation is unstable with respect to its elements.

2. Referring back to the general chemical reaction we defined on page 400, an equilibrium state can exist for many different quantities of the gases A, B, D, and E. But when the pressure of each gas at equilibrium is substituted into the expression for K, the value of K must be the specific value defined by the $\Delta G°$ and T of the system for the equilibrium state. The pressure of each gas depends on the amount of gas in the system, but at equilibrium these pressures must be consistent with the relationship between K, $\Delta G°$, and T.

3. The amounts of liquid and solid components present at equilibrium do not

influence the pressures of the gaseous components at equilibrium. But the nature of the solid and liquid components do determine the value of $\Delta G°$ and therefore of K.

Consider the reaction:

$$CaCO_3(s) \rightleftharpoons CaO(s) + CO_2(g)$$

If a quantity of $CaCO_3(s)$ is heated to 1000 K, the pressure of $CO_2(g)$ at equilibrium is about 4×10^{-2} atm. The equilibrium pressure of $CO_2(g)$ is related to the difference between the standard free energy of the products, $CO_2(g)$ and $CaO(s)$, and the reactant, $CaCO_3(s)$, at this temperature. However, adding or removing either solid, $CaCO_3(s)$ or $CaO(s)$, does not change the equilibrium pressure.

4. In general, when $\Delta G° < 0$, the products of a reaction are favored. When $\Delta G° > 0$, the reactants tend to be favored. We can put the relationship between $\Delta G°$ and K in tabular form:

$\Delta G°$	<0	>0	0
K	>1	<1	1

As an example, consider the process:

$$H_2O(l) \rightleftharpoons H_2O(g)$$

We know that at 373 K, $\Delta G° = 0$. At that temperature, $K = 1$; or, $P_{H_2O} = 1$ atm. At temperatures lower than 373 K, liquid water at 1 atm is more stable than gaseous water at 1 atm; $\Delta G° > 0$ for the process, so $K < 1$. That is, at equilibrium at temperatures below 373 K, the vapor pressure of water is less than 1 atm. At temperatures above 373 K, $\Delta G° < 0$; gaseous water in its standard state at 1 atm is more stable—has a lower free energy—than liquid water at 1 atm. For $H_2O(l)$ and $H_2O(g)$ to be at equilibrium at temperatures above 373 K, the pressure must be greater than 1 atm.

Solution Equilibrium

We can broaden the relationship between $\Delta G°$, K, and the composition of the system to include dissolved substances. For a discussion of solution equilibrium, we need a standard state for solutes. The standard state for pure substances is $P = 1$. A concentration of $1M$ can be used as the standard state of a dissolved solute. We shall express solute concentrations as molarities. We can then derive a relationship, similar to Equation 13.22, which includes the molarities of solutes as well as the pressure of gases. The molarity of any substance in solution appears in the equilibrium constant expression. In the expression, the molarity is raised to the power of the coefficient of the solute in the chemical equation.

Thus, the equilibrium expression for the reaction in which gaseous CO_2 dissolves in water, $CO_2(g) \rightleftharpoons CO_2$, is:

$$K = \frac{[CO_2]}{P_{CO_2}}$$

where, by convention, the molarity of each species actually present in solution is indicated by placing the formula of the species in square brackets.

For the reaction in which solid lead chloride dissolves in water, $PbCl_2(s) \rightarrow Pb^{2+} + 2Cl^-$, the equilibrium expression is:

$$K = [Pb^{2+}][Cl^-]^2$$

We shall discuss solution equilibria in greater detail in later chapters.

Temperature Dependence of K

It is often important to know how K for a given reaction varies with temperature. Equation 13.23 gives this information, if the value of $\Delta G°$ at the temperature of interest is known. But in many cases, the data are not available for the calculation of $\Delta G°$ at all temperatures. However, while $\Delta G°$ varies considerably with temperature, $\Delta H°$ and $\Delta S°$ are relatively constant over a range of temperatures. On this basis, a relationship between K and T can be derived from Equations 13.15 and 13.23:

$$\frac{\Delta H°(T_2 - T_1)}{T_1 T_2} = 2.303R \log \frac{K_2}{K_1} \tag{13.24}$$

where K_1 and K_2 are the equilibrium constants at T_1 and T_2 for the reaction whose standard enthalpy change is $\Delta H°$. Equation 13.24 allows us to predict how K varies with T, if we know whether the reaction is exothermic or endothermic. Suppose that T_2 is greater than T_1 in Equation 13.24. Then $T_2 - T_1$ is positive. If the reaction is endothermic, $\Delta H°$ is positive. Therefore, $\log (K_2/K_1)$ is positive, and $(K_2/K_1) > 1$. The value of K for an endothermic reaction increases with temperature.

The reasoning is similar for an exothermic reaction. In this case, $\Delta H° < 0$, and the left side of Equation 13.24 is negative; $\log (K_2/K_1)$ is also negative, and $(K_2/K_1) < 1$. The value of K for an exothermic reaction decreases with temperature.

EXAMPLE 13.11

Calculate the value of K at 500 K for the reaction of Example 13.10, $N_2O_4(g) \rightleftharpoons 2NO_2(g)$, using the data in Appendix II.

SOLUTION

First we must calculate $\Delta H°$ for the reaction in the usual way, using $\Delta H_f°$ of $NO_2 = 34$ kJ/mol and $\Delta H_f°$ of $N_2O_4 = 10$ kJ/mol:

$\Delta H° = \Delta H_f°(\text{products}) - \Delta H_f°(\text{reactants})$
$\Delta H° = [(2 \text{ mol})(34 \text{ kJ/mol})] - [10 \text{ kJ/mol}] = 58 \text{ kJ} = 58\,000 \text{ J}$

From Example 13.10, we know that $K_{298} = 3.0 \times 10^{-1}$ atm. Equation 13.24 gives:

$$\frac{58\,000 \text{ J}(500 \text{ K} - 298 \text{ K})}{(500 \text{ K})(298 \text{ K})} = 2.303(8.31 \text{ J mol}^{-1} \text{ K}^{-1}) \log \frac{K_2}{3.0 \times 10^{-1}}$$

$$\log K_2 = 3.59$$

$$K_2 = 3.9 \times 10^3 \text{ atm}$$

As expected, when the temperature increases, the K of this endothermic reaction increases.

13.8 THERMODYNAMICS IN THE REAL WORLD

Chemical thermodynamics is more than a set of principles that are applied to idealized laboratory systems. Living things also obey the laws of thermodynamics. Because living things are complex systems, we must use approximations when we apply the laws of thermodynamics to them. But even this imperfect application teaches us some valuable lessons about the activity of living things, up to and including human beings—especially human beings who live in an industrialized society such as ours.

We can start with a basic thermodynamic study of an animal. The food that an animal eats goes through a long series of chemical reactions in the body. In these reactions, heat is released. Some of the heat is thrown off into the surroundings and some is used in the body of the animal.

Every process involving work, heat, and energy in the living organism must obey the first law of thermodynamics. Energy is neither destroyed nor created. In theory, we can do the same first-law bookkeeping for an animal as for an ideal gas. In practice, this bookkeeping is hard to do for an animal. The processes in a living organism are quite complex, and living systems are never really in an equilibrium state. In terms of thermodynamics, the net processes of life are spontaneous and irreversible, since living organisms are heading relentlessly toward an equilibrium state that can be reached only after the life process stops. All that lives must die. But as always, since thermodynamics does not give us information about time, we cannot predict when things will die.

The net result of the processes of life is an increase in the entropy of the universe. To demonstrate this increase rigorously, we must consider such parameters as the heat thrown off by the organism and the heat given off by the sun, which provides the energy for all life on earth. (It is assumed that any living organisms that may exist on other planets get their energy from stars similar to the sun.) There are so many of these factors to consider that a quantitative treatment of the thermodynamics of living systems is extremely difficult. But there are lessons to be learned from qualitative reasoning about the thermodynamics of living organisms. We know from the first law that no living thing can exist without food, since it cannot create energy. We know from the second law that the efficiency with which food is used to produce work is less than 100%. We know that all living things eventually die. These facts have always been known intuitively. In the age of science, we have learned that the laws of thermodynamics are also consistent with these facts about life.

The same laws can be applied to our society. A technological society needs not only food but also other sources of energy to run machines. These machines have given us a high standard of living. But we have used energy for our machines so prodigally that we have a crisis. Thermodynamic reasoning helps us to appreciate the nature and magnitude of the problem and to evaluate the possible solutions.

The first law tells us that we cannot create energy. We are limited to a fairly small number of energy sources: fossil fuels, direct use of solar energy, nuclear fission, nuclear fusion. Most of these sources depend on fuel supplies that are exhaustible.

The second law tells us that the work available from these energy sources is limited. We cannot achieve 100% conversion of heat to work. The consequences of the second law are important, because we must now regard an energy source as something that releases waste heat. Even in our most efficient available processes for the conversion of heat to work, a substantial fraction of the available heat is wasted. A coal- or oil-burning plant turns only 40% of the heat from its fuel into work. A nuclear fission plant turns only 33% of its heat into work. An automobile is even less efficient.

As our society uses more energy, it releases more waste heat. The result of this heat release is a change in the surroundings, the environment. The air over a city is several degrees warmer than the air over unoccupied land. The water of a river used to cool a generating plant is warmer than the water of an untouched river. We know that the "heat island" of a large city changes weather patterns, and that the "thermal discharges" of generating plants have some effect on living organisms in a river. While the long-term effects of such changes are difficult to establish, we have learned from hard experience in recent years to be cautious about any large-scale, unintentional, artificial change in the environment.

In recent years, human progress has been equated with an increase in the available supply of energy. There is a good deal to be said for this point of view, since more energy means better living standards for more people. But there is a

A solar energy research facility at Odeillo in the
French Pyrenees. Solar energy is one of the most
promising alternatives to the use of fossil fuels, whose
supply is dwindling (*Menzel, Stock, Boston*)

limiting factor that must be considered when we discuss the utopia of virtually
unlimited energy from such sources as nuclear fusion. That factor is waste-heat
release. It is possible that the generation of virtually unlimited energy from nuclear
fusion would cause the release of waste heat in quantities that could first melt the
polar ice caps and eventually make this planet unlivable.

Solar energy appears to offer a way out of this dilemma. The earth receives
energy from the sun and radiates almost the same quantity of energy back into
space. Only a small fraction of incoming solar energy is put to use, either by
photosynthesis in plants or by technological processes. Since the quantity of solar
energy that reaches the earth appears to be more than sufficient to satisfy
reasonable human needs, it would seem that capturing solar energy for human
use is harmless. However, attempts to use solar energy on a large scale could
have profound results—for example, a change in the albedo of the earth, the
percentage of sunlight that is reflected back into space. Such an occurrence
could cause a substantial change in the temperature of the earth, just as any
other energy conversion does, making the planet too hot for life. Even solar
energy, it seems, does not offer us the limitless, pollution-free energy source that
we are seeking.

EXERCISES

13.1 Indicate which of the following
are state functions and which are path
functions: (a) your elevation above sea
level, (b) your food consumption per
day, (c) your sign of the zodiac, (d) your
bank balance, (e) your expenditure of
effort to learn thermodynamics.

13.2 One mole of an ideal gas at STP
undergoes a change to a final state in
which its temperature becomes 300 K
and its pressure becomes 1.25 atm.
Find ΔP, ΔT, and ΔV for the process.

13.3 Indicate whether the work accompanying each of the following processes is positive, negative, or zero:
(a) the elevation of a weight (the sys-

tem), (b) $NH_4Cl(s) \rightleftharpoons NH_3(g) + HCl(g)$ in
a sealed container, (c) the formation of
liquid water from hydrogen and oxygen
initiated by a spark in an open container
at room temperature, (d) $H_2(g) +$
$Cl_2(g) \rightleftharpoons 2HCl(g)$.

13.4 Find the work associated with the
expansion of an ideal gas from an initial
volume of 22.4 liters to a volume of
44.8 liters against an opposing pressure
of 1.00 atm.

13.5 Find the work associated with the
evaporation of 2.0 kg of superheated
water against a pressure of 1.1 atm at a
temperature of 450 K.

13.6 Indicate whether the heat accom-

panying each of the following processes
taking place in a sealed container is positive, negative, or zero: (a) $2H_2(g) +$
$O_2(g) \rightarrow 2H_2O(g)$, (b) $Cl_2(g) \rightleftharpoons 2Cl(g)$.
(c) carbon dioxide sublimes.

13.7 Indicate whether the heat accompanying each of the following processes
is positive, negative, or zero: (a) an ideal
gas expands against an opposing pressure and its temperature remains constant, (b) an ideal gas is compressed
and its temperature decreases, (c) an
ideal gas expands into a vacuum and its
temperature remains constant. (*Hint:* The
internal energy of an ideal gas is directly
proportional to its temperature.)

13.8 To carry out a true adiabatic process we need a container made of a perfect insulating material. Predict the temperature change of the system when each of the following processes is carried out in such a container: (a) the amount of gas is doubled at constant volume, (b) $H_2(g) \rightarrow 2H(g)$ at constant pressure, (c) $2H_2(g) + O_2(g) \rightarrow 2H_2O(l)$.

13.9 A sample of an ideal gas expands against an opposing pressure of 0.94 atm from 2.0 liters to 4.2 liters. The system is also found to evolve 510 J of heat. Find ΔE.

13.10 One mole of NaI(s) is formed from its elements in their standard states at 298 K in a sealed container. The ΔH_f° of NaI is -287.9 kJ/mol. Find the value of ΔE.

13.11 An isothermal change in state is one for which $\Delta T = 0$. An ideal gas absorbs 1000 J of heat without undergoing any change in temperature. Find the work that accompanies this process. (*Hint:* The internal energy of an ideal gas is directly proportional to its temperature.)

13.12 For each of the following processes, indicate whether the quantity $\Delta H - \Delta E$ is positive, negative, or essentially zero: (a) the evaporation of a liquid, (b) the sublimation of a solid, (c) the combustion of methane to carbon dioxide and liquid water, (d) the dissociation of F_2 into F atoms.

13.13 The value of ΔH for the reaction $2SO_2(g) + O_2(g) \rightarrow 2SO_3(g)$ at 298 K is -198 kJ. Find ΔE.

13.14 The heat of combustion of liquid C_8H_{18}, octane, to carbon dioxide and liquid water at 298 K is 1303 kJ/mol. Calculate ΔE for this combustion reaction.

13.15 Derive a relationship between ΔH and ΔE for a process in which the temperature of a fixed amount of an ideal gas changes. (*Hint:* Follow a procedure similar to the one used to derive Equation 13.8).

13.16 The reaction $CH_4(g) + 2O_2(g) \rightarrow CO_2(g) + 2H_2O(l)$, $\Delta E = -885.4$ kJ/mol CH_4, is carried out in a calorimeter of the type shown in Figure 13.3. It is found that when a 5.347 g sample of methane burns, the temperature of the calorimeter increases from 296.14 K to 299.88 K. Find the heat capacity of the calorimeter.

13.17 Find the quantity of heat that must be liberated by a reaction to raise the temperature of a bomb calorimeter whose heat capacity is 75.4 kJ/K by 1.00 K.

13.18 Find ΔE for the combustion of C_2H_5OH, ethanol, from the following data: The heat capacity of the calorimeter is 34.65 kJ/K. The combustion of 1.765 g of ethanol raises the temperature of the calorimeter from 294.33 K to 295.84 K.

13.19 One tablespoon of peanut butter has a mass of 16 g. It is combusted in a calorimeter whose heat capacity is 120 kJ/K. The temperature of the calorimeter rises from 295.2 to 298.4 K. Find the caloric content of peanut butter. (*Hint:* 1 food calorie is 4.184 kJ.)

13.20 The ΔH of the reaction $2CO(g) + O_2(g) \rightarrow 2CO_2(g)$ is -482.3 kJ. The reaction is carried out at constant volume with a 2.359-g sample of CO in a calorimeter whose heat capacity is 15.44 kJ/K. The initial temperature of the calorimeter is 298.01 K. Find its temperature when the reaction is over.

13.21 A 51.42-g sample of ice is placed in a calorimeter whose heat capacity is 2.348 kJ/K. The temperature of the calorimeter drops from 282.47 K to 275.16 K as all the ice melts. Calculate the heat of fusion of ice.

13.22 Calculate the heat required to raise the temperature of a 5.46-g block of aluminum from 298.0 K to 317.6 K.[a]

13.23 Calculate the heat evolved when the temperature of 1.0 mol of graphite falls 100 K.[a]

13.24 The heat of combustion of methane is 890.4 kJ/mol. A tank of water is heated by combustion of methane. Calculate the temperature increase in 20.56 kg of water as the result of the combustion of 2.00 mol of methane, assuming no heat loss to the surroundings.[a]

13.25 The heat of combustion of octane is 1303 kJ/mol. Find the mass of octane that must be burned in order to raise the temperature of a 1-kg block of iron by 200 K, assuming no heat loss.[a]

*13.26 An ice cube of mass 9.0 g is added to a cup of coffee at 363 K, which contains 120 g of liquid. Assume the heat capacity of coffee is the same as that of water. The heat of fusion of ice is 6.0 kJ/mol. Find the temperature of the coffee after the ice melts.

13.27 A 1.0-mol amount of helium at STP absorbs 510 J of heat at constant pressure. (a) Calculate the temperature

of the helium. (b) Calculate the volume of the helium. (c) Calculate the work associated with the process. (d) Calculate ΔE. Assume ideal gas behavior.[a]

13.28 The temperature of 2.0 mol of helium drops at a constant pressure of 1.0 atm from 325 K to 300 K. Calculate q, w, ΔE, and ΔH for this process. Assume ideal gas behavior.[a]

13.29 The heat of vaporization of water at 1 atm and 373 K is 40.66 kJ/mol. Find the value of ΔS_{sys} for the change in state $H_2O(l) \rightleftharpoons H_2O(g)$ at 1 atm.

13.30 Find the ΔS of the system and the surroundings for the process of Exercise 13.29 at 374 K and at 372 K. Show that the results are consistent with the second law.

13.31 The ΔS_{sys} for the evaporation of HF is 0.0257 kJ mol^{-1} K^{-1} and the ΔH is 7.53 kJ/mol at the boiling point. Find the boiling point of HF.

13.32 The ΔS_{sys} for the fusion of platinum is 9.62 J mol^{-1} K^{-1} and $\Delta S_{sys} + \Delta S_{surr} = 0$ at 2043 K. Find the heat of fusion of platinum.

13.33 Indicate the sign of ΔS_{univ} for each of the following processes: (a) $2H_2(g) + O_2(g) \rightleftharpoons 2H_2O(l)$ at 298 K, (b) the electrolysis of water to form $H_2(g)$ and $O_2(g)$ at 298 K, (c) the growth of an oak tree from a little acorn.

13.34 Indicate the sign of ΔS_{sys} for each of the following phase changes: (a) evaporation, (b) condensation, (c) sublimation, (d) fusion, (e) freezing. Explain your answers.

13.35 Indicate the sign of ΔS_{sys} for each of the following changes in state: (a) the contraction of a gas at constant temperature, (b) a decrease in the temperature of a gas at constant volume, (c) the mixing of two gases at constant temperature and volume, (d) an increase in the pressure of a gas at constant temperature. Explain your answers.

13.36 Indicate the sign of ΔS_{sys} for the following chemical changes: (a) $Br_2(g) \rightleftharpoons 2Br(g)$, (b) $CaCO_3(s) \rightleftharpoons CaO(s) + CO_2(g)$, (c) $H_2(g) + CO_2(g) \rightleftharpoons H_2O(g) + CO(g)$, (d) $N_2(g) + 3H_2(g) \rightleftharpoons 2NH_3(g)$. Explain your answers.

13.37 Entropy is sometimes called "the arrow of time." Suggest an explanation for this phrase.

13.38 Calculate ΔS° of the following reactions at 298 K: (a) $H_2(g) + Cl_2(g) \rightleftharpoons 2HCl(g)$, (b) $H_2(g) + Br_2(g) \rightleftharpoons 2HBr(g)$, (c) $H_2(g) + Br_2(l) \rightleftharpoons 2HBr(g)$.

---

[a] The data for this exercise can be found in Table 13.1.

Discuss the similarities and differences between the three results.[b]

13.39 Calculate the ΔS_f° at 298 K of the following gases: (a) $CO_2(g)$, (b) $CO(g)$, (c) $O_3(g)$, (d) $N_2O_4(g)$.[b]

13.40 Calculate the ΔS_f° at 298 K of the following solids: (a) NaCl, (b) NH_4Cl, (c) diamond.

13.41 The boiling point of chloroform, $CHCl_3$, is 334.4 K. Use the values of S° in Appendix II to find the ΔH° of vaporization at the boiling point.

13.42 Find the ΔS° of combustion of methane at 298 K to carbon dioxide and liquid water.[b]

13.43 Discuss the effect of increasing temperature on the extent to which the following reactions proceed:
(a) $CO_2(g) \rightleftharpoons CO_2(s)$, (b) $2NO_2(g) \rightleftharpoons N_2O_4(g)$, (c) $2NH_3(g) \rightleftharpoons N_2(g) + 3H_2(g)$, endothermic, (d) $2SO_2(g) + O_2(g) \rightleftharpoons 2SO_3(g)$, exothermic.

13.44 Discuss the effect of temperature on the formation of the following substances from their elements in their standard states: (a) CO(g), (b) KCl(s), (c) NO(g).

13.45 The ΔH_f° of AgCl(s) at 298 K is -127 kJ/mol and its ΔS_f° is -58 J/mol K. Find ΔG_f° of AgCl(s) at 298 K.

13.46 Use the following data to calculate ΔG° of the reaction $2NO_2(g) \rightleftharpoons N_2O_4(g)$ at 500 K: $NO_2(g)$, $S^\circ = 261$ J/mol K, $\Delta H_f^\circ = 32.2$ kJ/mol; $N_2O_4(g)$, $S^\circ = 349$ J/mol K, $\Delta H_f^\circ = 8.67$ kJ/mol, all at 500 K. Predict the value of ΔG° at 600 K relative to your answer.

13.47 Calculate ΔG° at 298 K for the following reactions: (a) $NH_3(g) + HBr(g) \rightleftharpoons NH_4Br(s)$, (b) $CaCO_3(s) \rightleftharpoons CaO(s) + CO_2(g)$, (c) $CH_4(g) + 3Cl_2(g) \rightleftharpoons CHCl_3(g) + 3HCl(g)$.[c]

13.48 The salt NH_4NO_3 can follow three modes of decomposition: (a) to $HNO_3(g) + NH_3(g)$, (b) to $N_2O(g) + H_2O(g)$, or (c) to N_2, O_2, and $H_2O(g)$. Calculate the ΔG° for each mode of decomposition at 298 K.[c]

13.49 The following reactions are important ones in catalytic converters in automobiles. Find ΔG° for each at 298 K. Predict the effect of increasing temperature on the magnitude of ΔG°:
(a) $2CO(g) + 2NO(g) \rightleftharpoons N_2(g) + 2CO_2(g)$,
(b) $5H_2(g) + 2NO(g) \rightleftharpoons 2NH_3(g) + 2H_2O(g)$, (c) $2H_2(g) + 2NO(g) \rightleftharpoons N_2(g) + 2H_2O(g)$, (d) $2NH_3(g) + 2O_2(g) \rightleftharpoons N_2O(g) + 3H_2O(g)$.[c,d]

13.50 Calculate ΔG for the formation of the following pressures of $CO_2(g)$ at 298 K from its elements in their standard states: (a) 2.0 atm, (b) 0.10 atm.

13.51 Calculate ΔG for the formation of $NH_3(g)$ at 298 K at a pressure of 1.0 atm from $H_2(g)$ at a pressure of 10 atm and $N_2(g)$ at a pressure of 5 atm.[c]

13.52 Suppose we redefine the standard state as $P = 2$ atm. Find the new standard ΔG_f at 298 K of each of the following substances: (a) HCl(g), (b) $N_2O(g)$, (c) H.[c]

13.53 Explain the results of Exercise 13.52 in terms of the effect of increased pressure on the relative entropies of the reactants and products of each

reaction. (*Hint:* An increase in pressure is equivalent to a decrease in volume.)

13.54 Use the result of Exercise 13.47 to calculate K for the reactions listed, at 298 K.

13.55 Find the value of K at 298 K for the three modes of decomposition of $NH_4NO_3(s)$ in Exercise 13.48.

13.56 Find the value of K for the decomposition of $H_2(g)$ at the temperatures given in Table 13.3. Account for the trend with increasing temperature.

13.57 Use the result of Exercise 13.46 to find the K of the reaction $N_2O_4(g) \rightleftharpoons 2NO_2(g)$ at 600 K. Explain the difference between your result and that of Example 13.10.

13.58 Calculate the K for each of the reactions given in Exercise 13.49. Predict the effect of temperature on the magnitude of each K.

13.59 Find the ΔH° of a reaction whose equilibrium constant doubles when the temperature is raised from 300 K to 400 K.

13.60 Find the ΔH° of a reaction whose equilibrium constant decreases by a factor 2 when the temperature is raised from 300 K to 400 K.

13.61 The ΔH° of the reaction $\frac{1}{2}H_2(g) \rightleftharpoons H(g)$ at 298 K is 218 kJ and $K = 2.54 \times 10^{-36}$. Use these data to calculate K of this reaction at 1000 K. Compare your result to the one obtained in Exercise 13.56. Account for the difference.

13.62 Calculate K for the reaction $N_2(g) + 3H_2(g) \rightleftharpoons 2NH_3(g)$, the Haber process, at 500 K and at 800 K.[c,d]

---

[b] The values of S° that are necessary for this exercise can be found in Appendix II.

[c] The necessary values of ΔG_f° for the solution of this exercise can be found in Appendix II.

[d] The necessary values of ΔH_f° can be found in Appendix II.

CHEMICAL EQUILIBRIUM

14

The equilibrium state is the ultimate goal of every chemical system. A system that is in the equilibrium state does not change spontaneously. Only the application of an external influence, a stress of some sort, can cause a system that is at equilibrium to undergo a change. If a system is not at equilibrium, it undergoes spontaneous change and attempts to attain the equilibrium state.

All of us have some intuitive knowledge about spontaneous reactions and the equilibrium state. We know that sugar dissolves in water, but only up to a point. We know that if we evaporate some water from the saturated solution of sugar, or if we cool it, sugar crystallizes from the solution. We know that silver reacts with sulfur compounds in the air to form tarnish. We may not be so aware of the fact that a small amount of the tarnish on a silver object may decompose back to silver and sulfur. In this chapter and the next, we shall discuss the criteria that enable us to tell whether a reaction is spontaneous and the extent to which it can proceed. We shall also discuss the qualitative and quantitative features of the equilibrium state.

14.1   THE EQUILIBRIUM STATE AND THE EQUILIBRIUM CONSTANT

In principle, we can easily define an equilibrium state: A system is in an equilibrium state if it does not change spontaneously, even over a prolonged period of time. But in practice, it is often difficult to know whether a system is in an equilibrium state. Many systems reach equilibrium rapidly, but some proceed so slowly to equilibrium that they appear not to be changing at all. These systems appear to be at equilibrium, but they are not. For example, a sample of laughing gas, N_2O, should decompose spontaneously at room temperature until almost all of it is converted to N_2 and O_2. However, no detectable amount of N_2 or O_2 appears in a purified sample of N_2O, because the rate of decomposition is immeasurably slow. Since the equilibrium state in this system includes N_2O, N_2, and O_2, a mixture of N_2 and O_2 should spontaneously form a very small quantity of N_2O. This process also is not observed, because the rate is too slow. In our discussion, we shall not be concerned with the amount of time needed for a system to reach equilibrium. We shall discuss only the state of the system when it reaches equilibrium.

A system at equilibrium does not appear to undergo any change. This appearance is deceptive, even though any measurements we can make on the composition or characteristics of the system are consistent with its apparent static nature. In fact, a system at equilibrium is static only on the macroscopic level. On the atomic and molecular level, an equilibrium system is dynamic; an enormous number of changes occur continuously. But the net result of all these changes on the molecular level is no change on the macroscopic level. At equilibrium any reaction and the exact reverse of that reaction take place at the same rate.

We have already discussed equilibrium in a system consisting of a liquid and its vapor in a closed container. At any given temperature, such a system reaches an equilibrium state that is characterized by a certain vapor pressure above the liquid. As long as the system is left alone, it does not change macroscopically. However, it *is* changing constantly on the molecular level. Molecules continually evaporate from the liquid and condense from the vapor. There is no overall change because the rates of evaporation and condensation are equal. If there is an external stress, such as a change in temperature, the system changes, as the concentration of substance in the vapor phase increases or decreases. The system goes to a new equilibrium state, characterized by a different vapor pressure.

Now we shall consider equilibrium states of systems that contain substances which are related by simple chemical changes. We shall examine the composition of the equilibrium states of these systems, and we shall see how these systems behave in response to external stress. Such systems usually differ from the simple equilibrium system of a liquid and its vapor in a closed container, in which the pressure of vapor depends only on the temperature.

Consider the system described by the reaction:

$$H_2(g) + I_2(g) \rightleftharpoons 2HI(g)$$

Equilibrium is established rapidly at 700 K. The equilibrium pressure of each gas depends not only on the temperature, but also on the quantity of material in the system. The introduction of 10 mol of HI into a container results in a greater equilibrium pressure of HI than the introduction of 1 mol of HI into the same container. By the same token, the introduction of 10 mol each of H_2 and I_2 into a container results in a greater equilibrium pressure of HI than the introduction of 1 mol of each in the same container.

The composition of equilibrium states as a function of the quantities of the substances in a system has been studied for more than a century. The results of a typical study are shown in Table 14.1. The equilibrium pressures of the components of the system can be seen to vary considerably. However, the value of the fraction $P^2_{HI}/P_{H_2}P_{I_2}$, which we shall call K, can be seen to be relatively constant.

The data in Table 14.1 reveal two crucial aspects of chemical equilibrium. The value of the expression that we designate as K does not change significantly with changes in the quantities of the substances present in the system. The exact composition of the equilibrium state is determined by the quantities of the substances in the system, but *not* by the direction from which equilibrium is approached.

We shall deal with chemical systems at equilibrium by using the **equilibrium constant, K,** an expression whose form is based on the chemical equation that describes the chemical change. For a given system at a given temperature, the value of K is fixed. If we want to determine the value of K for a given system, the first and most important step is to *write a balanced chemical reaction that relates the components of the system*. We then obtain the expression for K by following these rules:

1. The form of the expression for K is influenced by the way in which we write the reaction for the system. In general, K is a fraction whose numerator includes the products of the reaction and whose denominator includes the reactants.

2. The expression for K includes terms for gases and substances in solution. The terms for gases usually express the pressure of the gases in atmospheres (atm), although they sometimes give concentrations of the gases in moles per liter (mol/liter). The terms for solutions usually are expressed in units of concentration, such as moles per liter (molarity) of the solutes. By convention, molarity is symbolized by placing the formula of a substance in square brackets, [ ], so [Br_2]

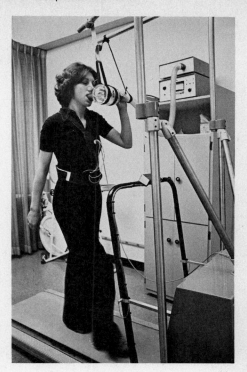

A period of exercise on a treadmill provides a rough analogy for the dynamic equilibrium of a chemical system. The chemical system remains unchanged on the macroscopic level despite a large number of changes on the molecular level. Vigorous movement on the treadmill produces no change in the position of the exerciser. (*Marmaras, Woodfin Camp*)

TABLE 14.1
Equilibrium in the System $H_2(g) + I_2(g) \rightleftharpoons 2HI(g)$

Initial Composition (atm)			Equilibrium Composition (atm)			$K = P^2_{HI}/P_{H_2}P_{I_2}$
P_{H_2}	P_{I_2}	P_{HI}	P_{H_2}	P_{I_2}	P_{HI}	
0.638	0.570	0	0.165	0.0978	0.945	55.3
0.610	0.686	0	0.103	0.179	1.013	55.7
0	0	0.612	0.0644	0.0654	0.482	55.2
0	0	0.256	0.0270	0.0274	0.202	55.2

means the molarity of bromine in solution and [Na$^+$] means the molarity of Na$^+$ ion in solution. Unless otherwise indicated, "solution" means an aqueous solution.

3. Terms for pure liquids and solids are not included in the expression for K, because the concentration of a pure liquid or a pure solid at a given temperature is essentially a constant—that is, the number of moles of solid per liter of solid or of liquid per liter of liquid is the same no matter what the quantity of solid or liquid.

4. The numerator of the expression for K is the product of all the terms written for substances on the right side of the chemical equation. The denominator of the expression for K is the product of all the terms written for substances on the left side of the chemical equation.

5. The coefficients of the chemical equation also have a direct influence on the value of K. The pressure of a gas or the concentration of a solute in the expression for K is raised to the power of its coefficient in the equation. If a gas has the coefficient 2 in an equation, the pressure of the gas is squared in calculating the value of K. If a substance in solution has the coefficient 3, its molarity is cubed in calculating the value of K.

EXAMPLE 14.1

Write the expressions for K for the following chemical equations, expressing gases as pressures:

(a) $C(s) + CO_2(g) \rightleftharpoons 2CO(g)$
(b) $NH_3(g) + HCN(l) \rightleftharpoons NH_4^+ + CN^-$
(c) $5Fe^{2+} + MnO_4^- + 8H^+ \rightleftharpoons 5Fe^{3+} + Mn^{2+} + 4H_2O$

SOLUTION

(a) Solids do not appear in the expression for K, and the coefficient of the CO appears as an exponent. The expression for K thus is:

$$K = \frac{P_{CO}^2}{P_{CO_2}}$$

(b) If a formula is not followed by a state symbol, the substance is understood to be in aqueous solution. Liquids do not appear in the expression for K. The expression thus is:

$$K = \frac{[NH_4^+][CN^-]}{P_{NH_3}}$$

(c) The concentration of water, the solvent, is not included in the expression, which is:

$$K = \frac{[Mn^{2+}][Fe^{3+}]^5}{[MnO_4^-][Fe^{2+}]^5[H^+]^8}$$

From these rules, you can see that it is necessary to indicate the state of each substance when writing a chemical equation. A substance is included in the expression for K if it is a gas or if it is in solution. It is not included if it is a solid or a liquid. If an equation does not indicate the states of all substances, it does not permit us to write the expression for the equilibrium constant, K. In Section 2.8, we discussed the need for indicating the states of substances when writing equations from word descriptions of chemical changes. *Review that section carefully*. To help you review, we shall repeat the rules to follow when writing a chemical equation.

1. First classify substances into two groups: those that are not in aqueous solution and those that are. Divide the substances that are not in aqueous solution into solids, liquids, and gases. The formula for each such substance is followed by a state symbol in the equation: (g) for gas, (l) for liquid, (s) for solid.

The state of any substance at atmospheric pressure depends directly on temperature. If the word description of a chemical reaction does not mention temperature, assume that the reaction occurs at room temperature, about 300 K, and that each substance is in its appropriate state at that temperature.

2. If a substance is in aqueous solution, its formula is not followed by a state symbol in the chemical equation that we write. Therefore, the absence of a state symbol is as meaningful as the presence of a state symbol.

3. Substances in aqueous solution can be subdivided into three categories (Section 2.6, page 37). A substance can be a strong electrolyte which exists entirely as dissociated ions in aqueous solution. It can be a weak electrolyte which is partially dissociated into ions in aqueous solutions. Or it can be a nonelectrolyte which does not form ions in aqueous solution.

If a substance is a strong electrolyte, its formula should be written to indicate complete dissociation into ions in aqueous solution. *The formula of an undissociated strong electrolyte should not appear in a chemical reaction without a state symbol. Only the dissociated ions exist in solution.*

There are three main types of strong electrolytes: salts, strong acids, and bases. A salt usually is a compound of a metal ion with either a nonmetal or a polyatomic ion. Table 2.4 lists some important polyatomic ions. All salts are solids at room temperature. Therefore, in reactions occurring at room temperature, a salt that is not in solution gets the state symbol (s). A salt that is in solution is represented as dissociated ions without state symbols.

Some common strong acids, which also are represented as dissociated ions when in solution, are listed in Table 2.6. As for strong bases, many of them are compounds of a metal and a hydroxide ion, OH^-, and are solids. Strong bases in solution are represented as dissociated ions. We shall discuss acids and bases in detail in Chapter 15.

The formulas for weak electrolytes and nonelectrolytes in aqueous solution are written to show them undissociated.

4. Any component of a system that does not change in any way is not included in the chemical reaction. Do not write a chemical reaction that includes the same substance, unchanged, on both sides of the reaction.

For example, when a solution of lead nitrate, which contains Pb^{2+} cations and NO_3^- anions, is mixed with a solution of sodium chloride, which contains Na^+ cations and Cl^- anions, a precipitate of lead chloride forms. No other chemical change accompanies this precipitation; the Na^+ ions and the NO_3^- ions remain in solution. The chemical equation for this reaction is thus:

$$Pb^{2+} + 2Cl^- \rightleftharpoons PbCl_2(s)$$

There is nothing in this equation to show that sodium chloride is the source of the Cl^- ion. The ion could have come from any soluble chloride, and the reaction would have been the same. Therefore, it is appropriate to omit the Na^+ from the equation. For our equilibrium and thermodynamic calculations, all we want to know is the amount of chloride in the solution, not the exact composition of the starting solution.

EXAMPLE 14.2
Write balanced chemical reactions and the corresponding expressions for K for the following processes:
(a) the sublimation of iodine
(b) the dissolution of silver sulfate in water
(c) the decomposition of calcium carbonate to calcium oxide and carbon dioxide
(d) the formation of a precipitate of silver chloride after a solution of hydrochloric acid and a solution of silver nitrate are mixed
(e) the conversion of nitrogen dioxide to dinitrogen tetroxide

SOLUTION

(a) Sublimation is the conversion of a solid to a gas. The equation for the sublimation of iodine is:

$$I_2(s) \rightleftharpoons I_2(g) \qquad K = P_{I_2(g)}$$

The form of the expression for the equilibrium constant shows that the pressure of $I_2(g)$ at equilibrium does not depend on the amount of solid present.

(b) Silver sulfate is a solid at room temperature, and it dissociates into ions when it is dissolved in water. The equation is thus:

$$Ag_2SO_4(s) \rightleftharpoons 2Ag^+ + SO_4^{2-} \qquad K = [Ag^+]^2[SO_4^{2-}]$$

Once again, the amount of solid does not determine the concentration of ions in the solution, so long as there is enough solid to form a saturated solution.

(c) The equation for the reaction is:

$$CaCO_3(s) \rightleftharpoons CaO(s) + CO_2(g) \qquad K = P_{CO_2(g)}$$

Neither solid appears in the expression for K.

(d) Hydrochloric acid is a strong acid that is a solution of H^+ and Cl^-. A solution of silver nitrate, a salt that dissociates into ions, contains Ag^+ ion and NO_3^- ion. Silver chloride forms from silver ion and chloride ion. The net equation is thus:

$$Ag^+ + Cl^- \rightleftharpoons AgCl(s)$$

The H^+ and NO_3^- ions are properly omitted because neither undergoes a change, and the equilibrium constant for this equation is $K = 1/[Ag^+][Cl^-]$. The numerator of the expression for K is 1 when no other terms appear.

(e) The balanced equation for this reaction can be written as:

$$2NO_2(g) \rightleftharpoons N_2O_4(g)$$

For this equation, $K = P_{N_2O_4}/P_{NO_2}^2$. If the equation is written as $NO_2(g) \rightleftharpoons \frac{1}{2}N_2O_4$, then $K = P_{N_2O_4}^{\frac{1}{2}}/P_{NO_2}$, and it has a different numerical value. Remember that the numerical value of K can be changed by changing the form of a chemical equation. We normally write equations with the smallest whole-number coefficients, for uniformity.

We can regard K as an expression that conveys a great deal of information about the behavior of chemical systems. We can obtain this information because K is the equilibrium *constant,* which is always the same for a given reaction at a given temperature. The value of K defines the equilibrium state and enables us to find the direction and extent of a reaction. That value depends on both the temperature and the nature of the components of the system.

Let us see what these statements mean for a specific system, the conversion of nitrogen dioxide to dinitrogen tetroxide. The equation is $N_2O_4(g) \rightleftharpoons 2NO_2(g)$, and $K = P_{NO_2}^2/P_{N_2O_4}$. If the temperature is 373 K, the value of K for this equation is found to be about 80 atm. This brief statement has several implications.

To begin with, when a system containing $NO_2(g)$ and $N_2O_4(g)$ at 373 is at equilibrium, the value of the pressure of NO_2 squared divided by the pressure of N_2O_4 is 80 atm. This statement might seem nothing but a verbal way of saying what the expression for K states symbolically. But let us turn the idea around. Suppose we measure the pressures of NO_2 and N_2O_4 in a system at 373 K and substitute the values we obtain into the expression for K. If the value obtained for K is 80 atm, we know that the system is in an equilibrium state. If we get a different value, the system is not at equilibrium. It is only when the system is at equilibrium that the expression $P_{NO_2}^2/P_{N_2O_4}$ is K. When the system is not at equilibrium, this expression is designated by Q and is called the *reaction quotient.*

The expression for Q has the same form as the expression for K, but it has a different numerical value, which changes as the reaction proceeds and becomes equal to K at equilibrium. Therefore, the value we get after substituting into this expression tells us something important about the system.

We can learn even more by a closer examination of the numerical value of the expression. Suppose we measure the pressures of NO_2 and N_2O_4 in a system, substitute those values into Q, and obtain a numerical value that is larger than 80 atm. Now we know that the system is not in an equilibrium state. But we know something more: The system is not in an equilibrium state because there is relatively too much NO_2 present. Why? Look at the expression for Q. The pressure of NO_2, which is on the right side of the chemical equation, is in the numerator. An increase in the value of the numerator means an increase in the value of the expression. We can make a general rule out of this observation: Whenever the value obtained by substituting numerical values into the expression for Q in a given system is larger than the value of K, there is an excess of the substances on the right side of the chemical equation.

Suppose that the opposite is true, that the numerical value of Q is smaller than the value of K. Now we know that the system is not at equilibrium because it has an excess of the substances on the left side of the equation, in the present case, $N_2O_4(g)$. The pressure of $N_2O_4(g)$ appears in the denominator of the expression for Q, and an increase in the value of the denominator means a decrease in the value of Q.

By comparing the values of Q and K, we can predict the direction of spontaneous change for a chemical system that is not at equilibrium. If Q is larger than K, there is a relative excess of the materials on the right side of the chemical equation and the reaction proceeds spontaneously from right to left toward equilibrium. If Q is smaller than K, there is a relative excess of the materials on the left side of the equation and the equation proceeds spontaneously from left to right toward equilibrium.

EXAMPLE 14.3

Predict the behavior of each of the following systems prepared from the given pressures of the two gases in atmospheres: (a) $P_{NO_2} = 0.4$, $P_{N_2O_4} = 0.002$; (b) $P_{NO_2} = P_{N_2O_4} = 0.004$; (c) $P_{NO_2} = 1.0$, $P_{N_2O_4} = 0.002$; (d) $P_{NO} = 0.001$, $P_{N_2O_4} = 0.0000$, at 373 K.

SOLUTION

We use the same method in each case: Substitute the data into the expression for Q and then compare the result with the value of K for the reaction $N_2O_4(g) \rightleftharpoons 2NO_2(g)$, which is given as 80 atm at 373 K.

(a) $\dfrac{P_{NO_2}^2}{P_{N_2O_4}} = \dfrac{(0.4 \text{ atm})^2}{0.002 \text{ atm}} = 80 \text{ atm}$

The system is at equilibrium, since $Q = K$.

(b) $\dfrac{(0.2 \text{ atm})^2}{0.004 \text{ atm}} = 10 \text{ atm}$

Since $Q < K$, the system is not at equilibrium. There is relatively too little of the product, $NO_2(g)$. The reaction will proceed spontaneously from left to right to produce $NO_2(g)$ and use up $N_2O_4(g)$ until $Q = K = 80$ atm.

(c) $\dfrac{(1.0 \text{ atm})^2}{0.002 \text{ atm}} = 500 \text{ atm}$

Since $Q > K$, the system is not at equilibrium. There is an excess of $NO_2(g)$. The reaction will proceed from right to left, consuming $NO_2(g)$ and producing $N_2O_4(g)$ until equilibrium is reached and $Q = K = 80$ atm.

(d)   Whenever one of the components that is present at equilibrium does not appear, the reaction will proceed in the direction which produces that substance, in this case from right to left to produce $N_2O_4(g)$.

One reason why K is so useful is that its value is independent of the quantities of materials in the system. *Once the composition of one equilibrium state of a system is known, the composition of any other equilibrium state of the same system containing different concentrations or pressures of the same substances at the same temperature can be found.*

EXAMPLE 14.4

An equilibrium state of a system containing $NO_2(g)$ and $N_2O_4(g)$ at an unknown temperature (not 373 K) is found to have $P_{NO_2} = 0.86$ atm and $P_{N_2O_4} = 0.12$ atm. Another system at the same temperature is also in equilibrium, and is found to have $P_{NO_2} = 0.98$ atm. Calculate the pressure of N_2O_4 in the second system.

SOLUTION

The value of the equilibrium constant can be found for the first system by using the given data:

$$K = \frac{(0.86 \text{ atm})^2}{(0.12 \text{ atm})} = 6.2 \text{ atm}$$

This value is K for any equilibrium state of these components at this temperature. For the second system, therefore:

$$6.2 \text{ atm} = \frac{(0.98 \text{ atm})^2}{P_{N_2O_4}}$$

$$P_{N_2O_4} = 0.15 \text{ atm}$$

There are dangers in comparing different systems that have different forms for K. But we can allow ourselves one useful generalization derived from observation. Suppose we begin with the pressures of all gases equal to 1 atm and the concentrations of all solutes equal to $1M$. In a system where the value of K is large ($K > 1$), the reaction proceeds from left to right, or toward completion. We can say that such a reaction is favorable. Conversely, for a system in which the value of K is small ($K < 1$), the reaction proceeds from right to left, and we can say that such a reaction is unfavorable.

It is also important to understand the role of liquids or solids in equilibrium systems. The composition of the equilibrium state of such systems does not change if we vary the amounts of liquid or solid in the system. The expression for K contains no terms for liquids and solids. Any amount of solid or liquid in such a system is enough to keep the system at equilibrium.

For example, the composition of the equilibrium state of a system containing a liquid and its vapor is described by the vapor pressure alone. We can change the value of the equilibrium constant by changing the temperature, but not by adding or removing liquid. As long as some liquid is present, the pressure of vapor is the equilibrium pressure. The pressure of gas can fall below the equilibrium pressure only if the liquid is removed (Figure 14.1).

Now look at a saturated solution, the equilibrium state of a system containing dissolved and undissolved material. If the solute is a gas, say $O_2(g)$, the reaction of interest is:

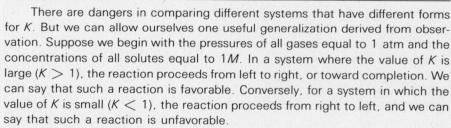

$$O_2 \rightleftharpoons O_2(g) \qquad K = P_{O_2}/[O_2]$$

The composition of the solution varies as the pressure of the undissolved solute varies. But if the solute is a pure liquid or a pure solid, a variation in the quantity of undissolved solute does not affect the composition of the system. Suppose we

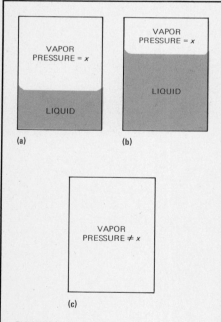

FIGURE 14.1
At a given temperature, the vapor pressure of a system at equilibrium is the same no matter how much or how little liquid is in the system. If the vapor pressure of the system in (a) is said to be x, then the vapor pressure of the system in (b), which contains more liquid, is also x. Only when the system contains no liquid, as shown in (c), can the pressure of the gas fall below the equilibrium vapor pressure.

have such a system, composed of solid glucose, $C_6H_{12}O_6$, dissolved glucose, and water. The equilibrium state of the system can be defined by the appropriate value of K for the reaction: $C_6H_{12}O_6(s) \rightleftharpoons C_6H_{12}O_6$; $K = [C_6H_{12}O_6]$. By looking at the expression for K, we can see that the quantity of solid glucose that is present at equilibrium does not influence the quantity of dissolved glucose.

The same is true for a reaction that does not occur in aqueous solution, such as the decomposition of magnesium carbonate. At 800 K, the value of the equilibrium constant for this process is about 0.95 atm. The reaction is $MgCO_3(s) \rightleftharpoons MgO(s) + CO_2(g)$ and $K = P_{CO_2}(g) = 0.95$ atm. The equilibrium state of this system at 800 K is any state that has carbon dioxide at a pressure of 0.95 atm and that also contains any amount of magnesium carbonate and magnesium oxide. Adding or removing even large amounts of either or both solids does not affect the pressure of carbon dioxide if the system is at equilibrium, because the "concentrations" of the solids do not change. The P_{CO_2} will fall below 0.95 atm at 800 K only if the $MgCO_3$ is removed completely.

14.2 THE PRINCIPLE OF LE CHATELIER

We have emphasized that a system in an equilibrium state will not change unless something is done in the surroundings to stress the system. We shall now examine the effects of stresses that cause a system to change from the equilibrium state—stresses that result in changes in the concentrations or pressures of the components of the system, or in the temperature of the system. We shall examine the relationship of the new state to the old equilibrium state, and we shall describe how to predict the way in which the system will change in reaching the new equilibrium state.

This problem was considered by Henri Le Chatelier (1850–1936), who in 1888 enunciated a remarkably simple rule that has come to be called the **principle of Le Chatelier:**

When a system in an equilibrium state undergoes a change because of an external stress, the system will reach a new equilibrium state in such a way as to minimize the external stress.

Let us look at this principle in action by studying the equilibrium state of the reaction for the fixation of nitrogen by the Haber process, a major source of nitrogen fertilizers. The reaction is:

$$3H_2(g) + N_2(g) \rightleftharpoons 2NH_3(g)$$

with the nitrogen coming from air. The expression

$$\frac{P_{NH_3}^2}{P_{N_2} P_{H_2}^3}$$

defines K and Q for this reaction.

Suppose a system containing these three gases is subjected to external stresses. By applying the principle of Le Chatelier, we can predict what the system will do in response to each external stress.

For our first example, let us add $N_2(g)$ to the system to try to increase the yield of NH_3. Now the pressure (or concentration) of $N_2(g)$ is greater than the equilibrium pressure and the system is no longer at equilibrium; $Q < K$. The principle of Le Chatelier tells us that the system will go to a new equilibrium in a way that minimizes the stress, in this case by removing some of the excess nitrogen. The reaction will proceed from left to right; some—but not all—of the added $N_2(g)$ will be converted to $NH_3(g)$. Why is the conversion incomplete? Because if all the added nitrogen were converted, the system would have an excess of ammonia and Q would be greater than K.

Now suppose that a quantity of $NH_3(g)$ is added to the system at equilibrium.

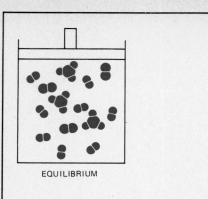

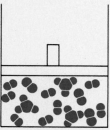

EQUILIBRIUM

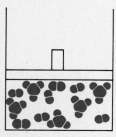

DECREASE IN VOLUME

NEW EQUILIBRIUM

FIGURE 14.2

If a system containing $H_2(g)$, $N_2(g)$, and $NH_3(g)$ is at equilibrium (top), and its volume is reduced (center), the system is no longer at equilibrium. The principle of Le Chatelier tells us that the system will move to a new equilibrium state by spontaneously undergoing reaction to decrease the number of molecules of gas (bottom).

We follow the same line of reasoning as before. This time, the stress is the addition of substances that are on the right side of the chemical equation. We find that the reaction proceeds from right to left to reach the new equilibrium state.

If we decrease the volume of the system, we find that the reaction proceeds in the direction that will produce the smallest number of moles of gas. We can see why by reasoning from the equilibrium constant. A decrease in the volume causes an increase in the pressure (or concentration) of each gaseous component. For this reaction, the numerator of the expression for Q includes a pressure term to the second power, while the denominator includes pressure terms to the fourth power. Since the pressure of each gas increases by a constant factor, the denominator of Q increases more than the numerator. The system is no longer at equilibrium as a result of the volume change. The reaction must proceed from left to right to reduce the value of the denominator of Q as the system moves to a new equilibrium state (Figure 14.2). The same line of reasoning enables us to predict that if the volume of the system increases, the reaction proceeds in the direction that increases the number of moles of gas. In this system, the reaction proceeds from right to left.

However, a change in volume does not always cause a system to move out of an equilibrium state. For example, the system represented by the reaction $2HCl(g) \rightleftharpoons H_2(g) + Cl_2(g)$ is not affected by changes in volume, because the number of moles of gas on each side of the chemical equation is the same.

When the temperature of a system at equilibrium is changed, the system usually is no longer at equilibrium. The value of K is influenced by the temperature. We can predict the way in which the system adjusts to a temperature change if we know how its equilibrium constant changes with temperature. That change depends on whether the reaction is endothermic or exothermic. Usually, the variation of K with temperature is related to the sign of $\Delta H°$. For an endothermic process, the value of K increases as the temperature increases. For an exothermic process, the value of K decreases as the temperature increases. The synthesis of ammonia by the Haber process is exothermic, so the value of K decreases as the temperature goes up. As the temperature increases, the relative amount of $NH_3(g)$ at equilibrium decreases. When the temperature of this system is raised, the reaction proceeds from right to left until the new equilibrium state is reached.

We can reach the same conclusion by using the principle of Le Chatelier. Let us write heat as part of the chemical reaction, as if it is a reagent. An exothermic reaction gives off heat, so the equation can be written as $3H_2(g) + N_2(g) \rightleftharpoons 2NH_3(g) + heat$. Heat is on the right side of the equation. To raise the temperature, we can add heat to the system. The effect is the same as adding $NH_3(g)$, the other reagent on the right. The reaction proceeds from right to left to reach equilibrium at the new temperature. Conversely, if we lower temperature we remove heat. The reaction then proceeds from left to right until equilibrium is restored. This reasoning is also consistent with the observed increase in the value of K for an exothermic process when the temperature is lowered. The same arguments in reverse can be made for an endothermic system by placing heat on the left of the reaction. Table 14.2 summarizes the changes we have discussed for the Haber process.

The reasoning we have used for systems that contain gases can also be applied to equilibrium systems that contain substances in solution. Consider the equilibrium system composed of a saturated solution of lead chloride and some solid lead chloride. The reaction is $PbCl_2(s) \rightleftharpoons Pb^{2+} + 2Cl^-$, and $K = [Pb^{2+}][Cl^-]^2$. Suppose the following things happen:

1. More solid lead chloride is added to the system. Result: The system is still at equilibrium, which is defined only by the concentrations of the dissolved ions. These concentrations are independent of the amount of solid in the system.

2. More water is added. Result: The concentration of all the dissolved

TABLE 14.2
The Principle of Le Chatelier and the Haber Process

Change in System	Subsequent Changes in Return to Equilibrium[a]		
	P_{NH_3}	P_{N_2}	P_{H_2}
add $NH_3(g)$	−	+	+
add $N_2(g)$ or $H_2(g)$	+	−	−
increase total pressure[b]	+	−	−
increase total volume[b]	−	+	+
increase temperature[b]	−	+	+

[a] Decrease is −, increase is +.
[b] The effect of a decrease is the opposite.

components is lowered, and the system is no longer at equilibrium. It returns to equilibrium in a way that minimizes the stress. More $PbCl_2(s)$ dissolves until the concentrations of the dissolved ions are back to the equilibrium value. That is, the equation proceeds from left to right.

3. The concentration of one or both of the dissolved ions is increased. Result: The system is no longer at equilibrium; $Q > K$. The reaction proceeds from right to left until the system returns to equilibrium; Some $PbCl_2(s)$ precipitates.

The concentration of the dissolved ions can be raised in several ways. The concentration of chloride ion can be raised by adding any salt that contains chloride ion, such as sodium chloride. The concentration of Pb^{2+} ion can also be increased by adding any salt that contains lead ion, such as lead nitrate. The source of the lead and chloride ions need not (and should not) be mentioned in the equation, since the accompanying sodium or nitrate ions do not take part in the reaction.

The composition of the equilibrium state is affected by the addition of either chloride ions or lead ions. If chloride ions are added, the new equilibrium state has a higher concentration of chloride ion and a lower concentration of lead ion than the original equilibrium state. The reverse is true if lead ions are added. But in either case, the value of the expression $[Pb^{2+}][Cl^-]^2$ is the same in the new equilibrium state as in the original equilibrium state.

4. The temperature of the system is raised. To predict the result, we must know the sign of $\Delta H°$. Lead chloride is observed to absorb heat when it dissolves in water. The reaction is endothermic and heat can be included on the left side of the equation. When the temperature is raised, the reaction proceeds from left to right. More solid dissolves, and $[Pb^{2+}]$, $[Cl^-]$, and K increase.

The principle of Le Chatelier permits qualitative predictions about the behavior of equilibrium systems. Now we shall see that we can also treat equilibrium systems quantitatively.

14.3 CALCULATION OF GAS PHASE EQUILIBRIUM CONSTANTS

The quantitative treatment of equilibrium systems usually requires one of two calculations. Either we must calculate the value of K at a given temperature from measurements of pressure or concentration, or we must calculate the composition of an equilibrium state, reached in a designated way, when the value of K is given. For some problems, a combination of these two kinds of calculations is necessary.

We shall start with gas phase systems for the sake of simplicity. Later, we shall try much the same calculations for aqueous solutions. We shall begin with the most basic type of equilibrium problem, in which we are given the pressure of each gas at equilibrium and are asked to calculate the value of K.

EXAMPLE 14.5

Automobile catalytic converters are designed to convert pollutants in exhaust gases to carbon dioxide and water. But it has been found that these devices can also convert sulfur dioxide in the air to sulfur trioxide, a more worrisome pollutant. A sealed bulb at 1000 K contains sulfur dioxide, $P = 0.15$ atm; sulfur trioxide, $P = 0.23$ atm; and oxygen, $P = 0.73$ atm. The system is at equilibrium. Calculate the value of K for this system.

SOLUTION

Before we begin any calculations, we must consider the first, and often the most important, part of a problem in chemical equilibrium: writing a relevant, balanced chemical reaction that relates the components of the system and describes the chemical change. The way in which the equation for the reaction is written will determine the form and value of K. Once the equation is written, we must then write the expression for K that corresponds to the equation. The third step is to substitute the data on the composition of the system at equilibrium into the expression for K and to calculate the numerical value of K.

An equation for the relevant chemical change in this example is:

$$2SO_2(g) + O_2(g) \rightleftharpoons 2SO_3(g)$$

The form of the equilibrium constant for this equation is therefore:

$$K = \frac{P^2_{SO_3}}{P^2_{SO_2}P_{O_2}}$$

Since the pressure of each component is given, we can calculate the numerical value of K:

$$K = \frac{(0.23 \text{ atm})^2}{(0.15 \text{ atm})^2(0.73 \text{ atm})} = 3.2 \text{ atm}^{-1}$$

If we had written the equation in reverse, as $2SO_3(g) \rightleftharpoons 2SO_2(g) + O_2(g)$, the answer would have been $K = (1/3.2)$ atm, the reciprocal of the answer given above, since reversing an equation inverts K. If we had written the equation with different coefficients, both the units and the value of K would be different.

Example 14.5 introduced the question of the units in which K is expressed. This is not always a simple question. We have seen that the numerical value of K depends on the way in which an equation is written. The units in which K is expressed also are determined by the equation. Example 14.6 looks at some of the possibilities.

EXAMPLE 14.6

Assuming that the quantity of each gas at equilibrium is expressed as a pressure in atmospheres, find the appropriate units of K for each of the following reactions:

(a) $N_2O_4(g) \rightleftharpoons 2NO_2(g)$
(b) $CO_2(g) + H_2(g) \rightleftharpoons CO(g) + H_2O(g)$
(c) $CO_2(g) + H_2(g) \rightleftharpoons CO(g) + H_2O(l)$
(d) $4NH_3(g) + 5O_2(g) \rightleftharpoons 4NO(g) + 6H_2O(g)$

SOLUTION

In each part, the essential first step is to write the expression for K that corresponds to the equation.

(a) $K = P^2_{NO_2}/P_{N_2O_4}$. Since the pressure of each component is expressed in atmospheres, let us substitute the symbol atm for each pressure P in the expression for K. We now have the expression $K = \text{atm}^2/\text{atm}$.

Reducing the fraction to lowest terms gives atm as the unit in which K is expressed for this reaction.

(b) $K = P_{CO}P_{H_2O}/P_{CO_2}P_{H_2}$. Here, substitution yields the fraction (atm atm)/(atm atm). This reduces to 1. For this reaction, K is dimensionless; it has no units.

(c) $K = P_{CO}/P_{CO_2}P_{H_2}$. In this reaction, H_2O is a liquid and does not appear in the expression for K. Substitution as before gives the fraction atm/(atm atm), or atm^{-1}, the units in which K is expressed for this reaction.

(d) $K = P_{NO}^4 P_{H_2O}^6/P_{NH_3}^4 P_{O_2}^5$. Substitution of the units into this expression gives the fraction (atm^4 atm^6)/(atm^4 atm^5) = atm, the units of K for this reaction.

You might encounter some problems such as part (c) of Example 14.6, where K is dimensionless. But it is more likely that you will encounter cases in which K has units. The value of K then depends on the units used to express the quantities of the components of the system. If units other than atmospheres are used, difficulties can be avoided by converting the units of such measurements to atmospheres before substituting into the expression for K. We shall follow this practice. The standard state and $\Delta G°$ refer to 1 atm; the equilibrium constant in the relationship $\Delta G° = -2.303RT$ log K is based on pressures of gases in atmospheres (Section 13.7, page 399).

EXAMPLE 14.7

A 7.24-g sample of IBr is placed in a container whose volume is 0.225 liter and is heated to 500 K. At this temperature, some of the IBr decomposes to I_2 and Br_2. All three substances are in the gas phase. The system is at equilibrium and the measured pressure of Br_2(g) in the system is 3.01 atm. Calculate the value of K.

SOLUTION

As always, we start by writing a balanced equation from the word description. One possibility is:

$$2IBr(g) \rightleftharpoons I_2(g) + Br_2(g)$$

The value of K can be calculated by finding the pressure of each gas present at equilibrium. Since the partial pressure of a gas is directly proportional to the amount of the gas, we can use the molar relationships expressed in the balanced equation. It states that bromine and iodine are formed in equal amounts. Therefore, $P_{Br_2} = P_{I_2} = 3.01$ atm.

The equation also states that 2 atm of product forms when 2 atm of IBr reacts. Therefore, the total pressure of I_2 and Br_2 that forms, $P_{I_2} + P_{Br_2} = P_{IBr}$ that reacts. The arithmetic is simple: 3.01 atm + 3.01 atm = 6.02 atm of IBr undergoing reaction.

In order to calculate the pressure of IBr present at equilibrium, we need the original pressure of IBr, which can be found by using the ideal gas equation and the data given:

$$P = \frac{(7.24 \text{ g IBr})(0.0821 \text{ liter atm mol}^{-1} \text{ K}^{-1})(500 \text{ K})}{(206.8 \text{ g IBr/mol IBr})(0.225 \text{ liter})}$$

$$= 6.39 \text{ atm}$$

The pressure of IBr at equilibrium is 6.39 atm − 6.02 atm = 0.36 atm.

This type of reasoning is quite common. If we know the quantity of a substance originally in the system and the quantity that undergoes reaction, we know the quantity of the substance that remains.

Now that we have the value of the pressure of each component at equilibrium, we can calculate the value of K:

$$K = \frac{P_{I_2}P_{Br_2}}{P_{IBr}^2} = \frac{(3.01 \text{ atm})(3.01 \text{ atm})}{(0.36 \text{ atm})^2} = 68.0$$

In Example 14.7, we found the value of K for a system starting with the pressures of the components of the system. The same kind of calculation can be done from measurements of the total pressure of the system. It is common to find the value of K in this way, since measurements of total pressure usually are the easiest that can be done on a system containing gases.

EXAMPLE 14.8

A container at 1000 K holds carbon dioxide, $P = 0.464$ atm. Graphite is added to the container, and some of the carbon dioxide is converted to carbon monoxide. At equilibrium, the total pressure in the container is 0.746 atm. Calculate the value of K.

SOLUTION

The equation of interest can be written as: $CO_2(g) + C(s) \rightleftharpoons 2CO(g)$, $K = P_{CO}^2/P_{CO_2}$. To calculate a value of K, we must know the pressures of both CO_2 and CO at equilibrium.

As this equation proceeds from left to right, two moles of CO(g) are produced for each mole of reactant CO_2(g). Therefore, the total pressure increases. If we let x equal the pressure of CO_2 in atmospheres which must react for the system to reach equilibrium, we can say from the stoichiometry of the reaction that $2x$ will be the pressure of CO at equilibrium. Thus, when the system proceeds to equilibrium, we can see that there is a decrease of x atm in the starting pressure of CO_2 and an increase of $2x$ atm in the pressure of CO, which started as zero.

	$CO_2(g) + C(s)$	$\rightleftharpoons 2CO(g)$
start	0.464 atm	0
equil	0.464 atm $- x$	$2x$

The initial conditions are expressed on the line labeled "start." The equilibrium conditions are given on the line labeled "equil." No entry is needed for the solid, since the solid does not appear in the expression for K.

The value for K can be found by solving for x, using the additional information that the total pressure is 0.746 atm. Total pressure is the sum of the partial pressures of the gases: $P_T = P_{CO_2} + P_{CO}$. If we express the values of these pressures in terms of x, we get:

$$0.746 \text{ atm} = (0.464 \text{ atm} - x) + 2x$$
$$x = 0.282 \text{ atm}$$

This value of x is used to find the pressures of the gases at equilibrium: $P_{CO_2} = 0.464$ atm $- 0.282$ atm $= 0.128$ atm, and $P_{CO} = 2(0.282 \text{ atm}) = 0.564$ atm.

We can calculate the value of K using these values for P:

$$K = \frac{(0.564 \text{ atm})^2}{(0.182 \text{ atm})} = 1.75 \text{ atm}$$

Many variations of this type of problem are possible, depending on the measurements made and the stoichiometry of the reaction. In addition, many methods can be used in the laboratory to find the data needed to calculate an equilibrium constant. One method is to calculate K by finding how much of a reactant is consumed.

EXAMPLE 14.9

When sulfur in the form of S_8 is heated to 900 K, at equilibrium the pres-

sure of S_8 falls by 29% from 1.00 atm because some of the S_8 is converted
to S_2. Find the value of K for this reaction.
SOLUTION
The chemical equation and the starting and equilibrium pressures are:

$$S_8(g) \rightleftharpoons 4S_2(g)$$

start   1.00 atm   0

equil   1.00 atm − 0.29 atm   4(0.29 atm)

The equilibrium pressures are found by this line of reasoning: The pressure of
S_8 at equilibrium is the starting pressure minus 29% of the original
1.00 atm of S_8, that is, 1.00 atm − 0.29 atm. For each mole or atmosphere of S_8 that reacts, four moles or four atmospheres of S_2 are formed. If
0.29 atm of S_8 reacts, 4(0.29 atm) of S_2 is formed.

Using these data, we can calculate the value for K:

$$K = \frac{P_{S_2}^4}{P_{S_8}} = \frac{[4(0.29 \text{ atm})]^4}{1.00 \text{ atm} - 0.29 \text{ atm}} = 2.6 \text{ atm}^3$$

14.4 CALCULATIONS FROM GAS PHASE EQUILIBRIUM CONSTANTS

The previous section dealt with the calculation of gas phase equilibrium constants
from experimental data. In this section we shall see how the equilibrium constant
that is calculated for one state of a system can be used to find the composition of
other equilibrium states of the system at the same temperature. Most of the
equilibrium calculations that you will encounter in general chemistry will require
you to find the composition of an equilibrium state from a set of starting conditions and a given value of an equilibrium constant. Here are some examples.

EXAMPLE 14.10
At 400 K, solid ammonium chloride decomposes to gaseous ammonia and
hydrogen chloride. For this process, $K = 6.0 \times 10^{-9}$ atm^2. Calculate the
equilibrium pressures for the two gases at this temperature.
SOLUTION
At first glance, it might seem that we cannot solve this problem because no
starting quantity of ammonium chloride is given. But ammonium chloride is a
solid. The quantity of ammonium chloride is therefore unimportant, as long
as there is enough present for the system to reach equilibrium. In problems
where the exact quantity of starting material is unimportant, we shall designate this quantity by the letter a. Thus, for this example, the equation and
starting conditions are:

$$NH_4Cl(s) \rightleftharpoons NH_3(g) + HCl(g)$$

start a 0   0

For the system to reach equilibrium, some ammonium chloride must decompose and some ammonia must form. Let us designate the pressure (in
atmospheres) of ammonia that forms by the letter x. From the stoichiometry
of the reaction, the pressure of hydrogen chloride that forms is also x. Now
we can describe the equilibrium conditions symbolically while ignoring the
quantity of ammonium chloride:

equil x x

We know that the expression for K includes only the equilibrium pressures of the two gases:

$$K = P_{NH_3}P_{HCl} = 6.0 \times 10^{-9} \text{ atm}^2$$

Expressing the pressures in terms of x gives:

$(x)(x) = 6.0 \times 10^{-9}$ atm^2

$x = 7.7 \times 10^{-5}$ atm $= P_{NH_3} = P_{HCl}$

Since the value of K is small, the equilibrium pressures of the two gases are also small.

The mathematics of the problem in Example 14.10 are relatively simple, because K has a relatively simple form. But you may encounter some more involved algebraic calculations in equilibrium problems. A knowledge of chemistry can enable you to simplify the calculations.

EXAMPLE 14.11

The value of K for the reaction in which two moles of nitrogen dioxide form one mole of dinitrogen tetroxide at 451 K is 5.33×10^{-3} atm^{-1}. A 0.460-g sample of nitrogen dioxide is heated to 451 K in a reaction vessel whose volume is 0.500 liter. Calculate the pressure of dinitrogen tetroxide that forms at equilibrium.

SOLUTION

We must first write the chemical equation and tabulate the starting quantities. But there is a problem. The starting quantity of NO_2 is given in units of mass, while K has units of pressure. We must therefore express the given mass of NO_2 as a pressure of NO_2 at 451 K. We substitute the data into the ideal gas equation:

$$P = \frac{(0.460 \text{ g } NO_2)(0.0821 \text{ liter atm mol}^{-1} \text{ K}^{-1})(451 \text{ K})}{(46.0 \text{ g } NO_2/\text{mol } NO_2)(0.500 \text{ liter})}$$

$$= 0.741 \text{ atm}$$

Now we can write the equation:

	$2NO_2$(g)	$\rightleftharpoons$	N_2O_4(g)
start	0.741 atm		0

To avoid fractions, let $2x$ be the pressure (in atmospheres) of NO_2 that is consumed to reach equilibrium. Then x is the pressure in atmospheres of N_2O_4 formed. The equilibrium pressures are:

equil   0.741 atm $- 2x$ x

When we substitute these equilibrium pressures into the expression for K, the result is a quadratic equation:

$$K = \frac{P_{N_2O_4}}{P_{NO_2}^2} = \frac{x}{(0.741 \text{ atm} - 2x)^2} = 5.33 \times 10^{-3} \text{ atm}^{-1}$$

This expression can be rearranged to:

$$(2.13 \times 10^{-2})x^2 - 1.0158x + 2.93 \times 10^{-3} = 0$$

and can be solved using the quadratic formula

$$x = \frac{-b \pm \sqrt{b^2 - 4ac}}{2a}$$

The value of x found in this way is 2.89×10^{-3} atm. In using the quadratic formula, two values are obtained for x. Only one makes chemical sense. In this case, the other value for x is larger than the initial pressure of NO_2. In other cases, one of the values for x may be negative, a chemical impossibility.

If you feel that the computation with the quadratic formula is tedious, you may be interested to learn that it is often unnecessary. A simpler way can be found by examining the problem and applying some chemical reasoning to it.

We start with the fact that the value for K is small. We learned earlier that a small value for K indicates that the reaction does not proceed from left to right to any appreciable extent in most circumstances; only a small amount of the starting material undergoes reaction. The unknown (x) in an equilibrium problem usually is assigned to a quantity of material that undergoes reaction. Therefore, the value of x is small relative to the quantity of starting material whenever the numerical value of K for the process is small. In this example, for instance, the value of x is quite small compared to the starting pressure of NO_2, so $2x$, the pressure of NO_2 that reacts, is small: 0.00288 atm, compared to 0.741 atm of NO_2.

In both mathematics and everyday life, the subtraction or addition of a very small number and a very large number makes no appreciable difference to the larger number. We can go on using the larger number as if no subtraction took place. We do this all the time. If you are in a traffic jam and estimate idly that you are surrounded by 800 cars (one significant figure), you will not change your estimate if three cars pull off the road. Rather than revising your estimate to 797 cars (three significant figures), you assume that 800 is still a good approximation. That approximation also holds if five cars pull onto the highway. An addition of five or a subtraction of three is small compared to the admittedly rough estimate of 800 cars. We can state this numerically as $800 \cong 800 - 3 \cong 800 + 5$. We shall now apply this reasoning to examples in chemical equilibrium.

Let us apply it to this system. We start with 0.741 atm of NO_2. Since K is small, we know that a very small amount of NO_2, designated as $2x$, undergoes reaction. Since $2x$ is much smaller than 0.741, we can say that $0.741 \cong 0.741 - 2x$. This approximation helps us greatly, since substituting into the expression for K gives a much simpler expression:

$$\frac{x}{(0.741 \text{ atm})^2} = 5.33 \times 10^{-3} \text{ atm}^{-1}$$

The previous expression was a quadratic equation. This one is not. It can be solved directly, giving $x = 2.93 \times 10^{-3}$ atm. You will note that this value is quite close to the value obtained with the quadratic formula.

There are many problems in which the method of approximation used in Example 14.11 cannot be utilized. There are guidelines to help you decide when it is appropriate to use approximations.

To begin with, approximations can be used only for sums and differences, never for products or quotients. Any time a multiplication or division must be performed, approximation is ruled out. Thus, we can say $2 + x \cong 2$ or $2 - x \cong 2$ if x is suitably small, but we cannot say $2x \cong 2$ or $2/x \cong 2$.

How small must x be? We shall say that the approximation is valid when the error in the value of a concentration that is introduced by an approximation is less than 5%. When the value of an unknown is calculated by dropping the unknown from a difference or a sum, we check on the approximation by comparing the approximated value of the term with the value that includes the calculated value of the unknown. To give a specific instance, in Example 14.11 we used the approximation that $0.741 - 2x = 0.741$ and found that $x = 2.93 \times 10^{-3}$, or 0.00293. If this value is used to verify the approximation, we find that we have said that: $0.741 - 2(0.00293) = 0.735 \cong 0.741$. The error—that is, the difference between 0.735 and 0.741—is less than 1%.

The value of K often provides a useful clue that helps us to decide whether an approximation is reasonable in an equilibrium calculation. We usually assign an unknown to represent a quantity of material that reacts in order for a system to reach equilibrium. We shall develop methods for assigning unknowns that will generally allow us to make approximations when K is either small (less than 10^{-2}) or large (greater than 10^2). If the value of K for a system is either large enough or small enough, it *may* be possible to make approximations.

In any problem where an approximation has been made, it is crucial to go back and check the approximation using the original data and the value of the unknown that has been calculated on the basis of the approximation. Only by checking in this way can you be sure that it was appropriate to make an approximation.

EXAMPLE 14.12

When phosgene, $COCl_2$, is heated to 600 K, it partially decomposes to form carbon monoxide and chlorine. The value of K for this process is 4.10×10^{-3} atm. Calculate the equilibrium pressures of the components of the system produced by a starting pressure of phosgene of 0.124 atm.

SOLUTION

The balanced chemical equation and starting conditions are:

$$COCl_2(g) \rightleftharpoons CO(g) + Cl_2(g)$$

start	0.124 atm	0	0

Let x be the pressure (in atmospheres) of $COCl_2$ that must react for the system to reach equilibrium. The equilibrium line is:

equil	0.124 atm − x	x	x

Substituting these pressures into the expression for K gives:

$$K = \frac{P_{CO}P_{Cl_2}}{P_{COCl_2}} = \frac{(x)(x)}{(0.124 \text{ atm} - x)} = 4.10 \times 10^{-3} \text{ atm}$$

To avoid the use of the quadratic formula, we turn to the approximation 0.124 atm − $x \cong$ 0.124 atm. The equation becomes:

$$\frac{x^2}{0.124 \text{ atm}} = 4.10 \times 10^{-3} \text{ atm}$$

$$x = 2.25 \times 10^{-2} \text{ atm}$$

Checking the approximation, we find that we have said that $0.124 - 0.0225 = 0.102 \cong 0.124$. This error of 20% is unacceptable. We must either use the quadratic formula for an exact solution or find a procedure for improving the approximation. Such a procedure, called the method of successive approximations, is outlined in Appendix IV.

Solution of the equation:

$$\frac{(x)(x)}{(0.124 \text{ atm} - x)} = 4.10 \times 10^{-3}$$

using the quadratic formula gives $x = 2.06 \times 10^{-2}$ atm.

So far we have dealt with chemical reactions that have relatively small equilibrium constants. Let us now consider chemical reactions with large equilibrium constants. If the equilibrium constant is large enough, the methods of approximation can also be used to simplify calculations.

In general, we can make useful approximations only when the unknown, x, is much smaller than the starting quantities of the components of the system. But when a reaction with a large equilibrium constant proceeds to equilibrium from left to right, most of the reactants are consumed. If we let the unknown be the quantity of reactant that is consumed, the value of the unknown would be relatively large. But if a reaction has a large equilibrium constant, the reverse reaction has a small one. If the reaction with the large K could be made to proceed from right to left to reach equilibrium, only a relatively small fraction of the materials on the right would be consumed, and the value of the unknown would be relatively small.

Now we have a rule to help us make approximations. The equilibrium state of the system should be approached from the direction that corresponds to the small K. Stated another way, it should be approached from the side of the equation that is closer to the final equilibrium state. When K is large, equilibrium should be approached from right to left. The final answer will be the same no matter how we approach equilibrium, since the equilibrium state is independent of the direction of approach.

To make use of this rule, we introduce an extra step into calculations for systems where K is large (usually larger than 10^2) and the starting conditions are such that the system reaches equilibrium from left to right. In the extra step, we take the reaction past the equilibrium state to completion, and then we approach the equilibrium state from right to left. In this way, the unknown that we assign will be relatively small. It should be clearly understood that the completion step in which the system passes equilibrium is entirely imaginary. It bears no relation to the actual behavior of the system. Its only purpose is to simplify algebra for us. The completion step is simply a stoichiometric calculation, sometimes of the limiting reagent type that we discussed in Section 3.3.

Now let us see this idea in action.

EXAMPLE 14.13

At 700 K, sulfur dioxide is converted almost completely to sulfur trioxide by oxygen. For a reaction of two moles of sulfur dioxide, $K = 8.24 \times 10^4$ atm^{-1}. Calculate the pressures at equilibrium when 0.490 atm of SO_2 and 0.245 atm of O_2 are brought together at 700 K.

SOLUTION

We write the relevant balanced equation and tabulate the starting quantity:

	$2SO_2(g)$	$+ O_2(g)$	$\rightleftharpoons$	$2SO_3(g)$
start	0.490 atm	0.245 atm		0

Since K is large, almost all these starting quantities of SO_2 and O_2 are consumed when the system reaches equilibrium. Let us take the reaction past equilibrium to completion. We now need to tabulate the pressures of all the substances in the system at completion. The reaction indicates that 2 mol (or 2 atm) of SO_2 reacts with 1 mol (or 1 atm) of O_2 to produce 2 mol (or 2 atm) of SO_3. The starting pressure of SO_2 in this system is exactly twice that of O_2. Therefore:

complete	0	0	0.490 atm

Now we allow the system to approach equilibrium from right to left. If $2x$ is the pressure (in atmospheres) of SO_3 that reacts when the system proceeds to equilibrium, the equilibrium line is:

equil	$2x$	x	0.490 atm $- 2x$

Since x is relatively small, 0.490 atm $- 2x \cong 0.490$ atm. Using this approximation, substitution into the expression for K gives a manageable equation:

$$K = \frac{P_{SO_3}^2}{P_{SO_2}^2 P_{O_2}} = \frac{(0.490 \text{ atm})^2}{(2x)^2(x)} = 8.24 \times 10^4 \text{ atm}^{-1}$$

$$4x^3 = \frac{(0.490 \text{ atm})^2}{8.24 \times 10^4 \text{ atm}^{-1}}$$

$$x = 0.00900 \text{ atm}$$

The validity of the approximation is checked by using the calculated value of x. We find that 0.490 atm $-$ 2(0.00900 atm) = 0.478 atm $\cong$ 0.490 atm, about a 2% error. Using the value of x gives: $P_{SO_2} =$ 0.018 atm, $P_{O_2} =$ 0.0090 atm.

If you wish, you may try to solve this problem by approaching equilibrium from left to right. This approach leads to a cubic equation that is virtually unsolvable and cannot be simplified by chemically valid approximations. By introducing a purely imaginary completion step, we have greatly simplified our calculations. But it must again be emphasized that the completion step is entirely imaginary. It tells us nothing about the way in which the system proceeds to equilibrium.

We started with a system whose components were present in stoichiometric amounts. Now it is time for a problem in which a reaction with a large equilibrium constant starts with nonstoichiometric quantities of reactants.

EXAMPLE 14.14

For the reaction in which two moles of hydrogen bromide form from hydrogen and bromine at 700 K, $K = 5.5 \times 10^8$. A mixture of 0.34 mol of hydrogen and 0.22 mol of bromine is heated to 700 K. Calculate the composition of the system at equilibrium.

SOLUTION

We write the balanced chemical reaction and tabulate the starting conditions:

	$H_2(g)$	$+ Br_2(g)$	$\rightleftharpoons$ 2HBr(g)
start	0.34 mol	0.22 mol	0

Since K is dimensionless, there is no need to convert from moles of gas to atmospheres of gas. And since K is large for the reaction as written, we introduce the extra completion step.

The reaction shows that one mole of hydrogen reacts with one mole of bromine to produce two moles of hydrogen bromide. Less bromine than hydrogen is present, so bromine is the limiting reagent. Therefore, 0.22 mol of Br_2 and 0.22 mol of H_2 are consumed and 2(0.22) mol of HBr is formed, giving the completion line:

| complete | 0.34 mol $-$ 0.22 mol | 0 | 0.44 mol |

Now we allow the system to reach equilibrium from right to left. The amount of HBr (in moles) that reacts as the system approaches equilibrium from the right is designated as $2x$. The equilibrium line is:

| equil | 0.12 mol $+ x$ | x | 0.44 mol $- 2x$ |

Since x is relatively small, we can make the approximations 0.12 mol $+ x \cong$ 0.12 mol and 0.44 mol $- 2x \cong$ 0.44 mol. The expression for K then gives:

$$K = \frac{(0.44 \text{ mol})^2}{(0.12 \text{ mol})(x)} = 5.5 \times 10^8$$

$$x = 2.9 \times 10^{-9} \text{ mol}$$

Because x is exceedingly small compared to the other quantities in the system, the approximations are valid. The value of x is now used to calculate the composition of the equilibrium state: $n_{H_2} = 0.12$ mol; $n_{Br_2} = 2.9 \times 10^{-9}$ mol; $n_{HBr} = 0.44$ mol.

The previous examples dealt with systems that start in a nonequilibrium state and move to equilibrium. The same sort of calculations can be used to deal with systems that are in equilibrium, are perturbed, and then return to equilibrium.

EXAMPLE 14.15

At 700 K, carbon dioxide and hydrogen react to form carbon monoxide and water. For this process, $K = 0.11$. A mixture of 0.45 mol of CO_2 and 0.45 mol of H_2 is heated to 700 K. (a) Find the amount of each gas at equilibrium. (b) After the system reaches equilibrium, another 0.34 mol of CO_2 and 0.34 mol of H_2 are added to the system. Find the composition of the new equilibrium state.

SOLUTION

(a) The relevant chemical equation with the data tabulated is:

	$CO_2(g)$	$+ H_2(g)$	$\rightleftharpoons$ $CO(g) +$	$H_2O(g)$
start	0.45 mol	0.45 mol	0	0

Since K for this reaction is dimensionless, we need not convert moles to units of pressure. Let x be the amount of CO_2 that undergoes reaction when the system reaches equilibrium.

| equil | 0.45 mol $- x$ | 0.45 mol $- x$ | x | x |

Substitution into the expression for K gives:

$$K = \frac{n_{CO}n_{H_2O}}{n_{CO_2}n_{H_2}} = \frac{(x)(x)}{(0.45 \text{ mol} - x)(0.45 \text{ mol} - x)} = 0.11$$

If the equation is rewritten as:

$$\frac{x^2}{(0.45 \text{ mol} - x)^2} = 0.11$$

we can see that an easy method of solution is available. Take the square root of both sides:

$$\frac{x}{(0.45 \text{ mol} - x)} = 0.33$$

and $x = 0.11 \text{ mol} = n_{CO} = n_{H_2O}$; 0.45 mol $-$ 0.11 mol $= 0.34$ mol $= n_{CO_2} = n_{H_2}$.

(b) When more CO_2 and H_2 are added to the system, we have a new set of starting conditions which can be tabulated in the usual way:

	$CO_2(g)$	$+ H_2(g)$	$\rightleftharpoons$ $CO(g)$	$+ H_2O(g)$
	0.34 mol	0.34 mol	0.11 mol	0.11 mol
start	+ 0.34 mol	+ 0.34 mol		

The principle of Le Chatelier tells us that the reaction proceeds from left to right. Once again we let x be the amount of CO_2 that must undergo reaction for the system to reach equilibrium.

| equil | 0.68 mol $- x$ | 0.68 mol $- x$ | 0.11 mol $+ x$ | 0.11 mol $+ x$ |

We again use the expression for K to find x:

$$\frac{(0.11 \text{ mol} + x)^2}{(0.68 \text{ mol} - x)^2} = 0.11$$

Taking the square root of both sides of the equation gives $x = 0.09$ mol; 0.11 mol + 0.09 mol = 0.20 mol $= n_{CO} = n_{H_2O}$. 0.68 mol $-$ 0.09 mol = 0.59 mol $= n_{CO_2} = n_{H_2}$. As a check, we can substitute the calculated amounts back into the expression for K to see if we obtain the correct value of K:

$$\frac{(0.20 \text{ mol})(0.20 \text{ mol})}{(0.59 \text{ mol})(0.59 \text{ mol})} = 0.11$$

EXAMPLE 14.16

An equilibrium mixture contains NO_2, $P = 0.28$ atm, and N_2O_4, $P = 1.1$ atm, at 350 K. The volume of the container is doubled. Calculate the equilibrium pressures of the two gases when the system reaches a new equilibrium.

SOLUTION

The reaction of interest and the conditions are:

$$N_2O_4(g) \;\rightleftharpoons\; 2NO_2(g)$$
equil   0.28 atm   1.1 atm

We use the data to calculate the value of K:

$$K = \frac{P_{NO_2}^2}{P_{N_2O_4}} = \frac{(1.1 \text{ atm})^2}{(0.28 \text{ atm})} = 4.3 \text{ atm}$$

According to Boyle's law, doubling the volume of the container of a gas decreases the pressure of each gas by a factor of two. The system is now out of equilibrium. We find the pressure of each gas by dividing the equilibrium pressure by two, and tabulate the data as starting conditions:

start $\dfrac{0.28 \text{ atm}}{2}$ $\dfrac{1.1 \text{ atm}}{2}$

We can easily verify that these are no longer equilibrium pressures by substitution into the expression for K.

The principle of Le Chatelier tells us that the reaction in a system whose pressure is lowered proceeds in the direction that produces more moles of gas. For this reaction, the direction is from left to right: N_2O_4 is consumed. We let x be the pressure of N_2O_4 consumed and write a new equilibrium line:

$$N_2O_4(g) \;\rightleftharpoons\; 2NO_2(g)$$
equil $\dfrac{0.28 \text{ atm}}{2} - x$ $\dfrac{1.1 \text{ atm}}{2} + 2x$

We use the expression for K to solve for x:

$$K = \frac{(0.55 \text{ atm} + 2x)^2}{(0.14 \text{ atm} - x)} = 4.3 \text{ atm}$$

We must use the quadratic formula to find x; $x = 0.045$ atm. Thus, the new equilibrium pressures are $P_{NO_2} = 0.55$ atm $+ 2x = 0.55$ atm $+ 2(0.045 \text{ atm}) = 0.64$ atm and $P_{N_2O_4} = 0.14$ atm $- x = 0.14$ atm $- 0.045$ atm $= 0.095$ atm.

We can check these values by substituting into the expression for K:

$$K = \frac{(0.64 \text{ atm})^2}{0.095 \text{ atm}} = 4.3 \text{ atm}$$

You now have been exposed to many problems in gas phase equilibrium. The lessons learned in solving these problems can be applied to the next step, problems dealing with equilibrium systems in aqueous solutions. Most of these problems have features in common. A review of these features provides a guide to some of the best methods of solving such problems.

The invariable first step is the writing of a balanced chemical equation that expresses the chemical changes that occur in the system.

Next, to keep the calculation orderly, we tabulate the initial conditions. The statement of the problem may give the quantities of some or all of the substances in the system. After making any necessary unit conversions, we list each value under the formula of the appropriate substance in the equation.

The statement of the problem will say whether the system starts in an equilibrium state. If it does, we look to see whether a numerical value can be listed under every substance in the equation that appears in the expression for K, remembering that only gases and dissolved substances appear in the expression. If there are no unknowns on this line, the data can be substituted into the expression for K and the value of K can be determined.

If the initial conditions given for the system are not equilibrium conditions, we must determine the direction in which the reaction proceeds to reach equilibrium. If one of the components is initially absent, the reaction must proceed in the direction to produce that component. In other cases, the relative values of Q and K indicate the direction.

We then assign an unknown. The unknown usually is related to the quantity of one of the substances that is consumed when the system moves from the initial nonequilibrium state to equilibrium. The quantity is expressed in units appropriate to the problem.

Once the initial conditions of a nonequilibrium state are tabulated on a "start" line and an unknown is assigned, the composition of the equilibrium state is tabulated on an "equil" line. Below each substance in the chemical equation that appears in the expression for K, we write a term for its pressure or concentration at equilibrium. The form of the equilibrium line—specifically, the number of unknowns on the line—can provide clues to the solution of the problem.

To find the values of the unknowns on the equilibrium line, we must be given an equal number of bits of data about the composition of the equilibrium state. For example, when the equilibrium line has one unknown and we are given one bit of data, the value of K, we can find the value of the unknown. In the next example, we develop an "equil" line with two unknowns, and we are given two bits of data about the composition of the equilibrium state. We use both of them to find the values of the two unknowns.

EXAMPLE 14.17

For the decomposition of one mole of carbon tetrachloride vapor to graphite and chlorine at 700 K, $K = 0.76$ atm. Calculate the starting pressure of carbon tetrachloride that will produce a total pressure of 1.0 atm at equilibrium.

SOLUTION

The only information given about the initial conditions is that we start with an unknown quantity of CCl_4. Let us represent this starting quantity by an unknown, y, the required starting pressure of CCl_4 in atmospheres. The relevant reaction and the starting conditions are:

$$CCl_4(g) \rightleftharpoons 2Cl_2(g) + C(s)$$

start y 0

The system now proceeds to equilibrium as an unknown quantity of CCl_4 reacts. Let x be this quantity in atmospheres. The equilibrium line is:

equil $y - x$ $2x$

The equilibrium line has two unknowns, so we must look for two bits of data that are given about the composition of the equilibrium state. They are the value of K and the total pressure at equilibrium. First we eliminate one of the unknowns from the "equil" line by using the total pressure:

$$P_{\mathbf{T}} = P_{CCl_4} + P_{Cl_2} = (y - x) + 2x = 1.0 \text{ atm}$$

which can be rearranged to $y = 1.0 \text{ atm} - x$. Substituting the expression for y in terms of x on the "equil" line gives:

$$\text{equil} \quad \begin{array}{ccc} CCl_4 & \rightleftharpoons & 2Cl_2(g) + C(s) \\ 1.0 \text{ atm} - 2x & & 2x \end{array}$$

Now that the equilibrium line has only one unknown, we can substitute pressure terms into the expression for K to solve for x:

$$K = \frac{(2x)^2}{(1.0 \text{ atm} - 2x)} = 0.76 \text{ atm}$$

Since K is neither particularly large nor particularly small, an approximation cannot be used. Applying the quadratic formula gives $x = 0.29$ atm. Since $y = 1.0$ atm $- x$, then $y = 1.0$ atm $- 0.29$ atm $= 0.71$ atm, the starting pressure of carbon tetrachloride needed to produce a total pressure of 1.0 atm at equilibrium.

14.5 SOLUTIONS OF SPARINGLY SOLUBLE SUBSTANCES: THE SOLUBILITY PRODUCT

We can use the principles of chemical equilibrium to get quantitative information about any equilibrium system that consists of water and a sparingly soluble ionic solid. The reasoning that was used in Sections 14.3 and 14.4 for gas phase equilibrium can also be used for equilibrium systems of these solutions.

We limit ourselves to sparingly soluble substances because our treatment of chemical equilibrium is based on the ideal solution approximation. The concentration of ions in saturated solutions of appreciably soluble ionic substances is so great that the ideal solution approach gives poor results. Such systems can be studied quantitatively, but only by methods that are beyond our present scope.

As you might expect, the first step in treating equilibrium systems of solutions of sparingly soluble substances is to write the chemical equation describing the appropriate reaction. The second step is to find the equilibrium constant for the reaction. Sometimes the equilibrium constant must be calculated from the data that are given. But in other cases, the equilibrium constant can be found in standard reference works. Once the equilibrium constant is known, it can be used to find the composition of equilibrium states that are reached from various starting conditions.

All the systems we are considering here consist of aqueous solutions of sparingly soluble ionic solids. The chemical equations for such systems all have the same form: The undissolved solid is on the left and the dissociated ions are on the right. Some examples are:

silver chloride: $AgCl(s) \rightleftharpoons Ag^+ + Cl^-$; $K_{sp} = [Ag^+][Cl^-]$
barium fluoride: $BaF_2(s) \rightleftharpoons Ba^{2+} + 2F^-$; $K_{sp} = [Ba^{2+}][F^-]^2$
ferric hydroxide: $Fe(OH)_3(s) \rightleftharpoons Fe^{3+} + 3OH^-$; $K_{sp} = [Fe^{3+}][OH^-]^3$
calcium phosphate: $Ca_3(PO_4)_2(s) \rightleftharpoons 3Ca^{2+} + 2PO_4^{3-}$; $K_{sp} = [Ca^{2+}]^3[PO_4^{3-}]^2$

None of the expressions for the equilibrium constants of these reactions has a denominator, since the left sides of the equations have only solids, and solids are not included in the expression for K. The expression for K includes the product of the concentrations of at least two ions, because the right side of each equation contains at least two ions, an anion and a cation. Each concentration has an exponent equal to the coefficient of the ion in the equilibrium.

The equilibrium constant for the dissolution of a sparingly soluble ionic solid in water is called the **solubility product**. It is abbreviated K_{sp}. The expression for the K_{sp} for each solid is shown next to the equilibrium above. Values have been

TABLE 14.3
Solubility Products of Some Sparingly Soluble Ionic Solids at Room
Temperature

Substance	Solubility Product[a]	Substance	Solubility Product[a]
AgBr	5.2×10^{-13}	CuCN	1.0×10^{-11}
AgCN	1.2×10^{-16}	CuCl	3.2×10^{-7}
Ag_2CO_3	8.2×10^{-12}	CuI	1.1×10^{-12}
AgCl	1.6×10^{-10}	Cu_2S	3×10^{-48}
Ag_2CrO_4	1.1×10^{-12}	CuS	6×10^{-36}
AgI	8.5×10^{-17}	$Fe(OH)_3$	4×10^{-38}
$AgIO_3$	3.0×10^{-8}	Hg_2Cl_2	1.3×10^{-18}
$AgNO_2$	1.6×10^{-4}	Hg_2I_2	2.5×10^{-26}
Ag_3PO_4	1.6×10^{-18}	HgI_2	8.8×10^{-12}
Ag_2S	6×10^{-50}	HgS	4×10^{-53}
Ag_2SO_3	1.5×10^{-14}	$MgCO_3$	1.6×10^{-6}
Ag_2SO_4	1.6×10^{-5}	$Mg(OH)_2$	8.9×10^{-12}
$Al(OH)_3$	2×10^{-32}	$Mg_3(PO_4)_2$	1.0×10^{-13}
$AlPO_4$	5.8×10^{-19}	$Mn(OH)_2$	1.9×10^{-13}
$Au(OH)_3$	5.5×10^{-46}	$Ni(OH)_2$	6.5×10^{-18}
$BaCO_3$	5.1×10^{-9}	NiS	3×10^{-19}
$BaCrO_4$	1.2×10^{-10}	$PbBr_2$	4.6×10^{-6}
BaF_2	1.0×10^{-6}	$PbCO_3$	3.3×10^{-14}
$BaSO_4$	1.3×10^{-10}	$PbCl_2$	1.6×10^{-5}
$CaCO_3$	4.8×10^{-9}	PbI_2	7.1×10^{-9}
CaF_2	1.7×10^{-10}	$Pb(IO_3)_2$	1.2×10^{-13}
$Ca(OH)_2$	5.5×10^{-6}	PbO_2	3.2×10^{-66}
$Ca_3(PO_4)_2$	2.0×10^{-29}	$PbSO_4$	1.2×10^{-8}
$CaSO_4$	1.2×10^{-6}	SnS	1×10^{-25}
$Cd(OH)_2$	5.9×10^{-15}	$SrCO_3$	7.0×10^{-10}
CdS	2×10^{-28}	SrF_2	7.9×10^{-10}
$Co(OH)_2$	2×10^{-16}	$SrSO_4$	3.2×10^{-7}
CoS	4×10^{-21}	$Zn(CN)_2$	2.6×10^{-13}
$Cr(OH)_2$	1.0×10^{-17}	$ZnCO_3$	1.4×10^{-11}
$Cr(OH)_3$	6×10^{-31}	$Zn(OH)_2$	1.8×10^{-14}
CuBr	5.2×10^{-9}	ZnS	2×10^{-24}

[a] All ion concentrations are in moles per liter.

determined for the solubility products of many substances. Table 14.3 lists some
of them.

The solution to problems that use solubility products begins with the writing
of correct chemical equations. To write the equation, you must know which ions
are formed when a given ionic substance dissolves in water. It is advisable at this
point to review the information on polyatomic ions in Table 2.4 and the procedure
for writing net ionic equations in Section 2.8. We shall use net equations
throughout.

Once the correct chemical equation is written, you must be able to find the
value of the equilibrium constant, usually from a table containing solubility
product data. Because our discussion is on an elementary level, some explanation
about the way in which we shall use the solubility product, K_{sp}, is necessary.

In addition to using the ideal solution approximation, we also make some
simplifying assumptions about the chemistry of solutions of ionic substances. We
shall assume that the only process of concern in a system containing water and a
sparingly soluble material is the one in which the material dissolves. Other
chemical reactions can occur in such a solution; an ion can react with water or
with other ions. Such reactions are especially common for cations of the heavy
metals in the center of the periodic table. For the moment, we shall ignore these
complications.

It is important not to confuse the term *solubility product* with the term *solubility*. The solubility product is the equilibrium constant of a specific reaction. The solubility is the quantity of a substance that dissolves in a given quantity of water. These two terms have different numerical values for a given reaction, but they are related. The relationship between the two numbers can be calculated.

EXAMPLE 14.18

A saturated solution of lead sulfate is prepared by shaking an excess of the solid with water until no more dissolves. By measuring the starting quantity of lead sulfate and the amount remaining undissolved, we find that 1.1×10^{-4} mol of $PbSO_4$ dissolves in 1 liter of water at 300 K to produce a saturated solution. Calculate the value of K_{sp} for lead sulfate.

SOLUTION

Simpler methods are available, but we shall use the same method that we have utilized for previous equilibrium calculations.

The relevant chemical equation is:

$$PbSO_4(s) \rightleftharpoons Pb^{2+} + SO_4^{2-}$$

start a 0   0

We use the symbol a for the starting quantity of $PbSO_4(s)$. The exact value is not needed, since it is not included in the expression for K_{sp}.

Let x be the amount (in moles) of lead sulfate that dissolves in 1 liter of water when the system comes to equilibrium, so that the concentrations of dissolved species will have units of moles per liter (M). We shall calculate and express all equilibrium constants for aqueous solutions using the molarities of the dissolved species.

equil x x

We need not list the quantity of solid lead sulfate remaining after the system reaches equilibrium, but we can see that it is $a - x$.

In this problem, we are given that $x = 1.1 \times 10^{-4}$ mol/liter (M). The value for K_{sp} can be determined by substituting the given value of x into the expression for K_{sp}:

$$K_{sp} = [Pb^{2+}][SO_4^{2-}] = (x)(x) = (1.1 \times 10^{-4} \text{ mol/liter})^2$$
$$= 1.2 \times 10^{-8} \text{ mol}^2/\text{liter}^2$$

You should note that the solution of the problem took for granted one of the key steps, the correct interpretation of the stoichiometry of the reaction. By this time, you should be able to discern that one mole of lead sulfate dissolves to produce one mole of Pb^{2+} ion and one mole of SO_4^{2-} ion. You must always pay close attention to stoichiometry in these calculations.

EXAMPLE 14.19

A saturated solution of calcium fluoride is prepared by shaking an excess of the solid with water until no more dissolves. By measuring the starting quantity of calcium fluoride and the amount remaining undissolved, we find that 7.0×10^{-4} mol of CaF_2 dissolves in 2 liters of water at 300 K to produce a saturated solution. Calculate the value of K_{sp} for calcium fluoride.

SOLUTION

The relevant chemical reaction and starting conditions are:

$$CaF_2(s) \rightleftharpoons Ca^{2+} + 2F^-$$

start a 0   0

To write the equilibrium line, we must keep in mind the fact that the expression for K_{sp} includes the molarities of the substances in the equation. By letting x, the unknown, stand for the solubility of CaF_2—that is, the number of moles of calcium fluoride that dissolve in 1 liter of water—we obtain expressions for the molarities of the dissolved species in terms of x. The chemical equation tells us that two moles of fluoride ion and one mole of calcium ion are formed when one mole of calcium fluoride dissolves. The equilibrium line reads:

$$CaF_2(s) \rightleftharpoons Ca^{2+} + 2F^-$$

equil $\qquad\qquad x \qquad\quad 2x$

The value of x can be found from the information given in the problem. If 7.0×10^{-4} mol of CaF_2 dissolves in 2 liters of water, then half that amount, x, dissolves in 1 liter. Thus, $x = (7.0 \times 10^{-4}/2)$ mol/liter $= 3.5 \times 10^{-4}$ mol/liter. Using this value of x, we can find the value of K_{sp}:

$$K_{sp} = [Ca^{2+}][F^-]^2 = (x)(2x)^2 = 4(3.5 \times 10^{-4} \text{ mol/liter})^3$$
$$= 1.7 \times 10^{-10} \text{ mol}^3/\text{liter}^3$$

It is easy to go wrong in this problem unless careful attention is paid to details. The chemical equation says that two moles of fluoride ion are formed when one mole of solid CaF_2 dissolves. If x mol/liter of solid dissolves, the concentration of F^- ion is $2x$ mol/liter. The expression for K_{sp} states that the concentration of F^- ion must be raised to the second power in the calculations; $2x$ to the second power is $(2x)^2$, or $4x^2$. Carelessness can give an erroneous value.

EXAMPLE 14.20
The solubility product of lead iodide is 7.1×10^{-9} mol^3/liter^3 at 298 K. Calculate the solubility of this salt in moles per liter and find the molarity of the ions in a saturated solution of PbI_2.
SOLUTION
The relevant equation and the starting conditions are:

$$PbI_2(s) \rightleftharpoons Pb^{2+} + 2I^-$$

start $\qquad a \qquad\qquad 0 \qquad\quad 0$

You can think of these starting conditions as describing an experimental procedure for finding the solubility of lead iodide. Some lead iodide is placed in water and the amount that dissolves when the system reaches equilibrium is measured. Let x be the amount (in moles) of solid lead iodide that dissolves in 1 liter of water. The equilibrium concentrations (in moles per liter) of the ions then are:

equil $\qquad\qquad\qquad\quad x \qquad\quad 2x$

The value of the unknown can be found by substituting the concentrations into the expression for K_{sp}:

$$K_{sp} = [Pb^{2+}][I^-]^2 = (x)(2x)^2 = 7.1 \times 10^{-9} \text{ mol}^3/\text{liter}^3$$
$$x = 1.2 \times 10^{-3} \text{ mol/liter}$$

The value of x is the solubility of lead iodide, the number of moles of solid PbI_2 that dissolve in 1 liter of water to form a saturated solution. The value of x can be used to find the molarity of each ion in the solution: $[Pb^{2+}] = x = 1.2 \times 10^{-3}$ mol/liter and $[I^-] = 2x = 2.4 \times 10^{-3}$ mol/liter.

Let us now suppose that the concentration of I^- ion in the equilibrium system of Example 14.20 is raised by adding I^- ions from a different source than PbI_2; for example, from a highly soluble iodide salt. The principle of Le Chatelier tells us that the system will return to equilibrium, with the reaction proceeding from right to left. Some solid lead iodide will precipitate from the saturated solution, removing part of the added I^- ion. Stated another way, the solubility of lead iodide is lower in a solution that contains some iodide ion from another source, than it is in pure water. We get the same result if Pb^{2+} ions from another source are in the solution; the solubility of lead iodide is lowered. The Pb^{2+} ion and the I^- ion from other sources are called *common ions* of lead iodide. These are the two ions produced when lead iodide dissolves in water. The *common ion effect* is a general rule that can be applied to sparingly soluble ionic solids: *The solubility of an ionic solid in a solution containing one or more of its common ions is lower than its solubility in pure water.* However, the presence of unrelated ions in water does not affect the solubility of an ionic substance.

If the value of the solubility product is known, it is possible to find how the presence of common ions in a solution affects the solubility of a sparingly soluble ionic solid.

EXAMPLE 14.21

Use the data given in Example 14.20 to calculate the solubility of lead iodide in a 0.10M solution of sodium iodide.

SOLUTION

In a 0.10M solution of sodium iodide, $[Na^+] = 0.10M$ and $[I^-] = 0.10M$. If lead iodide is added, the common ion is I^-. The relevant reaction and starting conditions are:

$$PbI_2(s) \rightleftharpoons Pb^{2+} + 2I^-$$

start a 0   0.10M

The reaction must proceed from left to right for the system to reach equilibrium. If we let x be the number of moles of solid PbI_2 that dissolve in 1 liter of the sodium iodide solution, the molarities of the ions in solution at equilibrium are:

equil x 0.10M + 2x

If we substitute the concentrations into the expression for K_{sp}, we get:

$$K_{sp} = (x)(0.10M + 2x)^2 = 7.1 \times 10^{-9}M^3$$

This equation is not easy to solve exactly. We can simplify the calculations by making an approximation. Since K_{sp} is small, the amount of lead iodide that dissolves is small, especially because of the common ion effect. The value of x is thus small, so we can make the approximation that 0.10M + 2$x \cong$ 0.10M. Now the equation becomes:

$$(x)(0.10M)^2 = 7.1 \times 10^{-9}M^3$$
$$x = 7.1 \times 10^{-7} \text{ mol/liter}$$

If you compare this value with the solubility of lead iodide in pure water that was calculated in Example 14.20, you can see that lead iodide is less soluble in a solution containing iodide ion.

You will note that Na^+ was omitted from the calculations, because it is not part of the net ionic equation; it is not a common ion in this solution.

We can also calculate the effect of the common ion Pb^{2+} in solution on the solubility of lead iodide.

EXAMPLE 14.22

Calculate the solubility of lead iodide in a 0.10M solution of lead nitrate.

SOLUTION

A solution of lead nitrate contains Pb^{2+} ions and NO_3^- ions. The common ion is Pb^{2+}, whose initial concentration is 0.10M. Following the same procedure as in Example 14.20 and 14.21, we write:

$$PbI_2 \rightleftharpoons Pb^{2+} + 2I^-$$

start a 0.10M 0

equil   0.10M + x 2x

Substitution into the expression for K_{sp} gives:

$$(0.10M + x)(2x)^2 = 7.1 \times 10^{-9}M^3$$

Making the approximation that $0.10M + x \cong 0.10M$, the equation becomes $(0.10M)(2x)^2 = 7.1 \times 10^{-9}M^3$ and $x = 1.3 \times 10^{-4}$ mol/liter.

Comparing this result to those of the two previous examples, we see that the presence of either Pb^{2+} ion or I^- ion reduces the solubility of lead iodide. But a given concentration of Pb^{2+} reduces the solubility less than the same concentration of I^- ion. Since two moles of I^- ions and only one mole of Pb^{2+} ions are formed when one mole of PbI_2 dissolves, there is a smaller effect from the Pb^{2+} ion.

Precipitation Reactions

The solubility product can be of great practical importance when it is applied to precipitation reactions, which are processes in which a sparingly soluble material comes out of solution. Both in the laboratory and in industrial processes, it is often necessary to know whether and to what extent a precipitation will occur. The solubility product can help us to obtain that information.

For example, the K_{sp} of AgCl is $1.6 \times 10^{-10}M^2$ at 298 K. Thus, a solution that contains Ag^+ ion and Cl^- ion, from whatever source, is at equilibrium only when $[Ag^+][Cl^-] = 1.6 \times 10^{-10}M^2$. In a solution in which $[Ag^+][Cl^-] < 1.6 \times 10^{-10}M^2$, the solution is not saturated and the system is not at equilibrium. If solid silver chloride is added, some of it dissolves until the product of the concentrations reaches the value of K_{sp}.

If the product of the two ion concentrations exceeds the value of K_{sp}, $[Ag^+][Cl^-] > 1.6 \times 10^{-10}M^2$, the system is not at equilibrium. It proceeds spontaneously to equilibrium by lowering the concentration of the dissolved ions. A precipitate of silver chloride forms, spontaneously and rapidly. Precipitation stops when the concentrations of the ions are reduced and the product of the concentrations equals $1.6 \times 10^{-10}M^2$.

The solubility product of lead iodate, $Pb(IO_3)_2$, at 291 K is $1.2 \times 10^{-13}M^3$. The expression $[Pb^{2+}][IO_3^-]^2$ must equal $1.2 \times 10^{-13}M^3$ at equilibrium. If a solution is prepared in which the concentration of Pb^{2+} ion multiplied by the square of the concentration of IO_3^- ion is greater than this value, lead iodate will precipitate.

A common method of preparing a solution in which the solubility product of a sparingly soluble ionic solid is exceeded is to mix two solutions. One solution contains a relatively soluble salt of the desired cation and the other solution contains a relatively soluble salt of the desired anion. For example, when solutions of appropriate concentrations of sodium chloride and silver nitrate are mixed, the value of the K_{sp} of silver chloride, $[Ag^+][Cl^-]$, is exceeded and a precipitate of AgCl forms almost immediately after mixing. The chemical equation we write to describe precipitation includes neither the nitrate ion nor the sodium ion, since neither undergoes change. They simply remain in solution and are not included in a net ionic equation.

So far, we have sought quantitative information about aqueous solutions of sparingly soluble ionic compounds. In many precipitation reactions, the information of interest is qualitative. Typically, we want to know whether a given precipitate forms when certain solutions of ions are mixed. If more than one precipitate is possible when solutions are mixed, we want to know which precipitate forms. As before, the information conveyed by solubility product data can simplify our calculations.

EXAMPLE 14.23

A 15-cm³ volume of a $0.050M$ solution of magnesium bromide is mixed with a 25-cm^3 volume of a 0.050M solution of lead nitrate at 298 K. Will a precipitate form?

SOLUTION

This question can be answered without a complete equilibrium calculation by some chemical reasoning.

The magnesium bromide solution contains two ions, Mg^{2+} and Br^-. The lead nitrate solution contains two ions, Pb^{2+} and NO_3^-. Two precipitates are possible by an exchange of ions: lead bromide, $PbBr_2$, and magnesium nitrate, $Mg(NO_3)_2$. We must determine if one of these precipitates will form and, if so, which one. In Section 12.3, page 357, we set forth some general solubility rules which you should review. Since all common nitrates are soluble, magnesium nitrate is not likely to precipitate. On the other hand, lead bromide is sparingly soluble. You will find a value for its K_{sp} in Table 14.3. If a precipitate forms in the system, it will be lead bromide, $PbBr_2(s)$.

At 298 K, K_{sp} for $PbBr_2$ is $4.6 \times 10^{-6}M^3$. Therefore, at equilibrium, $[Pb^+][Br^-]^2$ cannot exceed this value. If the concentrations of the ions in a solution are such that the value of K_{sp} is exceeded, a precipitate will form. We must therefore calculate the concentrations of the two ions of interest when the solutions are mixed and substitute those concentrations into the expression for K_{sp}. If the calculated value exceeds the given value of K_{sp}, a precipitate of lead bromide will form.

When two solutions are mixed, the volume of the resulting solution is approximately the sum of the volumes of the original solutions. That is, $V = V_1 + V_2$, where V is the volume of the resulting solution and V_1 and V_2 are the volumes of the original solutions. The increase in volume results in a decrease in the concentrations of the ions in the original solutions. The new molarity of an ion from solution 1, M_1', is:

$$M_1' = \frac{n_1}{V_1 + V_2}$$

where n_1 is the number of moles of the ion present. If M_1 is the original concentration of this ion, then $n_1 = M_1 V_1$. Therefore:

$$M_1' = \frac{M_1 V_1}{V_1 + V_2}$$

Similarly, the new molarity of an ion from solution 2, M_2', is given by:

$$M_2' = \frac{M_2 V_2}{V_1 + V_2}$$

Now we can find the concentrations of the ions of interest in the solution after the mixing. The original concentration of Br^- ion in a 0.050M solution of $MgBr_2$ is 2(0.050M) = 0.10M. Therefore:

$$M_1' = [Br^-] = \frac{(0.10M)(15 \text{ cm}^3)}{15 \text{ cm}^3 + 25 \text{ cm}^3} = 3.8 \times 10^{-2}M$$

The original concentration of Pb^{2+} ion in a 0.050M solution of $Pb(NO_3)_2$ is 0.050M. The concentration after mixing is:

$$M_2' = [Pb^{2+}] = \frac{(0.050M)(25 \text{ cm}^3)}{15 \text{ cm}^3 + 25 \text{ cm}^3} = 3.1 \times 10^{-2}M$$

Substituting these values into the expression for the solubility product of lead bromide gives:

$$[Pb^{2+}][Br^-]^2 = (3.1 \times 10^{-2}M)(3.8 \times 10^{-2}M)^2 = 4.29 \times 10^{-5}M^3$$

This value is larger than $4.6 \times 10^{-6}M^3$, the value of K_{sp}. The system is

not at equilibrium. It will proceed to equilibrium by forming a precipitate of
$PbBr_2$. The answer to the question is yes.

Solubility product data can tell us many things about precipitation reactions.
For example, it is possible to have a solution yield first one solid and then another
by careful addition of solutions of selected ions. This procedure is called the
method of *selective precipitation*. It is important in the synthesis of inorganic
substances and in both quantitative and qualitative analysis of mixtures of ions in
solutions. Solubility product data enable us to carry out selective precipitation
calculations.

EXAMPLE 14.24

A solution contains a mixture of silver ion, $[Ag^+] = 0.10M$, and mercury(I)
ion, $[Hg_2^{2+}] = 0.10M$. You are asked to separate the two ions. How would
you go about it?
SOLUTION
The method of selective precipitation can be used to separate the two ions.
The values of the solubility products indicate that both these cations form
iodide salts that are sparingly soluble. For AgI, $K_{sp} = 8.5 \times 10^{-17}M^2$ at
298 K. For Hg_2I_2, $K_{sp} = 2.5 \times 10^{-26}M^3$ at 298 K. (Note that the mer-
cury(I) cation in solution is unusual; it exists as a polyatomic ion
$[Hg—Hg]^{2+}$.) To use the method of selective precipitation successfully, we
must meet several conditions. We must first add iodide ion to the solution to
cause precipitation of either silver iodide or mercury(I) iodide. If the precipitation of one of these iodides is virtually complete before the other begins to
precipitate, the two ions can be separated.

To see whether the method of selective precipitation is feasible, we
must determine which iodide precipitates first. Then we determine the concentration of this cation that still remains in solution when the second iodide
begins to precipitate.

We can determine which iodide precipitates first by using appropriate
K_{sp} expressions to calculate the concentration of iodide ion that must be
reached for each solid to begin precipitating. For the sake of simplicity, we
shall assume that we can add a solution of a salt, such as sodium iodide,
that is concentrated enough so that its addition causes a negligible change
in the volume of the solution.

For silver iodide, $[Ag^+][I^-] = 8.5 \times 10^{-17}M^2$ at equilibrium. Since the
$[Ag^+] = 0.10M$, we can find the concentration of I^- ion at equilibrium:
$[I^-] = 8.5 \times 10^{-17}M^2/0.10M = 8.5 \times 10^{-16}M$.

As iodide ion is added gradually to the solution, its concentration increases until it reaches this value. The system is then saturated with respect
to silver iodide. If more iodide is added, silver iodide will precipitate.

Will mercury(I) iodide precipitate first? A similar calculation will tell us:

$$K_{sp} = [Hg_2^{2+}][I^-]^2 = (0.10M)[I^-]^2 = 2.5 \times 10^{-26}M^3$$
$$[I^-] = 5.0 \times 10^{-13}M$$

Precipitation of silver iodide begins at a lower concentration of iodide
ion than does precipitation of mercury(I) ion. Therefore, selective precipitation may be possible. But the procedure will be practical only if most of the
silver iodide has already precipitated when the concentration of iodide ion
becomes great enough to cause the precipitation of mercury(I) ion to begin.
We have determined that mercury(I) iodide begins precipitating when the
concentration of I^- exceeds $5.0 \times 10^{-13}M$. Now we must calculate the
concentration of silver ion remaining in solution when the iodide ion concentration reaches this value. We can find out by substituting this value of $[I^-]$
into the expression for K_{sp} of silver iodide:

SOLUTION, PRECIPITATION, AND CAVES

Vast, majestic underground cave systems, such as Carlsbad Caverns and Mammoth Cave, with all their incredibly beautiful and varied decorative formations, are created by a process of solution and precipitation that takes place over many centuries. If we strip away the complexities, we find a process that is simple in principle.

Caves are created because rocks are soluble in natural waters. Limestone, dolomite, gypsum, and anhydrite, in particular, dissolve rather readily under the proper conditions. Limestone, for example, is made up primarily of calcium carbonate, $CaCO_3$. Pure, running water always has some carbon dioxide from the atmosphere dissolved in it. When water comes in contact with limestone, it can dissolve the calcium carbonate to form a solution of calcium bicarbonate:

$$CaCO_3(s) + CO_2 + H_2O \rightleftharpoons Ca^{2+} + 2HCO_3^-$$

The extent to which this reaction takes place is extremely sensitive to the partial pressure of CO_2. Caves can form rapidly because the partial pressure of CO_2 is much greater in air trapped in soil than in the atmosphere. As water trickles through the soil, it is exposed to the higher partial pressure of CO_2, and its ability to dissolve calcium carbonate is increased manyfold. The less saturated the water and the faster it flows, the more rapidly solution proceeds. Thus, as calcium carbonate goes into solution, more water flows into the resulting space, and the cave grows more rapidly.

As more calcium carbonate is dissolved, the natural solution comes closer to saturation. If the water should emerge into an air-filled chamber, where the partial pressure of CO_2 is lower than in the soil, precipitation can occur to form stalactites, stalagmites, and other familiar features of caves. The precipitation often occurs because the partial pressure of CO_2 within the cave is lower than the partial pressure in the soil. In addition, the air in the cave may be warmer than ambient air above. The higher temperature reduces the solubility of CO_2 in water. As CO_2 comes out of solution, the aqueous solution of $Ca(HCO_3)_2$ becomes saturated, and $CaCO_3$ precipitates.

If the water drips from a cave ceiling, a precipitate of $CaCO_3$ can form both on the ceiling and on the floor below. The cylindrical deposit that forms on the ceiling as water falls drop by drop is called a stalactite. The bulbous deposit that forms on the floor where the splashing drops fall is called a stalagmite. Given enough time, the stalactite and stalagmite can meet and form a column. Changing conditions over long periods of time produce the fantastic variety of formations found in many caves. Standing pools of water can become saturated with $CaCO_3$ as CO_2 is released, precipitating limestone in shapes resembling lily pads. "Cave coral" consists of similar precipitates formed where moisture adheres to cave floors or walls. Water running down cave walls can lead to the growth of delicate mineral traceries called draperies or flowstone. In some

caves, precipitation can occur on such a scale that large passages that were formed by dissolution of rock are closed again by the deposition of minerals from solution.

$K_{sp} = [Ag^+][I^-] = [Ag^+](5.0 \times 10^{-13}M) = 8.5 \times 10^{-17}M^2$
$[Ag^+] = 1.7 \times 10^{-4}M$

We now know that when there is finally enough iodide ion in the solution to cause the precipitation of mercury(I) iodide, almost all the Ag^+ has precipitated as AgI. Only $(1.7 \times 10^{-4}M/0.10M) \times 100\% = 0.17\%$ of the original Ag^+ remains in solution. In other words, 99.83% of the silver ion is removed from solution as a precipitate of silver iodide before mercury(I) iodide begins to precipitate. If the silver iodide is separated just at this point, we are left with a solution of mercury(I) ion that is only slightly contaminated by a trace impurity of silver ion.

In this case, selective precipitation works.

As you saw in Example 14.24, it is sometimes necessary to think about more than one solubility product at a time. When a system contains several species, it is at equilibrium only when the concentration of the dissolved ions satisfies all the relevant solubility product expressions. Consider an aqueous solution that contains ions from two sparingly soluble ionic solids. If the two substances have an ion in common, the interdependence of the solubility products can be used to calculate the composition of the system at equilibrium, as in the next example.

EXAMPLE 14.25

A solution is saturated with respect to strontium carbonate, $SrCO_3$, $K_{sp} = 7.0 \times 10^{-10}M^2$, and strontium fluoride, SrF_2, $K_{sp} = 7.9 \times 10^{-10}M^3$. The $[CO_3^{2-}]$ is found to be $1.2 \times 10^{-3}M$. Calculate $[F^-]$.

SOLUTION

Since we are given a value for the K_{sp} of strontium carbonate and we are also given the concentration of carbonate ion, we can calculate the equilibrium concentration of strontium ion in the solution:

$K_{sp} = [Sr^{2+}][CO_3^{2-}] = [Sr^{2+}](1.2 \times 10^{-3}M) = 7.0 \times 10^{-10}M^2$
$[Sr^{2+}] = 5.8 \times 10^{-7}M$

Any other equilibrium expression that includes the strontium cation must also be satisfied by this concentration. The solution is also saturated with respect to strontium fluoride:

$K_{sp} = [Sr^{2+}][F^-]^2 = 7.9 \times 10^{-10}M^3$

Substituting the equilibrium concentration of Sr^{2+} into this expression gives:

$(5.8 \times 10^{-7}M)[F^-]^2 = 7.9 \times 10^{-10}M^3$
$[F^-] = 3.7 \times 10^{-2}M$

14.6 QUALITATIVE ANALYSIS BY SELECTIVE PRECIPITATION

It has been a traditional practice in introductory chemistry laboratory courses to have students perform a qualitative analysis of an unknown mixture of cations in solution, using methods based on solubility differences. In recent years, instruments have been developed to do such analyses almost automatically. However, a review of the traditional method of qualitative analysis is still quite useful and instructive.

In the ideal method of separating and identifying two or more cations in solution, we must first find an anion that forms a sparingly soluble solid with only one of the cations. A source of this anion is added to the solution, and the precipitate that forms is removed from the solution. Then a second anion, which forms an insoluble solid with only one of the remaining cations, is added, and the second precipitate is removed. This procedure is repeated until all the ions have been separated and identified.

Let us suppose that we have a solution that may contain any or all of the cations Ag^+, Cd^{2+}, and Ba^{2+}. The first step in the analysis is to consult a table of solubility product data, such as Table 14.3. A complete table of K_{sp} values will show that of the three cations, only silver forms a sparingly soluble chloride, since no K_{sp} values will be listed for either cadmium chloride or barium chloride.

If a solution containing chloride ion is added to the solution to be tested, a precipitate of silver chloride forms if Ag^+ ion is in the solution. We remove the precipitate from the solution. The small K_{sp} indicates that almost all the silver ion will precipitate if a reasonable amount of Cl^- ion is added.

Again, we consult a complete table of K_{sp} values. It shows that cadmium sulfide is sparingly soluble. There is no listing for barium sulfide; it is soluble. We add a solution containing sulfide ion to the solution that may contain Cd^{2+} and Ba^{2+}. If Cd^{2+} is present, cadmium sulfide precipitates. The solid is removed and we are left with a solution that may contain only the barium ion. Another consultation with the K_{sp} table suggests the addition of a solution containing carbonate anion to precipitate barium carbonate to test for Ba^{2+} in solution.

Careful selection of the sequence in which the solutions are added is vital for the success of this procedure. For example, silver sulfide is insoluble. If the concentration of silver ion is not greatly reduced by the chloride precipitation of the first step, a precipitate will form when sulfide ion is added in the second step to test for cadmium, even if cadmium is absent. That precipitate would be silver sulfide.

You might have to test a solution containing more than three ions. It is usually not possible to find an anion that forms a sparingly soluble solid with only one of the cations. In such a situation, you will find it convenient to group the cations by the order in which they are precipitated.

Group I includes cations that are precipitated by chloride ion. It includes Ag^+, Hg_2^{2+}, and Pb^{2+}. When a source of chloride ion is added to a solution that may contain one or more of these cations, a precipitate that may include $AgCl$, Hg_2Cl_2, and $PbCl_2$ forms. This precipitate is separated from the solution and appropriate procedures are used to separate and identify the mixture of precipitates.

Group II includes cations that form sulfides of extremely low solubility. Suitable procedures are available for keeping $[S^{2-}]$ very low and for precipitating only those cations that form very insoluble sulfides. Some cations in group II are Cu^{2+}, Cd^{2+}, Hg^{2+}, and Sn^{4+}.

Group III includes cations that form less insoluble sulfides than group II. The concentration of S^{2-} ion is increased by making the solution alkaline, so this group of cations also includes those that form insoluble hydroxides. Some important cations in group III are Zn^{2+}, Ni^{2+}, and Co^{2+}, which precipitate as sulfides, and Al^{3+}, Fe^{3+}, and Cr^{3+}, which precipitate as hydroxides.

Group IV includes cations that form insoluble carbonates. These cations derive from alkaline-earth metals. They include Mg^{2+}, Ca^{2+}, Sr^{2+}, and Ba^{2+}.

Group V includes all the cations that remain after all the others have precipitated. The important members of group V are NH_4^+, Na^+, and K^+.

There are procedures that allow separation and identification of all the cations in each of the groups. These procedures are based in part on solubility differences. They also are based on other types of aqueous equilibria, which we shall discuss in Chapters 15 and 18.

EXERCISES

14.1 On the basis of nonlaboratory observations, classify as many of the following processes as you can as spontaneous or nonspontaneous. Indicate which processes cannot be classified solely on the basis of simple observations. (a) $CH_4(g) + 2O_2(g) \rightarrow CO_2 + 2H_2O$, (b) a shrub blossoms in the spring, (c) $2NO(g) \rightarrow N_2 + O_2$, (d) ice melts at 298 K, (e) a weight loss occurs after not eating.

14.2 Write the expressions for the equilibrium constants of the following processes: (a) $2SO_2(g) + O_2(g) \rightleftharpoons 2SO_3(g)$, (b) $Br_2(l) + F_2(g) \rightleftharpoons 2BrF(g)$, (c) $Ca_3(PO_4)_2(s) \rightleftharpoons 3Ca^{2+} + 2PO_4^{3-}$, (d) $3HNO_2 \rightleftharpoons NO_3^- + H^+ + H_2O + 2NO(g)$.

14.3 The following processes differ only in the states of the reactants and products. Write the expression for the K of each: (a) pure ammonia and pure hydrogen chloride form ammonium chloride, a salt, (b) pure ammonia and hydrochloric acid form a solution of ammonium chloride, (c) a solution of ammonia and pure hydrogen chloride form a solution of ammonium chloride, (d) ammonia in solution and hydrochloric acid form ammonium chloride in solution.

14.4 Write the expression for K for each of the following phase changes: (a) the sublimation of carbon dioxide, (b) the condensation of steam, (c) the melting of ice.

14.5 Using atmospheres (atm) as the units for quantity of gas and moles per liter (M) as the units for concentration of solutes, find the units of the expressions for K of the following processes: (a) $2NO(g) \rightleftharpoons N_2(g) + O_2(g)$, (b) $Cl_2(g) \rightleftharpoons Cl_2$, (c) $H_2(g) + O_2(g) \rightleftharpoons H_2O_2(l)$, (d) $5Fe^{2+} + MnO_4^- + 8H^+ \rightleftharpoons 5Fe^{3+} + Mn^{2+} + 4H_2O$.

14.6 The reaction $2NO(g) + Cl_2(g) \rightleftharpoons 2NOCl(g)$ takes place readily at 700 K. The equilibrium constant $K = 0.26$ atm^{-1} at this temperature. Predict the behavior of each of the following systems, which contain the given pressures of the three gases in atmospheres: (a) $P_{NO} = P_{Cl_2} = 0.10$, $P_{NOCl} = 0.016$, (b) $P_{NOCl} = P_{Cl_2} = 2.4$, $P_{NO} = 3.0$, (c) $P_{NO} = P_{Cl_2} = P_{NOCl} = 1.0$, (d) $P_{NO} = 12$, $P_{NOCl} = 0.011$, $P_{Cl_2} = 0.00$.

14.7 The equilibrium constant for the reaction $Br_2(g) \rightleftharpoons 2Br(g)$ at 1400 K is 3.03×10^{-2} atm. The pressure of $Br_2(g)$ in a system at equilibrium at this temperature is 2.48 atm. Find the pressure of $Br(g)$ in the system.

14.8 The equilibrium constant for the reaction $CO(g) + Cl_2(g) \rightleftharpoons COCl_2(g)$ is 22.7 atm^{-1} at 670 K. At equilibrium the pressure of $COCl_2$ is found to be 1.05 atm and the pressure of CO is found to be twice the pressure of Cl_2. Find the pressure of CO at equilibrium.

14.9 When the system $H_2(g) + CO_2(g) \rightleftharpoons H_2O(g) + CO(g)$ is at equilibrium at 1260 K it is found to contain 0.19 mol of CO_2, 22.6 mol of H_2, and 2.6 mol of CO and of H_2O. Find the value of K.

14.10 At 1000 K the composition of the system $2CO(g) \rightleftharpoons CO_2(g) + C(s)$ is found to be 0.420 atm of CO, 0.101 atm of CO_2, and 1 kg of C. Calculate the value of K for the reaction.

14.11 Propose an explanation for Henry's law (Equation 12.9) using the principle of Le Chatelier.

14.12 Predict the direction of chemical change in the system $H_2(g) \rightleftharpoons 2H(g)$, which is originally at equilibrium, when it is stressed in each of the following ways: (a) H_2 is added, (b) H is added, (c) the temperature is increased, (d) the volume of the container is decreased.

14.13 The reaction between H_2 and CO_2 to form CO and H_2O in the gas phase is exothermic. Predict the changes that take place when this system, originally at equilibrium, is stressed in each of the following ways: (a) CO_2 is removed, (b) CO is removed, (c) the temperature is decreased, (d) the pressure of the system is increased, (e) the volume of the system is increased.

14.14 At a given temperature a system containing $O_2(g)$ and some oxides of nitrogen can be described by the following reactions: $2NO(g) + O_2(g) \rightleftharpoons 2NO_2(g)$, $K = 10^4$ atm^{-1}, and $2NO_2(g) \rightleftharpoons N_2O_4(g)$, $K = 0.10$ atm^{-1}. A pressure of 1 atm of N_2O_4 is placed in a container at this temperature. Predict which, if any, component will be present at a pressure greater than 0.2 atm at equilibrium.

14.15 Both of the reactions in Exercise 14.14 are exothermic. The temperature of the system at equilibrium is raised. Predict the change in the relative amount of each of the four components of the system when it again reaches equilibrium.

14.16 The container of the system of Exercise 14.14 is compressed to half its original volume. Predict the change in the relative amount of each of the components of the system when it again reaches equilibrium.

14.17 Diamonds are denser than graphite, as we discussed in Section 9.2. Use the principle of Le Chatelier to account for the use of high pressures for the synthesis of artificial diamonds.

14.18 The reaction $N_2(g) + O_2(g) \rightleftharpoons 2NO(g)$ is endothermic. The formation of NO in automobile engines is a serious cause of air pollution. Another cause of air pollution is incomplete combustion of gasoline to CO and soot particles. The combustion of gasoline is much cleaner at very high temperatures. Suggest possible drawbacks to the operation of automobile engines at very high temperatures.

14.19 A system at equilibrium at 1300 K contains HCN, $P = 0.0011$ atm, and N_2 and H_2 each with $P = 5.5$ atm. Write a suitable reaction for the system and calculate the value of K.

14.20 At 1300 K, a system at equilibrium contains $BrF_5(g)$, $P = 5.7$ atm; $Br_2(g)$, $P = 0.21$ atm; and $F_2(g)$, $P = 0.31$ atm. Write a suitable reaction for the system and evaluate its K.

14.21 The value of K for the reaction $2NO(g) + Cl_2(g) \rightleftharpoons 2NOCl(g)$ at 500 K is 52.0 atm^{-1}. At equilibrium the $P_{NO} = 0.095$ atm and the $P_{Cl_2} = 0.171$ atm. Find the pressure of NOCl.

14.22 The equilibrium constant for the reaction $H_2(g) + S(g) \rightleftharpoons H_2S(g)$ at 1200 K is 23.1 atm^{-1}. Equimolar amounts of H_2 and S are mixed in a container and the system is allowed to reach equilibrium. The pressure of H_2S is 3.51 atm. Find the pressure of H_2 and of S at equilibrium.

14.23 The equilibrium constant for the reaction $NO_2Cl(g) \rightleftharpoons NO_2(g) + \frac{1}{2}Cl_2(g)$ at 410 K is measured by placing 2.50 atm of NO_2Cl into a container and allowing the system to come to equilibrium. The pressure of Cl_2 is then found to be 1.20 atm. Find the value of K for the reaction.

14.24 For each of the following decompositions indicate the relationship, if any, between the change in the total pressure of the system and the extent of decomposition of the reactant.
(a) $2NO_2(g) \rightleftharpoons 2NO(g) + O_2(g)$,
(b) $2HBr(g) \rightleftharpoons H_2(g) + Br_2(g)$,
(c) $2N_2O_5(g) \rightleftharpoons 4NO_2(g) + O_2(g)$.

14.25 An initial pressure of 1.000 atm of NO_2 is placed in a container at 1000 K. At equilibrium the total pressure is 1.463 atm. Find K for the reaction $2NO_2(g) \rightleftharpoons 2NO(g) + O_2(g)$.

14.26 A mixture of 0.373 atm of $NO(g)$ and 0.310 atm of $Cl_2(g)$ is prepared at 500 K. The reaction $2NO(g) + Cl_2(g) \rightleftharpoons 2NOCl(g)$ takes place. The total pressure at equilibrium is 0.544 atm. Find K for the reaction.

14.27 A pressure of 1.00 atm of H_2 and 1.00 atm of P_2 is placed in a container at 800 K. When the equilibrium $P_2(g) + 3H_2(g) \rightleftharpoons 2PH_3(g)$ is established, the total pressure is 1.91 atm. Calculate K for the reaction.

14.28 Sometimes equilibrium constants for gas phase reactions are given in units of concentration, moles per liter, instead of in atmospheres. Such an equilibrium constant is sometimes designated as K_c. Find the units of K_c for each of the following reactions:
(a) $N_2(g) + 3H_2(g) \rightleftharpoons 2NH_3(g)$,
(b) $2SO_3(g) \rightleftharpoons 2SO_2(g) + O_2(g)$,
(c) $CO_2(g) + H_2(g) \rightleftharpoons CO(g) + H_2O(g)$.

*__14.29__ Assuming ideal gas behavior for all the gases in an equilibrium system, derive a relationship between K (in atmospheres) and K_c (in moles per liter) as a function of the temperature and the change in the number of moles of gas as a result of the reaction.

14.30 A 0.252-mol amount of $NO_2(g)$ is placed in a sealed 2.00-liter container and heated to 700 K. At equilibrium it is found that there remains 0.179 mol of $NO_2(g)$. Calculate K_c for the reaction $2NO_2(g) \rightleftharpoons 2NO(g) + O_2(g)$.

14.31 The equilibrium constant for the reaction $S_2(g) + C(s) \rightleftharpoons CS_2(g)$ is 9.40 at 900 K. Calculate the pressure of the two gases at equilibrium when 1.42 atm of S_2 and excess C(s) come to equilibrium. Repeat the same calculation for the equilibrium state reached when 1.42 atm of CS_2 comes to equilibrium at this temperature.

14.32 The equilibrium constant for the reaction $NH_4CO_2NH_2(s) \rightleftharpoons 2NH_3(g) + CO_2(g)$ is 2.5×10^{-4} atm^3 at 310 K. Calculate the equilibrium pressures of the two gases.

14.33 The equilibrium constant for the dissociation of chlorine: $Cl_2(g) \rightleftharpoons 2Cl(g)$ at 1200 K is 2.48×10^{-5} atm. Find the pressure of Cl at equilibrium when the pressure of Cl_2 is 1.00 atm. Find the pressure of Cl at equilibrium when the initial pressure of Cl_2 is 1.00 atm.

14.34 A sample of $CaCO_3(s)$ is introduced into a sealed container of volume 0.654 liter and heated to 1000 K until equilibrium is reached. The equilibrium constant for the reaction $CaCO_3(s) \rightleftharpoons CaO(s) + CO_2(g)$ is 3.9×10^{-2} atm at this temperature. Calculate the mass of CaO that is present at equilibrium.

14.35 At 300 K the equilibrium constant for the reaction $2NH_3(g) \rightleftharpoons N_2(g) + 3H_2(g)$ is 1.7×10^{-6} atm^2. (a) Calculate the composition of the equilibrium state reached from an original pressure of NH_3 of 1.0 atm. (b) Calculate the fraction of NH_3 that decomposes.

*__14.36__ The equilibrium constant for the reaction $2NO(g) + I_2(g) \rightleftharpoons 2NOI(g)$ at 500 K is 7.48×10^{-5} atm^{-1}. Calculate the pressure of NOI at equilibrium from initial pressures of 0.688 atm of NO and of I_2.

14.37 A mixture of $NH_3(g)$, $P = 0.50$ atm, and $H_2(g)$, $P = 0.40$ atm, is placed in a reaction vessel at 300 K. Using the data given in Exercise 14.35, find the pressure of $N_2(g)$ at equilibrium.

14.38 The equilibrium constant for the reaction $H_2(g) + CO_2(g) \rightleftharpoons CO(g) + H_2O(g)$ is 1.59 at 1260 K. Determine the composition of the equilibrium state when 0.449 atm of H_2 and 0.449 atm of CO_2 are brought together at this temperature.

*__14.39__ At 700 K the equilibrium constant for the decomposition of two moles of sulfur trioxide to two moles of sulfur dioxide and one mole of O_2 is 1.21×10^{-5} atm. Find the composition of the equilibrium state that forms when 0.098 atm of SO_3 is kept at this temperature.

14.40 At 1000 K the equilibrium constant for the formation of one mole of $Br_2(g)$ from gaseous bromine atoms is 28 000 atm^{-1}. Find the equilibrium pressure of bromine atoms remaining from an initial pressure of 0.0024 atm of Br at this temperature.

*__14.41__ When one mole of carbon dioxide and excess graphite are heated to 1000 K, some carbon monoxide forms; $K = 1.74$ atm. Calculate the total pressure at equilibrium formed from an initial pressure of carbon dioxide of 1.00 atm and excess graphite.

14.42 The equilibrium constant for the formation of two moles of ClF_3 from Cl_2 and F_2 at 700 K is 6.25×10^9 atm^{-2}. A mixture of 0.20 atm of Cl_2 and 0.60 atm of F_2 at 700 K is prepared. Determine the composition of the equilibrium state.

14.43 Determine the composition of the equilibrium state of the system described in Exercise 14.42 when the initial pressures of Cl_2 and F_2 are each 1.00 atm at 700 K.

14.44 A system at equilibrium at 700 K contains 0.18 atm of H_2, 0.29 atm of methane, and some graphite. An additional 0.11 atm of hydrogen is introduced. Calculate the composition of the system when it returns to equilibrium.

14.45 The equilibrium constant for the reaction $SO_2(g) + NO_2(g) \rightleftharpoons SO_3(g) + NO(g)$ is 3.0. Find the amount of NO_2 that must be added to 2.4 mol of SO_2 in order to form 1.2 mol of SO_3 at equilibrium.

14.46 Write solubility product expressions for the following solids: (a) $BaCO_3$, (b) CaF_2, (c) Ag_2SO_4, (d) $Fe(OH)_3$, (e) Hg_2Cl_2.

*__14.47__ Indicate the units of each of the solubility products in Exercise 14.46.

14.48 Find the solubility (in moles per liter) in water of the following solids: (a) AgI, (b) BaF_2, (c) Cu_2S.[a]

*__14.49__ Find the molarity of each ion in a saturated solution of $Mg_3(PO_4)_2$.[a]

14.50 The $[Bi^{3+}]$ in a saturated solution prepared from $BiPO_4$ is $3.6 \times 10^{-12} M$. Find the K_{sp} of bismuth phosphate.

[a] The data necessary for this exercise can be found in Table 14.3.

14.51 The $[Pb^{2+}]$ in a saturated solution prepared from lead fluoride is $1.9 \times 10^{-3}M$. Find the K_{sp} of lead fluoride.

14.52 The solubility of magnesium fluoride in water is 1.2×10^{-3} mol/liter. Find its solubility product.

14.53 The solubility of calcium arsenate, $Ca_3(AsO_4)_2$, in water is 9.0×10^{-5} mol/liter. Find its solubility product.

14.54 A system at equilibrium contains solid silver phosphate and PO_4^{3-} in solution. The $[PO_4^{3-}] = 0.048M$. Find the $[Ag^+]$.[a]

14.55 Find the solubility of $CaCO_3$ (a) in pure water and in the following solutions: (b) $0.20M$ calcium nitrate, (c) $0.20M$ sodium carbonate.[a]

14.56 Find the solubility of BaF_2 (a) in pure water and in the following solutions: (b) $0.10M$ barium nitrate, (c) $0.10M$ sodium fluoride, (d) $0.10M$ sodium nitrate.[a]

14.57 Find the solubility of Ag_3PO_4 in (a) pure water and in the following solutions: (b) $0.10M$ silver nitrate, (c) $0.10M$ sodium phosphate.[a]

14.58 Find the $[Ag^+]$ necessary to just begin precipitation of AgBr from a solution in which the $[Br^-] = 8.9 \times 10^{-3}M$.[a]

14.59 Find the $[S^{2-}]$ necessary to just begin precipitation of Cu_2S from a solution in which the $[Cu^+] = 0.10M$.[a]

14.60 A 10.0-cm^3 volume of $0.0110M$ lead nitrate solution is mixed with a 10.0-cm^3 volume of $0.000823M$ sodium iodate solution. Describe the chemical change that takes place.[a]

14.61 A volume of 250 cm^3 of $0.20M$ silver nitrate solution is mixed with a volume of 250 cm^3 of $0.30M$ sodium carbonate solution. Calculate the amount of precipitate that forms and the concentrations of the ions in solution at equilibrium.[a]

* 14.62 You have 1.00 liter of a solution of $0.10M$ magnesium nitrate from which you wish to precipitate 99.0% of the Mg^{2+} ion. Calculate the volume of $0.10M$ sodium carbonate solution required for this purpose.[a]

14.63 A solution contains a mixture of chloride ion, $[Cl^-] = 0.10M$, and iodide ion, $[I^-] = 0.010M$. Evaluate the feasibility of a separation of the ions from the solution by selective precipitation carried out by the slow addition of concentrated silver nitrate solution. (Neglect volume changes.) Be sure to calculate the remaining concentration in solution of the first ion that precipitates when the second one just begins to precipitate.[a]

14.64 A solution contains a mixture of lead(II) ion, $[Pb^{2+}] = 0.10M$ and copper(I) ion, $[Cu^+] = 0.010M$. Repeat the evaluation of Exercise 14.63 for a separation of these ions by a selective precipitation by addition of concentrated sodium iodide solution.[a]

14.65 A solution is saturated with respect to both magnesium carbonate and silver carbonate. The $[Mg^{2+}] = 2.2 \times 10^{-2}M$. Find the $[Ag^+]$.[a]

14.66 A solution is saturated with respect to both barium fluoride and barium sulfate. The $[F^-] = 7.5 \times 10^{-4}M$. Find the $[SO_4^{2-}]$.[a]

14.67 Write net ionic equations for the precipitation reactions of the group III cations.

14.68 Using only the data in Table 14.3, propose a procedure for the separation of the following cations: Na^+, Ag^+, Pb^{2+}, Ba^{2+}, Zn^{2+}, and Co^{2+}.

14.69 Using only the data in Table 14.3, propose a procedure for the separation of the following anions: NO_3^-, NO_2^-, F^-, Cl^-, CN^-, and SO_4^{2-}.

ACIDS AND BASES

15

Whether or not you have studied chemistry before, you almost certainly have picked up some knowledge about acids and bases. Most of us know that an acid is something that is put into automobile batteries, that causes heartburn, and that is neutralized by products advertised on television. Folk wisdom adds that an acid is something that tastes sour, and that a ''strong'' acid will burn the skin. Bases are not as well-defined in everyday life, but you may have heard that a base feels soapy, that it can also burn the skin, that it can be used to unclog drains, and that it has something to do with washing. The words ''alkali'' and ''alkaline'' also are associated with bases in some way. This chapter will expand your knowledge of acids and bases as they are defined and used in chemistry.

15.1 DEFINITIONS OF ACIDS AND BASES

When we first mentioned acids and bases in Chapter 2, page 38, we used a temporary working definition: An acid is a substance that releases hydrogen ions (H^+) in aqueous solution, and a base is a substance that produces hydroxide ions (OH^-) in aqueous solution. This definition was first proposed in 1890 by Svante Arrhenius (1859–1927). It was generally accepted for 30 years and was used to develop many important quantitative relationships in the chemistry of acids and bases.

The Arrhenius definition was a major advance that helped to correlate a great deal of experimental data, in a field not previously noted for precise definitions. Originally, ''acid'' was used to refer to substances with a sour taste; the word itself, in English and other languages, is derived from words meaning ''sour.'' A base, or alkali, was any substance that destroyed an acid and formed a salt and water. Given this background, the Arrhenius definition was an important step forward.

If you assume that an acid gives rise to H^+ ions and a base gives rise to OH^- ions, the reaction of one with the other is explained by the reaction

$$H^+ + OH^- \rightleftharpoons H_2O$$

However, as time passed, the Arrhenius definition began to seem less satisfactory. For one thing, there seems to be two kinds of bases. Metal hydroxides such as sodium hydroxide produce hydroxide ion in water by ionic dissociation. Bases such as ammonia, NH_3, produce hydroxide ions in aqueous solution by undergoing a reaction with water:

$$NH_3 + H_2O \rightleftharpoons NH_4^+ + OH^-$$

But more important, the Arrhenius definition is narrow. It defines acids and bases only in aqueous solution.

The Brønsted Theory

A more general definition of acids and bases was proposed in 1923, primarily by J. N. Brønsted (1879–1947). According to Brønsted, **an acid is a substance that can donate a proton, and a base is a substance that can accept a proton.** This definition can be represented by the general chemical reaction:

$$A \rightleftharpoons B + H^+$$

which does not attempt to show electrical charge balance. In this equation, A is the acid, B is the base, and H^+, a hydrogen atom without an electron, is a proton. Together, A and B are called a **conjugate acid-base pair.** There is a close relationship between A and B. Remove a proton from A and you have B. Add a proton to B and you have A. We can call B the conjugate base of A, and we can call A the conjugate acid of B. You should note that in this general representation, we do not show any electrical charges that A and B may bear. For example, if A is a neutral molecule, then B will be an anion of charge -1. If A is a cation of charge $+1$, then B will be a neutral molecule. In a reaction of specific species we must always include the necessary charges.

The Brønsted definition is more general than the Arrhenius definition. It does not refer to any specific solvent, and it does not require any differentiation between the various classes of substances defined as either acids or bases. We can simply say that any substance that can lose a proton is an acid. Among the acids are electrically neutral substances such as HCl, HCN, and H_2SO_4; anions such as HSO_4^-, HCO_3^-, $H_2PO_4^-$, and HPO_4^{2-}; and cations such as NH_4^+ and $CH_3NH_3^+$. Similarly, the bases include electrically neutral substances such as NH_3 and CH_3NH_2, anions such as OH^-, HCO_3^-, and HPO_4^{2-}; and cations such as $Mg(OH)^+$ and $Al(OH)^{2+}$.

While the relationship $A \rightleftharpoons B + H^+$ is a very general definition of an acid and a base, there is a problem in applying this definition to acid-base systems in solution. The problem arises because free protons, H^+, cannot exist in solution to any great extent. In many cases, the protons interact with the solvent, which acts as a base, by accepting protons.

The same reasoning can be applied to a base in solution. Since there are no free protons in solution, the base must obtain a proton from a proton source. Very often, the proton source is the solvent.

We can see that acid-base reactions in solution are not simply processes in which an acid loses a proton or a base gains a proton. Instead, we must regard all acid-base reactions in solution as proton transfer reactions. Any such reaction includes two acid-base conjugate pairs—the original acid-base pair, plus another pair to accept the proton from the acid or donate the proton to the base. Often, this second acid-base pair is derived from the solvent. Table 15.1 lists some conjugate acid-base pairs.

Each substance in the acid column is converted to its partner in the base column by removal of a proton. Each substance in the base column is converted to its partner in the acid column by the addition of a proton. Any two conjugate acid-base pairs can be combined in an acid-base reaction. For example, in the equation

$$HCl + CN^- \rightleftharpoons Cl^- + HCN$$

TABLE 15.1
Conjugate Acid-Base Pairs

Name of Acid	Acid	Base	Name of Base
hydrochloric acid	HCl	Cl^-	chloride
hydrocyanic acid	HCN	CN^-	cyanide
hypochlorous acid	$HClO$	ClO^-	hypochlorite
sulfuric acid	H_2SO_4	HSO_4^-	hydrogen sulfate
hydrogen sulfate	HSO_4^-	SO_4^{2-}	sulfate
ammonium	NH_4^+	NH_3	ammonia
ammonia	NH_3	NH_2^-	amide
water	H_2O	OH^-	hydroxide
oxonium or hydronium	H_3O^+	H_2O	water
carbonic acid	H_2CO_3	HCO_3^-	bicarbonate
phosphoric acid	H_3PO_4	$H_2PO_4^-$	dihydrogen phosphate
dihydrogen phosphate	$H_2PO_4^-$	HPO_4^{2-}	hydrogen phosphate
hydrogen phosphate	HPO_4^{2-}	PO_4^{3-}	phosphate
methyl ammonium	$CH_3NH_3^+$	CH_3NH_2	methyl amine

HCl and Cl⁻ are one conjugate acid-base pair and CN⁻ and HCN are the other pair. The proton donated by the acid HCl is accepted by the base CN⁻ in the forward reaction and the proton accepted by the Cl⁻ is donated by the HCN in the reverse reaction. In the equation

$$HClO + NH_3 \rightleftharpoons ClO^- + NH_4^+$$

HClO and ClO⁻ are one conjugate acid-base pair, and NH_3 and NH_4^+ are the other pair. The proton donated in the forward reaction by the acid HClO is accepted by the base NH_3, and the proton accepted in the reverse reaction by the ClO⁻ is donated by the NH_4^+.

You will note that some substances appear in both columns of Table 15.1. These substances are both acids and bases. Each can donate a proton or can accept a proton. Such substances are said to be *amphoteric*. Bicarbonate ion is an amphoteric substance. In the reaction $HCO_3^- + CN^- \rightleftharpoons CO_3^{2-} + HCN$, bicarbonate ion acts as an acid, donating a proton to cyanide ion, the base. In the reaction $HCO_3^- + HCN \rightleftharpoons H_2CO_3 + CN^-$, bicarbonate ion acts as a base, accepting a proton from hydrogen cyanide, the acid. Bicarbonate ion can even react with itself:

$$HCO_3^- + HCO_3^- \rightleftharpoons H_2CO_3 + CO_3^{2-}$$

In this reaction, the transfer of a proton from one bicarbonate to the other forms both CO_3^{2-}, the conjugate base, and H_2CO_3, the conjugate acid. A reaction in which a substance reacts with itself is called a *disproportionation*.

It is of special importance that water is amphoteric. Water appears in both columns of Table 15.1. Water can act as an acid, donating a proton to a base, as in $CN^- + H_2O \rightleftharpoons HCN + OH^-$ or $NH_3 + H_2O \rightleftharpoons NH_4^+ + OH^-$. Water can also act as a base, accepting a proton, as in $H_2O + HCN \rightleftharpoons H_3O^+ + CN^-$.

As an amphoteric substance, water can react with itself, by a process that can be represented as

$$2H_2O \rightleftharpoons H_3O^+ + OH^-$$

This process, in which one water molecule gains a proton from the other, is called the *autoprotolysis* of water. We shall discuss it in more detail later.

Since water is by far the most important solvent for inorganic materials, aqueous solutions will dominate our discussion of acid-base reactions. But it is worth mentioning that NH_3, ammonia, which is also used as a solvent for inorganic materials at low temperatures, also appears in both columns of Table 15.1. Ammonia, like water, is amphoteric and can undergo autoprotolysis:

$$2NH_3 \rightleftharpoons NH_2^- + NH_4^+$$

If you look back over the last few pages, you will see that the Brønsted definition of acids and bases is more general and more useful than the Arrhenius definition. The Brønsted definition can be summarized in a few sentences. An acid is a proton donor, and a base is a proton acceptor. We emphasize the relationship by designating conjugate acid-base pairs. Acid-base reactions in solution require two conjugate acid-base pairs, because free protons do not exist in solution.

The Lewis Theory

The Brønsted definition has its shortcomings. Since it defines an acid as a proton donor, it excludes from the category of acids substances that have no protons to donate. This limitation was overcome by a more general definition of acids and bases, based on electronic structure, which was proposed by G. N. Lewis.

By the Brønsted definition, a substance must accept a proton to be classified as a base. In other words, the base must form a chemical bond with a proton. Such a bond requires two electrons. Since the proton has no electrons, the base must have an electron pair available to form the bond. In the Lewis definition, therefore, a base is a substance that has a nonbonding valence electron pair that can be used to form a chemical bond. More simply, *a Lewis base is an electron pair donor*.

The Lewis definition does not greatly expand our ideas about bases. But it does significantly broaden the category of substances that can be classified as acids. When a base accepts a proton, the base donates electrons to form the bond with the proton. Therefore, the proton accepts the electrons. *A Lewis acid is a substance that is an electron pair acceptor*. A proton is the simplest Lewis acid. Many other substances fit the definition, including a number of cations.

A cation has an electron deficiency, and it usually can accept electrons to form bonds. Almost any reaction in which a metal cation forms a bond can be described as a reaction in which the metal cation acts as a Lewis acid, accepting electrons from a Lewis base. In the reaction

$$Ag^+ + 2NH_3 \rightleftharpoons Ag(NH_3)_2{}^+$$

the silver cation is a Lewis acid and NH_3 is a Lewis base. In

$$Zn^{2+} + 2OH^- \rightleftharpoons Zn(OH)_2(s)$$

the zinc cation is a Lewis acid and hydroxide ion is the Lewis base.

In organic chemistry it is especially common to find cations that behave as Lewis acids, accepting electrons. The reaction

$$C_2H_4 + Br_2 \rightleftharpoons C_2H_4Br_2$$

can proceed in three steps. In the first step, the Br_2 forms Br^+ and Br^-. Then Br^+, a Lewis acid, joins to C_2H_4, a Lewis base. In this step, C_2H_4 donates the π electron pair of the $C{=}C$ double bond; this electron pair is bound relatively loosely.

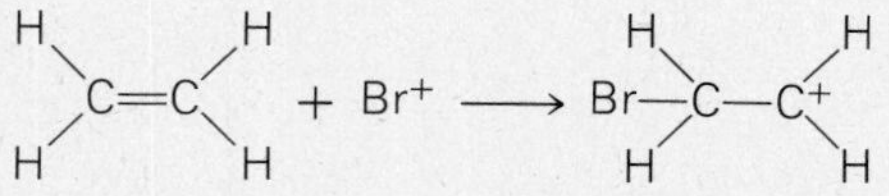

In the third step, Br^-, a Lewis base, donates an electron pair and joins to the cation formed in the first step. The cation is a Lewis acid:

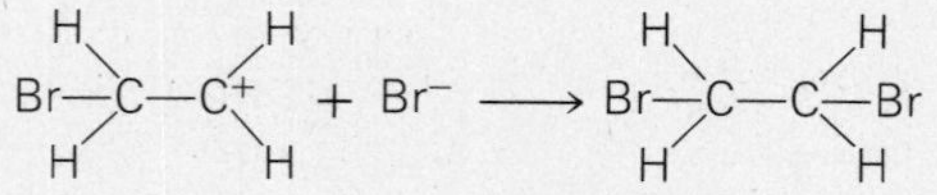

A great many reactions can be understood as the combination of a Lewis acid and a Lewis base. In fact, a common method of classifying chemical reagents uses the Lewis acid and Lewis base definition. A Lewis acid, a substance that seeks electrons, is called an *electrophile*, from the Greek for "lover of electrons." A Lewis base, which is an electron donor or a substance that seeks substances that are electron-deficient, can be classified as a *nucleophile*, from the Greek for "lover of nuclei." In the reaction between Br_2 and C_2H_4, the cation Br^+ is an electrophile and the anion Br^- is a nucleophile. When an electrophile and a nucleophile find each other, a reaction occurs.

Many Lewis acids are substances that are electrically neutral. These substances can accept electrons and form additional bonds either because they have incomplete octets or because octet expansion is possible. Many compounds of the elements in column IIIA of the periodic table are Lewis acids. The best known of them are the halides of boron and aluminum, such as BF_3, BCl_3, $AlCl_3$, and $AlBr_3$. In all compounds of this type, the central atom has an incomplete octet and can form another bond. We classify these compounds as electrophiles because they seek more electrons to achieve complete octets. These compounds react with Lewis bases:

$$BF_3 + :NH_3 \rightleftharpoons F_3B—NH_3$$
$$AlCl_3 + Cl^- \rightleftharpoons AlCl_4^-$$

Many compounds of the transition metals, other heavy metals, and some nonmetals are Lewis acids because of octet expansion. Examples are

$$PtCl_4 + 2Cl^- \rightleftharpoons PtCl_6^{2-}$$
$$PF_5 + :NH_3 \rightleftharpoons F_5P—NH_3$$

We shall discuss some of these reactions in greater detail in Chapter 18.

The Lewis definition is useful because it allows us to classify so many substances as acids or bases, and because it helps provide an understanding of the details of many chemical reactions. But the Brønsted definition is quite adequate for the chemistry of acids and bases in aqueous solution. Since the rest of this chapter deals primarily with aqueous equilibria, we shall use the Brønsted definition.

15.2 STRENGTHS OF ACIDS AND BASES

We listed some weak and strong acids in Chapter 2, page 38. The words "weak" and "strong" are commonplace in talking about acids and bases, but the exact meaning of these words is often blurred in everyday use. By using the Brønsted definition, we can formulate a simple yet precise description of the strength of acids and bases.

The description starts with the equilibrium $A \rightleftharpoons B + H^+$. We define the relative strength of the acid A and the base B by determining the amount of each substance present when the system is at equilibrium. The stronger the acid A, the further to the right is the position of equilibrium and the larger the value of the equilibrium constant K. The weaker the acid, the further to the left is the position of equilibrium and the smaller the value of K. In the same way, the stronger the base B, the further to the left is the position of equilibrium; the weaker the base B, the further to the right is the position of equilibrium. You can see that the strength of an acid and the strength of its conjugate base are interdependent. The stronger the acid, the weaker its conjugate base, and vice versa.

The definition of acid and base strengths that starts with this simple equilibrium is fundamental, but it is not practical. Free protons are included in the equilibrium reaction. Since free protons usually do not exist in solution, the equilibrium reaction does not occur in solution, so we cannot use it to measure acid and base strength. A practical definition of acid and base strength requires an equilibrium that we can make measurements on. As we have said, such an equilibrium must include another acid-base conjugate pair to accept the proton from the acid and to donate the proton to the base. Thus, when we define the strength of an acid or a base we must also specify the nature of the other conjugate acid-base pair in the equilibrium.

Since most of our discussion will deal with aqueous solutions, it is convenient to have water as one of the components of the other conjugate acid-base pair. The equilibrium reaction that we shall use to define the relative strengths of acids is thus:

$$A + H_2O \rightleftharpoons B + H_3O^+$$

Once again you should note that electrical charge balance is not shown. The value of K, the equilibrium constant, for this reaction expresses the strength of the acid A quantitatively. The larger the value of K, the stronger the acid. For example, we can measure the value of K for the reaction:

$$HClO + H_2O \rightleftharpoons ClO^- + H_3O^+$$

and find that it is larger than the value of K for the reaction:

$$NH_4^+ + H_2O \rightleftharpoons NH_3 + H_3O^+$$

Therefore, hypochlorous acid is a stronger acid than the ammonium ion.

You should also note that the value of the equilibrium constant for this general type of reaction also gives us information about the relative strengths of the two acids that appear in the equilibrium, A and H_3O^+. The larger the value of K, the stronger is acid A relative to H_3O^+, the conjugate acid of water.

The form of the equilibrium constant for the general reaction is:

$$K = \frac{[B][H_3O^+]}{[A][H_2O]}$$

As always, the expression for the equilibrium constant includes terms that designate the molarities of the dissolved substances in the system. But since water is the solvent, its concentration is very large (55M) and relatively constant. For this reason, it is customary to omit the term [H_2O] from the expression for K, especially in dilute solutions. The expression for K then becomes:

$$K = \frac{[B][H_3O^+]}{[A]}$$

This is an important expression. We shall use it whenever possible to express the strength of acids in water. This expression is often called the **acid dissociation constant** of the acid A, and usually is represented by the symbol K_A. However, you should remember that this "constant" is an actual constant only in dilute solutions, under conditions in which the ideal solution approximation can be used. We shall assume that the aqueous solutions are dilute enough so that K_A is a constant.

There is another reason why the expression for K_A is not exact. The H_3O^+ ion, like most cations, does not exist in aqueous solution because it is further hydrated. There is evidence to indicate that a particularly important form of the hydronium ion has the composition $H_9O_4^+$, with the H_3O^+ ion associated with three more water molecules. The structure can be represented as

We shall not concern ourselves with this complication in discussing acid-base equilibria.

The expression for K_A can also be used to indicate the relative strength of the conjugate base B of the acid. It is convenient to use the reciprocal of the expression:

$$\frac{1}{K_A} = \frac{[A]}{[B][H_3O^+]}$$

A large value of $1/K_A$ means that a given base is relatively strong. A small value of $1/K_A$ means that a given base is relatively weak.

In practice, the use of K_A to define the strengths of acids is limited to acids that are weaker than H_3O^+. If an acid is stronger than H_3O^+, it exists almost completely as its conjugate base when it dissolves in water. An example is nitric acid, which is a stronger acid than H_3O^+. The equilibrium is

$$HNO_3 + H_2O \rightleftharpoons H_3O^+ + NO_3^-$$

This system contains essentially no undissociated HNO_3 at equilibrium. In fact, it is difficult to measure anything but $[NO_3^-]$. We can get a measurable concentration of undissociated HNO_3 and thus a measure of K only by beginning with very high concentrations of nitric acid. But the resulting high concentrations of ions on the right side of the equation cause a complete breakdown of the ideal solution approximation. Indeed, in any solution that has water and a strong acid in the usual range of concentrations, K is too large to measure, since there is no detectable concentration of undissociated acid. For this reason, strong acids in solution are always written to show complete dissociation of the acid into ions (Section 2.6, page 38).

But if the equilibrium between any strong acid and water lies completely to the right, we cannot evaluate relative strengths of strong acids by comparing them to H_3O^+. In aqueous solutions, all acids stronger than H_3O^+ appear equally strong. This phenomenon is called the *leveling effect*, because water is said to level the strengths of all these acids, making them appear identical.

But it is possible to assess the relative strengths of acids that appear strong in water by using a solvent that is less basic than water. One such solvent is methanol. Methanol can accept a proton to form the methyl oxonium ion:

$$CH_3OH + H^+ \rightleftharpoons CH_3OH_2^+$$

The methyl oxonium ion is a stronger acid than the hydronium ion because methanol is a weaker base than water. When nitric acid is dissolved in methanol, the transfer of a proton from nitric acid to methanol is not complete, since nitric acid is a weaker acid than $CH_3OH_2^+$. It is possible to obtain a value for the equilibrium constant of the reaction:

$$HNO_3 + CH_3OH \rightleftharpoons NO_3^- + CH_3OH_2^+$$

Perchloric acid, $HClO_4$, is another strong acid. The equilibrium in water:

$$HClO_4 + H_2O \rightleftharpoons H_3O^+ + ClO_4^-$$

lies completely to the right, as is true of an aqueous solution of nitric acid. We cannot tell whether perchloric acid or nitric acid is stronger in water. But the evaluation can be made in methanol by studying the equilibrium between perchloric acid and methanol:

$$HClO_4 + CH_3OH \rightleftharpoons ClO_4^- + CH_3OH_2^+$$

This equilibrium lies completely to the right. The corresponding equilibrium for nitric acid does not. Methanol differentiates between the two acids; perchloric acid is the stronger of the two. This evaluation is another reminder that the terms ''strong'' and ''weak'' are relative. In water, both of these acids are strong. In methanol, perchloric acid is strong and nitric acid is weak.

Only a few common acids are stronger than H_3O^+. Many of them can be found in Table 2.5. As a working rule, you should assume that any acid that is not specifically described as being strong is weak. Since the value of K_A for strong acids cannot be measured by any practical method, we shall list values of K_A for weak acids only. Generally, any acid that is listed in a table of acid dissociation constants is a weak acid. Table 15.2 lists the acid dissociation constants of some common weak acids.

The leveling effect of a solvent can also be observed with bases. We can write reactions between a base and water in which the water acts as the acid and donates a proton to the base. One such reaction is:

$$NH_3 + H_2O \rightleftharpoons NH_4^+ + OH^-$$

TABLE 15.2
Dissociation Constants of Weak Acids at Room Temperature

Neutral Inorganic Acids	$K_A(M)$
hydrogen peroxide, H_2O_2	2.4×10^{-12}
hypoiodous acid, HIO	2.3×10^{-11}
hydrocyanic acid, HCN	4.93×10^{-10}
hypobromous acid, HBrO	2.06×10^{-9}
hypochlorous acid, HClO	3.2×10^{-8}
hydrofluoric acid, HF	3.53×10^{-4}
nitrous acid, HNO_2	4.5×10^{-4}
chlorous acid, $HClO_2$	1.1×10^{-2}
periodic acid, H_5IO_6	2.3×10^{-2}
iodic acid, HIO_3	1.69×10^{-1}

Inorganic Cations	
ammonium, NH_4^+	5.59×10^{-10}
hydrazinium, $N_2H_5^+$	5.9×10^{-9}
hydroxylammonium, NH_3OH^+	9.1×10^{-7}

Neutral Organic Acids	
saccharin	2.1×10^{-12}
phenol	1.28×10^{-10}
butyric acid	1.54×10^{-5}
acetic acid	1.76×10^{-5}
acrylic acid	5.6×10^{-5}
uric acid	1.3×10^{-4}
lactic acid	1.37×10^{-4}
formic acid	1.77×10^{-4}
sulfanilic acid	5.9×10^{-4}
chloroacetic acid	1.40×10^{-3}
dichloroacetic acid	3.32×10^{-2}
trichloroacetic acid	2×10^{-1}

Protonated Organic Bases	
methyl amine	2.70×10^{-11}
ephedrine	7.26×10^{-11}
codeine	6.15×10^{-9}
morphine	6.16×10^{-9}
nicotine	9.55×10^{-9}
imidazole	1.11×10^{-7}
aniline	2.34×10^{-5}

in which ammonia is the base and ammonium ion is its conjugate acid, while water is the acid and hydroxide ion is its conjugate base. We shall not usually use such reactions to compare the strengths of bases, but we shall use them here to illustrate the leveling effect.

If a base that is stronger than hydroxide ion is placed in water, virtually all the base becomes protonated. For example, the amide ion is a stronger base than the hydroxide ion. When the amide ion is dissolved in water:

$$NH_2^- + H_2O \rightleftharpoons NH_3 + OH^-$$

no detectable amount of NH_2^- is present at equilibrium; it is all converted to NH_3. The same is true of the equilibrium between water and a base such as CH_3^-, which is also much stronger than hydroxide ion. In water, therefore, the strengths of these bases and of all bases stronger than hydroxide are leveled. We need a solvent that is less acidic than water to differentiate between the strengths of these bases. One such solvent is liquid ammonia, which is useful at temperatures below 240 K. In liquid ammonia, CH_3^- is still completely protonated to CH_4.

$$CH_3^- + NH_3 \rightleftharpoons CH_4 + NH_2^-$$

Therefore, CH_3^- is a stronger base than NH_2^-, the conjugate base of ammonia.

There are solvents that are more strongly acidic than water. One is acetic acid, $HC_2H_3O_2$, which we shall represent by the formula HOAc. Water differentiates between NH_3, a weak base in aqueous solution, and OH^-, a strong base in aqueous solution. In HOAc, both these bases are strong. Because acetic acid is more acidic than water, it levels the strength of the two bases. Both equilibria lie far to the right:

$$HOAc + NH_3 \rightleftharpoons NH_4^+ + OAc^-$$
$$HOAc + OH^- \rightleftharpoons H_2O + OAc^-$$

In this chapter, we shall divide acids into two groups, strong and weak, based on their reaction with water. The formula for any strong acid in solution will be written to show complete dissociation, and we shall not use a K_A for a strong acid. An acid dissociation constant K_A will be used to show the extent of dissociation of a weak acid in solution. We shall assume that weak acids in solution are essentially undissociated, and that the major species in solution is the un-ionized weak acid. We shall derive the strengths of bases from the values of K_A for the related acids. The conjugate bases of strong acids are so weak that they do not participate in acid-base reactions in aqueous solution.

The strong base that is most frequently found in water is the hydroxide ion. Any compound that yields a relatively substantial amount of hydroxide ion in solution is defined as a strong base. Two such compounds are sodium hydroxide, NaOH, and potassium hydroxide, KOH. Both are soluble in water. They are strong bases because they dissociate completely to cations and free hydroxide ions. The hydroxides of many metals behave this way, but many of them are not very soluble. Calcium hydroxide, $Ca(OH)_2$ for example, dissociates completely in solution and is a strong base, but is relatively insoluble. Very little $Ca(OH)_2$ dissolves, so a solution of this strong base is not very alkaline.

The conjugate bases of the weak acids are all relatively weaker bases than OH^- ion. Most of them are anions. A common exception is NH_3, the conjugate base of NH_4^+, which is the most important weak base that is electrically neutral.

EXAMPLE 15.1
Write the chemical equations for the reactions that occur when the following substances are mixed:

(a) A solution of hydrochloric acid and a solution of sodium hydroxide.
(b) A solution of hydrochloric acid and a solution of ammonia.
(c) A solution of hydrogen cyanide and a solution of sodium hydroxide.
(d) A solution of hydrogen cyanide and a solution of ammonia.
(e) Hydrogen chloride and ammonia, not in solution.

SOLUTION (See Section 2.8, page 41)
(a) Since hydrochloric acid is a strong acid, the equation is written to show its complete dissociation into ions, H_3O^+ and Cl^-, in aqueous solution. Sodium hydroxide is a strong base, so in solution it is written as dissociated ions, Na^+ and OH^- ions. In the reaction that occurs, H_3O^+ transfers a proton to OH^-:

$$H_3O^+ + OH^- \rightleftharpoons 2H_2O$$

The Cl^- ion is too weak a base to take part in any reaction that could compete with this process. The same is true of the Na^+ cation, which is too weak an acid to compete.

(b) The strongest acid in this mixture is H_3O^+ from the hydrochloric acid in solution. The base is NH_3, which is weak. The reaction is H_3O^+ + $NH_3 \rightleftharpoons H_2O + NH_4^+$, in which a proton is transferred from the hydronium ion to ammonia, forming water, the conjugate base of hydronium ion, and ammonium ion, the conjugate acid of ammonia.

(c) Hydrocyanic acid is a weak acid, and so does not dissociate appreciably in solution. Sodium hydroxide is a strong base, so it is dissociated into sodium ions and hydroxide ions. The reaction is:

$$HCN + OH^- \rightleftharpoons CN^- + H_2O$$

in which a proton is transferred from the HCN to the OH^-, forming cyanide ion, the conjugate base of HCN, and water, the conjugate acid of OH^-. As in part (a), the sodium cation is so weak an acid that it does not participate in acid-base reactions in water.

(d) Here both the acid and the base that are mixed are weak. Neither is written as dissociated ions. The reaction is:

$$HCN + NH_3 \rightleftharpoons CN^- + NH_4^+$$

(e) When not in solution, both hydrogen chloride and ammonia are gases. An acid-base reaction, the transfer of a proton, does occur. The product is ammonium chloride, a salt that is a solid. The reaction is

$$HCl(g) + NH_3(g) \rightleftharpoons NH_4Cl(s)$$

in which a proton is transferred from the hydrogen chloride to the ammonia, forming chloride ion, the conjugate base of HCl, and ammonium ion, the conjugate acid of NH_3. These are the component ions of the salt ammonium chloride.

15.3 ACID-BASE REACTIONS IN AQUEOUS SOLUTIONS

Let us review some of the general features of acid-base systems that were outlined in the preceding discussion. An acid-base reaction is a proton transfer from an acid to a base to form the conjugate base of the acid and the conjugate acid of the base. An acid-base reaction therefore contains two conjugate acid-base pairs. The relative strengths of the acid and the base can be expressed by the numerical value of the equilibrium constants for the appropriate acid-base reactions. The strength of an acid is given by the value of K_A, the equilibrium constant for the reaction in which the acid donates a proton to water. The K_A of a strong acid is too large to measure. The strength of a base is given by $1/K_A$, where K_A is the acid dissociation constant of the conjugate acid of the base.

We can now make some simplifications for convenience. One simplification has already been mentioned. Since the concentration of water does not change in dilute solutions, water can be omitted from the expression for the equilibrium constant. In addition, H_3O^+ is often written as H^+, as we did earlier in the text. Thus, water can be subtracted from both sides of the equation for acid-base reactions, which include the H_2O/H_3O^+ conjugate pair. But it must be understood that H^+ does not exist in solution. The symbol H^+ is used as an abbreviation for all hydrated forms of the proton in these systems.

The reaction in which an acid transfers a proton to water, the base, can again be written $A \rightleftharpoons B + H^+$, which is the fundamental equation of the Brønsted theory. The equilibrium constant K_A of this reaction is written

$$K_A = \frac{[H^+][B]}{[A]}$$

Using this simplification, let us write equations for the reactions that correspond to some of the dissociation constants K_A given in Table 15.2. For acetic acid, the reaction is $HOAc \rightleftharpoons H^+ + OAc^-$. For bicarbonate ion, the reaction is $HCO_3^- \rightleftharpoons H^+ + CO_3^{2-}$. For ammonium ion, the reaction is $NH_4^+ \rightleftharpoons H^+ + NH_3$. Water does not appear in these equations. But it should be understood that water is the base which accepts the proton in these reactions, and that H^+ represents hydrated forms of the proton.

The Water Equilibrium

The equilibrium for the autoprotolysis of water is thus written $H_2O \rightleftharpoons H^+ + OH^-$. This equation shows only one water molecule, which acts as an acid. A second water molecule, which acts as a base, is omitted from the equation. The equilibrium constant for this reaction is $K = [H^+][OH^-]/[H_2O]$. In a dilute solution, we can omit the term for the concentration of water. The resulting equilibrium expression. $[H^+][OH^-]$, called the *ion product of water* and represented by the symbol K_W, is extremely important. At room temperature (298 K), its value is $1.0 \times 10^{-14}M^2$.

You should memorize this value. It gives you some useful information about any aqueous solution at equilibrium. No matter what else occurs in the solution, the concentration of hydrated protons multiplied by the concentration of hydroxide ion is $1.0 \times 10^{-14}M^2$ at 298 K. That is, for any aqueous solution at equilibrium

$$[H^+][OH^-] = 1.0 \times 10^{-14}M^2$$

In pure water—water that contains no added bases or acids—$[H^+] = [OH^-] = x$, so at room temperature:

$$x^2 = 1.0 \times 10^{-14}M^2 \quad \text{and} \quad x = 1.0 \times 10^{-7}M$$

A solution in which both the $[H^+]$ and the $[OH^-]$ are $1.0 \times 10^{-7}M$ is called a neutral solution. In practice, a neutral solution is almost impossible to obtain. When water is exposed to the atmosphere, some carbon dioxide dissolves in it. Since carbon dioxide forms carbonic acid, H_2CO_3, in water, the solution is no longer neutral. If water contains a dissolved acid, the $[H^+]$ has a value greater than $1.0 \times 10^{-7}M$ and the $[OH^-]$ therefore must have a value less than $1.0 \times 10^{-7}M$. Such a solution is said to be acidic. If water contains a dissolved base, the value of $[OH^-]$ is greater than $1.0 \times 10^{-7}M$ and the $[H^+]$ must be less than $1.0 \times 10^{-7}M$. Such a solution is said to be basic or alkaline. Aqueous solutions can be classified as acidic, alkaline, or neutral according to the values of $[H^+]$ or $[OH^-]$.

EXAMPLE 15.2
Calculate the $[H^+]$ and the $[OH^-]$ in a solution prepared from 0.030 mol of HI and enough water to form 0.50 liter of solution.
SOLUTION
Hydriodic acid, HI, is a strong acid which dissociates completely in solution. The solution has 0.030 mol HI/0.50 liter = $0.060M$ in HI. Since one mole of HI forms one mole of H^+ in solution, the $[H^+] = 0.060M$. From the expression for K_W:

$$[H^+][OH^-] = (0.060M)[OH^-] = 1.0 \times 10^{-14}M^2$$

$$[OH^-] = \frac{1.0 \times 10^{-14}M^2}{0.060M} = 1.7 \times 10^{-13}M$$

EXAMPLE 15.3
Calculate the $[H^+]$ and the $[OH^-]$ in a solution that is $1M$ in potassium hydroxide.

SOLUTION

Although the concentration of potassium hydroxide is high, we shall use the ideal solution approximation. Potassium hydroxide is a strong base which dissociates completely in solution, and thus $[OH^-] = 1M$. From the expression for the ion product of water:,

$$[H^+][OH^-] = [H^+][1M] = 1.0 \times 10^{-14}M^2$$
$$[H^+] = 1.0 \times 10^{-14}M$$

In these examples, we see that the concentration of hydroxide ion is low when the solution is acidic, and that the concentration of protons is low when the solution is basic. As the acidity of a solution increases, the $[H^+]$ increases and the $[OH^-]$ decreases. As the basicity of a solution increases, the $[OH^-]$ increases and the $[H^+]$ decreases.

The Symbol p

In chemistry, the symbol p is often used as a convenient method of representing small numbers. The symbol p precedes a quantity. It means ''take the negative logarithm of the quantity,'' a definition that requires further explanation.

Most calculations of acid-base equilibria in water use small numbers, less than 1, because most equilibrium constants, such as the K_A of a weak acid and K_W itself, are small numbers. The $[H^+]$ and $[OH^-]$ are also small numbers, except in relatively concentrated solutions of strong acids or strong bases. Any small number is written as the product of two numbers (Appendix IV). One is an exponential, 10 raised to a negative power. The other is an ordinary number, usually with one integer to the left of the decimal point and as many integers to the right of the decimal point as are warranted by the accuracy of the measurement. One example of this notation, taken from the preceding section, is 1.0×10^{-14}.

The concept of the negative logarithm is useful in dealing with small numbers (Appendix IV). The logarithm of 10 raised to an exponent is the exponent. The negative logarithm is the negative of the exponent. If 10 is raised to a negative exponent, then the negative logarithm will be a positive number, since the negative of a negative number is a positive number. For example, $\log 10^{-7} = -7$, and $-\log 10^{-7} = 7$. Similarly, $-\log 10^{-23} = 23$ and $-\log 10^{-2} = 2$.

It is rather rare to deal only with an exponential. More commonly, we find a two-term expression in which the exponential is multiplied by the other number. There are several ways to obtain a logarithm. You can find the logarithm of a number with one integer to the left of the decimal point in a table of logarithms, you can calculate it on a slide rule, or you can obtain it from an electronic calculator. Such a logarithm is a decimal between 0 and 1. The negative logarithm is the negative of this decimal. For example, $\log 2.05 = 0.312$, and $-\log 2.05 = -0.312$; $\log 7.37 = 0.867$, and $-\log 7.37 = -0.867$.

We can multiply numbers by adding their logarithms. (Logarithms were developed because they simplify multiplication.) The logarithm of the two-term expression, number times exponential, is the sum of their two logarithms. The negative logarithm of the two-term expression is the sum of their negative logarithms. For example, the negative logarithm of $2.05 \times 10^{-14} = -0.312 + 14 = 13.688$. The negative logarithm of $1.0 \times 10^{-14} = 0.000 + 14 = 14$ ($\log 1.0 = 0.000$). We shall be dealing primarily with numbers smaller than 1 in this chapter, so the negative logarithms will be positive and greater than 1.

EXAMPLE 15.4

Find the negative logarithms of the following numbers: (a) 1.0×10^{-7}, (b) 3.64×10^{-12}, (c) 0.25, (d) 1.3.

SOLUTION

In each case, we find the negative logarithms of two terms and add them.

(a) $-\log 1.0 = 0.00$; $-\log 10^{-7} = 7$; $-\log(1.0 \times 10^{-7}) = 7.00$.

(b) $-\log 3.64 = -0.561$; $-\log 10^{-12} = 12$; $-\log(3.64 \times 10^{-12}) = 12 - 0.561 = 11.439$.

(c) First convert 0.25 to an exponential form, 2.5×10^{-1}. Then $-\log 2.5 = -0.40$; $-\log 10^{-1} = 1$; $-\log(2.5 \times 10^{-1}) = -0.40 + 1 = 0.60$.

(d) 1.3 can be written 1.3×10^0. $-\log 1.3 = -0.11 + 0 = -0.11$.

At the beginning of this discussion, we said that when the symbol p is placed before any quantity, it stands for the negative logarithm of that quantity. Thus, pH means the negative logarithm of the $[H^+]$, pOH means the negative logarithm of the $[OH^-]$, pK_A means the negative logarithm of K_A, the acid dissociation constant, and so on. These negative logarithms are used because they are easier to handle than very small numbers and because addition and subtraction replace multiplication and division. But some adjustment in thinking is needed.

The smaller a negative logarithm, the larger the quantity it represents. A pH of 7 represents a higher $[H^+]$ than a pH of 9. A pH of 2 represents a $[H^+]$ of 1.0×10^{-2}, while a pH of -1 represents a $[H^+]$ of 1.0×10^1, a higher concentration. The pK_A of acetic acid is 4.74 and the pK_A of HCN is 9.40. Acetic acid has the smaller negative logarithm, and therefore has the larger K_A and is the stronger acid.

Another fact to keep in mind is that logarithms represent an exponential scale. Each integer represents a factor of 10. The $[H^+]$ of a solution whose pH is 7 is 10 times greater than the $[H^+]$ of a solution of pH 8; it is 100 times (10^2) greater than the $[H^+]$ in a solution of pH 9, and 1000 (10^3) times greater than the $[H^+]$ in a solution of pH 10. A solution of pH 2 has a $[H^+]$ that is 10^5 times greater than that in a solution of pH 7.

The symbol p is especially convenient for expressing the acidity of basicity or aqueous solutions. Let us take the negative logarithms of the terms on both sides of the equation for the ion product of water. Since we can multiply numbers by adding their logarithms, we can write

$$[H^+][OH^-] = K_W = 1.0 \times 10^{-14}$$
$$pH + pOH = pK_W = 14$$

The sum of the pH and the pOH in any aqueous solution at room temperature is 14. The lower the pH, the higher the pOH. Given one, the other is easily found by subtraction from 14. Acidic solutions can be described as having either a low pH or a high pOH. Basic solutions have high pH and low pOH. In a neutral solution, $pH = pOH = 7$. We can use either pH or pOH for any solution. In practice, pH is almost always used. These relationships are shown in Table 15.3.

EXAMPLE 15.5

Calculate the pH and the pOH of a solution prepared from 0.0330 mol of perchloric acid and enough water to form 250 cm^3 of solution.

SOLUTION

Since $HClO_4$ is a strong acid, it dissociates completely in solution. Therefore, there is 0.0330 mol of H^+ in the 250 cm^3 of solution.

$$[H^+] = \frac{0.0330 \text{ mol } H^+}{250 \text{ cm}^3} \times \frac{1000 \text{ cm}^3}{1 \text{ liter}} = 0.132 M$$

The pH is the negative logarithm of this number.

$$pH = -\log 1.32 \times 10^{-1} = 1 - 0.121 = 0.879$$
$$pOH = 14 - pH = 14 - 0.879 = 13.121$$

TABLE 15.3
The pH Scale

pH	$[H^+]$ (M)	$[OH^-]$ (M)	pOH	
−1	10	1×10^{-15}	15	acid
0	1	1×10^{-14}	14	
1	1×10^{-1}	1×10^{-13}	13	
2	1×10^{-2}	1×10^{-12}	12	
3	1×10^{-3}	1×10^{-11}	11	
4	1×10^{-4}	1×10^{-10}	10	
5	1×10^{-5}	1×10^{-9}	9	
6	1×10^{-6}	1×10^{-8}	8	
7	1×10^{-7}	1×10^{-7}	7	neutral
8	1×10^{-8}	1×10^{-6}	6	
9	1×10^{-9}	1×10^{-5}	5	
10	1×10^{-10}	1×10^{-4}	4	
11	1×10^{-11}	1×10^{-3}	3	
12	1×10^{-12}	1×10^{-2}	2	
13	1×10^{-13}	1×10^{-1}	1	
14	1×10^{-14}	1	0	
15	1×10^{-15}	10	−1	alkaline

EXAMPLE 15.6
Calculate the pH and the pOH of a solution prepared from 0.00135 mol of
barium hydroxide and enough water to form 100 cm³ of solution.
SOLUTION
Since each mole of $Ba(OH)_2$ in solution forms two moles of OH^-,

$$[OH^-] = \frac{(2)(0.00135 \text{ mol } Ba(OH)_2)}{100 \text{ cm}^3} \times \frac{1000 \text{ cm}^3}{1 \text{ liter}} = 2.70 \times 10^{-2} M$$

$$pOH = -\log 2.70 \times 10^{-2} = 2 - 0.431 = 1.569$$

$$pH = 14 - pOH = 14 - 1.569 = 12.431$$

An instrument found in almost any chemistry laboratory is the pH meter,
which measures the pH of a solution electrochemically. When you use a pH
meter, you will sometimes have to convert a given value of pH into a $[H^+]$.

EXAMPLE 15.7
The pH of a solution is found to be 8.78. Calculate the $[H^+]$ and the $[OH^-]$.
SOLUTION
This problem asks us to find the number whose negative logarithm is 8.78.
We must often work with negative logarithms in chemistry, so we need a
general method for this type of calculation. We first write 8.78 as the sum
of an integer and a decimal: $8 + 0.78$. We can also write 8.78 as the difference between the next highest integer and a decimal: $9 - (1 - 0.78) =
9 - 0.22$. The conversion to a negative logarithm is easiest when the num-
ber is written in the latter form, in which the integer becomes the negative
exponent of 10. Here, the integer is 9, so we write 10^{-9}. We find the number by which this exponential term is multiplied by finding the antilog of
0.22, using a table of logarithms, a slide rule, or an electronic calculator.
Any of these methods shows that the antilog of 0.22 is 1.66. Thus:

$$\text{antilog } (-8.78) = 1.66 \times 10^{-9}$$

$$[H^+] = 1.66 \times 10^{-9} M$$

$$[OH^-] = \frac{1.00 \times 10^{-14} M^2}{[H^+]} = \frac{1.00 \times 10^{-14} M^2}{1.66 \times 10^{-9} M} = 6.02 \times 10^{-6} M$$

ACID RAIN AND OUR ENVIRONMENT

An unintentional experiment on a global scale is providing dramatic evidence of the sensitivity of living organisms and ecosystems to even slight changes in pH.

For the past two decades, the acidity of rain and snow falling on large areas of the United States and Europe has been increasing, apparently because more and more pollutants such as sulfur dioxide and nitrogen oxides are being produced from the combustion of fossil fuels. The acid rain has caused a marked decline in the numbers of animals of some sensitive species, such as salmon and trout, and severe damage to plants.

Some acidity is normal in precipitation, since gaseous carbon dioxide in the atmosphere dissolves in water to produce a slightly acid solution. The minimum pH value expected for water in equilibrium with atmospheric CO_2 is about 5.6. But the pH of rain and snow falling on much of Northern Europe and the eastern United States in recent years has often been as low as 5 and has sometimes been as low as 3. The increased acidity of the rain is due to the presence of sulfuric acid and nitric acid.

The effects of acid precipitation have been documented particularly well in Scandinavia. Pollutants from England and other industrialized nations drift north and east with prevailing winds and are deposited on the Scandinavian nations. About 5000 lakes in Sweden are estimated to have pH values of 5.0 or below, and fish populations have been seriously affected. Studies have shown that the mean acidity of precipitation falling on southern Norway has decreased to a pH value of 4.6, and that the number of lakes without salmon and trout has increased alarmingly.

The same effect has been observed in the lakes of New York's Adirondack mountains. A study in the 1930s found that only 4% of 217 mountain lakes had a pH of 5.0 or under and had no fish populations. In the early 1970s, a survey found that half of the 217 lakes had pH values below 5.0, and that 90% of the lakes with low pH contained no fish.

Other effects of acid precipitation are more difficult to define. It is believed that a lower pH reduces the diversity of aquatic life forms, as acid-sensitive species are eliminated. It is also believed that acid precipitation harms trees and crops, but studies of such possible damage are only beginning. As for direct damage to property, it has been estimated that corrosion caused by acid precipitation could cost more than $1 billion a year in the United States alone.

The search for a solution to the problem has begun. One suggested approach is to add limestone or another alkaline substance to affected lakes in an effort to decrease the acidity. However, it appears that the only truly effective long-term solution is a drastic decrease in the amounts of pollutants that are added to the air. A reduction in pollutants will require not only the burning of low-sulfur fuels but also changes in combustion processes to reduce

emission of nitrogen oxides. These measures would cost billions of dollars annually, but there seems no alternative way of stopping the damage done by acid rain and snow.

Common Types of Acid-Base Reactions

Chemists are often asked to write balanced equations for reactions that occur in aqueous solutions of acids and bases. Solutions of acids and bases are also commonly encountered in disciplines such as the health sciences and the agricultural sciences. In writing equations for acid-base reactions, it is helpful to remember that reactions between acids and bases are proton transfers from acids to bases.

We usually start with a word description of an experiment or a problem. One key to writing the relevant equation for the experiment or problem is to go from the word description to a set of reactants. The relevant equilibrium usually is the reaction between the strongest acid and the strongest base that are present in appreciable concentration. While more than one equilibrium reaction takes place in any aqueous solution, we generally are interested only in the reaction whose equilibrium constant K has the largest value.

Let us start with a simple word description: ''A solution of acetic acid.'' This statement defines the reactants in the system: acetic acid, HOAc, and water, H_2O. The system has two acids, HOAc and H_2O. The acetic acid is the stronger, and it is included in the equilibrium of interest. The system has one base, H_2O. Therefore, the acid-base reaction is the transfer of a proton from the strongest acid in the system, HOAc, to the only base, H_2O: $HOAc + H_2O \rightleftharpoons OAc^- + H_3O^+$ or, more simply, $HOAc \rightleftharpoons H^+ + OAc^-$. Another equilibrium that occurs in this system is the autoprotolysis of water, $H_2O \rightleftharpoons H^+ + OH^-$. The K_W for this reaction is much smaller than the K_A for the dissociation of acetic acid, because acetic acid is a stronger acid than water (Table 15.2).

What about the other species in the solution, such as OH^- and OAc^-, which are both stronger bases than water? Should we write reactions that include these ions as reactants if the equilibrium constants of the reactions are larger than the K_A of acetic acid? The answer is no. The reaction we are seeking, the one that best describes the equilibrium state reached by the system when we prepare a solution of acetic acid in water, must include as reactants only those substances that are initially present in appreciable concentration. *No substance may appear as a reactant in an equation unless it is present in appreciable concentration when the system is initially constituted.* Generally, these substances are mentioned in the word description of the problem.

There are two footnotes to this rule. ''Solution'' means ''aqueous solution.'' Also, any reaction we write must be consistent with the rules described in Section 2.8, page 43, for the correct representation of solutions of ionic substances.

If a strong acid is included in the set of reactants, the acid is written as dissociated ions. The equation for a reaction of a solution of hydrobromic acid may include H^+ and Br^-. Either or both of these ions may appear as reactants. But HBr cannot be a reactant, because it does not exist in any appreciable concentration in solution. Similarly, if the problem mentions a strong base such as sodium hydroxide, in solution, the equation may include the ions Na^+ or OH^-, which may appear on the left side of the equation as reactants. But NaOH undissociated in solution may not be a reactant. For salts in solution, for example NH_4Cl, the salt is written as dissociated ions. The equation for a reaction of a solution of ammonium chloride may include NH_4^+ or Cl^-, but not NH_4Cl as a reactant.

Thus, the acid-base system described by the phrase ''a solution of potassium acetate'' contains K^+ ion, which does not enter into acid-base equilibria; OAc^- ion, which is a base that is stronger than water; and water itself, which is both an acid and a base. The system does *not* initially contain appreciable concentrations of acetic acid, H^+, OH^-, or undissociated KOAc. None of these materials may appear as reactants. The relevant reaction that we shall use to find the composition of the equilibrium state of this system is the reaction between H_2O, the strongest acid present in appreciable concentration, and OAc^-, the strongest base. As usual, a proton is transferred from the acid to the base:

$$H_2O + OAc^- \rightleftharpoons OH^- + HOAc$$

With these considerations in mind, we can define some of the important categories of acid-base systems, and then write the relevant equilibrium reaction for each category. Once we do this, we can write the relevant equilibrium for any system that is identified as belonging to one of these categories. The categories are:

1. A solution of a weak acid. The relevant equilibrium for problem-solving in this system is the reaction between water acting as a base and the weak acid. The equilibrium constant for this reaction is K_A. One such system is a solution of hydrogen cyanide, $HCN + H_2O \rightleftharpoons CN^- + H_3O^+$ or in the short form, $HCN \rightleftharpoons H^+ + CN^-$. Another is a solution of ammonium ion. $NH_4^+ \rightleftharpoons NH_3 + H^+$.

2. A solution of a weak base. The relevant equilibrium for problem-solving in this system is the reaction between water acting as an acid and the weak base. We must be careful to represent the solution of the weak base correctly. Many weak bases are anions that are the conjugate bases of weak acids. The sources of these anions, which include CN^-, OAc^-, ClO^-, and $H_2PO_4^-$, are usually salts, and a solution of such an anion is obtained by dissolving the appropriate salt in water. Solutions of salts are written to show dissociated ions, but the cations of these salts usually do not take part in the relevant reaction. Neutral bases such as ammonia do not present this problem. Some examples of reactions that could be used to solve problems related to solutions of weak bases are:

$$ClO^- + H_2O \rightleftharpoons HClO + OH^-$$
$$H_2PO_4^- + H_2O \rightleftharpoons H_3PO_4 + OH^-$$
$$NH_3 + H_2O \rightleftharpoons NH_4^+ + OH^-$$

3. A solution of an acid mixed with a solution of a base. The reaction that takes place is called a **neutralization reaction** when water does not appear as a reactant in the relevant equilibrium reaction. We shall discuss neutralization reactions in Section 15.5.

4. A solution of a weak acid mixed with a solution of its conjugate base. These components form what is called a **buffer solution**, which we shall discuss in Section 15.6. Buffer systems are best handled by using the reaction that defines the K_A of the weak acid. An example of a buffer system is the buffer solution formed by mixing a solution of acetic acid with a solution of sodium acetate. The salt is a source of acetate ion, the conjugate base of acetic acid. The reaction between the strongest acid (HOAc) and the strongest base (OAc^-) results in no change. The relevant reaction is between the strongest acid and the next-strongest base (H_2O):

$$HOAc \rightleftharpoons H^+ + OAc^-$$

Another buffer solution is formed by mixing a solution of a salt such as ammonium chloride with a solution of ammonia. Again a convenient equilibrium is the one that defines the K_A for the weak acid, NH_4^+:

$$NH_4^+ \rightleftharpoons NH_3 + H^+$$

Now that we have described some of the most common acid-base systems, we can solve some problems relating to the composition of the equilibrium states that we obtain when we prepare these systems.

15.4 WEAK ACIDS AND WEAK BASES

Many inorganic compounds are weak acids, having acid dissociation constants small enough to measure in water. A number of them are called oxyacids, because they are composed of hydrogen, oxygen, and a third element, usually a nonmetal or a semimetal. The general formula for an oxyacid can be written as H_mXO_n, where X is the third element. The subscript m indicates the number of hydrogen atoms in a molecule of the oxyacid, and the subscript n indicates the number of oxygen atoms.

The strength of an oxyacid can often be predicted from the value of $n - m$. Generally, the larger the value, the stronger the oxyacid. For $HClO_4$, $n - m = 3$, and $HClO_4$ is a very strong acid. For H_2SO_4, $n - m = 2$, and H_2SO_4 is a strong acid. For H_3PO_4, $n - m = 1$, and H_3PO_4 is a weak acid ($K_A \cong 10^{-2}$). For $HClO$, $n - m = 0$, and $HClO$ is a very weak acid ($K_A \cong 10^{-8}$). Additionally, there are a number of inorganic acids that do not contain oxygen; some of the weak ones are listed in Table 15.2.

Among organic acids, the vast majority are weak acids. The acidic proton in an organic acid is generally bonded to an oxygen atom, but there are many exceptions to this rule.

Another important group of weak acids are cations that are the conjugate acids of weak bases. The ammonium ion is the most common inorganic cation that is a weak acid. You will find others in Table 15.2.

The most common inorganic weak bases are the anions that are the conjugate bases of weak acids and can be obtained as salts. A small number of neutral inorganic molecules are also weak bases; ammonia is the most common. A very large number of neutral organic molecules are weak bases. Such weak bases are often related to $:NH_3$. They contain a nitrogen atom with a nonbonding valence electron pair that can form a bond with a proton.

To successfully use experimental data obtained for aqueous solutions of any of these substances, you must be able to carry out the appropriate calculations. The best way to approach these calculations is by example. Let us go through some typical problems.

EXAMPLE 15.8
While hydrofluoric acid etches glass and causes severe burns, it is correctly described as a weak acid. You will find its K_A in Table 15.2. Calculate the pH of a solution prepared from 0.245 mol of HF and enough water to form 1.00 liter of solution.
SOLUTION
Acid-base equilibrium problems of this type are attacked in the same way as problems in other types of equilibrium systems described in Chapter 14. We first write the appropriate balanced chemical equation, the dissociation of the acid. For hydrofluoric acid, the reaction is

$$HF \rightleftharpoons H^+ + F^-$$

We get the equilibrium constant K_A for this reaction from Table 15.2: $K_A = 3.53 \times 10^{-4}M$.

As we did in Chapter 14, we first write a starting line that lists the initial concentrations of all the materials appearing in the equilibrium reaction. Except for special cases which we shall specify, the concentration of H^+ (or of OH^-) from the autoprotolysis of water is not included on the start line. It can be omitted because it is negligibly small compared to the H^+ (or OH^-) that will be present at equilibrium. For this system, the start line is:

$$HF \rightleftharpoons H^+ + F^-$$

start $0.245M$ 0 0

The reaction proceeds to equilibrium by the transfer of protons from HF to water or, in this abbreviated version of the acid-base reaction, by the dissociation of some quantity of HF. Let x represent the [HF] that dissociates. The equilibrium line is

equil $0.245M - x$ x x

The expression for the equilibrium constant of the reaction gives:

$$K_A = \frac{[H^+][F^-]}{[HF]} = \frac{(x)(x)}{(0.245M - x)} = 3.53 \times 10^{-4}M$$

Since K_A is small (as it usually is in these problems) we can use the approximation $0.245 - x \cong 0.245$ to solve for x. We find $x = 9.30 \times 10^{-3}M$. A second approximation (Appendix IV) gives a better value, $x = 9.12 \times 10^{-3}M$, which is the value of both the [H$^+$] and the [F$^-$]. The pH is the negative logarithm of the [H$^+$].

$$pH = -\log 9.12 \times 10^{-3} = 3 - \log 9.12 = 3 - 0.960 = 2.040$$

This problem could have been phrased in a different way. You might have been asked to calculate the fraction of HF that dissociates in a solution that is originally $0.245M$ in HF. The fraction that dissociates, f, is the amount of HF that dissociates divided by the original amount: $f = x/0.245M = (9.12 \times 10^{-3}M)/0.245M = 0.0372$. Only a small fraction of HF dissociates because HF is a weak acid.

EXAMPLE 15.9

Solutions of ammonia, a weak base, suitable for cleaning windows can be bought in any supermarket. One such solution is 10% ammonia by mass and has a density of 0.99 g/cm. Calculate the pH of the solution.

SOLUTION

We must first convert the concentration from units of mass percent to units of molarity, since all equilibrium constant data are given in units of molarity.

One liter, or 1000 cm^3, of this solution has a mass of 990 g. Since the solution is 10% ammonia, it contains 99 g of NH_3, or (99 g)/(17 g NH_3/mol NH_3) = 5.8 mol of NH_3, so the [NH_3] = $5.8M$.

The relevant reaction is between the ammonia, a base, and water, an acid:

$$NH_3 + H_2O \rightleftharpoons NH_4^+ + OH^-$$

Before going further, we need the value of the equilibrium constant for this reaction. Table 15.2 does not list it, but it does list the value of K_A for the ammonium ion, NH_4^+, the conjugate acid of ammonia. There is a simple relationship between the K_A of NH_4^+ and the K of the relevant reaction. The relationship can be found by manipulating reactions with known equilibrium constants to obtain the reaction between NH_3 and H_2O.

We start by writing in reverse the reaction that defines the K_A of NH_4^+:

$$NH_3 + H^+ \rightleftharpoons NH_4^+$$

The equilibrium constant K for this reaction is the reciprocal of the K_A for NH_4^+, whose value we get from Table 15.2:

$$\frac{1}{K_A} = \frac{1}{5.59 \times 10^{-10}M}$$

We obtain the reaction of interest by combining the equation above with the equation for the water equilibrium:

$$NH_3 + H^+ \rightleftharpoons NH_4^+$$
$$\underline{H_2O \rightleftharpoons H^+ + OH^-}$$
$$NH_3 + H_2O \rightleftharpoons NH_4^+ + OH^-$$

The equilibrium constant of this reaction is the product of the equilibrium constants of the two reactions:

$$K = K_W \times \frac{1}{K_A} = \frac{K_W}{K_A} = \frac{1.00 \times 10^{-14}M^2}{5.59 \times 10^{-10}M^2} = 1.79 \times 10^{-9}M$$

The equilibrium constant for the reaction in which water transfers a proton to a base is K_W divided by the acid dissociation constant of the conjugate acid of the base. This is an important relationship to remember. This equilibrium constant is sometimes given the symbol K_B and is used to express relative strengths of bases. We shall not make use of the symbol K_B.

The equilibrium constant for the relevant equation is now known, and the rest of the calculation follows the pattern outlined in Example 15.8.

	NH$_3$	+ H$_2$O $\rightleftharpoons$	NH$_4^+$ +	OH$^-$
start	5.8M		0	0
equil	5.8M − x		x	x

Substituting into the expression for the equilibrium constant gives

$$K = \frac{(x)(x)}{5.8M - x} = 1.79 \times 10^{-5}M$$
$$x = 1.02 \times 10^{-2}M$$

The value of x is found by making the approximation that $5.8M - x \cong 5.8M$. Since $x = [OH^-] = 1.02 \times 10^{-2}M$, pOH = $2 - \log 1.02 = 2 - 0.009 = 1.99$. The pH = $14 - 1.99 = 12.01$.

Since this solution has a high concentration of NH$_3$, the answer is only approximate. The ideal solution approximation does not serve very well in such a concentrated solution.

We can often determine the dissociation constant of a weak acid by using a pH meter to measure the pH of a solution of a known concentration of the acid. Once this value is known, the pH is converted into K_A, the acid dissociation constant, by a calculation that is a reverse of the kind done in Examples 15.8 and 15.9.

EXAMPLE 15.10

The parent compound of the barbiturates, drugs that are widely used as central nervous system depressants, is barbituric acid, an organic acid. Aside from its practical use, barbituric acid is interesting because its acidic proton is bonded to a carbon atom, rather than an oxygen atom as is usually the case. We shall symbolize barbituric acid as HBar.

A solution is prepared in which the initial concentration of HBar is 0.25M. The pH of this solution is found to be 2.31. Calculate K_A of HBar.

SOLUTION

The relevant equilibrium and the tabulated concentrations are

	HBar	$\rightleftharpoons$ H$^+$ +	Bar$^-$
start	0.25M	0	0
equil	0.25M − x	x	x

The value of x, which is the $[H^+]$, can be found by converting the pH to a $[H^+]$ by the method used in Example 15.7.

$$2.31 = 3 - (1 - 0.31) = 3 - 0.69$$
$$[H^+] = \text{antilog}\,[-(3 - 0.6)] = 4.9 \times 10^{-3}M = x = [Bar^-]$$

The equilibrium concentrations can be found by using this value of x. We can then substitute the equilibrium concentrations into the expression for K_A.

$$K_A = \frac{[H^+][Bar^-]}{[HBar]} = \frac{(4.9 \times 10^{-3}M)^2}{0.25M - 4.9 \times 10^{-3}M} = 9.80 \times 10^{-5}M$$

The strength of a weak base can be measured by the same kind of experiment. However, the result is expressed as the K_A of the conjugate acid of the weak base.

EXAMPLE 15.11

Lactic acid gets its name from milk, in which it is found. Salts of this organic acid are called lactates. One salt, calcium lactate, is often used as a source of calcium for rapidly growing animals.

Calcium lactate can be represented as $Ca(Lac)_2$. A saturated solution of $Ca(Lac)_2$ contains 0.26 mol of this salt in 1 liter of solution. The pOH of such a solution is found to be 5.60 at 373 K. Assuming the ideal solution approximation and complete dissociation of the salt, calculate K_A of lactic acid.

SOLUTION

The relevant equilibrium that is established in a solution of lactate ion and the tabulated concentrations are:

	Lac^-	$+ H_2O \rightleftharpoons$	$HLac$	$+ OH^-$
start	$0.52M$		0	0
equil	$0.52M - x$		x	x

Since each mole of calcium lactate forms two moles of lactate ion, the starting concentration of lactate is $2 \times 0.26 = 0.52M$. The concentration of hydroxide ion and of lactic acid at equilibrium can be found from the pOH

$$[OH^-] = \text{antilog}\,(-5.60) = 2.5 \times 10^{-6}M = x = [HLac]$$

Since $x = 2.5 \times 10^{-6}M$, the value of the equilibrium constant for the reaction can be found by using the expression for the equilibrium constant. We use the approximation that $0.52M - x \cong 0.52M$.

$$K = \frac{(2.5 \times 10^{-6}M)^2}{0.52M} = 1.2 \times 10^{-11}M$$

In Example 15.9, we saw that the relationship between the equilibrium constant for the reaction of a base with water and the acid dissociation constant of the conjugate acid is

$$K = \frac{K_W}{K_A}$$

Thus, K_A for lactic acid can be found from the value of K calculated above:

$$1.2 \times 10^{-11} = \frac{1.0 \times 10^{-14}M}{K_A}$$

$$K_A = 8.3 \times 10^{-4}M$$

The common ion effect was mentioned in our discussion of solubility products. It can also be important in acid-base systems. The dissociation of a weak acid can be repressed by the presence of a different acid in the solution, and the reaction of a base with water can be repressed by the presence of a different base.

EXAMPLE 15.12

The artificial sweetener saccharin is a weak organic acid that can be represented as HSac. The concentration of Sac$^-$ ions at equilibrium in an acidic solution is lower than in a neutral solution. Since lemon juice is acidic, less ionization of HSac occurs in tea with lemon than in plain tea.

A 2.8×10^{-4} mol amount of saccharin is added to a glass of tea and lemon whose volume is 150 cm^3 and pH is 2.0. Calculate the [Sac$^-$] at equilibrium.

SOLUTION

This problem differs from most of those presented until now because there is a significant starting concentration of H$^+$. A pH of 2.0 corresponds to [H$^+$] = $1.0 \times 10^{-2}M$. The starting [HSac] = 2.8×10^{-4} mol/0.150 liter = 0.0019M.

The relevant equation is the dissociation of the weak acid:

	HSac	$\rightleftharpoons$ H$^+$	+ Sac$^-$
start	0.0019M	0.01M	0
equil	0.0019$M - x$	0.01$M + x$	x

The value of K_A is listed in Table 15.2 as $2.1 \times 10^{-12}M$. Since this value is small, and since the ionization of saccharin is repressed by the common ion, we can expect x to be very small. Thus, we approximate that $0.01M - x \cong 0.01M$, and that $0.0019M - x \cong 0.0019M$. Using these approximations and substituting the equilibrium concentrations into the expression for K_A gives

$$K_A = \frac{(0.01M)(x)}{0.0019M} = 2.1 \times 10^{-12}M$$

and

$$x = 4.0 \times 10^{-13}M = [\text{Sac}^-] \text{ at equilibrium}$$

This value is substantially lower than [Sac$^-$] in a neutral solution. The [Sac$^-$] in a solution of 0.0019M saccharin can be calculated to be $6.3 \times 10^{-8}M$.

The pH of a solution can affect the position of an acid-base equilibrium. We can take advantage of this phenomenon by adjusting experimental conditions in advance to obtain a given composition at equilibrium.

EXAMPLE 15.13

Aniline, $C_6H_5NH_2$, is a weak organic base. Its conjugate acid is $C_6H_5NH_3^+$, which is called the anilinium ion. The acidic proton is on the nitrogen atom in the anilinium ion. Aniline is a major intermediate in many important chemical processes, especially the manufacture of synthetic dyes. The success of a process often depends on careful control of the relative concentrations of the base and its conjugate acid.

In one such process, it is necessary to keep the concentration of anilinium ion no higher than $1.0 \times 10^{-9}M$ in a solution that is 0.10M in aniline. Find the necessary concentration of sodium hydroxide for this process.

SOLUTION

Let the desired [OH$^-$] = x. The relevant equilibrium for a solution of a weak base is

	$C_6H_5NH_2$	+ H$_2$O $\rightleftharpoons$ $C_6H_5NH_3^+$	+ OH$^-$
start	0.10M	0	x
equil	0.10$M - y$	y	$x + y$

Here y is the concentration of aniline that reacts when the system reaches equilibrium. Since the problem states that the equilibrium concentration of anilinium ion is $1.0 \times 10^{-9}M$, this is the value of y. Since y is so small, $0.10M - y \cong 0.10M$ and $x + y \cong x$. Table 15.2 gives the value of the K_A of the anilinium ion as $2.34 \times 10^{-5}M$. Therefore, K for this reaction is:

$$K = \frac{K_W}{K_A} = \frac{1.0 \times 10^{-14}M^2}{2.34 \times 10^{-5}M} = 4.3 \times 10^{-10}M$$

Substituting the concentrations on the "equil" line into the expression for K gives

$$K = \frac{(1.0 \times 10^{-9}M)(x)}{0.10} = 4.3 \times 10^{-10}$$

and $x = 0.043M$, the concentration of hydroxide ion needed to keep the concentration of anilinium ion at the desired level.

15.5 NEUTRALIZATION

Using the Brønsted definitions, the term "neutralization" can be used to describe the reaction of any acid with any base. But we shall use it in a more precise sense and say that a neutralization reaction occurs when an aqueous solution of an acid stronger than water is mixed with an aqueous solution of a base stronger than water. Neutralization procedures are generally carried out with stoichiometric quantities of acid and base. The correct molar quantities of acid and base are mixed in such a neutralization. In this section, we shall discuss some quantitative aspects of neutralization procedures.

Since both acids and bases can be classified as either strong or weak, there are four classes of neutralization reactions: (1) a strong acid with a strong base, (2) a strong acid with a weak base, (3) a weak acid with a strong base, and (4) a weak acid with a weak base.

1. Strong Acid-Strong Base. A solution of hydrochloric acid is mixed with a solution of sodium hydroxide. The strongest base in the system is OH^- and the strongest acid is H^+. The reaction occurs between the strongest acid and the strongest base:

$$H^+ + OH^- \rightleftharpoons H_2O$$

This is the net reaction that occurs when a solution of any strong acid is mixed with a solution of any strong base. It is the equation for the autoprotolysis of water, written in reverse, so its equilibrium constant is the reciprocal of K_W.

$$K = \frac{1'}{K_W} = \frac{1}{1.0 \times 10^{-14}M^2} = 1.0 \times 10^{14}M^{-2}$$

We can predict that this equilibrium constant will have a large value, since a vigorous exothermic reaction usually occurs when a strong acid and a strong base are mixed. In fact, the equilibrium constants of most neutralization reactions have large values.

2. Weak Acid-Strong Base. A solution of sodium hydroxide is mixed with a solution of formic acid. The strongest base in the system is OH^- and the strongest acid in the system is HCOOH, formic acid, which is stronger than water. (Table 15.2 lists the K_A of formic acid as $1.77 \times 10^{-4}M$.) The neutralization reaction is

$$OH^- + HCOOH \rightleftharpoons H_2O + HCOO^-$$

The equilibrium constant for this reaction is related to the K_A of the weak acid and to K_W. If we combine the equation that defines K_A with the reverse of the equation that defines K_W we get

$$HCOOH \rightleftharpoons H^+ + HCOO^- \qquad K_A$$

$$H^+ + OH^- \rightleftharpoons H_2O \qquad \frac{1}{K_W}$$

$$\overline{HCOOH + OH^- \rightleftharpoons H_2O + HCOO^- \qquad K_A\left(\frac{1}{K_W}\right)}$$

The resulting equation is the neutralization reaction for this process. Therefore, its equilibrium constant $K = K_A/K_W$. In this case,

$$K = \frac{1.77 \times 10^{-4}M}{1.0 \times 10^{-14}M^2} = 1.77 \times 10^{10}M^{-1}$$

The expression K_A/K_W defines the equilibrium constant for the neutralization of any weak acid with any strong base in solution. This kind of neutralization reaction is the reverse of the reaction of the conjugate base of a weak acid with water. The equilibrium constant for that reaction is K_W/K_A, the reciprocal of the equilibrium constant of this reaction.

    3. Strong Acid-Weak Base. A solution of nitric acid is mixed with a solution of sodium formate. The strongest acid in the system is H^+ and the strongest base is the formate ion, $HCOO^-$. The reaction for the neutralization is

$$H^+ + HCOO^- \rightleftharpoons HCOOH$$

This reaction is the reverse of the reaction that defines K_A for the system. Therefore, its equilibrium constant $K = 1/K_A$. In this case, $K = 1/(1.77 \times 10^{-4}M) = 5.65 \times 10^3 M^{-1}$. Again, the equilibrium constant for the neutralization reaction is large. The expression $1/K_A$ defines the equilibrium constant for the neutralization of any strong acid with any weak base in solution, where K_A is the acid dissociation constant of the conjugate acid of the weak base.

    4. Weak Acid-Weak Base. A solution of ammonium chloride is mixed with a solution of potassium cyanide. The ions that are formed by these salts in solution are NH_4^+, the weak acid; Cl^-, which is the conjugate base of a strong acid and therefore has no appreciable base strength in water; K^+, which does not enter into acid-base reactions in water; and CN^-, the conjugate base of a weak acid. The strongest acid in the system is the ammonium ion, which is a stronger acid than water. The strongest base is the cyanide ion, which is a stronger base than water. The neutralization reaction occurs between these two ions:

$$NH_4^+ + CN^- \rightleftharpoons NH_3 + HCN$$

To find the equilibrium constant, we must combine reactions of species with known equilibrium constants to give this reaction:

$$NH_4^+ \rightleftharpoons NH_3 + H^+ \qquad K_A(NH_4^+)$$

$$CN^- + H^+ \rightleftharpoons HCN \qquad \frac{1}{K_A(HCN)}$$

$$\overline{NH_4^+ + CN^- \rightleftharpoons HN_3 + HCN \qquad K_A(NH_4^+) \times \frac{1}{K_A(HCN)}}$$

Thus, from the data in Table 15.2,

$$K = \frac{K_A(NH_4^+)}{K_A(HCN)} = \frac{5.59 \times 10^{-10}M}{4.93 \times 10^{-10}M} = 1.13$$

The equilibrium constant for the neutralization of a weak acid by a weak base is the acid dissociation constant of the weak acid divided by the acid dissociation constant of the conjugate acid of the weak base. Such an equilibrium constant often is not very large. For this reason, this type of neutralization is the least common.

This four-part classification will be useful to you in calculations relating to neutralization reactions. You will find it helpful to decide which of the four types of neutralization is occurring, and therefore which acid dissociation constants are needed to calculate the relevant equilibrium constant.

Titration

Neutralization reactions are quite common in chemistry laboratories. Because the equilibrium constants of most neutralization reactions are large, these reactions can be treated as if they proceed to completion. Neutralization reactions often are used for quantitative analysis of acidic or basic substances.

Most often, a neutralization is carried out by using a technique called *titration,* in which a measured volume of one solution is added gradually to a measured amount of another solution. Figure 15.1 shows a typical titration apparatus. The addition of the solution is usually regulated by a buret, which is a graduated tube with a stopcock at the bottom. The stopcock is used to control the rate of flow of the solution out of the buret, while the graduations on the buret indicate the volume of solution that has been added at any point during the titration.

For a successful neutralization, we need a method of indicating when the desired stoichiometric quantity of the solution in the buret is added to the other solution. There are a number of methods, all based on measuring a property of the system that changes as one solution is added to another. Either instruments or colored substances called *indicators* can be used to make these measurements. We shall discuss indicators later in this chapter.

We shall consider two quantitative factors in a neutralization: the stoichiometry of the reaction and the composition of the system when the neutralization is complete.

We approach the stoichiometry of a neutralization reaction in the same way as we approach the stoichiometry of any other chemical reaction. We start with a balanced chemical equation that gives the molar ratios of the substances in the reaction. We may then consider the composition of the solutions of acids and bases after they have been mixed and equilibrium is reached.

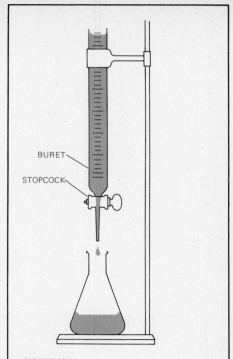

FIGURE 15.1
Titration, as it is usually done in the laboratory. The flask contains a measured volume of solution. A measured volume of another solution is added gradually. The rate of flow is controlled by the stopcock in the buret.

EXAMPLE 15.14

A volume of 24.4 cm³ of 0.117M nitric acid solution is needed to neutralize 50.1 cm³ of a sodium hydroxide solution. What is the concentration of the solution of sodium hydroxide?

SOLUTION

The equation for the neutralization is $H^+ + OH^- \rightleftharpoons H_2O$. One mole of protons neutralizes one mole of hydroxide ions. Therefore, the amount of protons in 24.4 cm³ of nitric acid solution is equal to the amount of hydroxide ion in 50.1 cm³ of sodium hydroxide solution. From the definition of molarity we can write

moles (n) = molarity (M) × volume (V)

where the volume is measured in liters.

Using this relationship and the given data, we get:

$$n_{H^+} = \left(0.117 \, \frac{\text{mol } HNO_3}{\text{liter}} \right) \times \frac{1 \text{ mol } H^+}{1 \text{ mol } HNO_3} \times 24.4 \text{ cm}^3 \times \frac{1 \text{ liter}}{1000 \text{ cm}^3}$$

$$= 2.85 \times 10^{-3} \text{ mol } H^+$$

From the stoichiometry, the amount of OH^- in the 50.1 cm³ of sodium hydroxide solution must also be 2.85×10^{-3} mol OH^-. Therefore, the concentration of the sodium hydroxide solution is:

$$M = \frac{n_{OH^-}}{V} = \frac{(2.85 \times 10^{-13}\,\text{mol OH}^-)}{50.1\,\text{cm}^3} \times \frac{1\,\text{mol NaOH}}{1\,\text{mol OH}^-} \times \frac{1000\,\text{cm}^3}{\text{liter}} = 0.0569\,M$$

which is the concentration of NaOH.

The solution of Example 15.14 demonstrates a useful generalization about the neutralization of any strong acid with any strong base: The number of moles of protons must equal the number of moles of hydroxide ions. There are several ways in which this generalization can be expressed, including

$$n_{H^+} = n_{OH^-} \tag{15.1}$$

or

$$(M_{H^+})(V_{H^+}) = (M_{OH^-})(V_{OH^-}) \tag{15.2}$$

EXAMPLE 15.15

What volume of a solution of 0.129M hydrochloric acid is required to neutralize 0.237 g of barium hydroxide?

SOLUTION

Since the mass of the barium hydroxide is given, the number of moles of barium hydroxide can be found by using the table of atomic weights:

$$n_{Ba(OH)_2} = \frac{0.237\,\text{g Ba(OH)}_2}{171.4\,\text{g Ba(OH)}_2/\text{mol Ba(OH)}_2} = 1.38 \times 10^{-3}\,\text{mol Ba(OH)}_2$$

Since each mole of $Ba(OH)_2$ is a source of two moles of hydroxide ion, $n_{OH^-} = (2\,\text{mol OH}^-/\text{mol Ba(OH)}_2)(1.38 \times 10^{-3}\,\text{mol Ba(OH)}_2) = 2.76 \times 10^{-3}\,\text{mol OH}^-$.

We can combine Equations 15.1 and 15.2 to derive a relationship suitable for this example:

$$(M_{H^+})(V_{H^+}) = n_{OH^-}$$

Since HCl has only one proton to donate, the molarity of H^+ is the same as the molarity of the hydrochloric acid. Using the data and the relationship above gives:

$$(0.129\,\text{mol/liter})(V_{H^+}) = 2.76 \times 10^{-3}\,\text{mol}$$

and

$$V_{H^+} = 0.0214\,\text{liter} = 21.4\,\text{cm}^3$$

Equations 15.1 and 15.2 can be generalized even further. Even in the case of weak acids or weak bases, neutralizations proceed essentially to completion. Therefore, the symbol n_{H^+} need not mean only the number of dissociated protons in a solution. It can also be taken to mean the number of moles of available protons from the acid in the neutralization.

For example, acetic acid is a relatively weak acid. In solution, acetic acid is largely undissociated. But if a stoichiometric quantity of a strong base such as hydroxide ion is present, essentially all of the protons of the acetic acid will be transferred to the hydroxide ion. For neutralization calculations, a solution of 0.1M acetic acid has 0.1 mol of protons available in each liter of solution.

We can use the same reasoning for solutions of acids that have more than one available proton per molecule. In a neutralization of sulfuric acid, each mole of H_2SO_4 can be a source of two moles of protons. Similarly, one mole of phosphoric acid, H_3PO_4, can be a source of three moles of protons. Such polyprotic acids, as they are called, will be discussed in more detail in Section 15.7.

As we generalize the meaning of the symbol n_{H^+}, we can also generalize the meaning of the symbol n_{OH^-}. We can take it to mean either the total number of moles of available hydroxide ion in the solution or the total number of moles of protons that can be accepted by the base in the solution. Just as some acids can donate more than one proton per molecule, some bases can accept more than one proton per molecule. Each mole of the weak base CO_3^{2-}, carbonate ion, can accept two moles of protons and form carbonic acid, H_2CO_3. Each phosphate anion, PO_4^{3-}, can accept up to three protons to form H_3PO_4.

This generalized definition of n_{H^+} and n_{OH^-} simplifies stoichiometric calculations of neutralization reactions between polyprotic acids and bases that can accept more than one proton.

Neutralization procedures have many applications other than the analysis of the composition of mixtures. A classic type of experiment is illustrated in the next example.

EXAMPLE 15.16
You are asked to calculate the molecular weight of a solid weak acid of unknown composition. The acid donates one proton per molecule. When a 1.02-g sample of the acid is dissolved in water, the resulting solution requires 48.0 cm^3 of a 0.241M solution of sodium hydroxide for neutralization.
SOLUTION
The number of moles of acid is equal to the number of moles of hydroxide ion:

$$n_{OH^-} = (M_{OH^-})(V_{OH^-}) = \left(0.241\ \frac{\text{mol OH}^-}{\text{liter}}\right)(48.0\ \text{cm}^3)\left(\frac{1\ \text{liter}}{1000\ \text{cm}^3}\right)$$

$$= 0.0116\ \text{mol OH}^- = n_{H^+}$$

We now know the mass and the number of moles of the acid, and the molecular weight of the acid can thus be calculated:

$$MW = \frac{1.02\ \text{g acid}}{0.0116\ \text{mol acid}} = 88.2$$

In Example 15.16, you were given the information that the unknown acid has only one proton to donate. This kind of information about an unknown acid usually is not available, because the composition of the acid is unknown. In such cases, it is more common to measure what is called the equivalent weight of the acid, the molecular weight divided by the number of protons that one molecule of the acid donates.

$$\text{equivalent weight} = \frac{\text{molecular weight}}{\text{available protons per molecule}}$$

In the case of a monoprotic acid, which has only one proton per molecule to donate, the molecular weight and the equivalent weight are identical. For a diprotic acid, which has two protons per molecule to donate, the equivalent weight is half the molecular weight. For an acid that has three protons per molecule to donate, the equivalent weight is one-third of the molecular weight.

Just as molecular weight can be defined as the weight of one mole of a substance, we can define the equivalent weight as the weight of one equivalent, that quantity of acid which furnishes one mole of protons to a base, or that quantity of a base which can accept one mole of protons. An experiment of the kind described in Example 15.16 can give only the equivalent weight of an acid, not the molecular weight if the composition of the acid is unknown. Only if the number of acidic protons per molecule is known, does the experiment give us the molecular weight.

The Equivalence Point

A neutralization procedure usually is carried out by titration. A quantity of a solution in a buret is added gradually to another solution. When the exact volume of the solution in the buret required for neutralization has been added to the other solution, the *equivalence point* has been reached.

A major object of many neutralization procedures is the determination of the equilibrium composition of the system at the equivalence point. Equivalence-point calculations use the same procedure as other equilibrium calculations. The pH of the solution at the equivalence point is of special interest, since most methods of monitoring a neutralization measure the pH.

The pH at the equivalence point in a neutralization is determined by the nature of the acid and the base. In the neutralization of any strong acid with any strong base, the pH at the equivalence point is 7.00 and the solution is neutral. In the neutralization of a weak acid by a strong base, the pH at the equivalence point is greater than 7 and the solution is alkaline; the exact pH depends on the concentrations of the solutions and on the nature of the weak acid. In the neutralization of a strong acid by a weak base, the pH of the solution at the equivalence point is less than 7 and the solution is acidic.

EXAMPLE 15.17

Calculate the pH at the equivalence point when a solution of $0.10M$ HF is titrated with a solution of $0.10M$ NaOH. Calculate the pH at the equivalence point when a $0.10M$ solution of NaF is titrated with a $0.10M$ solution of HCl.

SOLUTION

The two parts of the problem are related. In each part, the first step is to write the relevant reaction, the starting conditions, and the equilibrium line. In both parts, we must be careful to correct the concentrations of the original solutions to account for the increased volume of the solution at the equivalence point.

The solutions of the acids and of the bases have the same initial concentrations. Therefore, equal volumes of each solution are required for neutralization. As a result, the volume of the solution at the equivalence point is twice the original volume of either solution, and the concentrations are half those of the original solutions.

$$HF \quad + \quad OH^- \quad \rightleftharpoons \quad F^- + H_2O$$

start $\qquad \dfrac{0.10M}{2} \qquad \dfrac{0.10M}{2} \qquad 0$

K is large for this reaction, so a completion step is helpful:

complete $\quad$ 0 $\qquad\quad$ 0 $\qquad\qquad$ 0.050M
equil $\qquad\;\; x$ $\qquad\qquad x$ $\qquad\qquad$ 0.050$M - x$

As usual, the concentration of water is not included.

The value of K for this reaction is

$$K = \frac{K_A}{K_W} = \frac{3.53 \times 10^{-4}M}{1.0 \times 10^{-14}M^2} = 3.53 \times 10^{10}M^{-1}$$

which is found by using the data in Table 15.2.

From the concentration terms on the equilibrium line and the expression for K, we obtain:

$$K = \frac{(0.050M - x)}{x^2} = 3.53 \times 10^{10}M^{-1}$$

Making the approximation that $0.050M - x \cong 0.050M$, we find:

$$x = 1.19 \times 10^{-6}M = [OH^-]$$
$$pOH = 6 - \log 1.19 = 5.92$$
$$pH = 14 - pOH = 14 - 5.92 = 8.08$$

The second part of the problem is handled in the same way. The relevant equation and tabulated data are:

	F^-	$+ \quad H^+$	$\rightleftharpoons$	HF
start	$\dfrac{0.10M}{2}$	$\dfrac{0.10M}{2}$		0
complete	0	0		$0.050M$
equil	x	x		$0.050M - x$

$$K = \frac{1}{K_A} = \frac{1}{3.53 \times 10^{-4}M} = \frac{(0.050M)}{x^2}$$

Since x is small, we can again make the approximation that $0.050M - x \cong 0.050M$.

$$x = 4.20 \times 10^{-3}M = [H^+]$$
$$pH = 3 - \log 4.20 = 2.38$$

Indicators

Clearly, a neutralization procedure needs some method of detecting the equivalence point, when the addition of solution from the buret should be stopped. The most direct method is to use an instrument called a pH meter to measure the pH as the solution from the buret is added. But a simpler and cheaper method of detecting the equivalence point is to use a substance called an **indicator.** An indicator usually is an organic molecule of complex structure, whose color in aqueous solution changes as the pH of the solution changes.

Tea is an everyday example of an indicator. Its color changes as an acid, lemon juice, is added. Many indicators are used in the laboratory. We can represent any one of them as a weak acid of the general formula HIn, where In represents the indicator. Like any weak acid, an indicator dissociates in aqueous solution:

$$HIn \rightleftharpoons H^+ + In^-$$

If HIn and its conjugate base, In^-, have different colors, the relative quantities of each acid and conjugate base determine the color of a solution that we see. Since these relative quantities depend on the $[H^+]$, the color of the solution changes as the pH changes.

Perhaps the best-known laboratory indicator is litmus, whose acid form, HIn, is pink, and whose base form, In^-, is blue. According to the principle of Le Chatelier, when the $[H^+]$ is high, the dissociation of HIn is repressed. The relatively high [HIn] causes the solution to appear pink. When the $[H^+]$ is low, the dissociation of HIn is enhanced, and the relatively high $[In^-]$ causes the solution to appear blue.

Suppose we are titrating a solution of an acid with a solution of a base, using litmus as an indicator. The pH increases, until at some point we see a change in the color of the indicator from pink to blue. We can define the exact point of the color change as the pH at which $[HIn] = [In^-]$. (However, our eyes are not

TABLE 15.4
Acid-Base Indicators[a]

Common Name	pK_{ind}	pH Interval[b]	Color Acid	Alkaline
cresol red[c]		0.2–1.8	red	yellow
thymol blue[c]	1.6	1.2–2.8	red	yellow
tropeoline OO		1.3–3.0	red	yellow
methyl yellow	3.3	2.8–4.0	red	yellow
bromphenol blue	3.8	3.0–4.6	yellow	purple
methyl orange	3.5	3.1–4.4	red	yellow
bromcresol green	4.7	3.8–5.4	yellow	blue
methyl red	5.0	4.2–6.2	red	yellow
chlorphenol red	6.0	4.8–6.4	yellow	red
bromcresol purple	6.1	5.2–6.8	yellow	purple
bromthymol blue	7.1	6.0–7.6	yellow	blue
phenol red	7.8	6.4–8.0	yellow	red
neutral red	6.8	6.8–8.0	red	yellow-brown
cresol red[d]	8.1	7.2–8.8	yellow	red
cresol purple[d]	8.3	7.4–9.0	yellow	purple
thymol blue[d]	8.9	8.0–9.6	yellow	blue
phenolphthalein	9.3	8.0–9.8	colorless	red-violet
thymolphthalein		9.3–10.5	colorless	blue
alizarin yellow		10.1–12.0	yellow	violet

[a] From Gilbert H. Ayres, *Quantitative Chemical Analysis*, Second Edition, New York, Harper and Row, 1968.
[b] Will vary with the observer.
[c] Acid range; the indicator has two color change intervals.
[d] Alkaline range; the indicator has two color change intervals.

sensitive enough to detect very subtle color changes. What we actually see is a color change over a pH interval that depends on the ability of our eyes to detect one color in the presence of another.) The value of the pH at the point where [HIn] = [In⁻] depends on the value of the K_A for the indicator equilibrium. This equilibrium constant is sometimes symbolized by K_{In}.

$$K_{In} = \frac{[H^+][In^-]}{[HIn]}$$

At the $[H^+]$ where the color change occurs, $[In^-] = [HIn]$, and this expression becomes

$$K_{In} = [H^+] \quad \text{or} \quad pK_{In} = pH$$

In other words, if the pK_{In} of an indicator is equal to the pH at the equivalence point, the indicator changes color at the equivalence point. Table 15.4 lists some indicators and their pK_{In} values.

There are many advantages to the use of an indicator. No instruments are needed. You need only look to detect the equivalence point in a neutralization, if the proper indicator is used. In Example 15.17, a suitable indicator for the titration of HF by NaOH is one with $pK_{In} = 8.08$, the pH of the solution at the equivalence point. Similarly, a suitable indicator for the titration of NaF with HCl is one with $pK_{In} = 2.38$.

But there are also some possible problems in the use of indicators. It may not be possible to find an indicator with a pK_{In} that coincides exactly with the pH at the equivalence point. The experimenter's color perception may not be perfect, so the color change is not detected exactly at the equivalence point. Given these sources of imprecision, will there be a large error in the measured volume at the equivalence point?

This question can be answered by studying how the pH of a solution changes as the titration proceeds. It is possible to calculate the pH at any point during a titration and to present the results graphically. Such a plot of pH against volume of added solution is called a *titration curve*. The simplest such curve, shown in Figure 15.2, describes the titration of a strong acid such as HCl with a strong base such as NaOH. The most significant feature of this curve is the steepness of the slope around the equivalence point. The fact that the curve is so steep means that very large changes in pH are caused by small volumes of added solution. Therefore, a large error in monitoring the pH will cause only a small error in measuring the volume of added solution.

Titration curves for the neutralization of a weak acid with a strong base (Figure 15.3) or of a weak base with a strong acid are not as steep around the equivalence point. In these titrations, therefore, the use of an indicator whose pK_{In} is relatively close to the pH at the equivalence point is desirable.

EXAMPLE 15.18

Ephedrine, $C_{10}H_{15}NO$, is a weak organic base. It is a central nervous system stimulant that is often used in nasal sprays because it is also a decongestant. Table 15.2 lists the K_A of the conjugate acid of ephedrine as $7.26 \times 10^{-11}M$. To monitor the composition of nasal sprays, solutions of ephedrine in water whose concentrations are around $0.2M$ can be titrated with hydrochloric acid solutions of about the same concentration. Which of the indicators listed in Table 15.4 is most suitable for use in this titration?

SOLUTION

The most suitable indicator for the titration is one whose $pK_{In} = $ pH at the equivalence point. We can calculate the pH at the equivalence point, using Eph to represent ephedrine and EphH$^+$ to represent its conjugate acid.

	Eph	+ H$^+$	$\rightleftharpoons$ EphH$^+$
start	$\dfrac{0.2M}{2}$	$\dfrac{0.2M}{2}$	0
complete	0	0	$0.1M$
equil	x	x	$0.1M - x$

Since $K = 1/K_A = 1/(7.6 \times 10^{-11}M) = 1.4 \times 10^{10}M^{-1}$, $0.1M - x \cong 0.1M$. Therefore

$$\frac{0.1M}{x^2} = 1.4 \times 10^{10}M^{-1}$$

and

$x = 2.7 \times 10^{-6}M = [H^+]$
pH $= 6 - \log 2.7 = 5.6$

The indicator whose pK_{In} is closest to this pH is chlorphenol red, with $pK_{In} = 6.0$.

15.6 BUFFERS

While pure water has a pH of 7, it is almost impossible to obtain a sample whose pH is even close to 7. Even small traces of dissolved acidic or basic impurities can cause the pH of water to vary substantially, and such impurities are present in virtually every sample of water.

When water is exposed to air, some of the carbon dioxide in the atmosphere dissolves. When such a solution is saturated, $[CO_2]$ is about $1 \times 10^{-5}M$. As we shall see in Section 15.7, carbon dioxide acts as a weak acid when it is dissolved

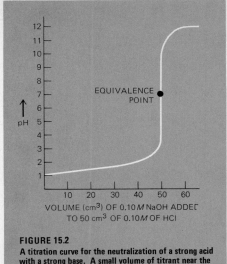

FIGURE 15.2
A titration curve for the neutralization of a strong acid with a strong base. A small volume of titrant near the equivalence point causes a large change in pH.

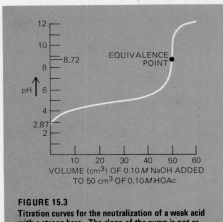

FIGURE 15.3
Titration curves for the neutralization of a weak acid with a strong base. The slope of the curve is not as steep at the equivalence point as the slope of the curve for the neutralization of a strong base with a strong acid.

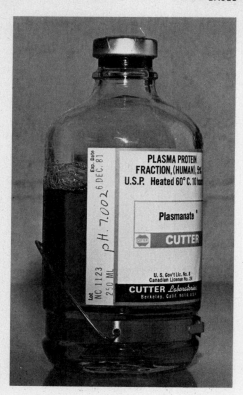

The importance of pH in physiological solutions is demonstrated by the prominent display of the pH of a unit of blood plasma destined for transfusion. (*Lester Bergman & Assoc.*)

in water. The pH of water that is saturated with respect to the carbon dioxide in the air is about 5.7.

Since pH is the negative logarithm of the concentration of dissolved protons, changes in pH reflect exponential changes in [H⁺]. Many chemical processes, including a large number of reactions in aqueous solutions, are sensitive to variations in [H⁺]. For example, there are many reactions that proceed rapidly at a given pH but much more slowly if the pH is several units higher or lower than the optimum. In some reactions, variations in the pH of a solution cause changes in the products.

Many reactions in living organisms are pH-dependent, and proceed normally only in a narrow pH range. A change in the pH of the aqueous solutions in a living organism can be fatal. Living organisms thus need a mechanism to keep the pH of their physiological solutions relatively constant.

In this respect, a living organism has the same problem as a chemist who must carry out pH-sensitive reactions. In fact, the method used by a chemist to prevent large variations in pH even when relatively large amounts of acid or base are added to a system is essentially the same as the method used in biological systems. In a living organism, as in a laboratory, the way to minimize the decrease in the pH of a solution when acid is added to a system is to have some base already present in the system to neutralize the acid. Similarly, the way to minimize the increase in the pH when base is added is to have some acid in the system. In other words, a system will keep a relatively constant pH if it already contains a quantity of acid and a quantity of base.

A solution that is relatively resistant to changes in pH is called a buffer solution. We can also say that the solution is buffered. The simplest way to achieve substantial concentrations of both acid and base in the same solution is to use an acid and its conjugate base. The most common form of buffer solution is one that contains a weak acid and its conjugate base. *The chemical reaction that is most conveniently used to calculate the equilibrium composition or chemical behavior of such buffer solutions is the equation that defines K_A, the dissociation constant of the weak acid.*

For example, a solution can be prepared from formic acid, a weak acid, and sodium formate, a salt of its conjugate base. Transfer of a proton from formic acid to formate ion does not result in an overall chemical change. $HCOOH + HCOO^- \rightleftharpoons HCOO^- + HCOOH$. Therefore the composition of an aqueous solution of formic acid and formate ion at equilibrium—or, more generally, of any aqueous solution of a weak acid and its conjugate base—depends primarily on the quantities of each that are originally added to the solution.

The buffer solution consisting of formic acid and sodium formate is most conveniently described by the reaction that defines K_A of formic acid:

$$HCOOH \rightleftharpoons HCOO^- + H^+$$

The pH of a buffer solution is determined mainly by two factors: the value of K_A of the weak acid, and the ratio of the concentrations of the weak acid and its conjugate base in the buffer solution. The action of a buffer can be understood qualitatively by applying the principle of Le Chatelier to the acid dissociation equilibrium reaction that we used to describe the buffer. If acid (H⁺) is added to the formic acid-formate buffer, the pH of the solution does not change appreciably, since the acid dissociation equilibrium reaction of formic acid proceeds from right to left, consuming most of the added H⁺. If the amount of added H⁺ is not too large, the ratio of formic acid to formate does not change very much. If base is added to the buffer, some of the H⁺ in the solution is consumed by reaction with the base. But the acid dissociation reaction proceeds from left to right to restore most of the lost H⁺ and return the system to equilibrium. Again, if the amount of added base is not too large, the ratio of formic acid to formate does not change appreciably.

EXAMPLE 15.19

Calculate the pH of a buffer solution prepared by dissolving 40.2 g of NH_4Cl and 20.3 g of NH_3 in enough water to form 1 liter of solution.

SOLUTION

Since NH_4Cl is a salt that is a source of NH_4^+ cations, the relevant equilibrium is:

$$NH_4^+ \rightleftharpoons NH_3 + H^+$$

From the data in Table 15.2:

$$K_A = \frac{[NH_3][H^+]}{[NH_4^+]} = 5.59 \times 10^{-10}M$$

From data given in the problem, and the appropriate molecular weights:

$$[NH_4^+] = \frac{\dfrac{40.2 \text{ g } NH_4Cl}{53.5 \text{ g } NH_4Cl/\text{mol } NH_4Cl}}{1 \text{ liter}} = 0.751M$$

$$[NH_3] = \frac{\dfrac{20.3 \text{ g } NH_3}{17.0 \text{ g } NH_3/\text{mol } NH_3}}{1 \text{ liter}} = 1.19M$$

The $[H^+]$ can be calculated by substituting this concentration data into the equilibrium constant expression for K_A of NH_4^+:

$$\frac{(1.19M)[H^+]}{(0.752M)} = 5.59 \times 10^{-10}M$$

$$[H^+] = 3.53 \times 10^{-10}M$$

$$pH = 9.452$$

A full-fledged equilibrium calculation is not necessary in a problem of this type. The unknown in this kind of calculation—in this case the $[H^+]$ that is formed or the $[NH_4^+]$ that must react for the system to reach equilibrium—is negligibly small compared to the concentrations of the components of the buffer. The direct substitution is sufficiently accurate.

Example 15.19 illustrates how the pH of a buffer solution depends on the *ratio* of the concentrations of the acid and the base, not on their absolute quantities. In Example 15.19 the ratio of the concentration of ammonium to ammonia is 1.19:0.752, and the pH of the buffer is 9.45. Any solution containing these two substances in this ratio will have a pH = 9.45, assuming the ideal solution approximation is valid.

The relationship between the pH of the solution, the pK_A of the weak acid used to prepare it, and the ratio of the concentrations of the weak acid, HA, and its conjugate base, A^-, can be expressed in convenient form by converting the equilibrium expression that defines K_A into negative logarithmic form:

$$K_A = \frac{[H^+][A^-]}{[HA]}$$

or, rearranging terms:

$$[H^+] = \frac{K_A[HA]}{[A^-]}$$

Taking the negative logarithms of both sides of this equation gives

$$-\log [H^+] = -\log \left(\frac{K_A[HA]}{[A^-]} \right) = -\log K_A + \log \frac{[A^-]}{[HA]}$$

BREATH, BLOOD, AND pH

"Take a deep breath" is a common bit of advice for someone in trouble. But there are circumstances in which deep breathing can cause trouble. Hyperventilation—breathing too deeply for too long—can cause symptoms that mimic heart disease and that can bring sudden death to divers.

When a swimmer takes a number of deep breaths before diving into the water, several things happen inside the body. More air flows into the lungs, and the composition of the gases in the lungs changes. The partial pressure of O_2 goes up and the partial pressure of CO_2 goes down. Breathing is controlled by the nerve cells in the carotid body, in the carotid artery of the head, which monitors blood O_2, CO_2, and pH. As the blood content of O_2 goes up and the CO_2 concentration goes down, the pH rises, since a loss of CO_2 decreases the carbonic acid content of the blood plasma.

The effect of this alkalosis, as it is called, can be a constriction of the blood vessels, including those in the brain. At the same time, the reduced CO_2 blood level and increased O_2 level affect the carotid body so that the swimmer can wait much longer before breathing. Too long, in fact, because the O_2 pressure may fall below the level needed to sustain consciousness. There have been several cases recorded of divers who have suddenly lost consciousness in this way. Unless immediate help is available, the swimmer may simply sink to the bottom and drown.

Even for persons on dry land, a small change in blood pH can cause problems. The normal pH of arterial blood is about 7.40. An increase to only 7.45 can be produced by the wrong kind of breathing habits. If an individual breathes too often and too deeply, CO_2 is lost from the body and arterial pH rises. The resulting symptoms can include light-headedness, agitation, a burning or prickling sensation in the limbs, fainting, and severe chest pains.

In some cases, the pains are so severe that the individual is diagnosed as having heart disease. The usual diagnosis is angina pectoris, a condition that causes excruciating chest pains. In true angina, the pain results from partial blockage of the coronary arteries, which reduces the flow of blood to the heart muscle. In false angina, the pain can be due to hyperventilation.

If hyperventilation is known to be the problem, the cure may be as simple as having the patient breathe into a paper bag. Exhaled breath has a higher pressure of CO_2 than ordinary air, so exhaled air helps bring the blood CO_2 level back to normal. But for individuals who hyperventilate by force of habit, weeks or even months of training may be needed to restore a normal breathing pattern that will keep blood pH within the normal range.

or:

$$pH = pK_A + \frac{\log [A^-]}{[HA]} \tag{15.3}$$

Equation 15.3 is sometimes called the Henderson-Hasselbalch equation. It is used extensively in biology. It is a valid approximation when the ratio of weak acid to its conjugate base is neither very large nor very small, that is, when the pH of the solution does not differ greatly from the pK_A of the acid.

Since a buffer solution must maintain a relatively constant pH when either acid or base is added, it is usually desirable for the buffer to have roughly comparable concentrations of the weak acid and its conjugate base. In the ideal case, when $[HA] = [A^-]$, $\log [A^-]/[HA] = 0$, and $pH = pK_A$.

In the buffer system of Example 15.19, for instance, if $[NH_4^+] = [NH_3] = 0.1M$, then $K_A = (0.1M)[H^+]/(0.1M) = 5.64 \times 10^{-10}M$, $[H^+] = 5.64 \times 10^{-10}M$, and $pH = 9.25$.

The Henderson-Hasselbalch equation is important to biologists because their experimental work often requires the use of solutions that are buffered at or close to a given pH. For a buffer to contain roughly comparable concentrations of the weak acid and its conjugate base, a weak acid with a pK_A close to the pH desired for the buffer solution is needed.

EXAMPLE 15.20

A buffer solution of pH = 3.00 is needed. Using the data in Table 15.2, suggest an appropriate weak acid-conjugate base pair, and determine the ratio of the concentrations of the two substances in the solution at this pH.

SOLUTION

A pH of 3.00 corresponds to $[H^+] = 1.0 \times 10^{-3}M$. We should select a weak acid with a K_A as close to $1.0 \times 10^{-3}M$ as possible. Chloroacetic acid, $ClCH_2COOH$, has $K_A = 1.40 \times 10^{-3}$, $pK_A = 2.85$, and is therefore suitable.

The expression for K_A with the desired $[H^+]$ is:

$$K_A = \frac{[ClCH_2CO_2^-](1.0 \times 10^{-3}M)}{[ClCH_2CO_2H]} = 1.40 \times 10^{-3}M$$

This equation can be rearranged to solve for the desired ratio:

$$\frac{[ClCH_2CO_2^-]}{[ClCH_2CO_2H]} = \frac{1.40 \times 10^{-3}M}{1.0 \times 10^{-3}M} = 1.40$$

If we select a weak acid whose pK_A differs substantially from the pH desired for the buffer solution, then one of the two components used to prepare the buffer will have to be present in great excess. Working with buffer solutions that have disparate concentrations of the two components usually is quite inconvenient.

The *capacity* of a buffer refers to the amount of acid or base that can be added to the buffer solution without changing its pH appreciably. The capacity of the buffer solution is related to the quantities of the weak acid and its conjugate base that are used to prepare the solution. A solution of $0.1M$ HCOOH and $0.1M$ HCOO$^-$ has roughly one-tenth the capacity of a buffer solution of $1M$ HCOOH and $1M$ HCOO$^-$, even though both solutions have the same pH. The pH of the buffer solution depends on the *ratio* of the two components, their relative amounts, while the capacity of the buffer solution depends on the actual *concentration* of the two components, their absolute amounts.

It is often necessary to calculate the change in the pH of a buffer solution caused by the addition of a given amount of an acid or a base.

EXAMPLE 15.21

Proteins play an important role in buffering the blood, lymph, and other physiological fluids of animals. Proteins are large molecules, portions of which act as weak acids or weak bases. A weak organic nitrogen-containing base called imidazole is found in various parts of protein molecules, and often plays a role in their buffering action. Imidazole can accept a proton to form the imidazolium ion, in the same way that ammonia can accept a proton to form the ammonium ion. The acid dissociation equilibrium can be symbolized as

$$ImidNH^+ \rightleftharpoons ImidN + H^+$$

where $ImidNH^+$ is the imidazolium ion and $ImidN$ is imidazole. Table 15.2 gives K_A as $1.11 \times 10^{-7}M$.

A study of the complex protein buffer system could start with a study of the relatively simple imidazole system. A buffer prepared from the imidazolium ion—imidazole acid-base system ideally has equal concentrations of both substances. A simple way to equalize the concentrations is to start with a solution of imidazole in water and to neutralize half of the dissolved imidazole with a strong acid. If half of a quantity of imidazole is converted to the imidazolium ion by the strong acid, one obtains the ideal buffer solution, containing equal concentrations of acid and base.

A solution of 1.00 mol of imidazole and 0.50 mol of HNO_3 in a 1.00-liter volume of solution is prepared. Calculate the pH of this solution. A sample of 4.00 g of NaOH is dissolved in the solution. Calculate the pH of this new solution.

SOLUTION

Since the amount of strong acid, HNO_3, is half the amount of imidazole, half the imidazole is converted to the imidazolium ion, and at equilibrium the concentration of imidazole is equal to the concentration of the imidazolium ion: $[ImidN] = [ImidNH^+] = 0.50M$. Therefore, the pH of the solution is equal to the pK_A of the imidazolium ion:

$$-\log(1.11 \times 10^{-7}) = 6.95$$

When NaOH sodium hydroxide, is added to the solution, a reaction occurs between the imidazolium ion, the acid of the buffer, and the hydroxide ion:

$$ImidNH^+ + OH^- \rightleftharpoons ImidN + H_2O$$

This reaction proceeds virtually to completion, so essentially all the added hydroxide ion is neutralized. As a result, the ratio of imidazole to imidazolium ion changes. The amount of hydroxide ion added is:

$$\frac{4.00 \text{ g NaOH}}{40.0 \text{ g NaOH/mol NaOH}} = 0.10 \text{ mol}$$

Since 0.10 mol of OH^- consumes 0.10 mol of imidazolium ion to form 0.10 mol of imidazole, the concentration of imidazolium ion after it reacts with the hydroxide ion is $0.50M - 0.10M = 0.40M$, and the concentration of imidazole is $0.50M + 0.10M = 0.60M$.

Equation 15.3 gives

$$pH = 6.95 + \log \frac{[ImidN]}{[ImidNH^+]}$$

$$= 6.95 + \log \frac{(0.60)}{(0.40)} = 7.13$$

Because the solution is buffered, there is an increase of only 0.18 pH units despite the addition of a substantial amount of a strong base. The same quantity of sodium hydroxide would change the pH of a liter of pure water from 7 to 13.

Two relatively simple buffer systems are also found in the fluids of living organisms. The more important of these systems is based on dissolved carbon dioxide and bicarbonate ion, HCO_3^-. Carbon dioxide might not appear to be an acid at first glance, since it has no proton to donate. But when CO_2 is dissolved in water, a small portion of it forms carbonic acid, H_2CO_3, by the reaction

$$CO_2 + H_2O \rightleftharpoons H_2CO_3$$

whose equilibrium constant, K, is estimated to be 3.7×10^{-3}.

The conjugate base of carbonic acid is the bicarbonate ion, HCO_3^-:

$$H_2CO_3 \rightleftharpoons HCO_3^- + H^+$$

The overall reaction relating carbon dioxide and bicarbonate ion is the sum of these two reactions:

$$CO_2 + H_2O \rightleftharpoons HCO_3^- + H^+$$

This equilibrium is used for calculations in this buffer system. The value of the equilibrium constant, K, for this reaction is $4.30 \times 10^{-7} M$; $pK = 6.37$. The pK of the equilibrium for this buffer system as well as the pK for the imidazole equilibrium are both close to 7.

A second class of physiological buffers is based on the dihydrogen phosphate anion, $H_2PO_4^-$, which is a weak acid, and HPO_4^{2-}, its conjugate base. The acid dissociation equilibrium of the dihydrogen phosphate anion is:

$$H_2PO_4^- \rightleftharpoons HPO_4^{2-} + H^+$$

$K = 6.23 \times 10^{-8}$ and $pK = 7.21$. Again, the pK is close to physiological pH.

Both of these buffer systems are complex, because their components are part of acid-base systems that can transfer more than one proton. Such systems are the subject of the next section.

15.7 POLYPROTIC ACIDS

The *polyprotic acids* are those which have more than one proton per molecule to donate. They include sulfuric acid, H_2SO_4, a diprotic acid; carbonic acid, H_2CO_3, also a diprotic acid; and phosphoric acid, H_3PO_4, a triprotic acid.

One way to study equilibria of polyprotic acids is to assume that the acid molecule donates one proton at a time to the base. That is, a polyprotic acid is assumed to transfer its first proton completely before any transfer of the second proton begins. An example is H_2S, hydrogen sulfide, which behaves as a diprotic acid in water. Two equilibria, each one for the transfer of a single proton, are written to describe its behavior:

$$H_2S \rightleftharpoons HS^- + H^+$$
$$HS^- \rightleftharpoons S^{2-} + H^+$$

The equilibrium constant for the first equilibrium, the transfer of the first proton from H_2S, is designated K_1. The equilibrium constant for the second reaction, the transfer of the second of the H_2S protons, is designated K_2. Thus, K_2 of H_2S is the equilibrium constant of a reaction that does not include H_2S. Instead, the reactant is HS^-, the conjugate base of H_2S. The hydrogen sulfide anion, HS^-, is *amphoteric*. It can accept a proton and return to H_2S, so it is a base. It can donate a proton and form S^{2-}, so it is an acid. All anions formed from the transfer of some of the acidic protons of a polyprotic acid are amphoteric.

Phosphoric acid is a triprotic acid. Thus there are three equilibria and three equilibrium constants to consider in describing its behavior and the behavior of anions derived from it:

$$H_3PO_4 \rightleftharpoons H_2PO_4^- + H^+ \qquad K_1$$
$$H_2PO_4^- \rightleftharpoons HPO_4^{2-} + H^+ \qquad K_2$$
$$HPO_4^{2-} \rightleftharpoons PO_4^{3-} + H^+ \qquad K_3$$

Both $H_2PO_4^{2-}$ and HPO_4^- are amphoteric.

Sulfuric acid is unusual among polyprotic acids because it is a strong acid with respect to the loss of its first proton. A solution of sulfuric acid starts as H^+ and HSO_4^-. But the HSO_4^- anion is a relatively weak acid:

$$HSO_4^- \rightleftharpoons SO_4^{2-} + H^+ \qquad K_2 = 1.20 \times 10^{-2} M$$

Table 15.5 lists the acid dissociation constants for some polyprotic acids. Although the symbols for these constants have numerical subscripts, each one is actually an acid dissociation constant and thus can be regarded as a K_A. For all polyprotic acids, K_1 is greater than K_2. For a triprotic acid, K_2 is greater than K_3. The loss of a proton by the acid creates a negative charge. The loss of a second proton adds more negative charge. It is unfavorable to add negative charge to an entity that is already negatively charged. Thus, each successive proton is more difficult to remove.

Equilibrium calculations for systems including polyprotic acids generally are more complicated than those for monoprotic acids. But the calculations are manageable if suitable approximations are made.

EXAMPLE 15.22

A 0.20-mol amount of $H_2S(g)$ is dissolved in water to form a solution of volume 2.00 liter. Calculate the concentrations of H_2S and all the ions in the solution at equilibrium, using the data in Table 15.5.

TABLE 15.5
Dissociation Constants of Polyprotic Acids at Room Temperature

Inorganic Acids	$K(M)$	
carbon dioxide, CO_2 or	K_1	4.30×10^{-7}
carbonic acid, H_2CO_3	K_2	5.61×10^{-11}
chromic acid, H_2CrO_4	K_1	1.8×10^{-1}
	K_2	3.20×10^{-7}
hydrogen sulfide, H_2S	K_1	1.1×10^{-7}
	K_2	1.0×10^{-12}
phosphoric acid, H_3PO_4	K_1	7.52×10^{-3}
	K_2	6.23×10^{-8}
	K_3	4.7×10^{-13}
phosphorous acid, H_3PO_3	K_1	1.0×10^{-2}
	K_2	2.6×10^{-7}
selenic acid, H_2SeO_4	K_1	large
	K_2	1×10^{-2}
sulfuric acid, H_2SO_4	K_1	large
	K_2	1.20×10^{-2}
sulfurous acid, H_2SO_3	K_1	1.54×10^{-2}
	K_2	1.02×10^{-7}
Organic Acids		
adipic acid	K_1	3.80×10^{-5}
	K_2	3.89×10^{-6}
citric acid	K_1	8.4×10^{-4}
	K_2	1.8×10^{-5}
	K_3	5.0×10^{-7}
oxalic acid	K_1	5.36×10^{-2}
	K_2	5.42×10^{-5}

SOLUTION

At equilibrium, the solution, like any aqueous solution, contains H^+ and OH^- ions in addition to dissolved H_2S. Since H_2S is an acid, we can predict that the $[H^+]$ is greater than $1.7 \times 10^{-7}M$ and that the $[OH^-]$ is smaller than $1.7 \times 10^{-7}M$. The concentrations of these ions at equilibrium are related to the acid strength of H_2S—that is, to the magnitudes of K_1 and K_2. In addition, the transfer of protons from the H_2S molecules to water molecules results in the formation of HS^- and S^{2-} anions. The concentrations of these anions are also related to the magnitudes of K_1 and K_2.

As always, the first step in the equilibrium calculation is to write the relevant equation. There are several possibilities. We could use the equilibrium that defines K_1, the equilibrium that defines K_2, or an equilibrium that combines them, such as:

$$H_2S \rightleftharpoons S^{2-} + 2H^+ \qquad K = K_1K_2$$

Unfortunately, no single one of these equilibrium reactions includes all the species that interest us. We must choose between using a number of equilibrium reactions simultaneously or working with the reaction that is most important, leaving the others for later consideration. Whenever possible, we shall use the second approach.

The equilibrium that seems most important is the one that has the largest equilibrium constant. For H_2S in solution, it is the equilibrium that defines K_1. The relevant reaction is

	H_2S	$\rightleftharpoons$	HS^-	$+$	H^+
start	$\dfrac{0.20 \text{ mol } H_2O}{2.00 \text{ liter}}$		0		0
equil	$0.10M - x$		x		x

The expression for K_1 gives

$$K_1 = \frac{[H^+][HS^-]}{[H_2S]} = \frac{(x)(x)}{(0.10 - x)} = 1.1 \times 10^{-7}M$$

Making the usual approximation for a weak acid that $0.10M - x \cong 0.10M$, and solving, $x = 1.0 \times 10^{-4}M = [H^+] = [HS^-]$; $[H_2S] = 0.10M - 1.0 \times 10^{-4}M = 0.10M$.

We now have the equilibrium concentrations of all the species appearing in the K_1 equilibrium. To calculate the concentrations of the other ions in the solution we can use the other equilibria that are established in the system. For example, to find the $[OH^-]$, we substitute the calculated $[H^+]$ into the expression for K_W. Since the product of these two concentrations in any aqueous solution at equilibrium is $1.0 \times 10^{-14}M^2$, $[OH^-]$ is fixed once $[H^+]$ is found:

$$[OH^-](1.0 \times 10^{-4}M) = 1.0 \times 10^{-14}M^2$$
$$[OH^-] = 1.0 \times 10^{-10}M$$

The concentration of sulfide ion, $[S^{2-}]$, is found by an analogous procedure. A reaction that includes sulfide ion and has a known equilibrium constant is needed. The equation for K_2 is most convenient:

$$HS^- \rightleftharpoons S^{2-} + H^+$$

$$K_2 = \frac{[S^{2-}][H^+]}{[HS^-]} = 1.0 \times 10^{-12}M$$

Substitution of the calculated concentrations into this expression gives:

$$\frac{[S^{2-}](1.0 \times 10^{-4}M)}{(1.0 \times 10^{-4}M)} = 1.0 \times 10^{-12}M$$

and

$$[S^{2-}] = 1.0 \times 10^{-12}M$$

The method described in Example 15.22 can be used for any solution prepared from a polyprotic acid. *The relevant reaction for calculations of the equilibrium composition of solutions of polyprotic acids is the reaction that defines* K_1. The first step in such calculations is to find the concentrations of all the substances that appear in the equation for this reaction, using the customary procedure for equilibrium calculations. The next step is to find the concentrations of the substances that are present in the system at equilibrium but do not appear in this equation. To find these concentrations, we substitute the concentrations found in the first step into the equilibrium constant expression for other equilibria, called secondary equilibria, that occur in the solution. An approximation is implicit in this procedure. It is assumed that the concentrations determined in the first step are not changed by the less important equilibria.

This approximation was true for the system in Example 15.22. By using K_1, we found that the $[HS^-]$ in a solution of $0.10M$ H_2S is $1.0 \times 10^{-4}M$. The transfer of the second proton does not really change this value. The value of K_2 is so small that only an inconsequential fraction of HS^- undergoes reaction in the K_2 equilibrium. This fraction is the $[S^{2-}]$ or the $[H^+]$ which forms in K_2, divided by the $[HS^-]$ present from K_1:

$$\frac{1.0 \times 10^{-4}M}{1.0 \times 10^{-12}M} = 1.0 \times 10^{-8}$$

Similarly, the H^+ that is formed by the second ionization of H_2S is inconsequential compared to the H^+ formed by the first ionization.

The lesser importance of the secondary equilibria helps to simplify the calculations. The full-fledged equilibrium calculation must be done to find only the concentration of species that appear in the most important relevant equilibrium. To calculate the concentration of species that appear only in the secondary equilibria, we simply substitute concentrations directly into the expression for K.

EXAMPLE 15.23

The importance of the carbon dioxide-bicarbonate buffer in physiological fluids was mentioned earlier. Blood is one such fluid. A sample of blood plasma is found to have $[HCO_3^-] = 2.4 \times 10^{-2}M$ and $[CO_2] = 1.2 \times 10^{-3}M$. The value of K_1 for carbon dioxide under physiological conditions is $7.9 \times 10^{-7}M$ and the value of K_2 is $1.0 \times 10^{-10}M$. Calculate the pH of the blood plasma sample and the $[CO_3^{2-}]$ in the solution.

SOLUTION

The relevant equilibrium is:

	CO_2	$+ H_2O \rightleftharpoons$	HCO_3^-	$+ H^+$
start	$1.2 \times 10^{-3}M$		$2.4 \times 10^{-2}M$	0
equil	$1.2 \times 10^{-3}M - x$		$2.4 \times 10^{-2}M + x$	x

Making the approximation that $1.2 \times 10^{-3}M - x \cong 1.2 \times 10^{-3}M$ and that $2.4 \times 10^{-2}M + x = 2.4 \times 10^{-2}M$, the expression for K gives:

$$\frac{(2.4 \times 10^{-2}M)(x)}{(1.2 \times 10^{-3}M)} = 7.9 \times 10^{-7}M$$

and $x = [H^+] = 4.0 \times 10^{-8}M$, and pH $= 7.40$.

To find the $[CO_3^{2-}]$, we use a secondary equilibrium in which it is included, such as K_2. Using the concentrations that we have just found, the K_2 expression gives:

$$K_2 = \frac{[CO_3^{2-}][H^+]}{[HCO_3^-]} = \frac{[CO_3^{2-}](4.0 \times 10^{-8}M)}{(2.4 \times 10^{-2}M)} = 1.0 \times 10^{-10}M$$

and

$$[CO_3^{2-}] = 6.0 \times 10^{-5}M$$

Like the carbon dioxide-bicarbonate buffer that is the subject of Example 15.23, many other physiological buffers are composed of substances related to polyprotic acids. For these buffers, *the relevant equilibrium is the reaction that most simply relates the two substances of which the buffer is composed*. We can apply this rule to another important system, one composed of substances related to phosphoric acid, a triprotic acid. Consider the buffer system composed of the dihydrogen phosphate anion, $H_2PO_4^-$, an acid, and its conjugate base, the hydrogen phosphate dianion, HPO_4^{2-}. The relevant equilibrium for calculations on this system is the simplest equilibrium that relates these two anions:

$$H_2PO_4^- \rightleftharpoons HPO_4^{2-} + H^+$$

This is the equation that defines K_2 for phosphoric acid. Under physiological conditions, the value of K_2 is $1.6 \times 10^{-7}M$ and $pK_2 = 6.8$. As you might expect, this value is close to the range of physiological pH.

Simultaneous Equilibria

The complete description of solutions that contain species related to polyprotic acids often requires us to consider a number of different equilibrium reactions. A quantitative approach to such systems can be quite complex, and a number of methods have been developed to carry out the calculations. Special methods of calculation may also be needed for other equilibrium systems, such as those that include relatively insoluble materials as well as ions in solution. In Appendix V, we outline a method for solving problems related to the composition of systems in which a number of simultaneous reactions take place.

EXERCISES

15.1 Classify each of the following species as a Brønsted acid or base: (a) HI, (b) HNO_2, (c) NH_4^+, (d) NH_2^-, (e) HCO_3^-, (f) CH_3^+, (g) SO_4^{2-}, (h) $H_2AsO_4^-$.

15.2 Write equations for the reactions between the following species: (a) hypobromous acid and ammonia, (b) hydrocyanic acid and methyl amine, (c) dihydrogen phosphate ion and excess cyanide ion, (d) hydrogen sulfate ion and phosphate ion.

15.3 Write equations for the autoprotolysis of the following species: (a) dihydrogen phosphate ion, (b) methyl amine.

15.4 Indicate which of the following species are Lewis acids: (a) NH_4^+, (b) CH_3^+, (c) BCl_3, (d) Fe^{2+}, (e) H_2S. Justify your choices.

15.5 Write equations for the reactions between the following species: (a) ammonia and aluminum chloride,

(b) phosphorus pentachloride and chloride ion, (c) ethylene, C_2H_4, and hydrogen chloride.

15.6 Write reactions to show that aluminum hydroxide is an amphoteric hydroxide.

15.7 Classify each of the following species as a weak or strong acid or base: (a) H_2SO_4, (b) HSO_4^-, (c) KOH, (d) CH_3NH^-, (e) HCO_3^-, (f) $HBrO_4$, (g) CN^-, (h) OH^-.

15.8 Designate the stronger acid in each of the following pairs: (a) $HClO_3$ and $HClO_2$, (b) $CH_3NH_3^+$ and CH_3NH_2, (c) $CH_3OH_2^+$ and H_3O^+.

15.9 Designate the stronger base in each of the following pairs: (a) ClO_2^- and ClO_3^-, (b) $CH_3CH_2^-$ and CH_3NH^-, (c) NH_3 and NO_3^-.

*15.10 When HNO_3 is dissolved in concentrated sulfuric acid, it can act as

a base and accept a proton from the sulfuric acid. Draw the Lewis structure of the conjugate acid of HNO_3 and write an equation for the acid-base reaction.

15.11 Four solutions are prepared from 1 mol of NH_3 in 0.1 liter of: methyl amine, sulfuric acid, acetic acid, and water. List the solutions in order of increasing concentration of NH_4^+ ion at equilibrium.

15.12 Write net ionic equations for the reactions that occur when the following solutions are mixed: (a) a solution of nitric acid and a solution of potassium hydroxide, (b) a solution of perchloric acid and a solution of barium hydroxide, (c) a solution of sodium hydroxide and a solution of ammonia, (d) a solution of ammonia and a solution of hydrochloric acid, (e) a solution of methyl amine and a solution of hydriodic acid.

15.13   Write net ionic equations for the reactions that occur when the following solutions are mixed: (a) a solution of nitrous acid and a solution of potassium hydroxide, (b) a solution of hypochlorous acid and a solution of methyl amine, (c) a solution of acetic acid and a solution of hydrochloric acid, (d) a solution of calcium hydroxide and a solution of chlorous acid.

15.14   Write net ionic equations for the reactions that take place when aqueous solutions of the following substances are mixed: (a) sodium cyanide and nitric acid, (b) ammonium chloride and sodium hydroxide, (c) sodium cyanide and ammonium chloride, (d) potassium hydrogen sulfate and lithium acetate, (e) sodium hypochlorite and ammonia.

15.15   Write the net ionic equation that defines the K_A of each of the following species listed in Table 15.2: (a) hydrogen peroxide, (b) periodic acid, (c) hydroxylammonium ion, (d) formic acid, (e) methyl ammonium ion.

15.16   Calculate the $[H^+]$ and $[OH^-]$ in each of the following solutions: (a) $2M$ NaOH, (b) $0.5M$ HNO_3, (c) $0.0003M$ $Ca(OH)_2$, (d) $2.4 \times 10^{-5}M$ HCl.

15.17   Calculate the pH and pOH of each of the following solutions: (a) $0.002M$ HCl, (b) $0.0002M$ KOH, (c) $5.2 \times 10^{-4}M$ $Ba(OH)_2$.

15.18   Calculate the $[H^+]$ in solutions of the following pH: (a) 7.30, (b) 6.70, (c) −0.68.

15.19   Calculate the $[OH^-]$ in solutions of the following pH: (a) 4.35, (b) 9.89, (c) 14.52.

15.20   Arrange the following oxyacids in order of increasing strength: (a) $HMnO_4$, (b) H_3BO_3, (c) H_2SeO_4, (d) H_3PO_4.

15.21   Arrange the following bases in order of increasing strength: (a) ClO_4^-, (b) $H_3SiO_4^-$, (c) HSO_3^-, (d) HSO_4^-.

15.22   Calculate the pH of a solution prepared from 0.14 mol of formic acid and enough water to make 1.0 liter of solution.[a]

15.23   Calculate the pH and pOH of a solution prepared from 0.25 mol of NH_4Cl and enough water to make one liter of solution.[a]

15.24   Calculate the difference in pH between solutions prepared from 1.0 mol of acetic acid and 1.0 mol of dichloroacetic acid, each in enough water to make one liter of solution.[a]

15.25   Write the equation and calculate the value of K, the equilibrium constant, for the reaction in which the conjugate base of each of the following acids accepts a proton from water: (a) $HClO_2$, (b) H_5IO_6, (c) hydrazinium ion, (d) phenol (PheOH), (e) protonated codeine (CodNH$^+$).

15.26   Calculate the $[H^+]$ and $[OH^-]$ in a solution prepared from 0.86 mol of ammonia and enough water to make 0.25 liter of solution.[a]

15.27   Calculate the pH of a solution prepared from 0.56 mol of NaCN and enough water to make 0.40 liter of solution.

15.28   A solution prepared from 0.13 mol of an organic acid and enough water to form one liter of solution is found to have a pH of 5.45. Calculate K_A of the acid. The acid is listed in Table 15.2. Identify the acid.

15.29   A solution of 0.23 mol of the chloride salt of protonated quinine, a weak organic base used in the treatment of malaria, in enough water to form 1.0 liter of solution has a pH of 4.58. Calculate K_A of protonated quinine.

15.30   A solution of the sodium salt of ascorbic acid (AcbOH), commonly known as vitamin C, is prepared from 0.025 mol of the salt in enough water to prepare 0.50 liter of solution. The $[OH^-]$ is found to be $2.5 \times 10^{-6}M$. Calculate K_A of ascorbic acid.

*15.31   The pH of a $1.00M$ solution of urea, a weak organic base, is 7.050. Calculate K_A of protonated urea.

15.32   Calculate the amount of sodium acetate that must be dissolved in 0.25 liter of water to produce a solution of pH = 8.90.[a]

15.33   Calculate the amount of ammonium chloride that must be dissolved in 0.50 liter of water to produce a solution of pH = 4.79.[a]

15.34   Calculate the $[ClO^-]$ in a solution prepared from 0.010 mol of HCl and 0.020 mol of HClO in enough water to make 1.0 liter of solution.[a]

15.35   Calculate the fraction of HNO_2 that dissociates in a solution that has an initial $[HNO_2] = 0.115M$ at the following pHs: (a) 7, (b) 4, (c) 1.[a]

15.36   Calculate the amount of protonated nicotine in 1.0 liter of solution prepared from 1.0 mol of nicotine and 0.00010 mol of NaOH.[a]

15.37   Calculate the minimum amount of nitric acid that must be added to 1.0 liter of a solution prepared from 1.0 mol of iodic acid in order to keep $[IO_3^-]$ below $0.1M$.[a]

15.38   Calculate the amount of ammonia that must be dissolved in 1.0 liter of

a solution prepared from 0.087 mol of ammonium sulfate to produce a solution with pH = 8.88.[a]

15.39   Write the relevant equilibrium and find the value of its K for the reaction that takes place when each of the following pairs of solutions are mixed: (a) calcium hydroxide and hydriodic acid, (b) sodium hydroxide and hydrofluoric acid, (c) hydroxylamine and hydrochloric acid, (d) iodic acid and ammonia.[a]

15.40   Write the relevant equilibrium and find the value of its K for the reaction that takes place when each of the following pairs of solutions are mixed: (a) sodium formate and nitric acid, (b) hydrazinium chloride and sodium hydroxide, (c) sodium cyanide and methyl ammonium chloride, (d) ammonia and sodium nitrate.

15.41   A volume of 18.5 cm^3 of $0.274M$ HCl solution is required to neutralize 25.0 cm^3 of a solution of KOH. Calculate the concentration of the KOH solution.

15.42   A volume of 23.9 cm^3 of $0.106M$ sulfuric acid solution is required to neutralize a sample of solid NaOH dissolved in some water. Calculate the mass of the NaOH.

15.43   A sample of fruit juice requires 12.3 cm^3 of $0.125M$ sodium hydroxide solution for neutralization. Find the total amount of available protons in the sample of fruit juice.

15.44   Calculate the volume of a $0.318M$ nitric acid solution necessary to neutralize 10.2 cm^3 of a solution prepared from 7.24 g of sodium nitrite and enough water for 1.0 liter of solution.

*15.45   A 25.0-cm^3 volume of a sodium hydroxide solution requires 19.6 cm^3 of a $0.189M$ hydrochloric acid solution for neutralization. A 10.0-cm^3 volume of a phosphoric acid solution requires 34.9 cm^3 of the sodium hydroxide solution for complete neutralization. Calculate the concentration of the phosphoric acid solution.

15.46   A solution is prepared from 0.364 mol of ammonium sulfate and enough water to make 1.00 liter of solution. Calculate the volume of $0.0464M$ sodium hydroxide solution required for its neutralization.

15.47   A 0.867-g sample of an unknown acid requires 32.2 cm^3 of a $0.182M$ barium hydroxide solution for neutralization. Calculate the equivalent weight of the acid. Assuming that the acid donates two protons per mole, calculate the molecular weight of the acid.

---

[a] The necessary data for this exercise are in Table 15.2.

15.48   A 1.24-g sample of ephedrine, a base that accepts only one proton per mole, requires 21.7 cm³ of a 0.347M solution of hydrochloric acid for neutralization. Calculate the molecular weight of ephedrine.

*15.49   Morphine, the well-known narcotic, accepts one proton per molecule. Its composition is $C_{17}H_{19}NO_3$. The major source of morphine is opium. A 0.682-g sample of opium is found to require 8.92 cm³ of a 0.0116M solution of sulfuric acid for neutralization. Assuming that morphine is the only acid or base present in opium, calculate the percentage of morphine in the sample of opium.

15.50   Calculate the pH at the equivalence point in the titration of 0.20M acetic acid solution with 0.20M sodium hydroxide solution.[a]

15.51   Calculate the pH at the equivalence point in the titration of solid sodium cyanide with 0.115M hydrochloric acid solution.[a]

15.52   Calculate the pH at the equivalence point in the titration of 50.1 cm³ of a 0.880M solution of methyl amine by a 0.254M solution of lactic acid.[a]

*15.53   A 100-cm³ sample of 0.100M nitric acid solution is titrated with a 0.100M solution of potassium hydroxide. Calculate the pH after the addition of the following volumes of the solution of the base: (a) 50.0 cm³, (b) 90.0 cm³, (c) 99.0 cm³, (d) 99.9 cm³, (e) 100.1 cm³.

*15.54   Repeat the calculation of Exercise 15.53 for the titration of a 0.100M solution of ammonium chloride with a 0.100M solution of sodium hydroxide.[a]

15.55   Select a suitable indicator from those listed in Table 15.4 for each of the following titrations: (a) perchloric acid with barium hydroxide, (b) 0.10M chloric acid with 0.10M ammonia, (c) 0.10M nitrous acid with 0.10M potassium hydroxide.[a]

15.56   Select a suitable indicator from those listed in Table 15.4 for each of the following titrations: (a) 0.10M nitric acid with 0.10M sodium chloroacetate, (b) 0.10M butyric acid with 0.10M ammonia, (c) 0.10M hydrazinium chloride with 0.10M sodium hydroxide.

15.57   Calculate the pH of a buffer solution prepared from 0.25 mol of formic acid and 0.35 mol of sodium formate in enough water to make 1.0 liter of solution.[a]

15.58   Calculate the pH of a buffer solution prepared from 0.16 mol of methyl ammonium chloride and 0.46 mol of

methyl amine in enough water to make 1.0 liter of solution.[a]

15.59   Calculate the quantity of sodium acetate that must be dissolved in 1.0 liter of a 1.4M acetic acid solution to prepare a buffer solution of pH = 4.36. (Assume no volume changes.)[a]

15.60   Calculate the relative masses of phenol (C_6H_5OH) and sodium phenoxide (C_6H_5ONa) necessary to prepare a buffer solution of pH = 9.89.[a]

15.61   Explain, using appropriate equilibrium reactions and the principle of Le Chatelier, how a buffer solution of ammonia and ammonium chloride acts to keep the pH of the solution relatively constant when small amounts of hydrochloric acid or sodium hydroxide are added.

15.62   Select a suitable acid-conjugate base pair from those listed in Table 15.2 for the preparation of a buffer solution of pH = 2.74. Find the necessary molar ratio of the two substances.

15.63   Calculate the pH of a buffer solution prepared by mixing 100 cm³ of a 0.36M hydroxylamine solution with 50 cm³ of a 0.26M hydrochloric acid solution.[a]

15.64   Calculate the pH of 1.0 liter of a buffer solution prepared from 1.0 mol of lactic acid and 1.0 mol of sodium lactate before and after the addition of 0.10 mol of HCl.[a]

15.65   Repeat the calculation of Exercise 15.64 for a buffer solution prepared from 0.10 mol of each component. Discuss your result in terms of buffer capacity.

15.66   Discuss the titration curve of Figure 15.3 from the point of view of the buffer solutions that are formed during the titration.

15.67   Calculate the molar ratio of formic acid to sodium formate necessary to prepare a buffer solution of pH = 2.89. Discuss the disadvantages of such a buffer.

15.68   Calculate the concentration of all species at equilibrium in a solution prepared from 1.00 mol of H_2SO_4 and enough water to make 1.00 liter of solution.[b]

15.69   The formula of oxalic acid can be written as $H_2C_2O_4$. Both protons are acidic. Calculate the concentration of all species at equilibrium in a solution prepared from 0.76 mol of oxalic acid and enough water to make one liter of solution.[b]

15.70   Buffer solutions are often prepared using salts of anions formed from dissociation of phosphoric acid. Calculate the pH of the buffer solution prepared from equimolar quantities of each of the following salts: (a) sodium dihydrogen phosphate and sodium hydrogen phosphate, (b) sodium hydrogen phosphate and sodium phosphate.[b]

15.71   A solution of 0.65M sodium carbonate is alkaline. Calculate the pH of the solution and $[HCO_3^-]$ and $[H_2CO_3]$ at equilibrium.

*15.72   Since the anions formed from partial dissociation of polyprotic acids are amphoteric, they can react with themselves. In a solution of sodium bicarbonate, the most important equilibrium is the one in which the bicarbonate ion undergoes just such a reaction. Calculate the equilibrium constant of the reaction and the pH of 0.10M sodium bicarbonate solution.[b]

15.73   When strong acid is added to a solution of sodium carbonate, the evolution of a gas is observed. Identify the gas and account for its evolution using the principle of Le Chatelier.

*15.74   The solubility product of $Fe(OH)_3$ is 4×10^{-38}. Calculate the solubility of $Fe(OH)_3$ at pH = 7 and at pH = 1.

15.75   A solution is 0.30M in HCl and a 0.20-mol amount of $H_2S(g)$ is dissolved in the solution. Calculate the $[S^{2-}]$ at equilibrium.[b]

15.76   Use the data in Example 15.23 to find the $[HCO_3^-]$ and $[CO_2]$ when the pH of the blood decreases to 7.30. Assume that the total of the two concentrations does not change.

15.77   Using only water or HS^- or both as reactants, list all the equilibria that are established in a solution of sodium hydrogen sulfide, and evaluate the equilibrium constant of each.[b]

15.78   Calculate the solubility of $Au(OH)_3$ in neutral solution. The K_{sp} is 5.5×10^{-46}. (Hint: pOH = 7; consider the common ion effect.)

15.79   Find the solubility of $Au(OH)_3$ in a solution that is 1M in nitric acid. (Hint: the relevant equilibrium includes the solid and H^+ as reactants.)

*15.80   The K_{sp} of calcium carbonate is 4.8×10^{-9}. Find the K for the reaction $CaCO_3(s) + CO_2 + H_2O \rightleftharpoons Ca^{2+} + 2HCO_3^-$. If the $[CO_2]$ in a saturated solution is 0.042M, find the solubility of calcium carbonate in a saturated solution of CO_2 in water.[b]

[b] The necessary data for this exercise are in Table 15.5.

OXIDATION-REDUCTION

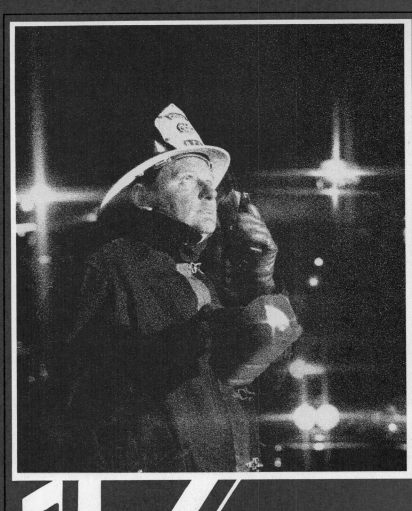

16

Any ordinary chemical change is accompanied by a change in electronic structure. Even a simple process such as the dissolution of a solute in a solvent causes a change in the electronic structures of both the solute and the solvent. In the more complex proton transfer reactions described in Chapter 15, a bonding electron pair of an acid becomes a nonbonding pair, and a nonbonding electron pair of a base becomes a bonding pair, forming a bond with the proton that the base accepts.

But all the aqueous equilibrium processes that we have discussed so far have one feature in common: There is no change in the number of electrons associated with each participant in the reaction. When H^+ is transferred from an acid to a base, there is no change in the number of electrons of either the acid or the base, since H^+ has no electrons.

Now we shall discuss processes that cause changes in the number of electrons of some or all of the participants in the reaction. These changes can be visualized as the result of real or imagined electron transfer between the substances that undergo reaction. **A chemical change that occurs as the result of an electron transfer is called an oxidation-reduction, or redox, process.**

Oxidation-reduction reactions occur all around us—and, in fact, within us. The energy that enables you to read these words comes from biological oxidation-reduction reactions. Most of the energy that powers our technological society comes from oxidation-reduction reactions. In short, oxidation-reduction reactions are a very important class of chemical processes. Let us examine them in detail.

16.1 OXIDATION-REDUCTION REACTIONS

In Section 7.8 (page 223), we discussed the concept of an oxidation state or oxidation number. We described how every atom in a molecule or an ion can be assigned an oxidation state, a positive or negative integer related to the electronic configuration of the atom in the molecule or ion. If you do not have a good understanding of these ideas, a review of Section 7.8 is strongly recommended, since you must know how oxidation numbers are assigned if you are to grasp the material in this chapter.

An oxidation-reduction reaction can be defined as a chemical change in which there are changes in oxidation states. In an oxidation-reduction process, the oxidation state of one or more atoms will increase and the oxidation state of one or more atoms will decrease. If the oxidation state of an atom increases, that atom is said to be oxidized. If the oxidation state of an atom decreases, that atom is said to be reduced.

When we talk about a change in oxidation state, we are really talking about a gain or loss of electrons. The oxidation state of an atom increases when it loses electrons; the oxidation state of an atom decreases when it gains electrons. We can see the relationship most clearly by studying changes in a single atom. In Mn^{2+}, the oxidation state of manganese is $+2$. If the Mn^{2+} cation *loses* an electron, its charge increases by $+1$, and it becomes Mn^{3+}. The oxidation state of Mn^{3+} is $+3$, and the Mn^{2+} cation is said to have been oxidized to Mn^{3+}. If Mn^{2+} *gains* two electrons, its charge is changed by -2; Mn^{2+} becomes Mn. We call this a reduction, because the charge has been reduced. When Mn^{2+} is reduced to Mn, its oxidation state is reduced from $+2$ to 0.

Oxidation is the loss of electrons; reduction is the gain of electrons. Under ordinary chemical conditions, one of these processes cannot occur to any appreciable extent unless the other process also occurs.

491

Oxidation-reduction processes are visualized as electron transfer processes. Electrons are transferred from the substance that is oxidized to the substance that is reduced. A substance that loses electrons, and is thus oxidized, transfers those electrons to another substance, which thus is reduced. The substance that is oxidized is sometimes called a **reducing agent,** because it is the agent that brings about the reduction of another substance. Similarly, a substance that is reduced by gaining electrons causes the oxidation of the substance that loses the electrons. The substance that is reduced is therefore called an **oxidizing agent.** The terminology may be confusing at first, but the basic idea is summed up simply: A substance that is oxidized is a reducing agent; a substance that is reduced is an oxidizing agent.

For example, the reaction:

$$Cl_2 + 2Br^- \longrightarrow Br_2 + 2Cl^-$$

is an oxidation-reduction reaction. The oxidation state of chlorine decreases from 0 to -1, while the oxidation state of bromine increases from -1 to 0. The Cl_2 is reduced and the Br^- is oxidized. The Cl_2 is an oxidizing agent that causes the oxidation of Br^-. The Br^- is a reducing agent that causes the reduction of Cl_2.

The Half-Reaction Concept

An oxidation cannot occur without an accompanying reduction. But it is often convenient to divide oxidation-reduction processes in half artificially to help in the analysis of these processes. This artificial division gives us two half-reactions. You can distinguish a half-reaction from a normal reaction easily. The half-reaction will include one or more electrons, whose symbol is e^-.

As you might expect, there are two kinds of half-reactions, one for oxidations and one for reductions. In an oxidation half-reaction, the electrons appear on the right side of the chemical equation as products. Electrons are lost in an oxidation. The half-reaction for the oxidation of manganese from the $+2$ to the $+3$ state is

$$Mn^{2+} \longrightarrow Mn^{3+} + e^-$$

In a reduction half-reaction, the electrons appear as reactants on the left side of the chemical equation. Electrons are gained in a reduction. The half-reaction for the reduction of manganese from the $+2$ to the 0 oxidation state is

$$Mn^{2+} + 2e^- \longrightarrow Mn$$

Every half-reaction must be balanced in two ways. We must balance not only the mass of all the elements but also the total charge. To balance the charge in a half-reaction, we take the charge of an electron to be -1. The two-half-reactions that we used as examples are balanced in both respects.

An oxidation half-reaction cannot occur unless a reduction half-reaction occurs simultaneously. A complete oxidation-reduction reaction consists of a suitable combination of an oxidation half-reaction and a reduction half-reaction. *The equation for an oxidation-reduction reaction is written correctly only if it is balanced with respect to both mass and charge and if it contains no electrons.*

Half-reactions have several uses. They are used to help balance oxidation-reduction reactions. They are used to compare the strength of oxidizing agents or reducing agents. They also help to clarify some aspects of the relationship between chemical energy and electrical energy.

Balancing Oxidation-Reduction Reactions

The methods for writing balanced chemical equations that were presented in Section 2.5 (page 33) should be applied whenever possible to the writing of equations for oxidation-reduction reactions. In the simplest case, an expression for a chemical change can be balanced by inspection, with no need for more formal techniques. Inspection alone will suffice to balance some oxidation-reduction equations, even though the charges as well as the masses must be balanced. One example is the reaction in which silver cation in solution reacts with zinc metal to form zinc cation in solution and silver metal:

$$Ag^+ + Zn(s) \longrightarrow Ag(s) + Zn^{2+}$$

The mass in this expression is balanced. The charge is not, since the left side of the expression has a net charge of $+1$ while the right side has a net charge of $+2$. We can balance the charge by giving the Ag^+ ion the coefficient 2. Now the mass is no longer balanced. The mass balance can be restored by giving the $Ag(s)$ the coefficient 2. The balanced equation is

$$2Ag^+ + Zn(s) \longrightarrow 2Ag(s) + Zn^{2+}$$

However, it is not possible to write balanced equations for many oxidation-reduction reactions by inspection alone. Several formal procedures have been developed to guide the writing of balanced oxidation-reduction equations when the reactants and products of the process are known. We shall discuss the two methods that are most commonly used. In this discussion, we shall follow the usual conventions for writing chemical equations in net ionic form. State symbols will be written after substances that are not in aqueous solution; the absence of a state symbol will indicate that a substance is in aqueous solution. The formulas of strong electrolytes such as salts, strong acids, and strong bases will be written to show that they dissociate into ions in solution. The formulas for weak electrolytes and nonelectrolytes will be written to show that they do not dissociate in solution. Only substances that undergo change will appear in the chemical equation. Coefficients will be in their simplest integral form.

A. The Method of Half-Reactions. This approach divides the oxidation-reduction process into two half-reactions. Both these half-reactions are first balanced with respect to mass and charge. They are then combined in a way that eliminates the electrons, giving the overall balanced equation for the oxidation-reduction process.

To begin, we must identify the reactant that is oxidized and the product that it forms as well as the reactant that is reduced and the product that it forms. The description of the oxidation-reduction process usually gives enough information to make the identification possible. We shall then have two pairs of substances that provide the skeletons for the two half-reactions.

Suppose we are told that an acidic solution of sodium dichromate, $Na_2Cr_2O_7$, is mixed with a solution of potassium bromide, KBr; that a reaction occurs; and that the products are identified as bromine, Br_2, and chromium(III) cation, Cr^{3+}, in solution. The two starting materials are salts in solution, so we write their formulas to show their dissociation into ions: $Na^+ + Cr_2O_7^{2-}$ and $K^+ + Br^-$. We can see that the bromide ion is oxidized, since its oxidation state increases from -1 in the ion to 0 in Br_2. The dichromate ion is reduced, since the oxidation state of Cr in $Cr_2O_7^{2-}$ is $+6$ and in the Cr^{3+} cation, the product, it is $+3$. We can now write the two skeleton half-reactions:

$$Br^- \longrightarrow Br_2$$
$$Cr_2O_7^{2-} \longrightarrow Cr^{3+}$$

We do not include the sodium and potassium ions, which are unchanged in the process.

Each skeleton half-reaction must now be converted into a half-reaction balanced with respect to both mass and charge. This is done by following a fixed sequence of steps.

a. Balance the number of atoms of all elements except H and O, using coefficients in the usual way. In this case, we write

$$2Br^- \longrightarrow Br_2$$
$$Cr_2O_7{}^{2-} \longrightarrow 2Cr^{3+}$$

b. Balance the oxygen by adding the necessary number of oxygen atoms, in the form of H_2O, to the oxygen-deficient side of the expression. In this case, only the second expression must be balanced for oxygen:

$$Cr_2O_7{}^{2-} \longrightarrow 2Cr^{3+} + 7H_2O$$

c. Balance the hydrogen by adding the appropriate number of hydrogen atoms, in the form of H^+, to the hydrogen-deficient side of the expression. Again in this case, only the second expression must be balanced for hydrogen:

$$14H^+ + Cr_2O_7{}^{2-} \longrightarrow 2Cr^{3+} + 7H_2O$$

d. Balance each half-reaction for charge by adding the appropriate number of electrons to the more positive side of the expression. The number of electrons added should equal the difference in charge between the two sides of the expression. In this case, the left side of the first expression has a charge of -2 and the right side has a charge of 0. We add two electrons to the right side:

$$2Br^- \longrightarrow Br_2 + 2e^-$$

The left side of the second expression has a charge of $+12$ and the right side has a charge of $+6$. We add six electrons to the left side:

$$6e^- + 14H^+ + Cr_2O_7{}^{2-} \longrightarrow 2Cr^{3+} + 7H_2O$$

This gives us two balanced half-reactions, one for the reduction and one for the oxidation. We must add the two to get the overall equation for the process. The overall equation cannot include electrons. Therefore, we must make sure that both half-reactions have the same number of electrons. If they do, the electrons will be eliminated from the overall equation when the half-reactions are added. When, as in this case, the half-reactions do not have the same number of electrons, we multiply one or both of the half-reactions by the necessary factor. Inspection shows that both half-reactions will include six electrons if the oxidation half-reaction is multiplied by three. The half-reactions can then be combined:

$$6Br^- \longrightarrow 3Br_2 + 6e^-$$
$$\underline{6e^- + 14H^+ + Cr_2O_7{}^{2-} \longrightarrow 2Cr^{3+} + 7H_2O}$$
$$6Br^- + 14H^+ + Cr_2O_7{}^{2-} \longrightarrow 3Br_2 + 2Cr^{3+} + 7H_2O$$

There is a procedure that we can use to find the factors by which the half-reactions are multiplied to give them both the same number of electrons. First find the smallest number that is divisible by the number of electrons in each half-reaction. Then divide the number of electrons in each half-reaction into this number. The division gives the factor by which each half-reaction must be multiplied.

For example, suppose we wish to add two half-reactions which have, respectively, $2e^-$ and $5e^-$. In this case, the lowest number divisible by both is 10. The factor for the first half-reaction is $10/2 = 5$, and the factor for the second half-reaction is $10/5 = 2$. After multiplication by these factors, each half-reaction has $10e^-$. When the half-reactions are combined, the electrons will be canceled in the addition.

Two steps remain after the half-reactions are combined. The overall reaction must be checked to see whether its coefficients are in the lowest terms. If not, they should be reduced. The reaction must also be checked for the presence of substances that appear unchanged on both sides of the equation. If there are such substances, they should be eliminated from the equation. Finally, there should be one last check to be sure that the equation is indeed balanced with respect to both mass and charge.

An additional step is needed for systems that are designated as alkaline. When an oxidation-reduction reaction occurs in alkaline solution, there is no appreciable concentration of H^+, so an extra step is needed to replace H^+ by OH^- in the equation. One OH^- ion is added to each side of the equation for every H^+ that is present. One side of the equation will then have an equal number of H^+ and OH^- ions. These are combined to form H_2O, thus eliminating the H^+ from the equation. The equation now has H_2O on both sides, so it must be made net by appropriate subtraction of H_2O. This procedure has no chemical or physical significance, but it gives the right answer.

EXAMPLE 16.1

When an alkaline solution of potassium permanganate is mixed with an alkaline solution of sodium sulfide, a yellow precipitate of sulfur and a brown precipitate of manganese dioxide form. Write a balanced equation for this oxidation-reduction reaction.

SOLUTION

The two skeleton half-reactions are

$$S^{2-} \longrightarrow S(s)$$
$$MnO_4^- \longrightarrow MnO_2(s)$$

First we balance these skeleton half-reactions. The first one, the oxidation, already is balanced with respect to mass, but electrons must be added:

$$S^{2-} \longrightarrow S(s) + 2e^-$$

For the second half-reaction, the reduction, H_2O and H^+ must be added to balance the mass:

$$MnO_4^- + 4H^+ \longrightarrow MnO_2(s) + 2H_2O$$

and then electrons must be added to balance the charge:

$$MnO_4^- + 4H^+ + 3e^- \longrightarrow MnO_2(s) + 2H_2O$$

The oxidation half-reaction has $2e^-$ and the reduction half-reaction has $3e^-$. The smallest number that is divisible by both 2 and 3 is 6. Therefore, the oxidation half-reaction is multiplied by $6/2 = 3$ and the reduction half-reaction is multiplied by $6/3 = 2$. The two half-reactions are then combined:

$$3S^{2-} \longrightarrow 3S(s) + 6e^-$$
$$\underline{2MnO_4^- + 8H^+ + 6e^- \longrightarrow 2MnO_2(s) + 4H_2O}$$
$$3S^{2-} + 2MnO_4^- + 8H^+ \longrightarrow 3S(s) + 2MnO_2(s) + 4H_2O$$

Since this is an alkaline solution and $8H^+$ are present, we add $8OH^-$ to each side of the equation. The $8H^+$ on the left side of the equation are combined with the $8OH^-$ to form $8H_2O$. The equation is now:

$$3S^{2-} + 2MnO_4^- + 8H_2O \longrightarrow 3S(s) + 2MnO_2(s) + 4H_2O + 8OH^-$$

This equation is not net, because H_2O appears on both sides. We get the final net balanced equation by subtracting $4H_2O$ from both sides:

$$3S^{2-} + 2MnO_4^- + 4H_2O \longrightarrow 3S(s) + 2MnO_2(s) + 8OH^-$$

B. The Method of Change in Oxidation Numbers. The second method approaches the writing of balanced equations for oxidation-reduction reactions by determining the changes that occur in oxidation numbers. In an oxidation-reduction process, one or more elements will have an increase in oxidation number and one or more elements will have a decrease in oxidation number. *In any balanced* *equation for an oxidation-reduction process, the total increase in oxidation numbers must equal the total decrease in oxidation numbers.* The method of change in oxidation numbers uses this principle to write correct equations by balancing the changes in oxidation numbers.

We start by using the description of the process to write a skeleton expression that includes the elements which undergo a change in oxidation number. In this method, we write a single overall skeleton expression.

Let us take the process in which an acidic solution of potassium permanganate, $KMnO_4$, is added to a solution of sodium chloride. The products are chlorine and manganese(II) cation. The two starting materials are salts in solution, so their formulas are written to show them as dissociated ions, $K^+ + MnO_4^-$ and $Na^+ + Cl^-$. Since neither Na^+ nor K^+ undergoes change in the process, they are left out of the equation. The skeleton expression includes the two ions that undergo reaction and the given products:

$$MnO_4^- + Cl^- \longrightarrow Mn^{2+} + Cl_2$$

a. The first step is to identify those elements whose oxidation numbers increase and those whose oxidation numbers decrease. We see that the oxidation number of Mn decreases from $+7$ to $+2$, a change of five units for each Mn atom, and that the oxidation number of Cl^- increases from -1 to 0, a change of one unit for each Cl atom.

b. Now we must make the total increase in oxidation number equal the total decrease in oxidation number. To do this, we find the element that undergoes the smaller total change in oxidation number. We then find the number of moles of this element that are needed to balance the change undergone by one mole of the other element. In this process, one mole of Mn undergoes a change of five units and one mole of Cl undergoes a change of one unit. Therefore, the increase and decrease in oxidation numbers will be balanced if there are five moles of Cl for each mole of Mn. We add the appropriate coefficient to Cl, while at the same time we balance all the elements in the expression except H and O:

$$MnO_4^- + 5Cl^- \longrightarrow Mn^{2+} + \tfrac{5}{2}Cl_2$$

c. The next step is to balance for charge by adding H^+ cations to the more negative side of the expression. We find the difference in charge between the two sides of the expression and add this number of H^+ cations to the more negative side of the equation. In this expression, the charge on the left is -6 and the charge on the right is $+2$. The difference is $+8$, so we add $8H^+$ to the left side, the more negative side of the expression:

$$MnO_4^- + 5Cl^- + 8H^+ \longrightarrow Mn^{2+} + \tfrac{5}{2}Cl_2$$

d. We now balance for oxygen by adding H_2O to the oxygen-deficient side of the expression:

$$MnO_4^- + 5Cl^- + 8H^+ \longrightarrow Mn^{2+} + \tfrac{5}{2}Cl_2 + 4H_2O$$

The equation should now be balanced. Check to be sure that it is net and that the coefficients are in lowest terms. If there is a fractional coefficient, as there is in this case, you may multiply to eliminate the fraction. Here, we multiply by two:

$$2MnO_4^- + 10Cl^- + 16H^+ \longrightarrow 2Mn^{2+} + 5Cl_2 + 8H_2O$$

If the solution is described as alkaline, step c is different. We still use H_2O as a source of hydrogen, but we add OH^- to the more positive side of the expression to balance for charge. For example, in alkaline solution the products of the reaction of potassium permanganate and sodium chloride are Cl_2 and $MnO_2(s)$. The skeleton expression is:

$$MnO_4^- + 3Cl^- \longrightarrow MnO_2(s) + \tfrac{3}{2}Cl_2$$

Balancing charge gives:

$$MnO_4^- + 3Cl^- \longrightarrow MnO_2(s) + \tfrac{3}{2}Cl_2 + 4OH^-$$

Balancing the oxygen gives the correct equation:

$$MnO_4^- + 3Cl^- + 2H_2O \longrightarrow MnO_2(s) + \tfrac{3}{2}Cl_2 + 4OH^-$$

16.2 OXIDATION-REDUCTION PROCESSES IN AQUEOUS SOLUTION

Many oxidation-reduction processes proceed differently in aqueous solution than the aqueous equilibria that we discussed in Chapters 14 and 15. Proton transfer reactions and many precipitation reactions occur very quickly, virtually at the instant that the solutions are mixed. These systems rapidly reach the equilibrium state. By contrast, many oxidation-reduction processes proceed slowly in aqueous solution. We cannot always predict what will happen on the basis of equilibrium constants alone. A given system may take so long to reach its equilibrium state that equilibrium is never achieved. In a multistep process, one step may occur so slowly that the system stops at this intermediate state. We usually must know the products that are formed under a given set of conditions to know the outcome of an oxidation-reduction process.

But there are also oxidation-reduction processes in aqueous solution that do proceed smoothly and rapidly. The equilibrium constants of these processes are often much larger than those of proton transfer or precipitation reactions. For example, the value of K at 298 K for the reaction

$$10I^- + 2MnO_4^- + 16H^+ \longrightarrow 2Mn^{2+} + 8H_2O + 5I_2$$

is about 3×10^{164}. We can treat this reaction as if it proceeds to completion.

Oxidation-Reduction Titrations

Oxidation-reduction processes in aqueous solution are extremely useful in analytical chemistry. If an element exists in more than one oxidation state, a procedure based on an oxidation-reduction process can be used to measure the amount of that element in a sample. One widely used method is oxidation-reduction titration. The technique is similar to that of the acid-base titrations discussed in Section 15.5 (page 471): A solution of known concentration is added slowly from a buret to a solution of the sample.

In an oxidation-reduction titration, the solution in the buret, called the titrant, is a solution of an oxidizing agent or a reducing agent. The equivalence point occurs when the stoichiometric quantity of the reagent in the buret has been added to the solution of the sample, that is, when the oxidation-reduction reaction is complete. There are a number of ways to determine when the equivalence point is reached. Some are based on the electrical properties of the solution and some on a change in color of a substance in the solution.

The choice between an oxidizing agent and a reducing agent as the titrant depends on the oxidation state of the element that is being analyzed. If the element is in a low oxidation state, a solution of an oxidizing agent is the titrant. If the element is in a high oxidation state, a solution of a reducing agent is the titrant. The oxidizing agent or reducing agent that is used must be powerful enough to oxidize or reduce the sample completely, and the reaction must occur rapidly. Preliminary treatment often is needed to ensure that all of the element that is being analyzed is in the same oxidation state at the start. In addition, it is important to ensure that the titrant reacts only with the element of interest.

In practice, the titrant usually is an oxidizing agent. Some common oxidizing

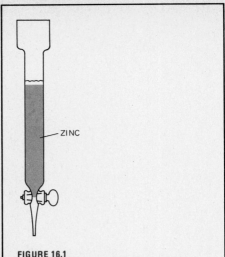

FIGURE 16.1
A Jones reductor. Iron in the +3 oxidation state is reduced to the +2 oxidation state by passing it through a column of specially treated zinc.

ZINC

agents are solutions of potassium permanganate, MnO_4^-; potassium dichromate, $Cr_2O_7^{2-}$; cerium(IV) sulfate, Ce^{4+}; potassium bromate, BrO_3^-; and iodine, I_2. Many of these substances are colored and so can be used as self-indicators. Some common reducing agents are solutions of iron(II) cation, Fe^{2+}; sodium arsenite, AsO_3^{3-}; sodium oxalate, $C_2O_4^{2-}$; oxalic acid, $H_2C_2O_4$; and sodium thiosulfate, $S_2O_3^{2-}$.

One common oxidation-reduction titration is the determination of the quantity of iron in a sample of iron ore. The first step is the dissolution of the iron ore in a mixture of perchloric acid and phosphoric acid. In this solution, the iron ore exists as a complex ion:

$$Fe_2O_3(s) + 2H^+ + 2H_3PO_4 \rightleftharpoons 2Fe(HPO_4)^+ + 3H_2O$$

To carry out an oxidation-reduction titration using an oxidizing agent, the iron must first be reduced from its +3 oxidation state in this ore to the +2 oxidation state. This reduction is accomplished by passing the solution through a column filled with specially treated zinc, a so-called Jones reductor (Figure 16.1). The reaction can be represented as:

$$2Fe^{3+} + Zn(s) \rightleftharpoons 2Fe^{2+} + Zn^{2+}$$

We are now ready for the titration, which is carried out by oxidizing the Fe^{2+} cation. The oxidizing agent is a solution of potassium permanganate, and the reaction between the two substances is rapid and essentially complete:

$$MnO_4^- + 5Fe^{2+} + 8H^+ \rightleftharpoons Mn^{2+} + 5Fe^{3+} + 4H_2O$$

According to the stoichiometry of this process, 0.2 mol of permanganate ion is needed to oxidize one mole of Fe^{2+} ion in the sample. The equivalence point in the titration is reached when this relative quantity of MnO_4^- is added from the burette. The oxidizing agent is the indicator for the titration. A solution of $KMnO_4$ is purple, while its reduction product, the Mn^{2+} ion, is pale pink. As the titrant is added to the sample, the color of the titrant changes. When all the Fe^{2+} has been oxidized to Fe^{3+}, the permanganate no longer changes color. The addition of the first drop of titrant solution past the equivalence point causes an easily detected purple coloration. The amount of iron in the sample is determined by measuring the amount of MnO_4^- added at that point.

Oxidation-reduction titrations require essentially the same sort of calculations used for neutralizations. But oxidation-reduction calculations may seem more difficult, because we encounter more complex equations and more complicated stoichiometry. Shortcuts can be developed, but if we start with a balanced chemical equation and use the molar relationships that arise from the equation, we can do any calculation without introducing extra concepts.

EXAMPLE 16.2
A sample of a substance whose only oxidizable material is tin in the +2 state is titrated with a dichromate solution that is prepared by dissolving 1.226 g of $K_2Cr_2O_7$ in enough water to give a total volume of 0.250 liter. A 0.0821-g sample of the substance requires a volume of 23.9 cm^3 of the titrant to reach the equivalence point. The product of the oxidation of the Sn^{2+} is the Sn^{4+} cation. The reduction product of the dichromate is the Cr^{3+} cation. Calculate the percentage of tin in the substance.
SOLUTION
The half-reactions for this oxidation-reduction are

$$Sn^{2+} \longrightarrow Sn^{4+} + 2e^-$$
$$Cr_2O_7^{2-} + 14H^+ + 6e^- \longrightarrow 2Cr^{3+} + 7H_2O$$

We combine the half-reactions to get the overall reaction for the process:

$$3Sn^{2+} + Cr_2O_7^{2-} + 14H^+ \longrightarrow 3Sn^{4+} + 2Cr^{3+} + 7H_2O$$

This equation tells us that one mole of dichromate oxidizes three moles of tin. We find the concentration of the solution of dichromate:

$$[Cr_2O_7^{2-}] = 1.226 \text{ g} \times \frac{1 \text{ mol K}_2\text{Cr}_2\text{O}_7}{294.2 \text{ g K}_2\text{Cr}_2\text{O}_7} \times \frac{1}{0.250 \text{ liter}} = 0.001667M$$

The amount of dichromate required to reach the equivalence point was

$$\text{amount Cr}_2\text{O}_7^{2-} = 0.001667 \frac{\text{mol}}{\text{liter}} \times 23.9 \text{ cm}^3 \times \frac{1 \text{ liter}}{1000 \text{ cm}^3}$$
$$= 3.98 \times 10^{-5} \text{ mol Cr}_2\text{O}_7^{2-}$$

From the stoichiometry of the reaction, the amount of tin in the sample is

$$\text{amount Sn}^{2+} = \frac{3 \text{ mol Sn}^{2+}}{1 \text{ mol Cr}_2\text{O}_7^{2-}} \times 3.98 \times 10^{-5} \text{ mol Cr}_2\text{O}_7^{2-}$$
$$= 1.20 \times 10^{-4} \text{ mol Sn}^{2+}$$

The mass of the tin in the sample is

$$\text{mass Sn} = 1.20 \times 10^{-4} \text{ mol Sn} \times 118.7 \frac{\text{g Sn}}{\text{mol Sn}} = 1.42 \times 10^{-2} \text{ g Sn}$$

and the percentage of tin in the sample is

$$\text{percentage Sn} = \frac{1.42 \times 10^{-2} \text{ g Sn}}{0.0821 \text{ g sample}} \times 100\% = 17.3\%$$

If you examine this problem, you will see that once we write the balanced equation for the oxidation-reduction reaction on which the titration is based, all the remaining steps are familiar from neutralization problems.

16.3   ELECTROLYSIS AND FARADAY'S LAWS

Electrolysis is a chemical change that results from the interaction between matter and an electric current. By logic, we expect this chemical change to be a reduction. A current is a flow of electrons. In an oxidation-reduction process, the substance that is reduced gains electrons. Therefore, if we have a process in which electrons are transferred to a substance, it seems logical that the substance will be reduced.

But no reduction can occur without an accompanying oxidation. If electrons flow into a chemical system, the charge of the system will increase steadily. As the negative charge of the system builds up, it becomes increasingly difficult for electrons to enter the system. An electric current can start a reduction, but only a small amount of material will actually be reduced unless there is an accompanying oxidation that provides a way for electrons to leave the system, maintaining its neutrality.

We must therefore alter our definition and say that electrolysis is an oxidation-reduction process in which the transfer of electrons is brought about by the application of an electric current from an external power source. In practice, electrolysis usually is carried out on liquids or liquid solutions. The electrons enter and depart through two electrodes, which usually are rods of inert substances that conduct electricity, such as graphite or platinum. The current enters the liquid at the *cathode,* and the transfer of electrons to the system occurs here. Electrons leave the system at the *anode.* Figure 16.2 shows a simplified electrolysis apparatus.

The chemical changes of an electrolysis are the result of the way in which an

FIGURE 16.2
Electrolysis. Electric current enters the system at the cathode. Electrons are transferred to the system at the cathode and leave the system at the anode, completing the circuit.

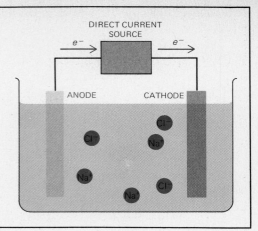

ionic substance conducts electricity. In a metal, an electric current is the flow of the electrons of the metal, so a metallic conductor need not undergo a chemical change as the result of an electric current. In an ionic substance, an electric current requires electron transfer from the circuit to the system at the cathode and from the system to the circuit at the anode. These two electron transfers result in an oxidation-reduction process. They maintain the current but also result in chemical changes.

The electrolysis of molten sodium chloride (Figure 16.3) is an example. Electrons enter the liquid at the cathode. They are accepted by the sodium cation, the only species in the system that is capable of accepting electrons. The process which can be represented by a half-reaction:

$$Na^+ + e^- \longrightarrow Na$$

is the reduction of sodium from the $+1$ oxidation state to the 0 oxidation state. As the process proceeds, Na^+ cations migrate to the cathode to accept the electrons that enter the system. Meanwhile, the electrical circuit is completed by the release of electrons from Cl^- anions to the anode, a process in which chlorine is oxidized from the -1 oxidation state to the 0 state:

$$Cl^- \longrightarrow \tfrac{1}{2}Cl_2 + e^-$$

As this process proceeds, Cl^- anions migrate to the anode to release electrons that leave the system, preventing the buildup of negative charge in the system. Thus, we see that the current is maintained in the system by the movement of ions, not of electrons. Cations move to the cathode, anions move to the anode, and the charge passes through the medium as a result.

In an electrolysis, *the half-reaction that occurs at the cathode is a reduction and the half-reaction that occurs at the anode is an oxidation.* As in any oxidation-reduction process, the chemical change of the entire system is obtained by adding the two half-reactions, eliminating the electrons in the addition. In this case the overall reaction is

$$Na^+ + Cl^- \longrightarrow Na + \tfrac{1}{2}Cl_2$$

We can think of an electrolysis as the conversion of electrical energy to chemical energy. In the electrolysis of sodium chloride, for example, a chemical system in an equilibrium state is changed to one in a nonequilibrium state by an external stress, the application of an electric current. In the system described in the reaction above, equilibrium lies far to the left; at equilibrium, the sodium and chlorine exist as ions. The electric current supplies the energy that forces the system to the right. If the current is turned off, the system does not return to equilibrium immediately, because the sodium metal and Cl_2 that form are kept

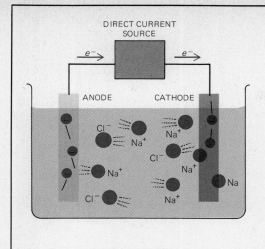

FIGURE 16.3
Electrolysis of molten sodium chloride. The Na^+
cations migrate to the cathode, where they gain
electrons. The Cl^- anions migrate to the anode,
where they release electrons. This flow of electrons
maintains the current.

apart by the distance between the electrodes. But when the Na and Cl_2 are brought together, they react violently, forming NaCl and releasing the large amount of energy that it took to form Na and Cl_2 from NaCl by the original electrolysis. We can imagine that the electrical energy that went into the electrolysis was stored in the nonequilibrium system of Na and Cl_2 and was then released as chemical energy.

Electrolysis has many practical uses because it is a good way to force chemical systems into nonequilibrium states. Electrolysis is often used to prepare materials that would not otherwise be available—reactive metals, for example. Without it, industry would have to use more complicated and more expensive procedures to obtain pure samples of reactive metals.

Electrolysis can occur in any liquid ionic medium—not only molten salts but also aqueous solutions, even those with relatively low concentrations of ions. In aqueous solutions, the flow of charge through the medium is due to the movement of ions, but the processes at the electrodes often differ from those that occur in molten salts. Water molecules can accept electrons from the cathode and become reduced, so the reaction that occurs at the cathode in aqueous solutions depends on the nature of the dissolved cations. Some cations are readily reduced; others are not. If a cation accepts electrons readily, its reduction is likely to be the cathode process. If a cation does not accept electrons readily, the reduction of water may instead be the cathode process.

In the electrolysis of a solution of sodium chloride, for example (Figure 16.4), the reduction of water is the cathode process:

$$2H_2O + 2e^- \longrightarrow H_2(g) + 2OH^-$$

ELECTROPLATING

The silver-plated spoon and the chromium-plated automobile bumper are familiar objects that are made by the well-known process of electroplating, in which a thin layer of metal is deposited on an electrically conducting surface in an electrolytic cell. Electroplating is more than a century old, but some of its most inventive and useful applications have been developed relatively recently.

For example, most of us know that the "tin can" is actually a steel can that has been plated with a protective layer of tin. But an increasing percentage of tin cans now are made of corrosion-resistant "tin-free steel," which is actually plated with chromium. The tinless tin can became possible in the 1960s, with the development of a method of putting an extremely thin coating of chromium on steel very quickly. Ordinary chromium plating is about 2×10^{-4} mm thick and takes several minutes of plating time to produce. The chromium plating on a tinless tin can is about 1.5×10^{-6} mm thick and is applied in a plating process that takes one-third of a second.

Metal-plated plastics are also a development of the 1960s, which saw the introduction of acrylonitrile-butadiene-styrene (ABS) plastic, which is readily plated. The plastic is etched chemically, usually by being dipped into an acid solution. After further treatment, it is coated with copper or nickel and then is placed in an electrolytic cell for the final plating. The process produces lightweight parts that can bear moderate stress and are useful in many applications where weight reduction is desired.

A specialized branch of electroplating called electrotyping is commonly used to print books and magazines. To make a book plate, an impression of the type is made by using wax, a soft plastic, or a lead sheet. This mold is coated with graphite to make it electrically conducting and is then put into an electroplating bath of copper sulfate or a similar salt. A thin layer of copper, nickel, or another metal is deposited on the surface of the mold, which is then removed. The thin metal shell is backed with a thicker layer to give it strength. A million or more pages can be printed from a single such plate.

Continual improvements are being made in electroplating technology, and there is an ongoing effort to develop new plating methods that will serve specialized needs. For example, the successful replacement of worn or damaged joints in the human body by artificial hips, knees, and the like depends on the existence of materials that are strong, light, and capable of withstanding the strong corrosive effects of saline body fluids. Several alloys for artificial joints have been developed, but the attempt to produce better materials goes on. One material that has shown promise in experiments is made by plating a lightweight alloy with a thin coat of tantalum, a metal that is unusually resistant to corrosion by body fluids and that is completely nonirritating.

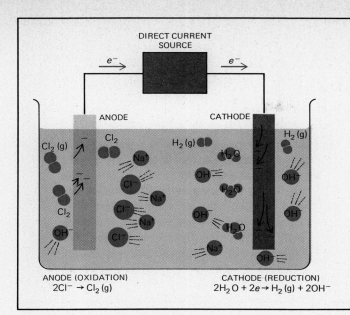

DIRECT CURRENT
SOURCE

e^- e^-

ANODE   CATHODE

Cl_2 (g) Cl_2 H_2 (g) H_2 (g)

Cl_2 Na^+ H_2O

Cl^- OH^-

Cl^- Na^+ H_2O

Cl_2 Cl^- OH^- OH^-

Na^+

OH^- Cl^- H_2O

Na^+

OH^-

ANODE (OXIDATION)   CATHODE (REDUCTION)
$2Cl^- \rightarrow Cl_2$ (g) $2H_2O + 2e \rightarrow H_2$ (g) $+ 2OH^-$

FIGURE 16.4
Electrolysis of a solution of sodium chloride. The
Cl^- anions migrate to the anode and release electrons,
but the presence of water prevents the Na^+ cations
from accepting electrons at the cathode. The cathode
process is, instead, the reduction of H_2O.

In solution, the Na^+ cation does not readily accept electrons. Sodium metal, the reduction product of Na^+, reacts violently with H_2O, forming OH^- anion and liberating $H_2(g)$. But in a solution containing Ag^+ cations, which readily accept electrons, the cathode process is

$$Ag^+ + e^- \longrightarrow Ag(s)$$

In acidic solutions with relatively high concentrations of H^+, the cathode process can be written in a simpler way as:

$$H^+ + e^- \longrightarrow \tfrac{1}{2}H_2(g)$$

Similarly, the oxidation that occurs at the anode in the electrolysis of aqueous solutions depends on the nature of the anions. In a solution of sodium chloride, the Cl^- anion is readily oxidized and the anode process generally is:

$$Cl^- \longrightarrow \tfrac{1}{2}Cl_2(g) + e^-$$

Chlorine gas usually is manufactured by the electrolysis of brine solutions, concentrated aqueous solutions of sodium chloride.

But if the anion is sulfate, SO_4^{2-}, or nitrate, NO_3^-, which are difficult to oxidize, the anode reaction usually is the oxidation of water:

$$2H_2O \longrightarrow O_2(g) + 4H^+ + 4e^-$$

If the solution is alkaline, the oxidation can be written as

$$4OH^- \longrightarrow O_2(g) + 2H_2O + 4e^-$$

In a solution where neither the anion nor the cation participates in the electrode processes (Figure 16.5), only the water molecules react. Although the ions Na^+ and NO_3^- do not undergo chemical change, their presence greatly facilitates the electrolysis of water because they carry the current. The electrolysis of water is much less expensive when electrolytes such as $NaNO_3$ are present. In such a solution, the electrolysis of water can be represented by the sum of the two half-reactions for the oxidation and reduction of H_2O:

$$6H_2O \longrightarrow 2H_2(g) + O_2(g) + 4H^+ + 4OH^-$$

In this reaction, H^+ is produced at the anode and OH^- at the cathode. These ions diffuse through the solution, neutralizing each other. If the solution is stirred,

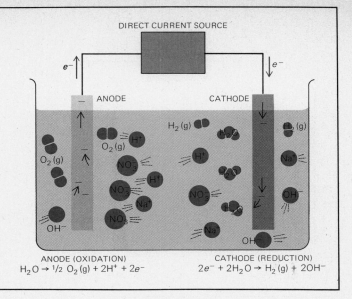

FIGURE 16.5
Electrolysis of water. At the anode, the H_2O molecule loses electrons to form oxygen gas and protons. At the cathode, the H_2O molecule gains electrons to form hydrogen gas and hydroxide ion.

ANODE (OXIDATION)
$$H_2O \rightarrow {}^{1}\!/_{2}\, O_2(g) + 2H^+ + 2e^-$$

CATHODE (REDUCTION)
$$2e^- + 2H_2O \rightarrow H_2(g) + 2OH^-$$

the neutralization occurs quickly. Since the ions in the reaction above form four H_2O molecules, the overall reaction for the electrolysis of water can be written as

$$2H_2O \longrightarrow 2H_2(g) + O_2(g)$$

Again, an electric current has caused a chemical system to go from equilibrium to nonequilibrium. In this reaction, the energy from the electrical current is stored in the mixture of hydrogen and oxygen. It can be released when a spark causes the two gases to combine explosively, forming water.

Faraday's Laws

In the early 1830s, Michael Faraday (1791–1867) systematically investigated the newly discovered process of electrolysis, discovering relationships that today are called Faraday's laws. These laws seem obvious to a generation that has a knowledge of the nature of electricity and matter. But Faraday's laws were a major scientific achievement in an era that had virtually no understanding of atomic and subatomic phenomena. In fact, Faraday's laws, with the atomic theory of Dalton, played a major role in the development of modern concepts of the structure of matter and the nature of electricity (Chapter 5).

One of Faraday's achievements was to demonstrate that the passage of an electric current through a given liquid causes reproducible chemical behavior related to the quantity of electricity and the nature of the substance. Faraday developed descriptions of the relationship between a given electric current and the mass of a substance that undergoes chemical change. In modern terms, these relationships, which are called Faraday's laws, are:

1. The mass of a substance formed or consumed in an electrolysis is proportional to the quantity of charge passing through it.
2. The mass of a substance formed or consumed in an electrolysis is also proportional to its atomic or molecular weight.
3. The mass of a substance formed or liberated in an electrolysis is inversely proportional to the number of electrons per mole needed to cause the indicated change in oxidation state.

To use these relationships, we need units of quantity for electricity. An electric current is a flow of electrons. Each electron has the same negative charge. The quantity of electricity is most simply designated as the total charge of the electrons in the current. The unit of current in the SI is the ampere, whose symbol is A. A current of 1 A is the flow of 1 coulomb, C, of charge, a specified number of electrons, for one second past a given point. This relationship between charge, current, and time usually is written as

$$q = it \qquad (16.1)$$

where q is charge, i is current, and t is time, measured respectively in coulombs, amperes, and seconds.

Once we know that a quantity of electricity (or charge) is a specific number of electrons, we can express the relationships of Faraday's laws in simple chemical terms. Thus, the half-reaction for the cathode process in the electrolysis of molten sodium chloride.

$$Na^+ + e^- \longrightarrow Na$$

can be described as the combination of one mole of sodium cations and one mole of electrons to form one mole of sodium metal. The half-reaction for the anode process:

$$Cl^- \longrightarrow \tfrac{1}{2}Cl_2 + e^-$$

can be described as the formation of one-half mole of chlorine and one mole of electrons from one mole of chloride ion.

Here we have an explanation for Faraday's first law. It is evident why a given quantity of charge produces a given quantity of sodium metal or of chlorine and consumes a given quantity of sodium cation or chloride anion. The transfer of one mole of electrons through molten sodium chloride causes the formation of one mole of sodium metal and one-half mole of chlorine. If charge equal to two moles of electrons is transferred through molten NaCl, two moles of sodium metal and one mole of chlorine are formed. By treating the symbol e^- in a half-reaction as if it were the symbol for an ordinary chemical reagent, we can understand electrolysis without introducing any new concepts.

We can also understand why the quantity of material formed or consumed in electrolysis is inversely proportional to the number of electrons per mole required to bring about the indicated change in oxidation state. One mole of electrons will cause the formation of one mole of sodium metal in the electrolysis of molten sodium chloride. But the cathode half-reaction in the electrolysis of a molten magnesium salt is:

$$Mg^{2+} + 2e^- \longrightarrow Mg$$

One mole of electrons forms only half of one mole of magnesium metal, because twice as many electrons per mole are needed to bring about the indicated change in oxidation state.

We can see this effect even more clearly in the electrolysis of metal salts that display more than one oxidation state. The reduction of tin in the +4 oxidation state requires four moles of electrons per mole of tin:

$$Sn^{4+} + 4e^- \longrightarrow Sn$$

Since the atomic weight of tin is 118.7, when one mole of electrons is passed through a system containing Sn^{4+}, a quantity of (118.7 g Sn/mol Sn)/ (4 mol e^-/mol Sn) = 29.68 g Sn forms. But in the electrolysis of a solution of Sn^{2+}, only two moles of electrons are required per mole of tin:

$$Sn^{2+} + 2e^- \longrightarrow Sn$$

A quantity of (118.7 g Sn/mol Sn)/(2 mol e^-/mol Sn) = 59.35 g Sn metal forms from one mole of electrons.

A unit other than the mole is convenient when we are dealing with electrons. One mole of a substance is 6.0221×10^{23}—Avogadro's number—of elementary entities of that substance. When we talk about "a mole," we usually indicate the mass of the substance: One mole of Na is 22.990 g, and one mole of Cl_2 is 70.906 g. It is more convenient to measure a mole of electrons as charge in coulombs, rather than as mass in grams, because the mass of electrons is so small. We usually express one mole of electrons as 96 500 C (the actual charge is 9.6487×10^4 C), a quantity that is called the **Faraday constant.** The charge on a single electron is 1.6022×10^{-19} C. If we multiply this charge by Avogadro's number, we get

$$(1.6022 \times 10^{-19} \ C/e^-)(6.0221 \times 10^{23} \ e^-/mol \ e^-)$$
$$= 9.6486 \times 10^4 \ C/mol \ e^-$$

which agrees closely with the value of Faraday's constant measured from mass-charge relationships in electrolysis.

The quantity of material formed by 96 500 C of charge is sometimes called an "equivalent." An equivalent is related to a mole, but the relationship depends on the number of electrons per mole that are needed for the desired change in oxidation state. We shall use moles, not equivalents, for our calculations.

EXAMPLE 16.3

When a solution of potassium iodide is electrolyzed, I_2 is produced at the anode and H_2 at the cathode. What mass of each substance is formed when a current of 5.20 A is applied for 46 minutes?

SOLUTION

The quantity of charge that flows through the solution determines the quantity of products in an electrolysis. The first step, therefore, is to convert the data we are given on current and time to a quantity of charge, using Equation 16.1:

$$q = it = (5.20 \ A)(46 \ min)\left(60 \ \frac{s}{min}\right) = 14 \ 400 \ C$$

Note that the number of minutes is multiplied by 60 to express time in seconds, the units of time for the coulomb, C.

Now the balanced half-reactions for the two electrode processes must be written. These half-reactions will give us the number of moles of electrons needed to bring about a change in oxidation state per mole of substance.

anode: $\qquad\qquad\qquad I^- \longrightarrow \frac{1}{2}I_2 + e^-$
cathode: $H_2O + e^- \longrightarrow \frac{1}{2}H_2(g) + OH^-$

In each process, 96 500 C will form a half-mole of the product of interest. Using this information, we can find the mass of each product that is formed by 14 400 C:

$$mass \ I_2 = 14 \ 400 \ C \times \frac{0.5 \ mol \ I_2}{96 \ 500 \ C} \times \frac{254 \ g \ I_2}{1 \ mol \ I_2} = 19.0 \ g \ I_2$$

The same sort of calculation can be used to find the mass of $H_2(g)$ that is formed, using the molecular weight of H_2, 2.02:

$$mass \ H_2 = 0.150 \ g \ H_2$$

In Example 16.3, we determined the quantity of material formed in an electrolysis by measuring the total quantity of charge. The reasoning can be reversed to get information about current or time by measuring the mass of products formed in an electrolysis.

EXAMPLE 16.4

The production of aluminum metal from bauxite ore, Al_2O_3, by electrolysis is the cornerstone of the aluminum industry. One step in this complex process is the reduction of aluminum from the $+3$ to the 0 oxidation state. Calculate the time needed for a current of 11.2 A to deposit 454 g of metallic aluminum.

SOLUTION

The half-reaction for the reduction of aluminum:

$$Al^{3+} + 3e^- \longrightarrow Al$$

shows that three moles of electrons, or 3(96 500) C of charge, are needed to form one mole of Al. The charge required to form the desired quantity thus is

$$454 \text{ g Al} \times \frac{1 \text{ mol Al}}{27.0 \text{ g Al}} \times \frac{3(96\ 500) \text{ C}}{1 \text{ mol Al}} = 4.87 \times 10^6 \text{ C}$$

To calculate the time needed to obtain this quantity of charge from a current of 11.2 A, we use Equation 16.1:

$$4.87 \times 10^6 \text{ C} = (11.2/A)(t)$$

and $t = 4.35 \times 10^5$ s or 121 hours.

16.4 GALVANIC CELLS

Oxidation-reduction processes can occur even when the reactants are separated from one another. This remarkable phenomenon occurs because an electron transfer from one reactant to another can take place through an electrical conductor, usually a metal wire.

An apparatus for carrying out an oxidation-reduction process in this way is called a **galvanic** or **voltaic cell** (Figure 16.6). You will notice a resemblance to the electrolytic cells mentioned in Section 16.3. Each cell has two electrodes, an anode and a cathode, in contact with liquid and with an electrical connection between them. But you will also notice differences. Unlike an electrolytic cell, the galvanic cell has no external source of current. The current between the electrodes in a galvanic cell is created by the electron transfer of an oxidation-reduction process within the cell. The galvanic cell shown in Figure 16.6 has a barrier, not found in an electrolytic cell. The barrier prevents the solutions in the anode and cathode compartments from mixing but allows electrical contact between the two solutions.

We can best regard the galvanic cell as a combination of two half-cells, each consisting of an electrode in contact with a solution. A half-reaction occurs in each half-cell. Such a half-reaction is often called an *electrode process*, which gives a different meaning to the term "electrode." Instead of referring only to the bar of solid attached to an external electrical connection, "electrode" in a galvanic cell is often assumed to mean an entire half-cell.

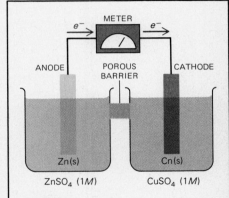

FIGURE 16.6
A galvanic cell. An oxidation process occurs in the cell on the left and a reduction process occurs in the cell on the right. The electrons travel through the wire at top, creating an electric current. The meter measures the amount of current. The porous barrier allows movement of ions to preserve neutrality but prevents the solutions from mixing.

Operation of a Galvanic Cell

If we build the galvanic cell of Figure 16.6, we will observe distinct chemical changes when the electrodes are connected. The zinc electrode shrinks and the copper electrode grows. At the same time, the $[Zn^{2+}]$ becomes greater than $1M$ and the $[Cu^{2+}]$ becomes less than $1M$. These changes can be described by a pair of half-reactions:

$$Zn(s) \longrightarrow Zn^{2+} + 2e^-$$
$$Cu^{2+} + 2e^- \longrightarrow Cu(s)$$

The first half-reaction is an oxidation that takes place in the anode compartment of the cell. The second half-reaction is a reduction that occurs in the cathode compartment. Until now, we have described electrolysis, in which the electrons come from an external source. In this galvanic cell, the electrons originate from a transfer within the reacting chemical system. The reduction at the cathode occurs because electrons liberated by the anodic oxidation travel through the wire to the cathode, where they are accepted by the Cu^{2+} ion. Since the electrodes take part in these processes, the electrolysis must stop when the anode is consumed.

It seems that as the cell operates and more Zn^{2+} is formed, the positive charge of the anode compartment should increase. At the same time, the negative charge of the cathode compartment should increase as more Cu^{2+} ion is removed from solution. But such a buildup of charge cannot occur. Electrical neutrality must be maintained if the cell is to continue to function: Negative charge must enter the anode compartment and leave the cathode compartment. This transfer of charge does, indeed, occur. The SO_4^{2-} anion passes from the cathode compartment to the anode compartment, through the porous barrier, so that neutrality is preserved as the reaction proceeds. The SO_4^{2-} anions do not undergo any chemical change; their migration serves only to preserve electrical neutrality.

We can describe the overall operation of this galvanic cell by writing an equation that is the sum of the two half-reactions for the electrode processes of the two half-cells:

$$Zn(s) + Cu^{2+} \rightleftharpoons Zn^{2+} + Cu(s)$$

The description given earlier indicates that the reaction will proceed from left to right when each cation has a concentration of $1M$. A galvanic cell obeys the laws of thermodynamics. Since the reaction proceeds spontaneously from left to right, in the equilibrium state there is a preponderance of the substances on the right side of the equation. The value of the equilibrium constant, K, of the system is large, so the $[Zn^{2+}]$ will be much greater than the $[Cu^{2+}]$ at equilibrium.

While proceeding spontaneously to an equilibrium state, a galvanic cell produces an electric current that can be used to perform work. In other words, a galvanic cell is a device that converts chemical energy into electrical energy. This capability is put to practical use on an enormous scale. The batteries that power our automobile electrical systems, our flashlights, tape recorders, portable radios, golf carts—even our spacecraft—are galvanic cells. The wide range of applications for galvanic cells—batteries for short—helps explain why so much effort has gone into inventing different kinds of galvanic cells. All these cells are similar in essence. Every galvanic cell has a set of components that are not at equilibrium and that proceed spontaneously to equilibrium by an electron transfer. The electron transfer occurs through an electrical connection between the electrodes, rather than directly.

Notation for Galvanic Cells

A conventional notation has been developed to simplify the task of describing the large variety of galvanic cells. In this notation, the cell pictured in Figure 16.6 is represented as:

$$Zn(s)|Zn^{2+}(1M)\|Cu^{2+}(1M)|Cu(s)$$

The single vertical lines indicate boundaries between phases. The double vertical line indicates a barrier that allows the movement of ions but not the mixing of solutions. By convention, the anode half-cell is on the left, the cathode on the right. The notation for this system does not include the SO_4^{2-} anions, since they serve only to maintain neutrality.

Many different kinds of electrodes and many different kinds of barriers can be used in galvanic cells. Figure 16.7 shows a galvanic cell that can be represented as

Pt(s) | Sn^{2+}(1M), SN^{4+}(1M) ‖ Fe^{2+}(1M), Fe^{3+}(1M) | Pt(s).

In this system, the oxidation half-reaction that occurs in the anode half-cell is

$$Sn^{2+} \longrightarrow Sn^{4+} + 2e^-$$

and the reduction half-reaction that occurs in the cathode is

$$Fe^{3+} + e^- \longrightarrow Fe^{2+}$$

The overall cell reaction is obtained by adding the half-reactions so as to cancel the electrons from the equation:

$$Sn^{2+} + 2Fe^{3+} \longrightarrow Sn^{4+} + 2Fe^{2+}$$

In this cell, ions move between the two solutions through a barrier called a *salt bridge*, a tube filled with a concentrated solution of a salt, such as potassium chloride, that does not participate in the cell reaction. A salt bridge provides a means of adding negative ions to the anode compartment and positive ions to the cathode compartment. It allows ions to pass but prevents the solutions from mixing.

Types of Galvanic Cells

The electrodes of the two galvanic cells described thus far are metals. Many other kinds of electrodes can be used. A galvanic cell can have a gas electrode, consisting of a strip of an inert conductor that is in contact with both the liquid solution and a stream of gas whose pressure is kept constant. Other cells can have a metal-insoluble salt electrode, in which the metal of the electrode is coated with one of the sparingly soluble salts of the metal. The salt is in contact with the solution containing the anion of the metal salt.

Figure 16.8 shows a galvanic cell whose anode is a gas electrode and whose cathode is a metal-insoluble salt electrode. The cell can be represented as

Pt(s) | H$_2$(g)(1 atm) | H$^+$(1M), Cl$^-$(1M) | AgCl(s) | Ag(s)

The anode half-reaction is

$$H_2(g) \longrightarrow 2H^+ + 2e^-$$

in which a stream of hydrogen bubbles, at a pressure of 1 atm, is in contact with a 1M solution of H$^+$ and is passed over a platinum strip. This system is called the *standard hydrogen electrode*. We shall discuss it in greater detail later in this chapter.

The metal of the cathode is silver, and the salt is therefore a silver salt. Since the anion of the silver salt is chloride, the solution contains chloride ion. The half-reaction is

$$AgCl(s) + e^- \longrightarrow Ag(s) + Cl^-$$

This cell does not contain a porous barrier or salt bridge because both half-cells contain the same solution, 1M hydrochloric acid. The overall reaction of the cell is

$$H_2(g) + 2AgCl(s) \longrightarrow 2H^+ + 2Ag(s) + 2Cl^-$$

Galvanic cells that have the same solution at both electrodes have practical advantages and are quite common.

EXAMPLE 16.5

Write the two half-cell processes and the overall reaction that takes place in the cell:

Zn(s) | NH$_3$(1M), Zn(NH$_3$)$_4$$^{2+}$(1$M$) ‖ Br$^-$(1$M$) | Br$_2$(l) | Pt(s)

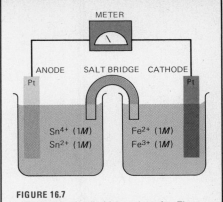

FIGURE 16.7
A tin-iron galvanic cell with inert electrodes. The oxidation occurs in the tin electrode at left, the reduction in the iron electrode at right. The salt bridge prevents the solutions from mixing.

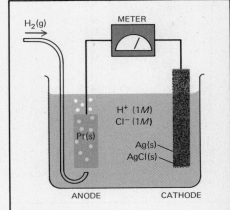

FIGURE 16.8
In this galvanic cell, the anode at the left has a gas electrode, consisting of a stream of hydrogen bubbles in contact with both the solution of H$^+$ and the platinum strip. The cathode at right has an electrode consisting of silver and one of its insoluble salts, silver chloride.

SOLUTION

By convention, the half-cell on the left is the anode, in which the oxidation takes place. The material undergoing a change in oxidation state is zinc, which goes from the 0 oxidation state in the metal to the $+2$ state in the complex ion. The balanced half-reaction is

$$Zn(s) + 4NH_3 \longrightarrow Zn(NH_3)_4^{2+} + 2e^-$$

The cathode is on the right, separated from the anode by the porous barrier that is represented by the double vertical line. In the cathode, bromine is reduced from the 0 oxidation state in $Br_2(l)$ to the -1 state in the bromide ion. The half-reaction is

$$Br_2(l) + 2e^- \longrightarrow 2Br^-$$

The overall cell reaction is obtained by adding the half-reactions so as to cancel the electrons from the equation:

$$Zn(s) + 4NH_3 + Br_2(l) \longrightarrow Zn(NH_3)_4^{2+} + 2Br^-$$

Batteries

Batteries have several unique advantages. A galvanic cell converts chemical energy to electrical energy and then to work with much greater efficiency than a steam engine or a diesel engine. The energy of a galvanic cell can be stored until it is needed, simply by keeping the electrodes unconnected. The most familiar galvanic cell is the dry cell (Figure 16.9), which is available in several forms, including the flashlight battery. The anode, the source of electrons, is the zinc of the cell wall. The oxidation process at the anode is

$$Zn(s) \longrightarrow Zn^{2+} + 2e^-$$

The cathode is a rod of graphite surrounded by a paste of MnO_2 and carbon. The cathode reaction is complex but can be represented as

$$2NH_4^+ + 2MnO_2(s) + 2e^- \longrightarrow 2MnO(OH) + 2NH_3$$

Both electrodes are in contact with a wet paste of ammonium chloride, zinc chloride, water, and an inert filler, so the "dry cell" is not really dry.

The lead storage battery used in automobiles is really a number of galvanic cells in series. It differs from the dry cell in one important respect: The galvanic cells are designed so that the cell reactions that produce current can be reversed by applying an external current. This design feature makes the battery rechargeable and thus long-lived.

The chemistry of the average automobile battery is based on changes in the oxidation state of lead. In one half-cell, the oxidation state of lead is changed from 0 to $+2$. In the other half-cell, the oxidation state is changed from $+4$ to $+2$. The lead in the 0 oxidation state is in the form of a spongy alloy. The lead in the $+2$ oxidation state is in the form of $PbSO_4$, a white insoluble salt that can sometimes be seen as a coating on dead batteries. Lead in the $+4$ oxidation state is in the form of PbO_2, lead dioxide, which is also insoluble.

The lead storage battery consists of two kinds of plates with different fillings (Figure 16.10). The anodes are filled with the spongy lead alloy. The cathodes are filled with lead dioxide. Both electrodes are in contact with a solution of H_2SO_4, sulfuric acid, in water. The solution is about 38% sulfuric acid by mass, and its density is 1.30 g/cm^3. The sulfuric acid participates in the cell reaction. Its role is to ensure that lead in the $+2$ oxidation state is produced in the form of the insoluble $PbSO_4$.

The chemical system of the lead storage battery is not at equilibrium, and an

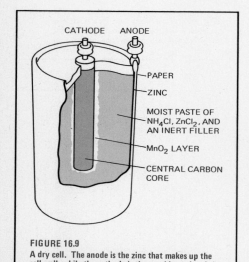

FIGURE 16.9
A dry cell. The anode is the zinc that makes up the cell wall, while the cathode is the graphite rod and the moist paste that surrounds it.

CATHODE   ANODE

PAPER

ZINC

MOIST PASTE OF NH_4Cl, $ZnCl_2$, AND AN INERT FILLER

MnO_2 LAYER

CENTRAL CARBON CORE

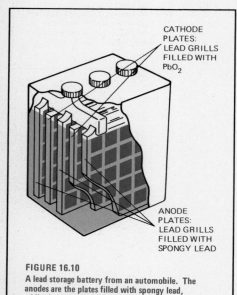

CATHODE PLATES: LEAD GRILLS FILLED WITH PbO_2

ANODE PLATES: LEAD GRILLS FILLED WITH SPONGY LEAD

FIGURE 16.10
A lead storage battery from an automobile. The anodes are the plates filled with spongy lead, while the cathodes are plates filled with lead dioxide. The plates are immersed in a solution of sulfuric acid.

electric current will flow when the plates of lead and of lead dioxide are connected. The oxidation process that occurs in the anode is

$$Pb(s) + SO_4{}^{2-} \longrightarrow PbSO_4(s) + 2e^-$$

The reduction that occurs when the lead dioxide plate, the cathode, receives the
electrons released by the oxidation of lead is

$$PbO_2(s) + SO_4{}^{2-} + 4H^+ + 2e^- \longrightarrow PbSO_4(s) + 2H_2O$$

You will note that both these processes produce $PbSO_4$, in which lead is in
the $+2$ oxidation state. The overall reaction for the cell is

$$Pb(s) + PbO_2(s) + 2SO_4{}^{2-} + 4H^+ \underset{\text{charge}}{\overset{\text{discharge}}{\rightleftharpoons}} 2PbSO_4(s) + 2H_2O$$

As the cell discharges electricity and proceeds toward equilibrium, insoluble
lead sulfate is deposited on both plates and sulfuric acid is consumed. The
consumption of H_2SO_4 reduces the density of the solution in the battery. The
extent to which a battery has been discharged can be determined by measuring
the density of the solution with an instrument called a hydrometer.

Lead storage batteries have long lifetimes. An automobile battery will last for
several years. Its long life is based on the ability of the battery to function not only
as a galvanic cell but also as an electrolytic cell. When an automobile engine
runs, it drives an electrical generator or alternator, a source of electric current.
This current flows into the battery in a direction that reverses the two half-cell
reactions and thus the overall cell reaction. Lead sulfate is converted to lead and
lead dioxide, while the sulfate ion regenerates sulfuric acid. In other words, the
system is moving away from equilibrium, back toward its starting conditions. As a
galvanic cell, the lead storage battery converts chemical energy into electrical
energy. As an electrolytic cell, the battery converts electrical energy into chemical
energy. The cycle can be repeated many times.

Despite their advantages, galvanic cells provide only a small fraction of the
electrical energy we use. The direct conversion of chemical energy to electrical
energy that takes place in galvanic cells may be highly efficient. But much more
electricity is generated by the indirect conversion that takes place in power plants,
where heat produced by burning fossil fuels is used to make steam that drives
generators. At best, fossil-fuel plants are only about 40% efficient, which means
that most of their energy is lost to the surroundings. But even in these days of high
energy prices, coal, oil, and natural gas are much cheaper than the reactants
used in galvanic cells—so much cheaper that batteries are economically competitive only in specialized applications.

In recent years, a great deal of research has gone into development of the
fuel cell, which potentially combines the high efficiency of the galvanic cell and
the low cost of fossil-fuel plants. The fuel cell uses the combustion reaction as the
oxidation-reduction process of a galvanic cell. In a typical fuel cell, hydrogen
combines with oxygen, producing electricity with high efficiency and low pollution. Unfortunately, such practical problems as the extremely high cost of
electrodes have limited the use of fuel cells to applications where cost is not an
important factor, notably in manned spacecraft such as the Apollo modules.

The Bacon hydrogen-oxygen fuel cell, diagramed in Figure 16.11, has been
used on manned space flights. The two half-reactions are

$$2H_2(g) \longrightarrow 4H^+ + 4e^-$$
$$O_2(g) + 4H^+ + 4e^- \longrightarrow 2H_2O$$

and the overall reaction is the familiar formation of water:

$$2H_2(g) + O_2(g) \longrightarrow 2H_2O(l)$$

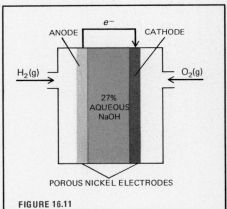

FIGURE 16.11
A fuel cell. Hydrogen gas is oxidized at the anode
and oxygen gas is reduced at the cathode, causing a
flow of electrons and an electric current. The only
product is pure water.

THE SEARCH FOR A SUPERBATTERY

The lead-acid batteries that dominate today's market for rechargeable electric sources are not good enough for the projected needs of tomorrow. To a surprising extent, current efforts to reduce pollution and develop inexhaustible energy sources depend on the success of research programs whose aim is to produce a "superbattery."

More efficient batteries could enable electric power plants to operate at constant levels. The batteries could store electricity generated at hours of low consumption for use during periods of peak consumption. Better batteries are needed to store the electricity from solar or wind-powered generators. More efficient batteries could also help make quiet-running, nonpolluting electric automobiles competitive with gasoline-powered vehicles.

Today's lead-acid batteries have a power density of 30 watt-hours per kilogram and a life of about 700 charge-discharge cycles. A typical 1700-kg electric car today needs 500 kg of lead-acid batteries to achieve a range of 30 km at a steady speed of 11 km an hour, a performance that is markedly inferior to that of gasoline-powered vehicles. The government is funding research to develop a battery that would achieve a fourfold increase in automobile driving range with half the present weight of batteries.

One candidate is the lithium-sulfur battery, which operates at about 700 K. A lithium alloy is the anode, a metal sulfide is the cathode, and a molten salt is the electrolyte. An energy density of 75 watt-hours per kilogram has been achieved, enough to give an automobile a range of 45 km at a speed of 25 km an hour.

Another major research effort is being devoted to the sodium-sulfur battery, which uses molten sodium for the anode, a mixture of molten sulfur and sodium polysulfide as the cathode, and a solid ceramic material, which is an excellent conductor of sodium ions, as the barrier between the electrodes. Sodium-sulfur batteries operate at 575–625 K and have achieved power densities of more than 120 watt-hours per kilogram.

A major difficulty in both these systems is the inconveniently high operating temperature, which enhances the rate of corrosion. For such batteries to be commercially feasible, materials that are highly resistant to corrosion and are relatively cheap must be available in large quantity. A third battery that is being tested is a zinc-chlorine system that has an aqueous electrolyte and operates at ordinary temperatures. One unusual feature is the method used to store chlorine when the battery is charged. The chlorine gas is chilled in the presence of the water to form solid chlorine hydrate. To discharge the battery, the frozen chlorine hydrate is heated. The chlorine evaporates and can then take part in the reaction.

If government plans are fulfilled, a fleet of more than 5000 electric automobiles will be operating on such superbatteries in the early 1980s. The results of this test program will help determine whether the electric automobile can make a comeback and will also be a major factor in the development of solar energy sources.

Pollution is an inevitable byproduct of plants which generate electricity by burning coal and other fossil fuels. The fuel cell and other methods of converting chemical energy directly to electrical energy are potentially more efficient and less polluting. (*Brody, Stock, Boston*)

Porous electrodes and other techniques are used to make the cell processes occur rapidly.

If the technical and practical problems that keep the cost of fuel cells so high can be surmounted, they could have a vast number of uses. Electric utilities are trying to develop hydrogen-oxygen fuel cells suitable for large-scale power generation. Biomedical researchers envisage fuel cells that are powered by the oxidation of food as the energy source for an artificial heart. So far, these possibilities lie in the future.

16.5 CELL POTENTIAL

It is the difference in potential energy between the two electrodes of a galvanic cell that causes an electric current to flow. This energy difference is most conveniently described as **potential**, which is the *potential energy per unit charge*. The unit of charge is the coulomb, C; the unit of potential energy is the joule, J; and the unit of potential is the volt, V:

$$1 \text{ V} = 1 \text{ J/C}$$

Charge flows only if there is a difference in potential between two points. In a galvanic cell, electrons flow from the anode to the cathode because the negative potential at the anode is higher than at the cathode. The potential at any one point cannot be measured. But it is possible to measure the difference in potential between two points by using a voltmeter. As Figure 16.6 shows, a voltmeter can be placed in the line between the electrodes of a galvanic cell to measure the potential difference between the half-cells. The reading obtained in this way is called the potential of the cell; it is represented by the symbol $\mathcal{E}$.*

---

* The value of the voltage measured in this way varies with the current of the cell. Ideally, we want the potential of the cell when the current is 0. This potential can be measured with an instrument called a potentiometer. The cell potential when the current is 0 is called the electromotive force (emf) of the cell. We shall use ''potential'' synonymously with ''emf.''

Voltmeter measurements on different galvanic cells show that the cell potential is affected by a number of factors: the nature of the substances that make up each half-cell, the concentrations of dissolved ions and molecules, the pressures of gases, and the temperature. These factors are the same ones that determine both the free energy difference between substances and the value of the equilibrium constant, K, for a chemical system.

A direct and useful relationship between free energy and potential can be derived. Free energy (Section 13.7) gives us information about the equilibrium state of a chemical system. It also gives us qualitative and quantitative information about the behavior of a mixture of substances, and about the amount of work that can be obtained from a chemical system that is proceeding to an equilibrium state. We cannot measure the free energy of a system easily, but it is easy to measure the potential of a galvanic cell. Since voltage is proportional to free energy, our measurements of the potential of galvanic cells will give us the same information that we would get by measuring free energy.

Standard Potential

To make comparisons of potentials of different cells, we must define a standard state for a galvanic cell. This standard state is defined in terms of the concentration of dissolved substances and the pressure of gases, which together determine the value of the cell potential. The standard state for cell potential is the same as the standard state for free energy. In each, the concentration of all dissolved species is $1M$,* and the pressure of all gases is 1 atm. When these conditions of concentration and pressure are met, the cell displays what is called **standard potential**, whose symbol is $\mathcal{E}°$. Since the $\mathcal{E}°$ of a galvanic cell is temperature-dependent, we shall assume that all measurements are made at 298 K, unless otherwise stated. The cells shown in Figures 16.6, 16.7, and 16.8 are all in the standard state, and their measured potential is the standard potential.

Potential and Free Energy

The relationship between $\mathcal{E}°$ for a galvanic cell and $\Delta G°$, the standard free energy change, for the chemical reaction of the cell is simple, important, and easily derived. It is

$$\Delta G° = -nF\mathcal{E}° \tag{16.2}$$

where $\Delta G°$ is measured in joules, $\mathcal{E}°$ is measured in volts, F is the Faraday, the charge on one mole of electrons, 9.6487×10^4 C, and n is the number of moles of electrons transferred by the cell reaction.

Calculating n for a chemical reaction is not always easy. In the reaction for the cell of Figure 16.6,

$$Zn(s) + Cu^{2+} \longrightarrow Zn^{2+} + Cu(s)$$

we can see that the reaction occurs by the transfer of two moles of electrons. But the reaction of Example 16.1:

$$3S^{2-} + 2MnO_4^- + 4H_2O \longrightarrow 3S(s) + MnO_2(s) + 8OH^-$$

requires more analysis. The best approach is to break down the overall reaction into its two constituent half-reactions, in effect reversing the procedure by which the balanced overall reaction is obtained. The reaction of Example 16.1 is the sum of these half-reactions:

$$2S^{2-} \longrightarrow 3S(s) + 6e^-$$
$$2MnO_4^- + 4H_2O + 6e^- \longrightarrow 2MnO_2(s) + 8OH^-$$

We can see that six electrons are transferred by each half-reaction, so $n = 6$.

---

*The standard state is actually unit activity, which we are approximating as unit molarity.

EXAMPLE 16.6

It is found that $\mathcal{E}°$ for the cell

$Cu(s)\,|\,Cu^{2+}(1M)\,\|\,Ag^+(1M)\,|\,Ag(s)$

is 0.4594 V. Calculate $\Delta G°$ for the cell reaction.

SOLUTION

The best approach is to write the two half-cell reactions. Since the anode half-cell is by convention, the left side of the expression, the oxidation half-reaction is

$Cu(s) \longrightarrow Cu^{2+} + 2e^-$

and the reduction is

$Ag^+ + e^- \longrightarrow Ag(s)$

Each half-reaction must have the same number of electrons so that the two half-reactions can be combined into an overall oxidation-reduction equation. In this case, we multiply the reduction half-reaction by two, each half-reaction then has two electrons. When the half-reactions are combined, the electrons cancel and the overall reaction is

$Cu(s) + 2Ag^+ \longrightarrow Cu^{2+} + 2Ag(s)$

The number of electrons in each half-reaction is the number of moles of electrons transferred in the overall reaction. That is, $n = 2$. We now calculate $\Delta G°$, using the relationship expressed in Equation 16.2:

$\Delta G° = -2 \text{ mol } e^-(96\,490 \text{ C/mol})(0.4594 \text{ V}) = -88\,660 \text{ J}$

Note that the unit of free energy is the joule, since 1 V = 1 J/C.

In Chapter 13 we discussed the meaning of the standard free energy, $\Delta G°$, for a chemical system. The sign of the free energy change gives the direction of spontaneous change, while the magnitude of $\Delta G°$ gives the extent to which the system must change from the standard state to reach equilibrium. The value of $\Delta G°$ also gives the amount of work that can be obtained from a reacting chemical system. In fact, the magnitude of the equilibrium constant K is directly related to the magnitude of the standard free energy change of the system:

$\Delta G° = -2.303RT \log K$

We can get the same kind of information from the standard potential by using the relationship between the standard free energy change and the standard potential that is expressed in Equation 16.2. We know that the operation of a galvanic cell corresponds to a chemical reaction, and that the measured standard potential, $\mathcal{E}°$, of the galvanic cell gives us information about this reaction. To start with, note that Equation 16.2 includes a negative sign, which affects our interpretation of $\mathcal{E}°$. *The direction of spontaneous change gives a negative value for $\Delta G°$ but a positive value for $\mathcal{E}°$.*

In the expression for a galvanic cell, the anode half-cell is on the left. If the measured $\mathcal{E}°$ of the cell is positive, we know that the system changes spontaneously from the standard state to equilibrium when the oxidation occurs in the half-cell on the left. If the measured $\mathcal{E}°$ of the cell is negative, we know that we have written the expression for the cell incorrectly. Two examples will demonstrate. For the cell

$Zn(s)\,|\,Zn^{2+}(1M)\,\|\,Cu^{2+}(1M)\,|\,Cu(s)$

$\mathcal{E}° = 1.1030$ V. The positive value indicates that the system proceeds to equilibrium by an oxidation in the zinc half-cell and a reduction in the copper half-cell. The overall reaction

$$Zn(s) + Cu^{2+} \rightleftharpoons Zn^{2+} + Cu(s)$$

proceeds from left to right to reach the equilibrium state when all the components of the system are in their standard states.

But when we construct the cell

$$Zn(s)|Zn^{2+}(1M)\|Mg^{2+}(1M)|Mg(s)$$

we find that the standard potential of the cell is -1.612 V. The oxidation occurs in the magnesium half-cell and the reduction in the zinc half-cell. The negative value of the potential tells us that it is best to rewrite the expression for the cell as

$$Mg(s)|Mg^{2+}(1M)\|Zn^{2+}(1M)|Zn(s)$$

to indicate that the oxidation takes place in the left half-cell. In its rewritten form, the cell has $\mathcal{E}° = 1.612$ V, and the overall reaction is

$$Mg(s) + Zn^{2+} \rightleftharpoons Mg^{2+} + Zn(s)$$

The reaction now proceeds from left to right to reach equilibrium when all the components of the system start in their standard states.

There are some rare cases where a galvanic cell has $\mathcal{E}° = 0$. In such a case, the system is at equilibrium when all its components are in their standard states. The equilibrium constant, K, for such a cell reaction has a numerical value of 1. There is no potential difference between the electrodes of such a cell when the concentration of each dissolved species is $1M$ and the pressure of each gas is 1 atm, so there is no net transfer of charge.

We can say that the sign of $\mathcal{E}°$ indicates the direction of spontaneous change and the magnitude of $\mathcal{E}°$ indicates the extent to which the system must change from the standard state to reach equilibrium. The magnitude of $\mathcal{E}°$ can also be used to find the value of K for the cell reaction. We can derive this relationship by combining the relationship between $\Delta G°$ and K with the relationship between $\Delta G°$ and $\mathcal{E}°$:

$$-nF\mathcal{E}° = -2.303RT \log K$$

or

$$\mathcal{E}° = \frac{2.303RT}{nF} \log K \tag{16.3}$$

Equation 16.3 is used to calculate the equilibrium constant at a given temperature of any reaction for which $\mathcal{E}°$ can be measured. Generally, we assume that the temperature is 298 K. We can obtain a simpler version of Equation 16.3 by using 298 K for T and calculating the numerical value of the term $2.303RT/F$:

$$\mathcal{E}° = \frac{0.05913}{n} \log K \tag{16.4}$$

You should note that the units of $2.303RT/F$ are joules/coulombs, or volts.

EXAMPLE 16.7

The cell

$$Pt(s)|Fe^{2+}(1M),\ Fe^{3+}(1M)\|MnO_4^-(1M),\ Mn^{2+}(1M),\ H^+(1M)|Pt(s)$$

is found to have a potential of 0.721 V at 298 K. Calculate K for the cell reaction.

SOLUTION

Since all the components of the system are in their standard states, the measured potential is $\mathcal{E}°$. Therefore, we can use Equation 16.4 to calculate the value of K.

To be sure of the cell reaction and to help find the value of n, it is best to write the half-reactions.

At the anode:

$$Fe^{2+} \longrightarrow Fe^{3+} + e^-$$

At the cathode:

$$MnO_4^- + 8H^+ + 5e^- \longrightarrow Mn^{2+} + 4H_2O$$

We multiply the oxidation half-reaction by five to equalize the number of electrons. The electrons then cancel when the half-reactions are combined to obtain the overall reaction:

$$5Fe^{2+} + MnO_4^- + 8H^+ \longrightarrow 5Fe^{3+} + Mn^{2+} + 4H_2O$$

Since each half-reaction has $5e^-$, $n = 5$. Substituting the data into Equation 16.4 gives

$$0.721 \text{ V} = \frac{0.0591 \text{ V}}{5 \text{ mol } e^-} \log K$$

$$\log K = 61.0$$

$$K = 1.0 \times 10^{61}$$

Electrochemical measurements usually are needed to evaluate equilibrium constants of such large magnitude. Potential measurements often are used for this purpose.

16.6 ELECTRODE POTENTIALS

While the potential difference between two electrodes or half-cells of a galvanic cell can be measured, no one has yet devised a way to measure the potential of a single electrode, and it seems probable that no one ever will. If such a method were available, it would be an easy matter to use a list of electrode potentials to find the potentials of cells that have never been constructed; we would simply compare the potentials of different half-cells.

But the same thing can be done indirectly. We can use the approach that enables us to evaluate thermodynamic functions such as ΔH and ΔG without knowing the values of H and G. Cell potential is the *difference* between two electrode potentials. If we know the relative potentials of the two electrodes, we can find the cell potential. The problem is comparable to the question of finding the distance between a person on the fifth floor of a building and someone on the eighth floor. If we know that the fifth floor is 18 m above the ground floor and the eighth floor is 27 m above the ground floor, we know that the two persons are 9 m apart. The absolute height of either floor above sea level can be ignored.

We can use the same procedure to assign potentials to electrodes and to calculate the potential difference between electrodes in a galvanic cell. By convention, the ''ground floor'' is the *standard hydrogen electrode* whose electrode potential is defined as exactly 0 V. This electrode consists of H_2 gas at a pressure of 1 atm, in contact with both a platinum strip and a solution where $[H^+] = 1M$. The half-reaction for this electrode process is

$$2H^+ + 2e^- \longrightarrow H_2(g)$$

Suppose we want to find the potential of the $Zn(s)\,|\,Zn^{2+}(1M)$ half-cell relative to the standard hydrogen cell. We must assign a numerical value to the potential of the half-reaction:

$$Zn^{2+} + 2e^- \longrightarrow Zn(s)$$

We can make this assignment by constructing a galvanic cell of these two half-cells and measuring the magnitude and sign of its potential.

To start with, note that both half-reactions are written as reductions. It is convenient to write all half-reactions in the same way, since we are comparing their potentials, and there has been an international agreement to write all electrode processes as reductions when expressing their relative potentials. In a galvanic cell, of course, only one of the half-cells will proceed as a reduction. The half-reaction at the other electrode will be an oxidation, the reverse of what is written.

The galvanic cell made of these two half-cells is

$$Zn(s)|Zn^{2+}(1M)||H^+(1M)|H_2(g)(1\ atm)|Pt(s)$$

It is found to have a standard potential of +0.7628 V, with the zinc half-cell as the anode. The half-reaction for the zinc electrode in this cell is an oxidation, the reverse of the half-reaction written above. The cell reaction is

$$Zn(s) + 2H^+ \longrightarrow Zn^{2+} + H_2(g)$$

By observing the spontaneous direction of the cell reaction and measuring the magnitude of the cell potential, we can assign a value to the relative potential of the $Zn(s)|Zn^{2+}(1M)$ electrode. The H^+ is reduced, which means that there is a greater tendency for the reduction of H^+ to proceed than for the reduction of Zn^{2+} to proceed. We can say that the free energy change, $\Delta G°$, for the reduction of Zn^{2+} is more positive (or less negative) than $\Delta G°$ for the reduction of H^+.

Equation 16.2 shows that $\mathcal{E}°$ is negative when $\Delta G°$ is positive. Since $\Delta G°$ for the reduction of Zn^{2+} is more positive than $\Delta G°$ for the reduction of H^+, then $\mathcal{E}°$ is more negative for the reduction of Zn^{2+} than for the reduction of H^+. Since $\mathcal{E}°$ for the reduction of H^+ is zero, the $\mathcal{E}°$ for the reduction of Zn^{2+} must be negative, and its value must be -0.7628 V.

The measured potential of a galvanic cell is the difference in potential between its two half-cells. By the convention we are using, *the potential of the anode half-reaction must be subtracted from the potential of the cathode half-reaction, when both are written as reductions*. That is

$$\mathcal{E}_{cell} = \mathcal{E}_{cathode} - \mathcal{E}_{anode} \qquad (16.5)$$

For this cell, $\mathcal{E}_{cathode}$ is zero. Therefore,

$$0.7628 = 0 - \mathcal{E}_{anode}$$
$$\mathcal{E}_{anode} = -07628\ V$$

for the half-reaction written as a reduction.

You can see why the anode potential is subtracted from the cathode potential in Equation 16.5 by considering how the half-reactions combine to give the overall reaction. The anode half-reaction is an oxidation, not a reduction. The reduction half-reaction which defines the potential must be reversed to give the correct anode half-reaction. If a half-reaction is reversed, the sign of its $\mathcal{E}°$ is also reversed. For the cell we have been discussing:

$$Zn^{2+} + 2e^- \longrightarrow Zn(s) \qquad \mathcal{E}° = -0.7628$$

or

$$Zn(s) \longrightarrow Zn^{2+} + 2e^- \qquad \mathcal{E}° = +0.7628$$

which is the half-reaction that we combine with a reduction half-reaction to get an overall reaction.

To summarize briefly: An electrode process corresponds to a half-cell or a half-reaction that we write as a reduction. When the substance being reduced in this half-reaction is more difficult to reduce than H^+, the potential of the half-reaction is negative. The negative value of the potential increases as the difficulty of reducing the substance in the half-cell increases.

We shall also find substances that are more easily reduced than H^+. The same sort of relationship applies: If a substance is easier to reduce than H^+, the potential of the half-reaction will be positive, and the positive value of the potential will increase with the ease of reducing the substance.

By making a series of such comparisons, we can build up a tabulation of standard potentials. For example, the cell

$$Cr(s)|Cr^{3+}(1M)\|H^+(1M)|H_2(g)(1\ atm)|Pt(s)$$

is found to have $\mathcal{E}° = +0.74$ V. The cell reaction is

$$2Cr(s) + 6H^+ \longrightarrow 2Cr^{3+} + 3H_2(g)$$

We find that H^+ is more easily reduced than Cr^{3+}, the substance in the other half-cell. Therefore, the potential of the reduction half-reaction

$$Cr^{3+} + 3e^- \longrightarrow Cr(s)$$

is negative; $\mathcal{E}° = -0.74$ V. This value is not as negative as that for the zinc electrode, so we can see that Cr^{3+} is more easily reduced than Zn^{2+}. We now have standard potentials for three half-reactions, which we can list in numerical order:

$$Zn^{2+} + 2e^- \longrightarrow Zn(s) \qquad \mathcal{E}° = -0.7628\ V$$
$$Cr^{3+} + 3e^- \longrightarrow Cr(s) \qquad \mathcal{E}° = -0.74\ V$$
$$2H^+ + 2e^- \longrightarrow H_2(g) \qquad \mathcal{E}° = -0.0000\ V$$

Again, the cell

$$Pt(s)|H_2(g)(1\ atm)|H^+(1M)\|Ag^+(1M)|Ag(s)$$

is found to have $\mathcal{E}° = +0.7996$ V with the standard hydrogen electrode as the anode. The cell reaction is

$$2Ag^+ + H_2(g) \longrightarrow 2Ag(s) + 2H^+$$

Therefore, Ag^+ is more easily reduced than H^+, and the reduction half-reaction for the silver electrode process has a positive potential. Its value is $\mathcal{E}° = +0.7996$ V, the same as that of the cell. We can also get this result by substitution into Equation 16.5:

$$0.7996 = \mathcal{E}_{cathode} - 0$$
$$\mathcal{E}_{cathode} = +0.7996\ V$$

Another example is the cell

$$Pt(s)|H_2(g)(1\ atm)|H^+(1M)\|Fe^{3+}(1M),\ Fe^{2+}(1M)|Pt(s)$$

for which $\mathcal{E}° = +0.770$ V with the hydrogen electrode as the anode. The cell reaction is

$$H_2(g) + 2Fe^{3+} \longrightarrow 2H^+ + 2Fe^{2+}$$

Again, Fe^{3+}, the substance in the cathode, is more easily reduced than H^+. The potential of the reduction half-reaction

$$Fe^{3+} + e^- \longrightarrow Fe^{2+}$$

is positive; it has the value of the cell potential, $\mathcal{E}° = +0.770$ V.

The emf Series

We can continue our list of half-reactions and their standard potentials with these two electrode processes:

$$Zn^{2+} + 2e^- \rightleftharpoons Zn(s) \qquad \mathcal{E}° = -0.7628\ V$$
$$Cr^{3+} + 3e^- \rightleftharpoons Cr(s) \qquad \mathcal{E}° = -0.74\ V$$
$$2H^+ + 2e^- \rightleftharpoons H_2(g) \qquad \mathcal{E}° = 0.0000\ V$$
$$Fe^{3+} + e^- \rightleftharpoons Fe^{2+} \qquad \mathcal{E}° = 0.770\ V$$
$$Ag^+ + e^- \rightleftharpoons Ag(s) \qquad \mathcal{E}° = 0.7996\ V$$

TABLE 16.1
Standard^a Reduction Potentials at 298 K

Half-Reaction	$\mathcal{E}°$ (V)	Half-Reaction	$\mathcal{E}°$ (V)
$Li^+ + e^- \rightleftharpoons Li(s)$	-3.045	$I_2(s) + 2e^- \rightleftharpoons 2I^-$	0.535
$Rb^+ + e^- \rightleftharpoons Rb(s)$	-2.925	$MnO_4^- + e^- \rightleftharpoons MnO_4^{2-}$	0.564
$K^+ + e^- \rightleftharpoons K(s)$	-2.924	$O_2(g) + 2H^+ + 2e^- \rightleftharpoons H_2O_2$	0.682
$Cs^+ + e^- \rightleftharpoons Cs(s)$	-2.923	$Fe^{3+} + e^- \rightleftharpoons Fe^{2+}$	0.770
$Ba^{2+} + 2e^- \rightleftharpoons Ba(s)$	-2.90	$Hg_2^{2+} + 2e^- \rightleftharpoons 2Hg(l)$	0.7961
$Sr^{2+} + 2e^- \rightleftharpoons Sr(s)$	-2.89	$Ag^+ + e^- \rightleftharpoons Ag(s)$	0.7996
$Ca^{2+} + 2e^- \rightleftharpoons Ca(s)$	-2.76	$2NO_3^- + 4H^+ + 2e^- \rightleftharpoons N_2O_4(g) + 2H_2O$	0.81
$Na^+ + e^- \rightleftharpoons Na(s)$	-2.7109	$NO_3^- + 3H^+ + 2e^- \rightleftharpoons HNO_2 + H_2O$	0.94
$Mg^{2+} + 2e^- \rightleftharpoons Mg(s)$	-2.375	$NO_3^- + 4H^+ + 3e^- \rightleftharpoons NO(g) + 2H_2O$	0.96
$Al^{3+} + 3e^- \rightleftharpoons Al(s)$	-1.706	$HNO_2 + H^+ + e^- \rightleftharpoons NO(g) + H_2O$	0.99
$Be^{2+} + 2e^- \rightleftharpoons Be(s)$	-1.70	$AuCl_4^- + 3e^- \rightleftharpoons Au(s) + 4Cl^-$	0.994
$Zn(NH_3)_4^{2+} + 2e^- \rightleftharpoons Zn(s) + 4NH_3$	-1.04	$Br_2(l) + 2e^- \rightleftharpoons 2Br^-$	1.065
$Mn^{2+} + 2e^- \rightleftharpoons Mn(s)$	-1.029	$Cu^{2+} + 2CN^- + e^- \rightleftharpoons Cu(CN)_2^-$	1.12
$Cr^{2+} + 2e^- \rightleftharpoons Cr(s)$	-0.91	$MnO_2(s) + 4H^+ + 2e^- \rightleftharpoons Mn^{2+} + 2H_2O$	1.208
$V^{3+} + 3e^- \rightleftharpoons V(s)$	-0.89	$O_2(g) + 4H^+ + 4e^- \rightleftharpoons 2H_2O$	1.229
$Zn^{2+} + 2e^- \rightleftharpoons Zn(s)$	-0.7628	$Au^{3+} + 2e^- \rightleftharpoons Au^+$	1.29
$Cr^{3+} + 3e^- \rightleftharpoons Cr(s)$	-0.74	$Cr_2O_7^{2-} + 14H^+ + 6e^- \rightleftharpoons 2Cr^{3+} + 7H_2O$	1.33
$Ag_2S(s) + 2e^- \rightleftharpoons 2Ag(s) + S^{2-}$	-0.7051	$Cl_2(g) + 2e^- \rightleftharpoons 2Cl^-$	1.3583
$S(s) + 2e^- \rightleftharpoons S^{2-}$	-0.508	$Au^{3+} + 3e^- \rightleftharpoons Au(s)$	1.42
$Ni^{2+} + 2e^- \rightleftharpoons Ni(s)$	-0.23	$MnO_4^- + 8H^+ + 5e^- \rightleftharpoons Mn^{2+} + 4H_2O$	1.491
$Sn^{2+} + 2e^- \rightleftharpoons Sn(s)$	-0.1364	$Mn^{3+} + e^- \rightleftharpoons Mn^{2+}$	1.51
$Pb^{2+} + 2e^- \rightleftharpoons Pb(s)$	-0.1263	$HClO + H^+ + e^- \rightleftharpoons \frac{1}{2}Cl_2(g) + H_2O$	1.63
$Fe^{3+} + 3e^- \rightleftharpoons Fe(s)$	-0.036	$HClO_2 + 2H^+ + 2e^- \rightleftharpoons HClO + H_2O$	1.64
$2H^+ + 2e^- \rightleftharpoons H_2(g)$	0	$MnO_4^- + 4H^+ + 3e^- \rightleftharpoons MnO_2(s) + 2H_2O$	1.679
$AgBr(s) + e^- \rightleftharpoons Ag(s) + Br^-$	0.0713	$N_2O(g) + 2H^+ + 2e^- \rightleftharpoons N_2(g) + H_2O$	1.77
$Sn^{4+} + 2e^- \rightleftharpoons Sn^{2+}$	0.15	$H_2O_2 + 2H^+ + 2e^- \rightleftharpoons 2H_2O$	1.776
$Cu^{2+} + e^- \rightleftharpoons Cu^+$	0.158	$O_3(g) + 2H^+ + 2e^- \rightleftharpoons O_2(g) + H_2O$	2.07
$Hg_2Cl_2(s) + 2e^- \rightleftharpoons 2Hg(l) + 2Cl^-$	0.2676	$F_2(g) + 2e^- \rightleftharpoons 2F^-$	2.87
$Cu^{2+} + 2e^- \rightleftharpoons Cu(s)$	0.3402		

Alkaline Solution

Half-Reaction	$\mathcal{E}°$ (V)	Half-Reaction	$\mathcal{E}°$ (V)
$Al(OH)_4^- + 3e^- \rightleftharpoons Al(s) + 4OH^-$	-2.35	$ClO_4^- + H_2O + 2e^- \rightleftharpoons ClO_3^- + 2OH^-$	0.17
$Mn(OH)_2(s) + 2e^- \rightleftharpoons Mn(s) + 2OH^-$	-1.47	$2ClO^- + 2H_2O + 2e^- \rightleftharpoons Cl_2(g) + 4OH^-$	0.40
$Cr(OH)_3(s) + 3e^- \rightleftharpoons Cr(s) + 3OH^-$	-1.3	$MnO_4^- + 2H_2O + 3e^- \rightleftharpoons MnO_2(s) + 4OH^-$	0.58
$Zn(OH)_4^{2-} + 2e^- \rightleftharpoons Zn(s) + 4OH^-$	-1.216	$ClO_2 + H_2O + 2e^- \rightleftharpoons ClO^- + 2OH^-$	0.59
$2H_2O + 2e^- \rightleftharpoons H_2(g) + 2OH^-$	-0.8277	$ClO_3^- + 3H_2O + 6e^- \rightleftharpoons Cl^- + 6OH^-$	0.62
$O_2(g) + 2H_2O + 2e^- \rightleftharpoons H_2O_2 + 2OH^-$	-0.146	$O_3(g) + H_2O + 2e^- \rightleftharpoons O_2(g) + 2OH^-$	1.24
$NO_3^- + H_2O + 2e^- \rightleftharpoons NO_2^- + 2OH^-$	0.01		

^{a} All species without state symbols are in aqueous solution at $1M$ concentration.

A tabulation of this kind, listing reduction half-reactions in order of increasing standard potential, is sometimes called an **emf series** or an **electrochemical series**. It can be used to predict and correlate a large body of data on chemical behavior. Table 16.1 is a more extensive list of this kind. Before we illustrate the uses of an emf series by working with the smaller list above, we shall summarize some of the conventions we have used or implied about these half-reactions and their standard potentials.

1. The potentials in this series are standard potentials, since all gases in the systems are at a pressure of 1 atm and all dissolved species are at a concentration of $1M$.
2. By definition, the standard hydrogen electrode has a standard potential of zero. All other electrode potentials are measured relative to this electrode, directly or indirectly.

3. All the half-reactions in the series are written as reductions. But it is understood that when two half-reactions are coupled in a cell or in a chemical system, one will reverse and proceed as an oxidation.

4. The magnitude of the standard electrode potential is a measure of the extent to which the half-reaction proceeds from left to right. The more positive the value of $\mathcal{E}°$, the greater the tendency of the reaction to proceed from left to right. The more negative the value of $\mathcal{E}°$, the smaller is the tendency of the reaction to proceed from left to right.

5. The potential of a half-reaction is not related to the coefficients of the equation, which means that the value of $\mathcal{E}°$ does not change if both sides of a half-reaction are multiplied by a factor. It also means that the number of electrons in the half-reaction does not affect the value of $\mathcal{E}°$; potential is potential energy *per charge*. The half-reactions

$$Zn^{2+} + 2e^- \longrightarrow Zn(s)$$
$$\tfrac{1}{2}Zn^{2+} + e^- \longrightarrow \tfrac{1}{2}Zn(s)$$
$$2Zn^{2+} + 4e^- \longrightarrow 2Zn(s)$$

all have $\mathcal{E}° = -0.7628$ V.

6. When the direction of a half-reaction is reversed, the sign of its potential is reversed. Thus, an unfavorable reduction half-reaction with a large negative potential is a favorable oxidation half-reaction with a large positive potential. The more negative $\mathcal{E}°$ is, the greater the tendency is for the reaction to reverse.

Cell Potentials from Electrode Potentials

Going back to the list on page 519, any two half-reactions can be selected from this series to be the half-cells of a galvanic cell. We can find the standard potential of this cell without constructing it and without making any measurements, simply by using the values of the standard potentials of the two half-reactions. We must first decide which of the half-reactions that we have selected is the one that reverses in the galvanic cell. The positions of the half-reactions in the series are the key to making this decision. Consistent with point 4 above, *the higher a half-reaction is in the series, the more easily it is reversed*. In this small series, the $Zn^{2+}|Zn$ half-reaction at the top of the series is the easiest to reverse and the $Ag^+|Ag$ half-reaction at the bottom of the series is the most difficult to reverse.

The half-reaction that reverses becomes the anode of the galvanic cell. Once the anode is chosen, the standard potential of the cell can be found by using Equation 16.5. For example, the galvanic cell made of the half-cells at the top and bottom of the series is:

$$Zn(s)|Zn^{2+}(1M)\|Ag^+(1M)|Ag(s)$$

Its standard potential is given by

$$\mathcal{E}°_{cell} = (0.7996) - (-0.7628) = 1.5624 \text{ V}$$

Since these two half-reactions are the farthest apart in the series, the cell they form has the greatest standard potential of any galvanic cell we could construct using the half-cells in this series.

Chemical Behavior from Electrical Potential

The emf series has more important applications than simply predicting the standard potential of galvanic cells. It can also be used to predict chemical behavior.

There is a cell reaction associated with every galvanic cell. The reaction can occur either in the cell or in an ordinary chemical system, where electron transfer occurs directly, rather than through an external connection. The data in an emf series can be used to obtain a great deal of information, qualitative and quantitative, about the behavior of chemical systems.

Each half-reaction in the series gives a relationship between two substances—an element in two oxidation states. The substance on the left of the reduction half-reaction is the element as an oxidizing agent; the substance on the right of the half-reaction is the element as a reducing agent. The standard potential of the half-reaction is a measure of the relative strengths of the oxidizing agent and the reducing agent. The greater the value of $\mathcal{E}°$, the greater is the strength of the oxidizing agent and the weaker is the strength of the reducing agent.

An emf series has the reduction half-reaction with the most negative $\mathcal{E}°$ at the top. Therefore, the half-reactions that stand highest in the emf series have the greatest tendency to reverse, proceeding from right to left as oxidations. We can say more: A substance whose half-reaction is toward the top of the emf series is a strong reducing agent, or a weak oxidizing agent. The closer a substance is to the bottom of the series, the more powerful an oxidizing agent (or weaker a reducing agent) it is. In our small series, the weakest oxidizing agent is Zn^{2+} at the top. As we go down the list, we encounter progressively stronger oxidizing agents; Ag^+, at the bottom, is the most powerful oxidizing agent of the group. Conversely, $Zn(s)$ is the strongest reducing agent and $Ag(s)$ is the weakest reducing agent in the series. Table 16.2 shows these relationships.

Since an overall chemical reaction is a combination of two half-reactions, we can use the relative positions of the two half-reactions in the emf series to predict the spontaneous direction of any reaction that starts with all reactants in their standard states. The half-reaction with the relatively stronger oxidizing agent proceeds spontaneously as a reduction. The half-reaction with the relatively stronger reducing agent reverses and proceeds spontaneously as an oxidation. Again, we see that the reaction that is closest to the top of the series is the one that reverses—the characteristic that we used in selecting the anode for the galvanic cell.

Consider a solution with $[Zn^{2+}] = 1M$, $[H^+] = 1M$, and $P_{H_2} = 1$ atm, in contact with zinc metal. The spontaneous direction of the reaction

$$Zn(s) + 2H^+ \rightleftharpoons Zn^{2+} + H_2(g)$$

is from left to right. The zinc half-reaction is higher in the emf series, so it reverses and proceeds as an oxidation. We can say the same thing in a different way: When zinc metal is in contact with an acidic solution, the reaction shown above will proceed substantially to the right, so that the zinc metal can dissolve until all the H^+ is consumed.

The same reasoning can be applied to the combination of the next half-reaction in our series and the standard hydrogen half-reaction. The spontaneous direction of the reaction

$$2Cr(s) + 6H^+ \rightleftharpoons 2Cr^{3+} + 3H_2(g)$$

is from left to right; chromium can dissolve in acid solution.

TABLE 16.2
Relative Strengths of Oxidizing Agents and Reducing Agents

Oxidizing Agent		Reducing Agent		Standard Potential (V)
$Zn^{2+} + 2e^-$	$\rightleftharpoons$	$Zn(s)$		-0.7628
$Cr^{3+} + 3e^-$	$\rightleftharpoons$	$Cr(s)$		-0.74
$2H^+ + 2e^-$	$\rightleftharpoons$	$H_2(g)$		0.0000
$Fe^{3+} + e^-$	$\rightleftharpoons$	Fe^{2+}		0.770
$Ag^+ + e^-$	$\rightleftharpoons$	$Ag(s)$		0.7996

(left margin, vertical: increasing strength ↓)   (middle, vertical: increasing strength ↑)

But the combination of the silver and hydrogen half-reactions gives the reaction

$$2Ag^+ + H_2(g) \rightleftharpoons 2Ag(s) + 2H^+$$

whose spontaneous direction is from left to right. Silver metal will not dissolve to any appreciable extent in acid solution.

We can generalize from these examples. Any metal that appears in a half-reaction above the standard hydrogen half-reaction in the emf series will dissolve in acid solution. Any metal that appears below the standard hydrogen half-reaction in the emf series will not dissolve in acid to any appreciable extent.

A number of chemical correlations can be made from the relative positions of substances in the emf series. In general, we can say that any oxidizing agent will oxidize any reducing agent, if the reducing agent is above the oxidizing agent in the emf series. Similarly, any oxidizing agent will be reduced by any reducing agent that appears above it in the emf series. But this generalization does not tell us anything about the kinetics of these processes. A process that is predicted to be favorable or spontaneous may occur very slowly. In using the emf series, all we can predict is the composition of the final equilibrium state.

These qualitative arguments can be made quantitatively. We can calculate a numerical value for the standard potential of any reaction that is a combination of two half-reactions of known standard potential. And if we know the standard potential, we can calculate both the standard free energy change, $\Delta G°$, and the equilibrium constant, K, for the reaction. We can also use the sign of $\mathcal{E}°$ to determine the spontaneous direction of the reaction when all the substances in it are in their standard states. The reaction is spontaneous from left to right when $\mathcal{E}°$ is positive and from right to left when $\mathcal{E}°$ is negative.

Reaction Potential from Electrode Potential

We find $\mathcal{E}°$ for a reaction by the same procedure used to find $\mathcal{E}°$ for a cell. The first step is to select the reduction half-reaction that reverses and proceeds as an oxidation. Once again, the half-reaction that is highest in the emf series is the one to reverse. The difference between the two potentials, taken directly from the series, gives the value of $\mathcal{E}°$. No other arithmetic operations are necessary. It is important to remember that the value of the standard potential is independent of the number of electrons in the half-reactions. When calculating $\mathcal{E}°$ for a reaction, we can ignore the number of electrons in each half-reaction.

EXAMPLE 16.8

Calculate the standard potential, the standard free energy, and the equilibrium constant for the reaction

$$2Cr^{3+} + 3Zn(s) \rightleftharpoons 2Cr(s) + 3Zn^{2+}$$

SOLUTION

We must first determine the two half-reactions that are combined to make this overall process. The reduction is

$$Cr^{3+} + 3e^- \rightleftharpoons Cr(s)$$

and the oxidation is

$$Zn(s) \rightleftharpoons Zn^{2+} + 2e^-$$

Both these half-reactions are listed in the series. The standard potential of the overall reaction is the listed potential of the half-reaction that occurs as a reduction minus the listed potential of the half-reaction that occurs as the oxidation. (We subtract, rather than add, the potentials because the series lists only reduction half-reactions. One of these reductions, the one higher in the series is reversed to obtain the overall reaction.)

$$\varepsilon^{\circ}_{\text{reaction}} = \varepsilon^{\circ}_{\text{reduction}} - \varepsilon^{\circ}_{\text{oxidation}} \qquad (16.6)$$

$$\varepsilon^{\circ} = (-0.74) - (-0.76) = 0.02 \text{ V}$$

Equation 16.2 is used to find the value of ΔG°. The value of n is found by inspection, which indicates that when the two half-reactions are combined in the overall reaction, the total number of electrons in each is six. Thus,

$$\Delta G^{\circ} = -(6 \text{ mol } e^-)(96\ 500 \text{ C/mol } e^-)(0.02 \text{ V}) = -12\ 000 \text{ J}$$

The value of K can be found by using Equation 16.4:

$$0.02 \text{ V} = \frac{0.059 \text{ V}}{6 \text{ mol } e^-} \log K$$

$$K = 1 \times 10^2$$

EXAMPLE 16.9

Using the data in Table 16.1, predict whether Cl_2 disproportionates in alkaline solution.

SOLUTION

Chlorine can exist in many oxidation states. Chlorine in the 0 oxidation state, as in Cl_2, can act as an oxidizing agent and can be reduced to the -1 oxidation state, as in Cl^-. Chlorine can also act as a reducing agent and be oxidized to one of several positive oxidation states. So Cl_2 can react with itself in an oxidation-reduction reaction, which means that it is possible for chlorine to disproportionate.

For the half-reaction

$$Cl_2 + 2e^- \rightleftharpoons 2Cl^-$$

Table 16.1 lists $\varepsilon^{\circ} = +1.3583$ V. This large positive reduction potential shows that Cl_2 is a relatively powerful oxidizing agent. The table also lists the half-reaction

$$2ClO^- + 2H_2O + 2e^- \rightleftharpoons Cl_2 + 4OH^-$$

with ε° of 0.40 V. The reverse of this half-reaction is an oxidation of Cl_2. When the two half-reactions are combined, the overall reaction is

$$2Cl_2 + 4OH^- \rightleftharpoons 2Cl^- + 2ClO^- + 2H_2O$$

or, more simply

$$Cl_2 + 2OH^- \rightleftharpoons Cl^- + ClO^- + H_2O$$

The standard potential of this reaction is positive: 1.36 V $-$ 0.40 V $=$ 0.96 V. Therefore, the reaction proceeds spontaneously from left to right. In other words, Cl_2 is unstable in alkaline solution; it is converted to chloride anion and hypochlorite anion. Chlorine does disproportionate, in a reaction that is put to practical use to keep swimming pools clear of algae. The reaction can actually proceed further forming ClO_3^-, if we wait long enough.

It might seem that equilibrium constants can be calculated from an emf series only for oxidation-reduction reactions. But there are data that allow us to calculate the equilibrium constants of reactions that are not oxidation-reductions, if they are the combination of two half-reactions listed in the emf series. Some examples of such reactions are the dissociation of complex ions into the metal ion and its ligands, and the dissolution of a sparingly soluble solid in water. For example, Table 16.1 lists a standard potential for the half-reaction

$$Zn(NH_3)_4^{2+} + 2e^- \rightleftharpoons Zn(s) + 4NH_3$$

which is the reduction of zinc from the $+2$ oxidation state in a complex ion to zinc metal in the 0 oxidation state. Table 16.1 also lists the half-reaction

$$Zn^{2+} + 2e^- \rightleftharpoons Zn(s)$$

in which zinc again is reduced from the $+2$ oxidation state to zinc metal in the 0 oxidation state. When the first half-reaction is reversed and these two half-reactions are combined to give an overall reaction, we find that not only the two electrons but also the $Zn(s)$ cancels out:

$$Zn^{2+} + 4NH_3 \rightleftharpoons Zn(NH_3)_4^{2+}$$

The overall reaction thus is a change from one form of zinc in the $+2$ oxidation state to another form of zinc in the $+2$ oxidation state. This process is not an oxidation-reduction reaction in the usual sense, although the reaction does occur as an oxidation-reduction if the cell

$$Zn(s)|NH_3(1M), Zn(NH_3)_4^{2+}(1M)\|Zn^{2+}(1M)|Zn(s)$$

is constructed. A transfer of two electrons can be visualized, since two electrons in each half-reaction do cancel when the half-reactions are combined. We have the standard potential for each half-reaction, so we can find the standard potential of the overall reaction:

$$\mathcal{E}° = (-0.76 \text{ V}) - (-1.04 \text{ V}) = 0.28 \text{ V}$$

Since we have $n = 2$, we can also find $\Delta G°$ and K.

EXAMPLE 16.10

Using the data in Table 16.1, calculate the solubility product of silver bromide.

SOLUTION

The reaction of interest that defines the K_{sp} is

$$AgBr(s) \rightleftharpoons Ag^+ + Br^-$$

One of the half-reactions in Table 16.1 includes solid silver bromide:

$$AgBr(s) + e^- \rightleftharpoons Ag(s) + Br^- \qquad \mathcal{E}° = 0.0713 \text{ V}$$

This half-reaction is the reduction of silver from the $+1$ to the 0 oxidation state. To obtain an overall reaction, we must combine it with another half-reaction that includes the same change in oxidation state. We find in Table 16.1 the half-reaction

$$Ag^+ + e^- \rightleftharpoons Ag(s) \qquad \mathcal{E}° = 0.7996 \text{ V}$$

The overall reaction that we are trying to form must have Ag^+ on the right. This half-reaction has Ag^+ on the left, so it is the one that we must reverse. Therefore, we change the sign of the $\mathcal{E}°$ of this half-reaction. In tabular form:

$AgBr(s) + e^- \rightleftharpoons Ag(s) + Br^-$	$\mathcal{E}° = 0.0713 \text{ V}$
$Ag(s) \rightleftharpoons Ag^+(s) + e^-$	$\mathcal{E}° = -0.7996 \text{ V}$
$AgBr(s) \rightleftharpoons Ag^+ + Br^-$	$\mathcal{E}° = -0.7283 \text{ V}$

The number of electrons transferred, n, is 1, the number of electrons that cancel when the half-reactions are combined. We calculate K by using Equation 16.4:

$$-0.7283 \text{ V} = \frac{0.05913 \text{ V}}{1 \text{ mol } e^-} \log K$$

$$K = 4.82 \times 10^{-13}$$

The small equilibrium constant is consistent with the negative $\mathcal{E}°$.

Half-Reaction Potentials from Electrode Potentials

Even more information may be obtained from the emf series. Many elements exist in a number of different oxidation states. The emf series does not list the values of all the standard potentials of every possible half-reaction for these elements. But we can obtain these values, as long as each oxidation state of interest appears in a half-reaction in the series. For example, the emf series in Table 16.1 lists the half-reaction for the reduction of tin from the $+4$ to the $+2$ oxidation state. It also lists the half-reaction for the reduction of tin from the $+2$ to the 0 oxidation state. The series thus gives us the information needed to calculate the standard potential of the half-reaction in which tin is reduced directly from the $+4$ to the 0 oxidation state.

To obtain this kind of information, we must use a procedure in which half-reactions in the series are combined to make a new half-reaction that is not in the series. However, the standard potential of the new half-reaction cannot be found simply by combining the potentials of the original half-reactions. We could convert the $\mathcal{E}°$ of each half-reaction to the corresponding $\Delta G°$, combine the $\Delta G°$ values to find the $\Delta G°$ of the new half-reaction, and then convert this $\Delta G°$ back to $\mathcal{E}°$. But we shall use a different method, in which the $\mathcal{E}°$ of each half-reaction is multiplied by the number of electrons in that half-reaction. The resulting values then are combined. Finally, the resulting potential is divided by the number of electrons in the new half-reaction to get its potential. The procedure is symbolized by

$$\mathcal{E}_T° = \frac{n_1\mathcal{E}_1° + n_2\mathcal{E}_2° + \cdots}{n_T} \tag{16.7}$$

where n_1 is the number of electrons in the first half-reaction and $\mathcal{E}_1°$ is its standard potential, n_2 is the number of electrons in the second half-reaction and $\mathcal{E}_2°$ is its standard potential (and so on for as many half-reactions as are being combined), and n_T is the number of electrons that appear in the resulting half-reaction. Equation 16.7 is derived from Equation 16.2. In using this relationship, care must be taken to make the sign of each $\mathcal{E}°$ consistent with the way in which its half-reaction is combined to make the new half-reaction.

EXAMPLE 16.11

Using the data in Table 16.1, find the standard potential of the half-reaction for the reduction of iron from the $+2$ to the 0 oxidation state.

SOLUTION

We begin by writing the desired half-reaction:

$$Fe^{2+} + 2e^- \rightleftharpoons Fe(s)$$

We then inspect the table for half-reactions that include these oxidation states of iron. Listed are

$$Fe^{3+} + 3e^- \rightleftharpoons Fe(s) \qquad \mathcal{E}° = -0.036 \text{ V}$$
$$Fe^{3+} + e^- \rightleftharpoons Fe^{2+} \qquad \mathcal{E}° = 0.770 \text{ V}$$

Applying the reasoning used earlier to combine reactions, we see that the second half-reaction (and the sign of its $\mathcal{E}°$) should be reversed and added to the first to get the desired half-reaction. The $\mathcal{E}°$ of the new half-reaction can be calculated by using the relationship expressed in Equation 16.7:

$$\mathcal{E}_T° = \frac{(3 \text{ mol } e^-/\text{mol})(-0.036 \text{ V}) + (1 \text{ mol } e^-/\text{mol})(-0.770 \text{ V})}{2 \text{ mol } e^-/\text{mol}}$$

$$= -0.439 \text{ V}$$

In more complex examples, it is sometimes difficult to choose the half-reactions to combine for the desired half-reaction. A line of reasoning based on oxidation states often is helpful in such cases.

16.7 POTENTIAL AND CONCENTRATION. THE NERNST EQUATION

So far, we have discussed standard potentials—that is, potentials when the concentrations of all dissolved substances are $1M$ and the pressures of all gases are 1 atm. By defining the standard state, we are able to make meaningful comparisons of many different galvanic cells and chemical systems. But in practice, we often encounter galvanic cells or chemical systems whose components are not in the standard state. We know that changes in concentration and pressure influence the potential of a system. What we need to know is just how variations from the standard state of a system affect its potential. We could then convert the standard potential of any given system to the potential under non-standard conditions, obtaining valuable information about the behavior of chemical systems in nonstandard states.

There is a known relationship between the potential, $\mathcal{E}$, of a nonstandard system, its standard potential, $\mathcal{E}°$, and its concentration and pressure terms. It was described by Walter Nernst (1864–1941) and is usually called the Nernst equation:

$$\mathcal{E} = \mathcal{E}° - \frac{2.303RT}{nF} \log Q \qquad (16.8)$$

At 298 K, Equation 16.8 can be simplified to

$$\mathcal{E} = \mathcal{E}° - \frac{0.05913}{n} \log Q \qquad (16.9)$$

where n has its usual meaning as the number of moles of electrons transferred in the process and Q is called the reaction quotient. The expression for Q has exactly the same form as the expression for the equilibrium constant, K. We use a different symbol because the concentrations and pressures in the expression for Q are not those of the equilibrium state, as they are in the expression for K.

The Nernst equation can be derived from relationships between the free energy and the composition of a system and the expression $\Delta G = -nF\mathcal{E}$.

The Nernst equation clarifies the relationship between $\mathcal{E}°$ and K. Equation 16.3 gave this relationship as $\mathcal{E}° = (2.303RT/nF)\log K$. The relationship between equilibrium constant and standard potential can be described quite simply. A system in which electron transfer occurs is at equilibrium only if there is no net transfer of electrons. Net electron transfer occurs because of a potential difference. Therefore, when net electron transfer stops and the system is at equilibrium, there is no potential difference; $\mathcal{E} = 0$. When this happens, $Q = K$, the Nernst equation becomes

$$0 = \mathcal{E}° - \frac{2.303RT}{nF} \log K$$

which is Equation 16.3.

The Nernst equation can be used to find the potential of a galvanic cell or any chemical system if we know (1) the standard potential of the system and (2) the concentrations and pressures of the relevant constituents of the system. The following procedure can be used in many calculations of this kind:

1. Write the reaction that is occurring in the cell or the chemical system.
2. Break the overall reaction into two half-reactions.
3. If possible, find the $\mathcal{E}°$ for each of these half-reactions in the emf series, and calculate the $\mathcal{E}°$ for the overall reaction.

4. Find n, the number of moles of electrons transferred in the reaction. This number is also the number of electrons that are cancelled when the half-reactions are combined to give the overall reaction.
5. Write the expression for Q from the overall reaction, keeping in mind the fact that Q and K have the same form.
6. Substitute all the data, including that on the composition of the system, into the Nernst equation.

EXAMPLE 16.12

Calculate the potential of the cell

$V(s)\,|\,V^{3+}(0.0011M)\,\|\,Ni^{2+}(0.24M)\,|\,Ni(s)$

SOLUTION

The cell reaction is

$2V(s) + 3Ni^{2+} \rightleftharpoons 2V^{3+} + 3Ni(s)$

The two half-reactions can be found in Table 16.1:

$V^{3+} + 3e^- \rightleftharpoons V(s) \qquad \mathcal{E}° = -0.89$ V
$Ni^{2+} + 2e^- \rightleftharpoons Ni(s) \qquad \mathcal{E}° = -0.23$ V

The first half-reaction reverses, and $\mathcal{E}°$ for the overall reaction is:

$\mathcal{E}_T° = (-0.23 \text{ V}) - (-0.089 \text{ V}) = 0.66$ V

The vanadium half-reaction is reversed and multiplied by two, and the nickel half-reaction is multiplied by three in order to obtain the overall reaction. The number of electrons that cancel when the two half-reactions are combined is six, so $n = 6$.

The expression for Q for the overall reaction is

$$Q = \frac{[V^{3+}]^2}{[Ni^{2+}]^3}$$

Using the calculated value of $\mathcal{E}°$ and the given data on concentration, the potential of the cell can be calculated with the help of the Nernst equation:

$$\mathcal{E} = 0.66 \text{ V} - \frac{0.059 \text{ V}}{6 \text{ mol } e^-} \log \frac{(0.0011)^2}{(0.24)^3}$$

$$\mathcal{E} = 0.70 \text{ V}$$

The values of $\mathcal{E}°$ and of $\mathcal{E}$ tell us a great deal about the behavior of the chemical system that corresponds to the cell reaction. The large positive value of $\mathcal{E}°$ means that the reaction has a large equilibrium constant and proceeds substantially to completion as written. The small concentration of V^{3+} ion in the cell relative to the concentration of Ni^{2+} means that the system is further from equilibrium than it would be at standard concentrations; therefore, $\mathcal{E}$ is even larger than $\mathcal{E}°$. Since the cell delivers electric current until its potential falls to zero, the magnitude of $\mathcal{E}$ can give us an idea of the amount of electrical work that can be obtained from a given amount of material in a galvanic cell.

The Nernst equation can also be used to convert measured potentials to standard potentials.

EXAMPLE 16.13

In investigating the properties of rare and expensive metals such as rhenium, element 75, it is both impractical and uneconomical to prepare cells with standard concentrations. To find the standard potential of the $Re^{3+}\,|\,Re(s)$ electrode, this cell is constructed:

$Pt(s)\,|\,Re(s)\,|\,Re^{3+}(0.0018M)\,\|\,Ag^+(0.010M)\,|\,Ag(s)$

The potential of this cell is found to be 0.42 V. Calculate the standard potential of the half-reaction $Re^{3+} + 3e^- \rightleftharpoons Re(s)$.

SOLUTION

Table 16.1 lists the $\mathcal{E}°$ for the $Ag^+|Ag(s)$ electrode as 0.80 V. Therefore, we can find the $\mathcal{E}°$ of the other electrode by finding the $\mathcal{E}°$ of the cell and using a relationship such as Equation 16.5.

The cell reaction is

$$Re(s) + 3Ag^+ \rightleftharpoons Re^{3+} + 3Ag(s)$$

The two half-reactions are

$$Re^{3+} + 3e^- \rightleftharpoons Re(s)$$
$$Ag^+ + e^- \rightleftharpoons Ag(s)$$

The overall reaction is obtained by reversing the rhenium half-reaction and multiplying the silver half-reaction by three. Thus, $n = 3$. The expression for Q is

$$Q = \frac{[Re^{3+}]}{[Ag^+]^3}$$

We now can solve for the standard potential of the cell:

$$0.42\ V = \mathcal{E}° - \frac{0.059\ V}{3\ mol\ e^-} \log \frac{(0.0018)}{(0.010)^3}$$

$$\mathcal{E}° = 0.48\ V$$

Since $\mathcal{E}°_T = \mathcal{E}°_{cathode} - \mathcal{E}°_{anode}$,

$$0.48\ V = 0.80\ V - \mathcal{E}°_{anode}$$

and $\mathcal{E}°_{anode} = 0.32\ V$, which is the standard reduction potential of the $Re^{3+}|Re(s)$ half-reaction.

Although a single half-reaction cannot occur alone, the Nernst equation can still be used to calculate the potential of a half-reaction at nonstandard conditions.

EXAMPLE 16.14

The potential of a Cu^+, Cu^{2+} electrode is 0.188 V. What is the ratio of these two ions in solution?

SOLUTION

The half-reaction of interest is

$$Cu^{2+} + e^- \rightleftharpoons Cu^+$$

Table 16.1 gives its standard potential as 0.158 V. The electrode could be: $Pt(s)|Cu^+, Cu^{2+}$. The number of electrons in this half-reaction is $n = 1$. The expression for the reaction quotient is $Q = [Cu^+]/[Cu^{2+}]$. Therefore, the value of Q at the given potential is the ratio of $[Cu^+]$ to $[Cu^{2+}]$, the ratio we are seeking.

$$0.188\ V = 0.158\ V - \frac{0.059\ V}{1\ mol\ e^-} \log Q$$

$$Q = 0.310 = \frac{[Cu^+]}{[Cu^{2+}]}$$

An answer of this sort is predictable, since the potential of the electrode of interest is greater than the standard potential of the half-reaction. The system is further from the equilibrium state in which Cu^+ predominates.

16.8 SOME IMPORTANT ELECTRON TRANSFER PROCESSES

Electron transfer processes play major roles in biological systems and in technology. While the electron transfer processes of living organisms and technology are considerably more complex than those of the simple chemical systems we have discussed in this chapter, the general principles are the same. We can apply these principles to understand some common electron transfer processes.

Corrosion

Corrosion, the deterioration of metals caused by chemical reactions on their surfaces, costs our society billions of dollars each year. The most widely used metals—iron, aluminum, copper, nickel—all undergo corrosion when in contact with the air, unless they are protected in some way. Only a few so-called "noble metals," such as gold and platinum, do not undergo corrosion; their resistance to corrosion is a major reason why these metals are so sought after and therefore so expensive.

Corrosion processes are electron transfers, oxidation-reduction processes that can occur when the surface of a metal is in contact with the atmosphere. A potential difference exists on the surface of a metal because of small differences in the composition of the metal, such as lattice defects, impurities, or even partial oxidation. This potential difference makes it possible for a corrosion process resembling the operation of a galvanic cell to take place on the metal surface. The site of the oxidation and the site of the reduction can be separated in space. Electrons can flow between these sites through the metal in the way that electrons flow through the external electrical connection of a galvanic cell. Water vapor from the air that condenses on the metal surface provides the solution through which the ions flow. Corrosion usually occurs faster near the sea, because droplets of water in the air contain some dissolved sodium chloride, forming an ionic solution that conducts electricity better than pure water does.

As might be expected, the anode reaction in corrosion is the oxidation of the metal. This process forms metal ions that dissolve in the ion-conducting medium, the atmospheric moisture in contact with the metal surface. In the corrosion of iron, for example, the oxidation is

$$Fe(s) \longrightarrow Fe^{2+} + 2e^-$$

We can predict that metals that are high in the emf series—called *active metals*—should be most susceptible to corrosion, since the potential of the oxidation half-reaction of an active metal is relatively positive. Cesium and rubidium, which are toward the top of the emf series, corrode very quickly in moist air. Other metals near the top of the emf series usually are stored away from air, to prevent the swift corrosion that occurs when they come in contact with the atmosphere for any period of time. There are exceptions to this rule, however; some active metals corrode quite slowly.

Given an oxidation, there must be a reduction to consume the electrons released in the anode process. Some of the important cathode, or reduction, half-reactions in corrosion are:

$$2H^+ + 2e^- \longrightarrow H_2(g) \qquad \text{(acid solution)}$$
$$O_2 + 4H^+ + 4e^- \longrightarrow 2H_2O \qquad \text{(acid solution)}$$
$$O_2 + 2H_2O + 4e^- \longrightarrow 4OH^- \qquad \text{(neutral or alkaline solution)}$$

Figure 16.12 shows schematically how iron corrodes, with the last half-reaction as the reduction process.

The standard potentials of the half-reactions in which dissolved O_2 is reduced are positive, so these reductions are more favorable than those involving H^+ or

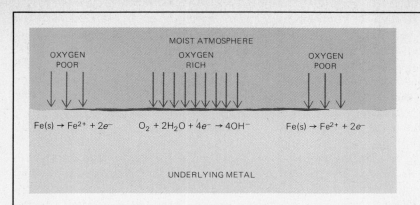

FIGURE 16.12
The corrosion of iron in moist air. The reduction
half-reaction occurs in the oxygen-rich area in the
center. The oxidation process occurs in the oxygen-
poor areas at the right and the left. The net result
is a combination of iron with oxygen to form rust.

H_2O alone. Metals corrode faster at higher pressures of oxygen, because more oxygen then dissolves into the moisture that is in contact with the metal. There is an interesting aspect of corrosion called the "principle of differential aeration": If a part of a metal surface is exposed to a relatively high concentration of O_2, corrosion occurs in another region of the metal. The fact that the potential difference that leads to corrosion requires the physical separation of the oxidation and reduction processes, and the analogy between a corroding metal and a galvanic cell, helps us to understand this principle. The reduction process in corrosion, which is the reduction of O_2, occurs in the region where the concentration of O_2 is highest. Therefore the oxidation half-reaction, which does the real damage, takes place elsewhere.

In the case of iron, the Fe^{2+} cations that form migrate toward the cathodic regions of the metal surface. There they react with water or OH^- to form $Fe(OH)_2$, which undergoes further oxidation to form $Fe(OH)_3$, the familiar reddish material called rust. Meanwhile, the anodic region of the metal surface undergoes the real damage: Holes appear, the surface is eaten away, and the metal's structural strength is weakened. The worst damage seems to occur when the reduction process liberates $H_2(g)$, apparently because the gas penetrates below the surface and further weakens the metal.

The corrosion of an automobile is a familiar demonstration of the principle of differential aeration. When some of the paint that protects the metal of the automobile chips off, corrosion does not occur at the site of the chipping. Rust does form at this spot, because it is the place where the reduction half-reaction occurs. The real damage is done at the anode region, which is a site near the exposed area (Figure 16.13).

Corrosion by differential aeration can also be demonstrated by embedding metal rods in sand, leaving the upper part of the rods in water. As Figure 16.14 shows, it is the segment of the rod in the sand that corrodes, not the segment in the water. The sand prevents O_2 from reaching the metal, but the part of the rod in the water is exposed to dissolved O_2. The reduction occurs in the water and the damaging oxidation process occurs in the sand.

The driving force of corrosion is the same as the driving force of the galvanic cell: a potential difference. More specifically, a metal can corrode because the potential of the half-reaction for its oxidation is relatively more positive than any other oxidation that can occur when the metal is exposed to moisture in the atmosphere. This concept is the basis of several methods for preventing corrosion.

The most widely used method is to coat the metal surface with paint or a similar substance to prevent O_2 and H_2O from coming in contact with the metal surface. This method has had varying success. Another group of methods use what are called corrosion inhibitors, substances that interfere with the flow of

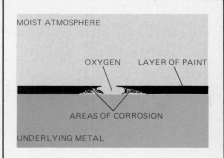

FIGURE 16.13
How an auto body corrodes. While the driving force
for corrosion is the reduction process that occurs at
the spot where the auto's paint has chipped and
metal is in contact with moist air, the structural
damage is done at the sites where the oxidation
process occurs.

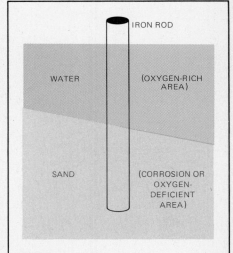

FIGURE 16.14
In the corrosion of a metal bar half-buried in sand,
the reduction process occurs in the segment of the
bar exposed to oxygen-carrying water. The segment
of the bar that corrodes is under the sand, where the
oxidation process occurs.

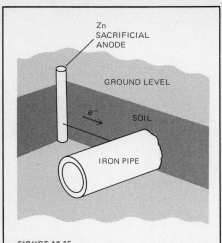

FIGURE 16.15
Cathodic protection. The iron pipe is connected to a "sacrificial anode" made of zinc, a metal that is higher in the emf series than iron. The zinc corrodes and the iron does not.

charge needed for corrosion to occur by forming films on the surface of the metal. There are anodic inhibitors, which act at the anode and interfere with the metal dissolution reaction. These include inorganic salts such as chromates, phosphates, and carbonates, as well as compounds containing organic nitrogen or sulfur. There are also cathodic inhibitors, which interfere with the reduction process that occurs at the cathodic region of the metal surface. These include salts of magnesium, zinc, or nickel.

Another corrosion-control method is cathodic protection in which the relative potential of the metal is changed. Cathodic protection prevents a metal from corroding by connecting it to another metal that is more active. The more active metal, which is higher in the emf series, is the one that corrodes. An iron pipe can be protected from corrosion by connecting it to a bar of zinc, as shown in Figure 16.15. Since the potential for the oxidation of zinc is more positive than the potential for the oxidation of iron, it is the zinc bar that corrodes. The zinc is called the "sacrificial anode." It prevents corrosion by pumping electrons into the iron, thus preventing the electron-releasing oxidation of iron from occurring. We could accomplish the same result by connecting the iron to an external power supply, using an inert electrode to complete the electrical circuit.

Surprisingly, corrosion can also be inhibited by pumping electrons out of the metal. This method is based on a rather complex phenomenon called *passivation*. The basic idea is to force the corrosion process to occur so rapidly that an oxide film with special properties forms on the metal surface and prevents further corrosion. This method requires less current than cathodic prevention, so it is used to protect large objects such as oil storage tanks from corrosion.

Passivation can occur naturally, without an applied electric current. Some metals that are high in the emf series are relatively corrosion-resistant because of passivation. Iron corrodes completely in dilute nitric acid but it does not corrode in concentrated nitric acid because of passivation. The rapid reaction that takes place in the concentrated acid causes formation of an oxide film that protects the iron from further attack.

Biological Oxidations

Living things eat food to get energy that enables them to perform work on the surroundings and to maintain the nonequilibrium state called life. The major source of this energy is the oxidation of food, a process that, for convenience, can be represented by the overall reaction for the oxidation of glucose:

$$C_6H_{12}O_6 + 6O_2 \longrightarrow 6CO_2 + 6H_2O + \text{energy}$$

The $\Delta G°$ for this process is very large. Its value is -2870 kJ/mol, and 24 mol of electrons is transferred for each mole of glucose that is oxidized. If such a large quantity of free energy were made available suddenly as heat—as it is when we set fire to a lump of sugar—most of it would be wasted by the organism, and it even could cause damage. In living cells, the oxidation of glucose and other molecules that are biological energy sources does not occur in one step. Biological oxidations take place in a complex, multistep pathway, in which each step releases only a small quantity of the total free energy of the process. This free energy is not released primarily as heat, most of which would be wasted. Rather, the free energy is used to drive a chemical system away from equilibrium.

We can compare this biological process to the process by which electrical energy is converted to chemical energy in an electrolytic cell. In electrolysis, an applied current drives a chemical system away from equilibrium. When the system is allowed to return to equilibrium, the stored energy becomes available. In a living organism, energy is also used to drive a system away from equilibrium. While the energy in the organism comes from a chemical reaction rather than from an electric current, the principle is the same. When the system is allowed to return to equilibrium, the energy is released at a convenient moment and in a convenient quantity.

The free energy from each of many steps in the oxidation of glucose is used to synthesize an unstable, energy-rich molecule called adenosine triphosphate (usually abbreviated ATP) from constituents in the cell. The formation of ATP is the process proceeding away from equilibrium. Under physiological conditions, ΔG for the synthesis of ATP is about $+40$ kJ/mol. Ideally, each step in the oxidation of glucose should release roughly this amount of energy.

Most of the details of the oxidation of glucose in the living cell have been worked out by biochemists. Each intermediate, partially oxidized form of glucose has been identified, and we can write the entire sequence of steps, starting with glucose and ending with CO_2 and H_2O, for a number of biological oxidation pathways. These pathways are quite complex and have many intermediates. We shall not attempt to describe them in any detail. Rather, we shall focus on one typical step, which will illustrate some characteristics of these pathways. In this step, malic acid, $C_4H_6O_5$, which is formed after many steps from the partial oxidation of glucose, is further oxidized to another organic acid, oxaloacetic acid, $C_4H_4O_5$. The overall reaction for this step is

$$C_4H_6O_5 + \tfrac{1}{2}O_2 \longrightarrow C_4H_4O_5 + H_2O$$

This overall reaction can be broken into two half-reactions. We can write both these half-reactions as reductions, giving their potentials in physiological conditions:

$$C_4H_4O_5 + 2H^+ + 2e^- \longrightarrow C_4H_6O_5 \qquad \mathcal{E} = -0.17 \text{ V}$$
$$\tfrac{1}{2}O_2 + 2H^+ + 2e^- \longrightarrow H_2O \qquad \mathcal{E} = 0.82 \text{ V}$$

The number of electrons transferred in this process is $n = 2$. Generally, two electrons are transferred in each step of the oxidation of glucose. We can calculate $\mathcal{E}$ for the overall reaction in the usual way, by noting that the half-reaction with the negative potential reverses. Thus,

$$\mathcal{E} = 0.82 \text{ V} - (-0.17 \text{ V}) = 0.99 \text{ V}$$

Since we know the value of the potential and the number of electrons transferred, we can calculate the free energy change for the reaction:

$$\Delta G = -nF\mathcal{E} = -(2 \text{ mol } e^-/\text{mol})(96\,500 \text{ C/mol } e^-)(0.99 \text{ V})$$
$$= -190\,000 \text{ J/mol}$$

Thus, the quantity of energy that is released when two electrons are transferred from malic acid to O_2 is 190 kJ/mol. But this quantity is more energy than a biological system can use efficiently. If this electron transfer occurred directly, most of the energy would be wasted. Detailed studies of biological oxidations have found that such electron transfers from intermediates in the oxidation of glucose do not occur directly. It has been shown that there is an electron transfer chain of at least seven steps, in which the two electrons are transferred from one substance to another as they travel from malic acid to O_2. Many of these transfer steps release a quantity of energy that is small enough to be used efficiently to synthesize ATP.

We could say that combining the malic acid oxidation half-reaction directly with the O_2 reduction half-reaction is like jumping from the roof of a building. Sending the electrons along the electron transfer chain, in this analogy, is like walking down a flight of stairs. The same amount of energy is released; in the one case impractically, in the other case in usable form. The large positive potential of the O_2 reduction half-reaction results in the large potential of the overall reaction and the consequent large free energy release. To "walk" this energy downstairs via the electron transport chain, the malic acid oxidation half-reaction is combined with a reduction half-reaction of much lower potential than that of the O_2 reduction. This half-reaction can be represented as

$$oxid_1 + 2e^- \longrightarrow red_1$$

where $oxid_1$ represents the oxidized form of the first substance in the electron transport chain and red_1 represents the reduced form. The reaction is

$$malic\ acid + oxid_1 \longrightarrow oxaloacetic\ acid + red_1$$

Because the potential of this reduction half-reaction is lower than that of the O_2 reduction half-reaction, the potential of this reaction is lower than that of the overall reaction with O_2, and the free energy release is lower.

At this point, malic acid and oxaloacetic acid have fulfilled their function, and the pair of electrons passes on to the next step, which can be represented as

$$oxid_2 + 2e^- \longrightarrow red_2$$

where $oxid_2$ and red_2 are, respectively, the oxidized and reduced forms of the next substances in the chain. The potential of this reduction half-reaction must be more positive than that of the first reduction half-reaction, so that the overall reaction for the second step:

$$oxid_2 + red_1 \longrightarrow red_2 + oxid_1$$

can occur spontaneously. This process reverses the first reduction by transferring the two electrons from red_1, the reduced form of the substance in the first step, to $oxid_2$. To put it another way, when the first and second half-reactions are combined, the first one reverses because it has the less positive potential. The effect of this reversal is the regeneration of $oxid_1$, which allows another electron transfer from malic acid.

Each subsequent step of the electron transfer chain is roughly similar to this one, although there are some differences in detail in the later steps. In each step, a half-reaction has a more positive potential than the preceding half-reaction. The preceding half-reaction reverses, transfers two electrons, and a small, manageable quantity of free energy is released. At the end of the chain, the electrons are transferred to O_2 in a process that can be represented as

$$\tfrac{1}{2}O_2 + 2H^+ + 2oxid_6 \longrightarrow H_2O + 2red_6$$

Since the potential of the half-reaction

$$2red_6 + 2e^- \longrightarrow 2oxid_6$$

is 0.5 V, the potential of the last overall reaction is

$$\mathcal{E} = 0.82 - 0.5 = 0.3\ V$$

and the corresponding free energy change is

$$\Delta G = -(2\ mol\ e^-/mol)(96\ 500\ C/mol\ e^-)(0.3\ V) = -60\ kJ/mol$$

Thus, the electron transfer chain utilizes a series of substances that can be oxidized and reduced reversibly. Each substance starts in its oxidized form, accepts electrons, and is converted to its reduced form. The reduced form then transfers the electrons to the next substance in the chain and returns to its oxidized form. The sequence in which the substances are utilized is important, since each reduction half-reaction must have a more positive potential than its predecessor in the chain.

Not all the details of these processes have been worked out. The substances in the electron transport chain have complex molecular structures that require a great deal of study. But it is known that the electron transfers appear to occur primarily in cell structures called mitochondria. Apparently, all the substances in the electron transfer chain are held rather rigidly in line, so that the electron transfers occur in the proper sequence. It appears that one function of the mitochondria is to hold the substances in line so that electron transfer will occur in the most effective way.

EXERCISES

16.1 Indicate for each of the following descriptions of chemical change whether an oxidation-reduction reaction takes place:
(a) Silver chloride forms when solutions of silver nitrate and sodium chloride are mixed.
(b) Sodium phosphate and sulfuric acid form phosphoric acid and sodium sulfate.
(c) Sodium bromide and chlorine form sodium chloride and bromine.
(d) Hydrogen peroxide forms water and oxygen.
(e) Nitric acid and nitric oxide form nitrous acid.
(f) Sodium chromate and nitric acid give sodium dichromate.

16.2 Classify each of the following changes as an oxidation or a reduction, and write a balanced half-reaction for the change: (a) $MnO_2(s)$ to MnO_4^-, (b) I_2 to I^-, (c) $HClO$ to ClO_4^-, (d) $Cr_2O_7^{2-}$ to Cr^{3+}, (e) $Fe(s)$ and Cl^- to $FeCl_4^-$.

16.3 Balance each of the following skeleton expressions of chemical change taking place in acid solution:
(a) $Sn^{2+} + Cu^{2+} \rightarrow Sn^{4+} + Cu^+$,
(b) $H_2S + Hg_2^{2+} \rightarrow S(s) + Hg(l)$,
(c) $I^- + MnO_4^- \rightarrow I_2(s) + Mn^{2+}$,
(d) $Cr_2O_7^{2-} + NO(g) \rightarrow Cr^{3+} + NO_3^-$.

16.4 Balance each of the following skeleton expressions of chemical change taking place in acid solution: (a) $Fe^{2+} + H_2O_2 \rightarrow Fe^{3+}$, (b) $O_3(g) + I^- \rightarrow IO_3^-$, (c) $Pt(s) + NO_3^- + Cl^- \rightarrow PtCl_6^{2-} + NO_2(g)$.

16.5 Balance each of the following skeleton expressions of chemical change taking place in alkaline solution:
(a) $CN^- + ClO_2^- \rightarrow CNO^- + Cl^-$,
(b) $MnO_4^- + H_2O \rightarrow MnO_2(s) + O_2(g)$,
(c) $Zn(s) + BrO_4^- \rightarrow Zn(OH)_4^{2-} + Br^-$.

16.6 Balance each of the following skeleton expressions of chemical change taking place in alkaline solution:
(a) $O_3(g) + Cl^- \rightarrow O_2(g) + ClO_4^-$,
(b) $F_2(g) + H_2O \rightarrow O_2(g) + F^-$,
(c) $PH_3(g) + CrO_4^{2-} \rightarrow Cr(OH)_4^- + P_4(s)$.

16.7 Write balanced reactions for each of the following disproportionations:

(a) $MnO_4^{2-} \rightarrow Mn^{2+} + MnO_4^-$ (acid),
(b) $MnO_2(s) \rightarrow Mn(s) + MnO_4^-$ (alkaline), (c) $ClO^- \rightarrow ClO_2^- + Cl_2(g)$ (alkaline), (d) $Fe^{2+} \rightarrow Fe^{3+} + Fe(s)$ (acid).

16.8 Find the amount of Fe^{2+} ion in a sample that requires 19.7 cm^3 of a 0.0843M solution of $KMnO_4$ to reach the equivalence point.

16.9 A 10.1-cm^3 sample of a solution of Cl^- requires 10.8 cm^3 of 0.0834M $KMnO_4$ solution to reach the equivalence point for its oxidation to ClO^- in alkaline solution. The MnO_4^- forms $MnO_2(s)$. Find the concentration of Cl^- ion in the original solution.

16.10 A sample of impure zinc metal of mass 2.54 g is analyzed by titration with $KBrO_3$ solution which oxidizes the zinc metal to the +2 oxidation state. The BrO_3^- is reduced to Br_2. The sample requires 50.4 cm^3 of 0.274M potassium bromate solution. Find the percentage of zinc metal in the sample, assuming that it does not contain any other reducing agents.

16.11 The analysis of chrome steels is carried out by an initial reaction of the steel with perchloric acid, which converts the metallic chromium to dichromate ion. The perchloric acid itself is reduced to Cl_2. The dichromate is then reduced by the addition of a measured excess of Fe^{2+} solution. The remaining ferrous ion is then titrated with Ce^{4+} ion in solution, which is reduced to Ce^{3+} ion. Write balanced equations for the reactions of this procedure.

16.12 Sodium oxalate, $Na_2C_2O_4$, in solution is oxidized to $CO_2(g)$ by MnO_4^-, which is reduced to Mn^{2+}. A 50.1-cm^3 volume of a solution of MnO_4^- is required to titrate a 0.339-g sample of sodium oxalate. Find the concentration of the MnO_4^- solution.

16.13 A volume of 32.5 cm^3 of the MnO_4^- solution of Exercise 16.12 is used to titrate a 4.62-g sample of a uranium-containing material. The oxidation of the uranium can be represented by the change $UO^{2+} \rightarrow UO_2^{2+}$. Calculate the percentage of uranium in the sample. (*Hint:* the answer to Exercise 16.12

is not necessary for this calculation; the stoichiometric relationships between oxalate and UO^{2+} are enough.)

16.14 Write equations for the anode and cathode processes and the overall reactions that occur in the electrolysis of: (a) molten magnesium bromide, (b) an aqueous solution of cesium iodide, (c) an aqueous solution of potassium sulfate, (d) an aqueous solution of silver nitrate.

16.15 Aluminum is manufactured by the electrolysis of aluminum oxide under special conditions, using graphite electrodes. The products of the electrolysis are aluminum metal and CO_2. Write equations for the anode and cathode processes and the overall reaction.

16.16 Calculate the quantity of charge required for the reduction of 0.80 mol of Fe^{3+} to Fe^{2+}, of 0.80 mol of Fe^{3+} to $Fe(s)$, and of 0.80 mol of Fe^{2+} to $Fe(s)$.

16.17 Calculate the quantity of charge necessary to produce 10 liters of $H_2(g)$ measured at STP from the electrolysis of water.

16.18 A current of 12.3 A is available for the electrolysis of molten sodium chloride. Calculate the length of time the electrolysis must be carried out to form 1.00 kg of sodium metal.

16.19 One method of manufacturing $KClO_3$ is based on the electrolysis, with stirring, of solutions of KCl in water. Find the mass of $KClO_3$ that can be formed by the application of a current of 6.42 A for 23.5 minutes.

16.20 The atomic weight of a metal is 50.9, and it forms a chloride of unknown composition. Electrolysis of the molten chloride produces 1.0 g of metal for every 2.8 g of $Cl_2(g)$. Find the empirical formula of the chloride.

16.21 A metal forms the fluoride MF_3. Electrolysis of the molten fluoride by a current of 3.86 A for 16.2 minutes deposits 1.25 g of the metal. Calculate the atomic weight of the metal.

16.22 A current of 10.0 A is applied to 1.0 liter of 1.0M HCl solution for 1.0 hour. Calculate the pH of the solution at this time.

*16.23 A thin layer of gold can be applied to another material by an electrolytic process. The surface area of an object to be gold plated is 49.8 cm^2 and the density of gold is 19.3 g/cm^3. A current of 3.25 A is applied to a solution that contains gold in the +3 oxidation state. Calculate the time required to deposit an even layer of gold 1.00 × 10^{-3} cm thick on the object.

16.24 Find the number of electrons required to reduce 1.00 g of Br$_2$ to Br$^-$ ion.

16.25 Suppose a galvanic cell is constructed with a solution of silver nitrate in contact with a silver electrode in one compartment (the cathode) and a solution of copper(II) nitrate in contact with a copper electrode in the other compartment. Write the two half-cell reactions and the overall cell reaction. Suppose that the barrier between the two compartments is one that allows the passage of cations but not anions from one compartment to the other. Describe the movement of ions as the cell operates.

16.26 Write the two half-cell reactions and the overall cell reaction for the following cells:
(a) Sn(s)|Sn^{2+}(1M)||Pb^{2+}(1M)|Pb(s)
(b)Cr(s)|Cr^{3+}(1M)||Br$^-$(1M)|AgBr(s)|Ag(s)
(c)Pt(s)|NO$_3^-$(1M), HNO$_2$(1M)||Mn^{2+}(1M), Mn^{3+}(1M)|Pt(s)

16.27 Use cell notation to describe the following cells:
(a) The anode is the standard hydrogen electrode and the cathode is the standard. Br$_2$(l),|Br$^-$ half-cell.
(b) The anode is the silver metal, solid silver bromide half-cell and the cathode is the silver metal, silver cation half-cell; all concentrations are 1M.
(c) The anode is Zn(s)|Zn(OH)$_4^{2-}$ and the cathode is Cl$_2$(g)|Cl$^-$ with a platinum electrode; all concentrations are 1M.

16.28 Calculate the mass of zinc consumed if a dry cell delivers a current of 0.28 A for 15 minutes.

16.29 Find the quantity of charge required to convert PbSO$_4$(s) into 50 g of lead in the lead storage battery.

16.30 Determne $\Delta G°$ for the cell: Ni(s)|Ni^{2+}(1M)||Fe^{3+}(1M)|Fe(s), for which $\mathcal{E}° = 0.19$ V.

16.31 The $\Delta G°$ of the reaction 2I$^-$ + Cl$_2$(g) → I$_2$(s) 2Cl$^-$ is −159 kJ.
(a) Calculate $\mathcal{E}°$ of the corresponding cell.
(b) Find $\Delta G°$ and $\mathcal{E}°$ for the reaction: I$^-$ + $\frac{1}{2}$Cl$_2$(g) → Cl$^-$ + $\frac{1}{2}$I$_2$(s).

16.32 The $\Delta G°$ of the reaction Cu^{2+} + Au$^+$ → Cu(s) + Au^{3+} is 183 kJ.

(a) Calculate the $\mathcal{E}°$ of the reaction.
(b) Write the cell made from the components in the reaction. (c) Calculate $\mathcal{E}°$ for the cell.

16.33 The $\mathcal{E}°$ for the cell Zn(s)|Zn^{2+}(1M)||Sn^{2+}(1M)|Sn(s) is 0.6264 V. Calculate the K of the cell reaction at 298 K.

16.34 The $\mathcal{E}°$ for the diproportionation of MnO$_2$(s) to MnO$_4^-$ and Mn^{2+} in acid solution is −0.471 V. Calculate K for the reaction at 298 K.

16.35 The K for the reaction 2V(s) + 6H$^+$ ⇌ 3H$_2$(g) + 2V^{3+} is 2.27 × 10^{90}. Calculate $\mathcal{E}°$ for the cell V(s)|V^{3+}(1M)||H$^+$(1M)|H$_2$(g)(1 atm)|Pt(s).

*16.36 The K_{sp} of CuI is 1.1 × 10$^{-12}M^2$. Find the $\mathcal{E}°$ for the cell: Cu(s)|CuI(s)|I$^-$(1M)||Cu$^+$(1M)|Cu(s).

16.37 Two cells are constructed in which the standard hydrogen electrode is one of the half-cells. When the other half-cell is the Be(s)|Be^{2+} electrode, the $\mathcal{E}°$ is measured as 1.70 V with the hydrogen electrode as the cathode. When the Au(s), Au^{3+} electrode is the other half-cell, the $\mathcal{E}°$ is measured as 1.29 V with the hydrogen electrode as the anode. Calculate $\mathcal{E}°$ for the cell made of the two metal half-cells and designate the anode and the cathode.

16.38 Two cells are constructed in which the Ni(s)|Ni^{2+} electrode is one of the half-cells. When the other half-cell is Cu(s)|Cu^{2+}, the $\mathcal{E}°$ is measured as 0.570 V with the nickel electrode as the anode. When the Mg(s)|Mg^{2+} electrode is the other half-cell, the $\mathcal{E}°$ is measured as 2.145 V with the magnesium electrode as the anode. Calculate $\mathcal{E}°$ for the cell made of the magnesium and copper electrodes and designate the anode and the cathode.

16.39 Calculate the standard potential of the following cells:
(a) Pb(s)|Pb^{2+}(1M)||Ag$^+$(1M)|Ag(s)
(b) Pt(s)|Sn^{4+}(1M), Sn^{2+}(1M)||Fe^{3+}(1M), Fe^{2+}(1M)|Pt(s)
(c) Pt(s)|Cl$_2$(g)(1 atm)|Cl$^-$(1M)||H$^+$(1M), H$_2$O$_2$(1M)|Pt(s).^{a}

16.40 Calculate the standard potential of the following alkaline cells:
(a) Al(s)|Al(OH)$_4^-$(1M), OH$^-$(1M)|H$_2$(g)(1 atm)|Pt(s)
(b) Cr(s)|Cr(OH)$_3$(s)|OH$^-$(1M), Ba^{2+}(1M)|Ba(s).^{a}

16.41 List the following species in order of increasing strength as oxidizing

agents: Zn(NH$_3$)$_4^{2+}$, Na$^+$, Cl$_2$(g), Zn^{2+}, H$^+$, H$_2$O$_2$, MnO$_4^-$, Au^{3+}.^{a}

16.42 List the following species in order of increasing strength as reducing agents: Mn^{2+}, Cl$^-$, H$_2$, Ba, Cu(CN)$_2^-$, F$^-$, I$^-$, Sn^{2+a}.

16.43 Predict whether there should be a substantial extent of reaction when the following substances are brought together in their standard states: (a) Cl$_2$(g) and MnO$_4^-$, (b) Ag(s) and Fe^{3+}, (c) Cr(s) and Ni^{2+}, (d) Sn and Sn^{4+}.^{a}

16.44 Predict whether there should be a substantial extent of reaction when the following substances are brought together in their standard states in solution of pH = 14: (a) MnO$_4^-$ and Cl$_2$, (b) Zn and NO$_3^-$.^{a}

16.45 Indicate which of the following metals should dissolve to an appreciable extent in a solution of pH = 0: (a) V, (b) Ag, (c) Pt, (d) Zn.^{a}

16.46 Indicate which of the following metals should dissolve to an appreciable extent in a solution of pH = 14: (a) Zn, (b) Mn, (c) Ag.^{a}

16.47 We can predict from the emf series that Cr should dissolve in 1M acid solution, yet very often we do not observe such a reaction. Suggest an explanation for this behavior of Cr.

16.48 Predict which of the following metal cations should disproportionate appreciably in 1M aqueous solution: (a) Cu$^+$, (b) Sn^{2+}, (c) Fe^{2+}, (d) Au$^+$.^{a}

16.49 Predict which of the following species should disproportionate in 1M aqueous solution: (a) HNO$_2$, (b) ClO$_3^-$ (alkaline), (c) ClO$^-$ (alkaline).^{a}

16.50 Find the $\mathcal{E}°$ of the following reactions: (a) Cr(s) + Fe^{3+} ⇌ Fe(s) + Cr^{3+} (b) 2V(s) + 3Cu^{2+} ⇌ 2V^{3+} + 3Cu(s). (c) Li(s) + H$^+$ ⇌ Li$^+$ + $\frac{1}{2}$H$_2$(g).^{a}

16.51 Calculate $\Delta G°$ for the reactions in Exercise 16.50.

16.52 Calculate K for the reactions in Exercise 16.50.

16.53 Find the $\mathcal{E}°$ of the following chemical changes in aqueous solution: (a) sodium dichromate and bromide ion form bromine and Cr^{3+} ion, (b) nitrous acid and potassium permanganate form nitric acid and Mn^{2+} ion, (c) H$_2$(g) and O$_2$(g) form hydrogen peroxide.^{a}

16.54 Calculate $\Delta G°$ for the reactions in Exercise 16.53.

16.55 Calculate K for the reactions in Exercise 16.53.

16.56 The reaction: 2AgBr(s) + S^{2-} ⇌ Ag$_2$S(s) + 2Br$^-$ is not an oxidation-reduction reaction. Nevertheless find $\mathcal{E}°$,

---

^{a} The necessary data for this exercise are found in Table 16.1.

$\Delta G°$, and K using data from Table 16.1.

16.57 Calculate the K_{sp} for Hg_2Cl_2.[a]

*__16.58__ The K_{sp} of $Zn(OH)_2$ is $1.8 \times 10^{-14}M^3$. Find $\mathcal{E}°$ for the half-reaction $Zn(OH)_2(s) + 2e^- \rightleftharpoons Zn(s) + 2OH^-$.[a]

*__16.59__ Use the result of Exercise 16.58 to find the solubility of $Zn(OH)_2$ in $1M$ OH^- solution. (*Hint:* consider the half-reaction for the reduction of $Zn(OH)_4{}^{2-}$ listed in Table 16.1.)

16.60 Calculate $\Delta G°$ and K for the disproportionation of the following species: (a) $MnO_2(s)$ (acid solution) to Mn^{2+} and $MnO_4{}^-$, (b) $Cl_2(g)$ to $HClO$ and Cl^-.[a]

16.61 Find $\mathcal{E}°$ of the following half-reactions. (a) $Sn^{4+} + 4e^- \rightleftharpoons Sn(s)$, (b) $Cu^+ + e^- \rightleftharpoons Cu(s)$, (c) $Mn^{3+} + 3e^- \rightleftharpoons Mn(s)$.[a]

*__16.62__ Calculate $\mathcal{E}°$ of the following half-reactions: (a) $MnO_4{}^-$ to Mn^{3+}, (b) $Cr_2O_7{}^{2-}$ to Cr^{2+}. (c) $ClO_2{}^-$ to $ClO_3{}^-$.[a]

*__16.63__ Find $\mathcal{E}°$, $\Delta G°$, and K for the reaction in which $HClO_2$ and HNO_2 react in acid solution to form Cl^- and N_2O_4.[a]

16.64 Calculate $\mathcal{E}$ for the cell: $Pb(s)|Pb^{2+}(0.0046M)\|Ag^+(0.10M)|Ag(s)$.[a]

16.65 Calculate $\mathcal{E}$ for the cell: $Pt(s)|I_2(s)|I^-(1.1M)\|Cl^-(0.0011M)|Cl_2(g)$ $(0.52 atm)|Pt(s)$.[a]

16.66 Find $\mathcal{E}$ for the given cell at (a) pH = 0, (b) pH = 3, (c) pH = 7: $Ni(s)|Ni^{2+}(1.0M)\|H^+|H_2(1 atm)|Pt(s)$.[a]

16.67 The potential of the cell: $In(s)|In^{3+}(0.0085M)\|H^+(1M)|H_2(1 atm)|Pt(s)$ is found to be 0.28 V. Calculate $\mathcal{E}°$ of the cell and of the $In|In^{3+}$ electrode.

16.68 The potential of the cell: $U(s)|U^{3+}(0.0022M)\|Zn^{2+}(0.50M)|Zn(s)$ is 1.00 V. Calculate $\mathcal{E}°$ of the $U(s)|U^{3+}$ electrode.[a]

16.69 Find the concentration of Fe^{3+} in solution necessary to raise the potential of the $Fe(s)|Fe^{3+}$ electrode to 0.00 V.[a]

16.70 Calculate the reduction potential of the half-cell: $Pt(s)|H_2(g)(1 atm)|H^+$ at pH = 2, pH = 4, and pH = 7.

16.71 A concentration cell is a galvanic cell in which both half-cells are composed of the same substances, but in different concentrations. Since $\mathcal{E}$ is determined by concentrations, a potential difference and a current are established. Consider a cell in which both half-cells contain zinc metal electrodes and Zn^{2+} in solution. The $[Zn^{2+}]$ is $1.0M$ in one half-cell and $0.10M$ in the other. Identify the anode and the cathode and caluculate $\mathcal{E}$ of the cell.

16.72 Determine the $\mathcal{E}$ of a concentration cell in which the electrodes are $Sn(s)$ and the $[Sn^{2+}]$ in the two half-cells is $0.34M$ and $0.0088M$.

16.73 Find the $[Fe^{3+}]$ in a half-cell of a concentration cell with $Fe(s)$ electrodes whose $\mathcal{E} = 0.020$ V and whose other half-cell is the standard electrode.

16.74 Write the anode and cathode processes and the overall reaction for the corrosion of tin under alkaline conditions.

16.75 Write the anode and cathode processes and the overall reaction for the corrosion or iron in dilute nitric acid. Write the anode and cathode processes and the overall reaction for the passivation of iron in concentrated nitric acid. Assume that $NO_2(g)$ is one of the products.

16.76 Show the anode and cathode processes and the overall reaction that takes place when the corrosion of iron is prevented by cathodic protection with magnesium. The conditions are acidic and oxygen rich.

16.77 Find the potential that corresponds to a ΔG of -40 kJ/mol for a one-electron transfer and for a two-electron transfer.

*__16.78__ An inorganic electron transport chain can be constructed by using the couples: Sn^{2+}, Sn^{4+}; Fe^{2+}, Fe^{3+}; Cu^{2+}, Cu^+; and Au^{3+}, Au^+ suitably arranged. Suppose we wish to use this chain to transfer 2 mol of electrons from $H_2(g)$ to $O_2(g)$ in acid solution. List the couples in the order in which they should be used. Write the overall reaction for each step and the overall reaction for the entire process. Calculate $\mathcal{E}°$ and $\Delta G°$ for each step of the process.[a]

CHEMICAL KINETICS

17

Our discussion of chemical reactions in previous chapters has left out one crucial factor: time. We have considered, qualitatively and quantitatively, the relative stability of different substances and the direction of change for chemical systems. With the knowledge we now have, and given the appropriate data, we can predict whether a chemical system should change, the nature of the change, and the composition of the system at the equilibrium state. But the omission of the time factor means that we cannot say whether any chemical system will indeed change in the way we predict.

Until now, we have studied thermodynamics, which tells us whether or not a chemical change is possible. We shall now study chemical kinetics, which tells us how much time is needed for a chemical change to occur. We cannot get complete information about a chemical change without applying both chemical thermodynamics and chemical kinetics.

Thermodynamics, for example, tells us that H_2O does not spontaneously dissociate to form H_2 and O_2 under ordinary conditions. The position of equilibrium in the reaction

$$H_2O(l) \longrightarrow H_2(g) + \tfrac{1}{2}O_2(g)$$

lies far to the left. If we want this change to occur, we must make it happen—for example, by passing an electric current through the water. This sort of nonspontaneous change will not take place unless some work is done on the system.

Thermodynamics also tells us that the reverse process, in which H_2 and O_2 form water, is spontaneous; all we have to do is wait, and water will be formed. But thermodynamics does not tell us that it will be a very long wait, because the reaction occurs very slowly. Kinetics gives us this information. Kinetics also tells us that H_2 and O_2 will react very quickly if a spark is applied to the mixture of the two gases—so quickly that an explosion may occur.

There are many such examples in everyday life. Thermodynamics tells us that the contents of an egg will change from liquid to solid spontaneously. Kinetics tells us that the best way to make this change occur is to put the egg in boiling water. Thermodynamics tells us that nitrous oxide, the propellant in a can of whipped cream, is unstable and forms N_2 and O_2 spontaneously. Kinetics tells us that the nitrous oxide dissociates so slowly that we need not rush to use a can of whipped cream.

We can sum up by saying that chemical kinetics deals primarily with the answers to three questions:

1. At what rate does a chemical system undergo change under a given set of conditions?
2. What effect will a change in conditions have on the rate at which a chemical change occurs?
3. Given the answers to the first two questions, what information is available about the details of the chemical change?

17.1 RATE OF REACTION

Expressions that define the rate at which events occur are commonplace in everyday life. An individual drives a car at 90 km an hour, reads 50 pages of a book in an hour, loses one kilogram a week on a diet, drinks half a glass of milk a minute. All these expressions of the rate at which a process occurs have certain features in common. Each of them is the expression of a change that occurs in a given interval of time. The speed of an automobile, for example, expresses a change in position, measured in kilometers, in an interval of one hour. If an individual drives at 90 km an hour, the position of the automobile in one hour changes by 90 km. Similarly, if someone reads a book for one hour, the number of pages that have been read increases by 50. A dieter's weight decreases by one kilogram in a week; a person drinks half a glass of milk in a minute, and the glass is half-emptied in that time interval.

The rate of a chemical reaction can also be expressed as a given change in a given interval of time. The rate usually is expressed in terms of the change in the concentration or the pressure of one component. It may be expressed as a decrease in the concentration of a reactant or an increase in the concentration of a product. The rate of change of a chemical reaction is always expressed as a positive quantity. The time interval usually is one second (although other time intervals are sometimes more convenient).

For reactions in solutions, the rate is the change in concentration (usually the molarity) of a component per second. For reactions in the gas phase, the rate is the change in the pressure of a component per second. The symbol Δ is used to express change, in the way in which we have used it previously: Δ means the final value minus the initial value.

For example, the rate at which the reaction

$$Cl_2 + 2I^- \longrightarrow 2Cl^- + I_2$$

occurs in solution can be expressed as the change in concentration of any of the four components of the system. If we write $\Delta[Cl_2]$, we mean the concentration of Cl_2 at the end of a given time interval minus the concentration of Cl_2 at the beginning of the time interval:

$$\Delta[Cl_2] = [Cl_2]_{final} - [Cl_2]_{initial}$$

The time interval can also be represented as $\Delta t = t_{final} - t_{initial}$. Thus, the rate at which Cl_2 disappears in this reaction is $-\Delta[Cl_2]/\Delta t$. The negative sign is necessary because Cl_2 is disappearing, so the final concentration is lower than the initial concentration. By writing the negative sign, we fulfill the requirement that the rate must be a positive quantity. The units of the rate at which Cl_2 disappears in solution are moles per liter per second (mol liter^{-1} sec^{-1}). (We shall use "sec" as the abbreviation for second in this chapter, although the SI abbreviation is "s.")

We could also express the rate of this reaction as the rate of disappearance of I^-, which is $-\Delta[I^-]/\Delta t$. The rate of disappearance of I^- is not the same as the rate of disappearance of Cl_2 in this reaction. The stoichiometry of the reaction shows that two moles of I^- are consumed for each mole of Cl_2 that is consumed. Therefore, the rate of disappearance of I^- is twice that of Cl_2. That is,

$$\frac{-\Delta[I^-]}{\Delta t} = 2\left(\frac{-\Delta[Cl_2]}{\Delta t}\right)$$

We could also express the rate of this process as the rate of appearance of one of the products. If we use this method to express the rate, the form of the expression is the same. But a negative sign is not necessary, since the final concentration of the product is greater than the initial concentration. Again, the stoichiometry of the reaction indicates that Cl^- will appear twice as fast as I_2. Thus

$$\frac{\Delta[Cl^-]}{\Delta t} = 2\left(\frac{\Delta[I_2]}{\Delta t}\right) = 2\left(\frac{-\Delta[Cl_2]}{\Delta t}\right) = \frac{-\Delta[I^-]}{\Delta t}$$

In the case of a reaction of gases:

$$2O_3(g) \longrightarrow 3O_2(g)$$

the changes in the quantities of the components can also be expressed as changes in pressure, and the rate can be given in units of atmospheres per second (atm/sec). For this reaction, the rate expressions are

$$3\left(\frac{-\Delta P_{O_3}}{\Delta t}\right) = 2\left(\frac{\Delta P_{O_2}}{\Delta t}\right) \quad \text{or} \quad \tfrac{1}{2}\left(\frac{-\Delta P_{O_3}}{\Delta t}\right) = \tfrac{1}{3}\left(\frac{\Delta P_{O_2}}{\Delta t}\right)$$

The pressure of O_2 increases faster than the pressure of O_3 decreases, so the total pressure of the system increases as the reaction proceeds.

For a general reaction

$$a\text{A} + b\text{B} \longrightarrow c\text{C} + d\text{D}$$

the rates of appearance and disappearance of the components in the reaction are given by

$$\frac{1}{a}\left(\frac{-\Delta \text{A}}{\Delta t}\right) = \frac{1}{b}\left(\frac{-\Delta \text{B}}{\Delta t}\right) = \frac{1}{c}\left(\frac{\Delta \text{C}}{\Delta t}\right) = \frac{1}{d}\left(\frac{\Delta \text{D}}{\Delta t}\right)$$

where the lowercase letters are the coefficients of the reaction.

EXAMPLE 17.1

When ammonia is treated with O_2 at elevated temperatures, the rate of disappearance of ammonia is found to be 3.5×10^{-2} mol liter^{-1} sec^{-1} during a measured time interval. Calculate the rate of appearance of nitric oxide and water.

SOLUTION

Since the relative rates of appearance of the products are governed by the coefficients of the chemical equation, the first step is to write the equation for the reaction:

$$4NH_3(g) + 5O_2(g) \longrightarrow 4NO(g) + 6H_2O(g)$$

Since NO and NH_3 have the same coefficient, the rate of appearance of nitric oxide is the same as the rate of disappearence or ammonia:

$$\frac{\Delta[\text{NO}]}{\Delta t} = 3.5 \times 10^{-2} \text{ mol liter}^{-1} \text{ sec}^{-1}$$

The rate of appearance of H_2O is $6/4$ or 1.5 times as fast:

$$\frac{\Delta[\text{H}_2\text{O}]}{\Delta t} = 1.5(3.5 \times 10^{-2} \text{ mol liter}^{-1} \text{ sec}^{-1}) = 5.3 \times 10^{-2} \text{ mol liter}^{-1} \text{ sec}^{-1}$$

To know the numerical value of a rate expression, we must know the concentration or the pressure of a substance at two different times during the course of a reaction. This information can be obtained experimentally, if we have (1) a method of measuring time, (2) a method of measuring concentration or pressure, and (3) a method of keeping the conditions—especially the temperature—constant.

Table 17.1 lists some kinetic data for the reaction

$$CH_3Cl + I^- \longrightarrow CH_3I + Cl^-$$

taking place in aqueous solution at 298 K.

The data in Table 17.1 can be used to calculate the rate of the reaction for the time intervals that are given. For example, the rate of disappearance of I^- from the start of the reaction to 180 minutes later is given by

$$\frac{-\Delta[\text{I}^-]}{\Delta t} = \frac{-(0.45 \text{ mol/liter} - 0.50 \text{ mol/liter})}{(180 \text{ min} - 0)(60 \text{ sec/min})} = 4.6 \times 10^{-6} \text{ mol liter}^{-1} \text{ sec}^{-1}$$

TABLE 17.1
Disappearance of I$^-$

$[\text{I}^-](M)$	0.50	0.45	0.41	0.35	0.27
t (min)	0	180	360	720	1440

with the factor 60 included in the denominator to convert the time, given in minutes, to seconds. This calculation can be done for any time interval. For example, the rate of disappearance of I^- in the interval from 360 minutes after the start of the experiment to 1440 minutes after the start is given by

$$\frac{-\Delta[I^-]}{\Delta t} = \frac{-(0.27 \text{ mol/liter} - 0.41 \text{ mol/liter})}{(1441 \text{ min} - 360 \text{ min})(60)} = 2.2 \times 10^{-6} \text{ mol liter}^{-1} \text{ sec}^{-1}$$

In many kinetics experiments, we do not make a direct measurement of concentration. Instead, we measure a property that can be related to the concentration of one substance in the system. In a gas phase system, the total pressure of the gases may be the property that is measured. A total-pressure measurement usually is a good way to determine the rate of a process that causes a change in the number of moles of gas in a system.

EXAMPLE 17.2

At high temperatures, nitrous oxide decomposes to N_2 and O_2. If the rate of the reaction is monitored by measuring the change in total pressure with time, the following data are obtained from an initial pressure of N_2O of 0.29 atm at 970 K:

P_T (atm)	0.33	0.36	0.39	0.41
t (sec)	300	900	2000	4000

Find the rate of disappearance of N_2O and the rate of appearance of O_2 for the first 300 sec and the last 2000 sec of this process.

SOLUTION

The rate of disappearance of nitrous oxide can be found by using the total pressure to find the pressure of nitrous oxide at any given time. The relationship between P_T and P_{N_2O} is derived from the equation for the reaction:

$$2N_2O(g) \longrightarrow 2N_2(g) + O_2(g)$$

start   0.29 atm   0   0

Let $2x$ be the pressure in atmospheres of N_2O that reacts in the time interval of interest. The pressure of each gas at the end of this interval can be tabulated:

time   0.29 atm $- 2x$ $2x$ x

The measured total pressure is the sum of these pressures:

$$P_T = P_{N_2O} + P_{N_2} + P_{O_2} = 0.29 \text{ atm} - 2x + 2x + x$$
$$= 0.29 \text{ atm} + x$$

The total pressure after 300 sec is given as 0.33 atm. Therefore

$$0.33 \text{ atm} = 0.29 \text{ atm} + x$$

and $x = 0.04$ atm. At time $= 300$ sec,

$$\Delta P_{N_2O} = 0.29 \text{ atm} - 2x = 0.29 \text{ atm} - 2(0.04 \text{ atm}) = 0.21 \text{ atm}$$

This value can be used to find the rate of disappearance of N_2O:

$$\frac{-\Delta P_{N_2O}}{\Delta t} = \frac{-(0.21 \text{ atm} - 0.29 \text{ atm})}{300 \text{ sec}} = 2.7 \times 10^{-4} \text{ atm/sec}$$

There are a number of ways to find the rate of appearance of O_2 in this time interval. The stoichiometry of the reaction shows that the rate of appearance of O_2 is half the rate of disappearance of N_2O. Thus

$$\frac{\Delta P_{O_2}}{\Delta t} = \frac{1}{2}(2.7 \times 10^{-4} \text{ atm/sec}) = 1.4 \times 10^{-4} \text{ atm/sec}$$

Another method finds the rate of appearance of O_2 directly from the P_{O_2} at the start and end of the given time interval. At the start, no O_2 is present. After 300 sec, the $P_{O_2} = 0.04$ atm

$$\frac{\Delta P_{O_2}}{\Delta t} = \frac{0.04 \text{ atm} - 0}{300 \text{ sec} - 0} = 1.3 \times 10^{-4} \text{ atm/sec}$$

Similarly, we find the rate for the last 2000 sec by finding the pressures at the start and finish of this interval.

At 2000 sec,

$$P_T = 0.39 \text{ atm} = 0.29 \text{ atm} + x, \text{ and } x = 0.10 \text{ atm} = P_{O_2}$$

At 4000 sec,

$$P_T = 0.41 \text{ atm} = 0.29 \text{ atm} + x, \text{ and } x = 0.12 \text{ atm} = P_{O_2}$$

The rate of appearance of O_2 can be calculated directly:

$$\frac{\Delta P_{O_2}}{\Delta t} = \frac{0.12 \text{ atm} - 0.10 \text{ atm}}{4000 \text{ sec} - 2000 \text{ sec}} = 1.0 \times 10^{-5} \text{ atm/sec}$$

The rate of disappearance of N_2O is twice as fast:

$$\frac{-\Delta P_{N_2O}}{\Delta t} = 2.0 \times 10^{-5} \text{ atm/sec}$$

Average Rates and Instantaneous Rates

In both Example 17.1 and 17.2, the rates of the processes become slower with time. Most chemical reactions become slower as they proceed. This characteristic can be explained quite simply. *The rate of most chemical reactions depends in some way on the concentration of one or more of the reactants.* Since the reactants are consumed as the reaction proceeds, their concentrations decrease and the rate of the reaction decreases.

Since the quantities of the reactants in a chemical reaction decrease continuously as the reaction proceeds, the rate of the reaction also decreases continuously. *The rate measured for a time interval is only the average rate for that interval.* Therefore, the rate during a measured time interval is called the *average rate;* the rate at any one instant during the interval is called the *instantaneous rate.* The average rate and the instantaneous rate are equal for only one instant in any time interval. Over the entire interval, the instantaneous rate changes continuously. At first, the rate is higher than the average rate. At the end of the interval, the rate is lower than the average rate. Thus, with the exception of one instant, the average rate is different from the instantaneous rate during the entire time interval.

We can illustrate the relationship between average rate and instantaneous rate by looking at data for a reaction that takes place in polluted air. This process is the reaction between ozone, O_3, and the hydrocarbon ethylene, C_2H_4, in the gas phase, which can be represented as

$$O_3(g) + C_2H_4(g) \longrightarrow O_2(g) + C_2H_4O(g)$$

The rate of disappearance of O_3 can be followed easily at 303 K. Table 17.2 lists the data.

TABLE 17.2
Disappearance of O_3

$[O_3]$(mol/liter)	3.20×10^{-5}	2.42×10^{-5}	1.95×10^{-5}	
t (sec)	0	10.0	20.0	
$[O_3]$(mol/liter)	1.63×10^{-5}	1.40×10^{-5}	1.23×10^{-5}	1.10×10^{-5}
t (sec)	30.0	40.0	50.0	60.0

CHLOROFLUOROCARBONS, THE OZONE LAYER, AND KINETICS

The debate about the extent to which the earth's ozone layer is damaged by chlorofluorocarbons that are used as propellants in some aerosol spray cans and as the working liquid in refrigerators is essentially a debate about chemical kinetics. Decisions on banning the chlorofluorocarbons are based on knowledge about the rates at which a number of chemical reactions occur in the atmosphere.

The chlorofluorocarbons are inert gases often sold under the trade name Freon. The two most widely used are Freon 11, $CFCl_3$, and Freon 12, CF_2Cl_2. Early in the 1970s, it was found that these gases drift slowly to the upper atmosphere, where ultraviolet radiation from the sun initiates their dissociation. The chlorine atoms produced by this reaction can bring about the conversion of O_3 to O_2:

$$Cl + O_3 \longrightarrow ClO + O_2$$
$$ClO + O \longrightarrow Cl + O_2$$

A single chlorine atom can cause the destruction of tens of thousands of O_3 molecules.

The extent to which the ozone layer is destroyed depends on the rate at which chlorofluorocarbons reach the upper atmosphere, the region of the ozone layer. If the chlorofluorocarbon molecules participate in other reactions in the lower atmosphere, only negligible amounts of chlorine atoms may be freed in the ozone layer. The chlorofluorocarbons can participate in more than seven different reactions in the lower atmosphere. A major research effort has been mounted to determine how many of these reactions actually occur and the rates at which they take place.

One example illustrates the complexity of the research. It was proposed that the chlorine atoms freed from the chlorofluorocarbons eventually would react with hydrogen to form hydrochloric acid. Estimates were made of the concentration of hydrochloric acid to be found in the atmosphere if the theory of ozone destruction were accurate. One balloon flight failed to detect the predicted concentration of hydrochloric acid. This finding led to the theory that the chlorine atoms might be combining with oxygen and nitrogen in the lower atmosphere to form chlorine nitrate, $ClONO_2$. The rate at which chlorine nitrate is formed in the atmosphere therefore became a critical test of the theory.

Further studies found that the concentration of chlorine nitrate in the lower atmosphere is very low. Therefore, its rate of formation must be very slow. At the same time, it was found that the failure to detect hydrochloric acid was due to a technical mistake; a technician had put the wrong amounts of two reagents into the instrument carried aloft by the balloon.

The task of verifying the chlorofluorocarbon-ozone theory was complicated by our limited knowledge of atmospheric physics and chemistry, and by the extremely small concentrations of the species being sought in the atmosphere. Many of them are present in parts per billion, and some of the most important are present only in parts per trillion.

Let us say we want to know the instantaneous rate of this reaction 30 sec after the start; or, to put it another way, the instantaneous rate when $[O_3] = 1.63 \times 10^{-5}M$. We can approximate this instantaneous rate by calculating an average rate from the data. For example, the average rate of disappearance of O_3 for the entire time interval is

$$\frac{-\Delta[O_3]}{\Delta t} = \frac{-(1.10 \times 10^{-5} \text{ mol/liter} - 3.20 \times 10^{-5} \text{ mol/liter})}{60.0 \text{ sec} - 0 \text{ sec}}$$

$$= 3.50 \times 10^{-7} \text{ mol liter}^{-1} \text{ sec}^{-1}$$

This value is an approximation of the instantaneous rate. But it is not a very good approximation. A better value for the instantaneous rate is the average rate between 10 sec and 50 sec:

$$\frac{-\Delta[O_3]}{\Delta t} = \frac{-(1.23 \times 10^{-5} \text{ mol/liter} - 2.42 \times 10^{-5} \text{ mol/liter})}{50.0 \text{ sec} - 10.0 \text{ sec}}$$

$$= 2.98 \times 10^{-7} \text{ mol liter}^{-1} \text{ sec}^{-1}$$

And a still better approximation is the average rate for the interval between 20 sec and 40 sec:

$$\frac{-\Delta[O_3]}{\Delta t} = \frac{-(1.40 \times 10^{-5} \text{ mol/liter} - 1.95 \times 10^{-5} \text{ mol/liter})}{40.0 \text{ sec} - 20.0 \text{ sec}}$$

$$= 2.75 \times 10^{-7} \text{ mol liter}^{-1} \text{ sec}^{-1}$$

As the time interval becomes smaller, the average rate becomes a better and better approximation of the instantaneous rate. Carrying this observation to a logical conclusion, the average rate will be the same as the instantaneous rate when the time interval is zero. While we cannot measure two concentrations (or anything else) during a time interval of zero seconds, we can say that the average rate gets closer and closer to the instantaneous rate as the time interval gets closer and closer to zero:

$$\text{instantaneous rate} = \lim_{\Delta t \to 0} \frac{-\Delta[A]}{\Delta t}$$

That is, the instantaneous rate is the limit of the average rate as Δt approaches 0. Using the notation of the calculus, the expression for this limit is written as $-d[A]/dt$.

The relationship between instantaneous rate and average rate can be shown graphically. Figure 17.1 shows a plot of $[O_3]$ against time. The solid curve shows the continuous decrease in O_3 as the reaction proceeds. The slopes of the dashed lines show the average rates for the time points they connect. The slope of a dashed line is $-\Delta[O_3]/\Delta t$. The instantaneous rate at any time is the slope of the tangent to the curve at the point corresponding to that time. The graph shows that as the time interval represented by an avarage rate line grows smaller, the slope of the line comes closer to the slope of the tangent to the curve at the 30-sec mark.

The curve for the disappearance of O_3 shows that the rate of the process

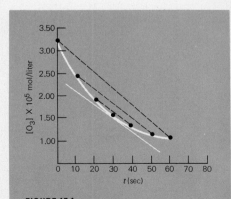

FIGURE 17.1

The solid curve shows the rate of disappearance of ozone with time in the reaction with C_2H_4. The instantaneous rate of the reaction at any given time is found by finding the tangent to the curve at that time. The slope of each dashed line, $-\Delta[O_3]/\Delta t$ shows the average rate of the reaction for a given interval. As the intervals are shortened, the slopes of the dashed lines come closer to the slope of the tangent to the curve.

slows with time. At the start, the curve falls sharply. As the reaction proceeds, it levels off toward the horizontal. Tangents to the beginning segment of the curve have a steeper slope than the tangents to the end segments of the curve.

17.2 REACTION RATE AND CONCENTRATION

We have seen that the rate of a reaction is related to the concentration of one or more of the substances in a system. One important goal of a kinetic investigation is to establish the exact relationship between concentration and the rate of reaction for a given system. If we know how the concentration affects the rate of reaction for a specific system, we can get practical information about the best way to conduct the reaction and valuable insights into the way that the chemical change occurs.

The relationship between rate and concentration for a given reaction can be established only by experiment. Fortunately, a few relatively simple relationships describe most reactions. The equation that describes such a relationship is called a **rate law.** For example, the rate of the reaction

$$2H_2O_2 \longrightarrow 2H_2O + O_2$$

has been found to be directly proportional to the concentration of hydrogen peroxide. We can write the rate law for the reaction as:

$$\text{rate} = k[H_2O_2]$$

where "rate" is the instantaneous rate of disappearance or appearance of a specified substance and k is the proportionality constant between the rate and the concentration and is called the **specific rate constant.** The units of the specific rate constant are not the same as the units of rate, and you should be careful not to confuse the specific rate constant with the rate. In this rate law, for example, the units of rate are mol liter^{-1} sec^{-1}, and the units of concentration are mol liter^{-1}. Substituting these units into the rate law gives

$$\text{mol liter}^{-1} \text{ sec}^{-1} = k \text{ mol liter}^{-1}$$

The units of mol liter^{-1} can be canceled from both sides of the equation, so that the units of k are sec^{-1} for this rate law.

The rate law for the gas phase decomposition of nitrosyl chloride

$$2NOCl(g) \longrightarrow 2NO(g) + Cl_2(g)$$

is found to be

$$\text{rate} = k[NOCl]^2$$

Here, the rate of the reaction is proportional to the square of a concentration term. The units of the rate constant k in this rate law can be found by substitution:

$$\text{mol liter}^{-1} \text{ sec}^{-1} = k(\text{mol liter}^{-1})^2$$

$$k = \frac{\text{mol liter}^{-1} \text{ sec}^{-1}}{\text{mol}^2 \text{ liter}^{-2}} = \text{liter mol}^{-1} \text{ sec}^{-1}$$

Comparing the two rate laws above, you can see that the only difference between them is that they have different exponents on the concentration terms. These exponents are quite important, because they define a central characteristic, the **order** of a rate law or a reaction. In the first case, the exponent on the term for the concentration of hydrogen peroxide is 1, so we say that this is a *first-order reaction*. In the second rate law, the exponent on the term for the concentration of nitrosyl chloride is 2, so we say that this is a *second-order reaction*.

The exponents of the concentration terms also are used to define the order of more complex reactions, those whose rate laws include the products of two or more concentration terms. The gas phase reaction

$$NO_2(g) + CO(g) \longrightarrow NO(g) + CO_2(g)$$

follows the rate law

$$\text{rate} = k[NO_2][CO]$$

Each of the two concentration terms has the exponent 1. The sum of the exponents on the concentration terms is 2, so the reaction is said to be second order. We can be more precise and say that the reaction is first order in nitrogen dioxide, first order in carbon monoxide, and second order overall. In other words, the overall order of a reaction is the sum of all the exponents on the concentration terms in the rate law. We note that the units of the specific rate constant in this two-term second-order rate law are liter mol^{-1} sec^{-1}, the same units as in the one-term second-order rate law for the NOCl reaction above.

The gas phase reaction

$$2NO(g) + O_2(g) \longrightarrow 2NO_2(g)$$

follows the rate law

$$\text{rate} = k[NO]^2[O_2]$$

The sum of the exponents is 3, so this is a third-order rate law; it is second order in NO and first order in O_2. The units of k are $liter^2$ mol^{-2} sec^{-1}.

Fractional exponents appear in some rate laws. Under certain conditions, the rate law for the reaction

$$H_2(g) + Br_2(g) \longrightarrow 2HBr(g)$$

is

$$\text{rate} = k[H_2][Br_2]^{\frac{1}{2}}$$

This can be called a three-halves-order reaction. It is first order in hydrogen and half-order in bromine.

The rate law of a reaction cannot be found by inspection of the equation. The rate law can be found only by experiment. In simple cases, the dependence of the rate of a reaction on concentration can be found by measuring rates at different concentrations. One method is based on the measurement of initial rates. An *initial rate* is the average rate during a relatively short time interval beginning at $t = 0$. The order of the reaction can often be found by obtaining values of the initial rate at different concentrations. This method is especially useful for reactions that are complicated by secondary reactions involving the products. Table 17.3, which lists data for the reaction between nitric oxide and ozone at 340 K, an important reaction related to the formation of smog, can be used to demonstrate this method.

In the first three entries in the table, the [NO] is held constant and the [O_3] is varied. The rate can be seen to vary as the [O_3] changes. Specifically, the rate changes by the same factor as the [O_3]. When the [O_3] is doubled, the rate

TABLE 17.3
Initial Rates of the Reaction $O_3(g) + NO(g) \rightarrow O_2(g) + NO_2(g)$

[O_3](mol liter^{-1})	[NO](mol liter^{-1})	Initial Rate (mol liter^{-1} sec^{-1})
2.1×10^{-6}	2.1×10^{-6}	1.6×10^{-5}
4.2×10^{-6}	2.1×10^{-6}	3.2×10^{-5}
6.3×10^{-6}	2.1×10^{-6}	4.8×10^{-5}
6.3×10^{-6}	4.2×10^{-6}	9.5×10^{-5}
6.3×10^{-6}	6.3×10^{-6}	14.3×10^{-5}

doubles. When the $[O_3]$ increases by a factor of 1.5, as it does from the second entry to the third, the rate increases by the same factor of 1.5. The rate of this reaction, therefore, is proportional to the first power of the $[O_3]$.

In the last three entries in the table, the $[O_3]$ is held constant while the $[NO]$ is varied. Again, the rate also varies. When $[NO]$ doubles from 2.1×10^{-6} mol/liter to 4.2×10^{-6} mol/liter, the rate also doubles, from 4.8×10^{-5} mol liter^{-1} sec^{-1} to 9.5×10^{-5} mol liter^{-1} sec^{-1}. Therefore, the rate of this reaction is also proportional to the first power of the $[NO]$. The rate law for this reaction is: rate $= k[O_3][NO]$.

Note that when each concentration changes by a factor, the rate changes by the product of the factors. For example, each concentration changes by a factor of three between the first and last entries in Table 17.3. We see that the rate changes by a factor of nine, from 1.6×10^{-5} mol liter^{-1} sec^{-1} to 14.3×10^{-5} mol liter^{-1} sec^{-1}.

EXAMPLE 17.3

The oxidation of manganate ion to permanganate ion by periodate ion in alkaline solution can be represented by the equation

$$2MnO_4{}^{2-} + H_3IO_6{}^{2-} \longrightarrow 2MnO_4{}^- + IO_3{}^- + 3OH^-$$

The data for the initial rate of the reaction measured as the rate of appearance of $MnO_4{}^-$ as a function of concentration at 310 K are:

$[MnO_4{}^{2-}]$ (mol liter^{-1})	$[H_3IO_6{}^{2-}]$ (mol liter^{-1})	Initial Rate (mol liter^{-1} sec^{-1})
1.6×10^{-4}	3.1×10^{-4}	2.6×10^{-6}
3.2×10^{-4}	3.1×10^{-4}	1.0×10^{-5}
6.4×10^{-4}	3.1×10^{-4}	4.1×10^{-5}
1.6×10^{-4}	4.7×10^{-4}	2.6×10^{-6}
1.6×10^{-4}	6.2×10^{-4}	2.6×10^{-6}

Find the rate law for this reaction.
SOLUTION
In the first three entries in the table, only the $[MnO_4{}^{2-}]$ changes. When the $[MnO_4{}^{2-}]$ changes by the factor $3.2/1.6 = 2$, the initial rate changes by the factor $10/2.6 = 4$. When the $[MnO_4{}^{2-}]$ changes by the factor $6.4/3.2 = 2$, the initial rate changes by the factor $4.1/1.0 \cong 4$. We see that when the concentration of $MnO_4{}^{2-}$ doubles, (a factor of 2), the rate quadruples (a factor of 2^2). When the concentration is increased by a factor of $6.4/1.6 = 4$, the rate therefore increases by the factor $4^2 = 16$. The rate is proportional to the $[MnO_4{}^{2-}]$ raised to a power. If we let x be the power, then $2^x = 4$ and $x = 2$. The rate is proportional to $[MnO_4{}^{2-}]^2$.

Is the rate also proportional to the $[H_3IO_6{}^{2-}]$? In the last two entries of the table, the $[MnO_4{}^{2-}]$ is constant and the $[H_3IO_6{}^{2-}]$ varies. The rate does not vary; therefore, the rate does not depend on the $[H_3IO_6{}^{2-}]$. We can say that the reaction is zero order in periodate ion. The rate law of the reaction is

$$\text{rate} = k[MnO_4{}^{2-}]^2$$

It is not unusual to find that the rate law does not reflect the overall stoichiometry of an equation, and that one or more of the reactants does not appear in the rate law.

We cannot always find the rate law by measuring rates at different concentrations. For example, the reaction

$$CH_3I + H_2O \longrightarrow CH_3OH + H^+ + I^-$$

follows the rate law

rate $= k[CH_3I][H_2O]$

This reaction is carried out in aqueous solution. The concentration of H_2O does not drop to any appreciable extent as the reaction proceeds. Under these conditions, we can regard the concentration of H_2O as a constant, and we can define a new specific rate constant, k', which is equal to $k[H_2O]$. The rate law then becomes

rate $= k'[CH_3I]$

A rate law of this kind is called a pseudo-first-order rate law. While it appears to be first order, it actually is second order. We encounter such rate laws when a large excess of one of the reactants leads to an apparent simplification of the rate law.

Rate Equations

Since the rate law defines the relationship between reaction rate and concentration, it can be used to calculate the course of a reaction with time. To perform such calculations, it is convenient to convert the rate law into a form that gives the relationship between the initial concentration of a reactant (that is, the concentration at $t = 0$) and its concentration at any time during the course of the reaction. For the simpler rate laws, this conversion is readily accomplished by the methods of the integral calculus. For this reason, the forms of the rate laws that give the relationships between concentration and time are often called *integrated rate equations*. We shall use the two simplest integrated rate equations, the equation for the first-order rate and the equation for the one-term second-order rate.

First Order

Consider a general first-order reaction, which can be represented as

A $\longrightarrow$ B

It follows the rate law

rate $= k[A]$

Assume the reaction begins at time $t = 0$, with a concentration of A that can be represented as c_0. Then the relationship between the concentration of A at any other time, the specific rate constant k, and the initial concentration c_0 is given by the integrated first-order equation

$$2.303 \log \frac{c}{c_0} = -kt \qquad\qquad (17.1)$$

where c is the concentration of A at time t.*

The integrated first-order rate equation defines the relationship between the specific rate constant k, the time t, and the *ratio* of the concentration c at this

---

*Those familiar with the methods of the calculus will be able to follow the derivation of this equation. The first-order rate law is actually a differential equation of the form $-dc/dt = kc$. This equation can be rearranged to $dc/c = -kdt$. Both sides of the equation can be integrated from $t = 0$ to t:

$$\int_0^t \frac{dc}{c} = -k \int_0^t dt \quad \text{or} \quad \frac{\ln c}{c_0} = -kt$$

The factor 2.303 appears in Equation 17.1 to convert from natural logarithms to logarithms whose base is 10.

time to the initial concentration, c/c_0. A ratio is dimensionally independent—that is, it has no units—and so the values of k and t do not depend on the units of concentration. Therefore, Equation 17.1 can be used with any measurable property of a system that is proportional to concentration. Data on the decrease in pressure of a reactant, on the decrease in color intensity of a reactant, or even, as we shall see, on the decrease in radioactivity of a reactant, can be used directly in Equation 17.1. Thus, c in the integrated first-order rate equation means more than just concentration. It refers to any property that is proportional to concentration.

EXAMPLE 17.4

The decomposition of Cl_2O_7 at 400 K in the gas phase to Cl_2 and O_2 follows first-order kinetics.

(a) After 55 sec at 400 K, the pressure of Cl_2O_7 falls from 0.062 atm to 0.044 atm. Calculate the specific rate constant.*

(b) Calculate the pressure of Cl_2O_7 after 100 sec of decomposition at this temperature.

(c) Calculate the time required for the pressure of Cl_2O_7 to fall to one-tenth of its original value.

SOLUTION

(a) The overall reaction is

$$2Cl_2O_7(g) \longrightarrow 2Cl_2(g) + 7O_2(g)$$

Since values are given for three of the four terms in the first-order rate equation, the value of the fourth can be found:

$$2.3 \log \frac{0.044 \text{ atm}}{0.062 \text{ atm}} = -k(55 \text{ sec})$$

$$k = 6.2 \times 10^{-3} \text{ sec}^{-1}$$

(b) Now that the value of the specific rate constant is known, it can be used to find the pressure of reactant that remains at any given time. Thus:

$$2.3 \log \frac{c}{0.062 \text{ atm}} = -(6.2 \times 10^{-3} \text{ sec}^{-1})(100 \text{ sec})$$

This equation can be solved more easily if we use the relationship $log\ (a/b) = \log a - \log b$ to rewrite it as

$$2.3 \log c - 2.3 \log 0.062 = -(6.2 \times 10^{-3} \text{ sec}^{-1})(100 \text{ sec})$$

$$c = 0.033 \text{ atm}$$

(c) When the pressure falls to one-tenth of its original value, it will be 0.0062 atm. We could substitute this value, along with the values of c_0 and k, into the rate equation to find t. However, if we recognize that the fraction c/c_0 has the value 0.1 when the pressure falls to one-tenth its original value, and that $\log 0.1 = -1$, the problem is more easily solved. We can write

$$2.3(-1) = -(6.2 \times 10^{-3} \text{ sec}^{-1})t$$

and

$$t = 370 \text{ sec}$$

---

* In practice, an experiment designed to find the value of the specific rate constant would not rely on only one data point, but would use a number of concentrations at different times. Often the value of k is then found graphically. A plot of $2.303 \log c/c_0$ against t is a straight line whose slope is $-k$ for a first-order reaction.

Half-life

In Example 17.4(c), we calculated the time in which a given fraction of the starting quantity of a reactant is consumed. Such a calculation is done quite often for first-order reactions. For first-order reactions only, *the time in which a given fraction of a reactant is consumed is independent of the starting quantity of reactant.* We can show why this is so.

Suppose we want to know the time interval in which 20% of the starting quantity of Cl_2O_7 in Example 17.4 will decompose. When 20% decomposes, there remains $c_0 - 0.2c_0 = 0.8c_0$. Substituting this value into the rate equation gives:

$$2.3 \log \frac{0.8c_0}{c_0} = -kt$$

or

$$2.3 \log 0.8 = -kt$$

Since c_0 does not appear in the equation, the time t is independent of c_0. The value of t is determined only by k.

Very often, we want to find the time t required for half the starting quantity of a reactant to be consumed, the *half-life* of the reaction. The expression for finding the half-life is based on the simple observation that when half the starting quantity of material is gone, the other half remains. Using the first-order rate equation, we find that

$$2.303 \log 0.5 = -kt_{1/2}$$

where $t_{1/2}$ is the half-life. This is usually written as

$$t_{1/2} = \frac{0.693}{k} \tag{17.2}$$

Given this simple relationship between the half-life and the specific rate constant, we can use the half-life as a convenient expression for the rate of a first-order process. If necessary, Equation 17.2 can also be used to calculate k from the half-life.

In many cases, the time elapsed from the start of a process is given in terms of the half-life. We can easily calculate that the half-life of the process in Example 17.4 is

$$t_{\frac{1}{2}} = \frac{0.693}{6.2 \times 10^{-3} \text{ sec}^{-1}} = 110 \text{ sec}$$

We can use the half-life to describe the extent and duration of a first-order reaction. For example, if we say that the reaction has proceeded for three half-lives, we mean that $t = 3(110) = 330$ sec. During the period of time equal to the first half-life, one-half of the starting quantity of material is consumed; that is, after 110 sec, $\frac{1}{2}$ the original quantity remains. At the end of the second half-life, at 220 sec, $\frac{1}{2}$ of $\frac{1}{2}$, or $\frac{1}{4}$ of the original quantity of material remains. At the end of the third half-life, at 330 sec, $\frac{1}{2}$ of $\frac{1}{4}$, or $\frac{1}{8}$ of the original quantity of material remains. Thus, after three half-lives, the pressure of Cl_2O_7 in Example 17.4 is $\frac{1}{8} \times (0.062 \text{ atm}) = 0.0078$ atm. For the general case, after n half-lives, $(\frac{1}{2})^n$ of the original quantity of material remains.

The half-life concept is commonly used to describe the rate at which radioactive decay processes occur (Section 20.2). Radioactive decay is a first-order process. The half-life of a radioactive element is the time in which the quantity of the element is reduced by half, by a process in which the element decays either to a different element or to a different isotope of the same element. The easiest way to measure the rate of decay is to count the number of particles (usually alpha or beta particles) emitted by the sample during a given time

An artist's recreation of Copan, a Mayan ceremonial site in what is now Honduras. Radiocarbon dating has helped archeologists interpret the Mayan calendar, so that a detailed chronology of the history of this site has been worked out. (*Peabody Museum, Harvard University*)

interval, using instruments such as the Geiger counter or the scintillation counter. The value of the count is directly proportional to the amount of radioactive substance present, and it can be used directly in the first-order rate equation for c_0 and c. For practical purposes, the half-life of a radioactive element is assumed to be the time in which the measured count rate of particles emitted by a given sample falls by one-half.

Radiocarbon Dating

An isotope of carbon that is of great value in several fields of science is ^{14}C, which is radioactive and decays with a half-life of 5760 years. Carbon-14 is produced continually in the atmosphere by the bombardment of nitrogen by cosmic rays. The rate at which ^{14}C decays is balanced by the rate at which it forms, giving a fairly constant concentration of ^{14}C in the atmosphere for relatively long periods of time. It was once believed that the atmospheric concentration of ^{14}C never changes. Recent evidence indicates that the concentration does vary over very long time periods, apparently because of changes in solar activity. The ^{14}C in the atmosphere forms $^{14}CO_2$, which is incorporated in plants and is transferred to animals that eat the plants. As a result, all living things contain this unstable isotope of carbon. As long as an organism is alive, it continually incorporates ^{14}C to replace the ^{14}C lost by radioactive decay. When the organism dies, the incorporation of ^{14}C stops. The ratio of ^{14}C to ^{12}C, the stable isotope of carbon, decreases steadily after death. The length of time since the death of an organism can be determined by measuring the ratio of ^{14}C to ^{12}C. Radiocarbon dating, the measurement of the ratio of ^{14}C to ^{12}C in organic matter, is a basic method of measuring the age of any carbon-containing matter—wood, bone, and so on—

that has been dead for periods ranging from several hundred years to about 50 000 years.

However, the discovery that the atmospheric concentration of ^{14}C apparently has not been constant over some time periods has required some revision of radiocarbon dates. The discovery was made because of discrepancies between ages determined by radiocarbon dating and those arrived at by dendrochronology, a method in which ages can be measured by analysis of tree ring patterns. It has now been established that the tree ring chronology is accurate and the radiocarbon dates are in error, apparently because ^{14}C is less abundant now than it was several thousand years ago. The resulting revision of radiocarbon dates is having a major impact on archeology.

One practical use of radiocarbon dating is the authentication of works of art that are suspected of being forgeries. Example 17.5 shows how radiocarbon dating can be used by art historians.

EXAMPLE 17.5

A museum curator, doubtful about the authenticity of an Egyptian papyrus painting that purports to be from the twenty-first century B.C., asks for a radiocarbon test. A small piece of the papyrus is burned to CO_2, which is collected. A Geiger counter measures 14.7 counts per minute (c.p.m.) per gram of carbon, compared to 15.3 c.p.m. from ^{14}C in a living organism. It is assumed that the ^{14}C content of the rushes from which the papyrus was made was 15.3 c.p.m. when they were alive. Is the painting genuine?

SOLUTION

The specific rate constant for the first-order radioactive decay of ^{14}C can be found from the half-life:

$$k = \frac{0.693}{5760 \text{ yr}} = 1.20 \times 10^{-4} \text{ yr}^{-1}$$

Since k, c_0, and c are known, the time elapsed since the death of the rushes can be found by using the first-order rate equation:

$$2.30 \log \frac{14.7 \text{ c.p.m.}}{15.3 \text{ c.p.m.}} = -(1.20 \times 10^{-4} \text{ yr}^{-1})t$$

$$t = 330 \text{ yr}$$

This measurement shows clearly that the painting is a forgery, since the papyrus is only 330 years old, rather than 3000 years old as it should be. Assuming a steady-state concentration of ^{14}C of 15.3 c.p.m., the radioactive count of a 3000-year-old papyrus should be

$$2.30 \log \frac{c}{15.3 \text{ c.p.m.}} = -(1.20 \times 10^{-4} \text{ yr}^{-1})(3000 \text{ yr})$$

$$c = 10.7 \text{ c.p.m. per gram of carbon}$$

This expected value is substantially lower than the measured value, confirming the attempt at forgery.

Second Order

Now let us consider reactions that follow the one-term second-order rate law. The general reaction can be represented as A → B. The rate law that it follows is: rate = $k[A]^2$. The integrated rate equation for this rate law is

$$\frac{1}{c} - \frac{1}{c_0} = kt \tag{17.3}$$

where the symbols have the same meaning as in the first-order rate equation: The initial concentration is c_0, the concentration at time t is c, and k is the specific rate constant. Usually, c and c_0 will be expressed in units of mol/liter, so k, as we have seen, has the units liter mol^{-1} sec^{-1}.

Calculations using the second-order rate equation are similar to those using the first-order equation, but there is one important difference. In the second-order equation, the units—and therefore the numerical value of k—depend on the units in which c and c_0 are expressed. The value of the specific rate constant for an equation that follows a second-order rate law can be found by measuring the concentration of a reactant with time. Once the specific rate constant is known, it can be used either to find the concentration of a reactant with time or to find the time needed to reach a given concentration of a reactant.

EXAMPLE 17.6

The decomposition of nitric oxide to N_2 and O_2 in the gas phase at elevated temperatures has been studied extensively, because it plays a role in atmospheric chemistry. It has been found that the reaction is second order in nitric oxide at 1370 K.

(a) Over a period of 2000 sec, the concentration of NO falls from an initial value of 2.8×10^{-3} mol/liter to 2.0×10^{-3} mol/liter. Find the value of the specific rate constant.

(b) Find the concentration of NO after the reaction proceeds for another 2000 sec.

SOLUTION
The reaction is

$$2NO(g) \longrightarrow N_2(g) + O_2(g)$$

(a) The specific rate constant is found by using the second-order rate equation:

$$\frac{1}{2.0 \times 10^{-3} \text{ mol/liter}} - \frac{1}{2.8 \times 10^{-3} \text{ mol/liter}} = k(2000 \text{ sec})$$

$$k = 7.1 \times 10^{-2} \text{ liter mol}^{-1} \text{ sec}^{-1}$$

Again, in practice more than one measurement of concentration would be made. If the reaction follows this rate law, a plot of $1/c$ against t is a straight line whose slope is k.

(b) The value of the specific rate constant found in part (a) can be used to determine the concentration at any given time, using the rate equation:

$$\frac{1}{c} - \frac{1}{2.8 \times 10^{-3} \text{ mol/liter}}$$

$$= (7.1 \times 10^{-2} \text{ liter mol}^{-1} \text{ sec}^{-1})(2000 \text{ sec} + 2000 \text{ sec})$$

$$c = 1.6 \times 10^{-3} \text{ mol/liter}$$

The half-life of a first-order reaction is not affected by the starting concentrations. A second-order process is different. In a second-order process, the half-life depends on the value of the starting concentrations as well as the value of the specific rate constant. The rate equation shows this relationship.

The concentration at $t_{1/2}$ is $c_0/2$. Thus

$$\frac{1}{\dfrac{c_0}{2}} - \frac{1}{c_0} = kt_{1/2}$$

Solving for $t_{1/2}$ gives

$$t_{1/2} = \frac{1}{kc_0} \tag{17.4}$$

In Example 17.6, the half-life of the NO when it has an initial concentration of 2.8×10^{-3} mol/liter is:

$$t_{1/2} = \frac{1}{(7.1 \times 10^{-2} \text{ liter mol}^{-1} \text{ sec}^{-1})(2.8 \times 10^{-3} \text{ mol liter}^{-1})} = 5000 \text{ sec}$$

Thus, after 5000 sec the concentration of NO is $(2.8 \times 10^{-3}$ mol/liter$)/2 = 1.4 \times 10^{-3}$ mol/liter. But this concentration of NO will not be reduced by half after another 5000 sec. By Equation 17.4, the time required to reduce this concentration by half is:

$$t_{1/2} = \frac{1}{(7.1 \times 10^{-2})(1.4 \times 10^{-3})} = 10\,000 \text{ sec}$$

The second-order rate slows down faster than the first-order rate. Therefore, the half-life of a reactant that follows a second-order rate law increases as the concentration decreases.

17.3 RATE LAWS AND REACTION MECHANISMS

A reaction mechanism is a description on a molecular level of all the changes that reactants undergo during their transformation to products in a chemical reaction. At the most sophisticated level, a reaction mechanism will describe the movement of the electrons and nuclei of the reactants, giving a continuous description of the changes in chemical bonding. But we must have a more elementary description before we can develop such a detailed reaction mechanism.

Elementary Reactions

Many chemical reactions proceed in a number of steps. These steps are called elementary reactions. For example, it has been proposed that the conversion of ozone to oxygen:

$$2O_3(g) \longrightarrow 3O_2(g)$$

proceeds in two steps:

(1) $\quad\quad O_3 \rightleftharpoons O_2 + O$
(2) $\quad O_3 + O \longrightarrow 2O_2$

as shown in Figure 17.2. Each of the two steps is an elementary reaction. One of the products of the first elementary reaction is atomic oxygen, O(g), which is an *intermediate* and is not included among the final products of the overall reaction. Since the overall reaction takes place in more than one step, we can describe it as a complex reaction.

It is important to understand the distinction between the overall reaction, which gives the stoichiometry of the process, and the elementary reactions, which together provide a two-step description of a possible reaction mechanism for the conversion of O_3 to O_2. While the overall reaction shows two O_3 molecules as reactants, the reaction does not occur by a collision between the molecules. The products are not formed directly by a reaction between two O_3 molecules, so the overall reaction does not occur exactly as written. But the two *elementary reactions occur exactly as written*. In the first step, one O_3 molecule does break into two parts, an O_2 molecule and an O atom. In the second step, an O_3 molecule and an O atom come together to form two O_2 molecules.

Molecularity

Molecularity is a term that applies only to elementary reactions. *The molecularity of an elementary reaction is generally the number of reactant molecules (or atoms*

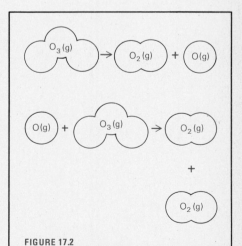

FIGURE 17.2
The two-step mechanism by which ozone is converted to molecular oxygen. In the first step, an O_3 molecule decomposes to an O_2 molecule and an O atom. In the second step, another O_3 molecule reacts with the O atom to form two O_2 molecules.

or ions) that come together to form the products. It is the sum of the coefficients of the reactants in the elementary reaction. The first elementary reaction in the conversion of ozone to oxygen is said to be *unimolecular,* since the reactant is one O_3 molecule. The second elementary reaction is said to be *bimolecular,* since the reactants are one O_3 molecule and one O atom. Most elementary reactions are unimolecular or bimolecular. There are a few termolecular elementary reactions, in which three species are reactants, but these are rare. Elementary reactions with molecularity of four or higher are unknown. The reason is simple. An elementary reaction describes an actual collision. The simultaneous collision of three or more molecules (or atoms or ions) is extremely improbable. In thermo-dynamic terms, ΔS for a termolecular (or higher) collision is quite negative.

Rate Laws of Elementary Reactions

We said earlier that the rate law and reaction order cannot be predicted from the stoichiometry of a complex reaction. Now you can understand why. The rate law is determined by the characteristics of the elementary reactions that make up the complex reaction. The stoichiometry of a complex reaction gives no information about its elementary reactions.

However, we can write the rate law for an elementary reaction directly from its molecularity. A unimolecular reaction follows a first-order rate law, and a bimolecular reaction follows a second-order rate law. Thus, the rate law for the first elementary reaction in the conversion of O_3 to O_2 is

rate $= k_1[O_3]$

and the rate law for the second elementary reaction is

rate $= k_2[O_3][O]$

The order and the molecularity of an elementary reaction are equal. The rate law for any elementary reaction is written as the product of the reactants raised to the power of their coefficients. For example, the rate law for the elementary reaction

$2CH_3(g) \longrightarrow C_2H_6(g)$

is

rate $= k[CH_3]^2$

While an overall chemical change usually consists of several elementary reactions, some ordinary chemical processes occur essentially in a single step. One of them is the reaction

$CH_3Cl + I^- \longrightarrow CH_3I + Cl^-$

which occurs in solution. The process can be pictured as an elementary reaction in which an iodide ion collides directly with the methyl chloride molecule, forming methyl iodide and displacing a chloride ion, as shown in Figure 17.3. Since this reaction is a bimolecular process, it should follow a second-order rate law if it does indeed proceed as a single elementary step. The rate law for the reaction has been found by observation to be:

rate $= k[CH_3Cl][I^-]$

This rate law is consistent with the picture in which the reaction occurs in a single elementary step. But it does not prove that the reaction occurs in this way. The only way to determine that this reaction occurs in more than one step is to find by experiment that it is not second order.

In the case of another reaction:

$2NO(g) + Cl_2(g) \longrightarrow 2NOCl$

which takes place in the gas phase, the reaction is believed to occur in a single

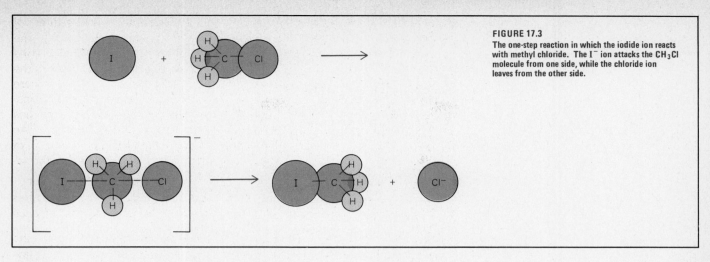

FIGURE 17.3
The one-step reaction in which the iodide ion reacts with methyl chloride. The I^- ion attacks the CH_3Cl molecule from one side, while the chloride ion leaves from the other side.

step: The chlorine atom collides with the two nitric oxide molecules to form the products directly, as shown in Figure 17.4. Experiments have found that this reaction occurs by a third-order rate law:

rate $= k[NO]^2[Cl_2]$

which is consistent with the single-step mechanism.

The rate law of a complex reaction is related to the rate laws of its elementary steps. Therefore, we can use an experimentally determined rate law for an overall reaction to propose elementary reactions by which the chemical change comes about. Kinetic measurements thus are important in formulating reaction mechanisms, since a proposed mechanism must be consistent with the observed kinetics. Kinetics is one of many experimental approaches to the investigation of reaction mechanisms. However, it is usually not possible to "prove" that a proposed mechanism is correct. The best we can do is to prove that a proposed mechanism is incorrect.

In the reactions between CH_3Cl and I or between Cl_2 and NO, for example, the observed rate laws are consistent with single-step mechanisms. But the rate laws do not prove that these are the actual mechanisms. While the reaction:

$H_2(g) + I_2(g) \longrightarrow 2HI$

follows a second-order rate law:

rate $= k[H_2][I_2]$

which is consistent with a single-step mechanism, there is other experimental evidence that proves that the single-step mechanism is incorrect.

Rate Laws From Reaction Mechanisms

It is usually quite difficult to formulate a possible mechanism from a rate law. It can also be difficult to derive the rate law required by a proposed mechanism. But the required rate law can be found easily for a mechanism that includes what is called a **rate-determining step.**

When one of the elementary reactions in a complex mechanism proceeds at a much slower rate than any of the others, the slow elementary reaction is the rate-determining step. The phrase means exactly what it says: The rate of this step determines the overall rate of the complex reaction, no matter whether the rate-determining step comes first, last, or in the middle of the series of elementary reactions. As a crude analogy, we can think of a mountain-climbing team composed of three experts and a beginner, all roped together. The beginner must climb at a slower rate than any of the experts. Therefore, the speed of the party is

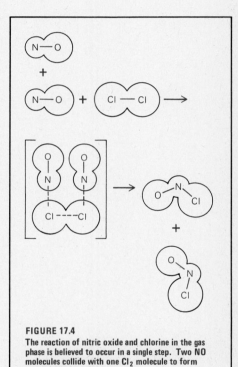

FIGURE 17.4
The reaction of nitric oxide and chlorine in the gas phase is believed to occur in a single step. Two NO molecules collide with one Cl_2 molecule to form two NOCl molecules.

determined by the speed of the beginner, no matter whether he leads, brings up the rear, or is in the middle.

We find a simple example of a mechanism with a rate-determining step in the gas phase reaction

$$NO_2(g) + CO(g) \longrightarrow NO(g) + CO_2(g)$$

which is believed to occur by a two-step mechanism below 500 K:

(1) $\quad NO_2 + NO_2 \xrightarrow{k_1} NO_3 + NO \qquad$ slow

(2) $\quad NO_3 + CO \xrightarrow{k_2} NO_2 + CO_2 \qquad$ fast

In this mechanism, the designations "slow" and "fast" indicate the relative rates of the steps. We follow custom by writing the specific rate constant of each elementary step above the arrow, with a subscript indicating the number of the step.

The first step is rate-determining, which means that the rate of the overall reaction is the rate of step (1). The rate law for step (1) is:

$$rate = k_1[NO_2]^2$$

which is also the rate law for the overall reaction. The observed rate law for the overall reaction is second order and is therefore consistent with the proposed mechanism. Because the unusual species NO_3 is a reactive intermediate, we expect that step (2) will be faster than step (1).

The process of obtaining the overall rate law from the mechanism of a reaction is simplified if we realize that *the rate law for the overall reaction is not influenced by steps that occur after the rate-determining step*. In the simplest case, the rate-determining step comes first, as in the reaction

$$H_2O_2 + 3I^- + 2H^+ \longrightarrow 2H_2O + I_3^-$$

which is believed to occur by the mechanism:

(1) $\qquad H_2O_2 + I^- \longrightarrow H_2O + IO^- \qquad$ slow

(2) $\qquad IO^- + H^+ \rightleftharpoons HOI \qquad$ fast

(3) $\quad HOI + H^+ + I^- \rightleftharpoons H_2O + I_2 \qquad$ fast

(4) $\qquad I^- + I_2 \rightleftharpoons I_3^- \qquad$ fast

The sum of all the steps gives the overall stoichiometry of the reaction. But steps (2), (3), and (4) do not influence the overall rate because they occur after the rate-determining slow step. The overall rate law is the rate law of step (1):

$$rate = k[H_2O_2][I^-]$$

which can be verified experimentally.

It is not as easy to derive the overall rate law for a mechanism in which the first step is not the rate-determining step. Such a mechanism often begins with what is called a fast preequilibrium, the rapid and reversible formation of an intermediate. The intermediate is then consumed in the rate-determining step. We find such a mechanism in the decomposition of nitric oxide, the least stable of all the oxides of nitrogen. While NO is stable at ordinary pressures at room temperature, it decomposes at pressures in excess of 100 atm by the reaction

$$3NO(g) \longrightarrow N_2O(g) + NO_2(g)$$

The proposed mechanism for this reaction is:

(1) $\qquad 2NO(g) \rightleftharpoons (NO)_2(g) \qquad$ fast

(2) $\quad (NO)_2(g) + NO(g) \xrightarrow{k_2} N_2O(g) + NO_2(g) \qquad$ slow

The rate law for the overall reaction is the rate law of step (2), the rate-determining step:

rate $= k_2[(NO)_2][NO]$

The usefulness of this rate law is limited because the concentration of $(NO)_2$ is not easily measured. A rate law is most useful when it contains only concentrations that can be measured. It is not possible to make a direct measurement of the concentration of NO dimer, as $(NO)_2$ is called, because it is an intermediate with a very short lifetime. Therefore, we cannot readily carry out an experimental test of the rate law. But there is a way out of this impasse. You can see that the value of $[(NO)_2]$ is related to the value of the starting concentration of NO. Since step (1) of the mechanism occurs relatively quickly, equilibrium is established between NO and $(NO)_2$. As usual, the composition of the equilibrium state is expressed by an equilibrium constant:

$$K_1 = \frac{[(NO)_2]}{[NO]^2}$$

Solving for the $[(NO)_2]$:

$$[(NO)_2] = K_1[NO]^2$$

We now have expressed the concentration of the NO dimer in terms of a concentration that can be measured. Substituting this expression into the rate law for the rate-determining step gives:

rate $= k_2[NO]K_1[NO]^2$

Since both k_2 and K_1 are constants, we can let $k' = k_2K_1$, which gives a simple one-term third-order rate law for the overall reaction:

rate $= k'[NO]^3$

This rate law has been verified experimentally. While a single-step mechanism is also consistent with this rate law, other experimental evidence led to the proposal of this two-step mechanism.

EXAMPLE 17.7

The oxidation of iodide ion by hypochlorite ion:

$$ClO^- + I^- \longrightarrow Cl^- + IO^-$$

has been postulated to occur by the three-step mechanism:

(1) $ClO^- + H_2O \rightleftharpoons HClO + OH^-$ fast

(2) $I^- + HClO \xrightarrow{k_2} HIO + Cl^-$ slow

(3) $OH^- + HIO \rightleftharpoons H_2O + IO^-$ fast

What rate law is required by this mechanism?

SOLUTION

The rate-determining step is step (2). Its rate law is

rate $= k_2[I^-][HClO]$

This rate law includes the concentration of HClO, a substance that is not one of the original reactants. To convert the rate law to a more useful form, we must find the relationship between the concentration of HClO and the concentration of substances originally present in the solution. Step (1) controls the [HClO], and step (1) is described by the equilibrium constant

$$K_1 = \frac{[HClO][OH^-]}{[ClO^-]}$$

Solving for [HClO] gives

$$[HClO] = \frac{K_1[ClO^-]}{[OH^-]}$$

Substituting this expression into the rate law gives

$$rate = \frac{k_2 K_1 [I^-][ClO^-]}{[OH^-]}$$

This is the overall rate law for the reaction. The product $k_2 K_1$ is a constant. The reaction is first order in iodide ion, first order in hypochlorite ion, and minus first order in hydroxide ion—that is, as the concentration of OH^- increases, the rate of the reaction decreases. This decrease in rate can be attributed to the decrease in the concentration of HClO as the solution becomes more alkaline. The rate-determining step proceeds more slowly as the concentration of HClO decreases. This sort of information can help us to determine the best conditions for carrying out this reaction.

One of the most interesting examples of the relationship between rate law and mechanism is provided by the reaction

$$H_2(g) + I_2(g) \longrightarrow 2HI(g)$$

We mentioned earlier that this reaction is first order in each reactant and follows a second-order rate law:

$$rate = k[H_2][I_2]$$

This rate law presents something of a problem: Is this reaction elementary or complex? While the rate law for the reaction between hydrogen and iodine is consistent with an elementary reaction, we may still find that the reaction is complex. We encountered this situation earlier. The high-pressure reaction of NO has a third-order rate law that is consistent with an elementary reaction, but the reaction is complex.

The reaction between H_2 and I_2 was the subject of one of the first kinetic investigations ever performed. Because second-order kinetics were observed, it was proposed that the reaction proceeds by a bimolecular single-step mechanism. This proposal, made toward the end of the nineteenth century, was accepted for decades. Most chemistry textbooks included this reaction as the standard example of a simple bimolecular process. However, this mechanism was disproven in the 1960s. The mechanism that is now accepted as being consistent with recent experimental work is:

(1) $\quad\quad I_2 \rightleftharpoons 2I \quad\quad$ fast
(2) $\quad I + H_2 \rightleftharpoons H_2I \quad\quad$ fast
(3) $\quad H_2I + I \longrightarrow 2HI \quad\quad$ slow

with step (3) as the rate-determining step. It seems probable that there will never be a definitive proof of this mechanism. At best, we can say that if the mechanism is correct, no one will ever be able to disprove it.

One of the simplest reactions imaginable:

$$H_2(g) + D_2(g) \longrightarrow 2HD$$

in which D represents deuterium, the heavy isotope of hydrogen, follows a second-order rate law:

$$rate = k[H_2][D_2]$$

But there is an ongoing debate about whether the reaction is elementary. The weight of the evidence indicates that the reaction is complex. These examples illustrate some of the difficulties encountered in the investigation of reaction mechanisms.

17.4   REACTION RATE AND TEMPERATURE

We know that chemical reactions require time to occur. A study of the reasons why time is needed for a chemical reaction to take place can give us valuable information about the way in which chemical changes come about.

Let us consider a simple bimolecular elementary reaction that occurs in the gas phase at 273 K:

$$NO(g) + O_3(g) \longrightarrow NO_2(g) + O_2(g)$$

This reaction is very favorable thermodynamically. Why doesn't it occur instantaneously? One answer is that the reaction takes place when two molecules collide, and that time is needed for the molecules to find each other. Thus, the rate of formation of products is limited by the rate at which collisions occur between molecules of NO and molecules of O_3.

We know that a gas consists mostly of empty space, and that molecules in the gas phase are relatively far from one another. We can calculate the exact rate of collision by using the kinetic theory of gases that was discussed in Chapter 4. This rate depends on the size of the molecules, their concentration, and their velocity. Their velocity, in turn, depends on the mass of the molecules and the temperature of the system. We can calculate that in this system, 10^{31} collisions between molecules occur per liter per second at standard conditions.

If we compare the rate at which collisions occur with the rate at which the reaction occurs, we find that only a small fraction of the collisions result in the formation of products. One reason is that the reaction will not take place unless the two reactant molecules collide in a way that allows the chemical change to occur, as Figure 17.5 shows in a simplified way. The formation of products requires the formation of a bond between the N atom of the NO molecule and an O atom of the O_3 molecule. Therefore, the collision must be one in which the O_3 molecule collides with the nitrogen end of the NO molecule. The fraction of the collisions in which the molecules have the proper orientation is called the *steric factor,* and it varies from reaction to reaction.

But there is an even more important reason why most collisions do not result in a chemical change. When two molecules approach, there is a natural repulsion between them; they tend to bounce away without any change. For a reaction to occur, this repulsion must be overcome. The colliding molecules usually must achieve relatively close contact for a change in bonding to occur. If this change is to happen, the molecules must have relatively high kinetic energies. Once this kind of collision occurs, even more energy is needed so that the electronic changes that lead to the formation of products can take place. When a collision takes place that is energetic enough to allow the formation of products, the system is in a state called the **activated complex** or **transition state**. The terms are synonymous.

The difference between the mean energy of all the collisions of the reactants and the average energy of collisions in which reaction takes place is called the **activation energy,** usually represented by E_a. The energy relationship between reactants, products, and the activated complex for a spontaneous process is shown schematically in Figure 17.6. For most reactions, the activation energy can be thought of as a barrier to the occurrence of the reaction. The greater the activation energy, the slower the rate of the reaction. The activation energy gives the highest energy state that the system must pass through to form products.

A relationship between the activation energy and the specific rate constant was proposed by Arrhenius in 1889. It has since been confirmed by many

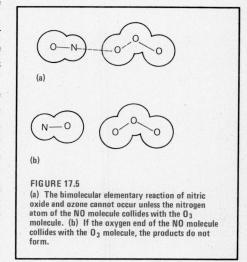

FIGURE 17.5
(a) The bimolecular elementary reaction of nitric oxide and ozone cannot occur unless the nitrogen atom of the NO molecule collides with the O_3 molecule. (b) If the oxygen end of the NO molecule collides with the O_3 molecule, the products do not form.

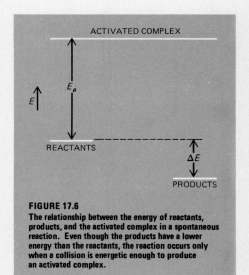

FIGURE 17.6
The relationship between the energy of reactants, products, and the activated complex in a spontaneous reaction. Even though the products have a lower energy than the reactants, the reaction occurs only when a collision is energetic enough to produce an activated complex.

experiments. The relationship is the **Arrhenius law**, which can be written as:

$$\log k = \log A - \frac{E_a}{2.303RT} \qquad (17.5)$$

where k is the specific rate constant, T is the temperature, and A is a constant called the *frequency factor*, which is related to the steric factor and the collisional frequency. The frequency factor is characteristic of the reaction and has the same units as the specific rate constant. The Arrhenius law assumes that E_a and A do not change with temperature.

If we look at the Arrhenius law term by term, we can see that it expresses much of what we have learned about reaction rates. The specific rate constant k is a convenient expression of relative rate; a reaction with a fast rate has a specific rate constant k whose value is high. Equation 17.5 shows that the value of k is related directly to the value of A, the frequency factor. The value of A increases if the frequency of collisions increases or if the steric factor increases. In other words, a reaction will be faster if there are more collisions or if a higher percentage of the collisions have the orientation needed for product formation.

The second term in the Arrhenius law is even more important because of the information it gives us about reaction rate. In general, the term $E_a/2.303RT$ has a positive value because E_a is almost always positive. Because a negative sign precedes this term in the Arrhenius law (Equation 17.5), the value of k goes down as the numerical value of this term goes up. An increase in the value of E_a increases the numerical value of this term, thus decreasing the value of k and the rate of the corresponding reaction. An increase in the value of T, which is in the denominator of this term, decreases the numerical value of the term and increases the value of the specific rate constant, k. The increase in the value of k with an increase in temperature is consistent with the observation that the rates of almost all chemical reactions increase as the temperature goes up.

It can be shown that the sensitivity of the rate of a reaction to changes in temperature depends on the value of E_a. If the value of E_a, the activation energy, is high, a reaction is more sensitive to temperature changes. The usual estimate is that the rate of a reaction in solution approximately doubles for every increase of 10 K in temperature. This generalization is roughly accurate around room temperature, although there are many exceptions to the rule. The generalization is not very useful for gas phase reactions, where we find a much wider range of values of E_a.

The easiest way to measure E_a is to measure the value of k at two different temperatures and substitute the values into the Arrhenius law. Better values of E_a can be obtained by making a number of such measurements. A plot of log k against $1/T$ gives a straight line whose slope is $-E_a/2.303R$. Figure 17.7 shows such a plot for the reaction between ozone and nitric oxide. The value of A, the frequency factor in the Arrhenius law, can be found if we know the value of E_a and of k at any temperature.

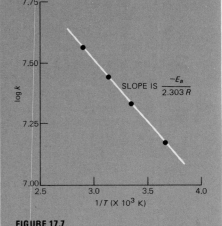

FIGURE 17.7
The activation energy of a reaction can be determined by measuring the specific rate constant k at several different temperatures, substituting the values of k into the Arrhenius law, and than making a plot of log k against $1/T$. The slope of the line gives the activation energy. Here we show such a plot for the reaction between O_3 and NO.

EXAMPLE 17.8

One of the earliest reactions to be investigated kinetically was the hydrolysis of sucrose in acidic solutions. This reaction is known as the inversion of cane sugar and is:

$$\underset{\text{sucrose}}{C_{12}H_{22}O_{11}} + H_2O \longrightarrow \underset{\text{glucose}}{C_6H_{12}O_6} + \underset{\text{fructose}}{C_6H_{12}O_6}$$

The reaction follows a second-order rate law:

$$\text{rate} = k[C_{12}H_{22}O_{11}][H^+]$$

At 300 K, $k = 2.12 \times 10^{-4}$ liter mol^{-1} sec^{-1}, and at 310 K, $k = 8.46 \times 10^{-4}$ liter mol^{-1} sec^{-1}.

(a) Calculate the value of E_a and A.

(b) Calculate the value of the specific rate constant at 320 K.

SOLUTION

(a) At 300 K, the Arrhenius law for this reaction is:

$$\log 2.12 \times 10^{-4} = \log A - \frac{E_a}{(2.30)(8.31 \text{ J mol}^{-1} \text{ K}^{-1})(300 \text{ K})}$$

At 310 K, it is:

$$\log 8.46 \times 10^{-4} = \log A - \frac{E_a}{(2.30)(8.31 \text{ J mol}^{-1} \text{ K}^{-1})(310 \text{ K})}$$

We can eliminate one of the two unknowns by subtracting the first equation from the second to obtain:

$$\log \frac{8.46 \times 10^{-4}}{2.12 \times 10^{-4}}$$

$$= \frac{E_a}{(2.30)(8.31 \text{ J mol}^{-1} \text{ K}^{-1})(300 \text{ K})} - \frac{E_a}{(2.30)(8.31 \text{ J mol}^{-1} \text{ K}^{-1})(310 \text{ K})}$$

$$= \frac{E_a}{(2.30)(8.31 \text{ J mol}^{-1} \text{ K}^{-1})} \left[\frac{1}{300 \text{ K}} - \frac{1}{310 \text{ K}} \right]$$

$$= \frac{E_a}{(2.30)(8.31 \text{ J mol}^{-1} \text{ K}^{-1})} \left[\frac{310 \text{ K} - 300 \text{ K}}{(300 \text{ K})(310 \text{ K})} \right]$$

$$E_a = 107\ 000 \text{ J/mol}$$

This value can then be substituted into either of the equations to find the value of A:

$$\log 2.12 \times 10^{-4} = \log A - \frac{107\ 000 \text{ J mol}^{-1}}{(2.30)(8.31 \text{ J mol}^{-1} \text{ K}^{-1})(300 \text{ K})}$$

$$\log A = 14.987$$

$$A = 9.71 \times 10^{14} \text{ liter mol}^{-1} \text{ sec}^{-1}$$

(b) Once the frequency factor A and the activation energy E_a are known, the Arrhenius law can be used to find the specific rate constant at any other temperature:

$$\log k = \log (9.71 \times 10^{14}) - \frac{107\ 000 \text{ J mol}^{-1}}{(2.30)(8.31 \text{ J mol}^{-1} \text{ K}^{-1})(320 \text{ K})}$$

$$= -2.507$$

$$k = 3.11 \times 10^{-3} \text{ liter mol}^{-1} \text{ sec}^{-1}$$

The algebraic manipulation carried out in part (a) of Example 17.8 can be used to derive a relationship between the activation energy E_a and the specific rate constants that are measured at any two temperatures. If the temperatures are T_1 and T_2 and the specific rate constants are k_1 and k_2, the relationship is:

$$\log \frac{k_2}{k_1} = \frac{E_a}{2.303R} \left[\frac{T_2 - T_1}{T_1 T_2} \right] \tag{17.6}$$

or, solving for E_a:

$$E_a = 2.303R \left[\frac{T_1 T_2}{T_2 - T_1} \right] \left[\log \frac{k_2}{k_1} \right] \tag{17.7}$$

The existence of an activation energy helps us to understand the way in which temperature affects the rate of a reaction. We know that the rate of many reactions increases by about a factor of two for every temperature increase of 10 K. This increase in rate is not due to an increase in the frequency of collisions.

When the temperature increases from 300 K to 310 K, the frequency of collisions increases by a factor of only 1.015. The rate increase is also not due to an increase in the mean energy of the system, which goes up by a factor of only 1.03 for a 10-K temperature increase.

To understand the reason for the increase in rate, we must consider the distribution of molecular speeds that we discussed in Section 4.8 (page 106). A distribution of molecular speeds also means a distribution of molecular energies. In any system, the energies of most molecules are close to the mean molecular energy. But there are a few molecules whose energies are much lower or much higher than the mean.

The activation energy of a reaction is usually much higher than the mean energy of the reacting system. Only a small fraction of the molecules in the system have energies sufficiently above the mean to react. When the temperature of the system increases by 10 K, the mean energy does not increase very much. But the number of molecules that are energetic enough to react is increased substantially, as Figure 17.8 shows.

If we accept the idea that only the relatively small number of molecules whose energy is substantially higher than the mean are converted to products, we are faced with an apparent difficulty. If only high-energy reactant molecules are converted to product, the system logically will soon have only reactant molecules whose energies are close to the mean. Lacking energetic molecules, the reaction will stop. But the reaction does not stop. The energy released when the activated complex forms product is distributed among the other molecules in the system by collisions. As a result, other reactant molecules become sufficiently energetic to undergo reaction.

Until now, we have discussed bimolecular processes. The same reasoning can be applied with little change to the relatively small number of termolecular processes. In a termolecular process, the activated complex contains three species rather than two. In these processes, collisions are much less frequent; we find that termolecular processes must have low activation energies to occur at reasonable rates.

A different line of reasoning would seem necessary for unimolecular processes, in which a single reactant molecule goes to the product. But the Arrhenius law applies to unimolecular processes, and activation energies can be measured for them. In a unimolecular process, the activated complex has only one molecule. When a molecule attains the energy of the activated complex, the products can form. The molecule attains this energy by collisions with other molecules in the system, as in bimolecular and termolecular processes. A unimolecular process must be preceded by collisions between the molecule that is to react and other molecules in the system, in order to raise the energy of the reacting molecule to the level of the activated complex.

This discussion helps us give a more precise definition of molecularity. We can say that the molecularity of an elementary reaction is the number of molecules in the activated complex.

We have said that the activation energy E_a that is found experimentally using the Arrhenius law can be regarded as an energy barrier which must be surmounted before the reactants can combine to form products. This interpretation is often given in graphical form, as in the energy diagram for the reaction of ozone and nitric acid that is shown in Figure 17.9. Such a diagram shows how the potential energy changes as reactants proceed to products; the "reaction coordinate" can be interpreted as the extent to which the reaction occurs.

An energy diagram gives only a crude picture of a chemical reaction, but it does give useful graphical information about some important characteristics of a reaction. Figure 17.9 shows that the reaction of ozone and nitric acid is highly exothermic, and that the activation energy of the reaction is relatively low.

Energy diagrams can be drawn for complex reactions with any number of

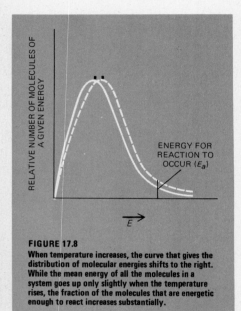

FIGURE 17.8
When temperature increases, the curve that gives the distribution of molecular energies shifts to the right. While the mean energy of all the molecules in a system goes up only slightly when the temperature rises, the fraction of the molecules that are energetic enough to react increases substantially.

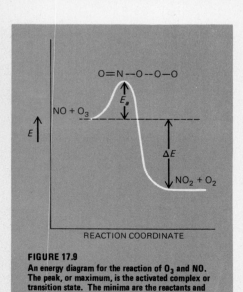

FIGURE 17.9
An energy diagram for the reaction of O_3 and NO. The peak, or maximum, is the activated complex or transition state. The minima are the reactants and the products.

steps. Figure 17.10 shows some examples. Each elementary reaction has an energy barrier. The rate-determining step has the energy barrier that is highest above the energy of the original reactants. The energy diagram for a complex reaction will have a number of maxima, with one maximum for each elementary reaction. There is a minimum between each two maxima. Each minimum corresponds to the existence of an intermediate. In the energy diagram, these minima are shown with higher energies than the reactants or products, consistent with the great reactivity and short lifetimes that are typical of reactive intermediates.

17.5 REACTION RATE AND EQUILIBRIUM

At first glance, it would seem that there is no need to discuss the kinetics of a system that is at equilibrium, since the overall composition of the system does not change. But a system at equilibrium is undergoing very many changes at the molecular level, so its kinetics are of interest to us.

There is an important relationship between reaction rate and equilibrium. It is not a simple, direct relationship, in which any reaction with a large equilibrium constant proceeds at a fast rate. We have already encountered a contrary example, the reaction in which water is formed from H_2 and O_2. The equilibrium constant of this reaction is large, but the reaction can be exceedingly slow.

To find the relationship between reaction rate and equilibrium, let us consider the process

$$2NOCl(g) \underset{k_{-1}}{\overset{k_1}{\rightleftharpoons}} 2NO(g) + Cl_2(g)$$

When the system is at equilibrium, the concentrations of the three gases do not change with time. However, the forward reaction in which NOCl is consumed occurs constantly, as does the back reaction in which NOCl is formed. Since the concentration of NOCl does not change, its rate of disappearance by the forward reaction must equal its rate of appearance by the back reaction. Both of these reactions have been found to be elementary reactions, and we can therefore write their rate laws.

The rate law for the forward reaction is

rate $= k_1[NOCl]^2$

and the rate law for the back reaction is

rate $= k_{-1}[NO]^2[Cl_2]$

(The symbol k_{-1} is commonly used for the specific rate constant of a reaction that is the reverse of a reaction whose specific rate constant is k_1.)

At equilibrium, the two rates are equal. That is:

$$k_1[NOCl]^2 = k_{-1}[NO]^2[Cl_2]$$

This equation can be rearranged to give:

$$\frac{k_1}{k_{-1}} = \frac{[NO]^2[Cl_2]}{[NOCl]^2}$$

The concentration expression on the right side of this equation is identical to the expression for K. Therefore, for an elementary reaction:

$$K = \frac{k_1}{k_{-1}} \tag{17.8}$$

Equation 17.8 defines the relationship between the equilibrium constant and the specific rate constants of a system at equilibrium. If a reaction is elementary and the specific rate constants of the forward and back reactions are known,

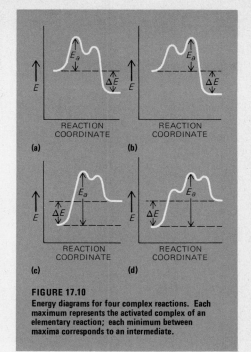

FIGURE 17.10
Energy diagrams for four complex reactions. Each maximum represents the activated complex of an elementary reaction; each minimum between maxima corresponds to an intermediate.

Equation 17.8 can be used to calculate the equilibrium constant. If the equilibrium constant and one specific rate constant are known, Equation 17.8 can be used to calculate the other specific rate constant.

EXAMPLE 17.9

The specific rate constant for the bimolecular decomposition of NOCl at 473 K is found to be 7.8×10^{-2} liter mol^{-1} sec^{-1}. The specific rate constant for the third order reaction of nitric oxide with chlorine at this temperature is found to be 4.7×10^2 $liter^2$ mol^{-2} sec^{-1}. Calculate the equilibrium constant for the reaction: $2NOCl(g) \rightleftharpoons 2NO(g) + Cl_2(g)$.

SOLUTION

Since we are given the specific rate constants for the forward and back reactions, Equation 17.8 can be used to find the value of K:

$$K = \frac{7.8 \times 10^{-2} \text{ liter } mol^{-1} sec^{-1}}{4.7 \times 10^2 \text{ liter}^2 mol^{-2} sec^{-1}} = 1.7 \times 10^{-4} \text{ mol/liter}$$

In this kind of calculation, care must be taken to ensure that the units are consistent.

The relationship between the equilibrium constant and the specific rate constants of complex reactions differs somewhat from the relationship for elementary reactions. The relationship for complex reactions is based on a general principle that is called *the principle of detailed balancing* when it is applied to large-scale systems and *the principle of microscopic reversibility* when it is applied on the molecular level. It can be stated as: *When equilibrium is reached in a reaction system, any elementary process and the reverse of that process must occur, on the average, at the same rate.*

We can show how this principle is used to find the relationship between the equilibrium constant and the relevant specific rate constants by analyzing the reaction system:

$$2NO_2(g) + F_2(g) \rightleftharpoons 2NO_2F(g)$$

It has been suggested that the mechanism for the formation of NO_2F includes two elementary reactions:

(1) $NO_2 + F_2 \xrightarrow{k_1} NO_2F + F$ slow

(2) $NO_2 + F \xrightarrow{k_2} NO_2F$ fast

The initial rate law for the reaction is:

rate $= k_1[NO_2][F_2]$

Figure 17.10(a) shows the energy diagram for these two elementary reactions. Until now, we have read such energy diagrams only from left to right. But at equilibrium, the steps in this mechanism proceed both forward and backward, since the NO_2F decomposes back to NO_2 and F_2. On the energy diagram, the system can be visualized as proceeding in both directions along the same energy pathway. The principle of detailed balancing tells us that the forward rate equals the back rate in each step at equilibrium. Since these are elementary processes, we can write the rate laws for each step:

step (1) forward rate $= k_1[NO_2][F_2]$
step (1) back rate $= k_{-1}[NO_2F][F]$

and since these rates are equal:

(a) $k_1[NO_2][F_2] = k_{-1}[NO_2F][F]$

step (2) forward rate $= k_2[NO_2][F]$
step (2) back rate $= k_{-2}[NO_2F]$

and since these rates are equal:

(b) $k_2[NO_2][F] = k_{-2}[NO_2F]$

To find the relationship of the equilibrium constant K to these specific rate constants, equations (a) and (b) can be combined. After the equations are multiplied and rearranged we have:

$$\frac{k_1 k_2}{k_{-1} k_{-2}} = \frac{[NO_2F][F][NO_2F]}{[NO_2][F_2][NO_2][F]} = \frac{[NO_2F]^2}{[NO_2]^2[F_2]} = K$$

For this mechanism, the equilibrium constant is the product of the specific rate constants of the forward reactions divided by the product of the specific rate constants of the back reactions.

We encounter a different aspect of the relationship between kinetics and equilibrium in systems where more than one product can form from a given set of reactants. This situation can be represented by the general scheme:

(a) $A + B \rightleftharpoons C$
(b) $A + B \rightleftharpoons D$

In such reaction systems, the product that forms fastest is not necessarily the most stable. Let us assume that product C forms fastest and product D is more stable, a situation represented by the energy diagram in Figure 17.11, and that reactions (a) and (b) are reversible. It is then possible to form selectively either C or D by controlling the way in which the reaction is carried out.

The product that forms faster, C, is called the *product of kinetic control*. It is obtained by allowing the reaction to proceed for only a relatively short time, or by removing C as it forms, thus preventing reaction (a) from reversing. The more stable product, D, is called the *product of thermodynamic control*. It is obtained by allowing the reaction to proceed for longer periods of time, so that reaction (a) can reverse and reaction (b) can take place. In other words, the system is allowed to proceed to equilibrium. Since D is more stable, it will predominate at equilibrium.

The buildup of product D, which is more stable than product C but forms more slowly, can be understood by examining the relative rates of the reverse reactions. While C forms faster, it also decomposes faster when conditions permit the reaction to reverse. When D finally forms, it decomposes to the reactants more slowly than C. Thus, when D forms, it tends to remain. The energy diagram in Figure 17.11 shows that the energy barriers for the reverse reactions are consistent with this explanation.

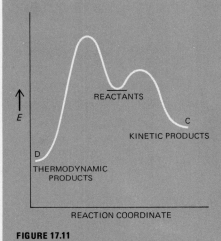

FIGURE 17.11
An energy diagram for a reaction in which one product, C, forms faster and another product, D, is more stable. Product C can be formed by stopping the reaction after a short time. Product D can be formed by allowing the reaction to proceed to equilibrium.

17.6   CATALYSIS

A catalyst is a substance that speeds up the rate of a reaction without being consumed. A catalyst enters into a reaction but emerges from it unchanged. It may, for example, form an intermediate with one or more of the reactants. The intermediate then decomposes to form the products and to regenerate the catalyst.

Catalysts are important in many reactions, organic and inorganic. We know the general mode of action of catalysts. It is the height of the energy barrier along the pathway from reactants to products that controls the rate of a reaction. As Figure 17.12 shows, a catalyst quickens the rate of a reaction by providing a different pathway with a lower energy barrier between reactants and products.

From Figure 17.12, we see that a catalyst does not change the relative energy of the reactants and the products, which means that *a catalyst has no effect on the position of equilibrium*. We also see that the system can go either way on the lower-energy pathway opened up by the catalyst. Therefore, *a catalyst*

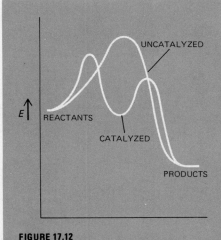

FIGURE 17.12
A catalyst makes a reaction go faster by providing a pathway with a lower energy barrier (bottom line) between reactants and products than the pathway that is normally available (top line).

speeds up the rate of a reaction in both the forward and reverse directions. By so doing, a catalyst causes a system to reach equilibrium faster.

We can divide catalyzed reactions into two main categories: *homogeneous reactions*, which occur entirely in one phase, usually in the gas phase or in solution, and *heterogeneous reactions*, which occur at the interface between two phases—for example, on the surface of a solid that is in contact with a solution or a gas. We shall study a few of the many processes in which catalysis can play a role.

Homogeneous Catalysis

Several catalytic processes in the gas phase have come under close scrutiny in recent years because of their apparent role in the unintentional alteration of the atmosphere caused by human activities. The mass of the substances released into the atmosphere by the human race is small compared to the total mass of the atmosphere. But some of these substances can cause relatively large changes because they act as catalysts.

One process by which technological activity produces damage is the oxidation of sulfur dioxide to sulfur trioxide in the lower atmosphere. This process eventually results in the formation of sulfates from the sulfur dioxide that is released when fossil fuels are burned. These sulfates are believed to be extremely harmful to human health. The oxidation reaction is

$$2SO_2(g) + O_2(g) \longrightarrow 2SO_3(g)$$

Normally, the rate of this reaction is slow. A termolecular collision is needed for the reaction to occur directly in a single step, and termolecular collisions generally are unfavorable (with the exception of some nitric oxide reactions). But it has been found that oxides of nitrogen, which also are produced by burning fossil fuels, catalyze the oxidation by the following steps:

$$2NO(g) + O_2(g) \longrightarrow 2NO_2(g)$$
$$NO_2(g) + SO_2(g) \longrightarrow NO(g) + SO_3(g)$$

The sulfur dioxide is oxidized by the NO_2, a faster process than its direct oxidation by O_2. The SO_3 then combines with H_2O in the atmosphere to produce sulfuric acid, H_2SO_4, and other sulfates.

Homogeneous catalysis in solution is also quite common. One important kind is acid-base catalysis, which often occurs when a substance reacts through its conjugate acid or conjugate base as an intermediate. The hydrolysis of nitramide, NH_2NO_2, for example, is believed to occur by the mechanism:

$$NH_2NO_2 + OH^- \rightleftharpoons H_2O + NHNO_2^-$$
$$NHNO_2^- \longrightarrow N_2O(g) + OH^-$$

The rate of the reaction increases as the concentration of hydroxide ion increases. You can see that the hydroxide ion qualifies as a catalyst because it is not consumed in the overall reaction.

Another reaction for which catalysis is important is the decomposition of hydrogen peroxide in aqueous solution:

$$2H_2O_2 \longrightarrow 2H_2O + O_2(g)$$

Even though the position of equilibrium lies far to the right for this system, the reaction is slow if the H_2O_2 is very pure. But it can be catalyzed in a number of different ways, one of which is based on the ability of H_2O_2 to be both an oxidizing agent and a reducing agent. The rate of decomposition is observed to increase substantially if bromide ion is in solution with the hydrogen peroxide. First the H_2O_2 oxidizes the bromide ion to bromine:

$$H_2O_2 + 2Br^- + 2H^+ \longrightarrow Br_2 + 2H_2O$$

THE CATALYSTS IN YOUR CAR

The catalytic converters that help reduce the emissions of carbon monoxide, hydrocarbons, and nitrogen oxides from many of today's automobiles are unusual because they actually are two converters in one: a reducing converter and an oxidizing converter, working in coordination to change pollutants into harmless gases.

The catalysts in the converters are primarily platinum and palladium, deposited in a thin layer on a porous material that has a very large surface area. Both converters use the same catalysts, but a wholly different set of reactions occurs in each.

The control system really begins at the carburetor, which is set for a rich fuel-to-air mixture that has two effects: It lowers the combustion temperature, so that less NO is produced, and it produces an abundance of hydrocarbons (HC), CO, and H_2. These gases, all reducing agents, are needed as reactants in the reducing converter, the first on line. Some of the reactions that occur in the reducing converter are:

$$2CO + 2NO \longrightarrow N_2 + 2CO_2$$
$$5H_2 + 2NO \longrightarrow 2NH_3 + 2H_2O$$
$$2H_2 + 2NO \longrightarrow N_2 + 2H_2O$$
$$CO + H_2O \longrightarrow CO_2 + H_2$$

For the oxidizing catalyst to do its job, the oxygen-poor output of the reducing converter must be made oxygen-rich. A stream of air is pumped into the exhaust as it enters the reducing converter, where these reactions take place:

$$2CO + O_2 \longrightarrow 2CO_2$$
$$4HC + 5O_2 \longrightarrow 4CO_2 + 2H_2O$$
$$2H_2 + O_2 \longrightarrow 2H_2O$$

Ideally, the exhaust gases that emerge from the oxidizing converter contain only CO_2, N_2, H_2O, and O_2. In reality, the undesirable gases are not removed completely. One of the products of the reducing converter is ammonia, NH_3. Not all of this ammonia is converted in the reducing converter. In the oxidizing converter, ammonia is converted to nitrogen oxides by the reactions:

$$4NH_3 + 5O_2 \longrightarrow 4NO + 6H_2O$$
$$2NH_3 + 2O_2 \longrightarrow N_2O + 3H_2O$$

There are further complications. When the engine starts, it takes several minutes for the oxidizing converter to reach its operating temperature of 650 K. The system thus must be engineered so that the reducing converter can bring about the oxidizing reactions temporarily. And since the tetraethyl lead used in gasoline can coat the catalyst and "poison" it, automobiles with catalytic converters can use only lead-free gasoline.

and then it reduces the Br_2 back to Br^-:

$$H_2O_2 + Br_2 \longrightarrow 2Br^- + 2H^+ + O_2(g)$$

The sum of the two reactions is

$$2H_2O_2 \longrightarrow 2H_2O + O_2(g)$$

which is the only net chemical change that occurs. The bromide ion is not consumed and so does not appear in the overall reaction.

There are a number of other oxidation-reduction couples that catalyze the decomposition of hydrogen peroxide. One group of catalysts that has been studied extensively is ions of transition metals that display more than one oxidation state. The catalysis of hydrogen peroxide decomposition by these ions is an important biological process.

In particular, the catalysis of the decomposition of H_2O_2 by cations of iron is of great interest. The overall changes are similar to those of catalysis by bromide ion:

$$H_2O_2 + 2H^+ + 2Fe^{2+} \longrightarrow 2H_2O + 2Fe^{3+}$$
$$H_2O_2 + 2Fe^{3+} \longrightarrow O_2(g) + 2H^+ + 2Fe^{2+}$$

The sum of these reactions is simply

$$2H_2O_2 \longrightarrow 2H_2O + O_2(g)$$

The details of these steps are rather complex, and have not yet been worked out completely.

Enzyme Catalysis

The phenomenon we call life would be impossible without catalysis. Most of the reactions that are essential to life take place very slowly outside the living organism. But inside living cells, biological catalysts called *enzymes* speed the rates of these reactions, often by enormous factors.

Enzymes are large proteins—indeed, they are the most important and numerous class of proteins. More than a thousand enzymes, grouped into six main classifications, have been identified. The name of an enzyme group, or of an individual enzyme, consists of a stem indicating the reaction that is catalyzed and the suffix -*ase*. For example, the major groups include the oxidoreductases, which catalyze oxidation-reduction reactions; the lyases, which catalyze addition reactions; and the hydrolases, which catalyze hydrolysis reactions.

Generally, enzymes can function as catalysts only within a narrow temperature range, so there is a need for temperature control in living organisms. In addition, enzymes can function only in a narrow pH range, so there is a need for buffering and close control of pH in most living organisms.

An enzyme often is associated with a smaller nonprotein molecule, called a *coenzyme* or prosthetic group, which is needed for the enzyme to perform its catalytic function. Undoubtedly the best-known group of coenzymes are the molecules called vitamins, which take part in a number of essential processes in the human body.

An enzyme is a very large molecule. But only a small region of the molecule, the *active site,* participates in the enzyme's catalytic action. In a sense, the structure of the entire molecule is built around the active site. The enzyme normally maintains a fixed three-dimensional structure that keeps the appropriate portions of the molecule at the active site, so that the enzyme can act as a catalyst.

The reactant in an enzyme-catalyzed reaction is called the substrate. There is a high degree of specificity in both the types of reactions that enzymes will catalyze and the substrates whose reactions they will catalyze. Often, an enzyme will catalyze a reaction for one substrate but not for another that is chemically

A model of the enzyme lysozyme, whose structure was determined by X-ray analysis. The vertical open groove in front, just left of the middle, is the active site of the enzyme. The black object at top center represents a molecule that can fit into the active site. (D. C. Phillips, *Proc. Roy. Soc. B167* (1967) p. 380.)

quite similar. In many such cases, it is possible to note three-dimensional structural differences between the substrates, which presumably explain why the enzyme catalyzes the reaction of one substrate but not the other.

The first step in an enzyme-catalyzed reaction is the rapid formation of an enzyme-substrate complex. Since this complex is easily broken up, it is assumed that the substrate is held to the active site of the enzyme by weak interactions, such as hydrogen bonding or ion-dipole attractions. The reaction can be represented as

$$E + S \rightleftharpoons ES$$

where E is the enzyme, S is the substrate, and ES is the complex. The equilibrium constant for this reaction usually is large, between 10^3 and 10^6. After the enzyme-substrate complex is formed, there are reactions that form the product and regenerate the enzyme. We can represent such reactions schematically as

$$ES \longrightarrow P + E$$

where P is the product.

The formation of the ES complex is often described as the insertion of a "key," the substrate, into a "lock," the active site of the enzyme, as shown in Figure 17.13. The point of this lock-and-key analogy is that the enzyme's activity is quite specific, since only a substrate with precisely the right shape will "unlock" the enzyme's catalytic capabilities. The precise fit is essential for the rapid formation of products, since it allows several catalytic groups that are part of the enzyme to work on the substrate simultaneously.

The structure of the active site and the nature of the enzyme-substrate complex have been the subject of much recent research. Investigations of this kind are unusually demanding, because of the complexity of enzyme structures. One of the most remarkable achievements of the past few decades has been the determination of the complete structure of several enzymes by X-ray crystallography, a feat that makes possible a reasonably good understanding of the way in which an enzyme works. Research on enzyme catalysis is one of the most active areas in chemistry today, and it promises to remain so for many years.

Research thus far has pointed up the remarkable efficiency of many enzymatic processes. For example, one of the most efficient enzymes known is catalase, which catalyzes the familiar reaction

$$2H_2O_2 \longrightarrow 2H_2O + O_2$$

One molecule of catalase will convert more than one million molecules of hydrogen peroxide to water and oxygen in one minute at room temperature. Since hydrogen peroxide is a necessary by-product of many biological oxidation steps and is also quite poisonous, its rapid removal is essential for the well-being of living organisms.

The details of the mode of action of catalase have not been worked out. But it is known that iron in the $+3$ oxidation state is part of the catalase molecule and takes part in the catalytic process, presumably in a way similar to the inorganic catalysis we have discussed.

There are measurements that give us a rough idea of the effectiveness of this biological catalysis. The uncatalyzed decomposition of H_2O_2 is reported to have $k = 1 \times 10^{-7}$ sec^{-1} and $E_a = 75$ kJ/mol at room temperature. For the decomposition brought about with catalase, $k = 4 \times 10^7$ sec^{-1} and $E_a = 8$ kJ/mol. Thus, for a given set of conditions, the presence of catalase increases the rate of decomposition of H_2O_2 by a factor of more than 10^{14}.

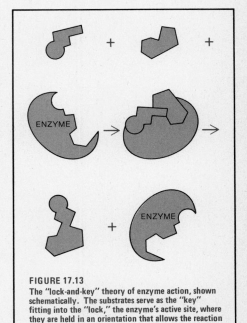

FIGURE 17.13
The "lock-and-key" theory of enzyme action, shown schematically. The substrates serve as the "key" fitting into the "lock," the enzyme's active site, where they are held in an orientation that allows the reaction to occur swiftly.

Heterogeneous Catalysis

Solids that increase the rate of chemical reactions because of their surface properties are called heterogeneous catalysts. There is a wide variety of heterogeneous catalysts, including metals, metal oxides, metal sulfides, and salts. Heterogeneous catalysts are widely used to catalyze reactions of gases or liquids on the surfaces of solids, including many processes that are essential to modern chemical industry. For example, most of the reactions by which petroleum is refined are catalyzed by solids such as alumina, Al_2O_3; silica, SiO_2; and various metal salts. The manufacture of ammonia by the Haber process:

$$N_2(g) + 3H_2(g) \longrightarrow 2NH_3(g)$$

is catalyzed by iron, molybdenum, or other metals. There is a whole group of reactions, called Fischer-Tropsch reactions, in which organic products are produced from H_2 and CO. One of these reactions, catalyzed by zinc oxide, is used for the production of methanol:

$$CO(g) + 2H_2(g) \longrightarrow CH_3OH(g)$$

The reaction of H_2 with CO can also be catalyzed with nickel to form methane, CH_4, or with cobalt to form more complex hydrocarbons. Another process in which heterogeneous catalysis is used is the oxidation of ammonia to nitric oxide, the first step in the Ostwald process (Section 9.3, page 274) for the manufacture of nitric acid:

$$4NH_3(g) + 5O_2(g) \longrightarrow 4NO(g) + 6H_2O(g)$$

which is catalyzed by an alloy of platinum and rhodium. The oxidation of SO_2 to SO_3, a step in the contact process for the manufacture of sulfuric acid, is catalyzed by V_2O_5. A list of industrial processes that use heterogeneous catalysis could run for many pages.

We can visualize the mechanism of reactions that take place on the surfaces of solids as a sequence of steps:

1. The reactants find their way to the surface of the catalyst.
2. The reactants are adsorbed on the surface.
3. A chemical reaction between the adsorbed substances takes place.
4. The products of the reaction are desorbed from the surface.
5. The products go back into the gas or liquid phase.

The first and last steps usually occur quickly and are not related to the reaction mechanism; we shall not consider them. The second step, adsorption, is the process by which a substance becomes attached to the surface of a solid. There are two ways in which a molecule can be adsorbed on the surface of a solid catalyst. In *physical adsorption*, relatively weak forces, typical of nonbonding interactions, hold the molecule on the surface. Physical adsorption is not a major factor in catalytic action. The second kind of adsorption, *chemisorption*, has most of the characteristics of a chemical reaction between the reactant molecule and the solid. Chemisorption evolves quantities of heat comparable to those of chemical reactions. The strength of the forces holding the reactant molecule to the surface is as great as that of chemical bonds, and chemisorption often changes the reactant molecule in a way that makes it easier for the desired reaction to occur. An appreciable energy of activation is associated with chemisorption in many cases, so reactions on solid surfaces must often be carried out at high temperatures.

The overall rate of a surface-catalyzed reaction usually is controlled by the rate at which reactions occur between the adsorbed molecules on the surface of the catalyst. But, depending on the reaction conditions, the overall rate can also be controlled by the rate of adsorption of the reactants or the rate of desorption of the products. We can see the controlling effect of adsorption and desorption in the reversible reaction for the synthesis of ammonia, which takes place on a metal surface:

$$N_2(g) + 3H_2(g) \rightleftharpoons 2NH_3(g)$$

At relatively high pressures and temperatures between 700 and 900 K, the rate of the forward reaction is controlled by the rate of chemisorption of N_2, while the rate of the reverse reaction is controlled by the rate of desorption of N_2.

The role of the solid catalyst in this reaction is of interest. The rates of reaction in this system in the absence of catalyst are extremely slow. It is not until the temperature rises above 1600 K that the homogeneous decomposition of NH_3 occurs, and then only at a moderate rate. The uncatalyzed thermal reaction of N_2 and H_2 has never been observed. The reaction is exothermic, so the position of equilibrium moves further to the left as the temperature goes up. In fact, the uncatalyzed reaction between N_2 and H_2 can occur at a measurable rate only at a temperature that is so high that no detectable NH_3 is found at equilibrium.

The major action of the catalyst in this reaction appears to be the activation of N_2. We have said that the $N\equiv N$ triple bond is one of the strongest bonds known, and that reactions with N_2 are extremely slow often because so much energy is needed to break, or even to weaken, the bond. A suitable catalyst weakens or even breaks the bond but does not adsorb the N_2 molecule too tightly for other reactions to occur. The beginning of the catalyzed synthesis of ammonia can be represented by the sequence:

$$N_2(g) \rightleftharpoons N_2 \text{ (adsorbed)}$$
$$N_2 \text{ (adsorbed)} \rightleftharpoons 2N \text{ (adsorbed)}$$
$$H_2(g) \rightleftharpoons H_2 \text{ (adsorbed)}$$
$$H_2 \text{ (adsorbed)} \rightleftharpoons 2H \text{ (adsorbed)}$$

When these chemisorption steps have taken place, the surface of the metal is covered with the equivalent of atomic nitrogen and atomic hydrogen, which can react with each other readily. An overall change in which three atoms of H and one atom of N combine can occur in steps. The NH_3 that is formed on the surface desorbs and goes into the gas phase, allowing more NH_3 to form:

$$N \text{ (adsorbed)} + 3H \text{ (adsorbed)} \rightleftharpoons NH_3 \text{ (adsorbed)}$$
$$NH_3 \text{ (adsorbed)} \rightleftharpoons NH_3(g)$$

Some surface-catalyzed reactions are among the few known examples of reactions that can be zero order overall. One of them is the decomposition of nitrous oxide gas on a hot platinum surface:

$$2N_2O(g) \longrightarrow 2N_2(g) + O_2(g)$$

When the concentration of N_2O is high enough, the surface of the metal is covered completely with adsorbed gas. An increase in the pressure of the gas above the solid does not affect the rate, since the surface is already saturated with N_2O and the reaction cannot proceed any faster. Since an increase in concentration of the reactant has no effect on the rate, the reaction is zero order.

EXERCISES

17.1 Write expressions that can be used to express the rate of the following reaction; show the relationship between these expressions: $2N_2O_5(g) \rightarrow 4NO_2(g) + O_2(g)$.

17.2 Write expressions that can be used to express the rate of the following reaction that takes place in aqueous solution; show the relationship between these expressions:
$16H^+ + 2MnO_4^- + 10I^- \rightarrow 2Mn^{2+} + 8H_2O + 5I_2$.

17.3 The rate of appearance of $NO_2(g)$ from $NO(g)$ and $O_2(g)$ is found to be 4.8×10^{-4} atm/sec during a measured time interval. Find the rate of disappearance of nitric oxide and of oxygen during this interval.

17.4 The rate of disappearance of ozone in the reaction $2O_3(g) \rightarrow 3O_2(g)$ is found to be 8.9×10^{-3} atm/sec during a measured time interval. Find the rate of appearance of $O_2(g)$ during this interval.

17.5 The rate of appearance of I_2 in aqueous acid solution as the result of the oxidation of I^- ion by hydrogen peroxide is found to be 3.4×10^{-5} mol liter^{-1} sec^{-1} during a measured time interval. Find the rate of disappearance of each of the reactants.

17.6 The reaction between $Cr_2O_7^{2-}$ ion and HNO_2 in acidic aqueous solution forms nitrate ion and Cr^{3+} ion. The rate of disappearance of the dichromate ion is found to be 2.8×10^{-4} mol liter^{-1} sec^{-1} during a measured time interval. Find the rate of disappearance of nitrous acid and the rate of appearance of each of the products during this interval.

17.7 The following data are obtained in a study of the rate of disappearance of H$^+$ ion in the reaction: $CH_3OH + H^+ + Cl^- \rightarrow CH_3Cl + H_2O$.

[H$^+$] (M)	1.85	1.67	1.52	1.30	1.00
t (min)	0	79.0	158	316	632

Find the average rate of disappearance of H$^+$ for the time interval between each measurement and for the total time interval.

17.8 The following data are obtained in a study of the rate of appearance of $NO_2(g)$ in the reaction: $NO(g) + CO_2(g) \rightarrow NO_2(g) + CO(g)$.

P_{NO_2} (atm)	0	0.217	0.370	0.482	0.556
t (sec)	0	29.9	60.1	90.0	119.8

Find the average rate of appearance of $NO_2(g)$ for the time interval between each measurement and for the total time interval.

17.9 The following data are obtained in a study of the rate of the reaction $2CO(g) \rightarrow CO_2(g) + C(s)$ by total pressure measurements:

P_T (atm)	0.329	0.313	0.295	0.276
t (sec)	0	401	998	1795

Find the average rate of disappearance of CO for the time interval between each measurement. Find the average rate of appearance of CO_2 for the total time interval.

17.10 The gas phase decomposition of N_2O_5 to NO_2 and O_2 is monitored by measurements of total pressure. The following data are obtained:

P_T (atm)	0.154	0.215	0.260	0.315	0.346
t (sec)	0	52	103	205	309

Find the average rate of disappearance of N_2O_5 for the time interval between each measurement and for the total time interval.

17.11 Before beginning a kinetic study of a reaction, an investigator might make some guesses about the rate law of a reaction. For example, for the reaction $2NO(g) + Br_2(g) \rightarrow 2NOBr(g)$ some simple rate laws suggest themselves. Write the rate law that corresponds to each of the following descriptions: (a) first order in nitric oxide, (b) second order in nitric oxide, (c) first order in nitric oxide, first order in bromine, (d) second order in nitric oxide, first order in bromine. Indicate the overall order of each rate law.

17.12 The rate law for the reaction $2Br^- + 2H^+ + H_2O_2 \rightarrow Br_2 + 2H_2O$ is third order overall and first order in each reactant. Write the equation for the rate law.

17.13 Find the units of the specific rate constant for a reaction in solution for (a) a rate law that is first order in each of two reactants, (b) a third-order rate law.

17.14 The following data are obtained for the decomposition of $N_2O_3(g)$ to nitric oxide and nitrogen dioxide:

$P_{N_2O_3}$ (atm)	1.2×10^{-3}	1.8×10^{-3}	2.7×10^{-3}
initial rate (atm/sec)	7.3×10^{-3}	1.1×10^{-2}	1.7×10^{-2}

Find the rate law of the decomposition from these data.

17.15 At 400 K oxalic acid decomposes according to the following reaction:

$$H_2C_2O_4(g) \longrightarrow CO_2(g) + HCOOH(g)$$

The rate of this reaction can be studied by measurements of the total pressure. Find the rate law of the reaction from the following measurements, which give the total pressure reached after 20 000 sec from the indicated starting pressures of oxalic acid, $H_2C_2O_4$.

$P_{H_2C_2O_4}$ (atm)	6.58×10^{-3}	9.21×10^{-3}	1.11×10^{-2}
P_T (atm)	9.46×10^{-3}	1.32×10^{-2}	1.60×10^{-2}

17.16 The rate of the reaction $H_2PO_2^- + OH^- \rightarrow HPO_3^{2-} + H_2(g)$ can be measured by a number of techniques. Suggest two possible techniques that can be used for this purpose.

17.17 The following measurements of initial rate are made for varying concentrations of OH$^-$ ion and $H_2PO_2^-$ ion:

[OH$^-$] (M)				
0.21	0.28	0.35	0.21	0.21

[$H_2PO_2^-$] (M)				
0.35	0.35	0.35	0.52	0.69

initial rate (mol liter^{-1} sec^{-1})				
5.2×10^{-5}	9.3×10^{-5}	1.5×10^{-4}	7.8×10^{-5}	1.0×10^{-4}

Find the rate law for the reaction.

17.18 The following data are obtained in a study of the rate of the reaction $SO_2Cl_2(g) \rightarrow SO_2(g) + Cl_2(g)$:

$P_{SO_2Cl_2}(g)$ (atm)	1.00	0.947	0.895	0.848	0.803
t (sec)	0	2500	5000	7500	10 000

The reaction follows a first-order rate law. Calculate the value of the specific rate constant.

*17.19 The rate of decomposition of $N_2O_3(g)$ to $NO_2(g)$ and $NO(g)$ is followed by measuring the [NO_2] at different times. The following data are obtained:

[NO_2] (mol/liter)	0	0.193	0.316	0.427	0.784
t (sec)	0	884	1610	2460	50 000

The reaction follows a first-order rate law. Calculate the specific rate constant.

17.20 The specific rate constant for the first-order decomposition of hydrogen peroxide to water and $O_2(g)$ is found to be 4.9×10^{-6} sec^{-1} at a given temperature. A solution is prepared in which the [H_2O_2] = 0.986M.

(a) Find the length of time required for the [H_2O_2] to fall to 0.329M.

(b) Find the length of time required for the [H_2O_2] to fall to 0.0986M.

(c) Find the [H_2O_2] after 40 hours.

*17.21 The specific rate constant for the first-order decomposition of $N_2O_5(g)$ to NO_2 and O_2 is 7.48×10^{-3} sec^{-1}, at a given temperature.

(a) Find the length of time required for the total pressure in a system containing N_2O_5 at an initial pressure of 0.100 atm to rise to 0.145 atm.

(b) To 0.200 atm.

(c) Find the total pressure after 100 sec of reaction.

17.22 Under certain conditions the rate of the reaction $I_2 + H_2O \rightarrow I^- + HIO + H^+$ is first order in I_2. Measurements of the [I_2] can be carried out indirectly by measuring the optical density of the brown solution of I_2 in water. The value of the optical density is proportional to the [I_2]. The optical density of a solution of I_2 is found to be 0.52 when the solution is first prepared. It drops to 0.42 after 100 sec and to 0.34 after 200 sec.

(a) Calculate the specific rate constant.

(b) Calculate the time required for the optical density to drop to 0.23.

17.23   Derive a general expression for the time required for the concentration of a reactant in a first-order reaction to fall by one-third.

17.24   Find the fraction of the starting concentration of a reactant that remains when a first-order reaction proceeds for 10 half-lives.

17.25   Find the half-life of the reaction in Exercise 17.21.

17.26   The half-life for radioactive decay of ^{239}Pu, produced from uranium in nuclear reactors, is 24 400 years.

(a) Find the specific rate constant for the decay.

(b) Find the number of years required for 99% of a sample of plutonium to decay.

17.27   A bit of vegetable dye scraped from a cave painting is converted to CO_2. The radioactivity of the CO_2 is measured as 0.968 c.p.m. Find the age of the cave painting, using the data in Example 17.5.

17.28   The reaction $CH_3CHO(g) \rightarrow CH_4(g) + CO(g)$ is second order in CH_3CHO at 700 K. The following data are obtained from a kinetic study of the reaction:

[CH_3CHO] (mol/liter)	0.022	0.020	0.017	0.013
t (sec)	0	1000	3000	7000

(a) Find the specific rate constant.

(b) Find the [CH_3CHO] after 15000 sec.

(c) Find the time required for [CH_4] to reach 0.012 mol/liter.

17.29   The reaction $I^- + ClO^- \rightarrow IO^- + Cl^-$ is second order overall at high [OH^-] and first order in each reactant. To simplify the treatment of the data, a kinetic study is carried out on a solution in which the [I^-] = [ClO^-] = 0.096M.

(a) Indicate why the treatment of the data will be simplified in this case.

(b) Use the following data to find the specific rate constant:

[I^-] (M)	0.096	0.049	0.032	0.020
t (sec)	0	29	63	146

(c) Find the [IO^-] after 200 sec.

17.30   Derive a general expression for the time required for the concentration of a reactant in a one-term second-order reaction to fall by one-third.

17.31   The following data are obtained for the reaction $A(g) \rightarrow B(g)$:

P_A (atm)	0.470	0.405	0.356	0.319	0.286
t (sec)	0	250	500	750	1000

Determine whether the reaction is first or second order and then calculate k.

17.32   Consider the reaction $A(g) \rightarrow B(g)$. A starting pressure of A = 1 atm falls to 0.5 atm after 100 sec. Calculate the pressure of A after another 100 sec of reaction if the reaction is (a) first order in A, (b) second order in A, (c) zero order in A.

17.33   Write rate laws for the following elementary reactions: (a) $N_2O_4(g) \rightarrow 2NO_2(g)$, (b) $2NO_2(g) \rightarrow N_2O_4(g)$, (c) $Cl_2(g) + O_2(g) \rightarrow 2ClO(g)$, (d) $2NO(g) + O_2(g) \rightarrow 2NO_2(g)$.

17.34   The reaction $C_4H_9I + Cl^- \rightarrow C_4H_9Cl + I^-$ is believed to take place in polar solvents by the mechanism:

(1) $C_4H_9I \xrightarrow{k_1} C_4H_9^+ + I^-$ slow

(2) $C_4H_9^+ + Cl^- \xrightarrow{k_2} C_4H_9Cl$ fast

Write the tare law that is consistent with this mechanism.

17.35   The reaction in the gas phase in which NO_2Cl decomposes to form Cl_2 and NO_2 is believed to take place by the following mechanism:

(1) $NO_2Cl(g) \xrightarrow{k_1} NO_2(g) + Cl(g)$ slow

(2) $NO_2Cl(g) + Cl(g) \xrightarrow{k_2} NO_2(g) + Cl_2(g)$ fast

Write the overall reaction and the rate law that are consistent with this mechanism.

17.36   One proposed, although probably incorrect, mechanism for the oxidation of nitric oxide is:

(1) $NO(g) + O_2(g) \rightleftharpoons NO_3(g)$ fast

(2) $NO_3(g) + NO(g) \xrightarrow{k_2} 2NO_2(g)$ slow

Write the overall reaction and the rate law that are consistent with this mechanism. An alternate mechanism is a one-step process. Write the rate law consistent with this mechanism.

17.37   The proposed two-step mechanism for the decomposition of $O_3(g)$ is

(1) $O_3(g) \rightleftharpoons O_2(g) + O(g)$

(2) $O(g) + O_3(g) \xrightarrow{k_2} O_2(g)$

(a) Write the rate law consistent with the first step being rate determining.

(b) Write the rate law consistent with the second step being rate determining.

(c) Suggest an experiment to distinguish between these rate laws.

17.38   The proposed mechanism for the formation of hydrogen bromide can be written in a simplified form as:

(1) $Br_2(g) \rightleftharpoons 2Br(g)$ fast

(2) $Br(g) + H_2(g) \xrightarrow{k_2} HBr(g) + H(g)$ slow

(3) $H(g) + Br(g) \xrightarrow{k_3} HBr(g) + Br(g)$ fast

This mechanism is consistent with a fractional-order rate law. Write the rate law.

17.39   The gas phase reaction $NO_2 + F_2 \rightarrow NO_2F + F$ is an elementary reaction. It is found that only a very small fraction of the collisions between reactant molecules lead to the formation of the products. Discuss in terms of molecular structure some of the factors that result in so many nonreactive collisions.

17.40   The activation energy of the reaction $NO(g) + Cl_2(g) \rightarrow NOCl(g) + Cl(g)$ is 84.9 kJ/mol, and the frequency factor is 3.98×10^9 liter mol^{-1} sec^{-1}. Find the specific rate constant at 500 K.

17.41   The specific rate constant for the reaction $C_4H_8(g) \rightarrow 2C_2H_4(g)$ is 6.07×10^{-8} sec^{-1} at 600 K, and the activation energy is 262 kJ/mol. Find the frequency factor.

17.42   The reaction $2N_2O_5 \rightarrow 2N_2O_4 + O_2(g)$ takes place around room temperature in solvents such as CCl_4. The specific rate constant at 293 K is found to be 2.35×10^{-4} sec^{-1}. At 303 K, it is found to be 9.15×10^{-4} sec^{-1}. Calculate the activation energy for the reaction.

17.43 The following data are obtained for the reaction
$H_2C_2O_4(g) \rightarrow CO_2(g) + HCOOH(g)$:

k (sec^{-1})	8.09×10^{-5}	1.87×10^{-4}	3.39×10^{-4}
T (K)	407.3	419.6	428.8

Calculate the activation energy of the reaction.

17.44 The activation energy of the reaction $NO_2(g) + CO(g) \rightarrow NO(g) + CO_2(g)$ is 132.0 kJ/mol, and the specific rate constant at 500 K is 2.02×10^{-6} liter mol^{-1} sec^{-1}. Find the specific rate constant at 550 K.

17.45 The specific rate constant of a reaction doubles when the temperature is raised from 298 K to 308 K. Calculate the activation energy.

*17.46 The rate of the reaction $2NO(g) + O_2(g) \rightarrow 2NO_2(g)$ decreases with increasing temperature.
(a) Account for this observation, assuming the reaction is an elementary process, by considering the sign and magnitude of the activation energy.
(b) Suggest an explanation in terms of electronic spins for the unusual activation energy.
(c) Suggest an explanation, based on ΔS, for the temperature effect.

17.47 Draw energy diagrams for the following reactions:
(a) a one-step nonspontaneous process, (b) a two-step spontaneous process in which the first step is rate determining, (c) a three-step nonspontaneous process in which the second step is rate determining and the third step is faster than the first step.

17.48 At 500 K the K for the reaction
$NO(g) + NO_2Cl(g) \rightleftharpoons NOCl(g) + NO_2(g)$ is 6.40×10^2, and the specific rate constant for the reaction between NO and NO_2Cl is 7.51×10^5 liter mol^{-1} sec^{-1}. Calculate the specific rate constant for the reaction between NOCl and NO_2 at this temperature.

17.49 The formation of phosgene, $COCl_2$, from chlorine and carbon monoxide is believed to proceed by the mechanism:

(1) $Cl(g) + CO(g) \underset{k_{-1}}{\overset{k_1}{\rightleftharpoons}} COCl(g)$

(2) $COCl(g) + Cl_2(g) \underset{k_{-2}}{\overset{k_2}{\rightleftharpoons}} COCl_2(g) + Cl(g)$

Show that this mechanism is consistent with the overall reaction $Cl_2(g) + CO(g) \rightleftharpoons COCl_2(g)$. Derive an expression for the K of this reaction in terms of the specific rate constants.

*17.50 Special techniques have been developed which allow the measurement of specific rate constants of very fast reactions. With these techniques, the k at 298 K of the reaction $H_3O^+ + OH^- \rightarrow 2H_2O$ has been found to be 1.4×10^{11} liter mol^{-1} sec^{-1}. Calculate the k for the reverse reaction. (*Hint:* Use $[H_2O] = 55.5M$ to correct K_w.)

17.51 At 400 K the K for the reaction $2SO_2(g) + O_2(g) \rightleftharpoons 2SO_3(g)$ is 1.02×10^{16} atm^{-1}, and the K for the reaction $2NO(g) + O_2(g) \rightleftharpoons 2NO_2(g)$ is 3.48×10^2 atm^{-1}. Calculate the K for the reaction $NO_2(g) + SO_2(g) \rightleftharpoons NO(g) + SO_3(g)$ at 400 K.

17.52 Under appropriate conditions in the mechanism shown on page 568 for the hydrolysis of nitramide, the second step is rate determining. Find the rate law for the reaction in terms of the specific rate constants of each step.

17.53 A material that is structurally similar to a substrate in a reaction catalyzed by an enzyme may act as an enzyme inhibitor, interfering markedly with the functioning of the enzyme. Explain this observation in terms of the lock and key analogy.

17.54 Suggest a reason why powdered charcoal rather than graphite is used in water and air purifiers.

17.55 Propose a mechanism for the decomposition of nitrous oxide to N_2 and O_2 catalyzed by platinum by analogy with the mechanism for the formation of NH_3. Predict the extent to which nitrous oxide can form from N_2 and O_2 at elevated temperatures in the presence of a hot platinum surface.

17.56 The chemisorption of $H_2(g)$ on a metal surface is found to liberate 50 kJ/mol. Estimate the strength of the metal-to-H bond.

18

COORDINATION CHEMISTRY

The study of coordination compounds is one of the most active and exciting areas of chemical research today. Coordination compounds have a broad range of interesting theoretical and practical features. In industry, they are used as catalysts in a number of processes, and the search for still better catalysts is being pressed. In biology, they play central roles in a number of life processes, and the study of their functions in living organisms is being pursued.

A coordination compound includes a metal atom surrounded by a set of other atoms or groups of atoms called **ligands.** To a large extent, the metal atom and its surrounding ligands act as a single chemical entity. In this chapter, we shall discuss the structure, the reactions, and the uses of coordination compounds.

18.1 STRUCTURE OF COORDINATION COMPOUNDS

The best way to examine most coordination compounds is to begin with the single entity represented by the metal atom and its coordinated ligands. This entity, called a coordination complex, may be either neutral or electrically charged. It is usually possible to assign a charge or oxidation number to the metal atom. This assignment is based on the nature of the ligands and the net charge on the complex. For example, the complex ion $[Co(NH_3)_6]^{3+}$ is built around cobalt in the $+3$ oxidation state. Since the ligands are neutral NH_3 molecules, the charge of the complex gives the oxidation state of the metal. The iron atom in $[Fe(CN)_6]^{4-}$ is in the $+2$ oxidation state, since each of the six ligands has a charge of -1. The net charge of -4 on the complex ion is the sum of -6 and $+2$. In the ion $[Fe(CN)_6]^{3-}$, the net charge of -3 on the complex ion reveals that the oxidation state of the iron is $+3$.

One of the puzzling features of coordination compounds to early investigators was the association of the metal atom with more groups than the number corresponding to what they called the "valence" of the metal. For example, it was believed that iron in the $+2$ oxidation state should have a valence of only two, as it does in a compound such as $FeCl_2$ or $Fe(OH)_2$. Therefore, the structure of a species such as $[Fe(CN)_6]^{4-}$ was a mystery.

The great historical figure in coordination chemistry is Alfred Werner (1866–1919), who was awarded the Nobel Prize in 1913. Werner provided much of the framework for understanding the structures and behavior of coordination compounds. His original coordination theory provided a basis for formulating the structures of coordination compounds, by proposing that the metal has two kinds of valences.

We can understand in modern terms the difference between his two kinds of valence by considering a coordination compound that includes the complex ion $[Co(NH_3)_6]^{3+}$, in which cobalt is in the $+3$ oxidation state. In a neutral compound this cation must be combined with anions of total charge -3, possibly three Cl^- ions. The formula of such a compound is written $[Co(NH_3)_6]Cl_3$. By convention, the complex is placed within brackets to emphasize that it acts as a single chemical unit. The structure of this compound is represented in Figure 18.1.

The six NH_3 ligands are closely associated with the metal, and are said to be in the *coordination sphere* of the metal. The three Cl^- anions are outside the coordination sphere and can be regarded as being less important. The *"ionic valence"* of the cobalt in this compound is simply its oxidation state; it is $+3$. This valence is satisfied by the three Cl^- anions that are necessary for electrical neutrality. The "coordination valence" of the metal is 6. This valence is satisfied by the six closely associated NH_3 ligands. In modern terms, the coordination valence of the metal is its *coordination number*.

The identity of the anions that satisfy the ionic valence of the metal and bring about electrical neutrality is not important. Only their charge matters. These

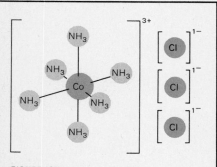

FIGURE 18.1
The coordination compound $[Co(NH_3)_6]Cl_3$ consists of the complex ion $[Co(NH_3)_6]^{3+}$ and three Cl^- anions. The three Cl^- ions are not as closely associated with the central cobalt atom as are the six NH_3 ligands.

anions are called *counter ions* when they are not in the coordination sphere of the metal.

The difference between the ligands in the coordination sphere and the counter ions often becomes apparent when the coordination compound is dissolved in water. The ligands tend to remain closely associated with the metal, while the counter ions dissociate. Thus, when $[Co(NH_3)_6]Cl_3$ is dissolved in water and a solution of silver nitrate is added, all the chloride is precipitated as $AgCl$. But when $[Co(NH_3)_5Cl]Cl_2$ is dissolved in water, only two-thirds of the chloride is precipitated as $AgCl$. In this compound one of the chlorides is in the coordination sphere of the metal and it does not dissociate in water.

There are many complexes in which some or all of the ligands in the coordination sphere are anions. These anionic ligands also satisfy some or all of the ionic valence of the metal. In such compounds the nature and number of the counter ions is determined simply by what is required for electrical neutrality. Thus the complex ion $[Co(NH_3)_5Cl]^{2+}$ of cobalt in the $+3$ oxidation state requires only two Cl^- counter ions for neutrality. The complex $[Co(NH_3)Cl_3]$ of cobalt in the $+3$ oxidation state is already neutral and requires no counter ions. Sometimes cationic counter ions are required to neutralize a complex ion with a net negative charge due to anionic ligands in the coordination sphere. An example is $K_2[PtCl_6]$, in which the platinum is in the $+4$ oxidation state, the complex ion bears a charge of -2, and two K^+ cations are the counter ions.

Our discussion of coordination chemistry will focus on the metal and the ligands in its coordination sphere. We shall be especially interested in the coordination number of the metal and how it affects the structure and behavior of the complex. We shall see that although the coordination number of a given metal may vary depending on its oxidation state and the nature of the ligands, most metals tend to have one, or at most a few, characteristic coordination numbers.

While all metals form coordination complexes, the most interesting—and often the most stable—complexes are those of the transition metals. The transition metals are found in groups IB and IIIB–VIII of the periodic table. A transition metal is defined as one that either has an incomplete d electronic subshell or readily forms a cation with an incomplete d electronic subshell. The d orbitals often play an important role in the behavior of transition metal complexes. The incompletely filled d orbitals can participate in the bonding of the ligands, allowing the formation of complexes that would not otherwise form. Transition metals can also achieve high oxidation states. The electrostatic attraction of highly charged metal cations with the negative charge of the ligands also favors the formation of complexes of transition metals.

Ligands

Many different molecules and ions can serve as ligands in coordination complexes. All are electron pair donors. Most ligands are anions or substances that contain nonbonding valence electron pairs. The common anionic ligands include the halide ions, F^-, Cl^-, Br^-, and I^-; conjugate bases of oxyacids, such as SO_4^{2-}, NO_2^-, NO_3^-, and OH^-; and a variety of other anions, such as S^{2-} and CN^-. The common neutral ligands include H_2O and other nonmetallic oxides in which oxygen, with its nonbonding valence electrons, is the electron donor. A nitrogen atom is an electron donor atom in many ligands; the simplest such ligand is NH_3. Two neutral ligands of biological importance are O_2 and N_2.

Ligands that have more than one donor atom in a single molecule can form complexes of special interest. One such molecule, which has two donor atoms, is ethylenediamine, whose structure is:

$$\begin{array}{cccc} H & H & H & H \\ | & | & | & | \\ :N\!-\!C\!-\!C\!-\!N: \\ | & | & | & | \\ H & H & H & H \end{array}$$

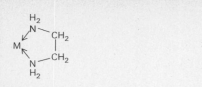

FIGURE 18.2
A chelate ring consisting of a metal atom and the bidentate ligand ethylenediamine. Each of the two nitrogen atoms in the ligand coordinates with the metal atom, forming a ring structure.

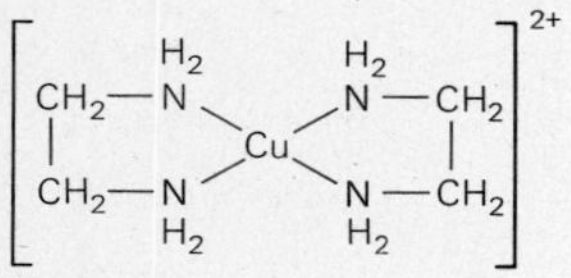

FIGURE 18.3
The complex ion $[Cu(en)_2]^{2+}$, consisting of a central copper(II) ion and two ethylenediamine ligands that form two chelate rings.

Each of the two nitrogen atoms in the molecule has a nonbonding valence electron pair, so each can coordinate with a metal. When both of these nitrogen atoms coordinate with the same metal atom, as shown in Figure 18.2, a ring structure called a *chelate* ring (from the Greek word for "claw") is formed. A ligand of this kind is called a *bidentate* (two-toothed) *ligand* or a *chelating agent*. In chemical formulas and equations, the abbreviation "en" is used to represent the ethylenediamine ligand. Thus, the complex ion with the structure shown in Figure 18.3 is written as $[Cu(en)_2]^{2+}$.

Complex ions with chelate rings often are unusually stable, as we can see from the equilibrium constant for the reaction

$$[Ni(NH_3)_6]^{2+} + 3en \rightleftharpoons [Ni(en)_3]^{2+} + 6NH_3$$

The complex ion on the right side of the equation has three chelate rings, while the complex ion on the left side has none. The equilibrium constant for the reaction is about 10^{10}, meaning that the equilibrium lies far on the side of the complex with the chelate rings. Both complex ions have six nitrogen atoms coordinated on a nickel atom, but the complex ion with the chelate rings is much more stable than the one with the NH_3 ligands.

This enhanced stability is known as the *chelate effect*. The large equilibrium constant for the formation of the complex with the chelate rings indicates that $\Delta G°$, the standard free energy change, is negative. The value of $\Delta G°$ is determined by the value of $\Delta H°$, the standard enthalpy change, and of $\Delta S°$, the standard entropy change. The total of the bond energies in each of the two complexes is about the same. Therefore, $\Delta H°$ for this reaction is not very large. It is $\Delta S°$ that is primarily responsible for the negative value of $\Delta G°$. The entropy change for the reaction is favorable from left to right because the number of particles increases from four to seven as the reaction goes from left to right. An increase in the number of particles means an increase in disorganization, and thus an increase in $\Delta S°$.

There are ligands that contain from three to six donor atoms in a single molecule. Some quadridentate ligands—that is, ligands with four donor atoms in a single molecule—play essential roles in the chemistry of living organisms. Each of these ligands has a structure that includes a large ring of atoms. A molecule containing such a large ring of atoms is called a macrocycle. Some organic macrocycles called porphyrins have rings that include four nitrogen atoms, which can all act as donors toward a single metal atom. Complexes with these macrocyclic ligands tend to be even more stable than complexes with simple chelate rings. Many important proteins include complexes between porphyrins and iron. One such coordination complex, heme, is found in many proteins. The heme-containing proteins include hemoglobin and myoglobin, which participate in biological oxygen transfer; cytochromes, which participate in biological energy transport; and enzymes such as catalase. We shall discuss biological complexes further in Section 18.6.

Geometry

The ligands in the coordination sphere around a metal atom are held rather rigidly in place. Because of this fixed orientation, we can describe the geometry of coordination complexes in the same way as we described the geometry of ordinary covalent compounds of nonmetals (Chapter 8). To describe the geometry of a given coordination complex, we specify points that correspond to the centers of the donor atoms that are complexed with the metal atom.

The geometry of metal complexes is conveniently related to the coordination number of the metal. There are ideal geometries associated with specific coordination numbers, and many complexes have geometries that are reasonably close to these ideals. However, the exact geometry is influenced by such factors as the nature of the metal and the ligand, and we find more than one geometry for some coordination numbers.

Complexes with metals whose coordination number is 2 are relatively uncommon. They are formed by the metal ions Cu^+, Ag^+, Au^+, and Hg^{2+}, each of which has a complete d electronic shell. The valence electronic configuration for Cu^+ is $3d^{10}$; for Ag^+, $4d^{10}$; for Au^+, $5d^{10}$; and for Hg^{2+}, $5d^{10}$. The complexes formed by these ions typically are linear—that is, the point at the center of each donor atom and the point at the center of the metal ion all lie on a straight line, as shown in Figure 18.4. Some typical complexes are $[CuCl_2]^-$. $[Ag(NH_3)_2]^+$, $[Au(CN)_2]^-$, and $HgCl_2$.

Complexes with a metal whose coordination number is 4 are more common. Two ideal geometries are possible when four ligands lie around a metal atom. One is the familiar tetrahedral geometry, encountered in covalent compounds of the nonmetals (Figure 8.7). In a coordination complex with this geometry, the points at the centers of the donor atoms of the four ligands lie at the corners of a regular tetrahedron. The center of the metal atom is at the center of the tetrahedron. Figure 18.5 shows the tetrahedral geometry of $[Zn(NH_3)_4]^{2+}$.

Not all complexes of metals with coordination number 4 have tetrahedral geometry. Some have square planar geometry. The points at the centers of the four donor atoms lie in a plane at the corners of a square. The center of the metal atom is in the same plane and at the center of the square. Figure 18.6 shows $[Pt(NH_3)_4]^{2+}$, which has square planar geometry.

Four-coordinate complexes of nontransition metals, such as $[AlCl_4]^-$, $[Zn(CN)_4]^{2-}$, and $SnBr_4$, almost always display tetrahedral geometry, sometimes in a distorted form. The geometry of four-coordinate metal complexes is influenced by several factors. Tetrahedral geometry minimizes repulsions between ligands. This geometry is especially favored when the ligands are relatively large and the metal is relatively small, as in $[MnO_4]^-$, VCl_4, $[FeCl_4]^-$, and $[NiBr_4]^{2-}$. But in some four-coordinate complexes with relatively small ligands, electronic factors may cause square planar geometry to be favored. Square planar geometry is most common for metals whose valence electronic configuration is d^8. Both the size of the ligands and the valence electronic configuration influence the geometry of the complex. Thus, complexes of Ni^{2+}, whose valence electronic configuration is $3d^8$, display square planar geometry only when the ligands are small. Square planar geometry is common for complexes of the larger metals, such as Pd^{2+} (valence electronic configuration $4d^8$), Pt^{2+} ($5d^8$), and Au^{3+} ($5d^8$).

The coordination number 6 is by far the most common for transition metal complexes. Some transition metal ions appear to form only six-coordinate complexes. These ions include Cr^{3+}, Co^{3+}, and Pt^{4+}. Aside from some rare exceptions, the geometry associated with the coordination number 6 is octahedral. The centers of the six donor atoms lie at the corners of an octahedron, either regular or asymmetrical (Figure 8.18). The center of the metal ion is at the center of the octahedron. Figure 18.7 shows the structure of $[Co(NH_3)_6]^{3+}$, which has regular octahedral geometry.

Because octahedral geometry is so common in transition metal complexes, you should know how to draw it.

The six corners of an octahedron lie at the ends of three lines of equal length that are mutually perpendicular, as Figure 18.8 shows. The six ligands are at the six corners of the octahedron. Because the lines are of equal length and mutually perpendicular, all six positions are equivalent. The three mutually perpendicular lines of Figure 18.8 are familiar to you as the x, y, and z coordinate axes in three dimensions.

There are coordination complexes with other coordination numbers, including 3, 5, and 7–12. But the structures of these complexes often are complicated and the complexes themselves are less often encountered, so we shall not discuss them.

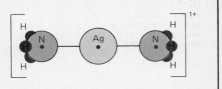

FIGURE 18.4
The complex ion $[Ag(NH_3)_2]^+$. The coordination number of the silver ion is 2, and the complex is linear, with the center of the metal ion and the centers of the two donor atoms of the ligands lying on a straight line. Linear geometry is typical of complexes of metals whose coordination number is 2.

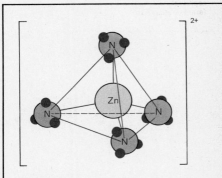

FIGURE 18.5
The coordination complex $[Zn(NH_3)_4]^{2+}$ has tetrahedral geometry. The four NH_3 ligands lie at the corners of a regular tetrahedron, with the metal atom at the center of the tetrahedron. Tetrahedral geometry is found in many complex ions of metals whose coordination number is 4.

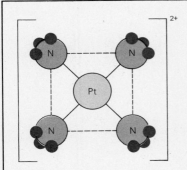

FIGURE 18.6
The coordination complex $[Pt(NH_3)_4]^{2+}$ has square planar geometry. The four ligands are at the corners of a square, with the Pt ion at the center of the square in the same plane as the ligands. Square planar geometry is found in some complex ions of metals whose coordination number is 4.

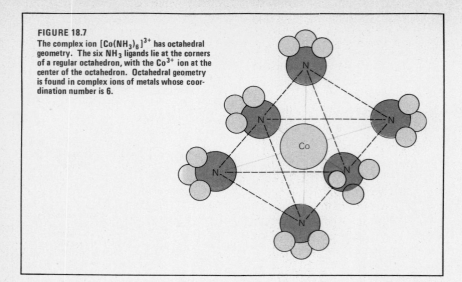

FIGURE 18.7
The complex ion $[Co(NH_3)_6]^{3+}$ has octahedral geometry. The six NH_3 ligands lie at the corners of a regular octahedron, with the Co^{3+} ion at the center of the octahedron. Octahedral geometry is found in complex ions of metals whose coordination number is 6.

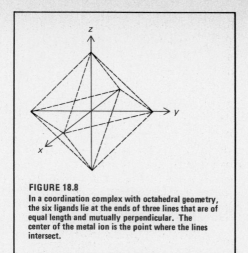

FIGURE 18.8
In a coordination complex with octahedral geometry, the six ligands lie at the ends of three lines that are of equal length and mutually perpendicular. The center of the metal ion is the point where the lines intersect.

18.2   ISOMERISM IN COORDINATION COMPOUNDS

Two chemical compounds that have the same molecular formula but different molecular structures are called **isomers**. We find several kinds of isomerism in coordination compounds. There is, for example, the isomerism that occurs when one of the ligands in the complex ion can be a counter ion and one of the counter ions can be a ligand. The molecular formula of the coordination compound does not change, but its structure does, in this *ionization isomerism*. The coordination compounds $[Co(NH_3)_5Cl]SO_4$ and $[Co(NH_3)_5SO_4]Cl$ are ionization isomers. They have the same composition but different structures. Both of these compounds are complexes of cobalt in the $+3$ oxidation state. In the first one, the Cl^- is the ligand and the SO_4^{2-} is the counter ion. In the second, the SO_4^{2-} is the ligand and the Cl^- is the counter ion.

Closely related to ionization isomerism is *hydrate isomerism,* which involves the water of hydration that is often included in ionic crystals (Chapter 11, page 337). Coordination complexes often include water molecules as ligands. Crystalline coordination compounds may also include water of hydration. Hydrate isomers have the same total number of water molecules, but these molecules are bonded differently in the isomers. Three hydrate isomers are $[Cr(H_2O)_6]Cl_3$, which is gray-blue; $[Cr(H_2O)_5Cl]Cl_2 \cdot H_2O$, which is light green; and $[Cr(H_2O)_4Cl_2]Cl \cdot 2H_2O$, which is dark green.

We find a different kind of isomerism in *ambidentate ligands*. An ambidentate ligand can complex with a metal atom in two or more ways. One well-known ambidentate ligand is NO_2^-. It can form nitrito complexes, in which one of the oxygen atoms is the donor:

$$\overset{..}{\underset{..}{O}} \quad \overset{..}{O} \longrightarrow M$$
$$\underset{..}{N}$$

or nitro complexes, in which the nitrogen atom is the donor:

$$\overset{..}{O}$$
$$N \longrightarrow M$$
$$\underset{..}{O}$$

Two compounds that differ only in the way that the NO_2^- ligand is attached are $[(NH_3)_5CoONO]Cl_2$, which is the red nitro complex, and $[(NH_3)_5CoNO_2]Cl_2$, which is the yellow nitrito complex. These compounds, and others that have the same kind of isomerism, are called *linkage isomers*.

Cyanide, $C\equiv N^-$, is another ambidentate ligand that forms linkage isomers. It is usually the carbon atom that coordinates with the metal, but there are complexes in which the nitrogen atom of the cyanide may instead coordinate with the metal. The thiocyanate anion, SCN^-, also gives rise to linkage isomers. Either the sulfur atom or the nitrogen atom can be the donor atom. There are some ligands, for example cyanate ion, NCO^-, that are known to be ambidentate although linkage isomers have not yet been isolated. The cyanate ion is known to use either the N atom or the O atom for bonding in different compounds, but no linkage isomer of this ligand is known.

An especially interesting kind of isomerism, found in many coordination compounds, is called **stereoisomerism.** Two compounds are said to be stereoisomers when they differ only in the arrangement of their atoms in space. Thus, two stereosiomers have both the same molecular formula and the same bonding and yet are different.

The most important type of stereoisomerism found in coordination complexes is called *geometric isomerism*. Two coordination complexes that are geometric isomers have the same overall geometry, composition, and bonding. They differ only in the spatial arrangement of the ligands around the metal ion. Whether geometric isomerism is possible for a coordination complex depends on the geometry of the complex and the nature of the ligands.

For example, geometric isomerism can occur in four-coordinate complexes with square planar geometry. The complex $[Pt(NH_3)_2Cl_2]$ of platinum in the $+2$ oxidation state, which has square planar geometry, exists as two geometric isomers. Each isomer has a central platinum ion surrounded by the same four ligands, which lie at the corners of a square. Yet the isomers are two different compounds with different properties. The differences are due to the spatial arrangement of the ligands. To see how this difference can come about, let us consider the various ways in which the two pairs of ligands can be distributed on the corners of a square.

Suppose we put a chloride ligand at one corner of an empty square, as shown in Figure 18.9(a). The second chloride can go either on the corner on the same edge of the square or on the corner diagonally opposite to the first chloride (Figure 18.9(b)). Because these two positions are not equivalent, two ligand arrangements are possible (Figures 18.9(c) and 18.9(d)). Each of these two arrangements corresponds to a different geometric isomer. The isomer with the two like ligands on the same edge of the square (Figure 18.9(c)) is called the *cis* isomer; the isomer with the two like ligands on opposite corners of the square (Figure 18.9(d)) is called the *trans* isomer. These isomers are different substances with different properties.

Geometric isomerism is possible only when no more than two positions of the square are occupied by identical ligands. If there are three identical ligands, as in the complex $[Pt(NH_3)_3Cl]^+$, all the arrangements are equivalent and geometric isomerism is impossible. Geometric isomerism cannot occur in a four-coordinate complex with tetrahedral geometry. After the first ligand is put at a corner of the tetrahedron, all three remaining positions are equivalent. No matter where the next ligand is placed, isomerism does not occur.

Geometric isomerism can occur in six-coordinate octahedral geometry. Consider the ion $[Pt(NH_3)_4Cl_2]^{2+}$, a complex of platinum in the $+4$ state. We first put a chloride ligand at one corner of the octahedron (Figure 18.10(a)). We now have a choice of nonequivalent positions for the second chloride ligand (Figure 18.10(b)). It can be placed on the same axis as the first chloride ligand, which puts it at the opposite corner of the octahedron, or it can be placed on one of the

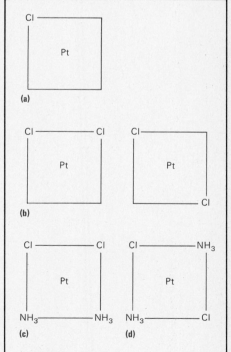

FIGURE 18.9
The geometric isomers of the complex $[Pt(NH_3)_2Cl_2]$, which has square planar geometry. With a chloride ligand in one corner of the square, the second chloride ligand can occupy either the opposite corner or an adjacent corner. Because the two positions are not equivalent, the complex has two isomers whose formulas are the same but whose structures are different.

PLATINUM AND CANCER

The *cis* isomer of $[Pt(NH_3)_2Cl_2]$ is a potent anticancer drug. The *trans* isomer is not. Many research groups have been working to explain why a relatively slight difference in structure is associated with a major difference in biological activity.

It is generally believed that the anticancer activity is due to the interaction of the compound with deoxyribonucleic acid (DNA), the genetic material of the cell. Cancer cells divide continually, eventually overwhelming the body by their sheer volume. In order for the cells to divide, the DNA of each cell must replicate itself. The DNA consists of two long, intertwined strands. In replication, the strands first separate and then form complementary strands. The *cis* isomer stops DNA synthesis; the *trans* isomer does not.

The original theory explained the activity of *cis*-$[Pt(NH_3)_2Cl_2]$ by saying that the complex is carried in the bloodstream to the tumor site, where hydrolysis takes place. The theory said that the hydrolysis product formed a link between the two strands of the DNA, preventing replication. According to this theory, the *trans* isomer was ineffective because it underwent hydrolysis at a faster rate than the *cis* isomer and so never reached the cancer. But further studies showed that the differences between the hydrolysis rates of the two isomers are too small to account for the major differences in their anticancer activity.

More recent work has shown that both isomers bind to the DNA, but in different ways. The *cis* isomer binds in a way that causes severe distortion of the DNA molecule while the *trans* isomer does not. Both isomers bind to DNA because their chloride ligands can be replaced by nitrogen-containing ligands that are part of the DNA molecule. It is believed that the inhibition of DNA synthesis by the *cis* isomer is due to replacement of both chloride ligands by ligands that are close to each other in one strand of the DNA molecule. For unknown reasons, only this mode of binding seems to result in marked inhibition of DNA synthesis. The *cis* isomer can form such bonds because the distance between its chloride ligands is about the same as the distance between binding sites on the DNA strand. The chloride ligands on the *trans* isomer are further apart, so it cannot form such a bifunctional linkage.

Early use of the compound to treat human cancer was highly promising. However, platinum compounds tend to cause severe kidney damage because they are reduced to metallic platinum in the kidney, so only limited doses could be given. It has been found that the damage can be reduced and the dosage increased substantially by giving a diuretic which speeds the passage of the compound through the kidney. Used in combination with other anticancer drugs, this platinum complex produced a 95% response rate in some groups of cancer patients. The search for other effective anticancer drugs is going on among coordination complexes of other group VIII metals such as palladium, ruthenium, and rhodium.

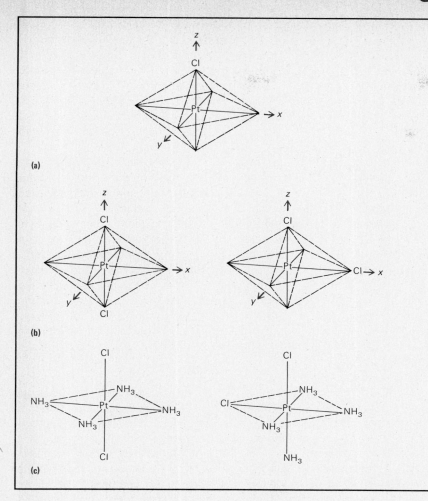

(a)

(b)

(c)

FIGURE 18.10
The geometric isomers of the complex $[Pt(NH_3)_4 Cl_2]^{2+}$, which has octahedral geometry. With a chloride ligand at one corner of the octahedron, the second chloride ligand can occupy either the opposite end of the same axis, creating a *trans* isomer, or one of the other two axes, creating a *cis* isomer.

other axes, which puts it at one of the four equivalent adjacent corners of the octahedron. Thus, two different ligand arrangements, and therefore two geometric isomers, are possible (Figures 18.10(c) and 18.10(d)). The isomer that has the two chloride ligands at opposite corners (Figure 18.10(c)) is called the *trans* isomer, while the isomer that has the two chloride ligands at adjacent corners (Figure 18.10(d)) is called the *cis* isomer. The *trans* isomer is green and the cis isomer is pink.

Two geometric isomers exist for the octahedral complex whose formula is $[Pt(NH_3)_3Cl_3]^+$ (Figure 18.11). One isomer has one chloride ligand on each of its three axes. The other isomer has two chloride ligands on the same axis, with the third chloride ligand on a different axis. More than two geometric isomers can exist when a complex has more than two different ligands. However, geometric isomerism in a six-coordinate octahedral complex is possible only when there are no more than four identical ligands.

Optical isomerism is another kind of stereoisomerism that is important in coordination compounds. It is also of great interest in organic chemistry and biochemistry.

Two optical isomers differ only in a property called *chirality* or *handedness*. There are many chiral objects around us. One of them is the human hand. You can see that your hands are not identical any time you try to put your right hand into a left-hand glove. If you hold your right hand before a mirror, the reflection you see appears to be a left hand. This is true of any chiral object: The object itself and its mirror image are different. One is left-handed, the other is right-handed,

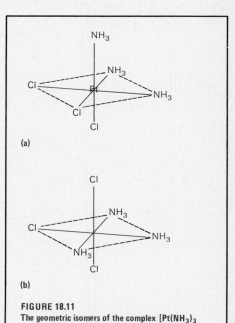

(a)

(b)

FIGURE 18.11
The geometric isomers of the complex $[Pt(NH_3)_3 Cl_3]^+$, which has octahedral geometry. The isomer in (a) has one chloride ligand on each of its three axes. The isomer in (b) has two chloride ligands on one axis, with the third on another axis.

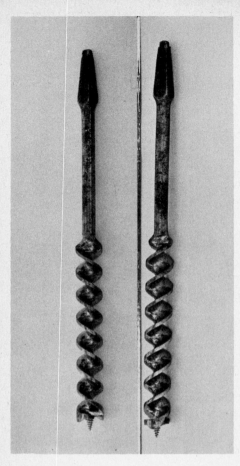

A drilling bit and its mirror image. The bit is a chiral object, as we can see by comparing it with the mirror image. The bit has a right-hand spiral, while the image has a left-hand spiral, and the two cannot be super-imposed. (*Beckwith Studios*)

and they cannot be superimposed. Wood screws are also chiral objects. A screw may have a thread that goes to the left or to the right; the screwdriver is turned clockwise or counterclockwise, depending on the chirality of the screw.

In any example of chirality, we have two different mirror-image objects, one left-handed and one right-handed. Chirality occurs on the molecular level, where we can have two isomers that are mirror images of each other, a left-handed molecule and a right-handed molecule. These are called optical isomers. Most simple molecules are not chiral and do not display optical isomerism, but there are many complex molecules that exist as optical isomers.

While two optical isomers generally have virtually identical physical and chemical properties, differences manifest themselves when the isomers interact with something that is also chiral. The usual method for distinguishing between two optical isomers uses plane-polarized light, which is chiral.

The most direct way to determine whether a specific coordination complex is chiral is to draw a picture of the complex and of its mirror reflection and compare the two. In Figure 18.12, we see such drawings for one of the geometrical isomers of a hypothetical complex, $[Co(NH_3)_2Cl_2Br_2]^-$. The complex and its mirror image are superimposable, and therefore are identical, so this complex does not display optical isomerism. Figure 18.13 shows a drawing of another geometric isomer of this complex and its mirror image. The two complexes are not superimposable. Therefore, they are optical isomers. Any time we recognize that a coordination complex has chirality, we know that there are two optical isomers of that complex.

A second way of determining whether a specific coordination complex is chiral is based on the relationship between chirality and symmetry. Most objects that are symmetrical can be superimposed on their mirror images and therefore do not display optical isomerism. Asymmetrical objects cannot be superimposed on their mirror images and therefore can display optical isomerism. Often, it is easier to determine whether a molecule is symmetrical than to draw its mirror image and test for superimposability. Some simple rules can be stated about the symmetry of complex ions of monodentate ligands. Linear complexes and square planar complexes never display optical activity. A tetrahedral complex displays optical isomerism only if it has four different ligands. Octahedral complexes of

FIGURE 18.12
One geometric isomer of the hypothetical complex $[Co(NH_3)_2Cl_2Br_2]^-$ and its mirror image. The two complexes can be superimposed, so they are identical.

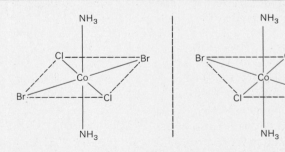

FIGURE 18.13
Another geometric isomer of the complex $[Co(NH_3)_2Cl_2Br_2]^-$ and its mirror image. Because the two complexes cannot be superimposed, they are optical isomers.

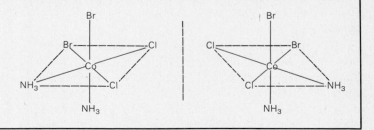

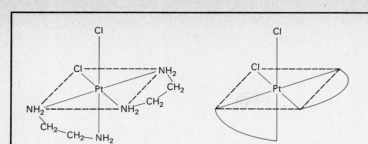

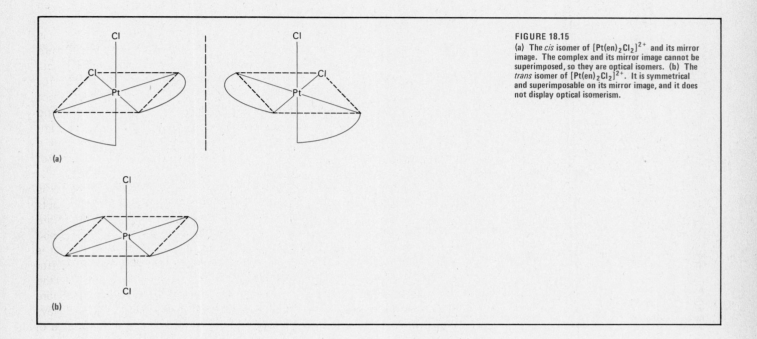

FIGURE 18.14
(a) The complex $[Pt(en)_2Cl_2]^{2+}$, which is formed by
replacing the four NH_3 ligands of the *cis* isomer of
the complex shown in Figure 18.10 (c) by two en
ligands. (b) A shorthand method of representing the
structure of a complex with en ligands.

FIGURE 18.15
(a) The *cis* isomer of $[Pt(en)_2Cl_2]^{2+}$ and its mirror
image. The complex and its mirror image cannot be
superimposed, so they are optical isomers. (b) The
trans isomer of $[Pt(en)_2Cl_2]^{2+}$. It is symmetrical
and superimposable on its mirror image, and it does
not display optical isomerism.

monodentate ligands are capable of optical isomerism only when they have at
least three different ligands.

Optical isomerism also occurs in octahedral complexes that include bidentate ligands. The *cis* isomer of the complex shown in Figure 18.10 is superimposable on its mirror image. If we replace its four NH_3 ligands with two en ligands,
we have the complex shown in Figure 18.14(a). Figure 18.14(b) shows a
shorthand representation of the structure of this complex, $[Pt(en)_2Cl_2]^{2+}$. Figure
18.15(a) shows that this complex is not superimposable on its mirror image.
Therefore, there are two optical isomers of this complex. The formation of two
chelate rings resulted in a loss of symmetry in the complex. But when two en
ligands are substituted for four NH_3 ligands in the *trans* isomer shown in Figure
18.10(c), the resulting complex, shown in Figure 18.15(b), is still symmetrical.
The new complex is superimposable on its mirror image and does not exhibit
optical isomerism.

We find an extreme example of the effect of chelate rings in the complex
$[Co(en)_3]^{3+}$. Even though there is only one ligand in the complex, and the complex
has only one geometric isomer, there are two optical isomers. Figure 18.16
shows the complex and its mirror image. You will note that the spiral arrangement
of the three ligands in the two optical isomers is analogous to the spiral arrangement of the grooves in right-handed and left-handed screws.

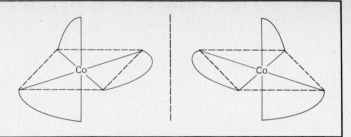

18.3 BONDING IN COORDINATION COMPOUNDS

We cannot use the terms developed in Section 7.3, page 196, to describe the bonding of nonmetal compounds for all aspects of the bonding in coordination complexes. Such a simple picture does not allow us to understand the geometry, the colors, and the magnetic properties of coordination compounds. Several different methods of describing the bonding in coordination compounds have been developed.

Valence Bond Theory

We start with the observation that the ligands in a coordination compound are electron donors. Both the electrons in the metal-ligand bond come from the ligand. This kind of bond was discussed in Section 7.3. It is called a *dative bond* or a *coordinate covalent bond;* the latter name comes from the existence of such bonds in coordination compounds.

In the traditional Lewis formulation, the existence of a coordinate covalent bond implies the existence of formal charges. In each of these bonds, there is a formal charge of $+1$ on the donor atom and a formal charge of -1 on the acceptor atom. In a complex, the metal ion is the acceptor. The complex $[Co(NH_3)_6]^{3+}$, for example, has six ligand-metal bonds. There are six formal positive charges, one on each ligand, and therefore six formal negative charges on the metal ion. Since the metal has an ionic charge of $+3$, the formal charge of -6 results in a net charge of -3 on the metal. However, such a net negative charge is contrary to chemical common sense, since metals rarely bear negative charges.

The metal does not bear a negative charge because the coordinate covalent bond has ionic character. The bonding electron pair is not shared equally between the ligand and the metal. Because the donor atom of the ligand usually is highly electronegative, the bonding electron pair is more closely associated with the ligand. Because of their ionic character, these bonds are polar. They are partially negative on the donor atom and partially positive on the metal atom. The polarity of these bonds is opposite to the polarity of the formal charges. The ionic bonding thus helps offset the formal charges.

A more detailed description of the bonding in coordination compounds, called the *valence bond* (VB) theory, has been developed by Linus Pauling (1901–   ). The VB theory is based on the classical principles of bonding theory. While it has several shortcomings, the VB theory does give valuable insights into the bonding in coordination compounds.

The theory assumes that the atomic orbitals of the metal atom form a number of hybrid orbitals equal to the coordination number of the metal. The hybrid orbitals can be used to form σ bonds with the ligands. Each hybrid orbital overlaps with an orbital of a ligand and holds the electron pair donated by the ligand, forming a coordinate covalent bond. The valence electrons of the metal atom are distributed among the remaining unhybridized orbitals of the metal in the standard way, consistent with the Aufbau principle.

The hybrid orbitals used to describe the bonding in a specific complex are selected primarily on the basis of the observed geometry and magnetic properties of the complex. Each of the ideal geometries observed in complexes corresponds to a type of hybridization. A six-coordinate octahedral geometry requires six symmetrically placed hybrid orbitals that can be formed from s, p, and d orbitals. Octahedral geometry can be accommodated by a hybridization of d^2sp^3. The six component atomic orbitals—one s orbital, three p orbitals, and two d orbitals— overlap to form six equivalent hybrid orbitals.

Tetrahedral geometry is consistent with four sp^3 hybrid orbitals, as we mentioned in our discussion of nonmetal compounds. The hybridization for square planar geometry calls for four dsp^2 orbitals, formed from one s orbital, two p orbitals, and one d orbital. The two kinds of four-coordinate geometry correspond to different types of hybridization. Therefore, we usually cannot formulate a valence bond picture of a four-coordinate transition metal complex without knowing its geometry. This difficulty is a major weakness of the valence bond theory.

Valence bond theory has other serious weaknesses. For example, it is usually difficult and sometimes impossible to account for the colors and many of the chemical properties of complexes with valence bond theory. While the valence bond theory was the best that was available for many years, it has recently been replaced by several new theories. In order of increasing complexity, these are the crystal field theory, the ligand field theory, and the molecular orbital theory. We shall discuss only the crystal field theory.

Crystal Field Theory

Crystal field theory differs from valence bond theory in its treatment of the ligand-metal bond. In valence bond theory, the emphasis is on the covalent character of the bond. Valence bond theory acknowledges that the bond is partially ionic, but it does not deal directly with the effects of the ionic character. The crystal field theory takes just the opposite approach. It treats the bond as completely ionic, omitting consideration of the bond's covalent character.

Crystal field theory assumes that the only interaction between the metal atom and the ligand is electrostatic, and it treats each ligand as a point of negative charge. The arrangement of the ligands around the metal atom that minimizes the repulsions between these negative point charges can be calculated by the equations of classical electrostatics. This concept is essentially the same approach as we used for compounds of nonmetals in Section 8.2, page 237, and the results are the same. Octahedral geometry is expected for a six-coordinate complex, tetrahedral geometry for a four-coordinate complex, and linear geometry for a two-coordinate complex.

But crystal field theory goes one important step further than a simple minimization of the repulsions between the negative ligands. The theory also considers the effect of the ligands on the relative energy of the d orbitals of the central metal atom. Since the six-coordinate octahedral geometry is the most common for transition metal complexes, we shall begin by discussing crystal field theory for this geometry.

Octahedral Complexes

When an atom of a transition metal is put into a spherical negative electrical field, the total energy of the five d orbitals of the atom increases. The increase is caused by electrostatic repulsions between the electrons in the d orbitals and the negative field.

We get a slightly different result when we study the d orbitals of a transition metal atom in a complex ion. Once again, the atom is in a negative electrical field, this one created by the negatively charged ligands. But in a complex with octahedral geometry, the individual d orbitals are affected differently by the electric field of the ligands, as we can see in Figure 18.17. The six ligands are on the x, y, and z axes (Figure 18.17(a)). The d orbitals can be oriented with respect to the same axes (Figure 18.17(b)). You can see that the d_{z^2} and $d_{x^2-y^2}$ orbitals point directly at the ligands, while the d_{xy}, d_{xz}, and d_{yz} orbitals have their lobes of maximum electron density between the ligands. The repulsion between the ligands and the two d orbitals that lie on the coordinate axes is greater than the repulsion between the ligands and the other three d orbitals. Thus, the energy of the d_{z^2} and $d_{x^2-y^2}$ orbitals is higher in the octahedral field than it would be in a spherical electric field of the same strength. But it can be shown that the total energy of the d orbitals is the same in the octahedral field as in the spherical field. Therefore, the energy of the d_{xy}, d_{xz}, and d_{yz} orbitals must be lower in the octahedral field than in the spherical field, as Figure 18.18 shows.

In an octahedral complex, therefore, the d orbitals of the metal atom are divided into two groups. There are two orbitals with relatively high energy and three orbitals with relatively low energy. Because the octahedron is symmetrical, the three low-energy orbitals all have the same energy, and the two high-energy orbitals have the same energy. Since the total energy of the five d orbitals in the octahedral field equals their total energy in a spherical field, the total increase in the energy of the d_{z^2} and $d_{x^2-y^2}$ orbitals must equal the total decrease in the energy of the d_{xy}, d_{xz}, and d_{yz} orbitals. The difference in energy between the high-level and the low-level orbitals is called the crystal field splitting energy, whose symbol is Δ_o.

The energy of each of the two high-energy orbitals is increased by $\frac{3}{5}\Delta_o$ above their energy in a spherical field, while the energy of each of the three low-energy orbitals is decreased by $\frac{2}{5}\Delta_o$ below their energy in a spherical field. The total increase is equal to the total decrease, so

$$(2)(\tfrac{3}{5}\Delta_o) = (3)(\tfrac{2}{5}\Delta_o)$$

The splitting of the d orbitals predicted by the crystal field theory not only helps explain a number of properties of complexes but also simplifies the description of the electronic structure of many complexes. All that is necessary is to assign the valence electrons of the metal cation to the d orbitals of the metal; the electron pairs of the ligands need not be considered.

The Aufbau principle can be applied in the usual way to formulate some electronic configurations, as Figure 18.19 shows. Thus, for a complex of a cation whose valence electronic configuration is d^1, such as Ti^{3+} ($3d^1$), V^{4+} ($3d^1$), or

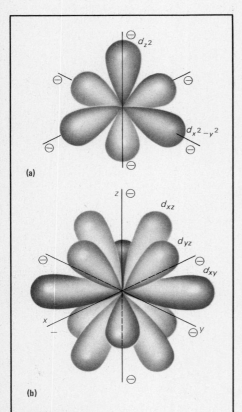

(a)

(b)

FIGURE 18.17
The effect of the octahedral field of a complex ion on the energies of the orbitals of the metal ion. Two of the d orbitals are on the x, y, and z axes while three of them are between the axes. The repulsion between the ligands and the two orbitals on the axes is greater, so these orbitals have higher energy than the other three orbitals.

FIGURE 18.18
When an isolated atom (a) is placed in a spherical electrical field, the energy of its orbitals increases equally (b). When the same atom is placed in the octahedral field of a complex ion, the energy increase is the same, but is distributed unequally among its orbitals (c). Three of the orbitals have lower energies than the other two.

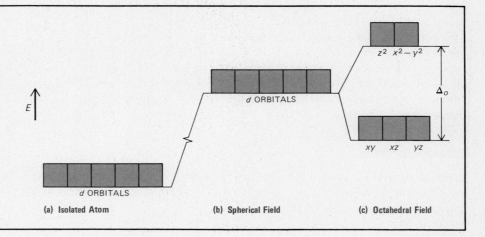

E

d ORBITALS

z^2 x^2-y^2

Δ_o

xy xz yz

d ORBITALS

(a) Isolated Atom   **(b) Spherical Field**   **(c) Octahedral Field**

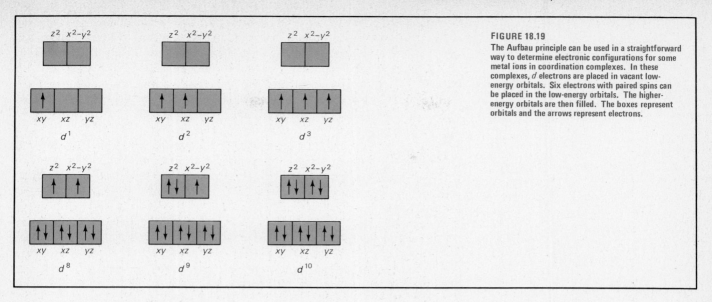

FIGURE 18.19
The Aufbau principle can be used in a straightforward way to determine electronic configurations for some metal ions in coordination complexes. In these complexes, d electrons are placed in vacant low-energy orbitals. Six electrons with paired spins can be placed in the low-energy orbitals. The higher-energy orbitals are then filled. The boxes represent orbitals and the arrows represent electrons.

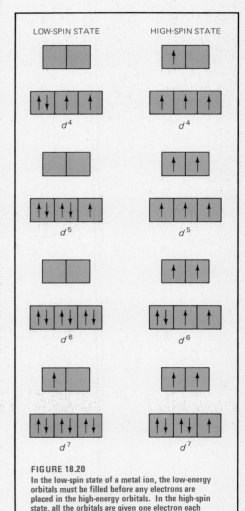

FIGURE 18.20
In the low-spin state of a metal ion, the low-energy orbitals must be filled before any electrons are placed in the high-energy orbitals. In the high-spin state, all the orbitals are given one electron each before additional electrons are placed in the low-energy orbitals.

Mo^{5+} ($4d^1$), the single d electron is placed in any one of the three low-energy d orbitals. For a complex of a cation with valence electronic configuration d^2, such as V^{3+} ($3d^2$), Nb^{3+} ($4d^2$), or W^{4+} ($5d^2$), the two d electrons are placed in two of the low-energy orbitals; in accordance with Hund's rules, their spins are parallel. Similarly, the three d electrons in d^3 cations such as V^{2+} ($3d^3$) or Cr^{3+} ($3d^3$) are in the three low-energy orbitals and have parallel spins.

The Aufbau principle can also be applied in the usual way to formulate the electronic structure of complexes of cations with electronic configuration d^8, such as Ni^{2+} ($3d^8$) or Au^{3+} ($5d^8$). As Figure 18.19 shows, the three low-energy orbitals are filled by three pairs of electrons, while the remaining two electrons are in the two high-energy orbitals and have parallel spin. Figure 18.19 also shows the electronic configurations of complexes of d^9 cations such as Cu^{2+} and d^{10} cations such as Zn^{2+}. Again, the Aufbau principle is used. With the exception of atoms whose configuration is d^{10}, octahedral complexes of cations with any of the other electronic configurations shown in Figure 18.19 have one or more unpaired electrons and thus are expected to be paramagnetic. This prediction is confirmed by experimental observation.

The electronic structure of complexes of metal cations whose electronic configuration is d^4, d^5, d^6, or d^7 cannot be formulated solely by applying the Aufbau principle. More information is needed because Δ_o, the energy difference between the high-energy and low-energy groups of orbitals, is relatively very small. In fact, Δ_o is much smaller than the energy difference between bonding and nonbonding orbitals in most molecules, or between electronic subshells in most atoms. Because the energy difference between orbitals is so small, we often have a choice of orbitals in which an electron can be placed.

Let us examine the electronic configuration of a d^4 cation in an octahedral complex. The first three d electrons are placed in the three low-energy orbitals, with parallel spins. Where should the fourth electron be placed? According to the Aufbau principle, we should place it in the unfilled orbital with the lowest energy, in this case one of the three low-energy orbitals. This placement would create the electronic configuration, shown in Figure 18.20, which has only two unpaired electrons and is called a *low-spin state*. But it is also possible to put the fourth electron in one of the two high-energy orbitals, as is also shown in Figure 18.20, keeping its spin parallel to that of the other three electrons. This arrangement, called the *high-spin state,* has four unpaired electrons.

As Figure 18.20 shows, we have the same choice in all the electronic configurations from d^4 to d^7. There is a low-spin state in which the relative energy of the three low-energy orbitals is dominant. These orbitals are filled before any electrons are placed in the high-energy orbitals. And there is a high-spin state, in which the dominant effect is to have the maximum number of electrons with parallel spins. In the high-spin state, electrons are placed in high-energy orbitals as soon as the low-energy orbitals are half-filled. The other low-energy orbitals are not filled until both the high-energy orbitals are half-filled.

The electronic configuration of any given complex is the one that is most stable. For each spin state, relative stability is a balance of two opposing effects. The low-spin state has many of its electrons in low-energy orbitals, which is a stabilizing effect. But many of its electrons have paired spins, which is a destabilizing effect. The high-energy spin state is stable because more of its electrons have unpaired spins but is unstable because one or two of its electrons are in high-energy orbitals. The relative stability of high-spin and low-spin states is determined primarily by the magnitude of Δ_o. If Δ_o is relatively large, the low-spin state is more stable because the difference in energy between the high-energy and low-energy orbitals is large. If Δ_o is small, the energy difference between high-energy and low-energy orbitals is small, and the high-spin state is more stable.

For any given metal cation, the magnitude of Δ_o depends on the nature of the ligands. The crystal field theory is not very helpful in determining the nature of the ligands, since it treats ligands only as point charges. But the effect of different ligands on the magnitude of Δ_o can be determined experimentally. Ligands that cause a large Δ_o are called *strong field ligands,* while those that cause a small Δ_o are called *weak field ligands*. Some common ligands can be listed in ascending order of Δ_o. The following list is a *spectrochemical series,* because it is based primarily on the study of the absorption of light by complexes.

strong field ligands		weak field ligands

$$CO > CN^- > NO_2^- > en > NH_3 > H_2O > OH^- > F^- > Cl^- > Br^- > I^-$$

Ligands at the left of the series are always strong field ligands, while those at the right are always weak field ligands. Those in the middle can be either strong field or weak field ligands, depending on the metal with which they are complexed. You should note that the order of ligands in the spectrochemical series cannot be predicted by electrostatic arguments alone. The order of the ligands is based on experiment and can only be understood by also considering the details of the covalent bonding between the ligands and the metal.

Magnetism

The behavior of a substance when it is placed in an external magnetic field can tell us a great deal about the electronic configuration of the substance (Section 7.6, page 217). A substance is *diamagnetic* (very weakly repelled by an external magnetic field) when all of its electrons have paired spins. A substance is *paramagnetic* (more strongly attracted by an external magnetic field) when it has one or more electrons with unpaired, or parallel, spins. All substances with an odd number of electrons are paramagnetic, but not all substances with an even number of electrons are diamagnetic.

Using the spectrochemical series, it is possible to understand, and in many cases even to predict, whether a coordination compound with an even number of electrons is diamagnetic or paramagnetic. Using the principle that strong field ligands cause low-spin states to be more stable and that weak field ligands cause high-spin states to be more stable, we can determine whether all the electronic spins are paired. For example, the low-spin state of d^6 ions such as Co^{3+}, Fe^{2+}, and Pt^{4+} is diamagnetic. In each of these ions, all six d electrons are paired in the three low-energy orbitals. The high-spin states of these same ions have four unpaired electrons and thus are paramagnetic. We expect the complexes of these ions with strong field ligands to be diamagnetic and complexes with weak field ligands to be paramagnetic. The experimental procedure shown in Figure 7.23 verifies this prediction.

A case in point is the complexes of Co^{3+}. The complex ion $[Co(NH_3)_6]^{3+}$ is diamagnetic, and the complex ion $[CoF_6]^{3-}$ is paramagnetic. The spectrochemical series shows that NH_3 is a relatively strong field ligand which causes a large Δ_o and favors low-spin states. The F^- ligand is shown to be a weak field ligand which favors the high-spin state. You can see by the magnetic properties of the two complex ions that the low-spin state, in which all the electron spins are paired, is favored in the complex of Co^{3+} with NH_3, while the high-spin state, in which four electrons have parallel spins, is favored in the complex of Co^{3+} with F^-.

Similarly, the complex $[Fe(CN)_6]^{4-}$ is diamagnetic, because the strong field ligand CN^- favors the low-spin state of the Fe^{2+} ion, whose valence electronic configuration is $3d^6$. But the complex $[Fe(NH_3)_6]^{2+}$ is paramagnetic. In complexes of Co^{3+}, NH_3 is a strong field ligand. But in complexes of Fe^{2+}, the NH_3 ligand behaves as a weak field ligand and favors the high-spin state. Some ligands can be either strong field or weak field, depending on the metal cation with which they complex.

High-spin and low-spin states can be detected experimentally by magnetic measurements in ions with other valence electronic configurations. While both the high-spin and low-spin states of atoms with valence electronic configuration d^4, d^5, and d^7 are paramagnetic, there are different numbers of unpaired electrons in the two states. It is possible to distinguish between high-spin and low-spin states for complexes of cations with these electronic configurations by determining the number of unpaired electrons. In general, these cations need stronger ligands—those further to the left in the spectrochemical series—to achieve low-spin states than do atoms whose configuration is d^6.

Colors

The splitting of d orbitals described in crystal field theory also helps account for some of the optical properties of coordination compounds. Color is one of those properties. Most pure chemical substances are colorless, which means that they do not absorb visible light. They usually absorb radiation in the ultraviolet region, from 100 nm to 400 nm. Most complexes of transition metals are colored, which means that they absorb some wavelengths of electromagnetic radiation in the visible region, from 400 nm to 750 nm.

We discussed the way in which atoms or molecules absorb radiation in Section 5.7. Absorbed radiation excites an electron from its ground state orbital to a higher-energy orbital that is either vacant or half-filled. The wavelength λ of the absorbed radiation is inversely proportional to ΔE, the difference in energy between the ground state orbital and the excited state orbital:

$$\lambda = \frac{hc}{\Delta E}$$

where h is Planck's constant and c is the speed of light.

Visible light has relatively long wavelengths, and therefore ΔE is small when visible light is absorbed. Thus, a molecule cannot absorb visible light unless it has a vacant or half-filled orbital whose energy is relatively close to that of the occupied orbitals. Most substances do not have such low-lying orbitals. But two low-lying orbitals can be created in a complex when the ligands split the d orbitals of the metal. The existence of these orbitals, the d_{z^2} orbital and the $d_{x^2-y^2}$ orbital, allows the absorption of visible light. Therefore, the compound is colored.

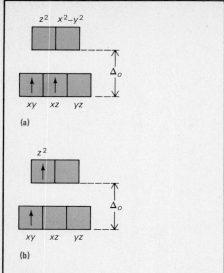

FIGURE 18.21
When a molecule absorbs visible light, one of its electrons moves from (a) a low-energy ground state orbital into (b) a higher-energy orbital. The difference between the energies of the two orbitals is ΔE.

Figure 18.21(a) shows the ground state electronic configuration of a d^2 ion. The absorption of light causes an electron to be promoted into a vacant higher-energy d orbital, so that an excited state is formed (Figure 18.21(b)).

The energy difference between the high-energy and low-energy d orbitals in complexes is Δ_o, which can be substituted for ΔE in the relationship above to give:

$$\lambda = \frac{hc}{\Delta_o}$$

The value of Δ_o is such that the wavelength of absorbed radiation usually is in the visible range. It is known that the absorption of a specific wavelength corresponds to the transition of an electron from a low-energy to a high-energy d orbital. The value of Δ_o can be found by measuring the wavelength of the absorbed radiation. This is the method used to construct the spectrochemical series.

Other Geometries

The crystal field theory can be applied to complexes with other coordination numbers and geometries, using the same basic approach as in the octahedral case. Figure 18.22 shows the relative splittings for the three most important geometries: tetrahedral, octahedral, and square planar. By examining the diagrams, you can see how the electronic configurations of some complexes allow the square planar geometry to be more stable than the tetrahedral geometry, although the square planar has more repulsions between ligands.

FIGURE 18.22
The relative splitting energies for tetrahedral, octahedral, and square planar coordination complexes, shown schematically.

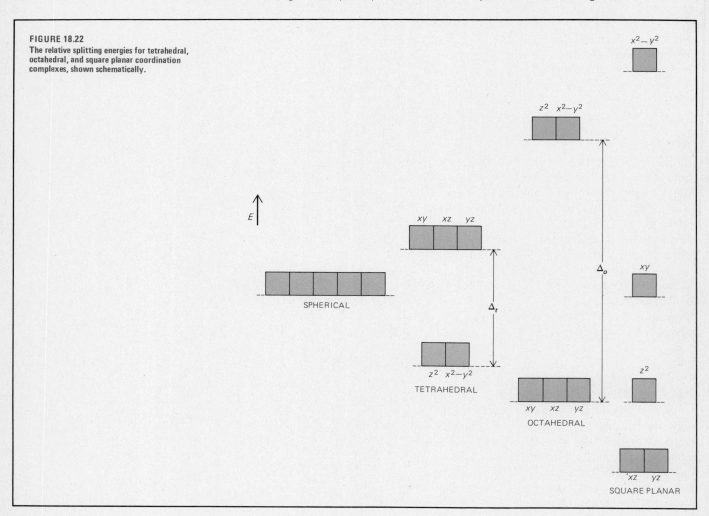

Valuable as it is, the crystal field theory has a weakness. It does not take into account the covalent interactions between the metal ion and the ligands. Both the ligand field theory and the molecular orbital theory include these interactions, as well as the differences in energy between the orbitals of the metal atom. These theories thus are more satisfactory, and considerably more complex, than the crystal field theory.

18.4 LABILITY AND STABILITY OF COORDINATION COMPLEXES

One of the most fundamental reactions of a coordination complex is the replacement of one ligand by another. In aqueous solution, for example, such a reaction is always possible because H_2O is a ligand. A complex can be classified as either *labile* or *inert* on the basis of the rate at which its ligands can be replaced. If ligand substitution is essentially complete in one minute at room temperature when the concentration of reactants is about $0.1M$, a complex is classified as labile. If ligand substitution takes longer under these conditions, the complex is classified as inert. It is often of practical importance to know whether a complex is labile. If the ligands of a complex can be replaced rapidly, precautions must be taken to prevent such reactions. Thus, the coordination chemist must understand the factors that influence the lability of a complex.

To start with, it is important to know that lability is a kinetic property. Lability refers to the rate at which a reaction takes place. Lability is not related directly to stability, which is a thermodynamic characteristic. The stability of a complex refers to the relative energy of the complex.

A classic example of this distinction between lability and stability is the behavior of the complex $[Co(NH_3)_6]^{3+}$. When this complex ion is dissolved in aqueous acid, the equilibrium

$$[Co(NH_3)_6]^{3+} + 6H^+ + 6H_2O \rightleftharpoons [Co(H_2O)_6]^{3+} + 6NH_4^+$$

lies far to the right, because the six ammonia ligands are neutralized by the six protons. The value of the equilibrium constant K is about 10^{25}. But this reaction may take weeks or even months to occur when a solution of the complex ion with the NH_3 ligands is acidified. The $[Co(NH_3)_6]^{3+}$ ion is relatively unstable in acid solution. But it is also inert, so the reaction occurs very slowly. Complexes of Cr^{3+}, of Co^{3+}, and of Pt^{2+} in low-spin states are also unstable but inert. Chemists were able to investigate the properties of these complexes in the early days of coordination chemistry because the complexes react slowly.

The three cyanide complexes $[Ni(CN)_4]^{2-}$, $[Mn(CN)_6]^{3-}$, and $[Cr(CN)_6]^{3-}$, all of which are extremely stable, also demonstrate the distinction between lability and stability. We cannot easily convert stable complexes to other complexes to measure their lability. Although a reaction such as

$$[Ni(CN)_4]^{2-} + 4H_2O \rightleftharpoons [Ni(H_2O)_4]^{2+} + 4CN^-$$

may have a fast rate, the position of equilibrium lies so far to the left that we cannot measure the rate, since no detectable amount of $[Ni(H_2O)_4]^{2+}$ forms. We need a method that enables us to measure ligand exchange without trying to convert a more stable complex to a less stable one.

One such method measures the rate at which CN^- ligands that are complexed to the metal ion are replaced by other CN^- ligands. Radioactive cyanide is used to make this measurement. Radioactive $^{14}CN^-$ is readily available. This radioactively labeled ion is added to an aqueous solution of $[Ni(CN)_4]^{2-}$. The reaction is

$$[Ni(CN)_4]^{2-} + 4\,^{14}CN^- \rightleftharpoons [Ni(^{14}CN)_4]^{2-} + 4CN^-$$

The equilibrium is established in less than a minute; the complex, while stable, is also quite labile. The same reaction with $[Mn(CN)_6]^{3-}$ takes several hours to occur. For the $[Cr(CN)_6]^{3-}$ ion, which is both stable and extremely inert, the reaction is not completed for several months. We thus have three complexes, all of which are quite stable but which differ in lability.

To some extent, the lability of complexes can be correlated with size and electronic characteristics:

1. A small metal atom tends to form relatively inert complexes, because it holds the ligands rather tightly.

2. Complexes which have electrons in high-energy d orbitals tend to hold ligands relatively weakly, and so are labile.

3. A complex with an empty low-energy d orbital tends to be labile, because an incoming ligand can approach the metal ion along this empty orbital without encountering much electrostatic repulsion from the electrons of the metal.

4. Complexes that do not have unfilled low-energy d orbitals or electrons in high-energy d orbitals are inert. Thus, octahedral complexes of metal cations of electronic configuration d^3 and low-spin states of d^4, d^5, and d^6 ions are inert.

18.5 COMPLEX IONS IN AQUEOUS EQUILIBRIA

When we deal with aqueous systems that contain dissolved metal cations, especially those of the transition metals and the heavier metals in the center of the periodic table, we must consider equilibria in which complex ions in solution are formed. In Chapter 11, we said that the most common neutral ligand in complex ions is the water molecule. You should assume that all metal cations are hydrated in aqueous solution. We shall not consider these water ligands in our discussion of aqueous equilibria.

Since many complex ions tend to be quite stable, they may form readily in systems containing the metal cation and appropriate ligands. The relevant equilibrium for a system often includes the formation of a complex ion. Therefore, some measure of the relative stability of complex ions in water is needed to evaluate the equilibrium constants of such reactions.

One approach in such cases is to assume that complex ions are in equilibrium with the free metal cation and the ligands, and to write reactions such as

$$Cu(NH_3)_4{}^{2+} \rightleftharpoons Cu^{2+} + 4NH_3$$
$$Al(OH)_6{}^{3-} \rightleftharpoons Al^{3+} + 6OH^-$$

We shall use this approach for the moment, but it must be noted that these reactions are not exactly correct on two counts: They do not include the water of hydration and they assume that all the ligands are separated from the metal cation simultaneously. In fact, most complex ions lose their ligands one at a time, forming other complex ions which also are stable. To be more accurate, we should write only stepwise reactions, such as

$$Al(OH)_6{}^{3-} \rightleftharpoons Al(OH)_5{}^{2-} + OH^-$$
$$Al(OH)_5{}^{2-} \rightleftharpoons Al(OH)_4{}^- + OH^-$$

We shall disregard these complications.

The equilibrium constant for the reaction in which all the ligands leave the metal cation is called K_i, where i stands for instability. The magnitude of K_i is a convenient measure of the stability of a complex ion. When K_i is small, there is very little tendency for the complex to dissociate, so it is relatively stable. If K_i is large, the complex is relatively unstable. Table 18.1 lists values of K_i for some complex ions.

Given the necessary K_i data, equilibrium calculations in which complex ion formation plays a role can be done by the same procedures that we have used previously.

TABLE 18.1
Instability Constants of Complex Ions at Room Temperature

Metal Ion	Ligand	Complex Ion	K_i
Ag^+	NH_3	$[Ag(NH_3)_2]^+$	6.3×10^{-8}
	Br^-	$[AgBr_4]^{3-}$	5.0×10^{-10}
	Cl^-	$[AgCl_4]^{3-}$	1.0×10^{-6}
	CN^-	$[Ag(CN)_2]^-$	7.9×10^{-22}
	en	$[Ag(en)_2]^+$	1.0×10^{-8}
	I^-	$[AgI_3]^{2-}$	1.3×10^{-14}
	$S_2O_3^{2-}$	$[Ag(S_2O_3)_2]^{3-}$	1.0×10^{-13}
Al^{3+}	F^-	$[AlF_6]^{3-}$	2.0×10^{-20}
	OH^-	$[AlOH]^{2+}$	1.3×10^{-9}
Cd^{2+}	NH_3	$[Cd(NH_3)_4]^{2+}$	1.0×10^{-7}
	Br^-	$[CdBr_4]^{2-}$	1.6×10^{-4}
	Cl^-	$[CdCl_3]^-$	5.0×10^{-3}
	CN^-	$[Cd(CN)_4]^{2-}$	1.6×10^{-19}
	en	$[Cd(en)_3]^{2+}$	6.3×10^{-13}
	I^-	$[CdI_4]^{2-}$	7.9×10^{-7}
Co^{2+}	NH_3	$[Co(NH_3)_6]^{2+}$	1.0×10^{-5}
	en	$[Co(en)_3]^{2+}$	1.3×10^{-14}
Co^{3+}	NH_3	$[Co(NH_3)_6]^{3+}$	1.0×10^{-34}
Cu^+	NH_3	$[Cu(NH_3)_2]^+$	1.6×10^{-11}
	Cl^-	$[CuCl_2]^-$	2.0×10^{-5}
	CN^-	$[Cu(CN)_2]^-$	1.0×10^{-24}
	en	$[Cu(en)_2]^+$	1.6×10^{-11}
	I^-	$[CuI_2]^-$	1.6×10^{-9}
Cu^{2+}	NH_3	$[Cu(NH_3)_4]^{2+}$	5.0×10^{-14}
	en	$[Cu(en)_2]^{2+}$	1.0×10^{-20}
Fe^{2+}	CN^-	$[Fe(CN)_6]^{4-}$	1.0×10^{-37}
	en	$[Fe(en)_3]^{2+}$	2.5×10^{-10}
Fe^{3+}	Cl^-	$[FeCl_4]^-$	1.0×10^{-2}
	CN^-	$[Fe(CN)_6]^{3-}$	1.0×10^{-44}
	SCN^-	$[Fe(SCN)_6]^{3-}$	8.0×10^{-10}
Hg^{2+}	NH_3	$[Hg(NH_3)_4]^{2+}$	5.0×10^{-20}
	Br^-	$[HgBr_4]^{2-}$	1.0×10^{-20}
	Cl^-	$[HgCl_4]^{2-}$	1.3×10^{-15}
	CN^-	$[Hg(CN)_4]^{2-}$	3.2×10^{-42}
	I^-	$[HgI_4]^{2-}$	1.6×10^{-30}
	SO_4^{2-}	$[Hg(SO_4)_2]^{2-}$	4.0×10^{-3}
Ni^{2+}	NH_3	$[Ni(NH_3)_6]^{2+}$	2.0×10^{-9}
	CN^-	$[Ni(CN)_4]^{2-}$	1.0×10^{-31}
	en	$[Ni(en)_3]^{2+}$	7.9×10^{-20}
Pb^{2+}	Br^-	$[PbBr_4]^{2-}$	1.0×10^{-3}
	Cl^-	$[PbCl_3]^-$	1.6×10^{-2}
	OH^-	$[PbOH]^+$	6.3×10^{-7}
	I^-	$[PbI_4]^{2-}$	1.3×10^{-4}
Sn^{2+}	Br^-	$[SnBr_3]^-$	5.0×10^{-2}
	OH^-	$[Sn(OH)_3]^-$	4.0×10^{-26}
Sn^{4+}	Cl^-	$[SnCl_6]^{2-}$	3.2×10^{-2}
	F^-	$[SnF_6]^{2-}$	1.0×10^{-18}
Zn^{2+}	NH_3	$[Zn(NH_3)_4]^{2+}$	4.0×10^{-10}
	CN^-	$[Zn(CN)_4]^{2-}$	6.3×10^{-21}
	OH^-	$[Zn(OH)_4]^{2-}$	2.5×10^{-16}

EXAMPLE 18.1

A 0.29-mol amount of NH_3 is dissolved in 0.45 liter of a 0.36M silver nitrate solution. Calculate the equilibrium concentrations of all species.

SOLUTION

The species present in appreciable concentration when the system is prepared are H_2O, NH_3, Ag^+, and NO_3^-. To select the relevant equilibrium for this system, we must consider that Ag^+ ion forms a complex ion with NH_3 ligands. The equilibria of interest are:

(a) $H_2O \rightleftharpoons H^+ + OH^-$ $\qquad\qquad K_W = 1.0 \times 10^{-14}M^2$

(b) $NH_3 + H_2O \rightleftharpoons NH_4^+ + OH^-$ $\qquad \dfrac{K_W}{K_A} = \dfrac{1.0 \times 10^{-14}M^2}{5.5 \times 10^{-10}M^2} = 1.8 \times 10^{-5}M$

(c) $Ag^+ + 2NH_3 \rightleftharpoons Ag(NH_3)_2^+$

Equation (c) is the reverse of the equation that defines K_i; it is the formation of the complex ion from the metal cation and the ligands. Therefore:

$$K = \frac{1}{K_i} = \frac{1}{6.3 \times 10^{-8}M^2} = 1.6 \times 10^7 M^{-2}$$

With its large equilibrium constant, reaction (c) is the relevant equilibrium. We shall use it in our calculations:

	Ag^+	$+ \ 2NH_3$	$\rightleftharpoons Ag(NH_3)_2^+$
start	0.36M	$\dfrac{0.29 \text{ mol}}{0.45 \text{ liter}}$	0
complete	0.36M − 0.32M	0.64M − 0.64M	0.32M
equil	0.04M + x	2x	0.32M − x

Assuming that x is very small relative to 0.04, and substituting into the equilibrium expression for the reaction:

$$K = \frac{(0.32M)}{(2x)^2(0.04M)} = 1.6 \times 10^7 M^{-2}$$

and

$$x = 3.5 \times 10^{-4}M$$

The concentrations of the species in the relevant equilibrium are:

$$[NH_3] = 2x = 7.0 \times 10^{-4}M$$
$$[Ag^+] = 0.04M$$
$$[Ag(NH_3)_2^+] = 0.32M$$

The concentrations of other ions in the solution can be found from the appropriate expressions for the secondary equilibria. For reaction (b)

$$K = \frac{[NH_4^+][OH^-]}{[NH_3]} = 1.8 \times 10^{-5}M$$

Since the concentrations of OH^- and NH_4^+ must be identical, according to the stoichiometry of reaction (b), and since the $[NH_3]$ is known from the first part of the calculation, we can write

$$\frac{(x)(x)}{(7.0 \times 10^{-4}M)} = 1.8 \times 10^{-5}M$$

where $x = [NH_4^+] = [OH^-]$ and $x = 1.1 \times 10^{-4}M$.

The remaining ion in solution is H^+. Its concentration can be found by using the $[OH^-]$ in the water equilibrium expression:

$$[H^+] = \frac{1.0 \times 10^{-14}M^2}{1.1 \times 10^{-4}M} = 8.9 \times 10^{-11}M$$

The major ion in this solution is the silver ammonia complex ion.

Solid hydroxides of metals that form stable complex ions with hydroxide as the ligand often are soluble in solutions of hydroxide, although they may be insoluble in water. These compounds are called amphoteric hydroxides, since they are also soluble in solutions of strong acids.

EXAMPLE 18.2

Calculate the solubility of $Zn(OH)_2$ in $1.0M$ NaOH solution. The K_{sp} of zinc hydroxide is $1.8 \times 10^{-14}M^3$.

SOLUTION

Table 18.1 shows that Zn^{2+} forms a stable complex ion, $Zn(OH)_4^{2-}$, with hydroxide as the ligand. Therefore, when $Zn(OH)_2(s)$ is added to a solution containing an appreciable concentration of hydroxide ion, the relevant equilibrium is the one that leads to formation of a stable complex ion:

$$Zn(OH)_2(s) + 2OH^- \rightleftharpoons Zn(OH)_4^{2-}$$

The equilibrium constant for this reaction is

$$K = \frac{K_{sp}}{K_i} = \frac{1.8 \times 10^{-14}M^3}{3.3 \times 10^{-16}M^4} = 5.5 \times 10^1 M^{-1}$$

The $[Zn(OH)_4^{2-}]$ is a convenient measure of the solubility of $Zn(OH)_2(s)$, since one mole of the complex ion is formed for every mole of the solid that dissolves. We determine $[Zn(OH)_4^{2-}]$ in the usual way:

	$Zn(OH)_2(s) + 2OH^-$	$\rightleftharpoons Zn(OH)_4^{2-}$
start	$1.0M$	0
equil	$1.0M - 2x$	x

Substitution into the expression for K gives

$$K = \frac{x}{(1.0M - 2x)^2} = 5.5 \times 10^1 M^{-1}$$

Using the quadratic formula to solve this equation gives $x = 0.46M$, which is the $[Zn(OH)_4^{2-}]$. Thus, the solubility of zinc hydroxide in $1.0M$ hydroxide solution is 0.46 mol/liter, substantially larger than the solubility of zinc hydroxide in water, which can be found to be 1.7×10^{-5} mol/liter.

18.6 SOME APPLICATIONS OF COORDINATION CHEMISTRY

Scientific interest in the chemistry of coordination compounds has increased greatly in recent years. The vital role they play in many biological processes has been recognized. New uses for coordination compounds in industry are being developed. We shall discuss some highlights of coordination chemistry in biology and in industry.

Catalysis

Coordination compounds are used as catalysts for many industrial processes. They have the major advantage of being homogeneous catalysts, which means that the entire reaction system, reactants and catalyst alike, is in the same phase. When a suitable homogeneous catalyst for a process can be found, it is better than a heterogeneous catalyst in several ways. More simple apparatus can be used with a homogeneous catalyst. Homogeneous catalysts generally are more efficient than heterogeneous catalysts, and their use allows the reaction to be carried out at lower temperatures.

The details of catalysis by coordination compounds tend to be complicated, as we can see by studying the catalysis of hydrogenation reactions by coordination compounds. In the hydrogenation of compounds with multiple bonds (Section 9.1, page 264), two hydrogen atoms are added to a double bond, converting it to a single bond. The reaction proceeds quite readily in the presence of a heterogeneous catalyst, such as finely divided platinum. The H_2 is chemisorbed (Section 17.6, page 572) on the surface of the metal. We can generalize such a reaction as:

$$H{-}H \; + \; \overset{\backslash}{\underset{/}{C}}{=}\overset{/}{\underset{\backslash}{C} } \longrightarrow H{-}\overset{\backslash}{\underset{/}{C}}{-}\overset{/}{\underset{\backslash}{C}}{-}H$$

Many transition metal complexes can act as homogeneous catalysts for hydrogenation. The catalysis can occur in several ways. For example, a cyanide complex of Co(II) reacts with H_2 to form a complex in which the cobalt is oxidized to Co(III) and an H^-, hydrido, ligand is added:

$$2[Co(CN_5]^{3-} + H_2 \longrightarrow 2[Co(CN)_5H]^{3-}$$

While the reactivity of H_2 is relatively low, the reactivity of the product complex is quite high, so the complex can donate the equivalent of H^- to a double bond. The formation of the hydrido complex thus provides a pathway for hydrogenation.

Some hydrogenation reactions are catalyzed by a chloride complex of rhodium, which can be represented as $RhClL_3$, where L represents a ligand of complicated structure. This homogeneous catalysis takes place in a different way. Under the conditions of the reaction, H_2 adds to the complex oxidatively to form $RhClH_2L_3$, which has two hydrido ligands. The next step in the reaction is the formation of a complex in which one of the ligands is replaced by the compound with the double bond that is to be hydrogenated. The loosely held π electrons of a double bond can act as donor electrons, so compounds with π electrons are often found as ligands in coordination compounds:

$$RhClH_2L_3 \; + \; \overset{\backslash}{\underset{/}{C}}{=}\overset{/}{\underset{\backslash}{C}} \longrightarrow \overset{C}{\underset{C}{\|}} \longrightarrow RhClH_2L_2 + L$$

The required changes in bonding that cause the H atom to be added to the double bond can now occur readily. This complex not only has the H atom in a reactive form but also brings the two reactants together.

A great deal of research currently is going into the effort to achieve nitrogen fixation through catalysis by coordination compounds. We have already described the great stability of the N_2 molecule, the drastic conditions required to make it react, and the importance of nitrogen fixation in the world economy. A method that makes nitrogen fixation possible under mild conditions will save a great deal of energy. Homogeneous catalysis through coordination complexes in which N_2 is a ligand seems to offer the most promise.

Complexes of the general form $M(N_2)_2(L)_4$, where M is either tungsten or molybdenum, and the other ligands, represented by L, are phosphorus compounds, have been found to react under mild conditions with acid to give ammonia:

$$M(N_2)_2(L)_4 + H^+ \longrightarrow NH_3 + N_2 + \text{other products}$$

Another reaction of current interest is

$$H_2O \longrightarrow H_2 + \tfrac{1}{2}O_2$$

This reaction is thermodynamically unfavorable, and can occur only if energy is supplied to the system. If energy in the form of light from the sun could be used to bring about the reaction, we would have a convenient way of producing hydrogen gas, a useful fuel, from solar energy. Since oceans and lakes do not evolve H_2 and O_2 when they are exposed to sunlight, we know that this reaction does not occur under ordinary conditions. Some coordination complexes, especially those of ruthenium, have been found to catalyze the reaction. If research can develop a catalysis process that is economically viable, ordinary water could become an important energy source.

Biological Complexes

We said at the beginning of this chapter that a number of biologically important substances contain a coordination complex of iron and porphyrin called heme, whose structure is shown in Figure 18.23. The major features of this complex structure are the central iron atom and the quadridentate macrocyclic ligand that complexes the iron. This portion of the heme molecule is flat. All the carbon atoms of the ring system, the coordinating nitrogen atoms, and the central iron atom lie in the same plane. The planar structure of heme influences the structure and activity of heme-containing proteins.

One of those heme-containing proteins is cytochrome c, which participates in the biological electron transport chain (Section 16.8, page 532). The function of cytochrome c is to donate or accept one electron. The iron atom of heme takes part in this electron transport, since it can exist in both the $+2$ and $+3$ oxidation states. (The ability of metal ions to exist in several different oxidation states, accepting or donating electrons as required, is one reason for their importance in biological processes.) But why should the iron, which performs what seems to be a simple function, be a part of a molecule as complex as cytochrome c (molecular weight 12 500)? There is no complete answer to this question as yet, but we can get a clue from the fact that the K_{sp} of ferric hydroxide is extremely small. Uncomplexed ferric ion will not dissolve in water at physiological pH. Enzyme-catalyzed reactions are homogeneous reactions. The reactants as well as the catalyst must be in solution. It appears that many metals that are important in biological processes cannot dissolve and serve as homogeneous catalysts unless they are part of a complex.

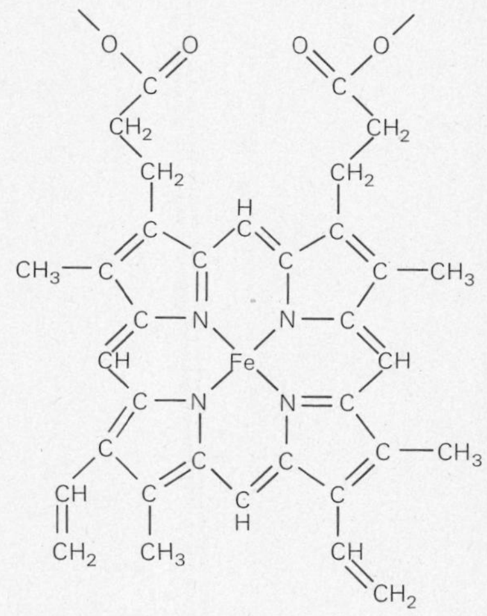

But we must also explain why the metal complex is part of a large, complicated molecule. We can get a partial explanation from the study of hemoglobin, the oxygen-carrying substance of the blood. Hemoglobin contains four heme groups embedded in a protein whose molecular weight is 64 000. The iron of each heme group is in the $+2$ oxidation state. Four of the six coordination sites of each iron are occupied by the quadridentate porphyrin ligand. A fifth site attaches the heme to the protein, and the sixth site is vacant. It is the vacant site to which O_2 molecules are attached as they are carried through the blood by hemoglobin; Figure 18.24 shows a possible schematic structure for this complex. Heme alone forms a complex with O_2, but this complex is not stable. It decomposes when it reacts with an iron cation in another molecule of heme. But when the heme that forms an oxygen complex is embedded in a large protein molecule, the bulk of the protein protects the heme from further attack. We can say that the iron cation is complexed to keep it in solution and that the complex is embedded in a protein to shield the relatively unstable oxygen complex from further attack.

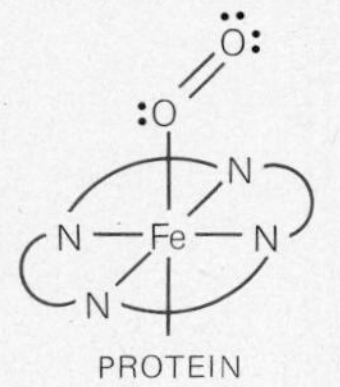

We find another reason why metals are complexed in biological molecules by studying vitamin B_{12}, a cobalt-containing substance that is a prosthetic group (Section 17.6, page 570) for a number of enzymes. The cobalt ion is complexed by a quadridentate ligand that is quite similar to the porphyrin ligand of hemoglobin. It appears that vitamin B_{12} is active only if cobalt is in the $+1$ oxidation state, a state that is normally not displayed by free cobalt. The potential for reducing

FIGURE 18.23
The structure of heme, a coordination complex of iron and porphyrin, which is biologically important. The iron atom, the nitrogen atoms, and the carbon atoms of the rings all lie in the same plane.

FIGURE 18.24
A schematic structure for the heme group of a hemoglobin molecule. Four of the six coordination sites of the iron are occupied by the quadridentate porphyrin ligand. The fifth site is occupied by a ligand belonging to the protein. The sixth site is available for the transport of O_2.

ESSENTIAL ELEMENTS: THE LIST GROWS

How many elements are essential to life? After more than a century of painstaking investigation, the question still cannot be answered with certainty. To demonstrate that an element is essential, all traces of the element must be kept from the organism. Many elements are essential only in exceedingly small quantities, so it is almost impossible to exclude them totally. In recent years, new techniques for keeping experimental animals in a completely isolated sterile environment have lengthened the list of essential elements.

Although only 10 elements account for more than 99% of the matter in living organisms, experiments have suggested that at least 29 other elements, including 21 metals, are essential to life. These elements are listed below in the Table of Essential Elements.

To determine trace-element requirements, rats or other small animals are kept in enclosures that are made of plastic to eliminate contaminants from metal, glass, and rubber. Air entering the enclosures is filtered to remove trace substances that might be present in dust. The animals are fed ultrapure amino acids, and their diet is carefully checked for metal contaminants. Controlled amounts of known essential elements are included in the diet. If a deficiency disease develops, small amounts of other elements can be added to determine which element corrects the deficiency.

For example, test animals on a diet that has no trace of vanadium suffer about a 30% growth retardation. The addition of one-tenth of a part per million of vanadium to the diet restores normal growth. One-half of a part per million of fluorine in the diet has been found to be essential for normal growth. Thirty parts per million of silicon and two parts per million of tin are needed for normal development.

The precise role of these trace elements in the body's metabolic processes is being investigated. It is presumed that in many cases coordination complexes of the trace metals are part of enzymes, but the nature of many of these metalloenzymes is unknown. It could be that the presence of one trace mineral in the body is necessary for the utilization of another. For example, it is known that copper is essential for the metabolism of iron. Ceruloplasmin, a copper-containing protein in the blood, promotes the release of iron from the liver. The iron complexes with transferrin, a protein in the blood serum, which enables the iron to be incorporated into the hemoglobin molecule. Similarly complex relationships may exist between the newly discovered essential trace elements and other elements in the body.

THE ESSENTIAL ELEMENTS

H																	He
Li	Be											B	C	N	O	F	Ne
Na	Mg											Al	Si	P	S	Cl	Ar
K	Ca	Sc	Ti	V	Cr	Mn	Fe	Co	Ni	Cu	Zn	Ga	Ge	As	Se	Br	Kr
Rb	Sr	Y	Zr	Nb	Mo	Tc	Ru	Rh	Pd	Ag	Cd	In	Sn	Sb	Te	I	Xe
Cs	Ba	La	Hf	Ta	W	Re	Os	Ir	Pt	Au	Hg	Tl	Pb	Bi	Po	At	Rn

Legend:
- Make up more than 99% of living matter
- Confirmed essential trace elements
- Suspected essential trace elements

higher oxidation states of cobalt to the +1 state is not favorable. Apparently, cobalt can be in the +1 oxidation state only when it is complexed. Thus, another reason for complexing metals is to achieve oxidation states that are not readily possible in the uncomplexed metals.

Many other metals are essential to biological processes, and these metals usually are in coordination complexes. Chlorophyll, the green plant pigment that plays a fundamental role in photosynthesis, is a complex of magnesium. Enzymes that participate in nitrogen fixation in bacteria contain both iron and molybdenum. Digestive enzymes contain zinc, and enzymes that catalyze electron transport processes contain copper. Other transition metals—vanadium, chromium, manganese, nickel—usually in the form of coordination complexes, are biologically important. The list of metals that are essential to life undoubtedly is not yet complete, and the details of the way in which these metals do their work are only beginning to be known. The field of bioinorganic chemistry, the biological chemistry of the metals and some of the heavier nonmetals, is one of the most active in chemical research today.

EXERCISES

18.1 Find the oxidation state of the metal in the following complexes: (a) $[AgF_4]^-$, (b) $[Ni(NH_3)_6]^{2+}$, (c) $[Cr(NH_3)_3(H_2O)_3]^{3+}$, (d) $[Hg(SO_4)_2]^{2-}$.

18.2 Find the oxidation state of the transition metal in the following compounds: (a) $K[Au(OH)_4]$, (b) $K[CrF_4O]$, (c) $[Co(ONO)(NH_3)_5]SO_4$, (d) $K[PtCl_3(C_2H_4)]$.

18.3 Find the coordination number of the metal in the following complexes: (a) $[Fe(CN)_5NO]^{2-}$, (b) $[Al(OH)(H_2O)_5]^{2+}$, (c) $[Ni(en)_2]^{2+}$, (d) $[Co(en)Cl_2Br_2]^-$.

18.4 The K for the reaction $Ni^{2+} + 6NH_3 \rightleftharpoons [Ni(NH_3)_6]^{2+}$ is 4.1×10^8. The K for the reaction $Ni^{2+} + 3$ en $\rightleftharpoons [Ni(en)_3]^{2+}$ is 1.9×10^{18}. Find K and $\Delta G°$ for the reaction $[Ni(NH_3)_6]^{2+} + 3$ en $\rightleftharpoons [Ni(en)_3]^{2+} + 6NH_3$.

18.5 Find the charge on a complex ion formed by iron(III) and the following ligands: (a) four CN^- and two H_2O, (b) three Cl^- and three NH_3, (c) one OH^- and five H_2O.

18.6 Write the formulas of four different neutral coordination compounds of Cr(III) with NH_3 and Cl^- ligands.

18.7 Predict the most likely geometry for the following complexes: (a) $Hg(CN)_2$, (b) $[HgI_3]^-$, (c) $[HgBr_4]^{2-}$, (d) $[Hg(en)_2]^{2+}$.

18.8 Predict the most likely geometry for the following complexes: (a) $[Cd(NH_3)_4]^{2+}$, (b) $[PbBr_4]^{2-}$, (c) $[NiF_4]^{2-}$, (d) $[Pd(NH_3)_2Cl_2]$.

18.9 Write the formulas of two ionization isomers of a complex with the composition $Cr(NH_3)_4BrCl_2$.

18.10 Write the formulas of all the ionization isomers of a complex with the composition $Pt(NH_3)_4Cl_2Br_2$.

18.11 Write the formulas of the hydrate isomers of a complex with the composition $Ir(H_2O)_6Br_3$.

18.12 Draw the Lewis structures of the two linkage isomers of the complex ion with the composition $[Co(NH_3)_5(NO_2)]^{2+}$.

18.13 Draw the Lewis structures of the two linkage isomers of the complex with the composition $PtCl_3(SCN)$.

18.14 Draw the structures of the geometric isomers of the complex $[Ni(CN)_2Cl_2]^{2-}$, which has square planar geometry. Use the method of representation shown in Figure 18.9.

18.15 Draw the structures of the geometric isomers of the complex $[Au(CN)_2ClBr]^-$, which has square planar geometry.

18.16 Indicate the number of geometric isomers that exist for each of the following complexes: (a) $[Pt(NH_3)_3Cl]^+$, square planar; (b) $[Zn(NH_3)_2Cl_2]$, tetrahedral; (c) $[PbClBrI]^-$, planar; (d) $[Pd(NH_3)_2ClF]$, square planar.

18.17 Draw the structures of the geometric isomers of the complex $[Cr(CN)_3Cl_3]^{3-}$, using the method of representation shown in Figure 18.10.

18.18 Draw the structures of the geometric isomers of the complex $[Fe(CN)_4(NH_3)_2]^-$.

18.19 Draw the structures of the geometric isomers of the complex $[Ru(H_2O)_2(NH_3)_2Cl_2]^+$.

18.20 Indicate which of the following objects are chiral: (a) a pair of scissors, (b) a propeller, (c) a drinking glass, (d) a snowflake.

18.21 Indicate which of the following hypothetical complexes display optical isomerism: (a) $[ZnClBrIF]^{2-}$, tetrahedral; (b) $[PtClBrIF]^{2-}$, square planar; (c) $[Hg(NH_3)Cl]^+$, (d) $[Cr(NH_3)_3(H_2O)_3]^{3+}$.

18.22 Indicate which of the geometric isomers of the complex in Exercise 18.19 are chiral. Draw the structures of the optical isomers.

18.23 Draw all the isomers of the complex with the composition $[Cr(en)_2ClBr]^+$. Indicate the relationship between the isomers.

18.24 Indicate the hybridization predicted by the valence bond theory for the metal in the following complexes: (a) $[Ag(NH_3)_2]^+$, (b) $[PbCl_3]^-$, (c) $[CdCl_4]^{2-}$, (d) $[NiF_4]^{2-}$, (e) $[Ni(CN)_5]^{3-}$, (f) $[RuCl_6]^{2-}$.

18.25 Indicate the number of electrons in the partially filled d subshells of the following ions: (a) Sc^{2+}, (b) Mo^{4+}, (c) V^{2+}, (d) W^{3+}, (e) Cu^{2+}, (f) Ni^{2+}, (g) Fe^{2+}, (h) Ir^{2+}.

18.26 Indicate the valence electronic configurations of the following isolated ions, showing electron spins: (a) Mn^{4+}, (b) Ru^{6+}, (c) Fe^{2+}, (d) Fe^{3+}.

18.27 Indicate the valence electronic configurations of the following ions in their low-spin states: (a) Mo^{2+}, (b) Mn^{2+}, (c) Os^{2+}, (d) Rh^{2+}.

18.28 Indicate the valence electronic

configurations of the following ions in their high-spin states: (a) Mn^{3+}, (b) Fe^{3+}, (c) Ir^{3+}, (d) Ni^{3+}.

18.29 Predict the magnetic properties of each of the following complexes. Justify your predictions: (a) $[Mo(CN)_6]^{4-}$, (b) $[MoF_6]^{4-}$, (c) $[Cu(NO_2)_6]^{4-}$, (d) $[Zn(OH)_4]^{2-}$.

18.30 Predict the magnetic properties of each of the following complexes. Justify your predictions: (a) $[AuF_6]^{3-}$, (b) $[RuCl_4(H_2O)_2]^-$, (c) $[OsCl_6]^{2-}$, (d) $[Co(NH_3)_6]^{2+}$.

18.31 Use the spectrochemical series to predict the magnetic properties of each of the following complexes. Justify your predictions: (a) $[Rh(en)_3]^{3+}$, (b) $[Fe(CN)_6]^{4-}$, (c) $[Fe(H_2O)_6]^{2+}$, (d) $[Re(CN)_6]^{5-}$.

18.32 The value of Δ_0 in $[Ti(H_2O)_6]^{3+}$ is 239 kJ/mol. Predict the frequency of the radiation that is absorbed by this complex ion.

18.33 The $[Cu(H_2O)_6]^{2+}$ ion absorbs radiation of wavelength 7.7×10^{-5} cm. Find the value of Δ_0 that corresponds to this wavelength.

18.34 Most organic substances are colorless because they do not have an empty or partly filled energy level sufficiently close to the highest occupied energy level to permit absorption of visible light. Predict the wavelength range of radiation absorbed by most organic compounds and justify your prediction.

18.35 Suggest an explanation for the observation that $[Fe(CN)_6]^{3-}$ is poisonous while $[Fe(CN)_6]^{4-}$ is not.

18.36 The $[Ni(H_2O)_6]^{2+}$ ion is bright green and the $[Ni(NH_3)_6]^{2+}$ ion is violet. When a solution of the latter ion in water is acidified, the color changes from violet to green. Write reactions for the changes that take place as a result of the addition of acid. Discuss the lability and stability of the two ions.

18.37 Halide complexes of metal M of the form $[MX_6]^{3-}$ are found to be stable in aqueous solution. Suggest an experiment to measure their lability which does not employ radioactive labels.

18.38 Suggest an explanation for the observation that d^6 metal cations in low-spin states are relatively inert.

18.39 A solution is prepared from 1.00 mol of $Cu(NO_3)_2$, 1.00 mol of NH_3, and enough water to make 1.00 liter of solution. Find the concentration of all species at equilibrium.[a]

18.40 Calculate the equilibrium concentrations of all species present in a solution prepared from 0.10 mol of SnF_4 and 0.20 mol of sodium fluoride and enough water to make 1.0 liter of solution.[a]

18.41 The K_{sp} of $Sn(OH)_2$ is 3.2×10^{-26}. Calculate the solubility of $Sn(OH)_2$ in a $2.0M$ NaOH solution.[a]

18.42 The K_{sp} of AgCl is 1.6×10^{-10}. Find the solubility of AgCl in a $1.0M$ solution of NH_3.[a]

18.43 The K_{sp} of $PbCl_2$ is 1.6×10^{-5}. Find the solubility of $PbCl_2$ in a $1.0M$ solution of NaCl.[a]

18.44 Indicate the changes in oxidation states that take place in the reaction: $2[Co(CN)_5]^{3-} + H_2 \rightarrow 2[Co(CN)_5H]^{3-}$.

18.45 Coordination compounds of copper have been found to be implicated in some important biological systems. Suggest some possible functions for the copper in such compounds.

---

[a] Some of the data necessary for the solution of this exercise can be found in Table 18.1.

METALS

19

The swiftest glance at the world around us shows the importance of metals to humanity. The rise of the human race to its present control of this planet has gone hand in hand with mastery over metals: from the Bronze Age to the Iron Age to the modern era of metallurgy. More recently, research has shown that metals are as essential to many biological processes as they are to technology.

Since more than three-quarters of all the elements are classified as metals, it is not surprising that we have encountered metals frequently in our previous discussions of such subjects as the chemistry of the nonmetals, the chemistry of aqueous solutions, and the chemistry of coordination compounds. In this chapter, we shall discuss the metallic state and the general characteristics of metals, with a closer look at the chemistry of two major groups of metals, those in groups IB and VIII of the periodic table.

19.1 THE METALLIC STATE

You will recall that we began by dividing the chemical elements into three groups: the nonmetals, which are found toward the right and top of the periodic table; the semimetals, which run in a diagonal band from boron at the upper left of the table to tellurium at the lower right; and the metals, which lie toward the left and the bottom of the periodic table (Figure 2.7). We can identify the metals by a number of chemical and physical properties that they have in common.

1. *Metals are good conductors of electricity*. Some metals conduct electricity better than others—the metals of group IB, copper, silver, and gold, are the best conductors, with aluminum and beryllium next—but every metal is a conductor. In metals, moreover, conduction of electricity takes place without the transfer of material or the chemical changes that occur in electrolysis or ionic conduction. Perhaps the most characteristic property of the metals is that their conductivity *decreases* as the temperature increases.

2. *Metals are good conductors of heat*. This property is evident when we compare metals to nonmetals, which tend to be thermal insulators. But there is a wide range of thermal conductivity among metals. The metals of group IB, which are the best electrical conductors, are also the best thermal conductors. Aluminum is the best electrical and thermal conductor of the other metals. We cook with metal pots and pans because of this property.

3. *Metals have luster*. Their surfaces, when clean, are shiny. The metallic luster results from the reflection of all wavelengths of visible light from the surface of most metals. (Gold and copper are exceptions because they absorb some frequencies of visible radiation, and thus have their characteristic colors.) The luster depends on the cleanliness of metal surfaces. When metals appear gray or black, it is because a chemical reaction has taken place on their surfaces.

4. *Metals are deformable and plastic*. The shape of a piece of metal can be changed considerably without disrupting the forces that hold the metal together. A piece of metal can be pounded into a thin sheet, so metals are said to be *malleable*. A piece of metal can be drawn into a fine wire, so metals are said to be *ductile*. Most nonmetallic solids do not have these properties. For example, if a piece of sodium chloride is hit with a hammer, it shatters. There are some nonmetallic solids that can be deformed: Rubber stretches, for instance. But generally, these nonmetallic solids are elastic; when the force that causes the deformation is removed, the solid returns to its original shape. Metals, by contrast, can be deformed well past the elastic range.

5. *Metals display the photoelectric effect*. When metals are exposed to radiation of short wavelength, they emit electrons (Section 5.6, page 132). A

related phenomenon, the thermionic effect, was observed in 1883 by Thomas Edison, who found that a metal wire emitted electrons when heated. The discovery of the thermionic effect led to the invention of electronic tubes and the birth of the electronics industry.

6. *All metals but mercury are crystalline solids at room temperature*. Metals generally have relatively high densities and high melting and boiling points. One reason why metals are so dense is that the atoms of their crystals are in close-packed arrangements. Cubic closest packing or hexagonal closest packing (Figures 10.22–10.24), in which each atom is surrounded by 12 identical and equidistant atoms, is common. In these crystal arrangements, we say that each atom has 12 nearest neighbors. Body-centered cubic packing (Figure 10.25) is found in about 20 metals, most of them in the top left part of the periodic table. In this arrangement, which is not as dense as cubic closest packing, each atom has only eight nearest neighbors. But six other atoms, called next-nearest neighbors, are almost as close.

7. *When a metal combines chemically with another element, the other element usually is a nonmetal and the metal generally is in a positive oxidation state*. The bonding between a metal and a nonmetal usually is ionic, with a substantial electron transfer from the metal to the nonmetal. Metals have relatively low electron affinities, so when they form covalent bonds, those bonds generally are not very strong. While nonmetals frequently exist as relatively small homonuclear molecules that are gases at room temperature, metals do not have any great tendency to form such molecules.

19.2   THE METALLIC BOND

While ordinary covalent bonding is not important between metal atoms, there are nonetheless strong forces between the atoms in a metallic crystal. Data on the heat of atomization, the amount of heat needed to form one mole of separated gaseous atoms, demonstrate the existence of these forces, as we can see by examining the data for sodium. Like many other metals, sodium in the gas phase can exist as a diatomic molecule with relatively weak covalent bonds. The strength of the sodium-sodium bond in the Na_2 molecule is 72.4 kJ/mol. The heat of atomization is the heat required to form one mole of atoms. For Na_2, it is thus $72.4/2 = 36.2$ kJ/mol, since 1 mol of Na_2 forms 2 mol of Na.

The heat of atomization for solid sodium is much higher. The heat needed to form one mole of sodium atoms from solid sodium is 108 kJ/mol. The forces holding the sodium atoms together in the solid are stronger than the covalent bonds in the Na_2 molecule. The same is true of other metals, as you can see in Table 19.1, which lists the heat of atomization of diatomic molecules and the heat of atomization at 298 K for a number of solid metals. In every case, the data show the existence of strong forces between atoms in a metallic crystal.

The force that holds metal atoms together in the crystal is known as the *metallic bond*. Any theory that attempts to describe the nature of the metallic bond must account for both the strength of this force and the unique observed properties of metals.

Compared to nonmetals, metals have a small number of *s* and *p* electrons in their valence shells. For example, the valence shell electronic configurations of the metals in the second row of the periodic table are: Na, $3s$; Mg, $3s^2$; Al, $3s^23p$, while those of the nonmetals are: P, $3s^23p^3$; S, $3s^23p^4$; Cl, $3s^23p^5$. The nonmetals can form only a limited number of covalent bonds, because of the relatively small number of electrons required to fill their valence shells. Thus, chlorine exists as Cl_2, with one covalent bond; sulfur as S_8, with two covalent bonds for each S atom; and phosphorus as P_4, with three covalent bonds for each P atom. In each case, the number of atoms that are bonded to a given atom is

TABLE 19.1
Heats of Atomization (kJ/mol) at 298 K

Metal	Diatomic[a]	Solid
lithium	52	162
potassium	24.7	90.0
copper	98	339
zinc	13	131
rubidium	22.6	81.6
silver	82	285
cadmium	4.4	112
tin	96	301
cesium	21.8	78.2
gold	109	368
lead	48	197

[a] The heat of atomization of a diatomic molecule is half the bond energy.

determined by the number of unpaired valence electrons that are available for bonding.

In metal atoms we find an entirely different situation, which can be described as ''delocalized bonding.'' There are more vacant orbitals than electrons in the valence shell of a metal atom. The availability of these vacant orbitals allows the formation of metal-metal bonds that are not localized between just two atoms. Instead, the bonding electrons are shared among the vacant orbitals in a number of neighboring atoms. We say that the bonds are *delocalized*.

For example, a sodium atom has one valence electron in a $3s$ orbital, but it has four valence orbitals, the $3s$ and three $3p$ orbitals. While the sodium atom can form only one bond because it has only one valence electron, that bond is not localized between two sodium atoms in the solid. The bond is spread out between the sodium atom and all its neighboring sodium atoms. Since solid sodium has a body-centered cubic crystal structure, we can say that the bond formed by a sodium atom is spread between this atom, its eight nearest neighbors, its six next-nearest neighbors and even, to some extent, more remote sodium atoms.

With some slight modifications, we get a similar picture for the other metals. Magnesium, for example, has a valence electron configuration of $3s^2$, which means that it has no unpaired valence electrons. But an excited state of magnesium whose electronic configuration is $3s3p$ can be used to form bonds. The energy needed to achieve the excited state is less than the energy gained by bond formation. Thus, each magnesium atom has two unpaired electrons and four valence orbitals available. Solid magnesium has a hexagonal closest-packed crystal structure, giving each atom 12 nearest neighbors. The two bonding electrons of each atom are delocalized between these 12 nearest neighbors and, to a lesser extent, between more remote magnesium atoms.

Similarly, aluminum has an electronic excited state that has three unpaired electrons, $3s3p^2$, which can be used for bonding. The crystal structure is cubic closest packed, giving each aluminum atom 12 nearest neighbors. The three bonding electrons are delocalized between the nearest neighbors and, to some extent, between more remote atoms in the crystal. The delocalization is possible because there are more valence orbitals (four) than valence electrons (three).

The most characteristic feature of metallic bonding is the delocalization of electrons permitted by the excess of vacant valence orbitals. In a rough, generalized way, many of the observed properties of metals can be explained by electron delocalization. The bonding electrons, which are not closely associated with one or two atoms, can move rather freely through the crystal. The high electrical conductivity of metals, the photoelectric effect, and the thermionic effect are all consistent with a picture in which the electrons are relatively free to move in the metallic crystal.

The strength of the metallic bond—or, to put it another way, the magnitude of the forces holding metal atoms together in a crystal—depends on a number of factors, including the geometry of the crystal, the ionization energy of the metal, and the electron affinity of the metal. But the most important factor by far is *the number of electrons that a metal atom has available for bonding, called the* **metallic valence** of the metal. The greater the metallic valence, the stronger the forces holding the atoms of the crystal together. The metallic valence of sodium is 1, of magnesium is 2, and of aluminum is 3. Therefore, atoms are held more tightly in solid aluminum than in magnesium, and more tightly in magnesium than in sodium.

It would seem that the heat of atomization is the most direct measurement of the forces holding metal atoms together in the solid, that is, of the strength of the metallic bond. For these three metals, the heats of atomization are as expected: greatest in aluminum (326 kJ/mol), less in magnesium (146 kJ/mol), lowest in sodium (108 kJ/mol). However, the picture is not as simple as it might seem. Each metal crystal has a different geometry. The differences in geometry affect the heats of atomization. In addition, an input of energy is required for both magnesium and aluminum to form the excited state that permits the maximum number of electrons to participate in bonding. Thus, the measured heats of atomization of magnesium and aluminum understate the magnitude of the forces holding nearest neighbors together, as the thermochemical diagram in Figure 19.1 shows.

Many properties of solids reflect the strength of the metallic bond. One such property is hardness, which is a measure of resistance to a change in shape when a force is applied. For a metal to change shape, some atoms must move away from one another. The hardness of a metal is thus related to the strength of the metallic bond. From their electronic structures, we would predict that aluminum is harder than magnesium and that magnesium is harder than sodium. The prediction is verified by observation.

The melting point of a metal is also related to the strength of the metallic bond: the stronger the bond, the higher the melting point. As expected, we find that aluminum has the highest melting point of the three metals, 933 K, with magnesium next at 923 K, and sodium last at 371 K. (The trend is not completely regular because of the influence of crystal structure and other factors.)

We find the same correlation between metallic valence and metallic properties in the transition metals. If we look at the third row of the periodic table, we find that potassium, with the smallest metallic valence, is a soft, low-density metal with a low melting point. The next metal is calcium, which is harder, denser, and higher melting than potassium. The trend of increasing hardness, density, and melting point continues with the first transition metal, scandium, through titanium, vanadium, and chromium. Table 19.2 summarizes the data for these six metals and other metals in the third row.

The increase in hardness, density, and melting point for these six metals can be attributed to an increase in the forces of attraction between nearest neighbors in the metallic crystal, which in turn is consistent with an increase in metallic valence. Each element in this row has one more valence electron than its predecessor. The metallic valences go up in a regular way; 1 for K, 2 for Ca, 3 for Sc, 4 for Ti, 5 for V, and 6 for Cr. The availability of d orbitals permits the use of as many as six electrons for metallic bonding. However, the metallic valence does not increase beyond six, which is the maximum number of orbitals that can be used effectively for bonding. Thus, the next few metals all have metallic valences of 6, and all have similar hardness, density, and melting points. The usefulness of transition metals such as chromium, manganese, iron, cobalt, and

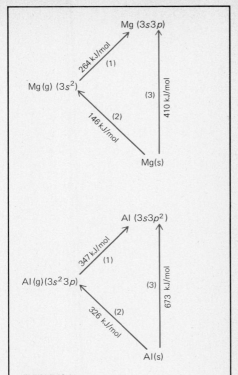

FIGURE 19.1
Thermochemical diagrams of the atomization of Mg and Al. Step (1) is the energy required to promote a $3s$ electron into a $3p$ orbital. Step (2) is the measured heat of atomization. Step (3) is the sum of steps (1) and (2) and is a better measure of the forces holding the atoms together in the metallic crystal than step (2) alone.

TABLE 19.2
Properties of Third-Row Metals

Metal	Melting Point (K)	Density (g/cm³)	Heat of Atomization (kJ/mol)
potassium	337	0.86	90.0
calcium	1110	1.55	178
scandium	1812	3.0	375
titanium	1941	4.51	469
vanadium	2173	6.1	515
chromium	2148	7.19	397
manganese	1518	7.43	285
iron	1809	7.86	416
cobalt	1768	8.9	428
nickel	1726	8.9	430
copper	1356	8.96	339
zinc	693	7.14	131
gallium	302	5.91	276

nickel is due to such properties as hardness and density, which are related to high metallic valences.

We find a decrease in hardness, density, and melting point starting with copper and continuing with zinc and gallium. The decrease reflects a drop in metallic valences that is caused by the filling of the *d* orbitals and the pairing of electrons. It should be noted that there are irregularities in all these trends, since the properties of metals are affected by such factors as crystal geometry.

The Electron Sea Model

Our picture of the metallic bond thus far has been rather crude. We can get a better picture by using what is called the *electron sea model,* in which the metal atom and all the electrons that do not participate in bonding are considered separately from the electrons that do participate in bonding. The electron sea model pictures the crystal lattice as a network of metal atoms, which bear positive charge because some of their electrons have been separated off for bonding. The lattice of positive metal ions is pictured as being immersed in a ''sea'' of negative electricity formed by the bonding electrons, which are completely delocalized, as Figure 19.2 shows. Thus, while the atoms maintain fixed positions, the bonding electrons lose all connection with their source atoms and move freely throughout the entire metal crystal, forming the electron sea.

This model can be used to account for the properties of metals. The photoelectric effect and the thermionic effect can be explained by saying that the ejected electrons jump off the surface of the electron sea because of an input of energy. The conductivity of electricity through a metal can be explained by saying that an electron enters the electron sea and creates a disturbance which is transmitted through the sea, until an electron leaves at the ''opposite shore.'' In this model, the overall neutrality of the metal is maintained, and the net result is the entrance of one electron at one end of the metal and the departure of another electron at the other end, as shown in Figure 19.3. This picture is consistent with the observed decrease in the electrical conductivity of metals with an increase in temperature. As the temperature rises, the thermal motion of ions in the lattice increases, interfering with the movement of a ''wave'' through the electron sea. The result is a decrease in electrical conductivity.

The electron sea model is also consistent with the plastic deformability of metals. To deform a metal, it is necessary to change the positions of some units of the crystal with respect to other units. Figure 19.4(a) shows an idealized picture of deformation: One plane of atoms slips with respect to another. Since the electron sea model pictures the bonds between atoms as being completely delocalized, this slippage does not create any bonding problems. There are no bonds to be broken, and since all the units of the crystal lattice have identical charges, any unit is an acceptable nearest neighbor for any other. Deformation is more difficult in an ionic crystal because of the nature of the bonding. When the units in an ionic crystal move with respect to each other, substantial electrical repulsions are created, as shown in Figure 19.4(b). In a sense, the movement of

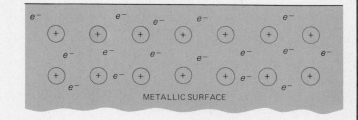

FIGURE 19.2
The electron sea model of the metallic bond. A metal crystal is pictured as a lattice of positively charged ions surrounded by a ''sea'' of completely delocalized bonding electrons.

METALLIC SURFACE

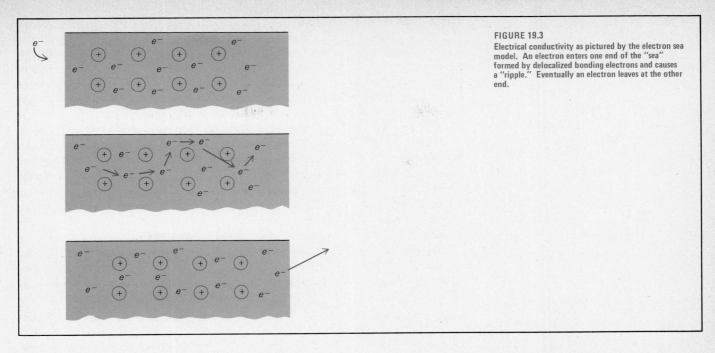

FIGURE 19.3
Electrical conductivity as pictured by the electron sea model. An electron enters one end of the "sea" formed by delocalized bonding electrons and causes a "ripple." Eventually an electron leaves at the other end.

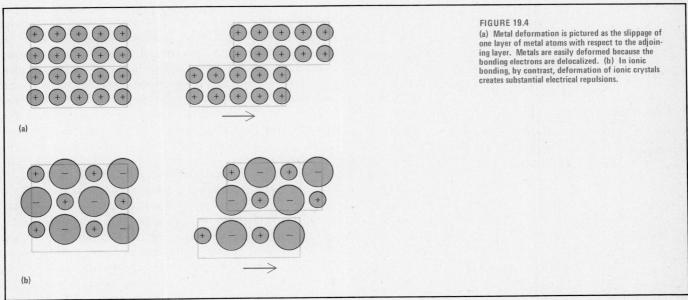

FIGURE 19.4
(a) Metal deformation is pictured as the slippage of one layer of metal atoms with respect to the adjoining layer. Metals are easily deformed because the bonding electrons are delocalized. (b) In ionic bonding, by contrast, deformation of ionic crystals creates substantial electrical repulsions.

the ions breaks some ionic bonds. Deformation is even more difficult in a crystal whose units of structure are covalently bonded. Diamond is one of the hardest substances known because it is a crystal of covalently bonded carbon atoms. Strong covalent bonds must be broken if those atoms are to be moved with respect to each other.

However, it is important to recognize that the electron sea picture is oversimplified and extreme because it overemphasizes delocalization of bonding electrons. If we followed the model to its logical conclusion, we would assume that metals have no mechanical strength because they have no resistance at all to applied stresses. But we know that metals are hard and resist deformation, and that some metals are harder than others. When we study the properties of metals, we must consider the picture we used originally, in which there are attractions between nearest neighbors in the metal crystal lattice.

19.3 THE BAND THEORY

For another description of bonding in metals, we must focus on the atomic orbitals that participate in the metallic bond and on the nature of the molecular orbitals formed by overlap of the atomic orbitals.

An orbital description of covalent bonding was developed in Section 7.5. Two atomic orbitals overlap to form two molecular orbitals: a bonding orbital, of lower energy than either of the two original nonbonding atomic orbitals; and an antibonding orbital, of higher energy than either of the two original orbitals. The two electrons of the covalent bond are found in the bonding orbital. The decrease in energy of the bonding orbital relative to the atomic orbital equals the increase in energy of the antibonding orbital.

Let us apply this description to the bonding of two lithium atoms. When the two atoms come together, the 2s orbital of each of the atoms overlaps with the other, forming two molecular orbitals that extend over both lithium atoms. The 2s electron of each atom is in the lower-energy bonding molecular orbital. If more than two atoms come together, more than two atomic orbitals can overlap to form molecular orbitals. In the case of lithium, as more atoms come together to form a crystal, more 2s orbitals overlap and form molecular orbitals that extend over more and more lithium atoms. As the number of atoms increases, the difference in energy between the molecular orbitals that form becomes steadily smaller, as Figure 19.5 shows schematically. When there is a very large number of atoms, as in a crystal, there are many molecular orbitals in a relatively narrow energy range. We can say that there is a *band of orbitals*.

We can see that the number of levels in an orbital band is the same as the number of atoms, and that a band can hold two electrons for each of its levels. For lithium, the 2s band is only half full. Since each orbital of the band extends over all the atoms whose orbitals overlap to form the band, the orbital band allows us to picture the electrons moving freely through the crystal, as in the electron sea model.

The orbital band concept can help us understand some properties of lithium, such as its luster. In general, metals are shiny because they easily absorb and emit all wavelengths of visible light. The orbital band in lithium, which is half-filled, includes a number of closely spaced levels, both occupied and unoccupied. All wavelengths of visible light can be absorbed and emitted because of electronic transitions between these levels. The band model also accounts for the electrical conductivity of lithium. When an electric current is applied, electrons enter vacant levels in the band. Since each level extends over the entire crystal, the vacant levels provide a route by which electric current may pass easily through the metal.

But we encounter a difficulty when we apply the band theory to beryllium, the next metal in the periodic table. The valence electronic configuration of beryllium is $2s^2$. Beryllium has twice as many valence electrons as lithium. Therefore, the 2s orbital band will be filled. It can be shown that a filled orbital band does not allow the conduction of electricity. Yet the electrical conductivity of beryllium is greater than that of lithium. How can this be so?

Conduction occurs because the 2s orbital is not the only atomic orbital that overlaps to form bands. Bands can be formed by the overlap of the 2p orbitals of the beryllium, among others. We can thus differentiate between two types of orbital bands. The valence band holds the valence electrons. If it is unfilled, as in lithium, it can be used to conduct electricity. If the valence band is filled, as in beryllium, it cannot be used to conduct electricity. We then turn our attention to the other orbital bands that are formed by the overlap of valence atomic orbitals. These orbital bands, which are higher in energy than the valence band, are called the *conduction bands*. They can be used to conduct electricity.

While the 2s and 2p atomic orbitals have different energies, the spread in energy that results from band formation makes it possible for parts of the bands

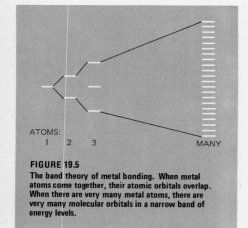

FIGURE 19.5
The band theory of metal bonding. When metal atoms come together, their atomic orbitals overlap. When there are very many metal atoms, there are very many molecular orbitals in a narrow band of energy levels.

ATOMS:
1 2 3   MANY

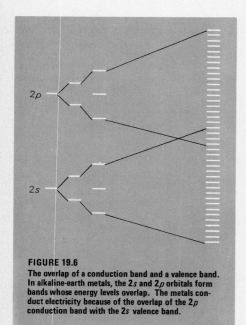

2p

2s

FIGURE 19.6
The overlap of a conduction band and a valence band. In alkaline-earth metals, the 2s and 2p orbitals form bands whose energy levels overlap. The metals conduct electricity because of the overlap of the 2p conduction band with the 2s valence band.

that they form to fall in the same energy range. In beryllium and the other alkaline-earth metals, the $2s$ and $2p$ bands have overlapping regions of energy, as Figure 19.6 shows. *If the valence band and the conduction band in a solid overlap, the solid is an electrical conductor* because the conduction band contains some electrons. All metals have either an unfilled valence band or overlap between the valence band and another band which can be the conduction band. In general, substances are electrical conductors when there are more valence orbitals than valence electrons. The only element that has more valence orbitals than valence electrons but does not have metallic properties is boron.

Band theory also gives an explanation of the properties of insulators, materials with very low electrical conductivity, and semiconductors, materials whose electrical conductivity is substantially larger than that of insulators but substantially smaller than that of metals. In both an insulator and a semiconductor, the energies of the conduction band and of the valence band are so different that there is no overlap. There is an energy gap, sometimes called a forbidden zone, between the bands. The difference between an insulator and a semiconductor can be traced to the magnitude of the energy gap, ΔE_0, which can be taken as the difference in energy between the top of the valence band and the bottom of the conduction band. Figure 19.7 is a schematic representation of the relative energies of the valence and conduction bands in insulators (a), semiconductors (b), and metals (c) and (d).

A material with a large ΔE_0 is an insulator because there is not sufficient energy available under normal conditions to promote enough electrons from the valence band to the conduction band to carry a current. But there are ways in which insulators can be made to carry current. An insulator can be placed in an intense electrical field, which may provide enough energy to raise some electrons into the conduction band. Some insulators may become conductors when subjected to extremely high pressure, which will decrease the distance between units of the crystal and squeeze the valence band and conduction band closer together. There are insulators that are photoconductors; that is, they conduct electricity when they are exposed to certain frequencies of ultraviolet radiation. The frequency of the radiation must be high enough to make $h\nu$ equal to or greater than ΔE_0, so that electrons are promoted into the conduction band when the radiation is absorbed.

In semiconductors, ΔE_0 is smaller than in insulators. The thermal energy available under ordinary conditions is enough to promote electrons out of the valence band into the conduction band. The semiconductor then conducts electricity for two reasons: There are electrons in the conduction band, and the valence band is no longer completely filled. The orbital band model predicts that an increase in temperature will promote more electrons to the conduction band, and that the electrical conductivity of a semiconductor will increase as the temperature rises. Indeed, the increase in conductivity with an increase in temperature is one of the features that distinguishes a semiconductor from a metal.

In general, the magnitude of ΔE_0 decreases in going down any given column of the periodic table. Table 19.3 gives the trend for group IVA. As we move down the column from a nonmetallic insulator to semimetallic conductors to metals, we find a decrease in the value of ΔE_0.

Semiconductors

There are really two kinds of semiconductors. Materials that are semiconducting because of a relatively small value of ΔE_0 are called intrinsic semiconductors. Materials that are semiconductors because they contain trace impurities are called extrinsic semiconductors. Silicon is a well-known extrinsic semiconductor. It is listed as an insulator in Table 19.3, but silicon must be of extreme purity to

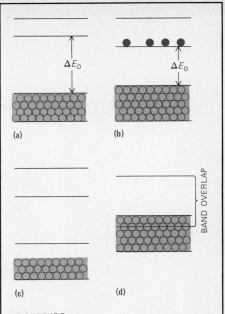

FIGURE 19.7
According to the band theory, the energy gap ΔE_0 between the energy of the valence band and that of the conduction band determines the electrical properties of a substance. In an insulator (a), the energy gap is so wide that no electrons can move into the conduction band. In a semiconductor (b), the gap is narrower and a few electrons can move into the conduction band. In a metal, the valence band is not filled (c) or the valence band and conduction band overlap (d), so there is no energy gap and the metal is a good conductor.

TABLE 19.3
Energy Gap of Elements in Group IVA

Substance	Conductivity	ΔE_0 (kJ/mol)
diamond	insulator	502
silicon	insulator	105
germanium	semiconductor	58
tin (gray)	semiconductor	8
tin (white)	metal	0
lead	metal	0

THE CALCULATOR EXPLOSION

In the 1960s, you could buy a hand-held calculator that would add, subtract, multiply, and divide for anywhere between $500 and $1000. In the early 1970s, as prices came down, there was speculation that a four-function calculator might someday be available for less than $100. Today, a basic four-function calculator can be had for $10 or less, and $100 will buy a programmable calculator whose capabilities approach those of a small computer.

The price revolution is the result of a technological revolution. The calculator of the 1960s used integrated electronic circuits that contained about a dozen transistors or similar components on a single chip. Today, mass-produced chips only a few millimeters square contain several thousand such components. This large-scale integration (LSI) has been achieved largely with metal-oxide-semiconductor (MOS) technology, which gets its name from the materials that are used.

A single MOS semiconductor consists of two semiconductor regions, called the source and the drain, that are made of silicon heavily doped with impurities. The silicon semiconductor is coated with silicon oxide, which acts as an electrical conductor. A metal is used as a "gate" electrode, which is placed between the source and the drain. When the voltage is high enough, a thin region of silicon under the gate becomes a conduction path for electrons, so the gate voltage controls the amount of current in the transistor.

Large-scale integration is possible because the method used to produce one transistor can be used to make many thousands of other components simultaneously on a single chip. In fact, hundreds of chips can be produced at once from a wafer about five centimeters square using a multistep chemical fabrication process that includes photoengraving, oxidation, controlled diffusion of impurities, ion implantation, and epitaxial growth, in which small amounts of substances are applied by exposing the wafer to an appropriate vapor.

In the first few steps of the process, a layer of silicon oxide is grown on the wafer and a layer of a photoresistant material, which hardens on exposure to ultraviolet radiation, is applied. Ultraviolet irradiation followed by acid etching creates the pattern of a circuit on the wafer. Phosphorus atoms then are diffused into exposed areas of silicon to form the source and drain areas. Perhaps a dozen more such steps are needed to produce a finished circuit.

Such a circuit is so small that the wavelength of the visible light used for some steps has become the limiting factor. In current technology, the smallest features are about five microns wide. Visible light cannot be used for features smaller than about two microns wide, so research has begun on the possible use of electron beams, which have produced features only one micron wide experimentally. Even further miniaturization might be possible using X rays, which have been used to fabricate features as small as 0.1 micron wide.

Continual advances in LSI have led to circuits of microscopic size and advanced capability. The past few years have seen the be an insulator. Even the tiniest traces of impurities increase its conductivity enough to make silicon a semiconductor. For example, 10 atoms of boron in 1 000 000 atoms of silicon increase its conductivity by a factor of 1000.

Impurities can increase the conductivity of both insulators and semiconductors in two different ways. By introducing impurities into a crystal, we create additional energy levels close to the existing bands. We do not create new bands; rather, the impurities add electronic levels, which are called impurity levels.

One way to increase conductivity is to add an element that has more electrons than the element in the pure crystal. For example, we can add a small quantity of arsenic or phosphorus to silicon, a process called "doping." Both phosphorus ($3s^23p^3$) and arsenic ($4s^24p^3$) have five valence electrons. Only four valence electrons bond a P or As atom into the lattice. As Figure 19.8(a) shows, the fifth electron resides in an impurity level that is fairly close in energy to the conduction band of silicon. Thermal energy can promote these electrons into the conduction band, so silicon doped with either arsenic or phosphorus is a semiconductor. Since the conductivity is due to the presence of negative electrons in the conduction band, silicon doped in this way is called an n-type semiconductor, where n stands for "negative."

We can also cause silicon to become a semiconductor by adding elements with fewer valence electrons than silicon, such as boron. In this case, the impurity levels are empty and are close to the valence band, as Figure 19.8(b) shows. Thermal energy can promote electrons from the valence band to the vacant impurity levels. When electrons are promoted out of the valence band, the band is no longer filled and the conduction of electricity is possible. Since the loss of an electron to the impurity level causes the silicon lattice to be positively charged, we can picture the lattice as containing positive holes. Electricity is conducted as electrons move into these positive holes, leaving new positive holes behind them. In a sense, we can visualize the current as being carried by a reverse flow of the positive holes. For this reason, an extrinsic semiconductor in which the impurity is relatively electron-deficient is called a p-type semiconductor, where p stands for "positive."

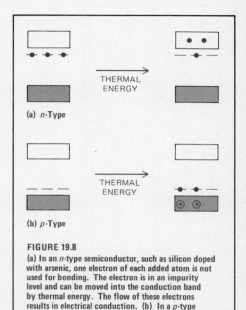

(a) n-Type

(b) p-Type

FIGURE 19.8
(a) In an n-type semiconductor, such as silicon doped with arsenic, one electron of each added atom is not used for bonding. The electron is in an impurity level and can be moved into the conduction band by thermal energy. The flow of these electrons results in electrical conduction. (b) In a p-type semiconductor, such as silicon doped with boron, electrons are promoted by thermal energy from the filled valence band into vacant impurity levels. Electricity is conducted by the movement of electrons into positive holes in the silicon lattice.

19.4 METALLURGY

Almost all the metals we use come from ores that are mined from the crust of the earth. We define an *ore* as a mineral or other naturally occurring material that is a usable source of a metal. The technology that uses the principles of physical science to produce metal in usable form from ore is *metallurgy*. Generally, metal cannot be extracted directly from ore. Ore usually undergoes some preliminary treatment, called *dressing*, which is followed by the actual extraction of the metal, called *winning*, and the purification of the metal, called *refining*. The details of the treatment, of course, depend on the nature of the ore and the metal.

Dressing; Preliminary Treatment

Ore usually is mined in the form of large chunks of metal-containing rock, mixed with unwanted materials such as sand or clay, collectively called gangue. The first step in dressing the ore is to crush the chunks into smaller particles. Next, as much as possible of the gangue is separated from the ore, most often by a process called flotation, in which air is blown through a mixture of the ore and water to form a foam that floats on top of the water. The particles of metal-containing material adhere to the foam and are skimmed off, while the gangue sinks to the bottom.

Other dressing methods are available. One technique that is familiar to movie-goers is panning, in which flowing water is used to wash away sand and clay from gold-containing particles. For ores that contain magnetic compounds of iron, magnets can be used for separation. For metals that have low melting points, heat will do the job. An ore that contains bismuth as the free metal can be purified by heating it above the melting point of bismuth. Liquid bismuth forms on the gangue and is poured off. Some metals combine readily with mercury to form solutions called *amalgams*. Mercury thus can be used as a solvent for ores that contain such metals. After the amalgam forms, it is poured off and subjected to further purification. Dressing with mercury is commonly used for ores of gold and silver.

Some ores undergo further dressing, either to remove impurities or to convert the metal-containing substances into compounds from which the metal is more easily won. Such a step is necessary in the production of aluminum. Iron oxide is a common contaminant of bauxite, the major ore of aluminum. Since iron cannot be easily separated from metallic aluminum, the iron oxide is removed before the metal is won from the ore. Bauxite, which is actually aluminum oxide, is soluble in sodium hydroxide solution, because aluminum forms a complex ion with hydroxide ligands. Iron does not, and iron oxides are insoluble in basic solution. To eliminate the iron oxides, the aluminum oxide is dissolved away from the rest of the ore by the process:

$$Al_2O_3(s) + 2OH^- + 3H_2O \longrightarrow 2Al(OH)_4^-$$

The solution is separated and then neutralized. The resulting precipitate of aluminum hydroxide is heated to drive off water. This leaves Al_2O_3, from which the metallic aluminum is produced.

Titanium is a metal whose mechanical properties are altered by only small amounts of impurities, so it is imperative to separate titanium from other metals that are found in the ore. To do this, the ore, which contains titanium as the oxide, is first heated with carbon to form the carbide:

$$TiO_2(s) + 2C(s) \longrightarrow TiC(s) + CO_2(g)$$

The titanium carbide is then treated with chlorine to form titanium tetrachloride:

$$TiC(s) + 4Cl_2(g) \longrightarrow TiCl_4(l) + CCl_4(l)$$

Titanium tetrachloride is relatively volatile, and so is easily separated from the remaining metallic impurities by distillation.

To win a metal, it is often desirable to have the metal as its oxide. In the case of carbonate ores, the oxides can be readily formed by heating:

$$ZnCO_3(s) \longrightarrow ZnO(s) + CO_2(g)$$

Sulfide ores can be converted to oxides by roasting, heating the ore in air:

$$2PbS(s) + 3O_2 \longrightarrow 2PbO(s) + 2SO_2(g)$$

Extraction

To some extent, the details of the process by which a given metal is won from its ore, as well as the nature of the compounds of the metal in the ore, can be

TABLE 19.4
Reduction Potential and Metallurgy

Metal	$\mathcal{E}°$ (V)	Occurrence	Extraction
lithium	−3.045	chlorides, carbonates,	electrolytic
rubidium	−2.925	sulfates, and	
potassium	−2.924	complex oxides	
barium	−2.90		
strontium	−2.89		
calcium	−2.76		
sodium	−2.711		
magnesium	−2.375	complex oxides,	electrolytic or
aluminum	−1.706	silicates, some	chemical
beryllium	−1.70	simple oxides	
		and carbonates	
manganese	−1.029	simple oxides and	chemical reduction
vanadium	−0.89	sulfides, some	with carbon
zinc	−0.7628	complex oxides	
chromium	−0.74		
iron	−0.44		
cobalt	−0.27		
nickel	−0.23		
tin	−0.1364		
lead	−0.1263		
bismuth	0.32	simple oxides and	chemical or physical
copper	0.3402	sulfides, free metal	
mercury	0.7961	(native)	
silver	0.7996		
platinum	1.23		
gold	1.42		

correlated with the reduction potential of the metal (Section 16.6, page 520, and Table 16.1). Table 19.4 summarizes this relationship.

Metals with large negative reduction potentials do not occur in uncombined form in nature. If a metal has a large negative reduction potential, its compounds have negative free energies of formation. Therefore, these compounds are stable with respect to the uncombined metal. The free alkali metals and most alkaline-earth metals, which all have large negative reduction potentials, react readily with oxygen in the air to form oxides. These metals are never found uncombined in nature. Interestingly, they are not found in the form of simple oxides, either. The simple oxides of these metals are so basic that they react readily with water and other acidic oxides, such as the carbon dioxide in the atmosphere. If some free sodium is exposed to the atmosphere, we can represent the reactions that take place as:

$$4Na(s) + O_2(g) \longrightarrow 2Na_2O(s)$$
$$Na_2O(s) + CO_2(g) \longrightarrow Na_2CO_3(s)$$

although the actual reactions in which the ores of such metals have been formed over geological time spans are considerably more complex. We do not find sodium oxide in the earth's crust, but we do find large deposits of sodium carbonate. The alkali metals and alkaline-earth metals are also found as chlorides, sulfates, and complex oxides.

Because compounds of these metals are readily soluble in water, their ores often are found where bodies of water once existed. Compounds of some metals, such as magnesium, potassium, and sodium can even be extracted from ocean water or from inland bodies of water, such as the Dead Sea, where the concentration of these compounds is relatively high.

Because these metals have high reduction potentials and form relatively stable compounds, they are not easily won from their ores by conventional chemical techniques. Thermal decomposition of the compounds is rarely practical. The usual method for winning metals with unfavorable reduction potentials is electrolysis. Since the metals react with water, the electrolysis generally must be carried out using the molten salts.

Metals such as magnesium and aluminum, which form less basic oxides than those of the alkali metals and the other alkaline-earth metals, can be found as simple oxides in nature. These metals are also found as silicates, in which they are combined with both silicon and oxide. We mentioned that the most common ore of aluminum is bauxite, which is hydrated aluminum oxide. Magnesium is often found as the silicate called asbestos, whose formula is $Mg_6Si_4O_{11}(OH)_6 \cdot H_2O$, or as talc, whose formula is $Mg_3Si_4O_{10}(OH)_2$.

Electrolysis is the standard method for winning aluminum from bauxite, although some nonelectrolytic processes have been developed recently. Magnesium is produced either by the electrolysis of molten magnesium chloride or by the ferrosilicon process, which is based on a chemical reduction of magnesium oxide. The source of the magnesium oxide is a plentiful mineral called dolomite, a magnesium calcium carbonate whose formula can be written $MgCO_3 \cdot CaCO_3$. Dolomite is converted to the oxides by heating. The oxides can be reduced by ferrosilicon, an alloy that contains roughly equal amounts of iron and silicon, in a reaction that can be written as:

$$2MgO \cdot CaO + Si(Fe) \longrightarrow 2Mg + Ca_2SiO_4 + Fe$$

and is carried out at about 900 K.

Metals with less negative reduction potentials than the alkaline-earth metals, such as manganese and all the metals below it in Table 19.4, usually are found as simple oxides or as sulfides. Compounds of these metals generally are less stable than compounds of metals with more negative reduction potentials. These metals thus are more easily won from the ores. Chemical reduction is used most often, although electrolysis is used for some metals.

Both methods can be employed to produce zinc. The most common ore of zinc is sphalerite, which is zinc sulfide, ZnS. The preliminary treatment of the ore is roasting:

$$2ZnS(s) + 3O_2(g) \xrightarrow{\Delta} 2ZnO(s) + 2SO_2(g)$$

In a method of chemical extraction called the distillation process, the zinc oxide is reduced at high temperature with carbon:

$$ZnO(s) + C(s) \longrightarrow Zn(g) + CO(g)$$

At this elevated temperature, the zinc metal is a gas that is collected in relatively pure form by condensation. This kind of reaction is also used to win other metals with reduction potentials close to that of zinc. Carbon is the reducing agent of choice. It is inexpensive and it forms gaseous oxidation products that are easily removed.

Zinc can also be formed from zinc oxide by electrolysis. The ZnO is first leached with sulfuric acid to form a concentrated solution of zinc sulfate:

$$ZnO(s) + H^+ + HSO_4^- \longrightarrow Zn^{2+} + SO_4^{2-} + H_2O$$

The solution is purified to eliminate other metal contaminants. It is then electrolyzed to produce zinc metal at the cathode and oxygen at the anode. The overall reaction is

$$2Zn^{2+} + 2SO_4^{2-} + 2H_2O \longrightarrow 2Zn(s) + 2HSO_4^- + O_2(g) + 2H^+$$

The electrolysis process used to win zinc, unlike those used for metals with more negative reduction potentials, is carried out in water. Zinc does not react with water; the other metals do.

Metals that have still less negative reduction potentials than zinc are even easier to win from their ores because their compounds are even less stable. Tin and lead are two such metals. Both are relatively inexpensive because they are easily won from their ores. Metals with positive reduction potentials can be found native—uncombined—in the earth's crust. While these metals are easily won, they are not always abundant. Gold, silver, and platinum are all native metals whose cost is high, in part because they are hard to find.

The most familiar metals, those that were known earliest in history, are those with positive or only slightly negative reduction potentials. Metals such as gold and copper are not abundant in the earth's crust, but the ease with which they can be won from their ores made them available to ancient civilizations. Indeed, abundance is generally no indication of a metal's role in history. Almost all the common metals, those most frequently encountered in our lives—tin, gold, lead, copper, silver, mercury—are not abundant in the earth's crust. The major exception is iron, which is both abundant and easily won, and which has played a prominent role in history.

Refining

Many methods are used to refine metals. Some of these are physical methods, based on a difference in phase transition temperature or in solubility between a metal and its impurities. Zinc, which has a relatively low boiling point (1180 K), can be separated from iron (B.P. 3272 K) and lead (B.P. 2010 K), both common impurities, by distillation, a method that is useful for all metals with low boiling points.

A method that can be used to prepare extremely pure samples of metals and other solids that do not decompose when they melt is zone refining, which is shown schematically in Figure 19.9.

Refining by amalgamation is based on the ready solubility of both silver and gold in mercury. Once these metals are dissolved in mercury, the solution usually can be separated easily from other solids that may be present. The final step is removal of the mercury by distillation, leaving pure gold or silver.

Metals can also be refined by electrolysis, a process that is especially important in copper production. The smelting of copper ore produces impure copper that is formed into large anodes, weighing about 300 kg each, which then are suspended in a solution of copper sulfate and sulfuric acid. The cathodes are thin sheets of pure copper. When current is passed through the cell, the half-reactions are:

anode: $\quad\quad$ $Cu(s) \longrightarrow Cu^{2+} + 2e^-$
cathode: $Cu^{2+} + 2e^- \longrightarrow Cu(s)$

The pure copper that collects on the cathode is removed and recast. Electrolysis not only provides pure copper but also produces an appreciable amount of other metals that are present as impurities in the copper. Many of these impurities—generally metals of more positive reduction potential than copper—do not dissolve in the electrolyte solution. Instead, they sink to the bottom of the tank and form a slime from which they can be recovered. Other metals dissolve in the electrolyte but do not plate out at the cathode. Copper ores yield about 32% of the gold mined in the United States, 28% of the silver, and substantial amounts of molybdenum, selenium, tellurium, and metals of the platinum group as by-products of the electrolytic refining process.

Another technique used to prepare very pure metals is the conversion of the metal to a volatile compound that can be purified by distillation. This technique is used for the purification of nickel, with the volatile nickel carbonyl as the intermediate:

$$Ni \text{ (impure)} + 4CO(g) \longrightarrow Ni(CO)_4(g) \xrightarrow{\Delta} Ni \text{ (pure)} + 4CO(g)$$

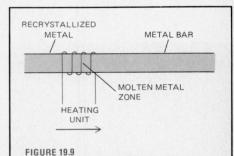

FIGURE 19.9
Zone refining. A small section of a metal bar is melted by the heating unit. As the heating unit moves along the bar, impurities move with it, since they are more soluble in liquid metal than in solid metal. The impurities are thus concentrated in one zone, which can be removed when the heating unit reaches the end of the bar. Successive passes can give metal of extremely high purity.

The cyanide process can be used to refine some precious metals that form relatively stable cyanide complexes. When impure gold is treated with cyanide solution through which air is bubbled, a complex ion forms:

$$4Au(s) + 8CN^- + O_2(g) + 2H_2O \longrightarrow 4Au(CN)_2^- + 4OH^-$$

The solution of the complex ion is separated from the impurities, and the gold is recovered from the complex ion by treatment with a metal such as zinc:

$$2Au(CN)_2^- + Zn(s) \longrightarrow 2Au(s) + Zn(CN)_4^{2-}$$

Similar chemical techniques are used to refine platinum and metals of the platinum group.

19.5 THE GROUP VIII METALS

The metals of group VIII can be divided into two subgroups on the basis of their properties. One subgroup consists of iron, cobalt, and nickel, which are in the third row of the periodic table, the first row of the transition elements. Many properties of these three elements differ from the properties of the six metals in the other subgroup, which are in the second and third rows of the transition metals. The six group VIII metals in these rows are called the platinum metals. They are ruthenium (Ru), rhodium (Rh), palladium (Pd), osmium (Os), iridium (Ir), and platinum (Pt).

With the exception of platinum and palladium, the group VIII metals have ground state valence electronic configurations in which the outermost s subshell is filled and the outermost d subshell is partly filled. All the group VIII metals display the $+2$ oxidation state in their chemistry. While the $+2$ oxidation state plays the dominant role, the $+3$ state is also common, especially in iron and cobalt. The higher oxidation states are more important in the chemistry of the platinum metals. For example, ruthenium is found in all the oxidation states from $+1$ to $+8$, with the $+3$, $+4$, $+6$, and $+8$ states the most important. For osmium, the principal oxidation state is $+8$.

All the elements of group VIII have enough valence electrons for the metallic valence of 6. The resulting strong metallic bonding is reflected in the strength and hardness that makes these metals so valuable. The six platinum metals are much more inert than the first three group VIII metals—so much so that they are often called "noble metals."

The group VIII metal that has had the widest range of uses is iron, the fourth most abundant element in the earth's crust (after oxygen, silicon, and aluminum). Modern civilization is built on a foundation of iron—but not iron in its pure form, since the mechanical properties of iron are greatly improved by the addition of small amounts of other elements, most notably carbon.

Metallurgy of Iron

Extracting iron from its ores and producing iron alloys (an alloy is a mixture of two or more metals) is one of the central activities of any industrial society. An industrialized nation requires an ample supply of iron ores—the oxides hematite, Fe_2O_3, and magnetite, Fe_3O_4, and the carbonate siderite, $FeCO_3$, are chief among them. Once the ores are mined, they are roasted. The iron is then won from the ores in a blast furnace (Figure 19.10). If the ores are oxides, the iron is liberated by reduction with carbon, in the form of coke, at high temperatures. The reactions are:

$$2C(s) + O_2(g) \longrightarrow 2CO(g)$$
$$3CO(g) + Fe_2O_3(s) \longrightarrow 2Fe(l) + 3CO_2(g)$$

The high temperatures required to win iron from its ores in a blast furnace are evident in this scene. (*Inland Steel*)

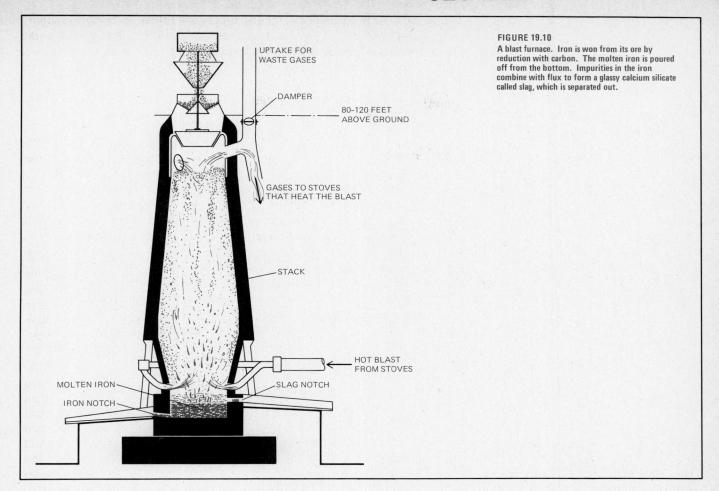

UPTAKE FOR
WASTE GASES

DAMPER

80-120 FEET
ABOVE GROUND

GASES TO STOVES
THAT HEAT THE BLAST

STACK

HOT BLAST
FROM STOVES

MOLTEN IRON

SLAG NOTCH

IRON NOTCH

FIGURE 19.10
A blast furnace. Iron is won from its ore by
reduction with carbon. The molten iron is poured
off from the bottom. Impurities in the iron
combine with flux to form a glassy calcium silicate
called slag, which is separated out.

The gases formed in the reactions escape through the top of the blast furnace. The liquid iron spills to the bottom of the furnace, where it is poured off.

The removal of substantial quantities of impurities in iron ore is accomplished by introducing a material called *flux*, which causes the impurities to fuse. The fused impurities combine with the flux to form a glassy substance called *slag*. The slag, which is less dense than liquid iron, forms a layer atop the iron that can be removed easily. The type of flux used in a blast furnace depends on the nature of the impurities in the ore. If the ore contains sand or clay, limestone, $CaCO_3$, is used as the flux. If the ore contains limestone as an impurity, sand or clay— actually silica, SiO_2—is used as the flux. In either case, the slag which forms is primarily a calcium silicate whose formation can be represented by the reactions:

$$CaCO_3(s) \longrightarrow CaO(s) + CO_2(g)$$
$$CaO(s) + SiO_2(s) \longrightarrow CaSiO_3(l)$$

The molten iron that collects at the bottom of a blast furnace is not pure. It contains about 3% or 4% carbon and a number of other impurities, including small amounts of silicon and sulfur. This iron is cast into bars that are called pigs. Most pig iron is processed further to wrought iron or steel. While pig iron is the cheapest form of iron, its usefulness is limited because it is brittle. The impurities also lower the melting point of pig iron more than 300 K below that of pure iron.

Wrought iron is essentially pig iron that has been further purified. It has more desirable mechanical properties than pig iron. Most of the impurities are removed by heating the pig iron to high temperatures in a furnace in which a bed of iron oxide causes the oxidation of impurities and is itself reduced to relatively pure iron; carbon becomes carbon monoxide, and sulfur becomes sulfur dioxide. Gases such as these escape from the iron, while impurities such as silicon and phosphorus form a slag that is poured off. The end product, wrought iron, is 99.5% pure.

Steel

Steel is manufactured by purifying the iron even further. A stream of heated air is blown into the iron to oxidize the impurities, while the furnace is lined with an oxide such as lime (CaO) or silica to absorb other oxidation products. Once the iron is purified, carefully measured amounts of carbon can be added to form carbon steel. There are several kinds of carbon steel, classified by the quantity of carbon they contain and their mechanical properties. Mild steels, which contain relatively low concentrations of carbon, have mechanical properties similar to those of wrought iron and are used to make wire, chains, pipes, and sheet iron. Medium steels, containing from 0.2% to 0.6% carbon, are structural steels, used for buildings, bridges, and railroad rails. High-carbon steels, containing up to 1.5% carbon, are used for razor blades, tools, and cutting instruments.

Steels with special properties can be produced by adding small amounts of elements other than carbon. Stainless steels, for example, are mild steels that contain at least 12% chromium and 7%–9% nickel. The resistance of stainless steel to corrosion makes it useful for cutlery and other instruments. Manganese steel is quite wear-resistant and is used for such things as safes and grinding machinery. Chromium-vanadium steels are both strong and elastic, and are used for, among other things, automobile axles. Other specialty steels contain tungsten, molybdenum, cobalt, and titanium.

Cobalt and Nickel

Cobalt and nickel are constituents of many alloys. Together, they are used in alnico alloys, which also contain aluminum and iron and which are used to make powerful magnets. Cobalt is added to tungsten steels and other steels designed to be highly resistant to oxidation and corrosion. Nickel is a constituent of stainless steel and a number of other alloy steels that are both tough and ductile. Nickel is often alloyed with copper. The ordinary five-cent piece, the ''nickel,'' is such an alloy; other nickel-copper alloys have many uses based on their exceptional resistance to corrosion. Nickel is often used to plate other metals, making them more attractive and corrosion-resistant. Nickel plating is done by electrolysis.

Ferromagnetism

The unusual magnetic properties of iron, cobalt, and nickel are perhaps their most characteristic feature. When we discussed magnetic properties in Section 7.6, page 217, we divided substances into two groups; those that are paramagnetic and are attracted by an external magnetic field, and those that are diamagnetic and are repelled by an external magnetic field. The difference reflects a difference in electronic configuration. All the electronic spins in a diamagnetic substance are paired, while a paramagnetic substance has one or more unpaired electrons. For most substances, the magnitude of the attraction or repulsion depends on the strength of the magnetic field. But a relatively small group of paramagnetic substances display an unexpectedly large attraction when they are exposed to a relatively weak external magnetic field. These substances have another unusual characteristic—they remain magnetized when the external field is removed. Such substances are said to be *ferromagnetic*. Iron, cobalt, and nickel are ferromagnetic at room temperature. Only one other pure element, gadolinium, is ferromagnetic at room temperature, although there are alloys and compounds that are ferromagnetic.

Only solids are ferromagnetic. The phenomenon occurs when all the units of structure in a small region of the crystal are lined up so that their magnetic moments are parallel. Such a region is called a *domain*. A typical domain is about 0.01 mm across and contains about 10^6 units. Ordinarily, the moments of domains are randomly oriented throughout a crystal, so that the solid has no net magnetic polarization. When even a weak external magnetic field is applied, the direction of magnetization of many of the domains becomes aligned with the field. With so many of the units of structure in the crystal aligned, the substance has a strong magnetic polarization. Even when the external magnetic field is removed, the domains retain their aligned direction of magnetization. To make the solid lose its strong magnetization, the alignment of the domains must be returned to their random orientation. The solid can be melted, or just heated, or even more simply, pounded with a hammer to destroy the ordered configuration. The solid then loses its magnetization.

Compounds of Iron, Cobalt, and Nickel

We mentioned that the +2 oxidation state is most often encountered in compounds of iron, cobalt, and nickel, although there are exceptions. The +3 oxidation state is found in many compounds of iron. Cobalt generally can be in the +3 oxidation state only when it is stabilized by surrounding ligands, while only a few compounds of nickel in the +3 oxidation state are known. There are many compounds of these three metals, and they have a broad range of practical applications. We shall examine a few of them.

The oxides of iron, cobalt, and nickel are of great interest. Iron forms three important oxides, FeO, Fe_2O_3, and Fe_3O_4. The last, named magnetite, is a black solid that was the first ferromagnetic substance to be discovered. It is called a mixed oxide, and is best thought of as a combination of FeO and Fe_2O_3. The iron(II) oxide is unstable; it disproportionates to iron and magnetite:

$$4FeO(s) \longrightarrow Fe(s) + Fe_3O_4(s)$$

The ease with which Fe(II) is oxidized to Fe(III) by air is a major feature of the chemistry of iron. It also creates practical difficulties for anyone who wants to work with iron in the +2 oxidation state.

For example, if a piece of iron(II) hydroxide, $Fe(OH)_2$, a white solid, is left standing on a table, it rapidly changes color, first to dark green and then to black, as O_2 in the air oxidizes the Fe(II) to Fe(III). The dark colors seen in this compound and in magnetite are typical of many solids that contain two oxidation states of the same element; iron hydroxide and magnetite each contain iron in both the +2 and +3 oxidation states. If the hydroxide is left standing long enough, its continued reaction with O_2 in the air eventually causes the color to change to red-brown, as the Fe(II) is converted completely to Fe(III). The formula $Fe(OH)_3$ is often written for the +3 hydroxide of iron, although no such compound has ever been isolated. The reddish brown substance formed in this way, or by precipitation from alkaline solutions of Fe^{3+}, actually has a composition close to $FeO(OH)$.

Cobalt also forms three oxides whose composition and behavior resemble those of the oxides of iron. The colors of these oxides make them useful as pigments in ceramics and pottery. If a solution of Co^{2+} is treated with a base, it precipitates $Co(OH)_2$, which can be either pink or blue, depending on the conditions. Solid $Co(OH)_2$ also is air-oxidized, although not as readily as the comparable hydroxide of iron. The brown solid that forms as Co(II) is oxidized to Co(III) and has a composition close to $CoO(OH)$.

The nickel(II) oxide, NiO, and the hydroxide, $Ni(OH)_2$, are easily prepared. They do not oxidize in air, since higher oxidation states of nickel are much more difficult to form than those of iron and cobalt. Both NiO and $Ni(OH)_2$ are green.

Iron, cobalt, and nickel all react directly with carbon monoxide to form compounds called carbonyls, in which the metal is formally in the 0 oxidation state. Carbonyls are interesting chemically and have a number of practical uses. Iron pentacarbonyl, $Fe(CO)_5$, is a yellow liquid with a trigonal bipyramidal structure, with the Fe atom at the center. It is soluble in organic solvents and is used widely as a homogeneous catalyst in industry. The most important carbonyl of cobalt, $Co_2(CO)_8$, has a bond between the two cobalt atoms. It is used extensively to catalyze organic reactions. Nickel tetracarbonyl, $Ni(CO)_4$, is an extremely toxic substance that is used to prepare high-purity nickel powders and coatings. It is an intermediate in the refining of nickel.

Other compounds of iron, cobalt, and nickel—halides, sulfates, cyanides, sulfides, and so on—are well known and have many applications. Among them are $FeSO_4$, which is used in water purification and ink manufacturing; $FeCl_3 \cdot 6H_2O$, which is used medically to treat anemia and to stop bleeding; $CoCl_2 \cdot 6H_2O$, which can be used as an invisible ink, since a small amount of heat causes it to lose some water of hydration and to change from colorless to blue; and $NiSO_4$, which is used in nickel-plating baths.

The Platinum Metals

Ruthenium, rhodium, palladium, osmium, iridium, and platinum are grouped together as the platinum metals because their properties are so similar. All six metals are dense, have high melting points, and are highly resistant to chemical attack, particularly oxidation by O_2 in the air. The combination of chemical inertness, desirable metallic properties, low abundance, and difficulty of refining makes these ''noble metals'' extremely expensive. Table 19.5 lists the prevailing 1977 prices of the platinum metals and some other metals in pure form.

The platinum metals occur in native form, so there is no great difficulty in winning them from their ores. But their abundance is so low that there are major difficulties in concentrating the ores. An even greater difficulty arises from the fact that all six metals occur together. Since they have similar properties and are unreactive chemically, they are not easily separated. Complex multistep procedures have been developed for their separation.

The platinum metals have many applications in pure form, alloyed with one another, or in combination with other metals. Platinum and palladium are used extensively as catalysts in the petroleum industry and in many industrial processes. Both metals are used for switches and relays in telecommunications systems. Alloys of Pt and Pd are used in dentistry, while an alloy of gold and platinum, ''white gold,'' and an alloy of platinum and iridium, are used in jewelry.

The four less common platinum metals have found a variety of specialized uses. Any of them can be alloyed with platinum. The resulting alloys are quite wear-resistant, and are used for such parts as electrical contacts, fountain pen tips, and phonograph needles, in which the high cost of the alloy is offset by the small quantity that is needed. All four metals are used as catalysts in chemical manufacturing. Jewelry makers plate silver objects with rhodium; medical supply companies use iridium for hypodermic needles and other medical accessories. A compound of osmium and oxygen, osmium tetroxide, OsO_4, is widely used as a mild oxidizing agent and as a stain for tissue samples in biology and medicine. Osmium tetroxide is highly toxic, and so must be handled with great care.

TABLE 19.5
Cost of Metals (dollars/g) in 1977

Metal	Cost
ruthenium	14
rhodium	46
palladium	20
osmium	30
iridium	60
platinum	27
copper	0.025
silver	1.60
gold	25

19.6 THE GROUP IB METALS

Group IB contains the three coinage metals: copper, silver, and gold. Even though the neutral atoms of these three elements have completed d subshells, all three are classified as transition metals because they can form cations with incomplete d subshells. The electronic configuration of these metals resembles that of the group IA metals, since both have a half-filled s shell. The configuration of copper is $4s3d^{10}$; that of silver is $5s4d^{10}$; and that of gold is $6s5d^{10}$. In the group IB metals, however, the electron in the s shell lies outside a completed d subshell, while in the IA metals it lies outside a completed noble gas electronic shell. A group IA metal that loses one electron becomes a monocation with a noble gas electronic configuration. A group IB metal that loses an electron does not have a noble gas electronic configuration, so the physical and chemical properties of the two groups of metals are quite different.

The Metals and Alloys

Copper, silver, and gold are among the best-known metals. They have been familiar from the earliest times—not because they are abundant metals, but because they were easily found and removed from the earth's crust. Gold may have been the first metal discovered and used. Copper, whose distinctive red color makes it readily recognizable, did not come far behind. Copper objects more than 10 000 years old have been found by archeologists. As early as 5000 B.C., copper ores were being smelted on a fairly wide scale in the Middle East. The mining and refining of copper and the manufacture of copper articles played a prominent role in biblical times. Silver, unlike copper and gold, is rarely found in native form and so was not used as widely in early civilizations. In fact, silver was so rare that some ancient cultures prized it more than gold.

Gold and silver have kept their hold on the minds of people down to the present day. In ancient times, gold and silver were considered not only precious but also, in some cases, sacred. They were used for sacramental purposes, as well as for money and jewelry. A good part of modern history, and of the history of science, is built around gold and silver. The quest for gold and silver was a driving force behind the exploration and conquest of the New World. The attempt to convert base metals into gold was a central activity of alchemy, the medieval pseudoscience that left a legacy of findings for modern chemistry.

Aside from the superstitious value placed on gold and silver, the group IB metals have properties that make them valuable for many practical uses. Pure copper is an excellent electrical and thermal conductor, and is widely used in electronics and in cookware. If pure copper is first heated and then cooled, it is soft enough to be worked easily. It can be drawn into wire or hammered into shape. Conveniently, the metal hardens as it is worked, and can be softened again by reheating.

Alloys in which copper is the chief constituent are used more widely than pure copper, because the alloys usually cost less and have more desirable properties. One major technological advance of early civilizations was the discovery of bronze, an alloy of copper and tin that is easier to cast, is harder, and is less malleable than pure copper. Bronze dominated early technology before iron came into use. Brass is an alloy of copper and zinc with multitudinous uses. Alloys of copper, nickel, and zinc, called nickel silver or German silver, have high resistance to corrosion and wear and are used as bases for silver plating and in costume jewelry. Among the many other alloys of copper used today are aluminum bronzes, alloys of copper and aluminum; silicon bronzes, alloys of copper, silicon, and traces of other metals; and phosphor bronzes, alloys of copper, tin, and phosphorus.

Silver, until recently, was used primarily in coins, in the form of coinage silver, or sterling silver, an alloy of 90% silver and 10% copper. Perhaps 10^{10} g of silver was used annually for coinage. In recent years, industrial demand for silver has raised its price to the point where its use for coinage is impractical. Silver has a number of desirable properties. In pure form, silver has the highest electrical and thermal conductivity of any metal, and is more ductile and malleable than all metals but gold and palladium. Silver is even valuable as a germicide.

A solution of silver that is harmless to higher organisms can kill microbes on contact.

Gold today remains important primarily as an international monetary standard, a role it has held since the late eighteenth century. Although governments and economists have worked diligently to change the situation, gold still is regarded by many as the ultimate medium of exchange—an opinion that gains strength in times when paper monies deteriorate in value. A large fraction of the world's gold production finds its way into government-controlled stores, such as Fort Knox in the United States. A large portion of the remainder is used for decorative purposes, in jewelry, religious objects, books, and ceramics. But as a metal, gold has properties that have led to its increasing use in electronics and in spacecraft. Gold is the most malleable and ductile of all the metals; it can be beaten into gold leaf only 1×10^{-6} cm thick. The electrical and thermal conductivity of gold is exceeded only by that of silver and copper. Thin films of gold reflect a high percentage of infrared radiation and thus are excellent heat shields; you may remember pictures of spacecraft wrapped in gold leaf. Many heat-sensitive electronic devices are electroplated with gold. For other electronic uses, gold can be drawn into extremely fine wire. One gram of gold can form a wire about 3 km long.

Pure gold is too soft to be usable in jewelry or coinage, so it usually is alloyed with copper, silver, nickel, zinc, palladium, or other metals. The composition of a gold alloy is usually expressed as the number of parts of gold in 24 parts of alloy. Each part is called a carat. Thus, pure gold is 24 carat. An alloy of 75% gold and 25% copper is customary in jewelry. It is 18 carat gold.

While copper, silver, and gold all have the same valence electronic configuration and are similar in some physical properties, there are many differences in their chemical behavior that are not explained simply. We must look at the details of their electronic configurations to understand the behavior of the three metals.

All three metals are ''noble,'' in the sense that their reduction potentials are positive, that they resist corrosion, and that they are not readily attacked by air. As we mentioned earlier, silver tarnishes when exposed to air that contains compounds of sulfur. When copper is exposed to moist air that contains sulfur compounds, it slowly forms a green patina that is a combination of $CuCO_3$, $Cu(OH)_2$, and some $CuSO_4$. The Statue of Liberty, which is made of copper, is covered by such a patina.

Compounds

Each of the three metals has a half-filled s electronic subshell, so we would not ordinarily expect them to be unreactive. Their relatively low reactivity is due to their relatively high ionization energies, which can be explained by a comparison with the group IA metals. In the group IA metals, the single s electron is shielded from the nucleus by a completed electronic shell. In the group IB metals, the shielding of the single electron from the nucleus is by a completed d subshell, which does a less effective job. Therefore, the negatively charged electron in a group IB metal is more tightly held by the positively charged nucleus. More energy is required to remove the electron, so the ionization energy is high compared to that of the comparable group IA metal.

There is also a contrast in second ionization energies of metals in the two groups. The second ionization energy of a group IA metal is very high, since an electron must be removed from a completed electronic shell. The second and third ionization energies of a group IB metals are much lower, since the electrons are removed from the d subshell. Thus, higher oxidation states of the group IA metals are unknown, while $+2$ and $+3$ oxidation states of the group IB metals are common.

The most common oxidation states are $+2$ for copper, $+1$ for silver, and

TABLE 19.6
Ionization Energies of Group IB Metals (kJ/mol)

Metal	Electronic Configuration	First	Second	Third
copper	$[Ar]3d4s^{10}$	745	1958	3554
silver	$[Kr]4d5s^{10}$	731	2074	3361
gold	$[Xe]4f^{14}5d6s^{10}$	890	1978	2940
potassium	$[Ar]4s$	419	3050	4410

+3 for gold. However, the +1 state is well known for copper, as are the +2 state for silver and the +1 state for gold. There is an interplay of factors that determines the relative stability of the oxidation states of each metal.

The electronic configuration associated with the +1 state would seem to be best for all three metals, since it has a closed electronic subshell: $3d^{10}$ for Cu^+, $4d^{10}$ for Ag^+, and $5d^{10}$ for Au^+. But these monocations may not be as stable as dications or tications in environments where there is electrostatic stabilization caused by solvation, or where they have anionic neighbors in a crystal. Therefore, we must consider the nature of the environment and the magnitude of the second and third ionization energies to determine the most stable oxidation state of the group IB metals.

In the case of copper, the Cu^+ cation is more stable than the Cu^{2+} cation in the gas phase, but the opposite is true in aqueous solutions or in crystalline salts. For silver, the Ag^+ cation is the most stable under most conditions, probably because the second ionization energy of silver is relatively high, at least in comparison with copper. Gold has an unusually low third ionization energy that is consistent with the stability of the +3 ionization state. Table 19.6 lists the ionization energies of the three metals, with those of potassium, a group IA metal, shown for comparison.

There are many compounds of copper in both the +1 and the +2 oxidation states. In compounds of the +1 oxidation state, the copper is usually covalently bonded or complexed. Thus, the copper halides, CuCl, CuBr, and CuI, can be regarded as covalent compounds. Copper(I) is present in biological systems, stabilized by complexation with ligands in protein molecules. Compounds of copper in the +2 oxidation state are ionic solids or complex ions with coordination numbers of four, five, or at most, six. In aqueous solution, the Cu^+ ion is unstable with respect to the Cu^{2+} ion. The equilibrium

$$2Cu^+ \rightleftharpoons Cu^{2+} + Cu(s)$$

lies far to the right.

Some compounds of copper with important practical uses are CuCl, which is a catalyst and desulfurizing agent in the petroleum industry; $CuCl_2$, which is a textile dye and a crop fungicide; Cu_2O, which is used in marine paints and electronic components; and CuO, an insoluble black solid which is used as an oxidizing agent. The most widely used copper compound is blue vitriol, hydrated copper(II) sulfate, $CuSO_4 \cdot 5H_2O$, which is a soil additive, a fungicide, a wood preservative, a component of electric cells, and a starting material for the manufacture of other copper compounds.

Many simple compounds and complexes of silver(I) are known. There are only two known simple compounds of silver(II), AgF_2 and AgO, although a number of complexes of silver(II) have been identified. One useful compound of silver(I) is silver nitrate, $AgNO_3$, which is unusual among simple silver salts because it is soluble in water; almost all the others are not. Silver nitrate is a starting material for the production of other silver salts. Another silver(I) compound, silver oxide, Ag_2O, is used as a mild oxidizing agent.

Gold compounds are not nearly as useful as those of copper and silver. Pure

PICTURES IN COLOR

The first color photographs were taken in the middle of the nineteenth century by what is called the additive color system. Three black-and-white negatives were made of a scene, each exposed through a filter of a different primary color—red, blue, and green. Black-and-white positive transparencies were made of the three negatives and were projected simultaneously through three color filters to produce a single image in color. The process produced pictures of good quality, but it was too cumbersome for widespread use.

The modern era of color photography began in the 1930s with the introduction of films that use what is called the subtractive color process. This process uses dyes that absorb all but the desired part of the visible spectrum. For example, a red image is obtained by using two dyes that absorb, respectively, the blue and green wavelengths of light. Only red light is not absorbed, so a red image is seen.

A color film has three layers of emulsion, each consisting of grains of silver halide dispersed in gelatin. Each layer is made sensitive to a different primary color by the addition of a sensitizer. A sensitizer is an organic substance that absorbs light of certain wavelengths and transfers the energy to the silver halide. When a color film is developed, the grains of silver halide in each emulsion layer are reduced by the developer to produce an image, just as in the processing of black-and-white film. The oxidized developer reacts with compounds called color couplers. The products of the reaction are dyes, which form in the areas where a silver image has been produced. The amount of dye that is generated is proportional to the density of the image. If an appropriate combination of couplers is used, any color can be reproduced on film. In the first modern color film, the couplers were in the developer, and the film had to go through three developing processes, one for each emulsion. Most color films today incorporate the couplers in the film, next to the emulsion layers, allowing one-step developing.

The instant color films that have been introduced in recent years work on the same basic principle, except that all the chemicals necessary for developing are incorporated in the film packet. Instead of using dye couplers, instant film uses dyes that are chemically linked to the developer. One system now in use has a layer that contains a dye-and-developer compound next to each of the three emulsion layers. When the film is exposed, the dye-developer molecules migrate into the emulsion layers, developing the silver halide and releasing their dye components to form a color image. A process that once took long and careful work in the darkroom now occurs automatically within seconds.

gold is needed for most applications. However, the gold cyanides, such as $KAu(CN)_2$ are used for electroplating, and some gold compounds are used medically to treat arthritis and tuberculosis—uses that are limited by the toxicity of gold. Most gold compounds are unstable thermally, decomposing to gold metal if they are heated. The only oxide of gold that is well characterized is Au_2O_3, which must be prepared indirectly because gold is the only metal that does not react with O_2, even at high temperatures.

Photography

Unquestionably the most important silver compounds are the silver halides, $AgCl$, $AgBr$, and AgI, which are the light-sensing materials in photographic film. More than 30% of the industrial silver in the United States goes into photography, so the chemistry of photography is of major interest.

When the silver halides are exposed to light, the silver is reduced from the $+1$ to the 0 oxidation state, as in the reaction

$$AgBr \longrightarrow Ag + \tfrac{1}{2}Br_2$$

This process is called a photoreduction, that is, a reduction brought about by light. In photography, the photoreduction occurs in a photographic emulsion, which consists of small grains of a silver halide (usually silver bromide) suspended in gelatin and coated on a strip of cellulose acetate. Dyes are added to the emulsion to make the silver bromide sensitive to a greater range of wavelengths of light.

When film is exposed, light strikes the photographic emulsion and a small number of the molecules on the surface of the grain of silver halide undergo photoreduction. Grains in which this occurs are said to be sensitized. They can be reduced much more easily than grains of silver halide in the emulsion that are not struck by the light. Photosensitization is a highly effective process. Reduction of only five atoms of silver to the 0 oxidation state on the surface of a grain of silver halide with 10^{10} atoms is enough to sensitize the grain. When the exposed film is treated with a mild reducing agent, all the sensitized grains of the halide are reduced to silver, while the grains that were not struck by light remain unchanged. This exposure to a reducing agent is commonly called developing. Organic reducing agents such as hydroquinone are the most widely used developers.

The next step is to remove the undeveloped grains of silver halide by formation of a complex ion:

$$AgBr + 2S_2O_3{}^{2-} \longrightarrow Ag(S_2O_3)_2{}^{3-} + Br^-$$

The process is called fixing. The most common fixer is a solution of sodium thiosulfate, $Na_2S_2O_3 \cdot 5H_2O$, customarily called hypo. When the silver(I) complex is washed away, all that remains on the film is grains of metallic silver, which are most numerous in areas where the incident light was strongest. The film is then called a negative. A positive print is made by passing light through the negative to strike another photographic emulsion on print paper. The pattern on the picture is just the reverse of the pattern on the negative; it is light where the negative is dark and dark where the negative is light.

The sensitivity of silver halides to light is not the reason why they are uniquely suited for photography. Many other substances are equally light-sensitive. The silver halides are unique because a very small amount of light sensitizes a large quantity of the silver compound to further reduction. In developing, the silver metal image formed by the initial exposure to light is intensified by as much as a factor of 10^{11}.

EXERCISES

19.1 List six characteristic properties of a material that allow us to classify it as a metal.

19.2 Use the data in Table 19.1 to find the energy released when $K_2(g)$ forms $K(s)$.

19.3 The heat of atomization of $Bi(s)$ is 207 kJ/mol, while the heat absorbed when $Bi(s)$ is converted to $Bi_2(g)$ is 125 kJ/(mol of $Bi(s)$). Calculate the bond energy of the Bi—Bi bond in $Bi_2(g)$.

19.4 Consider a five-atom portion of a crystal of sodium, one sodium atom and four nearest neighbors. Draw ordinary contributing structures with localized bonds, that in sum describe the bonding in this portion of the metallic crystal.

19.5 The ΔH_f° of metal atoms in the gas phase is often used as an estimate of metallic bond strength. Indicate for each of the following metals whether the estimate is a good one or is substantially in error: (a) lithium, (b) silver, (c) zinc, (d) vanadium.

19.6 Propose an explanation for the fact that a greater energy is required for aluminum to reach an excited state for metallic bonding than is required for magnesium.

19.7 Compare the metallic valence of the main group metals to their most common oxidation states.

19.8 The ideal density of a metal is defined as the density it would have if its atomic weight were 50. Thus the ideal density is 50 divided by the volume of one mole of the metal. Use the data in Table 19.2 to make a plot of ideal density on the y-axis against number of valence electrons on the x-axis.

19.9 Predict the trend in the melting point of the metals from rubidium to palladium in the periodic table.

19.10 Indicate which of the following metals have an incompletely filled valence band: (a) K, (b) Ag, (c) Cu, (d) Zn, (f) Ba.

19.11 Propose an explanation for the thermal conductivity of metals based on the band theory.

19.12 Graphite, although a nonmetal, is a good conductor of electricity. Propose an explanation for this observation.

19.13 Tin exists in two allotropic forms. Gray tin has a diamond structure and white tin has a close-packed structure. Predict which allotrope is (a) denser, (b) a conductor of electricity. Predict the valence and the electronic configuration of tin in each allotrope.

19.14 Calculate the minimum frequency of radiation required to promote an electron from the valence band of diamond into a conduction band.

19.15 The conductivity of a p-type semiconductor increases with temperature in a fairly narrow temperature range above room temperature and then levels off. Explain this observation.

19.16 Write balanced reactions for the following processes: (a) the roasting of molybdenum(IV) sulfide, (b) formation of the oxide from aluminum carbonate, (c) reduction of tin(IV) oxide with carbon, (d) reduction of beryllium oxide with ferrosilicon.

19.17 Cobaltite, CoAsS, is a mineral that is a source of cobalt. Propose a chemical procedure to win cobalt from this mineral. Discuss some of the hazards of the procedure.

19.18 The major ore of uranium is pitchblende, U_3O_8. The uranium can be won from the ore by reduction with carbon, calcium, or aluminum. Write equations for these three processes.

19.19 The K_i for $[Ag(CN)_2]^-$ is 7.9×10^{-22} and that of $[Cu(CN)_2]^-$ is 1.0×10^{-24}. The standard reduction potential, $\mathcal{E}^\circ$, is 0.80 V for Ag^+ and 0.52 V for Cu^+. Calculate K at 298 K for the reaction $[Ag(CN)_2]^- + Cu(s) \rightleftharpoons [Cu(CN)_2]^- + Ag(s)$.

19.20 The first ionization energies of iron, cobalt, nickel, and the first three platinum metals are all about the same; the ionization energies of osmium, iridium, and platinum are substantially higher. Suggest an explanation for this observation.

19.21 Discuss the formation of slag in the metallurgy of iron in terms of the Lewis theory of acids and bases.

19.22 Write chemical equations for the following processes: (a) the air oxidation of Fe(II) hydroxide, (b) the mixing of a solution of cobalt(II) nitrate with a solution of sodium hydroxide, (c) the formation of the carbonyl of cobalt from the metal and carbon monoxide.

19.23 Find the formal oxidation state of the group VIII metal in the following compounds, all of which have been prepared: (a) $K_4[Ni_2(CN)_6]$, (b) $Na_2[Ni(CO)_6]$, (c) $BaFeO_4$, (d) $Na_2[Fe(CO)_4]$.

19.24 One of the few substances that attacks platinum metal is aqua regia, a mixture of nitric acid and hydrochloric acid. The platinum goes into solution as the $[PtCl_6]^{2-}$ ion while a brown gas is evolved. Write the reaction for this process.

19.25 Predict the geometry of OsO_4 and RuO_4 and discuss the hybridization and bonding in these oxides.

19.26 Although both the group IA and IB metals have a half-filled s subshell, the IB metals have markedly higher first ionization energies, densities, and melting points and markedly lower second and third ionization energies than the corresponding IA metals. Explain these observations.

19.27 The standard reduction potential, $\mathcal{E}^\circ$, for the $Cu^{2+}|Cu(s)$ couple is 0.34 V. For the $Cu^{2+}|Cu^+$ couple it is 0.16 V. Find the value of K for the reaction $2Cu^+ \rightleftharpoons Cu^{2+} + Cu(s)$.

19.28 The K_i of $[Cu(NH_3)_2]^+$ is 1.6×10^{-11} and the K_i of $[Cu(NH_3)_4]^{2+}$ is 5.0×10^{-14}. Use these data and the result of Exercise 19.27 to find the K for the reaction $2[Cu(NH_3)_2]^+ \rightleftharpoons [Cu(NH_3)_4]^{2+} + Cu(s)$.

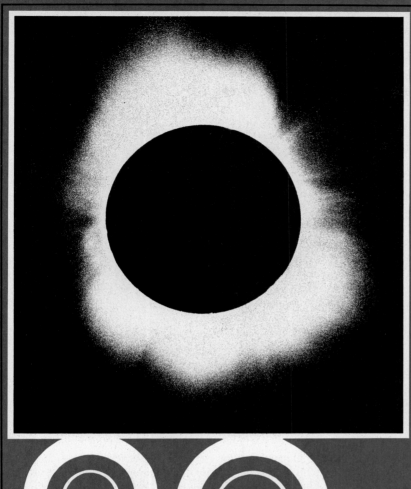

20

NUCLEAR CHEMISTRY

When the twentieth century began, it was believed that the transmutation of elements was impossible. The atom was regarded as the basic and unchangeable unit of matter. The idea that one element could be transformed into another was regarded with the same scorn as the efforts of medieval alchemists to prepare gold from base metals. That belief was to change rapidly.

The discovery of the electron (Section 5.2) demonstrated that atoms are made up of smaller particles. The nuclear model of the atom (Section 5.4) provided a detailed view of atomic structure. The discovery of natural radioactivity demonstrated that at least some atoms could change spontaneously. A whole new branch of chemistry, the field of nuclear chemistry, was opened.

Nuclear chemistry is concerned with changes in which the nucleus of the atom participates in a major way. Until now, we have discussed only ordinary chemical changes, the kind that most chemists deal with. In an ordinary chemical change, the nuclei of the atoms play only the most minor roles and have virtually no effect on the course of the reactions. There is no transmutation of elements (Section 2.5, page 35); indeed, expressions of ordinary chemical changes are written on the assumption that each side of a balanced equation has the same number of atoms of each element. Ordinary chemical changes are the result of electronic changes. Their study is primarily the study of the behavior of electrons. One other point is worth noting: Ordinary chemical changes generally are sensitive to external conditions, such as temperature and pressure. Most of the reactions in nuclear chemistry are not, under ordinary conditions.

In nuclear chemistry, the transmutation of elements is a normal event. Such chemical changes are called *nuclear reactions*. As early as 1903, not long after the original research of Becquerel and the Curies on natural radioactivity, Ernest Rutherford and others concluded that transmutation occurs in radioactive elements. By 1919, Rutherford had achieved the first artificial transmutation of an element, bombarding nitrogen atoms with alpha particles to produce oxygen atoms and protons. Since then, the study of nuclear reactions has become a major area of research. The nuclear technology that has grown out of this research has had an incalculable impact on mankind. In this chapter, we shall discuss the nature of the atomic nucleus, the nature of nuclear reactions, and some of their applications.

20.1 THE ATOMIC NUCLEUS

The atomic nucleus contains protons and neutrons, fundamental particles that are called *nucleons* (Section 5.4, page 122). Every atomic nucleus contains protons. A proton has a positive charge that is equal in magnitude to the negative charge of the electron; its value is 1.60219×10^{-19} C. The mass of the proton is 1.67251×10^{-24} g, which is about 1800 times the mass of the electron (9.1083×10^{-28} g). The mass of the proton is 1.007280 relative to the mass of an atom of ^{12}C, which has a mass of exactly 12. The number of protons in a nucleus is the atomic number of the element. The positive charge of the nucleus is the number of protons in the nucleus.

With the exception of atoms of ^{1}H, atomic nuclei also contain one or more neutrons. The neutron is an electrically neutral particle with a relative mass of 1.008665, slightly greater than that of a proton. In the notation that has been developed to describe atoms, the number of neutrons in the nucleus usually is represented by the symbol N. The number of protons in the nucleus is represented by the symbol Z, which is the atomic number. The symbol A represents the mass number of the nucleus. The mass number has a value close to that of the relative isotopic mass, and is defined as the sum of protons and neutrons in the nucleus:

$$A = Z + N \qquad (20.1)$$

To indicate the number of protons and the mass number of any atomic nucleus, we write the mass number as a superscript and the atomic number as a subscript to the left of the elemental symbol X:

$$^{A}_{Z}X$$

Two different atomic species with the same number of protons but different numbers of neutrons are said to be **isotopes** of the same chemical element. Isotopes can be characterized as stable or unstable. A stable isotope does not undergo radioactive decay. An unstable isotope undergoes radioactive decay and eventually is transformed into a stable isotope, usually of another element.

Most elements of atomic number 83 and under have more than one stable isotope. The average is three stable isotopes per element. Tin has the largest number of stable isotopes, ten. In general, the stable isotopes of an element occur in nearly constant proportions in samples taken from different parts of the earth, but there are many exceptions to this rule.

The word "isotope" generally is used when discussing two or more different nuclear species. When we are talking about a single nuclear species with a given value of A and Z, the term **nuclide** is preferred. Thus, either we can say that $^{16}_{8}O$ and $^{18}_{8}O$ are isotopes of oxygen, or we can say that $^{16}_{8}O$ is a nuclide and $^{18}_{8}O$ is a nuclide.

We can define isotopes as nuclides that have the same Z but different A. **Isobars** are nuclides with the same A but different Z; that is, $^{235}_{92}U$, $^{235}_{93}Np$, and $^{235}_{94}Pu$ are isobars. **Isotones** are nuclides with the same N but different Z. For example, $^{14}_{6}C$, $^{15}_{7}N$, and $^{16}_{8}O$ are isotones, since each nucleus contains eight neutrons.

While almost all the mass of the atom is concentrated in the nucleus, the size of the nucleus is quite small compared to that of the whole atom. Atoms can be regarded as spheres whose radii are in the range of 10^{-8} cm. The nucleus, which can be regarded as approximately spherical, has a radius a little less than 10^{-12} cm. It has been found that the magnitude of the nuclear radius R is directly proportional to the cube root of A, the mass number of the nucleus:

$$R = r_0 A^{1/3} \qquad (20.2)$$

where r_0 is a proportionality constant whose value is about 1.3×10^{-13} cm.

Since nuclear volumes are determined only by the nuclear mass, all nuclei have virtually the same density. If we assume that the nucleus is spherical, we can calculate its density from its mass and radius. The calculation gives a value for the nuclear density of about 1.8×10^{14} g/cm^3. This value is very large compared to the density of everyday materials. For example, osmium, the densest element, has a density of only 22.6 g/cm^3. We can get a better idea of the density of the nucleus by saying that the mass of a one-centimeter cube of nuclear matter would be greater than the combined mass of all the automobiles in the United States. In recent years, astronomers have discovered stars, called neutron stars, whose density is about the same as the density of the atomic nucleus. A neutron star with perhaps twice the mass of the sun would have a radius of about 10 km, less than the size of Manhattan Island.

From what we know about the electrical charges of nucleons, a special sort of attractive force is needed to explain the existence of the atomic nucleus. A nucleus consists of positively charged protons and neutral neutrons, packed closely together. The repulsive coulombic forces between the protons would make the nucleons fly apart, but there are forces of attraction stronger than the repulsive coulombic forces in the nucleus. The forces of attraction between nucleons that hold a nucleus together are very strong at small distances but they fall off rapidly as distance increases.

The magnitude of the attractive force that holds the nucleus together can be found by applying what is perhaps the most well-known equation in science, the equivalence between mass and energy given by Einstein's theory of special relativity:

$$E = mc^2 \qquad (20.3)$$

where E is the energy, m is mass, and c is the velocity of electromagnetic radiation. A convenient unit of energy to use in applying this relationship on the nuclear scale is the electronvolt, eV, whose magnitude is approximately 1.6022×10^{-19} J. Common multiples of the electronvolt are:

kiloelectronvolt (keV) = 10^3 eV
megaelectronvolt (MeV) = 10^6 eV
gigaelectronvolt (GeV) = 10^9 eV

A convenient unit of mass on the nuclear scale is the unified atomic mass unit (amu), which is exactly one-twelfth the mass of an atom of ^{12}C, or $1.6605655 \times 10^{-24}$ g.

We can find the energy equivalent of this mass if we use the value 2.998×10^{10} cm/sec as the velocity of electromagnetic radiation. By Equation 20.3, the energy equivalent of one unified atomic mass unit is 931 MeV; that is, 1 amu = 931 MeV.

Careful measurements of relative atomic masses show that their values are slightly less than those calculated by summing the masses of their nucleons. This deficiency can be regarded as the amount of energy associated with bringing nucleons together in the nucleus, the *nuclear binding energy*. The existence of nuclear binding energy is of enormous importance in the universe.

The mass of stars consists mostly of hydrogen nuclei, or protons. Stars also contain alpha particles, or helium nuclei. An alpha particle, $^4_2$He^{2+}, has a mass of 4.00151 amu. If we add up the relative masses of the two protons and two neutrons in the alpha particle, we get:

protons: 2 × 1.007277 amu = 2.014554 amu
neutrons: 2 × 1.008665 amu = 2.017330 amu
total = 4.031884 amu

The difference between the calculated mass and the measured mass is:

4.03189 amu − 4.00151 amu = 0.03038 amu

The energy equivalent of this mass is:

$$E = 0.03038 \text{ amu} \times \frac{931 \text{ MeV}}{1 \text{ amu}} = 28.3 \text{ MeV}$$

The binding energy in a single nucleus of $^4_2$He thus is 28 MeV. This binding energy is the amount of energy needed to break the helium nucleus into its constituent nucleons. It is also the energy that is liberated when the four nucleons come together in a reaction to form the helium nucleus:

$$2^1_1\text{H} + 2^1_0 n \longrightarrow {}^4_2\text{He}$$

The symbol n represents the neutron, whose $Z = 0$ and $A = 1$.

There is a resemblance between the binding energy of the nucleus and the bond energy of a chemical bond. But the magnitude of the nuclear binding energy is enormous compared to chemical bond energy. We are accustomed to measuring the bond energy of chemical bonds in molar amounts. To convert from the binding energy of a single nucleus to molar quantities of nuclei, we must use Avogadro's number, 6.022×10^{23}. Thus, one mole of helium nuclei has a binding energy of (28.3 MeV/nucleus) × (6.022×10^{23} nuclei/mol) = 1.70×10^{25} MeV/mol. In more familiar energy units, the binding energy is

2.72×10^9 kJ/mol, roughly the bond energy of 10 million moles of a fairly strong chemical bond. In other words, the binding energy of the nucleus is 10 million times greater than the bond energy of a chemical bond.

As the mass number of the nucleus increases, *total* binding energy increases. However, of more interest is the average binding energy—the binding energy *per nucleon*—which does not vary over a large range. We determine the value of the average binding energy per nucleon by dividing total binding energy by the mass number. For helium, average binding energy is $28.3/4 = 7.1$ MeV. With the exception of the lightest nuclei, the average binding energy per nucleon of a nucleus is in the range from 7.4 to 8.8 MeV.

The binding energy per nucleon reaches a peak for nuclei whose mass numbers are near 60, nuclei of iron and nickel. Nuclei with mass numbers near 60 tend to be the most stable. The stability of these nuclei helps account for the relatively high abundance of iron and nickel in the core of the earth and in meteorites. As Figure 20.1 shows, the average binding energy per nucleon falls off rapidly from this maximum.

A number of factors are responsible for the reduction in average binding energy per nucleon with increasing Z. A major factor is the coulombic repulsion between protons in the nucleus. Since these repulsions act over a much greater distance than the attractive forces between nucleons, coulombic repulsions become more important as more protons are added to the nucleus—that is, as Z increases.

The destabilizing effect of coulombic repulsions can be offset to some extent by the presence of neutrons, which are electrically neutral. The additional binding energy of the neutrons, which is not accompanied by any increase in coulombic repulsions, can confer stability on nuclei with many protons. Indeed, all stable nuclei with $Z > 20$ contain more neutrons than protons.

Another factor that affects binding energy is the relative number of protons and neutrons in a nucleus. It has been observed that nuclei in which the number of protons is the same, or nearly the same, as the number of neutrons have more binding energy than nuclei in which there is a large difference between the number of protons and the number of neutrons. In calculating binding energy, an empirical correction must be made for the difference between the number of protons and the number of neutrons, $N - Z$.

There is another factor that has a major effect on the binding energy. The most stable nuclei are those which have even numbers of both protons and neutrons. Even-even nuclei, as these are called, tend to be more stable than nuclei that have an even number of protons and an odd number of neutrons, and

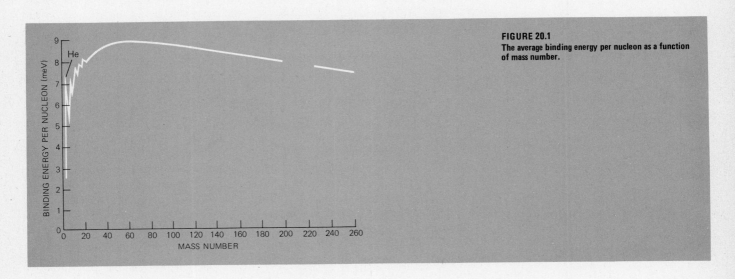

FIGURE 20.1
The average binding energy per nucleon as a function of mass number.

which are called even-odd nuclei. These even-odd nuclei, in turn, are more stable than odd-even nuclei, which have an odd number of protons and an even number of neutrons. Least stable of all are odd-odd nuclei, which have odd numbers of both protons and neutrons. The observed abundance of the known stable nuclides is consistent with these relationships. Of the known stable nuclides, 164 are even-even, 55 are even-odd, 50 are odd-even and only 4 are odd-odd. The four stable odd-odd nuclei are all of very light elements: ^1_1H, ^6_3Li, $^{10}_5B$, and $^{14}_7N$.

An analogy can be made between the pairing of electrons and the pairing of nucleons. Just as increased stability is associated with pairs of electrons of opposite spin, increased stability is associated with pairs of like nucleons. It has also been observed that nuclei with certain "magic numbers" of protons or neutrons seem to be especially stable. Stability is associated with 2, 8, 20, 28, 50, 82, and 126 of either protons or neutrons. The unusual stability of some of the light nuclei first called attention to this phenomenon. For example, the ^4_2He nucleus, with two protons and two neutrons, is exceptionally stable. So are the $^{16}_8O$ nucleus (eight protons, eight neutrons), and the $^{40}_{20}Ca$ nucleus (20 protons, 20 neutrons).

The stable isotopes of the heavier elements also provide evidence for the existence of magic numbers of nucleons. Elements for which Z is a magic number tend to have a relatively large number of stable isotopes. For example, we find the largest number of stable isotopes for tin, $Z = 50$. The natural radioactive decay of all heavy elements eventually ends at lead, $Z = 82$. As for neutrons, there are an unusually large number of isotones in which $N = 50$ and $N = 80$. Also, $N = 126$ for the two heaviest stable nuclides known, $^{208}_{82}Pb$ and $^{209}_{83}Bi$. In addition, the course of many nuclear reactions can be described in terms of the formation of nuclides with magic numbers of protons and neutrons, and of their relative stability. (It should be noted that while there are magic numbers for Z and N, there are no magic numbers for A.)

The nuclear shell model was developed in 1949 to explain the existence of magic numbers and other observations about nuclei. In many ways, the description of the nucleus that is given by the nuclear shell model is similar to our description of the electronic structure of the atom. Electrons are placed in energy levels on the basis of their spins and energies. Groups of electrons with given energy levels are arranged in shells. Extra stability is associated with a completely filled shell, the so-called noble gas configuration. Interactions between the individual electrons are not considered in this model.

In the same way, the nuclear shell model places nucleons into energy levels that are grouped into shells. Special stability is associated with closed shells which have "magic numbers" of nucleons. The interactions between individual nucleons are not considered. Nucleons are also known to be subject to the same quantum mechanical rules as electrons. The nuclear shell model does not agree very well with some of the measured magnetic and electric properties of nuclei, but it is quite helpful in rationalizing many aspects of nuclear reactions.

20.2 RADIOACTIVITY

A radioactive decay process is the transformation of a relatively unstable nuclide to a relatively stable nuclide with the accompanying emission of particles or electromagnetic radiation. In such a process, the unstable or *parent* nuclide decays into the *daughter* nuclide, which may itself be either radioactive or stable.

Radioactive decay reactions, unlike ordinary chemical reactions, are not affected by ordinary changes in temperature and pressure. With minor exceptions, their rates do not depend on the chemical or physical state of the reacting substance. The rate of a nuclear decay reaction depends only on the nature of the nucleus. Thus, as we mentioned in Section 17.2, a radioactive decay reaction is an ideal example of a first-order rate process. The half-life concept, which applies to first-order reactions, is commonly used to express the rate of a radioactive decay process and the relative stability of any given nuclide.

A nucleus decays because it is unstable. A major source of nuclear instability is coulombic repulsion between protons. As we have seen, this destabilizing force is counterbalanced by the presence of neutrons, which supply additional binding energy without additional coulombic repulsion.

The relationship between nuclear stability and nuclear composition can be conveniently represented by a graph such as that of Figure 20.2, a plot of Z against N for all the stable nuclides. When Z is small, all the stable nuclides lie close to the $Z = N$ line. As Z increases, the stable nuclides fall in a zone above the $Z = N$ line, reflecting the increase in the ratio of neutrons to protons that is necessary to offset the coulombic repulsions of the protons. The stable nuclides end at $Z = 83$. In a plot of this sort, the region in which the stable nuclides are found can be regarded as a "zone of stability." The relative stability of a nucleus and its mode of decay often can be predicted from its location relative to the zone of stability on a graph of the kind in Figure 20.2.

As we have noted, the binding energy per nucleon reaches a maximum in nuclei whose mass numbers are about 60, as Figure 20.1 showed. The most stable nuclei have mass numbers in this range. The most stable nucleus of all is believed to be $^{56}_{26}Fe$, which can thus be regarded as the thermodynamic equilibrium state for all nuclear matter. We can even say that all lighter and heavier nuclei are being transformed spontaneously into $^{56}_{26}Fe$. However, the rate at which these transformations occurs is so slow that these processes are of no practical importance for most nuclei.

The phenomenon of radioactive decay was discovered through the detection of the radiation emitted during decay. It was found that three kinds of radiation can be emitted. The three kinds of emitted "rays" are named α rays, β rays, and γ rays.

α **decay:** An α particle is $^{4}_{2}He^{2+}$. The loss of an α particle from a nucleus therefore lowers the mass number A of that nucleus by four, and its atomic number Z by two, and increases the ratio of neutrons to protons. Alpha decay is observed particularly often in nuclei with large mass numbers. At least one nuclide of every element beginning with lithium undergoes α decay. The lightest naturally occurring nuclide in which α decay has been observed is $^{144}_{60}Nd$. The process occurs very slowly in this nuclide. The half-life of $^{144}_{60}Nd$ is 2.4×10^{15} years.

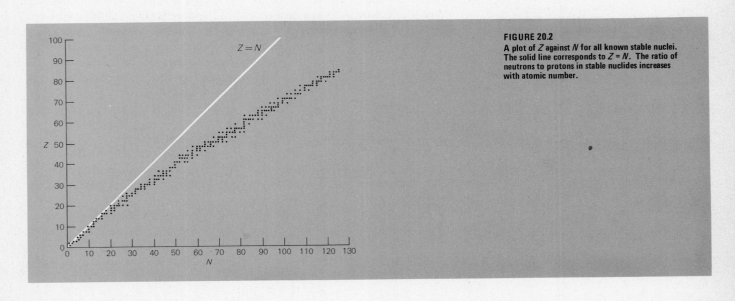

FIGURE 20.2
A plot of Z against N for all known stable nuclei. The solid line corresponds to $Z = N$. The ratio of neutrons to protons in stable nuclides increases with atomic number.

A typical α decay can be represented by an equation such as:

$$^{238}_{92}U \longrightarrow \; ^{234}_{90}Th + \; ^4_2He$$

The parent nuclide, uranium, element 92, decays to the daughter nuclide, thorium, element 90. You should note that equations for nuclear reactions must be balanced. The sum of the mass numbers on both sides of the equation are equal. The sum of the nuclear charges on both sides of the equation are equal. Thus, both A and Z are conserved in a nuclear reaction.

Why should a heavy nucleus decay by emitting a four-nucleon fragment, rather than a single nucleon? The answer lies in the unusually high binding energy of the ^4_2He nucleus compared to other light nuclides.

We can calculate the energy changes in α-decay reactions from the atomic masses of the parent and daughter nuclides. There is a loss of mass, ΔM, in any decay process. The loss in mass has an energy equivalent of 1 amu = 931 MeV. Atomic masses rather than nuclear masses are used in these energy calculations because the mass and binding energy of the electrons must be taken into account. Table 20.1 lists the atomic masses of some important nuclides.

The energy change in the decay of $^{238}_{92}U$ into $^{234}_{90}Th$ is:

$$\Delta M = (\text{mass } ^{238}_{92}U) - (\text{mass } ^{234}_{90}Th + \text{mass } ^4_2He)$$
$$= 238.0508 - (234.0436 + 4.0026)$$
$$= 0.0046 \text{ amu}$$
$$E = 0.0046 \text{ amu} \times 931 \; \frac{\text{MeV}}{\text{amu}} = 4.3 \text{ MeV}$$

Most of this energy appears as the kinetic energy of the emitted α particle. A small amount, about 0.1 MeV in this reaction, appears as the recoil energy of the nucleus.

γ **emission:** In many radioactive decay processes, γ rays, electromagnetic radiation of higher energy and shorter wavelength than X rays, are also emitted. The emission of γ rays in radioactive decay processes is due to the existence of excited nuclear states that are analogous to excited electronic states. A parent nuclide does not always decay to the ground state of the daughter nuclide. Instead, it may decay to an excited nuclear state of the daughter nuclide. After the excited daughter nuclide forms, it decays to its ground state with the emission of γ rays. Figure 20.3 diagrams these energy relationships for the decay of $^{226}_{88}Ra$ to $^{222}_{86}Rn$.

In Section 5.5, page 127, we described how an atom of any element can be identified by its emission spectrum. There is a resemblance between atomic spectra and the γ-ray spectra of excited state nuclides. In each, only certain wavelengths of radiation are seen. Each wavelength corresponds to the difference between allowed energy levels. Just as the electronic energy level structure of an atom can be found by studying its emission spectrum, the energy level structure of a nucleus can be found by studying its γ-ray spectrum.

The emission of γ rays accompanies not only α decay but also other nuclear transformations. The emission of γ rays represents an energy change within the nucleus, a decay from a high-energy state to a lower-energy state. By itself, the emission of γ rays does not result in any change in either A or Z of the nucleus. We find γ-ray emission only in nuclear transformations in which ground state daughter nuclides are not formed directly. Most radioactive decay processes are accompanied by some sort of γ-ray emission.

β **decay:** There are many radioactive decay processes in which there is no change in the value of the mass number A but there is a change in the value of Z, and therefore in the value of N. Such a process is called β decay. Beta decay can occur in several different ways, but only one type of β decay is observed in

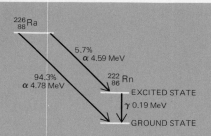

FIGURE 20.3
The decay of $^{226}_{88}Ra$ to $^{222}_{86}Rn$ occurs in two steps. The parent nuclide first emits an α particle and decays to an excited state of the daughter nuclide. The excited daughter nuclide then decays to the ground state with the emission of γ rays. As the diagram shows, most of the energy release in the process takes place in the first step.

TABLE 20.1

Relative Atomic Masses of Particles and Nuclides

Nuclide	Percent Abundance	Mass[a]	Nuclide	Percent Abundance	Mass[a]	Species	Mass[a]
$^{1}_{1}H$	99.985	1.0078252	$^{37}_{17}Cl$	24.47	36.965896	e^-	0.000548580
$^{2}_{1}H$	0.015	2.0141022	$^{40}_{18}Ar$	99.600	39.962384	$^{1}_{1}p^+$	1.007276470
$^{3}_{1}H$		3.0160494	$^{39}_{19}K$	93.10	38.963714	$^{1}_{0}n$	1.008665012
$^{3}_{2}He$	trace	3.0160299	$^{40}_{19}K$	0.0118	39.964008		
$^{4}_{2}He$	100	4.0026036	$^{40}_{20}Ca$	96.97	39.962589		
$^{6}_{3}Li$	7.42	6.015126	$^{55}_{25}Mn$	100	54.938054		
$^{7}_{3}Li$	92.58	7.016005	$^{56}_{26}Fe$	91.66	55.93493		
$^{10}_{5}B$	19.61	10.012939	$^{59}_{27}Co$	100	58.933189		
$^{11}_{5}B$	80.39	11.0093051	$^{58}_{28}Ni$	67.88	57.93534		
$^{12}_{6}C$	98.893	12 (exactly)	$^{60}_{28}Ni$	26.23	59.93078		
$^{13}_{6}C$	1.107	13.003354	$^{87}_{37}Rb$	27.85	86.90918		
$^{14}_{6}C$		14.0032419	$^{87}_{38}Sr$	7.02	86.9089		
$^{14}_{7}N$	99.634	14.0030744	$^{127}_{53}I$	100	126.90447		
$^{15}_{7}N$	0.366	15.000108	$^{129}_{53}I$		128.90498		
$^{16}_{8}O$	99.759	15.9949149	$^{206}_{82}Pb$	23.6	205.97446		
$^{17}_{8}O$	0.0374	16.999133	$^{207}_{82}Pb$	22.6	206.97590		
$^{18}_{8}O$	0.2039	17.9991598	$^{208}_{82}Pb$	52.3	207.97664		
$^{18}_{9}F$		18.000950	$^{209}_{83}Bi$	100	208.98042		
$^{20}_{10}Ne$	90.92	19.9924404	$^{232}_{90}Th$	100	232.03821		
$^{23}_{11}Na$	100	22.989773	$^{234}_{90}Th$		234.0436		
$^{24}_{12}Mg$	78.70	23.985045	$^{234}_{92}U$	0.0056	234.0409		
$^{27}_{13}Al$	100	26.981535	$^{235}_{92}U$	0.7205	235.04393		
$^{28}_{14}Si$	92.21	27.976927	$^{238}_{92}U$	99.274	238.0508		
$^{31}_{15}P$	100	30.973763	$^{237}_{93}Np$		237.04803		
$^{32}_{16}S$	95.0	31.972074	$^{239}_{94}Pu$		239.05216		
$^{35}_{17}Cl$	75.53	34.968854					

[a] Relative to ^{12}C, which has a mass of exactly 12.

naturally occurring radioactive nuclides. It is called β^- or negatron decay, and it generally occurs in nuclides that are unstable because the ratio of neutrons to protons is too high. Nuclides that lie to the right of the zone of stability in Figure 20.2 undergo β^- decay.

The most efficient way to correct an excess of neutrons over protons in a nucleus is to convert a neutron to a proton. This conversion is β^- decay. It can be represented as:

$$n \longrightarrow p^+ + e^-$$

where n represents the neutron and p^+ the proton. Electrons cannot exist in a nucleus, so the electron is created at the instant it is emitted, just as a proton that is emitted as the result of an electronic change in an atom is created at the instant of emission.

Thus, negatron decay is the emission of an electron with an increase of one in the value of Z and a decrease of one in the value of N of the nucleus. You will note that β^- decay causes an *increase* in atomic number; the daughter nuclide is one element higher in atomic number than the parent nuclide. A typical reaction is:

$$^{227}_{89}\text{Ac} \longrightarrow {}^{227}_{90}\text{Th} + {}^{0}_{-1}\beta$$

By giving the β particle (the electron) a subscript of -1, the equation is balanced with respect to both A and Z. A β-decay process changes the ratio of neutrons to protons in the nucleus. Beta decay often changes an odd-odd nucleus to an even-even nucleus.

In addition to the electron, another particle is emitted during β^- decay. The existence of such a particle was postulated in 1931 by Pauli to account for the observation that the electrons emitted in β decay have a continuous range of energies. Since β decay, like α decay, is a transition between two discrete energy states of the parent and daughter nuclides, the β particles should have only a few allowed energy values. Pauli proposed that part of the energy of β decay went into the formation of another fundamental particle, which he called the *neutrino*. A neutrino has no electrical charge, no measurable mass when at rest, negligible magnetic properties, and minimal interactions with other particles. Experimental evidence for the existence of the neutrino was not obtained until 1955.

A complete energy accounting of a β-decay process must include the energy carried away by the neutrino. Thus, the transformation of a neutron to a proton can be represented by:

$$^{1}_{0}n \longrightarrow {}^{1}_{1}p^{+} + {}^{0}_{-1}\beta + \nu$$

where ν represents the neutrino. Equations for β decay can be written to include the neutrino as a decay product, even though it is extremely difficult to detect. For example, the β decay of the unstable nuclide ${}^{14}_{6}\text{C}$, which is formed in the atmosphere by cosmic rays and is used in radiocarbon dating (Section 17.2, page 552), is represented by:

$$^{14}_{6}\text{C} \longrightarrow {}^{14}_{7}\text{N} + {}^{0}_{-1}\beta + \nu$$

Artificially produced unstable nuclides can undergo another type of β decay, called β^+ decay or positron decay. Nuclides in which the ratio of neutrons to protons is too low—that is, nuclides that lie to the left of the zone of stability of Figure 20.2—are converted to more stable daughter nuclides by such a process. In β^+ decay, a proton is converted to a neutron, with the formation of two other particles: a positron, which is identical to an electron except that its charge is positive; and a neutrino, similar to but not identical to the neutrino that is formed in β^- processes. The process is:

$$^{1}_{1}p \longrightarrow {}^{1}_{0}n + {}^{0}_{1}\beta + \nu$$

In positron decay, there is no change in mass number. There is an increase in N, and a decrease in Z. An example of β^+ decay is:

$$^{17}_{9}\text{F} \longrightarrow {}^{17}_{8}\text{O} + {}^{0}_{1}\beta + \nu$$

The positron is called the antiparticle of the electron. When the two collide, they both disappear. Their masses are converted to energy, in a process called annihilation. We can represent the process as:

$$e^{-} + e^{+} \longrightarrow \gamma$$

Another mode of β decay is called electron capture. In electron capture, the conversion of a proton to a neutron occurs when the nucleus captures one of the electrons of the atom:

$$p^{+} + e^{-} \longrightarrow n + \nu$$

The only radiation emitted from the nucleus in electron capture is in the form of the elusive neutrino. Electron capture is a mode of decay available to many nuclides in which the ratio of neutrons to protons is too low, and it competes with

β^+ decay. An example of electron capture is:

$$^{57}_{27}\text{Co} + {}^{0}_{-1}e^- \longrightarrow {}^{57}_{26}\text{Fe} + \nu$$

The electron that is captured by the nucleus usually is one in an orbital of low principal quantum number, since these orbitals lie closest to the nucleus. The loss of the electron creates a vacancy that is filled by an electron from an orbital of higher energy. When the electron drops from the higher-energy orbital, the atom emits electromagnetic radiation in the X-ray part of the spectrum. This emission occurs after the nuclear transformation, and the energy emitted by the atom is not part of the energy change of the decay process.

The energy changes associated with β-decay processes can be calculated from the atomic masses of parent and daughter nuclides, by procedures similar to those used for α decay. The use of atomic masses, which include the masses and binding energies of the electrons, eliminates the problem of keeping track of the electronic changes that accompany the nuclear change except in β^+ decay.

EXAMPLE 20.1

The unstable nuclide $^{40}_{19}\text{K}$, which occurs naturally and has been used in radioisotope dating, can undergo all three types of β decay that we have discussed. The relative atomic masses of the three isobars of $A = 40$ that are related by β decay are $^{40}_{19}\text{K} = 39.964008$; $^{40}_{18}\text{Ar} = 39.962384$; and $^{40}_{20}\text{Ca} = 39.962589$. Calculate the energy change for each of the three types of decay.

SOLUTION

The equation for β^- decay is:

$$^{40}_{19}\text{K} \longrightarrow {}^{40}_{20}\text{Ca} + {}^{0}_{-1}\beta$$

$$\Delta M = (\text{mass } {}^{40}_{19}\text{K}) - (\text{mass } {}^{40}_{20}\text{Ca})$$

$$= 39.964008 \text{ amu} - 39.962589 \text{ amu}$$

$$= 0.001419 \text{ amu}$$

$$E = 0.001419 \text{ amu} \times 931 \frac{\text{MeV}}{\text{amu}}$$

$$= 1.32 \text{ MeV}$$

The equation for electron capture is:

$$^{40}_{19}\text{K} + {}^{0}_{1}e^- \longrightarrow {}^{40}_{18}\text{Ar}$$

$$\Delta M = (\text{mass } {}^{40}_{19}\text{K}) - (\text{mass } {}^{40}_{18}\text{Ar})$$

$$= 39.964008 \text{ amu} - 39.962384 \text{ amu}$$

$$= 0.001624 \text{ amu}$$

$$E = 0.001624 \text{ amu} \times 931 \frac{\text{MeV}}{\text{amu}}$$

$$= 1.51 \text{ MeV}$$

The equation for β^+ decay is:

$$^{40}_{19}\text{K} \longrightarrow {}^{40}_{18}\text{Ar} + {}^{0}_{1}\beta$$

In β^+ decay, because of the use of atomic masses, the mass of two more electrons that are included in the mass of the daughter nuclide must be included in our calculation for mass balance.

$$\Delta M = (\text{mass } {}^{40}_{19}\text{K}) - [\text{mass } {}^{40}_{18}\text{Ar} + (2) \text{ mass } e^-]$$

$$= 39.964008 \text{ amu} - [39.962384 \text{ amu} + (2)(0.0005486 \text{ amu}]$$

$$= 0.000527 \text{ amu}$$

$$E = 0.000526 \text{ amu} \times 931 \frac{\text{MeV}}{\text{amu}}$$

$$= 0.490 \text{ MeV}$$

TABLE 20.2
Radioactive Decay Processes

Process	Nuclear Condition	Emitted Radiation	Change in A	Change in Z	Change in N
α decay	excess mass	$^4_2He^{2+}$ (alpha particle)	-4	-2	-2
β^- decay	neutron-to-proton ratio too high	$^0_{-1}e^-$ (electron)	0	$+1$	-1
β^+ decay	neutron-to-proton ratio too low	$^0_1e^+$ (positron)	0	-1	$+1$
Electron capture	neutron-to-proton ratio too low	none (except neutrino)	0	-1	$+1$
γ emission	excited nuclear state	γ rays	0	0	0

Table 20.2 summarizes all the radioactive decay processes that we have discussed.

Natural Radioactivity

The radioactive nuclides that occur naturally on earth are divided into three categories. One small group, of which $^{14}_6C$ is the most important, consists of nuclides that are formed by the action of cosmic rays on stable nuclides. A second group consists of nuclides with half-lives on the same order of magnitude as the life of the earth. Among these nuclides are $^{40}_{19}K$ (half-life 1.3×10^9 years), $^{87}_{37}Rb$ (half-life 5×10^{11} years), $^{232}_{90}Th$ (half-life 1.39×10^{10} years), $^{235}_{92}U$ (half-life 7.13×10^8 years), and $^{238}_{92}U$ (half-life 4.51×10^9 years).

The other naturally occurring radioactive nuclides have shorter half-lives and are formed by radioactive decay processes that begin with one of the long-lived nuclides. Such a decay process results in what is called a *radioactive family*, or *series*. Three such series are found in nature. Each series starts with a parent nuclide that decays through a series of daughter radioactive nuclides to an ultimate stable end product. Table 20.3 summarizes the starting and ending points of these three series.

The pathway by which the end product forms from the parent is similar in all three series. We shall discuss only the thorium series.

From Table 20.3, we see that the net change as a result of the entire decay process in the thorium series is a reduction of $232 - 208 = 24$ in the mass number A. The only mechanism for reduction of mass number is the emission of α particles, with $A = 4$. Therefore, the total reduction in A must be divisible by four and the reduction in mass must occur in steps of four mass units each. In the thorium series, there is a total of six such steps, each an α decay.

However, other processes must also be occurring. Six α decays should reduce the value of Z by 12. In fact, Z is only reduced by eight in going from Th to Pb. Four β^--decay processes must also occur to account for the observed reduction in the value of Z.

TABLE 20.3
Natural Radioactive Decay Series

Parent	Half-life (years)	End Product	Name
$^{238}_{92}U$	4.51×10^9	$^{206}_{82}Pb$	uranium series
$^{235}_{92}U$	7.13×10^8	$^{207}_{82}Pb$	actinium series
$^{232}_{90}Th$	1.39×10^{10}	$^{208}_{82}Pb$	thorium series

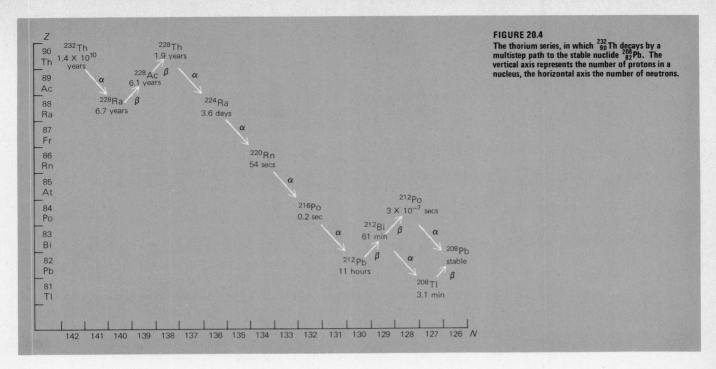

FIGURE 20.4
The thorium series, in which $^{232}_{90}$Th decays by a multistep path to the stable nuclide $^{208}_{82}$Pb. The vertical axis represents the number of protons in a nucleus, the horizontal axis the number of neutrons.

At several points, there are branches in the series. These branches occur when a parent nuclide can follow different paths to form a daughter product. In each branch, the intermediate nuclide is different. For example, $^{212}_{83}$Bi decays to $^{208}_{82}$Pb. It may do so by first forming $^{208}_{81}$Ti by α decay, which then forms $^{208}_{82}$Pb by β^- decay. Or $^{212}_{83}$Bi may first form $^{212}_{84}$Po by β^- decay, which then forms $^{208}_{82}$Pb by α decay.

Figure 20.4 shows a convenient way of diagraming the thorium series. All the nuclides are shown on a grid, on which each square corresponds to a value of Z and N. On this grid, α decay moves the path two boxes down and two boxes to the right, while β decay moves the path one box to the right and one box up. A grid of this type is useful for showing relationships between nuclides. Each horizontal row of boxes represents a series of isotopes of the same element, while each vertical column of boxes represents a series of isotones with the same number of neutrons. A diagonal of boxes is a series of isobars with the same mass number.

In addition to the naturally occurring radioactive nuclides, more than a thousand others have been created artificially. At least one unstable isotope of every element is known.

20.3 NUCLEAR REACTIONS

Many nuclear reactions are processes in which a nucleus is converted to one or more different nuclei as the result of an interaction with another nucleus or with an elementary particle. A nuclear reaction can be represented by an equation similar to those used for ordinary chemical reactions. The equation for a nuclear reaction includes all relevant elementary particles, those which interact with the reactant and those which are emitted with the products.

The first nuclear reaction carried out in the laboratory, Rutherford's formation of oxygen from nitrogen by bombardment with α particles, is represented by:

$$^{14}_{7}N + ^{4}_{2}He \longrightarrow ^{17}_{8}O + ^{1}_{1}H$$

Another nuclear reaction is the fusion reaction of deuterium:

$$^2_1H + ^2_1H \longrightarrow ^3_2He + ^1_0n$$

Both mass number A and atomic number Z are conserved in nuclear reactions. That is, the total mass number of the reactants is equal to the total mass number of the products, and the total Z of the reactants is equal to the total Z of the products, as can be seen in both these reactions.

The energy changes that accompany nuclear reactions are of great interest. Just as ordinary chemical reactions are described as exothermic or endothermic, so nuclear reactions are described as *exoergic*, energy-releasing, or *endoergic*, energy-absorbing. The energy released or absorbed by a nuclear reaction usually is represented by the symbol Q. It should be noted that Q refers to the energy associated with one nuclear event. To calculate the energy for nuclear reactions for molar quantities of material, we must multiply Q by Avogadro's number N. These energies are enormous compared to those of ordinary chemical reactions.

The value of Q can be found by several methods. Experimentally, we can measure the energy of the bombarding particle and of the products. The difference between these energies is the value of Q. We can also calculate the value of Q by the method used to calculate the energy associated with a radioactive decay process; that is, by finding the difference in mass between the reactants and the products.

For convenience, the atomic masses of the nuclides are used to calculate mass and energy changes for nuclear reactions, even though the reactions may include charged species. The excess or deficiency of electrons usually is the same on both sides of a nuclear reaction, so errors tend to cancel. The transmutation of nitrogen to oxygen, written to include charges, is:

$$^{14}_7N + ^4_2He^{2+} \longrightarrow ^{17}_8O^+ + ^1_1H^+$$

However, to calculate ΔM, the change in mass, we can use the atomic masses of these nuclides. The atomic masses include the masses of enough electrons to ensure neutrality.

$$\begin{aligned}
\Delta M &= (\text{mass } ^{17}_8O + \text{mass } ^1_1H) - (\text{mass } ^{14}_7N + \text{mass } ^4_2He) \\
&= (16.999133 \text{ amu} + 1.007825 \text{ amu}) - (14.003074 \text{ amu} \\
&\quad + 4.002604 \text{ amu}) \\
&= 0.001280 \text{ amu}
\end{aligned}$$

The increase in mass indicates that this reaction is endoergic. The magnitude of Q is found in the usual way:

$$\begin{aligned}
Q &= 0.001280 \text{ amu} \times 931 \frac{\text{MeV}}{\text{amu}} \\
&= 1.19 \text{ MeV}
\end{aligned}$$

For this endoergic reaction to occur, energy must be supplied in the form of the kinetic energy of the bombarding α particles. However, the reaction will not occur unless the kinetic energies of the bombarding α particles are substantially greater than the 1.19 MeV gained in this reaction. There is a barrier to the reaction of a nucleus with a positively charged particle. It is a coulombic barrier that is due to the repulsion between the positively charged nucleus and a positively charged particle that approaches it. There is a comparable situation in ordinary chemical reactions, in which there are activation energies that are substantially higher than the energy change of the overall reaction.

For the short-range nuclear forces to be effective, the bombarding α particle must come very close to the nucleus. To do so, it must have enough energy to surmount the coulombic barrier. The magnitude of the barrier increases with both the positive charge of the bombarding particle and the charge of the nucleus. For the heaviest elements, the barrier is about 25 MeV for the dipositive α particle and about 12 MeV for the monopositive proton or the deuteron, an isotope of hydrogen that has one proton and one neutron.

Even exoergic reactions in which a positively charged particle must interact with the nucleus have substantial barrier energies. The reaction for the fusion of deuterium:

$$\ce{^2_1H + ^2_1H -> ^3_2He + ^1_0 \textit{n}}$$

is exoergic:

$$\Delta M = (\text{mass } \ce{^3_2He} + \ce{^1_0\textit{n}}) - (2 \text{ mass } \ce{^2_1H})$$
$$= (3.0160299 \text{ amu} + 1.0086654 \text{ amu}) - (2)(2.0141022 \text{ amu})$$
$$= -0.0035091 \text{ amu}$$
$$Q = -0.0035091 \text{ amu} \times 931 \frac{\text{MeV}}{\text{amu}}$$
$$= -3.27 \text{ MeV}$$

But there is a coulombic barrier, of about 0.45 MeV. This barrier is relatively small because the two colliding particles have a charge of only $+1$ each. However, even this small barrier has caused major practical difficulties in the effort to use fusion reactions to generate electricity.

By the standard of ordinary chemical reactions, the kinetic enery needed for a positively charged particle to penetrate the coulombic barrier of a nucleus is huge. An average kinetic energy of 0.01 MeV corresponds to a temperature of about 100 000 000 K. The coulombic barrier can be overcome with machines called accelerators, which use electromagnetic fields to accelerate charged particles. Accelerators were developed mainly to study nuclear reactions of the heavy nuclides.

There is no coulombic barrier for the reaction of nuclei with neutrons, because neutrons have no electric charge. "Thermal" neutrons, so called because their kinetic energies are roughly the same as those of gas molecules at ordinary temperatures, react readily with nuclei. These slow-moving thermal neutrons take part in a class of nuclear reactions that are of great practical importance, the reactions that occur in the core of a nuclear reactor.

There are a number of models that describe nuclear reactions. In the compound nucleus model, the reaction starts when the bombarding particle is absorbed by the target nucleus. The result is the formation of a compound nucleus, analogous in many ways to the activated complex of ordinary chemical reactions. The complex nucleus is in a highly excited state. It decays quickly, forming the products and emitting one or more particles. The transmutation of nitrogen to oxygen by α-particle bombardment may proceed through a compound nucleus:

$$\ce{^{14}_7N + ^4_2He -> ^{18}_9F^* -> ^{17}_8O + ^1_1H}$$

A compound nucleus in this reaction is an excited state of a fluorine nuclide. The lifetime of a compound nucleus is extremely short, between 10^{-14} and 10^{-18} s.

At higher bombarding energies, greater than 5 MeV per nucleon, a different reaction mechanism begins to dominate. This mechanism is described by a second model, called the direct interaction model. These nuclear reactions appear to occur by the collision of the bombarding particle with one or several nucleons in the target nucleus. An intermediate compound nucleus is not formed, and the product particles are ejected directly from the target nucleus.

Varieties of Nuclear Reactions

Nuclear reactions have been initiated in many ways: by γ rays, X rays, electrons, and neutrons, and by a series of positively charged particles with low masses, such as protons, deuterons, and α particles. The development of high-energy accelerators has also made it possible to use heavy nuclei with positive charges, such as $^{10}_{5}B$, $^{12}_{6}C$, $^{16}_{8}O$, or even heavier nuclides as bombarding particles. The emissions in nuclear reactions include X rays, neutrons, and protons. Many nuclear reactions result in the emission of more than one kind of particle. In one class of nuclear reactions called *fission reactions,* the reactant nuclide forms two or more product nuclides of smaller mass number.

There is a simple notation for representing nuclear reactions that is based on the bombarding particle used for the reaction and the particle that is emitted. In this system, the symbols of the bombarding particle and of the emitted particle are placed in parentheses between the symbols of the reactant nucleus and the product nucleus. The expression for a given reaction will list, from left to right, the reactant nucleus, the bombarding particle, the emitted particle, and finally, the product nucleus. The symbols used in this system include n for the neutron, p for the proton, d for the deuteron, α for the alpha particle, e for the electron, and γ for the gamma ray. Thus, the reaction in which nitrogen is transmuted into oxygen by bombardment with α particles is represented as:

$$^{14}_{7}N(\alpha,p)^{17}_{8}O$$

Unstable nuclides that do not occur in nature can be prepared by nuclear reactions. Many of these nuclides are used in chemistry, biology, medicine, and technology.

A number of radioactive isotopes are prepared by bombarding stable nuclei with slow neutrons. Reactions in which slow, or thermal, neutrons are the bombarding particles are the lowest-energy nuclear reactions. The most common kind of slow-neutron reaction is the (n,γ) reaction, also called *neutron capture,* whose product is an isotope of the parent nuclide. The product nuclide contains one neutron more than the target nuclide. The product nuclide thus is often unstable, because the ratio of neutrons to protons is too high, so the nuclide usually undergoes subsequent β^{-} decay. A radioisotope of almost any element can be prepared by bombarding a stable isotope of the element with neutrons to cause a (n,γ) reaction. One such reaction is $^{59}_{27}Co(n,\gamma)^{60}_{27}Co$, which also can be represented as:

$$^{59}_{27}Co + {}^{1}_{0}n \longrightarrow {}^{60}_{27}Co^{*} \longrightarrow {}^{60}_{27}Co + \gamma$$

The compound nucleus is an excited state of the $^{60}_{27}Co$ nucleus that decays to the ground state by emitting energy in the form of γ rays. The product nuclide is an artificial radioisotope that undergoes β^{-} decay with a half-life of 5.26 years:

$$^{60}_{27}Co \longrightarrow {}^{60}_{28}Ni + {}^{0}_{-1}\beta$$

Radioactive cobalt is often used in radiation therapy against cancer.

The (n,γ) nuclear reaction is also one basis for a method of chemical analysis that is especially useful for finding the elemental composition of small samples. In the method, called neutron activation analysis, the sample to be analyzed is exposed to thermal neutrons. The nature of the radiation emitted by the products of the resulting nuclear reactions provides extremely accurate information on the constituent elements of the starting material. Neutron activation analysis is widely used in such fields as archeology and space exploration. A quantity of material as small as 10^{-12} g can be detected in a sample by neutron activation analysis.

The (n,γ) reaction is the most important one caused by bombardment with slow neutrons. However, nuclides of low atomic number can also undergo (n,p) and (n,α) reactions when bombarded with slow neutrons. One important (n,p) reaction is responsible for the formation of $^{14}_{6}C$ from $^{14}_{7}N$ in the atmosphere. The reaction can be written as $^{14}_{7}N(n,p)^{14}_{6}C$ or as:

$$^{14}_{7}N + {}^{1}_{0}n \longrightarrow {}^{15}_{7}N^{*} \longrightarrow {}^{14}_{6}C + {}^{1}_{1}H$$

The $^{14}_{6}C$ nuclide is unstable and reverts to the starting nuclide by β^{-} decay; with a half-life of 5760 years:

$$^{14}_{6}C \longrightarrow {}^{14}_{7}N + {}^{0}_{-1}\beta$$

Another group of nuclear reactions are produced by bombardment with medium-energy particles. These particles can be faster neutrons or positively charged particles such as protons, deuterons, α particles, or nuclei of higher mass number. The higher energy of the bombarding particles leads to a much wider range of reactions. The compound nuclei formed in these reactions tend to be highly excited and can decay in many different ways.

A medium-energy (α,n) reaction led to the discovery of the neutron in 1932. The reaction is $^{9}_{4}Be(\alpha,n)^{12}_{6}C$, which can be expressed as:

$$^{9}_{4}Be + {}^{4}_{2}He \longrightarrow {}^{13}_{6}C^* \longrightarrow {}^{12}_{6}C + {}^{1}_{0}n$$

This reaction still is used as a neutron source for nuclear experiments.

Simple alpha-capture (α,γ) reactions have also been observed for light nuclides. One such reaction is $^{7}_{3}Li(\alpha,\gamma)^{11}_{5}B$, which can also be written as:

$$^{7}_{3}Li + {}^{4}_{2}He \longrightarrow {}^{11}_{5}B^* \longrightarrow {}^{11}_{5}B + \gamma$$

A large number of different reactions are possible when the proton is the bombarding particle. Examples of (p,α), (p,n), (p,γ), and even (p,d) reactions are known. There is a similarly large variety of nuclear particles emitted in nuclear reactions in which the deuteron, $^{2}_{1}H$, is the bombarding particle. The exoergic reactions in which deuterium or tritium is bombarded with deuterons are of special interest in the effort to use thermonuclear power as an energy source.

Synthetic Elements

In the past few decades, nuclear reactions have been used to prepare elements that are not found on earth. One such element is technetium, element 43, whose most stable isotope, $^{97}_{43}Tc$, has a half-life of only 2.6×10^6 years, much less than the age of the earth. Technetium was first made artificially as the product of a (d,n) nuclear reaction,

$$^{96}_{42}Mo + {}^{2}_{1}H \longrightarrow {}^{98}_{43}Tc^* \longrightarrow {}^{97}_{43}Tc + {}^{1}_{0}n$$

Technetium can be prepared by several different reactions. It is also obtained as a by-product of the operation of nuclear reactors.

Element 61, promethium, also is not found on earth. All the isotopes of promethium are unstable. The least unstable is $^{145}_{61}Pm$, whose half-life is 17.7 years. Promethium is prepared from a stable isotope of samarium in two steps. The first step produces an unstable isotope of samarium:

$$^{144}_{62}Sm + {}^{1}_{0}n \longrightarrow {}^{145}_{62}Sm + \gamma$$

In the second step, this unstable isotope decays by electron capture to form $^{145}_{61}Pm$.

The only other member of the first 92 elements that is not found naturally occurring on earth is astatine. It was first prepared artificially by bombarding bismuth with α particles in the reaction $^{209}_{83}Bi(\alpha,2n)^{211}_{85}At$, which can also be written as:

$$^{209}_{83}Bi + {}^{4}_{2}He \longrightarrow {}^{211}_{85}At + 2{}^{1}_{0}n$$

This isotope of astatine is one of the longer-lived ones. Its half-life is about 7.2 hours.

All of the known elements whose atomic number is greater than 92, the transuranium elements, have been produced artificially by nuclear reactions. As yet, there is no definite evidence that any of these elements exists on earth, although they are known to exist in stars.

The first transuranium element to be produced artificially was neptunium,

Np, whose most stable isotope, $^{237}_{93}$Np, has a half-life of about 2.2×10^6 years. Neptunium is formed when $^{238}_{92}$U is bombarded by fast neutrons; the product of this reaction forms $^{237}_{93}$Np by β^- decay:

$$^{238}_{92}U + ^1_0n \longrightarrow {}^{237}_{92}U + 2^1_0n$$

$$^{237}_{92}U \longrightarrow {}^{237}_{93}Np + {}^0_{-1}\beta$$

This isotope of Np is the parent of a radioactive decay series similar to the three decay series of naturally occurring radioisotopes mentioned above. Neptunium decays through a series of steps that ends at the stable $^{209}_{83}$Bi nuclide. This series is not found in the earth's crust because the half-lives of the parent nuclide and other nuclides in the series are much shorter than the age of the earth. However, the series has been well established in the laboratory, and it probably existed on earth during the early years of the planet.

The transuranium element of greatest practical importance is plutonium, which is produced in large quantities in nuclear reactors. The most important isotope of plutonium is $^{239}_{94}$Pu, which can be used as fuel for a nuclear reactor or as explosive material. The widespread use of plutonium as nuclear reactor fuel has been the subject of intense debate in recent years. It is feared that nuclear weapons could become available to any nation, or even that plutonium could be stolen and fashioned into a bomb by terrorist groups if reprocessing of spent fuel rods to extract plutonium becomes routine.

Plutonium is prepared by the reaction of $^{238}_{92}$U with a neutron, followed by two β^--decay reactions:

$$^{238}_{92}U + ^1_0n \longrightarrow {}^{239}_{92}U + \gamma$$

$$^{239}_{92}U \longrightarrow {}^{239}_{93}Np + {}^0_{-1}\beta$$

$$^{239}_{93}Np \longrightarrow {}^{239}_{94}Pu + {}^0_{-1}\beta$$

This isotope of plutonium decays by α-emission with a half-life of 24 360 years. Another isotope, $^{244}_{94}$Pu, which has a half-life of 7.6×10^7 years, is believed to be the real parent nuclide of the thorium series. However, since the half-life of $^{244}_{94}$Pu is relatively short compared to the life of the earth, all of it that may once have existed on earth is believed to have decayed to the presently observed parent, $^{232}_{90}$Th, by a series of α-emission and β-emission reactions.

Still heavier elements can be made by bombarding suitable targets with charged particles. Nuclei of light atoms sometimes are used for these reactions. For example, curium, element 96, can be prepared by the reaction:

$$^{232}_{90}Th + ^{12}_6C \longrightarrow {}^{240}_{96}Cu + 4^1_0n$$

which can also be expressed as $^{232}_{90}$Th$(^{12}_6$C, $4n)^{240}_{96}$Cu. Fermium, element 100, can be prepared by the reaction:

$$^{238}_{92}U + ^{16}_8O \longrightarrow {}^{249}_{100}Fm + 5^1_0n$$

The number of transuranium elements that have been made in the laboratory continues to increase. In 1974, scientists from both the Soviet Union and the United States claimed to have synthesized element 106. The United States report said that the element was formed by the reaction $^{249}_{98}$Cf$(^{18}_8$O, $4n)^{263}_{106}$X. The Soviet report said the element had been created by several reactions, including some in which several isotopes of lead were bombarded with nuclei of $^{54}_{24}$Cr. In 1976, the Soviet Union claimed to have formed element 107 by bombarding $^{209}_{83}$Bi with nuclei of $^{54}_{24}$Cr. Attempts to create superheavy elements continue today.

20.4 THE CHEMISTRY OF THE UNIVERSE

One of the most challenging areas of scientific research is the investigation of the origin and evolution of the universe. A central part of this research deals with the

This cloud of gas and interstellar dust is the raw material for new stars. Gravitational collapse of such a cloud leads to the formation of increasingly dense regions. When the density and temperature are high enough to ignite fusion reactions, a star is born. (*Hale Observatories*)

chemical composition of the stars and the origin of the elements. Only limited experimental data are available. Theories are tentative and complex. Nevertheless, there seems to be a broad understanding of the nuclear reactions that occur on the cosmic scale, and that have created the chemical elements.

As we said earlier, by far the most abundant element in the universe is hydrogen. It is currently accepted that the universe as we know it originated in a "big bang" some 18 billion years ago. According to this theory, all the matter in the universe was packed into one mass, which somehow exploded to send matter streaming out. The expansion of the universe which began with the big bang still continues.

It is believed that during this expansion, clouds of gas condensed to form stars that were composed primarily of hydrogen. Such a star is called a *first-generation star*.

The prevailing picture of stellar evolution begins with the contraction of a cloud of gas, whose molecules are pulled together by gravitational attraction to form a star. As the gas cloud contracts, its temperature and density increase, until they are great enough to cause nuclear reactions. These nuclear reactions, which we shall discuss later in detail, release energy. The release of this energy offsets the contraction caused by gravitation, keeping the star in a stable state. When the fuel for these nuclear reactions is exhausted, the shrinkage of the star begins again, with a new increase in temperature and density that causes a new set of nuclear reactions to begin. This cycle is repeated a number of times, until the star reaches a stage where it no longer has fuel for any nuclear reactions. When this point is reached, events occur that are crucial in the formation of the heavier chemical elements. But let us discuss the story from the very beginning.

When the temperature in a first-generation star rises to about 10^7 K, hydrogen nuclei can begin reacting with each other. The reaction is:

$$_1^1H + {}_1^1H \longrightarrow {}_1^2H + {}_1^0\beta + \nu$$

You can see that the products of the reaction of two protons are a deuteron, a positron, and a neutrino. The coulombic barrier for this process is relatively low. However, this is a relatively slow process that requires surprisingly high temperatures. In a much faster nuclear reaction, the deuteron that is formed in the proton-proton reaction combines with another proton:

$$^2_1H + {}^1_1H \longrightarrow {}^3_2He + \nu$$

After enough ^3_2He accumulates, this nuclide can react with itself:

$$^3_2He + {}^3_2He \longrightarrow {}^4_2He + 2{}^1_1H$$

The overall process, therefore, is the fusion of four protons into a single α particle, as can be seen by summing these three equations. This process is called *hydrogen burning*. As all the hydrogen is converted to helium, the energy output of the star decreases and gravitational contraction begins again. When the temperature at the core of the star rises to 10^8 K, the helium nuclei that have become the dominant part of the core can begin to undergo fusion reactions. The basic helium reaction is:

$$3{}^4_2He \longrightarrow {}^{12}_6C$$

Some of the $^{12}_6C$ formed by this helium-burning process can react with α particles to form oxygen nuclei:

$$^{12}_6C + {}^4_2He \longrightarrow {}^{16}_8O + \gamma$$

When helium burning is complete, the core of the star contains carbon and oxygen nuclei. Shrinkage begins again, until the core temperature reaches 6×10^8 K. At this point, carbon-burning reactions begin. Two of these reactions are:

$$^{12}_6C + {}^{12}_6C \longrightarrow {}^{23}_{11}Na + {}^1_1H$$
$$^{12}_6C + {}^{12}_6C \longrightarrow {}^{20}_{10}Ne + {}^4_2He$$

The products of these reactions undergo further reactions very quickly at the elevated temperatures in the cores of the stars. Two processes that occur are:

$$^{23}_{11}Na + {}^1_1H \longrightarrow {}^{24}_{12}Mg + \gamma$$
$$^{20}_{10}Ne + {}^4_2He \longrightarrow {}^{24}_{12}Mg + \gamma$$

Many other reactions are possible. When carbon burning is finished, the core of the star consists of $^{16}_8O$, $^{20}_{10}Ne$, $^{24}_{12}Mg$, and other nuclides whose mass numbers are in this range.

Gravitational collapse then continues and the temperature of the star increases once more. The nuclear reactions that occur lead to the formation of many different nuclei, the heaviest being $^{32}_{16}S$. However, there is a limit to this progression. Even though the temperature of the core of the star reaches 4×10^9 K—temperatures at the surface are still much lower—reactions between heavy nuclei become difficult. Alternate multistep reaction paths become important in the formation of still heavier nuclei.

The interior of the star proceeds toward a condition called nuclear statistical equilibrium, which has a superficial resemblance to ordinary chemical equilibrium. Reactions occur quickly, and the species that comes to be predominant in the interior of the star is the nucleus with the greatest binding energy per nucleon, $^{56}_{26}Fe$. At this stage, the interior of the star can no longer release energy by nuclear reactions, and nuclei with mass numbers greater than 56 cannot form.

The star itself becomes unstable. It may undergo what is called a *supernova explosion,* expelling much of its material in a high-temperature shock wave that is accompanied by a number of nuclear reactions. At this point, the regions toward the surface of the star are of interest. Because the surface of the star is cooler than the core, nuclear reactions do not occur there. Even though the innermost part of the star may consist entirely of ^{56}Fe, there are cooler outer regions that contain the lighter elements, even hydrogen. When the supernova explosion occurs, all these elements are expelled. Much of the evidence used to develop

THE MISSING NEUTRINOS

The most unusual observatory in the history of astronomy is a 400 000-liter tank of dry-cleaning fluid 1500 m underground in a mine at Lead, South Dakota. This observatory has given astronomers their best view of the processes occurring deep in the interior of the sun. The results obtained from the observatory indicate that there could be basic flaws in the accepted theory of the nature of the solar interior.

The observatory was built to detect neutrinos produced in certain of the solar reactions, such as steps in the carbon-nitrogen cycle. The neutrinos emitted in a solar fusion reaction such as:

$$^2_1H + ^1_1H \longrightarrow ^3_2He + \nu$$

pass readily through the sun. Most of them also pass through the earth without reacting with other particles. Even though solar neutrinos are emitted in enormous numbers, only a few interactions occur. These interactions can easily be hidden by interactions due to cosmic rays. The observatory was placed underground to screen out all but solar neutrino interactions.

The tank in the mine contains perchloroethylene, C_2Cl_4. Every so often, a neutrino will react with a chlorine atom to form a radioactive argon atom:

$$\nu + ^{37}_{17}Cl \longrightarrow ^{37}_{18}Ar + ^{0}_{-1}\beta^-$$

Once every hundred days, inert gas is bubbled through the tank, and the number of neutrino interactions is measured by counting the number of ^{37}Ar atoms that are swept out by the gas. The chemical technique has been so perfected that even a few radioactive argon atoms can be isolated and counted.

If the existing model of the solar interior is correct, about one ^{37}Ar atom a day should be produced in the tank. The results obtained over several years are well below that level. Apparently, one ^{37}Ar atom is produced only every two and a half or three days.

This result has led to many speculative theories about the sun and the neutrino. It has been suggested that the sun is a variable star, whose changing energy output is responsible for the discrepancy between the predicted number of neutrinos and the observed number. Another theory says that there may be a black hole, a superdense energy-emitting body, at the center of the sun. A third speculative theory is that the sun may have formed in two steps, with the outer solar layer being added relatively late in the history of the sun. It has also been proposed that neutrinos behave differently over large distances from the way they do in laboratory observations, or that the neutrino, which is assumed to be massless, may indeed have a small mass. Each of these theories could explain the mystery of the missing neutrinos, but none of them has been widely accepted.

Several additional experiments have been proposed. One proposed experiment would use 200 000 liters of nearly saturated aqueous lithium chloride solution. The ^7Li atoms capture an occasional neutrino to form ^7Be atoms. About 30 such interactions would this model of stellar evolution comes from observations of the distribution of the elements in stars, interstellar clouds, and planets.

The most abundant nuclide is $^{1}_{1}H$. The second most abundant is $^{4}_{2}He$. Other light nuclides, such as deuterium, $^{3}_{2}He$, and isotopes of lithium, beryllium, and boron, which we believe are only transient intermediates in the evolution of a first-generation star, are not abundant. There is a relatively great abundance of the nuclides $^{12}_{6}C$ and $^{16}_{8}O$, which we believe form from hydrogen-burning reactions, and of $^{56}_{26}Fe$, the most stable nuclide.

Our evolutionary picture of a first-generation star does not account for the existence of any nuclide heavier than ^{56}Fe. The formation of the heavier nuclides takes place in second-generation stars, those that have formed from interstellar gas that includes the products of first-generation stars.

Because second-generation stars contain these elements in reasonable abundance, a whole new range of nuclear reactions in which these elements are energy sources is possible. These reactions are believed to play an important part in the energy-producing budget of stars like the sun.

One crucial sequence of reactions is called the *carbon-nitrogen cycle*. In this sequence, four protons are converted to an α particle, with an energy release of about 26 MeV. The presence of carbon and nitrogen makes this fusion process faster than the direct proton-proton reactions in first-generation stars. The important steps are:

$$^{12}_{6}C + {}^{1}_{1}H \longrightarrow {}^{13}_{7}N + \gamma$$
$$^{13}_{7}N \longrightarrow {}^{13}_{6}C + {}^{0}_{1}\beta + \nu \ (\beta^+ \text{ decay})$$
$$^{13}_{6}C + {}^{1}_{1}H \longrightarrow {}^{14}_{7}N + \gamma$$
$$^{14}_{7}N + {}^{1}_{1}H \longrightarrow {}^{15}_{8}O + \gamma$$
$$^{15}_{8}O \longrightarrow {}^{15}_{7}N + {}^{0}_{1}\beta + \nu \ (\beta^+ \text{ decay})$$
$$^{15}_{7}N + {}^{1}_{1}H \longrightarrow {}^{12}_{6}C + {}^{4}_{2}He$$

In a sense, the $^{12}_{6}C$ acts as a catalyst. It opens up a faster pathway for the reaction and it is regenerated at the end of the sequence. Some further branching steps in the sequence have the effect of converting $^{12}_{6}C$ to $^{14}_{7}N$. Thus, when hydrogen burning by this sequence is completed, the core of the star contains a considerable quantity of $^{14}_{7}N$, as well as helium. As a result, a relatively great abundance of $^{14}_{7}N$ is observed.

Helium burning in second-generation stars includes a sequence that begins with a reaction between He and N:

$$^{14}_{7}N + {}^{4}_{2}He \longrightarrow {}^{18}_{9}F + \gamma$$
$$^{18}_{9}F \longrightarrow {}^{18}_{8}O + {}^{0}_{1}\beta + \nu$$
$$^{18}_{8}O + {}^{4}_{2}He \longrightarrow {}^{22}_{10}Ne + \gamma$$
$$^{22}_{10}Ne + {}^{4}_{2}He \longrightarrow {}^{25}_{12}Mg + {}^{1}_{0}n$$

The notable point about this sequence and others like it that take place at this stage of stellar evolution is the production of a fairly large number of neutrons. The formation of many nuclides with mass number greater than 56 is believed to occur by neutron capture on a relatively slow time scale. By "slow," we mean that any unstable neutron-capture product has enough time to β^- decay before being struck by another neutron. It is the β^- decay that leads to an increase in Z. Starting with $^{56}_{26}$Fe, heavier nuclides are believed to form step by step through neutron capture and β^- decay. The sequence is believed to end with the formation of lead and bismuth, which cannot react to form nuclides of higher mass number.

This model of relatively slow nuclear reactions is consistent with many of the observations of the relative abundance of heavier nuclides. However, it is known that several heavier elements, most notably uranium and thorium, cannot form by such reactions. These nuclides are believed to be formed in supernova explosions of stars that are second generation and later. Such explosions are believed to result in a very rapid release of enormous numbers of neutrons. Nuclei can react with neutrons very rapidly during such an explosion to form nuclides of mass number up to 270. Support for this theory comes from the observation of rapid multiple neutron-capture reactions in the explosion of thermonuclear bombs, where $^{254}_{98}$Cf is formed through neutron capture by $^{238}_{92}$U.

The matter expelled by the supernova explosion of a second-generation star can serve as the raw material for the formation of a third-generation star. Table 20.4 outlines the major steps in stellar evolution. Uranium and thorium exist in our solar system, so our sun is believed to be at least a third-generation star. The known half-lives of the isotopes of uranium and thorium that are found in our solar system enable us to say that the supernova explosion that provided the raw material for the formation of the sun and the planets could not have occurred much more than five billion years ago.

We have only indirect or speculative evidence for many of the steps in this picture of stellar evolution. Of necessity, our information about events in the interior of a star must be indirect. However, we have fairly firm data on the relative abundance of the elements. The principles that govern nuclear reactions and the gravitational behavior of large masses are also believed to be well understood. Our understanding of the essential processes occurring within stars is firm enough to support an effort to make practical use on earth of the fusion reactions that produce energy in stars.

TABLE 20.4
Stellar Evolution

1. Contraction of interstellar hydrogen to form a first-generation star.
2. Nuclear reactions to form nuclides up to $^{56}_{26}$Fe.
3. Supernova explosion; the nuclides in the star are expelled and become interstellar material.
4. Contraction of interstellar material to form a second-generation star.
5. Nuclear reactions to form nuclides up to $^{56}_{26}$Fe; neutrons are released by some of these reactions.
6. Formation of many more nuclides as heavy as lead and bismuth as the result of slow reactions with slow neutrons.
7. Supernova explosion or other catastrophic event in which nuclides as heavy as thorium and uranium are formed by rapid multiple neutron-capture reactions. The nuclides are expelled from the star and become interstellar material.
8. Contraction of interstellar material to form a third-generation star.
9. Repeated cycles of nuclear reactions and eventual supernova explosions to produce succeeding generations of stars.

20.5 ENERGY FROM NUCLEAR REACTIONS; FISSION AND FUSION

Because nuclear reactions release much larger quantities of energy than ordinary chemical reactions, they have enormous potential for both military and peaceful applications. However, there are several barriers in the way. Nuclear processes are relatively improbable events. The probability that a bombarding particle will interact with a target nucleus is rather low because the nucleus is a small target. Furthermore, it is often difficult to produce enough bombarding particles with the energy needed to cause a reaction. And it is not always easy to obtain enough of the target nuclide to produce a large amount of energy. Finally, it is not always easy to create the conditions in which a desired nuclear reaction will occur.

Many of these difficulties have been overcome in one way or another in the past few decades. The major impetus for the scientific and technological advances in this field has been the military demand for more powerful weapons of destruction. Weapons research has been followed by the effort to develop peaceful uses of nuclear reactions, primarily for the generation of electricity.

The most important nuclear reaction for commercial energy production is the fission reaction, in which a nucleus splits into two nuclei of more or less equal mass with the accompanying release of energy. The existence of fission reactions was first recognized by German physicists in the late 1930s. The discovery led to the Manhattan Project of World War II, in which usable fission weapons—atomic bombs—were developed in less than four years by an all-out wartime effort.

In nature, fission can be a mode of spontaneous radioactive decay. However, only the fission of the very heaviest elements occurs rapidly enough to be detectable. Compared to other modes of decay, fission is a relatively improbable process for most elements. Thus, the half-life of $^{232}_{90}$Th is 1.39×10^{10} years with respect to α decay and 10^{21} years with respect to fission. The half-life of $^{238}_{92}$U is 7.13×10^8 years with respect to α decay and 6×10^{15} years with respect to fission. The half-life with respect to fission does decrease rapidly with increasing atomic number, which accounts in part for the difficulty in obtaining some of the heavier transuranium elements. Thus, the half-life of $^{244}_{96}$Cm is 1.4×10^7 years with respect to fission. The half-life with respect to fission of $^{254}_{100}$Fm is only 100 days. For element 104, it is believed to be 0.3 s, although this result is still open to question.

The data on average binding energy per nucleon in different nuclides shown in Figure 20.1 indicate that nuclei with mass numbers greater than about 60 are unstable with respect to two smaller nuclei. These heavier nuclei should thus undergo spontaneous fission. They generally do not undergo fission because there is an energy barrier to fission that is analogous to a large energy of activation in ordinary chemical reactions. This energy barrier can be overcome spontaneously or by supplying nuclei with a suitable amount of energy. Usually, the energy is supplied by bombarding the nuclei with particles to form compound nuclei. In some heavier elements, the bombarding particles do not have to be of very high energy. The most important fission reactions are those produced in heavy elements by thermal neutrons, which are not high-energy particles. The nuclides that undergo fission when they absorb thermal neutrons include the $^{235}_{92}$U used in the atomic bomb dropped on Hiroshima and the $^{239}_{94}$Pu used in the atomic bomb dropped on Nagasaki.

When such a heavy nucleus absorbs a thermal neutron, it undergoes fission after forming a compound nucleus that is sufficiently excited to overcome the energy barrier to fission. When fission occurs, the average binding energy per nucleon in the two fission product nuclei increases substantially compared to the average energy per nucleon in the original nucleus. As a result, fission is accompanied by a large release of energy, in the range of 200 MeV per fission.

The fission of $^{235}_{92}$U demonstrates this release of energy. When the nucleus absorbs a thermal neutron, it forms the compound nucleus $^{236}_{92}$U. The average binding energy per nucleon of a nucleus of this mass number is about 7.6 MeV. Assume that the fission products are two nuclei of mass number 118 each. In a nucleus of this mass number, the average binding energy per nucleon is about 8.5 MeV. The energy released by the fission thus is:

$$E = 2(118)(8.5 \text{ MeV}) - (236)(7.6 \text{ MeV}) = 210 \text{ MeV}$$

Most of this energy appears as kinetic energy of the fission products, but some energy is released in a different form in subsequent processes. The fission products generally have an excess of neutrons to protons, and are therefore unstable. Some of the energy of fission is released when the fission products undergo β^- decay because of their instability.

Energy calculations are somewhat complicated by the fact that fission reactions of a given nuclide usually do not result in two fission product nuclides of equal size. Instead, many product nuclides of unequal masses are formed. For example, the fission of $^{235}_{92}U$ forms products ranging from zinc ($Z = 30$) to terbium ($Z = 65$), with mass numbers from 72 to 161. Figure 20.5 shows the percentages of fission product nuclei of different mass number produced when $^{235}_{92}U$ fissions by thermal neutron absorption.

Neutrons and γ rays also are emitted when a heavy excited nucleus undergoes fission. In the thermal neutron fission of $^{235}_{92}U$, an average of 2.5 neutrons per nucleus is emitted. In the fission of $^{239}_{94}Pu$, an average of 2.9 neutrons per nucleus is emitted. Most of these neutrons are available to cause fission of other nuclei, which is one of the most important characteristics of these fission reactions.

If a nucleus absorbs one neutron and releases several neutrons, a chain reaction can occur. That is, the fission of a single nucleus leads to the production of neutrons that cause the fission of several nuclei. Each of these nuclei, in turn, produce several neutrons that cause the fission of more nuclei. Such a chain reaction is basic to both military and peaceful uses of fission energy. In an atomic bomb, the chain reaction is allowed to cascade, so that a large number of nuclei undergo fission almost simultaneously, releasing destructively large amounts of energy. In a nuclear reactor, the chain reaction is controlled, so that a manageable amount of energy is released slowly over a long period of time.

Chain reactions are rare in nature. For a chain reaction to occur, a single fission must release one or more neutrons that go on to initiate other fissions. The number of neutrons released per fission in naturally occurring samples must be larger than one, because some of the neutrons are absorbed by nuclei that cannot undergo fission. For example, naturally occurring uranium contains two isotopes, ^{238}U and ^{235}U, in a ratio of about 140:1. When a nucleus of ^{235}U absorbs a neutron, it undergoes fission. A nucleus of ^{238}U does not; instead, it undergoes β^- decay. A chain reaction is not possible in most deposits of naturally occurring uranium because most of the neutrons emitted when ^{235}U nuclei fission are absorbed by the ^{238}U nuclei. However, evidence that a natural chain reaction occurred two billion years ago in a uranium deposit in what is now the Gabon Republic in West Africa has been discovered. An unusual sequence of events apparently allowed the concentration of ^{235}U to increase, so that the chain reaction could occur. There are two major pieces of evidence for the existence of this natural fission reactor: The concentration of ^{235}U is lower than it should be, and the deposit contains elements, in particular xenon, that are characteristic products of fission reactions. As far as is known, this chain reaction was a unique event.

Even when the ^{235}U concentration is relatively high, a chain reaction may not occur if the sample of uranium is so small that many neutrons can escape without striking another nucleus. The minimum quantity of material required to sustain a chain reaction is called the *critical mass*.

One other requirement must be met for a chain reaction to occur. This

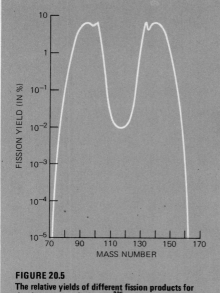

FIGURE 20.5
The relative yields of different fission products for the slow-neutron fission of ^{235}U, plotted as a function of mass number.

requirement is based on the fact that the neutrons emitted in fission reactions are not suitable for chain reactions because they are moving too fast. The probability that a nucleus will capture a bombarding particle and undergo a reaction is called the nuclear cross section. The unit in which nuclear cross sections are expressed is the barn, so called because it was first used to describe a nucleus whose cross section was "as wide as a barn door." One barn $= 10^{-28}$ m^2. The nuclear cross section depends not only on the nature of the nucleus but also on the nature of the bombarding particle. It has been found that nuclear cross sections are greater for the capture of slow-moving thermal neutrons than for fast neutrons. For example, the cross section of ^{235}U is 500 times greater for thermal neutrons than for fast neutrons. However, the neutrons released in fission reactions are fast neutrons. Therefore, fissionable material usually is mixed with a material called a *moderator,* which slows fast neutrons to thermal speeds. Light substances that do not undergo reactions with neutrons generally are used as moderators. The first nuclear reactor, or "pile," as it was called, used graphite as a moderator. Most nuclear generating plants in the United States use ordinary water as the moderator.

The uncontrolled chain reaction that takes place in an atomic bomb must occur with the fast neutrons released by fission, since there is no time to slow the neutrons with a moderator. An atomic bomb is set off by bringing together pieces of fissionable material very rapidly to form a critical mass. The fissionable material must be brought together in such a way that the beginning of the chain reaction does not blow the pieces apart again, causing a small premature explosion called a fizzle. A great deal of ingenuity has been devoted to producing nuclear explosives that use the minimum amount of fissionable material and have the maximum amount of destructive power.

In a nuclear reactor, the chain reaction must be controlled within fairly strict limits. The primary method of control is the use of rods made of a material that absorbs neutrons readily. Insertion or removal of these control rods can make the rate of the chain reaction slower or quicker. Control rods often are made of $^{10}_{5}$B, whose neutron-absorbing capabilities are excellent.

A typical nuclear power plant in the United States has a core consisting of about 40 000 fuel rods, weighing a total of 120 000 kg, made of uranium whose ^{235}U content has been enriched to 3%. The core is immersed in water, which acts as the moderator as well as a coolant (temperatures in the core can reach 600 K). Control rods are interspersed among the fuel rods. The core is enclosed in a steel vessel perhaps 12 m high and 5 m in diameter, with steel walls 30 cm thick. A number of safety features designed to prevent overheating of the core and escape of radioactive material are essential parts of a nuclear power plant. Despite these features, serious criticism has been leveled at the safety of nuclear power plants. Figure 20.6 shows one type of reactor design.

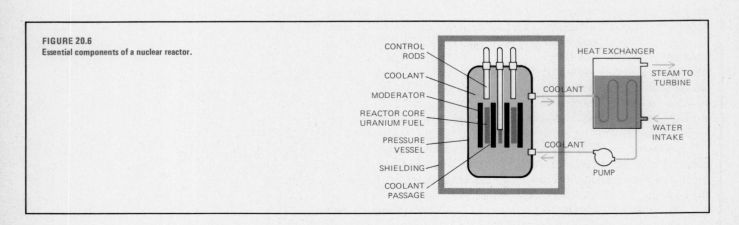

FIGURE 20.6
Essential components of a nuclear reactor.

THE NUCLEAR WASTE PROBLEM

When coal is burned to generate electricity, its waste products include sulfur oxides and ash. When nuclear fuel is "burned" to generate electricity, its waste products include a host of radioactive elements, some of them potentially useful and many of them quite dangerous. An intense debate about the most desirable way to handle nuclear wastes has been going on for several years.

One proposal is that the nuclear fuel be processed to remove fissionable material, which could then be used in new fuel rods. A typical fuel rod contains about 3.3% ^{235}U and 97.7% ^{238}U. As the fissionable ^{235}U is consumed, some of the ^{238}U is transformed by neutron absorption and β^- decay into transuranium elements, including ^{239}Pu and ^{240}Pu, which can be used as fissionable fuel. To obtain the plutonium, the fuel rod is dissolved in acid, and the resulting solution is processed chemically to isolate the fissionable fuels. Plutonium and ^{235}U are then fashioned into a so-called "mixed oxide" fuel for reactors.

The opposition to such fuel recycling is based partly on the fear that the plutonium obtained in this process could be used to make nuclear weapons, either by governments bent on aggression or by terrorist groups. In addition, fuel reprocessing leaves behind some liquid wastes of extremely high radioactivity that will be dangerous for many thousands of years.

For example, the fission products in the liquid wastes include ^{90}Sr, whose half-life is 29 years, and ^{137}Cs, whose half-life is 30 years. It will take 400 years for the radioactivity from these fission products to decrease to a reasonably safe level. But even after 400 years, there will be substantial danger from other radioactive elements such as ^{241}Am and ^{229}Th, whose half-lives are measured in thousands or even millions of years.

The most commonly discussed plan is to transform these liquid wastes into glassy ceramic rods, each about 3 m long and 30 cm in diameter, which would be put into salt beds, granite, or other geological deposits, where they presumably would remain untouched for the many thousands of years needed for the radioactive elements to decay. The wastes from a single 1000-MW nuclear reactor could be contained in 10 such rods. If all the electricity in the United States were generated by nuclear reactors, about 4000 fuel rods a year would be consumed. (At this writing, about 8% of U.S. electrical capacity is nuclear.)

Several studies have concluded that this method of waste disposal could be carried out safely using existing technology, and that safe geological deposits for the wastes can be found. However, other experts maintain that the risk of nuclear weapon proliferation makes fuel recycling undesirable. Doubts have also been raised about the long-term dangers of wastes with a high level of radioactivity. It has been pointed out that these wastes will have to be stored safely for a much longer time period than any human society has ever existed. It has been proposed that the fuel rods should be stored without reprocessing to lessen the danger of nuclear weapon proliferation.

Fusion Weapons

A major effort now is being made to obtain usable energy from fusion, the process that occurs in stars. Just as the maximum energy release in fission occurs in the heaviest elements, the largest energy release in fusion processes occurs in the lightest elements, where there are often substantial increases in binding energy per nucleon as the result of fusion. In principle, fusion reactions between the lightest elements are the easiest to bring about. The coulombic repulsion barrier that must be overcome for nuclei to collide is smallest for the lighter elements, whose nuclei have fewer protons.

Attention is being focused on the fusion reactions of heavier isotopes of hydrogen, such as deuterons, as sources of energy. A fusion reaction between two deuterons can take two different courses:

$$^2_1H + {}^2_1H \longrightarrow {}^3_2He + {}^1_0n + 3.3 \text{ MeV}$$
$$^2_1H + {}^2_1H \longrightarrow {}^3_1H + {}^1_1H + 4.0 \text{ MeV}$$

Such fusion reactions take place only at extremely high temperatures. They are called *thermonuclear reactions*. To start a thermonuclear reaction, the kinetic energy of the reacting nuclei must be raised to the equivalent of a temperature of 100 000 000 K. However, the average temperature can be lower. Most of the nuclei will have energies too low for fusion to occur, but because of the distribution of kinetic energy, a few nuclei will have the energy necessary for fusion. Once some appreciable fraction undergoes a fusion reaction, enough energy is liberated to raise the kinetic energy of other nuclei to the level needed for fusion, so the fusion reaction becomes self-sustaining.

So far, only uncontrolled thermonuclear reactions have been carried out on earth, using hydrogen bombs. The temperature necessary to begin the thermonuclear reaction in a hydrogen bomb is achieved by exploding an atomic bomb. This fission bomb is surrounded by deuterium, tritium, and ^6_3Li, all of which can undergo fusion reactions. The explosion of the fission bomb raises the temperature of these nuclides to 10^7 K. Fusion reactions occur, releasing energy that causes a rapid temperature increase and more fusion reactions. The reactions that occur in a hydrogen bomb include:

$$^2_1H + {}^3_1H \longrightarrow {}^4_2He + {}^1_0n + 17.6 \text{ MeV}$$
$$^6_3Li + {}^1_0n \longrightarrow {}^4_2He + {}^3_1H + 4.8 \text{ MeV}$$
$$^3_2He + {}^2_1H \longrightarrow {}^4_2He + {}^1_1H + 18.3 \text{ MeV}$$

The lithium reaction is important because it replenishes the supply of tritium, which is consumed in the fusion process.

To add explosive power, the material undergoing fusion is encased in a shell of ordinary uranium. The fusion reactions emit fast neutrons that can cause a fission chain reaction in the ^{238}U in the shell. This fission reaction cannot take place with slow neutrons. Thus, a thermonuclear bomb of the most modern design has three stages: nuclear fission of a few kilograms of $^{239}_{94}$Pu, nuclear fusion of about 150 kg of lithium, deuterium, and tritium, and nuclear fission of about 500 kg of $^{238}_{92}$U. The total mass of such a bomb is about 1500 kg. The energy released is equivalent to the explosion of 20 000 000 000 kg of TNT. There is no practical limit to the explosive yield of a modern thermonuclear weapon.

Fusion Reactors

In principle, thermonuclear reactions are a source of unlimited energy for our needs. Deuterium, the principal fuel proposed for a fusion generating plant, can be obtained in relatively large quantities and low cost from ordinary seawater. However, the task of creating the conditions under which a controlled thermonuclear reaction can occur are formidable. The control of thermonuclear reactions has been a high-priority research goal for more than two decades, but widespread use of peaceful thermonuclear energy still is believed to be decades away.

The principal problem is the difficulty of producing and maintaining matter in a physical state quite different from that which is encountered at ordinary temperatures. At the temperatures that exist in the interior of a star, there is virtually complete ionization of all atoms, forming a state of matter called *plasma*. Plasma is a homogeneous mixture of atomic nuclei and electrons, moving rapidly and randomly. At the very high temperatures necessary for fusion reactions, a plasma loses energy very quickly by emitting electromagnetic radiation. A fusion reaction cannot be self-sustaining unless the rate at which energy is produced by fusion exceeds the rate at which energy is lost from the plasma as radiation. A thermonuclear reactor will produce surplus energy only when the temperature of the plasma is at least 10^8 K.

An added complication is the loss of energy by the plasma through thermal conduction. Because of their free electrons, plasmas are extremely good conductors of heat. A plasma conducts heat more than 10^6 times better than a metal at room temperature. Therefore, the plasma must not be allowed to come in contact with the walls of a container, or it will lose energy so rapidly that no significant number of fusion reactions could occur. In most thermonuclear experiments, the plasma is kept from the walls of the container by magnetic fields, in what is called magnetic confinement. We can picture the plasma, a gas consisting of charged particles, as being enclosed in a cage whose bars are the lines of force of a magnetic field. Several different types of magnetic confinement are being tried. One is a toroidal (doughnut-shaped) device called the tokomak, originated by the Soviet Union and adapted by laboratories in the United States. Another is a so-called "magnetic mirror," in which the plasma can be pictured as washing back and forth between two "walls" of magnetic force.

A successful fusion reactor must satisfy what are called the Lawson criteria, concerning the product of n, the plasma density in particles per cubic centimeter, and τ, the confinement time of the plasma in seconds. For deuteron-deuteron reactions, $n\tau$ must exceed 10^{16}. For deuteron-tritium reactions, $n\tau$ must exceed 10^{14}. In 1977, leaders of the United States fusion research effort were predicting that the Lawson criteria would be met in a fusion reactor that achieved "scientific break-even," in which the amount of energy released by fusion reactions is equal to the energy required to cause fusion, by the early 1980s.

Magnetic confinement was the only approach to controlled thermonuclear fusion until the 1960s. In that decade, research began on a method called "inertial confinement," in which powerful beams of light from lasers would be used to compress tiny pellets of nuclear fuel, to produce fusion reactions. In essence, the inertial-confinement approach would produce a series of miniature thermonuclear explosions whose energy could be captured by the walls of the container in which the explosions occurred. The success of the inertial-confine-

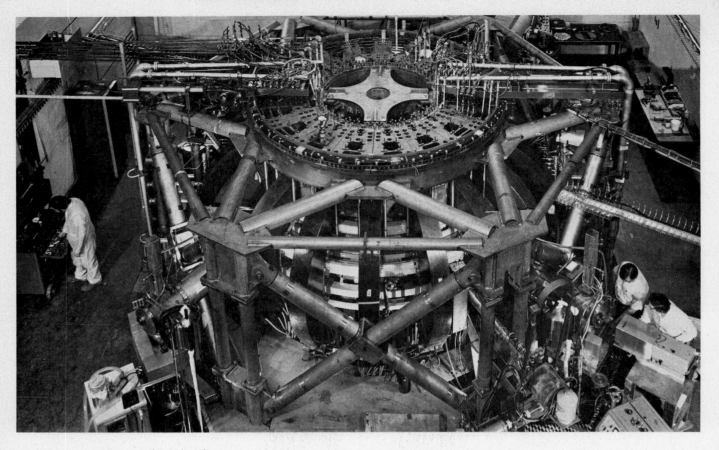

The Princeton Large Torus, one of the major U.S. fusion energy research facilities. Inside the torus is a plasma, which is prevented from touching the steel walls by intense magnetic fields. (*Princeton University Plasma Physics Laboratory*)

ment approach depends not only on the development of low-cost methods of making pellets containing fusion fuel but also on success in developing a new kind of laser of extremely high power and the required wavelength.

If either the magnetic-confinement approach or the inertial-confinement approach to harnessing thermonuclear fusion succeeds, we will have a source of energy essentially without limit. However, as we mentioned in Chapter 13, such an energy source could create problems as severe as any it solves. The challenge posed by the technology that has arisen from advances in nuclear science is one of the gravest that the human race has ever faced.

EXERCISES

20.1 Indicate the number of neutrons and protons in each of the following nuclides: (a) lithium-7, (b) neon-22, (c) plutonium-239, (d) platinum-194.

20.2 Use the conventional notation to list five isotopes of (a) carbon, (b) tin.

20.3 Use the conventional notation to list five isobars with (a) $A = 40$, (b) $A = 234$.

20.4 Use the conventional notation to list five isotones with (a) $N = 14$, (b) $N = 143$.

20.5 Find the radius of (a) a nucleus of ^{4}He, (b) a nucleus of ^{247}Bk, (c) the ratio of the two radii.

20.6 Find the mass of one mole of nuclei of ^{2}H.

20.7 Calculate the density of a nucleus of ^{12}C.

20.8 Calculate the mass of a one-centimeter cube of ^{12}C nuclei.

20.9 Find the energy equivalent of the mass of (a) an electron, (b) a proton.

20.10 Find the energy equivalent of one mole of ^{12}C.

20.11 The heat of combustion of gas-

oline is about 3×10^4 kJ/liter. Calculate the mass loss on combustion of one liter of gasoline.

20.12 The mass of ^{7}Li^{3+} = 7.014359. Use the relative masses of the proton and the neutron given in Table 20.1 to find the binding energy of this nuclide.

20.13 Find the binding energy in ^{3}He.[a]

20.14 Classify each of the following

[a]The necessary data for this exercise are listed in Table 20.1.

nuclides as even-even, even-odd, odd-even, or odd-odd: (a) ^{56}Fe, (b) ^{57}Fe, (c) ^{58}Co, (d) ^{59}Co.

20.15 Explain the observation that no stable nuclide of tin is known in which both Z and N are magic numbers.

20.16 Write reactions for the α decay of the following nuclides: (a) ^{8}Li, (b) ^{190}Pt, (c) ^{192}Pt, (d) ^{239}Pu.

20.17 Calculate the mass loss and energy release in the α decay of ^{239}Pu.^{a}

20.18 The relative mass of ^{242}Cm is 242.0588. Calculate the mass loss and energy release when this nuclide undergoes α decay to ^{234}U.

20.19 Find the energy change that corresponds to the emission of γ rays of wavelength 1.0×10^{-12} m.

20.20 Write reactions for the β^- decay of the following nuclides: (a) ^{28}Al, (b) ^{56}Mn, (c) ^{147}Pm, (d) ^{223}Fr.

20.21 Write reactions for the β^+ decay of the following nuclides: (a) ^{15}O, (b) ^{56}Co, (c) ^{94}Tc, (d) ^{206}Bi.

20.22 Write reactions for electron capture by the following nuclides: (a) ^{26}Al, (b) ^{60}Cu, (c) ^{209}Po.

20.23 Predict the most likely type of β decay for the following nuclides: (a) ^{17}F, (b) ^{21}F, (c) ^{197}Hg, (d) ^{206}Hg.

20.24 Calculate the mass loss and energy release for the β^- decay of ^{87}Rb.^{a}

20.25 Calculate the mass loss and energy release for the β^- decay of ^{14}C.^{a}

20.26 Find the mass loss and energy release when ^{10}C (relative mass = 10.016830) undergoes β^+ decay.^{a}

20.27 Find the mass loss and energy release when ^{15}O (relative mass = 15.003072) undergoes β^+ decay.^{a}

20.28 Find the mass loss and energy release when ^{209}Po (relative mass = 208.9829) decays by electron capture.

20.29 The nuclide ^{18}F decays by both electron capture and β^+ decay. Find the difference in the energy released by these two processes.^{a}

20.30 Find the energy released when 1.0 mol of ^{3}H undergoes β^- decay.^{a}

20.31 Write equations for each of the first three α-decay steps in the thorium series.

20.32 Write equations for three of the β^--decay steps in the thorium series.

20.33 Suggest an explanation for the absence of any isotopes of platinum or gold as intermediates in the thorium series.

20.34 Suggest an explanation for the observation that ^{236}Np (relative mass = 236.0466) undergoes electron capture but not β^+ decay to form ^{236}U (relative mass = 236.0457).

20.35 Calculate ΔM and Q for the reaction ^{59}Co$(n,\nu)^{60}$Co. Given that the relative mass of ^{60}Co is 59.93355.^{a}

20.36 Calculate ΔM and Q for the two-step process in which ^{59}Co is converted to ^{60}Ni by neutrons.^{a}

20.37 Find ΔM and Q for the formation of ^{14}C in the atmosphere.^{a}

20.38 Write the reaction and calculate ΔM and Q for the process ^{10}B$(n,\alpha)^7$Li.^{a}

20.39 Write the reactions, including the compound nuclei, for the following processes: (a) ^{6}Li$(n,\alpha)^3$H, (b) ^{7}Li$(n,\nu)^8$Li, (c) ^{45}Sc$(n,p)^{45}$Ca.

20.40 Write the reactions, including the compound nuclei, for the following processes. (a) ^{31}P$(d,p)^{32}$P. (b) ^{10}B$(p,\nu)^{11}$C, (c) ^{9}Be$(d,2p)^9$Li.

20.41 Write the reactions, including the compound nuclei, for the following processes: (a) ^{106}Pd$(\alpha,p)^{109}$Ag, (b) ^{6}Li$(^3$He$,n)^8$B, (c) ^{141}Pm$(^{12}$C$,4n)^{149}$Tb.

20.42 Suggest two possible pathways, using any bombarding particles you choose, for the conversion of ^{197}Au to ^{208}Pb.

20.43 The nuclide ^{241}Am has been made from the reaction between ^{238}U and ^{4}He. Suggest a pathway for the formation of this nuclide.

20.44 The nuclide ^{247}Es can be made by bombardment of ^{238}U in a reaction that emits five neutrons. Identify the bombarding particle.

20.45 Find Q for the overall hydrogen burning process in a star.^{a}

20.46 Write a series of steps by which ^{64}Zn could form from ^{56}Fe in a second-generation star.

20.47 Write a series of steps by which ^{238}U might form from ^{208}Pb in a supernova explosion.

20.48 Explain why a fission chain reaction does not ordinarily take place in natural uranium ores.

20.49 Explain the function of each of the following in a nuclear reactor: (a) critical mass, (b) moderator, (c) control rod, (d) coolant.

20.50 Predict the effect of the removal of each of the things mentioned in Exercise 20.49 on the operation of a nuclear reactor.

20.51 Account for the formation of plasma at high temperatures in terms of the second law of thermodynamics.

20.52 Given that the energy released in the fusion of two deuterons to a ^{3}He and a neutron is 3.3 MeV and in the fusion to tritium and a proton it is 4.0 MeV, calculate the energy change in the process ^{3}He$(n,p)^3$H. Suggest an explanation for the fact that this process occurs at much lower temperatures than either of the first two.

20.53 Discuss the possible effects of virtually unlimited energy from controlled fusion in terms of the conversion of heat to work and the second law of thermodynamics.

ORGANIC CHEMISTRY

21

Organic chemistry is the study of the compounds of carbon. Until the middle of the nineteenth century, the view of organic chemistry was much narrower. It was defined as the chemistry of living organisms and the substances isolated from them. Until that time, it was believed that anything alive possessed a ''vital force'' that was not present in nonliving things. All substances from living things—that is, all organic compounds—were believed to be unique in possessing this vital force. The downfall of the vital force theory was the result of many experiments that demonstrated that a sample of a substance from a living thing and a sample of the same substance from a nonliving source are indistinguishable. Today, organic chemistry deals with a very large number of substances that have nothing to do with living organisms.

Almost without exception, organic compounds contain hydrogen in addition to carbon. A very few contain one of the halogens instead of hydrogen. In addition, many organic compounds also contain one or more other elements, such as oxygen, nitrogen, sulfur, and phosphorus. A special class of compounds called organometallics contain a metal as well.

The number of organic compounds is impressive. More than three million of them are known, ten times more than all the compounds of the other elements. But aside from sheer numbers, organic compounds are important because much of the chemistry of life is organic chemistry. Almost every substance found in a living organism is an organic compound; H_2O is a notable exception. In addition, synthetic organic compounds are of enormous economic importance. A substantial part of the modern chemical industry is devoted to production of such compounds: plastics, pharmaceuticals, paints, detergents, fibers, and all the other products that are an essential part of our daily lives.

The existence of such a great number of organic compounds is due in large part to the ability of a carbon atom to form strong bonds with other carbon atoms, building chains or rings of carbon atoms. This property, called *catenation,* is not unique to carbon. Other elements near carbon in the periodic table can form rings or chains. But those formed by carbon are unusually stable and unreactive, in part because of the relative strength of the carbon-carbon bond. Not only is the typical carbon-carbon bond energy substantial, almost 350 kJ/mol, but it is also close to or larger than the bond energies of carbon with other elements. Chains or rings of carbon atoms are not easily broken.

In this chapter, we shall introduce some of the basic themes of organic chemistry: the systematic organization of the vast amount of information that is available about organic compounds; the simplified representations for the structures of organic molecules; the method for assigning an unambiguous name to any one of the enormous number of organic compounds. We shall first explore some of the structural relationships of organic compounds, and then discuss some characteristics of organic reactions.

The organic chemist tries to discern relationships between the structures of substances so that the chemical behavior of these substances can be understood as part of a general pattern. One of the most impressive achievements of organic chemists has been the development of a manageable theoretical framework for the almost innumerable experimental observations that are the body of knowledge of organic chemistry.

21.1 THE ALKANES; SATURATED HYDROCARBONS

Compounds that contain only carbon and hydrogen are called *hydrocarbons*. A hydrocarbon whose structure includes no double bonds and no rings is called an *alkane*. An organic compound with no multiple bonds is said to be *saturated,* and a compound with one or more multiple bonds is said to be *unsaturated*. All alkanes thus are saturated. The alkanes are the simplest group of organic compounds from the point of view of chemical behavior.

The simplest alkane, with one carbon atom, is methane. The alkane with two carbon atoms is ethane and the alkane with three carbon atoms is propane. Their structures are:

methane   ethane   propane

In these compounds, as in almost every other organic compound known, each carbon atom has a total of four bonds. We can use this fact to simplify the representation of organic molecules.

A first simplification is to eliminate the lines that indicate bonds between carbon and hydrogen atoms. Instead, all the hydrogen atoms that are bonded to a given carbon atom in a molecule are grouped together. The lines showing bonds between carbon atoms are retained. Thus, methane is represented as CH_4, ethane is represented as $CH_3—CH_3$, and propane is represented as $CH_3—CH_2—CH_3$. The Lewis structures can be written from this simplified representation by distributing the hydrogen atoms around the carbon atoms so that each carbon atom has a total of four bonds.

We can make a further simplification and eliminate the lines showing the carbon-carbon bonds. The structure of ethane then is written CH_3CH_3, and the structure of propane becomes $CH_3CH_2CH_3$. Since we know that each carbon atom has four bonds and each hydrogen atom has only one bond, an unambiguous Lewis structure can be written for each molecule from the simplified representation.

The relationship between the compositions of different alkanes can be seen even in the first three compounds in the group. Starting with methane, we can *imagine* that a CH_2 group is inserted into one of the C—H bonds of CH_4 to form ethane. We can also imagine a CH_2 group being inserted into one of the C—H bonds of ethane to form propane. In this way, we can construct an entire series of alkanes. Each alkane will have one more CH_2 group than its predecessor. Thus, the general formula for any alkane is C_nH_{2n+2}, where n is the number of carbon atoms.

But if we picture the formation of succeeding alkanes in this way, we quickly encounter a situation in which more than one structure is possible. Propane has two different kinds of hydrogens: six hydrogens bonded to the two carbon atoms at the ends of the three-carbon chain (the CH_3 groups) and two hydrogens bonded to the carbon atom in the middle of the chain (the CH_2 group). We can imagine the next alkane in the series being formed by inserting another CH_2 group into a C—H bond of propane. Since there are two types of C—H bonds, two alkanes are possible. They both have the composition C_4H_{10} and they are both called butane. Their structures are:

$$CH_3CH_2CH_2CH_3 \qquad CH_3CHCH_3$$
$$\qquad\qquad\qquad\qquad\qquad | $$
$$\qquad\qquad\qquad\qquad\qquad CH_3$$

n-butane   isobutane

As we mentioned earlier (Section 18.2, page 582), two compounds with the same composition are called **isomers**. Isomerism is a common phenomenon in organic chemistry. The type of isomerism observed in butane is frequently encountered. It is called *chain isomerism,* because the difference between the isomers of two butanes is in the chain of carbon atoms. The butane that has a straight chain of four carbon atoms is called *n*-butane, where *n* stands for "normal." The butane that has a chain of three carbon atoms with a one-carbon branch is called isobutane, where *iso-* stands for "different."

The number of isomers of an alkane increases rapidly as the number of carbon atoms increases. If we picture the insertion of a CH_2 group into a C—H bond of a butane, we find that there are three isomeric pentanes whose composition is C_5H_{12}. The three pentanes are:

$$CH_3CH_2CH_2CH_2CH_3 \qquad CH_3CHCH_2CH_3 \qquad CH_3CCH_3$$

$$\underset{\text{n-pentane}}{} \qquad \underset{\overset{|}{CH_3}}{\text{isopentane}} \qquad \underset{\overset{|}{CH_3}}{\overset{\overset{CH_3}{|}}{\text{neopentane}}}$$

You should note the names of the simple alkanes, because they illustrate some basic principles in naming organic compounds. A clear and unambiguous system of names is necessary because of the great number and complexity of organic compounds.

All the compounds we have discussed so far have names that end in the suffix *-ane*. This suffix conveys the information that the compound has no multiple bonds. You will find that other suffixes given to the names of organic compounds also convey specific information about their structures. The same is true of prefixes. The prefix *alk-* is used to name groups of compounds that contain only carbon and hydrogen atoms. The general prefix *alk-* is not used for individual compounds in the group, however. Instead, we use prefixes that indicate the number of carbon atoms in the compound. Table 21.1 lists some of these prefixes.

If necessary, additional prefixes and/or suffixes can be added to indicate additional details about the molecular structure of a compound. As the number of carbon atoms increases, it becomes a formidable task to write and name all the isomers of a given composition. There are 336 319 isomers with the composition $C_{20}H_{42}$.

How can we describe the properties of such an enormous number of compounds? Rather than studying the alkanes one by one, it is better to describe some properties that are common to all the alkanes. Then we can try to find correlations between properties and composition within the group.

The alkanes have a great deal in common. Every alkane contains only two kinds of bond, C—C bonds and C—H bonds. Both of these bonds are quite strong, so it is not surprising to find that the alkanes tend to be unreactive. When an alkane does undergo reaction, it generally does so at elevated temperatures or in other extreme conditions. The most important reaction for the alkanes is combustion, the rapid, high-temperature reaction with O_2.

We can group alkanes as well as other types of organic compounds into *homologous series* on the basis of their structures. A homologous series is a group of organic compounds that differ only in the number of CH_2 units in the longest chain. The simplest homologous series is that of the straight-chain alkanes. The first five members of the series are methane, ethane, propane, *n*-butane, and *n*-pentane. Each succeeding member of the series—*n*-hexane, *n*-heptane, and so on—has a chain that is longer by one carbon atom.

TABLE 21.1
Prefixes for Organic Compounds

Prefix	Number of C Atoms	Prefix	Number of C Atoms
meth	1	pent	5
eth	2	hex	6
prop	3	hept	7
but	4	oct	8

Another homologous series consists of alkanes that each have a one-carbon branch located on the next-to-last carbon atom of the chain. We can call this series the *iso* series. We already have mentioned isobutane and isopentane, the first two members of the series. Each succeeding member of the series has one more CH_2 group on the chain.

The members of a homologous series not only have similar chemical properties; even their physical properties change in a regular way as more CH_2 groups are added. One striking example of this regularity is the increase of 29 K in the boiling point of the alkanes with each added CH_2 group.

21.2 UNSATURATED HYDROCARBONS

We have mentioned that an organic compound that has one or more multiple bonds is said to be unsaturated. The simplest class of unsaturated compounds is the *alkenes*.

Like the alkanes, the alkenes are compounds that contain only carbon and hydrogen, as the prefix *alk-* indicates. An alkene has only one carbon-carbon double bond and does not contain any rings of carbon atoms. The suffix *-ene* means ''double bond.''

The two simplest alkenes are ethylene, C_2H_4, and propylene, C_3H_6. The location of the double bond is shown when writing the simplified structures of alkenes:

$$CH_2\!\!=\!\!CH_2 \qquad CH_2\!\!=\!\!CHCH_3$$
ethylene   propylene

Since each carbon atom can have only four bonds, the presence of a double bond between two carbon atoms requires the compound to have two fewer hydrogen atoms than the corresponding alkane. The general formula for an alkane is C_nH_{2n+2}, so the general formula for an alkene is C_nH_{2n}.

Positional Isomerism

The introduction of a double bond into a carbon chain creates more isomeric possibilities. Chain isomerism is found in the butenes, just as in the butanes. For example, two butenes with different chains are:

$$\underset{1}{CH_2}\!\!=\!\!\underset{2}{CH}\underset{3}{CH_2}\underset{4}{CH_3} \qquad CH_2\!\!=\!\!C\!\!\begin{array}{l} CH_3 \\ CH_3 \end{array}$$

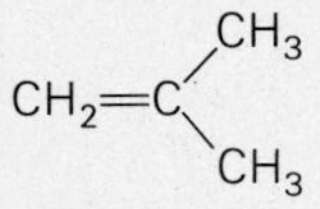

However, a double bond can be introduced into a four-carbon chain in one of two positions, giving rise to *positional isomers*. One of the positional isomers of the straight-chain butene is shown above. The other one is:

$$\underset{1}{CH_3}\underset{2}{CH}\!\!=\!\!\underset{3}{CH}\underset{4}{CH_3}$$

By numbering the carbon atoms of the chain in the indicated way, we can designate the location of the double bond and differentiate between the two positional isomers. Thus, one isomer is named 1-butene, because the double bond starts at the carbon atom numbered 1. The other isomer is named 2-butene, because the double bond starts at the carbon atom numbered 2. The chain isomer is called isobutene, to differentiate it from the two straight-chain positional isomers.

Geometric Isomerism

It has been found that there are actually four isomeric alkenes with the composition C_4H_8. There are two isomers of 2-butene that differ in the spatial arrangement of the substituents on the double bond. In one isomer, both CH_3 groups are on the same side of the double bond. This compound is called *cis*-2-butene. In the other isomer, the two CH_3 groups are on opposite sides of the double bond. This compound is called *trans*-2-butene. Their structures are:

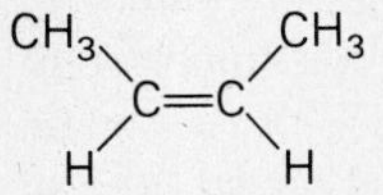

cis-2-butene 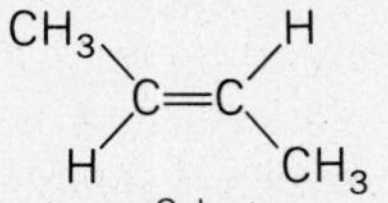   *trans*-2-butene

These two isomers of 2-butene are called *geometric isomers* (Section 18.2, page 583). They differ in the spatial arrangement of their atoms but not in the way that the atoms are attached to each other.

Geometric isomerism occurs in alkenes whenever each of the two carbon atoms connected by the double bond bears two different substituents. In 2-butene, for example, each of the two carbon atoms of the double bond bears one hydrogen atom and one CH_3 group, so geometric isomerism occurs. In 1-butene, however, the carbon atom that is numbered 1 bears two identical substituents, two hydrogen atoms, so there is no geometric isomerism in 1-butene.

The geometric isomer that has similar substituents on the same side of the double bond is called the *cis* isomer. The other one, in which the similar substituents are on opposite sides of the double bond, is called the *trans* isomer.

Polyenes

Many compounds have more than one double bond. Such compounds are called *polyenes*. The polyenes can be classified and named by the number of double bonds they contain. For example, the compounds that consist of only carbon and hydrogen, have no rings of carbon atoms, and have two double bonds are called *alkadienes*. You will note that this name includes both the prefix *alk-*, indicating that the compounds contain only carbon and hydrogen, and the prefix *di-*, which indicates the presence of two double bonds. Since an alkadiene has two double bonds, it has two fewer hydrogen atoms than the alkene with the same number of hydrogen atoms. The general formula for an alkadiene is C_nH_{2n-2}. The location of the double bonds in the chain is indicated by numbering the carbon atoms. For example, the compound whose name is 1,3-butadiene has the structure:

$$\underset{1}{CH_2}=\underset{2}{CH}-\underset{3}{CH}=\underset{4}{CH_2}$$

Chemistry of Alkenes

The alkenes are much more reactive than the alkanes. Even under mild conditions, they undergo a wide variety of reactions, most of which result in changes at the double bond. There are a number of reagents that add to double bonds but do not cleave the carbon chain. The π bond of the double bond becomes two σ bonds between the carbon atoms of the double bond and the species that are added. Some examples of this type of reaction, which is called an *addition reaction,* are shown in Figure 21.1(a). Other reactions of the alkenes result in the cleavage of the carbon chain at the site of the double bond. Figure 21.1(b) shows a reaction of this type. Still other reactions result in changes at carbon atoms adjacent to the double bond or changes in the location of the double bond in the chain.

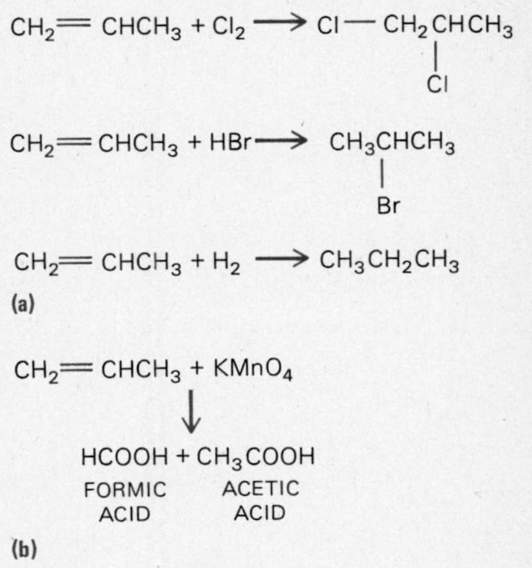

Functional Groups

The double bond in an alkene is one of the simplest examples of a **functional group**. A functional group can be defined as a small group of atoms in a molecule that differ in some way from an alkane structure. The most important generaliza-

FIGURE 21.1
Reactions of alkenes. In the first three reactions, the carbon skeleton remains intact. The addition is only to the π bond of the double bond. In the fourth reaction, the alkene is cleaved, forming two products.

tion in organic chemistry is the statement that *the chemical behavior of compounds is determined primarily by the behavior of their functional groups*. The overall structure of a compound is only of secondary importance in determining chemical behavior.

This generalization is a great simplifying principle in organic chemistry. Rather than trying to learn the chemistry of a huge number of individual compounds, we can learn the chemistry of a relatively small number of functional groups. For example, the chemistry of all the alkenes is similar because they all have the same functional group, a double bond.

However, the overall structure of a compound does influence its chemical behavior. There are differences in the chemical behavior of alkenes which are due to differences in their overall structures. One important part of organic chemistry is the study of the way such structural differences lead to observed differences in chemical behavior.

The presence of any atom other than carbon or hydrogen or of any multiple bond indicates the presence of a functional group. So far, we have encountered only the double bond. A number of other functional groups will be discussed in this chapter.

Alkynes

The suffix *-yne* means "triple bond." An alkyne, therefore, is a hydrocarbon with a triple bond. The simplest alkyne is acetylene, C_2H_2; the next is propyne, C_3H_4:

$$HC\equiv CH \qquad CH_3C\equiv CH$$
acetylene   propyne

The presence of the triple bond means that an alkyne has four fewer hydrogen atoms than the corresponding alkane. The general formula of the alkynes thus is C_nH_{2n-2}, which is identical to that of the alkadienes.

There are four compounds with a straight chain of four carbon atoms and the composition C_4H_6. They are shown in Figure 21.2. The compounds numbered **1** and **2** are positional isomers named, respectively, 1-butyne and 2-butyne. The compounds numbered **3** and **4** are also positional isomers which are named, respectively, 1,3-butadiene and 1,2-butadiene. But if we compare compound **1** with compounds **3** and **4**, or compound **2** with compounds **3** and **4**, we find a different kind of isomerism. These compounds are *functional isomers*. They have identical compositions but contain different functional groups. Compounds **1** and **2** contain a triple bond, while compounds **3** and **4** contain two double bonds.

There are similarities in chemical behavior in compounds that contain a double bond and those that contain a triple bond. Triple bonds undergo the addition reactions and cleavage reactions of the sort that double bonds undergo, although the details of the reactions may differ. However, there are some significant differences in the behavior of the two functional groups. The most important of these differences appears in what are called "terminal acetylenes," compounds such as propyne and 1-butyne, which have the triple bond at the end of the chain of carbon atoms. Such a triple bond bears a single hydrogen atom that is relatively acidic. These compounds are weak acids. The other hydrogen atoms in these hydrocarbons, as well as most hydrogen atoms bonded to carbon, are generally nonacidic and can be removed only by the strongest bases.

$CH_3CH_2C\equiv CH$
1

$CH_3C\equiv CCH_3$
2

$CH_2\equiv CHCH\equiv CH_2$
3

$CH_2\equiv C\equiv CHCH_3$
4

FIGURE 21.2
Some isomers of C_4H_6.

21.3 CYCLIC HYDROCARBONS

Carbon atoms can form rings as well as chains. The name of a compound that contains one or more rings of carbon atoms usually includes the prefix *cyclo-*. Compounds that contain only carbon and hydrogen and have at least one ring of carbon atoms are called *cyclic hydrocarbons*. The class of compounds called the *cycloalkanes* are those that have a ring of carbon atoms (*cyclo-*), contain only carbon and hydrogen atoms (-*alk*-), and have no multiple bonds (-*ane*). The cycloalkenes and the cycloalkadienes are just two of the many other classes of cyclic hydrocarbons.

We can imagine that a cycloalkane is formed by removing a hydrogen atom from each end of an alkane chain and then joining the two ends of the chain. The general formula of the cycloalkanes is C_nH_{2n}, the same as that of the alkenes. Thus, there are two compounds with the composition C_3H_6. One is propylene. The other, which has a ring of three carbon atoms, is called cyclopropane.

A number of methods have been developed to represent the compounds whose structures include rings of carbon atoms. Figure 21.3 shows several different representations of the simplest cycloalkane, cyclopropane. The first structure (Figure 21.3(a)), which shows all the bonds, is rarely used. In the second structure (Figure 21.3(b)), the practice of omitting the C—H bonds but including the C—C bonds is followed. The representation that is most commonly used is that of Figure 21.3(c), in which the cyclic structure is represented by a polygon. This greatly simplified representation can be used because it is understood that each carbon atom has four bonds.

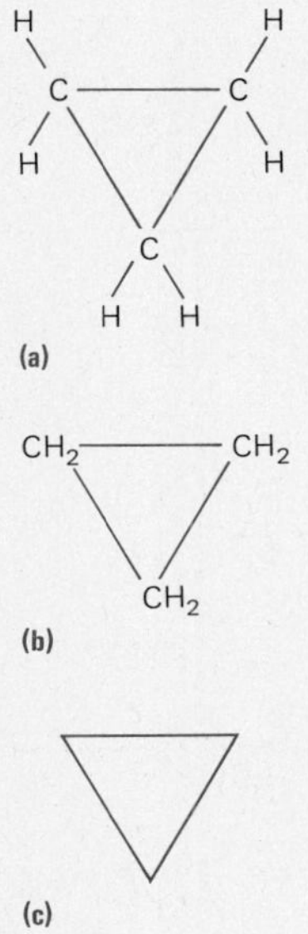

The sides of the polygon in Figure 21.3(c) represent carbon-carbon single bonds. Each corner represents a carbon atom. It is assumed that the molecule has just enough hydrogen atoms to allow each carbon atom to form four bonds. Substituent atoms or groups are shown only if they are not single hydrogen atoms. For example, the compound whose composition is C_4H_8 and whose name is cyclobutane is represented as:

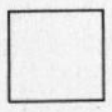

The representation indicates that cyclobutane has a ring of four carbon atoms, each of which has two singly bonded hydrogen atoms attached to it. The hydrogen atoms are not shown.

The name of an unsubstituted cycloalkane is derived by using the prefix *cyclo*- and the name of the alkane with the same number of carbon atoms. For example, the cycloalkane whose composition is C_6H_{12} is called cyclohexane. It has a six-carbon ring and its structure is represented as a regular hexagon.

Some of the principles that are followed in naming compounds whose rings bear substituents other than hydrogen atoms can be learned by studying specific compounds. A compound whose composition is C_4H_8 is:

—CH_3

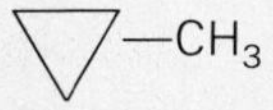

This compound is named methyl cyclopropane. Cyclopropane denotes the three-carbon-atom ring. The relationship between methane, CH_4, and the CH_3 substituent on the ring is reflected in the name *methyl*. The suffix -*yl* is a common one in chemistry. It designates a polyatomic group formed by the loss of one hydrogen atom from the molecule named in the stem. The loss of the hydrogen atom allows the attachment of the substituent, sometimes called a *radical*, to another atom, such as the carbon atom of a chain or ring. Many organic compounds are named by designating the type and location of radicals on a chain or ring of carbon atoms.

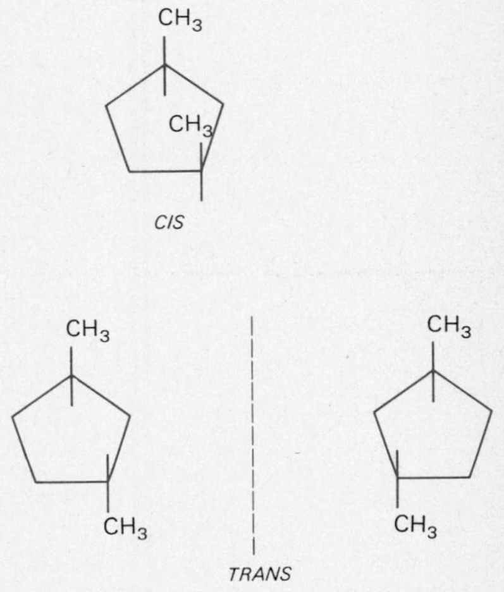

Compounds with ring structures that bear more than one substituent are quite common. Geometric isomerism, analogous to the geometric isomerism found in alkenes, can exist in compounds of this kind. If there are two substituents on the ring and both substituents are on the same side of the ring, we have the *cis* isomer. If the substituents are on opposite sides of the ring, we have the *trans* isomer. Figure 21.4 shows the *cis* and *trans* isomers of 1,3-dimethylcyclopentane.

Optical isomerism (Section 18.2, page 585) is also found in organic com-

FIGURE 21.3
Three ways of representing the simplest cycloalkane, cyclopropane.

FIGURE 21.4
The isomers of 1,3-dimethylcyclopentane.

pounds. If a structure has chirality, it will exist as two optical isomers. The *cis*-1,3-dimethylcyclopentane is not chiral. It has a plane of symmetry and therefore does not display optical isomerism. However, the *trans*-1,3-dimethyl-cyclopentane is an asymmetrical compound which is superimposable on its mirror image and which therefore exists as two optical isomers (Figure 21.4).

Like the alkanes, the cycloalkanes tend to be unreactive. However, cyclo-propanes and, to a lesser extent, cyclobutanes are more reactive than other cycloalkanes because of what is called *ring* or *angle strain*. When a straight chain of carbon atoms forms a three- or four-member ring, there must be considerable distortion of the tetrahedral geometry normally associated with sp^3 carbon atoms. It appears that the C—C—C bond angle in a three-member ring, which is geometrically similar to an equilateral triangle, must be only 60°, rather than the bond angle of 109°28' associated with tetrahedral geometry. Such a distortion of bond angles does not permit the kind of orbital overlap usually associated with σ bonding. As a result, there is a destabilization in the three-carbon ring. This destabilization is called ring or angle strain. A strained ring is more easily broken. Hence, cyclopropane is more reactive than other cycloalkanes. Angle strain is less important in cyclobutane and is relatively unimportant in larger rings.

In larger rings, a nonplanar arrangement of the carbon atoms minimizes angle strain. Although we represent cyclohexane as a regular hexagon, the six carbon atoms do not lie in the same plane. Instead, the ring has a nonplanar conformation in which all the C—C—C bond angles are 109°28'. Several such conformations are possible. Figure 21.5 shows two of them, called the "chair" and the "boat" because of their shapes. These two conformations are not stereoisomers. The "chair" and the "boat" convert rapidly to one another at room temperature; isomers, by contrast, are not interconvertible at room temperature. These two forms of cyclohexane are called *conformations,* or *conformers*. The study of the properties of conformers, especially those of cyclic molecules, called conformational analysis, is an important part of organic chemistry.

Some of the carbon atoms of a ring may be joined by double bonds. Even triple bonds are found in some rings, although they are rare. A cyclic hydrocarbon with one double bond is called a *cycloalkene*. To represent the structure of a cycloalkene, we indicate the location of the double bond in the polygon that represents the ring structure. For example, the structure of the simplest cyclo-alkene, cyclopropene, is represented as:

In part because of the reactivity of the double bond, cyclopropene is much more reactive than cyclopropane, the corresponding cycloalkane, and is quite difficult to prepare. Furthermore, the ideal bond angles associated with C=C double bonds are 120°, even further from the 60° angle of an equilateral triangle than the 109°28' angle associated with C—C single bonds. Therefore, cyclopropene is even more strained than cyclopropane.

Polycyclic Compounds

Many substances, called polycyclic compounds, have more than one ring as part of their structures. One major class of polycyclic compounds is the *steroids,* which are derivatives of the system of four rings shown in Figure 21.6. This type of ring system, in which the rings share some carbon atoms, is called a fused ring system. Although the representation in Figure 21.6 appears to be much more complex than that of a simple cycloalkane, it is interpreted in the same way.

A molecule with this ring system is a steroid. All steroids are physiologically active. A number of steroids occur naturally in plants and animals and many

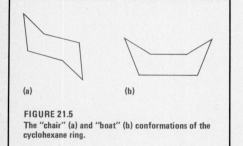

FIGURE 21.5
The "chair" (a) and "boat" (b) conformations of the cyclohexane ring.

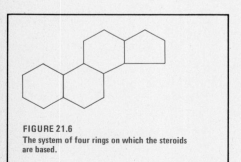

FIGURE 21.6
The system of four rings on which the steroids are based.

STEROIDS FROM PLANTS

One day in 1943, a chemist named Russell E. Marker walked into the office of a small Mexican pharmaceutical company named Laboratorios Hormona carrying two pickle jars wrapped in newspaper. In the jars were more than two kilograms of a steroid hormone called progesterone, worth about $160,000 at the going price and representing a substantial percentage of the existing annual world production of progesterone. Marker had produced the hormone on his own, by a method of his own devising, and that day in Mexico City was a significant milestone in modern pharmaceutical history.

At that time, almost all steroid hormones used in medicine were synthesized by a small number of European companies, using cholesterol from animal sources as a starting material. The European "hormone cartel" dominated the world market, keeping supplies limited and prices high.

Marker had been working on the synthesis of steroid hormones while he was a professor of steroid chemistry at Pennsylvania State University. He concentrated on steroid substances from plants, called sapogenins, whose structure resembles that of cholesterol. In 1939, he had worked out a way of approaching the structure of the steroid hormones by chemical treatment of the sapogenins. The following year, he found a way of synthesizing progesterone from one kind of sapogenin called diosgenin, taken from a plant of the genus *Dioscorea.* When American pharmaceutical companies refused to provide financial support for his research, Marker resigned his position and went to Mexico. He had found that a high yield of diosgenin could be obtained from the root of a yam called *cabeza de negro,* which grows wild in the jungles of the Mexican state of Veracruz.

Marker rented a rundown shack in Mexico City and went to work, helped only by unskilled laborers. Legend has it that he was so suspicious that he usually slept with a pistol next to his research notebook. Marker's efforts to keep his work secret were successful. When he appeared with his two kilograms of progesterone, it was a stunning surprise.

The owners of Laboratorios Hormona quickly recovered from the shock of their windfall and joined with Marker to found a new company to produce progesterone. The company, Syntex, today is one of the world's major pharmaceutical manufacturers. At that time, its main asset was Marker's knowledge. He quickly produced enough progesterone to break the cartel's hold on the world market. When Marker first appeared on the scene, progesterone was selling at $80 a gram. By 1945, the price was down to $18 a gram. In 1952, Syntex was able to fill a contract for more than 9000 kg of progesterone at 48 cents a gram.

But Marker, ever the individualist, stalked out of Syntex in 1945 after an argument with his partners, taking his knowledge with him. He did not reappear for two decades. Progesterone production at Syntex came to an abrupt but relatively brief halt. It took several months of research for others to work out the details of Marker's process, enabling production to be resumed.

others have been synthesized for research purposes or medical use. Perhaps the best-known and certainly the most abundant steroid in the human body is cholesterol, which makes up about one-sixth of the dry weight of the body. Cholesterol has been studied intensively because it has been implicated in diseases of the cardiovascular system.

Many other steroids play major roles in the human body. For example, the steroid sex hormones regulate sexual functions in both men and women. Female sex hormones are called estrogens. Male hormones are called androgens. A third type of steroid sex hormone, called progesterone, is active not only in regulating female sex function but also in regulating the many physiological changes that occur during pregnancy.

Oral contraceptives, the most widely used birth control method in the United States, include two synthetic steroids. One is a progestational hormone, the other is an estrogen. Another group of synthetic steroids which are frequently used in medicine are anti-inflammatory agents derived from the corticosteroids, hormones secreted by the adrenal gland.

Composition and Structure

From the information we already have, we can work out the relationship between the composition of a hydrocarbon and the kind of structure, or structures, that are possible for the compound. We can start with the alkanes, whose general formula is C_nH_{2n+2}. Any compound with this composition is a hydrocarbon with no rings and no multiple bonds. If the number of hydrogen atoms is reduced by two, the general formula becomes C_nH_{2n}. A compound with this composition can have either one ring or one double bond. If we reduce the number of hydrogen atoms by two again, the general formula becomes C_nH_{2n-2}. A compound with this composition can have two double bonds, or two rings, or one double bond and one ring, or one triple bond. Table 21.2 summarizes some relationships between the composition of a hydrocarbon and the number of multiple bonds and rings in the compound. You can see that the number of possible structures increases rapidly as the number of hydrogen atoms in the general formula decreases.

TABLE 21.2
Possible Structures of Hydrocarbons

Formula	Structure
C_nH_{2n}	one double bond or one ring
C_nH_{2n-2}	two double bonds; or one triple bond; or two rings; or one double bond and one ring
C_nH_{2n-4}	three double bonds; or one double bond and one triple bond; or three rings; or two rings and one double bond; or one ring and two double bonds; or one ring and one triple bond

EXAMPLE 21.1
Draw the structures of all the isomers of C_4H_6.
SOLUTION
We note from the formula that this compound has the general formula C_nH_{2n-2}. Table 21.2 gives the possible combinations of rings and multiple bonds.

We must also consider the possible arrangements of the four carbon atoms. They may be arranged in a straight chain of four atoms. They may form a straight chain of three atoms with a one-atom branch. They may form a four-membered ring. Or they may form a three-membered ring with a one-atom substituent.

The straight chain of carbon atoms can accommodate either one triple bond or two double bonds. Figure 21.7(a) shows the possible positional isomers of this arrangement. The branched-chain structure cannot accommodate two double bonds or one triple bond. There are no branched-chain isomers of C_4H_6.

Either a four-carbon ring or a three-carbon ring can accommodate a double bond. There is positional isomerism for the three-membered ring structure, since there are different ways of placing the double bond. Figure 21.7(b) shows the possible ring structures. Finally, there is one possible structure with two rings, shown in Figure 21.7(c).

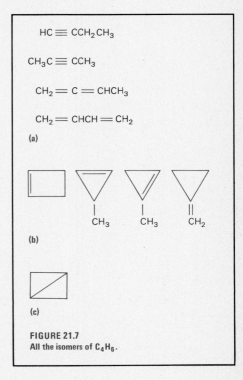

FIGURE 21.7
All the isomers of C_4H_6.

21.4 AROMATIC HYDROCARBONS

Benzene is a most unusual hydrocarbon. It has the composition C_6H_6, and the six carbon atoms form a six-membered ring. The rules summarized in Table 21.2 indicate that this ring should contain three additional π bonds and that a reasonable structure to write for benzene is:

But this structure is inconsistent with the experimental evidence on the structure of benzene. The six carbon atoms of the C_6H_6 molecule have been found to lie in the same plane and to have identical bonds whose length is midway between those of single and double bonds. All the carbon atoms of benzene are identical. So are all the hydrogen atoms and all the C—H bonds.

Benzene cannot be described by an ordinary structural formula. Rather, benzene is a resonance hybrid (Section 7.4, page 210) and is best described as a blend of two equivalent contributing structures:

Neither of these structures actually exists. We know that they are imaginary and write them to help us represent the structure of benzene conveniently.

The difference between these two contributing structures is in the distribution of their π electrons. Benzene has a total of six π electrons. The structure of benzene sometimes is written with a circle to represent the six π electrons:

As Figure 21.8(a) shows, each carbon atom of the benzene ring is sp^2 hybridized. The sp^2 hybrid orbitals form σ bonds with the three attached atoms. On each carbon atom, there is an unhybridized p orbital that is perpendicular to the plane of the carbon and hydrogen atoms. These six p orbitals are parallel to each other. They can overlap to form six molecular orbitals that extend over all of the six carbon atoms of the ring. The six π electrons of the benzene rings are in three of these orbitals, and can be regarded as circulating around the ring. The p

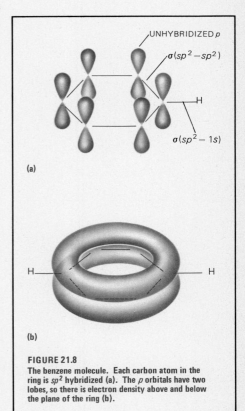

FIGURE 21.8
The benzene molecule. Each carbon atom in the ring is sp^2 hybridized (a). The p orbitals have two lobes, so there is electron density above and below the plane of the ring (b).

orbitals have two lobes, so there is electron density both above and below the plane of the ring, as Figure 21.8(b) shows.

Benzene has been found to be far more stable than would be expected on the basis of bond energies alone—even more stable than can be explained by simple resonance energy. A considerable body of experimental evidence indicates that unusually great stability is associated with the presence of six π electrons in a cyclic structure like that of benzene. This unusual stability is called *aromaticity*. The six-membered benzene ring is called an aromatic ring, and benzene can be regarded as the parent of all compounds containing such rings. A hydrocarbon whose structure has one or more benzene rings is called an *aromatic hydrocarbon* and is also unusually stable.

The chemical properties of aromatic compounds reflect their stability. Even though benzene appears to have double bonds, its chemistry is quite different from that of the alkenes. While alkenes readily add a variety of substances under mild conditions, benzene requires more extreme conditions for reaction. Furthermore, when reaction does occur, it is not addition but substitution. A typical substitution reaction of benzene takes place when it is treated with Cl_2. The products are HCl and chlorobenzene, C_6H_5Cl:

This reaction contrasts sharply with the reaction of propylene with Cl_2 (Figure 21.1), where the Cl_2 simply adds across the double bond.

Many compounds consist of a benzene ring with one or more substituents. These structures are represented in the same way as the structures of substituted cycloalkanes: The substituents are attached to the appropriate corners of the polygon that represents the ring. Figure 21.9(a) shows toluene, a simple substituted benzene.

Disubstituted benzenes—those with two substituents—can be named by using a special system of names developed for these compounds. Only three arrangements for two substituents on a benzene ring are possible. Figure 21.9(b) shows these arrangements for the compounds called xylenes, in which both substituents are methyl groups. The positions of the methyl groups are indicated by the prefixes ortho (*o*), meta (*m*), and para (*p*), as shown in Figure 21.9.

Many relatively simple substituted benzenes are physiologically active and are used as drugs or for other applications. Most of these substituted benzenes contain oxygen and/or nitrogen. Aspirin, acetylsalicylic acid, probably the most familiar of all drugs, is a substituted benzene. Many other simple substituted benzenes are used as painkillers. Benzocaine is a local anesthetic. Phenacetin is an analgesic. Oil of wintergreen (methyl salicylate) is applied externally for muscle aches. The amphetamines are a group of simple benzene derivatives that are central nervous system stimulants. A related compound is ephedrine, which is used as a nasal decongestant. A number of benzene derivatives are central nervous system depressants; the best known of them is phenobarbital. Mescaline, the hallucinogen found in peyote cactus, is a substituted benzene of fairly simple structure. So is adrenaline, a central nervous system stimulant found in the human body. Eugenol, or oil of cloves, is a toothache remedy; vanillin is a familiar flavoring agent. The list of substituted benzenes could go on for pages.

Polynuclear Aromatic Hydrocarbons

Compounds with more than one benzene ring are called polynuclear aromatic hydrocarbons. In most of these compounds, atoms are shared between rings. The simplest such compound is naphthalene, the familiar moth repellent, which is a resonance hybrid of three contributing structures:

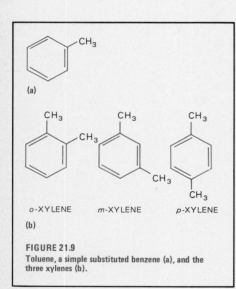

(a)

o-XYLENE   m-XYLENE   p-XYLENE

(b)

FIGURE 21.9
Toluene, a simple substituted benzene (a), and the three xylenes (b).

Coal tar contains a number of polynuclear aromatic hydrocarbons. Such compounds are also found in other materials, most notably in the "tars" of cigarette smoke. Many polynuclear aromatic hydrocarbons are carcinogens; that is, they cause cancer. There seems to be a correlation between the π electronic structure of a polynuclear aromatic hydrocarbon and its carcinogenicity. But despite intensive study, there is not yet a complete understanding of the way in which these compounds cause cancer.

Coal and petroleum consist primarily of hydrocarbons. When we burn coal or oil, we get heat from the combustion reactions of these hydrocarbons. By distilling crude petroleum, it is possible to obtain hydrocarbons that burn at different temperatures and that can be used for different purposes. The gas fraction of petroleum, which boils below 310 K, consists of C_1-C_5 hydrocarbons. The fraction that boils in the range from 310 K to 450 K is called gasoline, which consists of more than a hundred C_6-C_{10} hydrocarbons. Kerosine, which consists of C_{11} and C_{12} hydrocarbons, boils from 450 K to 500 K. Fractions of petroleum that boil at higher temperatures are called oils and lubricants. Vaseline is the last fraction to distill. It consists of $C_{26}-C_{38}$ hydrocarbons which boil in the range from 680 K to 790 K. The residue that remains after all these fractions are distilled from petroleum is the thick, black substance called asphalt.

21.5 OXYGEN FUNCTIONAL GROUPS

We have said that much of organic chemistry can be systematized by identifying the reactive portions of molecules, especially functional groups. By describing the chemistry associated with a specific functional group, we can in large measure describe the chemistry of compounds that have this functional group.

Many common functional groups include one or more oxygen atoms. We shall survey some of these oxygen-containing functional groups in this section. To do so, it is convenient to classify oxygen-containing functional groups on the basis of their oxidation states. In this classification, we note that formal oxidation numbers usually are not assigned in organic chemistry. Rather, the oxidation state is based on the number of oxygen atoms and hydrogen atoms in the functional group.

Alcohols and Ethers

There are two related functional groups in the lowest oxidation state associated with oxygen functional groups. One is the functional group —OH, which is found in *alcohols*. The other is the functional group —O—, which is found in *ethers*. Both these classes of compounds have the general formula $C_nH_{2n+2}O$.

Alcohols have a wide range of chemical reactions. The —OH group in an alcohol replaces a hydrogen atom on an alkane structure. The simplest alcohol, for example, has the composition CH_4O, which usually is written CH_3OH. This compound is called methyl alcohol, methanol, or wood alcohol.

The next alcohol in the series, the one that is most familiar to us, can be represented as CH_3CH_2OH and is called ethyl alcohol, ethanol, or grain alcohol. It is the alcohol that we drink. Alcohols, particularly those of higher molecular weight, tend to have similar properties.

We can understand many aspects of the chemical behavior of alcohols by studying the relationship between the alcohols and water. An alcohol can be regarded as an organic derivative of water in which one of the hydrogen atoms of the water molecule is replaced by an organic radical. An alcohol is both a weak acid and a weak base whose acid strength and base strength are somewhat less than those of water. The conjugate bases of the simple alcohols are the *alkoxides*. We can prepare salts of alkoxide anions and metal cations, such as CH_3CH_2ONa, which is called sodium ethoxide and is analogous to sodium hydroxide. Since its conjugate acid, ethanol, is a slightly weaker acid than water, sodium ethoxide is a slightly stronger base than sodium hydroxide.

Reactions of Alcohols

Alcohols undergo many different types of reactions. Substitution reactions, in which the —OH group is replaced by a different group, are easily carried out in many alcohols. A typical example is the reaction of an alcohol with hydrochloric acid:

$$\underset{\displaystyle CH_3\overset{\displaystyle |}{C}HCH_3}{\overset{\displaystyle OH}{|}} + HCl \longrightarrow \underset{\displaystyle CH_3\overset{\displaystyle |}{C}HCH_3}{\overset{\displaystyle Cl}{|}} + HOH$$

in which the —Cl group replaces the —OH group and water forms as the other product.

Dehydrations, in which an alcohol loses water, can be carried out with a dehydrating agent. When ethanol is treated with H_2SO_4, it loses H_2O to form ethylene:

$$CH_3CH_2OH \overset{H^+}{\longrightarrow} CH_2{=}CH_2 + H_2O$$

The H^+ can be written above the yield sign in this reaction to indicate that it is not consumed in the reaction.

Oxidation reactions can be carried out to convert an alcohol to a compound with an oxygen-containing functional group of higher oxidation state. Alcohols can also be converted to alkanes. This conversion is regarded as a reduction, because the alcohol loses an oxygen atom.

An especially interesting group of alcohols are those in which the —OH group is attached directly to a benzene ring. This group of compounds is called the *phenols*. The simplest compound of the group is called phenol and has the structure:

OH

Phenol was the first substance to be used as an antiseptic. Although phenol has been replaced by more effective substances, antiseptics still are rated by their "phenol number."

The chemical properties of the phenols are somewhat different from those of the alcohols. The most striking difference is in acidity. Phenols are much stronger acids than alcohols. The K_A of a typical phenol is 10^{-10}. The K_A of a typical alcohol is only about 10^{-18}.

The ethers are quite unreactive. They undergo chemical change only when treated with powerful reagents under vigorous conditions. In an ether, the —O— functional group is attached to two organic groups. The simplest ether has the structure CH_3—O—CH_3. It is called dimethyl ether. Dimethyl ether and ethyl alcohol have the same composition, C_2H_6O, and they are functional isomers.

Aldehydes and Ketones

The oxygen-containing functional groups in the next-higher oxidation state are the aldehydes and ketones. Both these classes of compounds have the *carbonyl* functional group, a carbon atom connected by a double bond to an oxygen atom:

$$-\overset{\overset{\displaystyle O}{\|}}{C}-$$

If the carbon atom of the carbonyl bears a hydrogen atom, the class of compounds is called the *aldehydes*, with the general formula:

$$R-\underset{\underset{\displaystyle O}{\|}}{C}-H$$

The simplest aldehyde, $CH_2{=}O$, is called formaldehyde. The next-simplest one is

$$CH_3C\overset{\diagup O}{\underset{\diagdown H}{}}$$

called acetaldehyde.

A class of compounds in which the carbonyl functional group bears two carbon substituents and no hydrogen atom is the *ketones*. The general formula of the ketones is:

$$R-\underset{\underset{\displaystyle O}{\|}}{C}-R$$

The simplest ketone is called acetone. Its structure is

$$CH_3-\overset{\overset{\displaystyle O}{\|}}{C}-CH_3$$

Functional isomerism exists between aldehydes and ketones. For example, a functional isomer of acetone is the three-carbon aldehyde called propionaldehyde,

$$CH_3CH_2\underset{\underset{\displaystyle O}{\|}}{C}-H$$

Both the aldehydes and the ketones have the general formula $C_nH_{2n}O$. By comparing this general formula to that of the alcohols and ethers, we can see that the aldehydes and the ketones are in a higher oxidation state. An aldehyde or ketone has two fewer hydrogen atoms than the corresponding alcohol or ether. Therefore, to convert an alcohol into a ketone or aldehyde, an oxidizing agent should be used. To convert a ketone or an aldehyde into an alcohol, a reducing agent should be used.

Aldehydes and ketones undergo a wide range of reactions, most of which are beyond the scope of our discussions. We shall describe one reaction that illustrates a number of general chemical principles.

Both aldehydes and ketones react readily with the halogens in the presence of hydroxide ion to form substitution products. A typical reaction is the bromination of acetone:

$$CH_3-\overset{\overset{\displaystyle O}{\|}}{C}-CH_3 + Br_2 \longrightarrow CH_3-\overset{\overset{\displaystyle O}{\|}}{C}-CH_2Br + HBr$$

The rate law for the bromination of acetone is first order in acetone and first order in hydroxide ion, and is independent of the concentration of bromine. The simplest mechanistic interpretation of this rate law is that the acetone and the hydroxide ion react in a slow rate-determining step to form an intermediate which then reacts rapidly with the Br_2.

Such a mechanism is consistent with a great deal of other evidence indicating that ketones and aldehydes are in equilibrium with other compounds called *enols*. The equilibrium for acetone and its corresponding enol is:

$$CH_3-\underset{\underset{O}{\|}}{C}-CH_3 \rightleftharpoons CH_3-\underset{\underset{OH}{|}}{C}=CH_2$$

The difference between these compounds is the shift of a hydrogen atom from a CH_3 group in acetone to the O atom in the enol, with the necessary accompanying change in the position of the double bond. These two compounds are functional isomers. The equilibrium is established rapidly at room temperature. A rapidly established equilibrium between two isomers is called a *tautomerism*. This one is called *keto-enol tautomerism*.

Generally, the equilibrium lies far on the side of the ketone. But even though the enol is a minor component at equilibrium, many reactions proceed by attack of the reagent on the enol, which is more reactive than the ketone. Halogenation is such a reaction. Even though the keto-enol equilibrium is established rapidly, the formation of the enol is the rate-determining step because the addition of bromine to the enol proceeds even faster.

Just as an alcohol loses H_2O to form an alkene, an aldehyde or a ketone can lose H_2O to form an alkyne. This change can be visualized more easily by including the appropriate enol as an intermediate. Thus, the conversion of acetone to propyne through the dehydration of the enol is:

$$CH_3\underset{\underset{O}{\|}}{C}CH_3 \rightleftharpoons CH_3\underset{\underset{OH\ \ H}{|\ \ |}}{C}=CH \longrightarrow CH_3C\equiv CH + H_2O$$

The reverse reaction can occur under appropriate conditions. Thus, aldehydes and ketones can be prepared from alkynes by the addition of H_2O, just as alcohols can be prepared from alkenes. The addition or loss of H_2O is not an oxidation-reduction reaction. Therefore, we can say that alcohols and alkenes are in the same oxidation state, and that aldehydes, ketones, and alkynes are in the same oxidation state.

Carboxylic Acids and Derivatives

The *carboxylic acid* functional group is the next-higher oxidation state of oxygen-containing functional groups. The carboxylic acid functional group appears to be a combination of the alcohol and the carbonyl groups. Its structure is:

$$-\underset{\underset{O}{\|}}{C}-OH$$

The two simplest carboxylic acids are formic acid and acetic acid:

$$\underset{\text{formic acid}}{H-\overset{\overset{O}{\|}}{C}-OH} \qquad \underset{\text{acetic acid}}{CH_3-\overset{\overset{O}{\|}}{C}-OH}$$

The hydrogen atom on the —OH portion of the carboxylic acid is an acidic proton. The conjugate base of a carboxylic acid is called a *carboxylate anion*. These anions are stabilized by resonance:

$$R-\underset{\underset{:O:}{\|}}{C}-\ddot{O}:^- \longleftrightarrow R-\underset{\underset{:O:^-}{|}}{C}=\ddot{O}$$

The resonance stabilization of their conjugate bases accounts in part for the enhanced acidity of carboxylic acids relative to alcohols.

The derivatives of the carboxylic acids are classes of compounds in which the —OH group of the acid is replaced by an atom or a group other than carbon or hydrogen. The *acid halides* are compounds in which the —OH group is replaced

TABLE 21.3
Oxidation State of Carbon Compounds

Increasing Oxidation State $\longrightarrow$

alkanes C_nH_{2n+2}	alkenes C_nH_{2n}	alkynes C_nH_{2n-2}	carboxylic acids and derivatives $C_nH_{2n}O_2$	carbon dioxide and derivatives CO_2
	alcohols and ethers $C_nH_{2n+2}O$	aldehydes and ketones $C_nH_{2n}O$		

by a halogen. A simple acid halide derived from acetic acid is acetyl chloride, whose structure is:

$$CH_3 - \underset{\underset{O}{\|}}{C} - Cl$$

The *amides* are compounds in which the —OH group is replaced by NH_2. A simple amide, derived from acetic acid, is acetamide, whose structure is:

$$CH_3 - \underset{\underset{O}{\|}}{C} - NH_2$$

The proteins, one of the most important classes of compounds in living organisms, are complex amides.

The most commonly encountered derivatives of carboxylic acids are the *esters*. An ester is formed when the H atom of the —OH part of the functional group is replaced by an organic radical. The general formula of an ester can be represented as:

$$R - \underset{\underset{O}{\|}}{C} - O - R'$$

where R and R' are two radicals. Two simple esters of acetic acid are:

$$CH_3 - \underset{\underset{O}{\|}}{C} - O - CH_3 \qquad CH_3 - \underset{\underset{O}{\|}}{C} - O - CH_2CH_3$$

methyl acetate   ethyl acetate

The formation of an ester from an alcohol and a carboxylic acid is called *esterification*. A typical esterification is the formation of ethyl acetate from acetic acid and ethanol:

$$CH_3 - \underset{\underset{O}{\|}}{C} - OH + CH_3CH_2 - OH \rightleftharpoons CH_3 - \underset{\underset{O}{\|}}{C} - O - CH_2CH_3 + H_2O$$

The other product is water. Virtually complete formation of ethyl acetate is assured by removing the water as it forms, thus driving the equilibrium to the right. This method is based on the principle of Le Chatelier.

The general formula of the carboxylic acids and the esters is $C_nH_{2n}O_2$. The presence of two oxygen atoms accounts for the increase in oxidation state relative to the ketones.

A still higher oxidation state than the carboxylic acid is found in CO_2 and its derivatives. Table 21.3 summarizes the oxidation states that we have discussed.

21.6 FUNCTIONAL GROUPS WITHOUT OXYGEN

Many nonmetallic elements other than oxygen are found in organic compounds. They include the halogens, nitrogen, sulfur, and phosphorus.

Organic Halides

A halogen atom that is attached to a carbon atom can be regarded as a functional group. Compounds with halogens as functional groups are called *organic halides*. Their properties depend both on the specific halogen and the type of carbon atom to which it is attached.

There are many differences in the properties of halides in which the halogen is attached to a benzene ring and those in which the halogen is on a nonaromatic carbon atom. Similarly, a halogen on the sp^2 carbon of a double bond has properties quite different from those of one on an sp^3 carbon. The properties of the halides can be influenced even by the number of hydrogen atoms bonded to the sp^3 carbon, by whether the carbon is part of a ring, and by the size of such a ring.

Nitrogen-Containing Functional Groups

Many functional groups contain one or more nitrogen atoms. The amides are nitrogen-containing carboxylic acid derivatives. The *amines* are a large class of compounds that are organic derivatives of ammonia, just as alcohols and ethers are organic derivatives of water. We can divide the amines into three groups on the basis of the number of hydrogen atoms in ammonia that are replaced by organic radicals.

In primary amines, which can be represented as RNH_2, one hydrogen atom is replaced by an organic radical. In secondary amines, which can be represented as

$$RNR'$$
$$|$$
$$H$$

two hydrogen atoms are replaced. In tertiary amines, which can be represented as

$$RNR'$$
$$|$$
$$R''$$

all three hydrogen atoms are replaced by organic groups. Compounds in which the nitrogen atom of the amine is attached directly to a benzene ring are called aromatic amines or *anilines*.

The amines generally are weak bases. We discussed some aspects of their acid-base chemistry in Chapter 15.

A group of amines, often of complex structure, that are found in many plants are the *alkaloids*. The alkaloids are physiologically active. Often, they are psychoactive—that is, they can influence the way in which the brain functions. Alkaloids often play important roles in medicine and in religion, in both primitive and advanced societies. The names and functions of many alkaloids are familiar to us.

The opium alkaloids, for example, are a group of more than 20 amines found in a resin that is exuded by the opium poppy, which is grown primarily in Turkey and Southeast Asia. The most abundant of the opium alkaloids is morphine, which is named for Morpheus, the Greek god of sleep. Morphine and many of its numerous derivatives, such as codeine, are extremely effective painkillers. Unfortunately, they are also physiologically addictive. For many decades, chemists have attempted to develop a synthetic morphine derivative that kills pain but is not addictive. All their efforts have failed. Heroin is one product of the effort to find a synthetic nonaddictive painkiller. It was found to be addictive only after it was used medically.

The ergot alkaloids are a group of alkaloids found in ergot, a fungus that

THE BRAIN'S OWN OPIATES

Why should human beings be so sensitive to opiates, substances that come from plants? The answer to this question has come from the discovery that there are natural opiates in the brain, a discovery that has provided a deeper understanding of how the brain works and of physiological drug addiction. It may also open the way to synthesis of new drugs to alleviate pain without the danger of addiction.

The first step toward the discovery was made in the early 1970s, when opiate receptors were isolated from the membranes of brain cells. Opiates such as morphine or heroin produce their effects by binding to these sites and altering the operations of the brain cells. The existence of these receptors suggested that the brain normally contains some molecule that is similar in shape and function to the opiates isolated from plants. Two different kinds of such molecules have been found in the brain.

One kind of molecule has been named enkephalin, from the Greek for ''in the head.'' The enkephalins that have been isolated to date are peptides, compounds related to proteins. They are quite different in overall composition from the alkaloids, but quite similar to morphine and other opiates in shape. It is believed that the enkephalins bind to receptors at the ends of nerve cells in the brain and inhibit the activity of the cells, thus deadening the sensation of pain.

The discovery of the enkephalins has led to a new theory about how two major features of drug addiction, tolerance and physical dependence, come about. Tolerance refers to the need for increasing amounts of an addictive drug to elicit the same response. Physical dependence refers to the severe withdrawal symptoms that develop when the drug is not administered. According to the theory, opiate receptors in the brain usually are exposed to a constant level of enkephalin. When heroin or another addictive drug is administered, these molecules bind to unoccupied opiate receptors. If the drug is taken habitually, enkephalin production decreases, so that larger amounts of the addictive drug are needed to produce the same effects. When administration of the drug is stopped, the withdrawal symptoms occur during the period in which the body has not yet resumed its normal production of enkephalins.

A second kind of natural opiates is the endorphins, a name that is an abbreviation for ''endogenous morphines.'' The endorphins are produced by the pituitary, a gland at the base of the brain. At this writing, the functions of the endorphins are not known.

Enkephalins have been found to have the same pain-killing potency as morphine. Attempts are being made to make analogs of the enkephalins that would ease pain but would not be addictive. The existence of these natural opiates has raised hopes that this long-sought goal may now be within reach.

grows on rye and other cereals. Carefully controlled doses of the ergot alkaloids have been used for centuries in obstetrics, for such purposes as the induction of labor. It is only in recent years that the ergot alkaloids have been generally replaced by other drugs.

The ergot alkaloids are deadly poisons. Epidemic poisonings called St. Anthony's fire, caused by the presence of ergot in rye bread, killed tens of thousands of persons in Europe in medieval times. The ergot alkaloids can cause severe physical and mental damage; the symptoms of ergotism include insanity and gangrene of the arms or legs.

In the past few years, one derivative of the ergot alkaloids that was prepared by an organic chemist seeking a nonaddictive painkiller has become quite well known. This synthetic compound is an amide derivative of lysergic acid, which is commonly called LSD, from the initials of its German name. LSD is the most powerful hallucinogen known.

The reserpine alkaloids were first isolated from the root of a plant that grows in India. Scientific interest in the plant was aroused by a tradition, perhaps thousands of years old, of using the roots to treat insanity. The alkaloids isolated from the roots of this Indian plant include some of the most powerful tranquilizers known.

The list of plant alkaloids is long. The quinine alkaloids include quinine itself, the drug of choice for treatment of several kinds of malaria. Drugs such as belladonna, nicotine, cocaine, and scopolamine are all alkaloids. In general, these drugs were discovered in the same way as the reserpine alkaloids, by modern scientists who followed the lead given by folk medicine. Today, the search for new drugs continues in relatively unexplored regions of the world, such as the headwaters of the Amazon River or the jungles of New Guinea, carried out by scientists who are called ethnobotanists or ethnopharmacologists.

Sulfur-Containing Functional Groups

There are a number of sulfur-containing functional groups that are of interest to both organic chemists and biologists. We can classify them according to the oxidation states of the sulfur atom in the functional group. They range from organic derivatives of H_2S, in which the sulfur is in the -2 oxidation state, to organic derivatives of sulfuric acid, in which the sulfur is in the $+6$ oxidation state.

Organic derivatives of H_2S can be regarded as the sulfur analogs of the alcohols and ethers. The mercaptans, whose general formula is RSH, correspond to the alcohols; the sulfides, whose general formula is RSR', correspond to the ethers. The disulfides, whose general formula is RSSR', are a class of organo-sulfur compounds in which the sulfur is in the -1 oxidation state.

Sulfur-containing functional groups are found in a number of physiologically active molecules. The mercaptans, which are also called thiols, are distinguishable by their smell. Butanethiol, $CH_3CH_2CH_2CH_2SH$, is the active principle of skunk odor, for example. Other organosulfur compounds are responsible for many of the unpleasant odors associated with industrial pollution.

The disulfide functional group is found in many proteins. The S—S group often acts as a bridge linking distant parts of the protein chain and thus plays a major role in maintaining the characteristic three-dimensional structures of proteins.

The chemistry of compounds that have more than one kind of functional group is beyond the scope of this discussion. The chemistry of many of these compounds is simply the chemistry expected from the individual functional groups. In other cases, the two functional groups may be close enough in the molecule to interact. In such compounds, the chemistry can be quite different from that expected from the simple combination of the functional groups.

21.7   WHAT ORGANIC CHEMISTS DO

A large fraction of practicing chemists consider themselves to be organic chemists. The research efforts of many organic chemists fall into a few well-defined categories.

Structural Determination

New substances with complex structures are continually being isolated from natural sources, primarily from plants. These compounds may help chemists learn more about biological processes; many of them are studied for possible use in medicine. The first question that is asked about a newly discovered compound is usually: What is its molecular structure? Answering that question is often a major challenge. More and more, organic chemists have come to rely on instruments to help them work out these structures (Section 8.1). Several different techniques can be used to obtain partial or complete answers to the question.

Infrared spectroscopy is used primarily to identify functional groups and, to a limited extent, to identify some features of the compound's carbon skeleton. Ultraviolet spectroscopy is helpful in establishing some structural details of compounds with aromatic rings or multiple bonds, such as the location of double bonds relative to each other. A technique called nuclear magnetic resonance spectroscopy is quite valuable to the organic chemist. It gives information not only about functional groups but also about the carbon skeleton of a molecule by identifying the chemically different types of hydrogen atoms in a molecule.

One of the most useful techniques in determining structures is mass spectrometry, in which a small sample of the unknown substance is bombarded with relatively high-energy electrons. These electrons cause the molecule to fragment into smaller pieces whose masses can be measured very accurately by the mass spectrometer. The correct structure of the unknown substance can be established by interpreting the fragmentation pattern of the molecule. Often, a computer is used to help in the analysis.

The most direct instrumental method of structural determination is X-ray crystallography (Section 10.5). However, X-ray crystallography requires such elaborate instrumentation that it cannot as yet be used routinely by organic chemists.

Chemical methods are also used to determine structure. The conversion of an unknown substance to a known substance by a well-defined reaction pathway is a good way to establish the structure of the unknown. The chemical behavior of an unknown substance can provide useful information about the nature of its functional groups or of its carbon skeleton. The chemical fragmentation of a complex molecule into smaller molecules, which are more easily identified, is another useful technique.

Despite the continued improvement in instrumentation, determining the structure of a natural product still can be a formidable task. It is therefore impressive to learn that the determination of exceedingly complex structures, such as that of strychnine, an alkaloid that is a well-known poison, was carried out without the use of modern instrumentation. Figure 21.10 shows the structure of strychnine. The molecule is even more complex than this two-dimensional structure indicates. In addition, there are many isomers, both geometric and optical, which are represented by this two-dimensional structure. It was a major achievement to single out strychnine from all the rest of the isomers.

Synthesis of Compounds

The synthesis of organic compounds of complex structure from simple and available starting materials is a major area of activity in organic chemistry. Some

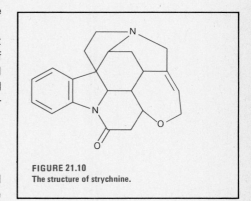

FIGURE 21.10
The structure of strychnine.

compounds prepared by synthetic organic chemists are only of theoretical interest. But many of these compounds are prepared with practical aims in mind, particularly in industrial research.

Most new drugs are substances made in the laboratory. Chemists sometimes design a molecule for a specific purpose, basing the design on existing molecules. More often, the synthesis of a new drug is a matter of trial and error, in which a large number of related compounds are prepared and tested to determine if any one of them is useful. In an effort to reduce the waste of the trial-and-error approach, a great deal of effort in organic chemistry is being devoted to the study of the relationship between molecular structure and physiological activity. There is a long way to go before this relationship is understood fully.

The pharmaceutical industry is not the only one that uses synthetic compounds. The plastics industry relies heavily on synthetic organic compounds. The list of commercial synthetic products is almost endless: solvents, paints, gasoline additives, insecticides, herbicides, fibers, and many more.

Several steps lie between the synthesis of a compound and its commercial use. The substance may have been prepared inadvertently. More often, some understanding of the relationship between properties and structure permits a chemist to prepare a compound for a specific purpose. If the substance has the desired properties, economically feasible methods of synthesis must be found. A considerable effort often is expended to maximize the yield of a compound. A difference of a few percent in the yield of the material may be the difference between marketing a product and abandoning it.

New Reactions

Organic chemists are constantly developing new kinds of reactions for bringing about chemical changes. The reactions of synthetic organic chemistry can be divided into two broad categories. One category consists of functional-group interconversion reactions, in which there is a change in the functional group but not in the carbon skeleton. The second category includes reactions in which a carbon skeleton is built up or changed.

Functional-group interconversion reactions usually are fairly easy to manage. We have described some of them—for example, the conversion of an alkene to a dichloride, a bromide, or an alkane (Figure 21.1), the conversion of a carboxylic acid to an ester, the oxidation of an alcohol to an aldehyde, the oxidation of an aldehyde to a carboxylic acid, and the reduction of a ketone to an alcohol.

There are a number of reasons for the ongoing search for new kinds of reactions that give high yields of the desired product and are accompanied by a minimum of side reactions. In many reactions, the organic chemist tries to achieve a change in just one functional group of a molecule that has several functional groups. A number of such specific reactions have been developed.

The second type of reaction, in which a carbon skeleton is built up, is generally much more difficult to conceive and develop. The synthetic chemist must not only construct a complex carbon skeleton, but the skeleton must also be constructed in a way that makes it possible to introduce functional groups at specific sites.

In general, a functional group cannot be introduced on a given carbon atom in a chain or ring unless there is a functional group already at, or close to, this site. Thus, in building up a carbon skeleton, the units must be connected so as to provide what are called functional-group ''handles'' at appropriate locations.

As a simple example, suppose we want to prepare the compound

$$CH_3CH_2CHCH_2CH_3$$
$$|$$
$$Cl$$

from compounds of fewer than five carbon atoms. One method is to join a three-carbon fragment and a two-carbon fragment in such a way as to leave a functional group at the desired position in the chain. Such a reaction could lead to the formation of the corresponding alcohol,

$$CH_3CH_2CHCH_2CH_3$$
$$\overset{|}{OH}$$

A reagent such as HCl could then be used to substitute for the OH in a functional-group interconversion reaction.

Another successful method is to join a one-carbon fragment and two two-carbon fragments to form the five-carbon chain with the functional group on the middle carbon. However, the compound cannot be formed by first preparing *n*-pentane and then trying to introduce the chlorine atom specifically at the middle atom of the chain. The methods for introducing chlorine substituents on alkane chains are not specific, so a mixture of products is formed in such reactions.

The existence of geometric and optical isomers creates another problem in forming carbon chains and rings. One of the most challenging areas in organic synthesis is to carry out stereospecific synthesis, that is, to make only the desired isomer. It is difficult to build up a carbon skeleton in a way that leads not only to the correct location of functional groups but also to their correct orientation in space.

An especially active area in organic synthesis is the preparation of natural products of complex structure from simple starting materials. Such syntheses may sometimes seem unnecessary to those outside the field, because they require an enormous expenditure of time and effort for preparation of a small amount of a compound that may be easily available from a natural source. However, these syntheses are invaluable in adding to chemical knowledge. The synthesis of a compound is sometimes the last and most elegant step in a structure determination. To prove that a structure proposed for a substance is correct, the compound with the proposed structure is synthesized. The identity of the synthetic compound and the natural compound provides the final proof that the proposed structure is correct.

Such syntheses are also pursued to clarify the chemical behavior of related compounds and to develop new reactions. But aside from these practical aims, a major motive for synthesizing a natural product of complex structure is pure intellectual curiosity. For example, morphine, the major alkaloid of opium, is available in large quantity from the natural source. Nevertheless, years of work by many excellent chemists were devoted to synthesizing morphine, whose structure is shown in Figure 21.11. Although a great deal of chemistry was learned along the way, the synthesis of morphine is primarily an impressive intellectual achievement. More than 20 steps, beginning with simple starting materials, were required for the synthesis of morphine.

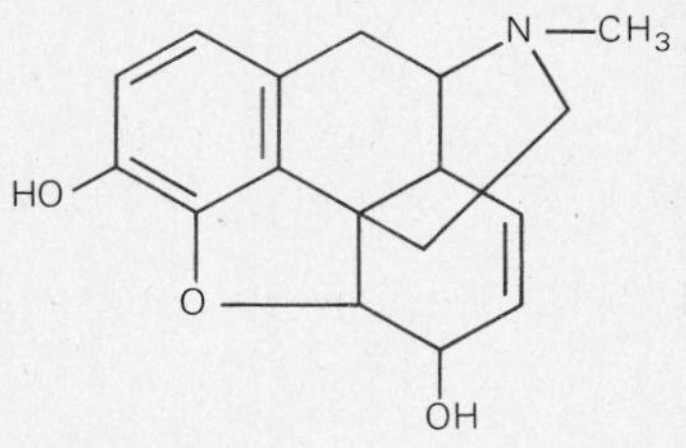

FIGURE 21.11
The structure of morphine.

Theory and Practice

Research in organic chemistry is also aimed at developing and refining the basic theoretical picture that allows chemists to rationalize, systematize, and predict the behavior of organic molecules. One approach is to apply the methods of quantum mechanics, with suitable simplifications, to formulating the structure of organic molecules. A detailed picture of the electrons and the orbitals in which they are found makes it possible to understand, and in some cases to predict, chemical behavior. This approach has been especially successful for molecules with π electrons.

Organic chemists also devote considerable attention to the study of reaction mechanisms. In organic chemistry, a reaction that appears to be relatively simple often proceeds by an involved pathway. The reaction mechanism is a detailed listing of all the changes that the system undergoes along the pathway from reactants to products. Kinetic methods, which we discussed in Chapter 17, are only some of the techniques used to elucidate reaction mechanisms. Great progress has been made in working out details of many reaction mechanisms. However, much remains to be done.

EXERCISES

21.1 Indicate which of the following compounds are alkanes: (a) C_6H_{12}, (b) C_6H_{14}, (c) $C_{100}H_{202}$, (d) $C_{20}H_{38}$.

21.2 Draw structures for: (a) a straight-chain alkane of seven carbons, (b) an alkane with a straight chain of five carbons and a two-carbon branch.

21.3 Draw structures for (a) *n*-octane, (b) isononane.

21.4 Draw structures for (a) *n*-hexane, (b) isohexane, (c) neohexane

21.5 Draw the structures of two chain isomers of C_6H_{14} in addition to the ones mentioned in Exercise 21.4.

21.6 Draw the structures of five isomers of C_7H_{16} that have a straight chain of five carbon atoms.

21.7 Draw the structures of the first three members of a homologous series of alkanes in which there is a CH_3 substituent two atoms from the carbon at the end of the chain.

21.8 The boiling point of *n*-hexane is 342 K. Predict the boiling point of *n*-octane.

21.9 Suggest an explanation for the observation that the boiling point of neopentane is 25 K lower than the boiling point of *n*-hexane.

21.10 Name and draw the structures of the positional isomers of *n*-octene, disregarding geometric isomerism.

21.11 Show that the name neohexene can reasonably correspond to only one alkene.

21.12 Find the number of compounds that are positional isomers of isooctene and draw their structures.

21.13 Indicate which of the following alkenes exist as geometric isomers: (a) propylene, (b) 1-heptene, (c) 2-hexene, (d) 1,3-butadiene.

21.14 Draw the structures of the two geometric isomers of 2-pentene.

21.15 $CH_3CH=CHCH=CHCH_3$ is a compound with three geometric isomers. Draw their structures and suggest names to differentiate between them.

21.16 $CH_3CH_2CH=CHCH=CHCH_3$ is a compound with four geometric isomers. Compare it to the compound in Exercise 21.15 and account for the different number of isomers of these two compounds.

21.17 A compound has a straight-chain structure, contains no triple bonds, and has the composition C_5H_8. Draw all the isomers with this composition and indicate the relationships between them.

21.18 Name the following compounds: (a) $CH_2=C=CHCH_3$, (b) $CH_2=CHCH=CHCH=CH_2$.

21.19 Write the products of the following reactions: (a) ethylene and Br_2, (b) ethylene and HCl, (c) isobutene and Cl_2, (d) *cis*-3-octene and H_2.

21.20 An alkene of unknown structure is found to yield only acetic acid on treatment with $KMnO_4$. Propose a structure for the alkene.

21.21 The addition of HCl to 1-butene can form two products while the addition of HCl to *cis*-2-butene can only form one product. Explain this observation using structural formulas.

21.22 Draw the structures of all the isomers of C_5H_8 that have a straight chain of carbon atoms. Name each isomer.

21.23 An alkyne such as 1-butyne reacts with Br_2 in excess to form a product that does not contain any multiple bonds. Write the product of the reaction of 1-butyne with Br_2.

21.24 Somewhat surprisingly it is found that alkynes are less reactive toward reagents such as Cl_2 than are alkenes. Predict the composition of the products that are isolated after treatment of one mole of 2-pentyne with one mole of Cl_2.

21.25 Draw and name all the cyclic isomers of C_5H_{10}, disregarding chirality.

21.26 Indicate which of the isomers that you found in Exercise 21.25 are chiral.

21.27 The *cis* isomer of 1,2-dimethylcyclopropane is found to be less stable than the *trans* isomer. Suggest an explanation for this observation.

21.28 Draw all the isomers of C_5H_8.

21.29 Find the composition of a hydrocarbon that has the steroid ring system, but no other substituents.

21.30 Find the number of geometric isomers that are possible for the unsubstituted steroid ring system. (*Hint:* Two rings may be fused in two ways.)

21.31 Draw the structures of the three isomers of dichlorobenzene ($C_6H_4Cl_2$) and name them.

21.32 Cyclooctatetraene, C_8H_8, is a compound with an eight-membered ring and four alternating double bonds, so that superficially its structure resembles benzene. It displays no special stability, and is not regarded as a resonance hybrid.
(a) Suggest a possible geometry for this compound.
(b) Suggest a reason for its dissimilarity to benzene.

21.33 The removal of a proton from cyclopentadiene forms the anion $C_5H_5^-$, which is surprisingly stable for such a species. Suggest an explanation for this stability.

21.34 There are two compounds that have three aromatic rings and the formula $C_{14}H_{10}$. Draw their structures.

21.35 Draw all the straight-chain isomers of $C_4H_{10}O$, indicating which are functional isomers and which are positional isomers.

21.36 Write the chemical reaction that takes place when solutions of ammonia and phenol are mixed.
(a) Name the product that forms.
(b) Use the data in Table 15.2 to calculate the K for this reaction.

21.37 Suggest a sequence of reactions that can be used to convert ethanol to $Br-CH_2CH_2-Br$.

21.38 Suggest a sequence of reactions that can be used to convert *n*-propyl alcohol, $CH_3CH_2CH_2OH$ to acetic acid.

21.39 Draw all the isomers with the composition C_4H_4O and a straight chain of four carbon atoms. Indicate which are functional isomers.

21.40 Suggest an explanation for the observation that aldehydes are quite easily oxidized, while drastic conditions are required to oxidize a ketone. (*Hint:* Consider the bonds that must be broken.)

21.41 The ketone that has a straight chain of four carbon atoms is called 2-butanone. Draw the structures of the enols that can be formed from this ketone.

21.42 The formation of an enol from a ketone in basic solution proceeds through the intermediacy of an enolate anion which is formed by the loss of a proton from the ketone. Draw the two contributing structures of this anion and use them to account for the reversibility of enol formation.

21.43 Draw the structures of all the monofunctional isomers with the composition $C_3H_6O_2$.

21.44 The carboxylic acid $CH_3CH_2CH_2COOH$ is called butyric acid and is responsible for the odor of rancid butter. Draw the structures of its derivatives.

21.45 An acid derivative called an anhydride is formed by the elimination of a water molecule, as two acid molecules join through an oxygen atom. The anhydride of acetic acid is called acetic anhydride. Draw its structure.

21.46 Draw the structures of all the amines with the composition $C_4H_{11}N$. (*Hint:* There are eight of them.)

21.47 Either from your general knowledge or on the basis of their names, suggest possible physiological effects of the following alkaloids: (a) nicotine, (b) caffeine, (c) emetine, (d) mescaline.

21.48 Suggest a general approach that can be used to construct a six-membered ring from smaller noncyclic compounds.

21.49 Find the composition of morphine from the structural formula given in Figure 21.11.

21.50 Alkanes react with Cl_2 when exposed to light. A Cl atom replaces a hydrogen on an alkane chain. Draw the structures of all the monochloride products that will form from the reaction of isopentane with Cl_2.

BIOCHEMISTRY

22

Biochemistry is the study of the chemistry of living organisms. The idea that life can be described as the sum of a series of chemical reactions that take place in living organisms is one of the most meaningful concepts of modern science. The number of chemical reactions that occur in any living organism is very large, and the molecules that participate in these reactions often are extremely large and complex. But no magical or mysterious properties are attributed to the chemical reactions of living organisms, and they can be described in the way that any chemical reaction is described.

There has been a huge growth of knowledge about biochemical processes in the past few decades, a period during which biochemistry has become one of the dominant fields of chemistry. For an increasing number of chemists, the study of general chemistry and of organic chemistry is merely the prelude to the study of biochemistry.

Despite the rapid growth of biochemical knowledge, much remains to be learned. Even the simplest living organism is composed of an enormously large number of different compounds. One important part of biochemistry is the effort to identify the constituent molecules of living organisms and to determine the structure of these molecules. This work is difficult, because there are so many molecules with complex structures that must be studied. An equally important part of biochemistry is the effort to find the mechanisms that coordinate the chemical reactions in which these molecules participate so that the processes of life are carried out most efficiently.

Biochemistry has helped to create the discipline called molecular biology, in which the composition and structure of the molecules in living organisms are related to the growth of the organisms, the synthesis of their myriad chemical components, the conversion of the matter that enters the cell into usable forms of energy, and the unique ability of living things to replicate themselves. The knowledge gained in these studies now is being applied in such fields of medicine as immunology and genetics.

We have discussed some aspects of biochemistry throughout this text. In this chapter, we shall give a detailed description of some of the more important compounds in living organisms. We shall also describe some of the chemical pathways associated with the processes of life.

22.1 PROTEINS

One of the most remarkable features of the molecules of living organisms is the basic simplicity beneath their apparent complexity. This underlying simplicity is one of the major themes of biochemistry. We find it when we study the proteins, the compounds that are the major organic constituents of living organisms. Proteins are important structural components in living things. Skin, hair, and muscle all consist of proteins, and proteins also play a vital role in almost all life processes as enzymes (Section 17.6). Proteins are large molecules, ranging in molecular weight from about 6000 to several million. All proteins contain carbon, hydrogen, nitrogen, oxygen, and sulfur. Some of them also include phosphorus and one or more metal atoms, most often iron. Proteins belong to the class of substances called *macromolecules,* a term that refers to their large size and complexity.

At first glance, an exact description of the structure of a macromolecule such as a protein seems almost hopeless. Even the simplest protein has several thousand atoms. Locating every one of these atoms in the molecular structure of a protein by a brute-force approach would be exceedingly difficult and tedious. However, the task is much simpler than it seems because the proteins—indeed, all the biological macromolecules—are found to consist of a large number of relatively simple units. A macromolecule that consists of such simple units of structure is called a *polymer*—in this case, a biopolymer.

In the simplest case, a polymer is a chain made up of identical units. For example, paraformaldehyde is a polymer of formaldehyde whose structure can be represented as $H—O—CH_2—(—O—CH_2—)_{n-2}—O—CH_2—O—H$, in which the repeating unit is $—O—CH_2—$. If n is the number of carbon atoms in the compound, then $n - 2$ is the number of repeating $—O—CH_2—$ units. While the value of n can be very large and a molecule of paraformaldehyde can be very long, the structure of this macromolecule is easy to represent because of its underlying simplicity. No matter how large the molecule, we know that it consists essentially of one basic unit, repeated over and over.

However, most biopolymers have more than one basic unit of structure. To find the structure of a biopolymer, we must identify the different basic units of structure, find the number of each of them in the polymer, and determine their sequence in the molecule. To relate the structure of the biopolymer to its chemical behavior in the organism, we must usually also know its three-dimensional structure. Determining the three-dimensional structure is the most difficult task, but it is a challenge that is being met with increasing success.

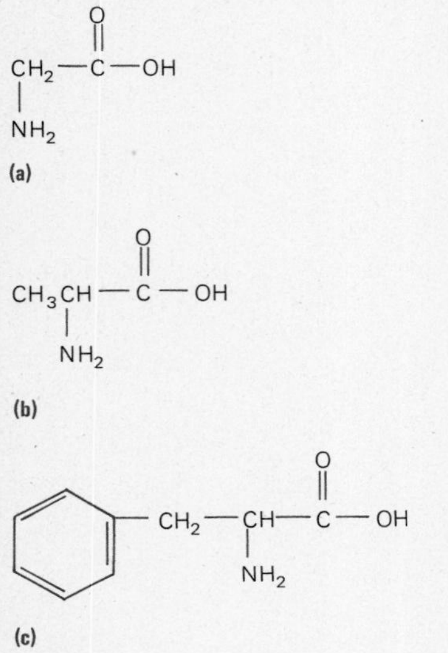

We have taken the first important step toward understanding the structures of proteins when we know that they should be viewed as polymers. We can then ask the first important question: What are the basic units of structure of proteins?

Remarkably, there are extraordinarily few basic units in the very large variety of proteins in living organisms on earth. It is estimated that there are more than 10^{10} different proteins to be found in the 10^6 species on this planet. Yet all of these proteins are built from about 20 different basic units of structure called the α-amino acids.

An α-amino acid is an organic compound that includes at least two functional groups, an amine group and a carboxylic acid group. The symbol α indicates that both the amine group and the carboxylic group are attached to the same carbon atom. Figure 22.1(a) shows the structure of glycine, the simplest α-amino acid that is found in proteins. Figures 22.1(b) and 22.1(c) show the structures of two other α-amino acids, alanine and phenylalanine. Other amino acids found in proteins have more carbon atoms and additional functional groups, including the —OH and —SH groups. However, none of these amino acids are very large or complex. The average molecular weight of these amino acids is 138.

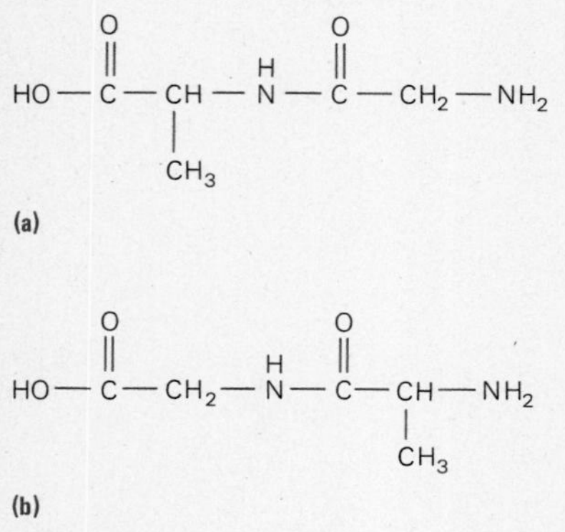

Amino acids can be joined together by what is called a *peptide bond*. In such a bond, the amino end of one amino acid is joined to the carboxyl end of the other amino acid. The —NH$_2$ group loses a hydrogen atom and the —COOH— group loses OH, so that the peptide bond is formed with the loss of a molecule of H$_2$O. The molecule that results from formation of a peptide bond is called a *peptide*. Figure 22.2 shows two peptides that can form between glycine and alanine. In one peptide (a), the amino of the alanine is joined to the carboxyl of glycine. In the other peptide (b), the amino of glycine is joined to the carboxyl of alanine.

Long chains of amino acids can be built up by the formation of peptide bonds. Such a chain always has a carboxylic acid at one end and an amine at the other end. A chain that contains many amino acids is called a *polypeptide*. Most proteins contain a number of polypeptide chains, which can be linked to each other through the other functional groups found in some of the amino acids. A small protein may contain between 100 and 200 amino acids. Large proteins may have tens of thousands of amino acid units. Almost every protein found on earth contains only members of the same group of 20 amino acids. Literally endless variety can be achieved by variations in the number and sequence of the amino acids in a protein.

We can specify the molecular composition or *primary structure* of a protein by specifying the sequence of the amino acids in the polypeptide chains and the

FIGURE 22.1
Three amino acids: (a) glycine, the simplest α-amino acid found in proteins; (b) alanine; (c) phenylalanine.

FIGURE 22.2
Two peptides that can form between alanine and glycine. (a) The amino group of the alanine is joined to the carboxyl group of glycine. (b) The amino group of glycine is joined to the carboxyl group of alanine.

SICKLE CELL AND THE GENE

Sickle-cell anemia is an inherited disease in which the normally round red blood cells assume a peculiar elongated shape that gives the condition its name. The oxygen-carrying red cells in individuals with the condition have been found to sickle when they release their oxygen and to regain the normal shape when they are oxygenated. The sickling phenomenon has been traced first to an inherited flaw in the hemoglobin molecule within the red cell and then to a specific change in the gene responsible for formation of the hemoglobin molecule.

Much of the work was done by Linus Pauling (1901–   ), who began research on sickle-cell anemia in 1945. Pauling first found that the hemoglobin from individuals with sickle-cell anemia carries a higher positive charge than the hemoglobin from normal individuals. Hemoglobin consists of a heme group (Section 18.4) and four polypeptide chains, two alpha chains and two beta chains. In experiments extending over many years, it has been found that in sickle hemoglobin, an uncharged amino acid called valine is substituted for a negatively charged amino acid called glutamic acid at the sixth position from one end of the beta chain.

It has been hypothesized that when the hemoglobin gives up its O_2, sickle-cell hemoglobin molecules tend to cling together in long, linear arrays. The hemoglobin rods that form in this way become long enough to push against the membrane of the red blood cell, distorting the cell into the familiar sickled appearance.

The reason why the sickle-cell hemoglobins stick together is still something of a mystery. The most widely held theory is that the loss of the oxygen somehow engenders a sticky site on the molecule which fits into a complementary site on the surface of a neighboring hemoglobin molecule in lock-and-key fashion. The valine on the beta chain is believed to form such a sticky site, while the glutamic acid which is normally present does not. At any rate, it is known that a change in a single nucleotide in a DNA chain can code for valine rather than glutamic acid.

Sickle-cell anemia has an unusually high incidence in some areas of Africa where falciparum malaria, a deadly form of the disease, is endemic. Somehow, the presence of sickled cells interferes with a successful malarial invasion of human cells. An individual who inherits only one gene for sickling obtains this protection and pays only a relatively low "price" in the form of sickled cells. An individual who inherits two sickling genes will have only sickled cells and will pay a much higher price. There is a comparable disease, thalassemia, which also causes damage to hemoglobin and provides protection against malaria. The gene for thalassemia has a relatively high incidence among Italians, Greeks, and other groups in the Mediterranean basin, where malaria was common. Again, the damage done by the disease can be offset by the protection it gives against a deadly malady.

way in which these chains are linked to each other. The first protein whose primary structure was specified in this way was insulin, a small protein consisting of two polypeptide chains with a total of 51 α-amino acids. The method by which the primary structure of insulin was determined was based on the use of acid hydrolysis to break the protein into smaller peptides by cleaving some peptide bonds. The experimenters were able to identify the structures of these small fragments, which contained two or three amino acids each. The structure of insulin then was worked out by determining how the small chains could fit together to form the complete molecule, in much the same way as a jigsaw puzzle is completed.

The structure of insulin was determined in 1953. Since then, the structures of many more complex proteins have been specified. Insulin and a number of other proteins have also been synthesized from their component amino acids.

When we specify the primary structure of a protein, we have taken only the first step toward a complete description of its structure. The biological functions of a protein depend on its characteristic three-dimensional structure. A polypeptide assumes a characteristic three-dimensional structure in part because of interactions between amino acids in the chain. In other words, the primary structure of a protein, its amino acid sequence, determines its secondary structure, part of its three-dimensional structure.

In 1951, Linus Pauling and Robert Corey proposed that a particularly stable three-dimensional structure for many polypeptide chains is an α helix, a spiral of a specific kind. Helices have chirality (Section 18.2, page 585); they may be right-handed or left-handed. The α helix structure of proteins proposed by Pauling and Corey is a right-handed spiral, as shown in Figure 22.3.

It has been found that the α helix is the most common secondary structure in proteins. At least part of the polypeptide chain of any given protein is in the shape of an α helix. Many proteins have large portions of their polypeptide chains in α helices. The α helix of polypeptide chains is unusually stable. This geometry is particularly favorable for the formation of hydrogen bonds between the H atoms of the amide NH group of one amino acid and the O atoms of the amide carbonyl group of a nearby amino acid.

Not all proteins exist as long, helical chains. In some, the polypeptide chains are folded into compact globular structures. This folding is described as the tertiary structure of a protein. The term "tertiary structure" describes the three-dimensional relationship between amino acids that may be quite distant in the polypeptide chain but are quite close spatially.

We have already discussed the importance of tertiary structure in the enzymatic activity of proteins. One of the most active areas of biochemical research is the investigation of the tertiary structure of proteins, and the effort to relate the tertiary structure to the primary structure and to the biological function of the protein. Using such methods as X-ray crystallography, it has been found that the amino acid sequence of the polypeptide chain determines the most stable tertiary structure for the protein. However, a great deal remains to be learned about the influence of primary structure on tertiary structure, and about the relationship between tertiary structure and biological function.

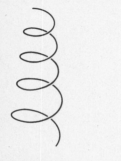

FIGURE 22.3
The right-handed spiral or α-helix, found in many proteins.

22.2   CARBOHYDRATES

Another major class of biological molecules are the carbohydrates, or saccharides. The carbohydrates are polyfunctional compounds. Each carbohydrate has a number of —OH functional groups and at least one aldehyde or ketone functional group. Any carbohydrate either has the empirical formula CH_2O or is derived from a compound that has this formula.

The carbohydrates come in many sizes. The smallest carbohydrates are the monosaccharides, most of which have from three to seven carbon atoms. The best-known monosaccharide is D-glucose, which is the sugar in candy. The prefix D- gives information about the stereochemistry of the molecule, the suffix -*ose* is found in the names of many carbohydrates.

The disaccharides are larger molecules in which two monosaccharides are linked. Sucrose, which is found in plants and honey and which is used as a sweetener, is a disaccharide. So is lactose, which is found in milk.

Some carbohydrates are biopolymers, that is, biological macromolecules in which the basic repeating unit of structure is a saccharide. These macromolecules are called polysaccharides. They are important constituents of cell walls and cell coats. The two most common polysaccharides are cellulose and starch.

D-Glucose and Other Monosaccharides

We can approach the chemistry of the monosaccharides by examining one of them, D-glucose, in detail. The composition of D-glucose is $C_6H_{12}O_6$. You will note that the ratio of hydrogen atoms to oxygen atoms is $2:1$, just as it is in water. D-glucose has six carbon atoms: it and all other saccharides with six carbon atoms are called hexoses. Chemically, D-glucose behaves as a polyhydroxy aldehyde.

The six carbon atoms of D-glucose are in a straight chain, an arrangement that is characteristic of all the saccharides. Each carbon atom of D-glucose bears an oxygen-containing functional group. There are other saccharides in which one carbon atom does not bear an oxygen-containing functional group. They are called deoxysaccharides.

One representation of the structure of D-glucose is:

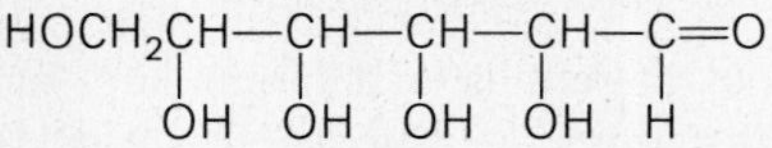

$$HOCH_2CH-CH-CH-CH-C=O$$
$$\qquad |\quad\ |\quad\ |\quad\ |\quad\ |$$
$$\qquad OH\ \ OH\ \ OH\ \ OH\ \ H$$

This structure is an oversimplification, for a number of reasons. There are actually 16 different stereoisomers that correspond to this two-dimensional representation. Only one of these 16 three-dimensional arrangements of the functional groups corresponds to that of the D-glucose molecule. The other 15 monosaccharides represented by this two-dimensional structure have all been characterized. Two of the most common are called galactose and mannose.

There is a great deal of experimental evidence indicating that this open chain is not an adequate representation of the structure of D-glucose in aqueous solution. It appears that there are two different isomeric forms of D-glucose in water. One is called α-D-glucose; the other is called β-D-glucose. The existence of these two isomers is due to interactions between functional groups of D-glucose. A carbonyl group and a hydroxyl group of the same molecule can react in water if the slightest amount of acid or base is present. When D-glucose is dissolved in water, there is a reaction in which a hydroxyl group is added across the C=O bond of the aldehyde.

The resulting compound is called a hemiacetal. Since the hydroxyl and aldehyde groups are in the same molecule, the reaction between them leads to the formation of a ring that includes an oxygen atom. In glucose, this reaction occurs in such a way as to form a six-membered ring consisting of five carbon atoms and an oxygen atom. Such a six-membered ring is the most stable and least strained that can form in this system.

The formation of this ring creates the possibility of geometric isomerism. There are two cyclic isomers of glucose, which are shown in Figure 22.4. The geometric isomerism occurs around the carbon atom just to the right of the oxygen atom in the ring, the carbon atom which previously was part of the aldehyde group. This carbon atom is called the *anomeric carbon,* and these two isomers of glucose are called *anomers.*

Ring formation reactions occur in all monosaccharides of five or more carbon

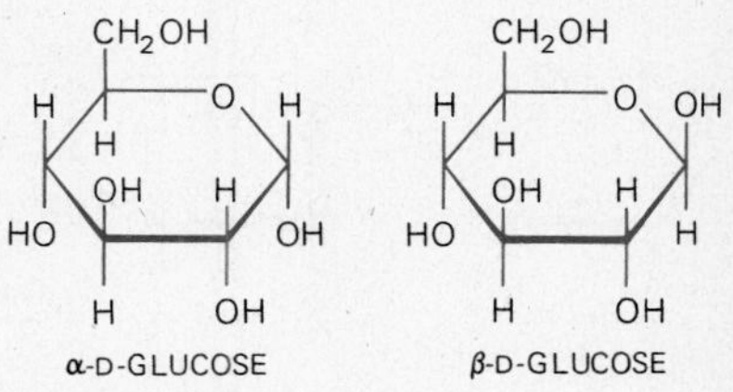

FIGURE 22.4
The two anomers of D-glucose.

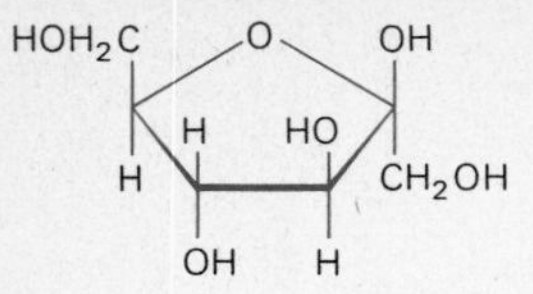

FIGURE 22.5
One of the anomers of D-fructose.

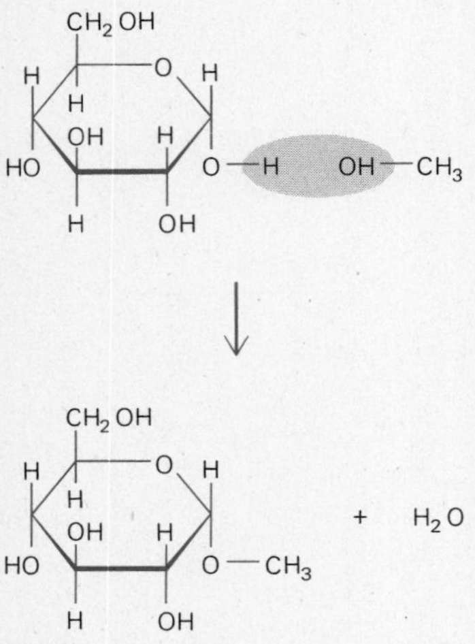

FIGURE 22.6
The reaction in which the methyl glycoside of α-D-glucose is formed.

FIGURE 22.7
Sucrose.

(a)

(b)

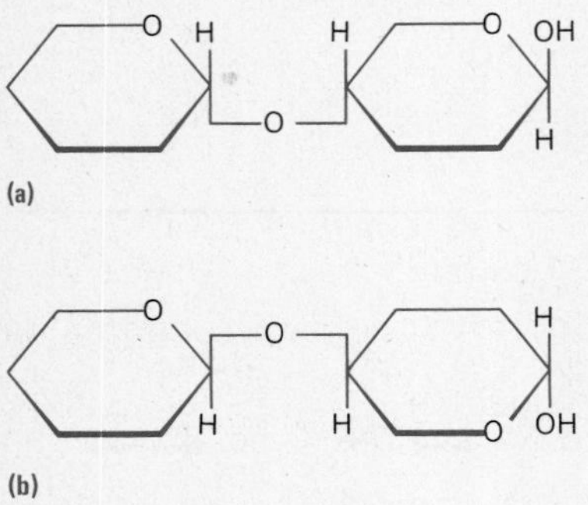

FIGURE 22.8
Maltose (a) and cellobiose (b), two disaccharides which differ only in the way their monosaccharide units are linked.

atoms. Six-membered rings, as in glucose, may form because of intramolecular reactions of the —OH group with the carbonyl group. Five-membered rings can form in monosaccharides with five carbon atoms or in hexoses that are ketones whose carbonyl group is one atom from the end of the chain. One such hexose is D-fructose, which is found in fruit juices and which forms a five-membered ring in solution. Figure 22.5 shows one of its anomers.

Saccharides often are attached to other molecules through the —OH group of the anomeric carbon. The —OH group of the anomeric carbon can also react with an ordinary —OH group to form a compound called a *glycoside*. Figure 22.6 shows the reaction of CH_3OH with α-D-glucose to form a methyl glycoside of this sugar.

Two saccharides can join together by a reaction between an ordinary —OH group of one and the anomeric —OH group of the other, or even by the reaction between two anomeric —OH groups. Not only disaccharides but also polysaccharides are formed by linkages of this kind.

To find the structure of any carbohydrate containing more than one monosaccharide, we must know the identity of the component monosaccharides, how they are joined, and whether each forms a five-membered or six-membered ring. We must also know the anomer that each monosaccharide forms in a larger molecule. We can identify most of the biologically significant saccharides by describing the component monosaccharides and the way in which the monosaccharides are joined.

For example, sucrose is a disaccharide made up of one α-D-glucose (Figure 22.4) and one β-D-fructose (Figure 22.5) which are joined by combining the anomeric —OH group of each and eliminating H_2O. Figure 22.7 shows the structure of sucrose.

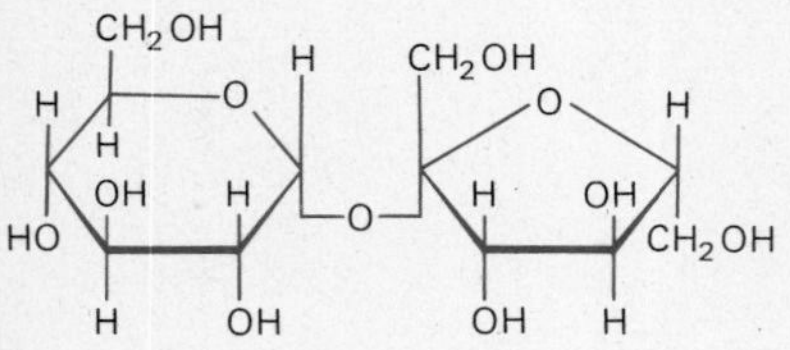

Maltose, obtained from the breakdown of starch, is a different kind of disaccharide. Both of its component monosaccharides are glucose. The —OH group at the α-anomeric carbon atom of one glucose is joined to an —OH group of the other. Maltose is said to have an α linkage. Figure 22.8(a) shows a schematic structure of maltose.

Cellobiose is a disaccharide that is formed from the breakdown of cellulose. Its structure is the same as that of maltose, except for the geometry about the linkage between the two monosaccharide units. In cellobiose as in maltose, the linkage is between the anomeric carbon atom of one glucose molecule and the carbon atom that is four places down the chain from the anomeric carbon in the other glucose molecule. However, the anomeric —OH group that connects the two monosaccharides in cellobiose is that of the β-anomer, and the connection is called a β-linkage. Figure 22.8(b) shows the schematic structure of cellobiose.

Polysaccharides; Cellulose and Starch

We mentioned that cellulose and starch are two of the most important polysaccharides. Cellulose, the structural component of fibrous and woody plants, is the most abundant biological compound. The paper of this book is one of many materials whose major component is cellulose. Starch, which is called a storage polysaccharide, is the material that a plant stores in its nutritional reservoir. Starch actually consists of two different polysaccharides, amylose and amylopectin, which have the same composition but somewhat different structures.

Starch and cellulose both consist of long chains of glucose molecules. The molecular weight of these chains ranges from 50 000 to 2 500 000. Cellulose is remarkably similar to amylose in molecular structure. Both polysaccharides consist of glucose units attached in a straight chain, with no branching. The connection is between the anomeric carbon atom of one glucose molecule and the carbon atoms four places down the chain of the other glucose molecule, just

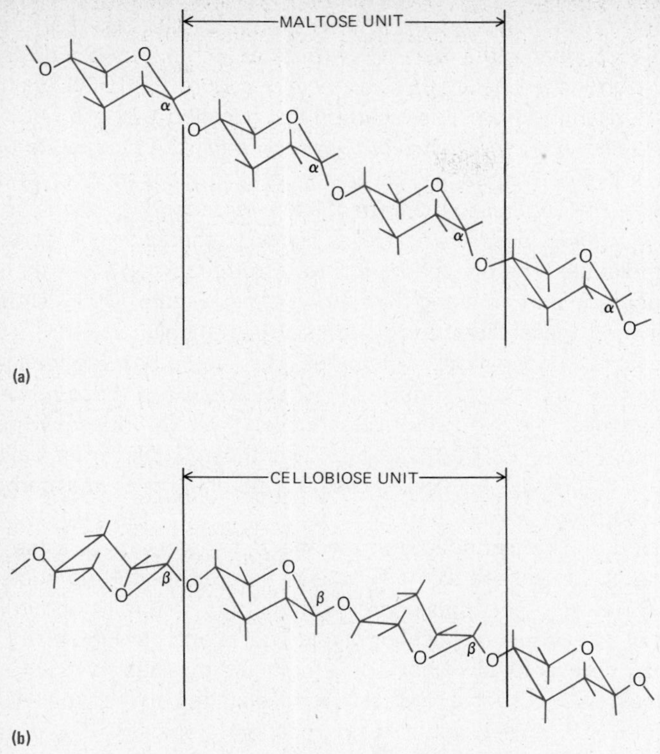

FIGURE 22.9
Amylose (a) and cellulose (b), two polysaccharides
which differ only in the way their monosaccharides
are linked.

as in maltose and cellobiose. In amylose (Figure 22.9(a)), the units are connected by α linkages. In cellulose (Figure 22.9(b)), they are connected by β linkage. There is no other difference between the two.

Most of the carbohydrates in our food are in the form of starch. When the body digests starch, the polysaccharide is broken apart to free its glucose units, which are used as an energy source. While cellulose resembles starch closely, we cannot use cellulose as a food because the human body does not have the enzymes needed to break it down to its component glucose units. Enzymes are quite specific in their action. The apparently small differences between the structure of starch and that of cellulose are enough to inactivate the enzymes that are active against starch. Most animals that use grass, bark, and other cellulose-rich plant parts as food really rely on enzymes provided by microorganisms living in their intestinal tracts to break the cellulose chains. These microorganisms are found in animals as different as cows and termites.

22.3 NUCLEOTIDES AND THE STRUCTURE OF NUCLEIC ACIDS

The nucleotides are found in all living things on earth. They play important roles in virtually every biochemical process. One nucleotide that we have already discussed, ATP (Section 9.4, page 282), is well known as the carrier of the chemical energy needed for a living organism to function. Other nucleotides are found in some important coenzymes. Still others take part in the regulation of many of the chemical reactions that occur as a living organism uses chemical energy. In addition—and perhaps most important—the nucleotides are the basic repeating units of the nucleic acids, RNA (ribonucleic acid) and DNA (deoxyribonucleic acid), which are the central elements that ensure the replication of living organisms from generation to generation.

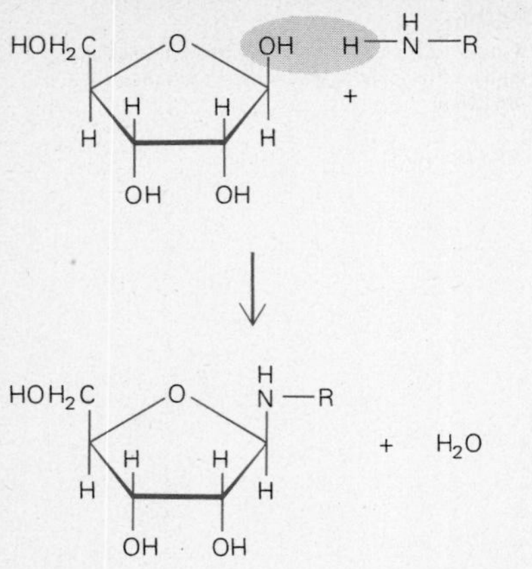

FIGURE 22.10
The formation of a glycosyl amine from D-ribose.

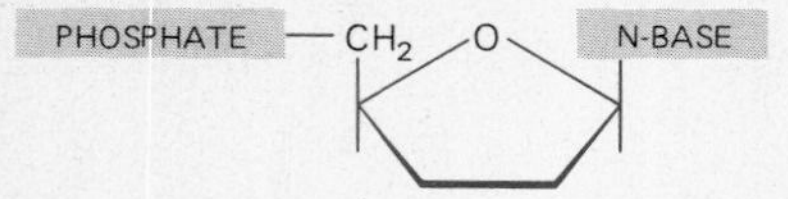

FIGURE 22.11
A schematic representation of the way in which a phosphate, a sugar, and an N-base combine to form a nucleotide.

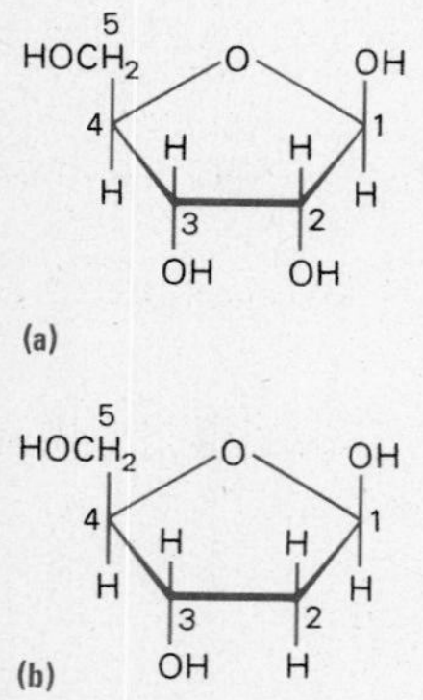

(a)

(b)

FIGURE 22.12
D-ribose (a) and D-2-deoxyribose (b).

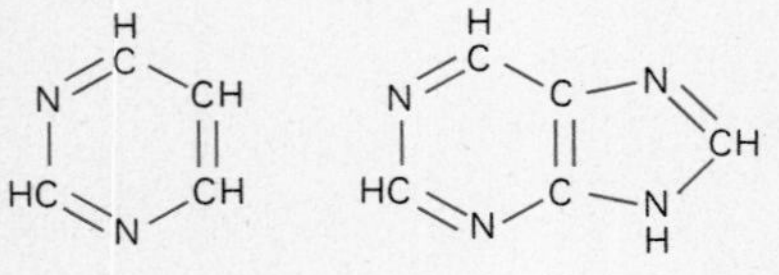

(a)   (b)

FIGURE 22.13
The pyrimidine ring system (a) and the purine ring system (b).

In Section 22.2, we mentioned that the anomeric—OH group of a monosaccharide is a reactive functional group. This —OH group can take part in a reaction with an amine, similar to the reactions with alcohol that form glycosides; the compounds thus formed are called *glycosyl amines*. Figure 22.10 shows the formation of a glycosyl amine from a five-carbon sugar called D-ribose.

A nucleotide is a phosphate derivative of a glycosyl amine. A nucleotide has three components: a five-carbon sugar, an amine, and a phosphate, or phosphoric acid equivalent. The sugar is in the form of a five-member ring. The amine, which is called a *nitrogenous base* or *N-base,* is linked to the anomeric carbon atom of the sugar molecule. The phosphate is linked to the carbon atom that is furthest from the anomeric carbon along the carbon chain. Figure 22.11 shows the relationship between these three components schematically.

One of two sugars is found in most nucleotides. The five-carbon sugar called D-ribose, whose structure is shown in Figure 22.12(a), is found in those nucleotides called *ribonucleotides*. The five-carbon sugar called D-2-deoxyribose, whose structure is shown in Figure 22.12(b), is found in those nucleotides called *deoxyribonucleotides*. Figure 22.12 also shows the numbering of the carbon atoms in these two sugars.

All naturally occurring nucleotides contain the β anomers of D-ribose or D-2-deoxyribose; that is, the linkage to the N-base in naturally occurring nucleotides is always a β linkage. The phosphate portion of a naturally occurring nucleotide consists of one or more phosphoric acid equivalents. In Figure 9.18, we showed how phosphoric acid can react with a molecule that has an OH group to form a phosphate ester. Just such a reaction can take place between the —OH group at carbon-5 of ribose or deoxyribose and phosphoric acid, H_3PO_4, or the pyrophosphoric acids, $H_4P_2O_7$ or $H_5P_3O_{10}$.

Five different amines, or N-bases, are commonly found in naturally occurring nucleotides. The N-bases in naturally occurring nucleotides are derived from either the pyrimidine ring system or the purine ring system, whose structures are shown in Figure 22.13. You should note the resemblance of these ring systems to the aromatic benzene ring (Section 21.4, page 673). Like benzene, both the purine and the pyrimidine ring systems have great stability, consistent with the presence of six π electrons.

The two major purine derivatives, called adenine and guanine (Figure 22.14), are found in both ribonucleotides and deoxyribonucleotides. The other three N-bases found in naturally occurring nucleotides are derivatives of the pyrimidine ring system. They are called cytosine, uracil, and thymine (Figure 22.14). Uracil is found only in ribonucleotides, thymine mainly in deoxyribonucleotides. Cytosine is found in both.

Each of these five N-bases has more than one nitrogen atom. The arrows in Figure 22.14 indicate the nitrogen atom that is attached to the anomeric carbon atom of the sugar in a nucleotide. Figure 22.15 shows the structure of a complete nucleotide, adenosine monophosphate, or AMP.

Nucleic Acids

The macromolecules built of long chains of nucleotide units are called the nucleic acids. Deoxyribonucleic acid (DNA) is built up of deoxyribonucleotides. Ribonucleic acid (RNA) is built up of ribonucleotides. In these chains, the nucleotides are linked by a bond between the phosphate of one nucleotide and carbon-3 of the sugar of the next nucleotide. Figure 22.16 shows this link for DNA.

To understand the nature of this link, we must remember that phosphoric acid can react with more than one alcohol molecule to form diesters or triesters. Two oxygen atoms of the parent phosphoric acid in such a diester bear organic residues rather than hydrogen atoms. The linkage in Figure 22.16 is called a *phosphodiester linkage*. It is between carbon-5 of one pentose and carbon-3

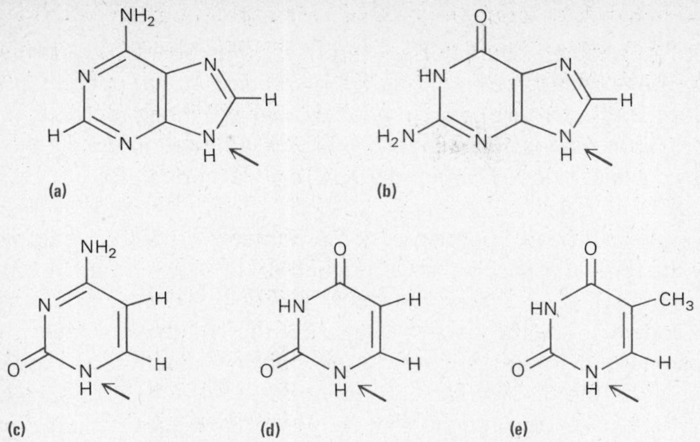

FIGURE 22.14
The two major purine derivatives, adenine (a) and guanine (b); the three major pyrimidine derivatives, cytosine (c), uracil (d), and thymine (e).

of the next pentose. This chain of alternating phosphodiesters and pentoses is the backbone of all nucleic acids. In DNA, the pentose is deoxyribose. In RNA, it is ribose. There is no other difference in the linkage found in RNA and DNA.

The N-bases of the nucleotides are not part of the nucleic acid chain. Rather, the N-bases can be regarded as substituents on the chain. But the N-bases play an extremely important role in the biological activity of nucleic acids. The differences between nucleic acids that come from different sources are due primarily to variations in the number and sequence of their N-bases.

The major functions of nucleic acids are the transmission of genetic information (which is done primarily by DNA) and the expression of the genetic information in terms of cell function (done primarily by RNA). Both these functions depend crucially on the sequence of N-bases in RNA and DNA molecules. One of the major achievements of modern biochemistry has been the discovery of the way in which the sequence of N-bases on nucleic acid chains controls both the replication and the functioning of living organisms.

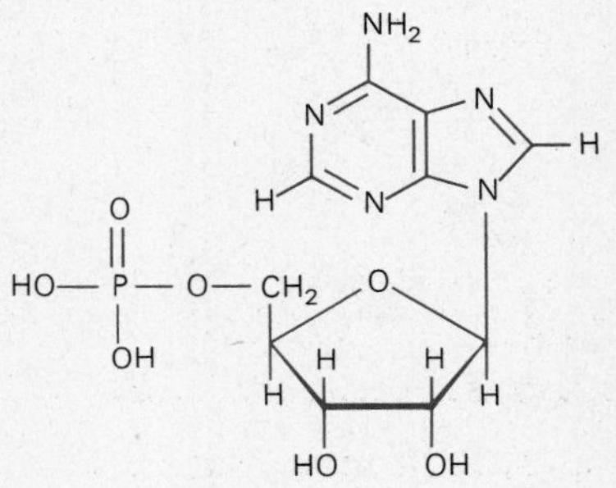

FIGURE 22.15
The structure of the nucleotide adenosine monophosphate.

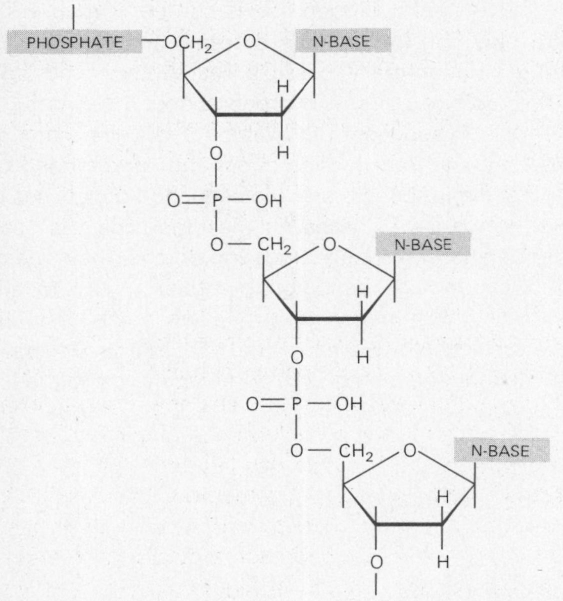

FIGURE 22.16
A portion of DNA, showing the linkages between the phosphates and sugars of succeeding nucleotides.

22.4 THE FUNCTIONS OF THE NUCLEIC ACIDS

One of the most remarkable features of a living thing is its ability to reproduce itself. This ability exists because every living thing contains stored genetic information which is transmitted from one generation to the next during reproduction. The genetic information then is interpreted and expressed through the synthesis of proteins. The nucleic acids, DNA and RNA, play a central role in these processes.

One of the best-publicized scientific accomplishments of recent years was the deduction of the three-dimensional structure of DNA from X-ray data in 1953 by James D. Watson (1928–   ) and Francis H. C. Crick (1916–   ). At that time, DNA was known to be the genetic material—that is, the molecule that stores and transmits the information necessary for reproduction. A correct description of the structure of DNA was important not only for its own sake but also because it made clear the mechanism by which DNA can replicate and transmit genetic information.

A typical DNA molecule consists of two very long polynucleotide chains. Both chains are right-handed helices, coiled around the same axis. The backbone of these chains consists of the sugar and phosphate portions of the polynucleotide; the N-bases are substituents on the chains. It is quite important to note that the N-bases are on the inside of the helix. It is also important to note that the purine and pyrimidine rings of these N-bases are planar, and that the plane of the rings is perpendicular to the axis around which the two helical chains are coiled. Figure 22.17 shows the structure of a portion of a DNA molecule schematically.

The polynucleotide chains of DNA are extremely long. A DNA molecule may have many millions of N-bases, and it may be millimeters, or even centimeters, long. Many DNA molecules have been visualized by electron microscopy.

Despite this extreme length, a DNA molecule contains only four N-bases: adenine, guanine, thymine, and cytosine. It is the sequence of these bases that constitutes a "genetic code" that enables DNA to store and transmit genetic information.

The relationship between the two polynucleotide chains of the double helix of DNA—specifically, the relationship between the bases of the two chains—is the crucial feature of its structure. The two polynucleotide chains are held together in the double helix by hydrogen bonding between their N-bases. Because of the sizes and shapes of the N-bases, only certain hydrogen bonds can be formed in the double helix. Adenine and thymine can form a pair; cytosine and guanine can form a pair. No other pairings are normally possible, as Watson and Crick deduced. Figure 22.18 shows the two pairs of hydrogen-bonded N-bases.

The specificity of N-base pairing suggested to Watson and Crick a mechanism for the replication of DNA and the transmission of genetic information from one generation to the next. If pairing is specific and the sequence of N-bases on one chain is known, then the sequence of bases on the other chain is easily determined. The presence of adenine at a specific location on one chain means that thymine must be at the corresponding location on the other chain. Similarly, the presence of guanine on one chain means that cytosine must be at the corresponding location on the other chain. We can write a sequence of N-bases for one small portion of a polynucleotide chain of a DNA molecule, using the first letters of the bases as abbreviations:

ACGGATCCTAAG

The corresponding sequence on the other chain must be:

TGCCTAGGATTC

because of the specificity of base pairing.

A conceptually simple sequence of steps by which a DNA molecule dupli-

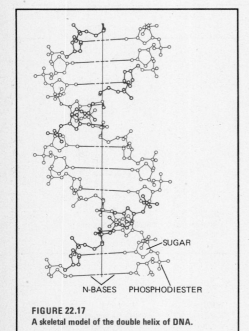

SUGAR

N-BASES   PHOSPHODIESTER

FIGURE 22.17
A skeletal model of the double helix of DNA.

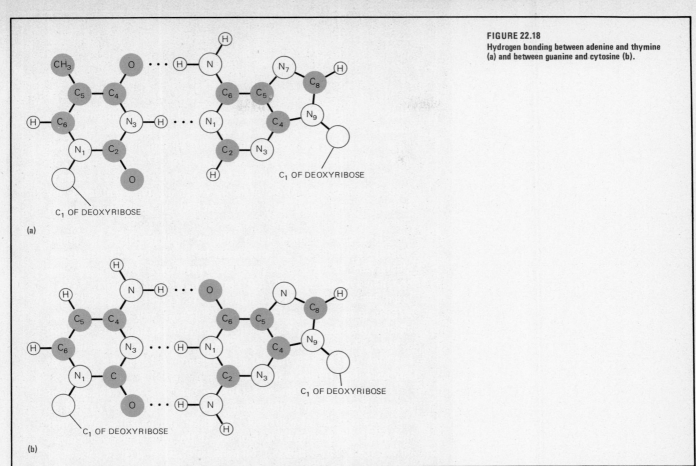

cates itself can now be described. In the first step, the hydrogen bonds holding the two chains together break, and the two chains separate from each other. In the next step, each chain acts as a template for forming a new matching chain. Because of the specificity of N-base pairing, the new chain has exactly the same sequence of bases as the original chain did. At the end of the process, two double helices exist, and two daughter DNA molecules have been formed from one parent. Each of the daughter molecules has one chain from the original DNA double helix and one new chain. Both new molecules should be identical to the parent molecule.

The specificity of N-base pairing is crucial not only to the transmission of genetic information during reproduction but also in the expression of the genetic information during the life of an organism. The sequence of N-bases determines the type of proteins made by a living cell. The proteins, in turn, play a central role in cell function.

However, DNA does not participate directly in protein synthesis. The information contained in a cell's DNA is transmitted to RNA molecules, which are involved in protein synthesis. At the actual sites of protein synthesis, other RNA molecules receive the information and participate directly in the chemical reactions by which proteins are synthesized. Although much remains to be learned, the basic steps in protein synthesis are known.

RNA molecules differ from DNA molecules in several ways. We have already mentioned that the sugar in RNA is D-ribose. There are other differences that are more important in studying protein synthesis. There is a difference in bases: RNA contains uracil; DNA does not. Uracil pairs with adenine as readily as thymine does. In addition, RNA molecules are almost always single-stranded, although

THE SIX MILLION DOLLAR MAN

Another annual cycle inevitably passed, and the pain was eased by a humorous birthday card from my daughter and son-in-law. The front bore the caption "According to BIO-CHEMISTS the materials that make up the HUMAN BODY are only worth 97¢" (Hallmark 25B 121-8, 1975). Before I could get to the birthday greeting I began to think that if the materials are only worth ninety-seven cents, my colleagues and I are really being taken by the biochemical supply companies.

I started by sitting down with my catalogue from the (name deleted) Biochemical Co. and began to list the ingredients. Hemoglobin was $2.95 a gram, purified trypsin was $36 a gram, and crystalline insulin was $47.50 a gram. Hyaluronic acid was $175 a gram, while bilirubin was a bargain at $12 a gram. Human DNA was $768 a gram, while collagen was as little as $15 a gram. Human albumin was down at $3 a gram, whereas bradykinin was $12,000 a gram. The real shocker came when I got to follicle-stimulating hormone, at $4,800,000 a gram—clearly outside the reach of anything that Tiffany's could offer.

I averaged all the constituents over the best estimate of percent composition of the human body and arrived at $245.54 as the average value of a gram dry weight of human being. With that fact burning in my head I rushed over to the gymnasium and jumped on the scale. There it was, 168 pounds, or, after a quick go-around with my pocket calculator, 76,364 grams. Remembering that I was 68% water, I calculated my dry weight to be 24,436 grams. The next computation was done with a great sense of excitement. I had to multiply $245.54 per gram dry weight by 24,436 grams. The number literally jumped out at me—$6,000,015.44. I was a Six Million Dollar Man—no doubt about it—and really an enormous upgrade to my ego after the ninety-seven cent evaluation.

We must still strike a balance between the ninety-seven cent figure and the six million dollar figure. The answer is at the same time very simple and very profound: information is much more expensive than matter. In the six million dollar figure I was paying for my atoms in the highest informational state in which they are commercially available, while in the ninety-seven cent figure I was paying for the informationally poorest form of coal, air, water, lime, bulk iron, etc.

This argument can be developed in terms of proteins as an example. The macromolecules of amino acid subunits cost somewhere between $3 and $20,000 a gram in purified form, yet the simpler, information-poorer amino acids sell for about twenty-five cents a gram. The proteins are linear arrays of the amino acids that must be assembled and folded. Thus we see the reason for the expense. The components such as coal, air, limestone, and iron nails are, of course, simple and correspondingly cheap. The small molecular weight monomers are of course much more complex and correspondingly more expensive, and so on for larger molecules.

A moment's reflection shows that even if I bought all the macromolecular components, I would not have purchased a human they have double-helical portions that result from folds in parts of the chain. In these folded portions, we find specific N-base pairing, as in DNA. Figure 22.19 shows a double-helical portion of an RNA chain, with only the sequence of N-bases indicated.

The synthesis of RNA in a living cell is controlled by the information contained in the N-base sequence of DNA. Just as one strand of DNA can serve as a template for the replication of a new strand of DNA, a portion of a DNA molecule can serve as a template for synthesis of a strand of RNA. For example, a DNA strand that has the sequence ACGGATCCTAAG will serve as a template for RNA whose base sequence is UGCCUAGGAUUC. Experiments indicate that the synthesis of RNA occurs in a multistep process, in which enzymes called RNA polymerases catalyze the addition of one nucleotide at a time to the growing RNA chain.

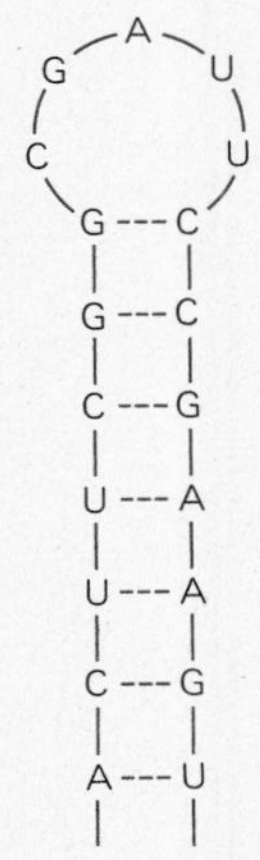

Three types of RNA, distinguished by their functions, are found in living cells. *Messenger RNA (mRNA)* serves as the template for protein synthesis. Messenger RNA contains the information, transmitted from DNA, that is needed to establish the sequence of amino acids in a protein. A typical mRNA molecule contains about 1200 nucleotides. *Transfer RNA (tRNA)* carries the amino acids necessary for protein synthesis to the mRNA template; tRNA also serves as a "label" that allows the mRNA to "recognize" the amino acids. There are many different tRNAs—at least one for each of the 20 amino acids found in proteins. Each tRNA carries only one kind of amino acid. The tRNA molecules are relatively small, containing about 75 nucleotides each. The third type of RNA is called *ribosomal RNA (rRNA)*, whose role in protein synthesis is not completely understood.

A detailed discussion of the chemical processes by which mRNA and tRNA convert the stored information of DNA into a protein is beyond our scope, but we can outline the general sequence of steps.

The portion of a DNA macromolecule that carries the information about a specific protein is called a gene. The mRNA for a particular protein is formed when the DNA in that gene acts as a template for RNA synthesis. The mRNA then

FIGURE 22.19
A schematic representation of base pairing caused by the folding of a single-strand RNA molecule. Only the sequence of bases is indicated.

carries this information to the region of the cell where small particles called ribosomes are found. Protein synthesis takes place in the presence of the ribosomes.

The process by which information is transferred from DNA to RNA is called *transcription*. The process by which this message is expressed in the ribosomes is called *translation*. In transcription, DNA is the template for the synthesis of mRNA. In translation, the mRNA is the template for the synthesis of proteins.

The genetic code that makes protein synthesis possible has been deciphered completely. The information carried by the mRNA consists of the sequence of bases in the mRNA molecule. It has been found that each three-base sequence in mRNA corresponds to an amino acid. Let us consider the genetic code as expressed in the language of mRNA. This language has only four letters, the four N-bases which we can abbreviate as U, C, A, and G. There are a total of $4 \times 4 \times 4 = 64$ three-letter combinations possible in this language. Each three-letter combination is called a *codon*. It has been found that three of these codons are "punctuation marks," which signal that the synthesis of a protein is at an end. Each of the other 61 codons corresponds to one amino acid. Since there are 20 amino acids and 61 codons, most amino acids have more than one codon. However, each codon corresponds to only one amino acid.

Thus, the sequence of codons along the mRNA chain determines the sequence of amino acids in the protein being synthesized. To illustrate, we can apply our knowledge of the genetic code to a small portion of a hypothetical mRNA chain, whose sequence of N-bases is:

—GGUGCGUUUUAA—

This sequence of N-bases consists of four codons. The first codon is GGU, which corresponds to glycine. Next comes GCG, which corresponds to alanine; then UUU, which corresponds to phenylalanine; finally UAA, which signals "stop." Thus, translation of this portion of the mRNA molecule codes for the synthesis of a portion of a peptide chain that has the amino acids glycine, alanine, and phenylalanine, in that order. These three amino acids are at the end of the peptide chain, since UAA is a "stop" codon.

This information in the mRNA molecule must be recognized and translated for protein synthesis to occur. The work of recognition and translation is accomplished primarily by tRNA molecules. A cell ordinarily contains a supply of free amino acids. During protein synthesis, each amino acid is joined to a tRNA molecule. Each tRNA is specific for one amino acid. In each tRNA chain, there is a sequence of three N-bases that is complementary to the three-N-base sequence of the codon for the amino acid in the mRNA. This sequence of three N-bases on the tRNA molecule is called an *anticodon*. The mRNA molecule recognizes the appropriate amino acid-tRNA combination at the appropriate point in protein synthesis because of the anticodon. The codon-anticodon relationship is once again governed by the specificity of N-base pairing; that is, by hydrogen bonding between the bases.

The message that the mRNA molecule has carried from the cell's DNA is read codon by codon, and the protein is assembled amino acid by amino acid at the ribosome. The process ends when the "stop" codon is read—not by a tRNA molecule, but rather by proteins called release factors. Figure 22.20 shows the process of protein synthesis schematically.

A great deal more is known about the process of protein synthesis than is shown in Figure 22.20. However, most of our knowledge relates to the primary structure of the nucleic acids. Thus, we know how the sequence of N-bases in a nucleic acid is related to the sequence of amino acids in the corresponding protein, and a good deal about the way in which this sequential relationship is expressed. But there is much more to the expression of genetic information than this straightforward one-to-one correspondence of nucleic acid bases and pro-

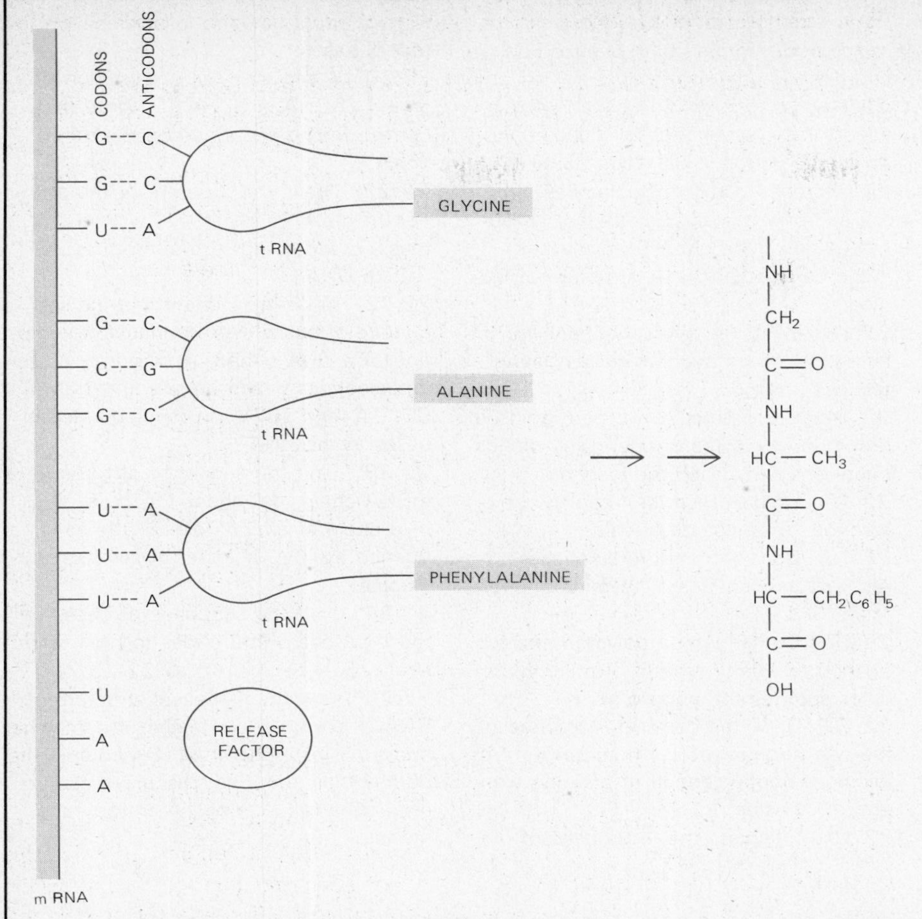

FIGURE 22.20
A schematic representation of the way in which the codons in a section of a messenger RNA molecule are read by transfer RNA molecules to synthesize a protein.

teins. Nucleic acids also contain information that controls the timing and manner of protein synthesis. Much of this information is in the secondary and tertiary structure of nucleic acids. The way in which this information is expressed seems to depend on interactions between nucleic acids and proteins in the cell.

A great deal of biochemical research is aimed at characterizing details of secondary and tertiary nucleic acid structure, and of nucleic acid-protein interactions. Success in these areas will give us a better understanding on the molecular level of such phenomena as protein synthesis and the relationship of specialized cells within an organism as complex as the human body. It seems reasonable to expect some major chemical advances in this field in the years ahead.

EXERCISES

22.1   Teflon is a polymer that consists of repeating units derived from tetrafluoroethylene, $CF_2=CF_2$. Draw the structure of a portion of the polymer, including one of the ends.

22.2   A tripeptide consists of three amino acids. Draw the structure of a tripeptide made up only of alanine.

22.3   Polyglycine is a polymer in which only one amino acid, glycine, is the repeating structural unit of the long chain. Draw a general representation of the structure of polyglycine showing the repeating unit and the structure of each end.

22.4   Find the number of tripeptides that can be made from glycine and alanine.

22.5 Distinguish between primary, secondary, and tertiary structure of a protein.

22.6 The interaction between various parts of a protein molecule and its aqueous surroundings in the cell play an important role in determining the three-dimensional structure of the protein. Some parts of the protein are called hydrophilic because they attract water, while other parts are called hydrophobic because they repel water. Suggest which parts are hydrophilic and which are hydrophobic and how the conformation of the protein can be influenced by their interaction with water.

22.7 Write the molecular formula of (a) a disaccharide, (b) a trisaccharide made up of glucose units.

22.8 Figure 22.5 shows one anomer of D-fructose. Draw the other one.

22.9 The hexose L-glucose is the mirror image of D-glucose. Draw the structures of the α and β anomer of L-glucose.

22.10 Write the structure of the glycoside that will form from the reaction of β-D-glucose with ethanol.

22.11 The structure of D-galactose, another important hexose, differs from that of D-glucose only in the orientation of the OH group on the carbon atom that is four places down the chain from the anomeric carbon. It forms a six-membered ring just as D-glucose does. Draw the structure of this ring.

22.12 Lactose, commonly called milk sugar, is a disaccharide of glucose and galactose. It has a β-linkage between the anomeric carbon of galactose and the carbon atom of glucose that is four places down the chain from the anomeric carbon. Draw the structure of lactose.

22.13 Write the overall reaction for the hydrolysis of a trisaccharide consisting of glucose units.

22.14 Use Figure 22.15 as a guide to draw the structures of adenosine diphosphate and adenosine triphosphate.

22.15 Draw the structures of two nucleotides that include guanine.

22.16 Draw the structures of the deoxyribonucleotides of thymine and cytosine.

22.17 Explain why a polyprotic acid is needed for the formation of macromolecules such as the nucleic acids.

22.18 Draw the complete structure of the two ribonucleotides that can form by joining a ribonucleotide of guanine with one of cytosine.

22.19 Calculate the mass percent of phosphorus in a strand of DNA that consists of equal amounts of each of the four N-bases.

22.20 A sequence of one portion of a strand of DNA is GCTAGTT. Write the corresponding sequence on the other chain.

22.21 Draw the structure of the portion of the RNA chain that will be synthesized according to the information GAT in DNA.

22.22 In order to determine the sequence of bases in a five-nucleotide portion of a DNA strand, it is hydrolyzed to dinucleotides which are identified as AT, CC, GA, and TC. Write the sequence of bases in the DNA.

22.23 Find the sequence of bases in a mRNA chain that codes for the synthesis of a chain of amino acids in the sequence alanine, alanine, glycine, phenylalanine.

22.24 Find the sequence of bases on the DNA chain that codes for the amino acid sequence in Exercise 22.23.

22.25 The double-helical structure of DNA is disrupted by heating it somewhat above room temperature. On cooling the double helix re-forms. Discuss these observations in terms of the entropy and enthalpy of each process.

Fundamental Constants

Quantity	Symbol	Value[a,b]
atomic mass unit	amu	$1.660\ 565\ 5 \times 10^{-27}$ kg
Avogadro constant[c]	N or N_A	$6.022\ 094\ 3 \times 10^{23}$ mol^{-1}
electronic charge	e	$1.602\ 189\ 2 \times 10^{-19}$ C
Faraday constant	F	$9.648\ 456 \times 10^4$ C mol^{-1}
gas constant	R	$8.314\ 41$ J mol^{-1} K^{-1}
		$0.082\ 056\ 8$ liter atm mol^{-1} K^{-1}
mass of electron	m_e	$9.109\ 534 \times 10^{-31}$ kg
mass of neutron	m_n	$1.674\ 954\ 3 \times 10^{-27}$ kg
mass of proton	m_p	$1.672\ 648\ 5 \times 10^{-27}$ kg
Planck constant	h	$6.626\ 176 \times 10^{-34}$ J s
Rydberg constant	R_H	$1.096\ 775\ 78 \times 10^7$ m^{-1}
speed of light in a vacuum	c	$2.997\ 924\ 58 \times 10^8$ m s^{-1}

[a] Values are from the *CODATA Bulletin*, December, 1973.
[b] A table of these constants to four significant figures can be found inside the back cover.
[c] The value of the Avogadro constant is from R. B. Deslattes et al., *Phys. Rev. Letters* **33**:463 (1974).

Conversion Factors

Quantity	Conversion Factor
energy	1 cal = 4.184 J
	1 liter atm = 101.3 J
	1 eV = 1.602×10^{-19} J
	1 eV = 9.648×10^4 J/mol
	1 cm^{-1} = 11.96 J/mol
pressure	1 atm = 1.013×10^5 Pa
	1 atm = 760 mmHg
	1 atm = 760 Torr
temperature	0°C = 273.15 K

APPENDIX II: VALUES OF THERMODYNAMIC PROPERTIES OF SUBSTANCES AT 298 K AND 1 ATM

Substance[a,b]	ΔH_f° (kJ/mol)	ΔG_f° (kJ/mol)	S° (J mol^{-1} K^{-1})
Ag(s)	0	0	43
Ag(g)	284	246	173
AgBr(s)	−100	−97	107
AgCl(s)	−127	−110	96
AgF(s)	−205	−187	84
AgI(s)	−62	−66	115
Ag$_2$O(s)	−31	−11	122
Al(s)	0	0	28
Al(g)	330	286	164
AlCl$_3$(s)	−704	−629	111
Al$_2$O$_3$(s)	−1670	−1576	51
As(g)	0	0	155
As(s) gray	0	0	35
As(g)	302	261	174
As$_4$(g)	144	92	314
AsCl$_3$(g)	−262	−249	327
As$_2$O$_3$(s)	−657	−578	107
As$_2$O$_5$(s)	−925	−782	105
Au(s)	0	0	48
Au(g)	366	326	180
Au(OH)$_3$(s)	−425	−317	190
B(s)	0	0	6.5
B(g)	573	529	153
BCl$_3$(g)	−395	−380	290
BF$_3$(g)	−1137	−1120	254
B$_2$H$_6$(g)	36	87	232
B$_2$O$_3$(s)	−1273	−1193	54
Ba(s)	0	0	67
Ba(g)	180	146	170
BaCl$_2$(s)	−860	−811	126
BaCO$_3$(s)	−1216	−1138	112
BaO(s)	−558	−529	70.3
Be(s)	0	0	9
Be(g)	324	287	136
BeCl$_2$(s)	−490	−445	83
BeO(s)	−610	−580	14
Bi(s)	0	0	57
Bi(g)	208	168	187
BiCl$_3$(s)	−379	−317	184
Bi$_2$O$_3$(s)	−574	−494	151
Br$_2$(l)	0	0	152
Br$_2$(g)	31	3	245
Br(g)	112	82	175
Br$^-$	−121	−104	83
BrF(g)	−59	−74	229
BrF$_3$(g)	−256	−229	292
BrF$_5$(g)	−429	−350	320
BrO$^-$	−94	−33	42
BrO$_3^-$	−67	2	163
BrO$_4^-$	13		
C(s) graphite	0	0	5.69
C(s) diamond	1.9	2.9	2.43
C(g)	718	671	158
C$_2$(g)	808	780	200

[a] Substances are listed alphabetically by formula, except the standard state of an element is listed first. Formulas are written in the usual way.
[b] Substances without state symbols are in aqueous solution at 1M concentration.

Substance[a,b]	ΔH_f° (kJ/mol)	ΔG_f° (kJ/mol)	S° (J mol^{-1} K^{-1})
$CBr_4(g)$	50	36	358
$CCl_4(g)$	−107	−58	310
$CH_4(g)$	−75	−51	186
$C_2H_2(g)$	227	209	201
$C_2H_4(g)$	52	68	220
$C_2H_6(g)$	−85	−33	230
$CHCl_3(g)$	−100	−69	296
$CHCl_3(g)^c$			303
$CHCl_3(l)$	−135	−74	203
$CHCl_3(l)^c$			215
$CH_2O(g)$	−116	−110	219
$CO(g)$	−110.5	−137.3	197.6
$CO_2(g)$	−393.5	−394.4	213.7
$CS_2(g)$	115	67	238
$Ca(s)$	0	0	42
$Ca(g)$	178	144	155
$CaCl_2(s)$	−795	−750	114
$CaCO_3(s)$	−1207	−1129	93
$CaF_2(s)$	−1220	−1167	69
$CaO(s)$	−636	−604	40
$Cd(s)$	0	0	51
$Cd(g)$	112	77	168
$CdCl_2(s)$	−389	−343	251
$CdO(s)$	−255	−225	55
$Cl_2(g)$	0	0	223
$Cl(g)$	121	105	165
Cl^-	−167	−131	57
$ClF(g)$	−51	−56	218
$ClF_3(g)$	−159	−123	281
$ClF_5(g)$	−240		
$ClO(g)$	109	98	227
$ClO_2(g)$	103	120	257
$Cl_2O(g)$	80	98	266
$Cl_2O_7(l)$	272		
ClO^-	−107	−37	42
ClO_2^-	−67	17	101
ClO_3^-	−104	−3	162
ClO_4^-	−128	−9	182
$Co(s)$	0	0	30
$Co(g)$	425	380	179
$CoCl_2(s)$	−318	−274	103
$CoO(s)$	−238	−214	53
$Cr(s)$	0	0	24
$Cr(g)$	397	352	173
$CrCl_2(s)$	−395	−356	115
$CrCl_3(s)$	−552	−481	123
$Cr_2O_3(s)$	−1140	−1058	81
$CrO_3(s)$	−590	−513	72
$Cs(s)$	0	0	85
$Cs(g)$	79	51	175
$CsCl(s)$	−447	−419	100
$Cs_2O(s)$	−318	−290	124
$Cu(s)$	0	0	33
$Cu(g)$	338	299	166
$CuCl(s)$	−137	−120	86
$CuCl_2(s)$	−220	−176	108
$CuO(s)$	−155	−127	44

^{c} At the boiling point, 334 K.

Values of Thermodynamic Properties of Substances at 298 K and 1 atm
(*continued*)

Substance[a,b]	ΔH_f° (kJ/mol)	ΔG_f° (kJ/mol)	S° (J mol^{-1} K^{-1})
$Cu_2O(s)$	−167	−146	101
$F_2(g)$	0	0	203
F^-	−333	−279	−14
$FSO_3H(l)$	−800		
$Fe(s)$	0	0	27
$Fe(g)$	418	372	180
$FeCl_2(s)$	−342	−302	118
$FeCl_3(s)$	−399	−334	142
$FeO(s)$	−272	−251	61
$Fe_2O_3(s)$	−824	−742	87
$Fe_3O_4(s)$	−1118	−1015	146
$Ge(s)$	0	0	3
$Ge(g)$	377	336	168
$GeCl_4(g)$	−496	−457	348
$GeH_4(g)$	91	113	217
$GeO_2(s)$	−551	−497	55
$H_2(g)$	0	0	131
$H(g)$	218	203	115
$H^+(g)$	1536	1517	109
$H^-(g)$	149	133	109
$HBr(g)$	−36	−53	198
$HBrO$	−113	−82	142
$HBrO_3$	−40	2	163
$HCl(g)$	−92	−95	187
$HClO$	−121	−80	142
$HClO_2$	−52	6	188
$HClO_3$	−98	3	162
$HClO_4(l)$	−41	84	188
$HClO_4$	−129	8.6	182
$HF(g)$	−273	−273	174
$HI(g)$	26	2	206
HIO	−138	−99	95
HIO_3	−230	−133	167
$H_5IO_6(s)$	−834		
HNO_2	−119	−56	153
$HNO_3(g)$	−135	−75	266
HNO_3	−207	−111	146
$H_2O(g)$	−242	−229	189
$H_2O(l)$	−286	−237	70
HH_2PO_2	−592		
H_2HPO_3	−955		
$H_3PO_4(s)$	−1260	−1126	110
$H_2S(g)$	−20	−34	206
H_2SO_3	−633	−538	232
$H_2SO_4(l)$	−814	−690	157
H_2SO_4	−908	−745	20
$He(g)$	0	0	126
$Hg(l)$	0	0	76
$Hg(g)$	61	32	175
$HgCl_2(s)$	−224	−179	146
$Hg_2Cl_2(s)$	−265	−211	192
$HgO(s)$ red	−91	−59	70
$I_2(s)$	0	0	116
$I_2(g)$	62	19	261
$I(g)$	107	70	181
I^-	−57	−52	107
$IF(g)$	−95	−118	236

Substance[a,b]	ΔH_f° (kJ/mol)	ΔG_f° (kJ/mol)	S° (J mol^{-1} K^{-1})
IF_3(g)	−485		
IF_5(g)	−840	−768	328
IF_7(g)	−961	−828	346
I_2O_5(s)	−158		
IO^-	−108	−38	−5
IO_3^-	−220	−128	118
IO_4^-	−145		
K(s)	0	0	65
K(g)	90	61	160
KBr(s)	−394	−379	97
$HBrO_4$(s)	−287		
KCl(s)	−437	−408	83
$KClO_4$(s)	−432	−300	151
KF(s)	−529	−533	67
KI(s)	−328	−322	104
KIO_4(s)	−461	−395	159
Kr(g)	0	0	164
Li(s)	0	0	29
Li(g)	161	128	138
LiCl(s)	−402	−377	59
Li_2O(s)	−596	−560	38
Mg(s)	0	0	33
Mg(g)	148	113	149
$MgCl_2$(s)	−642	−592	90
$MgCO_3$(s)	−1096	−1012	66
MgO(s)	−601	−570	27
Mn(s)	0	0	32
Mn(g)	281	238	174
$MnCl_2$(s)	−481	−441	118
Mn_2O_7(s)	−728		
N_2(g)	0	0	192
N(g)	473	456	153
NF_3(g)	−114	−83	261
N_2F_4(g)	−7	81	301
NH_3(g)	−46	−16	193
NH_3	−80	−27	111
N_2H_4(l)	50	149	121
NH_4Br(s)	−270	−175	110
NH_4Cl(s)	−314	−203	95
NH_4NO_3(s)	−365	−184	151
NH_2OH(s)	−107		
NO(g)	90	87	211
NO_2(g)	34	51	240
N_2O(g)	82	104	220
N_2O_3(g)	84	139	312
N_2O_4(g)	10	98	304
N_2O_5(s)	−42	114	178
Na(s)	0	0	51
Na(g)	109	78	154
NaBr(s)	−360	−347	84
NaCl(s)	−411	−384	72
NaF(s)	−574	−544	51
NaI(s)	−288	−282	91
Na_2O(s)	−416	−377	73
Ne(g)	0	0	146
O_2(g)	0	0	205
O(g)	249	232	161
O^+(g)	1567	1547	155
O^-(g)	110	89	158

Values of Thermodynamic Properties of Substances at 298 K and 1 atm (*continued*)

Substance[a,b]	ΔH_f° (kJ/mol)	ΔG_f° (kJ/mol)	S° (J mol^{-1} K^{-1})
O_2	−12	16	111
$O_2^+(g)$	1184	1166	205
$O_3(g)$	142	163	239
$OH(g)$	42	34	184
$OH^-(g)$	−133		
OH^-	−230	−157	−11
P(s) red	0	0	23
P(s) white	18	12	41
P(s) black	−17	−19	23
P(g)	334	290	163
$P_2(g)$	179	116	218
$P_4(g)$	77	36	280
$PBr_3(l)$	−167	−164	240
$PBr_5(s)$	−293		
$PCl_3(l)$	−300	−260	217
$PCl_5(s)$	−400		
$PH_3(g)$	23	25	210
$P_4O_6(s)$	−1593		
$P_4O_{10}(s)$	−2940	−2676	228
$POCl_3(l)$	−578	−509	222
Pb(s)	0	0	65
Pb(g)	194	163	175
$PbCl_2(s)$	−359	−314	136
PbO(s)	−217	−188	69
$PbO_2(s)$	−277	−217	69
Rb(s)	0	0	77
Rb(g)	86	64	170
RbCl(s)	−433	−404	92
$Rb_2O(s)$	−330	−297	110
$S_8(s)$ rhombic	0	0	32
S(g)	277	238	168
$S_2(g)$	128	79	228
$S_8(g)$	102	50	431
$S_2Cl_2(l)$	−60		
$SF_4(g)$	−780	−731	292
$SF_6(g)$	−1220	−1105	292
$SOCl_2(g)$	−210	−198	310
$SO_2Cl_2(g)$	−364	−320	312
SO(g)	6.9	−20	222
$SO_2(g)$	−297	−300	248
$SO_3(g)$	−396	−371	257
Sb(s)	0	0	46
Sb(g)	262	222	180
$SbCl_3(s)$	−382	−324	184
$SbH_3(g)$	155	148	233
$Sb_2O_5(s)$	−972	−829	125
Si(s)	0	0	19
Si(g)	450	395	168
$SiCl_4(g)$	−657	−617	331
$SiH_4(g)$	34	57	204
$SiO_2(s)$ quartz	−911	−856	41
Sn(s) white	0	0	52
Sn(s) gray	−2	0.1	44
Sn(g)	302	267	168
$SnCl_2(s)$	−325		
$SnH_4(g)$	163	188	228
SnO(s)	−285	−257	56

Substance[a,b]	ΔH_f° (kJ/mol)	ΔG_f° (kJ/mol)	S° (J mol^{-1} K^{-1})
$SnO_2(s)$	−581	−520	52
$Sr(s)$	0	0	52
$Sr(g)$	164	131	165
$SrCl_2(s)$	−829	−781	115
$SrCO_3(s)$	−1220	−1140	97
$SrO(s)$	−592	−562	54
$Xe(g)$	0	0	170
$XeF_2(g)$	−130	−96	260
$XeF_4(g)$	−215	−138	316
$XeF_6(g)$	−294		
$XeO_3(g)$	502	561	287
$Zn(s)$	0	0	42
$Zn(g)$	131	95	161
$ZnCl_2(s)$	−415	−369	111
$ZnO(s)$	−348	−318	44

Single Bonds[a]

Bond	Bond Energy (kJ/mol)	Bond	Bond Energy (kJ/mol)
H—H	436	O—Si	452
H—C	413	O—P	335
H—N	391	O—Cl	218
H—O	463	O—Br	201
H—F	563	O—I	201
H—Si	318	F—F	158
H—P	322	F—Si	586
H—S	368	F—P	503
H—Cl	432	F—S	327
H—Br	366	F—Cl	253
H—I	299	F—Br	249
C—C	346	F—I	280
C—N	305	Si—Si	176
C—O	358	Si—Cl	396
C—F	489	P—P	201
C—P	264	P—Cl	322
C—S	272	P—Br	264
C—Cl	328	P—I	184
C—Br	285	S—S	251
C—I	218	S—Cl	271
N—N	163	Cl—Cl	243
N—O	222	Cl—Br	219
N—F	275	Cl—I	211
N—Cl	192	Br—Br	193
O—O	146	Br—I	178
O—F	193	I—I	151

Double Bonds[a]

Bond	Bond Energy (kJ/mol)	Bond	Bond Energy (kJ/mol)
C=C	615	N=N	418
C=N	615	N=O	607
C=O	749	O=O	498
C=O	803[b]	O=P	504
C=S	536	O=S	498[c]

Triple Bonds[a]

Bond	Bond Energy (kJ/mol)
C≡C	812
C≡N	890
N≡N	946
P≡P	490

[a] Elements are listed in order of increasing atomic number.
[b] In carbon dioxide.
[c] In sulfur dioxide.

EXPONENTIAL NOTATION

The numbers we encounter in chemistry often are very large or very small. A convenient way to express these numbers uses exponential notation, in which a number is expressed as the product of a simple number and a power of 10:

$N \times 10^x$

where N is the simple number and x is the exponent, or power to which 10 is raised.

In general, we shall find it convenient for the value of N to be between 1 and 10, so that N has only one digit to the left of the decimal point. For numbers larger than 10, the value of x is positive; for numbers smaller than 1, it is negative.

To convert an ordinary number into exponential notation, the decimal point must be moved so that there is only one digit to the left of the decimal point. We move the decimal point to the left for large numbers and to the right for small numbers. The value of x in the exponential is equal to the number of places that the decimal point is moved. When the decimal point is moved to the left, the exponent is positive; when it is moved to the right, the exponent is negative. Some examples are:

Number	N	Number of Places Decimal Point Is Moved	x	Exponential Notation
4567	4.567	3, left	3	4.567×10^3
0.004567	4.567	3, right	-3	4.567×10^{-3}
100 000	1	5, left	5	1×10^5
0.000001	1	6, right	-6	1×10^{-6}

The same procedure can be used to convert from exponential notation back to an ordinary number or to move the decimal point in N. When a decimal point is moved one place to the left, the value of x is increased by one. Thus:

$45.67 \times 10^2 = 4.567 \times 10^3$

$456.7 \times 10^{-5} = 4.567 \times 10^{-3}$

When a decimal point is moved one place to the right, the value of x is decreased by one. Thus:

$0.1 \times 10^{-5} = 1 \times 10^{-6}$

$0.4567 \times 10^3 = 4.567 \times 10^2$

To convert from exponential notation to an ordinary number, the decimal point must be moved until the exponent $x = 0$, since $10^0 = 1$. For example, to convert 2.35×10^5 to an ordinary number, the exponent must be reduced from 5 to 0. Therefore, the decimal point must be moved five places to the right:

$2.35 \times 10^5 = 235\ 000$

Arithmetic operations with exponentials are carried out by the following rules:

1. When exponentials are multiplied, the exponents are added and the Ns are multiplied. Thus:

$(2.3 \times 10^3)(4.8 \times 10^2) = 11 \times 10^5 = 1.1 \times 10^6$

$(4.5 \times 10^{-5})(4.5 \times 10^{-3}) = 20 \times 10^{-8} = 2.0 \times 10^{-7}$

2. When exponentials are divided, the exponent in the denominator is subtracted from the exponent in the numerator, and the N in the numerator is divided by the N in the denominator. Thus:

$$\frac{(4.5 \times 10^{-5})}{(6.5 \times 10^{-3})} = 0.69 \times 10^{-2} = 6.9 \times 10^{-3}$$

$$\frac{(2.3 \times 10^{3})}{(4.8 \times 10^{2})} = 0.48 \times 10^{1} = 4.8$$

$$\frac{(4.5 \times 10^{-5})}{(6.5 \times 10^{3})} = 0.69 \times 10^{-8} = 6.9 \times 10^{-9}$$

3. When an exponential is raised to a power, N is raised to the power and x is multiplied by the power. Thus:

$$(3.3 \times 10^{7})^{2} = (3.3)^{2} \times 10^{14} = 1.1 \times 10^{15}$$
$$(6.8 \times 10^{-8})^{3} = (6.8)^{3} \times 10^{-24} = 310 \times 10^{-24} = 3.1 \times 10^{-22}$$

4. When a root of an exponential is extracted, the root of N is extracted in the usual way and the exponent x is multiplied by the fraction that corresponds to the root: one-half for the square root, one-third for the cube root, etc. To ensure that the exponent in the result is a whole number, the exponential must sometimes be rewritten so that it is divisible by the root. Thus:

$$\sqrt[2]{3.4 \times 10^{-5}} = \sqrt[2]{34 \times 10^{-6}} = \sqrt[2]{34} \times 10^{-3} = 5.8 \times 10^{-3}$$
$$(4.7 \times 10^{6})^{1/4} = (470 \times 10^{4})^{1/4} = (470)^{1/4} \times 10^{1} = 4.7 \times 10^{1}$$

5. When exponentials are added or subtracted, their exponents must have the same value. The exponent thus is unchanged, while one N is subtracted from or added to the other. Thus:

$$1.3 \times 10^{-2} + 4.2 \times 10^{-3} = 1.3 \times 10^{-2} + 0.42 \times 10^{-2} = 1.7 \times 10^{-2}$$
$$1.30 \times 10^{5} - 4.0 \times 10^{3} = 1.30 \times 10^{5} - 0.04 \times 10^{5} = 1.26 \times 10^{5}$$

LOGARITHMS

A logarithm is an exponent. A logarithm to the base 10 of a number is the exponent to which 10 must be raised to obtain the number. That is:

$$a = 10^{q} \quad \text{and} \quad \log a = q$$

where a is the number and $\log a$ is the logarithm to the base 10. We shall use logarithms to the base 10 as much as possible.

Another important system of logarithms uses the number e, whose value is 2.71828 . . . as its base. Logarithms to the base e are called natural logarithms. The natural logarithm of a number a is written as $\ln a$. The factor 2.303 converts between the two systems of logarithms:

$$\ln a = 2.303 \log a$$

A logarithm has two parts: the *characteristic*, which is the part of the logarithm to the left of the decimal point, and the *mantissa*, the part of the logarithm to the right of the decimal point. In the logarithm 2.456, 2 is the characteristic and 0.456 is the mantissa. In the logarithm 102.3, 102 is the characteristic and 0.3 is the mantissa.

One advantage of the form of exponential notation that we use is that there is a direct correspondence between an exponential and a logarithm to the base 10. The characteristic corresponds to the x, the exponent of 10. The mantissa is the logarithm of any simple number N whose value is between 1 and 10. This correspondence can be illustrated by some examples.

1. Taking the logarithm of a number:

$$\log (2.3 \times 10^{3}) = 3 + \log 2.3 = 3 + 0.36 = 3.36$$

The value of log 2.3 can be obtained from a table of logarithms or from a calculator with a "log" key.

$$\log 0.00047 = \log (4.7 \times 10^{-4}) = -4 + \log 4.7$$
$$= -4 + 0.67 = -3.33$$

2. Finding the antilogarithm, the number whose logarithm is given.

$$\text{antilog } (5.42) = \text{antilog } (0.42) \times 10^5 = 2.6 \times 10^5$$

Again, the value of the antilog can be obtained from a table of logarithms or from a calculator with an antilog, or "10^x" key.

$$\text{antilog } (-6.78) = \text{antilog } (-7 + 0.22)$$
$$= \text{antilog } (0.22) \times 10^{-7} = 1.7 \times 10^{-7}$$

Since a table of logarithms gives a mantissa as a positive number, the negative logarithm must be rewritten as the sum of a negative integer and a positive decimal (Examples 15.4–15.7).

Significant figures and units in the correct expression of numerical quantities are extremely important in chemistry. The value of a logarithm usually is determined by the units in which its antilogarithm is given, but the logarithm itself has no units. The number of significant figures in a logarithm can be taken as the number of figures in the mantissa. Thus, the logarithm 2.34 has two significant figures and the logarithm 102.3 has one significant figure.

Arithmetic operations using logarithms are carried out in the same way as the operations for any exponents. The relevant operations can be summarized by few relationships:

When two numbers are multiplied, their logarithms are added:

$$\log (ab) = \log a + \log b$$

When two numbers are divided, the logarithm of the denominator is subtracted from the logarithm of the numerator:

$$\log a/b = \log a - \log b$$

When a number is raised to a power, its logarithm is multiplied by the power:

$$\log a^n = n \log a$$

When a root is extracted from a number, its logarithm is multiplied by the fractional exponent corresponding to the root:

$$\log a^{1/n} \times (1/n) \log a$$

GRAPHS

Experimental data obtained in the laboratory are often presented by using a graph. When two variables are measured, it is convenient to show one of them along the vertical axis (the y axis) and the other along the horizontal axis (the x axis). Each pair of measurements is a point on the graph; the points are connected by a smooth curve.

The shape of the smooth curve sometimes reveals more clearly than the numbers alone the relationship between the two variables. For example, Figure 4.13 shows the relationship between the pressure P and the volume V of an ideal gas. The shape of the curve is consistent with an inverse proportionality; that is, the product of the two variables is a constant k: $PV = k$. We see similar examples in Section 17.2, where we discuss graphical methods for finding rate laws.

Straight-line plots are of special interest. A straight-line plot is observed when

FIGURE IV.1

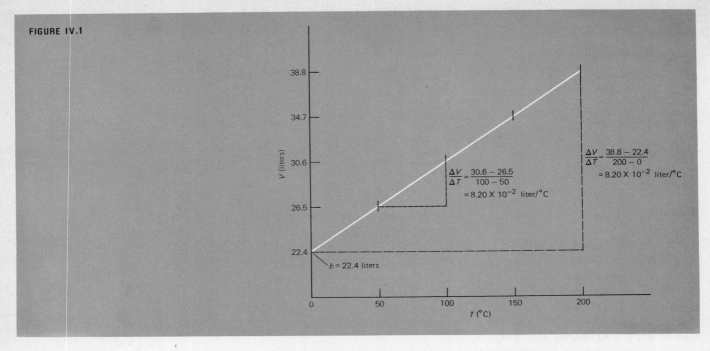

the two variables being studied are directly proportional. Figures 4.15 and 4.16 show the straight lines that are obtained when volume (V) and temperature (T) are the variables.

The general equation for a straight line usually is represented as:

$$y = mx + b$$

where x and y are the two variables, m is the slope of the line, and b is its intercept with the y axis. When the intercept with the y axis is at the origin, the equation becomes $y = mx$, which is the simple relationship of a direct proportionality. Such an equation is often encountered in chemical studies. For example, volume is directly proportional to absolute temperature: $V = kT$. The slope of the line gives the value of k.

To find the slope of a straight line from a graph, we pick two points on the line and find the change in each variable between the two points. The slope is then:

$$\frac{\Delta x}{\Delta y}$$

Figure IV.1 is a plot of the volume of one mole of an ideal gas at a pressure of 1 atm against the Celsius temperature. The value of b, the y intercept, can be seen to be 22.4 liters. The slope can be found from the graph, as shown in the figure.

THE METHOD OF SUCCESSIVE APPROXIMATIONS

In Example 14.11, we introduce a method for simplifying the algebra required for the solution of equations that are encountered in problems of chemical equilibrium. The basis of this method is to approximate sums or differences of a large number and a small number by using the larger number, unchanged. In Example 14.11, for instance, the equation:

$$\frac{x}{(0.741 \text{ atm} - 2x)^2} = 5.33 \times 10^{-3} \text{ atm}^{-1}$$

becomes

$$\frac{x}{(0.741 \text{ atm})^2} = 5.33 \times 10^{-3} \text{ atm}^{-1}$$

by making the approximation that $0.741 - 2x \cong 0.741$, since we know that x is very small.

We arbitrarily use the criterion that the approximation is valid if it introduces an error smaller than 5%. The value of x we obtain from the approximation is about 2.93×10^{-3}. Since we can show that $2x$ is less than 1% of 0.741, the approximation is valid. But sometimes the approximation introduces an unacceptably large error, as in Example 14.12. In the text, we use the quadratic formula for such a case. But another procedure is available. We can illustrate this procedures by using the calculation of Example 14.12.

The original equation is a difficult one:

$$\frac{x^2}{(0.124 \text{ atm} - x)} = 4.10 \times 10^{-3} \text{ atm}$$

We approximate that $0.124 - x = 0.124$ and find that $x = 0.0225$. The error of 25% in the approximation is unacceptably large. But we can use this admittedly erroneous value of x to improve the approximation. The correct value of $0.124 - x$ is closer to $0.124 - 0.0225$ than it is to 0.124 alone, since x is at least in the neighborhood of 0.0225. We now write the equation as:

$$\frac{x^2}{(0.124 - 0.0225)} = 4.10 \times 10^{-3}$$
$$x = 2.04 \times 10^{-2}$$

This value of x is closer to the correct one. We can repeat the approximation a second time, using this value of x.

$$\frac{x^2}{(0.124 - 0.0204)} = 4.10 \times 10^{-3}$$
$$x = 2.06 \times 10^{-2}$$

This last value is the same value obtained with the quadratic equation.

The method of successive approximations cannot be used indiscriminately. It is reserved for cases in which the first approximation is reasonably good, but where the error in the approximation is unacceptably large.

EXERCISES[a]

IV.1 Express the following numbers in exponential notation: (a) 9876, (b) 0.009 876, (c) 1 000 001, (d) 0.000 040 40, (e) 1.0004.

IV.2 Convert the following exponential expressions into ordinary numbers: (a) 2.345×10^5, (b) 2.345×10^{-5}, (c) 1.0×10^6, (d) 1.00×10^{-1}.

IV.3 Perform the following calculations and express your result in the usual exponential form: (a) $(4.5 \times 10^5)(6.5 \times 10^3)$. (b) $(4.5 \times 10^{-5})(6.5 \times 10^{-3})$, (c) $(4.5 \times 10^5)/(6.5 \times 10^3)$, (d) $(6.5 \times 10^{-3})(4.5 \times 10^{-5})$.

(e) $1/(4.5 \times 10^5)$, (f) $1/(6.5 \times 10^{-3})$.

IV.4 Find the square and the square root, the cube and the cube root of the following exponentials: (a) 4.5×10^{-5}, (b) 6.5×10^3.

IV.5 Using the logarithm table in Appendix VI, find the logarithms of the following numbers: (a) 2.33, (b) 101, (c) 0.0233, (d) 0.001 01.

IV.6 Using the logarithm table in Appendix VI, find the antilogarithms of the following numbers: (a) 0.289, (b) 3.289, (c) −2.67, (d) 1.000, (e) 101.1.

IV.7 Use logarithms to perform the fol-

lowing calculations: (a) (4.567)(7.654), (b) 8.765/5.678, (c) $\sqrt[3]{87.9}$, (d) $(0.00452)^4$.

IV.8 Find the slope and the y intercept of the straight line obtained by plotting the following data:

| P (atm) | 0.804 | 0.856 | 0.910 | 0.965 |
| T (°C) | 20 | 40 | 60 | 80 |

IV.9 Use the method of successive approximations to solve the following equation: $x/(0.30 - x)^2 = 0.62$.

IV.10 Use the method of successive approximations to solve the following equation: $x^2/(0.22 - x) = 0.024$.

[a] Do not use a calculator for Exer. 1–7, since they will then serve no purpose. Use a calculator to check your answers if you wish.

A number of different equilibrium reactions occur even in the simplest aqueous solutions. We have seen that there are several different equilibrium reactions in solutions of weak acids or weak bases. In such solutions, the equilibrium reaction between the strongest acid and the strongest base is chosen as the relevant one and is used for calculations. For example, in a solution of HCN, the relevant equilibrium is the one that defines K_A. The water equilibrium occurs simultaneously, but we ignore it until we wish to calculate [OH$^-$], which appears in K_W but not in K_A.

Many different equilibrium reactions also occur in an aqueous solution of a polyprotic acid. The procedure used for solutions of a polyprotic acid can be applied to any aqueous systems in which a large number of different equilibrium reactions take place simultaneously. We first choose a *main equilibrium reaction,* in polyprotic acids, the reaction that defines K_1, and we find the equilibrium concentrations of all the substances in this reaction. We then determine the concentrations of all the other species in solution by choosing secondary equilibria that include these substances and substituting the concentrations found from the primary equilibrium into the secondary equilibria.

Without such a procedure, calculations on the composition of these systems can require the solution of many simultaneous equations in many unknowns. A simple method is desirable, but we are faced with a problem. Out of all the simultaneous equilibria that occur in a system, how do we identify the one that qualifies as a main equilibrium? There are some guidelines that help us make the choice:

1. *To qualify as the main equilibrium, a reaction must have an equilibrium constant at least 100 times larger than the equilibrium constant of any other reaction that can be considered as a possible main equilibrium.* In a solution of phosphoric acid, for example, the ionization of the first proton, $H_3PO_4 \rightleftharpoons H_2PO_4^- + H^+$, is the main equilibrium because K_1 is more than 10^6 larger than K_2, more than 10^{10} larger than K_3, and more than 10^{11} larger than K_W.

2. *To qualify as the main equilibrium, a reaction must include as reactants only those substances that are present in appreciable concentrations when the solution is prepared.* In practice, this usually means that *substances that are not specifically mentioned in the description of the system of interest do not appear on the left side of any reaction that can be considered as a candidate for the main equilibrium.*

For example, many equilibria occur in a solution of hydrocyanic acid, HCN. Many of these equilibria have large equilibrium constants. The equilibrium constant for one of these reactions,

$$H_2O + CN^- \rightleftharpoons HCN + OH^-$$

is larger than K_A, the equilibrium constant for the main equilibrium. But this reaction is not a candidate for the main equilibrium because there is no appreciable concentration of CN$^-$ when a solution of HCN is first prepared. By rule 2, cyanide ion cannot appear on the left side of any reaction that can be used as the main equilibrium for this system. Only H$_2$O and/or HCN can appear on the left side of such a reaction. You will encounter many instances in which the application of this rule limits the number of reactions to consider in selecting a main equilibrium.

The procedure for selecting the main equilibrium is:

1. List all the substances that are present in appreciable concentrations when the solution is prepared. A careful reading of the description of the solution is highly desirable in this step.

2. List all the reactions that are possible if these substances are the only reactants. Each reactant is classified by its behavior in aqueous systems. Among the possible reactions are the transfer of a proton from an acid to a base, the dissolution of a sparingly soluble material, the precipitation of a sparingly soluble

material, the formation of a complex ion, or even an oxidation-reduction. Each reactant can therefore be identified as a proton donor, an acid; a proton acceptor, a base; or as both a donor and an acceptor, an amphoteric reactant; or as neither an acid nor a base. In addition, you must say whether each ion in solution is a possible component of a relatively insoluble material or of a complex ion. The solubility rules given in Chapter 12 and the information in Chapter 18 will help you make this identification. Use the information in Table 16.1 to decide if species that are oxidizing or reducing agents in water are present. Having made this classification, list all the reactions that are possible between the allowed reactants.

3. Use all the available data on equilibrium constants to obtain the equilibrium constant for each reaction that is listed as a possible main equilibrium.

4. Compare all these numerical values. If one equilibrium constant is at least 10^2 larger than any of the others, it identifies the desired main equilibrium reaction. The method described here can be used only if one equilibrium constant is at least 10^2 larger than any of the others.

If the method does identify a main reaction, the concentrations of all the substances present in the solution at equilibrium can be found by the methods described in the following examples.

EXAMPLE V.1

Calculate the concentrations of all species present at equilibrium in an aqueous solution that has a total volume of 1.0 liter and contains 0.10 mol of sodium bicarbonate.

SOLUTION

(1) Following the rules outlined above, we list all the substances present in appreciable concentration. In addition to H_2O, the solution of sodium bicarbonate, a salt, also contains the dissociated ions Na^+ and HCO_3^-.

(2) We now classify each of these substances according to its behavior in aqueous equilibria. All sodium salts are soluble, and Na^+ ion does not enter into acid-base reactions in water, it does not form complexes, and it is not an oxidizing agent in water. Therefore, the Na^+ ion can be disregarded. We can disregard all group IA metal cations in water; no possible main equilibrium will contain them as reactants. The HCO_3^- ion is amphoteric: It is an acid because it has a proton to donate and it is a base because it can accept a proton. Water is also amphoteric. We need consider only acid-base equilibria, that is, proton transfer reactions, and we can list all the possibilities:

(a) $H_2O \rightleftharpoons H^+ + OH^-$; water is both the acid and the base.
(b) $HCO_3^- + H_2O \rightleftharpoons H_2CO_3 + OH^-$; water is the acid and bicarbonate is the base.
(c) $HCO_3^- \rightleftharpoons CO_3^{2-} + H^+$; water is the base and bicarbonate is the acid. (As usual, water is not included in the equation when it acts as a base.)
(d) $HCO_3^- + HCO_3^- \rightleftharpoons H_2CO_3 + CO_3^{2-}$; bicarbonate is both the acid and the base.

(3) Table 15.5 lists the K_1 and K_2 for carbon dioxide, which we have represented as H_2CO_3 for the sake of simplicity. The value of K_W is also known, so we have sufficient data to evaluate the equilibrium constant for each of the equilibria.

(a) $K_W = 1.0 \times 10^{-14}$
(b) $K = K_W/K_1 = (1.0 \times 10^{-14})/(4.3 \times 10^{-7}) = 2.3 \times 10^{-8}$; note that K_1 is used here because the reaction includes carbonic acid and bicarbonate ion, as K_1 does.
(c) This reaction defines K_2, whose value is 5.6×10^{-11}.
(d) To find the equilibrium constant of this disproportionation, we must find a

combination of reactions with known equilibrium constants that will give the reaction. We see that HCO_3^- forms CO_3^{2-}, just as it does in the K_2 reaction. We also see that HCO_3^- forms H_2CO_3, which is equivalent to CO_2, in the reverse of the K_1 reaction. The combination of these reactions is

$$HCO_3^- \rightleftharpoons CO_3^{2-} + H^+ \qquad K_2$$

$$HCO_3^- + H^+ \rightleftharpoons H_2CO_3 \qquad \frac{1}{K_1}$$

$$\overline{2HCO_3^- \rightleftharpoons H_2CO_3 + CO_3^{2-} \qquad K = \frac{K_2}{K_1}}$$

$$K = \frac{5.6 \times 10^{-11}}{4.3 \times 10^{-7}} = 1.3 \times 10^{-4}$$

(4) Comparing the values obtained in step (3), we see that the equilibrium constant for the disproportionation is almost 10^4 larger than the second largest one, that for reaction (b). Therefore, reaction (d) is the main reaction. It can be used to carry out the usual equilibrium calculation.

	$2HCO_3^-$	$\rightleftharpoons$	H_2CO_3	$+$	CO_3^{2-}
start	0.10M		0		0
equil	0.10$M - 2x$		x		x

Substituting into the expression for K:

$$K = \frac{x^2}{(0.10M - 2x)^2} = 1.3 \times 10^{-4}$$

Taking the square root of both sides of this equation, we find that $x = 1.1 \times 10^{-3}M$.

Therefore,

$$[H_2CO_3] = [CO_3^{2-}] = 1.1 \times 10^{-3}M$$
$$[HCO_3^-] = 0.10M - (2)(0.0011M) = 0.10M$$

Both H^+ and OH^- are also present in the solution. To find the concentration of these ions, we use the secondary equilibria in which they appear, for instance

$$K_2 = \frac{[H^+][CO_3^{2-}]}{[HCO_3^-]} = 5.6 \times 10^{-11}M$$

The $[H^+]$ is found by substituting the concentration of bicarbonate ion and carbonate ion, calculated from the main equilibrium, directly into this expression:

$$\frac{[H^+](1.1 \times 10^{-3}M)}{(0.10M)} = 5.6 \times 10^{-11}M$$

and

$$[H^+] = 5.1 \times 10^{-9}M$$

The $[OH^-]$ is found in the usual way, by substituting into the expression for K_W:

$$[OH^-] = \frac{1.0 \times 10^{-14}M^2}{5.1 \times 10^{-9}M} = 2.0 \times 10^{-6}M$$

This approach can also be used for more complicated systems that include undissolved substances.

EXAMPLE V.2

Some magnesium hydroxide, $Mg(OH)_2(s)$, is added to a $1.0M$ solution of nitric acid. The K_{sp} of the magnesium hydroxide is 8.9×10^{-12}. Find the main equilibrium reaction for this system and evaluate its equilibrium constant.

SOLUTION

(1) List all the species present in appreciable quantity when the solution is prepared: $Mg(OH)_2(s)$, H_2O, H^+, and NO_3^-. No HNO_3 is present, since nitric acid, being a strong acid, is ionized completely in solution.

(2) Each substance is classified by its behavior in aqueous equilibria. The H_2O is amphoteric, and the H^+ is a strong acid—in fact, the strongest in the system. The NO_3^- is not part of any equilibrium of interest; it is too weak a base to react with water. Nitrate salts—including magnesium nitrate, which could form in this system—are all soluble. The $Mg(OH)_2(s)$ can dissolve in water.

We thus have a system that starts with an undissolved substance and ions in solution. In such a system, we must also consider the reactions between the ions (Mg^{2+} and OH^-) that form when the solid dissolves and the ions that are already in solution.

The simple equilibria are:

(a) $H_2O \rightleftharpoons H^+ + OH^-$ $\qquad K_W = 1.0 \times 10^{-14}$
(b) $Mg(OH)_2(s) \rightleftharpoons Mg^{2+} + 2OH^-$ $\qquad K_{sp} = 8.9 \times 10^{-12}$

We can assume that Mg^{2+} does not react further, but we must consider the reaction between the OH^-, produced when $Mg(OH)_2(s)$ dissolves, and H^+, the strongest acid in the system. The product of the reaction is H_2O. The reactants are H^+, which is present in the original system, and the added $Mg(OH)_2(s)$, which is the source of OH^-. The equation is:

(c) $Mg(OH)_2(s) + 2H^+ \rightleftharpoons Mg^{2+} + H_2O$

To evaluate the equilibrium constant for this reaction, we need a suitable combination of reactions with known equilibrium constants. Reaction (b), which defines K_{sp}, and reaction (c) both have solid magnesium hydroxide as a reactant and the magnesium cation as a product. Reaction (c) has water as a product and H^+ as a reactant, the reverse of reaction (a). But the coefficient is 2 in reaction (a) and 1 in reaction (c). To obtain the suitable equilibrium constant, we must therefore reverse reaction (a), multiply the resulting equation by 2, and raise its equilibrium constant to the second power. The combination is thus:

$$Mg(OH)_2(s) \rightleftharpoons Mg^{2+} + 2OH^- \qquad K_{sp} = 8.9 \times 10^{-12}$$
$$\underline{2H^+ + 2OH^- \rightleftharpoons 2H_2O \qquad\qquad \frac{1}{K_W^2} = \frac{1}{1.0 \times 10^{-28}}}$$
$$Mg(OH)_2(s) + 2H^+ \rightleftharpoons Mg^{2+} + 2H_2O$$

$$K = \frac{K_{sp}}{K_W^2}$$

$$= 8.9 \times 10^{16}$$

Reaction (c) has a very large equilibrium constant and is the main equilibrium.

This calculation tells us that magnesium hydroxide, which has a small K_{sp}, is relatively insoluble in water but is quite soluble in solutions of strong acids. If excess magnesium hydroxide is added to a solution of strong acid, it will dissolve until the strong acid is neutralized and the solution is slightly alkaline. The large value of K for the neutralization of a strong acid by a strong base accounts for the dissolution of magnesium hydroxide in the acid solution. The numerical value of the solubility of magnesium hydroxide in this solution can be found by using the main equilibrium to determine the $[Mg^{2+}]$ at equilibrium.

A more complex system is treated in the next example.

EXAMPLE V.3

Solid CuI is added to a 0.4M solution of HCN. The K_{sp} of CuI(s) = 1.1×10^{-12}, the K_{sp} of CuCN(s) = 1.0×10^{-11}. The K_A of HCN = 4.9×10^{-10} and the K_i of Cu(CN)$_2^-$ = 1.0×10^{-24}. Find the main reaction and evaluate its equilibrium constant.

SOLUTION

(1) List all the substances present in appreciable quantities when the system is prepared: CuI(s), HCN, and H$_2$O.

(2) Classify each substance according to its chemical behavior. The CuI(s) is a sparingly soluble ionic solid which can dissolve in water, the HCN is a weak acid, and the H$_2$O is an acid and a base. Also consider the Cu$^+$ ions formed when the solid dissolves; Cu$^+$ can form the sparingly soluble CuCN(s) or the complex ion Cu(CN)$_2^-$, if a source of CN$^-$ ion (such as HCN) is available.

(3) We list possible reactions, including not only the simple reactions but also those between the ions that form. The simple possibilities are:

(a) CuI(s) $\rightleftharpoons$ Cu$^+$ + I$^-$ $K_{sp} = 1.1 \times 10^{-12}$
(b) H$_2$O $\rightleftharpoons$ H$^+$ + OH$^-$ $K_W = 1.0 \times 10^{-14}$
(c) HCN $\rightleftharpoons$ H$^+$ + CN$^-$ $K_A = 4.9 \times 10^{-10}$

The more complex reactions are those between the Cu$^+$ from the CuI and the CN$^-$ from the HCN.

(d) CuI(s) + HCN $\rightleftharpoons$ CuCN(s) + H$^+$ + I$^-$

$$K = \frac{K_{sp}(CuI)K_A}{K_{sp}(CuCN)} = \frac{(1.1 \times 10^{-12})(4.9 \times 10^{-10})}{(1.0 \times 10^{-11})} = 5.4 \times 10^{-11}$$

(e) CuI(s) + 2HCN $\rightleftharpoons$ Cu(CN)$_2^-$ + 2H$^+$ + I$^-$

$$K = \frac{K_{sp}(CuI)K_A^2}{K_i} = \frac{(1.1 \times 10^{-12})(4.9 \times 10^{-10})^2}{1.0 \times 10^{-24}} = 2.6 \times 10^{-7}$$

Since the equilibrium constant for this reaction is more than 10^2 larger than the next largest one, it can be used as the main reaction for the purposes of calculation.

EXERCISES

V.1 Calculate the pH of a solution that is prepared from 0.10 mol of Na$_2$HPO$_4$ and enough water to make one liter of solution.[a]
V.2 Calculate the pH of a solution prepared from 0.10 mol of NaH$_2$PO$_4$ and enough water for one liter of solution.[a]
V.3 Calculate the solubility of Al(OH)$_3$

[a] The necessary data for this exercise can be found in Table 15.5.

in 1.0M nitric acid solution. The K_{sp} is 2×10^{-32}.
V.4 The K_{sp} of BaF$_2$ is 1.0×10^{-6}. Find its solubility in 0.10M hydrochloric acid solution. (K_A of HF = 3.53×10^{-4}.)
V.5 The K_i of [Zn(CN)$_4$]$^{2-}$ is 6.3×10^{-21}. Find the solubility of Zn(OH)$_2$, whose K_{sp} is 1.8×10^{-14}, in 0.20M NaCN solution.
V.6 Find the main reaction and evaluate its equilibrium constant in a system

prepared from solid AgCl, K_{sp} = 1.6×10^{-10}, and a solution that is 1.0M in NH$_3$ and 1.0M in NH$_4$Br. The K_i of [Ag(NH$_3$)$_2$]$^+$ is 6.3×10^{-8} and the K_{sp} of AgBr is 5.2×10^{-13}.
V.7 Use appropriate data from Tables 14.3 and 18.1 to find the main reaction and to evaluate its equilibrium constant in a system containing a 1.0M solution of mercury(II) nitrate and excess solid silver iodide.

	0	1	2	3	4	5	6	7	8	9
1.0	.0000	.0043	.0086	.0128	.0170	.0212	.0253	.0294	.0334	.0374
1.1	.0414	.0453	.0492	.0531	.0569	.0607	.0645	.0682	.0719	.0755
1.2	.0792	.0828	.0864	.0899	.0934	.0969	.1004	.1038	.1072	.1106
1.3	.1139	.1173	.1206	.1239	.1271	.1303	.1335	.1367	.1399	.1430
1.4	.1461	.1492	.1523	.1553	.1584	.1614	.1644	.1673	.1703	.1732
1.5	.1761	.1790	.1818	.1847	.1875	.1903	.1931	.1959	.1987	.2014
1.6	.2041	.2068	.2095	.2122	.2148	.2175	.2201	.2227	.2253	.2279
1.7	.2304	.2330	.2355	.2380	.2405	.2430	.2455	.2480	.2504	.2529
1.8	.2553	.2577	.2601	.2625	.2648	.2672	.2695	.2718	.2742	.2765
1.9	.2788	.2810	.2833	.2856	.2878	.2900	.2923	.2945	.2967	.2989
2.0	.3010	.3032	.3054	.3075	.3096	.3118	.3139	.3160	.3181	.3201
2.1	.3222	.3243	.3263	.3284	.3304	.3324	.3345	.3365	.3385	.3404
2.2	.3424	.3444	.3464	.3483	.3502	.3522	.3541	.3560	.3579	.3598
2.3	.3617	.3636	.3655	.3674	.3692	.3711	.3729	.3747	.3766	.3784
2.4	.3802	.3820	.3838	.3856	.3874	.3892	.3909	.3927	.3945	.3962
2.5	.3979	.3997	.4014	.4031	.4048	.4065	.4082	.4099	.4116	.4133
2.6	.4150	.4166	.4183	.4200	.4216	.4232	.4249	.4265	.4281	.4298
2.7	.4314	.4330	.4346	.4362	.4378	.4393	.4409	.4425	.4440	.4456
2.8	.4472	.4487	.4502	.4518	.4533	.4548	.4564	.4579	.4594	.4609
2.9	.4624	.4639	.4654	.4669	.4683	.4698	.4713	.4728	.4742	.4757
3.0	.4771	.4786	.4800	.4814	.4829	.4843	.4857	.4871	.4886	.4900
3.1	.4914	.4928	.4942	.4955	.4969	.4983	.4997	.5011	.5024	.5038
3.2	.5051	.5065	.5079	.5092	.5105	.5119	.5132	.5145	.5159	.5172
3.3	.5185	.5198	.5211	.5224	.5237	.5250	.5263	.5276	.5289	.5302
3.4	.5315	.5328	.5340	.5353	.5366	.5378	.5391	.5403	.5416	.5428
3.5	.5441	.5453	.5465	.5478	.5490	.5502	.5514	.5527	.5539	.5551
3.6	.5563	.5575	.5587	.5599	.5611	.5623	.5635	.5647	.5658	.5670
3.7	.5682	.5694	.5705	.5717	.5729	.5740	.5752	.5763	.5775	.5786
3.8	.5798	.5809	.5821	.5832	.5843	.5855	.5866	.5877	.5888	.5899
3.9	.5911	.5922	.5933	.5944	.5955	.5966	.5977	.5988	.5999	.6010
4.0	.6021	.6031	.6042	.6053	.6064	.6075	.6085	.6096	.6107	.6117
4.1	.6128	.6138	.6149	.6160	.6170	.6180	.6191	.6201	.6212	.6222
4.2	.6232	.6243	.6253	.6263	.6274	.6284	.6294	.6304	.6314	.6325
4.3	.6335	.6345	.6355	.6365	.6375	.6385	.6395	.6405	.6415	.6425
4.4	.6435	.6444	.6454	.6464	.6474	.6484	.6493	.6503	.6513	.6522
4.5	.6532	.6542	.6551	.6561	.6571	.6580	.6590	.6599	.6609	.6618
4.6	.6628	.6637	.6646	.6656	.6665	.6675	.6684	.6693	.6702	.6712
4.7	.6721	.6730	.6739	.6749	.6758	.6767	.6776	.6785	.6794	.6803
4.8	.6812	.6821	.6830	.6839	.6848	.6857	.6866	.6875	.6884	.6893
4.9	.6902	.6911	.6920	.6928	.6937	.6946	.6955	.6964	.6972	.6981
5.0	.6990	.6998	.7007	.7016	.7024	.7033	.7042	.7050	.7059	.7067
5.1	.7076	.7084	.7093	.7101	.7110	.7118	.7126	.7135	.7143	.7152
5.2	.7160	.7168	.7177	.7185	.7193	.7202	.7210	.7218	.7226	.7235
5.3	.7243	.7251	.7259	.7267	.7275	.7284	.7292	.7300	.7308	.7316
5.4	.7324	.7332	.7340	.7348	.7356	.7364	.7372	.7380	.7388	.7396
5.5	.7404	.7412	.7419	.7427	.7435	.7443	.7451	.7459	.7466	.7474
5.6	.7482	.7490	.7497	.7505	.7513	.7520	.7528	.7536	.7543	.7551
5.7	.7559	.7566	.7574	.7582	.7589	.7597	.7604	.7612	.7619	.7627
5.8	.7634	.7642	.7649	.7657	.7664	.7672	.7679	.7686	.7694	.7701
5.9	.7709	.7716	.7723	.7731	.7738	.7745	.7752	.7760	.7767	.7774
6.0	.7782	.7789	.7796	.7803	.7810	.7818	.7825	.7832	.7839	.7846
6.1	.7853	.7860	.7868	.7875	.7882	.7889	.7896	.7903	.7910	.7917
6.2	.7924	.7931	.7938	.7945	.7952	.7959	.7966	.7973	.7980	.7987
6.3	.7993	.8000	.8007	.8014	.8021	.8028	.8035	.8041	.8048	.8055
6.4	.8062	.8069	.8075	.8082	.8089	.8096	.8102	.8109	.8116	.8122
6.5	.8129	.8136	.8142	.8149	.8156	.8162	.8169	.8176	.8182	.8189
6.6	.8195	.8202	.8209	.8215	.8222	.8228	.8235	.8241	.8248	.8254
6.7	.8261	.8267	.8274	.8280	.8287	.8293	.8299	.8306	.8312	.8319
6.8	.8325	.8331	.8338	.8344	.8351	.8357	.8363	.8370	.8376	.8382
6.9	.8388	.8395	.8401	.8407	.8414	.8420	.8426	.8432	.8439	.8445

723

Table of Logarithms (*continued*)

	0	1	2	3	4	5	6	7	8	9
7.0	.8451	.8457	.8463	.8470	.8476	.8482	.8488	.8494	.8500	.8506
7.1	.8513	.8519	.8525	.8531	.8537	.8543	.8549	.8555	.8561	.8567
7.2	.8573	.8579	.8585	.8591	.8597	.8603	.8609	.8615	.8621	.8627
7.3	.8633	.8639	.8645	.8651	.8657	.8663	.8669	.8675	.8681	.8686
7.4	.8692	.8698	.8704	.8710	.8716	.8722	.8727	.8733	.8739	.8745
7.5	.8751	.8756	.8762	.8768	.8774	.8779	.8785	.8791	.8797	.8802
7.6	.8808	.8814	.8820	.8825	.8831	.8837	.8842	.8848	.8854	.8859
7.7	.8865	.8871	.8876	.8882	.8887	.8893	.8899	.8904	.8910	.8915
7.8	.8921	.8927	.8932	.8938	.8943	.8949	.8954	.8960	.8965	.8971
7.9	.8976	.8982	.8987	.8993	.8998	.9004	.9009	.9015	.9020	.9026
8.0	.9031	.9036	.9042	.9047	.9053	.9058	.9063	.9069	.9074	.9079
8.1	.9085	.9090	.9096	.9101	.9106	.9112	.9117	.9122	.9128	.9133
8.2	.9138	.9143	.9149	.9154	.9159	.9165	.9170	.9175	.9180	.9186
8.3	.9191	.9196	.9201	.9206	.9212	.9217	.9222	.9227	.9232	.9238
8.4	.9243	.9248	.9253	.9258	.9263	.9269	.9274	.9279	.9284	.9289
8.5	.9294	.9299	.9304	.9309	.9315	.9320	.9325	.9330	.9335	.9340
8.6	.9345	.9350	.9355	.9360	.9365	.9370	.9375	.9380	.9385	.9390
8.7	.9395	.9400	.9405	.9410	.9415	.9420	.9425	.9430	.9435	.9440
8.8	.9445	.9450	.9455	.9460	.9465	.9469	.9474	.9479	.9484	.9489
8.9	.9494	.9499	.9504	.9509	.9513	.9518	.9523	.9528	.9533	.9538
9.0	.9542	.9547	.9552	.9557	.9562	.9566	.9571	.9576	.9581	.9586
9.1	.9590	.9595	.9600	.9605	.9609	.9614	.9619	.9624	.9628	.9633
9.2	.9638	.9643	.9647	.9652	.9657	.9661	.9666	.9671	.9675	.9680
9.3	.9685	.9689	.9694	.9699	.9703	.9708	.9713	.9717	.9722	.9727
9.4	.9731	.9736	.9741	.9745	.9750	.9754	.9759	.9763	.9768	.9773
9.5	.9777	.9782	.9786	.9791	.9795	.9800	.9805	.9809	.9814	.9818
9.6	.9823	.9827	.9832	.9836	.9841	.9845	.9850	.9854	.9859	.9863
9.7	.9868	.9872	.9877	.9881	.9886	.9890	.9894	.9899	.9903	.9908
9.8	.9912	.9917	.9921	.9926	.9930	.9934	.9939	.9943	.9948	.9952
9.9	.9956	.9961	.9965	.9969	.9974	.9978	.9983	.9987	.9991	.9996

CHAPTER 1

1.1 (a) theory; (b) hypothesis; (c) observation; (d) law; (e) law; (f) observation. *1.5* (a) velocity, SI, coherent; (d) density, SI, not coherent. *1.8* (a) 1×10^{-3} kg/cm³; (b) 1×10^6 g/m³. *1.17* 3.1557×10^7 s.

CHAPTER 2

2.3 52 g. *2.4* 1.50 g. *2.10* (a) +4; (b) +7; (c) +5; (d) +8; (e) +3. *2.17* (a) SiH_4; (c) PH_3; (e) IF_7; (g) P_4O_{10}. *2.20* (a) H_2; (b) $GeCl_2$; (c) Cl^-; (d) I^-; (e) HPO_4^{2-}; (f) NO_2; (g) N_2O_4. *2.23* (a) $CCl_4(l) \rightarrow CCl_4(g)$; (e) $C(s) + O_2(g) \rightarrow CO_2(g)$; (g) $2C_4H_{10}(g) + 13O_2(g) \rightarrow 8CO_2(g) + 10H_2O(l)$. *2.27* (a) $Ag^+ + Br^- \rightarrow AgBr(s)$; (c) $H_2S + Zn^{2+} \rightarrow ZnS(s) + 2H^+$. *2.28* (a) $H^+ + OH^- \rightarrow H_2O$; (b) $HCN + OH^- \rightarrow H_2O + CN^-$; (c) $NH_3 + HCN \rightarrow NH_4^+ + CN^-$; (d) $HCl(g) + OH^- \rightarrow H_2O + Cl^-$; (e) $HCl(g) + NH_3 \rightarrow NH_4^+ + Cl^-$; (f) $NH_3(g) + H^+ \rightarrow NH_4^+$.

CHAPTER 3

3.1 50 pencils, 4.2 dozen pencils, 0.35 gross of pencils. *3.5* 2.3×10^{17} autos. *3.6* 4.5×10^{23} amu. *3.11* 36.947. *3.16* 42.081. *3.17* (a) 4.62 mol HCN. *3.19* 0.250 mol S_8, 1.00 mol S_2, 2.00 mol S. *3.21* 79.8, Br. *3.24* 8.91×10^{25} atoms U. *3.30* 2.54 mol N. *3.33* 88.81%. *3.39* C_2H_5N. *3.41* X_2Y_3. *3.42* C_3H_8. *3.45* $C_9H_{11}NO_2$. *3.47* 479 g Cl_2. *3.51* 0.60 mol B_5H_9. *3.52* 11.8 g O_2. *3.55* 128 g O. *3.57* 0.8 g W remains. *3.59* MnO_2, 96 g Mn. *3.61* 55.1%. *3.64* 92.90. *3.66* 6.86 mol HCl. *3.69* 6.39 liters solution. *3.71* 0.074 liter NaI solution. *3.74* 21 g Ba. *3.76* 3.32*M*. *3.78* 5.8 g S. *3.79* 24.0%. *3.80* 20.3 g S_2Cl_2. *3.81* 18.0 g $CrF_3(s)$, 8.59 g Cr^{3+}. *3.82* 1.6 g C_2H_2.

CHAPTER 4

4.4 $<2 \times 10^{-5}$ Pa. *4.7* 17 liters. *4.9* 16.4 m³. *4.13* 44.9 liters. *4.16* 19 g C_2H_2. *4.20* 1.19 g Zn. *4.23* 2.6×10^{-3} g. *4.27* 0.38 g/liter. *4.30* 0.019 mol H_2, 0.45 g NaH. *4.33* 7.3×10^{-3} mol CO_2. *4.35* 1.1 atm H.

4.36 0.42 CH_4. *4.37* 25% N_2H_4 by amount. *4.38* 1.5×10^8 m²/s². *4.41* 6.3 s for I_2, the slower. *4.45* 2.5×10^{-4}. *4.49* P correction = 0.013 atm.

CHAPTER 5

5.4 4.82241×10^7 C/kg. *5.5* 254.09. *5.9* $^{236}_{90}$Th, $^{237}_{91}$Pa, $^{239}_{93}$Np, $^{240}_{94}$Pu. *5.13* 127.054. *5.15* 3.0×10^{-24} m. *5.18* $n_f = 3$, 1.9×10^{-6} m, 1.3×10^{-6} m. *5.22* 8.11×10^{-26} J. *5.26* 4.84×10^{14} s⁻¹. *5.28* 7.39×10^5 m/s. *5.31* $3h/2\pi$, 4.77×10^{-10} m. *5.33* 2×10^{-13} m.

CHAPTER 6

6.1 (a) 4; (b) 5; (c) 0; (d) 5; (e) 25; (f) 1. *6.6* (b) 120; (c) 8. *6.8* Ti, Ni, Ge, Se. *6.11* $[Rn]5f^{14}6s^26p^66d^47s^2$ or $[Rn]5f^{14}6s^26p^66d^57s$. *6.14* (a) oxygen; (b) chlorine; (c) titanium; (d) sulfur. *6.15* (a) 1:4; (c) 1:4. *6.30* K_2O, Na_2O, CaO, TlO, Tl_2O_3, SiO_2, As_4O_{10}, SeO_3, I_2O_7, Br_2O_7.

CHAPTER 7

7.5 Na^+O^-, Na^+F^-, $Mg^{2+}F^-$, $Mg^{2+}O^{2-}$, $Al^{3+}O^{2-}$. *7.7* 2.6×10^{-9} N. *7.12* $2NaO(s) \rightarrow Na_2O(s) + \frac{1}{2}O_2(g)$. *7.14* ZnI_2, $FeCl_2$, $BaCl_2$, AlF_3, $CsCl$, LiF. *7.19* Six minus the number of *p* electrons equals the number of bonds.

7.23 (a)
$$:\!\overset{\displaystyle :\!\ddot{F}\!:}{\underset{\displaystyle :\!\ddot{F}\!:}{\ddot{F}\!-\!C\!-\!\ddot{F}}}\!:$$

7.25 (a) $\ddot{O}\!=\!C\!=\!\ddot{O}$

7.27 (a)
$$:\!\ddot{Cl}\!-\!\overset{\displaystyle \ddot{C}l}{\underset{\displaystyle \ddot{C}l\!:}{B}}$$

7.29 (a)
$$:\!\overset{\displaystyle :\!\ddot{O}\!:}{\underset{\displaystyle :\!\ddot{O}\!:}{\ddot{O}\!-\!Br\!-\!\ddot{O}\!-\!H}}$$

7.31 (a)
$$\left[\,H\!-\!\ddot{O}\!-\!\overset{\displaystyle :\!\ddot{O}\!:}{\underset{\displaystyle :\!\ddot{O}\!:}{P}}\!-\!\ddot{O}\!: \right]^{2-}$$

7.32 The eight carbon atoms lie at the corners of a cube; the edges of the cube are the C—C bonds. Each carbon atom has one hydrogen atom attached.

7.37 H—N—O:⁻, N—O—H (best)

(structures with H and lone pairs)

7.39 structures of N—Cl with O groups ⟷

7.41 :N═N═N: ⁻ ⟷

:N≡N—N:²⁻ ⟷ :N²⁻—N≡N:

7.43 (a) H(1s) + Cl(3p$_x$); (b) H(1s) + O(2p$_x$) and O(2p$_y$). **7.47** (a) $\frac{1}{2}$; (b) $\frac{1}{2}$; (c) 1; (d) 0. **7.55** (a) N +4, O −2; (b) Cl +7, O −2; (c) Al +3, S −2; (d) S +6, F −1. **7.59** 2 S atoms are −1 and 2 S atoms are 0. **7.62** Polarizability of Cl_2 is more important than the polarity of ClF.

CHAPTER 8

8.3 (b) C—H bond sp^3-1s, C—Cl bonds sp^3-3p. **8.10** (b) C—H bonds sp^2-1s, C—O σ bond sp^2-2p$_x$.

8.15 (diagram of C bonded to H, Cl, O with orbital labels σ(1s-sp²), π(2p$_z$-2p$_z$), σ(sp²-2p$_x$), σ(3p-sp²), π(2p$_z$-2p$_z$))

8.18 (b) linear; (d) pyramidal.

8.21 (structures) C═N═N: ⁻ or C—N≡N:

8.22 (c) sp^3d. **8.25** (c) distorted "T" shaped. **8.26** SnF_6^{2-}, SbF_6^-, IF_6^+. **8.28** There is no simple or obvious explanation that we can think of. It has been suggested that the inner electrons of the barium may influence the geometry. **8.35** −301 kJ/mol CH_4. **8.39** 51 kJ/mol. **8.43** 32 kJ/mol. **8.48** 2139 kJ/mol. **8.50** 536 kJ/mol. **8.52** 151 kJ/mol. **8.53** 108 kJ/mol. **8.55** −3318 kJ/mol = calculated ΔH_{comb}, 148 kJ/mol = resonance energy. **8.57** −2162 kJ/mol. **8.58** 127 kJ/mol.

CHAPTER 9

9.7

(diagram of B and H structure)

9.9 Two moles of bonds. ΔH_{obs} of sublimation = 716 kJ/mol, ΔH_{calc} of sublimation = 692 kJ/mol. The higher observed heat of sublimation may be due to van der Waals attractions between carbons that are close to each other, but not directly bonded, in the diamond crystal. **9.10** 5101 kJ/mol. **9.14** 365 kJ/mol. **9.17** H—O—N═N—O—H,

(structure with N, O, H)

9.20 $12NH_3(g) + 21O_2(g) \rightarrow 14H_2O(g) + 8H^+ + 8NO_3^- + 4NO(g)$. **9.24** 60°, orbital overlap does not occur well because this bond angle is much smaller than the bond angles associated with most hybrid orbitals or with unhybridized p orbitals. Poor orbital overlap results in weak bonds and therefore high reactivity. **9.29** 0.277. **9.31** One oxygen atom from 0 in O_3 to −2 in NO_2. **9.34** $Ba(s) + O_2(g) \rightarrow BaO_2(s)$, $BaO_2(s) + 2H^+ \rightarrow H_2O_2 + Ba^{2+}$.

9.40 :Cl—S—Cl: The gases are HCl and SO_2. (with :O: below)

9.43 −50 kJ/mol. **9.48** (a) ClF(g) + $H_2O \rightarrow$ HOCl + HF. **9.51** (a) Cl_2O_7. **9.53** (a) :O—Xe—O: (with :O: below)

Pyramidal, structures without formal charges can be obtained by expanding the octet of xenon.

CHAPTER 10

10.8 10.2 cm³/mol.
10.10 0.362 nm.
10.12 55.843. **10.17** $V = 1.126 \times 10^{-23}$ cm³, $r = 0.1391$ nm.
10.20 13.9 cm³. **10.22** 0.0749 nm.
10.24 0.303 nm. **10.26** 0.0586 nm.
10.34 (a) Vapor to solid at less than $P = 0.006$ atm; (b) vapor to solid at just below $P = 0.006$ atm, solid to liquid at $P = 1$ atm. **10.37** (b) Solid to triple point at 216 K to vapor; (c) solid to liquid, liquid to vapor.

CHAPTER 11

11.3 4°C is close to the temperature of maximum density. **11.8** 20 kJ/mol. **11.12** It is close to the tetrahedral angle of 109° 28', so that the tetrahedral geometry associated with maximum hydrogen bonding can form with only minor angle distortion. **11.16** Since Cs^+ is a relatively large cation, the lattice energies of the cesium salts are not as different as they are for the lithium salts. The hydrogen bonding of the fluoride ion by water is the major factor causing the solubility of CsF to be greater than that of CsCl. **11.19** Tetrahedral, octahedral. **11.21** 9.5×10^{21} g NaCl. **11.25** $Ca^{2+} + CO_3^{2-} \rightarrow CaCO_3(s)$. **11.28** 3.7 liter of O_2 for the NH_3 and 0.49 liter of O_2 for the NO_2^-.

CHAPTER 12

12.2 10.0 g sugar. **12.5** 0.298. **12.8** 0.0106. **12.11** 55.6m. **12.14** $1.08 \times 10^{-4}M$. **12.16** 0.821m. **12.19** 9.391M. **12.22** 40 cm³. **12.23** 2.50m, $X_{C_2H_5OH} = 0.0430$. **12.26** 6.3 g/100 cm³. **12.29** 1.9 liters of CO_2. **12.32** 0.2438 atm. **12.35** 267. **12.37** 0.067 atm n-propyl alcohol, 0.132 atm isopropyl alcohol. **12.39** 0.863. **12.41** 148 g sugar. **12.43** 0.230 atm. **12.45** 265.7 K. **12.50** 13.2 atm. **12.51** 272.3 K. **12.54** 0.0306 atm. **12.56** 1.1×10^{-3} g $Ca_3(PO_4)_2$. **12.58** MF_3. **12.60** 0.102. **12.62** 2.48.

CHAPTER 13

13.4 −2260 J. **13.8** (a) Temperature falls by one-half; (b) temperature falls by one-half; (c) temperature increases. **13.13** −196 kJ. **13.18** −1370 kJ/mol. **13.22** 96.4 J. **13.26** 357 K. (Neglect warming the melted ice.) **13.27** (a) $T = 297.7$ K; (b) 24.4 liters; (c) −205 J; (d) 305 J. **13.30** At 374 K, $\Delta S_T = 2.9 \times 10^{-4}$ J mol⁻¹ K⁻¹; at 372 K, $\Delta S_T = -2.9 \times 10^{-4}$ J mol⁻¹ K⁻¹. The negative ΔS_T at 372 K is consistent with the impossibility of boiling water at 372 K and 1 atm. **13.32** 19 700 J/mol. **13.35** (a) −; (b) −; (c) +; (d) −. **13.38** (a) 20 J/K. **13.41** 29 400 J/mol. **13.45** 110 kJ/mol. **13.47** (a) −106 kJ/mol. **13.51** −27 000 J/mol. **13.53** (a) $\Delta n = 0$, no change in equilibrium composition, therefore no change in $\Delta G°$; (b) $\Delta n = -\frac{1}{2}$, shift to right, K increases, so $\Delta G°$ decreases. **13.55** $NH_4NO_3(s) \rightarrow NH_3(g) + HNO_3(g)$, $K = 4.9 \times 10^{-17}$ atm², $NH_4NO_3(s) \rightarrow N_2O(g) + 2H_2O(g)$, $K = 6.6 \times 10^{29}$ atm, $2NH_4NO_3(s) \rightarrow 2N_2(g) + O_2(g) + 4H_2O(g)$, $K = 1.2 \times 10^{96}$ atm⁶. **13.57** 1.54×10^{-4} atm. The in-

creased importance of ΔS at high temperatures favors the products.
13.59 6.92 kJ/mol   *13.62* 1.2 × 10^{-1} atm^{-2}, 3.1 × 10^{-5} atm^{-2}.

CHAPTER 14

14.3 (a) $1/P_{NH_3}P_{HCl}$;
(b) $[NH_4^+]/P_{NH_3}[H^+]$;
(c) $[NH_4^+][Cl^-]/[NH_3]P_{HCl}$;
(d) $[NH_4^+]/[NH_3][H^+]$.   *14.5* (a) none;
(b) mol/liter^{-1} atm^{-1}; (c) atm^{-2};
(d) liter^4/mol^4.   *14.8* 0.30 atm.
14.10 0.573 atm^{-1}.   *14.12* (a) to the right; (b) to the left; (c) to the right; (d) to the left.   *14.15* N_2O_4 decreases, NO and O_2 increase, NO_2 decreases.
14.20 1.9 × 10^{-5} atm^4.
14.22 0.390 atm.   *14.24* (a) as P increases, decomposition decreases;
(b) none; (c) same as (a).   *14.26* $K =$ 50.1 atm^{-1}.   *14.28* liter^2/mol^2.
14.29 $K_c = K_P(1/RT)^{\Delta n}$.
14.31 $P_{CS_2} = 1.28$ atm, $P_{S_2} =$ 0.14 atm, in both cases.   *14.33* $P_{Cl} =$ 4.98 × 10^{-3} atm, $P_{Cl} = 4.97 × 10^{-3}$ atm.   *14.36* 4.94 × 10^{-3} atm.
14.39 $P_{O_2} = 2.9 × 10^{-3}$ atm, $P_{SO_2} =$ 5.9 × 10^{-3} atm, $P_{SO_3} = 9.2 × 10^{-3}$ atm.   *14.41* $P_T = 1.48$ atm.
14.44 $P_{CH_4} = 0.34$ atm, $P_{H_2} =$ 0.19 atm.   *14.47* (a) M^2; (b) M^3;
(e) M^3.   *14.49* $[Mg^{2+}] = 3.0 × 10^{-3}M$, $[PO_4^{3-}] = 2.0 × 10^{-3}M$.
14.51 2.7 × $10^{-8}M^3$.
14.53 6.4 × $10^{-19}M^5$.
14.55 (a) 6.9 × $10^{-5}M$;
(b) 2.4 × $10^{-8}M$; (c) 2.4 × $10^{-8}M$.
14.59 3.0 × $10^{-46}M$.   *14.61* $[Ag^+] =$ 9.1 × $10^{-6}M$, $[CO_3^{2-}] = 0.1M$,
mol $Ag_2CO_3(s) = 0.025$.   *14.62* 1.06 liters.   *14.64* CuI precipitates first at $[I^-] = 1.1 × 10^{-10}M$, PbI_2 precipitates when $[I^-] = 2.7 × 10^{-4}M$ at which point $[Cu^+] = 4.1 × 10^{-9}M$.
14.66 7.3 × $10^{-11}M$.   *14.69* (1) Add Zn^{2+} to precipitate $Zn(CN)_2$, (2) add Hg_2^{2+} to precipitate Hg_2Cl_2, (3) add Pb^{2+} to precipitate $PbSO_4$, (4) add Sr^{2+} to precipitate SrF_2, (5) add Ag^+ to precipitate $AgNO_2$.

CHAPTER 15

15.6 $Al(OH)_3(s) + OH^- \rightleftharpoons Al(OH)_4^-$, $Al(OH)_3 \rightleftharpoons Al^{3+} + 3OH^-$.

15.10 H—Ö—N̈⁺—Ö—H,
‖
:O:

$HNO_3 + H_2SO_4 \rightleftharpoons H_2NO_3^+ + HSO_4^-$.
15.12 (b) $H^+ + OH^- \rightleftharpoons H_2O$; (c) no reaction; (e) $CH_3NH_2 + H^+ \rightleftharpoons CH_3NH_3^+$.
15.17 (a) pH = 2.7, pOH = 11.3;

(c) pH = 11.02, pOH = 2.98.
15.21 a, d, c, b.   *15.23* 4.93.
15.27 11.73.   *15.29* 3.0 × $10^{-9}M$.
15.31 7.9 × $10^{-1}M$.   *15.33* 0.24 mol NH_4Cl.   *15.36* 9.7 × 10^{-4} mol.
15.39 (b) $OH^- + HF \rightleftharpoons H_2O + F^-$, $K = 3.53 × 10^{10}M^{-1}$;
(d) $HIO_3 + NH_3 \rightleftharpoons IO_3^- + NH_4^+$, $K = 3.02 × 10^8 M^{-1}$.   *15.42* 2.03 × 10^{-1} g NaOH.   *15.45* 0.172M.
15.49 8.66%.   *15.51* 5.12.
15.53 (a) 1.48; (b) 2.28; (c) 3.30;
(d) 4.30; (e) 9.70.   *15.54* (a) 9.25;
(b) 10.21; (c) 10.86; (d) 10.98;
(e) 10.99.   *15.55* (a) bromothymol blue; (b) chlorphenol red.
15.58 11.03.   *15.63* 6.29.
15.67 7.28.   *15.69* $[H_2C_2O_4] =$ 0.58M, $[HC_2O_4^-] = [H^+] = 0.18M$, $[C_2O_4^{2-}] = 5.4 × 10^{-5}M$.
15.70 (a) $pK_2 = 7.21$.
15.72 $K = 1.30 × 10^{-4}$, pH = 8.33.
15.74 4 × 10^{-17} mol/liter, 3.3 × 10^{-12} mol/liter.   *15.76* $[CO_2] =$ 1.5 × $10^{-3}M$, $[HCO_3^-] =$ 2.37 × $10^{-2}M$.   *15.78* 5.5 × 10^{-25} mol/liter.   *15.80* $K = 3.7 × 10^{-5}M^2$. 6.8 × 10^{-3} mol/liter.

CHAPTER 16

16.9 0.134M.   *16.13* 8.46%.
16.17 8600 C.   *16.20* MCl_4.
16.23 435 sec.   *16.26* (a) $Sn(s) \rightarrow$ $Sn^{2+} + 2e^-$, $Pb^{2+} + 2e^- \rightarrow Pb(s)$, $Sn(s) + Pb^{2+} \rightarrow Sn^{2+} + Pb(s)$;
(c) $HNO_2 + H_2O \rightarrow$ $NO_3^- + 3H^+ + 2e^-$, $Mn^{3+} + e^- \rightarrow$ Mn^{2+}, $HNO_2 + H_2O + 2Mn^{3+} \rightarrow$ $2Mn^{2+} + NO_3^- + 3H^+$.
16.30 −110 kJ   *16.33* 1.539 × 10^{21}
16.36 0.71 V.   *16.39* (a) 0.9259 V;
(c) 0.418 V.   *16.43* (a) yes; (d) yes.
16.48 (a) yes; (b) no; (c) no; (d) yes.
16.50 (a) 0.70 V; (d) 1.23 V.
16.55 (a) 2.5 × 10^{27},
(c) 1.17 × 10^{23}.   *16.58* −1.17 V.
16.59 0.45 mol/liter.
16.62 (a) 1.49 V; (c) −0.39 V.
16.63 $4HNO_2 + HClO_2 \rightarrow 2N_2O_4(g) +$ $2H_2O + Cl^- + H^+$, $\mathcal{E}° = 0.50$ V, $n = 4$, $\Delta G° = -190$ kJ, $K = 6.9 × 10^{33}$.   *16.65* 0.99 V.
16.69 67M.   *16.71* 0.10M anode, 1.0M cathode, $\mathcal{E} = 0.030$ V.
16.73 0.097M.   *16.78* (1) $Sn^{4+} +$ $H_2(g) \rightarrow Sn^{2+} + 2H^+$, $\mathcal{E}° = -0.15$ V, $\Delta G° = -30$ kJ; (2) $2Cu^{2+} + Sn^{2+} \rightarrow$ $2Cu^+ + Sn^{4+}$, $\mathcal{E}° = 0.01$ V, $\Delta G° = -2$ kJ; (3) $2Cu^+ + 2Fe^{3+} \rightarrow$ $2Cu^{2+} + 2Fe^{2+}$, $\mathcal{E}° = 0.61$ V, $\Delta G° = -120$ kJ; (4) $2Fe^{2+} +$

$Au^{3+} \rightarrow 2Fe^{3+} + Au^+$, $\mathcal{E}° = 0.52$ V, $\Delta G° = -100$ kJ; (5) $Au^+ + \frac{1}{2}O_2(g) +$ $2H^+ \rightarrow Au^{3+} + H_2O$, $\mathcal{E}° = -0.06$ V, $\Delta G° = 10$ kJ. Overall, $H_2(g) + \frac{1}{2}O_2(g)$ $\rightarrow H_2O$.

CHAPTER 17

17.1 $-\frac{1}{2}\Delta P_{N_2O_5}/\Delta t = \frac{1}{4}\Delta P_{NO_2}/\Delta t =$ $\Delta P_{O_2}/\Delta t$.   *17.4* 1.3 × 10^{-2} atm/sec.
17.7 interval 1, 2.28 × 10^{-3} mol liter^{-1} min^{-1}; total, 1.35 × 10^{-3} mol liter^{-1} min^{-1}.   *17.9* interval 1, 8.0 × 10^{-5} atm/sec; total, 3.0 × 10^{-5} atm/sec.   *17.12* Rate = $k[Br^-][H^+][H_2O_2]$.
17.15 $-dP_{H_2C_2O_4}/dt = kP_{H_2C_2O_4}$.
17.19 3.20 × 10^{-4} sec^{-1}.
17.21 (a) 47.8 sec; (b) 147 sec;
(c) 0.180 atm.   *17.25* 92.6 sec.
17.27 23 000 years.   *17.29* (a) Equal concentrations make this equivalent to a one-term second-order rate law;
(b) $k = 0.34$ liter mol^{-1} sec^{-1};
(c) 0.013M.   *17.30* $1/2c_0k = t$.
17.32 (a) 0.25 atm; (b) 0.33 atm;
(c) 0.   *17.37* (a) $-dP_{O_3}/dt = kP_{O_3}$;
(b) $-dP_{O_3}/dt = k_2KP_{O_3}^2/P_{O_2}$; (c) See if an increase in P_{O_2} affects the rate of disappearance of O_3.   *17.40* 5.34 liter mol^{-1} sec^{-1}.   *17.43* 96 700 J/mol.
17.46 (a) A negative E_a; (b) in the transition state, two unpaired electrons, one from each NO, start pairing in bonds;
(c) the $\Delta S^{\ddagger}$ is negative. As T increases, its contribution becomes more important, increasing $\Delta G^{\ddagger}$ and slowing the rate.   *17.50* 4.55 × 10^{-7} liter mol^{-1} sec^{-1}.

CHAPTER 18

18.2 (a) +3; (b) +5; (c) +3; (d) +2.
18.6 $[Cr(NH_3)_6]Cl_3$, $[Cr(NH_3)_5Cl]Cl_2$, $[Cr(NH_3)_4Cl_2]Cl$, $[Cr(NH_3)_3Cl_3]$.   *18.8* Tetrahedral; (b) tetrahedral; (c) square planar; (d) square planar.
18.10 $[Pt(NH_3)_4Cl_2]Br_2$, $[Pt(NH_3)_4Br_2]Cl_2$, $[Pt(NH_3)_4ClBr]ClBr$.
18.13

Cl—Pt—S—C≡N: , Cl—Pt—N̈=C=S̈

18.16 (a) 1; (b) 1; (c) 1; (d) 2.
18.19 Let A = H_2O, B = NH_3, and

C = Cl:

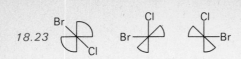

18.23

18.26 (a) $[Ar]3d^3(\uparrow\uparrow\uparrow)$; (b) $[Kr]4d^2(\uparrow\uparrow)$; (c) $[Ar]3d^6(\uparrow\downarrow\uparrow\uparrow\uparrow)$; (d) $[Ar]3d^5(\uparrow\uparrow\uparrow\uparrow\uparrow)$.
18.30 (a) 2 unpaired electrons; (b) 5 unpaired electrons; (c) 4 unpaired electrons; (d) 1 unpaired electron.
18.35 The $[Fe(CN)_6]^{4-}$ ion is not labile.
18.37 Mix two complexes such as $MCl_6{}^{3-}$ and $MBr_6{}^{3-}$ in solution and measure the rate of ligand exchange.
18.40 $[Sn^{4+}] = 4.2 \times 10^{-4}M$, $[F^-] = 2.5 \times 10^{-3}M$, $[SnF_6{}^{2-}] = 0.10M$.

CHAPTER 19

19.3 164 kJ/mol. *19.10* (a), (b), and (c). *19.13* (a) White tin; (b) white tin; four in gray, two in white, four $5sp^3$ in gray, $5s^25p^2$ in white. *19.17* Roast to the oxide, the arsenic oxides are relatively volatile and can be separated, reduce with coke. The liberated arsenic compounds are poisonous.
19.19 4.3×10^7. *19.23* (a) $+1$; (b) -2; (c) $+6$; (d) -2. *19.25* Tetrahedral, formally the metals are in the $+8$ oxidation state and can be regarded as having the noble gas electronic configuration. The hybridization can thus be sp^3 or sd^3. *19.28* $6.3 \times 10^{-3}M^{-1}$.

CHAPTER 20

20.3 (a) $^{40}_{17}Cl$, $^{40}_{18}Ar$, $^{40}_{19}K$, $^{40}_{20}Ca$, $^{40}_{21}Sc$.
20.6 2.0135536 g.
20.10 1.0785062×10^{16} J.
20.13 7.71 MeV. *20.16* (a) $^8_3Li \rightarrow$ $^4_2He + ^3_1H + ^1_0n$; (b) $^{190}_{78}Pt \rightarrow$ $^{186}_{76}Os + ^4_2He$. *20.18* 0.0127 amu, 11.8 MeV. *20.20* (a) $^{28}_{13}Al \rightarrow$ $^{28}_{14}Si + ^0_{-1}\beta + \nu$; (c) $^{147}_{61}Pm \rightarrow$ $^{147}_{62}Sm + ^0_{-1}\beta + \nu$. *20.22* (a) $^{26}_{13}Al +$ $^0_{-1}e \rightarrow ^{26}_{12}Mg + \nu$; (c) $^{209}_{84}Po + ^0_{-1}e \rightarrow$ $^{209}_{83}Bi + \nu$. *20.25* 0.0001675 amu, 0.156 MeV. *20.27* 0.001867 amu, 1.74 MeV. *20.34* Because the mass difference between the two nuclides is less than the total mass lost in the annihilation of an electron and a positron.
20.37 -0.000672 amu, -0.626 MeV.
20.39 (a) $^6_3Li + ^1_0n \rightarrow (^7_3Li) \rightarrow ^4_2He + ^3_1H$. *20.41* (a) $^{106}_{46}Pd + ^4_2He \rightarrow (^{110}_{48}Cd) \rightarrow$ $^{109}_{47}Ag + ^1_1H$. *20.45* -0.0265 amu, -24.7 MeV. *20.50* (a) Too many neutrons escape and the nuclear reaction slows; (b) too many fast neutrons, which would escape; the nuclear reaction slows (c) explosion; (d) explosion. *20.52* -0.7 MeV, no coulomb barrier for collision with a neutron.

CHAPTER 21

21.6 $CH_3CHCH_2CH_2CH_3$, $CH_3CH_2CHCH_2CH_3$, $CH_3CH_2CHCH_2CH_3$, with CH_3 and CH_2CH_3 substituents;
$CH_3CH-CHCH_2CH_3$, $CH_3CHCH_2CHCH_3$ with CH_3 CH_3 substituents.

21.9 The chainlike molecules of *n*-hexane become entangled with one another and are more difficult to separate into the vapor phase. *21.16* The pattern of substitution is not symmetrical as in Exercise 21.15. There are two *cis-trans* isomers.

21.19 (a) CH_2CH_2 (with Br, Br); (b) CH_3CH_2Cl; (c) CH_2CCH_3 (with Cl, Cl and CH_3); (d) *n*-octane.

21.22 $CH_2=C=CHCH_2CH_3$, $CH_2=CHCH=CHCH_3$, $CH_2=CHCH_2CH=CH_2$
1,2-pentadiene   1,3-pentadiene   1,4-pentadiene
$HC\equiv CCH_2CH_2CH_3$, $CH_3C\equiv CCH_2CH_3$, $CH_3CH=C=CHCH_3$
1-pentyne   2-pentyne   2,3-pentadiene

21.24 One-half mole of
$CH_3C-CCH_2CH_3$ (with Cl, Cl top and Cl, Cl bottom)
and one-half mole of unreacted starting 2-pentyne.

21.26

21.28 $HC\equiv C-CHCH_3$ (with CH_3); $CH_2=C=C$ (with CH_3, CH_3); $CH_2=CH-C$ (with CH_2, CH_3); plus numerous ring isomers — plus isomers of Exercise 21.22.

21.32 (a) Nonplanar, a tublike structure: ; (b) it has 8π electrons rather than 6.

21.33 It has 6π electrons.

21.35 *Alcohols:* $CH_3CH_2CH_2CH_2OH$, $CH_3CH_2CHCH_3$ (with OH);

Ethers: $CH_3-O-CH_2CH_2CH_3$. $CH_3CH_2-O-CH_2CH_3$.

21.37 Dehydration of ethanol with sulfuric acid to form ethylene followed by addition of Br_2. *21.40* Oxidation of an aldehyde breaks a C—H bond but forms a stronger O—H bond. In a ketone a C—C bond is replaced by a C—O bond of comparable strength.

21.43 CH_3CH_2COOH, CH_3COOCH_3, $HCOOCH_2CH_3$.

21.46

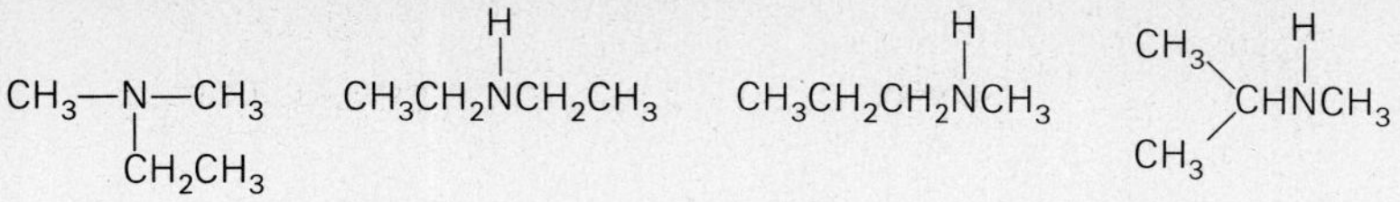

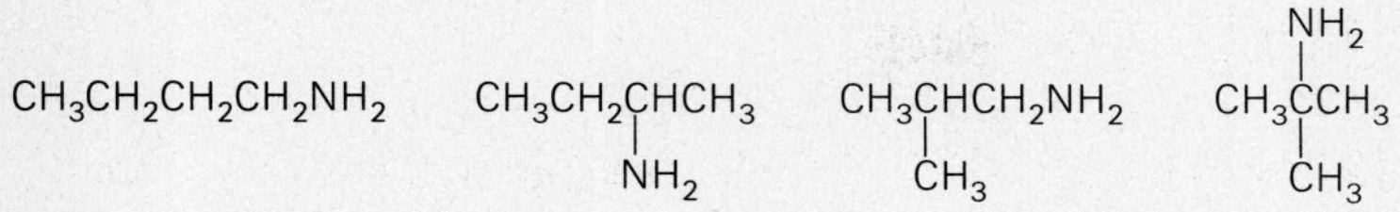

CH_3—$\overset{\overset{H}{|}}{N}$—$CH_3$ $CH_3CH_2\overset{\overset{H}{|}}{N}CH_2CH_3$ $CH_3CH_2CH_2\overset{\overset{H}{|}}{N}CH_3$ $\overset{CH_3}{\underset{CH_3}{\underset{|}{|}}}CHNCH_3$
$\underset{CH_2CH_3}{|}$

$CH_3CH_2CH_2CH_2NH_2$ $CH_3CH_2\underset{\underset{NH_2}{|}}{C}HCH_3$ $CH_3\underset{\underset{CH_3}{|}}{C}HCH_2NH_2$ $CH_3\overset{\overset{NH_2}{|}}{\underset{\underset{CH_3}{|}}{C}}CH_3$

21.50 $CH_3\underset{\underset{CH_3}{|}}{C}HCH_2CH_2Cl$ $CH_3\underset{\underset{CH_3}{|}}{C}H$—$\underset{\underset{Cl}{|}}{C}HCH_3$ $CH_3\overset{\overset{Cl}{|}}{\underset{\underset{CH_3}{|}}{C}}CH_2CH_3$ $ClCH_2\underset{\underset{CH_3}{|}}{C}HCH_2CH_3$

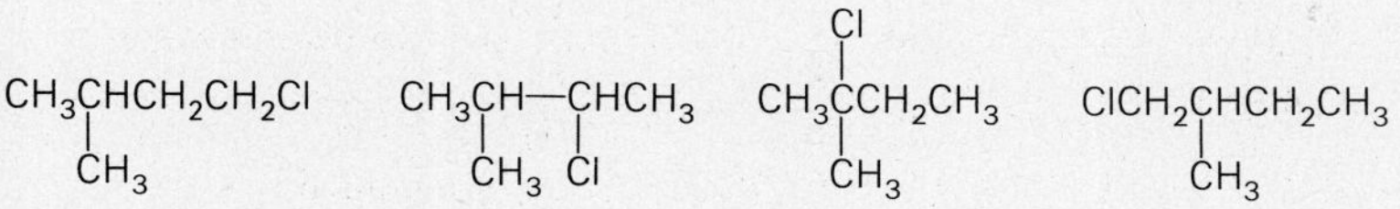

CHAPTER 22

22.3 $H_2NCH_2\underset{\underset{O}{\|}}{C}$—$\left[-NHCH_2\underset{\underset{O}{\|}}{C}-\right]_{n-2}$—$NHCH_2\underset{\underset{O}{\|}}{C}OH$

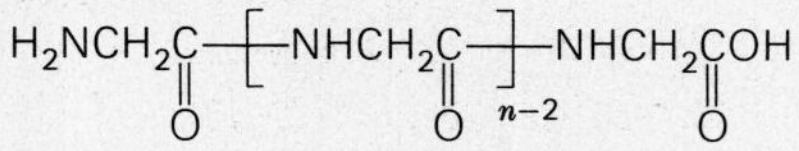

22.6 Hydrophobic: the nonpolar organic substituents on the α-carbon; hydrophilic, the amide groups, the polar organic substituents on the α-carbon. The protein folds so that the hydrophobic groups are in the interior shielded from water by the hydrophilic groups.

22.8

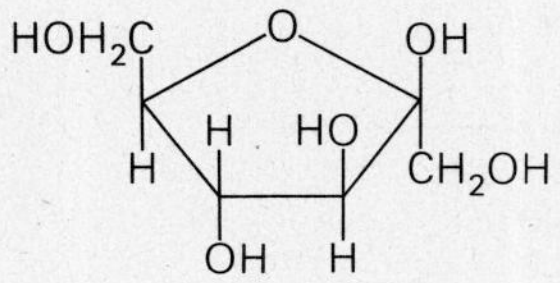

22.12

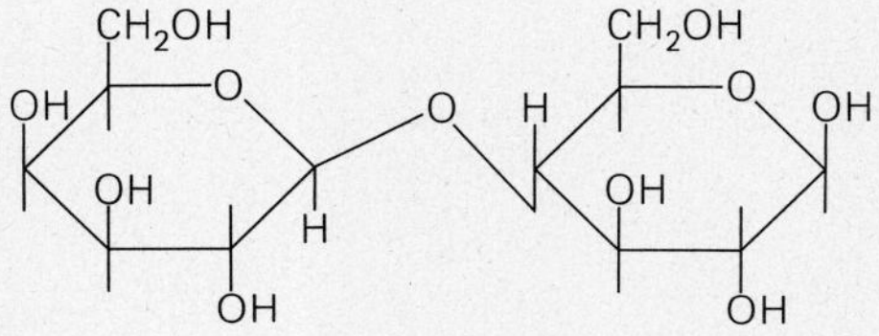

22.15

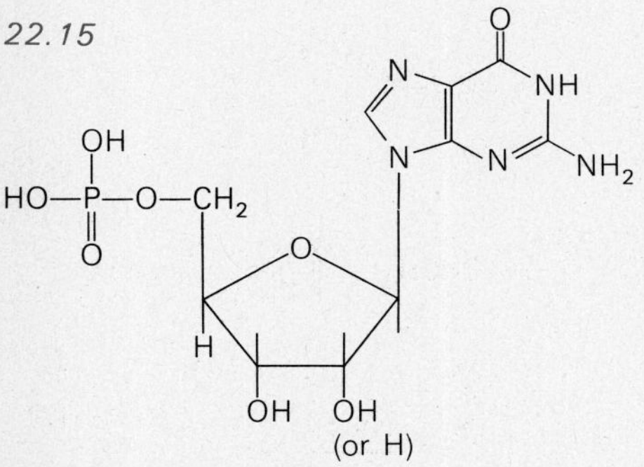

22.21

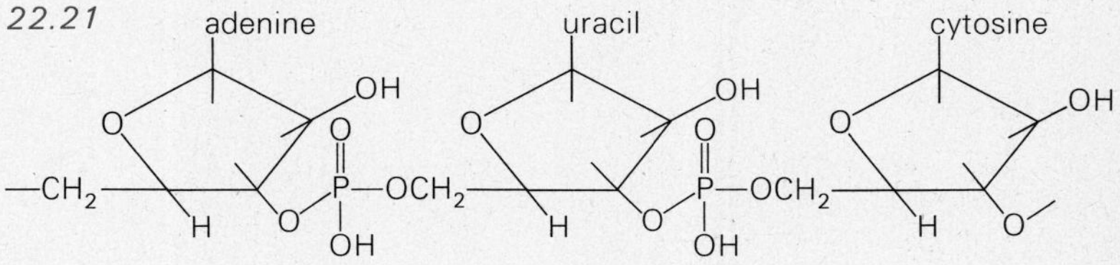

22.23 —GCGGCGGGUUUU—.

APPENDIX V

V.1 9.78.   V.4 0.046 mol/liter.   V.6 $AgCl(s) + Br^- \rightleftharpoons AgBr(s) + Cl^-$, $K = 3.1 \times 10^2$.

78 79 80 9 8 7 6 5 4 3 2 1

PHYSICAL AND CHEMICAL CONSTANTS

Quantity	Symbol	Value
Atomic Mass Unit	amu	1.661×10^{-27} kg
Avogadro Constant	N or N_A	6.022×10^{23} mol^{-1}
Electronic Charge	e	1.602×10^{-19} C
Faraday Constant	F	9.648×10^4 C mol^{-1}
Gas Constant	R	8.314 J mol^{-1} K^{-1}
		0.0821 liter atm mol^{-1} K^{-1}
Mass of Electron	m_e	9.110×10^{-31} kg
Mass of Neutron	m_n	1.675×10^{-27} kg
Mass of Proton	m_p	1.673×10^{-27} kg
Planck Constant	h	6.626×10^{-34} J s
Rydberg Constant	R_H	1.098×10^7 m^{-1}
Speed of Light in a Vacuum	c	2.998×10^8 m s^{-1}

UNITS AND CONVERSION FACTORS

Quantity	SI Unit	Symbol	Conversion Factors
Base Units			
Length	meter	m	1 cm = 10^{-2} m
			1 nm = 10^{-9} m
			1 Å = 10^{-10} m
			1 inch = 2.54×10^{-2} m
Mass	kilogram	kg	1 g = 10^{-3} kg
			1 mg = 10^{-6} kg
			1 lb = 0.454 kg
Time	second	s	1 day = 8.6×10^4 s
Temperature	kelvin	K	$0°$C = 273.15 K
Amount	mole	mol	
Electric current	ampere	A	
Derived Units			
Volume	cubic meter	m^3	1 liter = 10^{-3} m^3
			= 1000 cm^3
			1 ml = 1 cm^3
Energy	joule	J	1 cal = 4.184 J
			1 liter atm = 101.3 J
			1 eV = 1.602×10^{-19} J
Pressure	pascal	Pa	1 atm = 1.01×10^5 Pa
			= 760 Torr or mmHg
Force	newton	N	
Frequency	hertz	Hz	1 s^{-1} = 1 Hz
Electric charge	coulomb	C	
Electric potential difference	volt	V	